Fluorescence Microscopy in Life Sciences

Authored by

Juan Carlos Stockert

Department of Biology, Faculty of Sciences, Autonomous University of Madrid, Spain

Institute of Oncology Angel H. Roffo, Research Area, University of Buenos Aires, Argentina

Institute of Research and Technology in Animal Reproduction, Faculty of Veterinary Sciences, University of Buenos Aires, Argentina

&

Alfonso Blázquez-Castro

Aarhus Institute of Advanced Studies, Aarhus University, Aarhus, Denmark

Bentham Science Publishers Ltd.
Executive Suite Y - 2
PO Box 7917, Saif Zone
Sharjah, U.A.E.
Email: subscriptions@benthamscience.org

BENTHAM SCIENCE

CONTENTS

FOREWORD i

PREFACE iii

DEDICATION v

ACKNOWLEDGEMENTS vi

CHAPTER 1 INTRODUCTION 1
- **1. FLUORESCENCE METHODOLOGY: GENERAL ASPECTS** 1
- **2. COLOR AND ENERGY SPECTRUM** 2
- **3. BRIEF HISTORY OF STAINING AND FLUORESCENCE METHODS** 4
  - 3.1. Natural Dyes 4
  - 3.2. Synthetic Dyes 6
  - 3.3. Fluorescent Compounds 7
- **4. TYPES OF DYES AND FLUOROCHROMES: NOMENCLATURE** 9
- **5. MICROSCOPIC STAINING AND FLUORESCENCE** 11
- **6. ADVANTAGES AND APPLICATIONS OF FLUORESCENCE METHODOLOGY** 13
- **CONFLICT OF INTEREST** 17
- **ACKNOWLEDGEMENTS** 17
- **REFERENCES** 17

CHAPTER 2 PHYSICAL FUNDAMENTALS OF LUMINESCENCE 22
- **1. LIGHT/ELECTROMAGNETIC WAVES** 23
- **2. LUMINESCENCE** 26
  - 2.1. Grotrian Diagram 27
  - 2.2. Jablonski Diagram 28
- **3. ATOMIC SPECTROSCOPY** 29
  - 3.1. Electronic Orbitals 29
  - 3.2. Atomic Hydrogen 30
  - 3.3. Multielectronic Atoms 31
- **4. MOLECULAR SPECTROSCOPY** 32
  - 4.1. Kinetic Energy 33
  - 4.2. Molecular Rotation 34
  - 4.3. Molecular Vibration 34
  - 4.4. Potential Energy Curves 37
  - 4.5. Molecular Electronic Levels 38
  - 4.6. Franck-Condon Principle 40
  - 4.7. Molecular Multiplicity 42
  - 4.8. Potential Energy Curves Expanded 45
  - 4.9. Molecular Jablonski Diagram 47
- **5. NON-RADIATIVE DE-EXCITATION MECHANISMS** 50
  - 5.1. Internal Conversion 50
  - 5.2. Photodissociation 52
  - 5.3. Photoinduced Electron Transfer 53
  - 5.4. Photochemistry 56
  - 5.5. Energy Transfer 57
  - 5.6. Quantum Yields 57
- **CONFLICT OF INTEREST** 58
- **ACKNOWLEDGEMENTS** 59
- **REFERENCES** 59

CHAPTER 3 DYES AND FLUOROCHROMES 61
1. CLASSIFICATION AND CHEMISTRY OF DYES 61
1.1. General Features 61
1.2. Examples of Natural Dyes: Flavonoids 64
2. DYES AND COLOR THEORY 66
2.1. Traditional Color Theory 66
2.2. Free Base and Pseudo-Base 69
2.3. Conjugation 70
2.4. Double Bonds and Energy Absorption 73
2.5. Molecular Orbitals of Excited States 74
3. FLUOROCHROMES 79
3.1. Classification and Chemistry 79
3.2. Size: QSAR Parameters 81
3.3. Scintillators, NIR and Blinking Fluorochromes 82
4. PREDICTION OF FLUORESCENCE 84
4.1. General Aspects 84
4.2. Non-Radiative Energy Dissipation 87
4.3. Hematoxylin and Eosin as a Case Example 89
CONFLICT OF INTEREST 90
ACKNOWLEDGEMENTS 90
REFERENCES 90

CHAPTER 4 FLUORESCENT LABELS 96
1. TYPES OF FLUORESCENT LABELS 96
1.1. Reactive Dyes and Labeling Mechanisms 97
1.2. Photolabeling and Photoswitching 100
1.3. Mycotoxins and Phytotoxins 102
2. LIPID AND LIPOSOME LABELING 104
3. POLYSACCHARIDE LABELING 107
4. NUCLEIC ACIDS LABELING 109
4.1. Classical Labels 109
4.2. Labeled Peptide Nucleic Acids 109
4.3. Thymidine Analogues 110
5. PROTEIN LABELING 116
5.1. Avidin and Streptavidin 117
5.2. Lectins 117
5.3. Labeled Enzymes 118
6. NATURAL FLUORESCENT PROTEINS 118
6.1. Phycobiliproteins 118
6.2. Green Fluorescent Protein, Derivatives and Analogues 121
7. OTHER LABELING METHODS 126
CONFLICT OF INTEREST 127
ACKNOWLEDGEMENTS 127
REFERENCES 127

CHAPTER 5 SPECTRAL ANALYSIS 135
1. SPECTRAL RECORDING AND INTERPRETATION 135
2. INFLUENCE OF DYE CONCENTRATION 141
3. INFLUENCE OF SOLVENTS, CO-SOLUTES AND SUBSTRATES 143
3.1. Solvents 143
3.2. Co-solutes 145

3.3. Biological Substrates 148
4. MACROSPECTRA 151
5. ENERGY TRANSFER 152
CONFLICT OF INTEREST 154
ACKNOWLEDGEMENTS 154
REFERENCES 154

CHAPTER 6 FLUORESCENCE INSTRUMENTAL AND TECHNIQUES 159
1. BASIC ELEMENTS OF A FLUORESCENCE MICROSCOPE 160
1.1. Light Sources 160
Gas-Discharge Lamps 161
Lasers 163
Supercontinuum Sources 165
LEDs 166
Halogen Lamps 168
1.2. Optical Filters and Dichroic Mirrors 170
Excitation Filters 170
Dichroic Mirrors 171
Emission Filters 172
1.3. Microscope Objectives 172
1.4. Additional Optical Elements 175
Imaging Devices 175
Additional Filters 177
Optical Fibers 178
Other Optical Elements 179
2. CONFOCAL MICROSCOPY 180
3. TIME-RESOLVED FLUORESCENCE MICROSCOPY 184
3.1. Time-domain Resolved Fluorescence 185
3.2. Frequency-domain Resolved Fluorescence 188
4. LIGHT SHEET FLUORESCENCE MICROSCOPY 192
5. TOTAL INTERNAL REFLECTION FLUORESCENCE MICROSCOPY 198
6. FLUORESCENCE PHOTOBLEACHING TECHNIQUES 200
6.1. Fluorescence Recovery after Photobleaching (FRAP) 201
6.2. Fluorescence Loss in Photobleaching (FLIP) 203
6.3. Fluorescence Localization after Photobleaching (FLAP) 204
7. FLUORESCENCE CORRELATION SPECTROSCOPY 205
8. FLOW CYTOMETRY 211
CONFLICT OF INTEREST 213
ACKNOWLEDGEMENTS 214
REFERENCES 214

CHAPTER 7 GENERAL FLUORESCENCE REACTIONS 219
1. MICROSCOPIC STRUCTURES AND BIOLOGICAL SUBSTRATES 219
2. FLUORESCENCE PARAMETERS 221
2.1. General Features 221
2.2. Counterstaining and Mounting 223
2.3. Variations in the Emission Color 225
2.4. Chromo- and Fluorogenic Reactions 225
3. AUTOFLUORESCENCE 227
4. PHOTOCHEMICAL DAMAGE: BLEACHING AND FADING 232
5. EMISSION PHOTO-ENHANCEMENT 235
6. ABSORPTION-EMISSION CONVERSION 237

7. FLUORESCENCE OF CLASSICAL STAINING 239
8. EVALUATION OF FLUORESCENCE IMAGES 241
9. FALSE RESULTS AND MISLEADING INTERPRETATIONS 242
CONFLICT OF INTEREST 244
ACKNOWLEDGEMENTS 244
REFERENCES 244

CHAPTER 8 FLUORESCENT METAL COMPLEXES AND ION INDICATORS 252
1. IONIC REACTIONS 253
2. METAL COMPLEXES 256
3. FLUORESCENCE OF METAL COMPLEXES 257
3.1. Phenol, Naphthol, and Quinoline Ligands 258
3.2. Bipyridine and Phenanthroline 259
3.3. Flavonols 259
3.4. Hydroxy-Triarylmethanes 259
3.5. β-Diketones 260
3.6. Anthraquinones 261
3.7. Azo Dyes and Other Reagents 264
4. FLUORESCENT PH AND ION INDICATORS 266
4.1. pH Indicators 266
4.2. Ion Indicators 269
4.3. Calcium 271
4.4. Sodium, Magnesium, Zinc, and Phosphate 273
5. CALCIFICATION MARKERS 275
5.1. Fluorescent Labels 275
5.2. Polychromatic Fluorescence 277
CONFLICT OF INTEREST 277
ACKNOWLEDGEMENTS 277
REFERENCES 277

CHAPTER 9 LUMINESCENT TRANSITION METAL PROBES 283
1. TRANSITION METAL COMPLEX 283
1.1. d-block Transition Metals 285
3d Transition Metals 285
4d Transition Metals 285
5d Transition Metals 285
1.2. 4f-block Transition Metals: Lanthanides 286
1.3. 5f-block Transition Metals: Actinides 286
1.4. Ligands 287
Inorganic Ligands 288
Organic Ligands 288
Macrocycles 289
2. PHOTOPHYSICS AND PHOTOCHEMISTRY OF TMCS 290
2.1. Optical Transitions 291
d-d Transitions 293
Metal-to-Ligand Charge Transfer 293
Ligand-to-Metal Charge Transfer 293
Intra-ligand Transitions 294
2.2. Luminescence 295
2.3. Lanthanides 298
2.4. Actinides 303
3. BIOLOGICAL PROBES BASED ON LUMINESCENT TMCS 304

3.1. 3d-block 305
Chromium (Cr) 305
Manganese (Mn) 305
Iron (Fe) 305
Cobalt (Co) and Nickel (Ni) 305
Copper (Cu) 306
Zinc (Zn) 306
3.2. 4d-block 307
Zirconium (Zr) 308
Molybdenum (Mo) 308
Ruthenium (Ru) 308
Rhodium (Rh) 311
Palladium (Pd) 311
Silver (Ag) 312
3.3. 5d-block 312
Hafnium (Hf) 312
Rhenium (Re) 312
Osmium (Os) 314
Iridium (Ir) 315
Platinum (Pt) 317
Gold (Au) 319
3.4. 4f-block 321
Europium (Eu) 322
Terbium (Tb) 323
Other 4f-block Elements 326
3.5. 5f-block 327
CONFLICT OF INTEREST 329
ACKNOWLEDGEMENTS 329
REFERENCES 329

CHAPTER 10 LIPIDS 336
1. LIPID COMPONENTS AND STRUCTURES 336
2. FIXING REQUISITES AND DYE SOLUBILITY 339
3. LIPOCHROMATIC REACTIONS 341
4. OXIDATIVE DEAMINATION OF DYES: NILE RED 342
5. OTHER LIPOPHILIC STAINS 346
6. POLYENES 349
6.1. Natural Polyenes 349
6.2. Synthetic Polyenes 353
CONFLICT OF INTEREST 354
ACKNOWLEDGEMENTS 354
REFERENCES 354

CHAPTER 11 POLYSACCHARIDES 359
1. TYPES OF POLYSACCHARIDES AND HISTOCHEMICAL REACTIONS 359
2. BISAZO DYES AND BRIGHTENING AGENTS 361
3. REACTIVE DYES AND FLUOROCHROMES 365
4. CATIONIC DYES 368
4.1. Acridines 368
4.2. Carbocyanines, Indoles, Aryl-Oxazoles, and Other Dyes 371
4.3. Ruthenium Complexes 375
CONFLICT OF INTEREST 377

ACKNOWLEDGEMENTS ... 377
REFERENCES ... 377

CHAPTER 12 NUCLEIC ACIDS ... 381
1. GENERAL STRUCTURE AND REACTIVITY ... 381
2. FLUOROGENIC METHODS ... 383
2.1. Feulgen-Rossenbeck Reaction ... 383
2.2. Osmium Tetroxide Reaction ... 388
3. IONIC BINDING ... 390
3.1. Metal Cations and Fluorescent Lakes ... 390
3.2. Cationic Fluorochromes ... 391
4. INTERCALATION ... 393
4.1. Geometrical Parameters ... 394
4.2. Quantitative Parameters ... 395
4.3. Applications of Acridine Fluorochromes ... 397
4.4. Alkaloids ... 399
4.5. Psoralens ... 401
4.6. Anthraquinones ... 401
4.7. Porphyrins and Phthalocyanines ... 401
4.8. Naphthalimides ... 402
4.9. Bis-Intercalation ... 403
4.10. Intercalating Antibiotics ... 404
5. MINOR GROOVE BINDING ... 405
5.1. Pyrrole-Carboxamide Antibiotics ... 406
5.2. Isohelicity ... 407
5.3. Benzimidazoles ... 408
5.4. Aromatic Diamidines ... 410
5.5. Benzothiazoles ... 412
5.6. Other Linear Ligands ... 414
5.7. Bifunctional Minor Groove Binding ... 414
5.8. Minor Groove-Binding Antibiotics ... 415
6. SPECIAL NUCLEIC ACIDS STRUCTURES ... 416
6.1. Aptamers ... 417
6.2. Multistranded Polynucleotides ... 418
6.3. Three- and Four-Way Junctions ... 419
6.4. Chromosome Banding Patterns ... 419
CONFLICT OF INTEREST ... 419
ACKNOWLEDGEMENTS ... 420
REFERENCES ... 420

CHAPTER 13 PROTEINS ... 436
1. GENERAL STAINING AND IONIC REACTIONS ... 436
1.1. Isoelectric Point and Ionic Binding ... 437
1.2. Applications of Fluorochromes in Protein Histochemistry ... 438
1.3. Bisazo Dyes ... 440
1.4. Benzothiazole Fluorochromes ... 443
1.5. Hydrophobic and Reactive Fluorochromes ... 444
2. SELECTIVE FLUOROGENIC REACTIONS FOR AMINO ACIDS ... 446
2.1. Amino Groups ... 446
2.2. Carboxyl Groups ... 449
2.3. Guanidino Groups ... 449
2.4. Aromatic Amino Acids ... 450

2.5. Sulfhydryls 454
3. FLUOROGENIC REACTIONS FOR BIOGENIC AMINES AND NEUROTRANSMITTERS 455
CONFLICT OF INTEREST 457
ACKNOWLEDGEMENTS 457
REFERENCES 457

CHAPTER 14 ENZYMES 462
1. GENERAL ASPECTS OF ENZYMATIC HISTOCHEMISTRY 462
2. FLUORESCENCE METHODS FOR ENZYMATIC ACTIVITY 464
2.1. Esterases and Hydrolases 465
2.2. Dehydrogenases 465
2.3. Peroxidases 470
2.4. Phosphatases 473
2.5. Proteases 476
CONFLICT OF INTEREST 477
ACKNOWLEDGEMENTS 477
REFERENCES 477

CHAPTER 15 VITAL PROBES 1. CELL ORGANELLES 481
1. MECHANISMS OF UPTAKE AND ACCUMULATION 481
2. QUANTITATIVE STRUCTURE-ACTIVITY RELATIONSHIPS 484
2.1. Numerical Parameters 484
2.2. Prediction Rules 488
3. FLUORESCENT LABELING OF CELL ORGANELLES 489
4. NUCLEIC ACIDS-CONTAINING STRUCTURES 491
4.1. Chromatin DNA 492
4.2. Nucleoli and Basophilic Cytoplasm 497
5. MEMBRANOUS STRUCTURES 501
5.1. Plasma Membrane 501
5.2. Endoplasmic Reticulum 503
5.3. Lysosomes 504
5.4. Mitochondria 508
5.5. Golgi Apparatus 512
5.6. Lipid Droplets 514
5.7. Localization of Drugs 514
CONFLICT OF INTEREST 517
ACKNOWLEDGEMENTS 517
REFERENCES 517

CHAPTER 16 VITAL PROBES 2. CELL VIABILITY AND SIGNALING 532
1. CELL VIABILITY ASSAYS 532
1.1. Dye Exclusion 532
1.2. Enzymatic Assays and Redox Probes 536
1.3. Organelle Labeling 538
1.4. Protein Quantification 539
2. TYPES OF CELL DEATH 539
2.1. Apoptosis 539
2.2. Necrosis 543
2.3. Other Types of Cell Death 546
Mitotic Catastrophe 546
Autophagy 546

Regulated Necrosis 546
2.4. Senescence 547
3. GENOTOXICITY 547
3.1. TUNEL Assay 547
3.2. Comet Assay 548
3.3. Halo Assay 549
4. CELL SIGNALING 550
4.1. ROS Detection 550
4.2. NO Detection 554
5. LABELING OF CELL POPULATIONS 555
5.1. Phagocytosis and Cell Tracking 555
5.2. Retrograde Axonal Tracers 556
CONFLICT OF INTEREST 557
ACKNOWLEDGEMENTS 557
REFERENCES 557

CHAPTER 17 IMMUNOFLUORESCENCE AND FISH 568
1. IMMUNOFLUORESCENCE 568
1.1. Direct Immunofluorescence 575
1.2. Indirect Immunofluorescence 576
1.3. Counterstain Dyes and Anti-fading Agents 577
Counterstaining Dyes 577
Anti-fading Agents 579
1.4. New Uses and Related Techniques 582
Proteome Studies 582
Nanobodies 583
Flow Cytometry 585
Enzyme-linked immunosorbent assay (ELISA) 586
Western Blot 587
2. FLUORESCENT PROTEINS 587
3. FLUORESCENCE IN SITU HYBRIDIZATION (FISH) 589
CONFLICT OF INTEREST 600
ACKNOWLEDGEMENTS 601
REFERENCES 601

CHAPTER 18 LUMINESCENT SOLID-STATE MARKERS 606
1. BAND STRUCTURE MODEL 606
1.1. Valence and Conduction Bands 608
1.2. Semiconductors 610
1.3. Solid-state Luminescence 613
Dielectric and Semiconductor Luminescence 613
Conductor Luminescence 615
2. SEMICONDUCTOR LUMINESCENT MARKERS 618
2.1. Quantum Dots (Qdots) 618
2.2. Carbon and Silicon Nanoparticles 623
2.3. Semiconducting Polymers (P-dots) 627
3. METALLIC LUMINESCENT MARKERS 627
3.1. Gold Luminescent Nanoparticles 628
3.2. Silver Luminescent Nanoparticles 629
4. DIELECTRIC LUMINESCENT MARKERS 631
4.1. Polymer Particles 631
4.2. Nanodiamonds 632

4.3. Silica Nanoparticles 635
4.4. Lanthanide-doped Dielectrics 636
4.5. Other Materials 637
CONFLICT OF INTEREST 638
ACKNOWLEDGEMENTS 638
REFERENCES 638

CHAPTER 19 NON-LINEAR OPTICS 642
1. NON-LINEAR OPTICS 642
2. MULTIPHOTON MICROSCOPY 647
2.1. Fundamentals 647
2.2. Biphotonic Microscopy 651
2.3. Higher-order Microscopy 660
3. HARMONIC MICROSCOPY 661
3.1. Fundamentals 661
3.2. Harmonic Generation in Biological Materials 664
3.3. Harmonic Generation in Artificial Materials 665
4. PHOTON UPCONVERSION 665
4.1. Lanthanide-based Upconversion 666
4.2. Transition Metal Complex-based Upconversion 673
4.3. Triplet-triplet Annihilation Upconversion 673
5. RAMAN MICROSCOPY 677
CONFLICT OF INTEREST 681
ACKNOWLEDGEMENTS 681
REFERENCES 681

CHAPTER 20 SUPER-RESOLUTION MICROSCOPY 687
1. THE DIFFRACTION LIMIT 687
2. LIGHT COLLECTION ENHANCEMENT 690
2.1. Confocal 691
2.2. 4Pi 691
2.3. I5M 692
2.4. Superlenses and Hyperlenses 692
3. NEAR-FIELD SRM TECHNIQUES 693
3.1. TIRFM 694
3.2. SNOM 694
4. FAR-FIELD SRM TECHNIQUES 695
SRM linear techniques 695
4.1. Single Molecule Localization (PALM/STORM) 695
4.2. SIM 701
SRM non-linear techniques 704
4.3. S-SIM 704
4.4. STED 705
CONFLICT OF INTEREST 711
ACKNOWLEDGEMENTS 711
REFERENCES 711

SUBJECT INDEX 715

FOREWORD

Using fluorescence microscopy to study the components and processes of cells, tissues and organisms is not of course a novel investigatory approach, as it dates back to the early years of the twentieth century. However, the widespread use of this methodology had to wait for the commercial availability of fluorescence microscopes, in particular for commercialised confocal instruments. The parallel flowering of fluorescent probes for use with such microscopes, together with the continuing evolution of novel and powerful optical systems, further accelerated uptake of such methods within a wide variety of biomedical fields.

Consequently, many books — and even more chapters and reviews — have considered various specific aspects of fluorescence microscopy or, less often, provided more global overviews. Nevertheless, even in this crowded field, the present volume is ambitious, adopting, as it does, a rather different approach to most of its predecessors.

A key difference is its inclusive approach. Thus this book adopts a firmly didactic approach on the fundamentals and the applications of fluorescent microscopy. The aim is clearly to bring biologists and physicians up to speed with a basic knowledge of the physical chemistry, technical equipment, and practical manipulations underlying a functional understanding of fluorescence methods. However, the book also provides sufficient detail of practical applications, within an intelligible biomedical context, to allow readers whose backgrounds are chemistry or physics to understand the types of problem which are currently addressed by this approach. But more than that, to grasp possible future expansions of applications of fluorescence microscopy.

Another striking feature is the amount of critically selected information made accessible. Consider the physicochemical background. An account of colour theory – traditional, valence bond and molecular orbital – leads to a description of the molecular mechanisms of fluorescence. And so to the role of quenching, thereby answering the biologist's plaintive enquiry, "But if that is how fluorescence arises, why are **all** dyes not fluorescent?" The separate discussion of fluorescent dyes and reactive fluorescent dyes (*i.e.,* labels) will constitute a further useful clarification for many biomedical investigators. Perhaps, having read this account, readers will not be so likely to generate publications saying that a "rhodamine/Bodipy/fluorescein" dye was used – and imagine they are providing useful information. Indeed, there are explicit sections clarifying issues of nomenclature. Then there is extensive consideration given to the nuts and bolts of everyday practice – both regarding the instrumentation and the protocols used. Finally there are the many chapters addressing applications. These address the staining of all common biochemical components and organelles of cells. Again, the chosen examples of applications are broad based. Staining of both biopsy material and of living cells are separately discussed. Consequently, this book is just the place to send those, frequent, enquirers who ask, "Which dye should you use for … ?"

Another marked characteristic of this book, supporting and illustrating the features discussed above, are the numerous, informative and often visually attractive coloured figures. Presumably it is the use of an e-book format which makes such lavish illustration possible.

So do I think anything is missing? Well, there is no significant account of the burgeoning field of clinical diagnostic applications of fluorochromes – to detect or indicate the borders of tumours for instance. Nor of "smart probes" – those imaging compounds which exploit biochemical quirks to obtain selective staining, for instance of particular tumour cell lines. But no book can be all things to all people – and this book is certainly a remarkable fusion of information, explanation and beauty. In fact it's hard to imagine who wouldn't benefit from having this volume available for repeated consultation and, just as valuably, for mind and possibility-broadening browsing.

Richard W. Horobin
Chemical Biology and Medicinal Chemistry, School of Chemistry,
The University of Glasgow,
Glasgow G12 8QQ, Scotland,
UK

PREFACE

The current use of light for so many applications is without precedence in history. If the 20th century would probably be recalled as the "age of the electron", then the 21st century will be, undoubtedly, the "age of light". In the Life Sciences one of these light applications is at the very foundations to obtain knowledge: fluorescence microscopy. Fluorescence microscopy has offered, since its inception during the second half of the 20th century, a definitive enhancement in the signal-to-noise ratio in comparison to bright field microscopy. This increased contrast has done nothing but to grow as the years have passed by, accompanied by an unprecedented increase in resolution too. The improvement in resolving power has reached a level that empowers fluorescence microscopes to image nanoscopic objects with a detail that rivals some electron microscopy techniques. All of these without forgetting that fluorescence microscopy allows for real-time assessment of *live* samples, something at the reach of not so many analytical techniques.

Our goal with this book has been to present a straight, didactical work. We have strived to present all topics in a practical way, with as many real-world examples as possible. Also our aim has been to introduce concepts from an as intuitive as possible perspective, without neither being over simplistic nor eluding the theoretical foundations of the underlying processes. From our experience there has been a lack of a didactical work that presents fluorescence microscopy, as encompassing as the field is, to the broad audience in the Life Sciences. Therefore our intention has been to present a comprehensive yet user-friendly compendium of both basic and advanced topics in the field.

The book has been structured in three parts although it has not been explicitly divided into such. The first part (Chapters 1-7) presents the general concepts and processes to understand fluorescence microscopy, both from theoretical and applied points of view. The second part (Chapters 8-17) focuses on the main application areas of fluorescence microscopy. Our approach for this has been to present the principal biomolecules and organic molecules that make up living organisms and how fluorescence microscopy is employed in every case to obtain useful information. Finally in the third part (Chapters 18-20) we introduce to the reader some of the new application fields of fluorescence microscopy, many of which (*e.g.* luminescent nanoparticles or multiphoton microscopy) have evolved from interdisciplinary overlap with other scientific areas (solid-state physics and non-linear optics).

Fluorescence microscopy is a continuously evolving field, as we can testimony during this book writing process. As a consequence many techniques introduced in the book will transform over time. But our intention has been to make the fundamentals as clear and approachable as possible. Understanding the supporting concepts ensures the comprehension of the techniques, no matter their evolution and improvement. Finally there is, we must admit, an aesthetic side in fluorescence microscopy that makes this technique so appealing. The pure colours and intense contrasts commonly seen in fluorescence images put many of them on pair with art pieces. Truly an image speaks more than a thousand words when talking about fluorescence microscopy. In this sense, we coincide with Francis Bacon when he stated: "*The beholding of the light is itself a more excellent and fairer thing than all the uses of it*".

Juan Carlos Stockert
Department of Biology, Faculty of Sciences,
Autonomous University of Madrid,
Spain

Institute of Oncology Angel H. Roffo, Research Area,
University of Buenos Aires,
Argentina
Institute of Research and Technology in Animal Reproduction,
Faculty of Veterinary Sciences,
University of Buenos Aires,
Argentina

Alfonso Blázquez-Castro
Aarhus Institute of Advanced Studies,
Aarhus University, Aarhus,
Denmark

DEDICATION

To my family, especially Federico and Alejandro (my heroes), and Gloria (my muse).

Juan Carlos Stockert

To my wife, Bego, for her patience and her support during this long road.

Alfonso Blázquez-Castro

ACKNOWLEDGEMENTS

We thank the following colleagues and friends for encouraging support, fructiferous collaboration, and stimulating discussions:

María I. Abasolo, Ulises Acuña, Ümit Akbey, Marco A. Alvarez, José L. Bella, Gabriel Bertolesi, Thomas Breitenbach, Magdalena Cañete, Paul J. CaraDonna, Elisa Carrasco, Adriana Casas, Lucas L. Colombo, Pedro Del Castillo, Edgardo N. Durantini, Jürgen Elvert, Jesús Espada, Pedro Esponda, Juan M. Ferrer, Sergio Galaz-Leiva, Antonio Gómez, Jorge Herkovits, Richard W. Horobin, Amy M. Iler, Angeles Juarranz, Markus Kempf, John A. Kiernan, Claudia Lanari, José A. Lisanti, Daniel M. Lombardo, Isabel Lüthy, Maria Luiza S. Mello, Roberto Mezzanotte, Arturo Morales, Karen ní Mheallaigh, Santi Nonell, Claus Pelling, Peter R. Ogilby, Carolina Ollé, Cristina Ortega-Villasante, Viviana Rivarola, Antonio Romero, Steven E. Ruzin, Sergio H. Simonetta, Alberto J. Solari, Arnaldo T. Soltermann, Carlos Soñez, Cristina Soñez, Francisco Vicente Pedrós, Benedicto C. Vidal, Angeles Villanueva, and Clara I. Trigoso.

Finally, our apologies to many other people lacking explicit mention but also very interested in this e-book and always supporting our research work.

Our institutional acknowledgements go to the following institutions for their help and support:

JCS - Max-Planck-Institut für Biologie, Tübingen, Germany, Department of Biology, Faculty of Sciences, Autonomous University of Madrid (grant CTQ2013-48767-C3-3-R from the Ministerio de Economía y Competitividad, Spain), Institute of Research and Technology in Animal Reproduction, Faculty of Veterinary Sciences, University of Buenos Aires, and Institute of Environmental Sciences and Health, Buenos Aires, Argentina.

ABC - Aarhus Institute of Advanced Studies (AIAS), Aarhus University, Aarhus, Denmark; Marie Skłodowska-Curie actions - Research Fellowship Programme (AIAS-COFUND programme 609033).

CHAPTER 1

Introduction

Abstract: This book tries to be a guide for understanding fluorescence microscopy, and explaining its chemical and physical principles for workers in biomedical sciences, especially those with limited expertise in chemistry and physics. In contrast to early morphological studies, considerable background of physics and chemistry is at present necessary to make fluorescence microscopy a more fruitful technique. In this book we therefore attempt to simplify and make understandable the basis of fluorescence reactions and their biomedical applications. When possible, mechanistic approaches have been introduced regarding dye affinity and fluorescent selectivity. To make the text more didactic and amenable, cell and tissue pictures, diagrams, graphs and chemical structures are included. Obviously, overlapping of several issues concerning fluorochromes and fluorescence techniques will be found along chapters. As an example, consider binding of a fluorescent ligand to a biomolecule. This can be studied from the point of view of (1) ligand properties and binding, (2) substrate structure and affinity, (3) reaction methodology, mechanisms, instrumentation, *etc.* In turn, ligands can be simple molecules (fluorochromes, vital probes), macromolecules (phycobilins, fluorescent proteins), or multi-molecular complexes (labeled IgG, lectins, oligonucleotides), in which different fluorescent labels are used for visualizing a great variety of biological substrates.

Keywords: Amphoteric dyes, Dark-field illumination, Fluorescent labels, Hematoxylin, Hydrophobic dyes, Indigo, Ionic dyes, *Lignum Nephriticum*, Mordant dyes, Nile blue, Orcein, Photoactive dyes, Polarization microscopy, Reactive dyes, Solvent dyes.

1. FLUORESCENCE METHODOLOGY: GENERAL ASPECTS

At present, there is almost no biomedical field in which fluorescence methods are not applied. Research professionals have been increasingly interested in this methodology, and now new concepts, instruments and techniques have developed rapidly leading to constant progress in the design and application of fluorescence methods. The literature of fluorescence methodology is so extensive that only the most relevant references are included in each chapter. Likewise, no special emphasis is given to precise chemical description of biological substrates (*i.e.* lipids, polysaccharides, nucleic acids, proteins, *etc.*), because these are explained in detail in textbooks of biochemistry and cell biology.

Juan Carlos Stockert & Alfonso Blázquez-Castro

On the other hand, the number of fluorescent images is overwhelming [1], and it must be recognized that even in an E-book, the number of pictures must be limited. It is unfortunate that many beautiful and interesting colored images remained out of this work. Many of them are, however, available at www.probes.com, www.lifetechnologies.com/bioprobes), and other commercial sources.

Classic books should be consulted for more detailed accounts in the fields of light, color and chemistry of dyes [2 - 9], microscopical and histochemical staining [10 - 17], and fluorescence methods [18 - 29]. In addition, exhaustive descriptions on specific properties and biological uses of dyes, fluorochromes, and fluorescent probes are available [1, 30 - 34]. Relevant critical reviews, and historical articles on the same subjects should also be consulted [35 - 45].

Here the terms, fluorescent dye and fluorochrome apply indistinctly to any fluorescent compound used for staining fixed cells or labeling live cells. The terms, luminophores and fluorophores are also used, meaning a chromophore (part of a molecule) that absorbs UV or visible light and emits light of a longer wavelength (*i.e.* fluorescence). Fluorescent probes are vital fluorochromes that localize selectively in some cell or tissue structures. Sometimes they are referred to as "fluoroprobes". A fluorescent label indicates a fluorochrome that is covalently or otherwise strongly bound to a biomolecule. The generic term "staining" is often applied to denote the use of a fluorochrome in fluorescence microscopy, *i.e.* "acridine orange staining".

Some misleading concepts should be taken into account and corrected regarding fluorescence microscopy. Often it is claimed that to observe fluorescence, ultraviolet (UV) excitation is required. This is true for only some fluorescent compounds, because others need blue or green excitation. Fluorochromes are sometimes abbreviated as "fluors". "Immunofluorescence microscope" is a bad name, because there is no specific microscope to be used for this technique. Description of excitation filters by their applications instead wavelength (*i.e.* "Texas red excitation/emission filters", "DAPI channel", or "DAPI cube") is another ambiguous and imprecise practice. High brightness is often taken as a synonym of strong affinity for a given substrate, but the emission intensity is only based on the fluorochrome chemistry and binding mode. Therefore, some prejudices and myths in the fluorescence methodology should be corrected.

2. COLOR AND ENERGY SPECTRUM

Color is the visual impression produced by light emitted by a luminous source or reflected from a material. Sometimes, a color is often subjectively described by comparing it with the color of a familiar object. A perceived color is largely dependent on the type of illumination and light intensity (*e.g.* brown is not a spectral color but a very dark orange). Clearly defined assessment of colored objects is possible through objective measurements and chromaticity diagrams.

Cyan, yellow and magenta are the three primary colors. All the remaining solid (absorption) colors can be obtained by mixing the primary ones. Techniques for color restitution in trichromy are shown in Fig. (**1.1**). Therefore, colors derived from light absorption or emission correspond to different physical processes. When a given color (wavelength) is absorbed by a material illuminated with white light, then the complementary color is observed (blue, green, red, respectively). This feature, as well as the distribution of radiation energy over a large scale, is illustrated in the next chapter (see Fig. **2.2**).

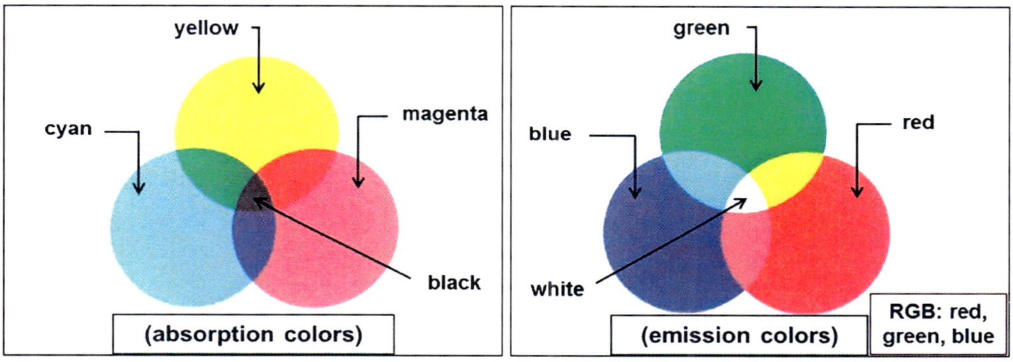

Fig. (1.1). Diagram of primary colors (absorption and emission), and trichromic RGB synthesis.

Depending on the energy (wavelength) of the light that is absorbed by a compound, several types of interactions occur between photons, electrons, atoms or the whole molecule, and can have different effects (*i.e.* heating, luminescence). According to the wavelength of photons their interaction with molecules results in electronic excitation (UV-visible), atomic vibration and rotation (near infrared, NIR), or movement of whole molecules (far infrared, FIR) (see Chapter 2.4). Luminescence consists in light emission from the excited matter. Photoluminescence is caused by light absorption that induces fluorescence and phosphorescence. When a chemical reaction takes place, the energy it generates can be released as light (chemoluminescence). In the case of enzymatic catalysis of chemiluminescent reactions, the process is known as bioluminescence. In contrast to the color of the world that surrounds us (due to the differential absorption of white light), luminescence color (*e.g.* fluorescence) has the true color (wavelength) of the emitted light.

3. BRIEF HISTORY OF STAINING AND FLUORESCENCE METHODS

First uses of dyes for general or histochemical staining reactions are listed in Fig. (**1.2**). It is only an approximate survey, because of doubts about the priority of some discoveries. As Lillie [31] validly stated: *One can never be certain how long any particular individual will remain credited with being the "first" to do something*. Unfortunately, current fashion is to quote only the most recent works, leaving the original sources in oblivion. Therefore, historical and cultural aspects of science and technology are often ignored in recent publications.

First uses of dyes for microscopical and histochemical staining

1714 Leeuwenhoek: Saffron (muscle fibers)

1807 Link: Iron sulfate (tannin)

1825 Raspail: Iodine (starch)

1849 Göppert and Cohn: Carmine (nuclei)

1865 Böhmer: Hematoxylin (hemalum) (nuclei)

1867 Perls: Ferrocyanide (iron)

1875 Ranvier: Quinoline blue (lipids, myelin)

1889 Gram: Crystal violet + iodine (bacteria)

1896 Daddi: Sudan III (lipids)

1924 Feulgen and Rossenbeck: Schiff's reagent (DNA)

1946 Hotchkiss and McMannus: PAS reaction (polysaccharides)

Fig. (1.2). Brief history of microscopical and histochemical staining (see [31, 34]).

3.1. Natural Dyes

The natural compound indigo (indigo blue, vat blue 1, CI 73000), known and applied as textile dye for 5000 years, is produced by fermentation of the glycoside "indican" from leaves of *Indigofera tinctoria*. Dimerization of the resulting indoxyl produces the insoluble blue dye, indigo. After alkaline reduction, indigo becomes soluble and colorless (indigo white, leuco-indigo). It is then applied to the fabric, and the blue color reappears by air oxidation. Now, the current commercial dye is synthetic. In the textile industry indigo is used mainly to dye blue jeans and other denim products.

Interestingly, although in the free form indigo undergoes photodegradation (fading), the dye is protected in some environments. In addition to faded indigo, Mayan artwork from the 6th century found in the Yucatan peninsula also contained the dye (produced from the leaves of the *añil* plant, *Indigofera suffruticosa*) included within channels of the clay palygorskite. This inclusion compound (Maya blue) conserved a bright blue color because of high chemical- and photo-stability of indigo when protected within the clay lattice [46, 47].

Tyrian purple (6,6'-dibromoindigo, the most precious indigoid dye, made in pre-Roman times by the Phoenicians), is obtained from the marine snails *Bolinus (Murex) brandaris, Hexaplex trunculus, Purpura pansa, Nucella lapillus,etc.* (Muricidae, Neogastropoda: Mollusca). The dye was greatly prized in antiquity because the color became brighter with weathering and sunlight. On account of their high cost, purple-dyed textiles became status symbols (*i.e.* royalty, prelates). Exploitation of Tyrian purple is known from as early as the 13th century B.C., and has attracted the interest of historians, artists, dyers, archeologists, chemists, biologists, and pharmacologists [48, 49]. Tyrian purple derivatives (as indirubin compounds) show important antiproliferative activity [50]. Histochemical use of indigogenic substrates is described in Chapter 14.

It is interesting to mention that murexide (ammonium purpurate) is a deep purple dye known since late the 18th century, which increased the fascination for the purple ("the color of power"). Its name (from *Murex* mollusks) evokes the Tyrian purple of the ancients used for the dresses, and was first prepared by heating uric acid with nitric acid, and then treating the dried solid with ammonia. Murexide has been used for textile dyeing, as product from the reaction to reveal uric acid, and for Ca^{2+} detection in tissues [10].

The first microscopic staining reaction using the natural dye saffron (crocin, CI 75100; from stigma of *Crocus sativus;* see Fig. **3.7**) is attributed to Leeuwenhoek in 1714 [51]. Other natural dyes were also applied to stain different cell and tissue components, notably cochineal (carmine, CI 75470) and hematoxylin (CI 75290). The latter compound is extracted from one of the trees with red wood [52], found by Spanish explorers in Yucatan, Mexico, in 1502. The generic name (*Hematoxylon; Haematoxylum* according to Linnaeus, 1753) derives from "blood-colored wood". Historical aspects of hematoxylin and its oxidized product were described [34, 53] (Fig. **1.3**) (see also Chapter 3.4.3).

A vigorous trade soon developed related to preparing hematoxylin for use in dyeing fabrics in Europe. In the mid 1800s, amateur microscopists applied hematoxylin to stain nuclei. Today, the aluminum complex of hematoxylin remains the most popular nuclear stain in histology and histopathology. It is interesting to note that the name Brazil comes from the term "brasa" (live coal) due to the occurrence in that country of a tree with red wood (brazilwood, *Caesalpinia echinata*), from which the hematoxylin-related dye brazilin (CI

75280) was obtained. This dye also occurs in other leguminous plants from Sumatra (*C. sappau*) and Central America (*C. brasiliensis, C. crista*).

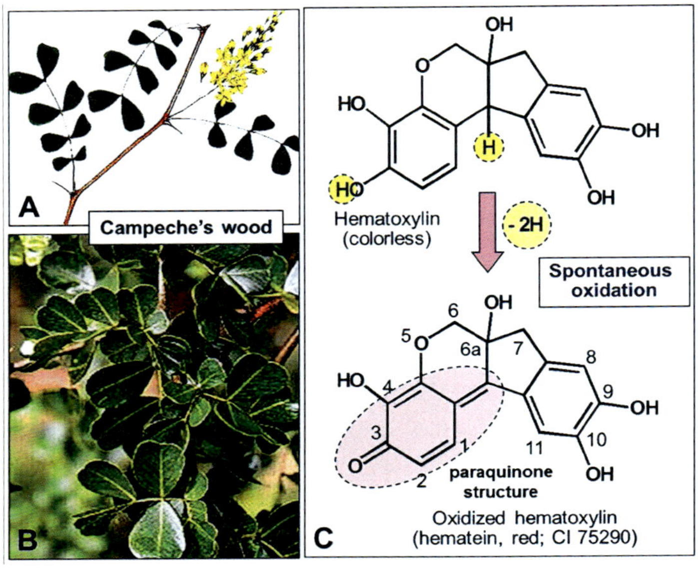

Fig. (1.3). Hematoxylin. A, B: The logwood *Haematoxylon* (*Caesalpinia*) *campechianum,* Leguminosae (Campeche's, wood). C: Oxidation process yielding the dye product, oxidized hematoxylin (hematein) showing the numbering of atoms.

3.2. Synthetic Dyes

The first synthetic dye, mauveine (mauve, aniline purple, Perkin´s violet, CI 50245), was obtained by the eighteen-year-old English chemist W.H. Perkin in 1856 using potassium dichromate and sulfuric acid on crude aniline. Its industrial application was very successful for dyeing silk and cotton fibers with higher brightness and fastness than natural dyes then in use. Mauveine is commonly mentioned in books on color chemistry on account of its great historical interest [54], but at present no application as a textile or histological dye is known. After examination, a sample of the original dye (λab: 548 nm) showed an exceptional high tinctorial value [30]. It is structurally related to safranine O (CI 50240) and thus red fluorescence would be expected to occur for mauveine.

Before Perkin's synthesis other dyes were also produced (reviewed by Lillie [31]). As early as about 1300, French purple (orcein, orseille; likely the first synthetic dye!) was synthesized by air oxidation of orcinol from the orchil *Rocella tinctoria* and other lichens in the presence of ammonia (from fermented urine). Indigocarmine was obtained in 1740 by treating indigo with sulfuric acid. Another artificial dye was murexide. Picric acid was prepared by Woulfe in 1771 [30] and produced commercially one hundred years later. Rosolic acid was synthesized by Runge in 1834, and produced commercially thirty years later. In contrast, the time interval between mauveine synthesis (1856) and industrial production (1857) by Perkin was only one year. Almost all dyes are now derived from raw materials– mainly benzene, toluene, naphthalene, anthracene – obtained from coal tar or petroleum. These aromatic hydrocarbons provide the molecular framework for the final structure of a dye, which is a substance that can confer color.

3.3. Fluorescent Compounds

On the other hand, fluorescence methods also have an interesting history (Fig. **1.4**) (for details see [42, 45]).

Historical perspective of the fluorescence techniques	
1565	**Monardes: First observation of fluorescence (*Lignum Nephriticum*)**
1833	**Brewster: Chlorophyll fluorescence**
1845	**Herschel: First emission spectrum (quinine)**
1852	**Stokes: Name and concept of fluorescence**
1867	**Göppelsröder: First fluorescence-based analysis (morin-Al^{3+})**
1871	**Baeyer: Fluorescein synthesis**
1877	**Baeyer: Link between Danube and Rhine (fluorescein)**
1941	**Richards and Miller: Use of auramine in microbiology (tuberculosis bacilli)**
1941	**Coons: Immunofluorescence (isocyanates: -N=C=O)**
1949	**Strugger: Use of acridine orange in fluorescence microscopy**
1955	**Bowman: First commercial fluorometer**
1955	**Minsky: Principles of the confocal scanning microscopy**
1958	**Riggs: Immunofluorescence (isothiocyanates: -N=C=S)**
1960	**Schawlow and Townes: LASER (Light Amplification by Stimulated Emission of Radiation)**
1967	**Ploem: Epifluorescence microscopy**

Fig. (1.4). Brief history of fluorescence methodology.

In 1565 the Spanish botanist and physician Nicolás Bautista Monardes (1508-1588) noted that a water extract of a Mexican wood exhibited an odd blue shimmer. Kircher (1601-1680) also referred to the blue tinge of water contained in a cup made from this Mexican wood. The medicinal wood, known in Europe in that time as *Lignum Nephriticum* from the Mexican tree *Eysenhardtia polystachya* (Leguminosae) had been used to treat kidney and bladder diseases; Aztec healers had already noticed the "blue" shine of infusions of slices of this wood.

Recent studies have revealed that at least one flavonoid derivative of this tree (the glycosyl dihydrochalcone, "coatline B") is rapidly converted by oxidation into the water-soluble fluorophore "matlaline", with an intense blue emission (λem: 465 nm) only at alkaline pH, with fluorescence quantum yield Φ_F: 1 (Fig. **1.5**) [55]. According to Boyle's (1627-1691) description, this pH dependence (blue shine or not at all at alkaline or acid pH, respectively) is possibly the first observation on a fluorescent pH indicator [42]. The same fluorophore seems to form also from infusions of similar trees such as *E. officinalis* and the Philippine *Pterocarpus indicus*.

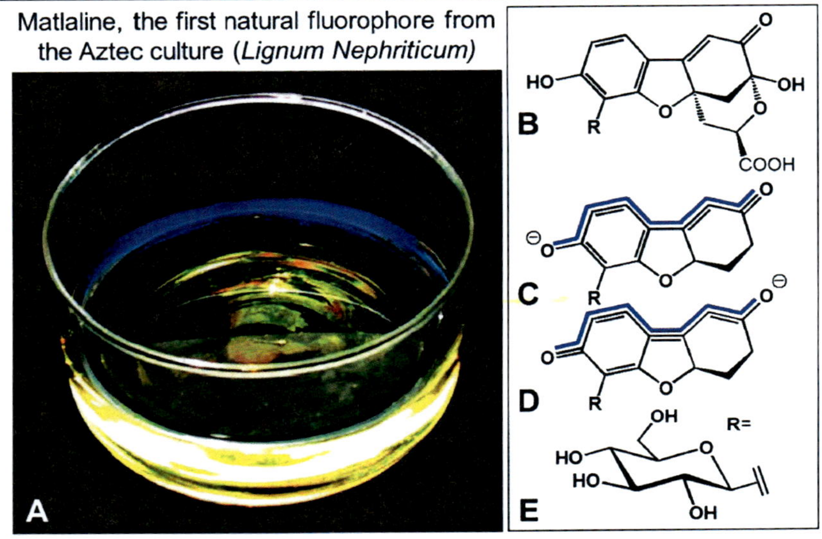

Fig. (1.5). **A**: Blue fluorescence of an aqueous infusion of *Lignum Nephriticum* under daylight illumination (reprinted and adapted with permission from [45], Copyright 2011, American Chemical Society). **B**: Chemical structure of matlaline in its acid form. **C**, **D**: Resonant forms of the ionized fluorophore at alkaline pH, showing the conjugated double bonds (blue) and charges, similar to oxonols, coumarins and xanthenes (see Figs. **3.6** and **3.18**). **E**: Chemical structure of the glycosyl residue at R.

Almost three hundred years later, the red emission of green leaf extracts (chlorophyll) and the blue emission of quinine were described by Brewster and Herschel, respectively. George G. Stokes and Adolf von Baeyer greatly contributed to establish the importance of fluorescence methods. Their potential was already understood in 1877, when Baeyer demonstrated a link between the headwaters of Danube (which flows into the Black Sea) and Rhine (which flows into the North Sea) [45, 56]. He suggested that 10 liters of concentrated uranin solution (disodium salt of fluorescein) be thrown into the Danube (near Immendingen). Fifty hours later, the characteristic green fluorescence of the dye was found in the river Aache (12 km to the south), then in Lake Constanz and finally in the Rhine. At present, fluorometric methods are used for tracking water pollution in rivers and lakes [57].

In contrast, some developments in fluorescence microscopy have only a short history. Immunofluorescence and epifluorescence, for example, are about 75 and 50 years old, respectively [58, 59], and now rapid developments occur in the case of new methods such as super-resolution fluorescence microscopy (see Chapter 20).

4. TYPES OF DYES AND FLUOROCHROMES: NOMENCLATURE

For practical purposes, the terms dyes, colorants and stains will be used here as synonyms and in a wide sense. Synthetic dyes were first used in the textile industry, and can be classified according to methods of application to fabrics. Today, dye classification is mainly based in the chemical structure of chromophores, although names and characteristics derived from industrial dyeing also appear in descriptions of dye properties. Dyes and fluorochromes can be classified according several criteria (Fig. **1.6**).

Types of dyes

(the classification is also valid for fluorochromes, as many dyes are fluorescent compounds)

SOURCE:

-- Natural
-- Synthetic

CONSTITUTION:

-- Organic compounds (conjugated double bonds)
-- Inorganic compounds (mixed valence)

CHARGE:

-- Anionic dyes (acid)
-- Cationic dyes (basic)
-- Neutral dyes (uncharged, hydrophobic)
-- Amphoteric dyes

BINDING MECHANISMS:

-- Ionic dyes
-- Hydrogen bonding dyes
-- Hydrophobic (lipophilic) dyes
-- Reactive dyes (forming covalent bonds)
-- Chromogenic and fluorogenic reagents

Fig. (1.6). General classification of dyes according to different criteria.

Anionic and cationic dyes are defined by the occurrence of negative or positive groups in the molecule, amphoteric dyes having both groups. Non-ionic, hydrophobic dyes have no charged groups. **Solvent dyes** are colored, nonpolar compounds that dissolve in lipids. **Reactive dyes** remain bound to the substrates by covalent linkages. **Direct dyes** are anionic or amphoteric compounds originally designed for staining cellulosic materials (see Chapter 11.2). **Mordant dyes** combines with metal ions to form metallic complexes (see Chapter 8.2). **Vat dyes** generally correspond to anthraquinone or indigoid dyes that are applied as leuco-compounds, and color is regenerated by oxidation. In general, names of dyes have initial capitals only for words that are proper nouns, as in Congo red, Nile blue, Texas red, *etc.*

On the other hand, trivial names or acronyms are often used to refer to more complex chemical denominations of dyes or well known staining procedures. Thus it is practice to employ the current name of H&E for the staining sequence aluminum ions (alum)-oxidized hematoxylin followed by eosin Y, the staining method most commonly used in histology and histopathology. Other examples are Bodipy (dipyrromethene-boron difluoride), DAPI (4',6-diamidino-2-phenyl indole), $DiOC_1(3)$ (3,3'-dimethyl-oxacarbocyanine), *etc.*

Regarding the chemical composition of natural and synthetic dyes, they belong to a great variety of chemical groups such as anthraquinone, azine, azo, di- and triphenylmethane, indigo, oxazine, phthalocyanine, polyene, thiazine, xanthene, *etc.* Several of these compounds are fluorescent, mainly those containing chemical groups such as acridine, aryloxazole, benzimidazole, benzoxazole, benzothiazole, bimane, carbocyanine, coumarin, dipyrrole-boron, flavone, indole, naphthalimide, phenanthridine, porphyrin, stilbene. More precise descriptions will be found in Chapters 3 and 4.

Unfortunately, dye nomenclature is variable and ambiguous. Some dyes have many synonyms and they can be easily confused. Therefore, in addition to the usual name, the Colour Index (CI) provides a precise way to identify dyes, providing a unique number for each chemical structure. Unfortunately, many fluorescent compounds are not included in the CI. In cases of contradictions in chemical numbering of chromophore rings, that of the Merck Index is here used [60]. Numerous names and acronyms of fluorochromes are cryptic and difficult to understand (*e.g.*, Alexa, Cy3, Cy5, FM 4-64, PKH 26, PKH 67). Some trivial names are unfortunate (indigocarmine is not red but blue, fluorone black is not black but red) or misleading (true blue, fast blue). Likewise, postfixes are generally unintelligible and therefore they are often omitted. This is a bad practice because postfixes define the type of dye. They often refer to the shade, solubility or other characteristics of the compounds (Y: yellowish; B: bluish; S: soluble).

Although at present numerous fluorescent probes are available, chemical structures are not always revealed by their manufacturers and vendors. Thus Molecular Probes and Ursa BioScience have produced fluorophores, some of which are their own, with non-revealed chemical structures. This secrecy hinders understanding of their properties and reaction mechanisms with biological substrates. Often fluorochromes are sold in kits with attractive or fantastic names, designed to promote their consumption: examples are AttoPhos®, GelGreen®, NanoOrange® protein reagent, OliGreen® ssDNA reagent, PicoGreen® dsDNA reagent, ProQ Diamond®, *etc.* Alexa fluorochromes have become fashionable and mainly correspond to coumarins, rhodamines and carbocyanines (all sulfonated). Advantages are mentioned for Alexa fluorochromes: high brightness and photostability, good solubility in water, insensitivity to pH changes, and multiple emission colors [1]. An attractive goal for usage is to design a gallery of dyes covering the full spectrum of colors.

5. MICROSCOPIC STAINING AND FLUORESCENCE

Without having differences in the refractive index or absorption characteristics between different cell structures, the need of staining to distinguish them becomes evident. Although staining and fluorescence reactions will be presented in detail in other Chapters, some examples are illustrated here. Ortho- and metachromatic reactions are very common in microscopy, and Fig. (**1.7**) shows representative bright-field images after staining with thiazine and xanthene dyes. In addition, fluorescence reactions induced in plastic semi-thin sections can be observed in Fig. (**1.8**). In the case of acridine orange, the differential orthochromatic (green) and metachromatic (red) fluorescence reactions of specific structures in live and dead (fixed) cells are shown in Fig. (**1.9**) (see Chapter 15.1).

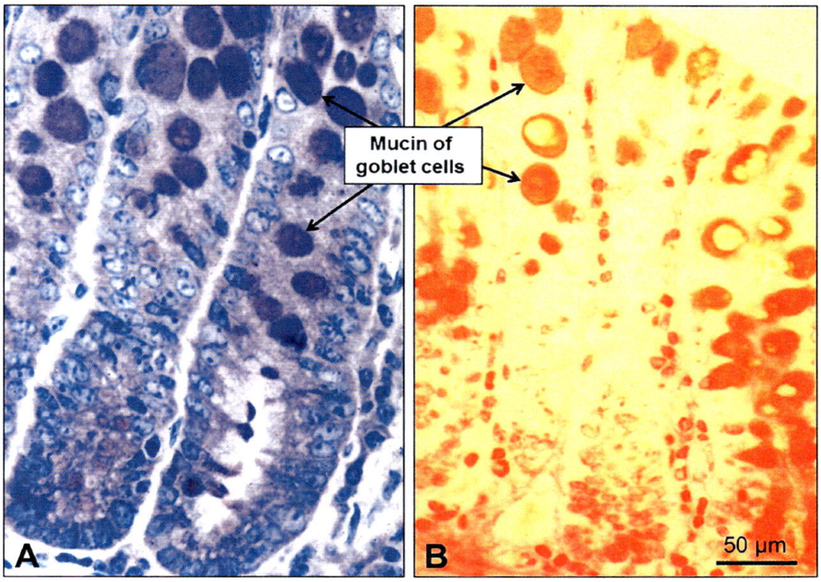

Fig. (1.7). Bright-field metachromatic reactions in Epon semithin sections of mouse large intestine stained by toluidine blue (**A**) and pyronine Y (**B**) (both dyes at 10^{-4} M, pH 5). Observe the violet (**A**) and orange (**B**) metachromatic reactions of mucin. Nuclei appear in orthochromatic blue and pink-red, respectively.

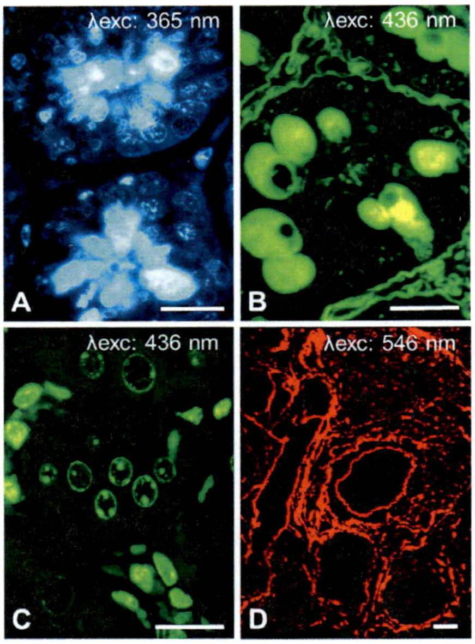

Fig. (1.8). Examples of fluorescence reactions from semithin Epon sections of mouse large intestine (**A**, **B**), kidney (**C**), and uterus (**D**) stained with 10^{-6} M DAPI (**A**), saturated morin in distilled water (**B**), 10^{-4} M berberine (**C**), and 0.2 mg/mL orcein (**D**). In **C** and **D**, dye solutions were in borate buffer pH 9.2. Mucin from goblet cells, chromatin, and reticular and elastic fibers are shown in **A** and **B**, **C**, and **D**, respectively. Scale bars: 30 μm.

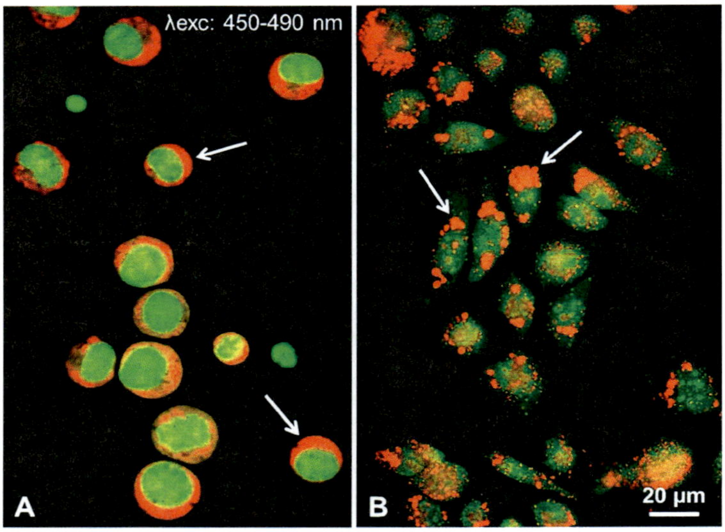

Fig. (1.9). Acridine orange staining of dead and living cells. **A**: Ehrlich ascites tumor cells fixed in methanol, showing metachromasia of cytoplasmic RNA (arrows). **B**: Living Pam212 keratinocytes in culture with metachromatic lysosomes (arrows). Acridine orange was applied as follows: 50 μg/mL, 1 min (**A**), and 5 μg/mL, 15 min (**B**).

It is well known that fluorescence methods offer advantages of enhanced sensitivity and contrast over transmitted light absorption methods. Fluorescence microscopy is no exception: fluorescent cell and tissue structures have higher visibility than those observed by bright-field microscopy and stand out against a black background. But not all bright structures contrasting on a dark background are necessarily fluorescent ones. Thus, this type of image pattern may be observed in dark-field illumination, and linear or circular polarization microscopy (Fig. **1.10**) [61]. In this context, an intriguing artifact was described in the histological technique. If paraffin is not adequately removed from sections, bright birefringent chromatin spots are seen within nuclei under polarization microscopy [62].

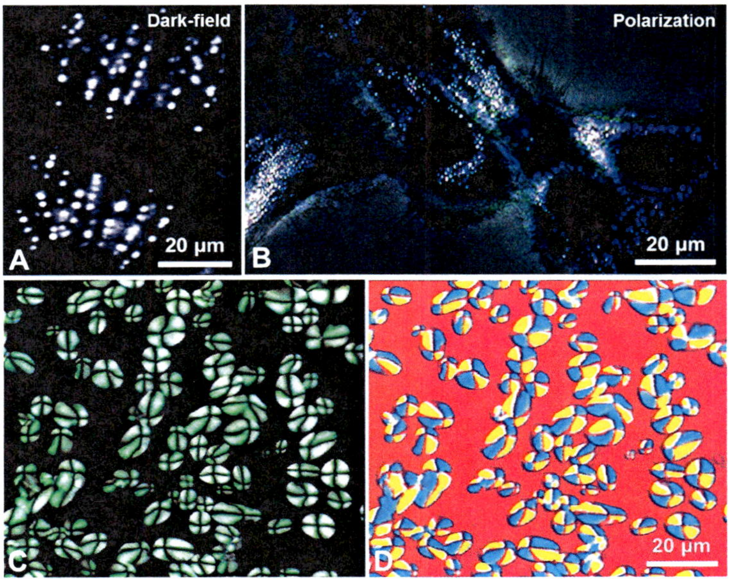

Fig. (1.10). Fluorescence-like images: dark-field and polarization microscopy. **A**: Dark-field image of a plant anaphase after silver staining showing bright prenucleolar bodies. **B**: HeLa cells treated with a sunflower oil emulsion showing birefringent lipid droplets in the cytoplasm. **C**: Starch granules of the arrowroot rhizome (*Maranta arundinacea*) showing typical Maltese crosses, polarization microscopy. **D**: Same image as **C**, but observed trough a λ-plate.

6. ADVANTAGES AND APPLICATIONS OF FLUORESCENCE METHODOLOGY

At present, luminescent molecules have many advantages and are very important research tools in biomedical sciences. Regarding sensitivity, conventional detectors (*e.g.* the eye) record the emitted light with intensity at least three orders of magnitude lower than the reflected light after specific absorption by a colored body. Therefore, the major advantage of fluorescence methods is their greater sensitivity (10^3 to 10^5 times higher than that achieved by light absorption). Very high fluorescence quantum yields exist in some natural fluorophores such as phycobiliproteins and those from the green fluorescent protein family, which show bright fluorescence (see Chapter 4.6).

On account of its high sensitivity, fluorescence is an extremely powerful tool for several analytical applications, including detection of biomolecules, ligand-receptor interactions, enzymatic activity, *etc.* [22, 56]. A survey of fluorescence methods that have important applications in biomedical fields include:

- Microscopical procedures in microbiology, genetics, histology, histochemistry, cell and molecular biology, developmental biology, toxicology, pathology (fluorescent vital probes, viability assays, fluorescent indicators, FISH, immunofluorescence).
- Immunological and image methods for medical diagnosis.
- Analytical (qualitative and quantitative) techniques in chemistry and biochemistry (electrophoresis, chromatography).
- Population analysis of cells (flow cytofluorometry).

A few examples will illustrate these issues. The diagnosis of some infectious diseases is now greatly improved by using fluorescent microscopic methods. The detection of *Mycobacterium tuberculosis* remains a diagnostic challenge in many resource-poor countries. Classical auramine O fluorescence is a useful method for mycobacteria [63, 64] (Fig. **1.11A**). Application of light-emitting diodes (LEDs, see Chapter 6.1) as excitation sources in fluorescence microscopy has resulted in rapid, efficient and low cost methods for mycobacteria screening in clinical specimens [65, 66]. Auramine O also allows easily recognizing kinetoplast DNA in trypanosomes (Fig. **1.11B**).

Likewise, acridine orange is also used for selective detection of mycobacteria, to which it imparts red-orange fluorescence [67, 68]. In addition, acridine orange is currently used for recognizing *Plasmodium falciparum* in human blood smears [69], providing a simpler and more rapid procedure for malaria diagnosis than traditional Giemsa staining. Comparable results can be achieved by means of DAPI staining. Examples of other diagnostic applications are the immunofluorescent detection of the CD-138 antigen in plasma cells from multiple myeloma, and the antigen pp65 from cytomegalovirus (pathogenic in fetuses and immuno-suppressed patients) in neutrophil leucocytes.

Fluorescent methods are also involved in the development of new fields in biomedical research. Examples are emerging areas such as systems biology (genomics, proteomics, lipidomics, glycomics), nanobiology (nanotoxicology, nanomedicine), redox biology, design of biosensors, cellular signaling, diagnostic studies, imaging technology, *etc.* Applications of fluorescence methodology in whole organisms are especially attractive, by using relatively innocuous probes and labels [38]. This is the case for the uptake and visualization of fluorescent photosensitizing dyes in living organisms such as the nematode *Caenorhabditis elegans* [70], one of the simplest and best known animal models (Fig. **1.12**).

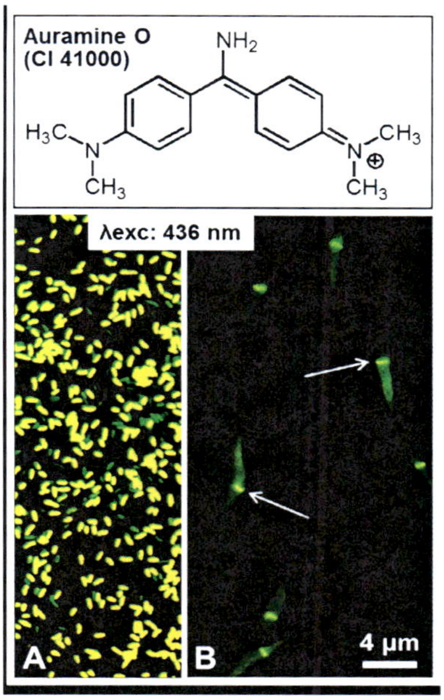

Fig. (1.11). Fluorescence of auramine O: uses in microbiology and parasitology. **A**: Visualization of *Mycobacterium tuberculosis* by auramine O fluorescence. **B**: Fluorescence of kinetoplast DNA (arrows) from *Trypanosoma cruzi* epimastigotes stained with auramine O.

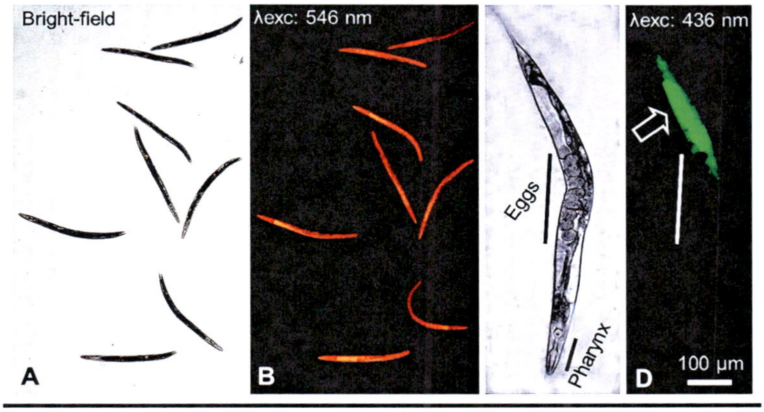

Fig. (1.12). *Caenorhabditis elegans* observed under bright-field (**A**, **C**) and fluorescence microscopy (**B**, **D**). Adult worms (SS104 strain) treated for one hour either with 10 µM rose Bengal (**B**) or 10 µM acridine orange (**D**). High fluorescence of the rostral (**B**) and caudal (**D**) digestive tract is seen (arrow). Pharynx and eggs are indicated (courtesy of J.I. Bianchi and S.H. Simonetta).

Fluorescence methods are very useful in morphological studies on bone development in whole organisms [71], an example being the use of calcein for visualizing skeletal calcification in living zebrafish embryos (Fig. **1.13**).

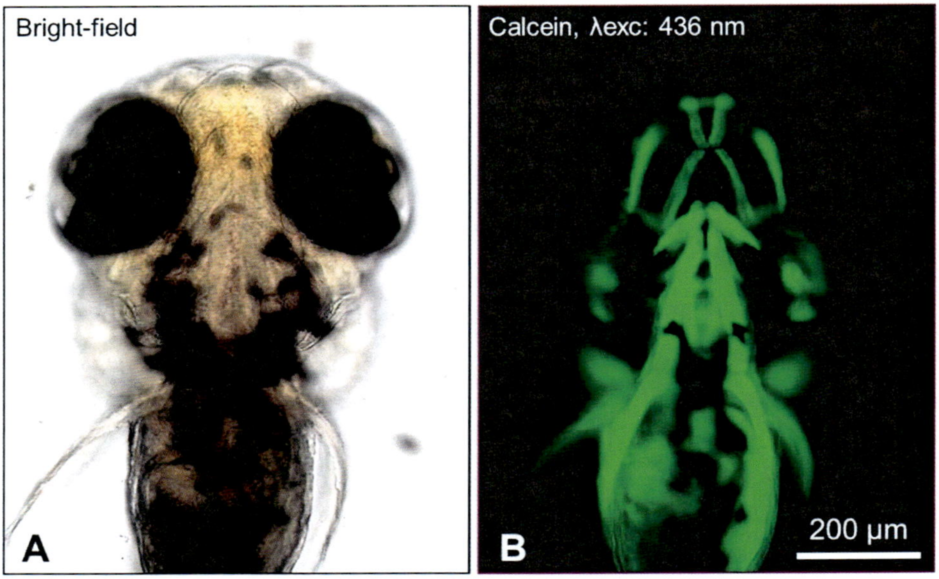

Fig. (1.13). Skeletal development in zebrafish embryo by calcein labeling. Ventral view of a zebrafish (*Danio rerio*) embryo at 5 day postfertilization showing normal bone development after treatment of living animals for 3 min with 0.2% calcein, followed by washing and observation by bright-field (**A**) and fluorescence microscopy (**B**). Calcified structures in the head emit green fluorescence (courtesy of M.A. Alvarez).

Diagnosis of tumors *in vivo* can be aided by clinical tests using fluorescent antineoplastic drugs used *e.g.* in photodynamic therapy. Although the dream of a "magic bullet" carrying a photo-active drug directed to the neoplastic tissue has not yet been achieved [72], fluorescent tumor imaging shows that some suitable drugs are selectively accumulated into neoplastic tissue after intravenous injection (Fig. **1.14**); this is an interesting prospect for diagnosis and therapy of cancer [73, 74].

In addition to applications in microscopic staining, fluorochromes and dyes are employed (1) in industry as pigments and lakes for dyeing paper, textiles, leather, and plastics, (2) as additives to food, cosmetics and cleaning products, (3) in pharmacy for coloring medicaments, and (4) in medicine as diagnostic and therapeutic agents [30, 34, 39, 75, 76]. Treatment of whole organisms with photoactive dyes can induce either useful therapeutic actions [73] or adverse toxic effects [77, 78].

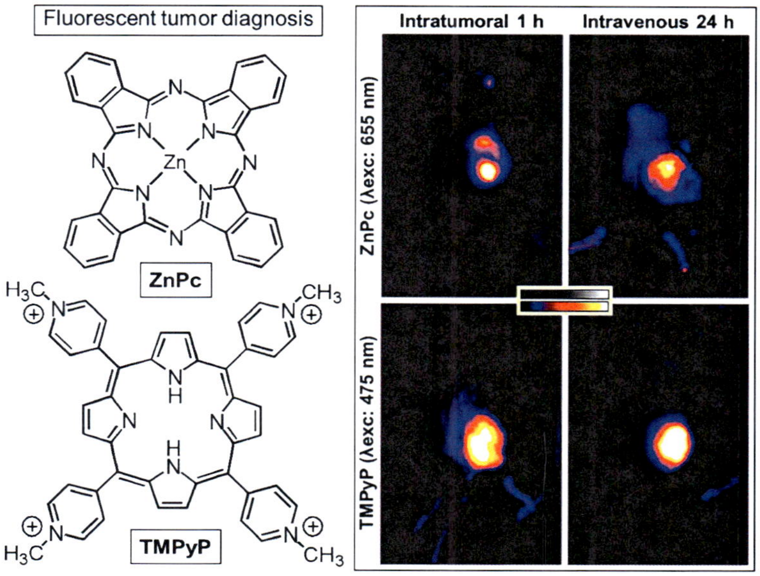

Fig. (1.14). Fluorescent localization of tumors by the photosensitizers, zinc(II)-phthalocyanine (ZnPc) and *meso*-tetra(4-N-methylpyridyl)porphine (TMPyP). C57BL/6 mice bearing a subcutaneous B78H1 amelanotic melanoma were subjected to intratumoral or intravenous injection of ZnPc in DPPC liposomes, and TMPyP in PBS (0.5 mg/kg and 4.1 mg/kg body weight, respectively). Fluorescence was recorded with an Aequoria MDS™ imaging system (Hamamatsu) and converted to pseudocolor images (ImageJ, LUT: fire).

CONFLICT OF INTEREST

The author (editor) declares no conflict of interest, financial or otherwise.

ACKNOWLEDGEMENTS

This work was supported by a grant from the Ministerio de Economía y Competitividad (CTQ2013-48767-C3-3-R), Spain.

REFERENCES

[1] Johnson I, Spence MT. The Molecular Probes Handbook. 11th ed., Grand Island, NY: Molecular Probes/Invitrogen 2010.

[2] Udenfriend S. Fluorescence Assay in Biology and Medicine. New York, London: Academic Press 1962.

[3] Seliger HH, McElroy WD. Light: Physical and Biological Action. New York, London: Academic Press 1965.

[4] Mason SF. Color and the Electronic States of Organic Molecules. Chemistry of the Synthetic Dyes. New York: J. Wiley and Sons 1970; pp. 169-221.

[5] Allen RL. Colour Chemistry. London: Thomas Nelson 1971. [http://dx.doi.org/10.1007/978-1-4615-6663-2]

[6] Gordon PF, Gregory P. Organic Chemistry in Colour. Berlin, Heidelberg: Springer 1987.

[http://dx.doi.org/10.1007/978-3-642-82959-8]

[7] Nassau K. The Fifteen Causes of Color: The Physics and Chemistry of Color. 2nd ed., New York: Wiley-Interscience 2001.

[8] Zollinger H. Color Chemistry: Synthesis, Properties and Applications of Organic Dyes and Pigment. Jhon Wiley & Sons 2003.

[9] Gupta AK, Ed. The Complete Book on Natural Dyes and Pigments National Institute of Industrial Research. Delhi: Asia Pacific Business Press 2005.

[10] Thompson SW. Selected Histochemical and Histopathological Methods 1966. http://www.ccthomas.com/

[11] Suvarna KS, Layton C, Bancroft JD, Eds. Bancroft's Theory and Practice of Histological Techniques. 7th ed., London: Elsevier 2013.

[12] Kasten FH. Methods for Fluorescence Microscopy. In: Clark G, Ed Staining Procedures Used by the Biological Stain Commission. 4th ed. Baltimore: Williams and Wilkins 1981; pp. 39-103.

[13] Horobin RW. Histochemistry An Explanatory Outline of Histochemistry and Biophysical Staining. Stuttgart, London: Gustav Fischer, Butterworths 1982.

[14] Gahan PB. Plant Histochemistry and Cytochemistry An Introduction. London, New York: Academic Press 1984.

[15] Horobin RW. Understanding Histochemistry: Selection, Evaluation and Design of Biological Stains. Chichester: Horwood 1988.

[16] Kiernan JA. Histological and Histochemical Methods: Theory and Practice. 5th ed., Banbury, UK: Scion Publishing 2015.

[17] Carson FL. 1993. https://www.amazon.ca/Histotechnology-Self-Instructional-Text-Carson/dp/0891895817

[18] Thaer AA, Sernetz M, Eds. Fluorescence Techniques in Cell Biology Berlin. Heidelberg: Springer Verlag 1973.
[http://dx.doi.org/10.1007/978-3-642-49204-4]

[19] Guilbault CG. Practical Fluorescence Theory, Methods and Techniques. New York: Marcel Dekker 1973.

[20] Wittekind D. Fluoreszenzmikroskopie.Moderne Untersuchungsmethoden in der Zytologie 2 Auflage. Baden-Baden, Köln, New York: Verlag Gerhard Witzstrock 1979; pp. 36-69.

[21] Beddart GS, West MA, Eds. Fluorescent Probes. London, New York: Academic Press 1981.

[22] Taylor DL, Waggoner AS, Murphy RF, Eds. Applications of Fluorescence in the Biomedical Sciences. New York: Alan R. Liss 1986.

[23] Taylor DL, Salmon ED. Basic fluorescence microscopy. Methods Cell Biol 1989; 29: 207-37.
[http://dx.doi.org/10.1016/S0091-679X(08)60196-X] [PMID: 2643761]

[24] Bach PH, Reynolds CH, Clark JM, Eds. Biotechnology Applications of Microinjection, Microscopic Imaging and Fluorescence. New York, London: Plenum Press 1993.
[http://dx.doi.org/10.1007/978-1-4615-2828-9]

[25] Rost FW. Fluorescence Microscopy. Cambridge: Cambridge University Press 1995.

[26] Herman B. Fluorescence Microscopy. 2nd ed., Oxford, UK: Bios Scientific Publishers 1998.

[27] Rettig W, Strehmel B, Schrader S, Seifert H. Applied Fluorescence in Chemistry, Biology and Medicine. New York, Berlin: Springer 1999.
[http://dx.doi.org/10.1007/978-3-642-59903-3]

[28] Mason WT, Ed. Fluorescent and Luminescent Probes for Biological Activity A Practical Guide to

Technology for Quantitative Real-Time Analysis. 2nd ed., San Diego, CA: Academic Press 1999.

[29] Ghiran IC. Introduction to fluorescence microscopy. Methods Mol Biol 2011; 689: 93-136. [http://dx.doi.org/10.1007/978-1-60761-950-5_7] [PMID: 21153789]

[30] Gurr E. Synthetic Dyes in Biology, Medicine and Chemistry. London, New York: Academic Press 1971.

[31] Lillie RD. Conn's Biological Stains. 9th ed., Baltimore: Williams and Wilkins 1977.

[32] Green FJ. The Sigma-Aldrich Handbook of Stains, Dyes and Indicators. Milwaukee, Wisconsin: Aldrich Chemical Company 1990.

[33] Slavick J. Fluorescent Probes in Cellular and Molecular Biology. London: Times Mirror International Publishers 1994.

[34] Horobin RW, Kiernan JA, Eds. Conn's Biological Stains A Handbook of Dyes, Stains and Fluorochromes for Use in Biology and Medicine. 10th ed., Oxford, UK: BIOS Scientific Publishers 2002.

[35] Kasten FH. The Origins of Modern Fluorescence Microscopy and Fluorescent Probes.Cell Structure and Function by Microspectrofluorometry. New York, San Diego: Academic Press 1989; pp. 3-50. [http://dx.doi.org/10.1016/B978-0-12-417760-4.50008-2]

[36] Christie RM. Fluorescent dyes. Rev Progr Color Relat Top 1993; 23: 1-18. [http://dx.doi.org/10.1111/j.1478-4408.1993.tb00092.x]

[37] Mullins JM. Overview of fluorophores. Methods Mol Biol 1994; 34: 107-16. [PMID: 7711838]

[38] Stockert JC. Reactive Staining Reagents and Fluorescent Labels.Conn's Biological Stains A Handbook of Dyes, Stains and Fluorochromes for Use in Biology and Medicine. 10th ed. Oxford: Bios Scientific Publishers 2002; pp. 77-88.

[39] Sugden JK. Photochemistry of dyes and fluorochromes used in biology and medicine: some physicochemical background and current applications. Biotech Histochem 2004; 79(2): 71-90. [http://dx.doi.org/10.1080/10520290412331292400] [PMID: 15513709]

[40] Dapson RW. Dye-tissue interactions: mechanisms, quantification and bonding parameters for dyes used in biological staining. Biotech Histochem 2005; 80(2): 49-72. [http://dx.doi.org/10.1080/10520290500219982] [PMID: 16195171]

[41] Coleman R. The long-term contribution of dyes and stains to histology and histopathology. Acta Histochem 2006; 108(2): 81-3. [http://dx.doi.org/10.1016/j.acthis.2006.04.001] [PMID: 16730052]

[42] Acuña AU, Amat-Guerri F. Early History of Solution Fluorescence: The *LignumNephriticum* of Nicolás Monardes.Fluorescence of Supermolecules, Polymers and Nanosystems Springer Series in Fluorescence. Berlin: Springer-Verlag 2008; Vol. 4: pp. 1-20.

[43] Hilderbrand SA. Labels and probes for live cell imaging: overview and selection guide. Methods Mol Biol 2010; 591: 17-45. [http://dx.doi.org/10.1007/978-1-60761-404-3_2] [PMID: 19957122]

[44] Stockert JC, Abasolo MI. Inaccurate chemical structures of dyes and fluorochromes found in the literature can be problematic for teaching and research. Biotech Histochem 2011; 86(1): 52-60. [http://dx.doi.org/10.3109/10520295.2010.489428] [PMID: 20500122]

[45] Valeur B, Berberan-Santos MN. A brief history of fluorescence and phosphorescence before the emergence of quantum theory. J Chem Educ 2011; 88: 731-8. [http://dx.doi.org/10.1021/ed100182h]

[46] Sánchez del Río M, Martinetto P, Reyes-Valerio C, *et al.* Synthesis and acid resistance of Maya blue pigment. Archaeometry 2006; 48: 115-30.

[http://dx.doi.org/10.1111/j.1475-4754.2006.00246.x]

[47] Giustetto G, Wahyudi O, Corazzari I, Turci F. Chemical stability and dehydration behavior of a sepiolite/indigo Maya Blue pigment. Appl Clay Sci 2011; 52: 41-50. [http://dx.doi.org/10.1016/j.clay.2011.01.027]

[48] Cooksey CJ. Tyrian purple: 6,6'-dibromoindigo and related dyes. Molecules 2001; 6: 736-69. [http://dx.doi.org/10.3390/60900736]

[49] Westley C, Benkendorff K. Sex-specific Tyrian purple genesis: precursor and pigment distribution in the reproductive system of the marine mollusc, *Dicathais orbita.* J Chem Ecol 2008; 34(1): 44-56. [http://dx.doi.org/10.1007/s10886-007-9402-2] [PMID: 18075711]

[50] Meijer L, Skaltsounis AL, Magiatis P, *et al.* GSK-3-selective inhibitors derived from Tyrian purple indirubins. Chem Biol 2003; 10(12): 1255-66. [http://dx.doi.org/10.1016/j.chembiol.2003.11.010] [PMID: 14700633]

[51] Lewis FT. The introduction of biological stains: Employment of saffron by Vieussens and Leeuwenhoek. Anat Rec 1942; 83: 229-53. [http://dx.doi.org/10.1002/ar.1090830204]

[52] Scarani P. [Trees with blood-colored wood]. Pathologica 2000; 92(4): 298-300. [PMID: 11029893]

[53] Titford M. The long history of hematoxylin. Biotech Histochem 2005; 80(2): 73-8. [http://dx.doi.org/10.1080/10520290500138372] [PMID: 16195172]

[54] Welham RD. The early history of the synthetic dye industry. (2) The industrial history (1856-1900). J Soc Dyers Colour 1963; 79: 146-52. [http://dx.doi.org/10.1111/j.1478-4408.1963.tb02544.x]

[55] Acuña AU, Amat-Guerri F, Morcillo P, Liras M, Rodríguez B. Structure and formation of the fluorescent compound of *Lignum nephriticum.* Org Lett 2009; 11(14): 3020-3. [http://dx.doi.org/10.1021/ol901022g] [PMID: 19586062]

[56] Bright FV. Bioanalytical applications of fluorescence spectroscopy. Anal Chem 1988; 60(18): 1031A-9A. [http://dx.doi.org/10.1021/ac00169a001] [PMID: 3067623]

[57] Madden CJ, Day JW. An instrument system for high-speed mapping of chlorophyll and physico-chemical variables in surface waters. Estuaries 1992; 15: 421-7. [http://dx.doi.org/10.2307/1352789]

[58] Coons AH. Labeling techniques in the diagnosis of viral diseases. Bacteriol Rev 1964; 28: 397-9. [PMID: 14244715]

[59] Ploem JS. The use of a vertical illuminator with interchangeable dichroic mirrors for fluorescence microscopy with incidental light. Z Wiss Mikrosk 1967; 68(3): 129-42. [PMID: 4882659]

[60] O'Neil MJ. The Merck Index 2013.https://www.amazon.ca/Merck-Index-Encyclopedia-Chemic-ls-iologicals/dp/1849736707/ref=sr_

[61] Frohlich MW. Birefringent objects visualized by circular polarization microscopy. Stain Technol 1986; 61(3): 139-43. [http://dx.doi.org/10.3109/10520298609110723] [PMID: 3523835]

[62] Nedzel GA. Intranuclear birefringent inclusions, an artifact occurring in paraffin sections. Q J Microsc Sci 1951; 92: 343-6.

[63] Kumar VA, Chandra PS. Auramine phenol staining of smears for screening acid fast bacilli in clinical specimens. J Commun Dis 2008; 40(1): 47-52. [PMID: 19127669]

[64] Hendry C, Dionne K, Hedgepeth A, Carroll K, Parrish N. Evaluation of a rapid fluorescent staining method for detection of mycobacteria in clinical specimens. J Clin Microbiol 2009; 47(4): 1206-8. [http://dx.doi.org/10.1128/JCM.02097-08] [PMID: 19244460]

[65] Anthony RM, Kolk AH, Kuijper S, Klatser PR. Light emitting diodes for auramine O fluorescence microscopic screening of Mycobacterium tuberculosis. Int J Tuberc Lung Dis 2006; 10(9): 1060-2. [PMID: 16964802]

[66] Hänscheid T. The future looks bright: low-cost fluorescent microscopes for detection of *Mycobacterium tuberculosis* and Coccidiae. Trans R Soc Trop Med Hyg 2008; 102(6): 520-1. [http://dx.doi.org/10.1016/j.trstmh.2008.02.020] [PMID: 18405928]

[67] Katila ML, Mäntyjärvi RA. Acridine orange staining of smears for demonstration of *Mycobacterium tuberculosis*. Eur J Clin Microbiol 1982; 1(6): 351-3. [http://dx.doi.org/10.1007/BF02019933] [PMID: 6186485]

[68] Smithwick RW, Bigbie MR Jr, Ferguson RB, Karlix MA, Wallis CK. Phenolic acridine orange fluorescent stain for mycobacteria. J Clin Microbiol 1995; 33(10): 2763-4. [PMID: 8567921]

[69] Keiser J, Utzinger J, Premji Z, Yamagata Y, Singer BH. Acridine Orange for malaria diagnosis: its diagnostic performance, its promotion and implementation in Tanzania, and the implications for malaria control. Ann Trop Med Parasitol 2002; 96(7): 643-54. [http://dx.doi.org/10.1179/000349802125001834] [PMID: 12537626]

[70] Bianchi JI, Stockert JC, Buzzi LI, Blázquez-Castro A, Simonetta SH. Reliable screening of dye phototoxicity by using a *Caenorhabditis elegans* fast bioassay. PLoS One 2015; 10(6): e0128898. [http://dx.doi.org/10.1371/journal.pone.0128898] [PMID: 26039060]

[71] Du SJ, Frenkel V, Kindschi G, Zohar Y. Visualizing normal and defective bone development in zebrafish embryos using the fluorescent chromophore calcein. Dev Biol 2001; 238(2): 239-46. [http://dx.doi.org/10.1006/dbio.2001.0390] [PMID: 11784007]

[72] Moser JG. Carrier and Delivery Systems.Photodynamic Tumor Therapy. 2nd and 3rd Generation Photosensitizers. Amsterdam: Harwood Academic Publishers 1998; pp. 127-36.

[73] Stockert JC, Juarranz A, Villanueva A, *et al.* Photodynamic therapy: selective uptake of photosensitizing drugs into tumor cells. Curr Top Pharmacol 2004; 4: 185-217.

[74] Acedo P, Stockert JC, Cañete M, Villanueva A. Two combined photosensitizers: a goal for more effective photodynamic therapy of cancer. Cell Death Dis 2014; 5: e1122. [http://dx.doi.org/10.1038/cddis.2014.77] [PMID: 24625981]

[75] Hughes C. The Additives Guide. Chichester: John Wiley & Sons 1987.

[76] Hayashi M, Kishi M, Sofuni T, Ishidate M Jr. Micronucleus tests in mice on 39 food additives and eight miscellaneous chemicals. Food Chem Toxicol 1988; 26(6): 487-500. [http://dx.doi.org/10.1016/0278-6915(88)90001-4] [PMID: 3169648]

[77] Stockert JC, Herkovits J. Photodynamic toxicity and its prevention by antioxidative agents in *Bufo arenarum* embryos. Toxicology 2003; 192(2-3): 211-8. [http://dx.doi.org/10.1016/S0300-483X(03)00334-2] [PMID: 14580787]

[78] Herkovits J, Pérez-Coll CS, Stockert JC, Blázquez A. The screening of photodynamic toxicity of dyes by means of a bioassay using amphibian embryos. Res J Chem Environ 2007; 11: 86-91.

CHAPTER 2

Physical Fundamentals of Luminescence

Abstract: The fundamental physical and chemical processes responsible for luminescence will be introduced and explained in this chapter. The classical division of luminescence into Fluorescence and Phosphorescence will be presented along the similarities and differences between these two processes. Starting with a simple introduction to light and electromagnetic waves and moving to atomic and molecular systems, the text will allow the reader to intuitively understand why physical systems absorb and emit light. This chapter represents the theoretical foundations on which the rest of the book is built upon, the platform from which all other chapters and processes explained therein can be adequately understood and apprehended.

Keywords: Atom, Electromagnetic wave, Electronic excitation, Excited state, Fluorescence, Grotrian diagram, Jablonski diagram, Light, Luminescence, Molecule, Non-radiative de-excitation, Phosphorescence, Photochemistry, Photon, Potential energy curve, Radiative de-excitation, Spin.

The fundamental processes that make luminescence possible are the absorption and emission of photons by some material substrate. Under the most elementary approach luminescence takes place after a photon is absorbed by a substrate which then emits another photon of a lower energy than the photon absorbed on the first place (see Fig. **2.1**).

To begin to understand the process of luminescence in general, and fluorescence in particular, the phenomenon of light itself must be understood. The question of the nature of light is not a trivial one, as History shows us [1]. Although few things are more natural and prosaic than the perception of light through the vision, nonetheless light in itself remains a difficult topic to explain. This is because the interaction between matter and light takes place at a subatomic level, where quantum mechanics rules impose over classical mechanics. We will approach light in this chapter as simple as possible, trying to keep physics to the least. However, some basic physical concepts are necessary to understand the nature and behaviour of light.

Juan Carlos Stockert & Alfonso Blázquez-Castro

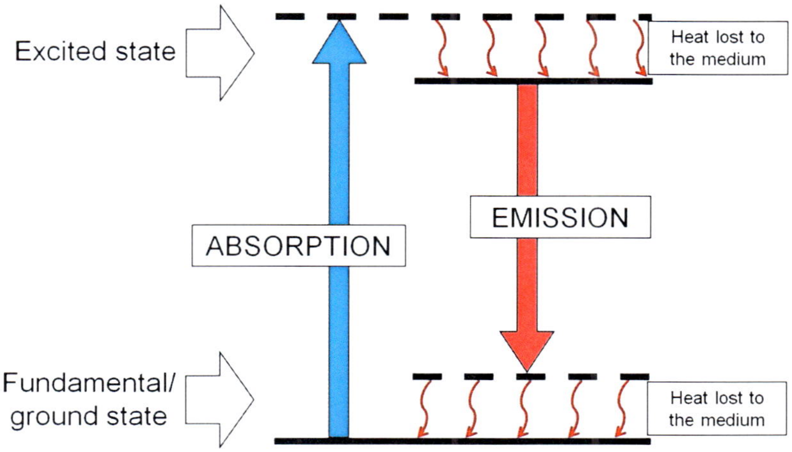

Fig. (2.1). Schematic of the basic absorption and emission of radiation by matter. Matter in the ground non-excited state absorbs a certain amount of energy (upward blue arrow) reaching some metastable excited state. After some amount of time the excited state deactivates emitting luminescence light (downward red arrow). Some energy is lost as heat (wavy arrows) to the medium.

1. LIGHT/ELECTROMAGNETIC WAVES

Light is an oscillation of the electromagnetic field. The electromagnetic force is one of the four fundamental forces (the other three being gravity, strong nuclear force and weak nuclear force) that support all physical interactions in the universe. Thus, the electromagnetic field permeates the entire universe. When energy is applied to the electromagnetic field, it deforms, giving rise to an oscillation. It is these oscillations that we perceive as light waves. These waves displace very quickly, being in fact the fastest entities in the universe. Their velocity in vacuum is 2.9979×10^8 m/s. When these waves move inside matter they reduce their velocity depending on the optical properties of the material they are traversing [2].

The electromagnetic field is itself composed of two aspects: an electrical field and a magnetic field. One goes with the other, because they are two faces of the same phenomenon. The usual depiction of an electromagnetic wave, displaying the orthogonal electric and magnetic vectors, is shown in Fig. (**2.2A**). As can easily be seen, when the electric vector increases so does the magnetic one. Because of Maxwell-Faraday law of induction, a changing electric field engenders a magnetic field and a changing magnetic field engenders an electrical one. In an electromagnetic wave, both fields are interlocked, one sustaining the other without a material supporting medium being necessary [3]. This is why electromagnetic waves can displace in a vacuum.

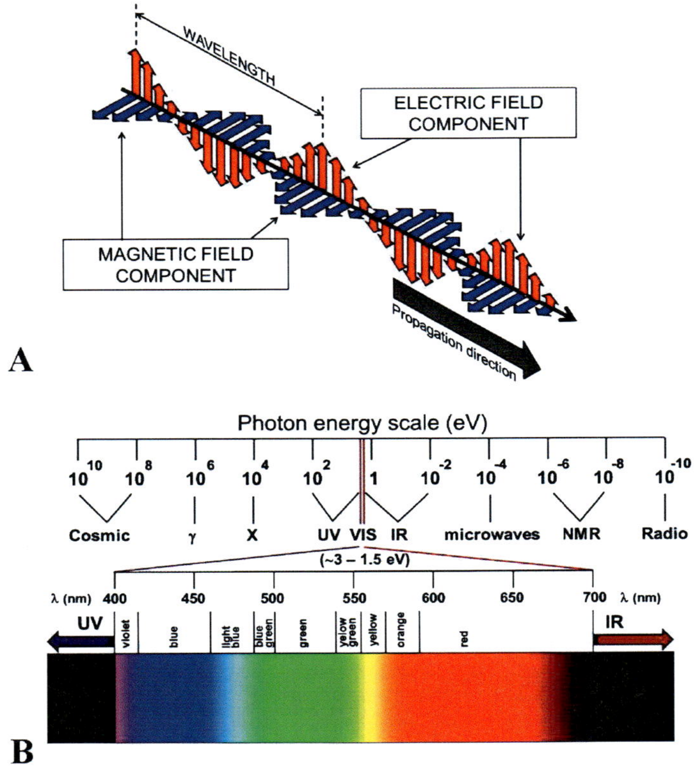

Fig. (2.2). **A**: Electromagnetic waves are composed of an electric field (red arrows) and a magnetic field (blue arrows), that oscillate synchronized. Both fields displace in a direction orthogonal (at 90°) with respect to each plane of oscillation. The distance between to equivalent points in the magnitude of one of the fields defines the wavelength of the wave. **B**: Diagram showing the many orders of magnitude covered by light energy. The energy carried by a photon in each portion of the electromagnetic spectrum is displayed in the logarithmic scale on top. The portion covered by the visible light spectrum is expanded at the bottom.

Electromagnetic waves are transversal waves. This means that the vibrating medium, the medium that stores the energy carried by the wave, oscillates in a direction that is orthogonal to the displacement vector. This is also represented in Fig. (**2.2A**): the electric and magnetic fields oscillations are at right angles to the direction of movement. The behaviour of electromagnetic waves is described by Maxwell equations, a set of four partial differential equations that account for their properties. It is not the purpose of this chapter to describe Maxwell equations, but the interested reader can expand his knowledge with deeper physical books [2, 4]. The number of times the electromagnetic field oscillates per second provides the frequency of the field. As waves, there exists a definite relationship between the speed at which such waves displace in a medium (or vacuum), their frequency and their wavelength. This relationship reads:

$$\lambda \cdot \nu = c \quad \textbf{(1)}$$

where λ is the wavelength, ν is the wave frequency (Hz) and c is the speed of light in the medium. The wavelength is the amount of distance it takes for the wave to oscillate one time. It is not the "size" of the wave.

Paying attention to the plane in which the electrical (and magnetic) field is oscillating will help us understand light polarization. In Fig. (**2.2A**) the electric field is oscillating in the vertical plane. We will not be dealing with the magnetic component, but remember that it always oscillates at right angles –orthogonal- to the electric field. If all the waves making up one light beam have their electric fields arranged in the same plane it is said that the light is polarized. If, on the other hand, each wave displays its own oscillation plane, the light is unpolarised and the average electric field will be pointing in random directions at different times. Polarized light has some uses in spectroscopy and fluorescence microscopy in particular (see Chapter 12.3).

The energy carried by an electromagnetic wave is directly related to its frequency. Said in another way, the more times per second the wave oscillates, the more energetic it is. The energy carried by an electromagnetic wave can be calculated through this equation:

$$E = h\nu \quad \textbf{(2)}$$

where E is the energy in Joules (J), *h* is Planck´s constant ($h = 6.63 \times 10^{-34}$ J · s) and ν is the oscillation frequency (Hz). However, the energy carried by an electromagnetic wave is very small when expressed in Joules. For this reason it is more usual to provide the energy, especially in the optical range of interest for fluorescence microscopy, in electron-volts (eV). One eV is equivalent to 1.602×10^{-19} J.

It is customary to represent the electromagnetic spectrum with respect to the frequency of the oscillation (see Fig. **2.2B**). As showed in the figure, the lowest energy is represented by the very low frequency fields that are usually employed to transmit radio signals. As we displace to higher and higher frequencies, the behaviour of light changes and becomes more and more energetic. In the end, very high frequency light waves (x-rays and gamma rays) carry so much energy that they are able to break chemical bonds, ionize matter and penetrate deeply into all objects.

Electromagnetic waves show wave-like nature at all frequencies, from radiofrequency to x-rays. However, the wave behaviour is better observed when the frequency is low (*i.e.* up to and including the optical range of frequencies). This does not mean that x-rays show no wave behaviour, but at these energies electromagnetic waves tend to display a more corpuscular behaviour. This is where the photon concept comes into play. To describe light interaction with matter above the terahertz range (10^{12} Hz) it is more feasible to think of light as pseudo-material corpuscles or particles, designed photons. But remember that the energy carried by a photon is always linked to a certain oscillating frequency of the electromagnetic wave [5, 6].

Most of the optical properties of light in regard to fluorescence microscopy can be better explained and developed by treating light as composed of photons instead of working from a wave point of view [1, 7]. Throughout the rest of this book the general trend will be to refer to light as composed of photons and only allude to its wave nature where strictly necessary to explain a concept or technique. Sometimes a misconception arises when the wavelength of a photon is interpreted as its size or volume. This is incorrect. The wavelength is the distance it takes for the associated electromagnetic wave to oscillate once. It is a wise strategy to remember always that light is a wave, and that to consider it composed of photon particles is a way to better interpret and understand its behaviour.

In Fig. (**2.2B**) the energy carried by a photon at several frequency ranges has been plotted. Less energetic photons correspond to the radiofrequency and microwave regions in the electromagnetic spectrum. Photons in the optical range, encompassing infrared (IR), visible and ultraviolet (UV) regions, show moderate energies per photon. These photons are unable to totally knock electrons from atoms and molecules (ionization) but they are capable of promoting them from the ground state to bounded higher energy levels. The properties of atoms and molecules usually change drastically when their electrons occupy an excited electronic state. After a certain time the electron decays to a lower excited state or the ground state. Several de-excitation pathways may be open for the electron to lose energy, one of them being the emission of a photon. We will explore these de-excitation mechanisms in following sections of this chapter. However, the most important de-excitation phenomenon from the point of view of this book is photon emission.

2. LUMINESCENCE

Luminescence is the emission of light, *i.e.* photons, by matter [8 - 10]. It has been traditionally assumed that this emission is in the visible part of the spectrum. Currently, with a whole range of photosensors available, this assumption can be relaxed to include the IR and UV too [11]. A key concept to understand luminescence is that for matter to emit light it must first get excited (energized) in some way. Luminescence is an energetic process so energy must be provided somehow in the first place [12]. Only after matter is excited it is able to de-excite by emitting a photon (see Fig. **2.1**).

The process shown in Fig. (**2.1**) is the simplest depiction of the absorption/luminescence phenomenon. However, in real life few luminescent processes are so simple. To study and display the energy changes that take place during such processes some graphical simplifications of the real steps have been implemented. Perhaps the most successful of these depictions are the Grotrian diagram for atomic transitions and the Jablonski diagram for molecular transitions [10].

2.1. Grotrian Diagram

This diagram was introduced by astrophysicist Walter Grotrian in 1928. In a Grotrian diagram the atomic energy levels are depicted as horizontal lines arranged in energy on a vertical scale. So, the fundamental (lowest energy) energy level is the line on the lower part of the diagram and more energetic levels are arranged above this fundamental according to their energies. An example of a Grotrian diagram for the hydrogen atom is shown in Fig. (**2.3**).

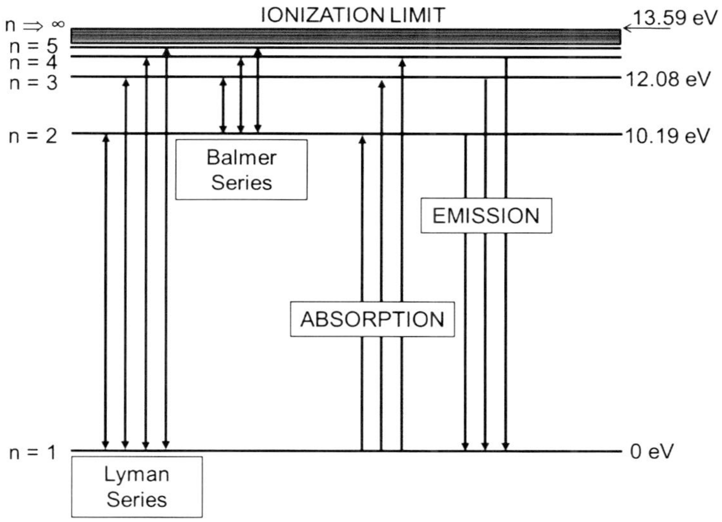

Fig. (2.3). Simplified atomic Hydrogen Grotrian diagram. The principal quantum numbers (n) are shown to the left and the energy (eV) to the right. The lowest level (n=1) is the ground electronic state. Photonic transitions from and to this level define the Lyman series in atomic spectroscopy. Transitions starting/finishing in the n=2 level are grouped as the Balmer series. Each electronic level has a sharply defined energy value above the ground state. Excited states tend to stack closer as they get higher in energy, finally converging at the atomic ionization level (13.59 eV for H).

The vertical axis always translates into energy. The horizontal axis shows the different electronic orbital configurations allowed for each transition (see section 3 below). An upward arrow (excitation) means that a photon is being absorbed to pump an electron to a higher orbit. Higher orbits are more energetic. A downward arrow denotes a de-excitation transition with emission of a photon, in other words a luminescent event. Other process, like collisional excitation or de-excitation, can also be included in a Grotrian diagram just by adopting some code that clearly states which kind of transition is taking place in each place (*e.g.*, straight arrows for photonic transitions and wavy arrows for collisional ones). Further details concerning the Grotrian diagram will be explored in section 3 (see also [2 - 4]).

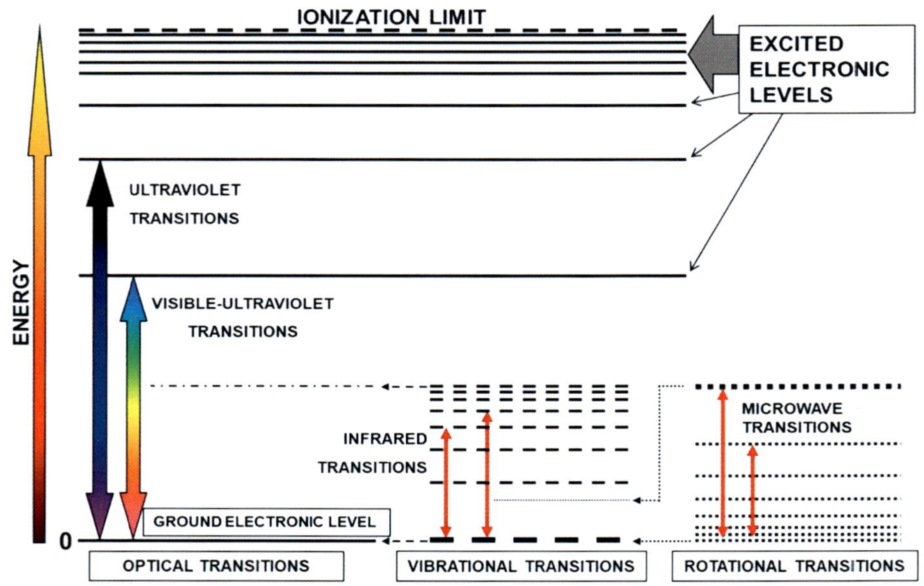

Fig. (2.4). Jablonski diagram schematic. Molecules can undergo optical (UV and visible), vibrational (IR) and rotational (microwave) transitions in order of decreasing energy per photon. Optical transitions connect electronic energy levels. Each electronic state (including the ground state) has a manifold of vibrational and rotational states. This rovibrational stacking makes molecular optical transitions appear as bands in contrast to atomic lines.

2.2. Jablonski Diagram

The Jablonski diagram was proposed by Prof. Alexander Jablonski in 1933. Its underlying fundamentals and actual depiction is very similar to that of the Grotrian diagram. However, it applies to molecular transitions [4, 13]. A simplified Jablonski diagram is shown in Fig. (**2.4**). As can be seen, it is quite similar in arrangement to a Grotrian diagram. Again, energy is represented in the vertical axis. The horizontal axis represents different molecular states. Straight arrows show absorption (upward) or emission (downward) of a photon by the molecule. As it happened with the Grotrian diagram other processes can be depicted by using other symbols, like wavy arrows usually showing non-radiative transitions. The scale of vibrational and rotational transitions/energies has been plotted as compared to optical transitions. The Jablonski diagram is a very useful tool to represent energy transitions in molecular spectroscopy and more details will be explained in section 4.

3. ATOMIC SPECTROSCOPY

The processes that make possible fluorescence microscopy most usually depend on molecular transitions. Nevertheless, it seems adequate to provide a short introduction to atomic spectroscopy first, for two reasons: 1) to introduce the phenomenon of light-matter interaction in a simpler system, that would make understanding of molecular luminescence easier, and 2) because some very important current applications of fluorescence microscopy depend on atomic transitions (as, for example, in QDots and lanthanide microscopy; see Chapter 18). This introduction to atomic spectroscopy will be simple and straightforward. Further specialized reading can be found in the bibliography [2, 4, 12, 13].

3.1. Electronic Orbitals

It is common knowledge that atoms are composed of an inner nucleus and orbiting electrons. The nucleus is composed of protons (which bear a positive electrical charge) and neutrons (which are electrically neutral). Both protons and neutrons are referred as nucleons and their masses are almost equal. The positive charge of the nucleus is compensated by an equal negative charge around the nucleus. However, while the nuclear charge is concentrated in the very small volume of the nucleus, the negative charge is spread out among the electrons that keep as much distance among them as possible. It is the electrical attraction between protons and electrons that keep these orbiting around the nucleus. The amount of protons (and compensating electrons) defines each of the known chemical elements.

The energy transitions giving rise to luminescence depend on the electrons. Although electrons are material particles they are so small and light that quantum mechanical rules dictate their behaviour in most cases. This means that the wave-like component of all material particles is strongly displayed by electrons. Because of this an electron cannot fill any arbitrary orbit around its nucleus. Only certain orbital values are allowed (for a simple but more technical introduction to this issue see [4]). The distance to the nucleus is directly related to the orbital energy. This is similar to a satellite which stores more potential energy the farther it is from Earth´s surface. But as already said only certain "distances" are allowed for the electron to orbit. Transitions between these orbits or energy levels are the reason of photon absorption/emission.

Taking again a look at Fig. (**2.3**) one can see that there is an orbit below which there are no other orbits. This minimum orbit is the ground or fundamental level. In the case of hydrogen this is the absolute ground level as this element has only one electron bound to a single nuclear proton. All the other elements have two or more protons with their corresponding two or more electrons. There are precise rules that state the order in which every electron fills higher and higher orbits as the elements get heavier and heavier [2, 13]. Again, looking for simplicity, when we talk about the fundamental/ground state of an atom or molecule we will be referring to the highest, less bonded electron. Transitions in the optical range most usually take place from this electron changing its orbit. Electrons moving in deeper orbits (*i.e.* closer to the nucleus) typically produce transitions in the extreme UV or X-ray range.

As already said transferring one electron from one level to another requires a precise amount of energy. It is custom to denote each electronic energy level by the letter n. The ground state is the n=1 state and n=2, 3, 4.... denote higher and higher energy levels. By substituting the value of n in relevant equations the energy value connecting two levels can be find out. Not only energy but other physical quantities (like orbital angular momentum or spin) define each electronic level. These levels are denoted by the letters s, p, d and f. For the sake of simplicity we are not going to enter in detail to explain the features of the electronic levels (see [2, 13] for more details).

3.2. Atomic Hydrogen

Now the utility of the Grotrian diagram appears clear for it allows the depiction of many energy levels and the energies involved (recall that knowing the transition energy one can obtain the wavelength and vice versa). In the case of atomic hydrogen transitions from the ground state to the excited states (1s → 2p, 3p, 4p, *etc.*) the wavelengths involved are in the far UV region. Not all transitions are allowed. Note, for example, that starting from the 1s level all transitions end into the p manifold but not in the s or d. This is because not only a specific energy must be provided in the transition, also there are restrictions regarding orbital angular momentum, orbital degeneracy and spin. So some transitions are said to be allowed and other are forbidden by the quantum mechanical rules.

As mentioned before the transitions from and to the ground state (n=1) of atomic hydrogen occur in the far UV. All these transitions are known as the Lyman series. Transitions that start or finish in the first excited state (n=2) are known as the Balmer series and mainly involve photons in the visible part of the spectrum.

Looking at Fig. (**2.3**) it can be noticed that energy levels tend to stack closer and closer the farther the electron is from the nucleus. In the end they coalesce into a final energy level above which the electron is no longer bonded to that particular atomic nucleus. This is known as ionization and gives rise to a free electron and an atom with excess positive charge (cation). Once an electron is above the ionization limit it can acquire any arbitrary amount of energy. In other words, it is no longer compelled to exchange precise quantities of energy. The energy an electron absorbs once it is free (essentially a continuum) is transformed in kinetic energy and not in potential energy as it was when the electron orbits the nucleus.

3.3. Multielectronic Atoms

Any other atom apart of hydrogen has, by definition, more than one electron. This fact complicates calculations enormously. Each electron "feels" the attraction to the positive charged nuclear core but, at the same time, "feels" the repulsion to all the other electrons in the different orbitals. As all electrons are moving this repulsive field changes from moment to moment. Given this complexity we are going to proceed with the most basic introduction to multielectron atomic spectroscopy.

First, we will concern ourselves only with the outermost electron(s) and ignore the rest of electronic population. This is justified because the energies involved in the transitions by the outer electron are of the order of magnitude of optical photons (≈ eV). On the other hand inner electrons deal with much higher energies in their transitions (tens of eVs to tens of keV). Those energies fall in the X-rays range and are employed in X-ray spectroscopy and analysis. These topics, however, are outside the scope of this book.

Fig. (**2.5**) shows the Grotrian diagram of atomic sodium. The level arrangement is very similar to that of hydrogen. It can be seen that there is a fundamental level (n=3), below of which there are a total of 10 inner electrons that do not interact with optical photons (these electrons are occupying levels n=1 and n=2). As with hydrogen the electron can make only certain allowed transitions to other n levels. For example, the electron can move from the 3s ground state level to any of the p levels. Because of orbital momentum considerations it cannot move directly (*i.e.* by photon absorption/emission) to the s, d or any other orbital from the 3s level. Also the energy levels get closer and closer the higher in energy we move, until they converge to a maximum energy value that defines the threshold for atom ionization. Notice that sodium ionization energy is 5.14 eV while hydrogen is 13.59 eV. Each element has its own ionization threshold although trends are found in the periodic table.

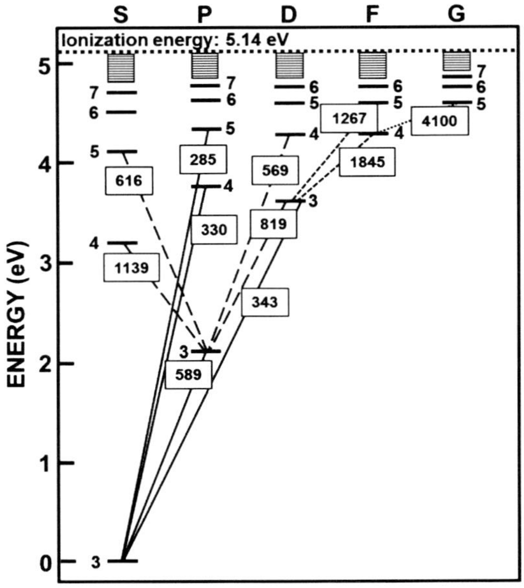

Fig. (2.5). Grotrian diagram of atomic Sodium. The different electronic states (n≥3) and different angular momentum orbitals (S, P, D, F, G) are shown. Some allowed transitions between levels are represented with lines. The wavelength (nm) associated with each transition is shown in a label tagged to the corresponding line. The transition linking the 3S (ground state) level with the 3P (first excited state) level is associated with light of 589 nm wavelength (broadly used in public lighting, for example).

An important difference between atoms and molecules is the modes in which they can store energy. Atoms can store energy basically in two ways: 1) as electronic excitation energy in the different electronic orbitals, and 2) as kinetic energy directly related to the speed at which the atom is moving (we purposefully ignore nuclear energy stored in the atomic nucleus). The kinetic energy can take any value, in other words, is not quantized. The electronic excitation energy, on the other side, is quantized and takes precise values as we have seen in the examples. Molecules display more freedom degrees to store additional energy. This will be our starting point to understand the interaction of molecules with photons.

4. MOLECULAR SPECTROSCOPY

When the material substrate interacting with the light is a molecule, both the absorption and emission spectra turn more complex than when dealing with atoms [13 - 15]. Molecules, as multiatomic entities, are capable of storing energy in more ways than bare atoms. In general terms a molecule might display the following different kind of energies:

- Kinetic energy (shared with atoms)
- Electronic excitation energy (shared with atoms)

- Vibrational energy (unique to molecules)
- Rotational energy (unique to molecules)

Each of these different kinds of molecular energies will be briefly explained in the following. We would like to recommend strongly at this point the classical but very enlightened work in molecular spectroscopy by Prof. Gerhard Herzberg [16 - 18].

4.1. Kinetic Energy

Kinetic energy can take continuous values, from 0 (no displacement at 0 °K) to any arbitrary value. The kinetic energy of a molecule has a minimum effect upon the luminescence behaviour of such a molecule, mainly through the Doppler effect taking place because of the relative movement between the molecule and the incoming/emitted photon. But a molecule has to be moving at really high speeds for the Doppler effect to have some impact on the absorption/emission when considering a fluorescence microscopy setup [13]. Also other subtle effects, like pressure broadening, will not be taken into account in this chapter because they show up under more energetic conditions than expected in an optical microscope [3]. The other three molecular energy reservoirs, electronic, vibrational and rotational, are quantized in their allowed energy values. Together they support the photonic (absorption and luminescence) processes that are at the basis of the most employed fluorescence techniques [10, 12, 15].

To introduce each type of energy just presented, the hydrogen molecule (H_2) will be employed to exemplify how molecules store and release energy. The H_2 is the simplest molecule and it is composed of two hydrogen atoms bonded by the interaction of their shared electrons. This chemical bond where atoms share "equally" their valence electrons is called covalent (see Chapter 3). As explained above for atoms, molecules store kinetic energy when they are moving. The kinetic energy of a molecule, as that of atoms, is a continuous quantity. This means that it can take any value from 0 eV upwards. At the microscopic scale kinetic energy is equal to temperature. However, for the range of temperatures usually found under fluorescence microscopy studies the kinetic energy of atoms and molecules can be neglected in regard to its influence on the luminescence events. An order of magnitude comparison between typical temperatures and energies in the optical range shows that a temperature of 25 °C (298 °K) corresponds to an average kinetic energy of 0.025 eV per atom/molecule. However photon energies in the optical range encompass roughly from 1 to 5 eV. It can be seen that kinetic energy at room temperature is at least two orders of magnitude lower than that carried by optical photons.

4.2. Molecular Rotation

A molecule is, by definition, composed of two or more bonded atoms. This means that molecules present a centre of masses that depends on all the atomic masses and distances between the atoms (see Fig. **2.6A**). This centre of masses is most usually at some point between the atoms. Molecules can spin around this centre of masses at different velocities and in different planes.

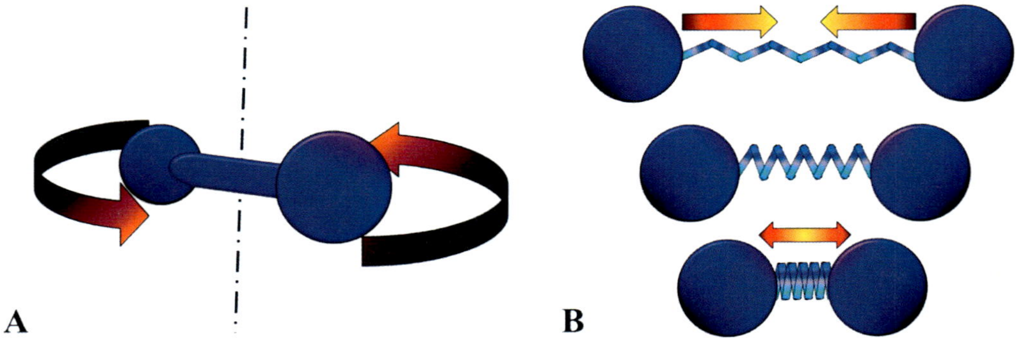

Fig. (2.6). A: Representation of molecular rotation. In the example two atoms are linked by a chemical bond. The bond acts as a rigid bar that defines a symmetry axis around of which the atoms can rotate. The absorption/emission of radiation makes the molecule rotate faster/slower in defined amounts. **B:** Molecular vibration. In the example, two atoms chemically bonded approach or recede periodically because of the spring-like properties of chemical bonds. The stronger the stretching/compression of the bond, the higher the vibrational energy involved. Absorption/emission of infrared photons makes this vibrational energy to increase or decrease, respectively.

The molecular rotational energy is quantized. This means that a molecule can spin only at certain velocities and not others. The energies involved in rotational transitions are of the order of 0.01-0.1 eV (wavelengths ≈ 10-100 μm, see Fig. **2.4**). A molecule that absorbs a photon at those energies will increase its spinning rate while a molecule emitting a photon in this range will spin at a lower rate or even stop rotating altogether. Rotational transitions will be of no more concern to us, as their energies are very low as compared to optical transitions and their influence in fluorescence is negligible. More interested readers can consult [4, 5] for deeper knowledge in rotational spectroscopy.

4.3. Molecular Vibration

Fig. (**2.6B**) represents the vibration mode of molecular hydrogen. It is clear that both atoms have the same mass, so they will approach each other at certain speed and then reverse their movement and recede until they stop and move again towards the other. A useful model is to assume that the atoms are connected by a "spring" (the "spring" being actually the molecular bond). Remember that a molecular bond is "forged" by electrons. When the atoms are far apart the electrons in the bond pull to make atoms closer (in this way the total energy of the system is lowered). However, when the atoms are too close, the bonding electrons start to repulse each other, practically exerting a pushing force that tends to put atoms apart.

Assembling the whole picture the atoms come to rest when they are at the most or the least distance from each other, and move at the highest speed when they are in between these two positions. However, they experience the strongest forces (attracting or repulsing) precisely at the turning points and almost no force at the moment they are moving faster. This behaviour is analogous to that displayed by a classical pendulum. When the pendulum is at rest it occupies the lowest part of the path it would cover if it was be moving. If energy is imparted to the pendulum it will swing sideward, reach a maximum point that depends on how much energy it stores, change its swinging sense and move in the opposite direction. The moving pendulum moves the slowest at the turning points of its trajectory. However, it is at these points that gravity exerts its maximum force on the pendulum. It can be said that acceleration (force) and velocity are dephased in a pendulum.

This example makes it easier to explain the vibrational behaviour of a molecule. For the sake of simplicity we will again deal with the hydrogen molecule, although the conclusions will be applicable to all diatomic molecules and, to a certain extent, to any other molecule. The situation is schematized in Fig. (**2.7A**). Here we represent the interatomic distance at any given moment in the x-axis and the system energy on the y-axis. One very important difference between a classical pendulum and molecular vibrations is that the latter ones always display some degree of movement. In other words, all molecules vibrate even at the lowest temperatures (0° K). This is a consequence of quantum mechanical laws and will not be further discussed [4, 16, 17]. Speaking more properly we can state that the fundamental or ground vibrational state stores some energy, even when it is the ground state.

The second point that deserves stress is that vibrational energy is quantized, as already mentioned. The vibrational modes of any molecule can accept or provide precise amounts of energy. If our H_2 molecule absorbs such a precise quantity in the ground state it will vibrate in a stronger way. The atoms will experience increased forces and expend more time away (to the right of the equilibrium position) and closer (to the left of its equilibrium position) to each other on average as compared with the ground state vibration. This statement is important, as we will see later, to understand the absorption/emission electronic molecular spectra. In the first excited vibrational state the two atoms expend on average more time on the turning points of their trajectories than in the ground state (see Fig. **2.7A**).

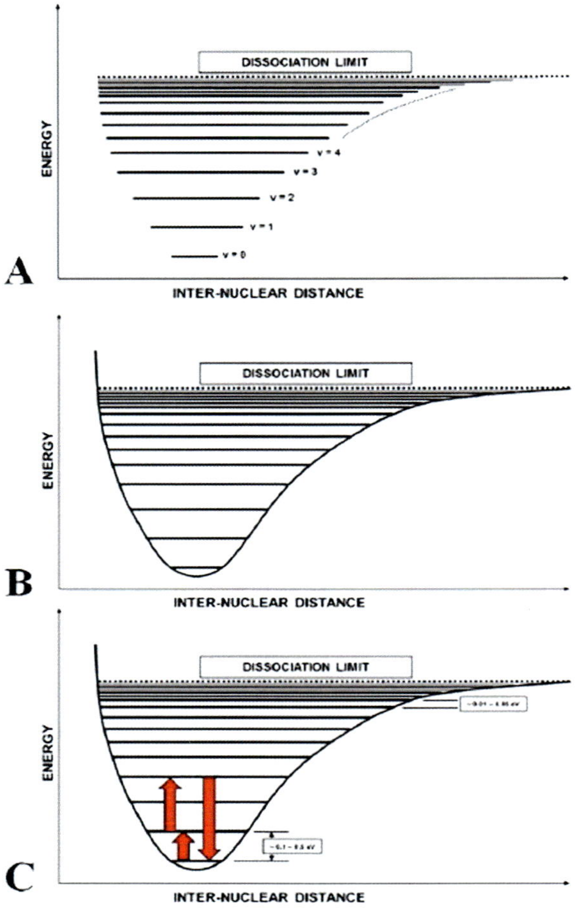

Fig. (2.7). **A**: Ordering of vibrational levels for a diatomic molecule. Plotting the allowed inter-nuclear distances (x-axis) for a given vibrational energy (y-axis) results in a series of vertically-ordered "lines". The limits of each of these lines represent the maximum compression or stretching the involved chemical bond permits for a certain amount of energy. The higher the vibrational energy the more inter-nuclear distance values are allowed. The compressed bond tends to converge to a certain unsurpassable distance (line limits to the left). However, the stretched bond sets no such a limit (line limits to the right) thereby allowing for its breaking after reaching some critical amount of vibrational energy (dissociation energy). Vibrational levels tend to stack one upon another as more energy is stored in the molecular vibration (vertical trend). **B**: Potential energy curve representation. Connecting the endpoints of each vibrational level to the lower and upper ones a curve is obtained. This curve (for a diatomic molecule in the figure) represents the potential energy for a molecule in a certain electronic level. The curve represents the limits for the distance between the atoms in a molecule. **C**: Optical transitions between vibrational levels within the same electronic level. Transitions are favoured between adjacent vibrational levels but jumps involving two or more steps are also allowed (arrows). Typical transition energies in eV are shown for low vibration and high vibration levels.

If the molecule keeps absorbing energy it will vibrate more and more, the distances between the atoms taking more extreme values. This trend could keep on and on, but other forces start at this point to shape the vibrational amplitude. Two things happen: 1) the repulsive force grows at a faster rate than the attractive force, and 2) above a certain energy level (y-axis) the attractive force in fact starts to diminish. The result of the first event is that atoms reach a minimum distance below which they cannot be pushed further together (they, in fact, can but much higher energies are be involved). The second event makes it possible to increase the vibrational amplitude to such a point where the atoms are really far apart and to reach a threshold energy which breaks the bond between the involved atoms (the "spring" is stretched to such a point that finally snaps). This is exemplified in Fig. (**2.7A**) for a prototypical diatomic molecule.

A final note regarding vibrational levels is adequate at this point. It was already said that no further development concerning rotational levels will be included in the chapter. It deserves mentioning, however, that each vibrational level is composed of many rotational levels, as each electronic level is composed of many vibrational levels. This is an important detail that will help the reader understand in a better way molecular absorption and emission spectra as they will be explained below.

4.4. Potential Energy Curves

With all this information we can approach now the issue of potential energy curves. Let´s take a look again at Fig. (**2.7A**). If we trace a curve that joins the positions at which the hydrogen atoms expend most of their time, we get a curve very similar to that draw in Fig. (**2.7B**). This curve provides information about how much potential energy, in the form of vibrational amplitude, a given molecule stores. In this kind of schemes the y-axis always means energy. The x-axis represents the distance between the two atoms that make up the molecule. It can be seen that above a certain energy level the molecule will break. Also, when atoms first form a bond by coming close enough, the just-formed molecule will be in a high vibrational mode. Only by losing part or all of its vibrational energy, can the molecule stabilize. This losing of energy is usually achieved by emitting a photon or by collisions with other atoms/molecules.

In the scheme (Fig. **2.7B**) it can also be appreciated that vibrational energy levels appear as rungs or steps in a ladder. Note that only certain energy levels are allowed. These levels become closer and closer the higher in vibrational energy the molecule goes. In the end, just before the bond is about to break at the dissociation level, these levels form almost a continuum. This is a very similar trend to that shown for the high energy electronic excited levels in atoms (see Fig.

2.3 and **2.5**, just below the ionization threshold). In summary, the more excited an atom or a molecule gets, the less quantized its energy levels are.

Potential energy curves can be drawn also for triatomic molecules. In this case the vertical axis still represents energy and the two horizontal axes represent the distances between the two different bonds that form between each pair of atoms. Sometimes the two bonds are equal, as in the case of H_2O (two O-H bonds) or CO_2 (two C-O bonds) molecules. Potential energy curves for triatomic molecules are more properly designed as potential energy surfaces, because the 3-dimensional plot translates effectively into a surface or sheet that defines, for each energy value, the different positions of the three atoms involved. Nevertheless, they usually are called potential energy curves for simplicity.

Higher order potential energy curves, for molecules with 4 or more atoms, are perfectible possible from the mathematical point of view. However, they cannot be represented because more than 3 dimensions are necessary to plot them. For this reason they are usually named potential energy hypercurves/hypersurfaces. In any case, the same principles apply to them: energy values *vs.* atomic spatial positions. As we will see later, 2-dimensional potential energy curves are also employed as simplified representations of hypercurves for molecules with many atoms. We stress that this is a simplification as the Jablonski diagram is a simplification of electronic energy levels.

Molecules can gain/loss vibrational energy by absorbing or emitting a photon (Fig. **2.7C**).

Transitions take place from one vibrational level to another in the vibrational ladder. The energies corresponding to these transitions are in general around 0.01-0.2 eV for most complex molecules. However, molecular hydrogen has the highest vibrational energy (0.516 eV or $\lambda \approx 2.4$ μm) when vibrational transitions between the v=0 and v=1 levels are involved. Transitions between vibrational levels are more probable between adjacent levels (0→1, 2→3, 2→1, *etc.*). But transitions involving changes in two or more steps at a time are not forbidden, although their probability is orders of magnitude lower than single step jumps [16, 17].

4.5. Molecular Electronic Levels

Now we have explained in a simplified way the rotational and vibrational energy in molecules. Now we introduce the electronic energy levels to complete the picture. In general terms, many concepts presented for atomic spectroscopy apply to molecular spectroscopy [12, 13]. The different electrons involved in the molecular bonds occupy different energy levels according to quantum mechanical laws. One electron will be the most distant from the surrounding atoms, at least for that region of the molecule if the molecule is big. For a basic understanding of electronic transitions we can ignore atomic electrons not involved in molecular bonding and also the inner electrons. In some instances these electrons absorb photons but that is usually in the far UV.

As was the case with atoms, an electron fills a higher energy electronic orbit when it absorbs a photon (see Fig. **2.8**).

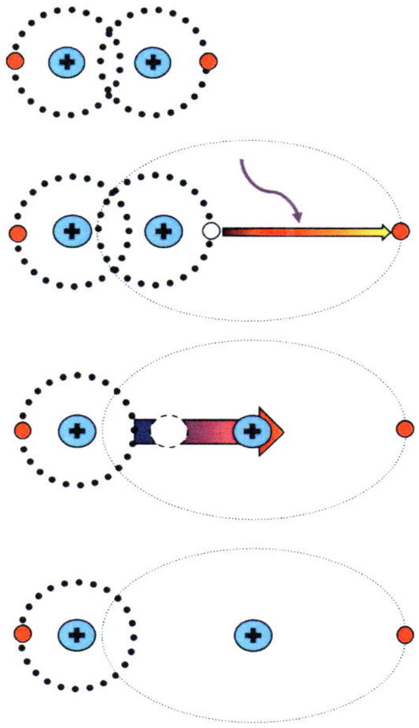

Fig. (2.8). Franck-Condon principle. A Hydrogen molecule (H_2) is represented in the ground state as two protons surrounded by two electrons orbiting the nuclei. Right after photon absorption, one electron transits to the next electronic energy level. This transit is fast enough so that the protons stay in the same positions they occupied before the absorption. After some delay the nuclei repel each other because the excited electron now shields less efficiently the positive nuclear charges. Consequently the protons move apart starting a vibrational swing in the molecule. Later the two protons come to equilibrium for the new electronic arrangement that balances the electrical charges all over the molecule.

This new orbital is farther from the molecule. This has two consequences: 1) the electron leaves behind a hole in the molecular orbital, and 2) as a consequence, the bond which the excited electron was part of is weakened. This can lead to several outcomes, like molecular rearrangement, photodissociation or redox chemistry, some of which will be introduced in section 5 in the final part of this chapter. In fluorescent molecules, the excited electron returns to the ground state before these other pathways take place, thereby restoring the molecule to its original arrangement. After some time, the excited electron will decay emitting a photon. However, both in absorption and emission the vibrational (and, to a certain extent, also the rotational) manifold of the molecule modulates the available transition levels [14].

4.6. Franck-Condon Principle

This modulation of the electronic transitions by the vibrational component is known as the Franck-Condon principle. It is an extension of the so called Born-Oppenheimer approximation [10, 13]. This approximation was employed by physicists Max Born and J. Robert Oppenheimer to try to simplify quantum calculations related to molecular spectroscopy. The approximation states that molecular kinetic, electronic, vibrational and rotational energy components can be treated separately in order to make easier the computation of the total energy stored by a molecule. One of the assumptions made by the Born-Oppenheimer approximation is that electrons, being much lighter than protons and neutrons, move faster than these nuclear particles when changing energy levels. In this simplification electrons move from one electronic level to another in a time frame where atomic nuclei are "frozen" in their places because of their much larger masses and inertia (see Fig. **2.8**). It is adequate to recall at this point that molecular vibrations depend on the distances between the nuclei. In this theoretical thinking frame the Franck-Condon principle states that electronic molecular transitions take place much faster than molecular vibrations can respond to such an electronic rearrangement.

Fig. (**2.9**) presents a simplified scheme of the Franck-Condon principle for potential energy curves to help understand the concept. In the figure two energy curves have been drawn: one (lower) corresponds to the fundamental electronic state of the molecule, the other (upper) to the first excited electronic state. One can appreciate several features in the scheme. Each curve displays its own set of vibrational levels. The vibrational ground state (the $v = 0$ level) in the lower and upper curves has different equilibrium positions. Electronic excited states usually have larger equilibrium positions. This is a consequence of the farther orbit of the excited electron which weakens the bonds between atoms.

Let's assume that we start with a diatomic molecule in the electronic and vibrational ground state ($v = 0$). This means we are on the lower curve and in the lower rung. As the molecule is barely vibrating in level $v = 0$ it is most probable for it to be at the middle of the rung when a photon arrives. At this moment an electron will be excited and the molecule "moves" to the upper curve. The molecule must "arrive" at a stable configuration. This means that the transition must end in one of the rungs of the upper curve. Recall that the molecule spends on average more time in the middle of the first ($v = 0$) rung but more time on the sides of vibrationally excited (v ≥ 1) levels. In our example the molecule will "arrive" to vibrational levels 2, 3 and 4 at the upper curve. This vertical transition from one curve to the other represents the electronic excitation. The atomic nuclei, on the other hand, remain at the same positions. It is clear now that the new potential energy curve dictates a change in the nuclei positions. As a consequence of the electronic transition the molecule that was barely vibrating on the ground state, starts to vibrate strongly in the excited state. In most cases the molecule will lose this additional vibrational energy by emitting IR photons and/or through collisions with third bodies. The result is that after a certain relaxation time, the molecule reaches the vibrational ground state (v' = 0) of the excited electronic state (upper curve).

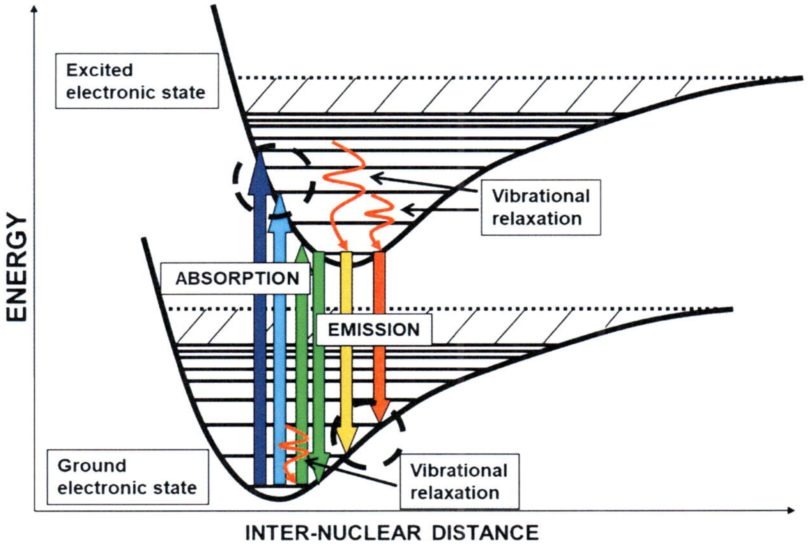

Fig. (2.9). Franck-Condon principle applied to potential energy curves and Stokes shift. Two electronic states, ground (lower) and first excited (upper), are plotted as energy curves in the figure. Because of the curves displacement the molecule is not in vibrational equilibrium after absorbing photons from the ground state (light and dark blue arrows). It takes some time for the molecule to lose part of this vibrational energy (vibrational relaxation) that ends up either as emitted IR radiation or as heat in the molecular environment. When the molecule emits a photon to return to the ground state (orange and red arrows) it again undergoes vibration excitation because the nuclear equilibrium positions in the excited state do not coincide with those in the ground state. Thus another round of vibrational relaxation takes place after photon emission. This chain of events leads to an emission spectrum that is less energetic –Stokes shifted- than the absorption spectrum.

This electronic configuration is only temporally stable (metastable). After some time the molecule will de-excite usually through different pathways. Non-luminescent mechanisms will be shortly reviewed in section 5. Assuming that the only available de-excitation pathway open for the molecule is photon emission, the process takes place in a very similar way as the excitation did. Now the excited electron emits a photon, losing energy and transitioning to the lower curve. It will "arrive" at any of the available points in the ground potential energy curve (see Fig. **2.9**). The nuclei are frozen in their positions while the electron decays. However, once in the ground state the nuclear positions no longer match an equilibrium distance (v = 1 and 2), giving rise to another cycle of vibration. As it happened after the initial excitation the molecule loses this excess of energy by IR photon emission and/or collisions. Here we are presenting the first hint of why molecules emit at longer wavelengths than those at which they absorb. A more complete picture will be provided shortly, but it can be seen that the arrangement of potential energy curves leads to some of the initial excitation energy to almost always been degraded to vibration and, finally, into heat. This is the origin of the so called Stokes shift effect in emission (see below and Chapter 5.1).

4.7. Molecular Multiplicity

Electronic transitions of molecular electrons also follow certain rules: not all empty higher energy levels are allowable for an electron to occupy [4, 10]. To simplify things for the purposes of this book, we are going to divide molecular electronic levels in two classes: singlet and triplet states. This division is a consequence of the molecule´s multiplicity. To understand the concept of multiplicity we must rely on what we know about molecular bonds. As already explained the outer electron in a molecular bond is the one excited by optical range photons. However, there are in fact two electrons involved in all molecular bonds. All electrons carry a spin which has a value of +1/2 (UP) or -1/2 (DOWN). Electronic bond couples usually arrange in such a way that one electron has its spin in the UP direction and the other electron points its spin in the DOWN direction (see Fig. **2.10A**).

Bonds can be formed only when the electrons couple their spins in this way. The multiplicity of a molecule is obtained through equation

$$M = 1 + 2 \cdot (\text{spin } 1 + \text{spin } 2) \qquad \textbf{(3)}$$

where M is the multiplicity, and spin 1 and spin 2 are the spin values of the two bond electrons. If we take a molecule in the ground state the two electrons forming a bond will have spin values +1/2 and (-1/2). Applying equation **(3)** we obtain a net spin sum of 0 = [+1/2 + (-1/2)]. The spin sum times 2 is also 0, plus 1 gives a total M = 1. Because the obtained multiplicity is 1, molecules in their ground state are designed as singlets (radicals and other molecular arrangements can produce M values of 2 and 4 but this is beyond this book´s objectives).

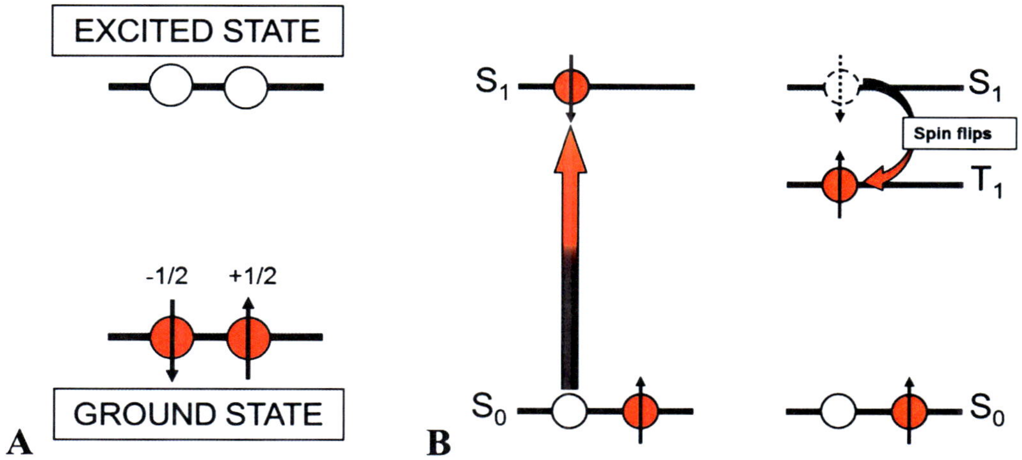

Fig. (2.10). **A**: Spin configuration in the ground state. The usual molecular multiplicity is a singlet state because the two outermost electrons (red circles) arrange their spins (arrows) in opposite directions. One is spinning UP (+1/2) and the other DOWN (-1/2). The first excited state is empty presenting "holes" that can be filled by the electrons when excited to this level (white spots). **B**: Spin configuration in the excited state. **A**: The absorption of a photon does not change the electronic spin. Therefore molecules display a singlet character after being excited. To distinguish between singlet states the ground state is S_0 and excited singlet sates are designed S_1, S_2, S_n, *etc.* The excited electron leaves a hole behind in the ground state orbital (white spot). **B**: Some events, like molecular collisions, can flip the excited electron spin. Under such circumstances both outermost electrons display the same spin (black arrows point in the same direction) and the obtained multiplicity is a triplet. In the scheme it is shown the first excited triplet that is designed as T_1.

When a bond is excited by light one of the two outer electrons fills a higher orbital leaving behind a hole in the bond and the other electron (Fig. **2.10B**). As a general rule photons are unable to flip the electron spin. Said in another way: the electron spin is conserved during the excitation process. This means that the excited state to which a molecule arrives immediately after excitation is also a singlet state. It can be seen in Fig. (**2.10B**) that the spin arrangement remains unchanged after the electronic excitation. Following this line of thought the molecular ground state is designed S_0 and the first singlet excited state as S_1. Upper excited electronic states with a singlet multiplicity are labelled following this order. However, above the S_2 excited levels tend to stack very closely one upon another. Collectively these levels are usually denoted by S_n.

While for the vast majority of molecules in the ground state the obligated option is to assume a singlet state (S_0), molecules in the excited state are not so constrained. It has already been established that a molecule transitions from the S_0 to the S_1 state whenever it absorbs a photon. This was a consequence of a photon´s inability to change or flip an electron´s spin. Once in the S_1 state the excited electron is no longer forced to stay in its current spin state because its orbital electron couple remains in the lower bonding orbital. Frequently (this depends on many variables, like phase, solvent, temperature, *etc.*) a molecule in the S_1 state undergoes a collision with a third body that induces the excited electron to flip its spin. Now both outer electrons, bonding and excited, have the same spin, either UP or DOWN (see Fig. **2.10B**).

This spin change brings additional molecular features. On first instance the multiplicity will change too. If equation **(3)** is applied the sum of spins is 1 (spin sign is a convention, the point is that now both spins are of equal sign so they add together to calculate M). When spins sum 1 the equation results in a value of 3. The excited molecule is now a triplet state. The first triplet state, which must be an excited state as already explained, is denoted by T_1. Higher energy triplets states will accordingly be designed T_2, T_3, *etc.* As it happened with the singlet states those levels above the T_1 or T_2 are usually pooled into the T_n manifold.

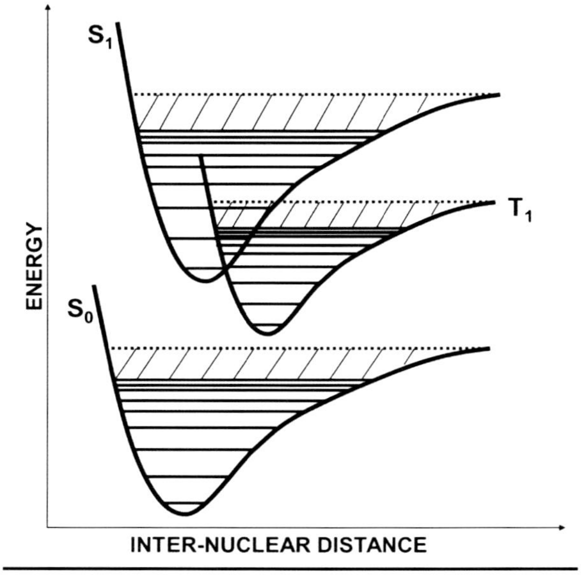

Fig. (2.11). Potential energy curves of different multiplicity. Shown is the usual arrangement for potential energy curves of different multiplicity. The T_1 curve is always localized at lower energy than its corresponding S_1 curve. At the same time the inter-nuclear equilibrium distances are also larger for T_1 than for S_0 and S_1. In many instances there is an overlap between the S_1 and the T_1 curves that allows for a transition to take place between these curves.

A triplet state always has less energy than its corresponding singlet state. Also equilibrium nuclear distances are relaxed which means that the triplet potential energy curve is displaced to the right in the graph of potential energy curves (see Fig. **2.11**). When the molecule "crosses" from the S_1 curve to the T_1 curve it is usually vibrationally excited (recall that electronic changes are too fast for nuclei to follow). After this crossing the molecule starts to vibrate and loses energy to the environment by IR photons or through collisions. In the end the triplet molecule reaches the equilibrium position at the bottom of the corresponding energy curve.

At this moment the molecule, we could say, finds itself in a blocked state. It cannot simply emit a photon to relax to the S_0 ground state because the spin of the excited state is the same as the unexcited state in the bond. And photons cannot change electronic spin when absorbed/emitted. The molecule cannot return to the S_1 state without absorbing energy. So, T_1 molecules find themselves locked in this state. This is the main reason explaining the very long time (in a molecular sense) of triplet molecules as compared to singlet states. The situation is mostly resolved when the T_1 molecule collisions with a third body and the electron changes its spin. Immediately the photonic transition becomes available and the molecule emits a photon returning to the S_0 state in the process. But the excited state lifetime depends now on the collision rate rather than to its own internal de-excitation rate.

4.8. Potential Energy Curves Expanded

With the knowledge of singlet and triplet states we come closer to describe the general picture of molecular absorption and emission bands. Fig. (**2.12A**) shows the greatly simplified two-dimensional potential energy curves scheme for an organic molecule.

Fluorescent organic molecules employed in microscopy usually have between 20 and 100 atoms. So it is important to keep in mind that the presented scheme is indeed an oversimplified representation [10]. Paying attention to Fig. (**2.12A**) it can be appreciated that photon absorption is more probable from the ground vibrational level than from higher energy vibrational levels. Transitions to the S_1 curve are more probable from the central region of the ground vibrational rung (see Franck-Condon principle above). These transitions end on the S_1 curve where they start a vibrational oscillation of the molecule.

During the vibrational relaxation the S_1 and T_1 curves may overlap for certain nuclear distance values. At this crossing point the molecule may transition from S_1 to T_1. This event is designed intersystem crossing (ISC). In our example some excited molecules out of a population will transfer to the T_1 curve (ISC) and some will remain in the S_1 curve.

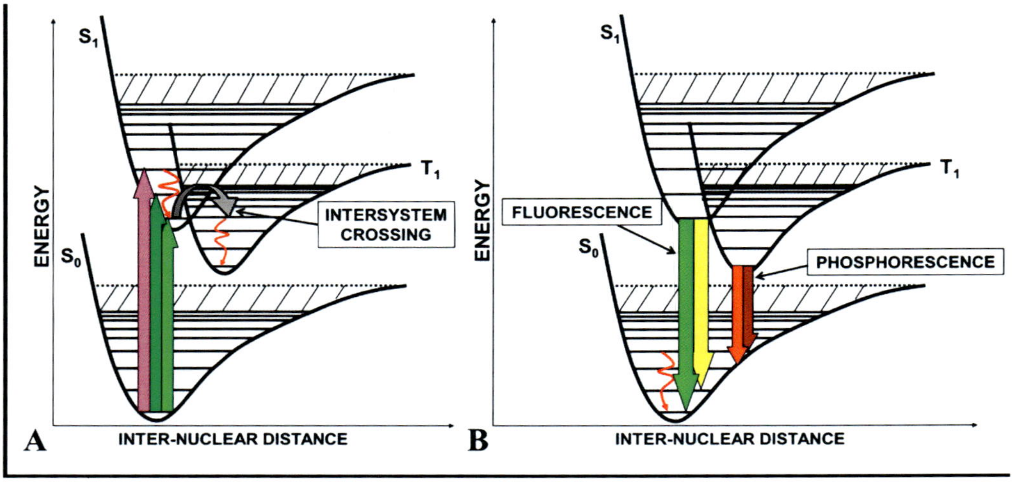

Fig. (2.12). **A**: Intersystem crossing (ISC). After absorbing a photon (coloured arrows) the molecule gets excited to the S_1 state. Vibrational relaxation (red wavy arrow) takes place very quickly after absorption and the molecule reaches its equilibrium position in the S_1 state. Under the right conditions (*e.g.* a collision) the molecule transits to the T_1 potential curve through the overlapping region between curves in the middle of the scheme (grey transition arrow). Afterwards a second round of vibrational relaxation takes place until the molecule approaches the equilibrium inter-nuclear distances (v=0 level) of the T_1 curve. **B**: Fluorescence and Phosphorescence. Fluorescence is the emission of light from an initial (excited) state that presents the *same* multiplicity as the final state (excited or fundamental). In the figure this is represented by photon emission (yellow and green arrows) from the S_1 curve to the S_0. Phosphorescence is the emission of light from an initial (excited) state that presents a *different* multiplicity to that of the final state (excited or fundamental). In the Figure this is represented by photon emission (light and deep red arrows) from the T_1 curve to the S_0.

After some relaxation time molecules in S_1 will reach the bottom of the potential curve. Assuming that non-radiative mechanisms (see below) do not compete efficiently with the photon emission process, a photon will be emitted after some characteristic time (concepts and techniques related to lifetimes and characteristics times will be treated more extensively in Chapter 6). *When a photon is emitted from a level with a certain multiplicity to another level of the same multiplicity ($S \rightarrow S$, $T \rightarrow T$) this kind of luminescence event is denominated* **FLUORESCENCE** (Fig. **2.12B**). In our current example, emission from S_1 to S_0, the light emission process is a fluorescent one. After the fluorescence photon has been emitted, with a Stokes shift due to less energy than the original absorbed photon, the molecule is vibrationally excited a relaxes back to the ground vibrational level of S_0 after a certain time. The energy difference between absorbed photons and emitted photons appears as vibrational energy that is finally degraded into heat in the medium surrounding the molecule.

Taking another look to Fig. (**2.12A**) we recall that some molecules have made a cross-over to the T_1 curve (ISC). When a collision changes the excited electron spin, a photon is emitted, returning the molecule to the S_0 level. *When a photon is emitted from a level with a certain multiplicity to another level of different multiplicity (S → T, T → S) this kind of luminescence event is denominated* **PHOSPHORESCENCE** (Fig. **2.12B**). Phosphorescence features two differential characteristics when compared to fluorescence: 1) the phosphorescent emission lifetime is usually orders of magnitude (~ $10^2 - 10^4$) longer than the fluorescent lifetime; 2) the Stokes shift (see Chapter 5.1) is larger for phosphorescence than for fluorescence. The first feature is due to the spin-flipping event that must take place before the T_1 molecule can emit the photon. The second characteristic is clear from Fig. (**2.12A** and **B**). A phosphorescent molecule loses more energy in the form of vibrational relaxation than a fluorescent molecule. ISC to the T_1 curve degrades part of the excited state energy into the vibrational manifold of the triplet level. Additionally, when the photon is emitted the molecule "arrives" at the S_0 curve at a higher energy point, which means more energy would be released as vibrations. The final result is that phosphorescent photons are almost always displaced to the red (Stokes-shifted) from fluorescent photons.

Molecules can absorb photons from the S_1 or T_1 levels. However, given the very short lifetimes of both levels, the circumstances and techniques that allow such transitions will be treated in more detail in Chapter 19 (Non-linear optics) [19]. At the same time it is conceivable for molecules to emit from levels above the S_1 or T_1. For reasons that will be treated shortly (see section 5), the most important levels related to luminescence (fluorescence and phosphorescence) are the S_1 and T_1 by far. Nevertheless, some examples exist of molecules emitting from the S_2 and various T levels. As these are exotic cases, they will be mentioned in the text as necessity arises to explain some process.

4.9. Molecular Jablonski Diagram

Going from the somewhat more complex to the simpler, we retrace our steps and now expand the concept of the Jablonski diagram with what has been explained about molecular potential energy curves and multiplicity. Towards this end Fig. (**2.13A**) will help in developing the molecular Jablonski diagram. In the diagram the nuclear positions are no longer taken into account. The x-axis has a meaning of spatial displacement but representing positions in a qualitative way only. The y-axis keeps on representing energy. Electronic levels are represented by bold lines. Vibrational levels are drawn as thin lines. The ground electronic state corresponds to energy level 0, so it usually overlaps with the x-axis. Additionally, vibrational ground levels (v = 0) for each electronic level also overlap with the corresponding bold line. Note how electronic levels stack upon each other as energy increases. The same trend is observed for the vibrational levels associated with each electronic level. Although partly incorrect singlet levels are allocated one over another in the vertical axis, while triplet levels are slightly displaced to the right but also allocated vertically.

The advantage of the Jablonski diagram, compensating its inaccuracies, is that it offers a straightforward way of presenting the different radiative and non-radiative mechanisms acting upon a molecule without going into excessive complexity. As can be seen in Fig. (**2.13A**) photonic transitions are represented by straight arrows (upward for absorption and downward for emission). Non-radiative processes are unified under wavy arrows symbols. The two phenomena of fluorescence and phosphorescence (previous ISC) are schematized in Fig. (**2.13A**). Please compare this figure to Figs. (**2.12A** and **B**) in other to appreciate the similarities and differences between both representations.

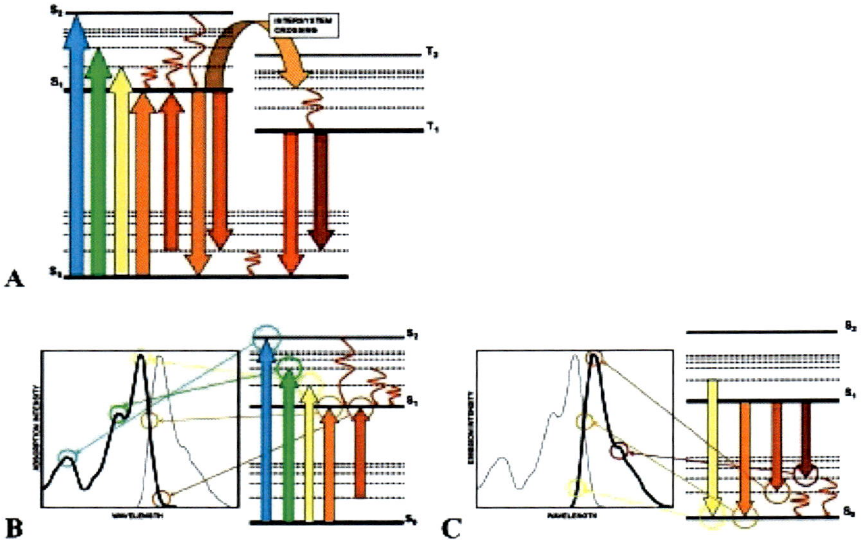

Fig. (2.13). **A**: Molecular Jablonski diagram. In the Jablonski diagram the x-axis no longer represents strictly inter-nuclear distances. Nevertheless molecular states with different multiplicities (singlets, triplets) are depicted as displaced in the diagram. Energy is represented in the vertical axis. Optical transitions are represented as upward/downward straight arrows. Vibrational relaxations appear as wavy downward red arrows and ISC is represented as the bridging orange arrow. Ground vibrational levels for each electronic state are represented as bolded lines and excited vibrational levels as dotted thinner lines. **B**: Molecular absorption band features explained through the Jablonski diagram. A general molecular absorption (thick line) and emission (dotted line) spectra are represented in the figure. Absorption features in the spectrum are linked to their counterparts in a Jablonski diagram. In this way the peak in the absorption spectrum corresponds to transitions involving yellow photons (~570 nm) that excite the molecule from the S_0 (v=0) level to the S_1 (v=1) level. C: Molecular emission band features explained through the Jablonski diagram. A general molecular absorption (dotted line) and emission (thick line) spectra are represented in the figure. Emission features in the spectrum are linked to their counterparts in a Jablonski diagram. The peak in the emission spectrum corresponds to transitions involving red photons (~640 nm) emitted from the S_1 (v=0) level to the S_0 (v=1) level. Therefore the emission peak is displaced to the red (see Fig. 2.20) or Stokes-shifted.

Using a Jablonski diagram the main features of the molecular absorption and emission spectrum will be introduced. Both absorption/emission spectra and Jablonski diagram are presented side by side in Fig. (**2.13B**) (absorption) and Fig. (**2.13C**) (emission). First we will focus on the absorption part of the spectrum and its correlate on the diagram. The absorption spectrum features a continuous curve with peaks at some wavelengths. This means that the molecule absorbs more efficiently at those wavelengths (see Chapter 5.1 for additional details). These peaks correlate with the allowed transitions from the electronic ground state (S_0) to some, usually the first, excited electronic state (S_1, S_2, *etc.*). This is represented in the Jablonski diagram by the upward straight arrows connecting the ground vibrational level (v = 0) of the S_0 state. Typical fluorescent molecules are composed of many atoms, so the amount of vibrational levels is really large. This is the reason for observing a smooth absorption curve and not discrete bands at some wavelengths. Additionally each vibrational level encompasses a manifold of rotational levels, which blurs further the absorption spectrum.

As can be seen in Fig. (**2.13B**) most upward transitions connect the v = 0 level of S_0 with vibrationally excited levels of S_1. At room temperature the most populated vibrational level will be the v = 0. Nevertheless at this temperature some molecules are, at any given moment, in the v = 1 vibrational level of S_0. For this reason a minority of transitions take place from the S_0 v = 1 level to the S_1 level. The energy required in these transitions is somehow lower than that required when starting from the v = 0 level. Thus these transitions are Stokes-shifted (= shifted to the red) with respect to the main transitions starting from v = 0. These Stokes-shifted transitions appear on the right side of the absorption spectrum (see figure).

On the blue side of the absorption spectrum (left side on figure) the photon energy is enough to pump the molecule from the S_0 level to the S_2. As it happened with the S_1 level, transitions to the S_2 level take place preferentially from the v = 0 vibrational level. This means that molecular absorption spectrum show, in fact, transitions to more than one excited electronic level. Of course, this depends on the width of the scanned wavelengths. The same rules apply to transitions to the S_2 as to the S_1 level. It is also true that more often than never the S_2 potential energy curve -usually hypercurve because so many atoms conform the molecule- overlaps with the high-lying S_1 vibrational levels and, as a consequence, the molecule transits very fast ($\sim 10^{-12} - 10^{-11}$ s) from the S_2 to the S_1 hypercurve (see section 5 below).

After internal relaxation the excited molecule population is mainly on the ground vibrational level (v = 0) of the S_1 state, with some molecules in the v = 1 level. The most probable transitions are to the high excited vibrational levels in the S_0

state. These are represented as straight downward arrows in Fig. (**2.13C**). Some energy is lost in vibrations which once again explains why the emitted photons are of less energy (=longer wavelength) than the absorbed photons on average. Some molecules emit from the v = 1 level of S_1 and these molecules register on the left side of the emission spectrum. Molecules finishing in high vibrational levels of the S_0 state register their emission on the right side of the spectrum.

These examples demonstrate the utility of the Jablonski diagram to explain and represent in a simple way the molecular transitions registered in molecular absorption and emission spectra. Nevertheless, we would like to stress once again at this point that the Jablonski diagram is an oversimplification of the real processes taking place when molecules absorb and emit light [5, 10, 13, 14].

5. NON-RADIATIVE DE-EXCITATION MECHANISMS

When a molecule absorbs or emits light some processes that dissipate energy in a radiationless way usually take place in parallel. These are known as non-radiative de-excitation mechanisms and they make up the final part of this chapter. These non-radiative processes occur because of the arrangement of potential energy curves already introduced in previous sections. So it must be understood that radiative and non-radiative processes are interlocked in a molecule´s response to light. We already have presented glimpses of some of these processes, as when a molecule transits from the S_1 to the T_1 energy curve or when it quickly de-excites from the S_2 to the S_1 curve after absorbing a high-energy photon. We will now introduce these mechanisms in more detail.

5.1. Internal Conversion

The first non-radiative mechanism is Internal Conversion (IC). The most general case in molecular spectroscopy is that in which the largest energy difference is between the S_0 and S_1 curves. Singlet curves above S_1 tend to stack one upon the other at closer and closer energy increments. This typical curve arrangement leads to the situation depicted in Fig. (**2.14A**).

It is easily understood that these curves overlap on both energy (y-axis) and spatial coordinates (x-axis). The consequence of these overlappings is that the molecule can transit from one curve to another just by exchanging electronic excitation energy with vibrational energy. In the depiction a molecule at the bottom of the S_2 energy curve (v = 0 vibrational level) transits to a vibrationally excited state (~v = 6) of the S_1 curve. From there the molecule relaxes to the v = 0 level of the S_1 state through collisions and IR photon emission. Finally the molecule "sits" at the bottom of the S_1 curve where no overlapping is .available to other energy curves and practically locks the molecule there until a photon is emitted.

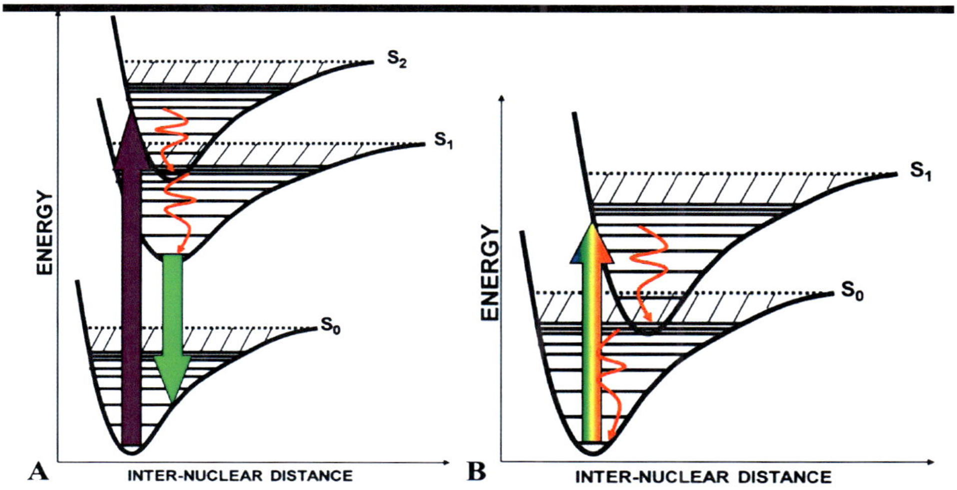

Fig. (2.14). **A**: Molecular internal conversion (IC). The IC process takes place when two potential energy curves overlap in the diagram (exemplify by S_2 and S_1 curves). Optical excitation to the S_2 curve (violet arrow) leads to fast vibrational relaxation to the bottom of the S_2 level. The v=0 level of S_2 overlaps with the high-energy vibrational manifold (v≥6) of S_1. Therefore before the molecule can emit a fluorescence photon from S_2 (v=0) it transits to the S_1 curve. A second vibrational relaxation takes place which de-excites the molecule to the bottom of the S_1 curve. From there the usual molecular fluorescence (green arrow) takes place. **B**: Internal conversion from the S_1 state. Many molecules show a potential energy curves arrangement such that the S_1 curve overlaps with S_0. In these cases optical excitation (blue arrow) is followed by very fast vibrational relaxation and crossing from S_1 to S_0 before the molecule has a chance to undergo photonic de-excitation. Molecules displaying this energy curves arrangement degrade photonic energy totally into heat.

This physical de-excitation mechanism is the reason behind the so called Kasha´s rule which states that an excited molecule emits from the lowest excited electronic state no matter at which wavelength it was initially excited. This is a general rule and it has its exemptions, but it is usually true for most molecules. What we would expect is for a molecule to emit from the S_1 curve no matter if it was initially excited to this level or to a higher energy one. This same principle, IC, applies to a molecule undergoing ISC to the triplet state manifold. If the molecule transits from an S_n level to a T_n level, curve overlapping usually warranties that the molecule ends up in the T_1 level through vibrational relaxation.

An extension of the IC mechanism explains why so many molecules do not fluoresce at all. Fig. (**2.14B**) shows a case where the S_1 energy curve overlaps some vibrational levels of the S_0 curve. It is easily seen that the molecule can transit from S_1 to S_0 without undergoing photon emission. In this case, all the energy initially provided by the exciting photon ends up in the form of vibrational excitation and, finally, appears as heat in the medium. This is the general case for the vast majority of molecules, in special when they are composed of many atoms. So molecules which feature a large enough energy difference between the S_0 and S_1 levels to allow for photon emission to take place are the exemption rather than the rule. This is also why we live in a world where things usually do not emit light when excited but get heated instead. The intersystem crossing process (ISC), already introduced above, is also an extension of IC. In this case one or more of the singlet curves overlap with a triplet curve. If a spin flipping event (*e.g.* a molecular collision) takes place at the right moment, the molecule passes over to the triplet curve releasing some energy as vibrations. After this the molecule relaxes to the bottom of the T_1 energy curve where it stands until it emits a phosphorescence photon or another non-radiative de-excitation process, like for example ISC to the S_0 curve, takes place.

5.2. Photodissociation

After photon absorption, a molecule can reach a potential energy curve which displays an anti-bonding character instead of a bonding one. This situation is shown in Fig. (**2.15A**). Anti-bonding curves do not feature a minimum potential well but an ever-increasing repulsive curve the closer the atoms come to one another. As a consequence, the molecule is unstable and stability is only achieved when the molecule breaks. In the figure the molecule reaches the dissociation curve after absorbing a photon. On first instance the molecule does not break because of the Franck-Condon principle. The curve indicates that the molecule has now some vibrational excitation and starts "moving" to the right of the curve. However, it never reaches another stable position and the atoms keep moving farther and farther apart until the chemical bond is broken. At this point the molecular fragments will be moving with kinetic energy equal to the absorbed photon energy minus the dissociation threshold energy (see Fig. **2.15A**).

In some instances the dissociation curve is not reached directly after photon absorption, but after IC takes place from a stable potential energy curve. This is depicted in Fig. (**2.15B**). The initial absorption event takes the molecule to a stable excited molecular level. After some amount of time the molecule transits to the dissociation curve by IC. At this moment the molecule breaks as a consequence on instability. When photodissociation does not take place immediately after photon absorption but as a consequence of IC it is usually denoted as delayed photodissociation.

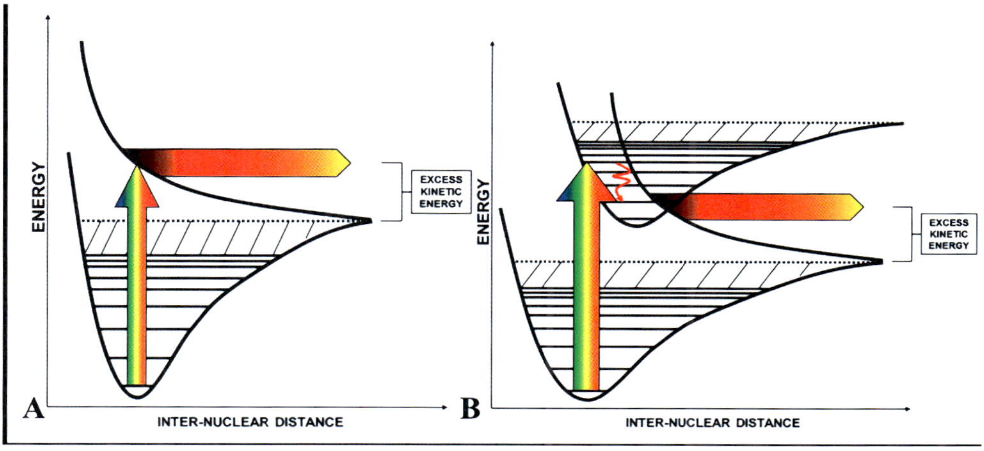

Fig. (2.15). **A**: Photodissociation. Some molecular excited levels display a repulsive character no matter the inter-nuclear distances (*i.e.* no equilibrium position is available). In the figure example a photon absorption event (coloured arrow) pumps the molecule to the upper repulsive energy curve. At this point the atoms start a molecular dislodgement that results in molecular fragments flying away from each other (reddish yellow arrow) with a total kinetic energy equal to the initial photon energy minus the molecular dissociation energy. **B**: Delayed photodissociation. A variant of the photodissociation process in this case the molecule reaches a stable excited state after absorbing a photon (coloured arrow). However this excited state overlaps with another repulsive excited state that was not originally available from the ground state because of the Frank-Condon principle. During the vibrational relaxation in the stable excited level there is a chance that the molecule transits to the repulsive energy curve. In such an event fast molecular fragmentation ensues (reddish yellow arrow) that is delayed in time in respect with the original photon absorption.

5.3. Photoinduced Electron Transfer

In virtue of its higher energy content when a molecule is in an excited electronic state it chemical reactivity changes, in most cases drastically. Photoinduced electron transfer (PET) is the mechanisms behind these phenomena. Chemical reactions involve the rearrangement of bonds and the way electrons are shared by the atoms [20 - 22]. An excited molecule has one of its electrons in a higher molecular orbital. Commonly this excited electron will move over to another part of the molecule, which brings about a molecular change. Due to the excited electron migration a certain portion of the molecule is left with a deficit in electronic population (positive charge) while the region where the excited electron finally migrates displays a net negative charge. This results in the establishment of a dipole configuration where originally there was a neutral molecule. This dipolar transient is known as the charge transfer complex (see Fig. **2.16A**). Note the changes in the chemical bonds, not only in regards to the nitrogen and oxygen atoms, but also the rearrangement within the two fused benzene moieties.

When the electron moves within the space of the original molecule we are referring to internal PET. Making use of the potential energy curves approximation the different molecular states can be represented by their corresponding curves as schematically shown in Fig. (**2.16B**). It is readily shown that the excited state curve position, which is unconnected to and above the ground state curve, would probably lead to a luminescent species. However, because of the PET state and associated potential curve, it "crosses over" to this charge transfer potential surface. Now this potential curve crosses with both the excited state and the ground state. This offers a faster deactivation way for the excited molecule as compared with photon emission. Therefore energy is lost mainly as bond rearrangements and vibrational emission [23]. Internal PET can be understood as a special case of ISC and IC. The charge transfer complex is unstable in regards to the ground state and the molecule recovers its original conformation at the end of the process. As a result the photon energy initially absorbed is lost to the medium as heat.

It also can occur that the charge transfer state connects to other ground state curves that are not necessarily the original molecular arrangement. In this case the molecule ends up in a different conformation leading to photochemistry (see below). Here we consider PET only as to giving rise to a transient energized state that reverses back to the original molecule which is therefore left unchanged. The net result of the photoreaction is heat production.

In other instances the electronic migration induces bond rearrangement that changes the molecule while keeping neutrality. This light-driven change is known as photoisomerization (see Fig. **2.16C** for an example with benzene). Photoisomerization can lead to fast transient changes that quickly revert to the original molecule (energy is lost as heat) or to more lasting or even permanent transformations. In this later case we again enter the realm of formal photochemistry.

The excited molecule can transiently react with a nearby species, most usually another molecule. This particular case is known as external PET. For example, it is quite common for fluorophores to engage in a transient energized state with a solvent molecule. The overall reaction is quite similar to the internal PET except that the excited electron promotes a charge transfer complex with the second molecule (see Fig. **2.16D**). The complex follows the same course as in the internal PET, with changes in charges and bonds. In the end these changes release vibrations and motions to the medium and the complex disassembles regenerating the absorbing molecule and the molecular partner unchanged. Again the light energy has been degraded to heat through this chain of molecular motions. However, in many cases the charge transfer complex does lead to more permanent chemical changes. When a net chemical transformation is taking place we speak of photochemistry (see below). But it is important to keep in mind that PET (both internal and external) is the common gateway for heat production or photochemistry. The final result does depend on many parameters of the reacting system: absorbing species, reacting species, solvent, photon energy, temperature, *etc.* [10]. The same chromophore under the same irradiation conditions can display internal PET, external PET and/or photochemistry simply by changing the solvent.

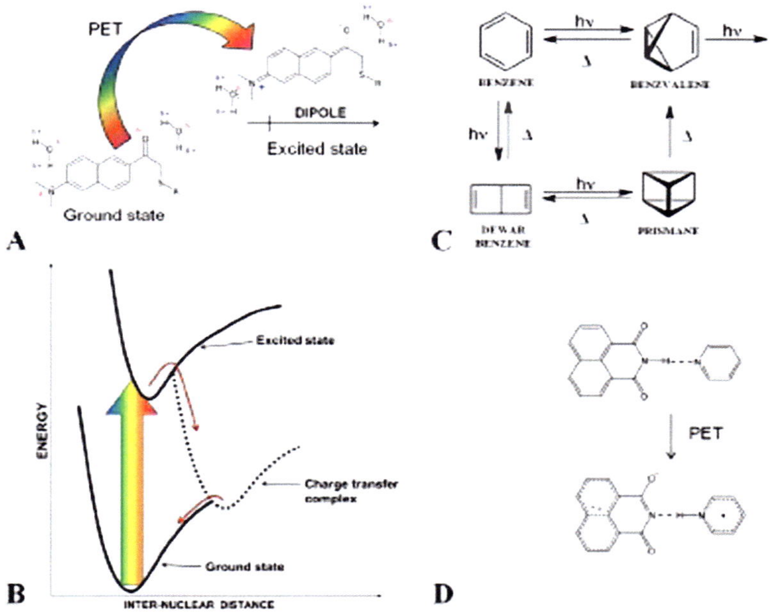

Fig. (2.16). Photoinduced electron transfer (PET). **A**: Schematics of internal PET. After photon absorption the excited molecule has one of its electrons relocalized on the oxygen atom. At the same time a positive charge develops on the nitrogen atom. This gives rise to a dipole configuration and the widespread rearrangement of chemical bonds within the molecule. This changes the way solvent molecules (water in this case) interact with the molecule. Reproduced with permission from Squidonius - https://commons.wikimedia.org/wiki/File:Badan.svg#/media/ File:Badan.svg). **B**: Potential energy curves of the molecular states involved in internal PET. In this particular example the molecule reaches an excited state after absorbing a photon. This high energy potential curve crosses with a potential curve that involves a charge transfer. Therefore there is a finite chance that the molecule "transits" to this curve losing energy in the process (reaching the bottom of the curve). Again the lower region of the charge transfer curve interacts with the ground state curve. This makes it possible for the molecule to shift back into the ground state conformation, releasing what energy was left from the original excitation as heat. **C**: Examples of photoisomerization brought about in benzene after light excitation (hv). After some time the isomer returns to the ground state configuration releasing heat in the process (Δ). For the light-mediated isomerization from benzvalene to fulvene the molecular rearrangement is permanent. **D**: Scheme of external PET whereas a 1,8 naphthalimide molecule absorbs a photon and undergoes PET to a solvent pyridine molecule. After PET the 1,8 naphthalimide is left with a defect of one electron (positive charge delocalized over the aromatic rings) which is transferred to the pyridine. This additional electron on the pyridine traps an hydrogen originally belonging to the 1,8 naphthalimide and bonds it to the nitrogen in the pyridine. Hydrogen bonding is represented by slashed line.

5.4. Photochemistry

Until now all the treated mechanisms have dealt with physical or physicochemical processes taking place in the molecule after absorbing a photon. However, an excited molecule is also capable of undergoing formal and long-lasting/permanent chemical reactions with other entities, atoms or molecules, during its lifetime. The mechanisms behind the photochemistry of molecules will be introduced in the simplest way. For textbooks treating this matter in more depth we recommend the following sources [20 - 22].

Photochemistry is based upon the following premise: when a molecule is excited one of the outer electrons is promoted to a high-energy farther orbital. This has two immediate consequences: 1) this electron is less bonded to the molecule, and 2) it leaves a hole in the former molecular orbital. As a result the excited molecule is more reducing (it tends to lose the excited electron easily) and more oxidizing (it tends to fill the created hole) at the same time. Its chemical reactions will depend, of course, on the nature of the available reactants. But it can be said that molecules in an excited state are more reactive than their ground state counterparts.

Photochemistry is a very important issue in fluorescence microscopy and in the fluorescence field in general. Because chemical reactions involve gaining/losing of electrons or bond rearrangement, the molecular makeup will change after photochemistry takes place. For this reason the molecule that initially absorbed light (*i.e.* chromophore) tends to disappear as it reacts from the excited state. Given that the chromophore photo-driven reaction usually involves a colour fading transition, this kind of reactions is designated photobleaching in a general sense. Photobleaching is usually irreversible, so molecules are chosen to avoid photobleaching in fluorescence analysis (which includes fluorescence microscopy).

One of the main environmental agents involved in photobleaching is molecular oxygen (O_2). O_2 is ubiquitous and it has a tendency to quickly react with other molecules. When organic molecules are in the excited state the reaction rate with O_2 increases many orders of magnitude. This is in part because of what explained in the previous paragraph and in part because of the possibility to transfer the excitation energy to O_2 (see ***Energy transfer*** below). This transfer leads to the generation of the first excited state of O_2 designed $O_2(a^1\Delta_g)$ or, more colloquially, singlet oxygen ($^1O_2$). This $^1O_2$ is very reactive and readily oxidizes the chromophore that transferred the energy on the first place [22, 24]. For this reason O_2 is to be excluded whenever possible from samples subjected to fluorescence analysis or, if not possible, quenchers or sacrificial reductants are commonly employed to delay as much as possible $^1O_2$ production and oxidation.

5.5. Energy Transfer

Another pathway for molecular de-excitation is the transfer of the excitation energy to another entity, molecule or atom. This is also a deep area of study so only a general introduction will be provided. Basically, excitation energy can be transferred through two mechanisms: 1) collision-mediated or 2) distance transfer. The first type is the most obvious and easily understandable. The excited molecule collides with a partner and transfers its excitation energy during the brief moment that both entities are very close and their electronic wave functions can interact. During these collisions multiplicity may change also as a result.

The second type of energy transfer, without any collision, takes place when the interacting entities are at some distance. We mean, of course, atomic/molecular distances that usually range in the nanometre scale. Distance energy transfer takes place when the excited electron interacts from some distance with the other electrons that make up the partner atom or molecule through the electromagnetic field. This subtle interaction may stir the excited electron to decay to a lower excited level or the ground state while at the same time transferring the excitation energy to the receiving electron. The crucial point here is that no real photon is emitted by the donor molecule: the whole interaction takes place through the electric interference of one electron upon another. The process can be visualized as the emission and immediate absorption of a virtual photon. Energy transfer is favoured when the interacting entities feature very similar energy levels. In this case, a matching energy transfer results. This phenomenon is described as resonance energy transfer and provides the fundamental mechanism for a very important fluorescent technique widely used in fluorescence microscopy denominated Fluorescence Resonance Energy Transfer (FRET). Because energy transfer of this type depends strongly on distance between partners ($\sim r^{-6}$) emission by one of them sensitized by absorption by the other can be employed as a very fine tool to measure nanometric distances (1-10 nm).

5.6. Quantum Yields

To finish the chapter the concept of quantum yield is introduced. The quantum yield is a measure of the different available de-excitation pathways open to an excited molecule. It provides the ratio between different de-excitation pathways after photon absorption [20, 22]. If a molecule is said to have a quantum yield of fluorescence of 1, this means that every time the molecule absorbs a photon it will emit a fluorescence photon after some time. In contrast, a quantum yield of IC of 1 will mean that the molecule does not fluoresce or phosphoresce at all, the photon absorbed being integrally degraded to vibration and heat. For most molecules such extreme behaviours are not observed. Most usually one finds that a molecule has, for example, a quantum yield for IC of 0.7, fluorescence quantum yield of 0.2 and a yield for ISC of 0.1.

The quantum yield is represented by the Greek letter phi Φ with a subscript that defines to which process the quantum yield refers. In the previous example the molecule will have Φ_{IC}=0.7, Φ_F=0.2 and Φ_{ISC}=0.1. Because of the way in which it is defined, the sum of all quantum yields for a molecule must equal 1. That the example molecule displays a Φ_F=0.2 does not mean that the energy of the emitted photon is 0.2 that of the absorbed photon. The correct interpretation of this value is that for every 10 photons absorbed by the molecule on average 2 fluorescence photons are expected. Or that for a molecular population absorbing photons on average we would expect 20% of the molecules to fluoresce afterwards. The Φ is a statistical concept. After a molecule has absorbed one photon it can only follow one of its available de-excitation pathways. Thus for each absorption event only one de-excitation pathway will be observed. Which one will occur, on a one-t--one basis, cannot be predicted. Only after many events have taken place a trend in the preferred pathways can begin to be established.

The Φ depends on many variables. Anything that can alter the electronic wave functions can, in principle, steer the Φ towards or away from a certain de-excitation pathway. Some variables that deeply affect the Φ include: phase (gas, liquid, solid), solvent, electric and magnetic fields, viscosity, temperature, pressure, *etc.* The exemplified Φ introduced above could fit for a molecule solubilized in water, for example. That very same molecule in an organic solvent, methanol, could display the following yields: Φ_{IC}=0.1, Φ_F=0.8 and Φ_{ISC}=0.1with Φ_{Ph}=0.1 for allowed phosphorescence. It is obvious that what was a mediocre fluorophore in water becomes an adequate fluorescent probe in methanol. Although success is not guaranteed some hints can be given beforehand to predict some trends in Φ. For example, viscosity hinders molecular vibrations and rotations. Given that IC and ISC depend strongly on electronic-vibrational/rotational coupling, it can be argued that increasing the medium´s viscosity will decrease both Φ_{IC} and Φ_{ISC}. This is usually the case, indeed. And, at the same time, Φ_F will most often increase because the IC and ISC pathways are narrowed so the molecule must get rid of the absorbed energy and increases its fluorescence.

CONFLICT OF INTEREST

The author (editor) declares no conflict of interest, financial or otherwise.

ACKNOWLEDGEMENTS

This work was supported by the Aarhus Institute of Advanced Studies (AIAS).

REFERENCES

[1] Nature Milestones-Photons. Nature milestones 2010.

[2] Band YB. Light and Matter Electromagnetism, Optics, Spectroscopy and Lasers. Chichester: John Wiley & Sons Ltd 2006.

[3] Saleh BE, Teich MC. Fundamentals of Photonics. New Jersey: Wiley-Interscience 2007.

[4] Banwell CN, McCash EM. Fundamentals of molecular spectroscopy. 4th ed., London: McGraw-Hill International Limited 1994.

[5] Parson WW. Modern Optical Spectroscopy. Berlin, Heidelberg: Springer-Verlag 2007. [http://dx.doi.org/10.1007/978-3-540-37542-5]

[6] Pedrotti FL, Pedrotti LS, Pedrotti LM. Introduction to optics. 3rd ed., San Francisco: Pearson Addison Wesley 2007.

[7] Nature Milestones-Light Microscopy. Nature milestones 2009.

[8] Methodology. In: Solomon EI, Lever AB, Eds. Inorganic electronic structure and spectroscopy. New York: Wiley-Interscience 1999; Vol. I.

[9] Applications and case studies. In: Solomon EI, Lever AB, Eds. Inorganic electronic structure and spectroscopy. New York: Wiley-Interscience 1999; Vol. II.

[10] Lakowicz JR. Principles of fluorescence spectroscopy. 3rd ed., New York: Springer Science + Business Media LLC 2010.

[11] In: Papkovsky DB, Ed. Live cell imaging Methods and protocols. New York: Humana Press. Springer Science+Business Media 2010. [http://dx.doi.org/10.1007/978-1-60761-404-3]

[12] Nassau K. The Fifteen Causes of Color: The Physics and Chemistry of Color. 2nd ed., New York: Wiley-Interscience 2001.

[13] Svanberg S. Atomic and molecular spectroscopy Basic aspects and practical applications. 4th ed., Berlin, Heidelberg: Springer-Verlag 2004.

[14] Valeur B. Molecular fluorescence Principles and applications. Weinheim: Wiley-VCH Verlag GmbH 2002.

[15] Schmidt W. Optical Spectroscopy in Chemistry and Life sciences. Weinheim: Wiley-VCH Verlag GmbH 2005.

[16] Herzberg G. Molecular spectra and molecular structure: I Spectra of diatomic molecules. 2nd ed., New York: Van Nostrand Reinhold Company-Prentice Hall 1950.

[17] Herzberg G. Molecular spectra and molecular structure: II Infrared and Raman spectra of polyatomic molecules. New York: Van Nostrand Reinhold Company 1945.

[18] Herzberg G. Molecular spectra and molecular structure: III Electronic spectra and electronic structure of polyatomic molecules. New York: Van Nostrand Reinhold Company 1966.

[19] König K. Multiphoton microscopy in life sciences. J Microsc 2000; 200(Pt 2): 83-104. [http://dx.doi.org/10.1046/j.1365-2818.2000.00738.x] [PMID: 11106949]

[20] Wayne CE, Wayne RP. Photochemistry. Oxford: Oxford University Press 1996.

[21] Klán P, Wirz J. Photochemistry of organic compounds: From concepts to practice. United Kingdom: John Wiley & Sons Ltd 2009.

[http://dx.doi.org/10.1002/9781444300017]

[22] Turro NJ, Ramamurthy V, Scaiano J. Modern molecular photochemistry of organic molecules. New Jersey: University Science Books 2010.

[23] Ceroni P, Balzani V. Photoinduced energy and electron transfer processes. In: Ceroni P, Ed. The Exploration of Supramolecular Systems and Nanostructures by Photochemical Techniques. Netherlands: Springer Science+Business Media 2012; pp. 2. 21-38.

[24] Ogilby PR. Singlet oxygen: there is indeed something new under the sun. Chem Soc Rev 2010; 39(8): 3181-209.
[http://dx.doi.org/10.1039/b926014p] [PMID: 20571680]

CHAPTER 3

Dyes and Fluorochromes

Abstract: This Chapter deals with the "molecular anatomy of dyes" as referred by Gurr [1], as well as with some chemical characteristics and properties. Obviously, structural aspects of dyes and fluorochromes are often shared and then they should be viewed as valid for both. In addition to shape and size, the molecular anatomy includes a collection of geometrical values such as planarity, bond lengths, dihedral and torsional angles, as well as the occurrence of substituents such as anionic, cationic, hydrophobic, hydrogen bonding, reactive, and metal complexing groups. Some of these physico-chemical parameters can be numerically formulated [2-4] (see section 3.2 and Chapter 15.2.1). The interdisciplinary character of color chemistry (*i.e.* staining and fluorescence methodology, modern microscopical techniques, industrial and medicinal applications, photosensitizing processes, *etc.*) is a fascinating research field of increasing importance.

Keywords: Atomic orbitals, Auxochromes, Bond order, Chromophores, Conjugated double bonds, Energy transfer, Eosin Y, Ethidium, Fluorescein, Fluorescence quantum yield, Mesomeric forms, Mixed valence, Molecular orbitals, Prediction of fluorescence, QSAR, Traditional color chemistry.

1. CLASSIFICATION AND CHEMISTRY OF DYES

1.1. General Features

The close relations between color and the chemical structure of organic molecules are known since long ago [5 - 9], but less known correlations occur for fluorescence [10, 11]. However, some practical rules and conceptual strategies can aid to predict or explain emission properties of dyes [12]. Those readers particularly interested in detailed chemical aspects of dyes and fluorochromes should consult other books [1, 4, 13 - 16]. A useful glossary on photochemistry is available [17].

Regarding terminology, "dye" is the term most used in biomedical applications. In bright-field (transmission) microscopy the name "diachrome" (in contrast with fluorochrome) has been sometimes employed. Dyes and fluorochromes have features in common, but emphasis in light - absorbing compounds (dyes) and

Juan Carlos Stockert & Alfonso Blázquez-Castro

light-emitting ones (fluorochromes) will be separately presented in this Chapter. The terms fluorescent dye and fluorochrome are used indistinctly.

It is worth noting that conventional chemical representation can result in misleading interpretations [18]. Among others, this occurs in the following cases:

a. Erroneous assumption of the double bond as two identical and equivalent single bonds.
b. Omission of the representation of H atoms.
c. Linear (‒N=, ‒CH=) instead angular bonds, and crossed instead tetrahedral configurations (*i.e.* sulfate, sulfonate, phosphate).
d. Structures lacking resonant forms: *i.e.* formal position of double bonds in aromatic compounds, sulfate, phosphate and carboxylate groups, *etc.*

Although commonly illustrated as simple 2D schemes, molecules have obviously a 3D structure. Computed-generated drawing and stereoscopic images (stereopairs) are far superior to any kind of 2D or perspective picture and can be easily viewed. At present, computer software and theoretical methods of computational chemistry are used as helpful tools for drawing and modeling molecular structures, and for analyzing electronic properties and chemical reactivity of dyes [18 - 21].

Effective light absorption occurs in molecules showing electron delocalization by the occurrence of either conjugated double bonds (organic dyes) or mixed valence (inorganic compounds). Several examples are given in Fig. (**3.1**). Prussian blue, ruthenium red, and phosphomolybdic blue are examples of important inorganic dyes. Fluorescence properties are related only with organic molecules, and therefore inorganic mixed valence dyes will be not taken into account. As stated in Chapter 1.4 (Fig. **1.6**), several criteria are used for classification of dyes. They can be natural, synthetic, charged (cationic, anionic), uncharged, lipophilic, reactive, *etc.* Some natural fluorescent dyes are morin, saffron, curcumin, berberine, carmine, *etc.* Natural fluorophores useful in microscopical work are also chlorophylls, nicotinamide and flavine coenzymes, fluorescent proteins, *etc.*, which will be described in forthcoming Chapters.

To illustrate well known and widely used cationic dyes in microscopical work, Fig. (**3.2**) shows several examples in which the common paraquinone structure is indicated. In these examples, dyes are also fluorescent. Thiazines (thionine, methylene and toluidine blue) require red excitation and fluoresce in the near infrared, whereas acridine orange, pyronine Y, and neutral red (acridine, xanthene, and azine dyes, respectively) fluoresce in the visible region under visible exciting light. Other currently used fluorescent dyes are described below.

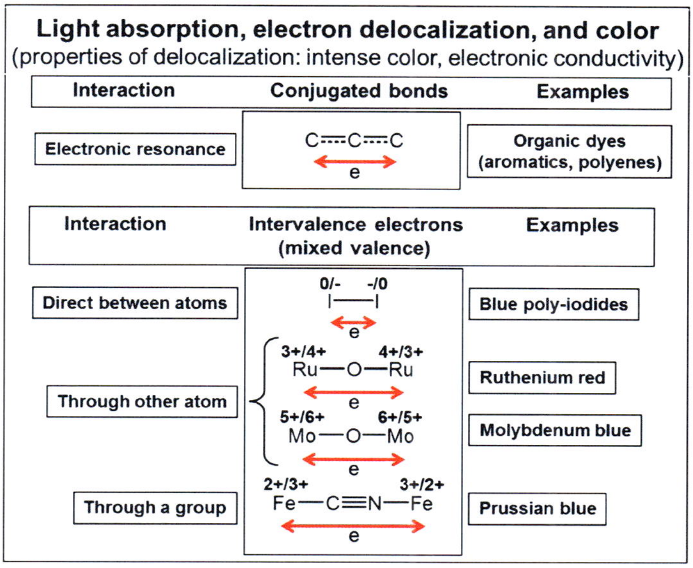

Fig. (3.1). Examples of different classes of dyes showing the principal mechanisms of electron delocalization in organic (conjugated) and inorganic (mixed-valence) compounds.

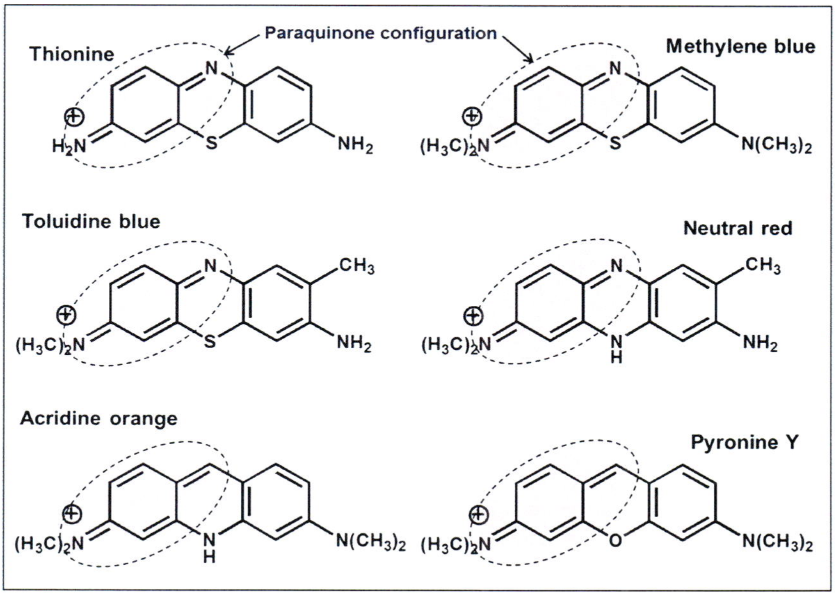

Fig. (3.2). Examples of tricyclic cationic dyes from the thiazine (thionine, methylene blue, toluidine blue), azine (neutral red), acridine (acridine orange), and xanthene (pyronine Y) groups. The usual paraquinone configuration is indicated.

1.2. Examples of Natural Dyes: Flavonoids

Important natural compounds in plants are the flavonoids [13, 16], which exist generally as glycosides with important properties and pharmacological effects (anti-oxidants, pesticides, fungicides, colored and fluorescent pigments, phyto-estrogens) (Fig. **3.3**). They derive from the chromane ring, and 2- and 4-keto chromone compounds are important derivatives such as vitamin E, and coumarins with anticoagulant (dicoumarol) and estrogenic activity (coumestrol). As expected, some coumarin derivatives are fluorescent (*i.e.* coumestrol, psoralidin). The term "bioflavonoids" refers to compounds that contribute to the decrease of capillary permeability and fragility. In addition, chelating capacity for metals is an important feature that often occurs in flavonols and neoflavones.

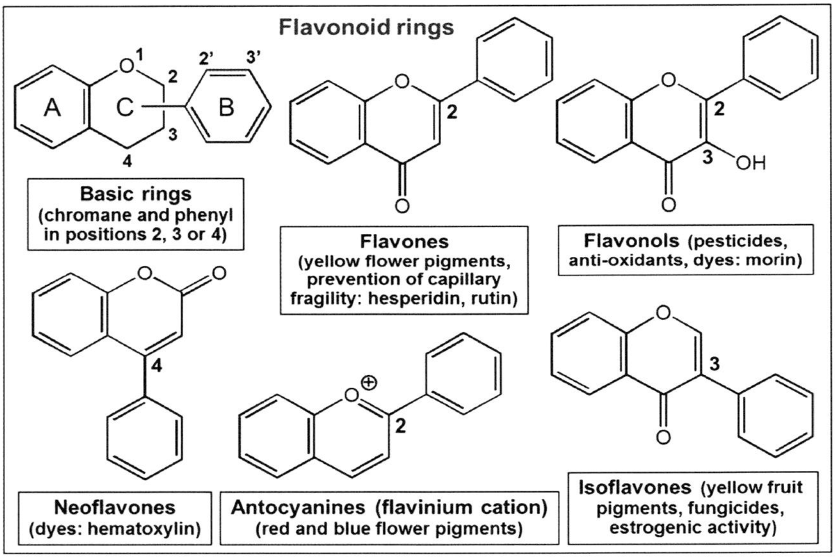

Fig. (3.3). Natural flavonoids showing the basic chromane ring and chromone derivatives, with examples of some compounds and properties.

Anthocyanins are flowers pigments that appear as glycosides with one or two glucose molecules, mainly existing as oxonium (pyrylium) salts with a high pH-dependent color [13]. Anthocyanins and flavonols have been employed for nuclear staining generally in presence of Al^{3+} salts. Color stabilization from pH changes takes place by formation of borate complexes with the catechol group of some flavonoids (see Chapter 8.3.3).

The flavonol morin (2',3,4',5,7-penta-hydroxyflavone, CI 75660; Fig. **3.4**) is extracted from fustic tree, moral fustete or yellow log (*Chlorophora tinctoria*), a wood type that is dear in cabinetmaking. Morin is fluorescent and like other flavonols it forms metal complexes that are useful for the detection of Al, Be, Ca, Zn, Ga, In, Zr, Sc, Ti, and Th ions [13, 22], using the O3, O4, and O5 atoms, which form H bonds (Fig. **3.4A**). As occurs in other flavonols, the H atom at O3 undergoes excited state proton transfer to O4. On account of its weakly acidic character, morin induces yellow fluorescence in acidophilic structures such as eosinophil and heterophil leucocyte granules (Fig. **3.4C**). This reaction is based on electrostatic forces (ionic binding) and does not depend on the presence of metal cations [23, 24].

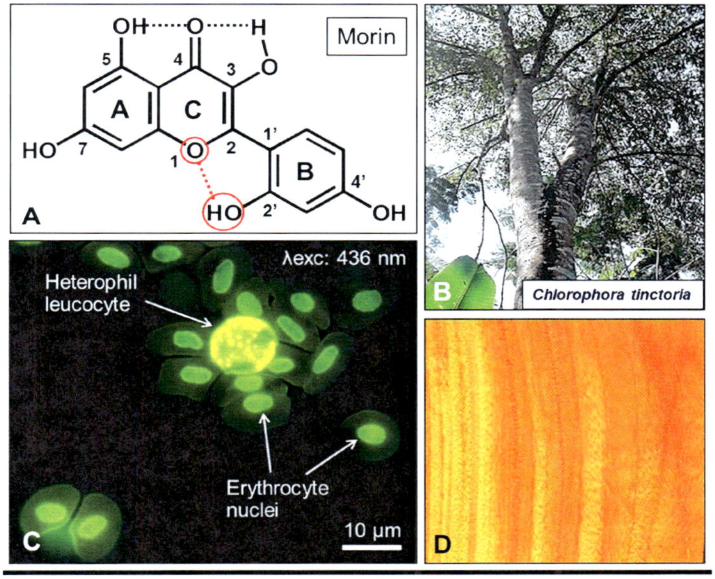

Fig. (3.4). A: Chemical structure of morin, showing the numbering system and the H bond between O1 and O2' (in red), and between O3, O4 and O5 atoms. **B:** Fustic tree. **C:** Methanol-fixed smear of chicken blood stained with morin (saturated aqueous solution) showing yellow and green fluorescence of heterophil leucocyte granules and erythrocyte nuclei, respectively. **D:** Shone wood of fustic.

In addition, chromatin, reticular fibers, mucin of goblet cells, and keratohyalin granules appear fluorescent after morin staining (see Fig. **1.8B**). Hydrophobic and/or H bonding interactions could be responsible for these reactions. Different emission colors of morin (green, yellow) are found using morin at low (10^{-7} M) or high (10^{-2} M) concentration in 10 mM borate buffer (pH 9.2), showing emission peaks at 530 and 550 nm, respectively. A suspension of 0.4 mg/mL morin in paraffin oil shifts the peak to 570 nm [23]. It is worth noting that the green fluorescence of morin-stained chromatin greatly increases after previous treatment with Al^{3+} or Zn^{2+} ions. Higher emission of morin is also observed when directly dissolved in borate buffer.

In flavonols lacking -OH on the 2' position, the ring B plane moves away considerably from the plane of rings A and C, the dihedral angle being about 60 °. The lack of coplanarity between rings B and C prevents transmission of resonant electronic effects from ring B to the C2-C3 double bond. As morin has a hydroxyl substituent in position 2' of ring B, this hydroxyl interacts with O1 of ring C forming an H bond (Fig. **3.4A**) that makes morin more coplanar, the dihedral angle diminishing to about 38 °. Thus singlet oxygen ($^1O_2$) can attack the site of high electron density (shortest C2-C3 double bond), and this feature explains the increased reactivity of morin towards $^1O_2$ [21] and its anti-oxidant effect. Fisetin, quercetin, rutin, isorhamnetin, *etc.*, are also fluorescent, and the possible applications of these and other flavonols in fluorescence microscopy should be investigated.

2. DYES AND COLOR THEORY

2.1. Traditional Color Theory

The first comprehensive theory on the color of organic compounds was developed by Witt in 1876 [5, 9]. Regarding the nomenclature of traditional color chemistry, the terms chromogen, chromophore and auxochrome originated in early color theory (Fig. **3.5**), the chromogen corresponding to an unsaturated (conjugated) linear or cyclic molecule. Today, the term chromophore is applied in a wide sense to define a light-absorbing conjugated compound, and shows a practical value in biological and industrial staining in relation to the chemical classification of dyes [15]. Therefore, a chromophore is the molecular region in which the electronic transition responsible for a given spectral absorption is located. Polar or ionizable auxochrome groups (anionic or cationic) are introduced to modify the absorption intensity and/or shift the spectral absorption [1], and to give water solubility and adequate affinity for substrates. Auxochromes are electron-donating (-NH_2, -OH) and electron acceptors (-COOH, NH_3^+) substituents.

The Witt's theory has served for a long time as a rough working frame in color chemistry, and some theoretical implications have became fundamental concerns. Significant advances in this field are the wave-mechanical treatment of atoms and molecules, leading to the notion of atomic orbitals (AOs) and molecular orbitals (MOs). Thus, quantum theories of color have been developed, based mainly in MOs methods. At present, ideas concerning unsaturation allow explaining energy absorption by the transition of electrons from ground to excited states. In the following sections, issues emerging from this approach to molecular structure, reactivity, light absorption and emission will be presented in a non-mathematical way.

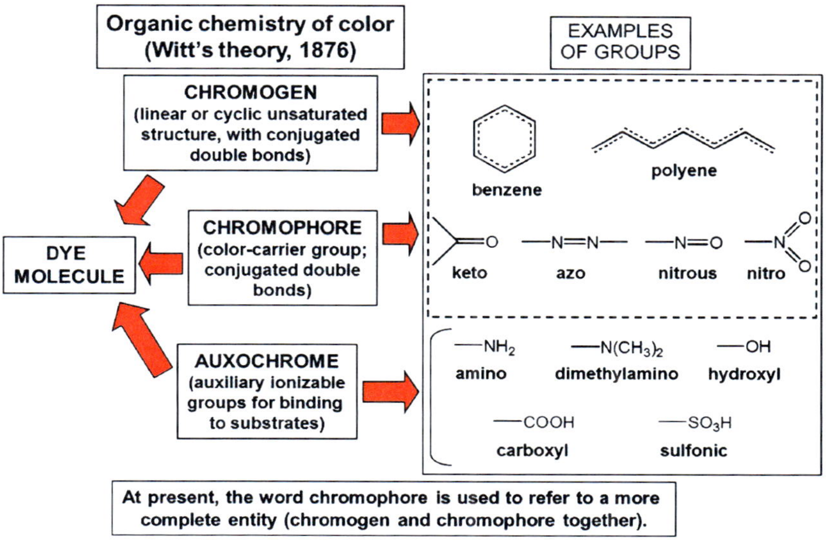

Fig. (3.5). Principal concepts on the color of dyes developed by Witt in 1876. Examples of chromophores and auxochrome groups are illustrated.

From a simplistic approach to explain the color chemistry of organic dyes, conjugated double bonds and resonance are the principal features to be considered (Fig. **3.6**). In both linear chains and cyclic structures the alternating double bond (-CH=CH-) generates a conjugated (mesomeric) C-C binding system in which the defined position of the double bond is only a formal representation. In fact, the delocalized double bonds can be more precisely described by dashed or dotted lines, which also are used to represent the conjugation of sulfonate, sulfate, phosphate and carboxylate groups. Mesomeric forms also involve the participation of O and N atoms, as well as charge resonance. Regarding unsaturated groups, light absorption is related to the number of double bonds, ethylene and benzene absorbing at 170 and 250 nm, respectively.

The use and properties of some chromophores will illustrate these concepts. The first natural dye used in light microscopy was probably saffron (crocin, CI 75100), applied by Leeuwenhoek in 1714 to stain muscle fibers, and now as collagen stain in polychrome staining procedures. Saffron is a glycoside of crocetin (dicarboxylic carotenoid) bound to the disaccharide gentiobiose and extracted from the dried stigma of *Crocus sativus* (Fig. **3.7**). Tin-mordanted polysaccharides give an orange color with saffron possibly due to the Sn union to the gentiobiose hydroxyls [13]. The bright green emission of this polyene dye (λexc: 436 nm, λem: 520 nm) was applied in fluorescence microscopy to reveal selectively eosinophil leucocyte granules [25]. The anionic crocetin (CBN: 19) is formed at the alkaline pH of the staining solution, and it is responsible for the fluorescence reaction with highly cationic proteins.

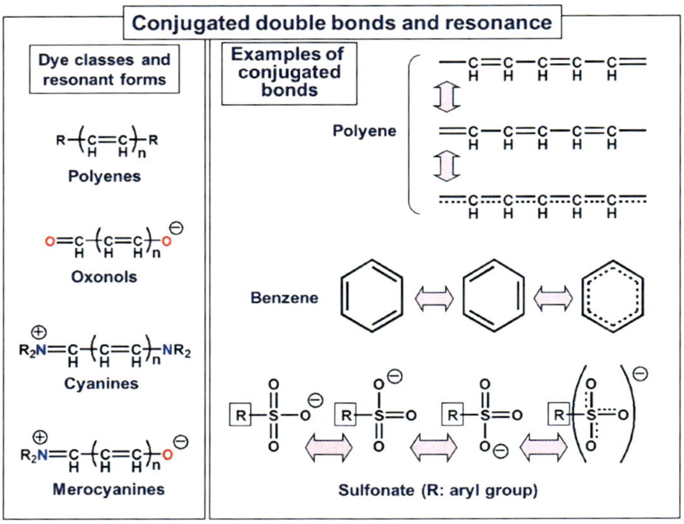

Fig. (3.6). Examples of resonant forms of chromophores and auxochromes. The conjugated double bond system of the polyene chain and benzene ring is shown at right, as well as the resonance of the sulfonate group. At left, the basic structure of polyenes, oxonols, cyanines, and merocyanines is illustrated.

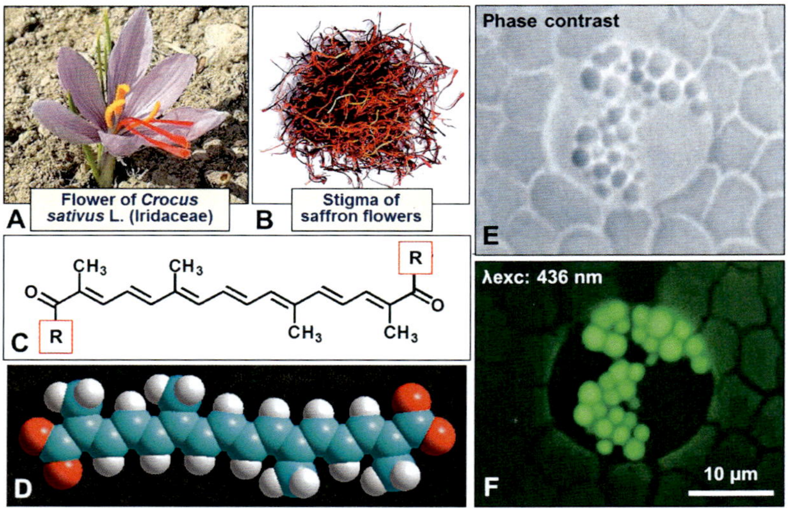

Fig. (3.7). Fluorescence of saffron, the first natural dye used in light microscopy. **A**, **B**, and **C**: Flower, stigma, and chemical structure of saffron. R represents the disaccharide gentiobiose. **D**: Molecular structure of the dicarboxylic carotenoid (polyene) dye, crocetin. **E**, **F**: Horse eosinophil leucocyte stained with saffron (saturated solution in borate buffer, pH 10) observed under phase contrast (**E**) and fluorescence microscopy (**F**).

2.2. Free Base and Pseudo-Base

Lipophilic (hydrophobic) dyes are specifically described in Chapter 10, but some modifications of dyes induced by pH variations result in hydrophobic derivatives and color changes. This is the case of the formation of the "free base" in cationic dyes (Fig. **3.8A**). The removal of one proton (H ion) at alkaline pH produces the uncharged hydrophobic free base with changed color, which finally precipitates. Common pH indicators are based on these color changes. Some cationic dyes contain quaternary nitrogen atoms that can not release any proton, and then at alkaline pH a hydroxyl ion attaches to an electropositive carbon forming a "pseudo-base" (carbinol) structure, typical examples being triphenylmethane dyes and ethidium bromide (Fig. **3.8B**). In these cases, cationic and pseudo-base forms are responsible for colored and colorless compounds, respectively.

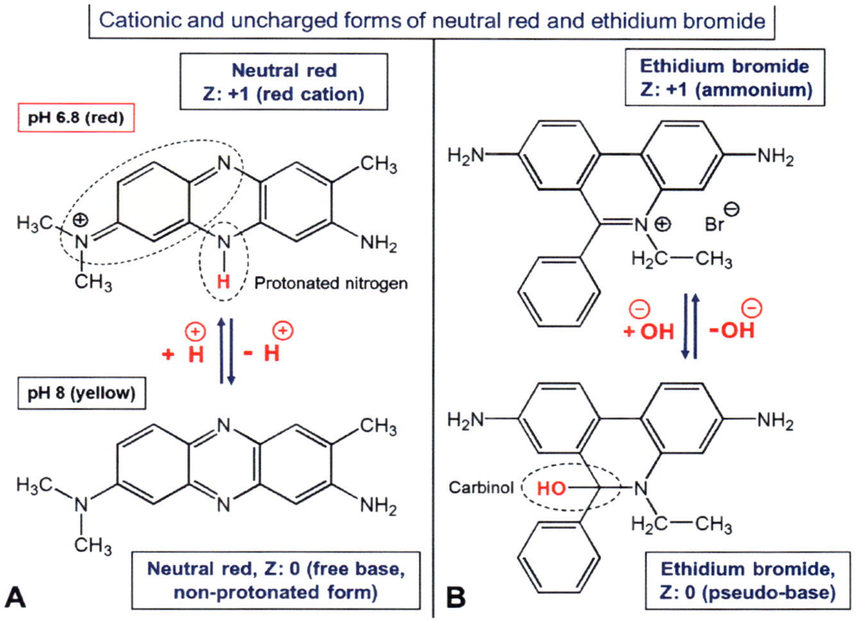

Fig. (3.8). Interconversions between cationic and uncharged forms from dyes. **A**: Free base (**A**) and pseudo-base (**B**) formation of neutral red and ethidium bromide, respectively.

An illustrative example regarding the relation between pseudo-base and color is that of hydrastinine, an oxidation product of the alkaloid hydrastine from *Hydrastis canadensis* with hemostatic activity (Fig. **3.9A**). Aqueous or alcoholic solutions are yellow and fluorescent, whereas those in non polar solvents (corresponding to the pseudo-base of the cationic compound) are colorless and do not fluoresce. The same occurs for cotarnine [26] and N-methyl-nicotinamide [27]). Likewise, the reaction product of fluorescamine and MDPF with primary amines gives the colorless and non-fluorescent pseudo-base at alkaline pH, which becomes fluorescent in the cationic form [18, 28] (see Chapter 13.2.1).

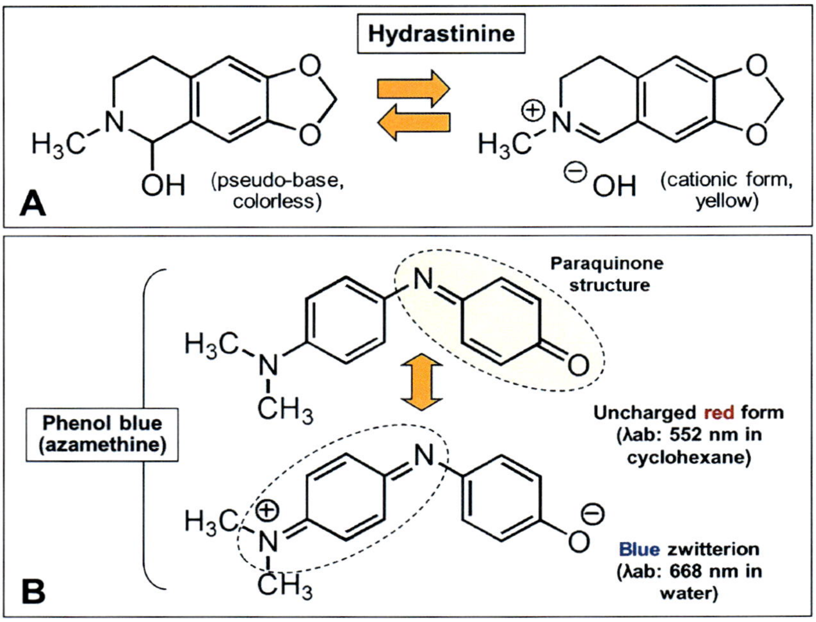

Fig. (3.9). Tautomeric forms of dyes. **A**: Hydrastinine, an example of interconversion between hydroxylated (pseudo-base) and ammonium forms associated to color and fluorescence changes. **B**: Resonant forms, properties, and spectral features of phenol blue.

On the other hand, the azamethine dye (merocyanine-type) phenol blue (N,N-dimethyl-indoaniline) can be formulated as nonionic or ionic structures [6] (Fig. **3.9B**), the absorption spectra depending on the solvent (solvatochromism) and light excitation. The ionic configuration (*zwitterion*, stabilized in polar media) corresponds to the excited state [29]. The absorption of phenol blue undergoes a large red shift from 552 nm in cyclohexane to 668 nm in water, indicating that the non-polar form has lower energy and makes the major contribution to the ground state. Other azamethine dyes such as indophenol blue, indonaphthol purple, and fat blue Z, show similar characteristics. Interestingly, tricyclic dyes as rhodols, thiazones, and oxazones have analogous mesomeric configurations as phenol blue (see Chapter 10.3).

2.3. Conjugation

Although conjugated double bonds appear as symmetrical structures in polyene and benzene (Figs. **3.5** and **3.6**), their distribution can vary in complex aromatic compounds depending on the molecular geometry. Thus there is a sort of high or low "double bond character". A simplistic description of this double bond character based on numeric methods was suggested long ago by Pauling [30]. The double bond index (DBI) is the ratio between the frequency of a given double bond and the total number of mesomeric structures (Fig. **3.10A-D**), and it describes in a simple way the lower or higher double bond character ($1/3 < 1/2 < 2/3 < 3/4$, *etc.*, or their fractions: 0.33, 0.5, 0.66, and 0.75, respectively). The DBI for ethylene ($H_2C{=}CH_2$) is obviously 1, whereas for benzene it is 1/2: 0.5.

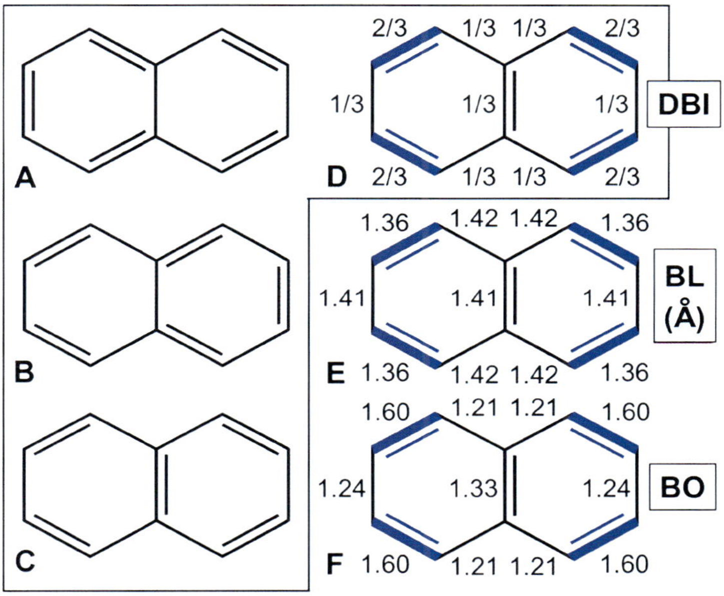

Fig. (3.10). Double bond parameters. **A, B, C:** Resonance forms of naphthalene according to the formal position of double bonds. In **D**, the resulting double bond index (DBI) is expressed as fractions (*i.e.* 2/3: 0.66). **E** and **F:** Bond lengths (BL, in Å) and bond orders (BO) of naphthalene.

But now more accurate approaches are available for expressing the double bound character. The bond length (BL) is a physical measure of the interatomic distance in Å (10^{-10} m) provided by X-ray and electron diffraction studies, and it reflects very precisely the occurrence of a double bond. Thus, the BL of a local CH=CH bond in ethylene is 1.32 Å, and for the conjugated (aromatic) double bonds of benzene the BL is 1.39. In the case of acetylene (CH≡CH), the BL of the triple bond is 1.19 Å. Therefore, the shorter BL, the higher the unsaturated character of the bond (Fig. **3.10E**). This is also valid for C=O and C=N bonds compared to C-O and C-N ones.

Other parameters are based on studies on the electronic configuration of molecules, which results from the application of the Schrödinger's wave function

(ψ) to each of their mesomeric structures. Readers interested in more detailed historical and physical descriptions of the double bond and aromaticity should consult specialized reviews and books [6, 8, 9, 31 - 34].

The widely used bond order (BO) is a quantity derived from ψ functions that describes the contribution of π electrons in addition to σ bonds [35] (see following section). Therefore, this parameter represents a "fractional bond character" between C-C and C=C bonds as a result of resonance among allowed structures (Fig. **3.10F**). The BO reflects in a simple and direct way the double bond character. The values for ethane (CH_3-CH_3) and ethylene (CH_2=CH_2) are 1.01 and 2, respectively. Benzene has a BO of 1.42. It is worth noting that in contrast with BL values, those of BO appear independent to the type of atom involved in the double bond. Thus while the C=O and C=C bonds of 1,4-benzoquinone have BLs of 1.21 Å and 1.33 Å, respectively, the BO values are 1.90 for both bonds, indicating a very high unsaturated character.

A good correspondence between BO and DBI can be envisaged by subtracting the value of 1 (equivalent to the single or σ-bond) from BO, then obtaining the value for the π-bond alone. Thus the BO for benzene is 1.42, but the π-BO is 0.42. Regarding all the mentioned double bond parameters, an illustrative comparison in the case of ethylene would be DBI, BL, and BO values: 1, 1.32 Å, and 2, respectively. The representation of mesomeric configurations, and comparative DBI values, BL, and BO for naphthalene is shown in Fig. (**3.10**). The correlation between the eleven DBI and BO values for this compound is high (R^2: 0.966), indicating that the simple DBI is adequate to show bonds with the highest double bond character.

Although a full conjugated (resonant) chromophore is clearly depicted by its mesomeric structures, the precise bond length and double bond character vary between different atoms. Both BL and BO values can be directly recorded by using the available molecular modeling software (*i.e.* ChemDraw, HyperChem, *etc.*). Similarly, other molecular characteristics such as bond angle, partial charge, dipolar moment, contour and surface electrostatic potential, QSAR properties (surface area, volume, log P), molecular orbitals of ground and excited states, total energy, *etc.*, are obtained after geometry optimization from computer-assisted modeling. Since long ago, the relationship between molecular structure (following quantum mechanical approaches) and biological activity has been a fascinating research field. The most relevant correlations were found between electronic configurations and chemical carcinogenesis [35], which led to use similar terms as in color theory, such as carcinogenophore (K-region) and auxocarcinogen.

2.4. Double Bonds and Energy Absorption

In the traditional representation of chemical structures, C-C single and double bonds are shown as one or two lines, respectively. But they are not equivalent (Fig. **3.11**): single bonds are due to interaction of either tetragonal sp^3 ($2s$-$2p_x$, $2p_y$, $2p_z$) or trigonal sp^2 ($2s$-$2p_x$, $2p_y$) electron hybridizations from two atomic orbitals (AOs) that generate a sigma (σ) bond centred along the C-C axis. The interaction between two $2p_z$ AOs originates a pi (π) bond.

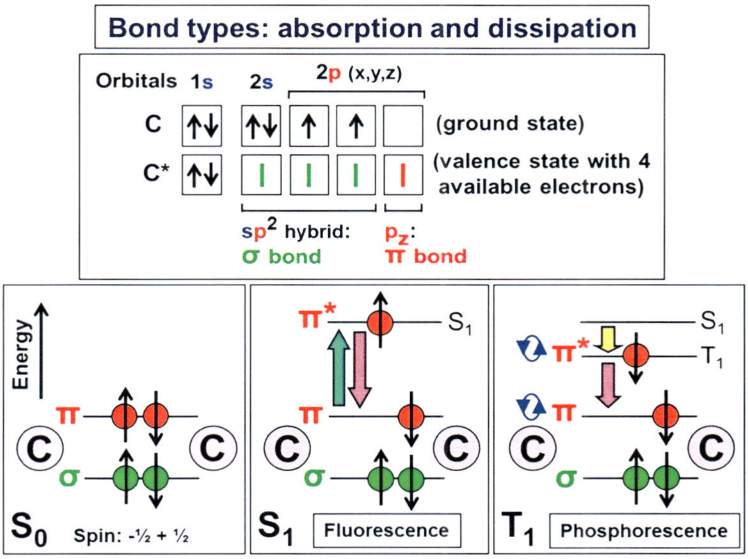

Fig. (3.11). Top: Distribution of electrons in the valence state of carbon. Hybridization of three valence electrons gives a trigonal sp^2 configuration with sigma (σ) bonds, whereas the remaining p_z electron gives a pi (π) bond. Antiparallel arrows indicate paired electrons with opposite spin. Bottom: Simplified scheme of the ground state (S_0) of a C=C bond. Light absorption promotes a π electron (cyan arrow) to a higher energy level (π^*), leading to an excited singlet state (S_1) and to energy decay by fluorescence. Spin inversion of a π^* electron (yellow arrow) results in an excited triplet state (T_1) giving phosphorescence. Red and green circles represent π and σ electrons, respectively.

When a molecule absorbs visible or UV light with an adequate wavelength, a valence (π) electron is promoted from a lower to a higher energy level. A simple and conventional approach illustrating the nature of the double bond and the related light absorption and emission process is shown in Fig. (**3.11**). The scheme shows the origin of the sp^2 hybridization of carbon, and after photon absorption, the promotion of a π electron from an aromatic compound to an excited π^* electron of S_1 ($^1\pi^*$) and T_1 ($^3\pi^*$) states, followed by fluorescence and phosphorescence emission. The excited singlet compound can cross over to T_1 by inverting the spin of the excited electron, and then again by spin inversion the molecule decays finally to the S_0 state. Other non-radiative decay mechanisms (thermal dissipation or radiationless deactivation, oxygen photosensitization) are not shown but a scheme accounting for these processes can be seen in Fig. (**7.11**).

It must be emphasized that the singlet $\pi \rightarrow \pi^*$ transition is the prototype of electronic processes in dye molecules. In relation to the molecular and electronic structure of a dye, there are two main sets of factors governing its color. The first and principal set relates to the size and shape of the π-electron system, including the number of conjugated atoms and the form of the bonding network between them, linear, branched, or cyclic (again an "anatomical" approach). The second set concerns substitution and atom-replacement effects related to the π-bonding and electronegativity of the new conjugated atoms.

2.5. Molecular Orbitals of Excited States

Historical perspectives and development of present concepts on excited molecules can be found in [6, 9, 32 - 34, 36]. Ethylene (Fig. **3.12A**) is the simplest example of ground-excited states and their molecular orbitals (MOs). The three trigonal sp^2 hybrids lie in a plane 120 ° apart, and when they interact a σ bond arises centered along the C-C axis. The overlapping of the remaining non-hybridized $2p_z$ AOs generates the π MO with positive (green) and negative (blue) lobes, that are located above and below the plane of the σ bond (Fig. **3.12B, C**). There is a given probability of electron density distribution between atomic nuclei: thus the electron cloud represents the charge density for the bond. In σ bonds, it is axially localized in the internuclear region. The π-bond pattern has zero electron density along the internuclear axis.

Bonding (π) and anti-bonding (π*) configurations result in different MOs, corresponding respectively to the highest occupied (HOMO) and lowest unoccupied (LUMO) patterns (Fig. **3.12D-F**). The molecule can absorb a photon when its energy is equal to the energy difference between the bonding HOMO and anti-bonding LUMO. In the simplest case of the H_2 molecule, it absorbs at a short wavelength in the far-UV region (160 nm), in agreement with the fact that $\sigma \rightarrow \sigma^*$ transitions require high-energy photons.

In the $\pi \rightarrow \pi^*$ transition of ethylene a nodal plane appears at the center of the excited C=C bond and perpendicular to it, and the corresponding transition dipole moment is in turn perpendicular to this plane. In comparison to ethylene, however, a more complex situation takes place in extensively unsaturated (aromatic) molecules, in which cyclic and planar structures are the main factors responsible for aromaticity. This is the typical case of benzene, the best known example of an organic molecule with delocalized electrons. In benzene, the overlapping of six $2p_z$ AOs forms a ring-shaped (doughnut-like) π MO with maximal electron density above and below the σ-bonds plane. The aromatic character extends to other rings such as pyrrole, furan, thiophene, oxazole, and cyclopentadiene.

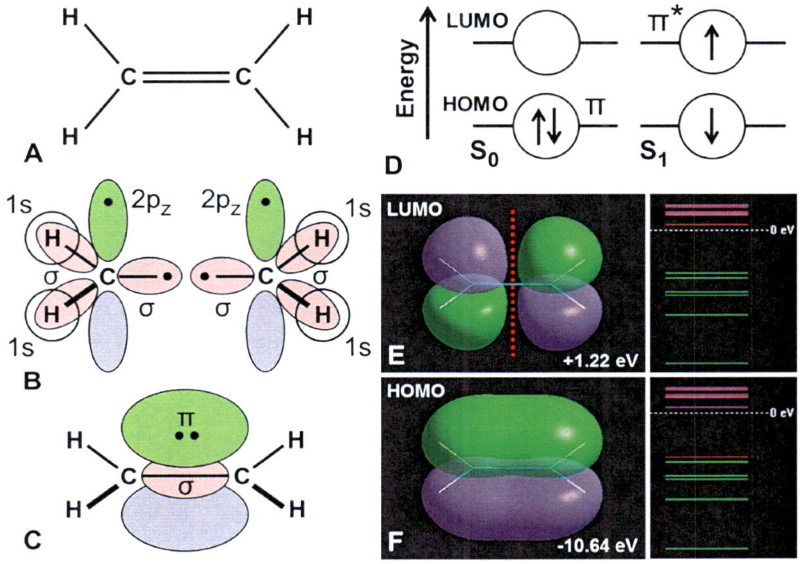

Fig. (3.12). A: Chemical structure of ethylene (ethene). **B:** Atomic orbitals (AOs) of carbon (sp^2) and hydrogen (1s) showing the σ bonds between C and H, and the remaining sp^2 and $2p_z$ AOs of carbon. **C:** The overlapping of the two $2p_z$ AOs forms a π bond, namely, a molecular orbital (MO) with positive (green) and negative (blue) lobes. The σ C-C bond is also shown. **D:** Scheme of the S_0 and S_1 states of ethylene with HOMO and LUMO orbitals. **F, E:** 3D isosurface plotting of the ethylene π (HOMO) and π* (LUMO) orbitals, respectively, with energy scale at right (red lines are the corresponding orbitals). Positive (green) and negative (blue) MOs are shown for the 15 ° tilted molecule from the frontal view (PM3, orbital contour: 0.05). The red dotted line in E represents the nodal plane between lobes of the anti-bonding π* orbital.

Regarding the electronic configuration of benzene, two formal (Kekulé) structures can be viewed as unexcited forms, whereas the three *para* (Dewar) structures would correspond to singly excited forms. The energy level and shape of the six π MO of benzene are represented in Fig. (**3.13**). In the ground state, the three anti-bonding π MOs remain unoccupied. The excited forms have anti-bonding π* MOs whose energy is determined by the number of nodal planes. High numbers of nodes represent greater levels of excited states.

Molecular excitation results when electronic transitions occur from low to high energy levels. The lower level may be either a bonding π-orbital, or a lone electron pair from a non-bonding (n) orbital provided by N or O atoms, and the higher level an unoccupied anti-bonding π* orbital. The corresponding transitions are π → π* and n → π*. When an aromatic ring carries a substituent with additional π orbitals such as $-NO_2$ and -COOH, π overlapping with an adjacent aromatic carbon is possible. The same effect takes place by substitution with atoms with a non-bonding n orbital (*i.e.* -OH, $-NH_2$), which can also overlap with π orbitals.

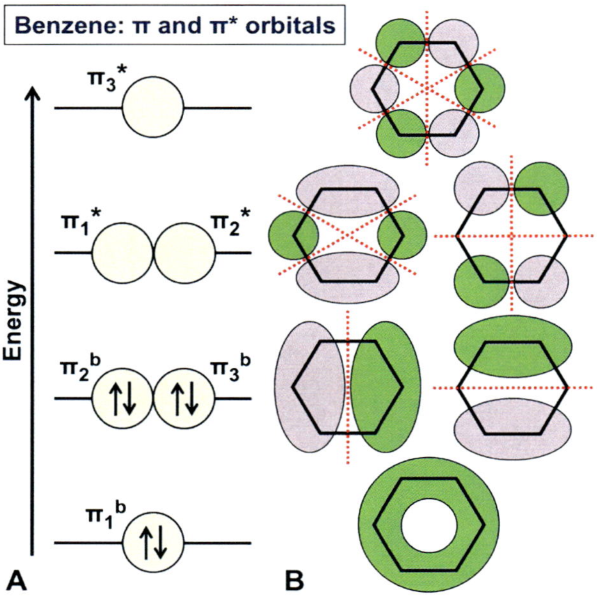

Fig. (3.13). Electronic configuration of benzene showing π-energy levels (A), and π-molecular orbitals (B), with one, two and three nodal planes indicated as red dotted lines. The three lower MOs correspond to ground states. Positive and negative MOs are green and blue, respectively.

The absorption (electronic) spectrum of a dye generally exhibits several bands, and each of these can be assigned to a electronic transition. Thus the absorption bands in simple molecules due to n → π* transitions (requiring lower energies) generally lie at longer wavelengths than those from aromatic π → π* transitions (Fig. **3.14A**). However, in complex dye molecules there are as many π-levels as are conjugated atoms, and the π-orbitals of both bonding and anti-bonding MOs are spread over a range of energies.

Generally, the higher-bonding π-orbitals have energies above those of the non-bonding n-level, so that the weak n → π* bands often remain submerged under the stronger π → π* bands and could not appear, although in simpler cases the former may be observed. Thus naphthalene shows the main absorption band due to π → π* transition between 250 and 300 nm, but in 1,2-diazanaphthalene (cinnoline) a new band at 380 nm results from the n → π* transition [6].

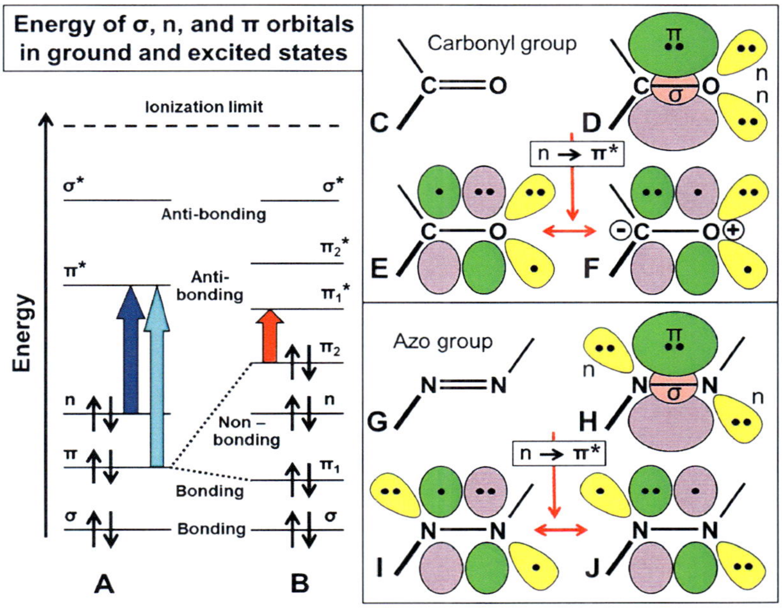

Fig. (3.14). A, B: Relative energies of the σ, π, n (lone pair), π*, and σ* orbitals in a simple (A) and complex (B) unsaturated molecule, showing n → π* and π → π* transitions. **C, D** and **G, H:** ground-state carbonyl and azo groups, with σ and π bonds. **E, F** and **I, J:** Pictorial description of mesomeric structures from the excited carbonyl and azo groups, respectively, showing positive (green) and negative (blue) π orbitals, and n orbitals in yellow. The charged carbonyl (F) is the predominant excited form.

In Fig. (**3.14 A,B**) two different energy levels are schematically shown for simple (**A**) and complex aromatic compounds (**B**). The molecule **A** possesses a pair of σ, π, and n electrons, while **B** has additional conjugated π electrons. Less energy (long wavelength) is required for the n → π* transition (blue arrow) in molecule **A** as compared to the π → π* transition (cyan arrow). In contrast, π levels are split in molecule B (π_1 and π_2) and the low energy (long wavelength) absorption band corresponds to a π → π* transition (red arrow). In Fig. (**3.14**) only one occupied σ- and n-level is shown for both molecules, but several energy levels can occur in dyes by further splitting of σ, σ*, π, and π* orbitals.

Based on the MOs involved, "pictorial descriptions" of electron distribution in the excited state of carbonyl (>C=O) and azo (-N=N-) groups have been published [20, 31]. In this context, mesomeric structures after n → π* transitions of these groups are represented in Fig. (**3.14 C-J**). As previously mentioned, n,π*-excited states can occur below the aromatic excitation. To illustrate this point, formazans are a good example. The formazan chromophore from reduction of tetrazolium salts (*e.g.,* MTT, see also Chapters 14.2.2 and 16.1.2) contain the π-bonding azohydrazone group (-N=N-C=N-NH-) attached to three aromatic rings. The broad absorption peaks of the formazan chromophore at 450-550 nm are due to π → π* transitions. In addition, peaks in the 300-350 nm region correspond to n → π* transitions of the -N=N- group, and peaks at 200-300 nm originate from n → π* transitions of the -C=N- group [37].

Ground and excited states of a cationic fluorophore may illustrate the shape of MOs, changes of partial charges, and features of the molecular structure. Pyridinium-phenyl-oxazoles (PyPO) are water soluble compounds based on the very efficient scintillator PPO (1,5-diphenyloxazole) [38]. In addition to lasing applications, the N-methyl derivative (mPyPO) is a bright fluorochrome useful for visualizing acid GAGs [39] (see Fig. **11.15**). The protonated HPyPO is used here to describe chemical changes during the excitation process. Fig. (**3.15**) shows the formal structure of the ground and excited state of HPyPO [38], as well as the changes in the partial charge distribution and MOs. In the ground state, most of the positive charge is localized on the pyridinium nitrogen (δ^+: 0.55), whereas in the excited state the charge is most dispersed on *ortho* and *para* carbons of the phenyl group. The partial charge in the >NH group of the excited paraquinone state appears now decreased to δ^+: 0.12.

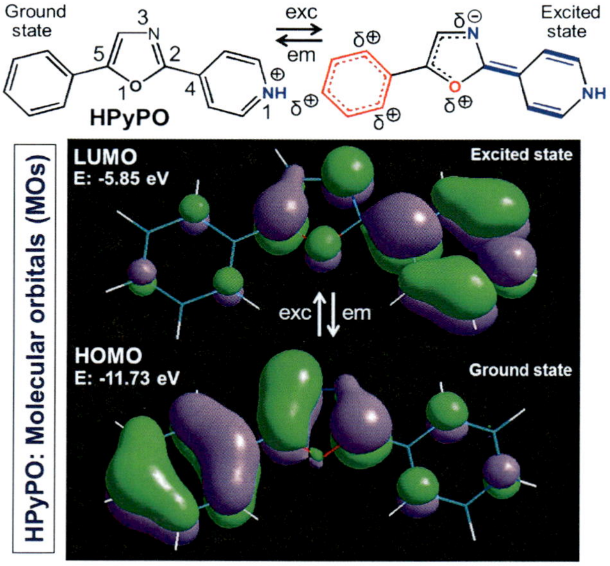

Fig. (3.15). Top: Molecular structure of ground (left) and excited (right) states of 4-(2-[5-phenyloxazol-l]-pyridinium cation (HPyPO) with atom numbering at left. In the excited dye the occurrence of preferential double bonds (blue) and partial charges (δ) are also indicated. Bottom: Molecular orbitals (HOMO and LUMO, PM3 geometry) of HPyPO showing energy values (orbital contour: 0.025).

3. FLUOROCHROMES

Although there are dyes that fluoresce in the UV or in the IR spectral region, most fluorescent dyes emit visible light, and thus are easily visualized. They are applied in biological staining, lasing, plastic and textile dyeing, *etc.* A number of dyes commonly used in bright-field microscopy also can be visualized because of their fluorescence signal (*i.e.* alcian yellow, acid fuchsine, carmine, eosin Y, fast green FCF, hematoxylin, mercurochrome, orcein, saffron, *etc.*) [25, 40 - 47]. These fluorescent dyes are described in Chapters 4, 8, 11, 12 and 13. Characteristics of an ideal fluorochrome are strong absorption, high fluorescence quantum yield (Φ_F), emission color different from autofluorescence, high photostability (low or no fading), good water solubility, very little triplet formation, and selective affinity for biological substrates.

3.1. Classification and Chemistry

Regarding dye and fluorochrome terminology, general concepts have been mentioned in Chapter 1.4. Although names as di-and triphenylmethanes are widely used, they correspond to misleading terms, since the central C atom of these dyes is not sp^3-, but sp^2-hybridized and therefore they must be called di- and triphenylmethine compounds [9]. In spite of that, the older name of "methane" will be conserved on historical grounds and because it refers to the uncharged parent chemical group from which the dyes derivate.

According to their origin, fluorochromes are natural or synthetic compounds. Some natural compounds, generally from plants and possessing biological activity (such as many alkaloids and toxins) are fluorescent. This characteristic is useful in botanic taxonomy and pharmacologic analysis, allowing the identification of these compounds [48]. Several of them are employed as fluorochromes (*i.e.* berberine, coumarin and quinine derivatives, flavonoids, polyenes, *etc.*). Synthetic fluorochromes are very varied regarding chemical constitution, generally containing fluorophores such as acridine, anthraquinone, arylmethane, azine, stilbene, pyrene, coumarin, benzofuran, benzimidazole, benzothiazole, flavonoid, indole, naphthalimide, phenanthridine, quinoline, oxazine, oxazole, xanthene, porphyrin and other aromatic groups [4, 13].

Size-related luminophores, arranged from large to small size, are illustrated in Fig. (**3.16**). In addition to the different size and shape, these compounds show distinct mechanisms of light excitation and emission, in all cases corresponding to electron transitions from low to high energy levels. In strict sense a true fluorescent emission only arises from the $^1\pi^*$ state, and accordingly photoluminescence and then luminophores are better names in order to include the remaining emission processes (from either conduction band-, $^3\pi^*$-, or f*-electrons).

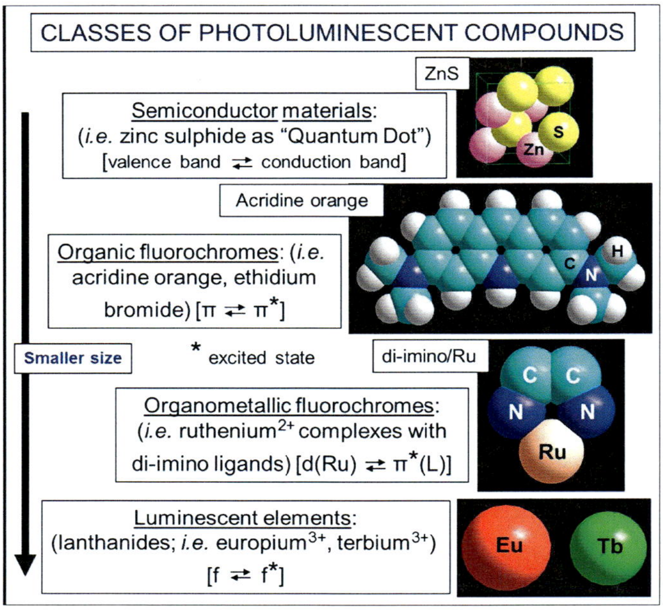

Fig. (3.16). Schematic classification of luminophores according to size, chemical constitution, and type of electron transitions.

Nanoparticles of crystalline semiconductors are important and promising photoluminescent solid-state materials, now called "quantum dots" or Qdots (see Chapter 18.2), most important constituents being cadmium selenide and zinc sulfide. Regarding other inorganic salts, it is interesting to mention that barium tetracyanoplatinite ($BaPt[CN]_4 \cdot 4H_2O$) is luminescent under UV and X photons, as well as when excited by radium and cathode rays, its principal use being for X-ray screens. The luminescence quantum yield of potassium tetracyanoplatinite is Φ_F: 1 in the crystalline state, and only 0.045 in water solution.

Ruthenium(II) forms luminescent complexes with diimino (-N=CH-CH=N-) ligands (*i.e.* bipyridine, *o*-phenanthroline). Light absorption induces a metal to ligand charge transfer (MLCT) of a d^6 electron from Ru to a ligand π^* orbital. The singlet ^{1}MLCT converts to the long-lived triplet ^{3}MLCT with Φ_T: 1, and in the absence of $^3O_2$ the excited ligand decays by phosphorescence. Ru-diimino complexes have been applied to visualize sulfated polyanions and DNA in fluorescence microscopy (see Chapters 11.4.3 and 15.4.1). Other metals (Ir, Os) also form luminescent complexes with these ligands.

Several lanthanide and actinide elements show light emission for themselves, although they are commonly employed as complexes with "antenna" ligands. Regarding lanthanides, the reduction of cerium(IV) by sugars gives Ce(III) that is highly fluorescent (λexc: 260 nm, λem: 350 nm) and has analytical applications [27, 49]. But most important in microscopy are terbium(III) and europium(III) ions complexed with suitable ligands (*i.e.* bipyridine, β-diketones), which shift the excitation toward longer wavelengths [50]. On the other hand, the luminescence of uranium salts is well known (*i.e.* the uranyl group, UO_2^{2+}, which is a precipitant reagent for phosphates and shows green emission). For crystalline uranyl sulfate, the Φ_F is 1, whereas it is 0.01 for the aqueous solution. It should be noted that the quantum yield value would be referred better as luminescence than as fluorescence.

Acridine orange (AO; Fig. **3.16**) is a useful fluorochrome with ortho- and metachromatic properties, applied since log ago in fluorescence microscopy. In this context, numerous fluorescent dyes and organic compounds are applied in research and diagnostic assays [4, 50 - 54]. Practical examples of most used fluorochromes in microscopy requiring UV (365 nm) excitation are the following: AMCA (coumarin), ANS (naphthalene), cascade blue (pyrene), DAPI (diamidine), Hoechst 33258 (bisbenzimidazole), fluorescamine- and MDPF-derivatives, with emission at 420-480 nm (white-blue). Those requiring violet-blue or blue excitation are AO and hydroxy-xanthenes (fluorescein, FITC, eosin), AO emitting in green, yellow, orange or red according to the substrate. Rhodamine, Texas red, and TRITC (amino-xanthenes), as well as ethidium and propidium (phenanthridines) are conveniently excited with green light (546 nm) and fluoresce in red.

3.2. Size: QSAR Parameters

Quantitative structure–activity relationships (QSAR) can be used to specify the physico-chemical characteristics of probes that accumulate preferentially within living cells. The QSAR approach requires describing relevant physico-chemical properties numerically. The overall size of dyes and fluorochromes can be precisely described by their projections on a planar area (Å^2). However, regarding the conjugated systems it is more easy and practical to record by simple inspection of the chemical structure some relevant parameters such as the conjugated bond number (CBN) for the size of the whole aromatic system, and largest conjugated fragment (LCF) for the size of the planar chromophore [3, 55]. These and other relevant QSAR parameters are described in Chapter 15.2.1.

Small cyclic fluorophores with low CBN are reduced nicotinamide (7), N-methylanthraniloyl (Mant) products (9), *o*-phthalaldehyde derivatives (10), bimane (11), indoxyl (11), azulene (11), dipyrromethene-boron difluoride (Bodipy, 12), dansyl (12), dicyano-dihydroxybenzene (12), hydroxy-coumarin (13), lumazine (13), nitrobenzoxadiazole (NBD, 14), hydroxy-quinoline, nitroso-naphthol and hydroxy-naphthaldehyde metal complexes (14), Ru-bipyridine (15), β-carboline (15), carbazole (15), *etc.* Among linear fluorophores, benziliden-imidazolones (green fluorescent protein, 15), stilbene (rhaponticin, 17), aryl-oxazoles (mPyPO, 19), and aryl-pyrrolone derivatives (20) also have a small molecular size. Generally, fluorophores with small CBN values show fluorescence at short wavelengths (blue), whereas long wavelength emission (red) occurs for fluorophores with high CBN.

On the other hand, there are also macrocyclic fluorophores, most important being porphyrins, porphycenes (structural isomers of porphyrins) and phthalocyanines. The CBN of unsubstituted porphyrins and phthalocyanines is 28 and 48, respectively. These fluorescent macrocycles may lack metal or contain diamagnetic metals (Mg^{2+}, Al^{3+}, Zn^{2+}). Porphyrin complexes with other metals (Pd^{2+}, Pt^{2+}) show phosphorescence emission if the $^3\pi^*$ state is not quenched by $^3O_2$. When complexed with Pd, Pt, Y, Lu, Zn and Sn, porphyrin phosphorescence is used for oxygen sensing, whereas Yb complexes are applied in IR-fluorescence tumor diagnosis [56]. For more details about macrocyclic complexes with transition metals see Chapter 9.

3.3. Scintillators, NIR and Blinking Fluorochromes

A number of fluorochromes have particular spectral or chemical characteristics that make them very suitable for specific uses such as diagnostic purposes and advanced microscopy. Although several of these especial fluorochromes will be described in next Chapters, it seems appropriate to introduce here some relevant features.

Scintillators are highly fluorescent compounds applied for scintillation counting methods [57, 58]. The basic process is that under high energy excitation (X- and γ-rays, α- and β-particles, *etc.*), free radicals, free electrons, and excited solvent molecules (toluene, xylene) are produced, which lead to solute excitation. Primary scintillator solutes are *p*-terphenyl and 2,5-diphenyloxazole (PPO), whereas secondary ones, used to shift the light emission to longer wavelengths, are diphenyl-hexatriene (DPH), 1,4-bis-2-(5-phenyloxazolyl)-benzene (POPOP), and its 4,4-dimethyl derivative (dmPOPOP). To measure radioisotopes (*i.e.* H^3, C^{14}, P^{32}, S^{35}, the natural K^{40}, Cr^{51}, I^{131}, *etc.*), scintillators are used either as crystals, incorporated into liquids (toluene, xylene) or plastics (polystyrene) as "solvent-like" energy donors. Several scintillator-derived cationic fluorochromes are applied in microscopical work to visualize selectively cell and tissue substrates, *i.e.* mPyPO and Q-dmPOPOP (see Figs. **5.16**, **11.15**, and **11.16**).

Near infrared (NIR) fluorescence (NIRF) methodology is now one of the most popular imaging modes in biomedical sciences for visualization of cells and tissues both *in vitro* and *in vivo* [59, 60]. Indocyanine green (cardio green, fox green) was developed for NIR photography by Kodak in 1955, and approved for clinical use already in 1956. It is a widely used NIRF dye, showing very little absorption in the visible range (λexc: 750 to 800 nm, λem: 835 nm). Obviously, as the emission of far red and NIR dyes is not directly visible by eye, infrared detectors are required to record the images. Indocyanine green binds selectively to plasma lipoproteins giving increased emission, and is currently applied in medicine for determination of hepatic function, cardiac output, and blood plasma volume [61, 62]. NIR laser emission of heptamethine pyrylium dyes (λexc: 1000 nm, λem: 1400 nm) [63] could have biological applications.

At present, NIRF imaging has increasing importance, and thus indocyanine green is used as fluorescent stain and marker for ophthalmology and angiography [64], but mainly to identify lymphatics and label sentinel lymph nodes in oncological surgery [65 - 67]. An interesting use of indocyanine green is for photodynamic and/or photothermic therapy of Kaposi's sarcoma and other tumors [68 - 73], and to trigger a PDT-induced tumor destruction in metastatic lymph nodes. A critical review on the photothermic and photodynamic effects of this dye has been recently published [74].

Disadvantages of indocyanine green are its high cost and instability of water solutions, and other NIR fluorochromes (*i.e.* Cy7, IRDye800CW, IR-783, *etc.*) are now available [66]. It is worth mentioning that sulfonated triphenylmethanes such as patent blue V (CI 42045), its isomer isosulfan blue, and methylene blue [4] may also be used as marker diachrome dyes to recognize metastatic lymph nodes without fluorescence technology. Methylene blue and carbon nanotubes (see Fig. **4.21**) are compounds with NIRF properties.

Finally, other especial types of synthetic or natural fluorochromes are noteworthy, namely photoswitching (blinking) fluorophores that are applied for subdiffraction-resolution fluorescence microscopy (super-resolution [75, 76] (see Chapter 20.4). The super-resolution methodology is based on the temporal control of fluorophore emission, using photochromic dyes that can be switched reversibly between a colored and fluorescent on- and a colorless non-fluorescent off-state. The spiropyran ↔ merocyanine system is an attractive optical switch to bind to rhodamine fluorophores, allowing super-resolution fluorescence imaging based on precise localization of individual emitters [77]. In addition, cyanines, expanded rhodamines, some fluorescent proteins, and Qdots have been used as blinking fluorophores [75, 78].

4. PREDICTION OF FLUORESCENCE

4.1. General Aspects

Attempts to correlate the chemical structure of compounds with fluorescence properties useful in microscopy are rather scarce [10, 12, 79]. Conjugated double bounds and planar structure are important factors but they do not explain all the fluorescence features nor allow precise predictions. Some practical rules based in numerical parameters such as CBN values (see Chapter 15.2.1) have been proposed to predict fluorescence of dyes. Using paraffin sections of tissues stained by a gallery of 90 dyes, 90% of them with CBNs of 29 or less were fluorescent, whilst 70% of dyes whose CBNs exceeded 30 were non-fluorescent [12].

Why is a compound fluorescent? All the compounds (colored) that are seen by the eye absorb light of a given wavelength, and we see a "subtraction spectrum" (light reflected by a body that has absorbed part of the white light). Then, all that we commonly see corresponds to matter that has absorbed energy (photons) and is in the excited state. It can decay to the ground state:

1) by photon emission in the visible range (fluorescence, phosphorescence), but most of all,

2) by internal conversion (heat generation and dissipation), energy transfer to other molecules (*i.e.* to $^3O_2$), photochemical reactions, or emission of photons that are not visible by the eye (λem < 400 nm [UV] or > 700 nm [NIR]).

Therefore, luminescence (light emission) is only one of the ways in which energy of the excited state is dissipated and the compound returns to its ground state. A typical feature for most fluorochromes is their rather planar and stiff molecular structure, which prevents the energy of the excited state from being lost by torsion (bond rotation). Conjugated double bonds and charge resonance between ionized and unionized atoms must also be taken into account as in the case of 1-naphthol (Fig. **3.17**). The same occurs for *p*-hydroxydiphenyl. In addition, fluorescence increases by nucleophilic (electron-donating) groups, typically -NH_2, -OH, -CN, and decreases by electrophilic (electron-accepting) substituents (*i.e.* -NO_2, -NO, and -$N^+{\equiv}N$).

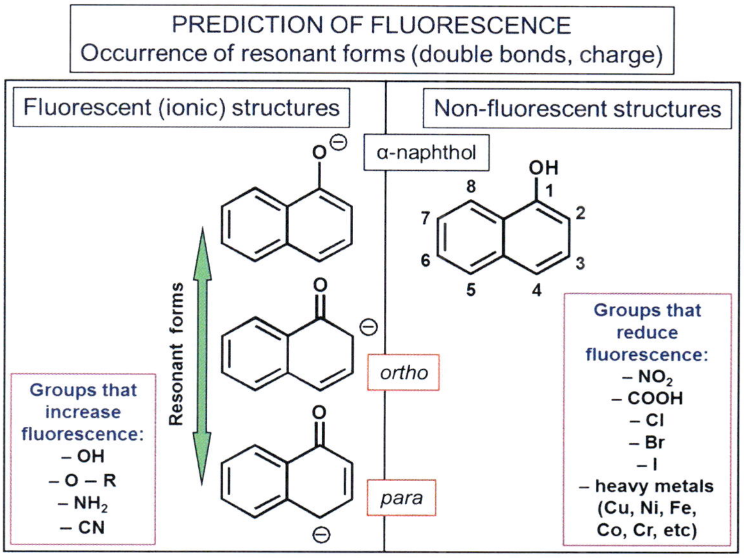

Fig. (3.17). An example of prediction of fluorescence. α-Naphthol with atoms numbered (right), and the possible resonance configurations of charged forms (left). Other resonances of negative charge (at C5 or C7) are possible. Substituents that increase or decrease the emission intensity are also shown.

Transition and heavy metals (Fe, Cu, Cr, Co, Ag Au, Pb, Hg, *etc.*) quench the fluorescence of organic compounds where they are incorporated. A well known example is that porphyrin complexes with these metals are not fluorescent [27]. However, an interesting exception is ferrocene, a 2:1 sandwich complex between two cyclopentadiene rings and Fe^{2+}. In this case the molecule is diamagnetic and its dipole moment is zero. Ferrocene solutions in nitric or sulfuric acid are red with blue fluorescence [13, 14]. On the other hand and due to a "heavy atom effect", the speed of forbidden transitions (singlet → triplet) increases when some heavy atoms (*i.e.* iodine, palladium) are present, and then the Φ_T also increases.

The occurrence of conjugated double bonds in aromatic molecules and charge resonance in O or N are important factors for appearance of fluorescence. Esculetin (6,7-dihydroxy-coumarin, the aglycone of esculin obtained from *Aesculus hippocastanum*) is an example of the correlation between resonant forms and fluorescence (Fig. **3.18**). Aqueous solutions of esculin above pH 5.8 are ionized and show blue fluorescence, and the same occurs for fraxin (fraxetin-8 - glycoside). Interestingly, these fluorescent coumarins are glycosylated at C6 (esculin) and C8 (fraxin), whereas in the non-fluorescent isomer cichorium, glucose is attached to C7, thus hindering the resonance between C2 and C7.

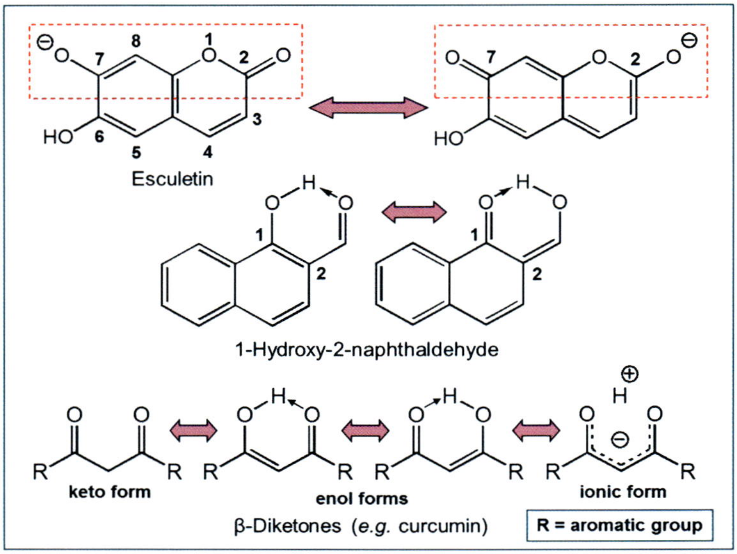

Fig. (3.18). Fluorescence mechanisms. Resonant forms (bonds, charges, H-bonding) of some fluorophores. Esculetin contains a coumarin ring (numbered atoms), with a chain of conjugated double bonds between oxygen atoms bound to C2 and C7. 1-Hydroxy-2-naphthaldehyde and β-diketones also show the resonance of H-bonds.

Other coumarin compounds are also strongly fluorescent, *i.e.*, hymecromone, umbelliferone (7-hydroxycoumarin, a fluorescent pH indicator, with blue emission at pH 6.2 or higher), and its 6-methoxy derivative, scopoletin [27]. 4-Methyl-esculetin) is also used as fluorogenic substrate [14]. The coumarin derivative coumestrol (an estrogenic factor from clovers and alfalfa, also related to isoflavones) shows blue fluorescence in neutral or acid solutions and green-yellow fluorescence at alkaline pH [26]. There are also coumarin antibiotics, such as novobiocin and coumermycin A_1, which are effective inhibitors of DNA gyrase.

On the other hand, lumazine tautomers have pH-dependent fluorescence. Aqueous solutions show green or blue fluorescence at alkaline or acid pH, respectively. Interestingly, this behavior also corresponds to that of more complex compounds such as riboflavine derivatives: lumiflavine (green emission) and lumichrome (blue emission) [26]. Mesomerism of the dark blue azine pigment pyocyanin (from *Pseudomonas aeruginosa*) is another example. In addition, the aqueous blue zwitterion at alkaline pH is easily reduced to a colorless compound.

Tautomeric H-bonding and resonance of keto-enol forms can be found in other fluorochromes (Fig. **3.18**). Curcumin (CI 75300) is a natural β-diketone (acetylacetonate) dye with chelating properties obtained from rhizomes of the Asian plants *Curcuma longa* (*C. domestica*), *C. tinctoria*, *etc.*, and mainly used as food additive (E-100). Curcumin shows anti-inflammatory, immuno-modulating, tumor-preventing and anti-atherogenic properties, generating a complex response because of its both anti- and pro-oxidative effects. Likewise, curcumin has photosensitizing activity [80]. In addition to its complexes with some metals (see Chapter 8.3.5), curcumin is fluorescent by itself, inducing a green emission in substrates containing polysaccharides (see Fig. **11.3**), and nucleic acids [81, 82]. Particularly interesting is the selective fluorescence observed in kinetoplast DNA (Fig. **3.19**). Carthamin (CI 75140, from leaves of the African saffron, *Carthamus tinctorius*, and used as surrogate for Spanish saffron) is similar to curcumin and fluorescence applications should be investigated.

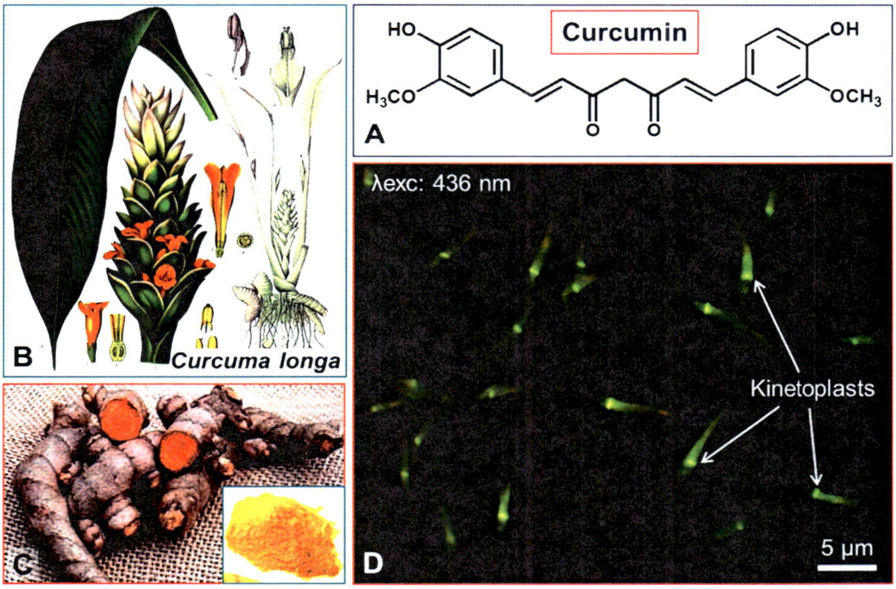

Fig. (3.19). **A**: Chemical structure of curcumin. **B**: The plant *Curcuma longa*. **C**: Rhizomes and powder of curcuma. **D**: *Trypanosoma cruzi* epimastigotes fixed in methanol and stained with aqueous 10 µg/mL curcumin for 5 min, showing strong fluorescence of kinetoplast DNA.

4.2. Non-Radiative Energy Dissipation

Many dyes that are non-fluorescent in aqueous solution become fluorescent or even phosphorescent when frozen, bond to suitable macromolecules, adsorbed to gels or solid surfaces, or dissolved in solvents of high viscosity [45]. This is due to lower solvent collisions and limitations of the vibrational and rotational decay from the excited state of the fluorophore, which result in thermal dissipation of energy. The emission increase of some azo and triphenyl-methane dyes in viscous solvents is a good example (see Figs. **5.10** and **5.11**).

Likewise, the non-coplanar conformation of dyes leads to energy dissipation by the occurrence of bonds with rotational freedom, generally abolishing photon emission. This radiation-less internal conversion involves rotation of different parts of the molecule resulting in a "free-rotor" effect. Thus the presence of a rigid planar ring is particularly favorable for fluorescence, which is well demonstrated by comparing phenolphthalein and fluorescein. The sodium salt of phenolphthalein (non-fluorescent in aqueous solution) is structurally very similar to the sodium salt of fluorescein, except for the oxygen (ether) bridge of the latter (Fig. **3.20**). Closing of the middle ring by oxygen is responsible for the fluorescence. However, on account of the tetrahedral (saturated) central C9 from dihydrofluorescin and fluorescein diacetate lactone these compounds are non-fluorescent. On the other hand, the diphenylmethane Michler's hydrol blue shows no fluorescence in solution, but bridging the two phenyl rings with a -NH- group converts it into the highly fluorescent acridine orange. In the case of fluorene and carbazole, in which planarity is stabilized by -CH_2- or -NH- bridges, respectively, fluorescence is higher than in biphenyl, in which the two phenyl groups can rotate about the single bond that joins them.

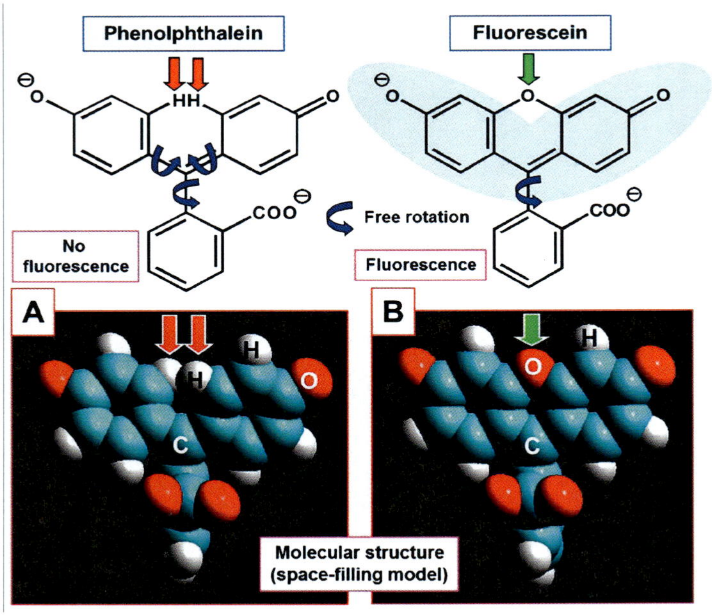

Fig. (3.20). Fluorescence requisites: coplanarity and rigidity. Comparison between the molecular structure of the pH indicator phenolphthalein (**A**) and the xanthene dye fluorescein (**B**). In the former, H atoms (red arrows) restrict coplanarity of phenyl rings due to steric hindrance. The rigid tricyclic fluorophore of fluorescein is formed through an oxygen bridge (green arrow) allowing fluorescence.

Under some conditions (*i.e.* viscous solvents, binding to substrates) the phenolphthalein structure becomes stabilized in a fluorescein-type planar configuration, in which case fluorescence is possible. This is actually observed for phenolphthalein in solid solutions of gelatin or sucrose. The same phenomenon occurs for solid or gel solutions of malachite green, crystal violet, fast green FCF, *etc.*, which in aqueous solution are non-fluorescent.

4.3. Hematoxylin and Eosin as a Case Example

Another correlative example of dyes in which bridging atoms make rigid a resonant hydroxy-diphenylmethane fluorophore is the couple hematoxylin-eosin Y. The natural and weakly acid neoflavone, hematoxylin (4-hydroxy-brazilin, CI 75290, see Figs. **1.3** and **7.6**), is a well known dye mainly used after oxidation to form metal (*i.e.* Al^{3+}, Cr^{3+}, Fe^{3+}) complexes, which selectively stain polyphosphate anions such as nucleic acids [4, 83, 84]. The oxidized dye has been named haematine, hemateïn, hämateïn, or hematein, but all these names are unfortunate, giving rise to confusion with the hemoglobin derivative pigment, hematin [13]. The name oxidized hematoxylin (hematoxylein), or simply hematoxylin (assuming that the dye is in the usual oxidized form) should be preferred. Likewise, brazilin (CI 75280) is a similar compound that after spontaneous oxidation gives the red dye, brazilein.

It is worth noting that the fluorescence properties of oxidized hematoxylin alone (lacking metals) have been overlooked, but its capacity for inducing selective fluorescence in several cell structures has been described [41] (Fig. **3.21**). On the other hand, the fluorescence of the hydroxy-xanthene dye, eosin Y, is also well known [46, 47]. In this connection, a rigid hydroxy-diphenylmethane is the common chromo-fluorophore of both hematoxylin and eosin Y [24]. The distribution of resonant double bonds, and some structural parameters are seen in Fig. (**3.21C**), where the boomerang-shape of conjugated atoms is indicated by shaded blue areas. The two mesomeric forms of hematoxylin ($-O_9^-$ / $=O_3$ $\leftrightarrow$ $=O_9$ / $-O_3^-$) has been indicated [24, 83].

In addition to hematoxylin and eosin Y, the following dyes have the same conjugated hydroxy-diphenylmethane pattern in which a rigid bridge avoids the free rotation of phenyl groups: fluorescein, gallein, iris blue, phenyl-fluorone, resorufin, and halogenated or mercurial fluorescein derivatives (erythrosin, phloxin B, rose Bengal, mercurochrome, *etc.*). Microscopic fluorescence reactions by all these anionic dyes are well known [13, 14, 24, 40, 85].

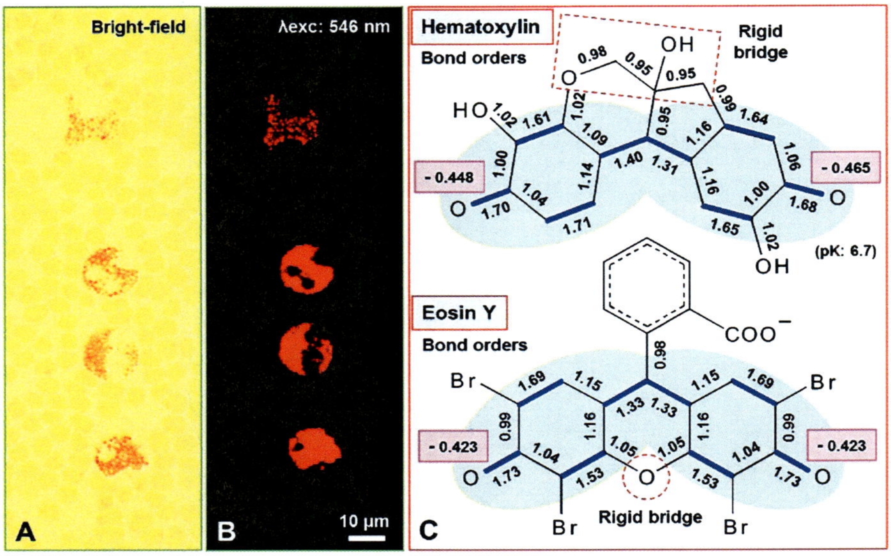

Fig. (3.21). **A**, **B**: Horse blood smear stained for 5 min with 0.1 mg/mL oxidized hematoxylin showing stained (**A**) and fluorescent eosinophils (**B**). **C**: Molecular correlations between oxidized hematoxylin and eosin Y, with the common hydroxy-diphenylmethane chromophore (shaded cyan areas) and rigid bridges. Bond orders, shortest double bonds (thick blue lines), and the partial charge of oxygen atoms (pink rectangles) are indicated (PM3 geometry).

CONFLICT OF INTEREST

The author (editor) declares no conflict of interest, financial or otherwise.

ACKNOWLEDGEMENTS

This work was supported by a grant from the Ministerio de Economía y Competitividad (CTQ2013-48767-C3-3-R), Spain.

REFERENCES

[1] Gurr E. Synthetic Dyes in Biology, Medicine and Chemistry. London, New York: Academic Press 1971.

[2] Horobin RW. Staining plastic sections: a review of problems, explanations and possible solutions. J Microsc 1983; 131(Pt 2): 173-86. [http://dx.doi.org/10.1111/j.1365-2818.1983.tb04244.x] [PMID: 6194302]

[3] Horobin RW. Uptake, distribution, and accumulation of dyes and fluorescent probes within living cells. A structure-activity modelling approach. Adv Colour Sci Technol 2001; 4: 101-7.

[4] Horobin RW, Kiernan JA, Eds. Conn's Biological Stains A Handbook of Dyes, Stains and Fluorochromes for Use in Biology and Medicine. 10th ed., Oxford, UK: BIOS Scientific Publishers 2002.

[5] Abrahart EN. Dyes and their Intermediates. Oxford, London: Pergamon Press 1968.

[6] Mason SF. Color and the Electronic States of Organic Molecules.Chemistry of the Synthetic Dyes. New York: J. Wiley and Sons 1970; pp. 169-221.

[7] Gordon PF, Gregory P. Organic Chemistry in Colour. Berlin, Heidelberg: Springer 1987. [http://dx.doi.org/10.1007/978-3-642-82959-8]

[8] Nassau K. The Fifteen Causes of Color: The Physics and Chemistry of Color. 2nd ed., New York: Wiley-Interscience 2001.

[9] Zollinger H. Color Chemistry.Syntheses, Properties, and Applications of Organic Dyes and Pigments. 3rd ed., Zürich, Weinheim: Wiley-VCH, VHCA 2003.

[10] Puchtler H, Sweat F, Gropp S. An investigation into the relation between structure and fluorescence of azo dyes. J R Microsc Soc 1967; 87(3): 309-28. [http://dx.doi.org/10.1111/j.1365-2818.1967.tb04513.x] [PMID: 5582815]

[11] Berlman IB. Handbook of Fluorescence Spectra of Aromatic Molecules. 2nd ed., New York, London: Academic Press 1971.

[12] Juarranz A, Horobin RW, Proctor GB. Prediction of in situ fluorescence of histochemical reagents using a structure-staining correlation procedure. Histochemistry 1986; 84(4-6): 426-31. [http://dx.doi.org/10.1007/BF00482974] [PMID: 3721913]

[13] Lillie RD. Conn's Biological Stains. 9th ed., Baltimore: Williams and Wilkins 1977.

[14] Green FJ. The Sigma-Aldrich Handbook of Stains, Dyes and Indicators. Milwaukee, Wisconsin: Aldrich Chemical Company 1990.

[15] Horobin RW. Histochemistry An Explanatory Outline of Histochemistry and Biophysical Staining. Stuttgart, London: Gustav Fischer, Butterworths 1982.

[16] Gupta AK, Ed. The Complete Book on Natural Dyes and Pigments National Institute of Industrial Research. Delhi: Asia Pacific Business Press 2005.

[17] Verhoeven JW. Glossary of terms used in photochemistry. IUPC Recommendations. Pure Appl Chem 1996; 68: 2223-86. [http://dx.doi.org/10.1351/pac199668122223]

[18] Stockert JC, Abasolo MI. Inaccurate chemical structures of dyes and fluorochromes found in the literature can be problematic for teaching and research. Biotech Histochem 2011; 86(1): 52-60. [http://dx.doi.org/10.3109/10520295.2010.489428] [PMID: 20500122]

[19] Stockert JC. Stereoscopy of computer-drawn molecular structures. Biochem Educ 1994; 22: 23-5. [http://dx.doi.org/10.1016/0307-4412(94)90160-0]

[20] Pischel U, Nau WM. Structure-reactivity relationships in the photoreduction of n,π*-excited ketones and azoalkanes: the effect of reaction thermodynamics, excited-state electrophilicity, and antibonding character in the transition state. Photochem Photobiol Sci 2002; 1(2): 141-7. [http://dx.doi.org/10.1039/b110108k] [PMID: 12659131]

[21] Morales J, Günther G, Zanocco A, Lemp E. Singlet oxygen reactions with flavonoids. A theoretical-experimental study. PLoS ONE 2012; 7(7): e40548.

[22] Feigl F. Spot Test in Inorganic Analysis. New York: Elsevier Publishing Co. 1958.

[23] Stockert JC, Colman OD, Cañete M. Fluorescence reaction of leukocyte granules by morin. Acta Histochem Suppl 1985; 31 (Suppl.): 243-52. [PMID: 2410949]

[24] Stockert JC. The horse eosinophil as a model leucocyte for morphological and cytochemical studies. Braz J Morphol Sci 2005; 22: 73-84.

[25] Trigoso CI, Stockert JC. Fluorescence of the natural dye saffron: selective reaction with eosinophil leucocyte granules. Histochem Cell Biol 1995; 104(1): 75-7.

[http://dx.doi.org/10.1007/BF01464789] [PMID: 7584563]

[26] Budavari S, Ed. The Merck Index. An Encyclopedia of Chemicals, Drugs and Biologicals. 11th ed.. Rahway, NJ: Merck and Co. 1989.

[27] Udenfriend S. Fluorescence Assay in Biology and Medicine. New York, London: Academic Press 1962.

[28] Stockert JC, Blázquez A, Galaz S, Juarranz A. A mechanism for the fluorogenic reaction of amino groups with fluorescamine and MDPF. Acta Histochem 2008; 110(4): 333-40. [http://dx.doi.org/10.1016/j.acthis.2007.10.011] [PMID: 18272208]

[29] Nagasawa Y, Watanabe A, Ando Y, Okada T. Solvent dependence of the ultrafast ground state recovery dynamics of phenol blue. J Mol Liq 2001; 90: 295-302. [http://dx.doi.org/10.1016/S0167-7322(01)00133-7]

[30] Pauling L. The nature of the chemical bond. 3rd ed. Ithaca, NY: Cornell University Press 1960; p. 139.

[31] Zimmerman HE. A new approach to mechanistic organic photochemistry. Adv Photochem 1963; 1: 183-208. [http://dx.doi.org/10.1002/9780470133316.ch6]

[32] Seel F. Atomic Structure and Chemical Bonding A Non-Mathematical Introduction. London, New York: Methuen, John Wiley 1965.

[33] Garratt PJ, Vollhardt P. Aromatizität. Stuttgart: Georg Thieme Verlag 1973.

[34] Christen HR. Atommodelle und chemische Bindung 3. Frankfurt am Main: Umschau Verlag 1976.

[35] Coulson CA. Electronic Configuration and Carcinogenesis.Advances in Cancer Research. New York: Academic Press 1953; Vol. 1: pp. 1-56.

[36] Seliger HH, McElroy WD. Light: Physical and Biological Action. New York, London: Academic Press 1965.

[37] Tunc T, Yildirim LT. Synthesis, crystal structure and spectroscopic studies of 1-(p-bromphenol)--,5-diphenylformazan. Open Cryst J 2010; 3: 54-8. [http://dx.doi.org/10.2174/1874846501003010054]

[38] Lee LA, Robb RA. Water soluble blue-green lasing dyes for flashlamp-pumped dye lasers. IEEE J Quantum Electron 1980; QE-16: 777-84. [http://dx.doi.org/10.1109/JQE.1980.1070567]

[39] Stockert JC. Cytochemistry of mast cells: new fluorescent methods selective for sulfated glycosaminoglycans. Acta Histochem 2000; 102(3): 259-72. [http://dx.doi.org/10.1078/S0065-1281(04)70034-3] [PMID: 10990064]

[40] Ferrer JM, González Garrigues M, Stockert JC. Mercurochrome: a fluorescent and electron opaque dye. Microsc Acta 1983; 87(3): 293-300. [PMID: 6195509]

[41] Stockert JC, Molero ML, Espelosin RH. Fluorescence patterns of chromatin and cytoplasm by hematoxylin solutions. Z Naturforsch C Biosci 1983; 38(1-2): 153-5. [PMID: 6845816]

[42] Molero ML, Del Castillo P, Stockert JC. Fluorescence reactions of orcein: new applications for an old stain. J Microsc 1986; 144(Pt 1): 45-53. [http://dx.doi.org/10.1111/j.1365-2818.1986.tb04672.x] [PMID: 2432271]

[43] Stockert JC, Del Castillo P, Armas-Portela R. Alcian yellow as a fluorescent dye. Acta Histochem 1989; 85(1): 59-63. [http://dx.doi.org/10.1016/S0065-1281(89)80099-6] [PMID: 2496573]

[44] Stockert JC, Llorente AR, Del Castillo P, Gómez A. Chromatin fluorescence after carmine staining. Stain Technol 1990; 65(6): 299-302.

[http://dx.doi.org/10.3109/10520299009105621] [PMID: 2080525]

[45] Stockert JC, Espada J, Trigoso CI, Del Castillo P. Fluorescence reaction of the anionic triaryl-methane dye fast green FCF. Appl Fluor Technol 1991; 3(6): 14-5.

[46] Espada J, Valverde P, Stockert JC. Selective fluorescence of eosinophilic structures in grasshopper and mammalian testis stained with haematoxylin-eosin. Histochemistry 1993; 99(5): 385-90. [http://dx.doi.org/10.1007/BF00717051] [PMID: 7687595]

[47] Apgar JM, Juarranz A, Espada J, Villanueva A, Cañete M, Stockert JC. Fluorescence microscopy of rat embryo sections stained with haematoxylin-eosin and Masson's trichrome method. J Microsc 1998; 191(Pt 1): 20-7. [http://dx.doi.org/10.1046/j.1365-2818.1998.00348.x] [PMID: 9723188]

[48] Frohne D. Anatomisch-mikrochemishe Drogenanalyse Ein Leitfaren 2. Stuttgart: Georg Thieme Verlag 1974.

[49] Rhys Williams AT. Fluorescence Detection in Liquid Chromatography. Buckinghamshire, UK: Perkin-Elmer 1980.

[50] Soini E, Hemmilä I. Fluoroimmunoassay: present status and key problems. Clin Chem 1979; 25(3): 353-61. [PMID: 400437]

[51] Taylor DL, Waggoner AS, Murphy RF, Eds. Applications of Fluorescence in the Biomedical Sciences. New York: Alan R. Liss 1986.

[52] Mullins JM. Overview of fluorophores. Methods Mol Biol 1994; 34: 107-16. [PMID: 7711838]

[53] Haugland RP. Handbook of Fluorescent Probes and Research Products. 9th ed., Eugene, OR: Molecular Probes, Inc. 2002.

[54] Hilderbrand SA. Labels and probes for live cell imaging: overview and selection guide. Methods Mol Biol 2010; 591: 17-45. [http://dx.doi.org/10.1007/978-1-60761-404-3_2] [PMID: 19957122]

[55] Horobin RW, Stockert JC, Rashid-Doubell F. Uptake and localisation of small-molecule fluorescent probes in living cells: a critical appraisal of QSAR models and a case study concerning probes for DNA and RNA. Histochem Cell Biol 2013; 139(5): 623-37. [http://dx.doi.org/10.1007/s00418-013-1090-0] [PMID: 23542926]

[56] Papkovsky DB. Luminescent porphyrin probes. Appl Fluor Technol 1991; 3(3): 16-24.

[57] Bell CG, Hayes FN, Eds. Liquid Scintillation Counting. New York, Oxford, Paris: Pergamon Press 1958.

[58] Birks JB. The Theory and Practice of Scintillation Counting. New York: Macmillan 1964.

[59] Marshall MV, Rasmussen JC, Tan IC, *et al.* Near-infrared fluorescence imaging in humans with indocyanine green: A review and update. Open Surg Oncol J 2010; 2(2): 12-25. [http://dx.doi.org/10.2174/1876504101002020012] [PMID: 22924087]

[60] Sevick-Muraca EM. Translation of near-infrared fluorescence imaging technologies: emerging clinical applications. Annu Rev Med 2012; 63: 217-31. [http://dx.doi.org/10.1146/annurev-med-070910-083323] [PMID: 22034868]

[61] Alander JT, Kaartinen I, Laakso A, *et al.* A review of indocyanine green fluorescent imaging in surgery. Int J Biomed Imaging 2012; 2012: 940585. [http://dx.doi.org/10.1155/2012/940585] [PMID: 22577366]

[62] Azhdarinia A, Ghosh P, Ghosh S, Wilganowski N, Sevick-Muraca EM. Dual-labeling strategies for nuclear and fluorescence molecular imaging: a review and analysis. Mol Imaging Biol 2012; 14(3): 261-76.

[http://dx.doi.org/10.1007/s11307-011-0528-9] [PMID: 22160875]

[63] Luo W, Zhu Z, Yao Y, *et al.* Some novel infrared laser dyes of superior photostability tunable from 1.30 to 1.52 µm. Laser Chem 1990; 10: 259-63.
[http://dx.doi.org/10.1155/1990/59157]

[64] Zarfati D, Harris A, Garzozi HJ, *et al.* A review on ocular blood flow measurement techniques. Neuroophthalmology 2000; 24: 401-9.
[http://dx.doi.org/10.1076/noph.24.3.401.7146]

[65] Hiratsuka M, Miyashiro I, Ishikawa O, *et al.* Application of sentinel node biopsy to gastric cancer surgery. Surgery 2001; 129(3): 335-40.
[http://dx.doi.org/10.1067/msy.2001.111699] [PMID: 11231462]

[66] Luo S, Zhang E, Su Y, Cheng T, Shi C. A review of NIR dyes in cancer targeting and imaging. Biomaterials 2011; 32(29): 7127-38.
[http://dx.doi.org/10.1016/j.biomaterials.2011.06.024] [PMID: 21724249]

[67] Polom K, Murawa D, Rho YS, Nowaczyk P, Hünerbein M, Murawa P. Current trends and emerging future of indocyanine green usage in surgery and oncology: a literature review. Cancer 2011; 117(21): 4812-22.
[http://dx.doi.org/10.1002/cncr.26087] [PMID: 21484779]

[68] Abels C, Karrer S, Bäumler W, Goetz AE, Landthaler M, Szeimies RM. Indocyanine green and laser light for the treatment of AIDS-associated cutaneous Kaposi's sarcoma. Br J Cancer 1998; 77(6): 1021-4.
[http://dx.doi.org/10.1038/bjc.1998.168] [PMID: 9528851]

[69] Urbanska K, Romanowska-Dixon B, Matuszak Z, Oszajca J, Nowak-Sliwinska P, Stochel G. Indocyanine green as a prospective sensitizer for photodynamic therapy of melanomas. Acta Biochim Pol 2002; 49(2): 387-91.
[PMID: 12362980]

[70] Shafirstein G, Bäumler W, Hennings LJ, *et al.* Indocyanine green enhanced near-infrared laser treatment of murine mammary carcinoma. Int J Cancer 2012; 130(5): 1208-15.
[http://dx.doi.org/10.1002/ijc.26126] [PMID: 21484791]

[71] Zheng X, Zhou F, Wu B, Chen WR, Xing D. Enhanced tumor treatment using biofunctional indocyanine green-containing nanostructure by intratumoral or intravenous injection. Mol Pharm 2012; 9(3): 514-22.
[http://dx.doi.org/10.1021/mp200526m] [PMID: 22332810]

[72] Barnes KD, Shafirstein G, Webber JS, Koonce NA, Harris Z, Griffin RJ. Hyperthermia-enhanced indocyanine green delivery for laser-induced thermal ablation of carcinomas. Int J Hyperthermia 2013; 29(5): 474-9.
[http://dx.doi.org/10.3109/02656736.2013.817615] [PMID: 23902340]

[73] Wang YG, Kim H, Mun S, Kim D, Choi Y. Indocyanine green-loaded perfluorocarbon nanoemulsions for bimodal (^{19}F-magnetic resonance/nearinfrared fluorescence imaging and subsequent phototherapy. Quant Imaging Med Surg 2013; 3(3): 132-40.
[PMID: 23833726]

[74] Giraudeau C, Moussaron A, Stallivieri A, Mordon S, Frochot C. Indocyanine green: photosensitizer or chromophore? Still a debate. Curr Med Chem 2014; 21(16): 1871-97.
[http://dx.doi.org/10.2174/09298673216661312180958021 [PMID: 24350844]

[75] Vogelsang J, Steinhauer C, Forthmann C, *et al.* Make them blink: probes for super-resolution microscopy. ChemPhysChem 2010; 11(12): 2475-90.
[http://dx.doi.org/10.1002/cphc.201000189] [PMID: 20632356]

[76] Flors C. Super-resolution fluorescence imaging of directly labelled DNA: from microscopy standards to living cells. J Microsc 2013; 251(1): 1-4.

[http://dx.doi.org/10.1111/jmi.12054] [PMID: 23700988]

[77] Seefeldt B, Kasper R, Beining M, *et al.* Spiropyrans as molecular optical switches. Photochem Photobiol Sci 2010; 9(2): 213-20.
[http://dx.doi.org/10.1039/b9pp00118b] [PMID: 20126797]

[78] Fernández-Suárez M, Ting AY. Fluorescent probes for super-resolution imaging in living cells. Nat Rev Mol Cell Biol 2008; 9(12): 929-43.
[http://dx.doi.org/10.1038/nrm2531] [PMID: 19002208]

[79] Puchtler H, Sweat Waldrop F, Meloan SN. Application of thiazole dyes to amyloid under conditions of direct cotton dyeing: correlation of histochemical and chemical data. Histochemistry 1983; 77(4): 431-45.
[http://dx.doi.org/10.1007/BF00495799] [PMID: 6190786]

[80] Bruzell EM, Morisbak E, Tønnesen HH. Studies on curcumin and curcuminoids. XXIX. Photoinduced cytotoxicity of curcumin in selected aqueous preparations. Photochem Photobiol Sci 2005; 4(7): 523-30.
[http://dx.doi.org/10.1039/b503397g] [PMID: 15986060]

[81] Stockert JC, Del Castillo P, Gómez A, Llorente AR. Fluorescence reaction of chromatin by curcumin. Z Natforsch C J Biosci 1989; 44(3-4): 327-9.
[PMID: 2663005]

[82] Stockert JC, Del Castillo P, Testillano PS, Risueño MC. Fluorescence of plastic embedded tissue sections after curcumin staining. Stain Technol 1989; 64(4): 207-9.
[http://dx.doi.org/10.3109/10520298909107001] [PMID: 2482998]

[83] Bettinger C, Zimmermann HW. New investigations on hematoxylin, hematein, and hematein-aluminium complexes. II. Hematein-aluminium complexes and hemalum staining. Histochemistry 1991; 96(3): 215-28.
[http://dx.doi.org/10.1007/BF00271540] [PMID: 1717413]

[84] Stockert JC. Observations on the chromatin staining by aluminium-hematoxylin. Z Naturforsch C Biosci 1979; 34(12): 1285-6.
[PMID: 94723]

[85] Bobadilla JR, Gutiérrez-Gonzálvez MG, Armas-Portela R, Stockert JC. 9-Fenil-2,3,7- trihidroxi-6-fluorona como colorante y fluorocromo. Acta Bioquim Clin Latinoam 1988; 22: 419-25.

CHAPTER 4

Fluorescent Labels

Abstract: This Chapter deals with the labeling of biomolecules with synthetic or natural fluorochromes. In spite of the development of enzyme and gold protocols for biomolecular labeling, the use of fluorescent tags keeps on being important procedures applied in biomedical research. Covalently attached fluorophores are the most general linking type and they can be either synthetic (FITC, TRITC) or natural (GFP and analogues, phycobiliproteins). Generally proteins are not fluorescent, but they can be labeled with suitable fluorophores. However, there are natural fluorescent proteins that are increasingly applied in biomedical studies. In addition to lipid labeling with synthetic dyes, there are also natural and synthetic fluorescent lipids that are more physiological for studies *in vivo* (*i.e.* pentaene derivatives). General aspects on the type and use of fluorescent labels can be found in [1-8]. Relevant issues for fluorescent labeling are the type of fluorophore, its reactive group and binding mode, and the biomolecule to be labeled. Increasing interest is now devoted to the development of new methods for fluorescent labeling [9], mainly using nanoparticles [10, 11].

Keywords: Bimane, Bio-orthogonal reactions, Biotin, Boronic acid, Caged fluorophores, Chlorophyll, *Cis-trans* isomerism, GFP, Isothiocyanate, Lectins, NBD-ceramide, Nucleic acids labeling, Phallotoxins, Phycobiliproteins, Polyenes, PRINS, Taxanes, Tyramide signal amplification.

1. TYPES OF FLUORESCENT LABELS

The first advances toward the development of modern fluorescent labeling methods are credited to the immunologist Albert Coons in the 1940s by developing fluorescein isothiocyanate (FITC) [12], which remains one of the most ubiquitous fluorescent labels for coupling to antibodies against specific antigens. Today, targeted fluorescent labels are among the most commonly employed imaging agents in microscopy.

Obviously, there are many types de fluorescent labels, depending on the fluorophore and the biomolecule (*i.e.* proteins, lipids, polysaccharides, lectins, nucleic acids, metabolites and compounds with biological activity). Labeling may also be achieved using phosphorescent complexes from transition metals (see Chapter 9.2), and quantum dots (see Chapter 18.2). It is important that the binding to fluorescent labels does not modify the biological activity of the biomolecule.

Juan Carlos Stockert & Alfonso Blázquez-Castro

Principal applications of proteins and nucleic acids labeling are described in Chapter 17.

1.1. Reactive Dyes and Labeling Mechanisms

Reactive dyes and fluorochromes have chemical groups that react with tissue substrates or isolated biomolecules and form covalent bonds with them, conferring fastness and washable features to stained structures. These dyes react preferentially with proteins, either microscopically, in electrophoretic gels, or as immobilized ligands for protein separation. They are widely used for labeling secondary antibodies and oligonucleotide probes for immunofluorescence and FISH methods [5, 7]. Some reactive groups are presented in Fig. (**4.1A**), most used being isothiocyanate, chloro-triazine, azido, sulfonyl chloride, and succinimidyl ester groups.

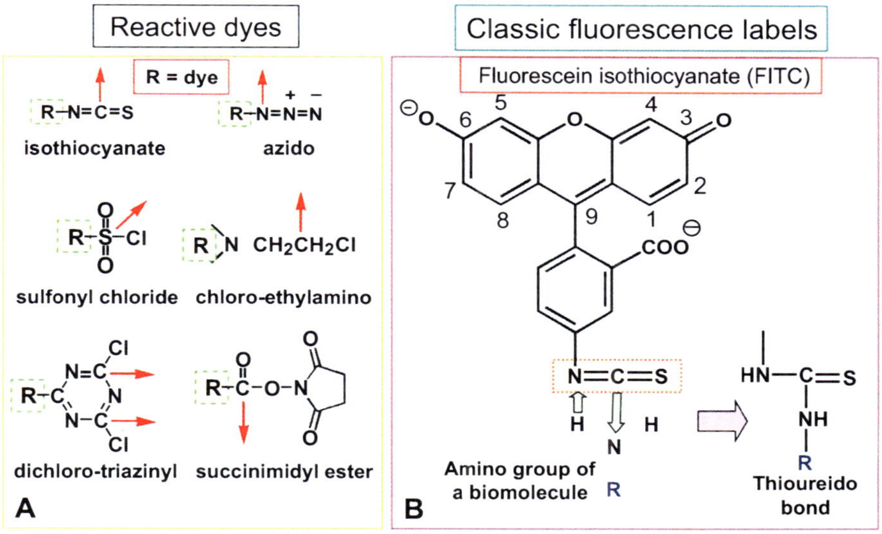

Fig. (4.1). **A**: Gallery of reactive dyes and covalent binding types. The red arrows indicate the new linkages to acceptor biomolecules. **B**: Chemical structure of FITC with numbering of the xanthene fluorophore, the isothiocyanate group, and the resulting thioureido bond.

Reactive dyes have advantages when they are followed by other staining procedures that could remove the former signal. In addition to FITC and tetramethyl-rhodamine isothiocyanate (TRITC) labels, examples of reactive dyes and fluorochromes are astrazon pink FG (CI 48015), cibacron blue F3G (CI 61211), Cy3, Cy5, MitoTracker red CMXRos, procion brilliant blue MR (CI 61205), procion yellow M4R, quinacrine mustard, *etc*. The main types of fluorescent labels are reviewed in [2, 3, 5, 7]. The relevant guide of fluorescent labels [6] should be consulted for more details.

At present, the reactive group isothiocyanate (-N=C=S) has fully replaced the formerly used isocyanate (-N=C=O). Main use of the classical FITC and TRITC is for secondary antibody labeling in immunofluorescence (Figs. **4.1B** and **4.2**).

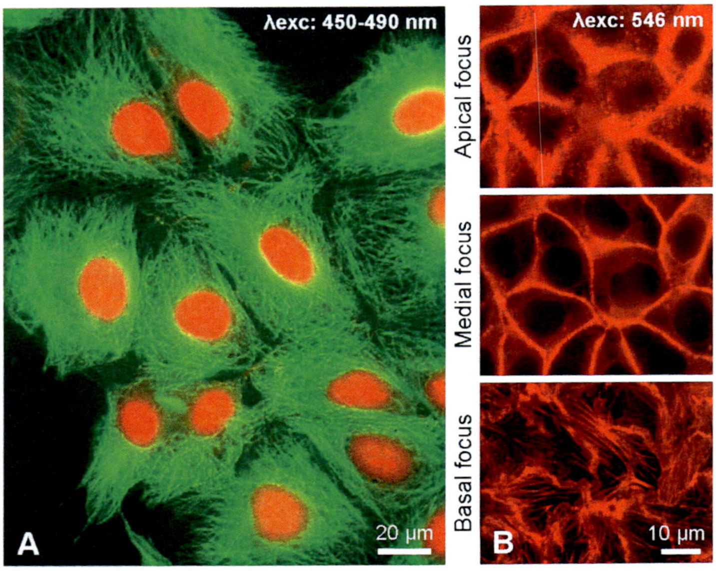

Fig. (4.2). Fluorescent labeling of proteins and peptides. **A**: Immunofluorescence labeling of α-tubulin in Pam212 mouse keratinocytes (FITC) showing the normal network of interphase microtubules. Nuclear counterstaining with the cationic porphyrin TMPyP. **B**: Three focal planes of Pam212 keratinocytes stained with phalloidin-TRITC, showing the different organization of actin microfilaments by conventional epifluorescence.

In both cases the thioureido bond is stable but cleaves at alkaline pH releasing free fluorescein or rhodamine derivatives that can give false positive results. The sulfonylchloride derivative of the red-emitting fluorophore sulforhodamine 101 (Texas red) has proved very useful for coupling to amino groups [13]. Texas red conjugates of antibodies and other biomolecules retain their biological activities and are strongly fluorescent, with spectral properties well separated from those of molecules labeled with FITC.

The incorporation of heavy halogens into dyes such as xanthene derivatives or some heavy metals such as Pd, Pt, *etc.*, to other compounds also diminishes fluorescence but increases the conversion of the labeling dye to the excited triplet state and then improves phosphorescence and photodynamic potential. Fluorescein is nearly inactive as a photodynamic sensitizer. It absorbs visible light but its excited singlet state fluoresces much faster than its intersystem crossing to the triplet state; thus, the quantum yield of fluorescein triplet (Φ_T) required for production of singlet oxygen, is only 0.03 in water. In contrast, fluorescein is of great value as fluorescent probe. Halogens dramatically increase the intersystem crossing and the extent of excited triplet state formation. In the case of xanthenes, the more halogenated the fluorescein ring, the higher the triplet yield. For example, 2,7-dibromo-fluorescein in water has Φ_T: 0.42; 2,4,5,7-tetrabrom--fluorescein (eosin Y, CI 45380) shows Φ_T: 0.57; and for 2,4,5,7-tetraiod--3',4',5',6'-tetrachloro-fluorescein (rose Bengal, CI 45440), Φ_T: 0.76. Relative efficiencies as photodynamic sensitizers parallel this trend, which has been profited by converting FITC-labeled antibodies into photosensitizers after iodination with Lugol reagent [14]. Likewise, iodination of fluorescein or rhodamine derivatives of taxol is another interesting possibility.

Bimane is one of the smallest fluorophores (Fig. **4.3**), and useful reactive fluorochromes for sulfhydryl groups are bromo-bimanes [15, 16]. In addition to FITC, TRITC, and Texas red, reactive indocyanines such as Cy3, Cy5 and Cy7, the pyrene derivative cascade blue, and reactive ruthenium complexes are also suitable fluorescent probes for macromolecular labeling. Likewise, several derivatives of rhodol, coumarin, Bodipy, dansyl chloride (dimethylamino-naphthalene-sulfonyl chloride), and NBD-Cl (7-chloro-4-nitrobenzo-2-o-a-1,3-diazole), are now currently employed as fluorescent labels. Diphenyl-furanones such as the known reagents fluorescamine and MDPF react with high selectivity with aliphatic primary amines (see Chapter 13.2.1), giving fluorescent diphenyl-pyrrolone products [17] useful for protein labeling.

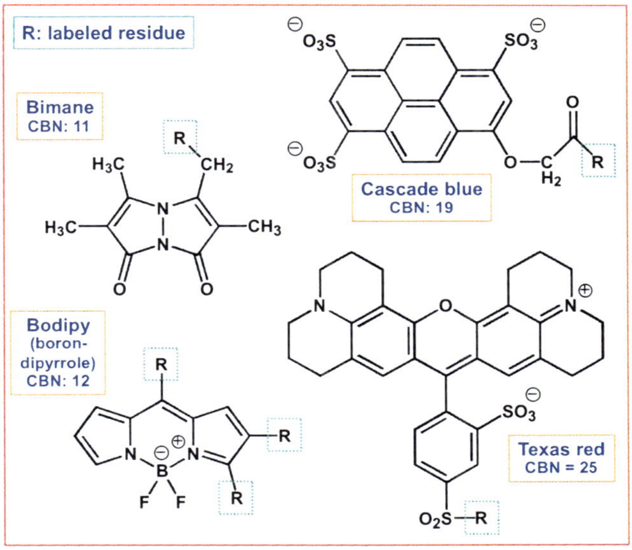

Fig. (4.3). New fluorescent labels for microscopy. Chemical structure of the labeling fluorophores bimane, Bodipy, Cascade blue, and Texas red, showing CBN values and linkages to residues (R) in biomolecules.

Other new fluorescent labels are reactive sulfonate derivatives of rhodamine dyes such as Alexa™ 488 and Alexa™ 546-tyramide, which are suitable for macromolecular labeling and tyramide signal amplification. At present, oligothiophenes (from the thiophene ring, C_4H_4S) are versatile fluorophores with intense and stable emission tunable over the entire visible range [18]. Typically, 2-hydroxyethyl-terthiophene was used as a fluorescent label for oligonucleotides (λexc: 356 nm, λem: 441 nm). Oligothiophenes can be derivatized with several groups such as N-hydroxy-succinimidyl, phosphoramidite, isothiocyanate and azide, allowing fluorescent labeling of proteins and antibodies, oligonucleotides, and magnetic Fe_2O_3-nanoparticles. When labeled with oligothiophenes, hairpin configurations of oligonucleotide probes can be also used as molecular beacons [18].

1.2. Photolabeling and Photoswitching

Many fluorophores can be modified to act as optical switches that are activated by exposure to UV or visible light. Caged fluorophores (masked, non-fluorescent, and photo-activatable) can be employed in dynamic imaging applications where specific temporal and spatial activation of fluorescent labels is required [19]. Short laser pulses of about 100 ms are provided through a pinhole to photoactivate only a restricted microscopic area. Caged fluorescein (2-nitrobenzyl-fluorescein lactone) is an example of masked fluorophore, which is activated by irradiation at 365 nm to cleave the *o*-nitrobenzyl group, releasing free fluorescein (Fig. **4.4A**). Incorporation of tubulin labeled with caged fluorescein into cell microtubules has allowed studying microtubule dynamics in the mitotic spindle [20].

In addition to caged fluorescein, caged resorufin and caged coumarin derivatives bound to proteins have been also applied to analyze dynamic cell processes. Likewise, tracking of cell descendants can be recorded by using dextrans bound to caged-fluorescein. Intranuclear movements of RNA molecules have been visualized after uptake of an anti-sense oligo(dT) labeled with caged fluorescein, which then hybridizes with nuclear poly(A) tails of RNA. Other caged compounds are neurotransmitters, metabolites, Ca^{2+} carriers, and proteins, allowing analysis of synapsis function, cell metabolism, ion fluctuations, and enzymatic activation [20]. Likewise, there are today numerous photoactivatable metal complexes that are increasingly applied in biotechnology and biomedical studies [21]. Photobiotin reacts with substrates under UV-blue irradiation in a simple and rapid way (Fig. **4.4B**), and is now widely used as labeling agent. Photobiotin binds to =CH–, $-NH_2$ or –SH groups, allowing the direct biotinylation of proteins and nucleic acids.

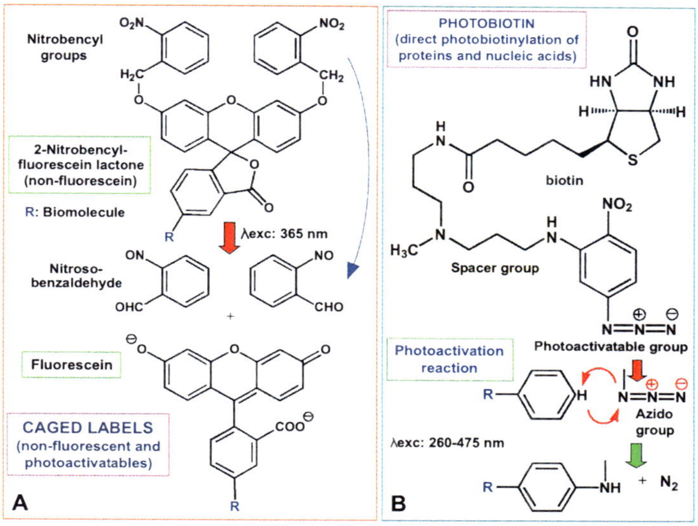

Fig. (4.4). Photo-activatable labeling. **A**: Caged labels: when subjected to 365 nm exciting light, the non-fluorescent 2-nitrobenzyl-fluorescein lactone yields fluorescein. **B**: Photobiotin labeling: excitation with UV-blue light results in the direct binding of the label to the biomolecule.

On the other hand, photoisomerism and photochromic reactions can be also applied to design photoswitchable drugs. The antimitotic alkaloid colchicine (from *Colchicum autumnale* Fig. **4.5A**) is a tropolone derivative that on account of its pharmacophore domain, binds with high affinity to tubulin [22]. Interestingly, colchicine is fluorescent only when bound to tubulin (λexc: 362 nm, λem: 435 nm, Φ_F: 0.03) [23]. A similar pharmacophore is also present in the antimitotic podophyllotoxin and derivatives. Tubulins have hydrophobic binding sites for small aromatic ligands such as GTP, colchicine and derivatives, *Vinca* alkaloids, griseofulvin, taxol, podophyllotoxin, nocodazole, chlorpromazine, rotenone, and other neuroleptic drugs [22].

Photostatins (*i.e.* PST-1) are new photoswitchable agents for tubulin binding that can reversibly change their conformations between *trans*-inactive and *cis*-active isomers [24], the latter showing the same pharmacophore domain as colchicine (Fig. **4.5A**). When cultured cells are treated with PST-1 and subjected to 390 nm light (pulses of 200 ms every 2 min), a striking breakdown of microtubule cytoskeleton and apoptotic fragmentation of nuclei are seen, while cells maintained in the dark show the normal structure of microtubules and nuclei (Fig. **4.5B, C**). PST-1 can modulate microtubule dynamics with subsecond response time, controlling mitosis with single-cell spatial precision. These features make this drug a very valuable tool for cell biology, and as promising chemotherapeutic agent.

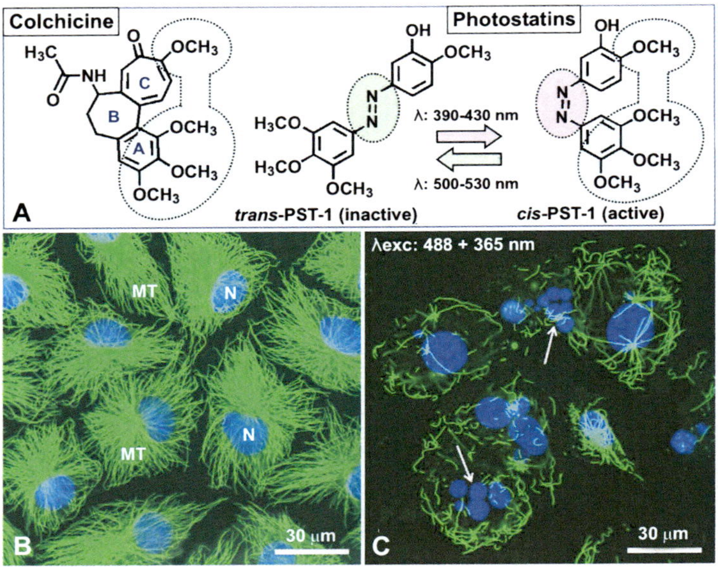

Fig. (4.5). A: Chemical structure of colchicine and photostatins *trans*- and *cis*-PTS-1, showing the pharmacophore domain (cyan areas). -N=N- isomers are *trans* (green) or *cis* (violet). Colchicine has the trimethoxy-benzene ring **A**, ring **B**, and methyl-tropolone ring **C**. Rings **A** and **C** are responsible for tubulin binding and fluorescence increase [24]. **B**, **C**: Confocal images of MDA-MB-231 human breast cancer cells treated for 20 h with 1.5 µM PST-1 in the dark (**B**) or under pulsed illumination with 390 nm light (**C**) [25]. Glutaraldehyde-fixed cells were stained with anti-α-tubulin-IgG and AlexaFluor 488-IgG, followed by 1 µg/mL Hoechst 33342 counterstaining. Images show intact nuclei (N) and microtubules (MT) under dark regime (**B**), and nuclear fragmentation (arrows) and breakdown of microtubules in apoptotic cells subjected to combined PST-1 plus 390 nm toxic regime (**C**) (courtesy of M. Borowiak, O. Thorn-Seshold, and D. Trauner).

1.3. Mycotoxins and Phytotoxins

There are also small molecules with biological activity that can be labeled with fluorescence. Although phallotoxins and taxanes are not antibodies, they bind with high affinity to some cytoskeleton proteins. Phalloidin is a fungal toxin from *Amanita phalloides* (Fig. **4.6A**) that binds polymeric actin very selectively, interfering with the function of actin microfilaments. Fluorescent FITC-, TRITC-, and coumarin-conjugates of phalloidin are extensively used to visualize filamentous actin, showing λexc/λem at 495/520 nm, 540/570 nm, and 385/470 nm, respectively [25 - 27] (Figs. **4.2B** and **4.6B**).

In addition to the well known phalloidin-TRITC (Fig. **4.6C**), F-actin can be identified by the similar toxin phallacidin (Fig. **4.6D**). In all cases, fluorescent labels are attached to the -OH group of the hydroxy-leucine. It is worth to note that the affinity of phalloidin-TRITC for F-actin is 6 times lower than that of the unlabeled toxin [26]. NBD has been also employed to label phalloidin [28].

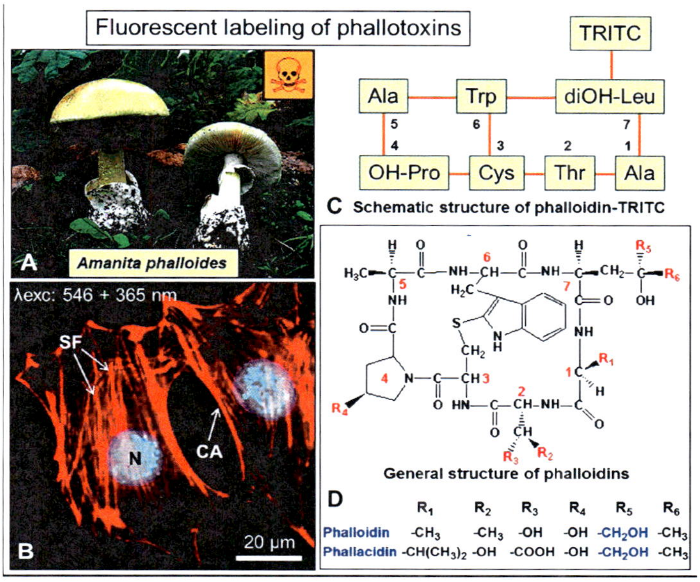

Fig. (4.6). **A**: *Amanita phalloides*. **B**: 3T3 mouse fibroblasts labeled with phalloidin-TRITC and counterstained with DAPI. SF: stress fibers, CA: cortical actin, N: nucleus. **C**, **D**: Schematic structure of phalloidin-TRITC and general structure of phalloidins, with numbering of amino acids and substituents.

Aflatoxins are a group of toxic mould metabolites (mycotoxins) produced by *Aspergillus flavus* and *A. parasiticus*. Chemically, they are coumarin derivatives (as well as psoralens) with extreme toxicity and carcinogenicity due to intercalation and inhibition of DNA synthesis [29]. Several types of aflatoxins (B, G, M, and R) have been reported to occur in peanuts, corn, *etc*. After metabolic activation, intercalation of aflatoxin B1-*exo*-8,9-epoxide forms a N7 adduct with guanine [30]. Aflatoxins are fluorescent, with λexc ~ 370 nm and λem between 405 and 430 nm, depending on the aflatoxin type [31], and their violet or blue emission allows to reveal and characterize them in food. Ochratoxins are also toxic metabolites from *Aspergillus ochraceous* and other molds that have been found as contaminant mycotoxins of corn, peanuts, storage grains, *etc*. Ochratoxins A and B exhibit green and blue fluorescence, respectively [29]. Fluorescence microscopy uses of both aflatoxins and ochratoxins seem to have been scarcely explored.

On the other hand, phytotoxins such as taxanes (taxoids) can be labeled with adequate fluorophores (Fig. **4.7**). Taxanes bind with high affinity to microtubules. In contrast with *Vinca* alkaloids, which prevent the assembly of microtubules, taxanes stabilize them and prevent their disassembly. Paclitaxel (formerly referred as taxol) is extracted from the bark of the yew tree from the Pacific Ocean (*Taxus brevifolia*), and from precursor compounds from leaves of the common yew

(*T. baccata*). It is used as chemotherapeutic drug for Kaposi sarcoma, mammary and ovary tumors. Fluorescent taxanes (Flutax-1, Flutax-2, Hexaflutax, FChitax-3, and Rotax) have been synthesized using the allowed -OH at C7 to bind fluorescent labels and employed to visualize microtubules [32 - 34]. Other fluorescent drugs such as TubulinTracker™ Green (Oregon 488® Taxol bis-acetate) and Bodipy-paclitaxel derivatives also induce microtubules stabilization, a G2/M cell cycle arrest, and cell death by apoptosis [35].

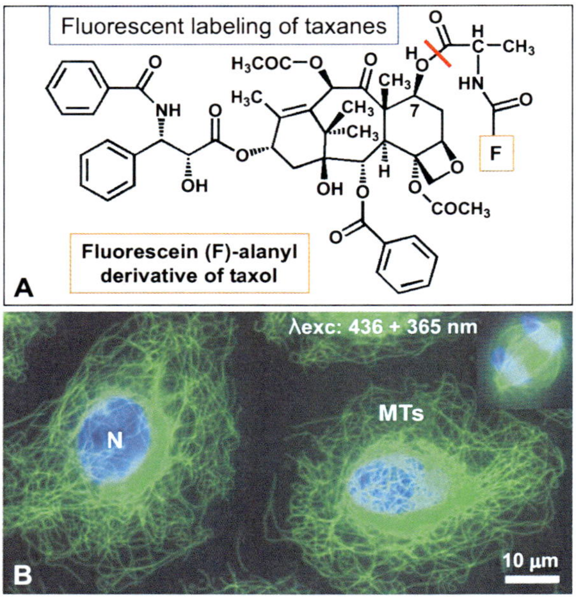

Fig. (4.7). Taxanes and fluorescent labeling. **A**: Chemical structure of the fluorescein (F)-alanyl derivative of taxol (Flutax), with the -OH and the ester fluorescein linkage (red line) to C7. **B**: A-549 human lung carcinoma cells showing green fluorescing microtubules (MTs). Unfixed cells were washed in a MT-stabilizing buffer, permeabilized with 0.5% Triton X-100 in PEM buffer for 90 s, washed again, and stained with 1 µM Flutax-2 for 5 min. Nuclear DNA (N) counterstained with Hoechst 33342. Observe the anaphase cell at the upper right corner (courtesy of I. Barasoain).

2. LIPID AND LIPOSOME LABELING

There are phospholipids, ceramides and sphingomyelins covalently labeled with several types of fluorophores (rhodamine, dansyl, NBD, Bodipy, pentaene, *etc*). Choline- and ethanolamine-dipalmitoyl phospholipids (*i.e.* DPPC and DPPE, respectively) are common substrates for labeling. Fluorescent tags can be attached either to acyl chains (*i.e.* NBD, pyrene), or polar head groups (*i.e.* dansyl, rhodamine). After binding of MDPF with the polar head of DPPE, its diphenyl-pyrrolone-labeled derivative should be useful as a fluorescent phospholipid derivative.

NBD-labeled ceramides and phospholipids are useful compounds for lipid visualization in membranous organelles of live and fixed cells (*i.e.* NBD-ceramide, NBD-phosphatidylcholine) (Fig. **4.8**) (see also Chapters 10 and 15). Regarding applications, incorporation of cholesterol has been visualized by using its Bodipy FL_{12} ester [36], and massive phospholipids accumulation due to treatment of cultured cells with antipsychotic drugs has been observed by staining with NBD-C_{12}-phosphatidylcholine [37].

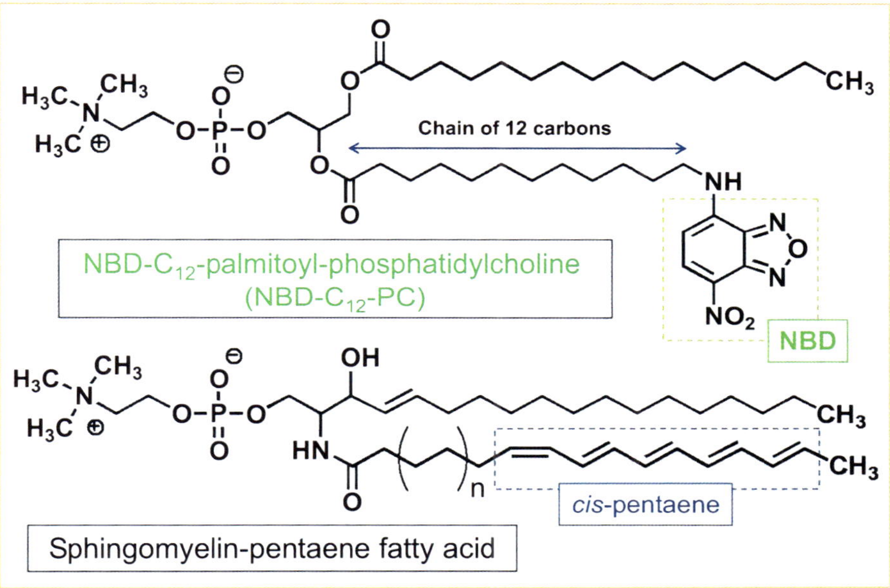

Fig. (4.8). Fluorescent derivatives of choline phospholipids: palmitoyl-C_{12}-NBD-phosphatidylcholine (C_{12}-NBD-PC) and sphingomyelin labeled with a *cis*-pentaene fatty acid.

On the other hand, fluorescent labeling can abolish the biological activity. A fluorescent NBD derivative of natamycin labels the plasma membrane of fungi, but it has no antifungal action. The labeled polyene antibiotic does not bind to ergosterol in the plasma membrane because NBD-tagged amine hinders the H bonding to the sterol. Interesting new developments for natural (physiological) fluorescent labeling of lipids are polyene fatty acids (Fig. **4.8**) (see Chapter 10.6.2).

Liposomes have long been used as experimental models for membranes in various biomedical fields [38], and there is great interest in using uni- and multilamellar liposomes to deliver macromolecules (DNA, RNA, oligonucleotides, proteins, peptides) to the interior of cells. Liposomes are also employed to deliver cytotoxic drugs (anthracyclines, camptothecins, vincristine, photosensitizers) within tumor cells [39]. Obviously, fluorophores covalently attached to phospholipids are very appropriate to form fluorescent liposomes [40], but liposomes simply loaded with drugs or fluorochromes such as acridine orange, neutral red, thiazine red, and eosin Y have been also used [41].

In addition to small liposomes, microscopical models of membrane-like lipid bilayers are the giant unilamellar vesicles (GUVs), which are cell-size membrane envelopes of increasing interest and applications in the transient electroporation for drug and gene delivery into cells [42]. GUVs composed of mixtures of palmitoyloleoyl-phosphatidyl-choline (POPC), palmitoyloleoyl-phosphatidy--glycerol (POPG), and dipalmitoyl-phosphatidyl-ethanolamine-rhodamine (DPPE-Rh), containing the fluorescent carboxyl SNARF-4F pH indicator are shown in Fig. (**4.9**). Both liposomes and GUVs are now useful models for information transfer in neoplastic tissues based on membrane protrusions, extracellular vesicles, and cell-cell fusion processes.

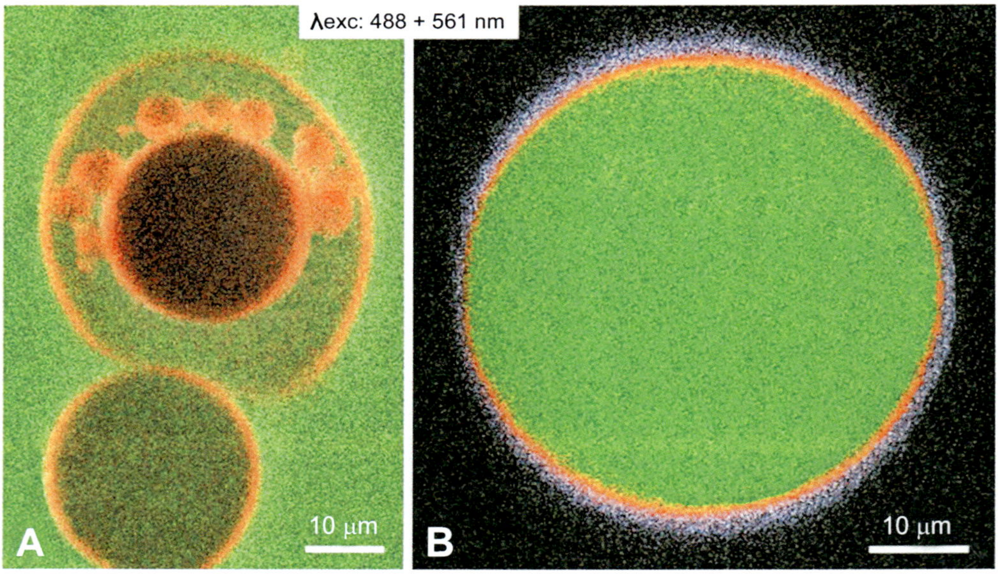

Fig. (4.9). Fluorescent images of giant unilamellar vesicles (GUVs) formed with 1-palmitoyl-2-ol-yl-phosphatidylcholine (POPC), labeled with dipalmitoyl-phosphatidylethanolamine-rhodamine (Rh-DPPE) (red), and containing the fluorescent pH indicator SNARF-4F (5-[and-6]-carboxylic acid) (green at acid pH). GUVs were placed in a chamber between two glass plates allowing observation [42]. **A**: GUVs with the pH indicator in the outside and internalized smaller vesicles. **B**: Fluorescent GUV image showing phase contrast (blue), Rh-DPPE (red), and acid pH indicator (green). Pixels are due to the maximum gain in confocal image capture under very low excitation (courtesy of N. Olaiz and R. Dimova).

On the other hand, some natural lipophilic fluorochromes are similar to polar lipids and can mimic the organization of lipid bilayers. Porphyrins and chlorophylls are fluorescent tetrapyrrole macrocycles with important functions in cell metabolism. The universal tetrapyrrole precursor is 5-aminolevulinic acid (ALA), which arises from condensation of glycine and succinyl-CoA in yeast and animals, but in cyanobacteria and algae it is formed by another route, using the intact carbon skeleton of glutamate [43]. Chlorophyll is the main photosynthetic pigment and occurs in plant chloroplasts associated to a complex lipoprotein organization. It is applied as deodorant and photographic sensitizer, to color liquors, soaps, oils, waxes, cosmetics, and to promote healing in skin lesions [29].

The deep red fluorescence of chlorophyll can be used to visualize chloroplasts (see Fig. **7.8A-C**) as well as for labeling of lipid micelles from chlorophyll-oil emulsions (Fig. **4.10A, B**). Treatments of cell cultures with chlorophyll extracted in ethanol from spinach leaves (mainly chlorophyll a), mixed (1:1, v/v) with sunflower oil, and then with DMEM (1:10, v/v) either alone or with 0.01% of DPPC show a selective localization of the red chlorophyll emission in the Golgi apparatus and lipid droplets (Fig. **4.10C, G**). On the other hand, it is worth to note that in chloroplasts, excited chlorophyll can act as a photosensitizer. When an inhibition of photosynthetic electron transport occurs, the chlorophyll triplet state will interact with $^3O_2$ to produce $^1O_2$ [44, 45], which finally can generate chlorophyll bleaching and chloroplasts damage.

3. POLYSACCHARIDE LABELING

Cell micro-injection with fluorescent dextrans are useful to identify cells that are co-injected with plasmids containing recombinant proteins [46]. Fig. (**4.11**) shows confluent patches of mouse MCA3D keratinocytes micro-injected with dextran-Texas red (mol wt 10,000), in which the immunofluorescence of E-cadherin expressed in cells with close contact between them is also observed. Likewise, intravenous injection of 25 μl dextran-fluorescein isothiocyanate (FITC-dextran, mol wt: 70,000; 250 mg/kg) in saline solution is commonly applied to label blood vessels [47].

In addition, other fluorescent polysaccharides have been used. Studies on the uptake of FITC-labeled heparin into smooth muscle cells showed that the fluorescent probe was localized around the cell nucleus and at the level of cell contacts [48]. Dextrans labeled with caged fluorescein have been employed for tracking embryonic cells during development [19]. Reactive fluorochromes with triazine chloride or boronic acid groups are also useful for polysaccharide labeling.

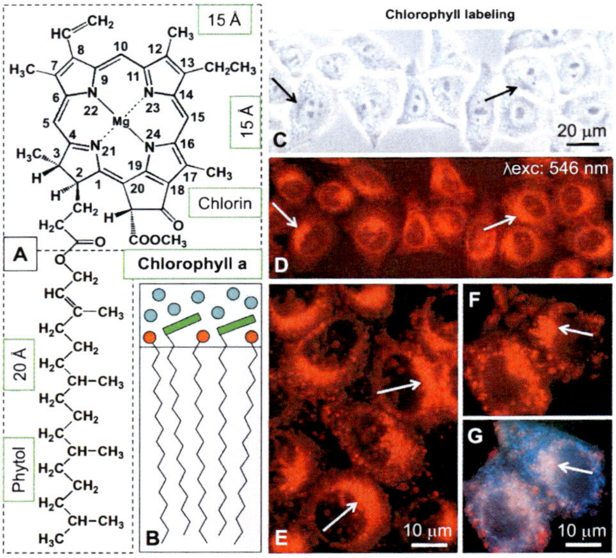

Fig. (4.10). **A**: Chlorophyll a showing the numbered chlorin macrocycle and the phytol tail (size in Å). **B**: Schematic model of an oil micelle loaded with chlorophyll. Chlorin (green rectangles), polar heads (red) and zig-zag chains of triglycerides and DPPC, phytol in the lipophilic phase (pink), and water (blue circles) in the aqueous phase (light blue) are seen. **C**, **D-G**: Phase contrast and red fluorescence images of living HeLa cells treated for 20 h with chlorophyll-sunflower oil emulsions in DMEM (1:10, v/v), either directly (C-E) or after mixing with DPPC (F,G), showing labeling of the Golgi apparatus and lipid droplets. **G**: Overlay of the chlorophyll image with the blue autofluorescence of mitochondria (λexc: 365 nm) (courtesy of A. Lázaro and A. Villanueva).

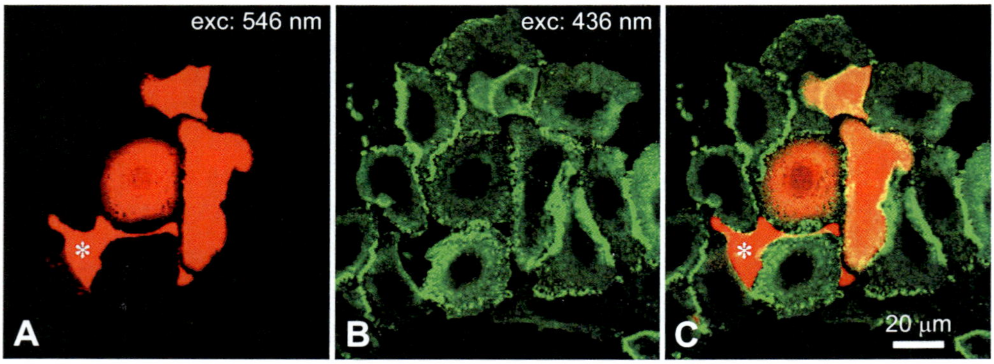

Fig. (4.11). Labeling by micro-injection of fluorescent dextran. **A-C**: Mouse MCA3D keratinocytes microinjected with dextran-Texas red (A), followed by immunolabeling of E-cadherin with a rat monoclonal anti-mouse E-cadherin antibody and FITC-conjugated IgG (B), and overlay of both images (C). Observe the extracellular localization of a small amount of labeled dextran (asterisk), and the low or null expression of the adhesion protein when no intercellular contacts are present (courtesy of J. Espada).

4. NUCLEIC ACIDS LABELING

DNA labeling has been performed by using tritiated (H^3)-thymidine, but at present the common labels are non-radioactive nucleosides. Anyway, DNA labeling methods involve the use of labeled precursors to study hybridization with complementary sequences in fixed cells or DNA replication in living cells. Thymidine analogues such as halogenated uridines are incorporated into replicating cells and then DNA becomes labeled, allowing differentiation between sister chromatids [49, 50], and identification of stem cell niches in adult eukaryotic organisms [51].

4.1. Classical Labels

FITC has found wide applications in fluorescence *in situ* hybridization methods (FISH, see Chapter 17.3) for chromosome classification and detection of chromosome aberrations [52]. Biotin-labeled oligonucleotide probes can be hybridized to target chromosomes and subsequently rendered fluorescent by treatment with fluorescein-labeled avidin. The sequence biotinylated anti-avidin antibody followed again by fluorescein-avidin is also possible, with signal amplification. Visualization of the label is detected easily at metaphase and in interphase cell nuclei. By using a 0.8-kilobase DNA probe specific for the Y chromosome, the characteristic XY and XYY sex chromosome pattern was visible in metaphase chromosomes and interphase amniocyte nuclei. Human 28S ribosomal RNA genes on metaphase chromosomes were also distinctly stained by using a 1.5-kilobase DNA probe [52].

In addition to direct fluorescent labels (*i.e.* Bodipy FL-14-dUTP), several non-fluorescent compounds can be used to label nucleic acids (either *in silico* or in live cells), revealing them by a fluorescence methodology. Examples are biotin- and digoxigenin-bound pyrimidines (*i.e.* digoxigenin-11-dUTP). In the case of purines, a non-fluorescent and stable ethylenediamino-platinum coordination complex attached by a linker to digoxygenin binds selectively to the N7 of guanine. The Ulysis® nucleic acids labeling kit is based on this strategy, but here the reagent is a Pt-linker-fluorochrome complex that directly reacts with guanine. Several fluorophores are available with λem from 440 to 700 nm.

4.2. Labeled Peptide Nucleic Acids

Instead ribo- or deoxyribonucleotide probes, peptide nucleic acids (PNAs) can be synthesized and labeled with fluorophores to perform FISH analysis. PNAs are oligonucleotide analogues in which the sugar-phosphate backbone is replaced by a pseudopeptide skeleton, composed of *N*-(2-aminoethyl)glycine units (Fig. **4.12 A**). Nucleobases are attached to the peptide chain through -CO-CH_2- links. PNAs bind complementary anti-parallel DNA and RNA chains with high affinity and selectivity, forming hybrid Watson–Crick base-paired duplexes with a so-called P-helix conformation [53, 54] (Fig. **4.12B**). Because of their thermal stability and resistance to nucleases, PNAs are ideal probes in FISH and as antigene therapeutic agents.

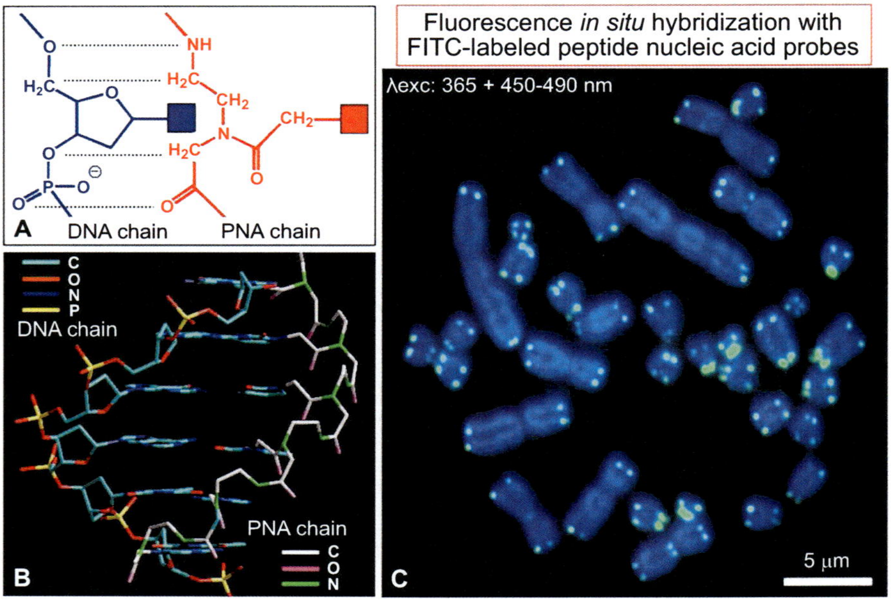

Fig. (4.12). A: Comparative chemical structure of DNA (blue) and PNA (red) chains; dashed lines indicate the corresponding atomic positions (squares: bases). **B**: Computer-drawn molecular model of a DNA (left)-PNA (right) hybrid duplex with the telomeric sequence TTAGGG. Hydrogens are omitted for clarity. **C**: FISH of *Akodon dolores* (Rodentia, Sigmodontinae) metaphase chromosomes with a telomeric FITC-labeled $(CCCTAA)_3$ PNA probe counterstained with 5 μg/mL DAPI. Additional centromeric signals in some chromosomes indicate chromosome rearrangements (see [Viera *et al.* 2001], courtesy of the authors).

The physical ends of vertebrate chromosomes consist of tandem repeats of the canonical hexamer $(TTAGGG)_n$. In many species, *in situ* hybridization allows the recognition of telomeric sequences located outside the telomere domains, which results from rearrangements involving the chromosomal ends [55]. An example of hybridization of telomeric sequences from metaphase chromosomes with PNA probes is shown in Fig. (**4.12C**). Obviously, FISH on telomeres within interphase nuclei is also possible.

4.3. Thymidine Analogues

Thymidine analogues with labeling groups at the C5 position of deoxyuridine (dU) can be: (1) bromine and other halogens, (2) an ethynyl residue (-C≡CH),

(3) aminoallyl group (-CH=CH-CH_2-NH_2), and (4) mercury (-Hg-OH). 5-Brom--2 ' -deoxyuridine, BrdU, and 5-ethynyl-2'-deoxyuridine (EdU) are the most used labels (Fig. **4.13A-C**).

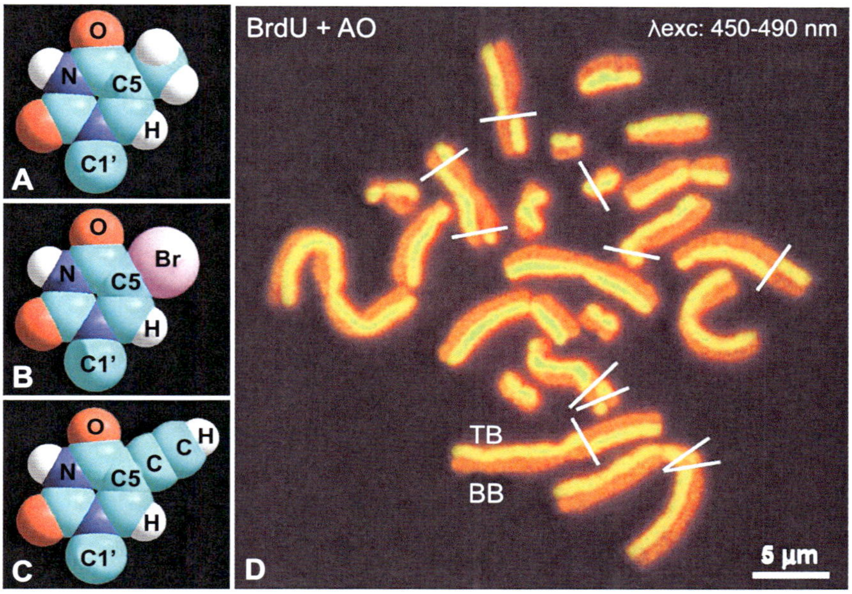

Fig. (4.13). Thymine analogues: C5-labeled derivatives of uracil. Molecular structure of thymine (**A**), and the labeled bases, 5-bromo-uracil (**B**) and 5-ethynyl-uracil (**C**), showing numbered atoms. **D**: CHO-K1 chromosomes from a cell grown for two cycles in the presence of BrdU. Fixed chromosomes (methanol-acetic acid) stained with 0.1% AO show sister chromatid differentiation: yellow-green emission of double-stranded DNA and red emission of single-stranded DNA, corresponding to TB and BB chromatids, which appear with concave and convex shape, respectively. White lines indicate sister chromatid exchanges (courtesy of R. Mezzanotte).

BrdU is widely employed for DNA labeling. When the thymine of DNA is replaced by 5-halogeno (Br, I)-uracil, a striking increase of absorption and photosensitivity appears at 300 nm, leading to dehalogenation, formation of 5,6-cyclobutane products between adjacent pyrimidines, and then genetic inactivation [56]. In the absence of light excitation, living cells with bromo-substituted DNA allow to perform studies on replication and genetic damage, but some cell toxicity including mutagenic and teratogenic effects occur for BrdU a[nd] edU analogues.

When BrdU is incorporated into DNA for two cell cycles, replication produces chromosomes with one bromo-substituted chromatid (thymidine-Br, TB) and another doubly substituted (Br-Br, BB). Subsequent treatments result in a differential behavior between TB/BB (and also TT/TB) chromatids, such as changes of fluorochrome emission or specific staining responses. Incorporation of

BrdU into DNA can be detected using several methods:

A. Direct staining with the cationic fluorochromes, Hoechst 33258 [57] and acridine orange [58], which decrease their emission intensity or shift it to red when bound to bromo-substituted DNA. DNA labeling after one or two rounds of replication in the presence of BrdU allows recognizing particular fluorescence features (sister chromatid differentiation) in each sister chromatid of metaphase chromosomes [50]. The differential emission by AO and the occurrence of sister chromatid exchanges (SCEs) are illustrated in Fig. (**4.13D**). Differentially stained sister chromatids with SCE have a harlequin appearance. Interestingly, chromatids that have no o less Br show a concave contour, whereas a convex contour is seen in those containing high content of Br. This curvature is due to the different chromatin compaction related to changes in chromosomal proteins (*i.e.* histone H2B) and their binding to DNA [49].
B. Exposition to UV light after staining with photosensitizing fluorochromes followed by Giemsa staining [59 - 61]. This is a routine technique, generally using Hoechst 33258. Likewise, exposition to UV light alone or incubation in hot saline solutions followed by Giemsa staining allows recognizing differentially bromated DNA [50].
C. Binding to a FITC-conjugated mouse monoclonal antibody against BrdU. This method requires a previous mild acid hydrolysis for epitope unmasking in chromatin DNA. At present, BrdU labeling followed by FITC-conjugated antibodies against bromo-substituted DNA [62, 63] is a well established method for cytochemical studies on DNA replication. BrdU labeling is equivalent to H^3-thymidine autoradiography, but problems related to management of radioactive samples are avoided and resolution is improved. Since single-stranded DNA is required for BrdU immunodetection but native DNA must be preserved for DNA counterstaining, two contradictory aims have to be reconciled when DNA and its bromo-derivative should be visualized simultaneously.

The identification of stem cells in adult eukaryotic organisms is an example of the use of DNA labeling. Stem cell niches can be tagged with BrdU and later analyzed for searching label-retaining cells [51]. After injection of BrdU into neonatal mice along days 10-13 after birth and observation 50 days later, low-proliferating (quiescent) stem cells from hair follicles are easily detected in the bulge region by a FITC-conjugated monoclonal antibody against bromo-substituted DNA (Fig. **4.14A**, **B**).

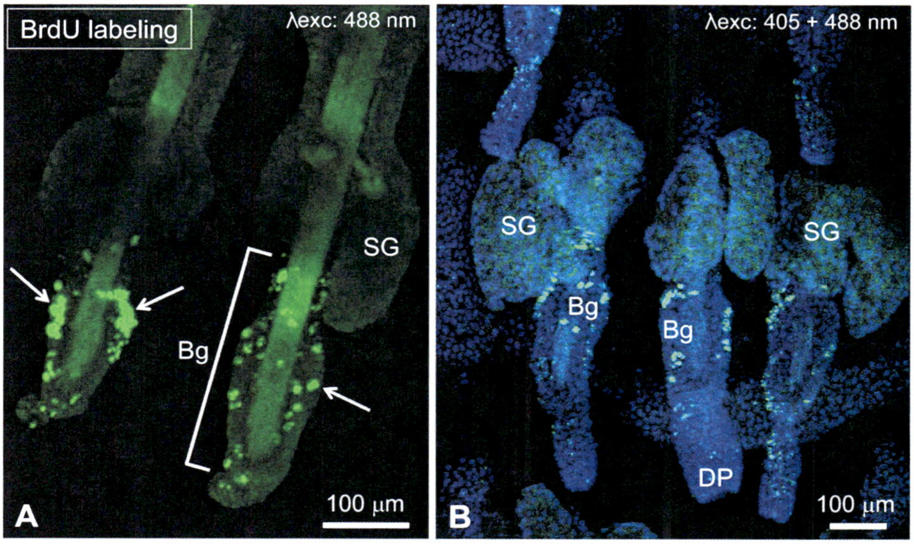

Fig. (4.14). Stacks of focal planes in whole mounts of BrdU-labeled mouse hair follicles. **A**: FITC-conjugated mouse monoclonal antibody against bromo-substituted DNA, showing green fluorescence of label-retaining cells in the bulge region (arrows). **B**: Overlay of BrdU-labeled nuclei (green) and DAPI-counterstained nuclei (blue). BG: bulge, SG: sebaceous gland, DP: dermal papilla (courtesy of J. Espada).

In the last decade, bio-orthogonal reactions [64] have been developed for labeling of biomolecules *in vivo*. Accordingly, the ethynyl group of EdU-substituted DNA is easily detected with fluorescent azides (*i.e.* AlexaFluor® 488 azide) [65] through a reaction known as "click chemistry", consisting in a cooper(I)-catalyzed alkaline cycloaddition [66, 67]. The corresponding chemical reactions and the replication banding pattern of metaphase chromosomes after EdU labeling followed by a fluorescent azide are shown in Fig. (**4.15**).

Incorporation of BrdU during DNA replication followed by UV irradiation (with or without photosensitization by Hoechst 33258) results in a method called strand break induction by photolysis (SBIP), which allows to recognize sites of DNA synthesis. SBIP consists in the induction of single- or double-strand DNA breaks in BrdU sites by UV or Hoechst 33258 + UV treatment, followed by formation of fluorescent oligonucleotide chains on the 3'-OH free ends of DNA using a terminal deoxynucleotidyl transferase and FITC- or Bodipy-labeled deoxynucleotides (Fig. **4.16A**). The SBIP method is currently applied as kits useful for cell proliferation assays, avoiding the denaturation step required for immunofluorescent labeling of Br-DNA.

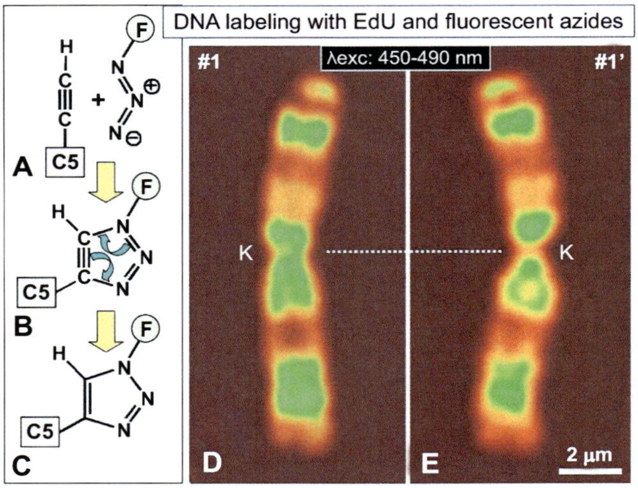

Fig. (4.15). A-C: Click reaction of EdU-labeled DNA with a fluorescent azide. The ethynyl residue (A, left) bound to the C5 atom of uracil and the azido group (A, right) react under Cu(I)-catalysis (B), yielding the triazene derivative (C). **D**, **E**: Chromosomes #1 and #1' from a human lymphocyte culture synchronized by 10^{-7} M methotrexate [49], treated wi^th edU, and then with AlexaFluor 488 azide and propidium iodide counterstaining. The late-replicating EdU-labeled DNA regions show green fluorescence and correspond to G bands.

On the other hand, primed *in situ* labeling (PRINS) is a fast and sensitive alternative to FISH for detection of repeat sequences using oligonucleotides or fragments of cloned probes as primers. Useful applications are the identification of chromosome aberrations in cancer and preimplantation genetic diagnosis, multicolor PRINS procedures for simultaneous visualization of several probes in different colors, detection of single-copy genes, and PRINS-painting procedures [68 - 71].

The strategy in PRINS (Fig. **4.16B**) is that a short unlabeled probe (about 18-35 nucleotides) hybridizes to matching DNA sequences in the chromatin, where it serves as a primer for chain elongation *in situ*. The reaction, commonly performed on a programmable temperature cycler, is catalyzed by Taq DNA polymerase that uses chromosomal DNA as template and deoxyribonucleotide triphosphates (some of which are labeled). Different non-fluorescent or fluorescent labels are usually employed, examples being biotin-16-dUTP (revealed by Texas red-streptavidin), digoxigenin-11-dUTP (revealed by immunofluorescence), or direct detection with fluorescein-12-dUTP and AMCA-11-dUTP. An example of PRINS labeling is shown in Fig. (**4.16C**), where the selective location of α-satellite in the human chromosome X is clearly visible.

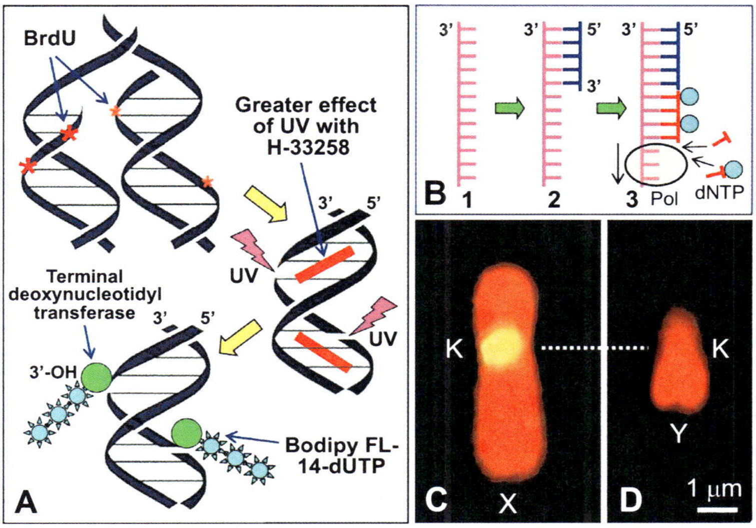

Fig. (4.16). A: Strand break induction by photolysis (SBIP) consists in BrdU incorporation into DNA, UV irradiation (with or without minor groove binding of Hoechst 33258), and synthesis of a fluorescent chain by a terminal deoxynucleotidyl transferase and labeled nucleotides. **B**: Schematic PRINS reaction involving (1) DNA denaturation (violet), (2) annealed primer (blue), and (3) synthesis of a polynucleotide (red) from labeled (blue circles) and unlabeled deoxyribonucleotide triphosphates (dNTP) by a DNA polymerase (Pol) along the arrow. **C**, **D**: Human chromosomes X and Y, showing the repetitive α-satellite sequences in the centromere of the chromosome X as revealed by PRINS labeling using digoxigenin-11-UTP, FITC-conjugated anti-Dig-IgG, and propidium bromide counterstaining. K: kinetochore.

Other labeling strategies are also possible. In the ARES™ DNA labeling kit, 5-(3-aminoallyl)-dUTP is enzymatically incorporated into DNA, and then a reactive fluorophore is used to label the aminoallyl group. Likewise, mercurated nucleic acid probes can be used for *in situ* hybridization [72]. The method is based on the reaction of 5-mercury pyrimidine residues of the hybridizing probe with the sulfhydryl group of a ligand containing a linker with two positively charged lysyl residues, and then an attached hapten (trinitrophenyl) or reporter group (fluorescein, biotin) is detected.

Fluorescent 2'(3')-O-(N-methylanthraniloyl) (Mant)-substituted nucleotides (modified on the ribose moiety) are a mixture of isomers widely used for the kinetic analysis of enzymes and signaling proteins [73]. GTP- and ATP-Mant derivatives are fluorescent analogues with λexc/λem: 355/448 nm. The Φ_F of Mant fluorophore is 0.2 in water, which increases significantly in nonpolar solvents or upon binding to most proteins. The compact nature of the Mant fluorophore and its attachment position result in nucleotide analogues that induce minimal perturbation of nucleotide-protein interactions. Because Mant fluorescence is sensitive to the pH environment of the fluorophore, these interactions may be directly detectable. Mant-nucleotides are valuable probes of the structure and enzymatic activity of nucleotide-binding proteins.

On account of structural similarities with purine bases, fluorescent analogues such as 8-azaguanine, lumazine and pteridine derivatives should be investigated in order to form the corresponding nucleoside analogues for enzymatic polymerization or incorporation in living cells. This approach could provide fluorescent polynucleotides. Likewise, it would be interesting to investigate the possible use of simple ruthenium complexes with purines for luminescence labeling.

5. PROTEIN LABELING

Conjugation of fluorochromes with proteins has allowed numerous studies [74], the most frequently used labels being the reactive dyes FITC and TRITC. In addition to proteins (antibodies), biomolecules with very high affinity for cell structures such as oligopeptides (phallotoxins) and phytotoxins (taxanes) can be labeled with suitable fluorophores, resulting in selective "immuno-like" fluorescence reactions. The immunolabeling using classical fluorophores conjugated to antibodies is illustrated in Figs. (**4.2A**) (see also Chapters 13 and 17). In the case of FITC labeling of antibodies, maximum brightness is attained with 2-4 fluorophores/antibody [75]. The combination of AMCA (7-amino-4-methylcoumarin-3-acetic acid), FITC, and TRITC remains the most common method for three-color analysis in fluorescence microscopy [27, 76]. New long wave-excitable coumarin compounds for protein labeling are available [77].

Non-covalent labeling of proteins is also suitable and frequently employed. Detection and imaging of individual cell-surface receptor molecules (*i.e.* low-density lipoprotein [LDL] receptor) is possible by using fluorescent labels and highly sensitive detectors. For instance, about 45 molecules of the fluorescent lipid analog DiI ($DiIC_{18}(3)$; 1,1'-dioctadecyl-3,3,3',3'-tetramethyl-indo-carbocyanine) can be inserted into the LDL molecule. This amount of fluorophores is sufficient to allow the visualization of an individual DiI-LDL particle, which has been applied to record the motion and trajectory of the DiI-LDL/LDL receptor complexes on the surface of live cells [78, 79]. Non-covalent labeling of albumin with trypan blue and similar bisazo dyes has been also applied in studies on retrograde axonal transport and protein transport from blood to tissues [80] (see Chapters 13.1.3 and 16.1.1).

On account of the indole ring, tryptophan is the dominant intrinsic fluorophore of most proteins, but as protein label its photophysics and fluorescence behavior is highly complex and environment-dependent [81]. Therefore an attractive approach is to replace the indole group by other similar but spectroscopically simpler fluorophores such as benzothiophene or azulene, the latter with $S_2 \rightarrow S_0$ emission [81]. However, drawbacks of these analogues are the lack of the N-H group of indole for H bonding, and difficulties in their biosynthetic incorporation into proteins. Replacement of the indole group of other biomolecules by these fluorescent analogues could be profitable (*i.e.* in the case of the actin-specific peptide derivative jasplakinolide [82]).

5.1. Avidin and Streptavidin

Avidin is a tetrameric glycoprotein from the white portion of eggs of birds and amphibia that is highly selective for biotin binding. As avidin is a cationic protein and can originate false positive results due to ionic binding to polyanionic substrates [83], it has been replaced by streptavidin (its analogue from *Streptomyces*), which does not suffer this drawback and is the most widely used compound to react with biotinylated biomolecules. Avidin and streptavidin are commonly conjugated with reactive fluorochromes or enzymes such as peroxidase and alkaline phosphatase, thus allowing signal amplification. Streptavidin nanoparticles are now available as intermediate labels.

5.2. Lectins

The term lectin (from the Latin *legere*, to choose) was originally applied to plant proteins that had the property of agglutinating erythrocytes and other mammalian cells. Lectins (also known as agglutinins) are "sugar-binding proteins or glycoproteins of non-immune origin that agglutinate cells and/or precipitate glycoconjugates" [29, 84]. This behavior mimics the selective reactivity of antibodies. Lectins bind to one or two glycoprotein determinants on the outside surface of cells, and thereby join the cells together. A few lectins are toxic but most are harmless, and their precise physiological role is still scarcely known.

Lectins can be conjugated with fluorescein and rhodamine fluorophores, the latter being more suitable because its red fluorescence is not easily confused with autofluorescence. An important practical issue is that alcoholic fixation must be preferred to that with aldehydes. Selective binding commonly requires Ca^{2+}, Mg^{2+}, or Mn^{2+} ions. Control procedures to confirm lectin positive reactions must be also performed (*i.e.* inhibition by binding with appropriate sugars, and chemical blocking such as methylation and acetylation of sugars, *etc.*).

Important members of the lectin family include *Phaseolus vulgaris* phytohemagglutinin (PHA, with strong mitogenic activity), as well as agglutinins from *Concavalia ensiformis* (concanavalin A, Con A), *Glycine max* (soybean lectin, SBA), *Triticum aestivum* (wheat germ lectin, WGA), *etc.* Examples of selective binding are Con A with α-D-mannose, α-D-glucose, and their *N*-acetylated derivatives, and PHA with galactose-β-1,4-*N*-acetylglucosamine-β-1, 2-mannose. In addition to plants, some lectins are obtained from animals (*i.e.* the snail *Helix pomatia*, and the horseshoe crab *Limulus polyphemus*). A survey of the selective affinity of lectins for sugars can be found in [85].

In spite of the low photodynamic action of fluorescein, fluoresceinated Con A binds to the outside of the plasma membrane and causes oxidative cross-linking of membrane proteins subjected to intense blue irradiation [86]. Similar effects are obtained by using TRITC-labeled Con A, FITC alone, or fluorescein mercuric acetate.

Specific lectins are important tools widely used for several applications such as mitogenic activation of lymphocytes, typing of human blood groups, analysis of lymphocyte subclasses, differentiation between normal and cancer cells, assessment of sugar-containing receptors, and studies on cell toxicity, migration, motility, and adhesion. AlexaFluor488 conjugated to GS-III lectin (from *Grifonia simplicifolia*) binds selectively to glycosylated proteins of the Golgi apparatus allowing its histochemical visualization by fluorescence microscopy.

5.3. Labeled Enzymes

Another original approach to identify macromolecular substrates is the use of labeled enzymes. The method is based on the specificity of particular enzymes toward their corresponding substrates, and requires the covalent labeling of the enzyme with a reactive fluorochrome avoiding the loss of enzyme affinity. A cellulase-FITC conjugate has been used to visualize cellulose in plant tissues [87]. Although interesting on account of its high selectivity, this method has found only scarce applications in fluorescence microscopy.

6. NATURAL FLUORESCENT PROTEINS

Some plant pigments (*i.e.* chlorophylls, carotenoids) exist as complexes with lipids or proteins. However, in this section emphasis will be given to more specific entities. Phycobiliproteins contain covalently bound fluorophores, and the same occurs in the case of fluorescent proteins (FPs, typically the green fluorescent protein, GFP), and all their derivatives and analogues from different species.

6.1. Phycobiliproteins

Phycobiliproteins are a family of chromoproteins derived from cyanobacteria (*Cyanophyceae*) and eukaryotic algae (*Rhodophyceae*) [6, 43, 88, 89]. They are also named simply as biliproteins due to the similarity of their chromophores with animal bile pigments (*i.e.* bilirubin, urobilin). Phycobiliproteins are deeply colored, highly fluorescent photoreceptor pigments composed of linear tetrapyrroles (phycobilins) and an apoprotein. They are classified as phycoerythrins (red), phycocyanins (blue), and allophycocyanins (pale blue). The two former compounds occur as supramolecular aggregates (phycobilisomes) closely linked to photosynthetic membrane systems for efficient energy transfer to chlorophyll molecules located in the photosynthetic reaction center. Interestingly, the related phytochromes are plant biosensors of environmental light signals that contain the linear phycocyanobilin-like tetrapyrrole, phytochromobilin [43, 90].

Linear tetrapyrroles were important chromophores for the evolution of life on earth [91]. About 3 x 10^9 years ago, anaerobic microorganisms evolved to cyanobacteria, the first life forms capable of performing photosynthesis. Instead of chlorophyll, these early cells did contain phycobiliproteins as photoreceptors. Later, phycobiliproteins became sensitizers of chlorophyll in the photosynthesis of algae. These ancient chromoproteins are specialized light harvesting molecular machines that act as fluorescent screens, converting the absorbed blue and blue-green light into orange and red light. This allows photosynthesis at a depth of 180 meters in sea water, where chlorophyll does not absorb any light because the 600-650 nm radiation is very scarce [91 - 94]. Absorption bands above 500 nm and at 350-400 nm of phycobiliproteins are attributed to the chromophores, whereas absorption at 270-280 nm are due to the apoprotein. Even UV energy absorbed by tyrosine and tryptophan in the phycocyanin molecule could contribute to its red fluorescence [92].

Thirty years ago a new class of fluorescent reagents, called phycofluors (conjugates of phycobiliproteins with molecules that confer biological specificity), were developed and applied as analytical and imaging reagents [89, 95]. B-phycoerythrin is an attractive tag because of its very high absorbance coefficient (2.4 x 10^6 cm^{-1} M^{-1} at 562 nm) due to 34 phycoerythrobilin molecules that are responsible for a very strong fluorescence quantum yield (Φ_F: 0.98).

Phycoerythrins are red-colored and orange-fluorescing chromoproteins while phycocyanins are blue in color with a red fluorescence. Phycobilins are covalently-bound fluorophores with chemical structures and emission illustrated in Fig. (**4.17A-C**) [43]. B- and R-phycoerythrins, and allophycocyanin are the best known phycobiliproteins, showing characteristic parameters (EC: extinction coefficient):

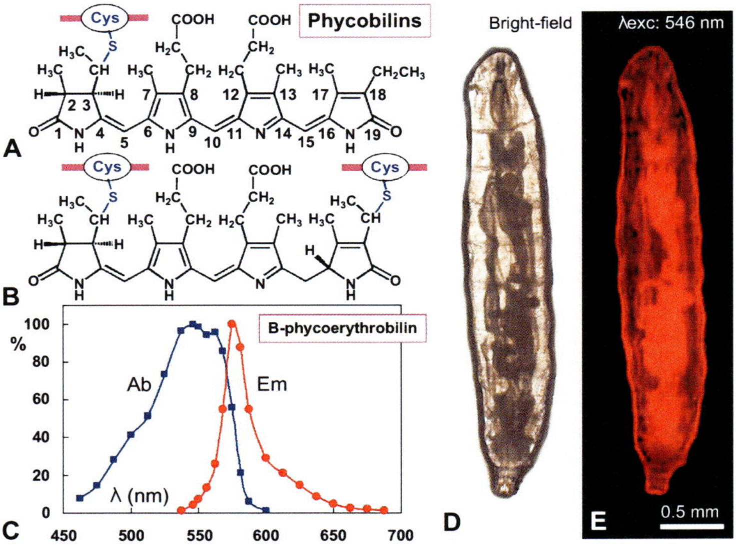

Fig. (4.17). Chemical structure of Cys-phycocyanobilin (**A**), and diCys-phycoerythrobilin (**B**), showing conventional bilin numbering in its extended conformation. Both bilins are bound to protein cystein residues by either one or two thioether bonds. **C**: Normalized absorption (blue) and emission (red) spectra of B-phycoerythrobilin. **D**, **E**: Bright-field and fluorescence images of a 3rd instar larva of *Drosophila melanogaster* after 72 h treatment with 2 mg/mL phycocyanin in the culture medium, washed in saline solution, and fixed in formaldehyde vapor. Fluorescent labeling corresponds mainly to the digestive tract (courtesy of S. Galaz-Leiva).

	Molecular weight	**λab (nm)**	**EC (cm^{-1} M^{-1})**	**λem (nm)**	**Φ_F**
B-phycoerythrin	240,000	546, 565	2,410,000	575	0.98
R-phycoerythrin	240,000	496, 546, 565	1,960,000	578	0.82
Allophycocyanin	104,000	650	700,000	660	0.68

The fluorescence of phycobiliproteins is less or not at all quenched either by internal energy transfer or by external factors (*i.e.* changes in pH, quenching molecules), because the fluorophores are protected by the covalently linked protein shell. A relevant consequence is the high Φ_F of phycoerythrins.

On account of its absorbance and Φ_F parameters, the fluorescence intensity of a phycoerythrin molecule is equivalent to at least 30 fluorescein molecules and 100 rhodamine molecules. The fluorescent signal arising from only three molecules of phycoerythrin was clearly observed [95]. More recently, single R-phycoerythrin molecules were imagined in both nucleus and cytoplasm of mammalian cells [96]. In practical applications such as flow cytometry and immunoassays, the sensitivity of phycoerythrin-conjugated antibodies is usually 5- to 10-times higher than fluorescein conjugates. Phycobiliproteins are efficient fluorescent labels due to stability, water solubility, and low nonspecific binding. In addition to their binding to antibodies, phycobiliproteins can be labeled with other fluorochromes (*i.e.* Alexa probes, Texas red), and used as tracers for studies on nucleo-cytoplasmic protein transport in living cells. Phycocyanins are suitable fluorescent labels (Fig. **4.17E**), and also show phototoxic activity [97].

6.2. Green Fluorescent Protein, Derivatives and Analogues

The prototype of fluorescent proteins (FPs) is the green fluorescent protein (GFP, 27 kDa) from light-emitting cells (photocytes) of the jellyfish *Aequorea victoria* (class Hydrozoa) (Fig. **4.18A**). When excited by violet or blue light (395 or 475 nm) GFP emits green light (peak: 509 nm, shoulder: 540 nm) (Fig. **4.18E**) with high efficiency (Φ_F: 0.8) and very low fading.

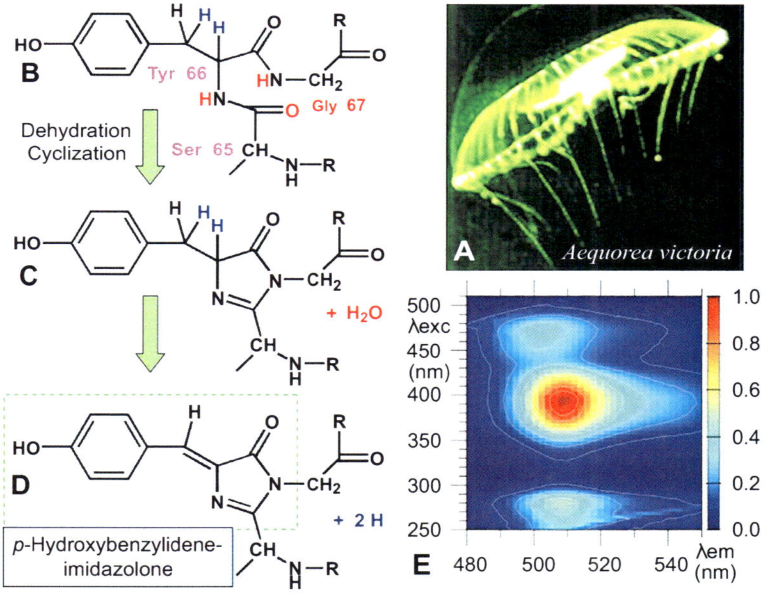

Fig. (4.18). "Living" dyes: green fluorescent protein (GFP). **A**: The jellyfish *Aequorea victoria* showing green fluorescence under blue excitation. **B**-**D**: Formation of GFP fluorophore, showing dehydration, cyclization and oxidation steps. **E**: 3D excitation-emission matrix recorded for the recombinant wt GFP dissolved in Tris-EDTA, pH 7.6 at 21 °C. The color map used for the intensity scale is set to 1 at the position λexc: 390 nm and λem: 510 nm (courtesy of Ursa BioScience, www.UrsaBioScience.com).

Other coelenterates (*i.e. Aequorea forskalea, Renilla reniformis, Phialidium gregarum*) also have similar GFPs. In nature, photoproteins such as the chemiluminescent Ca^{2+}-activated aequorin (λem: 470 nm) [98] excites the

Aequorea GFP by a radiation-less energy transfer that requires protein-protein interaction. In the case of *Renilla* GFP, λexc is at 498 with emission 6 times higher than for *Aequorea* GFP, and energy transfer occurs from the luciferase-coelenterazine system.

The small GFP fluorophore (CBN: 15) is 4-(*p*-hydroxybenzylidene)-5-imidazolone, which is formed by a spontaneous post-translational and oxygen-dependent auto-catalytic event (cyclization and oxidation) involving the sequence -Ser65-(or Thr65)-Tyr66-Gly67- (Fig. **4.18B-D**). GFP has 238 amino acids and a monomeric or dimeric structure when it occurs at low or high concentration, respectively. Relevant reviews on the GFP have been published [99, 100]. In 2008, the Nobel Prize of Chemistry was awarded to O. Shimomura, M. Chalfie and R.Y. Tsien for their studies on the GFP.

X-ray crystallography reveals that GFP structure consists of an 11-stranded barrel of folded β-sheets and a coaxial α-helix, with the fluorophore formed in the central helix, and loops and segments of α-helix closing the barrel ends [101]. The fluorophore appears well protected and without contact with the free solvent. Photosensitization of O_2 by the model chromophore of the EGFP) is very scarce (Φ_Δ: 0.004) [102]. N or C terminal groups at the same barrel end allow the fusion with other proteins.

The GFP fluorophore appears predominantly in its ionized (charged) form [101]. The ionization of the phenolic -OH group results in dramatic differences in the molecular geometry and charge distribution. A paraquinone benzene structure is most apparent in the ionized (anionic) form and is responsible for the absorption peak at 475 nm, whereas the peak at 395 nm corresponds to the unionized (neutral) phenolic form. The same fluorophore occurs in *Aequorea* and *Renilla* GFPs but their spectra are strongly dependent on the amino acid sequence within and outside the carrier peptide. GFPs fluorophores are considerably stable over a wide range of temperature and pH values. In "kindling fluorescent proteins" (and similar fluorophores) a *cis-trans* isomerism allows to change between KFP kindled (*cis*, fluorescent) and KFP not kindled (*trans*, non-fluorescent) states.

The remarkable diversity of coral reef coloration is mainly due to the presence of FPs and non-fluorescent chromoproteins that are related to the GFP. At present, red and yellow FPs from Anthozoa corals have attracted great interest on account of their novel fluorophores and spectral features, becoming important tools for *in vivo* imaging of cells and tissues [103, 104]. Similarly to GFP, the yellow zFP538 and DsRed form the fluorophore at the sequences -Lys66 - Tyr67 - Gly68-, and - Gln66-Tyr67-Gly68-, respectively [105].

Several GFP variants have been obtained by modifying the polypeptidic chain, in which different interactions of the fluorophore with polar or hydrophobic amino acids take place. Thus enhanced, cyan, yellow and red FPs, as well as FPs from other organisms are now available. Recently, a novel FP from the jellyfish *Rhacostoma atlantica* has been isolated (*Rhacostoma* GFP, R-GFP), which is similar to the popular E-GFP mutant. Spectral features of some FPs labels are the following:

Protein	λexc (nm)	λem (nm)
GFP	395 + 475	509 + 540
R-GFP	466	505
E-GFP	488	507
E-CFP	434	474
E-YFP	514	527
DsRed2	558	583
HcRed1	590	615
mCherry	585	615

E: enhanced (mutated FP), R-GFP: from *Rhacostoma atlantica*, DsRed2: from *Discosoma* sp. (Coelenterate), HcRed1: from the coral *Heteractis crispa*.

GFP and other similar FPs have become invaluable tools in cell and molecular biology. Instead direct application as fluorochromes, most relevant uses of FPs are intracellular synthesis after incorporation of their corresponding genes. The gene of GFP has been expressed in prokaryotic (*Escherichia coli*) or eukaryotic (*Caenorhabditis elegans*) cells [106]. Because exogenous substrates and cofactors are not required for GFP fluorescence, the fluorescent signal is physiological and related to gene expression in live cells. Thus GFP emission can be used to monitor gene expression and protein localization in living organisms, enabling to visualize fluorescent-tagged proteins without the need for chemical staining. The fluorescent response of transgenic GFP reporter strains of organisms can be also quantified [107]. It is worth to note that antibodies against natural GFP are now available to amplify weak fluorescent signals.

Different structural and functional effects can be visualized by intravital GFP fluorescence microscopy using a transparent plastic chamber implanted into the dorsal skinfold of mice [47]. Gene transfer is detected after injection of a GFP-containing plasmid into the skinfold chamber followed by electroporation and microscopic observation 5 days later (Fig. **4.19A**). On the other hand, the use of fusion proteins with FPs allows the visualization of specific structures in living cells, examples being the labeling of chromatin by histone1-GFP (Fig. **4.19B**) and the detection of the nuclear membrane by the fusion protein lamina-RFP (Fig. **4.20A, B**). Other fusion protein constructed with YFP stably bound to cytochrome c oxidase allows visualizing fluorescent mitochondria (Fig. **4.20C**).

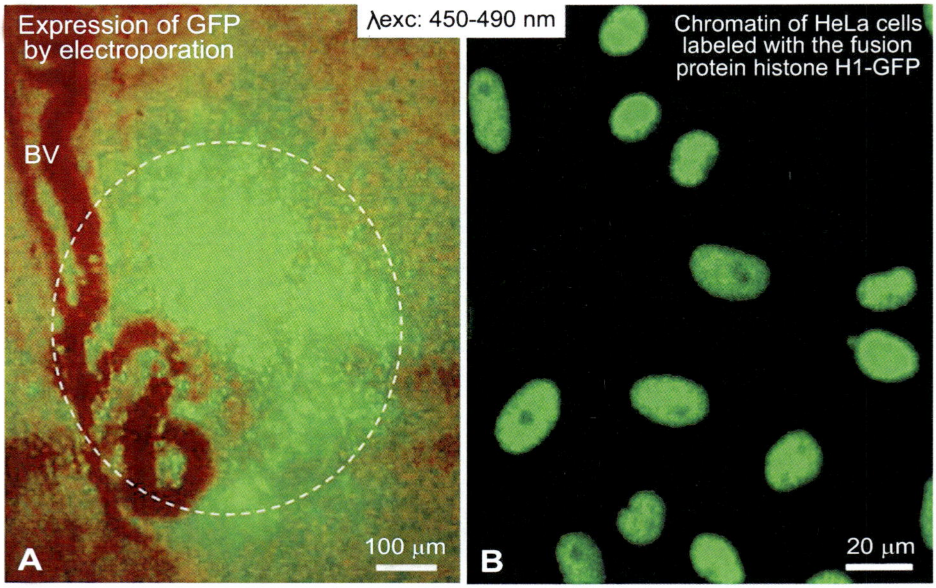

Fig. (4.19). Examples of GFP labeling. **A**: Gene electrotransfer visualized by GFP. Injection of 25 μL GFP-containing plasmid into a dorsal skinfold chamber of the mouse was followed by electroporation (10 pulses of 20 V and 20 ms at 1 Hz, using 1 mm-apart electrodes inserted through the skin), and microscopic observation 5 days later. BV: blood vessel (courtesy of N. Olaiz, L. Colombo and G. Marshall). **B**: Image from a video of HeLa cells showing the fusion protein histone H1-GFP. Observe the nucleoli without fluorescent signal (courtesy of J. Espada).

The preservation of GFP fluorescence after fixation is an important concern. Fixation with 1% paraformaldehyde followed by extensive washing with PBS generally gives good results. If the fluorescence decreases, a possible alternative is using an anti-GFP IgG and a fluorescent secondary IgG to amplify the signal. Another relevant point is the possible change of the fluorescent site. GFP, CFP, and YFP are considered to be inert proteins regarding possible binding to cell structures. After expression of the electroporated plasmids *pEGFP-C1*, *pECFP-C1*, and *pEYFP-C1* in cultured cells, FPs are localized homogeneously within living cells, but after paraformaldehyde fixation they appear associated to actin microfilaments [108]. This result is confirmed by colocalization of GFP with TRITC-phalloidin, which indicates that careful controls are necessary when observing FPs signals in fixed cells.

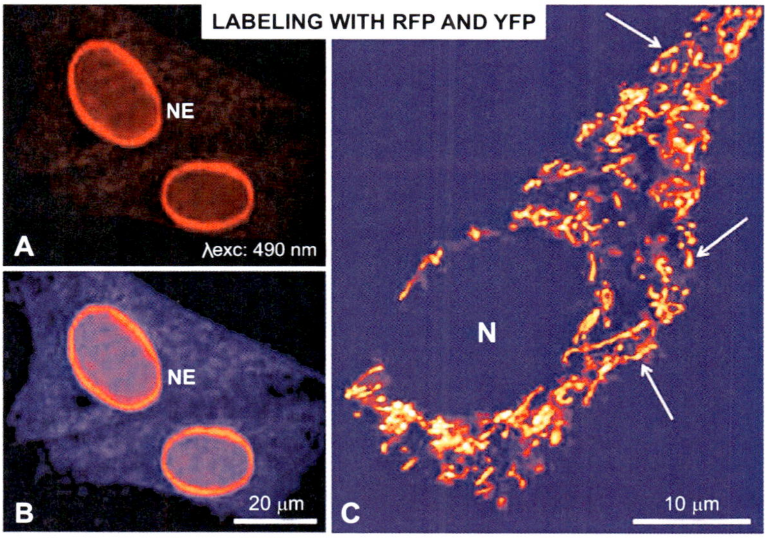

Fig. (4.20). **A**: HeLa cells labeled with the fusion protein lamina-RFP showing the fluorescence of nuclear envelope (NE) (confocal microscopy, courtesy of J. Espada). **B**: Same cells processed for color equidensities (ImageJ, LUT: fire). **C**: Ptk2 cell showing the mitochondrial fluorescence (arrows) after labeling with YFP stably bound to the MTS sequence of cytochrome c oxidase (subunit VIII). N: nucleus. ImageJ, LUT: fire.

In some particular cases, FP fluorophores are photo-active agents that represent endogenous, genetically-encoded photosensitizers for production of $^1O_2$ from $^3O_2$. Under green light, TagRFP contained in a pET20b plasmid induces cell death in *E. coli* cultures [109]. Another FP is the mini singlet oxygen generator (miniSOG, a non-covalent FMN-containing flavoprotein, which under blue excitation generates $^1O_2$) [110]. It must be noted that in both cases, the efficiency for $^1O_2$ production is very low (Φ_Δ: 0.004 for TagRFP and Φ_Δ: 0.03 for miniSOG). On the other hand, GFP, YFP and some derivatives can be used as blinking fluorophores for fluorescence super-resolution [111] (see Chapter 20.4).

Benzylidene reactions *in silico* with *p*-dimethylamino-benzaldehyde (Ehrlich's reagent, DMBA) also form GFP-like compounds. A chromophore similar to that of GFP is *p*-dimethylamino-benzylidene-rhodanine, which forms colored complexes with some heavy metals (particularly cooper), and is useful in histochemical work [80, 112]. Replacing the *p*-dimethylamino group for a nitrogen mustard residue (-N[$CH_2CH_2Cl]_2$) yields several reactive fluorochromes with cytostatic properties [113]. The hydantoin derivative is very similar to the GFP fluorophore. Likewise, condensation with DMBA with indole (tryptophan), imidazole (histidine), and pyrrole rings give red azomethine dyes that should be fluorescent.

In addition to GFP, imaging of gene expression can be achieved through luciferase bioluminescence [114, 115]. Luciferase from the firefly *Photinus pyralis* produces yellow-green light (λem: 560 nm, Φ_F: 0.88) by adenylating and then oxidizing the substrate luciferin in the presence of ATP, O_2, and Mg^{2+}. The enzyme shows no toxic effects on cell metabolism, and the corresponding reporter gene *luc* can be stably transfected providing bioluminescent images on gene activity. Bioengineered photoproteins from marine organisms (*i.e.* the bivalve *Pholas dactylus*, the barnacle *Vargula hilgendorfii*, and the coelenterate *Obelia*) could be also useful in studies of gene expression.

7. OTHER LABELING METHODS

The macromolecular structure itself can offer useful photoluminescence properties. This is the case of carbon nanotubes (see Chapter 18.2.2) and other nanoparticles such as Qdots (see Chapter 18.2.1). Carbon nanotubes are almost "one-dimensional" structures that can be envisioned as rolled single sheets of graphite (or more precisely graphene) with semiconductor, photovoltaic, and optic properties (*i.e.* fluorescence, electroluminescence).

Single-wall carbon nanotubes (SWNTs) emit near-infrared light upon suitable excitation. In the photoluminescence (PL) process an electron from the nanotube absorbs exciting light, creating an electron-hole pair (exciton). Both electron and hole rapidly relax (*via* phonon-assisted processes), and then they recombine with light emission, the position and intensity of emission peaks depending on the nanotube environment; no PL occurs in multi-wall carbon nanotubes. Isolated SWNTs lying on the substrate show weak PL. Efficiency can be modified by encapsulating linear dyes (carotene, lycopene, *etc.*) inside the tubes as inclusion compounds. Energy transfer occurs from the light absorbing dye and emitting SWNTs.

The fluorescence of carbon nanotubes are now used in biomedical fields for imaging and sensing applications. The structure of a SWNT with 14 carbon hexagonal units in a zig-zag configuration is shown in Fig. (**4.21A**). Although the emission spectrum of SWNT excited at 808 nm presents low intensity at 1300-1400 nm (Fig. **4.21B**), this NIR-IIa region is the most suitable for fluorescence detection due to little light scattering and no tissue autofluorescence [116]. After injection of fluorescent water-soluble SWNTs into the bloodstream of mice and NIR excitation with an 808 nm laser, the circulation inside the living brain can be visualized. NIR excitation by this non-invasive technique penetrates about three millimeters through intact scalp skin and skull to see vasculature details such as arterial and venous blood flow in the brain, useful for studying cerebral-vascular diseases (Fig. **4.21C, D**).

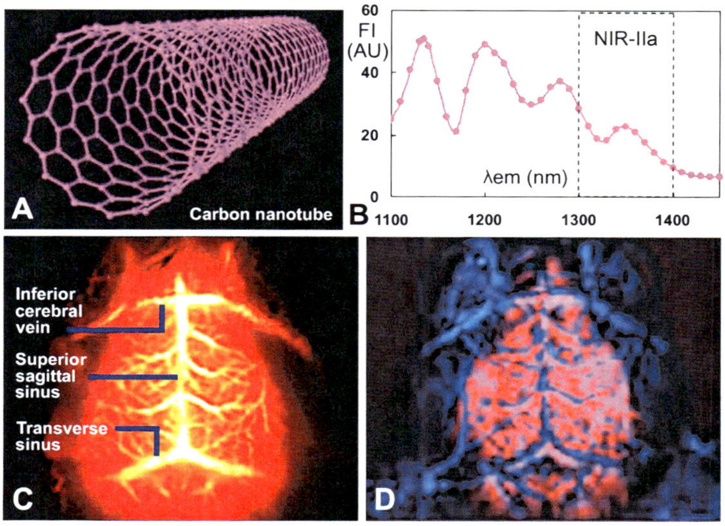

Fig. (4.21). NIR fluorescence by labeling *in vivo* with carbon nanotubes. **A**: Structure of a single wall nanotube (SWNT). **B**: Fluorescence spectrum of a solution of SWNTs showing the region NIR-IIa. **C**: NIR-IIa image of a healthy mouse brain vasculature through the intact scalp and skull, taken at 4.7 s postinjection of SWNTs (pseudocolor). **D**: Time-course NIR-IIa merged images of the mouse brain showing arterial (red) and venous (blue) vessels, 2.6 s and 18.5 s after SWNT injection, respectively [116] (courtesy of G. Hong and H. Dai).

CONFLICT OF INTEREST

The author (editor) declares no conflict of interest, financial or otherwise.

ACKNOWLEDGEMENTS

This work was supported by a grant from the Ministerio de Economía y Competitividad (CTQ2013-48767-C3-3-R), Spain.

REFERENCES

[1] Beddart GS, West MA, Eds. Fluorescent Probes. London, New York: Academic Press 1981.

[2] Waggoner AS. Fluorescent Probes for Analysis of Cell Structure, Function, and Health by Flow and Imaging Cytometry.Applications of Fluorescence in the Biomedical Sciences. New York: Alan R. Liss 1986; pp. 3-28.

[3] Koller E. Fluorescent labels for use in biology and medicine. Appl Fluor Technol 1989; 1(2): 1-8.

[4] Mullins JM. Overview of fluorophores. Methods Mol Biol 1994; 34: 107-16. [PMID: 7711838]

[5] Stockert JC. Reactive Staining Reagents and Fluorescent Labels.Conn's Biological Stains A Handbook of Dyes, Stains and Fluorochromes for Use in Biology and Medicine. 10th ed.., Oxford: Bios Scientific Publishers 2002.

[6] Haugland RP. Handbook of Fluorescent Probes and Research Products. 9th ed., Eugene, OR: Molecular Probes, Inc. 2002.

[7] Hilderbrand SA. Labels and probes for live cell imaging: overview and selection guide. Methods Mol Biol 2010; 591: 17-45.
[http://dx.doi.org/10.1007/978-1-60761-404-3_2] [PMID: 19957122]

[8] Sinkeldam RW, Greco NJ, Tor Y. Fluorescent analogs of biomolecular building blocks: design, properties, and applications. Chem Rev 2010; 110(5): 2579-619.
[http://dx.doi.org/10.1021/cr900301e] [PMID: 20205430]

[9] Bindman NA, van der Donk WA. A general method for fluorescent labeling of the N-termini of lanthipeptides and its application to visualize their cellular localization. J Am Chem Soc 2013; 135(28): 10362-71.
[http://dx.doi.org/10.1021/ja4010706] [PMID: 23789944]

[10] Beverloo HB, Tanke HJ. Inorganic phosphors: new luminescent immunolabels. Appl Fluor Tech 1991; 3(4): 27-9.

[11] Arap W, Pasqualini R, Montalti M, *et al.* Luminescent silica nanoparticles for cancer diagnosis. Curr Med Chem 2013; 20(17): 2195-211.
[http://dx.doi.org/10.2174/0929867311320170005] [PMID: 23458621]

[12] Coons AH, Creech HJ, Jones RN, Berliner E. The demonstration of pneumococcal antigen in tissues by the use of fluorescent antibody. 1942. J Immunol 1942; 45: 159-70.

[13] Titus JA, Haugland R, Sharrow SO, Segal DM. Texas Red, a hydrophilic, red-emitting fluorophore for use with fluorescein in dual parameter flow microfluorometric and fluorescence microscopic studies. J Immunol Methods 1982; 50(2): 193-204.
[http://dx.doi.org/10.1016/0022-1759(82)90225-3] [PMID: 6806389]

[14] Devanathan S, Dahl TA, Midden WR, Neckers DC. Readily available fluorescein isothiocyanate-conjugated antibodies can be easily converted into targeted phototoxic agents for antibacterial, antiviral, and anticancer therapy. Proc Natl Acad Sci USA 1990; 87(8): 2980-4.
[http://dx.doi.org/10.1073/pnas.87.8.2980] [PMID: 2109321]

[15] Kosower NS, Kosower EM, Newton GL, Ranney HM. Bimane fluorescent labels: labeling of normal human red cells under physiological conditions. Proc Natl Acad Sci USA 1979; 76(7): 3382-6.
[http://dx.doi.org/10.1073/pnas.76.7.3382] [PMID: 291011]

[16] Mirabelli F, Vairetti M, Bellomo G. Fluorescence analysis of intracellular soluble and protein thiols using bimane derivatives. Appl Fluor Technol 1992; 4(2 & 3): 9-13.

[17] Stockert JC, Blázquez A, Galaz S, Juarranz A. A mechanism for the fluorogenic reaction of amino groups with fluorescamine and MDPF. Acta Histochem 2008; 110(4): 333-40.
[http://dx.doi.org/10.1016/j.acthis.2007.10.011] [PMID: 18272208]

[18] Capobianco ML, Barbarella G, Manetto A. Oligothiophenes as fluorescent markers for biological applications. Molecules 2012; 17(1): 910-33.
[http://dx.doi.org/10.3390/molecules17010910] [PMID: 22258339]

[19] Politz JC. Use of caged fluorochromes to track macromolecular movement in living cells. Trends Cell Biol 1999; 9(7): 284-7.
[http://dx.doi.org/10.1016/S0962-8924(99)01585-8] [PMID: 10370245]

[20] Desai A, Mitchison TJ. Preparation and characterization of caged fluorescein tubulin. Methods Enzymol 1998; 298: 125-32.
[http://dx.doi.org/10.1016/S0076-6879(98)98014-4] [PMID: 9751877]

[21] Smith NA, Sadler PJ. Photoactivatable metal complexes: from theory to applications in biotechnology and medicine. Philos Trans A Math Phys Eng Sci 2013; 371(1995): 20120519.
[http://dx.doi.org/10.1098/rsta.2012.0519] [PMID: 23776303]

[22] Maccioni RB. Molecular Cytology of Microtubules.Cell Biology Reviews. Leioa-Vizcaya: Editorial Service, University of the Basque Country 1986.

[23] Bhattacharyya B, Wolff J. Promotion of fluorescence upon binding of colchicine to tubulin. Proc Natl Acad Sci USA 1974; 71(7): 2627-31. [http://dx.doi.org/10.1073/pnas.71.7.2627] [PMID: 4527949]

[24] Borowiak M, Nahaboo W, Reynders M, *et al.* Photoswitchable inhibitors of microtubule dynamics optically control mitosis and cell death. Cell 2015; 162(2): 403-11. [http://dx.doi.org/10.1016/j.cell.2015.06.049] [PMID: 26165941]

[25] Wulf E, Deboben A, Bautz FA, Faulstich H, Wieland T. Fluorescent phallotoxin, a tool for the visualization of cellular actin. Proc Natl Acad Sci USA 1979; 76(9): 4498-502. [http://dx.doi.org/10.1073/pnas.76.9.4498] [PMID: 291981]

[26] Faulstich H, Trischmann H, Mayer D. Preparation of tetramethylrhodaminyl-phalloidin and uptake of the toxin into short-term cultured hepatocytes by endocytosis. Exp Cell Res 1983; 144(1): 73-82. [http://dx.doi.org/10.1016/0014-4827(83)90443-3] [PMID: 6840214]

[27] Small JV, Zobeley S, Rinnerthaler G, Faulstich H. Coumarin-phalloidin: a new actin probe permitting triple immunofluorescence microscopy of the cytoskeleton. J Cell Sci 1988; 89(Pt 1): 21-4. [PMID: 2458368]

[28] Barak LS, Yocum RR. 7-Nitrobenz-2-oxa-1,3-diazole (NBD)--phallacidin: synthesis of a fluorescent actin probe. Anal Biochem 1981; 110(1): 31-8. [http://dx.doi.org/10.1016/0003-2697(81)90107-X] [PMID: 6894219]

[29] Budavari S, Ed. The Merck Index. An Encyclopedia of Chemicals, Drugs and Biologicals. 11th ed.. Rahway, NJ: Merck and Co. 1989.

[30] Gopalakrishnan S, Harris TM, Stone MP. Intercalation of aflatoxin B1 in two oligodeoxynucleotide adducts: comparative 1H NMR analysis of d(ATCAFBGAT).d(ATCGAT) and d(ATAFBGCAT)$_2$. Biochemistry 1990; 29(46): 10438-48. [http://dx.doi.org/10.1021/bi00498a002] [PMID: 2125491]

[31] Rhys Williams AT. Fluorescence Detection in Liquid Chromatography. Buckinghamshire, UK: Perkin-Elmer 1980.

[32] Souto AA, Acuña AU, Andreu JM, *et al.* New fluorescent water-soluble taxol derivatives. Angew Chem Int Ed Engl 1995; 34: 2710-2. [http://dx.doi.org/10.1002/anie.199527101]

[33] Evangelio JA, Abal M, Barasoain I, *et al.* Fluorescent taxoids as probes of the microtubule cytoskeleton. Cell Motil Cytoskeleton 1998; 39(1): 73-90. [http://dx.doi.org/10.1002/(SICI)1097-0169(1998)39:1<73::AID-CM7>3.0.CO;2-H] [PMID: 9453715]

[34] Barasoain I, Díaz JF, Andreu JM. Fluorescent taxoid probes for microtubule research. Methods Cell Biol 2010; 95: 353-72. [http://dx.doi.org/10.1016/S0091-679X(10)95019-X] [PMID: 20466144]

[35] Bicamumpaka C, Page M. *In situ* localization of paclitaxel binding structures: Labeling with a paclitaxel fluorescent analogue. Int J Mol Med 1998; 2(2): 161-5. [PMID: 9855683]

[36] Reaven E, Tsai L, Azhar S. Intracellular events in the "selective" transport of lipoprotein-derived cholesteryl esters. J Biol Chem 1996; 271(27): 16208-17. [http://dx.doi.org/10.1074/jbc.271.27.16208] [PMID: 8663101]

[37] Hope-Roberts M, Wainwright M, Horobin RW. A nitrobenzofuran-conjugated phosphatidylcholine (C_{12}-NBD-PC) as a stain for membrane lamellae for both microscopic imaging and spectrofluorimetry. Biotech Histochem 2008; 83(1): 25-8.

[http://dx.doi.org/10.1080/10520290801886224] [PMID: 18568674]

[38] Fendler JH. Membrane Mimetic Chemistry Characterizations and Applications of Micelles, Microemulsions, Monolayers, Bilayers, Vesicles, Host-guest Systems, and Polyions. New York: John Wiley & Sons 1982.

[39] Harrington KJ, Syrigos KN, Vile RG. Liposomally targeted cytotoxic drugs for the treatment of cancer. J Pharm Pharmacol 2002; 54(12): 1573-600. [http://dx.doi.org/10.1211/0022357002243] [PMID: 12542887]

[40] Morgan J. Application of fluorescent liposomes. Appl Fluor Technol 1990; 2(1): 14-6.

[41] van Rooijen N, van Nieuwmegen R. Fluorochrome staining of multilamellar liposomes. Stain Technol 1978; 53(6): 307-10. [http://dx.doi.org/10.3109/10520297809111951] [PMID: 89716]

[42] Riske KA, Knorr RL, Dimova R. Bursting of charged multicomponent vesicles subjected to electric pulses. Soft Matter 2009; 5: 1983-6. [http://dx.doi.org/10.1039/b900548j]

[43] Beale SI. Biosynthesis of phycobilins. Chem Rev 1993; 93: 785-802. [http://dx.doi.org/10.1021/cr00018a008]

[44] Owens TG. Processing of Excitation Energy by Antenna Pigments.Photosynthesis and the Environment. Dordrecht: Kluwer Academic Publishers 1996; pp. 1-23.

[45] Krieger-Liszkay A. Singlet oxygen production in photosynthesis. J Exp Bot 2005; 56(411): 337-46. [http://dx.doi.org/10.1093/jxb/erh237] [PMID: 15310815]

[46] Espada J, Pérez-Moreno M, Braga VM, Rodriguez-Viciana P, Cano A. H-Ras activation promotes cytoplasmic accumulation and phosphoinositide 3-OH kinase association of β-catenin in epidermal keratinocytes. J Cell Biol 1999; 146(5): 967-80. [http://dx.doi.org/10.1083/jcb.146.5.967] [PMID: 10477752]

[47] Olaiz N, Signori E, Maglietti F, *et al.* Tissue damage modeling in gene electrotransfer: the role of pH. Bioelectrochemistry 2014; 100: 105-11. [http://dx.doi.org/10.1016/j.bioelechem.2014.05.001] [PMID: 24925861]

[48] Bârzu T, Pascal M, Maman M, Roque C, Lafont F, Rousselet A. Entry and distribution of fluorescent antiproliferative heparin derivatives into rat vascular smooth muscle cells: comparison between heparin-sensitive and heparin-resistant cultures. J Cell Physiol 1996; 167(1): 8-21. [http://dx.doi.org/10.1002/(SICI)1097-4652(199604)167:1<8::AID-JCP2>3.0.CO;2-T] [PMID: 8698843]

[49] Ribas M, Korenberg JR, Peretti D, *et al.* Sister chromatid differentiation in 5-bromo-2′-deoxyuridine-substituted chromosomes: a study with DNA-specific ligands and monoclonal antibody to histone H2B. Chromosome Res 1994; 2(6): 428-38. [http://dx.doi.org/10.1007/BF01552865] [PMID: 7530578]

[50] Mezzanotte R, Nieddu M. A historical overview of bromo-substituted DNA and sister chromatid differentiation. Methods Mol Biol 2014; 1094: 89-98. [http://dx.doi.org/10.1007/978-1-62703-706-8_8] [PMID: 24162982]

[51] Carrasco E, Calvo MI, Espada J. DNA labeling *in vivo*: quantification of epidermal stem cell chromatin content in whole mouse hair follicles using Fiji image processing software. Methods Mol Biol 2014; 1094: 79-88. [http://dx.doi.org/10.1007/978-1-62703-706-8_7] [PMID: 24162981]

[52] Pinkel D, Straume T, Gray JW. Cytogenetic analysis using quantitative, high-sensitivity, fluorescence hybridization. Proc Natl Acad Sci USA 1986; 83(9): 2934-8. [http://dx.doi.org/10.1073/pnas.83.9.2934] [PMID: 3458254]

[53] Eriksson M, Nielsen PE. Solution structure of a peptide nucleic acid-DNA duplex. Nat Struct Biol

1996; 3(5): 410-3.
[http://dx.doi.org/10.1038/nsb0596-410] [PMID: 8612069]

[54] Menchise V, De Simone G, Tedeschi T, *et al.* Insights into peptide nucleic acid (PNA) structural features: the crystal structure of a D-lysine-based chiral PNA-DNA duplex. Proc Natl Acad Sci USA 2003; 100(21): 12021-6.
[http://dx.doi.org/10.1073/pnas.2034746100] [PMID: 14512516]

[55] Viera A, Ortiz MI, Pinna-Senn E, Dalmasso G, Bella JL, Lisanti JA. Chromosomal localization of telomeric sequences in three species of Akodon (Rodentia, Sigmodontinae). Cytogenet Genome Res 2004; 107(1-2): 99-102.
[http://dx.doi.org/10.1159/000079577] [PMID: 15305062]

[56] Bowen EJ, Ed. Recent Progress in Photobiology. New York: Academic Press 1965.

[57] Latt SA. Microfluorometric detection of deoxyribonucleic acid replication in human metaphase chromosomes. Proc Natl Acad Sci USA 1973; 70(12): 3395-9.
[http://dx.doi.org/10.1073/pnas.70.12.3395] [PMID: 4128545]

[58] Kato H. Spontaneous sister chromatid exchanges detected by a BUdR-labelling method. Nature 1974; 251(5470): 70-2.
[http://dx.doi.org/10.1038/251070a0] [PMID: 4137941]

[59] Wolff S, Perry P. Differential Giemsa staining of sister chromatids and the study of chromatid exchanges without autoradiography. Chromosoma 1974; 48(4): 341-53.
[http://dx.doi.org/10.1007/BF00290991] [PMID: 4141298]

[60] Hazen MJ, Villanueva A, Juarranz A, Cañete M, Stockert JC. Photosensitizing dyes and fluorochromes as substitutes for 33258 Hoechst in the fluorescence-plus-Giemsa (FPG) chromosome technique. Histochemistry 1985; 83(3): 241-4.
[http://dx.doi.org/10.1007/BF00953991] [PMID: 2412994]

[61] Stockert JC, Blázquez-Castro A, Horobin RW. Identifying different types of chromatin using Giemsa staining. Methods Mol Biol 2014; 1094: 25-38.
[http://dx.doi.org/10.1007/978-1-62703-706-8_3] [PMID: 24162977]

[62] Gratzner HG. Monoclonal antibody to 5-bromo- and 5-iododeoxyuridine: A new reagent for detection of DNA replication. Science 1982; 218(4571): 474-5.
[http://dx.doi.org/10.1126/science.7123245] [PMID: 7123245]

[63] Hemmer J. *In vitro* bromodeoxyuridine labeling and flow cytometry of solid tumors. Appl Fluor Technol 1991; 3(6): 2-4.

[64] Best MD. Click chemistry and bioorthogonal reactions: unprecedented selectivity in the labeling of biological molecules. Biochemistry 2009; 48(28): 6571-84.
[http://dx.doi.org/10.1021/bi9007726] [PMID: 19485420]

[65] Hoshi O, Ushiki T. Replication banding pattern in human chromosomes detected using 5-ethynyl-2′-deoxyuridine incorporation. Acta Histochem Cytochem 2011; 44(5): 233-7.
[http://dx.doi.org/10.1267/ahc.11029] [PMID: 22096263]

[66] Buck SB, Bradford J, Gee KR, Agnew BJ, Clarke ST, Salic A. Detection of S-phase cell cycle progression using 5-ethynyl-2′-deoxyuridine incorporation with click chemistry, an alternative to using 5-bromo-2′-deoxyuridine antibodies. Biotechniques 2008; 44(7): 927-9.
[http://dx.doi.org/10.2144/000112812] [PMID: 18533904]

[67] Salic A, Mitchison TJ. A chemical method for fast and sensitive detection of DNA synthesis *in vivo*. Proc Natl Acad Sci USA 2008; 105(7): 2415-20.
[http://dx.doi.org/10.1073/pnas.0712168105] [PMID: 18272492]

[68] Hindkjaer J, Koch J, Brandt C, Kølvraa S, Bolund L. Primed in situ labeling (PRINS). A fast method for in situ labeling of nucleic acids. Mol Biotechnol 1996; 6(2): 201-11.
[http://dx.doi.org/10.1007/BF02740774] [PMID: 8970173]

[69] Kubaláková M, Vrána J, Cíhalíková J, Lysák MA, Dolezel J. Localisation of DNA sequences on plant chromosomes using PRINS and C-PRINS. Methods Cell Sci 2001; 23(1-3): 71-82. [http://dx.doi.org/10.1023/A:1013193516001] [PMID: 11741145]

[70] Tharapel SA, Kadandale JS. Primed *in situ* labeling (PRINS) for evaluation of gene deletions in cancer. Am J Med Genet 2002; 107(2): 123-6. [http://dx.doi.org/10.1002/ajmg.10105] [PMID: 11807886]

[71] Mozdarani H, Pellestor F. The primed *in situ* (PRINS) technique: an alternative approach for preimplantation chromosomal diagnosis. Iranian J Biotechnol 2004; 2: 149-57.

[72] Hopman AH, Wiegant J, Tesser GI, Van Duijn P. A non-radioactive in situ hybridization method based on mercurated nucleic acid probes and sulfhydryl-hapten ligands. Nucleic Acids Res 1986; 14(16): 6471-88. [http://dx.doi.org/10.1093/nar/14.16.6471] [PMID: 3748817]

[73] Gille A, Seifert R. 2'(3')-O-(N-methylanthraniloyl)-substituted GTP analogs: a novel class of potent competitive adenylyl cyclase inhibitors. J Biol Chem 2003; 278(15): 12672-9. [http://dx.doi.org/10.1074/jbc.M211292200] [PMID: 12566433]

[74] Nairn RC. Fluorescent Protein Tracing. 4th ed. Edinburgh: Churchill Livingstone 1976.

[75] Hemmilä IA. Applications of Fluorescence in Immunoassays. New York: Wiley 1991.

[76] Nederlof PM, Robinson D, Abuknesha R, *et al.* Three-color fluorescence in situ hybridization for the simultaneous detection of multiple nucleic acid sequences. Cytometry 1989; 10(1): 20-7. [http://dx.doi.org/10.1002/cyto.990100105] [PMID: 2492920]

[77] Fauler G. New longwave-excitable coumarin-derived protein labels. Appl Fluor Technol 1989; 1(6): 14-5.

[78] Webb WW, Gross D. Patterns of Individual Molecular Motions Deduced From Fluorescent Image Analysis.Applications of Fluorescence in the Biomedical Sciences. New York: Alan R. Liss 1986; pp. 405-22.

[79] Gross D. Fluorescence labeled low density lipoprotein for studying cell receptors. Appl Fluor Technol 1989; 1(1): 13-4.

[80] Horobin RW, Kiernan JA, Eds. Conn's Biological Stains A Handbook of Dyes, Stains and Fluorochromes for Use in Biology and Medicine. 10th ed., Oxford, UK: BIOS Scientific Publishers 2002.

[81] Hudson BS, Harris DL, Ludescher RD, *et al.* Fluorescence Probe Studies of Proteins and Membranes.Applications of Fluorescence in the Biomedical Sciences. New York: Alan R. Liss 1986; pp. 159-202.

[82] Holzinger A. Jasplakinolide. An actin-specific reagent that promotes actin polymerization. Methods Mol Biol 2001; 161: 109-20. [PMID: 11190499]

[83] Tharp MD, Seelig LL Jr, Tigelaar RE, Bergstresser PR. Conjugated avidin binds to mast cell granules. J Histochem Cytochem 1985; 33(1): 27-32. [http://dx.doi.org/10.1177/33.1.2578142] [PMID: 2578142]

[84] Liener IR, Sharon N, Goldstein IJ. The Lectins Properties, Functions and Applications in Biology and Medicine. Orlando, Florida: Academic Press 1986.

[85] Kiernan JA. Histological and Histochemical Methods: Theory and Practice. Oxford, New York: Pergamon Press 1990.

[86] Sheetz MP, Koppel DE. Membrane damage caused by irradiation of fluorescent concanavalin A. Proc Natl Acad Sci USA 1979; 76(7): 3314-7. [http://dx.doi.org/10.1073/pnas.76.7.3314] [PMID: 291005]

[87] Seibert GR, Benjaminson MA, Hoffman H. A conjugate of cellulase with fluorescein isothiocyanate: a specific stain for cellulose. Stain Technol 1978; 53(2): 103-6. [http://dx.doi.org/10.3109/10520297809111450] [PMID: 80840]

[88] Haxo FT, Ó h Eocha C. Chromoproteins of Algae.Die CO_2 Assimilation. Encyclopedia of Plant Physiology. Berlin, Heidelberg: Springer-Verlag 1960; pp. 497-510.

[89] Oi VT, Glazer AN, Stryer L. Fluorescent phycobiliprotein conjugates for analyses of cells and molecules. J Cell Biol 1982; 93(3): 981-6. [http://dx.doi.org/10.1083/jcb.93.3.981] [PMID: 6749865]

[90] Smith H. Phytochromes and light signal perception by plants--an emerging synthesis. Nature 2000; 407(6804): 585-91. [http://dx.doi.org/10.1038/35036500] [PMID: 11034200]

[91] Zollinger H. Color Chemistry.Properties, and Applications of Organic Dyes and Pigments. 3rd ed. Zürich, Weinheim: Wiley-VCH, VHCA 2003.

[92] Udenfriend S. Fluorescence Assay in Biology and Medicine. New York, London: Academic Press 1962.

[93] Seliger HH, McElroy WD. Light: Physical and Biological Action. New York, London: Academic Press 1965.

[94] Dauvillier A. The Photochemical Origin of Life. New York, London: Academic Press 1965.

[95] Mathies RA, Stryer L. Single-Molecule Fluorescence Detection: A Feasibility Study Using Phycoerythrin.Applications of Fluorescence in the Biomedical Sciences. New York: Alan R. Liss 1986; pp. 129-40.

[96] Goulian M, Simon SM. Tracking single proteins within cells. Biophys J 2000; 79(4): 2188-98. [http://dx.doi.org/10.1016/S0006-3495(00)76467-8] [PMID: 11023923]

[97] Li B, Chu X, Gao M, Li W. Apoptotic mechanism of MCF-7 breast cells *in vivo* and *in vitro* induced by photodynamic therapy with C-phycocyanin. Acta Biochim Biophys Sin (Shanghai) 2010; 42(1): 80-9. [http://dx.doi.org/10.1093/abbs/gmp104] [PMID: 20043050]

[98] Tsuji FI, Ohmiya Y, Fagan TF, Toh H, Inouye S. Molecular evolution of the Ca($^{2+}$)-binding photoproteins of the Hydrozoa. Photochem Photobiol 1995; 62(4): 657-61. [http://dx.doi.org/10.1111/j.1751-1097.1995.tb08713.x] [PMID: 7480150]

[99] Chalfie M. Green fluorescent protein. Photochem Photobiol 1995; 62(4): 651-6. [http://dx.doi.org/10.1111/j.1751-1097.1995.tb08712.x] [PMID: 7480149]

[100] Tsien RY. The green fluorescent protein. Annu Rev Biochem 1998; 67: 509-44. [http://dx.doi.org/10.1146/annurev.biochem.67.1.509] [PMID: 9759496]

[101] Ormö M, Cubitt AB, Kallio K, Gross LA, Tsien RY, Remington SJ. Crystal structure of the *Aequorea victoria* green fluorescent protein. Science 1996; 273(5280): 1392-5. [http://dx.doi.org/10.1126/science.273.5280.1392] [PMID: 8703075]

[102] Jiménez-Banzo A, Nonell S, Hofkens J, Flors C. Singlet oxygen photosensitization by EGFP and its chromophore HBDI. Biophys J 2008; 94(1): 168-72. [http://dx.doi.org/10.1529/biophysj.107.107128] [PMID: 17766345]

[103] Verkhusha VV, Lukyanov KA. The molecular properties and applications of Anthozoa fluorescent proteins and chromoproteins. Nat Biotechnol 2004; 22(3): 289-96. [http://dx.doi.org/10.1038/nbt943] [PMID: 14990950]

[104] Miyawaki A, Shcherbakova DM, Verkhusha VV. Red fluorescent proteins: chromophore formation and cellular applications. Curr Opin Struct Biol 2012; 22(5): 679-88. [http://dx.doi.org/10.1016/j.sbi.2012.09.002] [PMID: 23000031]

[105] Remington SJ, Wachter RM, Yarbrough DK, *et al.* zFP538, a yellow-fluorescent protein from *Zoanthus*, contains a novel three-ring chromophore. Biochemistry 2005; 44(1): 202-12. [http://dx.doi.org/10.1021/bi048383r] [PMID: 15628861]

[106] Chalfie M, Tu Y, Euskirchen G, Ward WW, Prasher DC. Green fluorescent protein as a marker for gene expression. Science 1994; 263(5148): 802-5. [http://dx.doi.org/10.1126/science.8303295] [PMID: 8303295]

[107] Anbalagan C, Lafayette I, Antoniou-Kourounioti M, *et al.* Use of transgenic GFP reporter strains of the nematode *Caenorhabditis elegans* to investigate the patterns of stress responses induced by pesticides and by organic extracts from agricultural soils. Ecotoxicology 2013; 22(1): 72-85. [http://dx.doi.org/10.1007/s10646-012-1004-2] [PMID: 23081760]

[108] Schmitz HD, Bereiter-Hahn J. GFP associates with microfilaments in fixed cells. Histochem Cell Biol 2001; 116(1): 89-94. [PMID: 11479727]

[109] Ruiz-González R, White JH, Agut M, Nonell S, Flors C. A genetically-encoded photosensitiser demonstrates killing of bacteria by purely endogenous singlet oxygen. Photochem Photobiol Sci 2012; 11(9): 1411-3. [http://dx.doi.org/10.1039/c2pp25126d] [PMID: 22729069]

[110] Ruiz-González R, Cortajarena AL, Mejias SH, Agut M, Nonell S, Flors C. Singlet oxygen generation by the genetically encoded tag miniSOG. J Am Chem Soc 2013; 135(26): 9564-7. [http://dx.doi.org/10.1021/ja4020524] [PMID: 23781844]

[111] Fernández-Suárez M, Ting AY. Fluorescent probes for super-resolution imaging in living cells. Nat Rev Mol Cell Biol 2008; 9(12): 929-43. [http://dx.doi.org/10.1038/nrm2531] [PMID: 19002208]

[112] Lillie RD. Conn's Biological Stains. 9th ed., Baltimore: Williams and Wilkins 1977.

[113] Popov D, Thorell B. New reactive fluorescent dyes for cytofluorometry. Stain Technol 1978; 53(4): 246-7. [PMID: 734690]

[114] Sigworth LA, Chandler TR, Liao L, Geusz ME. Luciferase Imaging Reveals Distinct Patterns of Gene Regulation in Live Brain Slices.Bioluminescence and Chemiluminescence 2000. Singapore: World Scientific 2001. [http://dx.doi.org/10.1142/9789812811158_0046]

[115] Geusz M, Liao L. Imaging the dynamics of gene expression in live cells with luciferase bioluminescence. Spectrum 2001; 14: 15-20.

[116] Hong G, Diao S, Chang J, *et al.* Through-skull fluorescence imaging of the brain in a new near-infrared window. Nat Photonics 2014; 8(9): 723-30. [http://dx.doi.org/10.1038/nphoton.2014.166] [PMID: 27642366]

CHAPTER 5

Spectral Analysis

Abstract: Studies on the spectral (electronic) properties of dyes are often needed to know qualitative and quantitative aspects of their interactions with macromolecular substrates. This Chapter deals with the description and interpretation of absorption and emission spectra from some dyes and fluorochromes used in microscopical studies. General references on fluorescence spectroscopy can be found in [1-13]. For detailed information on microscopic photometry (*i.e.* ray paths, equipments, adjustments, and procedures), the book by Piller [14] should be consulted. At present, confocal spectral analysis generating 2D and 3D microscopical images of biological materials offers new possibilities for research applications, mainly in the localization of antitumor drugs within living cells [15]. Some key concepts and useful tips must be taken into account for a correct interpretation of spectra. Generally, emission spectra are corrected for the Raman scatter of the solvent and for harmonics. To record adequately far red emission, the detector sensitivity must be able to capture long wavelengths with fidelity.

Keywords: Acridine orange, Bathochromic shift, Binding mechanisms, Caffeine, Clay minerals, Co-solutes, Detergent micelles, Harmonics, Hypsochromic shift, Ionic strength, Macrospectra, Molecular rigidity, Raman band, Rayleigh-Tyndall band, Viscous solvents.

1. SPECTRAL RECORDING AND INTERPRETATION

Most common abbreviations are used along this and other Chapters. λ: wavelength; ν: wavelength number (ν: 1/λ in cm); UV: ultraviolet; IR: infrared; OD: optical density; FI: fluorescence intensity; AU: arbitrary units; AB, A: absorption; EM, E: emission; PS: photosensitizer; Φ_F, Φ_T, and Φ_Δ: fluorescence, triplet and $^1O_2$ quantum yield, respectively; λab and λem: maximum value of absorption and emission (peak); λexc: used exciting light. 1 and 3 as superscripts correspond to singlet and triplet states, respectively, *i.e.* $^3O_2$ and $^1O_2$ (triplet and singlet oxygen); * denotes an excited state (molecule, electron).

Spectral features of fluorochromes are successfully represented as 3D excitation-emission color maps using Y and X axes for λexc and λem, respectively (see Fig. **4.18**), but more conventional 2D curves showing absorption/emission intensity and wavelength along these axes are commonly employed. For illustrative purposes, recording of the emission spectrum of a given compound is shown in

Juan Carlos Stockert & Alfonso Blázquez-Castro

Fig. (**5.1**), with the position of Rayleigh-Tyndall and Raman scattering, emission signal, and the second higher harmonic of excitation. The Rayleigh-Tyndall band is due to molecular and particle scattering from the solvent, whereas the Raman band derives from elastic collisions between photons and chemical bonds (both with different energy) (see Chapter 19.5). Thus, when the population of chemical bonds in the solvent has a vibrational energy somewhat lower than the incident light, the photon yields energy to the bond and is scattered with a greater wavelength. On the contrary, scattering from high energy levels in bonds of the solvent increases the photon energy and diminishes its wavelength.

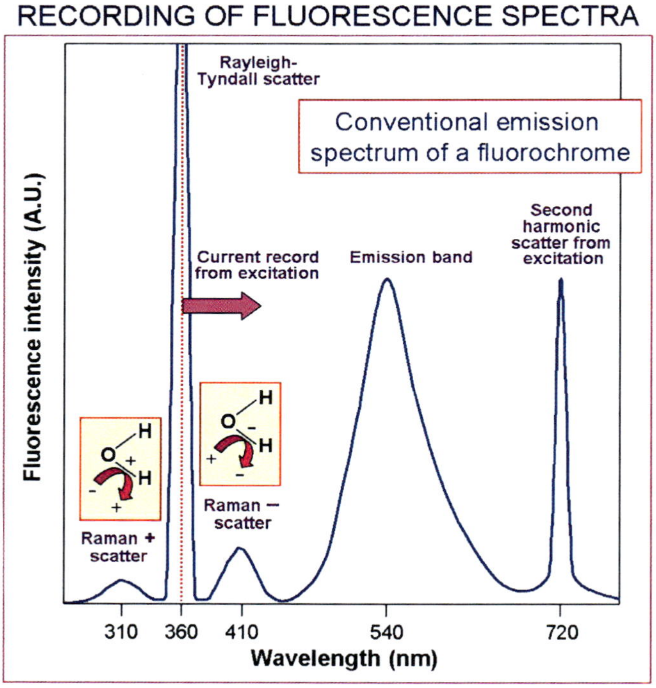

Fig. (5.1). Conventional emission spectrum of a given compound showing the excitation band (Rayleigh-Tyndall scatter), the high- and low-energy Raman scatter, the true emission band, and the higher second harmonic of exciting light.

Although the Raman band of the solvent commonly presents a considerable intensity when spectra are recorded at high instrumental sensitivity, the absorption of a dissolved fluorochrome reduces the Raman signal, which remains incorporated to the emission curve. This case is illustrated in Fig. (**5.2**), that shows the fluorescence increase of the natural dye hematoxylin depending on the degree of spontaneous oxidation [16, 17]. A similar feature of Raman hindering also occurs with increasing concentration of other fluorochromes.

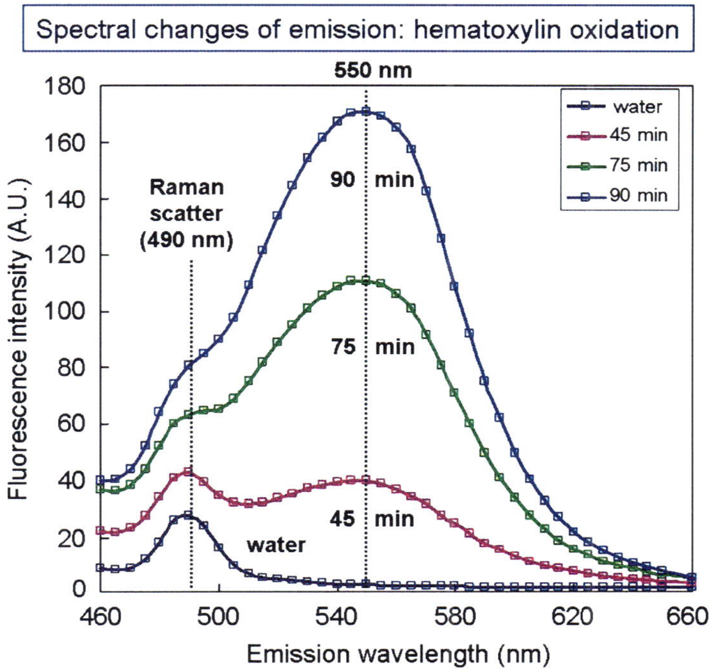

Fig. (5.2). Fluorescence increase of hematoxylin (0.25 mg/mL in distiller water) during spontaneous oxidation at the indicated times. λexc: 410 nm. Observe the higher emission at 550 nm with increasing oxidation and the decreased expression of the Raman scatter of water.

Because of their remote place, high and low harmonics of the excitation band are infrequently found in fluorescence spectra, but when recording extends to the red region, the second higher harmonic must be not confused with a far emission band. Harmonic components originate from the use of grating monochromators. An interesting example useful to avoid possible pitfalls in spectral recording is the fluorescence of oxidized hematoxylin solutions. When excited with blue light, microscopical emission of hematoxylin-stained samples is green-yellow, although a red color due to the barrier filter is observed under green excitation (see Chapter 7.2.3). The spectral analysis using λexc: 400 nm shows an emission peak at 550 nm, but with 500 nm excitation a clear red signal at 600 nm is also observed (Fig. **5.3**). This is an intriguing feature, because there is no barrier filters in fluorescence spectrophotometers. However, this puzzling effect is generated by the second lower harmonic of 500 nm excitation, which excites at 250 nm the still non-oxidized hematoxylin. The second higher harmonic of its emission at 300 nm corresponds to the 600 nm band (Fig. **5.3**). The simple strategy of placing a glass slide before the cuvette removes the UV excitation thus avoiding misleading results.

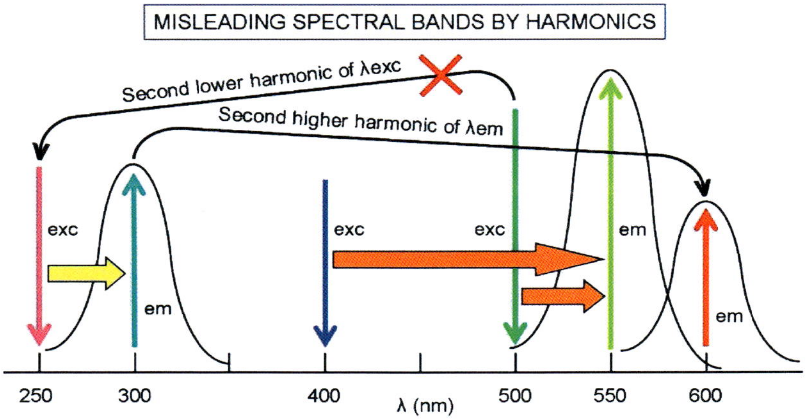

Fig. (5.3). Puzzling excitation and emission bands (color arrows). Excited at 400 nm (blue), the emission of oxidized hematoxylin occurs at 550 nm (yellow-green) that is the true fluorescence of oxidized hematoxylin. When using 500 nm (green), its lower harmonic at 250 nm (violet) excites the still present non-oxidized hematoxylin giving 300 nm emission (cyan) and its higher harmonic (red) at 600 nm. A glass filter placed before the cuvette avoids the low excitation harmonic reaching the sample (red cross) and thus abolishes the bands at 300 and 600 nm.

The emission wavelength of a fluorochrome is independent on the excitation wavelength and the fluorescence is only affected by changes of intensity. The dye orcein (a natural [and now synthetic] dye related to lacmoid and litmus, used as pH indicators) contains several components that are fluorescent. Lacmoid, litmus and orcein are variable and poorly defined oxazine and oxazone mixtures (α-amino-orcein, α-hydroxy-orcein, β-hydroxy-orcein, γ-amino-orcein, and γ-amino-orceimin). Typical emission features have been found in orcein solutions and after staining of several substrates such as nucleic acids, elastic and reticular fibers [18, 19]. The fluorescence intensity depends on the exciting wavelength, while the position of the peak at 587 nm remains constant (Fig. **5.4**). As expected, highly concentrated orcein solutions show greatly reduced and red shifted emission (λem: 620 nm at 20 mg/mL). In addition to orcein, other oxazone compounds show similar fluorescence (*i.e.* Nile red, brilliant cresyl blue oxazone, 7-amino-actinomycin D).

On the contrary, the position of other signals (Rayleigh-Tyndall harmonics and Raman scatter) depends on the excitation. This is an obvious fact for the first case, as harmonics correspond to the double or half of the scatter band. In the case of the Raman band, it is moved a variable distance depending on the position of excitation and the type of solvent. This shift is clearly seen when the correlation between excitation and emission is not expressed in nm but as wavelength number ($\nu = 1/\lambda$). Thus, the Raman scattering of unstained microscopical samples can be easily recorded showing a linear correlation (Fig. **5.5**). It must be noted that, although vulgarly described as emission, the Raman band does not correspond to a true emission.

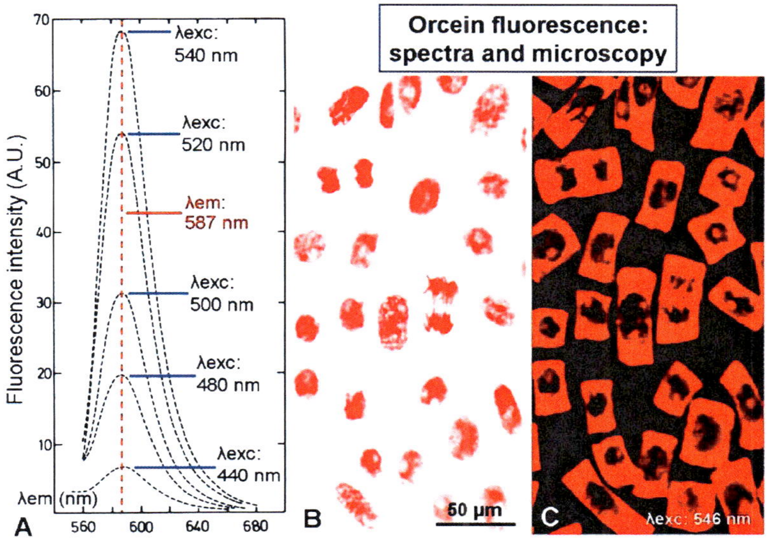

Fig. (5.4). A: Fluorescence spectra of synthetic orcein (2 µg/mL pH 6) excited at the indicated wavelengths. In all cases, the emission peak remains at 587 nm (dashed red line). **B**, **C**: Acetic orcein squash of *Allium cepa* root tip (2% orcein in 45% acetic acid) showing meristematic cells under bright-field **(B)** and fluorescence microscopy **(C)**. Observe the strong chromatin staining **(B)** and the fluorescence of nucleoli and cytoplasm **(C)**. Chromatin does not fluoresce due to the excess of dye concentration.

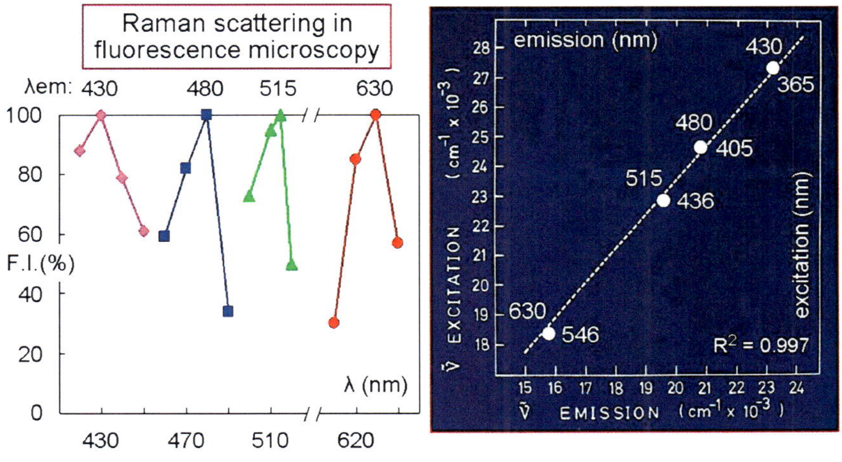

Fig. (5.5). Left: Microscopic fluorescence spectra of unstained chicken erythrocyte nuclei excited with 365, 405, 436, and 546 nm (violet, blue, green and red emission, respectively). **Right**: Raman correlation between the wavelength number ν: 1/λ (cm) of excitation and emission obtained from the spectra at left.

Recording the fluorescence spectra of microscopical samples at different excitations allows revealing the dependent and independent position of Raman scattering and emission peaks, respectively, in relation to the exciting light (Fig. **5.6**). Thus the different signals of both contributions in fluorescence microscopy can be easily distinguished.

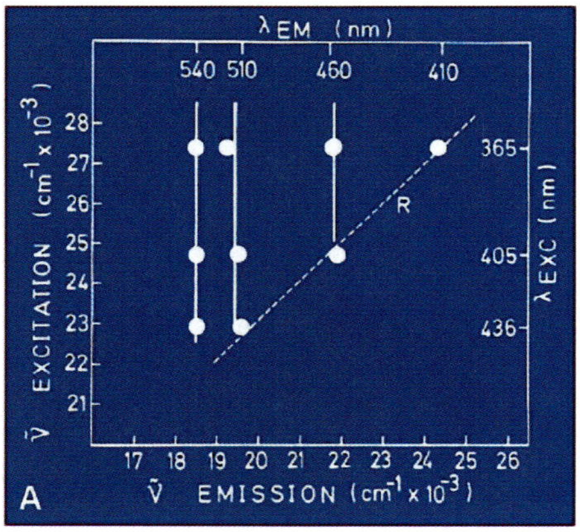

Fig. (5.6). Raman scatter and autofluorescence in microscopic samples. Correlation between excitation and emission wavelength numbers obtained from lipofuscin granules in macrophages of a mouse uterus (glutaraldehyde, Epon). Vertical lines correspond to autofluorescence emission, whereas the dashed line represents the contribution of the Raman scatter.

Commonly, absorption and emission curves are arranged according to mirror symmetry. The separation between wavelengths at which excitation and emission occurs is referred to as the Stokes shift (see Chapter 2.4). Several fluorochromes show small Stokes shift and thus an important overlapping of absorption and emission. In this respect, the absorbance of the sample should be less than 0.05 in the overlap region between absorption and emission to allow determination of Φ_F and to avoid quenching due to inner filter effects. A schematic representation of the correlation between monomeric and aggregated states of dyes and spectral features is seen in Fig. (**5.7**).

Higher and lower dye absorption (hyperchromic and hypochromic changes, respectively) is related to greater hydrophilic and hydrophobic interactions, caused by the separation or approximation between own or foreign aromatic surfaces. Likewise, hyper-emission and hypo-emission features are due to similar processes. Dye monomerization due to dye-solvent or dye-substrate interactions results in a bathochromic shift (orthochromatic color reaction), whereas dye aggregation by stacking is reflected by a hypsochromic shift (metachromatic color reaction). The counterpart in fluorescence spectra corresponds to red and blue shift, respectively. The absorption peak of cationic dyes often shifts to longer wavelengths (bathochromic shift) by using low concentration and ionic strength, acid pH, non-polar solvents, and high temperature, thus reflecting a greater contribution of dye monomers.

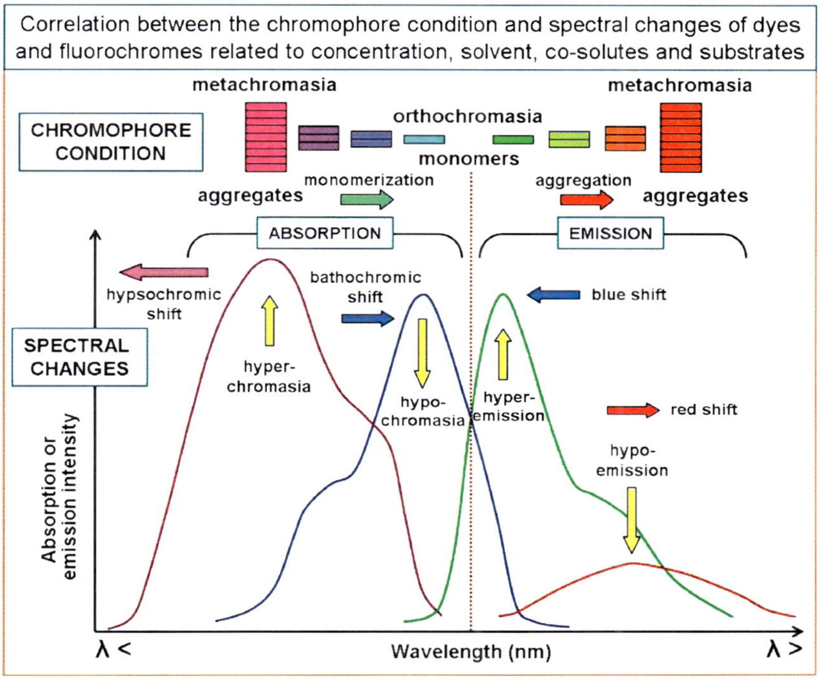

Fig. (5.7). Schematic correlation between chromophore condition and spectral characteristics. Aggregation is the cause of metachromatic absorption (hypsochromic shift) and emission (red shift). On the contrary, monomerization results in orthochromatic absorption (bathochromic shift) and emission (blue shift).

2. INFLUENCE OF DYE CONCENTRATION

Similarly with the absorption changes generated by dye concentration, a related effect is observed in stained biological materials, because of the local increase of dye amount imposed by the substrate. Thus, in addition to the different spectra induced by increasing toluidine blue concentration, also different colors appear using only one concentration of the dye for staining different cell structures, which allows recognizing ortho- and metachromatic responses of DNA (blue) and RNA (violet), respectively. To this respect, toluidine blue and acridine orange are the most used dyes to reveal clear metachromatic staining of anionic nucleic acids and polysaccharides.

Likewise, fluorescence features also reflect variations in the concentration of fluorochromes. In the case of the anionic dye, eosin Y, the emission peak at 540 nm is shifted to 568 nm when the concentration increases 3 magnitude orders (Fig. **5.8**), whereas first hyper-emission and then hypo-emission occurs. This spectral response is also observed in the concentration-dependent behavior of numerous fluorochromes, typically the cationic ones.

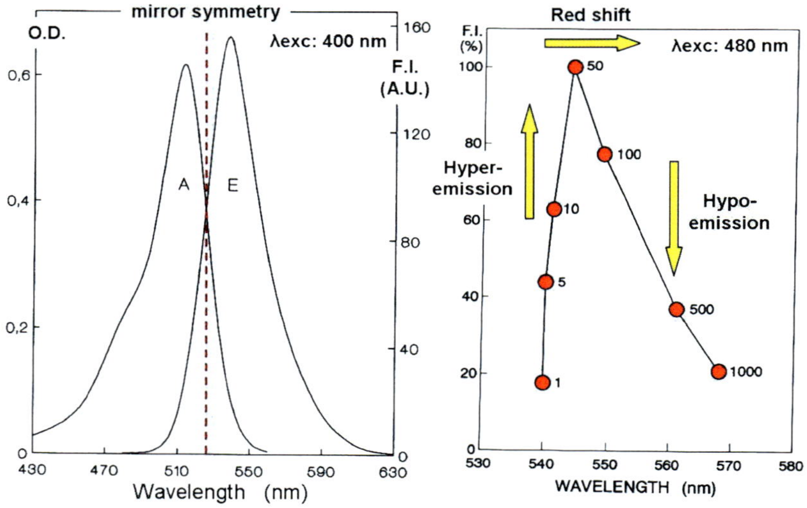

Fig. (5.8). Absorption and fluorescence spectra of eosin Y: emission changes related to dye concentration. **Left**: Absorption (A) and emission (E) spectra of 10 μg/mL eosin Y. **Right**: Effect of increasing concentrations of eosin Y (μg/mL) on the intensity and wavelength of emission. Hyper-emission, red shift and hypo-emission are observed as dye concentration increases.

The widely used cationic fluorochrome acridine orange also shows hypo-emission with red shift at increasing concentration, a feature which has shown interesting applications in fluorescence microscopy yielding ortho- and metachromatic colors in chromatin (DNA) and basophilic cytoplasm (RNA), respectively (Fig. **5.9**). This differential fluorescence reaction is highly appreciated in morphological and cytochemical studies. It is important to note that the hypo-emission and red shift correspond to a stacking effect of planar tricyclic dyes (see section 3.3). Quenching of fluorescence by concentration excess with formation of dye dimers and aggregates is a well known phenomenon [20], which is related to an "inner filter effect" induced by self-absorption.

An excited fluorophore may form a transient complex with a ground state molecule. Such a complex is called an excited dimer or excimer (between identical molecules) or an exciplex (between different molecules) [21]. This behavior is concentration-dependent. Fluorescence features of pyrene derivatives are unique due to the ability of the pyrene ring to form the excimer (MM*) between the ground state (M) and the excited (M*) monomer: $M^* + M \rightarrow MM^*$. Emission spectrum of monomers* (non-interacting pyrenes, λem: 376 nm) are different from that of the pyrene excimer (λem: 475 nm, 100 nm farther away!), allowing to study the concentration-dependent aggregation of pyrene-labeled glycolipids in cell membranes [22]. On the other hand, anthracene in the excited state forms an exciplex with *N,N*-diethylaniline in toluene. Spectral features of the anthracene monomer (λem: ~400 nm) and anthracene exciplex (λem: ~500 nm) are clearly different.

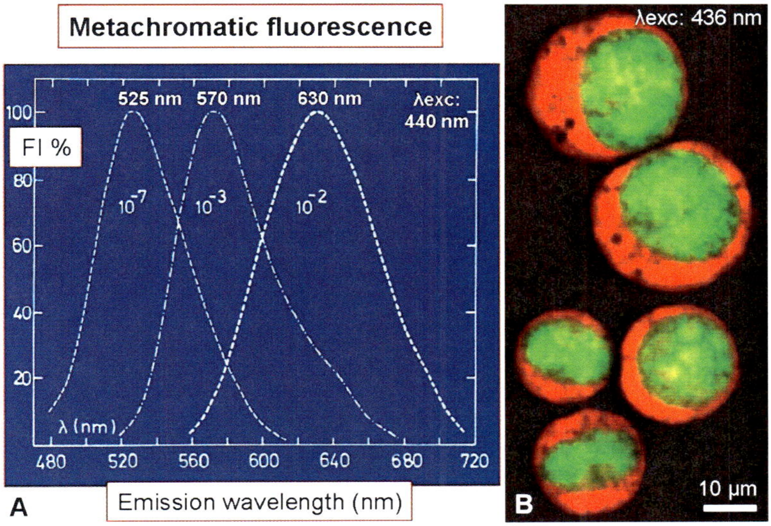

Fig. (5.9). Influence of dye concentration: metachromatic fluorescence. **A**: Normalized emission spectra of acridine orange at different molar concentrations in distilled water. **B**: Methanol-fixed Ehrlich ascites tumor cells stained by acridine orange (0.1 mg/mL, 5 min). Observe the orthochromatic (green) and metachromatic (red) fluorescence of DNA- and RNA-containing structures, respectively.

3. INFLUENCE OF SOLVENTS, CO-SOLUTES AND SUBSTRATES

3.1. Solvents

The influence of the environment on absorption and emission spectra is well known. Polar or apolar solvents, ionic strength, solvent viscosity, co-solutes (caffeine, tannin, dextrins, clays), detergent micelles and biological substrates are factors that can influence the fluorescence characteristics [23 - 29]. Generally, low ionic strength, organic or viscous solvents, detergent and phospholipid micelles, and presence of monomerizing substrates and co-solutes increase the fluorescence yield. In this respect, the addition of sucrose, glycerol, gelatin, *etc.*, to aqueous solvents, as well as absorption on rigid surfaces (cellulose, silica gel, aluminum hydroxide) are often used to increase the emission intensity [1]. In addition to low temperatures, these environmental conditions also improve the phosphorescence yield.

Spectrofluorometric studies reveal that when dissolved in water, several linear and/or non-rigid dyes are scarcely fluorescent or not at all. On the contrary, a striking enhancement of emission appears when these dyes are dissolved in viscous solvents such as water-glycerol solutions. The use of 90% glycerol results in a strong increase of fluorescence in linear, non-rigid dyes, typically bis-azo dyes such as Congo red, trypan blue, and Evans blue (Fig. **5.10**). This is the case for numerous indole, benzimidazole, benzothiazole, di- and triarylmethane dyes under conditions that increase their molecular rigidity such as viscous solvents (*i.e.* ethylene glycol, polyethylene glycol, sucrose solutions) or binding to suitable substrates [3 - 5]. The presence of increasing concentrations of sucrose causes a stronger and blue-shifted emission of the non-rigid fluorochrome BAO (bis-aminophenyl-oxadiazole). This compound induces a bright blue fluorescence of chromatin and basophilic cytoplasm of fixed cells on account of their nucleic acids content [30].

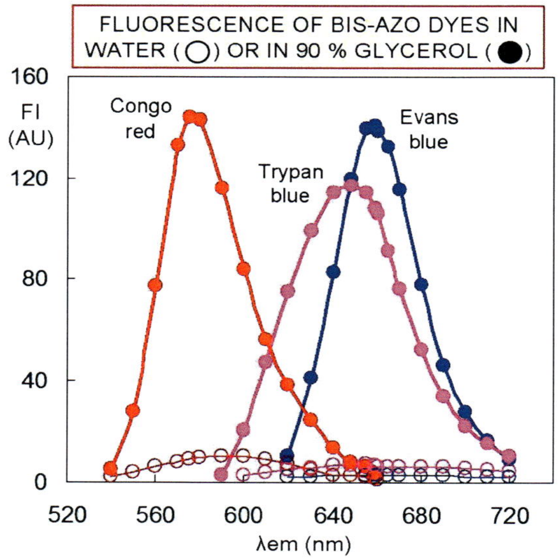

Fig. (5.10). Fluorescence increase of bis-azo dyes Congo red, trypan blue, and Evans blue (10 μg/mL, excited at 520, 540, and 540 nm, respectively) either in distilled water (empty circles) or in 90% glycerol (full circles). The Raman scattering of solvents were subtracted from all spectra.

Other illustrative example of viscosity-enhanced emission is the increase observed when the triarylmethane dye, fast green FCF is dissolved in 90% glycerol or applied for fluorescent staining of cell substrates (Fig. **5.11**). When fluorescent grade glycerol is not used for spectral recording, the second-order emission of this solvent excited by the lower harmonic of the main excitation beam can be entirely removed by setting a glass filter before the cuvette [31]. Thus, using λexc: 530 nm (lower harmonic at 265 nm), solvent emission at 300 nm (and higher harmonic at 600 nm) is avoided.

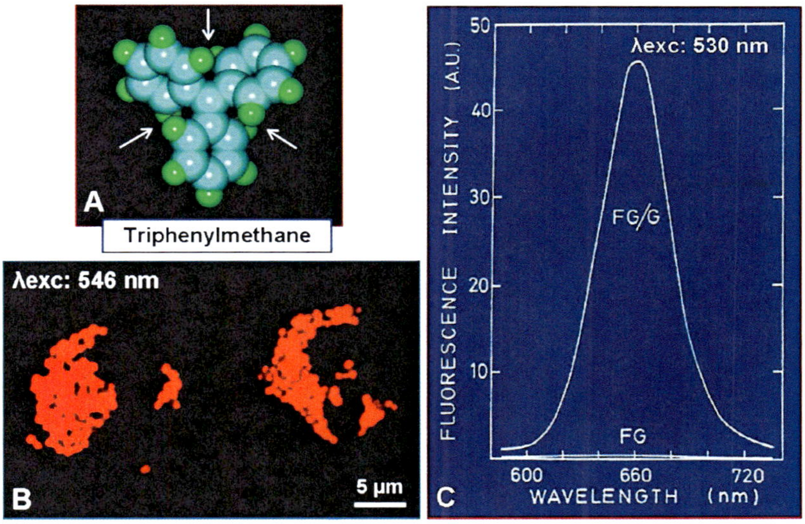

Fig. (5.11). Emission of non-rigid fluorochromes. **A**: Triphenylmethane chromophore. Arrows indicate steric hindrance for coplanarity. **B**: Fluorescence of horse eosinophil granules stained by fast green FCF (0.1 mg/mL, pH 9). **C**: Emission spectra of 25 µg/mL fast green FCF in distilled water (FG), and in 90% glycerol (FG/G).

3.2. Co-solutes

In addition to different solvents, the presence of small co-solutes is an adequate model system to explore some characteristics of the more complex interactions of dyes and small ligands with macromolecules (*i.e.* polysaccharides, nucleic acids, proteins) [28]. The addition of caffeine, *o*-phenanthroline and tannin to aqueous solutions of cationic dyes and fluorochromes induces strong monomerization with hyperchromic, bathochromic and hyper-emission changes (Fig. **5.12**).

Cyclodextrins are doughnut-shaped cyclic oligosaccharides containing 6, 7, or 8 glucose units (α-, β-, or γ-cyclodextrins, respectively) bound by α-1,4-glycosyl linkages and forming a toroidal structure. Small guest molecules can be trapped into the void cavity of cyclodextrins as inclusion compounds in the form of soluble or crystalline adducts [32, 33]. Inner channels of α-, β-, or γ-cyclodextrins have diameters of 5.7, 7.8, and 9.5 Å, respectively. The relatively hydrophobic internal wall of these rings is predominantly formed by H atoms from C3 and C5, and the glycosidic oxygen from C1.

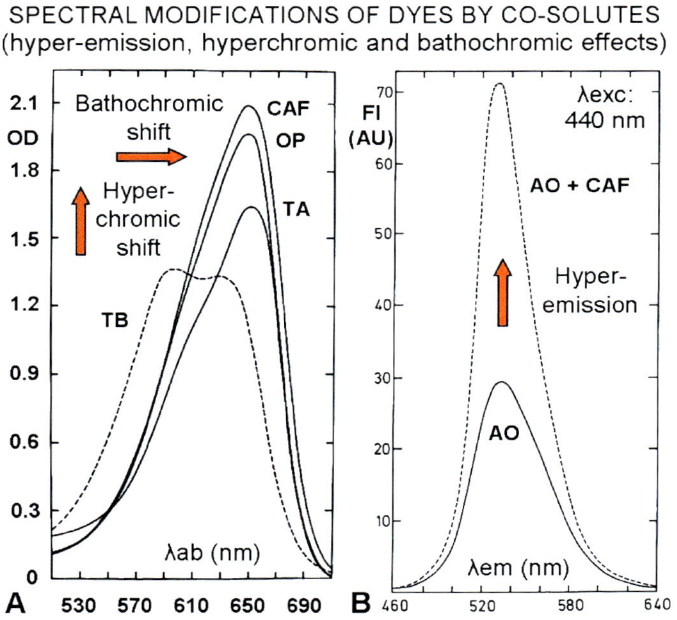

Fig. (5.12). Spectral changes induced by co-solutes. **A**: Absorption of free toluidine blue (TB, 5.5 x 10^{-5} M) and in the presence of tannin (TA, 3 mg/mL), caffeine (CAF), and *o*-phenanthroline (OP) (CAF and OP at 90% saturation). **B**: Emission increase of 10^{-4} M acridine orange (AO) in the presence of caffeine at 90% saturation.

Included compounds remain rigid and isolated from the solvent. The restricted motion and no access to the solvent avoid rotational and collisional decay of excited dyes. Cyclodextrins also stack to form cylinders that can accommodate long-chain guest ligands (channel inclusion compounds), in a similar way to the helically structured dextrins. Depending on the guest molecule, hydrogen bonding, van der Waals forces, hydrophobic interactions, and stereochemical fitting contribute to form and stabilize the adduct. Bathochromic and hyperchromic shifts occur for nitro, azo and thiazine dyes in the presence of α-cyclodextrin [32, 34]. Enhanced fluorescence is also observed for 1-aniline-8-naphthalene sulfonic acid (ANS) [33], berenil, curcumin, DAPI, Hoechst 33258 and other linear and non-rigid cationic and anionic dyes [29] (Fig. **5.13**). In addition, on account of their monomerizing and solubilizing properties, cyclodextrins are commonly employed as excipients for drug delivery [35, 36]. It is worth to note that iodine anions (linear chains of I_3^- and I_5^-) form inclusion compounds with cyclodextrins and starch, which are responsible for the known blue reaction [37].

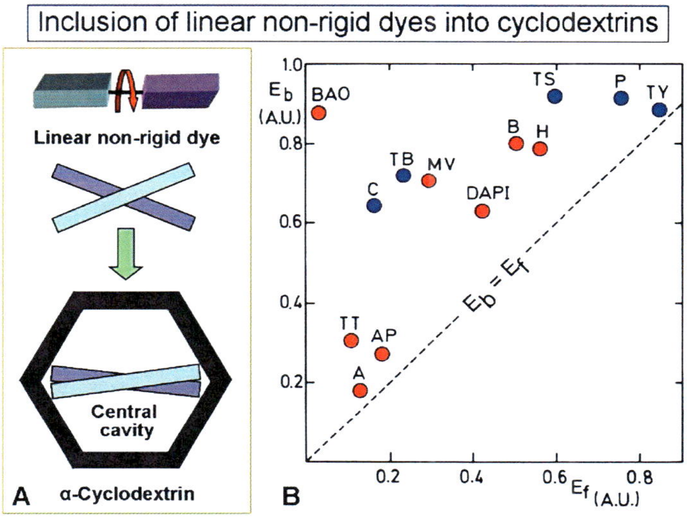

Fig. (5.13). A: Schematic representation of the inclusion of a linear and non-rigid dye into the central cavity of α-cyclodextrin. **B**: Emission of cationic (red) and anionic (blue) non-rigid fluorochromes either free (E_f) or in the presence (E_b) of 1.25% α-cyclodextrin. Concentrations are: BAO, berenil (B), curcumin (C), and Hoechst 33258 (H): 7.5 µg/mL; DAPI, methylviologen (MV), trypan blue (TB), thioflavine S (TS), primuline (P), astrazon pink FG (AP), and titan yellow GR (TY): 5 µg/mL; thioflavine T (TT), and auramine O (A): 10 µg/mL.

On the other hand, some microcrystalline clays can be co-solutes that change dye properties. Clay minerals are alternating 2- or 3-layered sheets of tetrahedral silicate and octahedral alumina groups [38], and inorganic cations and organic compounds can easily insert within interlamellar spaces forming interlayer complexes [39]. In the presence of the 3-layered silicate clay montmorillonite (Fig. **5.14**), diluted solutions of cationic thiazine dyes give an immediate metachromatic shift (from blue to reddish violet) in the absorption spectrum. Hypo-emission and red shift in the fluorescence spectra of cationic fluorochromes (*i.e.* acridine orange, pyronine Y and ethidium bromide) appear in the presence of montmorillonite.

Spectral changes indicate that this clay has a strong aggregating capacity, allowing formation of stacked chromophores between negatively charged silicate layers, and offering a suitable model substrate to study metachromatic reactions [40, 41]. Likewise, the inclusion of methylene blue within montmorillonite produces metachromasia and protects the dye from photodegradation [42]. On the contrary, only dye monomers can insert within void channels of reticular clays like palygorskite [43]. Methylviologen also binds to montmorillonite as monomers with blue shift and strong increase of emission [44]. Clay minerals have been also used as model substrates for photochemical reactions [45].

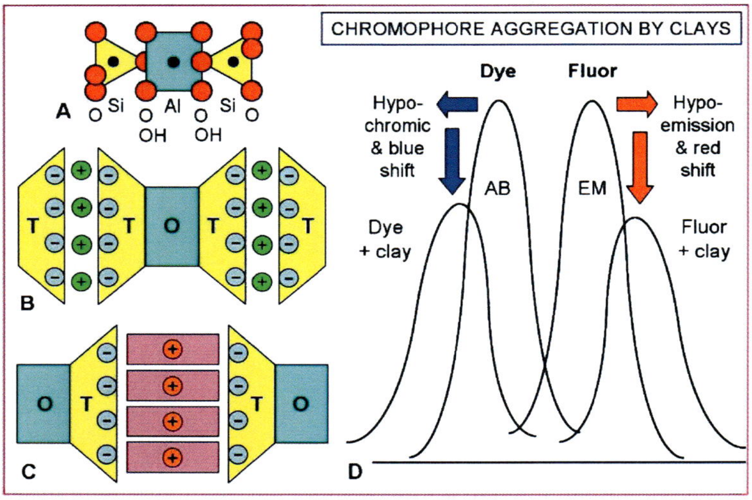

Fig. (5.14). A: Structure of montmorillonite clay showing silicate tetrahedra (T) and alumina octahedra (O). Some Al can be replaced for Mg. **B** and **C**: Schematic TOT layer structure with inorganic cations between negative charges at free T surfaces **(B)**, and stacked planar cationic chromophores **(C)** in the interlamellar space. **D:** Schematic absorption **(AB)** and emission (EM) spectra of an planar dye or fluorochrome (Fluor) at low concentration, either alone or in the presence of montmorillonite clay (0.2 mg/mL), showing the spectral changes.

Quenching processes due to the presence of some co-solutes (*i.e.* heavy atoms, aromatic chromophores) induce appreciable changes in the absorption and emission spectra of dyes and fluorochromes. The inclusion of fluorescent molecules into the cavity of cyclodextrins protects the former against collisional quenching by heavy ions (*i.e.* iodate, iodide) [46]. A similar effect is observed in the case of fluorochrome-substrate interactions.

3.3. Biological Substrates

Numerous cell and tissue components are microscopical substrates that bind and change the absorption and emission spectra of dyes and fluorochromes. Binding mechanisms generally involve ionic, hydrophobic, hydrogen bridges or covalent bonds (see Chapter 7.1). In the case of some tissue components (*i.e.* neutral polysaccharides, lignin, melanin, nucleic acids duplexes), bathochromic changes (red shift in the absorption spectrum) are related to the monomerization of dyes, which remain trapped between aliphatic chains or aromatic rings of the biopolymer [23, 25, 28, 34]. In addition, solutions of specific substrates *in silico* such as dextrans, dextrins, and nucleic acids also show similar changes to those observed *in situ*.

In the presence of polyanions, a bathochromic shift occurs at low "r" values (r: molar ratio between bound dye and binding site), whereas a hypsochromic change results at high ratios. The latter modification accounts for the metachromatic staining. Generally, the term orthochromatic staining is applied when λab is either similar or red shifted in relation to the free dye, although from the spectral point of view, the bathochromic shift has been viewed as a "negative metachromasia" [47]. An illustrative case is the interaction of pyronine Y with polyanions (Fig. **5.15**). The shoulder at 510 nm (corresponding to dye aggregates) reduces in the presence of DNA, and the peak shows hypochromic and bathochromic shifts due to intercalative binding of monomer to hydrophobic sites. On the contrary, the interaction of pyronine Y with heparin results in a clear metachromatic change.

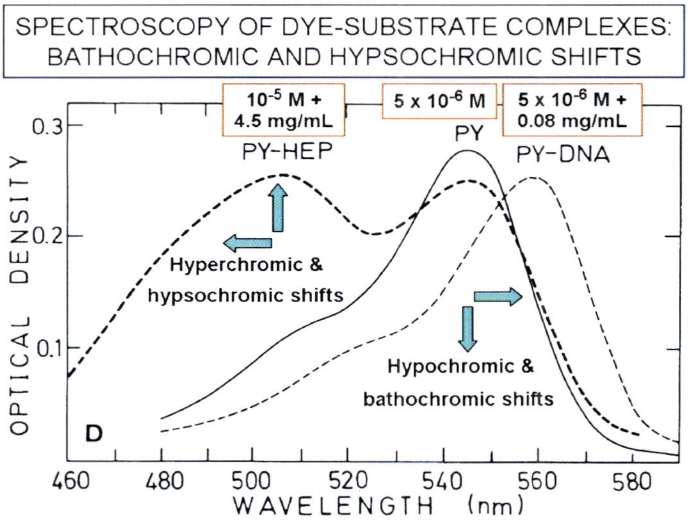

Fig. (5.15). Absorption spectra of pyronine Y alone and complexed to the polyanionic substrates, DNA and heparin, which induce dye monomerization (orthochromasia) and aggregation (metachromasia), respectively.

In the case of emission spectra, cationic fluorochromes that bind selectively to DNA at low r ratios generally induce a higher fluorescent signal on account of monomerizing intercalative or minor groove binding modes, which are responsible for the orthochromatic fluorescence. Typical examples are the interaction of DNA with intercalating (acridine orange, ethidium bromide) and minor-groove binding fluorochromes (DAPI, Hoechst 33258, nuclear yellow) (see Chapter 12).

Regarding the current views on metachromasia, dyes containing only tertiary amino groups are not metachromatic. This would be the case of methylene blue

[48], which appears blue at either low or high concentration. However, the strongly metachromatic acridine orange has full substituted (tetramethylated) amino groups and then should be not metachromatic. This puzzling effect is due to the fact that for dyes absorbing almost entirely in the red (methylene blue) or blue region (pyronine Y, safranine O, neutral red), metachromatic changes will be scarcely appreciated. On the contrary, for bordering red-blue dyes (toluidine blue, thionine) conspicuous ortho- (blue) and metachromatic (red) colors are easily recognized [49]. In all cases, microspectrophotometric methods permit unambiguously to achieve a clear characterization of color differences. Thus spectral data confirm the metachromatic response of methylene blue [42]. In addition to methylene blue, other dyes such as alcian blue 8G and astra blau do not show color differences to the naked eye, but their ortho- and metachromatic changes can be spectroscopically identified [24].

The mechanism of metachromatic fluorescence relates to electronic orientation of dyes and deserves a particular mention. It has long been known that in aqueous solution dyes generally do not obey the Lambert-Beer law closely, and the absorption spectra of dyes bound to polymers are often substantially changed. Deviations are due to the formation of dimers and high dye aggregates. The strong green emission of monomeric acridine orange is quenched and shifted to the red in the aggregated condition, in which antiparallel stacking of the transition dipole moment (a vector perpendicular to a nodal plane of the excited state) almost cancel mutually, and decay to a lower excitation level thus generating a spectrum equivalent to a ^{3}dye* emission. On the contrary to planar dyes, pseudoisocyanine (a linear fluorophores with parallel head-to-tail orientation) almost does not fluoresce as monomer but becomes strongly fluorescent on aggregation on a suitable counterionic substrate [50].

Binding to some substrates can either protect or quench the fluorescence of fluorophores. In the case of nucleic acids duplexes, as the access of solvent molecules to the intercalated ligands is hindered, no collisional quenching occurs and the emission generally increases. However, fluorescence quenching is expected to occur in the case of dark chromophores contained in macromolecules such as melanin, which absorb emission by an inner filter (screen) effect. The chromophore of melanin (actually a melano-protein) is an indol- and catechol-polymer [51] that strongly absorbs light in the order UV > violet > blue > green > yellow > red. Also fluorescence and phosphorescence quenching by photo-oxidation with $^1O_2$ and energy-transfer to $^3O_2$, respectively, can diminish the emission in dependence on the access of oxygen to the bound fluorophores.

4. MACROSPECTRA

The microscopic assessment of absorption and emission spectra is not often employed in analytical and applicative approaches. The use of an interference monochromator filter before a photometer head as light detector results in a rather tedious procedures, although numerous fluorescence microspectra have been recorded and published. The spectral diode array detector has been applied for light absorption analysis of samples in HPLC, and it would be also appropriate to perform the emission analysis in fluorescence microscopy.

On the other hand, homogeneous biological samples can be employed for spectral analysis of specific substrates (mainly cell smears, cultured cells, wood and cartilage sections, *etc.*) [8, 10, 23, 52]. Conventional spectrophotometers and spectrofluorometers can be applied for recording absorption and emission characteristics in cell populations and tissue preparations placed on glass slides.

Cationic derivatives of classical scintillating agents such as *p*-bis(phenyloxazole)-phenylene (POPOP) are useful in fluorescence microscopy studies. The cationic N-quaternary derivative of dimethyl-POPOP (Q-dmPOPOP) induces fluorescence in polyanionic substrates (*i.e.* heparin-containing granules of mast cells [53] and chromatin DNA [54]). The corresponding macrospectra from cell nuclei can be seen in Fig. (**5.16**).

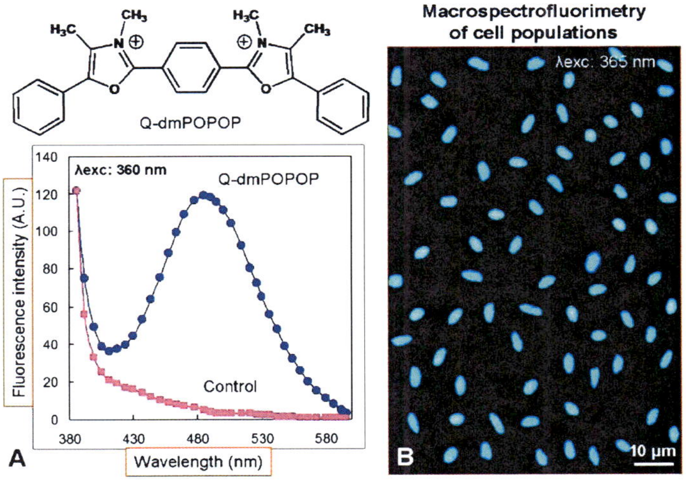

Fig. (5.16). A: Chemical structure of Q-dmPOPOP. **B**: Emission macrospectra: erythrocyte nuclei from a chicken blood smear stained by aqueous Q-dmPOPOP (10 μg/mL, 5 min), and the unstained smear as previous control for light scattering. **C**: Fluorescence of chicken erythrocyte nuclei stained with 10 μg/mL Q-dmPOPOP.

Macrospectra are also performed after staining with two fluorochromes simultaneously. Smears of Ehrlich ascites tumor cells stained with DAPI and pyronine Y (PY) show emission bands corresponding to both fluorochromes (Fig. **5.17**). The peaks of DAPI and PY at 445 and 580 nm, respectively, correlates very well with the nuclear (DAPI) and cytoplasmic (PY) emission observed under fluorescence microscopy. Likewise, the same correlation is found when tissue sections are subjected to macrospectral analysis: after staining with acridine orange, unembedded cartilage sections [52] show a fluorescence curve where the contribution of orthochromatic nuclei (green) and metachromatic matrix (red) are clearly distinguished (Fig. **5.17**).

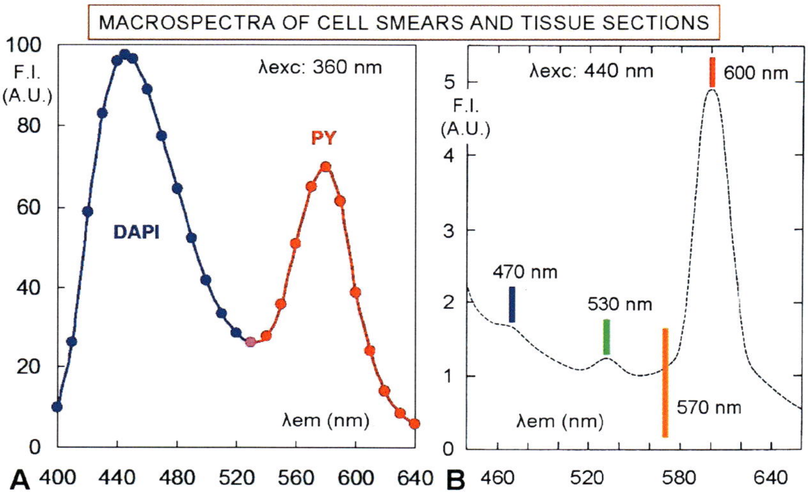

Fig. (5.17). A: Macrospectrum of Ehrlich ascites cells fixed in methanol and stained with 10^{-5} M DAPI at pH 8 for 5 min followed by 5 x 10^{-5} M pyronine Y (PY) for 10 min. **B**: Fluorescence macrospectrum of an unembedded cartilage section (chicken xiphoid) stained with acridine orange (1 mM, 5 min). Raman scattering (470 nm), green emission of chondrocyte nuclei (530 nm), and red emission of cartilage matrix (600 nm) are indicated. For comparison, the fluorescence peak at 570 nm of 1 mM solution of the free dye is also shown.

5. ENERGY TRANSFER

The interesting phenomenon of energy transfer can take place as intra- or intermolecular processes (see Chapter 2.5). Regarding biochemical systems, the inductive resonance type of singlet excitation energy transfer over 5 nm or more is a well known fact between aromatic amino acids in proteins [55]. In addition to aromatic residues, peptide groups (carboxamide, -CO-NH-) act as photon receptor at λab: 190 nm, leading to some energy leakage across them although they are separated by isolating tetrahedral carbons. The fluorescence of distamycin A would be related to the carboxamide coupling between methyl-pyrrole units (see

Chapter 12.5.1). Of course, H-bonds connecting close peptide groups would be the main pathway for migration of the excitation energy in proteins [55].

Illustrative examples of intramolecular energy transfer are shown in Fig. (**5.18**) for synthetic or natural compounds. In the first case, photon energy captured by an antenna ligand is directly transferred to a chelated luminescent lanthanide. In other cases (*i.e.* the linked chromophores benzoyl-biphenyl), photoexcitation localized in one group (>C=O) induces emission from the other group (biphenyl), the mechanism being a triplet excitation energy transfer [55].

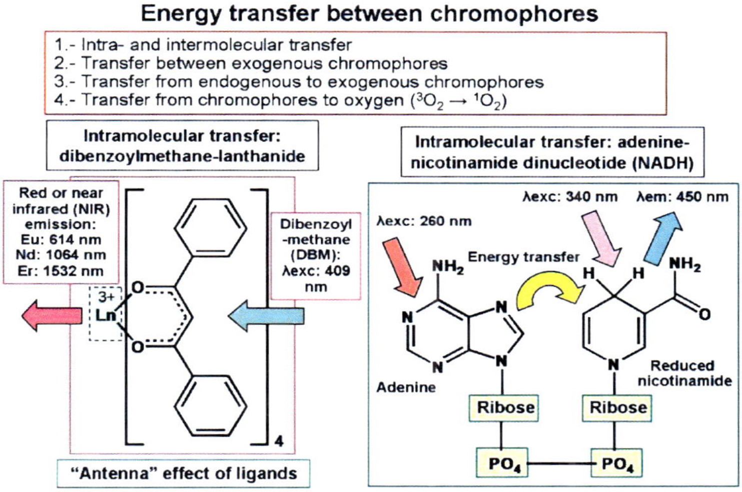

Fig. (5.18). Examples of energy transfer between chromophores. **Left**: Intramolecular transfer from an antenna ligand (dibenzoylmethane) to the luminescent lanthanide. **Right**: Intramolecular transfer from adenine to the reduced (fluorescent) nicotinamide of NADH coenzyme.

Another interesting case is the intramolecular energy transfer between the independent chromophores of the reduced nicotinamide-adenine-dinucleotide (NADH), in which they are connected without conjugation by isolated single bonds (Fig. **5.18**). Energy transfer from the excited adenine to nicotinamide occurs on account of the electronic overlap between the stacked chromophores. Thus the energy absorbed by adenine (260 nm) is transferred with high efficiency to the reduced nicotinamide and emitted as fluorescence (450 nm). A similar energy transfer occurs in flavine-adenine-dinucleotide (FAD) between adenine and isoalloxazine [55].

Intermolecular energy transfer also occurs between DNA bases and intercalated fluorochromes [56]. In the presence of ethidium bromide, UV-excited DNA emission at 360 nm is greatly abolished [57] and strong EB emission at 590 nm is induced by energy transfer between DNA bases and intercalated EB (Fig. **5.19**).

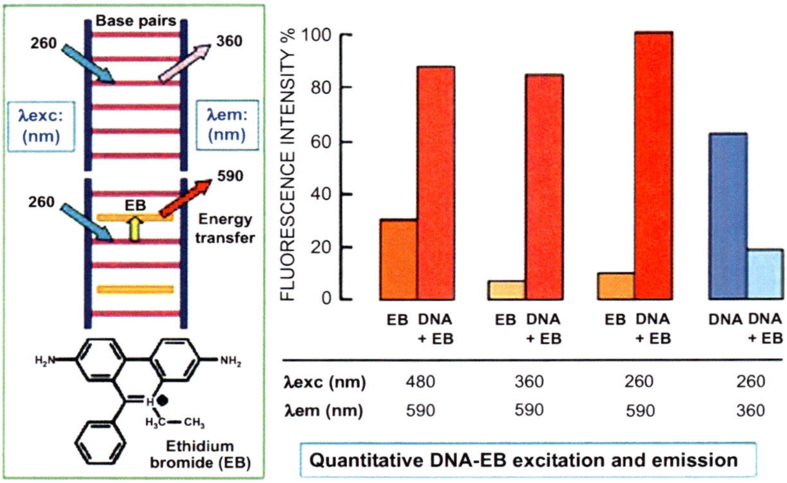

Fig. (5.19). Intermolecular energy transfer between chromophores. **Left**: Schematic diagram of DNA emission (alone and intercalated with ethidium bromide, EB) showing energy transfer. **Right**: Quantitative relationships between DNA-EB excitation and emission with fluorescence intensity (λem) of DNA solutions (8×10^{-5} M in bases), ethidium bromide (EB, 2×10^{-5} M), and DNA-EB complexes under different excitations (λexc). Observe the high EB fluorescence at 590 nm when the complexes are excited at 260 nm, and the low DNA emission at 360 nm in the presence of EB.

In turn, photoinduced electron transfer from intercalated EB to electron acceptors (*i.e.* DNA-bound methylviologen) can also occur [58]. Combination of the fluorochromes EB and mithramycin results in energy transfer from mithramycin to EB [59], a feature that must be taken into account when simultaneous DNA staining with these dyes is performed. Interestingly, fluorescence resonance energy transfer can be now applied to microscopic instrumentation [60].

CONFLICT OF INTEREST

The author (editor) declares no conflict of interest, financial or otherwise.

ACKNOWLEDGEMENTS

This work was supported by a grant from the Ministerio de Economía y Competitividad (CTQ2013-48767-C3-3-R), Spain.

REFERENCES

[1] Förster T. Fluoreszenz Organischer Verbindungen. Göttingen: Vandenhoeck and Ruprecht 1951.

[2] Udenfriend S. Fluorescence Assay in Biology and Medicine. New York, London: Academic Press 1962.

[3] Winkler MH. Some notes on fluorescence intensity. Biophys J 1967; 7(6): 719-25. [http://dx.doi.org/10.1016/S0006-3495(67)86618-9] [PMID: 19210994]

[4] Puchtler H, Sweat F, Gropp S. An investigation into the relation between structure and fluorescence of azo dyes. J R Microsc Soc 1967; 87(3): 309-28. [http://dx.doi.org/10.1111/j.1365-2818.1967.tb04513.x] [PMID: 5582815]

[5] Berlman IB. Handbook of Fluorescence Spectra of Aromatic Molecules. 2nd ed., New York, London: Academic Press 1971.

[6] Porumb H. The solution spectroscopy of drugs and the drug-nucleic acid interactions. Prog Biophys Mol Biol 1978; 34(3): 175-95. [PMID: 375301]

[7] Juarranz A, Stockert JC. Cytofluorometric and cytophotometric measurements by using the exposure control device in the automatic photomicroscope. Microsc Acta 1980; 83(4): 297-305. [PMID: 6158657]

[8] Villanueva A, Stockert JC, Armas-Portela R. A simple method for the fluorescence analysis of nucleic acid-dye complexes in cytological preparations. Histochemistry 1984; 81(1): 103-4. [http://dx.doi.org/10.1007/BF00495409] [PMID: 6469718]

[9] Latt SA, Lalande M, Kunkel LM, Schreck R, Tantravahi U. Applications of fluorescence spectroscopy to molecular cytogenetics. Biopolymers 1985; 24(1): 77-95. [http://dx.doi.org/10.1002/bip.360240108] [PMID: 3886034]

[10] Gutiérrez-Gonzálvez MG, Stockert JC. A simple method for the spectral analysis of dye-substrate complexes in cytologic preparations. Biol Zbl 1985; 104: 47-50.

[11] Bright FV. Bioanalytical applications of fluorescence spectroscopy. Anal Chem 1988; 60(18): 1031A-9A. [http://dx.doi.org/10.1021/ac00169a001] [PMID: 3067623]

[12] Wolfbeis OS, Ed. Fluorescence Spectroscopy New Methods and Applications. Berlin, Heidelberg: Springer-Verlag 1993. [http://dx.doi.org/10.1007/978-3-642-77372-3]

[13] Lakowicz JR. Principles of Fluorescence Spectroscopy. 3rd ed., New York: Springer Science + Business Media LLC 2010.

[14] Piller H. Microscope Photometry. Berlin, New York: Springer-Verlag 1977. [http://dx.doi.org/10.1007/978-3-642-66570-7]

[15] Sharonov S, Chourpa I, Valisa P, *et al.* Confocal spectral imaging analysis. Eur Microsc Anal 1994; 32: 23-5.

[16] Stockert JC, Molero ML, Espelosin RH. Fluorescence patterns of chromatin and cytoplasm by hematoxylin solutions. Z Naturforsch C Biosci 1983; 38(1-2): 153-5. [PMID: 6845816]

[17] Stockert JC. The horse eosinophil as a model leucocyte for morphological and cytochemical studies. Braz J Morphol Sci 2005; 22: 73-84.

[18] Molero ML, Hazen MJ, Stockert JC. Observations on the orcein fluorescence. Acta Histochem 1985; 76(1): 77-9. [http://dx.doi.org/10.1016/S0065-1281(85)80039-8] [PMID: 2409729]

[19] Molero ML, Del Castillo P, Stockert JC. Fluorescence reactions of orcein: new applications for an old stain. J Microsc 1986; 144(Pt 1): 45-53. [http://dx.doi.org/10.1111/j.1365-2818.1986.tb04672.x] [PMID: 2432271]

[20] Ghosh AK. Quenching of fluorescence of methylene blue by the dye dimer. Indian J Chem 1974; 12: 313-4.

[21] Galla HJ, Hartmann W. Excimer-forming lipids in membrane research. Chem Phys Lipids 1980; 27(3): 199-219.
[http://dx.doi.org/10.1016/0009-3084(80)90036-5] [PMID: 7418114]

[22] Masserini M. Pyrenyl-derivatives as a tool to study the behavior of glycolipids in the membranes. Appl Fluor Technol 1991; 3(1): 28-9.

[23] Stockert JC, Cañete M, Colman OD. Histochemical mechanism for the orthochromatic staining and fluorescence reaction of lignified tissues. Cell Mol Biol 1984; 30(6): 503-8.
[PMID: 6085282]

[24] Juarranz A, Stockert JC, Cañete M, Villanueva A. A micellar model for the disaggregating effect of detergents on phthalocyanin dyes. Cell Mol Biol 1985; 31(5): 379-84.
[PMID: 4084943]

[25] Stockert JC. Cytochemistry of nucleic acids: binding mechanisms of dyes and fluorochromes. Microsc Electron Biol Celular 1985; 9(2): 89-131.
[PMID: 2431250]

[26] Grases F, Genestar C, Gil JJ. Effect of surfactants on the fluorescence intensity of organic reagents. Microchem J 1985; 31: 44-9.
[http://dx.doi.org/10.1016/0026-265X(85)90007-4]

[27] Reinsborough VC, Holzwarth JF. Kinetics of the interactions between dyes and micelles. Can J Chem 1986; 64: 955-9.
[http://dx.doi.org/10.1139/v86-159]

[28] Stockert JC. Monomerizing effect of caffeine, *o*-phenanthroline, and tannin on cationic dyes: a model system to analyze spectral characteristics of the intercalative binding to nucleic acids. Acta Histochem 1989; 87(1): 33-42.
[http://dx.doi.org/10.1016/S0065-1281(89)80028-5] [PMID: 2513697]

[29] Stockert JC, Del Castillo P, Llorente AR, Rasskin DM, Romero JB, Gómez A. New fluorescence reactions in DNA cytochemistry. 1. Microscopic and spectroscopic studies on nonrigid fluorochromes. Anal Quant Cytol Histol 1990; 12(1): 1-10.
[PMID: 2302306]

[30] Stockert JC. 2,5-bis(4-aminophenyl)-1,3,4-oxadiazol: fluorescence reaction of chromatin and basophilic cytoplasm. Acta Histochem Cytochem 1983; 16: 66-9.
[http://dx.doi.org/10.1267/ahc.16.66]

[31] Stockert JC, Espada J, Trigoso CI, Del Castillo P. Fluorescence reaction of the anionic triarylmethane dye fast green FCF. Appl Fluor Technol 1991; 3(6): 14-5.

[32] Cramer F, Hettler H. Inclusion compounds of cyclodextrins. Naturwissenschaften 1967; 54(24): 625-32.
[http://dx.doi.org/10.1007/BF01142413] [PMID: 4298643]

[33] Cramer F, Saenger W, Spatz AC. Inclusion compounds. XIX. The formation of inclusion compounds of some α-cyclodextrin in aqueous solution. Thermodynamics and kinetics. J Am Chem Soc 1967; 89: 14-8.
[http://dx.doi.org/10.1021/ja00977a003]

[34] Armas-Portela R, Stockert JC. A reaction mechanism for the orthochromatic staining of polysaccharides: monomerizing interactions with basic dyes. Cell Mol Biol 1986; 32(4): 449-53.
[PMID: 3742556]

[35] Pitha J, Pitha J. Amorphous water-soluble derivatives of cyclodextrins: nontoxic dissolution enhancing excipients. J Pharm Sci 1985; 74(9): 987-90.

[http://dx.doi.org/10.1002/jps.2600740916] [PMID: 4067854]

[36] Moser JG. Carrier and Delivery Systems.Photodynamic Tumor Therapy. 2nd and 3rd Generation Photosensitizers.. Amsterdam: Harwood Academic Publishers 1998; pp. 127-36.

[37] Kim H. Normal modes and structure of iodine in the starch iodine complex. Biopolymers 1982; 21: 2083-96.
[http://dx.doi.org/10.1002/bip.360211102]

[38] Whittingham MS. Intercalation Chemistry. An Introduction.Intercalation Chemistry. New York, London: Academic Press 1982; pp. 1-18.
[http://dx.doi.org/10.1016/B978-0-12-747380-2.50006-7]

[39] Rausell-Colom JA, Serratosa JM. Reactions of Clays with Organic Substances. Chemistry of Clays and Clay Minerals. Essex: Longman Scientific and Technical 1987; pp. 731-422.

[40] Bujdák J. Effect of the layer charge of clay minerals on optical properties of organic dyes. A review. Appl Clay Sci 2006; 34: 58-73.
[http://dx.doi.org/10.1016/j.clay.2006.02.011]

[41] Stockert JC, Del Castillo P, Blázquez-Castro A. Induction of metachromasia in cationic dyes and fluorochromes using a clay mineral: a potentially valuable model for histochemical studies. Acta Histochem 2011; 113(6): 668-70.
[http://dx.doi.org/10.1016/j.acthis.2010.07.001] [PMID: 20674960]

[42] Samuels M, Mor O, Rytwo G. Metachromasy as an indicator of photostabilization of methylene blue adsorbed to clays and minerals. J Photochem Photobiol B 2013; 121: 23-6.
[http://dx.doi.org/10.1016/j.jphotobiol.2013.02.004] [PMID: 23474529]

[43] Sánchez del Río M, Martinetto P, Reyes-Valerio C, *et al.* Synthesis and acid resistance of Maya blue pigment. Archaeometry 2006; 48: 115-30.
[http://dx.doi.org/10.1111/j.1475-4754.2006.00246.x]

[44] Villemure G, Detellier C, Czabo AG. Fluorescence of clay-intercalated methylviologen. J Am Chem Soc 1986; 108: 4658-9.
[http://dx.doi.org/10.1021/ja00275a071]

[45] Madhavan D, Pitchumani K. Photoreactions in clay media: singlet oxygen oxidation of electron-rich substrates mediated by clay-bound dyes. J Photochem Photobiol Chem 2002; 153: 205-10.
[http://dx.doi.org/10.1016/S1010-6030(02)00293-9]

[46] De Korte A, Langlois R, Cantor CR. Fluorescence quenching of methyl-2-aminobenzoate in the presence of beta-cyclodextrin: a model for the dynamic quenching of chromophores in hydrophobic regions of biopolymers. Biopolymers 1980; 19: 1281-8.
[http://dx.doi.org/10.1002/bip.1980.360190704]

[47] Semmel M, Daune M. Etudes des complexes de colorants basiques avec le RNA. Biochim Biophys Acta 1967; 145(3): 561-76.
[http://dx.doi.org/10.1016/0005-2787(67)90116-5] [PMID: 6065691]

[48] Green FJ. The Sigma-Aldrich Handbook of Stains, Dyes and Indicators. Milwaukee, Wisconsin: Aldrich Chemical Company 1990.

[49] Bergeron JA, Singer M. Metachromasy: an experimental and theoretical reevaluation. J Biophys Biochem Cytol 1958; 4(4): 433-57.
[http://dx.doi.org/10.1083/jcb.4.4.433] [PMID: 13563551]

[50] Mason SF. Color and the Electronic States of Organic Molecules. Chemistry of the Synthetic Dyes. New York: J. Wiley and Sons 1970; pp. 169-221.

[51] Gupta AK, Ed. The Complete Book on Natural Dyes and Pigments National Institute of Industrial Research. Delhi: Asia Pacific Business Press 2005.

[52] Stockert JC, Del Castillo P. Direct sectioning of unembedded cartilage: a simple method for

microscopical and histochemical studies on chondrocytes and extracellular matrix. Z Mikrosk Anat Forsch 1990; 104(3): 377-84.
[PMID: 2238787]

[53] Espada J, Bertolesi GE, Trigoso CI, Gamallo C, Stockert JC. Fluorescence of mast cell granules in paraffin sections and cell smears induced by an N-quaternary oxazole scintillator. Histochem J 1995; 27(4): 318-22.
[http://dx.doi.org/10.1007/BF00398974] [PMID: 7635764]

[54] Stockert JC, Pelling C, Espada J. New cationic fluorochromes from diaryloxazole scintillators: fluorescence of chromatin DNA induced by N-quaternary POPOP derivatives. Acta Histochem 1997; 99(2): 195-205.
[http://dx.doi.org/10.1016/S0065-1281(97)80042-6] [PMID: 9248577]

[55] Bowen EJ, Ed. Recent Progress in Photobiology. New York: Academic Press 1965.

[56] Del Castillo P, Stockert JC. Fluorescencia del DNA: efecto de la desnaturalización y de la intercalación. Técn Laboratorio 1988; 11: 440-4.

[57] Brodie S, Giron J, Latt SA. Estimation of accessibility of DNA in chromatin from fluorescence measurements of electronic excitation energy transfer. Nature 1975; 253(5491): 470-1.
[http://dx.doi.org/10.1038/253470a0] [PMID: 1110800]

[58] Fromherz P, Rieger B. Photoinduced electron transfer in DNA matrix from intercalated ethidium to condensed methylviologen. J Am Chem Soc 1986; 108: 5361-2.
[http://dx.doi.org/10.1021/ja00277a060]

[59] Crissman HA, Tobey RA. Specific Staining of DNA with the Fluorescent Antibiotics, Mithramycin, Chromomycin, and Olivomycin.Techniques in Cell Cycle Analysis. Clifton: Human Press 1986; pp. 97-104.

[60] Cardullo RA, Parpura V. Fluorescence resonance energy transfer microscopy: theory and instrumentation. Methods Cell Biol 2003; 72: 415-30.
[http://dx.doi.org/10.1016/S0091-679X(03)72020-2] [PMID: 14719343]

CHAPTER 6

Fluorescence Instrumental and Techniques

Abstract: The use of fluorescence microscopy requires knowledge of the elements making up the fluorescence microscope. In this chapter the basic elements combining to make a fluorescence microscope will be introduced and its role explained. Afterwards an introduction to the most common fluorescence microscopic techniques will be made followed by more advanced and developed approaches. These will include total-internal reflection and time-resolved microscopies, to name a few. The chapter will end with a brief introduction to flow cytometry, a highly related fluorescence technique in the biomedical field.

Keywords: Confocal microscope, Filter, Flow cytometry, Fluorescence Correlation Spectroscopy (FCS), Fluorescence Lifetime Imaging Microscopy (FLIM), Fluorescence localization after photobleaching (FLAP), Fluorescence loss in photobleaching (FLIP), Fluorescence microscope, Fluorescence photobleaching, Fluorescence recovery after photobleaching (FRAP), Imaging system, Light-sheet microscope, Microscope objective, Time-resolved microscopy, Total Internal Reflection Fluorescence Microscopy (TIRFM).

Fluorescence techniques in general and fluorescence microscopy in particular need, like any other experimental branch of science, devices and instrumentation that allow for measuring. Thus, research employing fluorescence microscopy requires certain optical elements to produce pictures and data to support results. In this chapter the basic elements of a fluorescence microscope setup will be introduced and explained. Additional optical elements will also be introduced in connection to more specific microscopic techniques (*e.g.* confocal or total internal reflection). A brief introduction to some modern techniques will be provided in this chapter to set the stage for more specific developments in upcoming chapters (FRAP, FLIP, FLAP, *etc.*). A short explanation of fundamentals of time-resolved microscopy will be included too.

Nevertheless a point deserves clarification here. Because of their current widespread use microscopic techniques based upon non-linear optics or super-resolution deserve a deeper review and will be explained in their corresponding

Juan Carlos Stockert & Alfonso Blázquez-Castro

chapters (see Chapters 19 and 20 respectively). Of course, some optical instruments and devices employed for those techniques will be explained in this chapter (lasers as light sources, for example) and the corresponding reminder will be included whenever necessary to refresh a concept. Also some basics on flow cytometry have been included in this chapter. Although not strictly speaking an imaging fluorescence microscopy technique it nevertheless rests upon the foundations of fluorescence recognition. Thus brief introductory concepts regarding flow cytometry can be found at the end of chapter.

1. BASIC ELEMENTS OF A FLUORESCENCE MICROSCOPE

A fluorescence microscope can be considered on its simplest as an evolved version of a bright field microscope [1, 2]. As such it shares with its predecessor many of the optical elements that make up the basic setup for fluorescence microscopy (see Fig. **6.1A**). From the most basic point of view, and looking at Fig. (**6.1A**), a fluorescence microscope is composed of a *light source*, filtering elements (*optical filters* and *dichroic mirrors*) and imaging elements (*objectives* and *lenses*). The most widely spread fluorescence microscope design is that in which the sample's emission is collected by optics on top of the sample. Therefore emitted light that travels upwards from the sample is used to image it. This particular arrangement is known as an *epifluorescence* microscope (see Fig. **6.1B** and **C**). From this point on, when we refer to a fluorescence microscope it is assumed to be an epifluorescence microscope. In the following a brief introduction to the basic optical components that make up such a microscope will be outlined.

1.1. Light Sources

A fluorescence microscope is an imaging device. Therefore it requires some source of light that allows for an image to be obtained. Since its inception at the beginning of the 20th century several different light sources have been employed for fluorescence microscopes [3]. No ideal light source exists for all microscopic techniques so an overall balance must be found in each particular case. All in all, some general features can be detailed that account for the desirable properties a particular light source must have to be used in a fluorescence microscope: 1) emission lines instead of a continuous emission spectrum, 2) overlap of emission lines with the absorption spectra of desirable fluorophores, 3) high light output at those emission lines, 4) safe emission wavelengths (avoiding the UV specially below 350 nm), 5) high energy efficiency, 6) easy/economic maintenance, 7) che-apness. No light source complies with all these criteria, of course. In the following the most common in use light sources will be briefly introduced along with some particular features in each case. It is up to each user to decide which light source is most desirable in light of their specific research.

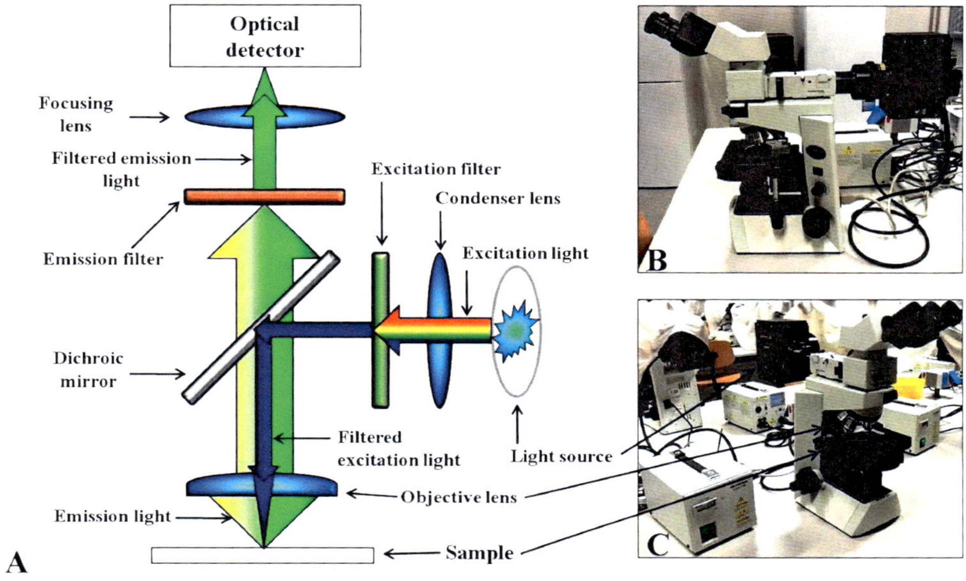

Fig. (6.1). A: General arrangement of common optical elements in an epifluorescence microscope. A multicolour light beam is first filtered by the excitation filter and this light is focused on the sample. The emitted fluorescent light is partly collected by the objective lens, traverses the dichroic mirror, is filtered by the emission filter and refocused by the focusing elements before reaching the corresponding optical detector. **B**: Example of a typical commercial epifluorescence microscope. **C**: Same epifluorescence microscope from a different angle. Some of the correspondences between components displayed schematically in Figure **6.1 A** and the real parts are indicated (arrows).

Gas-Discharge Lamps

These are, by far, the most common light sources currently in use for fluorescence microscopy [4]. Their working mechanism is based on the establishment of an electrical current inside a bulb filled with certain mixtures of gases and, sometimes, additional volatile elements/compounds. The gas inside the lamp becomes ionized under the action of an electrical current between the electrodes. The electrons flowing through the plasma collide with atoms, ions and molecules and transfer part of their kinetic energy to them. This pumps them to excited electronic states that finally radiate at several wavelengths (see Chapter 2.3 and 2.4), depending on the specific gas composition inside the bulb. The most employed discharge lamps in fluorescence microscopy are the high-pressure xenon (Xe) arc lamp (Fig. **6.2A**) and the high-pressure mercury (Hg) vapor lamp (Fig. **6.2B**).

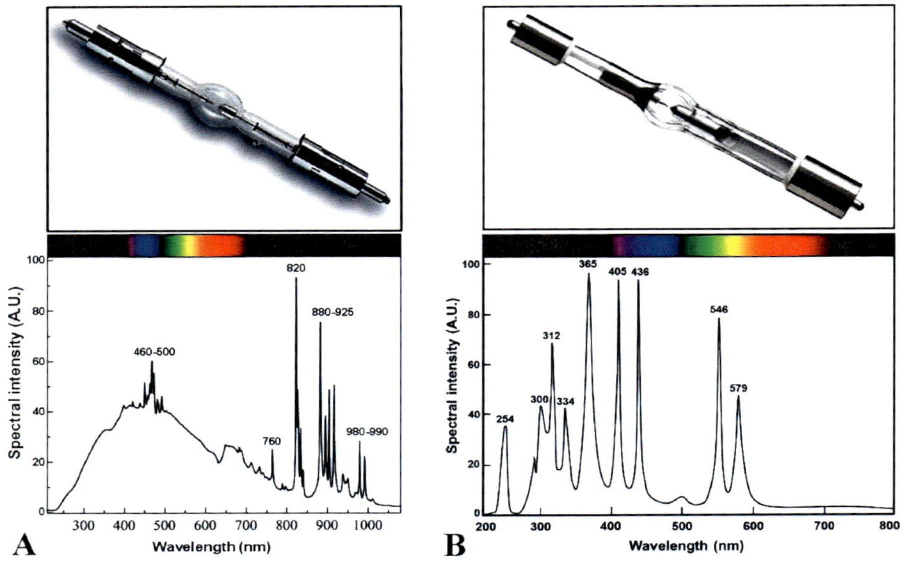

Fig. (6.2). A: Xenon lamp bulb (top, reprinted with permission from Osram) and its emission spectrum (bottom). It can be seen that the emission is quite broad with many peaks in the near infrared. **B**: Mercury (Hg) lamp bulb (top, reprinted with permission from Osram) and its emission spectrum. In this case the emission is made up of different peaks corresponding to atomic transitions of the Hg atom. Notice in both pictures the electrodes inside the bulb energizing the lamp. Numbers above the emission lines in the emission spectra are nanometres. The "rainbow" strip above the emission spectra is a guide to the eye in regards to the emission colours corresponding to the lines shown below.

The Xe arc lamp provides a continuous emission spectrum from the UV to the IR (250-1000 nm). The spectrum in the IR also displays many atomic lines (750-1000 nm). The Hg vapor lamp is a Xe-filled lamp with some additional liquid mercury in the bulb. Once the lamp is started it heats and vaporizes the mercury drops. Once in the gas phase mercury atoms enter the discharge zone and become electronically excited. Mercury atoms are very intense sources of light at certain wavelengths, mainly 184, 254, 365, 405, 436 and 546 nm. The lines at 184 and 254 nm make these lamps dangerous so special care must be taken not to expose oneself to the bare lamp emission (the same counts for Xe lamps, as they provide strong emission below 400 nm). Hg vapor lamps produce a line–dominated emission spectrum atop a continuous one produced mainly by the Xe filling. These two lamps types are very common nowadays a provide adequate excitation spectra –after correct filtering, see filters below- for most fluorophores currently in use.

Metal halide lamps belong to the gas discharge lamp group too. It is an Hg vapor lamp with additional metal halides (most usually bromides or iodides) that enhance the luminous output. Sodium and Scandium are routinely employed metals, while less common mixtures include Thallium, Indium, Dysprosium and other lanthanide elements. All these metallic elements increase the amount of light emitted by the lamp and allow a fine tailoring of the spectral output. Light is coupled to an optical fiber to make it reach the microscope stage. This setup is being adopted by more and more companies and users because of the enhanced spectral properties, the higher output and the longer lamp lifetime (1,000's of hours *vs.* 100's) as compared to classical discharge lamps (Xe and Xe-Hg).

Lasers

Lasers are another light source with its own specific characteristics (Fig. **6.3**) [5, 6]. Some of them make lasers really attractive as light sources for fluorescence microscopy, like spectral monochromacity, spatial and temporal coherence and pulsed operation for time-resolved microscopy and other advanced techniques (*e.g.* super-resolution microscope). Laser light emission can range from no pulsing at all (continuous wave or CW) to pulses of different lengths (from quasi-cw with 10^{-3} s pulses to femtosecond pulses of $\sim 10^{-14}$ s).

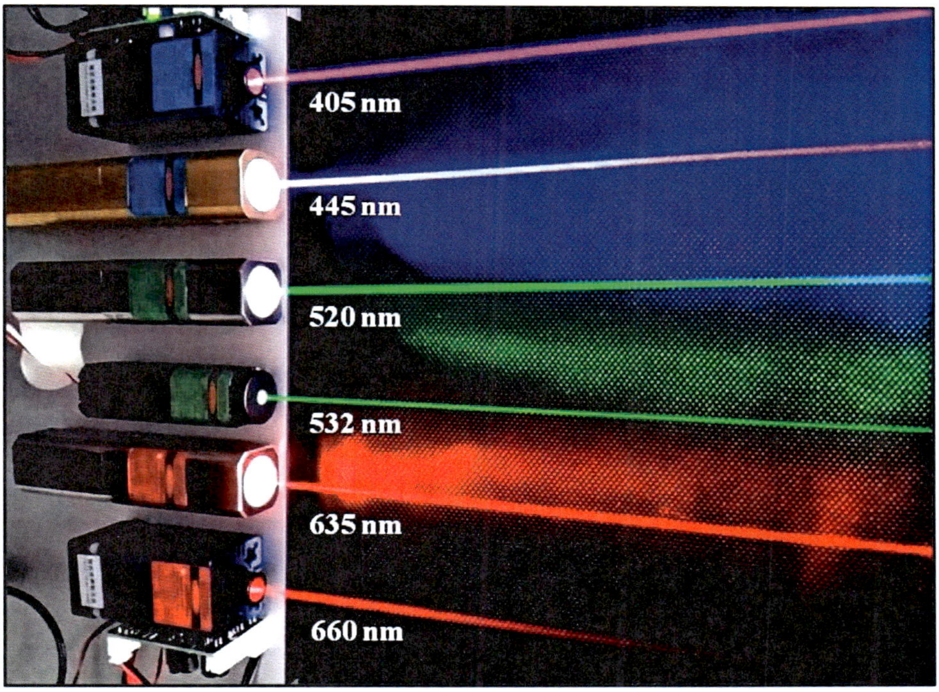

Fig. (6.3). Examples of different lasers emitting in the visible spectrum. From top to bottom the lasers emit at increasing wavelengths (see emission wavelength labelling for each case). Note the monochromaticity (single colour) and collimation (straightness) of each laser beam (reproduced with permission from https://commons.wikimedia.org/wiki/File:LASER.jpg).

Briefly a laser working principle is that of stimulating photon emission from an adequately prepared population of emitters. So a laser is based in the so called population inversion phenomenon. This means that, given an ensemble of emitting entities (ions, atoms, molecules), some excited state must be more populated than another state lying down in energy. This lower energy state can be the ground state or a less energetic excited state. Such situation is metastable: the system does not immediately undergo deactivation but is very eager to emit light at the first chance. If a photon with the right energy (this energy being equal to the difference between the upper and lower energy levels) interacts with this metastable inverted population it can stimulate light emission. The priming photon acts as a "template" for other photons being emitted so that, ideally, all photons emitted will be "clones" in terms of energy (= wavelength), polarization, travelling direction, *etc*. Therefore, a laser is a collection of photons very much alike in virtue of their emission process. This is, of course, a very simplistic way of presenting laser operation but provides a general idea. For much more technical detail we refer the reader to specific literature [5 - 7].

Because of its emission mechanism, laser light is monochromatic or single-color. Said in another way, all photons have the same energy/wavelength. This is very convenient for fluorescence microscopy as it was established at the beginning of the chapter that a high degree of monochromatism was one of the features a light source should have to qualify for this technique. This makes exciting light filtering much more easily than with broadband light sources (see filters below). Lasers usually emit one or several monochromatic lines, depending on each particular laser type and the wavelength tuning (see Table **6.1**). Again this is advantageous to excite different fluorophores. Also laser light is highly polarized which makes its manipulation by optical elements (*e.g.* polarizers, dichroic mirrors, *etc.*) much easier than with incoherent sources.

Table 6.1. Common lasers employed in fluorescence microscopy.

Laser	Laser type	Wavelengths (nanometres)
Excimer	Gas discharge	248 (KrF) / 308 (XeCl) / 353 (XeF)
Nitrogen	Gas discharge	337.1
Argon	Gas discharge	488 / 514.5
Copper	Metal vapour discharge	510.6 / 578.2
Krypton	Gas discharge	568.2 / 647.1 / 676.4
Helium-Neon	Gas discharge	632.8
Dye	Organic fluorophores	350 – 800+ (emission from UV to IR)
Diode	Semiconductor	365 – 1000+ (emission depends on composition)

(Table 6.1) contd.....

Laser	Laser type	Wavelengths (nanometres)
Titanium:sapphire	Solid-state	650 – 1100
Neodimium:YAG	Solid-state	1064 (frequency doubled or tripled to 532 and 355)
Raman	Fiber	650 – 1500+ (pumped by another laser)

Lasers are very intense light sources. In fact they are the most intense light sources when pulsed lasers are put to use. This feature is put to use in a variety of fluorescence microscope setups such as low signal to noise ratio, confocal microscopy (see section 2 below), time-resolved microscopy (see section 3 below), or non-linear techniques (see Chapter 19). However, this high intensity can also be a problem because of increased sample photobleaching (see Chapters 2.5 and 7.4) and even destruction by too high levels of light and/or thermal load. Therefore laser sources require adequate intensity filtering before reaching the fluorescent sample in order to avoid damage.

Perhaps the most important feature of laser light for fluorescence microscopy, along with its monochromatism, is its coherence. No other light sources employed in fluorescence microscopy are coherent. Coherence means that laser light keeps an almost constant phase shift between its different light waves (temporal coherence) and that the emission spot is very small (spatial coherence). These two characteristics make it possible to focus laser light to very small points in space because the different light waves do not cancel each other due to destructive interference, as happens with incoherent light sources. Therefore laser light can be focused to what is called the diffraction limit. This diffraction limit depends, not only on the laser wavelength, but also on the medium´s refraction index and the specific characteristics of the focusing element employed (most usually a microscope objective). As an example, diffraction limit focused laser light of λ=600 nm will converge to a focusing point of less than a micron. This account for the extraordinary ability of laser light to probe matter at sub-micrometric lengths (see Chapters 19 and 20 for further developments in this sense).

Lasers have also provided the basis for another new light source developed in recent years for fluorescence microscopy. This source, supercontinuum light, blends together some desired features of laser light with a wider emission spectrum typical of incoherent sources.

Supercontinuum Sources

A supercontinuum spectrum is obtained when light undergoes non-linear processes that make the overall spectrum to widen at the expense of intensity at each spectral region [8, 9]. Two reasons make lasers the almost unique choice to pump a supercontinuum: 1) non-linear optical processes commonly precise very high light intensities to start working and 2) because supercontinuum proceeds by depleting the pumping wavelength to produce other wavelengths an intense initial spectral source is mandatory to end up with bright enough light at other wavelengths to make them useful. The two prerequisites are perfectly fulfilled by lasers.

The supercontinuum output is obtained by launching laser light through some material medium where non-linear phenomena transform the original monochromatic wave into multi-wavelength light (see Fig. **6.4A**).

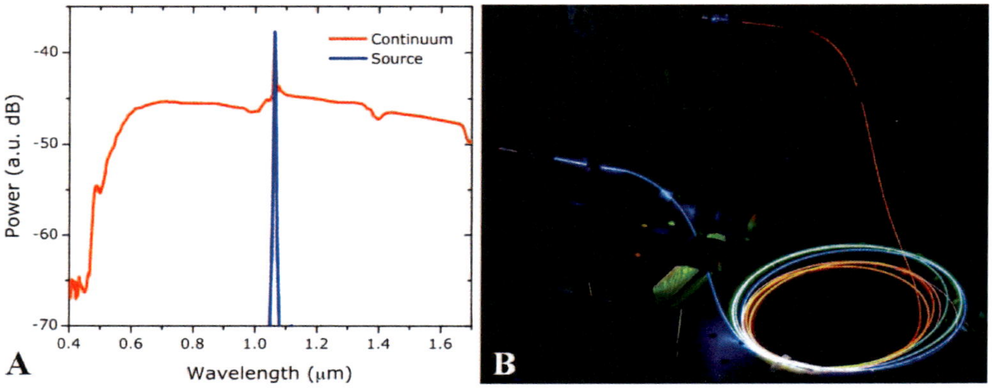

Fig. (6.4). A: Scheme of supercontinuum development (red spectrum) from a quasi-monochromatic light source (blue spectrum). Initially photon energies are restricted to a very narrow range typical of a laser source. With time this photons interact with the propagation medium and spread out in their spectral wideband, some photons gaining energy at the expense of others in the laser pulse (reproduced with permission from Burlyc - http://commons.wikimedia.org/wiki/File:Optics-SupercontinuumSpectrum.png#). **B**: Real example of supercontinuum generation in a non-linear optical fiber. An infrared laser pulse is pumped into the fiber in the upper part of the picture. As the pulse propagates inside the fiber all colours in the visible range are generated and, in virtue of internal scattering, some of this light propagates outside the fiber (reproduced with permission from Blinking Spirit - http://commons.wikimedia.org/wiki/File: Supercontinuum_in_a_microstructured_optical_fiber.PNG#).

The most employed material substrate for supercontinuum are optical fibers because optical losses are very low, the interaction length can be quite large (~kilometers) and their optical properties can be finely tailored to meet the expected output. Optical fibers are cheap and reliable, therefore they stand out as the preferred optical material to support supercontinuum generation (see Fig. **6.4B**). Supercontinuum light sources are finding more and more use in the same fields as lasers (confocal, time-resolved microscopy, *etc.*) but with the advantage that many wavelengths are available at the same time.

LEDs

Light Emitting Diodes or LEDs are solid-state optoelectronic devices that emit light when an electrical current flows through them (see Fig. **6.5A-B** for a scheme of the typical structure) [10].

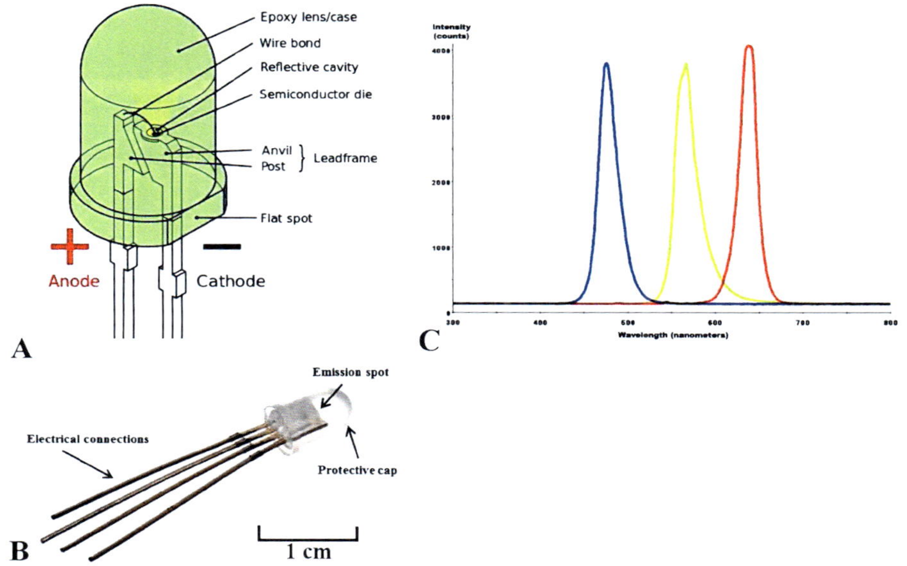

Fig. (6.5). **A**: Detailed structure of a LED showing its main parts. The semiconductor crystal giving rise to light emission is located in the cusp on top of the anvil (reproduced with permission from Inductiveload - https://commons.wikimedia.org/wiki/File:LED,_5mm,_green_(en).svg#). **B**: An example of an assembled LED displaying the light emitting dome and the sustaining electrical pins that provide the current (reproduced with permission from oomlout - http://commons.wikimedia.org/wiki/File:RGB_LED.jpg#). **C**: Emission spectrum from a white LED source that in fact includes emission from three different independent LEDs emitting in the blue, yellow-green, and red. Note the narrow emission bands from each LED that cover spectral bandwidths around 20-30 nm around the emission peak (reproduced with permission from Deglr6328 - http://commons.wikimedia.org/wiki/File:RedYellowGreenBlue_LED_spectra.png#).

It could be argued that, to some extent, a filament lamp could fit in this definition. The key difference, however, is that a LED emits light as a result of an electroluminescence process while incandescent lamps emit in virtue of a thermal mechanism (see halogen lamp below). LEDs have a precise composition and are structured in such a way as to facilitate the recombination of electrons and holes. This recombination gives rise to photon emission. A more detailed physical description of semiconductors, LEDs and their associated luminescence processes will be provided in Chapter 18.1. For the moment only the main lighting features of LEDs will be presented in order to compare them with other light sources currently employed in fluorescence microscopy.

As gas discharge lamps and halogen lamps LEDs are incoherent light sources. Nevertheless they show several features that make them very competitive ones

[11]. The first is their monochromatism. LED emission is probably the closest one can get to a mochromatic source without relying to a laser. The emission wavelength can be tailored by tuning the chemical composition and the internal structure of the LED (see Fig. **6.5C**). In this way the emission band offered by different LEDs spans from the UV to the IR (300-1000+ nm). The spectral bandwidth depends on several factors (composition, structure, quality) but it usually accounts for around 10-30 nm. For microscopy purposes this comes really close to practical monochromatism as such a narrow wavelength output is easily filtered (see optical filters below).

Other advantageous features are a very high electrical efficiency which makes them low-consuming devices. Although initially quite expensive, LEDs are becoming cheaper and cheaper to acquire, even at blue, violet or UV wavelengths. Their operational lifetimes are extremely long (tens of thousands of hours) and they show very good emission stability along this lifetime. This is in contrast to other light sources, especially gas discharge lamps, that display important changes in their luminous output (intensity, spectral quality, *etc.*) as their operational lifetime is approaching its end. Another attractive point is that the emission spot, although not getting to the limit of lasers, can be extremely small (10-100 μm). This makes focusing easier and imaging quality improves noticeably. Finally LED maintenance is usually minimal which adds to their other attractive features.

Halogen Lamps

Filament lamps have not been classically regarded as adequate excitation sources for fluorescence microscopy. This is because they display several disadvantageous features: broadband emission, very high thermal output (IR emission, conduction), large emitting surface, low energy efficiency to cite some. Many of these are the result of the light emission process that takes place in incandescent lamps, that is, blackbody thermal emission. In the simplest, an electrical current flows through a metallic conductor (tungsten) and atomic scale collisions between electrons and the metallic lattice produces large amounts of heat. The filament increases its temperature and emits electromagnetic radiation accordingly. The higher the temperature the more blue-shifted (towards the visible) is the emission spectrum. However this emission spectrum is continuous and very broad (see Fig. **6.6**).

This leads to large amounts of light being emitted in the IR and only a small fraction in the visible and in the UV (minimal). This broad spectrum makes filtering an arduous task and, at the same time, heat must be dissipated adequately too. To increase the filament lifetime, given the very high temperatures involved ($\geq$ 3000 K), some halogen element (*e.g.* iodine or bromine) is introduced in the lamp's bulb. Through different chemical reactions with tungsten they extend the filament and lamp's working service. This is the reason why these high-temperature filament lamps are named halogen lamps.

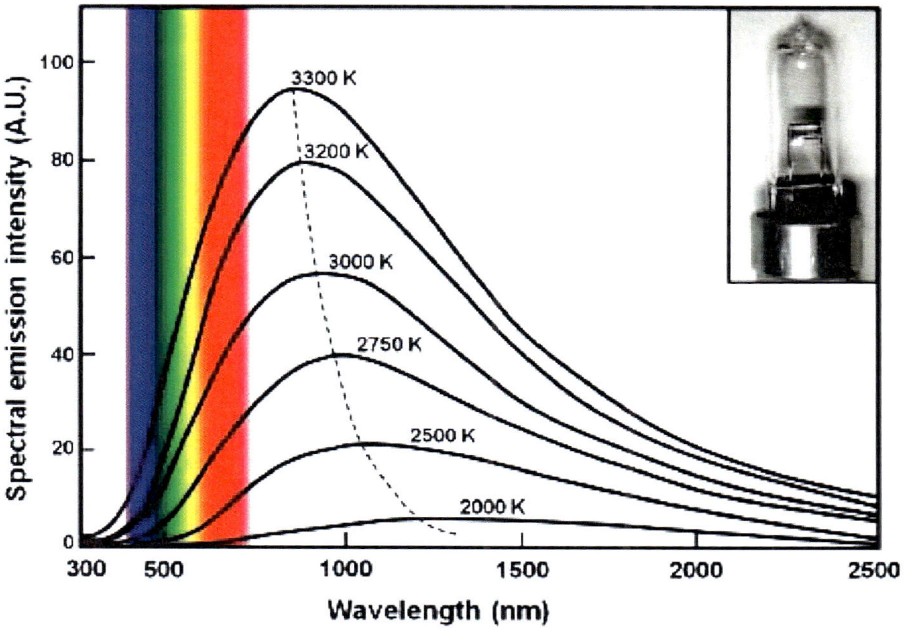

Fig. (6.6). Spectral emission curves of a blackbody at different temperatures (2000-3300 K). It can be appreciated that a hotter blackbody emits more energy at any wavelength than a colder one. Also the peak emission wavelength moves to shorter wavelengths with increasing temperature in accordance to Wien's law (dotted line). The rainbow strip is provided as a guide to the eye for the visible range. Inset: a halogen incandescent lamp in which the metal ribbon that gets incandescent when current flows can be seen inside the quartz bulb. This quartz bulb is filled with a special gas mixture to increase light output and lamp lifetime.

Taken all these disadvantages into account one is tempted to ask why one should even considering using filaments lamps. However halogen lamps are inexpensive, available and reliable. They have been proposed as alternatives to more advanced lighting systems and some recent literature points to their utility in this field [12]. Anyway halogen lamps are rarities in the fluorescence microscopy world and have been included only in the sake of comprehensiveness.

An important issue to obtain adequate and detailed fluorescent images is sample illumination by the exciting light. This illumination must be as smooth as possible to ensure equal amounts of light reaching the sample and to avoid lighting artifacts (*e.g.* shadowing, false positives, *etc.*). This even light exposure is achieved by making use of Köhler illumination [13]. To achieve Köhler illumination several optical elements are introduced in the exciting light beam path (lenses and condensers/pinholes) from the light source to the oculars. This setup ensures that the light source focusing plane is different from the sample's focusing plane. A very homogenous light spot excites the sample while at the same time avoiding light source focusing at the imaging plane. The sample emission is, on the other side, focused at the imaging plane therefore allowing a clear picture to be obtained.

1.2. Optical Filters and Dichroic Mirrors

The light source output needs filtering in order to obtain a fluorescent image. This is because the exciting light is orders of magnitude higher in intensity than the fluorescence emission intensity. Therefore exciting light photons must be reduced as much as possible (ideally being totally absorbed/reflected) to dismiss the possibility that they reach the imaging part of the microscope [2]. For this essential reason a filtering arrangement is deployed between the light source and the sample, and between the sample and the imaging system. Optical filters can selectively allow some wavelengths and not others by either light absorption or wave interference mechanisms. Modern filters work through interference as it leads to less problems in regard to filter heating and optical distortions because of thermally-derived changes in refraction indexes.

Basically filters can be of three types, attending to the wavelength range they filter out (see Fig. **6.7A**): short pass (SP) filters which remove light above a certain wavelength, long pass (LP) filters which remove light below a certain wavelength, and band pass (BP) filters which only allow a certain wavelength band to pass through, removing light above and below that band. Adequate use of these filtering options ensures the right wavelength combination to obtain microscopic fluorescent images. In the vast majority of fluorescence microscopes this arrangement is composed of 1) an excitation filter, 2) a dichroic mirror, and 3) an emission filter (see Fig. **6.1A**). These elements will be briefly explained below.

Excitation Filters

Because most exciting light sources emit a more or less wide band spectrum it is necessary to filter the output light so that it fits better the absorption spectrum of the fluorophore being used in the experiments and to make it easier to later filter the spurious excitation light before it reaches the imaging device. This is the reason for using an excitation filter. Excitation filters are usually SP filters or (infrequently) BP filters. This means that they allow light of a certain wavelength or shorter ones to pass through. Excitation filters are selected to match the excitation light to the specific fluorophore's absorption spectrum it is being used.

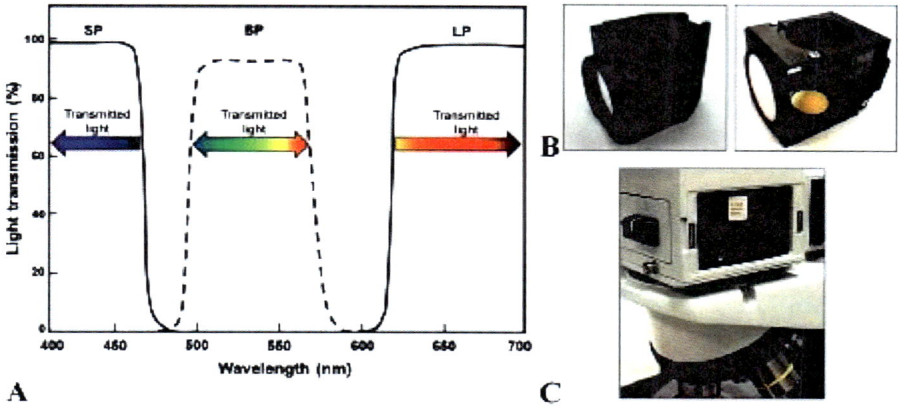

Fig. (6.7). **A**: Filtering features of the main filter types. Short-pass (SP) filters allow light of wavelengths below a certain threshold to pass. Band-pass (BP) filters allow transmission between two defined wavelengths. Long-pass filters block shorter wavelengths and transmit longer wavelengths. **B**: Pictures of two optical cubes (filter cubes) incorporated into epifluorescent microscopes. Excitation light is first filtered by an excitation filter (left sides), then reflects from the dichroic mirror (located at the diagonal) downwards to the sample through the bottom hole (right picture). The emission light is transmitted upwards through the dichroic mirror and filtered again by the emission filter (top hole) before reaching the oculars. **C**: Picture of the location where a filter cube is inserted in an epifluorescence microscope. The excitation light would arrive from the right at the same level as the cube. Then the filtered light would be directed downwards by the optical elements in the cube through the objectives (located below the cube). Fluorescent emission light would travel straight up from the objectives through the cube and to the oculars/imaging system. Notice the sliding cube mechanism (grey button with a vertical white strip) employed to change between two different cubes without modifying any other optical element.

Dichroic Mirrors

As already stated for a fluorescent microscope to function properly it is of the utmost importance to eliminate as much exciting light as possible after exciting the sample. Exciting light filtered by the excitation filter next reaches the dichroic mirror. This is a special interference filter/mirror that reflects light almost perfectly in some wave band (usually shortwave light) and allows light to pass through in another wave band (longwave light). In this sense, it behaves in a similar way to a LP or BP filter, except that selective reflection takes place in this case. Shortwave light coming from the light source/excitation filter combination is reflected by the dichroic mirror set at 45° and directed towards the microscope objective and sample (see Fig. **6.1**). Part of the fluorescence light emitted by the sample, with a longer wavelength than the exciting light, is collected by the objective and traces back the optical path towards the dichroic mirror. However, the dichroic allows longwave light to pass so emitted light takes, from this point, a different light path from that of the excitation light. The fluorescent light then reach the emission filter (see below).

The dichroic offers an additional advantage apart from splitting the excitation and fluorescence light paths: it reflects much of the scattered excitation light collected by the objective and sends it back to the light source. This property reduces further the signal to noise ratio for detecting the sample's fluorescence.

Emission Filters

To decrease as much as possible the amount of exciting light reaching the detection device, an additional filter is installed in the optical path: the emission filter. Emission filters are usually LP filters or (infrequently) BP filters. They allow longwave light to pass, therefore filtering out the small amount of excitation light that might have leaked through the dichroic mirror. An ideal emission filter should eliminate any leftover of excitation light while leaving emission light at its original level. In practice some emission light is also lost in the emission filter because of reflection, scattering, absorption, *etc*. Emission filters are sometimes named as barrier filters, as they act as a barrier for any excitation light.

In most modern fluorescence microscopes the excitation filter, the dichroic mirror and the emission (barrier) filter are assembled together in a structure called "optical cube" or just "cube" (see Fig. **6.7B**). These optical cubes display a preset arrangement of filters and dichroic that fits to a specific fluorophores probe or group of probes. They are easily exchanged by rotating or sliding mechanisms incorporated into the fluorescence microscope (see Fig. **6.7C**). In this way it is quick and error-proof to change from one probe to another while assessing the same sample.

1.3. Microscope Objectives

In contrast to their single imaging role in light microscopes, objectives in fluorescent microscopes play a dual role: providing the excitation light to induce sample's fluorescence and gathering/focusing the fluorescent light to produce a useful image. Usually the same objectives employed in light microscopy are employed for fluorescence microscopy (see Fig. **6.8A**).

For specialized applications unusual objectives with customized coatings and/or optical properties are used. In this sense there are objectives for detecting infrared or ultraviolet emission, for studying polarized light or objectives with a variable numerical aperture (see below) to be tuned with different samples. Objectives can be located above the studied sample (epifluorescence) or below it (traverse) as shown in Fig. (**6.8B**). However, most modern fluorescence microscopes commercially available are of the epifluorescence kind.

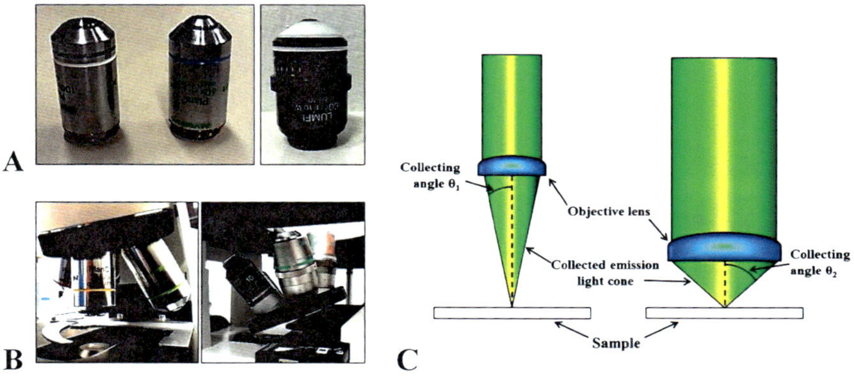

Fig. (6.8). **A**: Some examples of microscope objectives displaying different magnification (100x and 40x on the left, 60x on the right). **B**: Arrangement of microscopic objectives in an epifluorescence microscope (left) and a traverse fluorescence microscope (right). In the epifluorescence microscope the sample is located below the objectives (black platform and silvery grappling mechanism). In a traverse microscope excitation light still comes from above but the fluorescent light is emitted through the sample and collected by the objectives located below it. **C**: Geometrical relationships between the collecting angle and the amount of light collected by an objective. These relate to the numerical aperture (NA) of the objective (see main text for details).

Modern objectives are composed of several optical elements, mainly lenses, which control the way in which light travels through them and finally produce a magnified image of the studied object [14]. Objectives play a pivotal role in microscopy as they are the imaging forming components. Modern objectives are designed to deal with typical optical aberrations, like spherical and chromatic aberrations. Depending on the objective's quality these corrections apply in a narrower or wider light spectrum. Therefore it is important for a microscopist to know and understand the qualities of the objective being used to collect the fluorescence light. For example, because the refractive index changes with λ (chromatic aberration) it is important to know if the sample's fluorescent emission is going to be distorted by the objective by using it outside its working range.

The magnification is the increase in size a microscope produces. It usually ranges from 4x to 100x (Fig. **6.8A**) although special objectives for higher magnification are available (currently up to 250x). In order to inspect a microscopic sample, fluorescent or not, it seems obvious that it would be desirable to have very high magnifications to grasp as much details as possible. So, why is it that the current limit in magnification is around 250x. Why there are no 1,000x or 10,000x magnification objectives? The answer lies in the fundamental limits that light, especially its wavelength, imposes on magnification. If we want to image a very small spot two aspects come into play to limit the imaging process: 1) the smaller the spot the less light will come from it and 2) as the spot dimensions come to the same order of magnitude as the light wavelength we are using to inspect it, light diffraction ruins further magnification attempts. These two limiting aspects will be briefly treated below.

It comes clear that as the imaged spot becomes smaller and smaller the light emitted from it will be less and less. Therefore it is necessary for an objective to gather more and more light to make the spot visible in a meaningful amount of time. So to image very small spots it is unavoidable to collect more light than to image a larger spot (see Fig. **6.8C**). This is where the collecting angle of an objective comes into play. Microscopic limits to imaging takes into account three parameters: the light wavelength employed to image the sample (λ), the refractive index of the medium light traverses (n), and the cone angle over which the objective actually gathers light (θ). These parameters combine to give the theoretical diffraction limit to image a spot at a certain wavelength:

$$d = \lambda/[2n \cdot \sin\theta] \quad \textbf{(1)}$$

where d is the size of the limit-resolvable feature or spot, λ is the light wavelength employed to probe the sample, n is the refractive index of the medium, and sinθ is the sine of the light collecting angle θ. This equation gives the resolution for optical microscopy which relies on visible light to image objects. It is set between 1 μm and ~400 nm for most usual conditions and using air as the transmitting medium ($n_{air} \approx 1$).

The refractive indexes and the angles (part of the denominator in eq. **(1)**) can be combined in one quantity called numerical aperture (NA). The NA is a measure of the light collecting ability of a microscope. It is defined by:

$$NA = n \cdot \sin\theta \quad \textbf{(2)}$$

By combining n and sinθ it makes direct comparison possible between different objectives, no matter the refractive index of the transmitting medium (NA normalizes the collecting cone by the n in each case). Following this an objective with low magnification (imaging larger spots) will show a low NA (for a 10x objective NA≈0.25) while a higher magnification objective will have a larger NA(1.0 and higher for 60x and more).

As the imaged spot becomes smaller and smaller the collecting angle must become larger and larger. This practically means that the objective must be closer and closer to the sample (Fig. **6.8C**). If the transmitting medium is air there is a minimum distance beyond which the collecting angle should be so large that makes objectives impractical. However we can manage the refractive index n to obtain an enhanced magnification while still working with manageable objectives.

This is exactly what is done with immersion objectives where the medium between sample and objective is changed from air to some other higher refractive index fluid medium. The two most employed options are water (n≈1.333) and different types of oils (n≈1.2-1.6). A third option is open to increase magnification further which is to use a shorter light wavelength. In this way microscopic images can be obtained with UV light although dedicated detection devices must be employed because the human eye is blind to these wavelengths. Moreover, UV is dangerous and readily induces undesirable photochemistry in the sample, especially in biological ones. Because of this UV light is rarely employed to increase the magnification and relaying to the other two approaches, increasing collecting angle and medium's refractive index, is the widely strategy commonly employed today. Nevertheless new optical approaches have been developed recently that allow imaging beyond the classical diffraction limited established above. These approaches, known as super-resolution techniques, are revolutionizing fluorescence microscopy and deserve their own chapter to explain their principles, advantages, shortcomings and applications. Therefore they will be treated in detail in Chapter 20.

1.4. Additional Optical Elements

A light microscope is of course composed of more optical elements than those already described. They are important to understand how fluorescence microscopy is possible and to bear in mind when unexpected results or problems happen during investigation. These optical elements will be succinctly explained below.

Imaging Devices

Broadly speaking two kinds of imaging devices are nowadays employed in optical microscopy, including fluorescence microscopy: user's eyes and electronically-driven devices. Obviously the imaging device is the part in the system that receives and processes the optical signal generated by the sample. Here some basics about electronic detection devices will be introduced. The human eye provides a fine way to obtain fluorescence information at the qualitative level but some kind of objective data is required, banning direct human perception as an adequate source of scientific information, at least quantitatively. Also photographic film was the main imaging source in microscopy. However this role has diminished more and more with time as electronic detection and recording devices have been introduced in this research field. Today photographic film use can be said to be no longer in use.

The most widely employed electronic imaging devices are Charge-Coupled Devices (CCD) cameras. This imaging device is reliable, advanced and technology is currently cheap. The working mechanism is based upon a photoactive material, usually a semiconductor like silicon. The imaging system projects an image upon the photoactive material where, thanks to the photoelectric effect, electrical charges (usually electrons) are generated. The amount and distribution of charges therefore stores the image information in a latent pattern. Then arrays of tiny condensers set over the photoactive material move these charges to the sides to amplifiers there. These amplifiers read the signals and the electronic circuits transform the voltages into information that is displayed and/or stored. Working with a line or a surface array will produce the corresponding row or pattern of pixels. In its most basic form the CCD will produce a greyscale image (it measures levels of brightness).

This simple setup can be transformed into a colour palette by interposing a diffraction grating/prism and a Bayer filter between the incoming light and the CCD. The diffraction grating or prism splits the incoming light into the different colours that then project to slightly different areas on the CCD. Before reaching the photoactive material a Bayern mask filters the different primary colours (red-green-blue) so that only light of one specific colour excites a particular pixel. The incoming light still produces charge clouds read as brightness levels but the device, by knowing which colour each specific pixel responds to, can reconstruct the colour image assigning the adequate colour for each pixel. Of course this has been only an introductory glint to CCD technology. As with many other technologies and devices that conform and make fluorescence microscopy possible CCDs are in themselves a huge developing area.

There are some alternatives to the use of CCD technology for adequate imaging, of course. Active pixel sensor (APS) is probably the most encouraging and it is finding intense development and application in the last few years in digital cameras (cell phones, webcams, *etc.*). In a very basic way it can be said that APS are similar to CCDs in layout but each pixel has its own amplifier wired in. Therefore the acquisition time is faster because each pixel has not to wait for those "before" it in the row to be processed by the corresponding amplifier. These features make APS better suited for scalability and for reducing the signal-t--noise ratio.

A still widely employed photodetector is the Photomultiplier Tube or PMT. PMTs are a kind of vacuum tube photodetector. Although not widely employed for fluorescence microscopy anymore, nevertheless they a still play a key role for certain specialized techniques mainly time-resolved fluorescence measurements (see section 3 below). This is because PMTs display very adequate features in regards to time-resolved processes like very good signal response, low noise amplification, very good signal-to-noise ratio, very fast response and adequate spectral range [7, 15]. Additionally they are reliable and endurable, the technology for their use and making being mature for more than 70 years.

PMTs work on the principle of amplified current. The basic PMT setup corresponds to a photocatode, one or (most usually) several dynodes, and an anode. The photocathode is composed of a material that emits an electron after absorbing a photon (photoelectric effect). This initial electron or photoelectron is repulsed from the negatively charged photocathode and attracted to the positively charged dynodes. Dynodes are nothing more than intermediate charged plates positioned between the photocathode and the anode. Their role is to amplify the initial extremely low (usually a single electron) photocurrent. The photoelectron is accelerated in the potential from the photocathode and the first dynode. Therefore it impacts the dynode with a certain amount of energy. This energy results in more electrons being ejected by the dynode. This is a current-amplification step. The electrons are again accelerated in the electric potential towards the nest dynode in the chain and more electrons are ejected during the second collision event. It is clear that by the time the electron stream reaches the anode, it would be composed of thousands or millions of electrons. This is easily registered by the sensor. In essence a one photon-one electron event has been amplified to a level where the signal stands clearly above the noise. Given their working principle PMTs are most suited for time- and spectrally-resolved sensors, making them a much better option to CCDs in regards to time-resolved measurements.

Additional Filters

Apart from the main filtering system already explained previously, fluorescence microscopes have more filters that ensure additional features. On one side there are neutral density (ND) filters that reduce light in more or less the same amount throughout the whole spectrum. In this way light intensity can be adjusted in the interplay between sample adequate illumination and excessive excitation. In contrast to excitation and emission filters, neutral density filters aim at reducing light going through them in an even way to avoid enhancing light output at some spectral region.

Also the microscope has IR/UV safety filters. These filters reduce or totally absorb a range of harmful wavelengths, either because they are in the UV and can produce photochemical burns/sample photobleaching, or because they are in the IR and give rise to unwanted/dangerous temperature increases. IR filters are sometimes named heat filters. These filters provide a "cool" output light, devoid of harmful UV wavelengths below 360 nm.

Optical Fibers

Other optical elements that are becoming more and more common in fluorescence microscopy (especially when lasers are part of the system) are optical fibers. An optical fiber is a medium that acts as a waveguide for light. In this way the optical fiber keeps light from dispersing and provides a path for it so that it reaches its target with most of its initial properties (intensity, polarization, wavelength, *etc.*) unchanged. Optical fibers represent a huge and rapidly developing research area on itself, therefore only a brief introduction will be provided here. For those interested more information can be found in the bibliography [16, 17].

Broadly speaking an optical fiber is a medium that confines light propagation into certain allowed directions in virtue of the medium's refractive index (n). Most often light propagation is permitted only in one dimension therefore the optical fiber is a "tube" or "pipe" for light movement. Light confinement is obtained by tailoring the fiber n parameter and taking advantage of the total internal reflection phenomenon. Total internal reflection takes place when light is moving from a medium with a certain n to another medium with a lower n (see Fig. **6.9A**). For example, when light moves from water ($n \approx 1.333$) into air ($n \approx 1.0$) some light will be reflected at the interface between media. The amount of reflected light depends critically on the angle at which light approaches the interface. The more tilted the incoming angle with respect to the interface normal the more light will be reflected back into the higher n medium. There is a critical reflection angle below which all incoming light is reflected back into the higher index medium (this internal reflection phenomenon is also exploited in a microscopic technique known as total internal reflection fluorescence microscopy; see section 5 below for more details). Optical fibers have a higher n than their surrounding medium. All light injected within a certain approaching cone will be transmitted within the fiber with very low losses (Fig. **6.9B-C**). Some diffraction, fiber impurities acting as scattering centers and some unavoidable optical absorption will extinguish the light in the end. But very high transmission lengths are possible (~ hundreds of kms) with the current technology. These features make optical fibers essential elements of our present information-based civilization.

Nevertheless low-loss transmission is important in fluorescence microscopy use of optical fibers not because the light has to be transmitted large distances but because low loss means higher light source efficiency. At the same time it is very desirable to contain laser light within a non-scattering medium for safety reasons. Additionally, as stated previously when supercontinuum sources where introduced, the role of optical fibers in producing the supercontinuum is primordial in most applications. In summary, optical fibers represent a very adequate mode of making light reach the sample from any of the light sources already presented at the beginning of the chapter.

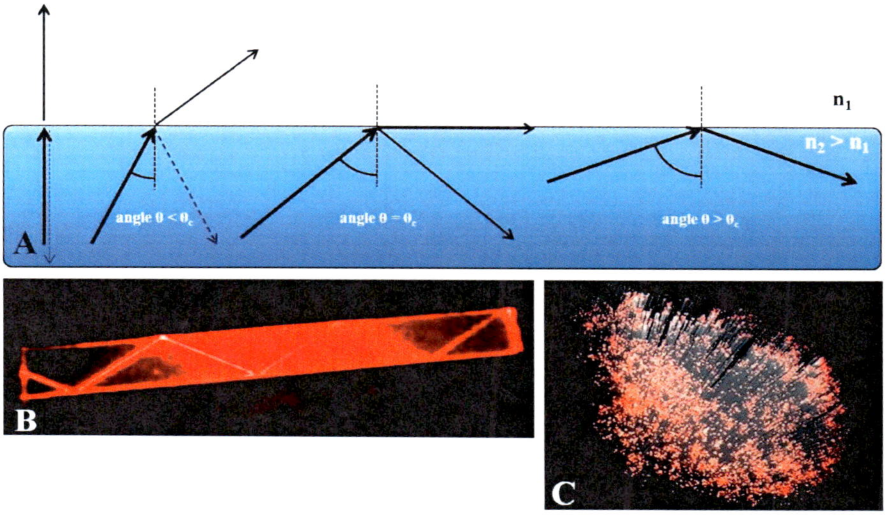

Fig. (6.9). **A**: Fundamentals of the total internal reflection (TIR) phenomenon. Light waves undergo reflection when transiting from a higher refraction index medium (n_2) to a lower refraction index medium (n_1). The amount of reflected light at the interface depends on the wave incidence angle (θ). For a normal incidence ($\theta = 0°$) the reflection is minimal. As the incidence angle increases (light arrives more oblique to interface) more light is reflected and propagates inside the n_2 medium. Above a certain critical angle of incidence (θ_c) all light is reflected, the interface behaving as a perfect mirror. Therefore a light beam is "trapped" inside the (n_2) medium. **B**: Example of TIR inside a plastic block (n_2) surrounded by air (n_1). A red laser beam can be seen reflecting repeatedly at the internal surface of the plastic block (reproduced with permission from Sai2020 - http://commons.wikimedia.org/wiki/File:TIR_in_PMMA.jpg). **C**: TIR allows very low loss light propagation inside optical fibers. No light leaks from the outside of the fibers. Only when the light reaches the fiber end it is able to transit to the n_1 medium, air in this case.

Other Optical Elements

To finish this section about additional optical elements some complementary devices will be introduced. Fluorescence microscopes have shutters that can stop or allow light to reach the sample and/or the imaging device. These shutters can be manual or motor-driven, depending on the speed at which the shutter must act. Shutters are important mainly for two reasons: they are part of the setup that allows for time-resolved fluorescence microscopy (see below), and they are an additional safety mechanisms to stop bright and/or dangerous light from reaching sensible equipment, the sample or the user. As an example a shutter is a mandatory feature in fluorescence microscopes that use discharge lamps to close the light output from the lamp when the lamp holder is manipulated. This avoids unwanted exposure to the harmful UV output from the lamp.

Pinholes are also necessary to collimate light beams and provide smooth illumination. As stated above the pinhole's role is critical for an adequate Köhler illumination of the sample (see light sources). A condenser (also related to Köhler illumination) is a lens coupled to a variable aperture/pinhole between the light source and the sample. By varying the lens focal distance and the pinhole's aperture the light cone reaching the sample can be controlled. The condenser is a piece more related to the bright field microscope but some fluorescence microscopes, especially transmitted light ones, make use of this optical element. Finally, for a useful short guide to fluorescent microscope use and common pitfalls, see the work published by North [18].

2. CONFOCAL MICROSCOPY

Once the most basic components and technologies for fluorescence microscopy have been introduced in the previous sections we go into describing some more advanced schemes that provide additional advantages as compared to the usual setup already presented. The first advanced setup is the confocal fluorescence microscope. Confocal microscopy is based upon the establishment of a focal point at the sample's plane that is conjugate with an excitation point at the light source plane (see Fig. **6.10A**).

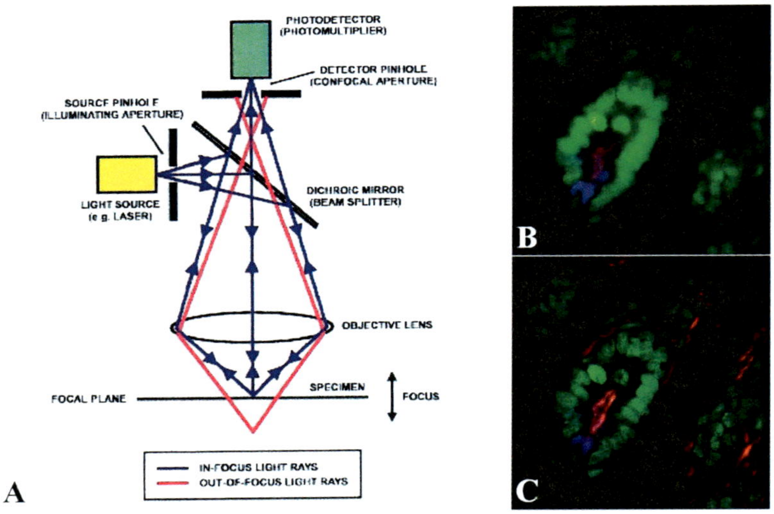

Fig. (6.10). **A**: Basic set up scheme of a confocal optical system. The main difference with a standard optical set up is the use of two pinholes in the light path. One is located after the light source and the other one right before the photodetector. Light coming from above and below the confocal plane is filtered out by this pinhole arrangement. **B-C**: Comparison of the resolution achievable by using a standard optical set up (**B**) and a confocal system (**C**) on the same biological sample (mouse small intestine crypts). Note the decreased image blurring and enhanced signal to noise ratio in the confocal picture (reproduced with permission from Leica Microsystems).

The aim of this arrangement is to illuminate the sample point by point, avoiding out of plane background excitation which gives rise to blurring and spurious noise [19, 20]. This setup greatly improves the contrast and resolution (signal-to-noise ratio) as compared to conventional wide-field fluorescence microscopy (see Fig. **6.10B-C**). Because of the increased resolution confocal microscopes are routinely employed to obtain high quality images from tissue slices too (see Fig. **6.11A-B**) [21, 22]. Nevertheless this increased resolution comes at the price of a much longer imaging time because the sample has to be scanned point by point to take advantage of this technique. In contrast, wide-field imaging illuminates all or most of the sample at the same time, imaging the whole field in a single scan.

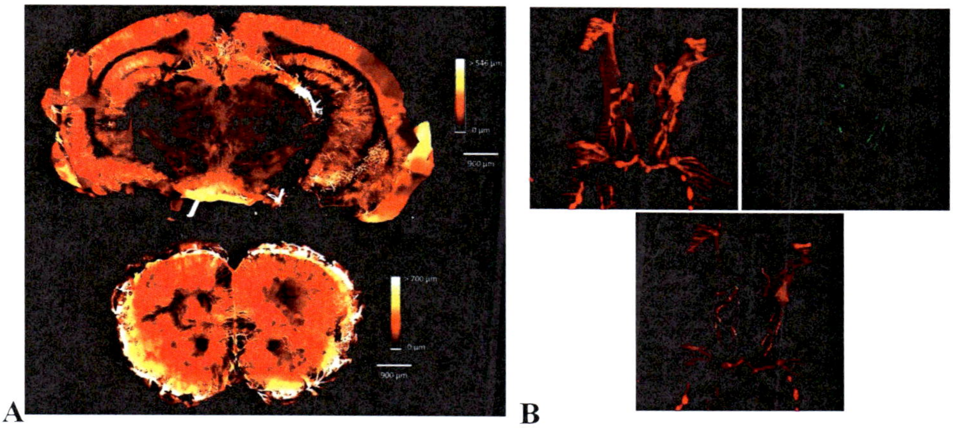

Fig. (6.11). Confocal images of thick biological samples. **A:** Mouse brain microvasculature imaged by a confocal system with scanning volumes of 850 µm (top) and 820 µm (bottom). The endothelial plasma membranes lining the vasculature were stained *in vivo* with the fluorescent lipophilic dye 1,1'-dioctadecyl-3,3,3',3'-tetramethylindocarbocyanine perchlorate (DiI) and then imaged with the confocal microscope. Several tissue stacks were merged using a computer program to obtain the final volume image shown. The whole organ microvasculature can be appreciated in the pictures. Reproduced with permission from [21]. **B:** Muscles (top left) and neuromuscular junctions (top right) in a *Drosophila melanogaster* abdomen as imaged with a confocal microscope. Cell membranes were stained in the muscles and the neuromuscular junctions were labelled through a structural protein in the active presynaptic neuron. Also shown is the merged image (bottom center) obtained from overlapping the previous pictures (images kindly provided by B. López-Arias).

Before the widespread use of lasers in microscopic systems, point sample excitation was obtained with the aid of a very small pinhole after the light source (see Fig. **6.10A**). This pinhole makes excitation light produce a point-like excitation spot after lens focusing given that most of this light is blocked by the aperture. It follows then that only that sample portion confocal with the focused light spot will be excited properly and emit fluorescence light. A second pinhole is positioned before the detection sensor. Only light that comes from the excited spot passes through to the detector (see Fig. **6.10A**). In this way most of the background is unable to reach the sensor and does not contribute with noise. By displacing the excitation spot, either by moving the sample or the excitation light through mirrors, a 2-dimensional scan of the sample is obtained (a fluorescent planar plot). Then the excitation spot depth is changed (again by moving the sample or the excitation light in the z-plane) and another scan is obtained in the x-y plane. Subsequently all of these x-y fluorescent planes at different z-heights can be stacked one upon each other to obtain a 3D image or reconstruction.

A definitive breakthrough came with the implementation of lasers as the confocal excitation source. Lasers offer the possibility to reduce the excitation spot to the (theoretically) diffraction limit (see section 1.3 above). This is because lasers are coherent light sources that can be strongly focused. This has the consequence that a better resolution is obtained as compared to other excitation light sources. At the same time the light intensity provided by lasers allows for a shorter scanning time as each sample spot needs to be excited for less time to provide the same amount of signal. Therefore laser confocal microscopy has more resolution, higher contrast and takes less time than other confocal approaches [23].

However lasers and the supporting equipment for confocal laser microscopy are expensive and require adequate maintenance. Especially light beam alignment and focusing must be checked regularly to assure the best performance. Not uncommon is also the possibility of sample photobleaching due to the very high irradiance of the focused laser. A compromise must be found between scanning times and sample bleaching all the most if the sample is to be conserved for later analysis.

The extremely high light intensities achievable with lasers open the possibility for non-linear optics and the so-called two- (multi-)photon confocal microscopy. This approach is very interesting as the sample excitation mechanism, by its multi-photonic nature, gives rise to a spontaneous confocal arrangement through a "virtual pinhole", without having to use a real pinhole neither for the light source nor for the optical detector. However, to fully appreciate this multi-photon confocal system, the particular features of non-linear optics must be introduced first. Given the numerous current applications of such non-linear optics, a full chapter will be dedicated to such issue. Therefore we derive the interested reader to Chapter 19 for a deeper explanation of this and many other non-linear processes. Also further increases in resolution, provided by super-resolution techniques, quite related to the confocal laser will be treated in Chapter 20.

Other confocal approaches put the stress on image acquisition speed instead of resolution. These approaches make use of a pinhole pattern that makes imaging in parallel possible. Therefore excitation light dwell time at each sample point is reduced and undesirable light-induced effects lessened. These approaches are especially well suited for live samples imaging where phototoxicity or sample heating are to be avoided. Below follows a description of the most notorious multi-pinhole confocal setups.

In the spinning disc confocal microscope an especially constructed disc with holes (Nipkow disc) regularly patterned on its surface is positioned between the exciting light and the sample [19]. During image acquisition the disc spins at a certain speed letting light through it only at discrete points one after another. In this way sample excitation is accomplished also at certain points right after the holes in a determined sequence. Each disc hole acts as the primary pinhole. Before the photoreceptor there is the second fixed pinhole that constrains light getting through. As already stated the spinning disc confocal has the advantage that scanning times at each particular sample point are much shorter than with classical confocal or laser confocal. This reduces adverse light-related effects. This approach shortcoming is that resolution is much lower than with laser confocal, for example, so that one trades resolution for imaging speed.

The dual spinning disk confocal system works on the same principle as the single spinning disc just introduced but a second disc is positioned before the pinhole disc. This secondary disc has micro lenses at the same positions as the pinholes in the primary disc. With the two discs spinning synchronized each micro lens focuses exciting light into each pinhole. This increases the amount of light reaching the sample and, as a consequence, it also increases the signal-to-noise ratio and the resolution. This partially resolves the setup shortcoming explained above for the single spinning disc approach [19].

Finally another approach to confocal systems is introduced with the programmable array microscope or PAM. This setup takes advantage of spatial light modulators (SLMs) which are screens that modulate some light property as it passes through. The most usual modulated feature is light transmission (brightness, colour, reflectivity, *etc.*) although phase or polarization can be controlled too. The advantage is that the SLM is made up of many pixels each one modulating the corresponding light property and being themselves controlled individually through an electronic circuit. This provides the basis for a very efficient tuning of the transmitted light with very fast responding times. It is easy to visualize that a SLM provides an outstanding platform, a programmable array, to "open" and "close" pinholes by controlling the device transmission, for example. In this way it acts in a similar way to the spinning disc confocal explained above but without having to introduce a mechanical setup, faster response times and much more system flexibility.

3. TIME-RESOLVED FLUORESCENCE MICROSCOPY

Most of the techniques and phenomena discussed so far deal with measuring the sample's fluorescence signal intensity, disregarding its time profile (recording the average emission during a certain time period). But indeed, fluorophores emit light with a certain time profile that provides much information when questioned adequately. Therefore time-resolved fluorescence techniques address the emission profile in relation to time and the fluorophore's environment. Time-resolved spectroscopy principles will be presented followed by specific fluorescence microscopy techniques that rely on those principles, specifically Fluorescence Lifetime Imaging Microscopy or FLIM [7, 24].

As was introduced in Chapter 2.4 each fluorophore spends a certain amount of time in the excited state before undergoing photon emission or relaxing by other mechanisms. This excited state time is known as the lifetime. The lifetime has a meaning as a statistical value. The amount of time a fluorophore spends in the excited state is random and depends of many factors (*e.g.* interactions with other molecules and/or solvent, *etc.*). It cannot be predicted exactly when a particular fluorophore will emit a photon. But when a large ensemble of fluorophores in a given environment is interrogated, a trend in the average time it takes for them to emit appears. This is the lifetime. One key aspect in the previous statement is the given environment. A fluorophore's lifetime depends on its environment. A plethora of factors modulate the lifetime, like phase (gas, liquid, solid), temperature, polarity, polarizability, viscosity, chemical reactivity, to name a few. Therefore, by measuring the lifetime a fluorophore informs in an enhanced way in respect to its environment as compared to just sampling the emission intensity [7, 15]. This is a clear advantage of time-resolved methods.

Time-resolved techniques and FLIM in particular, can be broadly divided in two approaches attending to the methodology employed to obtain the lifetime: Time-domain resolved (TD) and Frequency-domain resolved (FD) [15]. The fundamentals of each technique will be presented in what follows along with some examples applied to the life sciences. FLIM offers many advantages compared to other fluorescence microscopy techniques because of the way it obtains information from the sample. In this regard it can detect many fast molecular/cellular processes and is suitable for tissue/organ 3D imaging. It does not lead to the important photobleaching characteristic of confocal systems, it can interrogate several fluorophores at the same time because it can discern their different emission (spectral and lifetime) profiles, and it is a very adequate method to deal with the sample's autofluorescence noise. FLIM is most usually accomplished with PMTs because of the much better temporal response of these photodetectors as compared to CCDs.

3.1. Time-domain Resolved Fluorescence

Time-domain (TD) resolved fluorescence makes use of the most intuitive approach to fluorescent lifetime measurement: pulsed sample excitation. Basically the sample is subjected to the output from a short-pulse light source (laser, flashlamp, *etc.*) and its temporal emission profile is recorded (see Fig. **6.12A**). The corresponding photodetector does not record during the excitation pulse. It only starts measurements after the light pulse is over. Therefore this approach inherently avoids positive negatives induced by the excitation source [25].

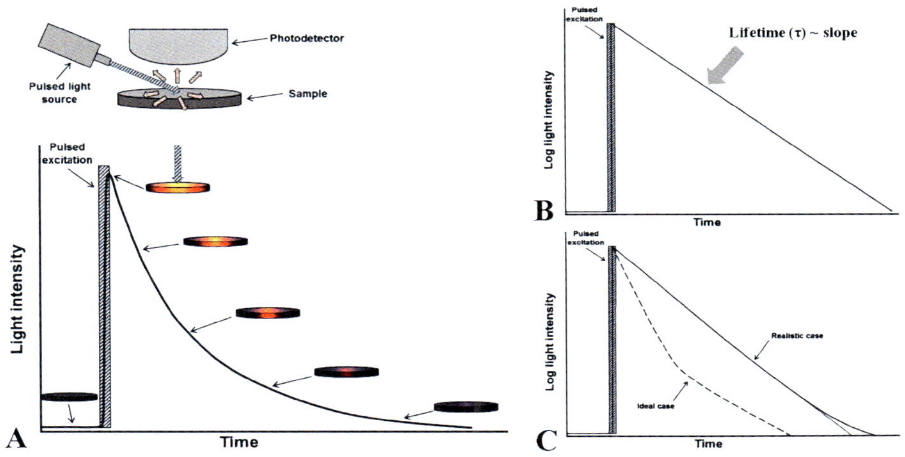

Fig. (6.12). Working principle behind time-domain FLIM. **A**: The technique is based on using a pulsed light source to excite a sample and then register the light emission decay with time. Before excitation the sample (disc) does not emit light. The sample is then exposed to a brief (as compared to the emitting excited species lifetime) light pulse (blue lines/arrow). Afterwards the photodetector registers the amount of light emitted by the sample with time. The emission gets dimmer with time and takes the form of the curve shown in the scheme. **B**: Changing the light intensity scale from linear (as shown in A) to logarithmic transforms the decay curve into a line. The slope of this line provides the lifetime of the emitting species. **C**: Decay plots when two emitting species with different lifetimes are involved in the light emission. In an ideal case (dashed line) the two species greatly differ in their lifetimes and the line shows two slopes. Under more realistic situations, both lifetimes will be quite similar and the slopes will be almost undiscernible except at very long times (small inflection near the right bottom part of the line).

In TD fluorescence the technique takes advantage of the finite time it takes for a fluorophore to emit a photon after it is pumped to an excited state. Therefore some time will pass between photon absorption ($\approx 10^{-15}$ - 10^{-14} s) and photon emission. As stated previously this process is statistical and a lifetime can only be assigned to a certain ensemble of fluorophores in a certain environment. Because the fluorescent lifetime is directly related to the fluorophore's environment, TD measurements provide information about this environment. For example, many fluorescent organic molecules reach a strained arrangement immediately after being excited (Franck-Condon principle, see Chapter 2.4). In a low viscosity environment (*e.g.* water) the molecule rearranges itself (nuclei readjust their distances, angles and vibrational frequencies) before it emits a photon. In many instances this molecular rearrangement precludes fluorescence at all, the excitation energy being lost as heat to the medium. In more viscous environments (*e.g.* glycerol) the same molecule will find itself much more constrained and molecular rearrangement is delayed. As a consequence the time for rearrangement and the time for fluorescence compete with each other and some molecules will emit light before there is an internal conversion. Many other factors affect lifetime apart from viscosity, like temperature, pressure, pH, dielectric constant, polarizability, co-solute interactions (both physical and chemical), *etc.*

In TD measurements a suitable pulsed light source excites the sample in a time shorter than the expected lifetime. From a kinetic point of view the amount of light emitted after pulsed excitation is given by equation **(3)**:

$$I(t) = I_0 \exp(-t/\tau) \quad \textbf{(3)}$$

where I(t) is the emission intensity after a certain amount of time t, I_0 is the initial light emission intensity after the pulsed excitation and τ is the fluorophore's lifetime. The fluorescence lifetime can be obtained from the slope of the log I(t) *vs.* t. This equation defines a single exponential decay (see Fig. **6.12B** for an ideal situation) which plots as a line instead of a curve (because of the logarithmic representation). However the most typical situation encountered in biological fluorescence studies and microscopy is that in which several decay kinetics are involved (same fluorophore in different environments and/or several different fluorophores being measured at the same time). Therefore TD measurements are more often than not multi-exponential (see Fig. **6.12C**) and require complex data processing to obtain the different τ involved in this situations [7]. Note in the figure how the logarithm of the lifetime is no longer a line but the overlap of two or more lines.

Some practical examples showing the use of TD-resolved fluorescence in cultured cells are displayed in Fig. (**6.13**). These examples take advantage of the autofluorescence signal from NADH (see Chapter 7.3). In this way no external label is necessary to assess cells. The first example regards the changes in NADH/NADPH fluorescence to measure changes in the metabolic state of the cell [26]. The fluorescence lifetime curves are displayed first (see Fig. **6.13A**). Then the comparison between control cultures and experimental conditions on fluorescence lifetime is shown in Fig. (**6.13B**). The second example assesses changes in the intracellular pH through differences of the NADH fluorescence due to those pH changes (see Fig. **6.13C**) [27]. From these examples it is clear that time-resolved analysis offers many advantages as compared to average-intensity fluorescence microscopy. TD approaches can work within a single spectral region (single channel) or at different wavelengths (multi-channel). This depends more on the photodetector employed. Working with multi-channel offers the opportunity to measure several different fluorophores and assess each one's environment.

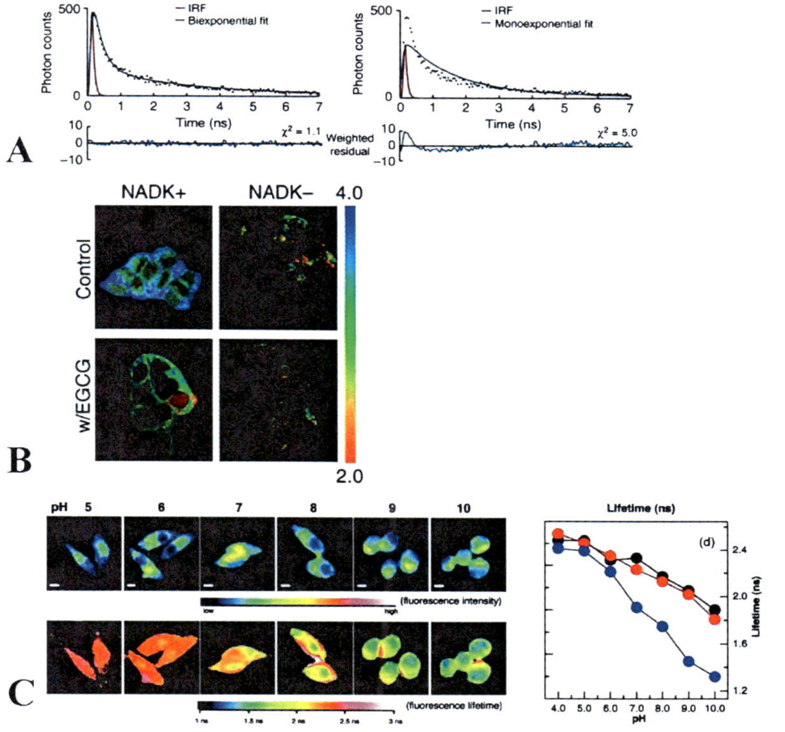

Fig. (6.13). Free *vs.* protein-bound NAD(P)H autofluorescence FLIM-TD measurements in live cells. A bi-exponential fit (left), with a short free lifetime and a longer bound lifetime, explains much better the light decay than a mono-exponential one (right) assuming a single emitting lifetime. Statistically this is demonstrated by checking the error with each assumption through a χ^2 test (shown below plots). The closer to 1the χ^2 value the better fit. IRF is the response time of the measuring device. Reproduced with permission from [26]. **B**: Measurement of the NAD(P)H lifetimes (false colour, right scale in ns) in cells overexpressing the NAD kinase (NADK+) *vs.* wild type (NADK-). The incubation with a selective inhibitor of NADPH binding (epigallocatechin gallate –EGCG-) results in a decrease in the lifetime (as NADPH is free in a greater proportion). This is particularly well appreciated in the NADK+ cells (almost no blue colour in the cells shown in lower left panel). Reproduced with permission from [26]. **C**: Effect of pH variation on NADH autofluorescence in living cells. The upper row shows the autofluorescence intensity and the lower one the autofluorescence lifetime as a false colour scale (same cells for both rows). Lifetimes are longer under acidic (low pH) conditions as compared to alkaline ones(high pH). The curves to the right show the lifetime *vs.* pH relationships at the whole cell (black), the mitochondria (red) and the cell nuclei (blue). Scale bar: 10 μm. Reproduced with permission from [27]. Copyright (2011) American Chemical Society.

TD studies are especially well-suited for long-lifetime probes. These include transition metal luminescent complexes and upconversion approaches (see Chapter 9 and 19) and luminescent solid-state markers (see Chapter 18) that offer the possibility to tailor the emitter lifetime by tightly regulating the emission environment.

3.2. Frequency-domain Resolved Fluorescence

Frequency-domain (FD) time-resolved fluorescence employs a different approach to obtain the fluorescence lifetime (τ). Instead of switching on and off the excitation light source, FD modulates the light intensity at a certain frequency [28]. The exciting light does not necessarily have to disappear during the excitation, merely it has to increase and reduce in brightness (see Fig. **6.14A**). The key point is that this light modulation has to be roughly on the same order of magnitude as the lifetime to be measured. Let's assume a typical fluorescence lifetime of 10 ns (10^{-8} s). In order to measure this lifetime with the FD approach the light has to be modulated around 10^8 Hz or 100 MHz, as the frequency is the inverse of time (f = 1/t).

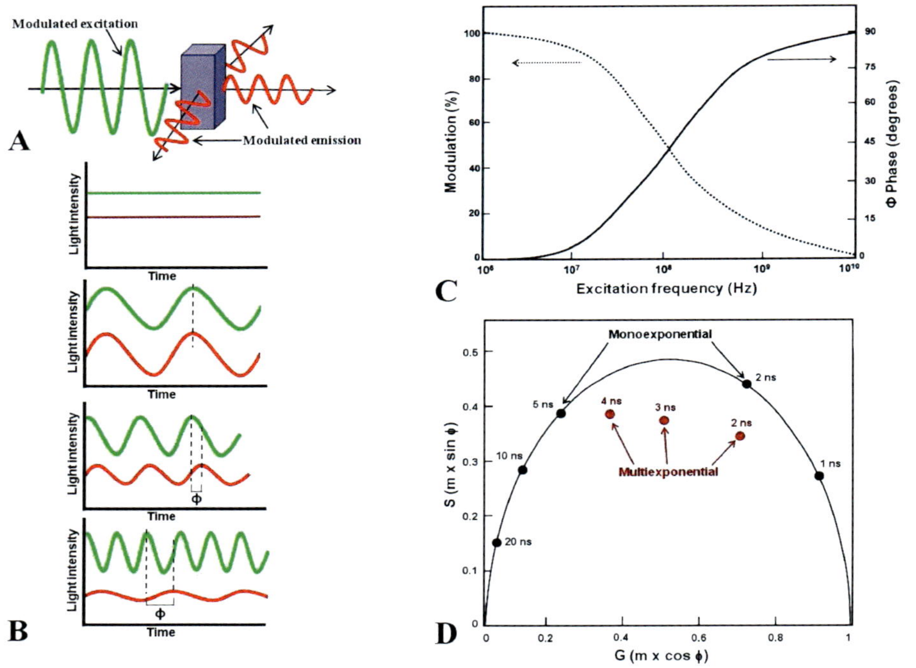

Fig. (6.14). **A**: Working principle behind frequency-domain FLIM. A modulated excitation light source (green light) intermittently illuminates the sample (cuvette). The fluorophore excited in this way emits light in a time-modulated way too (red light). **B**: Relationship between the excitation light modulation frequency and the emission light frequency. For a non-modulated excitation (upper graph) the emission is also continuous. The higher the excitation modulation the more is the emission light affected (sequence of lower graphs). This is seen primarily in two features: the emission light becomes dimmer (modulation) and there is a delay or "slip" (phase Φ) for the emission as compared to the excitation (compare the peaks and valleys for both excitation and emission light). Both changes provide information about the probe lifetime. **C**: Example of the expected FD-FLIM plot for a fluorophore with a nanosecond excited lifetime (10^{-8}-10^{-9} s). For low excitation frequencies ($<10^7$ Hz) the emission follows the excitation modulation (no intensity dimming –dotted line for left axis- and there is almost no dephasing-solid line for right axis-). Between 10^7 and 10^9 Hz there is a strong response by the fluorophore. For example, at a modulation frequency of 10^8 Hz the emission intensity drops to roughly half its initial value and the emission takes place with a diphase of $\Phi \approx 45°$. For 10^{10} Hz or higher the emission is almost lost and the diphase is $\Phi = 90°$. **D**: Phasor plot obtained by plotting the S and G parameters (see main text for their definition). When the emission kinetics follows a mono-exponential trend the function appears as a semicircle (solid line) on top of which the lifetimes can be found (black dots). When the kinetics are multi-exponential the lifetimes map below the solid line (red dots) in places that depend on the specific features of the probed system.

To understand how the modulated exciting light can provide a measure of the lifetime let's discuss what happens to an ensemble of fluorophores as the excitation light is modulated. When there is no modulation (steady light brightness) the fluorescence level is also steady (see Fig. **6.14B**). If the brightness level is modulated at increasing frequencies, but still slow enough as compared to the fluorophore emission lifetime, the fluorescence follows perfectly the ups and downs in excitation by emitting more or less light itself.

However as the modulated excitation approaches a time interval close to the fluorophore's lifetime (≈ modulation time around 10 times longer than τ) the molecules start to "lag" behind in their emission response (see Fig. **6.14B**). This is because it takes some finite time for the fluorophores to relax and reemit light. Two things happen at this point: 1) the intensity of the fluorescence light decreases (some molecules do not emit before the exciting light increases again), and 2) the fluorescence light is dephased (delayed) in respect to the exciting light. By measuring either the decrease in emission intensity or the dephasing in emission time the lifetime can be obtained (equations **(4)** and **(5)**):

$$\tan \phi(\omega) = \omega \cdot \tau \quad \textbf{(4)}$$

$$m(\omega) = 1 / \sqrt{(1 + \omega^2 \cdot \tau^2)} \quad \textbf{(5)}$$

where ω refers to the modulation frequency (Hz), ϕ is the dephasing angle, m is the ratio between the emission when there is no modulation in the excitation ω_0 (ω = 0 Hz) and the emission at the modulation frequency ω ($\omega > 0$ Hz), and τ is the fluorescence lifetime. The angle ϕ is not a geometrical angle but the measure of dephase between excitation and emission. If there is no delay between them (ω is much lower than $1 / \tau$) then $\phi \approx 0°$. As ω approaches $1 / \tau$, the delay increases leading to $0° < \phi < 90°$. Finally for $\omega > 1 / \tau$ the situation $\phi \approx 90°$ is reached (see Fig. **6.14C**).

Not only is the emission delayed in respect to the excitation, also the intensity of emission decreases as ω approaches $1 / \tau$. This is measured as a ratio m between the emission at the modulation frequency ω divided by the emission when there is no modulation (ω = 0 Hz or $\omega < 1 / \tau$) (Fig. **6.14C**). To determine τ the sample is scanned at different ω to obtain data. From the values of both ϕ and m as a function of ω, the lifetime τ can be obtained applying equations **(4)** and **(5)**. As τ varies depending on the particular fluorophore's environment FD provides a flexible tool to assess a sample simply by varying the excitation frequency. Nevertheless it is very common to obtain multiple signals from the different fluorophores/environments which require complex data analysis to isolate each fluorophore contribution [7].

Another way of interpreting time-resolved data, especially for FD-FLIM, is to transform the results to obtain a phasor plot. Knowing the modulation value $m(\omega)$ and the dephasing factor $\phi(\omega)$ at a certain frequency (ω), the following transformation is calculated:

$$G = m(\omega) \times \cos \phi(\omega) \quad \textbf{(6)}$$

$$S = m(\omega) \times \sin \phi(\omega) \quad \textbf{(7)}$$

The G values are usually represented on the x-axis and the S values on the y-axis [25, 28]. The typical phasor plot is displayed in Fig. (**6.14D**). Lifetimes can plot on the elliptical line or within the space between this line and the x-axis. Longer lifetimes appear closer to the y-axis. Those lifetimes on top of the line represent monoexponential decays, while those outside the line represent multiexponential ones.

An example of the use of FD-FLIM is shown in (see Fig. **6.15A**) [29]. In this particular case yeast cells –*Saccharomyces cerevisiae*- express fluorescent proteins which emit in the same spectral region (isospectral). However, the fluorescent proteins variants displayed different lifetimes. Therefore it is possible to detect cells expressing the proteins attending not to their spectral profile but to their emission lifetime. Furthermore, it was shown that a flow cytometer capable of working on the FLIM principle can sort very specifically cells expressing those isospectral proteins.

An example of the use of phasor plots for cellular analysis is shown in Fig. (**6.15B**) [28]. In this case the effect of FRET on a fluorescent Src kinase sensor protein is measured in MLO-Y4 osteocyte-like cells. When Src is inactive the sensor protein lifetime is shorter. After incubation with epidermal growth factor (EGF) for 10 min Src gets activated which reflects on a longer sensor fluorescence lifetime. This effect can be seen in the phasor plot as the average lifetime before and after exposure to EGF maps at different points in the plot. The false colour cell images serve as a guide to the eye. It is important to note that the colour code does not translate into brightness levels but different lifetimes (cold colours mean shorter lifetimes and warm colours longer ones).

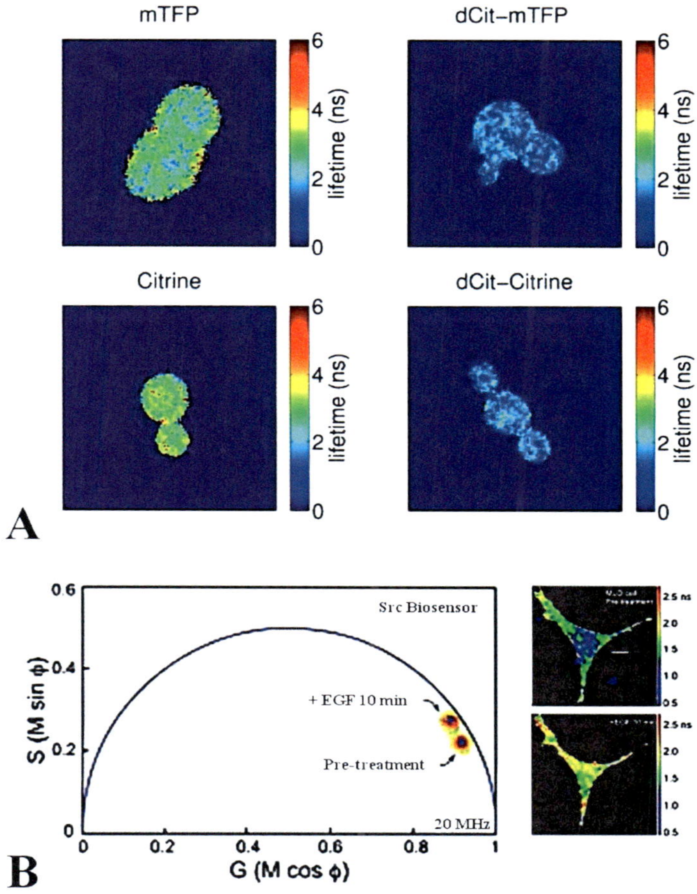

Fig. (6.15). Practical examples of FD-FLIM in biological samples. **A**: False colour images displaying the lifetimes of different fluorescent protein (shown on the left column) expressed by yeast cells (*S. cerevisiae*). They have different emission spectra (citrine *vs.* TFP). Additionally fusion proteins with a dCit (dark Citrine) domain were expressed which quenches the fluorescence and shortens the lifetime while leaving the emission spectra unchanged (shown on the right column). It can be clearly seen in the images that the dCit proteins show a shorter lifetime than their unquenched pair (more green and yellow false colour in the unquenched proteins *vs.* more blue in the quenched ones). Reproduced with permission from [29]. **B**: MLO-Y4 cell expressing a fluorescent Src kinase sensor show a different lifetime if Src is active or not. Before incubation with EGF (a known Src activator) the sensor lifetime is short on average (Pre-treatment spot on phasor plot and upper false colour cell image). After exposure to EGF, Src activates and the sensor fluorescence lifetime increases (+EGF 10 min on plot and lower cell image). False colour levels translate into lifetimes not to brightness under this approach. Reproduced with permission from [28].

As excitation light sources for FD diode lasers and LEDs are progressively substituting pulsed lasers and flash lamps, as they are cheaper and quite reliable with very long working lifetime. Their output can be directly modulated through the driving electrical current which bypasses problems with light modulators previously used.

Time-resolved techniques are increasingly being employed to study more and more biological aspects. Some examples are Ca^{2+}/ Na^{+}/ Cl^{-} levels, pH, pO_2, pCO_2, ROS levels, time-resolved FRET, fast photobiological processes (photosynthesis, vision). These techniques can use autofluorescence signals from endogenous fluorophores and turn them from a shortcoming into an information source. Different autofluorescence lifetimes are expected from different environments which provide critical information about cellular processes without having to perturb the system with exogenous probes.

4. LIGHT SHEET FLUORESCENCE MICROSCOPY

Light sheet fluorescence microscopy (LSFM or LSM) relies on illuminating the sample with a very thin plane of light and observing the fluorescent emission at an orthogonal plane [1, 30 - 32]. This setup can be conceived as an intermediate step between wide field fluorescence microscopy and confocal microscopy (see Fig. **6.16A**). To obtain very thin light sheets (≈ μm width) lasers are most usually employed taking advantage of their diffraction-limited focusing ability. The laser beam is focused into the sample through a cylindrical lens which produces the light sheet. The sampling objective is positioned at a 90° angle in respect to the exciting light. In this way a large sample area is scanned at the same time. Depending on the sample's thickness or the detail with which it wants to be studied the sample is moved towards/away the observing objective to sample different depths, obtaining "slices" of the material. A variant rotates the sample so that the light sheet impinges at slightly different angles but also exciting different slices.

Several variants exist in regards to LSFM mainly attending to different setup configurations (see common examples in Fig. **6.16B**). In orthogonal-plane fluorescence optical sectioning (OPFOS) the light sheet is generated exclusively by a cylindrical lens. A further refinement is achieved with selective plane illumination microscopy (SPIM) in which a focusing microscopic objective is placed between the cylindrical lens and the sample thereby producing microscopic images. In digital scanned laser light sheet microscopy (SLM or DSLM) the sample is held in place while the probing laser light scans different planes by use of a mirror [33]. By combining two or more excitation light sheets over the same sample a multidirectional selective plane illumination microscopy (mSPIM) is obtained which increases the images detail and helps to avoid the stripe artefact (see below). In multi-field of view selective plane illumination microscopy (MoViSPIM) the sample is excited by at least two light sheets and is monitored by at least two sensors.

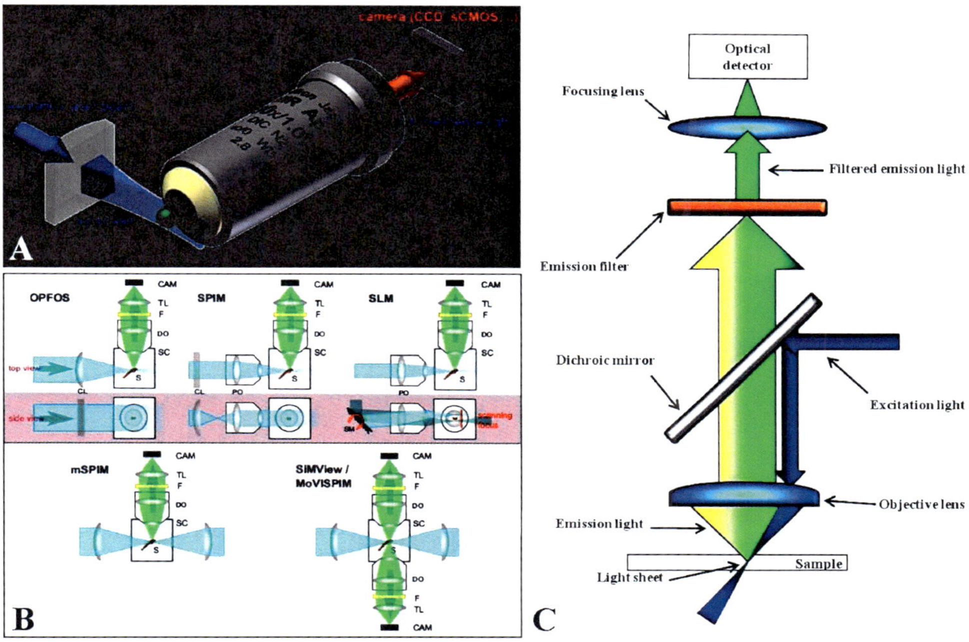

Fig. (6.16). **A**: Basic scheme of the light sheet fluorescence microscope (LSFM). The exciting laser beam (blue) is focused into a line on the sample thanks to a cylindrical lens. This produces a plane of light inside the sample from where emission light (red) is collected through an objective and processed (reproduced with permission from http://commons.wikimedia.org/wiki/File: Spim_prinziple_en.svg#). **B**: Schematics of several approaches to LSFM, depending on the number of exciting lasers and the arrangement of the optical elements. Approaches: OPFOS: orthogonal-plane fluorescence optical sectioning, SPIM: selective plane illumination microscopy, SLM: scanned laser light sheet microscopy, mSPIM: multidirectional selective plane illumination microscopy, MoViSPIM: multi-field of view selective plane illumination microscopy. Optical elements: Cl- cylindrical lens, PO- projection objective, S- sample, SC- sample chamber, DO- detection objective, F- emission filter, TL- tube lens, CAM- camera, SM- scanning mirror (reproduced with permission from http://commons.wikimedia.org/wiki/File:Lsfm_methods.svg). **C**: Scheme of oblique plane microscopy (OPM) setup, a LSFM variant. The excitation beam is focused into the sample through an oblique trajectory at the objective lens rim. The light sheet excites obliquely and the emission light, being isotropic, is collected from a much wider angle by the lens and directed to the imaging optical elements.

Additionally the oblique plane microscopy (OPM) approach has been developed in which the same microscopic objective is employed to focus the light sheet on the sample and then to collect the fluorescence light emitted [34]. This is accomplished by directing the excitation light through the objective lens side, in contrast to the usual LSFM in which the whole lens is employed to focus light (see Fig. **6.16C**). Afterwards fluorescent light is emitted at all angles from the sample and part of it is collected back by the objective (this time using the entire lens surface). Therefore it is possible to obtain an image for this excitation scheme. Including a dichroic mirror improves the signal-to-noise ratio.

One of the most common shortcomings of these techniques is the so-called stripe artefact. This happens because the observation point is now at some angle (usually 90°) with respect to the excitation light sheet. Therefore any sample feature that might affect light propagation along this sheet (diffraction, absorption) will result in some kind of optical aberration in the form of lines or stripes. DSLM deals with this problem by quickly changing the excitation angle so that different "point of views" are generated which minimize the stripes. Of course computer programs can be employed to recognize these stripes by comparing images from consecutive planes, correcting for them.

An example of an internal detailed structure obtained through LSFM is shown in Fig. (**6.17**) where the nuclear distribution of protein HP1α –a protein involved in heterochromatin stabilization- is displayed at different planes [35]. The detailed HP1α distribution highlights heterochromatin domains *vs.* euchromatin in a living cell, showing, for example, that heterochromatin is mainly localized close to the nuclear membrane while actively transcripting euchromatin organizes in the nucleus core region. LSFM provides very detailed images of cells in culture. These images can probe both the inside of a cell and their surfaces structure [36]. For example, the filopodia from a living HeLa cell have been imaged with great detail during time intervals of minutes using LSFM.

LSFM is routinely employed to study thicker biological samples like whole living embryos (see Fig. **6.18A**) [37, 38] or cellular spheroids composed of hundreds or thousands of cells (see Fig. **6.18B**) [39]. This technique allows for minimally invasive survey while producing high quality data.

Other biological samples, like small organisms (*Caenorhabditis elegans* in Fig. **6.19A**) [40] or vegetal tissues (*Arabidopsis thaliana* in Fig. **6.19B**) [41], have been successfully imaged *in vivo* using LSFM. Additionally LSFM has been used to image dynamic internal processes in the cell like cytoskeletal structure and rearrangements or mitosis/mitotic structures temporal evolution. Not only direct cellular processes can be imaged also cell-cell and cell-extracellular matrix interactions have been successfully captured using LSFM techniques [42]. These examples provide an idea of the impressive results LSFM can afford through a minimally invasive approach.

LSFM can be combined with other advanced techniques, with non-linear techniques and super-resolution standing out in their performances to capture details at a very high resolution spatial level. These techniques deserve their own deep explanation for fundamentals and applications and therefore will be treated in details in Chapters 19 (Non-linear optics) and 20 (Super-resolution).

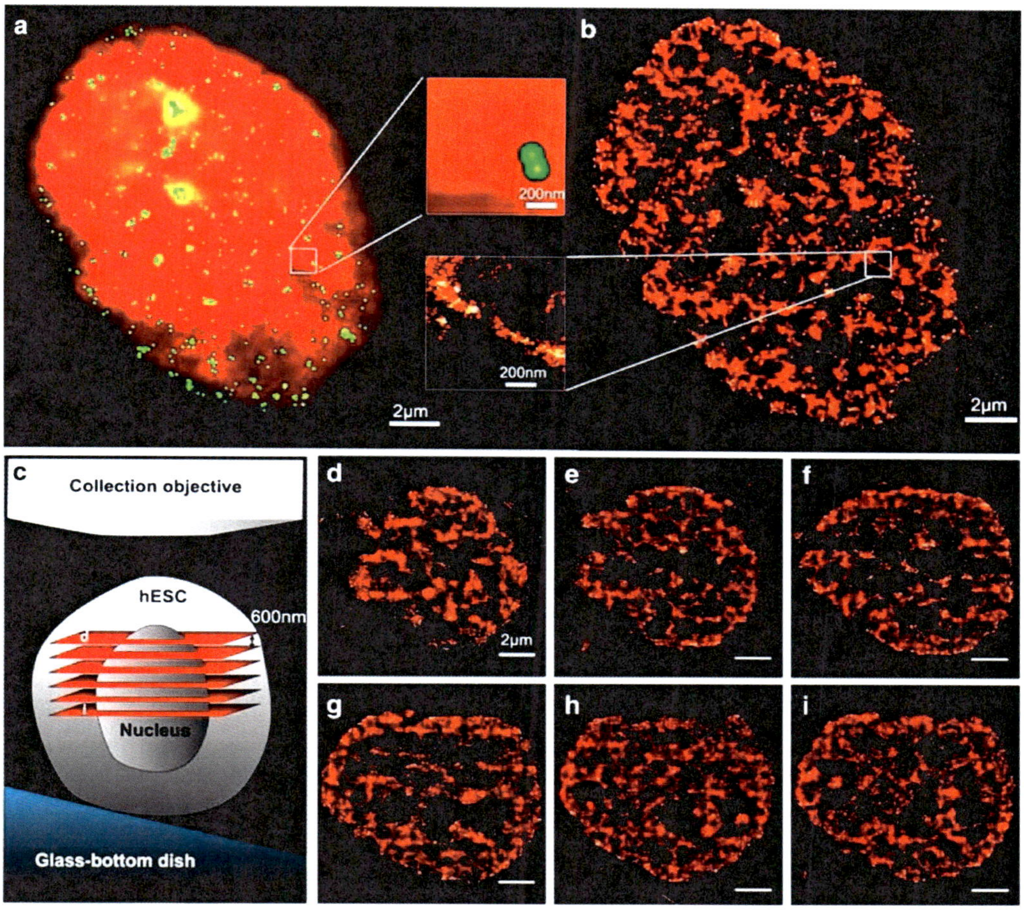

Fig. (6.17). Human stem cell heterochromatin images obtained by LSFM. Panels a and b (top) show the spatial distribution of heterochromatin-associated fluorescent-labelled protein HP1α. Panel a shows a conventional fluorescent microscope picture (blurred red-orange background) merged with a superresolution image (green dots obtained by PALM) which only acquires regions with a high concentration of HP1α. Panel b shows the same nucleus and reveals a much more detailed heterochromatin structure. Note the differences in resolution from the same sub-micrometric area using the different techniques (magnified squares). Panel c is a schematic of the sampling setup for the following panels. Panels d-i show LSFM ultrathin planes on the same nucleus which allow 3D reconstruction of the heterochromatin spatial arrangement. Reproduced with permission from [35].

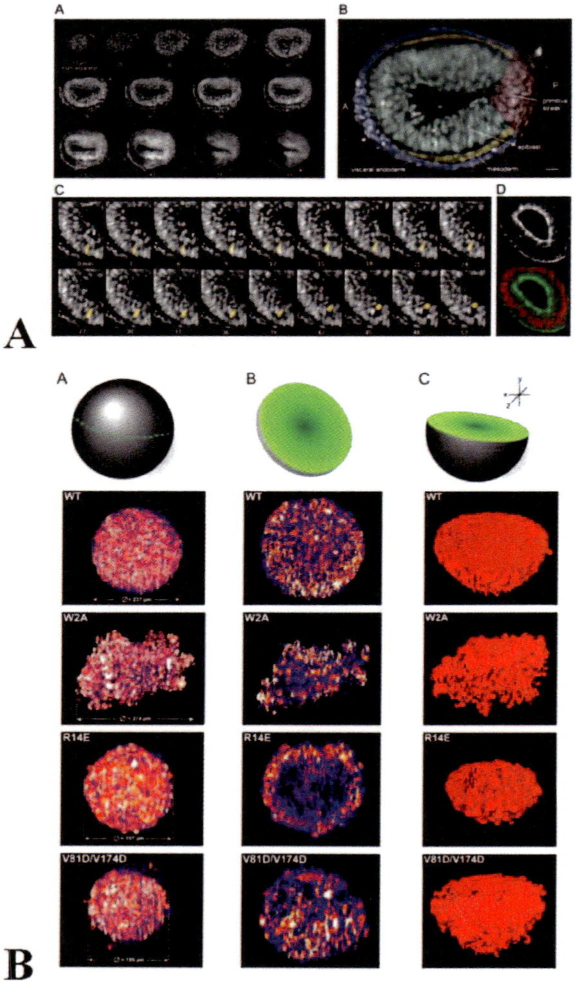

Fig. (6.18). A: Live imaging of a mouse embryo on day 6.5. Panel A shows Histone H2B was tagged with a GFP construction and images at different planes in the proximal-distal axis have been obtained through LSFM (numbers indicate plane, scale bar 20 µm). Panel B is a magnified image of plane 78 in panel A. Different embryonic tissues and structures are highlighted (A is anterior pole and P posterior pole, scale bar 20 µm). Panel C is a time series (time in min) where cell divisions (blue arrows) are observed and a migrating cell (yellow coloured) moving from the outside to the inside is clearly displayed. Panel D shows membrane-associated actin (top image) and both the actin-associated actin and a nuclear marker (bottom) in an embryo section revealing a high level of detail. Reproduced with permission from [38]. **B**: Impact of N-cadherin function on the spatial organization of cell spheroids. Fluorescently labelled N-cadherin was imaged with LSFM to obtain 3D images of wild type (WT) and N-cadherin mutant lines (W2A, R14E and V81D/V174D) cell spheroids. Column A shows the outer appearance of the spheroids. Column B shows a thin plane through the spheroids. Column C depicts reconstruction of half a spheroid for each condition. Notorious differences can be appreciated between the WT line and the mutants (especially W2A) in regards to spatial organization due to malfunctioning cell junctions (*e.g.*, W2A lacks a spherical shape and R14E is hollow). Reproduced with permission from [39].

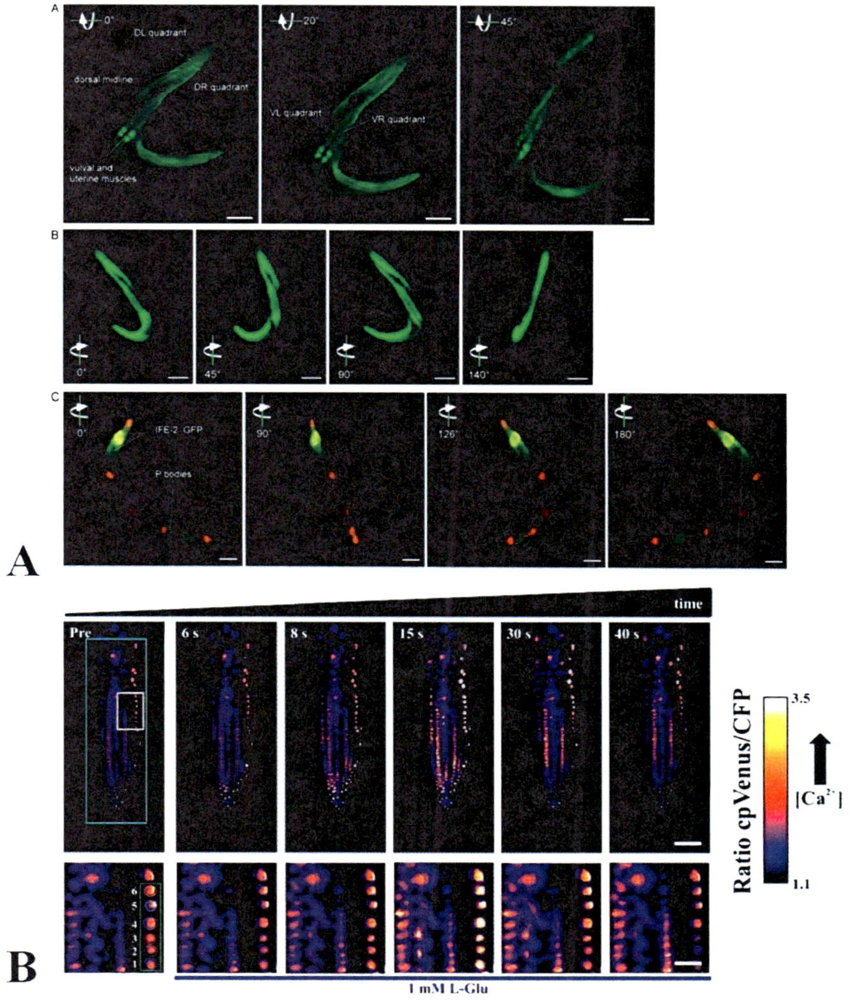

Fig. (6.19). Examples of LSFM in living organisms. **A**: Imaging of a living *C. elegans* specimen through LSFM. Panel A shows different observation angles of the same animal expressing a muscle-specific GFP (D=dorsal, V=ventral, R=right, L=left). Panel B shows the same specimen rotated on the vertical axis. Panel C shows a specimen co-expressing two fluorescent proteins for two different proteins. In this way protein level expression in different tissues and cells can be studied in a complex 3D animal. The worm is shown at four different angles. In all panels scale bar measures 100 μm. Reproduced with permission from [40]. **B**: Same approach to study a vegetal sample, in this case Ca^{2+} kinetics in *A. thaliana* roots. A fluorescent Ca^{2+} protein probe was expressed by root cells. The living root was incubated with 1 mM L-glutamate, a known activator of Ca^{2+} uptake. The top images show the kinetics of emission before and at several time points after L-glutamate exposure (false colour gradient indicated to the right; scale bar 50 μm). Bottom images are a magnified region taken from the top images (white rectangle). They display in further detail the emission changes and six cells (green rectangle) that were analysed in their emission response to L-glutamate (scale bar 10 μm). Reproduced with permission from [41].

5. TOTAL INTERNAL REFLECTION FLUORESCENCE MICROSCOPY

Total internal reflection fluorescence microscopy (TIRFM) is a technique that uses the phenomenon of total internal reflection of light at the interface of two media to probe events at very small scales (100-200 nm) in great detail [1]. The fundamentals of this total internal reflection were presented before in this Chapter (see Optical fibers in section 1 above). Light reflection is a consequence of refractive index differences between two media. In the particular case where light propagates from a medium A with a higher refraction index into a medium B with a lower index ($n_a > n_b$) the amount of reflected light depends on the angle at which light approaches the interface (see Fig. **6.9A** above). Importantly there is a certain angle, denoted critical reflection angle (ϕ_c), below which light is completely reflected back into the A medium. TIRFM is based on exciting the sample at smaller angles than this ϕ_c to take advantage of internal reflection to probe the sample.

When total internal reflection is taking place at an interface some fraction of the reflected light wave does indeed manage to appear beyond the interface in the medium with the lower n (see Fig. **6.20A**).

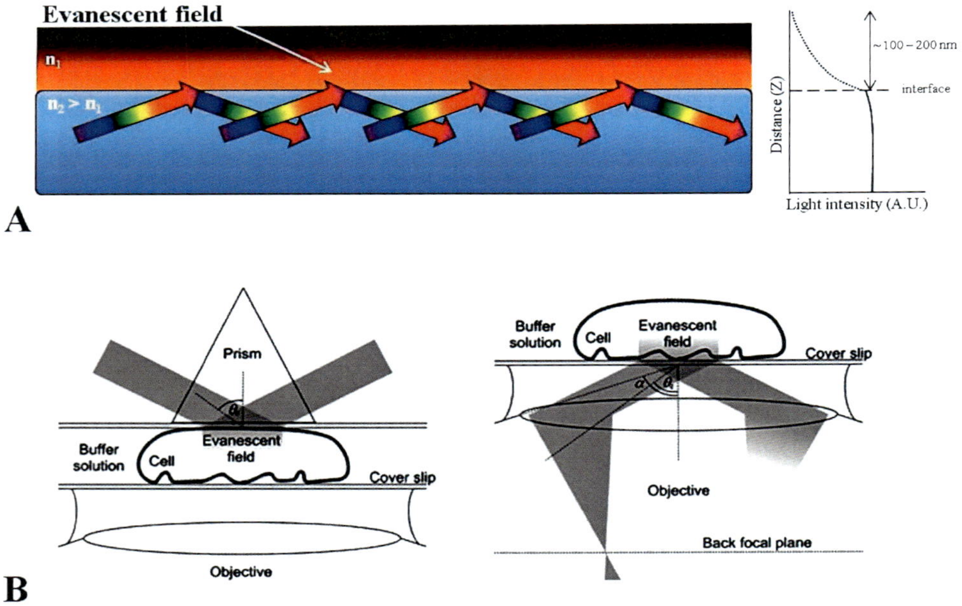

Fig. (6.20). **A**: Physical principle supporting total internal reflection fluorescence microscopy (TIRFM). During the total internal reflection process (see also Fig. **6.9**) a small amount of the energy carried by the electromagnetic wave (rainbow arrows) crosses (orange gradient) to the lower refractive index medium (n_1

medium in the scheme). This energy can be coupled to a fluorophore localized very close to the media interface. This "energy leakage" is represented in the graph on the right: some of the energy is available in the lower index medium at short distances from the interface. The energy available decreases exponentially the farther from the interface, with practically useful levels within the first 100-200 nm away from the interface. **B**: Two experimental setups to achieve TIRFM. On the left the TIRFM process takes place on the top of the cell and the emitted light is collected at the bottom. The prism helps to couple the excitation beam to the sample. On the right, both TIRFM and signal collection take place at the bottom. In both approaches the objective is optically coupled to the sample by an immersion medium that increases the numerical aperture. The grey areas on the schemes represent the light pathway. Reproduced with permission from [43].

This is required by electromagnetic energy conservation rules. However this fraction is very tenuous and diminishes exponentially with distance from the interface. This tenuous wave displacing close to the interface is named evanescent wave [43]. The spatial region inside which the intensity of the evanescent wave can be tapped for practical applications usually lies within one third of the associated reflected wavelength from the interface. This means that for optical wavelengths (400-700 nm) the evanescent wave can be put to use within 100-200 nm from the interface. Therefore within this volume the evanescent can couple with matter if an adequate interaction exists (*e.g.* excitation of excited electronic levels of a fluorophore).

For many applications this slim region where light interacts with matter is too thin for practical uses. However it is perfect to probe samples with spatial resolution better than the diffraction limit of the employed wavelength. TIRFM can use 500 nm light to probe a ~150 nm width space while its diffraction limit is around 250-300 nm [2, 44]. Therefore TIRFM is an excellent tool to study membrane-related and cell-substrate biological events. The main shortcoming, of course, is that sampling is restricted to this very limited spatial domain where the evanescent wave is sustained (see **6.20 B**) [43]. Images of the cytoskeletal microtubules of migrating fibroblasts obtained by TIRFM are displayed in Fig. (**6.21**) [45]. Although the imaged space is restricted, as noted above, TIRFM allows great detail within this sampled volume. Beck *et al.* report details below 100 nm when studying the plasma membrane and cytoskeleton of eukaryotic cells [46]. As it happens with many other fluorescence methodologies, TIRFM can be combined with other techniques to yield better results. For example, TIRFM has been implemented in conjunction with multiphoton excitation and super-resolution approaches. Very interesting results have been obtained by melding TIRFM with fluorophore photobleaching techniques (photobleaching techniques will be explained in section 6 below). In short photobleaching further increases the image details and provides a kinetic approach to cellular events.

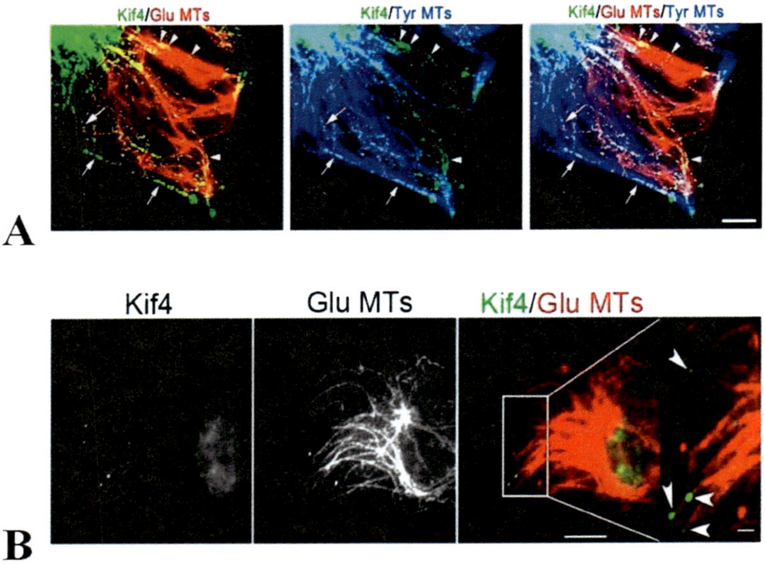

Fig. (6.21). A: TIRFM images of microtubules (MTs) and stabilizing protein Kif4 in migrating NIH3T3 fibroblasts at the basal level in contact with the substrate. Stabilized MTs have an exposed glutamate residue (Glu MTs) while dynamical MTs have a tyrosine residue (Tyr MTs). The fluorescence images show (left to right) Kif4+Glu MTs localization, Kif4+Tyr MTs localization and a merge of the two previous images. Arrowheads indicate Kif4-Glu MTs interaction places while arrows indicate Kif4-Tyr MTs interaction places (scale bar 5 μm). Reproduced with permission from [45]. **B**: Interactions between Kif4 and stabilized MTs (Glu MTs). The B/W images are TIR fluorescence images of the respective protein (Kif4 localizes both to the nucleus and the cytoplasm). The coloured image is a merge of the B/W images (scale bar 10 μm). The region inside the white rectangle has been enlarged (right) to better appreciate the interacting regions between Kif4 and stabilized MTs (arrowheads) at the cell edge (scale bar 2 μm). Reproduced with permission from [45].

TIRFM is currently being exploited within many research areas with success. Yamamoto *et al*. made use of TIRFM in combination with an optical tweezers setup to spatially manipulate a single cell in order to rotate it and expose parts of it to the evanescent wave [47]. This partially circumvents the main shortcoming of TIRFM making possible to study different parts of a cell or biological sample. Not only cells but other structures or even molecules can be studied by TIRFM as recently reviewed by Woods and Bain making use of the same principles for total internal reflection spectroscopy [48].

6. FLUORESCENCE PHOTOBLEACHING TECHNIQUES

These techniques encompass different approaches to measure chemical and biological issues but all have in common a photobleaching first step in order to carry out the measurements. The cells are first labelled with some fluorophore. Afterwards a bleaching light pulse is applied to a certain cell region and the spatiotemporal dynamics of the unbleached/bleached dye is measured. These measurements provide information in regards to diffusion, chemical/biochemical reactions and structural changes that take place in the cell. Depending on how the bleaching pulse is applied and which process is measured afterwards (*e.g.* diffusion of unbleached fluorophores into the initially bleached area) different photobleaching variants have been implemented. Extensions of these photobleaching techniques are employed in some super-resolution approaches (see Chapter 20).

6.1. Fluorescence Recovery after Photobleaching (FRAP)

In its simplest this technique selects a certain cell region or spot to bleach and then the time it takes for the fluorescence signal to recover to pre-bleaching levels is measured [24]. The mechanistic basis for FRAP is displayed in Fig. (**6.22A**).

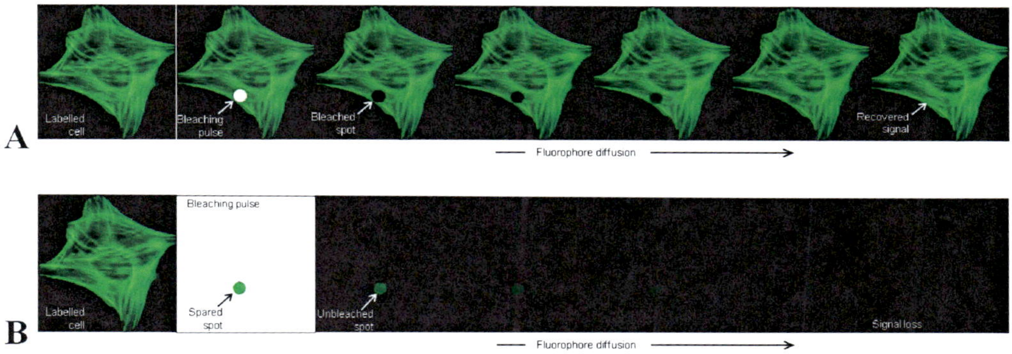

Fig. (6.22). **A**: Working principle in fluorescence recovery after photobleaching (FRAP). The process starts with a fluorophore-labelled sample (a cell in the images). A laser is employed to transiently illuminate a small spot of the cell (bleaching pulse). This pulse gives rise to a fluorophore-depleted area (bleached spot) from which no emission light is obtained under usual excitation conditions. With time unbleached fluorophores diffuse into the bleached spot until the original signal is recovered. **B**: In inverse FRAP (iFRAP) the whole cell is subjected to the bleaching pulse while a mask spares some small spot of it. Under these circumstances the unbleached fluorophore diffuses from this spot to the rest of the cell. With time there is a net loss of fluorophore signal from the spot which provides diffusional information.

The cell is labelled with a fluorescent dye. A particular cell area which wants to be studied is subjected to a short but bright light exposure (usually employing a pulsed laser). The fluorophore within the irradiated spot is bleached due to the high light intensity and, commonly, the presence of dissolved O_2 (see photobleaching mechanisms in Chapter 2.5.4). Afterwards bleached fluorophores diffuse outside the irradiated spot while unbleached luminescent fluorophores diffuse into the area. By measuring the time it takes for the spot to recover a certain brightness level diffusion constants can be obtained through FRAP.

In inverse FRAP (iFRAP) the surroundings of the spot under study are bleached while the spot is left intact. On many occasions the surroundings mean the whole cell. Once this large area has been bleached, functional fluorophores diffuse and/or move outside the spot recovering the emission of the surrounding area up to a certain point. The scheme behind iFRAP is shown in Fig. (**6.22B**).

Initially FRAP was developed and employed to study diffusion within cell membranes [49]. However it was soon realized that diffusion-mediated processes could be studied in regards to proteins and other cellular components [50]. Additionally, diffusion was measured not only in 2-dimensional structures like membranes but also in 3 dimensions regions like the cytoplasm or the nucleus. Figs. (**6.23A-B**) are examples of different FRAP uses in Cell Biology. Fig. (**6.23A**) shows a cell expressing a recombinant fluorescent virus membrane protein (pEGFP3A). The pictures display the photobleaching process and the signal recovery with time [51]. Fig. (**6.23B**) shows FRAP in action to photobleach chimeric fluorescent actin to study cytoskeleton restructuring [52]. The recovery kinetic of actin is clearly seen in the featured sequence.

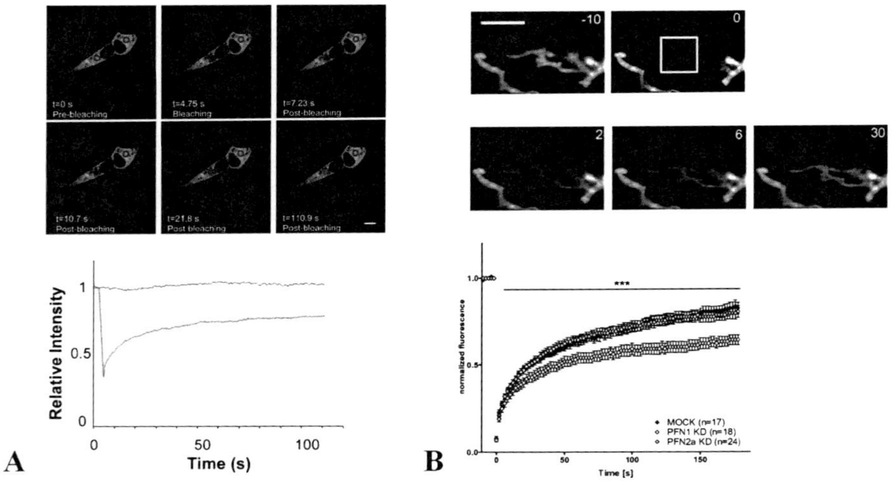

Fig. (6.23). Practical use of FRAP in living cells. **A**: Vero cells expressing a fluorescent viral protein were studied with FRAP to assess the protein dynamics. Two cells were studied, one subjected to a photobleaching pulse (left cell, red circle) while the other served as control (right, blue circle). The black area inside each cell is the nucleus where the viral protein does not enter (scale bar 20 μm). The fluorescent protein emission from each spot is shown in the lower graph where it can be seen how the bleached spot displays a marked decrease in emission which recovers with time although not completely. On the other hand the control spot shows a stable emission during the measuring period. Reproduced with permission from [51]. **B**: Actin-interacting factors reveal actin kinetics through FRAP. A murine astrocyte expressing a GFP-actin is pulse-bleached (t=0) in the area highlighted with a white rectangle. With time (2-30 s) the area recovers its emission (scale bar 5 μm). The emission recovery plot shows that Profilin1- PFN1 (open diamonds) interferes with actin dynamics as the emission recovery in these modified cells is significantly slower than the controls (solid diamonds) and cells expressing Profilin2a-PFN2a (grey diamonds). Reproduced with permission from [52].

6.2. Fluorescence Loss in Photobleaching (FLIP)

The approach employed with FLIP is a variant of FRAP [24]. Two are the main differences between these techniques. First, in FLIP a particular spot is bleached several consecutive times, as opposed to FRAP where bleaching usually takes place once. Second, the amount of fluorophore loss is measured *outside* the bleached region. Therefore FLIP is particularly well suited to measure the dynamic connections between different structures or regions. The mechanistic basis for FLIP is displayed in Fig. (**6.24A**). FLIP is mostly employed to study cell compartmentalization, cytoplasm-nucleus traffic and membrane dynamics. A use of FLIP to measure GFP kinetics is shown in Fig. (**6.24B**) [53]. It is clear how fluorescent emission is consistently dimmer the closer to the repeatedly bleached spot the measurements are done.

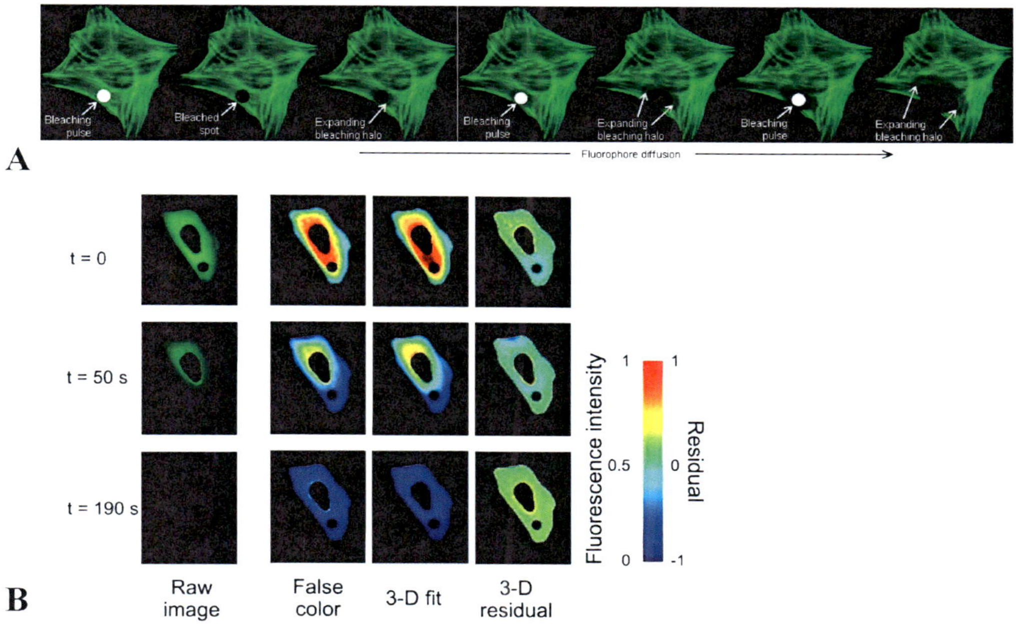

Fig. (6.24). A: In fluorescence loss in photobleaching (FLIP) several bleaching pulses are applied in succession instead of a single pulse as is used in FRAP. A certain time passes between pulses which allow some unbleached fluorophore to diffuse into the spot. With the next light pulse part of this replenishing fluorophore population is bleached too. The observed effect is that of an expanding bleaching halo centred on the illuminated spot. **B**: U2OS cell expressing GFP undergoes FLIP to measure intracellular diffusion. On the left real images of the cell are shown. The round smaller black spot is the repeatedly photobleached spot and the larger elongated black spot is the cell nucleus. By 190 s almost all GFP in the cell has been photobleached by irradiating the same spot. The three columns to the right are false colour images increasing the contrast and showing the diffusion fit and residual fit employed to model GFP diffusion. Reproduced with permission from [53].

6.3. Fluorescence Localization after Photobleaching (FLAP)

Finally fluorescence localization after photobleaching or FLAP allows the direct following of fluorophore populations. It makes use of two types of fluorophores that can be different molecules or be part of a larger molecular entity (like a fluorescent protein with two fluorescent moieties). The light pulse bleaches one of the two fluorophores but leaves the other unaffected. Now it is possible to measure the movement of the unbleached molecules (single emission profile) against the background of mixed emission by the unbleached mixture of fluorophores outside the irradiated spot [24]. This technique is very adequate to resolve fast molecular diffusion/transport and provides a ratiometric objective parameter by which to assess molecular movements.

Two developments of FLAP are Photoactivation and Photoconversion. In Photoactivation FLAP a single fluorophore is employed. This fluorophore features a low quantum yield of fluorescence. However under specific wavelength irradiation it undergoes a photochemical rearrangement that drastically increases the fluorescence yield. Therefore, after photoactivation, the high-fluorescence subpopulation can be tracked against the low-fluorescence ("dark") background. In Photoconversion FLAP the fluorophore already emits with a high quantum yield but exposure to a certain wavelength irreversibly changes its spectroscopic emission features, mainly the emission colour. After the photoconverting pulse the fluorophore subpopulation, emitting in different region of the spectrum, can be tracked *vs.* the background. An example of Photoconversion FLAP is shown in Fig. (**6.25**). The actin polymerization process is revealed by this FLAP technique in a murine macrophage cell line [54].

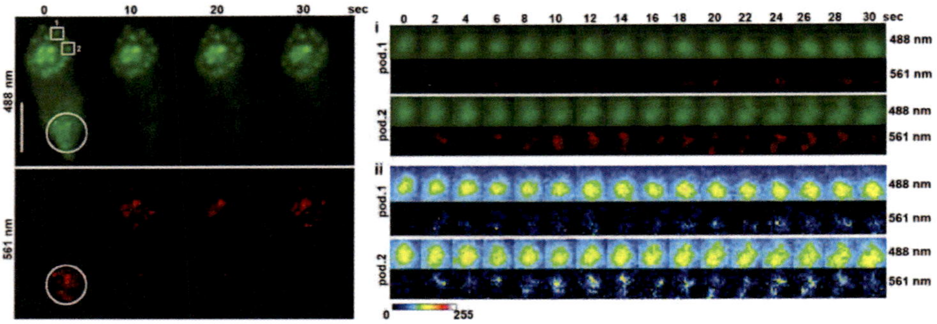

Fig. (6.25). Example of FLAP in a murine macrophage cell. On the left there are images of the same cell excited at two different wavelengths (488 and 561 nm). The white circle indicates the region of photoconversion (light pulse at t=0). The time evolution (0-30 s) of translocation of photoproduct to regions 1 and 2 (white squares) is displayed in detail at the top right. With time more signal registers in the 561 nm excitation channel (red emission). At bottom right false colour images of the same 1 and 2 regions to increase contrast and to quantify kinetics. Reproduced with permission from [54].

7. FLUORESCENCE CORRELATION SPECTROSCOPY

Photobleaching techniques as those described in section 6 allow the measurement of kinetic properties like diffusion constants. However, because of their inherent pump-probe nature the results obtained are related to non-equilibrium figures. Said in another way, the system under study is perturbed, brought out of equilibrium (pump), then the perturbation is switched off and the system relaxes back towards equilibrium. During this "equilibrium-regaining" period the system is interrogated (probe). It would be desirable to have a technique able to probe the system under equilibrium conditions too. This is possible by employing fluorescence correlation spectroscopy (FCS).

FCS measures the fluorescent signal from one or several fluorophores in a very small (~ femtoliter) excitation volume. By comparing the intensity of light coming from this volume at different times FCS provides the diffusion times of the fluorophores. Once the diffusion time is known, and given that the excitation volume dimensions are known too, it is relatively easy to arrive to the diffusion constants of the probed molecules [7]. Because of the mathematical basis of this technique, the less amount of fluorescent labels inside the analysed volume, the better. Therefore to obtain reliable results with FCS it is mandatory to sample a very small volume. This is achieved by using a confocal system (taking advantage of pinholes) or by employing non-linear fluorescence excitation (usually biphotonic excitation) with lasers. Under these conditions the probed volume is commonly near-diffraction, an ellipsoid around one micron in width and 2-3 microns in height. Also the fluorophore concentration in the studied medium is very low (1-100 nM) in order to reduce the number of molecules diffusing in and out the excitation volume.

FCS relies on complex mathematical assumptions, the detailed explanation of which is beyond the scope of this book (see chapter 24 of reference [7] for a deeper presentation of this topic). But we will provide a simple introduction to it in what follows. Assume a laser beam exciting a very small volume of a solution in which a certain amount of fluorescent molecules have been dissolved. At any particular moment some of the molecules may cross the excitation volume as they move around under the Brownian movement (see Fig. **6.26A**).

After a certain time some of these molecules move out of the excitation volume and other molecules move in. Sampling the emission intensity coming from the volume provides information about how many molecules reside within it during a certain time period. Measuring the emission intensity for very short times returns random-like light fluctuations but increasing the sampling time provides with an emerging pattern. If the molecular diffusion time is slow when compared to the measuring time the light intensity registered will change slowly with time (Fig. **6.26B**). On the other hand if the diffusion time is fast in regards to the measuring time, light intensity will change in a faster way (Fig. **6.26C**). Then the light intensity at a certain moment t_1 is compared with the intensity at a later time $t_1 + \Delta t$ and the rate of change between these two measures is recorded. Under slow diffusion conditions it would be expected that the rate change is to be small (*i.e.*, light intensity changes slowly with time) while a higher rate change is expected under fast diffusion.

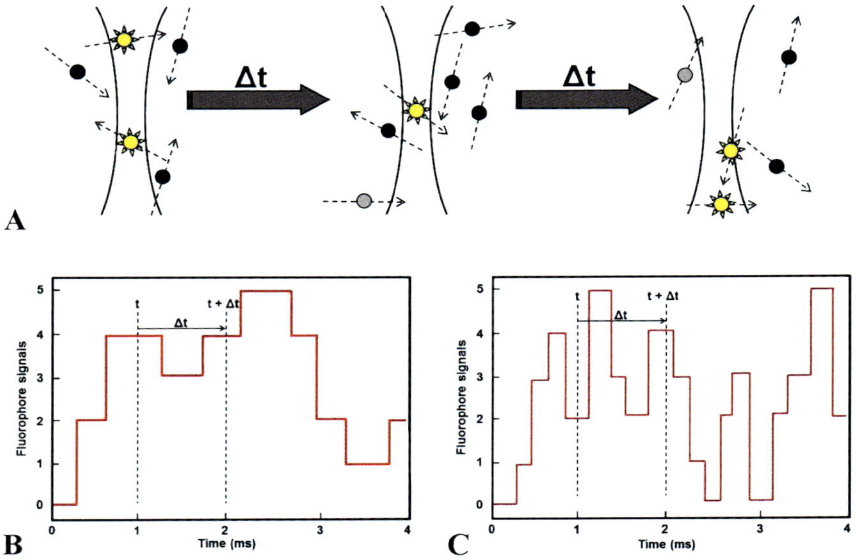

Fig. (6.26). Fluorescence correlation spectroscopy (FCS) working principle. **A**: In FCS a very small optical excitation volume is established in the medium. Initially fluorescent labels (black spheres) diffuse randomly into and out of this volume. While transiting the volume they emit light (bright spots). After some time (Δt) these emitting entities move outside the volume but others move into (grey spheres) repeating the process. **B**: Light intensity *vs.* sampling time for a medium with a slow diffusion. As fluorophores move slowly into and out of the excitation volume the light intensity measured (red line) changes slowly with time. **C**: Light intensity *vs.* sampling time for a medium with a fast diffusion. In this case fluorophores move quickly into and out of the excitation volume. As a consequence the light signal obtained (red line) changes faster for a certain amount of time.

On a statistical basis, assuming a certain diffusional behaviour that takes into account multiple parameters, FCS models provide a mathematical function of how the light intensity should change under the experimental conditions. The next step is to compare the experimental light intensity rate value obtained between t_1 and $t_1 + \Delta t$ with the expected rate values. By making this comparison between what is measured and what is expected multiple times an autocorrelation function is obtained. This function is usually denoted G(t). An example of how G(t) looks like is displayed in Fig. (**6.27A**). The function appears as a sigmoidal curve as a function of time. The time point where G(t) takes half its initial value gives the diffusional time (τ) of the molecule (red downward arrows in Fig. **6.27A**). The shorter is τ the higher the diffusional constant and the diffusion speed. Once τ is established and given that the linear dimensions of the excitation volume are known, or at least finely estimated, it is easy to obtain the diffusion constant that takes into account both quantities: time and distance. In the example depicted in Fig. (**6.27A**) fluorophore "A" has a higher diffusion constant than fluorophore "B", and these two much higher than fluorophore "C".

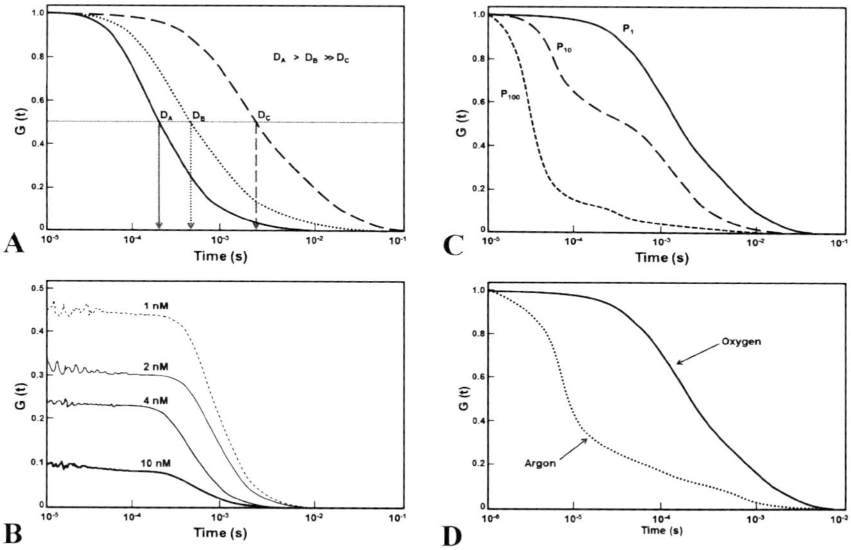

Fig. (6.27). **A**: Examples of autocorrelation curves for different diffusion coefficients. These curves show the relationship between the autocorrelation function (G(τ)) and sampling time. Fluorophore A has the highest diffusion coefficient (D_A) while fluorophore C has the lowest (D_C). The time points corresponding to half the initial value of G(τ) (thin horizontal line) indicate the diffusional times (red downward arrows) for each case. **B**: Effect of fluorophore concentration in G(τ). As fluorophore concentration increases more individual emitters populate the excitation volume. This smears out the fluctuations in light intensity upon which FCS is based. As a consequence the autocorrelation curves diminish and return less information the higher the fluorophore concentration (from 1 to 10 nM in the example shown). **C**: Increasing the excitation power (from P_1 to P_{100} in the example) can lead to a higher triplet population in the excitation volume. These triplets, in virtue of their long excited state lifetime, behave as "dark states" that provide no information employing FCS. The autocorrelation curves display a typical multiexponential trend when this triplet pumping process is taking place, making very difficult to obtain the diffusion time. **D**: The environmental composition can affect this triplet population. Molecular oxygen (O_2) is capable of accepting energy from a triplet molecule very efficiently. In the case where O_2 is in the medium (solid line) the triplet effect is not observed in contrast to a medium where argon displaces O_2 where triplets are present again (dotted line).

As it was already mentioned above FCS is critically dependent on the fluorophore concentration in the medium. A too high concentration eliminates the emission fluctuations by maintaining a more or less steady population of fluorophores within the excitation volume. The result on the correlation plots is shown schematically in Fig. (**6.27B**). As the fluorophore concentration is increased the correlation function returns lower levels. Probe concentrations above 100 nM should be avoided in order to obtain reliable data.

Another issue to take into account is that a fluorophore in the excited state can have more than one deactivation pathway (see Chapter 2.4 and 2.5). Most commonly this deactivation is the intersystem crossing (ISC) to the triplet state. Although triplets states can be luminescent, their very long lifetimes (in comparison to singlets) and their large Stokes shift typically make them unsuitable for FCS. Therefore when a fluorophore undergoes ISC it reaches a "dark-state" which does not provide information and, at the same time, makes the molecule insensitive to excitation because of the different absorption spectrum of the triplet as compared with the ground state. The practical consequence is that the correlation plot changes and returns incorrect diffusion values. Because the probability of pumping the triplet state depends on the excitation power, an increase in laser power usually leads to a larger error because of ISC taking place in the sample. This trend can be shown in Fig. (**6.27C**) where an increase of power of 10x or 100x leads to an alteration of the correlation curve. This should be avoided and a reduction in the excitation power decreases the triplet population. The environment can also modulate this phenomenon. In Fig. (**6.27D**) two functions are shown obtained under the same experimental conditions except the amount of dissolved gases. One depicts the function with atmospheric pressure oxygen in equilibrium with the sample and the other a sample in which argon purging was done before measurements. It is readily obvious that, excitation power and probe concentration being the same, the O_2 sample displays an adequate plot while the argon-purged sample shows clear signs of triplet population. As it was explained in Chapter 2.5.4, ground-state O_2 interacts very efficiently with triplet molecules which transfer their excitation energy to O_2. After this transfer O_2 is excited to its first excited electronic state commonly known as singlet oxygen ($^1O_2$). Therefore O_2 quenches triplets fast and relieves the sample from the ISC burden. Argon does not quench triplets which results in the different result. The disadvantage of using O_2 to quench triplets is that $^1O_2$ is a highly oxidative molecule which tends to oxidize irreversibly molecules, including fluorescent probes, and consequently deplete the sample of them. In conclusion, it is advisable to use a probe with the highest possible fluorescent quantum yield for good results in FCS.

Diffusion depends on many parameters (temperature, viscosity, pressure, molecular mass, molecular hydrodynamic radius, molecular interactions, *etc.*). FCS provides information directly from the probes (different probe sizes or probe interaction with the environment –"probe stickiness"-) or by attaching/labelling the probe to a target (different proteins, nucleic acids, *etc.*) [55, 56]. Two or more probes with different properties can be used in the same environment to assay different conditions. FCS can be used in 3-D environments (cuvettes, liquid medium, intracellular medium), 2-D environments (cellular organelles or membranes, liquid crystals) or 1-D environments (microcapillaries, microfluidic flow measurements). The obvious advantage of FCS is that it measures diffusion in the steady state, so it does not require to perturb the medium to return valuable information. However, it is perfectly possible to use a pump-probe approach and still use FCS to measure diffusion. FCS can be coupled to other techniques like TIRF [7], multiphoton excitation [57] or dual-colour probing [58]. An example taken from this last reference is shown in Fig. (**6.28**). In this case FCS was employed to measure the differences in diffusion of a mammalian cell receptor located in the plasma membrane when it is binding or not its ligand. The upper panel (a) shows the correlation plots for the receptor (emitting in red). Two fits are shown, one corresponding to the autocorrelation function (upper trace), and the other corresponding to a cross-correlation function between the measurements taken for the receptor in two different probing lines, scanned by the laser on the cell membrane (laser lines shown as yellow arrows in the cell picture upper right side of the figure). The same was done for the ligand (emitting in the green) as shown in middle panel (b). Finally, cross correlations between receptor signal (red) and ligand signal (green) have been computed in the lower panel (c) which returns the diffusion for ligand-receptor bind complex (see below). Diffusion times when ligand and receptor are bound are slightly larger than when they are unbound. Images to the right of the plots are standard confocal fluorescence images of the sampled cells.

These images allow us to introduce two evolutions of the FCS technique: Image Correlation Spectroscopy (ICS) and Image Cross-Correlation Spectroscopy (ICCS). In ICS the laser system exciting the fluorophores is moving, scanning the sample. This is opposite what happens in FCS, where the exciting system is static. By systematically scanning the sample, ICS imposes a "virtual movement" on the sample, which partly substitutes the real fluorophore movement/diffusion. This is an advantage when slow-moving molecular species are being imaged. Therefore, ICS compares different images of the same field at different times, from which it can obtain the correlation functions [59 - 61].

In ICCS the system excites and detects two fluorophores at the same time (see Fig. **6.28**). The protocol computes the autocorrelation functions for each fluorophore, and then obtains the cross-correlation between these functions [62 - 64]. The more similar the autocorrelations are the more probable is that the fluorophores are moving together. This is graphically represented by a cross-correlation function that appears below the autocorrelation function. The more similar is the cross-correlation curve to the autocorrelation curve, the higher the number of co-moving fluorophores. In other words, if the cross-correlation curve overlaps with the autocorrelation curve, then all fluorophores are moving together. If there is no overlap at all, the cross-correlation curve appears practically flat at the bottom of the graph, then no binding exists between any of the two types of fluorophores. In the examples shown in Fig. (**6.28**) one can notice that the cross-correlation curves are not flat but, nevertheless, they are far from overlapping with the autocorrelation curves. This can be interpreted as a low-t--moderate level of fluorophore binding, in this case interaction between a cell receptor (red fluorophore) and its ligand (green fluorophore).

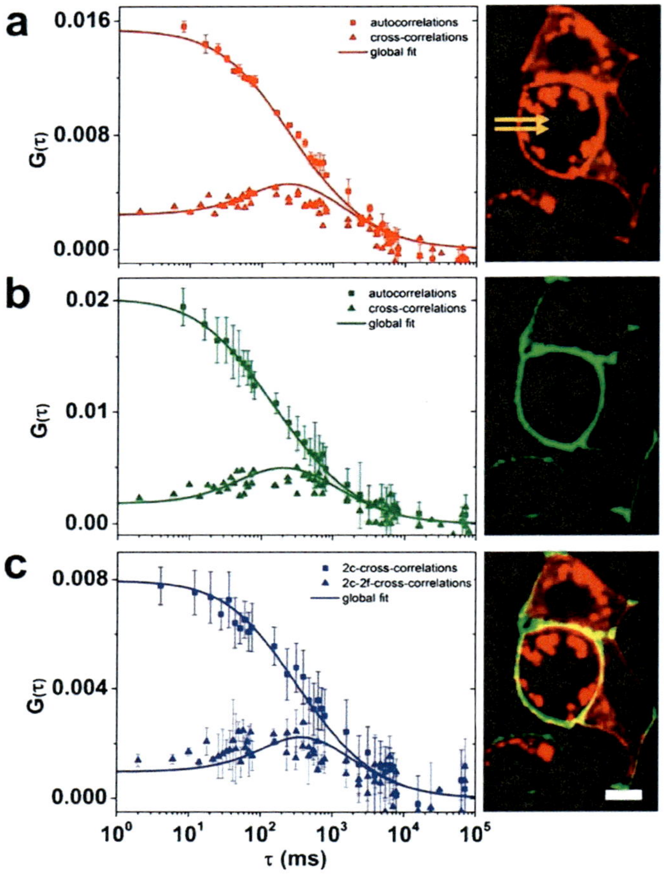

Fig. (6.28). Example of experimental use of FCS in cell biology. Panel (**a**) shows the autocorrelation curves for the plasma membrane diffusion of a red fluorescent cell receptor. Also shown are the cross correlation curves between the data obtained in two membrane scans (laser scans obtained along the two yellow arrows displayed in the confocal cell picture to the right). Panel (**b**) shows the same parameters but for a green fluorescent protein ligand specific for the cell receptor shown in (**a**). Panel (**c**) shows the results for these parameters when receptor and ligand are bound and, therefore, they result in a more massive molecular complex. The three autocorrelation curves show slightly different diffusion times especially for the data in panel (**c**). Displayed to the right of the plots there are confocal fluorescent pictures of the studied cells with their corresponding spectral emission (merged image in the case of panel (**c**)). Reproduced with permission from [58].

ICS and ICCS have clear advantages over FCS in regards to slowly-moving or static biological structures. For example, they are indicated to obtain movement information about large proteins, protein complexes, cellular membrane dynamics, or cell receptors, to cite a few examples [63, 64]. Some variations exist, like employing a dual-focus approach instead of imaging two fluorophores; or combining two fluorophores plus dual-focus for resolution enhancement [64]. Critical issues with these techniques are the establishment of very small imaging spots, use of fluorophores with emission spectra as less overlapping as possible, chromatic aberrations of the exciting/imaging systems, or precise positioning of the focal volumes in dual-focus approach [63, 64].

8. FLOW CYTOMETRY

Technically speaking flow cytometry is not a fluorescent imaging technique although it makes use of fluorescence to sort out cell sub-populations attending to a certain feature. Nevertheless we feel that flow cytometry is in the blurred border between fluorescent imaging techniques and fluorescent spectroscopic techniques. For this reason a brief introduction to flow cytometry has been included in this book although we do not intend to give an in depth explanation to the technique. Many works already exists that treat this topic with much more detail and the interested reader is invited to consult them for a more profound knowledge (introductory material can be found in [65, 66]).

The basic flow cytometry methodology starts by labelling a cell population with some fluorophore that will tag some subpopulation preferentially. Examples are markers for cell surface molecules, for internal organelles or DNA dyes that allow determination of the cell cycle point.

Once the cells are incubated with the corresponding fluorescent probe or probes, the cells are transferred to the flow cytometer (see Fig. **6.29A**). The device then takes a certain volume of the cell suspension and produces very small droplets each one, ideally, containing one cell. The droplets are directed through a laser system that illuminates them and photoreceptors analyse the fluorescence signal emitted by each droplet. The obtained data are expressed in typical flow cytometer plots that show cell subpopulations distribution in regards to the emitted signal (see Fig. **6.29B-C**) [67]. Other features can be obtained from the

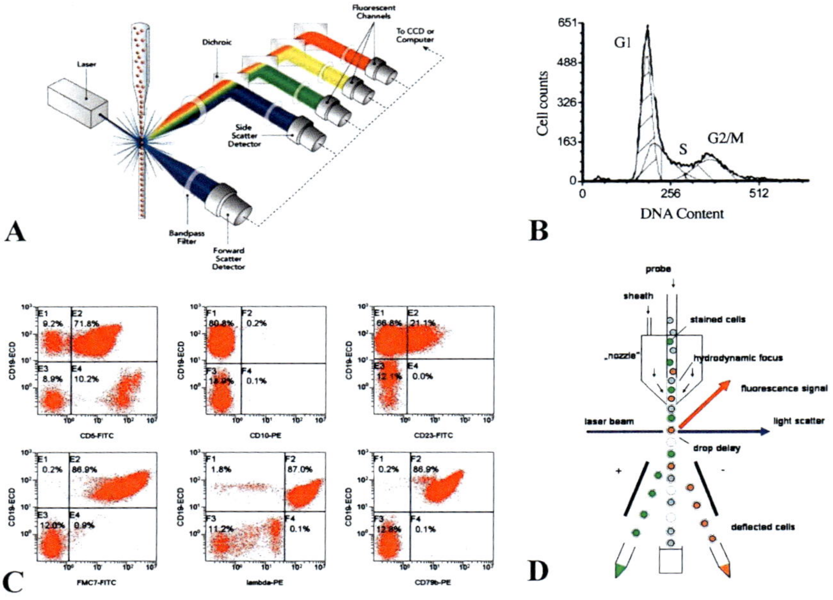

Fig. (6.29). **A**: Basic flow cytometer set up scheme. A light source (mostly a laser) excites sample droplets which contain one cell each coming from a capillary tube. The emitted light is registered sideways in different channels for different wavelengths providing information from each fluorophore in the sample. The laser excitation light is also scattered to some degree by each droplet and its contain. Registering the forward and sideward scattered light provides additional information without having to relay to fluorescent probes. **B**: Typical cytometry plot of the DNA contain in a cell population. Cellular DNA has been labelled with a fluorophore. The amount of fluorophore per cell depends on the amount of DNA. Most of the cells are in the G1 stage of the cell cycle (first and largest peak to the left). A small percentage of the cells are in the S phase and show a larger amount of DNA because they are in the process of replicating it (intermediate plateau). Finally cells in G2 or undergoing active mitosis (M) have double amount of DNA and map at double the signal of the G1 sub-population. This corresponds to the second shorter peak to the right. **C**: Typical plots from a flow cytometer analysis of a cell population. In this case the lymphocytes from a person diagnosed of chronic lymphocytic leukaemia were labelled with different lymphocyte cell surface markers (6 markers, one for each plot). In all cases each marker was compared with the amount of cells expressing the CD19 marker, a protein expressed by B-cells but not by mature plasma cells. Each red dot in each plot represents a measured event by the cytometer, a single cell registered. Most of the cells were B-cells (they map largely in the upper half of all plots) because there was a CD19 positive signal in most cases. Most cells were positive for both CD19 and CD5 (upper left plot region E2) but were negative for CD10 (upper middle plot region F1) for example. Reproduced with permission from [67]. **D**: Working principle of a Fluorescence-Activated Cell Sorting system (FACS). The device is basically a flow cytometer that, upon sampling each droplet, activates a sorting mechanism (usually electrostatic as shown in the scheme) that physically separates different samples. In this way, for example, a FACS could provide pure populations of CD19-positive *vs.* CD19-negative cells in regards to the cell analysis shown in **6.29 C** previously.

cell sample without having to label them previously to flow analysis. For example, the autofluorescence signal from some biomolecule or specific cell type can be obtained through flow cytometry. Also by measuring the interaction of the laser light with the cell an estimation of size distribution or cellular complexity (loosely equivalent to cell "roughness" in regards to how much light is scattered by the cells) can be studied without even having to rely on fluorescent emission.

Due to the ability of antibody-decorated beads to bind different biomolecules in a very efficient way, it is also possible to analyse these molecules attached to beads by means of a flow cytometer. This technique, named Cytometric Bead Array (CBA), employs the same principles as for cell analysis but substituting the former by beads [68]. Beads of different sizes displaying different antibodies allow for distinguishing different ligands. The amount of ligand is established through a second antibody reaction (secondary antibody) that is tagged with a fluorescent moiety (see Chapter 17.1 for further details in regards to immunofluorescence).

Flow cytometry systems are able to measure and analyse thousands and tens of thousands of particular events per second. Therefore a further step was taken in the early 1970s to use the freshly obtained signal from particular events to sort or physically separate those events in different chambers or receptacles. In this way the first Fluorescence-Activated Cell Sorting system (FACS) was implemented (see Fig. **6.29D**) [69]. The FACS has instructions to change a particular droplet trajectory after it has probed it and has obtained information about the sample's features. Therefore after confirming if a certain droplet contains a mitotic cell or an interphasic cell, for example, the device can redirect mitotic cells to one receptacle and interphasic cells to another one [70]. FACS is a very efficient and fast methodology to obtain characterized cell subpopulations.

Apart from their essential role in mammalian cell research, both flow cytometry and FACS are being employed in many other areas of Cell Biology. Some examples are the detection and sorting of rare bacterial strains [71], and their use for assessing environmental stress effects on vegetal cells and their developmental programs [72]. Additionally flow cytometry techniques can be fitted to other analytical techniques. For example, flow analysis can be implemented with non-linear optical approaches (see Chapter 19) to obtain fluorescence data by multiphoton sample excitation [73]. A very interesting merging of techniques comes with the use of flow cytometry in conjunction with microfluidics (the manipulation/analysis of fluids at the micrometer scale). A comprehensive review on this topic has been recently published by Wyatt Shields IV *et al.* [74].

CONFLICT OF INTEREST

The author (editor) declares no conflict of interest, financial or otherwise.

ACKNOWLEDGEMENTS

This work was supported by the Aarhus Institute of Advanced Studies (AIAS).

REFERENCES

[1] Stender AS, Marchuk K, Liu C, *et al.* Single cell optical imaging and spectroscopy. Chem Rev 2013; 113(4): 2469-527.
[http://dx.doi.org/10.1021/cr300336e] [PMID: 23410134]

[2] Sanderson MJ, Smith I, Parker I, Bootman MD. Fluorescence microscopy. Cold Spring Harb Protoc 2014; 2014(10): top071795.
[http://dx.doi.org/10.1101/pdb.top071795] [PMID: 25275114]

[3] Nature Milestones-Light Microscopy. Nature milestones 2009.

[4] Aswani K, Jinadasa T, Brown CM. Fluorescence microscopy light sources. Microscopy Today 2012; 7: 22-8.
[http://dx.doi.org/10.1017/S1551929512000399]

[5] Silfvast WT. Laser Fundamentals. 2nd ed., Cambridge: Cambridge University Press 2004.
[http://dx.doi.org/10.1017/CBO9780511616426]

[6] Milonni PW, Eberly JH. Laser physics. New Jersey: John Wiley & Sons Inc. 2010.
[http://dx.doi.org/10.1002/9780470409718]

[7] Lakowicz JR. Principles of fluorescence spectroscopy. 3rd ed., New York: Springer Science + Business Media LLC 2010.

[8] Dunsby C, Lanigan PM, McGinty J, *et al.* An electronically tunable ultrafast laser source applied to fluorescence imaging and fluorescence lifetime imaging microscopy. J Phys D Appl Phys 2004; 37: 3296-303.
[http://dx.doi.org/10.1088/0022-3727/37/23/011]

[9] Tu H, Boppart SA. Coherent fiber supercontinuum for biophotonics. Laser Photonics Rev 2013; 7(5): 628-45.
[http://dx.doi.org/10.1002/lpor.201200014] [PMID: 24358056]

[10] Fox M. Optical properties of solids. Oxford: Oxford University Press 2001.

[11] Albeanu DF, Soucy E, Sato TF, Meister M, Murthy VN. LED arrays as cost effective and efficient light sources for widefield microscopy. PLoS One 2008; 3(5): e2146.
[http://dx.doi.org/10.1371/journal.pone.0002146] [PMID: 18478056]

[12] Yamagata K, Iwamoto D, Terashita Y, *et al.* Fluorescence cell imaging and manipulation using conventional halogen lamp microscopy. PLoS One 2012; 7(2): e31638.
[http://dx.doi.org/10.1371/journal.pone.0031638] [PMID: 22347500]

[13] Abramowitz M, Davidson MW. Köhler illumination 2012. http://www.olympusmicro.com/primer/anatomy/kohler.html

[14] Spring KR, Keller HE, Davidson MW. Microscope objectives introduction 2012. http://www.olympusmicro.com/primer/anatomy/objectives.html

[15] Niesner R, Gericke KH. Fluorescence lifetime imaging in biosciences: technologies and applications. Front Phys China 2008; 3: 88-104.
[http://dx.doi.org/10.1007/s11467-008-0002-6]

[16] Band YB. Light and Matter Electromagnetism, Optics, Spectroscopy and Lasers. Chichester: John Wiley & Sons Ltd 2006.

[17] Saleh BE, Teich MC. Fundamentals of Photonics. New Jersey: Wiley-Interscience 2007.

[18] North AJ. Seeing is believing? A beginners' guide to practical pitfalls in image acquisition. J Cell Biol 2006; 172(1): 9-18.
[http://dx.doi.org/10.1083/jcb.200507103] [PMID: 16390995]

[19] Pawley J. Handbook of Biological Confocal Microscopy. 3rd ed., New York: Springer 2006.
[http://dx.doi.org/10.1007/978-0-387-45524-2]

[20] Papkovsky DB, Ed. Live cell imaging.Methods and protocols Methods in molecular biology. New York: Humana Press, Springer Science+Business Media 2010.
[http://dx.doi.org/10.1007/978-1-60761-404-3]

[21] Moy AJ, Wiersma MP, Choi B. Optical histology: a method to visualize microvasculature in thick tissue sections of mouse brain. PLoS One 2013; 8(1): e53753.
[http://dx.doi.org/10.1371/journal.pone.0053753] [PMID: 23372668]

[22] López-Arias B. Personal communication 2016.

[23] Dhawan AP, D'Alessandro B, Fu X. Optical imaging modalities for biomedical applications. IEEE Rev Biomed Eng 2010; 3: 69-92.
[http://dx.doi.org/10.1109/RBME.2010.2081975] [PMID: 22275202]

[24] Ishikawa-Ankerhold HC, Ankerhold R, Drummen GP. Advanced fluorescence microscopy techniques--FRAP, FLIP, FLAP, FRET and FLIM. Molecules 2012; 17(4): 4047-132.
[http://dx.doi.org/10.3390/molecules17044047] [PMID: 22469598]

[25] Collier BB, McShane MJ. Time-resolved measurements of luminescence. J Lumin 2013; 144: 180-90.
[http://dx.doi.org/10.1016/j.jlumin.2013.06.034]

[26] Blacker TS, Mann ZF, Gale JE, *et al.* Separating NADH and NADPH fluorescence in live cells and tissues using FLIM. Nat Commun 2014; 5: 3936.
[http://dx.doi.org/10.1038/ncomms4936] [PMID: 24874098]

[27] Ogikubo S, Nakabayashi T, Adachi T, *et al.* Intracellular pH sensing using autofluorescence lifetime microscopy. J Phys Chem B 2011; 115(34): 10385-90.
[http://dx.doi.org/10.1021/jp2058904] [PMID: 21776989]

[28] Hum JM, Siegel AP, Pavalko FM, Day RN. Monitoring biosensor activity in living cells with fluorescence lifetime imaging microscopy. Int J Mol Sci 2012; 13(11): 14385-400.
[http://dx.doi.org/10.3390/ijms131114385] [PMID: 23203070]

[29] Sands B, Jenkins P, Peria WJ, Naivar M, Houston JP, Brent R. Measuring and sorting cell populations expressing isospectral fluorescent proteins with different fluorescence lifetimes. PLoS One 2014; 9(10): e109940.
[http://dx.doi.org/10.1371/journal.pone.0109940] [PMID: 25302964]

[30] Santi PA. Light sheet fluorescence microscopy: a review. J Histochem Cytochem 2011; 59(2): 129-38.
[http://dx.doi.org/10.1369/0022155410394857] [PMID: 21339178]

[31] Becker K, Jährling N, Saghafi S, Dodt HU. Ultramicroscopy: light-sheet-based microscopy for imaging centimeter-sized objects with micrometer resolution. Cold Spring Harb Protoc 2013; 2013(8): 704-13.
[http://dx.doi.org/10.1101/pdb.top076539] [PMID: 23906921]

[32] Swoger J, Pampaloni F, Stelzer EH. Light-sheet-based fluorescence microscopy for three-dimensional imaging of biological samples. Cold Spring Harb Protoc 2014; 2014(1): 1-8.
[http://dx.doi.org/10.1101/pdb.top080168] [PMID: 24371323]

[33] Keller PJ, Stelzer EH. Digital scanned laser light sheet fluorescence microscopy. Cold Spring Harb Protoc 2010; 2010(5): top78.
[http://dx.doi.org/10.1101/pdb.top78] [PMID: 20439423]

[34] Dunsby C. Optically sectioned imaging by oblique plane microscopy. Opt Express 2008; 16(25): 20306-16.

[http://dx.doi.org/10.1364/OE.16.020306] [PMID: 19065169]

[35] Hu YS, Zhu Q, Elkins K, *et al.* Light-sheet Bayesian microscopy enables deep-cell super-resolution imaging of heterochromatin in live human embryonic stem cells. Opt Nanoscopy 2013; 2(1): 7. [http://dx.doi.org/10.1186/2192-2853-2-7] [PMID: 27795878]

[36] Planchon TA, Gao L, Milkie DE, *et al.* Rapid three-dimensional isotropic imaging of living cells using Bessel beam plane illumination. Nat Methods 2011; 8(5): 417-23. [http://dx.doi.org/10.1038/nmeth.1586] [PMID: 21378978]

[37] Keller PJ, Schmidt AD, Wittbrodt J, Stelzer EH. Reconstruction of zebrafish early embryonic development by scanned light sheet microscopy. Science 2008; 322(5904): 1065-9. [http://dx.doi.org/10.1126/science.1162493] [PMID: 18845710]

[38] Ichikawa T, Nakazato K, Keller PJ, *et al.* Live imaging of whole mouse embryos during gastrulation: migration analyses of epiblast and mesodermal cells. PLoS One 2013; 8(7): e64506. [http://dx.doi.org/10.1371/journal.pone.0064506] [PMID: 23861733]

[39] Bunse S, Garg S, Junek S, *et al.* Role of N-cadherin *cis* and *trans* interfaces in the dynamics of adherens junctions in living cells. PLoS One 2013; 8(12): e81517. [http://dx.doi.org/10.1371/journal.pone.0081517] [PMID: 24312555]

[40] Rieckher M, Kyparissidis-Kokkinidis I, Zacharopoulos A, *et al.* A customized light sheet microscope to measure spatio-temporal protein dynamics in small model organisms. PLoS One 2015; 10(5): e0127869. [http://dx.doi.org/10.1371/journal.pone.0127869] [PMID: 26000610]

[41] Costa A, Candeo A, Fieramonti L, Valentini G, Bassi A. Calcium dynamics in root cells of *Arabidopsis thaliana* visualized with selective plane illumination microscopy. PLoS One 2013; 8(10): e75646. [http://dx.doi.org/10.1371/journal.pone.0075646] [PMID: 24146766]

[42] Chen BC, Legant WR, Wang K, *et al.* Lattice light-sheet microscopy: Imaging molecules to embryos at high spatiotemporal resolution. Science 2014; 346: 1257998_1-1257998-12.

[43] Martin-Fernandez ML, Tynan CJ, Webb SE. A 'pocket guide' to total internal reflection fluorescence. J Microsc 2013; 252(1): 16-22. [http://dx.doi.org/10.1111/jmi.12070] [PMID: 23889125]

[44] Axelrod D. Total internal reflection fluorescence microscopy in cell biology. Traffic 2001; 2(11): 764-74. [http://dx.doi.org/10.1034/j.1600-0854.2001.21104.x] [PMID: 11733042]

[45] Morris EJ, Nader GP, Ramalingam N, Bartolini F, Gundersen GG. Kif4 interacts with EB1 and stabilizes microtubules downstream of Rho-mDia in migrating fibroblasts. PLoS One 2014; 9(3): e91568. [http://dx.doi.org/10.1371/journal.pone.0091568] [PMID: 24658398]

[46] Beck M, Aschwanden M, Stemmer A. Sub-100-nanometre resolution in total internal reflection fluorescence microscopy. J Microsc 2008; 232(1): 99-105. [http://dx.doi.org/10.1111/j.1365-2818.2008.02075.x] [PMID: 19017206]

[47] Yamamoto K, Ishimaru I, Fujii Y, *et al.* Method for measuring the three-dimensional distribution of a fluorescent dye in a cell membrane. Appl Phys Lett 2007; 90: 021110. [http://dx.doi.org/10.1063/1.2428457]

[48] Woods DA, Bain CD. Total internal reflection spectroscopy for studying soft matter. Soft Matter 2014; 10(8): 1071-96. [http://dx.doi.org/10.1039/c3sm52817k] [PMID: 24651911]

[49] Axelrod D, Koppel DE, Schlessinger J, Elson E, Webb WW. Mobility measurement by analysis of fluorescence photobleaching recovery kinetics. Biophys J 1976; 16(9): 1055-69. [http://dx.doi.org/10.1016/S0006-3495(76)85755-4] [PMID: 786399]

[50] Sprague BL, Pego RL, Stavreva DA, McNally JG. Analysis of binding reactions by fluorescence recovery after photobleaching. Biophys J 2004; 86(6): 3473-95. [http://dx.doi.org/10.1529/biophysj.103.026765] [PMID: 15189848]

[51] González-Magaldi M, Martín-Acebes MA, Kremer L, Sobrino F. Membrane topology and cellular dynamics of foot-and-mouth disease virus 3A protein. PLoS One 2014; 9(9): e106685. [http://dx.doi.org/10.1371/journal.pone.0106685] [PMID: 25275544]

[52] Schweinhuber SK, Meßerschmidt T, Hänsch R, Korte M, Rothkegel M. Profilin isoforms modulate astrocytic morphology and the motility of astrocytic processes. PLoS One 2015; 10(1): e0117244. [http://dx.doi.org/10.1371/journal.pone.0117244] [PMID: 25629407]

[53] Guo M, Gelman H, Gruebele M. Coupled protein diffusion and folding in the cell. PLoS One 2014; 9(12): e113040. [http://dx.doi.org/10.1371/journal.pone.0113040] [PMID: 25436502]

[54] Dovas A, Gligorijevic B, Chen X, Entenberg D, Condeelis J, Cox D. Visualization of actin polymerization in invasive structures of macrophages and carcinoma cells using photoconvertible β-actin-Dendra2 fusion proteins. PLoS One 2011; 6(2): e16485. [http://dx.doi.org/10.1371/journal.pone.0016485] [PMID: 21339827]

[55] Kim SA, Heinze KG, Schwille P. Fluorescence correlation spectroscopy in living cells. Nat Methods 2007; 4(11): 963-73. [http://dx.doi.org/10.1038/nmeth1104] [PMID: 17971781]

[56] Wachsmuth M, Conrad C, Bulkescher J, *et al.* High-throughput fluorescence correlation spectroscopy enables analysis of proteome dynamics in living cells. Nat Biotechnol 2015; 33(4): 384-9. [http://dx.doi.org/10.1038/nbt.3146] [PMID: 25774713]

[57] Guldbrand S, Kirejev V, Simonsson C, Goksör M, Smedh M, Ericson MB. Two-photon fluorescence correlation spectroscopy as a tool for measuring molecular diffusion within human skin. Eur J Pharm Biopharm 2013; 84(2): 430-6. [http://dx.doi.org/10.1016/j.ejpb.2012.10.001] [PMID: 23085333]

[58] Dörlich RM, Chen Q, Niklas Hedde P, *et al.* Dual-color dual-focus line-scanning FCS for quantitative analysis of receptor-ligand interactions in living specimens. Sci Rep 2015; 5: 10149. [http://dx.doi.org/10.1038/srep10149] [PMID: 25951521]

[59] Costantino S, Comeau JW, Kolin DL, Wiseman PW. Accuracy and dynamic range of spatial image correlation and cross-correlation spectroscopy. Biophys J 2005; 89(2): 1251-60. [http://dx.doi.org/10.1529/biophysj.104.057364] [PMID: 15923223]

[60] Hebert B, Costantino S, Wiseman PW. Spatiotemporal image correlation spectroscopy (STICS) theory, verification, and application to protein velocity mapping in living CHO cells. Biophys J 2005; 88(5): 3601-14. [http://dx.doi.org/10.1529/biophysj.104.054874] [PMID: 15722439]

[61] Wiseman PW. Image correlation spectroscopy: principles and applications. Cold Spring Harb Protoc 2015; 2015(4): 336-48. [http://dx.doi.org/10.1101/pdb.top086124] [PMID: 25834267]

[62] Bacia K, Schwille P. A dynamic view of cellular processes by *in vivo* fluorescence auto- and cross-correlation spectroscopy. Methods 2003; 29(1): 74-85. [http://dx.doi.org/10.1016/S1046-2023(02)00291-8] [PMID: 12543073]

[63] Bacia K, Kim SA, Schwille P. Fluorescence cross-correlation spectroscopy in living cells. Nat Methods 2006; 3(2): 83-9. [http://dx.doi.org/10.1038/nmeth822] [PMID: 16432516]

[64] Ries J, Petrášek Z, García-Sáez AJ, *et al.* A comprehensive framework for fluorescence cross-correlation spectroscopy. New J Phys 2010; 12: 113009. [http://dx.doi.org/10.1088/1367-2630/12/11/113009]

[65] Watson JV. Introduction to flow cytometry. Cambridge: Cambridge University Press 2004.

[66] Hawley TS, Hawley RG. Flow cytometry protocols, Methods in molecular biology. New York: Springer Science+Business Media 2011; 699.

[67] Rocha CK, Praulich I, Gehrke I, Hallek M, Kreuzer KA. A rare case of t(11;22) in a mantle cell lymphoma like B-cell neoplasia resulting in a fusion of IGL and CCND1: case report. Mol Cytogenet 2011; 4(1): 8.
[http://dx.doi.org/10.1186/1755-8166-4-8] [PMID: 21457541]

[68] Morgan E, Varro R, Sepulveda H, *et al.* Cytometric bead array: a multiplexed assay platform with applications in various areas of biology. Clin Immunol 2004; 110(3): 252-66.
[http://dx.doi.org/10.1016/j.clim.2003.11.017] [PMID: 15047203]

[69] Herzenberg LA, Parks D, Sahaf B, Perez O, Roederer M, Herzenberg LA. The history and future of the fluorescence activated cell sorter and flow cytometry: a view from Stanford. Clin Chem 2002; 48(10): 1819-27.
[PMID: 12324512]

[70] Juan G, Hernando E, Cordon-Cardo C. Separation of live cells in different phases of the cell cycle for gene expression analysis. Cytometry 2002; 49(4): 170-5.
[http://dx.doi.org/10.1002/cyto.10173] [PMID: 12454980]

[71] Lim SW, Tran TM, Abate AR. PCR-activated cell sorting for cultivation-free enrichment and sequencing of rare microbes. PLoS One 2015; 10(1): e0113549.
[http://dx.doi.org/10.1371/journal.pone.0113549] [PMID: 25629401]

[72] Carter AD, Bonyadi R, Gifford ML. The use of fluorescence-activated cell sorting in studying plant development and environmental responses. Int J Dev Biol 2013; 57(6-8): 545-52.
[http://dx.doi.org/10.1387/ijdb.130195mg] [PMID: 24166437]

[73] Tkaczyk ER, Tkaczyk AH. Multiphoton flow cytometry strategies and applications. Cytometry A 2011; 79(10): 775-88.
[http://dx.doi.org/10.1002/cyto.a.21110] [PMID: 21796772]

[74] Shields CW IV, Reyes CD, López GP. Microfluidic cell sorting: a review of the advances in the separation of cells from debulking to rare cell isolation. Lab Chip 2015; 15(5): 1230-49.
[http://dx.doi.org/10.1039/C4LC01246A] [PMID: 25598308]

CHAPTER 7

General Fluorescence Reactions

Abstract: Fluorescence analysis has become a powerful tool in Biology and Medicine. Fluorescence microscopy and image technology are now widely used in many research fields. In this Chapter, most important issues are the description of cell and tissue structures as model systems, reactivity of biological substrates, binding modes of fluorochromes, fluorescence parameters, counterstaining, emission color, fading, autofluorescence, photo-enhancement, evaluation of fluorescent images, fluorescence of classical stains, *etc*. Some of these issues are also referred in other Chapters. In the case of reporting experimental observations and when otherwise is not stated, dyes and fluorochromes are applied simply dissolved in distilled water. Generally excitation and/or emission values are indicated directly on the figures.

Keywords: Antifading agents, Autofluorescence, Binding mechanisms, Chromogenic reaction, Competing agents, Counterstaining, Fading, Fluorogenic reactions, H bonding, Hydrophobic forces, Image processing and analysis, Lipofuscin granules, Mounting, Quenching, Reactive oxygen species.

1. MICROSCOPIC STRUCTURES AND BIOLOGICAL SUBSTRATES

Microscopic and spectroscopic techniques constitute a useful approach for analyzing the affinity of dyes and fluorochromes for biological substrates, and indications on possible binding modes can be often put forward. In this respect, the fluorescence of drugs and biologically active compounds is advantageous because they can be detected and localized at the microscopic level.

Numerous biological structures are employed as model systems to study selectivity, sensitivity and mechanisms of fluorescence reactions. For general morphological studies, aldehyde-fixed, hematoxylin and eosin (H&E)-stained paraffin sections are usually employed. Advantages of embryo sections are that almost all tissues and organs can be found, although paraffin sections of specific tissue and organs are the most frequently used materials.

Model microscopic systems for fluorescence analysis in cell biology and histology are frozen, paraffin and epoxy-resin embedded sections from plant and animal tissues, blood smears, cell cultures (living or fixed cells), *etc*. More specific, well known and widely used biological structures are cell types such as

Juan Carlos Stockert & Alfonso Blázquez-Castro

eosinophils from horse blood and nucleated erythrocytes from avian blood, meristematic plant cells, polytene and meiotic chromosomes, yeasts, *Trypanosoma* epimastigotes, peritoneal exudates, ascites tumors, *etc.*

Monolayer cell cultures are useful *in vitro* systems mainly in cell biology studies and for the search of new experimental treatments. At present, 3D cell cultures have been successfully developed, particularly as tumor spheroids [1], and scaffold-based 3D cell cultures [2], in which nanofibers of self-assembling peptides form a 3D matrix (Fig. **7.1**). Fluorescence reactions induced and evaluated in whole organisms have also increasing use.

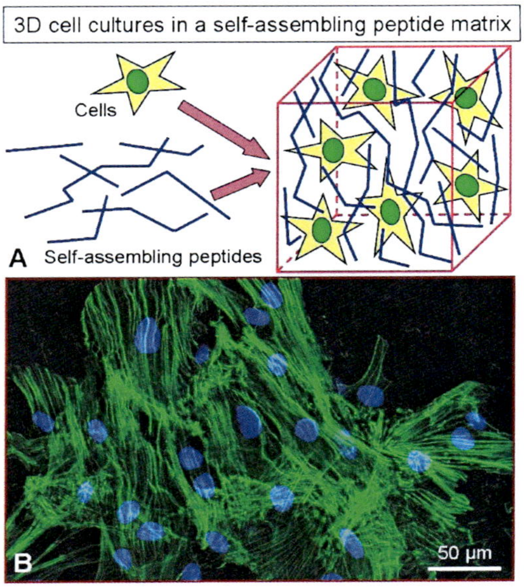

Fig. (7.1). A: Schematic formation of a 3D culture with cells embedded in a peptide matrix. B: 3D culture of normal human skin fibroblasts grown for 10 days in a self-assembling peptide scaffold (PuraMatrix™, AcN-[arginine-alanine-aspartic acid-alanine]$_4$-$CONH_2$) after 1% paraformaldehyde fixation, phalloidin-FITC labeling, and DAPI counterstaining (courtesy of M. Alemany-Ribes and C.E. Semino).

It must point out that before valid conclusions are drawn from microscopical procedures, differential staining or fluorescence have to be interpreted in terms of structure compaction (mass, dry weight). A structure with a high density will also show a high reaction intensity, and then staining or fluorescence degree, which must be borne in mind when comparing and interpreting microscopic results. However, highly compact structures can also result in diminished reactions due to a slowly access of dyes. A gallery of some relevant types of chemical groups in biological substrates, as well as binding modes to ions and ligands (dyes, fluorochromes) are shown in Fig. (**7.2**).

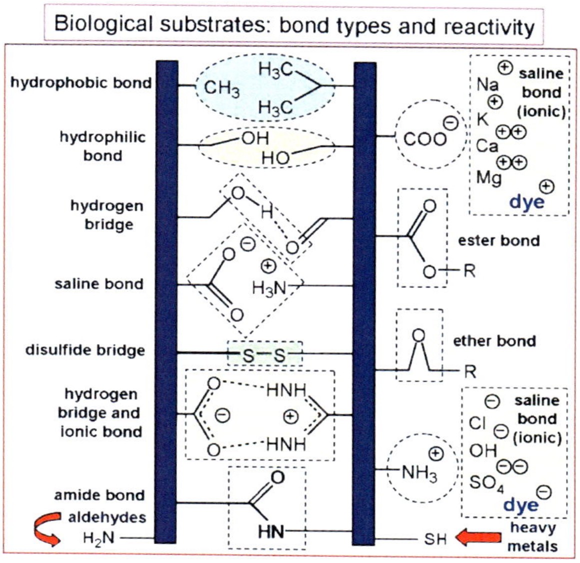

Fig. (7.2). Possible binding modes between chains of biopolymers, or with ions and small molecules. Observe the electrostatic binding mode of cationic and anionic ions and dyes (at right) and the blocking action of some fixatives (at bottom).

Ionic (electrostatic), hydrophobic, covalent, and hydrogen (H) bonding are important binding mechanisms between biopolymers, as well as between them and dyes and fluorochromes. The reactivity of some groups toward aldehydes and heavy metals used in fixing procedures is also indicated. Accordingly, fixing fluids containing heavy ions should be not used to avoid fluorescence quenching, and aldehyde fixation could modify the binding to cationic dyes.

Among the binding mechanisms between dyes and suitable groups of substrates the most relevant are: 1) electrostatic (ionic) linkages, 2) H bonding, 3) covalent bonds, 4) hydrophobic forces, 5) coordinative bonds, and 6) van der Waals attractions. In this case, dipoles are formed when an opposite distribution of positive and negative partial charges (δ) occur along a given bond. These uneven position of δ include Keesom (dipole-dipole: $\delta^+C{=}O\delta^-\ldots\delta^+CO{=}O\delta^-$), Debye (dipole-induced dipole: $\delta^+C{=}O\delta^-\ldots\delta^+C{=}C\delta^-$), and London (induced dipole-induced dipole) forces.

2. FLUORESCENCE PARAMETERS

2.1. General Features

General concepts on important fluorescence parameters such as affinity, selectivity, specificity, intensity, stability, and competitivity are graphically explained in Fig. (**7.3**). The claim that "... this dye is specific for..." should be avoided because almost no reaction has absolute specificity (this is the case for immunology [cross-reactions], enzymatic reactions [analogues, inhibitors], DNA replication [mutations], *etc.*). When specificity of a fluorochrome is claimed for a macromolecular substrate, surely no other substrates are present in the sample. Unfortunately, the designation of "specific" is currently used to signify "high selective". On the other hand, when poor fluorescence intensity is found, it could be due not only to a too small but a too high fluorochrome concentration applied, which leads to the quenching effect (see Chapter 5.2).

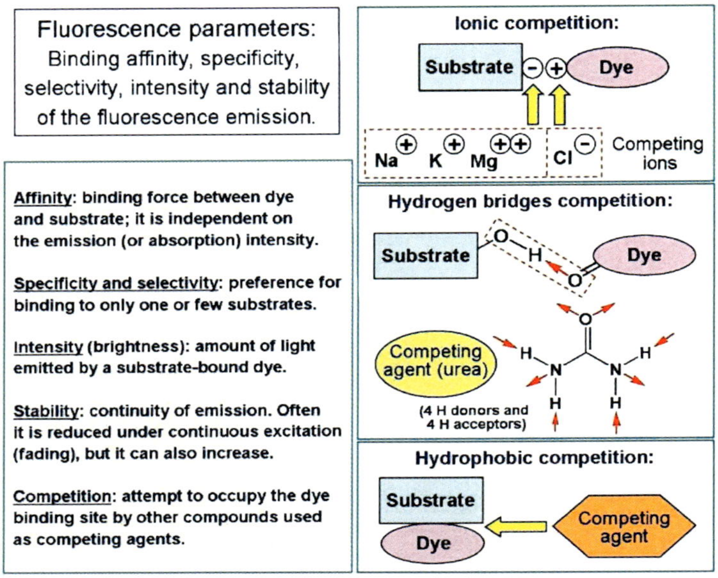

Fig. (7.3). Relevant fluorescence parameters: binding affinity, specificity, selectivity, intensity and stability of the fluorescence emission. The strategy of competition methods is also indicated at right.

The binding affinity of ligands (*i.e.* dyes, fluorochromes) to biological structures can be studied and quantitatively evaluated by competitive strategies. The concept of "critical electrolyte concentration" (CEC), in which salts (NaCl, $MgCl_2$) are used as competing agents to quantify the intensity of ionic binding, has been applied to study staining reactions involving cationic [3, 4] and anionic dyes [5] (see Chapter 8.1). Since in some cases staining and fluorescence are not abolished by 2 M or higher NaCl concentrations, residual non-ionic binding mechanisms (hydrophobic, van der Waals, and H bonding forces) can not be excluded. The type of cations (Al^{3+}, Ba^{2+}, *etc.*) is also relevant for binding to anionic substrates.

The preferential binding of inorganic cations to different anionic groups has found useful applications in histochemistry [6 - 8].

Other competing agents to analyze H bonding and hydrophobic binding can be used (Fig. **7.3**). On account of the very high capacity for H bonding, urea competes efficiently with dyes that use this binding mode. On the other hand, 8-hydroxyquinoline, 1,10-phenanthroline, and methylated oxypurines such as caffeine are planar aromatic compounds that exert effective competition for hydrophobic binding sites. This occurs in the case of DNA intercalation of dyes and drugs, which are susceptible of competition with these agents (see Chapter 12.4.1).

2.2. Counterstaining and Mounting

Lacking reference points provided by autofluorescence, counterstaining is often performed to analyze more precisely the localization of fluorescent signals. In the case of labeled cytoplasm, chromatin counterstaining with cationic fluorochromes allows recognizing the position of nuclei (Fig. **7.4A**). In addition to ethidium bromide and propidium iodide, the cationic porphyrins *meso*-tetra(4-N-methylpyridine)-porphine (TMPyP) and *meso*-tetra(N-trimethylaniline)- porphine (TMAP) are very convenient for nuclear counterstaining to green immunolabeled structures [9]. The wide spectral separation between TMAP excitation (436 nm) and emission (665 nm), as well as non-intercalative (and non-genotoxic) DNA binding [10] makes this fluorochrome suitable as counterstain of multiple fluorescence labeling. The autofluorescence under different exciting light can be used as a counterstaining procedure (Fig. **7.4B**, **C**). Diaminobenzidine-induced emission also provides useful counterstaining in sections of nervous tissue [11].

Obviously, the color of staining and counterstaining must be not the same. Some anionic triphenylmethane dyes such as light green and fast green FCF (Fig. **7.5**) are commonly used for cytoplasmic counterstaining of the Feulgen reaction for DNA in bright-field microscopy [12, 13]. As these dyes exhibit a red fluorescence similar to the Feulgen product, they should be not applied as fluorescent counterstaining. Likewise, methyl green is often used under bright-field illumination as nuclear counterstaining of the periodic acid-Schiff (PAS) reaction for polysaccharides. Under UV excitation (and also blue to a lesser extent), the heptamethylated triphenylmethane dye, methyl green, loses one methyl group and converts in the red-fluorescing crystal violet (Fig. **7.5**), which precludes the use of this counterstaining in fluorescence microscopy when the specific signal is also red.

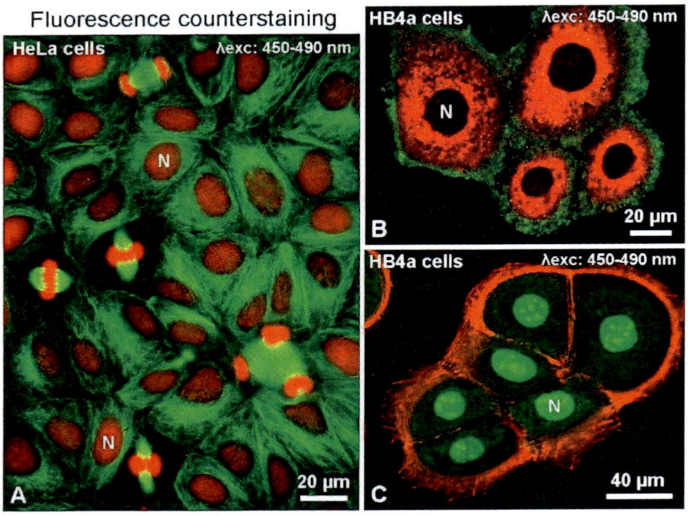

Fig. (7.4). A: Methanol-fixed Pam-212 cells, after immunofluorescence for α-tubulin (FITC, green) and DNA counterstaining (red) with the cationic porphyrin TMPyP (5 x 10^{-6} M). Observe several metaphases and a tripolar telophase (courtesy of A. Juarranz). **B**: Living HB4a cells treated with 0.5 mM verteporfin in liposomes for 3 h (red signal) showing green autofluorescent peripheral cytoplasm. **C**: HB4a cells fixed in 4% formaldehyde and treated with phalloidin-TRITC followed by Hoechst 33258 counterstaining of nuclei, which show green fluorescence under blue exciting light. Observe the red signal of F-actin and green autofluorescent cytoplasm. N: nuclei (courtesy of A. Casas).

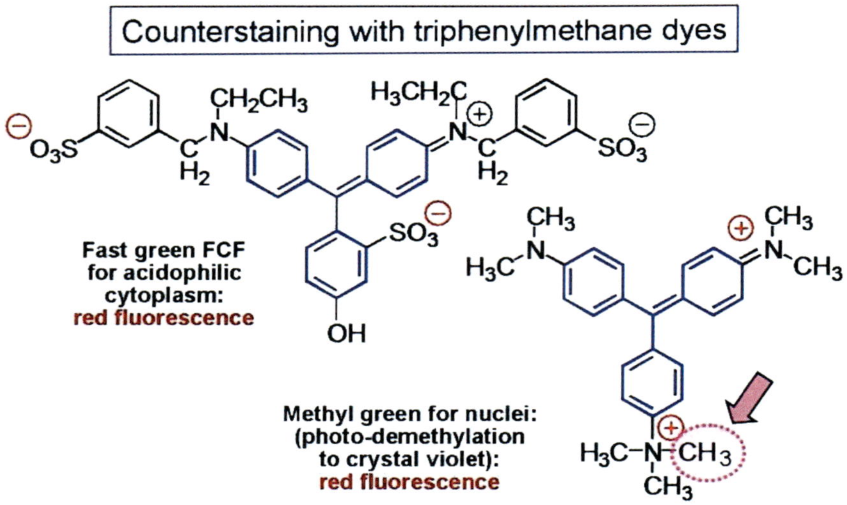

Fig. (7.5). Chemistry of counterstaining. The application of fast green FCF and methyl green counterstaining is adequate for PAS and Feulgen methods in bright-field microscopy, but these dyes are unsuitable for fluorescence observations.

Although there is a widely held belief that aqueous mounting is need for fluorescence studies, (*i.e.* transient mounting in most common aqueous media, either solidified or not, such as glycerol-water, arabic gum, Apáthy's gum syrup, Kaiser's gelatin, Mowiol 40-88, Vectashield), permanent anhydrous mounting does not necessarily hinder observation by fluorescence microscopy. Direct immersion oil, as well as definitive mounting media can be used (Crystal/Mount, Damar resin, Euparal, Eukitt, Entellan, Krystalon), the best results being achieved with DePeX [14], a non-fluorescent solution of polystyrene in xylene that preserves perfectly the result of fluorescence reactions. In this connection, it is worth noting that the fluorescence yield of most fluorophores employed in microscopical work increases in non-polar rigid media. DePeX mounting has the additional advantage that it corresponds to the final step of the routine technique used in pathology laboratories.

2.3. Variations in the Emission Color

The emission color induced by the same fluorochrome can be very different when applied to fixed (dead) or living cells. It is important to note that this phenomenon generally occur with several types of fluorochromes. In the case of acridine orange, staining of fixed cells reveals the distribution of orthochromatic DNA (green) and metachromatic RNA (red-orange) (see Fig. **5.9**), but when the dye is used as a probe for live cells, a red-orange emission is selectively seen in lysosomes whereas nuclei and remaining cytoplasm show green orthochromasia (see Chapter 15.1).

However, there are other causes of unexpected differences in the emission color of a given fluorochrome. Fluorochromes may fluoresce in different colors depending on the microscopic excitation. In the case of hydroxy-diphenylmethane fluorophores such as oxidized hematoxylin and eosin Y, different colors are observed when the same sample is observed under blue or green excitation. This leads to the doubt of which color of emission is the true, an important point especially when the microscopic color does not correspond to spectroscopic features. Thus, the emission color induced by aged hematoxylin is green-yellow under blue excitation [15], but it is red under green excitation (Fig. **7.6**). Spectroscopically, the peak is at 550 nm, which corresponds to the green-yellow visual impression, and the red color is caused by the influence of the limiting barrier filter on the emission spectrum leading to a red shift.

2.4. Chromo- and Fluorogenic Reactions

These reactions correspond to the formation of new chromophores and fluorophores that are generated by using both uncolored substrates and reagents, and therefore they are highly selective because the only color seen in bright-field or fluorescence microscopy is that of the reaction product. The blue reaction of the Lugol solution (containing potassium iodide and iodine) with starch described by Raspail in 1825 is considered the first histochemical reaction and a historical example of a chromogenic reaction. Important chromo- and fluorogenic reactions in histochemistry are the PAS and Feulgen methods. In many cases these reactions implicate the formation of covalent linkages between substrates and colored products. Therefore, they also correspond to reactive staining and fluorescence, making the product resistant to washing.

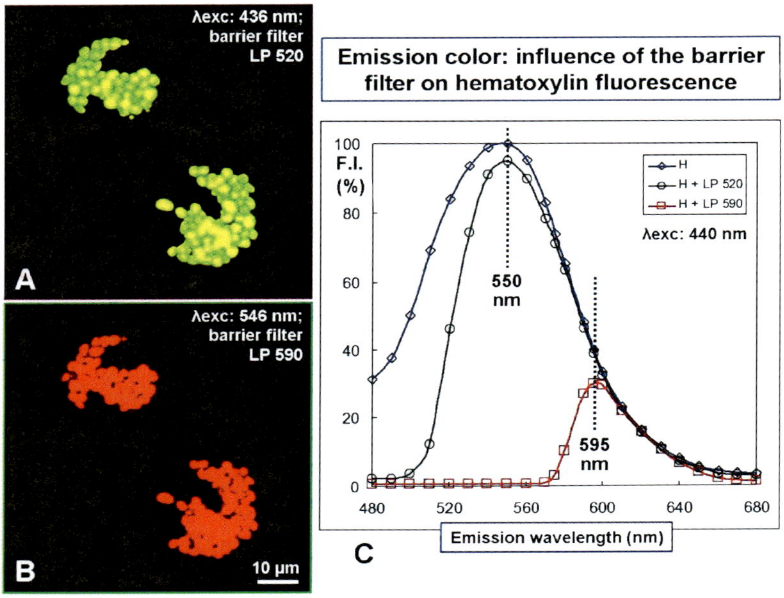

Fig. (7.6). Fluorescence of horse eosinophil leucocytes stained with 4 days aged 0.1 mg/mL hematoxylin and observed under excitation at 436 nm (**A**) and 546 nm (**B**). The type of barrier filter is indicated. **C**: Emission spectra of 4 days aged 50 μg/mL hematoxylin in distilled water recorded without any filter or through the barrier filters LP 520 and LP 590. Observe the peak shift from 550 to 595 nm.

Other examples of fluorogenic reactions are (1) aromatic amino acids with either N-bromosuccinimide or tetrazonium salts, (2) tryptophan with ethanolic isatin, (3) arginine residues with 9,10-phenanthrenequinone, (4) primary amines with either dansyl, furanones, nitrobenzoxadiazole, or 1,8-naphthalic anhydride derivatives, and (5) biogenic amines with formaldehyde. These reactions correspond mainly to proteins and will be described in more detail in Chapters 13.2 and 13.3. Likewise, there are a number of fluorogenic substrates for assessing enzymatic activities, the most important being those based on 4-methylumbelliferone fluorophores, fluorescein diacetate, and ELF substrate.

3. AUTOFLUORESCENCE

Several components of cells show autofluorescence (endogenous fluorescence, primary fluorescence). The case of nucleic acids and proteins, containing bases and aromatic amino acids, respectively, represents a special concern because of their dependence on both UV excitation and emission, which limits microscopic applications. Purines and pyrimidines strongly absorb at 260 nm and can emit UV photons [16 - 19]. On account of the extensive base stacking and hypo-emission, the main dissipation energy mode is predominantly non radiative.

However, some minor bases in RNA show considerable fluorescence, and also duplex denaturation produces unstacking and increase of both fluorescence and phosphorescence (Fig. **7.7**). Chromatin *in situ* was found to have emission from 330 to 480 nm when excited at 240-280 nm [20]. Interestingly, a possible biological significance for this DNA emission has been suggested [21]. Likewise, non-enzymatic glycosylation of single stranded nucleic acids, possibly related to aging, produces fluorescent derivatives of guanine and adenine with λexc: 315 nm and λem: 420 nm [22, 23].

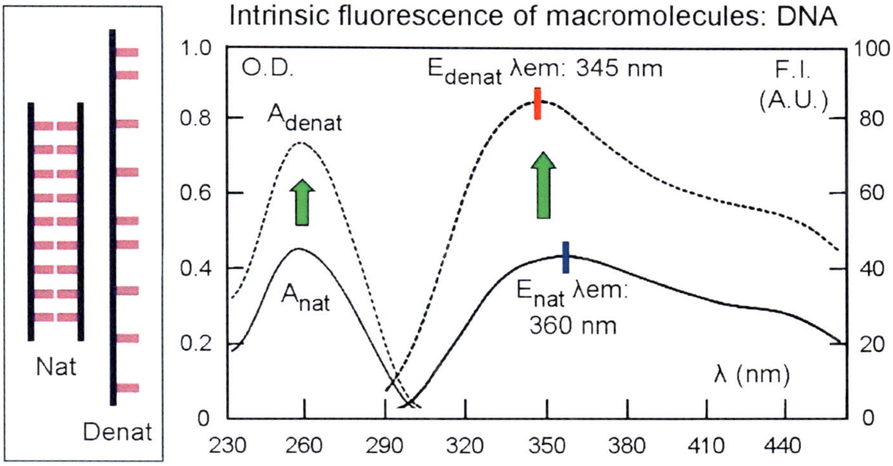

Fig. (7.7). DNA fluorescence. **Left**: scheme of the position of bases, stacked in double-stranded (native) DNA, and partially unstacked in single-stranded (denatured by heating) DNA. **Right**: Absorption (A) and emission (E) spectra of native and denatured DNA in distilled water (9 x 10^{-5} M in bases). Observe fluorescence peaks at 345 nm and 360 nm, corresponding to denatured and native DNA, respectively, and phosphorescence shoulders at 440 nm. Hyperchromic, hyper-emission and blue shift effects are clearly visible after denaturation (λexc: 260 nm).

The occurrence of autofluorescence is a very important subject in microscopy, which may cause misleading results. Although a low level of autofluorescence occurs in most cells and tissues, several structures have a conspicuous emission before fixing procedures. Examples of biological structures with autofluorescence due to endogenous (natural) fluorophores are elastic fibers, chloroplasts, mitochondria, lysosomes, xylem, lipofuscin granules, amyloid, ceroid, suberin, *etc.* Some autofluorescence characteristics are shown here:

Compounds and/or structures	**λexc**	**λem**
Chlorophyll (chloroplast)	blue	red
Porphyrins (no metal)	blue	red
Lignin (xylem)	blue	yellow-green
Elastin	blue	green
Lysosomes	blue	green
Eosinophil granules	blue	green
NADH (mitochondria)	UV	blue
Suberin	UV	yellow

In addition, unfixed frozen sections from mammalian tissues show pale green autofluorescence in keratin, collagen, and skeletal muscle under 436 nm excitation. A compilation of autofluorescent structures is reported in [24]. Natural compound such as chlorophyll, porphyrins, pteridines, flavonoids, carotenes, alkaloids, pyridine and riboflavine coenzymes, vitamins A, E and K, as well as oxidation products of phospholipids, unsaturated fatty acids, and proteins are responsible for the autofluorescence of cells and tissues. Microscopical autofluorescence in plant tissues is seen in Fig. (**7.8**). In this context, it is well known that chlorophyll shows strong red fluorescence (see Fig. **4.10**), allowing recognizing the chloroplasts in a simple way. However, this autofluorescence is a drawback when induced red emission by fluorochromes or immunofluorescent reactions must be recorded.

Regarding chlorophyll, it must be taken into account that in the live cell the main decay pathway of excited chlorophyll is not by fluorescence. In spite of this, chlorophyll emission has been used to roughly evaluate photosynthesis [16], with applications at the macroscopical (planetary) level. Recent advances in spectroscopy enables the space-based monitoring of sun-induced chlorophyll fluorescence from terrestrial plants, providing a global perspective on crop photosynthesis and agricultural productivity [25], but this approach could be controversial.

A low amount of protoporphyrin IX is found in its free form in erythrocytes endowing them with red autofluorescence, which increases in iron deficiency, erythropoietic response, and lead poisoning [16]. When the methine bridges of linear degradation products of natural tetrapyrrole macrocycles are reduced they are poorly fluorescent, but their oxidation by iodine in presence of zinc acetate yields the strongly green fluorescent urobilin-zinc complex [16].

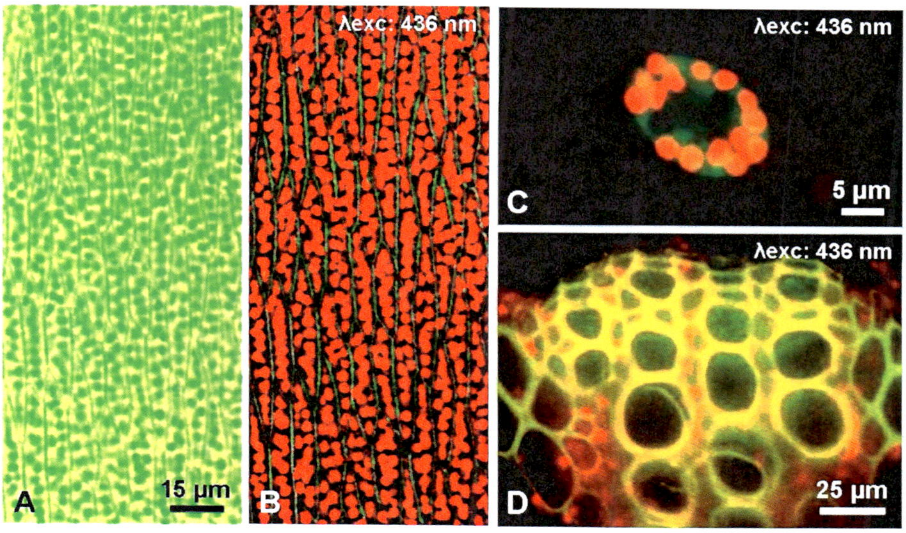

Fig. (7.8). Autofluorescence of plant structures. **A**, **B**: Moos leaf (*Funaria hygrometrica*) observed with bright-field illumination (**A**) and fluorescence microscopy (**B**), showing autofluorescence: red emission of chloroplasts and green fluorescence of cell walls. **C**: Unstained leaf epithelium from *Plectranthus australis* showing stomata guard cells with red fluorescing chloroplasts. **D**: Unstained section of a grape leaf, showing yellow-green fluorescing lignin in xylem.

A considerable but often overlooked autofluorescence occurs in mammalian eosinophil granules [26, 27]. They are modified lysosomes that contain some of their typical enzymes, highly basic proteins, phospholipids, unsaturated fatty acids, and ubiquinone. Some of these compounds (as well as the coenzyme FAD [28]) seem to be responsible for the greenish autofluorescence of eosinophil granules (see Fig. **15.15**). Likewise, lysosome remnants or age pigments could account for autofluorescence of cytoplasmic granules in chicken erythrocytes (Fig. **7.9A**).

It is known that the autofluorescence signal is positively correlated with tissue section thickness [29, 30], an important feature that should be taken into account when autofluorescence intensity is a concern in histological studies (Fig. **7.9B**). In addition, autofluorescence can be originated (or exacerbated) from the application of fixation procedures. Thus the increase of autofluorescence by aldehyde fixatives is a well known fact [31, 32]. Formaldehyde and glutaraldehyde give fluorescent derivatives of catecholamine and indolamine neurotransmitters (see Chapter 13.3), as well as bipyridine and pyridinium products from lysine residues in proteins that can contribute to autofluorescence [33].

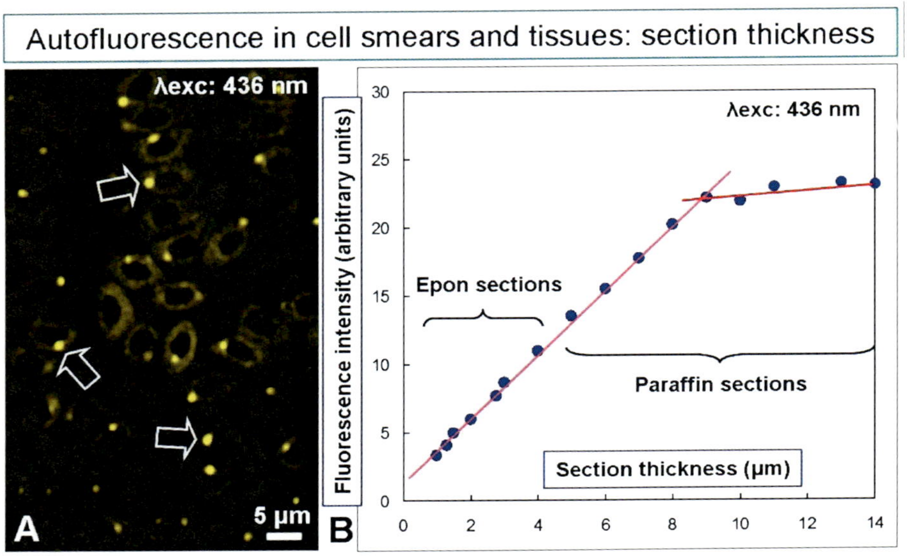

Fig. (7.9). A: Autofluorescent granules (arrows) in the cytoplasm of chicken erythrocytes. **B**: Correlation between thickness of unstained Epon and paraffin sections and autofluorescence intensity measured in glutaraldehyde-fixed, Epon-embedded cartilage (semithin sections of chicken trachea), and in glutaraldehyde-fixed paraffin sections (chicken xiphoid appendix). There is a linear correlation but the slope becomes lower in thicker sections due to limitations from excitation and emission.

Relevant autofluorescence is also found in mitochondria under UV excitation [34 - 38], which is due to the reduced nicotinamide-adenine dinucleotide, NAD(P)H. Under λexc: 365 nm the reduced coenzyme has λem: ~450 nm, and therefore this autofluorescence signal is very useful to visualize mitochondria in living cells (see Chapter 15.5.4). Mitochondria autofluorescence is useful as spontaneous counterstaining, but it can become a drawback when precise co-localization studies must be done using probes for other organelles [39].

A general biomarker of cellular ageing is the appearance over time of autofluorescent residual bodies called age pigments or lipofuscin granules [40]. A yellow-emitting fluorophore from the retinal pigment epithelium (RPE) was isolated and shown to be N-retinylidene-N-retinyl-ethanolamine [41]. This is a Schiff-base product that as cationic surfactant displays a lysosomotropic behavior producing cell damage and death. Although retinal lipofuscin consists in a complex mixture of pigments, the major component seems to be the fluorophore

A2E (λem: 565-570 nm) and its *cis*-isomer (Fig. **7.10A**), which arises from the reaction between all-*trans* retinaldehyde and ethanolamine from membrane phospholipids [42]. A2E accumulation in lysosomes is correlated with age-related macular degeneration.

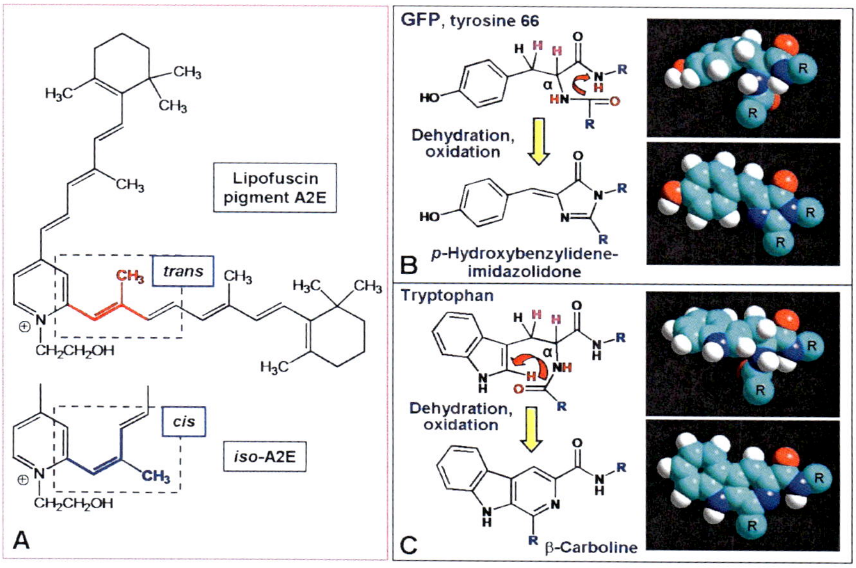

Fig. (7.10). Autofluorescence: chemistry of fluorophores. **A**: Structure of A2E fluorophore and its *cis*-isomer *iso*-A2E. Dashed rectangles indicate changes in the conformation of *trans* (red bonds) in A2E to *cis* (blue bonds) in *iso*-A2E. **B**: Formation *in vivo* of the GFP fluorophore. **C**: Possible formation *in situ* of fluorophores in proteins with aromatic amino acids through dehydration and oxidation, using the >C=O group of peptide bonds.

On the other hand, a great amount of lipofuscin components derives from malonaldehyde (the major product of lipid peroxidation [43]), which can react with amino groups from amino acids and proteins from live cells to yield fluorescent products. The chromophoric system of these products has an amino-iminopropene structure (R-N=CH-CH=CH-NH-R), like fluorescent conjugated Schiff bases [44]. It is interesting to note that spontaneous dehydration and oxidative processes originate the natural autofluorescent GFP (see Chapter 4.6.2). Similarly to the formation of GFP fluorophores, the reaction between aromatic residues in proteins and the carbonyl group of peptide bonds could induce the formation of autofluorescent protein derivatives (Fig. **7.10B, C**).

Some chemicals, physical or image procedures applied for mitigation of endogenous autofluorescence or that generated by fixation are digital processing [30], reducing treatment with sodium borohydride [45], treatment with ammonia/ethanol [46], Sudan black B staining [46, 47], and previous UV photobleaching followed by Sudan black B [48]. The effect of these reagents also depends on the type of tissue and processing techniques, mainly fixation. Therefore, no general recipe for the control of autofluorescence can be delineated. In some cases, results are even worse: although treatment of samples with the reducing agent sodium borohydride is used to quench the autofluorescence induced by glutaraldehyde fixation, in formaldehyde-fixed tissues sodium borohydride induces brilliant autofluorescence in erythrocytes [46].

4. PHOTOCHEMICAL DAMAGE: BLEACHING AND FADING

The reduction of color and fluorescence of dyes and fluorochromes after continuous illumination is known since long ago, and is mainly provoked by oxidation or reduction events (photodegradation, see Chapter 2.5.4). Commonly, the terms bleaching and fading are employed indistinctly in bright-field and fluorescence microscopy, but bleaching (or photobleaching) is more convenient to indicate loss of color in stained structures. A well known example in histology is the poor light-fastness of eosin Y [49], which also varies with the nature of the substrate. Chemical damage due to a self-sensitized photo-oxidation of dyes by singlet oxygen ($^1O_2$) is a common process resulting in the loss of color or fluorescence. This mechanism is responsible for the photofading of azo, indigoid, anthraquinone, triphenylmethane, quinophthalone, stilbene, and other dyes [50].

Photosensitizers (PSs, *i.e.* dyes, drugs) are able to absorb energy and become excited (without spin change) to the first electronically excited singlet state ($^1PS^*$) (Fig. **7.11**). Excited PSs can then decay to the background energy level by photon emission (fluorescence) or, after spin inversion, can convert to the excited triplet state ($^3PS^*$), which decays by phosphorescence. In both cases, energy loss by heating is also produced. When the excited $^3PS^*$ molecule interacts with molecular oxygen ($^3O_2$), the PS decays to its basal state whereas oxygen is promoted to its highly reactive singlet state ($^1O_2$).

In addition to $^1O_2$, other reactive oxygen species (ROS) are produced (*i.e.* hydroxyl and superoxide radicals, hydrogen peroxide, *etc.*). On account of their cytotoxicity, photosensitizing processes have applications in the photodynamic therapy (PDT) of cancer and in other biomedical fields [51 - 56]. In addition, $^1O_2$ can be formed in biological systems by a chemical mechanism that takes place in neutrophils for germicidal activity [57]. Although the photodynamic action has been described in 1900 [58, 59], the germicidal effect of light in the presence of oxygen and urine was just indicated in 1877 [60], and also a light-dark effect of some alkaloids (*i.e.* quinine, cinchonamine) on plants and eggs was anticipated in 1888 by Marcacci [61].

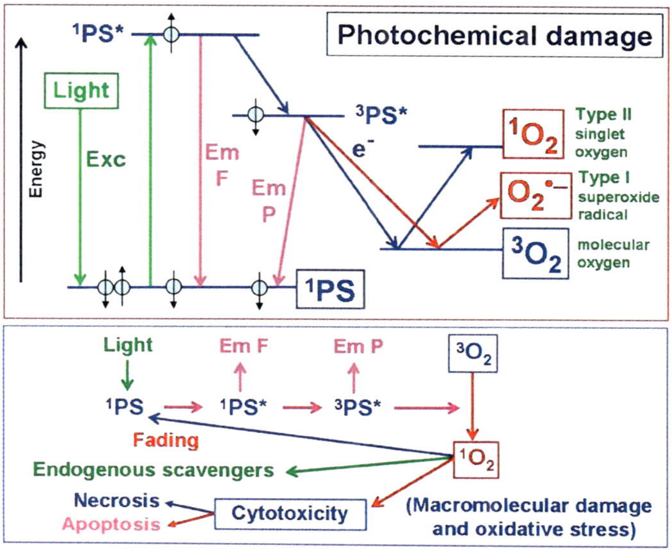

Fig. (7.11). Photochemical damage. **Top**: Excitation of a photosensitizer (PS) by light absorption (Exc) and energy (type II) or electron transfer (type I) to oxygen or neighbor molecules. Absorbed energy can also decay by fluorescence (Em F), phosphorescence (Em P), or heat dissipation (not shown). **Bottom**: Relation between photochemical generation of ROS and chemical (fading) and biological responses (cell damage and death). Endogenous (or exogenous) scavengers can mitigate the oxidative stress.

Recently, the implication of ROS in some human pathological conditions (*i.e.* cancer, inflammative disorders, cardiovascular diseases) has stimulated the interest on the imbalance between production and elimination of ROS. It is well known that strong (damaging) oxidative stress leads to adverse consequences for cells and organisms, whereas at very low concentration, ROS appear as signals for cell stimulation and proliferation [62, 63]. In this case, the use of anti-oxidants would lead to the paradoxical effect of inhibiting the beneficial response to a stimulating oxidative condition.

In the histochemical context, other photochemical reactions of fluorochromes and dyes can also occur: photoreduction of eosin Y and thionine, spontaneous and photochemical demethylation of methyl green to crystal violet, photo-activable caged labels, photo-isomerism, and photo-enhancement of fluorescence (see below). It is known that most dyes and fluorochromes have some degree of photo-activity and can generate $^1O_2$, which causes chemical damage [50]. In this context, most fading and bleaching processes are mainly due to ROS raised by the fluorochrome itself. A gallery of dye-peroxides produced by the action of $^1O_2$ is shown in Fig. (**7.12**). Anti-fading agents are anti-oxidative and reducing compounds that scavenge ROS and free radicals produced by photosensitization, and when added to the mounting medium they retard fluorescence fading. Fading reduction of fluorescein- and rhodamine-conjugates is relevant for immunofluorescence methods [64 - 66].

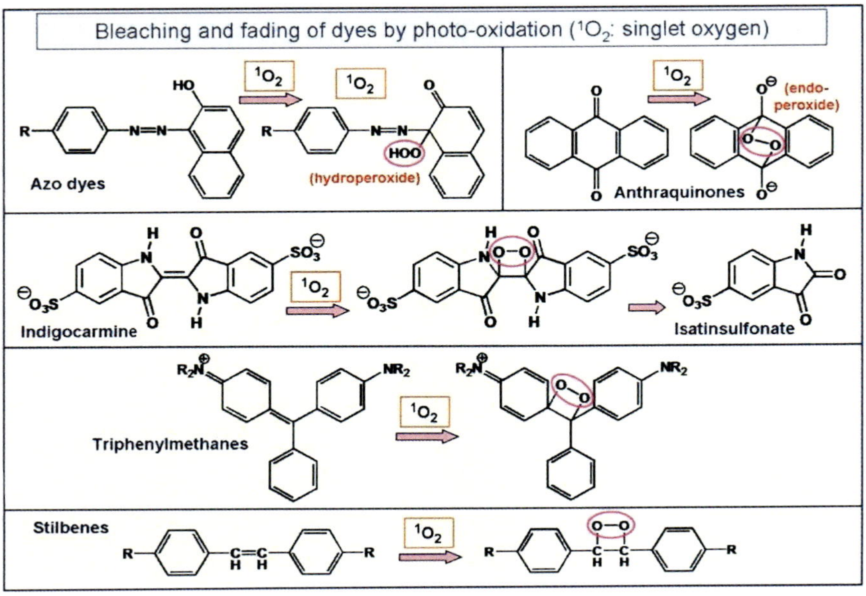

Fig. (7.12). Chemistry of bleaching and fluorescence fading, showing the attack of double bonds by singlet oxygen ($^1O_2$). Relevant examples of photodamage of several dye groups are shown.

DABCO (1,4-diazabicyclo[2.2.2]octane) is one of the most used anti-fading agent [67, 68]. The more trivial name of triethylenediamine (TED) should be preferable for DABCO. *p*-Phenylenediamine [67, 69, 70] and *n*-propyl gallate [65, 71] are other suitable anti-fading agents. When no anti-fading agents is used, typical kinetics of fluorochrome fading in stained materials are illustrated in Fig. (**7.13**).

At present, there are kits to avoid fluorescence fading (*i.e.* Prolong®), which are widely used for mounting. In addition to all the above mentioned agents, cheap and scarcely or not used reducing compounds such as hydroquinone, ascorbic acid, sodium dithionite, pyrogallol, hexamethylene-tetramine, and sodium azide should be useful as ROS scavengers and anti-fading agents. Likewise, some hydrophobic anti-oxidant additives such as butylated hydroxy-toluene (BHT) could be incorporated into organic mounting media (*i.e.* DePeX) to prevent fading.

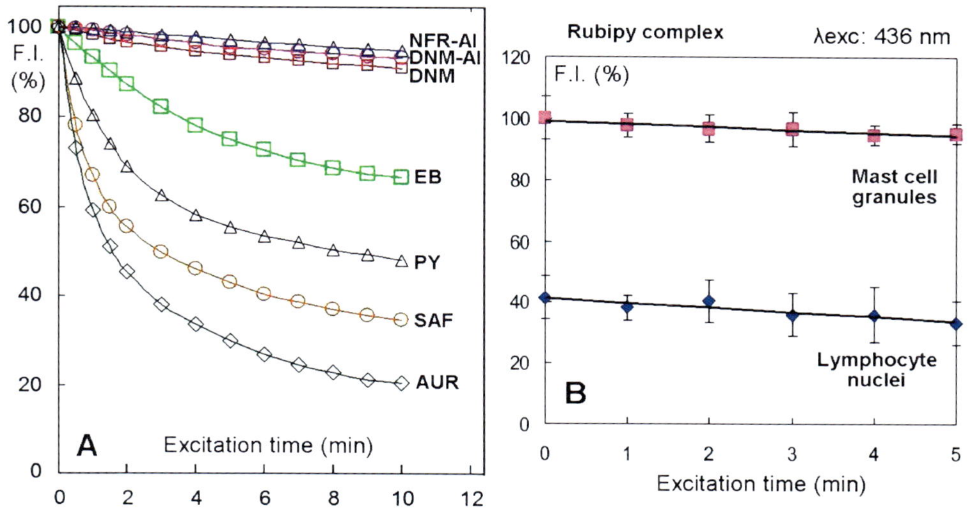

Fig. (7.13). Photofading curves of fluorochromes and fluorescent metal complexes. **A**: Fading rates of fluorescence in chicken erythrocyte nuclei stained with nuclear fast red-Al (NFR-Al), daunomycin-Al (DNM-Al), DNM, ethidium bromide (EB), pyronine Y (PY), safranine O (SAF), and auramine O (AUR). **B**: Fading of mouse mast cells and lymphocytes stained with Rubipy (25 µg/mL, pH 5). Mean values ± SD, *n*: 10.

Regarding fading of fluorescent probes in live cells, there are some natural anti-fading agents (anti-oxidants) that avoid the effects of ROS. Endogenous scavengers protect from oxidative stress in both lipid and aqueous domains, examples being enzymatic systems such as superoxide dismutase, peroxidases, catalases, as well as non-enzymatic anti-oxidants (*i.e.* glutathione, ascorbic acid, dihydrolipoic acid, estrogens, α-tocopherol, β-carotene, flavonoids, and other natural agents [72, 73]. Anti-oxidants also prevent oxidative damage induced by photosensitizers in whole living organisms [74]. On the other hand, to avoid oxygen photodamage, cells can be treated with the commercially available Oxyrase®, which is a suspension of bacterial (*E. coli*) membrane fragments containing an intact electron transport chain that in the presence of reducing sugars (glucose in the culture medium) reduce $^3O_2$ to water, avoiding the photogeneration of $^1O_2$ and improving the image quality [75]. Obviously, this produces an adverse hypoxic condition in live cells.

5. EMISSION PHOTO-ENHANCEMENT

In some cases, an unexpected process of increased emission with time occurs after staining or fluorescence reactions, and there are different mechanisms responsible for these phenomena. During excitation, the fading of highly accumulated fluorochromes in specific cell structures produced by photodamage results in monomerization, and then the fluorescence response increases with time [76, 77]. An interesting example is the case of mercurochrome (Fig. **7.14**).

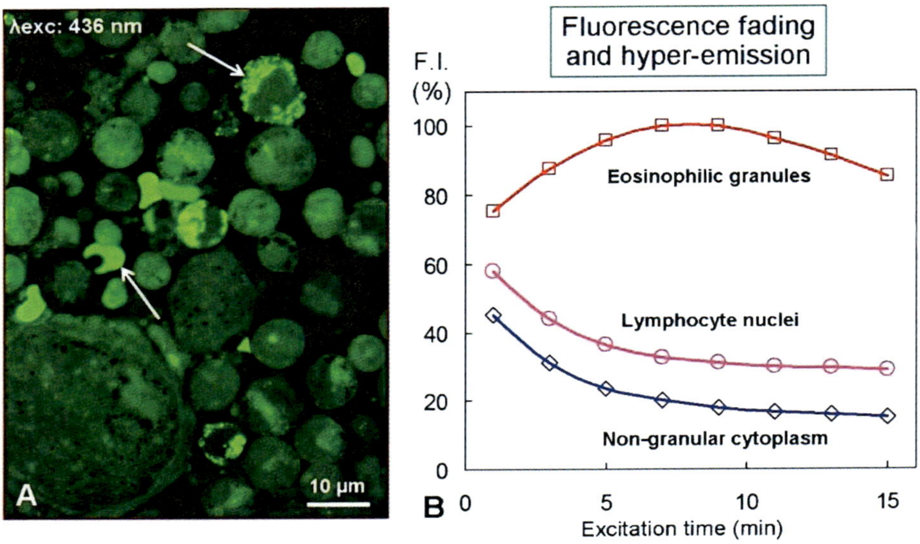

Fig. (7.14). A: Fluorescence in a semithin rat bone marrow section (glutaraldehyde, Durcupan) stained with 0.25 mg/mL mercurochrome in acetone 70%, showing high emission of eosinophil granules and erythrocytes (arrows). **B**: Fading rates of mercurochrome-stained structures under continuous 436 nm excitation.

On the other hand, the oxidizing reagent N-bromo-succinimide (NBS) reacts with aromatic amino acids yielding fluorescent products [78] (see Chapter 13.2.4). Under continuous UV excitation the emission rapidly increases along the first minutes (Fig. **7.15A**), and then a slow fading occurs. The precise mechanism of this process is not yet known.

The trypanocide drug berenil is a minor groove-binding DNA fluorochrome that also shows an increase of emission under continuous UV irradiation (Fig. **7.15B-D**). In this case, photoreduction of the dye would be possible giving a new conformation of the fluorophore (uncharged paraquinone form) with higher and blue shifted emission. Reduction of the slightly asymmetric dication containing the triazene group (-N=N-NH-) to the uncharged dye accentuates the paraquinone structure and stabilizes them. Interestingly, electrochemical reduction of berenil also occurs [79], and these changes could be similar to that of methylviologen under reducing conditions [80]. Other dyes such as eosin Y and thionine also show photoreduction processes [81]. Likewise, emission photo-enhancement phenomena are found for the nitro-naphthalimide product from primary amines

[82], the phenanthrene-imidazole product from arginine [15] (see Chapter 13.2.3), and the porphyrin photosensitizer TMAP [39]. An interesting example of fluorescence photogeneration is the UV-induced conversion of the non-fluorescent cannabinol (from *Cannabis sativa*) into a highly fluorescent phenanthrene derivative [83].

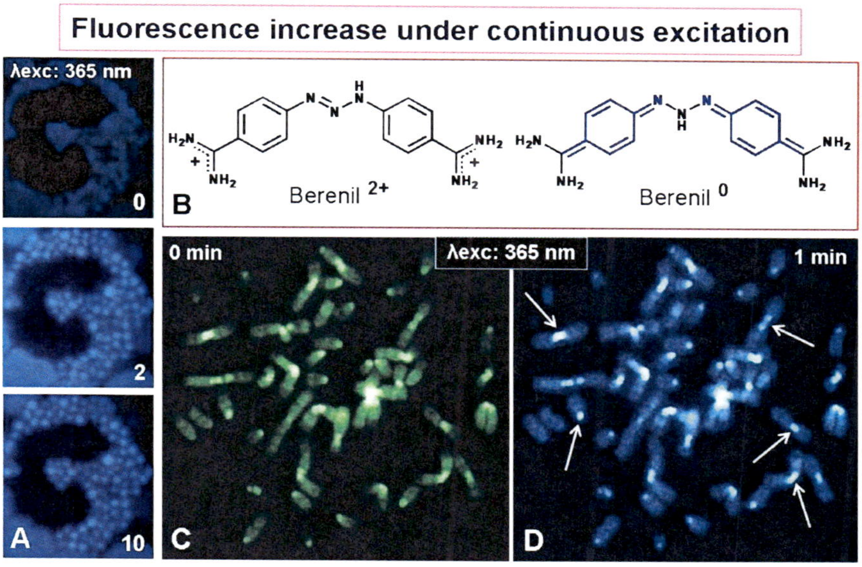

Fig. (7.15). **A**: Fluorogenic reaction of proteins in horse eosinophil granules treated with N-bromo-succinimide (10 mM, 10 min). A rapid increase in emission occurs after continuous UV irradiation (from 0 to 2 min) and then a slow fading is observed (10 min). **B**: Chemical structure of berenil dication showing the paraquinone configuration of a possible photoreduced product (blue). **C**: Metaphase chromosomes from a murine L929 cell stained with 5 µg/mL berenil immediately recorded after exposition to exciting light. **D**: The same field after 1 min excitation showing emission enhancement and blue shift. Arrows indicate the AT-rich centromeric heterochromatin.

6. ABSORPTION-EMISSION CONVERSION

Spontaneous (and more rapidly, photo-induced) demethylation of methyl green (heptamethylated dication) results in the formation of crystal violet (hexamethylated monocation) that in contrast with the former dye shows red fluorescence. In addition to this absorption-emission change, there are also other possibilities of converting bright-field microscopy images into fluorescence images and *vice versa*, a strategy that may be useful for specific purposes. Two examples will illustrate these absorption-emission conversions.

At alkaline pH, insoluble indigo (CI 73000) is easily reduced to give the colorless and water soluble fluorescent compound, leuco-indigo or indigo white (CI 73001). As occurs with indigo, the disulfonate derivative, indigocarmine (CI

73015), also converts into a green fluorescent leuco-derivative (λexc: 436 nm, λem: 528 nm) after reduction with lithium borohydride ($LiBH_4$) or sodium dithionite ($Na_2S_2O_4$) at pH 10 [84]. The fluorescence of indigocarmine-stained eosinophil leucocyte granules induced by alkaline reduction of the bound dye to its leuco derivative *in situ* is shown in Fig. (**7.16**). The fluorescence of leuco-indigocarmine has been exploited as a label for the determination of cationic proteins by capillary electrophoresis under laser excitation [85]. Recently, indigoid pigments from ancient textiles have been assessed as soluble leuco-indigoids by micro-extraction with $NaOH$-$Na_2S_2O_4$ [86].

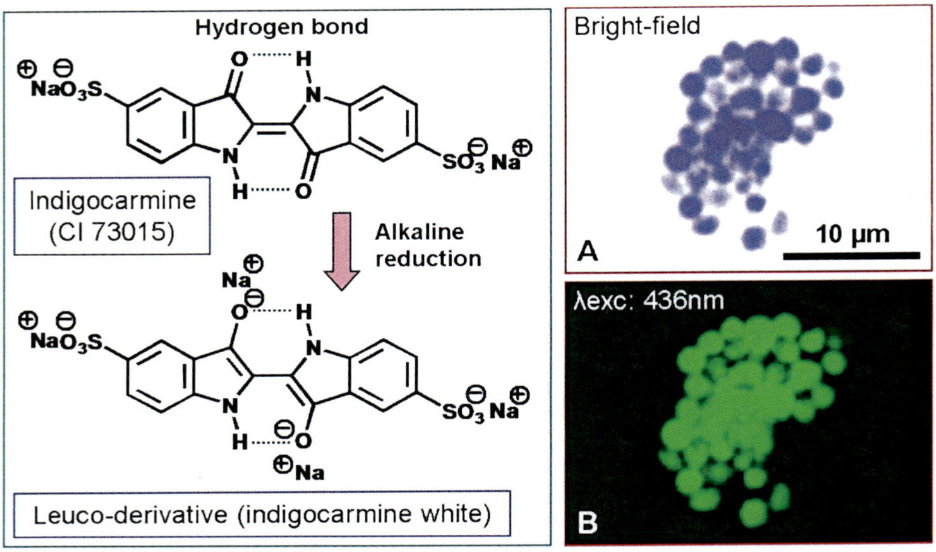

Fig. (7.16). Staining-fluorescence conversion: reduction of indigocarmine to its fluorescent leuco-derivative. **A**: Horse eosinophil stained with indigocarmine. **B**: The same cell after reduction with lithium borohydride.

On the other hand, photoconversion of fluorescent staining in the presence of diaminobenzidine (DAB) originates an oxidized DAB precipitate that is widely used for immunofluorescence and neuroanatomical studies. An important advantage of the approach is to make possible correlations between light and electron microscopy, because the DAB reaction product formed during photoconversion is osmiophilic and therefore, electron dense for electron microscopy [87 - 90]. Several photosensitizing fluorochromes are commonly employed (AO, DAPI, DiI, eosin Y, EB, Lucifer yellow CH, PI), one of the best photo-active labels being eosin Y-isothiocyanate [91, 92]. Interestingly, oxidation of DAB can also reveal and localize the photodynamic generation of ROS *in situ* [93].

It is worth noting that fluorescence and photosensitization are inverse processes. In the case of xanthene dyes, they are related to the amount of halogen substituents. Thus fluorescein, eosin Y and rose Bengal (no halogen, 4 bromine, and 4 iodine atoms in the planar xanthene chromophore, respectively) show an increasing quantum yield of the excited triplet state (Φ_T) in water: rose Bengal (0.76) > eosin Y (0.57) > fluorescein (0.03), with similar values of $^1O_2$ formation (Φ_Δ), but decreasing fluorescence intensity (Φ_F): rose Bengal (0.01) < eosin Y (0.15) < fluorescein (0.71). Likewise, substitution with more heavy halogens leads to a red-shift in the absorption spectrum. The chemical mechanism of this conversion from dye photosensitization to DAB oxidation is illustrated in Fig. (**7.17**). Eosin Y is especially attractive because it is both a good fluorochrome and photosensitizer, and its isothiocyanate derivative is available for antibody labeling.

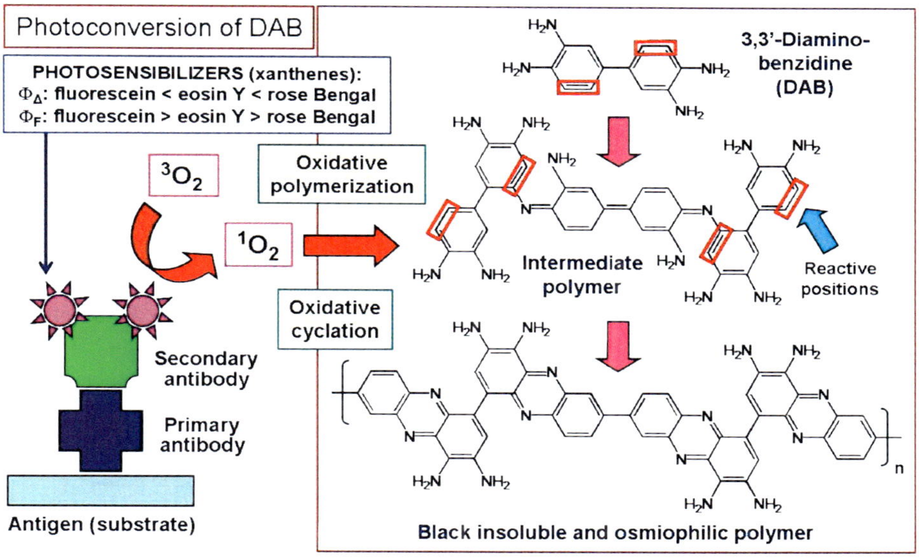

Fig. (7.17). Photoconversion of fluorescence to bright-field staining by using diaminobenzidine (DAB) in immunofluorescence methods. The position of the secondary antibody (labeled by a suitable fluorescent photosensitizer, *i.e.*, eosin Y) is revealed by generation of singlet oxygen ($^1O_2$) that oxidizes DAB to a black polymer.

7. FLUORESCENCE OF CLASSICAL STAINING

In addition to the color observed in bright-field microscopy, several stains also give fluorescence reactions. The use of orcein in bright-field microscopy is well known, but the emission of this dye has been scarcely explored. Orcein is a mixture of closely-related dyes, the predominant one being an amino-phenoxazone derivative, which is formed from the oxidation of orcinol in the presence of ammonia [49]. Both natural and synthetic orceins are widely used for visualization of nuclei, chromosomes, elastin, collagen, reticulin, *etc.* [13, 24, 49, 94]. Depending on the concentration and solvent, these substrates give selective red fluorescence reactions with orcein [95, 96] (see Fig. **5.4**). An example of orcein fluorescence in elastic fibers and reticulin is shown in Fig. (**7.18A**).

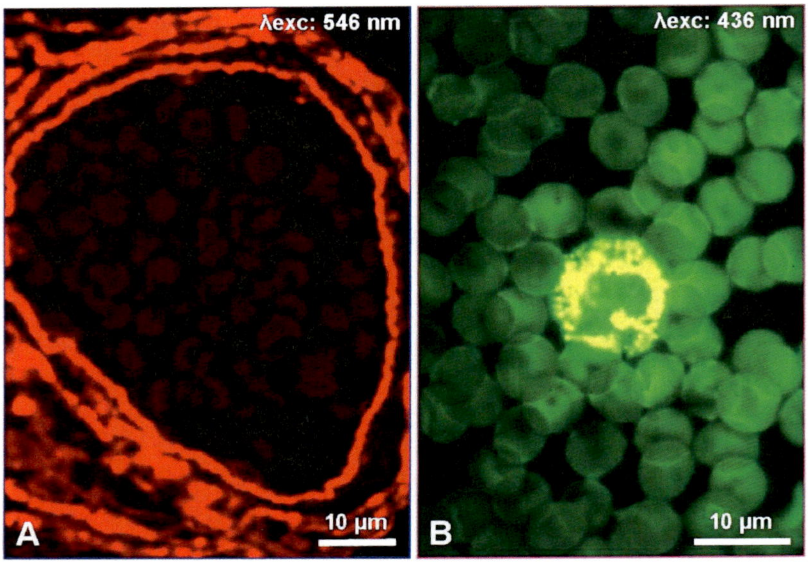

Fig. (7.18). Fluorescence of classical staining: orcein and Giemsa. **A**: Epon semithin section of a mouse uterus stained with 0.2 mg/mL orcein in borate buffer pH 9.2, showing the red fluorescence of elastic and reticular fibers in a blood vessel. **B**: Giemsa-stained human blood smear showing the bright emission of eosinophil granules and low fluorescence of erythrocytes.

H&E, Masson's trichrome and Giemsa staining are routinely used for cytological and histological studies [12, 97]. Eosin Y is a component of Giemsa stain and shows green emission under blue excitation (Fig. **7.18B**), which has been utilized in fluorescence microscopy [98 - 100]. In H&E-stained tissue sections, eosin Y emission is found in acidophilic (eosinophilic) tissue structures such as basal lamina, erythrocytes, eosinophil leucocytes, elastic fibers, pancreas zymogen granules, and muscle fibers [101]. Eosin Y is also a fluorescent probe for mitochondria labeling in living cells [102]. On the other hand, oxidized hematoxylin is fluorescent and allows visualizing acidophilic substrates [15, 103] (see Fig. **3.21**), but hematoxylin-metal complexes used in histology (*i.e.* with Al^{3+}, Cr^{3+}, Fe^{3+} ions) are non-fluorescent.

In the case of Masson's trichrome stain, several structures are stained by the anionic compounds xylidine ponceau and acid fuchsin, which are red fluorescing dyes when rigidly bound to acidophilic substrates and under green excitation [101]. All tissue structures stained with eosin Y appear also stained and fluorescing in red with Masson's trichrome. Obviously, the use of H&E and trichrome staining has the advantages of additional observation possibilities for analyzing archive tissue samples by fluorescence microscopy.

8. EVALUATION OF FLUORESCENCE IMAGES

Conventional, stereo, confocal, and multiple labeling fluorescence microscopy strategies are useful procedures in cell and molecular biology studies. In addition, several methods can be employed to process, improve and interpret images from fluorescence microscopy. Combined microscopy is generally used with esthetic intentions or to locate the object selectively marked in a general topographic context. The use of stained preparations, phase contrast, or differential interference contrast (DIC) combined with fluorescence images are particularly suitable procedures in microscopy, using simultaneous double illumination of the sample or merging by digital methods.

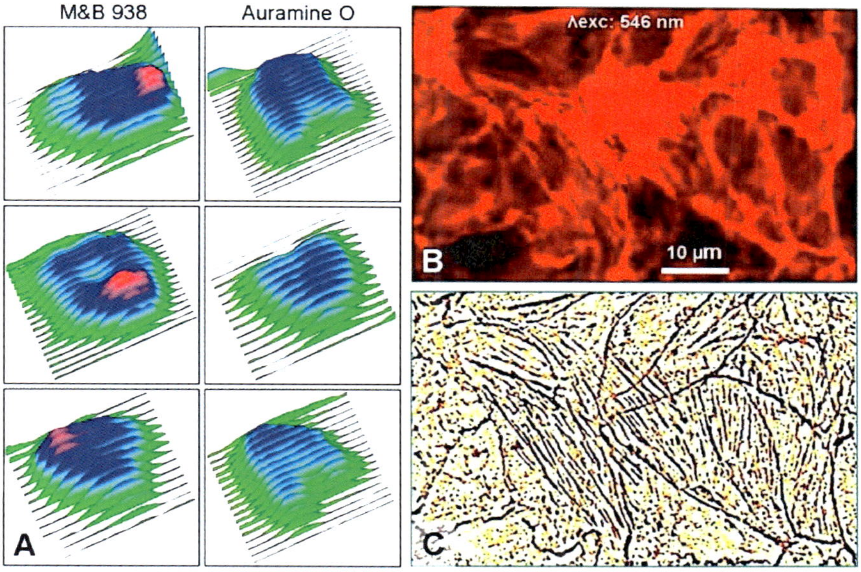

Fig. (7.19). Image processing and analysis. A: Densitometric surfaces from fluorescence images of murine metaphase chromosomes stained with M&B 938 and auramine O (see Chapter 12.5.4). The highest brightness of centromeric heterochromatin (C bands: red color) only appears after M&B 938 staining. B: Actin microfilaments of LM3 cells stained with phalloidin-TRITC showing stress and cortical fibers. C: Convolved image of the same field after white-black inversion and IPA processing (ImageJ 1.37; the convolve matrix filter of 25 points [5 columns and 5 rows] contains a central point with a +24 value, and remaining peripheral points each with a -1 value. LUT pseudocolor: fire).

Computer-assisted analysis of microscopy images can improve detection of structural details that are difficult to see by the naked eye. Image processing and analysis (IPA) methods are relevant for the study of fluorescence images; examples include the use of densitometric curves and surface plots [15, 104 - 106], accumulative densitometric profiles, RGB splitting and color subtraction from multicolor images [39, 107], and conversion of the continuous brightness gradient into pseudo-color equidensities [108, 109]. These techniques are useful in the study of chromosomes, cytoskeleton, and organelles (Fig. **7.19**). Additional examples of IPA applications will be found in other chapters. Public domain software packages are ImageJ 1.37 http://rsb.info.nih.gov/ij/) and Fiji (http://fiji.sc/Fiji).

Regarding fluorescence microscopy measurements, semi-quantitative or simplified quantitative methods [110, 111] are scarcely used, mainly due to the extensive application of photoelectrical devices and availability of easy IPA methods for quantitative measurement of brightness levels. Using computational methods, evaluation of the fluorescence intensity within three-dimensional cell structures is possible [112]. Specific procedures for image analysis are available for particular research designs. 3D imaging is an interesting complement in fluorescence studies on cytoskeleton-labeled cells [113, 114].

9. FALSE RESULTS AND MISLEADING INTERPRETATIONS

Obviously, false results can occur by the erroneous use of microscopic procedures, and here only some few examples will be described. The application of inconvenient methods (*i.e.* bad choice of fixatives, or incorrect handling of reagents and techniques) can produce false positive or negative results. On account of collateral affinities, the use of macromolecular reagents can also produce erroneous results. Avidin is a strongly cationic protein, and therefore, conjugated avidin binds to polyanions such as heparin from mast cell granules inducing a false positive result [115].

Fixation and preparative procedures can introduce a number of artifacts resulting in misleading results. Aldehydes bind to free amino groups and generally increase the basophilia of tissues. Heavy ions such as Pb^{2+} and Hg^{2+} bind with high affinity to -SH groups, making them undetectable. The removal of divalent cations (*i.e.* Mg^{2+}, Ca^{2+}, Zn^{2+}, Cu^{2+}) by treatment with chelating agents induces a striking acidophilia [116].

After glutaraldehyde fixation, remaining aldehyde groups in proteins will result in false PAS or Feulgen positive reactions (Fig. **7.20**). The same occurs by reacting free aldehydes with Schiff's reagent [117]. Likewise, after osmium tetroxide fixation, -CH=CH- bonds from unsaturated lipids and pyrimidines form glycols that oxidize and give PAS positivity in lipid- and nucleic acid-containing structures such as cell membranes, chromatin, and nucleoli. Selective cytochemical techniques (*i.e.* osmium tetroxide/*p*-phenylenediamine [118]) followed by the PAS reaction allows to visualize clearly the presence of osmiophilic structures. In all these cases, the red emission of recolored basic fuchsine is easily observable in fluorescence microscopy [119]. Critical conditions in some cytochemical methods can also produce erroneous results. Too short or too prolonged time of acid hydrolysis will give weakly positive Feulgen reactions or not at all (Fig. **7.20**). Also the attempt of detecting DNA below the methodological or instrumental limits will yield false negative results.

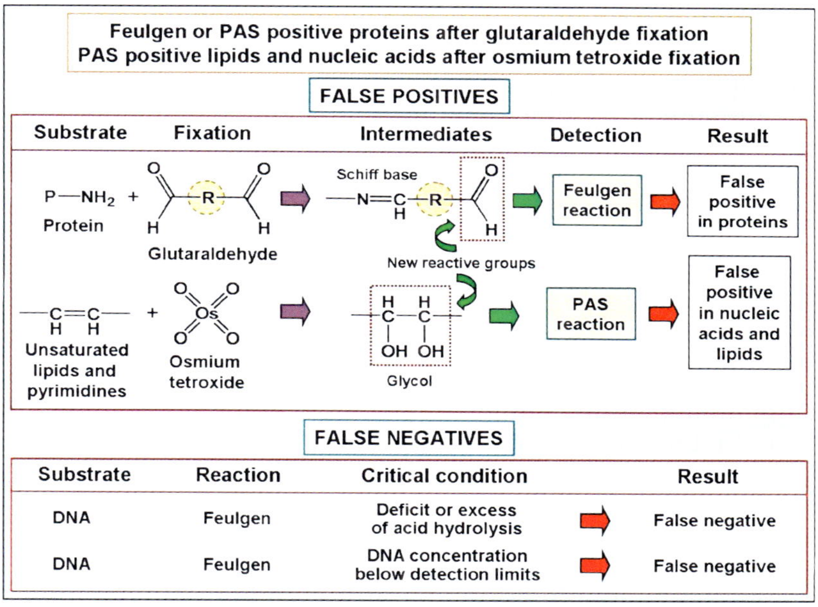

Fig. (7.20). A, B: A survey of false positive and false negative results in microscopical and histochemical fluorescence reactions.

Sometimes misleading interpretations result from unclear methodology, such as the intriguing green fluorescing image of the porphyrin XF73 in the corneous layer of frozen pig skin sections [120], applied to photoinactivate resistant *Staphylococcus aureus* strains. According to the authors, this green fluorescence (instead red) was due to an image capture in gray levels (λexc: 405 nm, barrier filter > 600 nm), followed by conversion of brightness to green color only for esthetic motives. As this explanation does not appear in the published work, the reader could be confused by this unexpected green emission.

CONFLICT OF INTEREST

The author (editor) declares no conflict of interest, financial or otherwise.

ACKNOWLEDGEMENTS

This work was supported by a grant from the Ministerio de Economía y Competitividad (CTQ2013-48767-C3-3-R), Spain.

REFERENCES

[1] Mehta G, Hsiao AY, Ingram M, Luker GD, Takayama S. Opportunities and challenges for use of tumor spheroids as models to test drug delivery and efficacy. J Control Release 2012; 164(2): 192-204. [http://dx.doi.org/10.1016/j.jconrel.2012.04.045] [PMID: 22613880]

[2] Alemany-Ribes M, García-Díaz M, Acedo P, *et al.* Why not introducing the third dimension in photodynamic therapy research? J Anal Bioanal Tech 2013; S1: 004.

[3] Scott JE. Affinity, competition and specific interactions in the biochemistry and histochemistry of polyelectrolytes. Biochem Soc Trans 1973; 1: 787-806. [http://dx.doi.org/10.1042/bst0010787]

[4] Stockert JC. Cytochemistry of nucleic acids: binding mechanisms of dyes and fluorochromes. Microsc Electron Biol Cellular 1985; 9(2): 89-131. [PMID: 2431250]

[5] Trigoso CI, Del Castillo P, Carlos Stockert J. Influence of inorganic salts on the staining reaction of eosinophil leukocyte granules by anionic dyes. Acta Histochem 1992; 93(1): 313-8. [http://dx.doi.org/10.1016/S0065-1281(11)80230-8] [PMID: 1382350]

[6] Heath ID. Observations on a highly specific method for the histochemical detection of sulphated mucopolysaccharides, and its possible mechanisms. Q J Microsc Sci 1962; 103: 457-75.

[7] Del Castillo P, Llorente AR, Gómez A, Gosálvez J, Goyanes VJ, Stockert JC. New fluorescence reactions in DNA cytochemistry. 2. Microscopic and spectroscopic studies on fluorescent aluminum complexes. Anal Quant Cytol Histol 1990; 12(1): 11-20. [PMID: 2302307]

[8] Stockert JC, Ibáñez N, Trigoso CI, Cañete M, Tato A. A barium method for the cytochemical detection of sulfated glycosaminoglycans in mast cells and basophilic leukocytes. Acta Histochem 1999; 101(4): 397-408. [http://dx.doi.org/10.1016/S0065-1281(99)80040-3] [PMID: 10611928]

[9] Juarranz A, Villanueva A, Cañete M, Stockert JC. Fluorescent porphyrin counterstaining of chromatin DNA in conjunction with immunofluorescence methods using FITC-labelled antibodies. J Microsc 1996; 182(Pt 1): 46-9. [http://dx.doi.org/10.1046/j.1365-2818.1996.114394.x] [PMID: 8632446]

[10] Carvlin MJ, Fiel RJ. Intercalative and nonintercalative binding of large cationic porphyrin ligands to calf thymus DNA. Nucleic Acids Res 1983; 11(17): 6121-39. [http://dx.doi.org/10.1093/nar/11.17.6121] [PMID: 6889138]

[11] von Bohlen und Halbach O, Kiernan JA. Diaminobenzidine induces fluorescence in nervous tissue and provides intrinsic counterstaining of sections prepared for peroxidase histochemistry. Biotech Histochem 1999; 74(5): 236-43. [http://dx.doi.org/10.3109/10520299909034659] [PMID: 10711503]

[12] Bancroft JD, Stevens A. Theory and Practice of Histological Techniques. 3rd ed., Edinburgh: Churchill Livingstone 1990.

[13] Horobin RW, Kiernan JA, Eds. Conn's Biological Stains A Handbook of Dyes, Stains and Fluorochromes for Use in Biology and Medicine. 10th ed., Oxford, UK: BIOS Scientific Publishers 2002.

[14] Espada J, Juarranz A, Galaz S, *et al.* Non-aqueous permanent mounting for immunofluorescence microscopy. Histochem Cell Biol 2005; 123(3): 329-34.
[http://dx.doi.org/10.1007/s00418-005-0769-2] [PMID: 15856278]

[15] Stockert JC. The horse eosinophil as a model leucocyte for morphological and cytochemical studies. Braz J Morphol Sci 2005; 22: 73-84.

[16] Udenfriend S. Fluorescence Assay in Biology and Medicine. New York, London: Academic Press 1962.

[17] Konev SV. Fluorescence and Phosphorescence of Proteins and Nucleic Acids. New York: Plenum Press 1967; pp. 147-72.
[http://dx.doi.org/10.1007/978-1-4684-0700-6_7]

[18] Morgan JP, Daniels M. Excited states of DNA and its components at room temperature. IV. Spectral, polarization and quantum yield studies of emissions from CpC and poly rC. Photochem Photobiol 1980; 31: 207-13.
[http://dx.doi.org/10.1111/j.1751-1097.1980.tb03709.x]

[19] Del Castillo P, Stockert JC. Fluorescencia del DNA: Efecto de la desnaturalización y de la intercalación. Tecn Laboratorio 1988; 11: 440-4.

[20] Matsubara H, Yamada M. Auto- and acquired fluorescence of chromatin. Cell Mol Biol 1986; 32(4): 411-8.
[PMID: 3742551]

[21] Popp FA, Nagl W, Li KH, Scholz W, Weingärtner O, Wolf R. Biophoton emission. New evidence for coherence and DNA as source. Cell Biophys 1984; 6(1): 33-52.
[http://dx.doi.org/10.1007/BF02788579] [PMID: 6204761]

[22] Bucala R, Model P, Cerami A. Modification of DNA by reducing sugars: a possible mechanism for nucleic acid aging and age-related dysfunction in gene expression. Proc Natl Acad Sci USA 1984; 81(1): 105-9.
[http://dx.doi.org/10.1073/pnas.81.1.105] [PMID: 6582469]

[23] Lee AT, Cerami A. *In vitro* and *in vivo* reactions of nucleic acids with reducing sugars. Mutat Res 1990; 238(3): 185-91.
[http://dx.doi.org/10.1016/0165-1110(90)90010-9] [PMID: 2111459]

[24] Thompson SW. Selected Histochemical and Histopathological Methods. Springfield, Illinois: Charles C. Thomas 1966.

[25] Guanter L, Zhang Y, Jung M, *et al.* Global and time-resolved monitoring of crop photosynthesis Proc Nat Acad Sci USA 2014.

[26] Fuerst DE, Jannach JR. Autofluorescence of eosinophils: a bone marrow study. Nature 1965; 205: 1333-4.
[http://dx.doi.org/10.1038/2051333a0]

[27] Stockert JC, Ferrer JM, González-Garrigues M. Autofluorescence of eosinophil leukocyte granules. Biol Zentralbl 1984; 103: 543-6.

[28] Mayeno AN, Hamann KJ, Gleich GJ. Granule-associated flavin adenine dinucleotide (FAD) is responsible for eosinophil autofluorescence. J Leukoc Biol 1992; 51(2): 172-5.
[PMID: 1431554]

[29] Del Castillo P, Llorente AR, Stockert JC. Influence of fixation, exciting light and section thickness on the primary fluorescence of samples for microfluorometric analysis. Basic Appl Histochem 1989; 33(3): 251-7.

[PMID: 2684138]

[30] Van de Lest CH, Versteeg EM, Veerkamp JH, Van Kuppevelt TH. Elimination of autofluorescence in immunofluorescence microscopy with digital image processing. J Histochem Cytochem 1995; 43(7): 727-30.
[http://dx.doi.org/10.1177/43.7.7608528] [PMID: 7608528]

[31] Collins JS, Goldsmith TH. Spectral properties of fluorescence induced by glutaraldehyde fixation. J Histochem Cytochem 1981; 29(3): 411-4.
[http://dx.doi.org/10.1177/29.3.6787116] [PMID: 6787116]

[32] del Castillo P, Molero ML, Ferrer JM, Stockert JC. Autofluorescence and induced fluorescence in Epon embedded tissue sections. Histochemistry 1986; 85(5): 439-40.
[http://dx.doi.org/10.1007/BF00982676] [PMID: 3781887]

[33] Hayat MA. Fixation for Electron Microscopy. New York, London: Academic Press 1981.

[34] Andersson H, Baechi T, Hoechl M, Richter C. Autofluorescence of living cells. J Microsc 1998; 191(Pt 1): 1-7.
[http://dx.doi.org/10.1046/j.1365-2818.1998.00347.x] [PMID: 9723186]

[35] Petrat F, Pindiur S, Kirsch M, de Groot H. NAD(P)H, a primary target of $^1O_2$ in mitochondria of intact cells. J Biol Chem 2003; 278(5): 3298-307.
[http://dx.doi.org/10.1074/jbc.M204230200] [PMID: 12433931]

[36] Cristóbal J, Stockert JC, Villanueva A, Rello-Varona S, Juarranz A, Cañete M. Caspase-2: a possible trigger of apoptosis induced in A-549 tumor cells by ZnPc photodynamic treatment. Int J Oncol 2006; 28(5): 1057-63.
[PMID: 16596221]

[37] Rello-Varona S, Stockert JC, Cañete M, Acedo P, Villanueva A. Mitotic catastrophe induced in HeLa cells by photodynamic treatment with Zn(II)-phthalocyanine. Int J Oncol 2008; 32(6): 1189-96.
[http://dx.doi.org/10.3892/ijo_32_6_1189] [PMID: 18497980]

[38] Stockert JC, Blázquez-Castro A, Cañete M, Horobin RW, Villanueva A. MTT assay for cell viability: Intracellular localization of the formazan product is in lipid droplets. Acta Histochem 2012; 114(8): 785-96.
[http://dx.doi.org/10.1016/j.acthis.2012.01.006] [PMID: 22341561]

[39] Stockert JC, Villanueva A, Cristóbal J, Cañete M. Improving images of fluorescent cell labeling by background signal subtraction. Biotech Histochem 2009; 84(2): 63-8.
[http://dx.doi.org/10.1080/10520290902804357] [PMID: 19267289]

[40] Colcolough HL, Helmy FM, Hack MH. Some histochemical observations on the lipofuscin of vertebrate liver, kidney and cardiac muscle. Acta Histochem 1970; 35(2): 343-56.
[PMID: 4986603]

[41] Eldred GE, Lasky MR. Retinal age pigments generated by self-assembling lysosomotropic detergents. Nature 1993; 361(6414): 724-6.
[http://dx.doi.org/10.1038/361724a0] [PMID: 8441466]

[42] Sparrow JR, Parish CA, Hashimoto M, Nakanishi K. A2E, a lipofuscin fluorophore, in human retinal pigmented epithelial cells in culture. Invest Ophthalmol Vis Sci 1999; 40(12): 2988-95.
[PMID: 10549662]

[43] Siakotos AN, Munkres KD. Recent Developments in the Isolation and Properties of Autofluorescent Lipopigments.Ceroid-Lipofuscinosis. New York: Elsevier 1982; pp. 167-78.

[44] Chio KS, Tappel AL. Synthesis and characterization of the fluorescent products derived from malonaldehyde and amino acids. Biochemistry 1969; 8(7): 2821-6.
[http://dx.doi.org/10.1021/bi00835a019] [PMID: 5808334]

[45] Clancy B, Cauller LJ. Reduction of background autofluorescence in brain sections following

immersion in sodium borohydride. J Neurosci Methods 1998; 83(2): 97-102. [http://dx.doi.org/10.1016/S0165-0270(98)00066-1] [PMID: 9765122]

[46] Baschong W, Suetterlin R, Laeng RH. Control of autofluorescence of archival formaldehyde-fixed, paraffin-embedded tissue in confocal laser scanning microscopy (CLSM). J Histochem Cytochem 2001; 49(12): 1565-72. [http://dx.doi.org/10.1177/002215540104901210] [PMID: 11724904]

[47] Oliveira VC, Carrara RC, Simoes DL, *et al.* Sudan Black B treatment reduces autofluorescence and improves resolution of *in situ* hybridization specific fluorescent signals of brain sections. Histol Histopathol 2010; 25(8): 1017-24. [PMID: 20552552]

[48] Viegas MS, Martins TC, Seco F, do Carmo A. An improved and cost-effective methodology for the reduction of autofluorescence in direct immunofluorescence studies on formalin-fixed paraffin-embedded tissues. Eur J Histochem 2007; 51(1): 59-66. [PMID: 17548270]

[49] Lillie RD. Conn's Biological Stains. 9th ed., Baltimore: Williams and Wilkins 1977.

[50] Kuramoto N, Kitao T. Contribution of singlet oxygen to the photofading os some dyes. J Soc Dyers Colour 1982; 98: 334-40. [http://dx.doi.org/10.1111/j.1478-4408.1982.tb03604.x]

[51] Stockert JC, Juarranz A, Villanueva A, *et al.* Photodynamic therapy: selective uptake of photosensitizing drugs into tumor cells. Curr Top Pharmacol 2004; 8: 185-217.

[52] Cañete M, Stockert JC, Villanueva A. Preclinical photodynamic therapy research in Spain. 3. Localization of photosensitizers and mechanisms of cell death *in vitro*. J Porphyr Phthalocyanines 2009; 13: 544-51. [http://dx.doi.org/10.1142/S1088424609000516]

[53] Agostinis P, Berg K, Cengel KA, *et al.* Photodynamic therapy of cancer: an update. CA Cancer J Clin 2011; 61(4): 250-81. [http://dx.doi.org/10.3322/caac.20114] [PMID: 21617154]

[54] Kharkwal GB, Sharma SK, Huang YY, Dai T, Hamblin MR. Photodynamic therapy for infections: clinical applications. Lasers Surg Med 2011; 43(7): 755-67. [http://dx.doi.org/10.1002/lsm.21080] [PMID: 22057503]

[55] Vera DM, Haynes MH, Ball AR, *et al.* Strategies to potentiate antimicrobial photoinactivation by overcoming resistant phenotypes. Photochem Photobiol 2012; 88(3): 499-511. [http://dx.doi.org/10.1111/j.1751-1097.2012.01087.x] [PMID: 22242675]

[56] Yoo JO, Ha KS. New insights into the mechanisms for photodynamic therapy-induced cancer cell death. Int Rev Cell Mol Biol 2012; 295: 139-74. [http://dx.doi.org/10.1016/B978-0-12-394306-4.00010-1] [PMID: 22449489]

[57] Stief TW. The physiology and pharmacology of singlet oxygen. Med Hypotheses 2003; 60(4): 567-72. [http://dx.doi.org/10.1016/S0306-9877(03)00026-4] [PMID: 12615524]

[58] Raab O. Über die Wirkung Fluoreszierender Stoffe auf infusorien. Zeitschr Biol 1900; 39: 524-46.

[59] Tappeiner Hv. Über die Wirkung Fluoreszierender Stoffe auf infusorien nach Versuchen von O. Raab. Munch Med Wochenschr 1900; 47: 5-7.

[60] Downes A, Blunt TP. Researches on the effect of light upon Bacteria and other organisms. Proc R Soc London Ser B 1877; 26: 488-500.

[61] Marcacci A. Sur l'action des alcaloïdes dans le règne végétal et animal. Arch Ital Biol 1888; 9: 2-4.

[62] Blázquez-Castro A, Carrasco E, Calvo MI, *et al.* Protoporphyrin IX-dependent photodynamic production of endogenous ROS stimulates cell proliferation. Eur J Cell Biol 2012; 91(3): 216-23. [http://dx.doi.org/10.1016/j.ejcb.2011.12.001] [PMID: 22293959]

[63] Blázquez-Castro A, Breitenbach T, Ogilby PR. Singlet oxygen and ROS in a new light: low-dose subcellular photodynamic treatment enhances proliferation at the single cell level. Photochem Photobiol Sci 2014; 13(9): 1235-40.
[http://dx.doi.org/10.1039/C4PP00113C] [PMID: 25051122]

[64] Picciolo GL, Kaplan DS. Reduction of fading of fluorescent reaction product for microphotometric quantitation. Adv Appl Microbiol 1984; 30: 197-234.
[http://dx.doi.org/10.1016/S0065-2164(08)70056-X] [PMID: 6442533]

[65] Valnes K, Brandtzaeg P. Retardation of immunofluorescence fading during microscopy. J Histochem Cytochem 1985; 33(8): 755-61.
[http://dx.doi.org/10.1177/33.8.3926864] [PMID: 3926864]

[66] Longin A, Souchier C, Ffrench M, Bryon PA. Comparison of anti-fading agents used in fluorescence microscopy: image analysis and laser confocal microscopy study. J Histochem Cytochem 1993; 41(12): 1833-40.
[http://dx.doi.org/10.1177/41.12.8245431] [PMID: 8245431]

[67] Johnson GD, Davidson RS, McNamee KC, Russell G, Goodwin D, Holborow EJ. Fading of immunofluorescence during microscopy: a study of the phenomenon and its remedy. J Immunol Methods 1982; 55(2): 231-42.
[http://dx.doi.org/10.1016/0022-1759(82)90035-7] [PMID: 6819318]

[68] Langanger G, De Mey J, Adam H. [1,4-Diazobicyclo-(2,2,2)-octane (DABCO) retards the fading of immunofluorescence preparations]. Mikroskopie 1983; 40(7-8): 237-41.
[PMID: 6361609]

[69] Storz H, Jelke E. Photomicrography of weakly fluorescent objects--employment of p-phenylene diamine as a blocker of fading. Acta Histochem 1984; 75(2): 133-9.
[http://dx.doi.org/10.1016/S0065-1281(84)80048-3] [PMID: 6441422]

[70] Small JV, Zobeley S, Rinnerthaler G, Faulstich H. Coumarin-phalloidin: a new actin probe permitting triple immunofluorescence microscopy of the cytoskeleton. J Cell Sci 1988; 89(Pt 1): 21-4.
[PMID: 2458368]

[71] Giloh H, Sedat JW. Fluorescence microscopy: reduced photobleaching of rhodamine and fluorescein protein conjugates by n-propyl gallate. Science 1982; 217(4566): 1252-5.
[http://dx.doi.org/10.1126/science.7112126] [PMID: 7112126]

[72] Meister A. Glutathione, ascorbate, and cellular protection. Cancer Res 1994; 54(7) (Suppl.): 1969s-75s.
[PMID: 8137322]

[73] Winkler BS, Orselli SM, Rex TS. The redox couple between glutathione and ascorbic acid: a chemical and physiological perspective. Free Radic Biol Med 1994; 17(4): 333-49.
[http://dx.doi.org/10.1016/0891-5849(94)90019-1] [PMID: 8001837]

[74] Stockert JC, Herkovits J. Photodynamic toxicity and its prevention by antioxidative agents in *Bufo arenarum* embryos. Toxicology 2003; 192(2-3): 211-8.
[http://dx.doi.org/10.1016/S0300-483X(03)00334-2] [PMID: 14580787]

[75] Mikhailov AV, Gundersen GG. Centripetal transport of microtubules in motile cells. Cell Motil Cytoskeleton 1995; 32(3): 173-86.
[http://dx.doi.org/10.1002/cm.970320303] [PMID: 8581974]

[76] De DN, Bhanja P. A phenomenon of photoenhancement of acridine orange-induced fluorescence. J Histochem Cytochem 1976; 24(5): 674-8.
[http://dx.doi.org/10.1177/24.5.58929] [PMID: 58929]

[77] Ferrer JM, González Garrigues M, Stockert JC. Mercurochrome: a fluorescent and electron opaque dye. Microsc Acta 1983; 87(3): 293-300.
[PMID: 6195509]

[78] Martínez-Ramírez A, Bella JL, Stockert JC. Fluorogenic reaction of blood cells induced by N-bromosuccinimide. Micron 2002; 33(4): 399-402.
[http://dx.doi.org/10.1016/S0968-4328(01)00030-0] [PMID: 11814878]

[79] Roig A, Peña MP, Vicente F, Stockert JC. Voltammetric behavior of berenil. J Pharm Sci 1993; 82(3): 251-3.
[http://dx.doi.org/10.1002/jps.2600820306] [PMID: 8450416]

[80] Fromherz P, Rieger B. Photoinduced electron transfer in DNA matrix from intercalated ethidium to condensed methylviologen. J Am Chem Soc 1986; 108: 5361-2.
[http://dx.doi.org/10.1021/ja00277a060]

[81] Stockert JC, Villanueva A, Cañete M, Juarranz A. Spectroscopic observations on photofading and fluorescence changes of eosin Y solutions in the presence of sodium azide. Spectrochim Acta 1994; 50A: 1197-200.
[http://dx.doi.org/10.1016/0584-8539(94)80041-3]

[82] Stockert JC, Trigoso CI, Braña MF. A new fluorescence reaction in protein cytochemistry: formation of naphthalimide fluorophores from primary amino groups and 1,8-naphthalic anhydride derivatives. Eur J Histochem 1994; 38(1): 29-39.
[PMID: 7517729]

[83] Rhys Williams AT. Fluorescence Detection in Liquid Chromatography. Buckinghamshire, UK: Perkin-Elmer 1980.

[84] Stockert JC, Trigoso CI. Selective fluorescence reaction of indigocarmine stained eosinophil leucocyte granules induced by alkaline reduction of the bound dye to its leuco derivative. Acta Histochem 1994; 96(1): 8-14.
[http://dx.doi.org/10.1016/S0065-1281(11)80003-6] [PMID: 7518177]

[85] McCorquodale EM, Piper J, Colyer CL. Capillary electrophoresis with laser-induced fluorescence detection of indigo carmine and indigo carmine-labeled proteins. J Capill Electrophor Microchip Technol 2002; 7(5-6): 95-101.
[PMID: 12546157]

[86] Platania E, Lofrumento C, Lottini E, Azzaro E, Ricci M, Becucci M. Tailored micro-extraction method for Raman/SERS detection of indigoids in ancient textiles. Anal Bioanal Chem 2015; 407(21): 6505-14.
[http://dx.doi.org/10.1007/s00216-015-8816-x] [PMID: 26082395]

[87] Sandell JH, Masland RH. Photoconversion of some fluorescent markers to a diaminobenzidine product. J Histochem Cytochem 1988; 36(5): 555-9.
[http://dx.doi.org/10.1177/36.5.3356898] [PMID: 3356898]

[88] Lübke J. Photoconversion of diaminobenzidine with different fluorescent neuronal markers into a light and electron microscopic dense reaction product. Microsc Res Tech 1993; 24(1): 2-14.
[http://dx.doi.org/10.1002/jemt.1070240103] [PMID: 7679591]

[89] Nikonenko I, Boda B, Alberi S, Muller D. Application of photoconversion technique for correlated confocal and ultrastructural studies in organotypic slice cultures. Microsc Res Tech 2005; 68(2): 90-6.
[http://dx.doi.org/10.1002/jemt.20239] [PMID: 16228980]

[90] Meisslitzer-Ruppitsch C, Röhrl C, Neumüller J, Pavelka M, Ellinger A. Photooxidation technology for correlated light and electron microscopy. J Microsc 2009; 235(3): 322-35.
[http://dx.doi.org/10.1111/j.1365-2818.2009.03220.x] [PMID: 19754726]

[91] Deerinck TJ, Martone ME, Lev-Ram V, *et al.* Fluorescence photooxidation with eosin: a method for high resolution immunolocalization and *in situ* hybridization detection for light and electron microscopy. J Cell Biol 1994; 126(4): 901-10.
[http://dx.doi.org/10.1083/jcb.126.4.901] [PMID: 7519623]

[92] Capani F, Deerinck TJ, Ellisman MH, Bushong E, Bobik M, Martone ME. Phalloidin-eosin followed

by photo-oxidation: a novel method for localizing F-actin at the light and electron microscopic levels. J Histochem Cytochem 2001; 49(11): 1351-61.
[http://dx.doi.org/10.1177/002215540104901103] [PMID: 11668188]

[93] Stockert JC, Blázquez-Castro A. Establishing the subcellular localization of photodynamically-induced ROS using 3,3′-diaminobenzidine: A methodological proposal, with a proof-of-concept demonstration. Methods 2016; 109: 175-9.
[http://dx.doi.org/10.1016/j.ymeth.2016.04.031] [PMID: 27154745]

[94] Kiernan JA. Histological and Histochemical Methods: Theory and Practice. Oxford, New York: Pergamon Press 1990.

[95] Molero ML, Hazen MJ, Stockert JC. Observations on the orcein fluorescence. Acta Histochem 1985; 76(1): 77-9.
[http://dx.doi.org/10.1016/S0065-1281(85)80039-8] [PMID: 2409729]

[96] Molero ML, Del Castillo P, Stockert JC. Fluorescence reactions of orcein: new applications for an old stain. J Microsc 1986; 144(Pt 1): 45-53.
[http://dx.doi.org/10.1111/j.1365-2818.1986.tb04672.x] [PMID: 2432271]

[97] Stockert JC, Blázquez-Castro A, Horobin RW. Identifying different types of chromatin using Giemsa staining. Methods Mol Biol 2014; 1094: 25-38.
[http://dx.doi.org/10.1007/978-1-62703-706-8_3] [PMID: 24162977]

[98] Goldstein DJ. The fluorescence of elastic fibres stained with eosin and excited by visible light. Histochem J 1969; 1(3): 187-98.
[http://dx.doi.org/10.1007/BF01081407] [PMID: 4113281]

[99] Espada J, Valverde P, Stockert JC. Selective fluorescence of eosinophilic structures in grasshopper and mammalian testis stained with haematoxylin-eosin. Histochemistry 1993; 99(5): 385-90.
[http://dx.doi.org/10.1007/BF00717051] [PMID: 7687595]

[100] Espada J, Stockert JC, Gamallo C. Fluorescencia selectiva de estructuras eosinofílicas en secciones teñidas con hematoxilina-eosina: mecanismos histoquímicos y aplicaciones histológicas. Patología 1994; 27: 335-41.

[101] Apgar JM, Juarranz A, Espada J, Villanueva A, Cañete M, Stockert JC. Fluorescence microscopy of rat embryo sections stained with haematoxylin-eosin and Masson's trichrome method. J Microsc 1998; 191(Pt 1): 20-7.
[http://dx.doi.org/10.1046/j.1365-2818.1998.00348.x] [PMID: 9723188]

[102] Rashid F, Horobin RW. Accumulation of fluorescent non-cationic probes in mitochondria of cultured cells: observations, a proposed mechanism, and some implications. J Microsc 1991; 163(Pt 2): 233-41.
[http://dx.doi.org/10.1111/j.1365-2818.1991.tb03175.x] [PMID: 1719208]

[103] Stockert JC, Molero ML, Espelosin RH. Fluorescence patterns of chromatin and cytoplasm by hematoxylin solutions. Z Naturforsch C Biosci 1983; 38(1-2): 153-5.
[PMID: 6845816]

[104] Buño I, Juarranz A, Cañete M, Villanueva A, Gosalvez J, Stockert JC. Image processing and analysis of fluorescent labelled cytoskeleton. Micron 1998; 29(6): 445-9.
[http://dx.doi.org/10.1016/S0968-4328(98)00023-7] [PMID: 10071868]

[105] Bertolesi G, Stockert J, Lauría de C. Microfluorometry and image analysis of peritoneal mast cells reveal differences in the morphology and content of sulfated polyanions between normal and tumor-bearing mice. Int J Oncol 1997; 11(6): 1221-5.
[PMID: 21528326]

[106] Espada J, Juarranz A, Villanueva A, Cañete M, Andrés I, Stockert JC. Recycling cultured cells for immunofluorescent labeling. Histochem Cell Biol 2001; 116(1): 41-7.
[PMID: 11479721]

[107] Casas A, Sanz-Rodríguez F, Di Venosa G, *et al.* Disorganisation of cytoskeleton in cells resistant to

photodynamic treatment with decreased metastatic phenotype. Cancer Lett 2008; 270(1): 56-65. [http://dx.doi.org/10.1016/j.canlet.2008.04.029] [PMID: 18513853]

[108] Stockert JC. Photographic equidensitometry of metaphase chromosomes. Histochemistry 1980; 66(3): 261-75.
[http://dx.doi.org/10.1007/BF00495739] [PMID: 7399971]

[109] Stockert JC, Pinna-Senn E, Bella JL, Lisanti JA. DNA-binding fluorochromes: correlation between C-banding of mouse metaphase chromosomes and hydrogen bonding to adenine-thymine base pairs. Acta Histochem 2005; 106(6): 413-20.
[http://dx.doi.org/10.1016/j.acthis.2004.09.001] [PMID: 15707650]

[110] Stockert JC. Cytofluorometric measurements by fluorescence extinction through a diaphragmed objective. Microsc Acta 1980; 82: 367-71.

[111] Juarranz A, Stockert JC. Cytofluorometric and cytophotometric measurements by using the exposure control device in the automatic photomicroscope. Microsc Acta 1980; 83(4): 297-305.
[PMID: 6158657]

[112] Caster AH, Kahn RA. Computational method for calculating fluorescence intensities within three-dimensional structures in cells. Cell Logist 2012; 2(4): 176-88.
[http://dx.doi.org/10.4161/cl.23150] [PMID: 23538475]

[113] Osborn M, Born T, Koitsch HJ, Weber K. Stereo immunofluorescence microscopy: I. Three-dimensional arrangement of microfilaments, microtubules and tonofilaments. Cell 1978; 14(3): 477-88.
[http://dx.doi.org/10.1016/0092-8674(78)90234-9] [PMID: 357010]

[114] Baschong W, Duerrenberger M, Mandinova A, Suetterlin R. Three-dimensional visualization of cytoskeleton by confocal laser scanning microscopy. Methods Enzymol 1999; 307: 173-89.
[http://dx.doi.org/10.1016/S0076-6879(99)07013-5] [PMID: 10506974]

[115] Tharp MD, Seelig LL Jr, Tigelaar RE, Bergstresser PR. Conjugated avidin binds to mast cell granules. J Histochem Cytochem 1985; 33(1): 27-32.
[http://dx.doi.org/10.1177/33.1.2578142] [PMID: 2578142]

[116] Gómez-Perretta C, Armas-Portela R, Stockert JC, Tato A, Ferrer JM. Increased acidophilia of eosinophil granules after EDTA treatment. Histochem J 1986; 18(1): 1-4.
[http://dx.doi.org/10.1007/BF01676191] [PMID: 3086258]

[117] Stockert JC. Feulgen positive nucleoli in Epon semithin sections: fact or artifact? Experientia 1977; 33(12): 1669-70.
[http://dx.doi.org/10.1007/BF01934064] [PMID: 73476]

[118] Stockert JC. Osmium tetroxide/*p*-phenylenediamine staining of nucleoli and Balbiani Rings in *Chironomus* salivary glands. Histochemistry 1977; 53(1): 43-56.
[http://dx.doi.org/10.1007/BF00511209] [PMID: 71287]

[119] Antonio C, González-García JM, Page J, Suja JA, Stockert JC, Rufas JS. The osmium tetroxide/ p-phenylenediamine procedure reveals the chromatid cores and kinetochores of meiotic chromosomes by light and electron microscopy. J Histochem Cytochem 1996; 44(11): 1279-88.
[http://dx.doi.org/10.1177/44.11.8918903] [PMID: 8918903]

[120] Maisch T, Bosl C, Szeimies RM, Love B, Abels C. Determination of the antibacterial efficacy of a new porphyrin-based photosensitizer against MRSA *ex vivo*. Photochem Photobiol Sci 2007; 6(5): 545-51.
[http://dx.doi.org/10.1039/b614770d] [PMID: 17487307]

CHAPTER 8

Fluorescent Metal Complexes and Ion Indicators

Abstract: Spectral changes of dyes related to pH and metal coordination have now increasing importance as analytical tools. Changes in the light absorption or emission that are characteristics of some dyes and reagents at different pH values or in the presence of metal ions, as well as their microscopical uses are described in this Chapter. Regarding histochemical and histophysiological detection, important metal cations are Mg, Ca, Fe, Zn, *etc*., with methodological procedures being spot tests, spectral studies and microscopic observations.

Keywords: Alizarin, Aluminum salts, Anthraquinones, Calcein, Calcification markers, Carmine, Chelating reagents, Crown ethers, Flavonols, Fluorescent pH probes, Intracellular pH, Polychromatic fluorescence, Pyranine, Ratio imaging.

At present, spectral changes of certain dyes in relation to variations of the pH and formation of metal complexes as analytical tools for recognizing metal cations have increasing relevance [1 - 9]. In this Chapter the following types and sequence of interactions of dyes with metal cations and hydrogen ions (H^+) will be considered:

1. Binding of metal cations with substrates to compete with fluorochromes (first cation, then fluorochrome or first fluorochrome, then cation). This later approach relates to the critical electrolyte concentration (CEC) method [10, 11].
2. Use of metal cations with ligands, to produce fluorescent complexes (first cation, then ligand; first ligand, then cation; preformed cation-ligand complexes). Both metal cations and ligands are administered as exogenous treatments. Ideal ligands should be non fluorescent or with emission features different enough to ligands complexed with metals. However, most ligands are already fluorescent, as occurs in the case of flavonols and anthraquinone dyes [4, 12].
3. Protonation and deprotonation (and the corresponding reactions involving hydroxyls) leading to free base or pseudo-base of dyes and fluorescent pH indicators.
4. Fluorescent probes for revealing physiological metal cations (*i.e.* Ca^{2+}) already present in living cells and organisms.

Juan Carlos Stockert & Alfonso Blázquez-Castro

1. IONIC REACTIONS

Ionization of substrate groups and dyes, as well as ionic binding processes are illustrated in Fig. (**8.1**). Obviously, inorganic ions bind to anionic and cationic sites on the substrate and can result in ionic competition to dye binding (see also Figs. **7.2** and **7.3**).

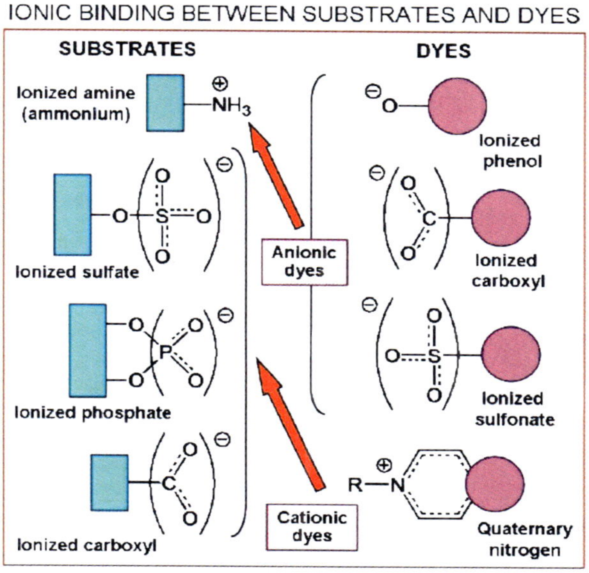

Fig. (8.1). Examples of ionic binding of anionic and cationic dyes to biological substrates. Ionizable groups are shown as resonant structures.

The concept of critical electrolyte concentration (CEC) introduced by Scott has been widely applied in staining reactions by cationic [10, 11] and anionic dyes [13, 14]. The use of inorganic salts such as NaCl and $MgCl_2$ as competing agents allows to obtain numerical data that are representative of the intensity of the electrostatic interaction between dyes and substrates (Fig. **8.2**). The parameter CEC is then expressed as the molar salt concentration that abolishes staining or fluorescence.

Since very often staining is not completely abolished at the highest sal concentration employed (which indicates that other non-ionic binding modes also occur), the CEC parameter could be improved. The minimal concentration of salt that leads to the first appreciable reduction in the staining intensity can be defined as MEC (minimal electrolyte concentration) for each dye, allowing a more precise evaluation of the strength of ionic interactions than CEC values [14].

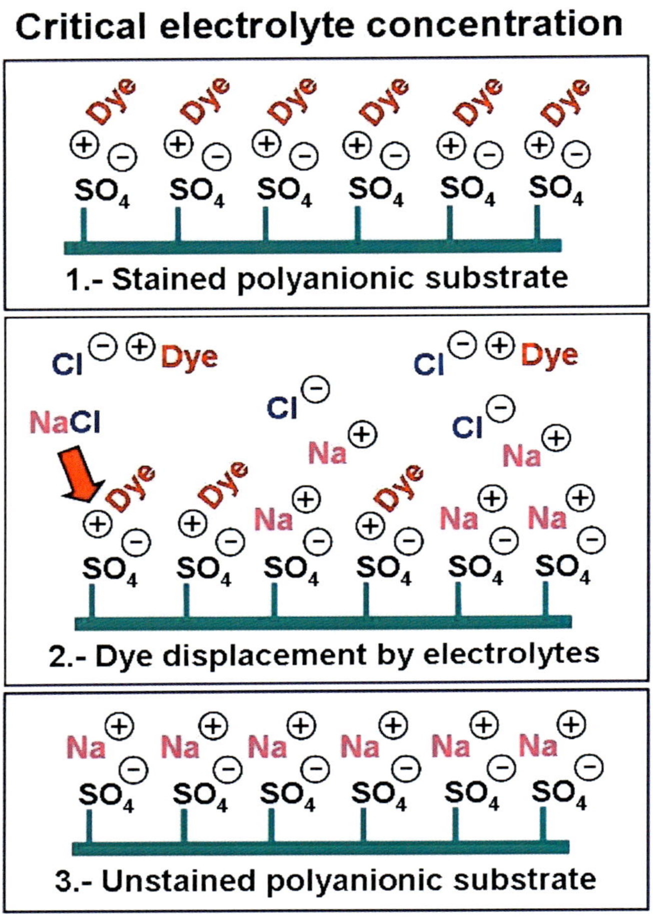

Fig. (8.2). Critical electrolyte concentration (CEC). The electrostatic binding strength can be expressed as the molar concentration of the competing salt that is required to abolish the staining reaction. Three steps are represented. 1: Normal staining with a cationic dye. 2: Competing effect and replacement of dyes by inorganic cations. Free dyes are now equilibrated by inorganic anions. 3: At CEC, the staining reaction is abolished.

There are several chemical strategies to reveal the influence of metal cations on staining and fluorescence reactions, examples being different competing salts and spectral analysis of dyes. In the first case, washing the safranine stained material with Mg^{2+} salt solutions leds to reduction of absorption and increase of emission (Fig. **8.3A**). Spectral studies confirm that previous treatment with Al^{3+} ions abolishes the chromatin staining by toluidine blue, whereas acid glycosaminoglycans remain metachromatically stained (Fig. **8.3B**, see also Fig. **11.10B**). Pre- or post-treatments with aluminum salts (*i.e.* 5% alum for 10 min) abolishes the basophilic and metachromatic staining and fluorescence reactions of chromatin by cationic dyes such as pyronine Y, thionine, safranine O, neutral red, and phenosafranine (10^{-4} M, 5 min) [15].

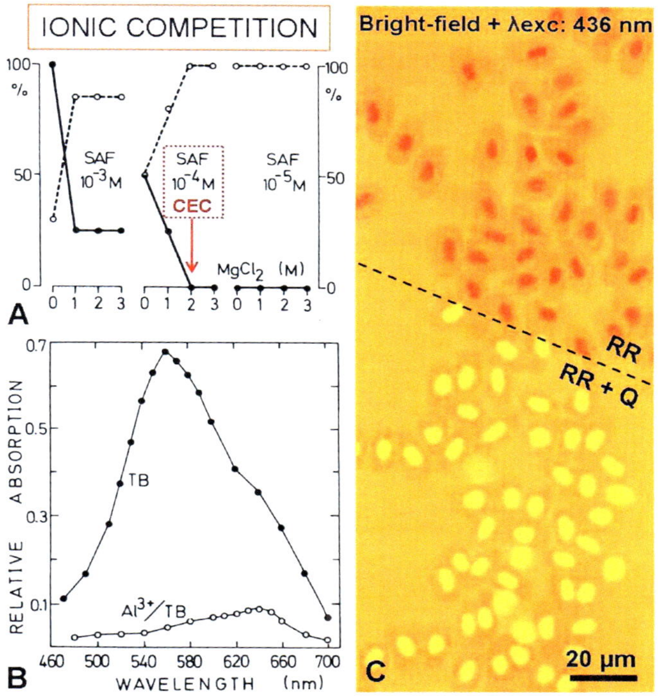

Fig. (8.3). **A**: Competing actions of different concentrations of electrolyte ($MgCl_2$) on the chromatin staining (solid lines) and fluorescence (dashed lines) of chicken erythrocyte nuclei by safranine O (SAF), showing the CEC value. **B**: Absorption spectra of chicken erythrocyte nuclei stained by toluidine blue alone (TB) or after previous treatment with 5% alum (Al^{3+}/TB). **C**: Metanol-fixed chicken blood smear showing erythrocyte nuclei stained by 0.1 mg/mL ruthenium red (RR, top), followed by 50 µg/mL quinacrine (RR+Q, bottom). Simultaneous bright-field and fluorescence microscopy.

Microscopic observation of double stained preparations offers illustrative examples of competing reactions between cationic dyes and fluorochromes. Fig. (**8.3C**) shows the effect of a fluorescent intercalating dye on chromatin staining by a polycationic dye. The inorganic dye ruthenium red ($Ru_3O_2[NH_3]_{14}Cl_6 \cdot 4H_2O$) is a linear and non-fluorescent mixed-valence complex used in bright-field microscopy for visualization of DNA, heparin, mucin, and pectin [16]. Ionic binding of ruthenium red to DNA phosphates in chromatin can be displaced by intercalating fluorochromes such as the antimalarial drug quinacrine. On the other hand, competition and differential binding affinity between Al-hematoxylin lake and luminescent Ru^{2+}-polypyridyl complexes (Rubipy, Ru(II)-bipyridine$_3$) for different polyanions are seen in Fig. (**11.9**) (see also Chapter 9.3.2).

Inorganic cations show preferential binding to different anionic groups. Generally monovalent metal cations bind to anionic substrates by means of weak ionic forces, whereas other cations have strong interactions that lead to insoluble reaction products (as the corresponding salts (*i.e.* barium with sulfate, *etc.*). Therefore, this differential affinity for specific anionic groups allows to recognize polyanions using metal complexes. As an example, chelate complexes of Al^{3+}, Cr^{3+}, and Fe^{3+} with hematoxylin show a strong affinity for phosphates, lower for carboxylates, and less still with sulfates [10]. This is an expected fact, because Al, Cr, and Fe sulfates are water soluble salts, whereas the corresponding phosphate salts are insoluble. The basophilia of DNA phosphates is severely abolished and only reduced by UO_2^{2+} and Ba^{2+} ions, respectively [17].

Differential affinity of metal cations for anionic groups (from [11, 15, 17]):

	K^+	Ba^{2+}	Al^{3+}
$-COO^-$	+	++	+
$-SO_4^-$	+	++++	+
$-PO_4^-$	+	+++	++++

2. METAL COMPLEXES

In the dye industry and biological staining the word lake (Sanskrit, *laksha*, a hundred thousand) usually means an adsorption compound formed by the union of a dye or ligand with a metal ion. In bi- or multidentate metal chelates the formation of new rings through ionic and coordinative bonds results in higher resonance and stability.

Ionic and chelating binding occur between anionic ligands and metal cations. Lake-forming cations are commonly referred as mordants (Latin, *mordere*, to bite), from an old idea that they attack the substrate causing dye binding. The word chelate (Greek, *chele*; Latin, *chela*, claw of crab) involves coordinative linkages to form stable 5- or 6-membered rings, with O and N atoms being the main electron donors. The chelating properties of dyes and reagents are applied to reveal the occurrence of inorganic constituents in biological systems, either physiological or xenobiotic elements.

Mordanted with different metal cations, the same chelating dye generates different colors. With the yellow-orange alizarin, red (Al), violet (Mg), purple-red (Ca), blue (Ba), violet-blue (Fe), and violet-brown (Cr) colors are produced. With the reddish oxidized hematoxylin, the main colors are dark blue (Al, Fe, Cr, In, Zn, Mo, Pb), blue-green (Cu), and purple-red (Sn) [4]. Preformed metal complex dyes, often termed "mordant dyes" are widely applied in "one-step" (as distinct from premordanted) staining methods in histology [18].

Preformed metal complexes such as Cr-gallocyanin, Cr- and Al-hematoxylin, Al-carminic acid, Cu-phthalocyanines, Fe-*o*-phenanthroline, *etc*., are used as selective stains for the visualization of polyanionic substrates [2, 4, 19]. Chromogenic reactions give colored complexes of ligands with several metal cations [3, 4]. As illustrative examples regarding biologically relevant cations, dithizone reveals Zn, Mn, Cd, Fe, Co, Cu, *etc*. Rhodizonic acid gives dark red of black precipitates with Ba, Sr, Fe, Zn, Sr, Hg, and Pb ions. The calcium reagent glyoxal-bis(2-hydroxyanil) (GBHA) produces a red precipitate with Ca^{2+} but also gives insoluble chelates with Mg, Sr, Ba, Cd, Cu, Co and Ni ions. No fluorescence seems to have been described for complexes of GBHA with Mg, Ca, Al, Zn or Ru ions.

3. FLUORESCENCE OF METAL COMPLEXES

The fluorescence of metal complexes finds wide application in analytical chemistry. Interestingly, the first fluorescence-based chemical analysis was performed in 1867 using the Al^{3+} chelate of morin, and this dye is still applied for detection of aluminum and calcium in tissue sections. Coordinative and ionic forces result in the formation of metal complexes with bi- or multidentate ligands, which contain resonance-stabilized chelate rings. Fluorescent complexes occur with anthraquinones, flavonols, tetracyclines, *etc*. [9], and are employed for different purposes.

The same metal cation originates diversely colored or fluorescent lakes with different chelating dyes, and diamagnetic metals often produce fluorescent complexes. For instance, the aluminun ion forms fluorescent complexes with morin (λem: 490 nm), 8-hydroxyquinoline (λem: 520 nm), curcumin (λem: 530 nm), purpurin (λem: 550 nm), nuclear fast red (λem: 565 nm), and alizarin red (λem: 580 nm) [12].

Several metal-dye complexes have useful histochemical applications for revealing polyanionic susbstrates in fluorescence microscopy. Fluorescent complexes of Al^{3+} with several ligands show high affinity binding to polyphosphates, allowing a selective visualization of nucleic acids-containing structures. The mechanism corresponds to the preferential binding of Al^{3+} to phosphate groups, whereas sulfate groups do not bind this cation. Examples are Al^{3+} complexes with quinolines, flavonols, β-diketones, anthraquinones, *etc*., which will be described in following sections. The complex (chromomycin $A_3)_2$-Mg^{2+} binds with high selectivity to the minor groove of DNA in guanine-cytosine rich regions, (see Chapter 12.5.8).

3.1. Phenol, Naphthol, and Quinoline Ligands

Aluminum complexes with small ligands such as nitroso-phenol (CBN: 9), nitroso-naphthol (CBN: 14), and 8-hydroxyquinoline (CBN: 12) are fluorescent and have histochemical applications. The fluorescent nitroso-phenol-Al^{3+} complex is the analogue product of the Millon's test for tyrosine (see Chapter 13.2.4). The complex of nitroso-naphthol with Fe^{3+} (naphthol green B, CI 10020) is a well known dye [4], and nitroso-naphthol is also a reagent for Co^{3+}. The Al^{3+} complex with this ligand induces fluorescence in nucleic acids-containing structures.

8-Hydroxyquinoline is a reagent applied for detecting several metal cations. Fluorescent metal complexes of 8-hydroxyquinoline and derivatives are well known, the most important cations being Be^{2+}, Mg^{2+}, Al^{3+} and Zn^{2+} [20, 21]. The preformed hydroxyquinoline-Al^{3+} complex induces blue-green fluorescence in chromatin DNA under UV excitation (Fig. **8.4**). The introduction of the Al ion with a residual positive charge, and unoccupied d orbitals involved in Al bonding contribute to the electron delocalization in the chelated quinoline ring [22]. The same fluorescence reaction is obtained after previous Be, Zn, Sn, In, and La mordanting followed by treatment with 8-hydroxyquinoline [12, 23].

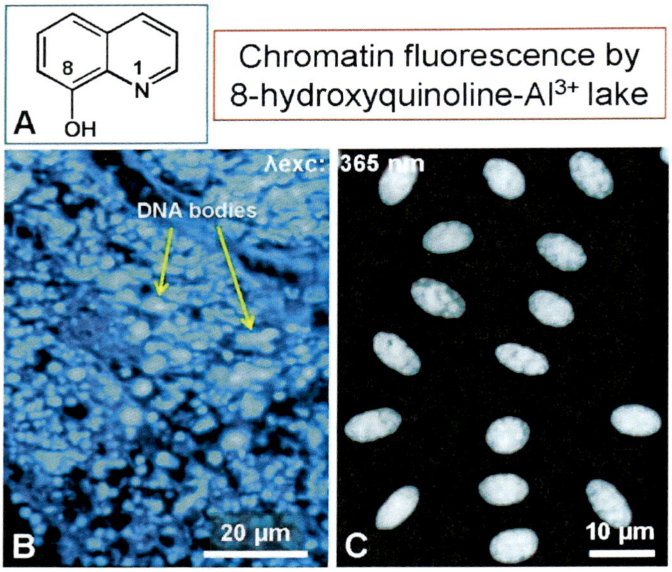

Fig. (8.4). **A**: Chemical structure of 8-hydroxyquinoline. **B**: Paraffin section of a human lymphatic node from an autoimmune disease containing both fluorescent DNA bodies (arrows) and nuclei after staining with the preformed 8-hydroxyquinoline-Al complex (0.01 mg/mL in saturated alum). **C**: methanol-fixed chicken blood smear stained with the preformed 8-hydroxyquinoline-Al complex, showing the fluorescence of erythrocyte nuclei.

3.2. Bipyridine and Phenanthroline

2,2'-Bipydirine and 1,10-phenanthroline are well known reagents that give red color with Fe^{2+} and blue color with Fe^{3+}. The red complex *o*-phenanthroline-Fe^{2+} has found microscopical use [19]. Complexes with Zn and Cu ions are also known. In the luminescent complexes of ruthenium(II) and iridium(II) with bipyridine and *o*-phenanthroline, the excitation process corresponds to a metal-t--ligand-charge-transfer (MLCT) mechanism (see Fig. **9.5**). The luminescence of the complex tris-(2,2'-bipyridine) ruthenium(II) (Rubipy) has microscopical and histochemical applications [24, 25]. These ligands also form cryptates with luminescent lanthanide ions.

3.3. Flavonols

The flavonol morin (2',3,4',5,7-penta-hidroxyflavone, CI 75660) (see Chapter 3.1.2 and Fig. **3.4**) is known to form fluorescent complexes with Al, Be, Ca, Zn, Ga, In, Sc, Zr, Sn, and Th [1, 4]. The metal complexes with Al^{3+} and In^{3+} have been employed as permanent stains in cytochemical studies on the chromatin structure by using fluorescence microscopy [26, 27]. In additon to these cations, premordanting with Mg^{2+}, Ca^{2+}, Zn^{2+}, and Sn^{2+} followed by morin induces clear fluorescence reaction of chromatin [28]. Hovewer, it must be noted that lacking metal cations, morin alone induce selective fluorescence in some cell and tissue structures such as basal membranes, mucin, granules from heterophil and eosinophil leucocytes, chromatin, *etc.* (see Figs. **1.8B**, **3.4** and **13.3**). In the presence of boric and oxalic acids, 5-hydroxy-flavones and flavonols give stable and strong yellow-green fluorescence due to the formation of borate complexes (Wilson-Tauböck detection test) [29]. The same fluorescence reaction also occurs in the case of anthraquinone and β-diketone derivatives.

3.4. Hydroxy-Triarylmethanes

Several hydroxy-triarylmethane dyes have carboxyl groups adjacent to hydroxy or ketone groups, which results in excellent metal chelating reagents [4]. A well known example is chromoxane cyanine R (CI 43820; solochrome cyanine R) [30] and its blue and red complexes with Fe and Al ions, respectively. The similar dye chromoxane pure blue B (CI 43830) give blue lakes with Be, Mn, Cu, Cr, Ni, Gd, and Pb ions, and pink lakes with Li, Al, Zn, Mo, and Bi ions [4]. Aluminum complexes of some of these dyes are fluorescent [3].

The tricarboxylic ammonium derivative of aurin (rosolic acid; CI 43800), known as aluminon (CI 43810), is a selective reagent for Al^{3+}, giving bright red color. Red fluorescence under green exciting light can be observed selectively in chromatin of fixed cells stained with the preformed aluminon-Al complex [12].

Other dyes from this group are used as calcification markers in living organims (see section 5). Naphthochrome green B (CI 44530) forms colored chelates with Be, Mg, Ca, Al, Fe, Cr, *etc*. [3], some of which should be fluorescent. Regarding dyes with the ortho hydroxy-ketone group, most known are the differently colored complexes of oxidized hematoxylin with numerous metal cations, which have histochemical applications.

3.5. β-Diketones

Curcumin is a β-diketone ligand dye from *Curcuma longa* (Figs. **8.5A**, and **3.19**). The natural dye is used as food aditive (E-100) and has anti-oxidant, anti-inflammatory and anti-tumoral properties. As occurs with other acetone-acetonate derivatives, curcumin can complex with several metal cations that enhance or modify fluorescence properties. The most common complexes are those with Be^{2+}, Mg^{2+}, Al^{3+}, and boric acid. The histological application of fluorescent curcumin-aluminum lake is illustrated in (Fig. **8.5B**). The complex induce bright emission in chromatin but the mucin from goblet cells remains non-fluorescent. It must be taken into account that the ligand alone also induce selective fluorescence reactions, for instance in the kinetoplast of *Trypanosoma* (see Fig. **3.19D**), and in callose walls (see Fig. **11.3**).

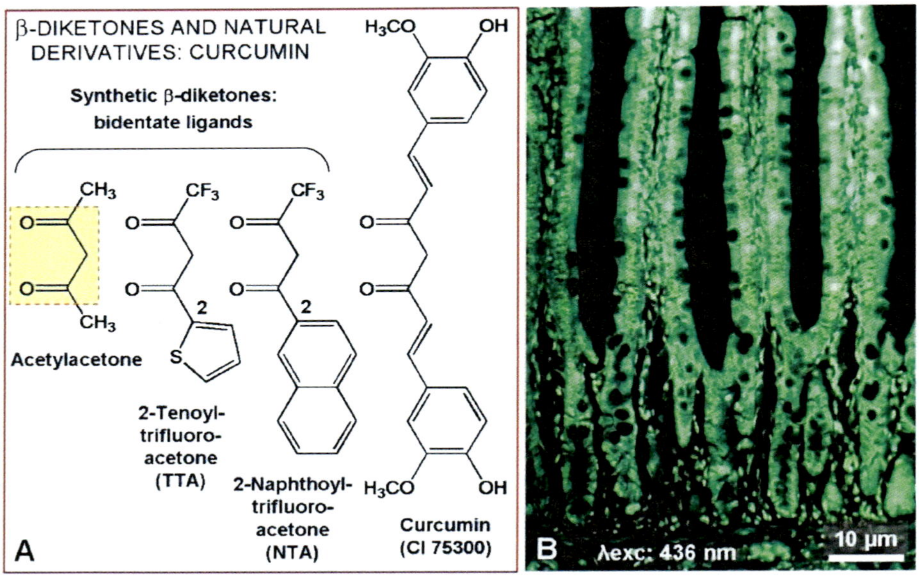

Fig. (8.5). **A**: Chemical structure of β-diketone (acetylacetonate) ligands. **B**: Paraffin section of formaldehyde-fixed rat small intestine stained for 10 min with the preformed curcumin-aluminum lake (0.01 mg/mL in saturated alum).

3.6. Anthraquinones

The most popular and widely used natural anthraquinone dye-metal complex is carmine (CI 75470) [4]. It is now accepted that carmine is a 2:1:1 complex between carminic acid, Al^{3+}, and Ca^{2+} ions [31] (Fig. **8.6A**). The dye is obtained from the dried bodies of the female cochineal insect *Dactylopius coccus* (*Coccus cacti* L.), growing on the tuna plant (*Opuntia ficus*, from Mexico, Peru, Ecuador and Bolivia). Carmine was employed by the Paracas culture (pre-Inca, about 2000 years B.C.) as a textil dye, giving different colors in wool fibers pre-mordanted with specific metals (Fig. **8.6B**). At present it is used as cosmetic and food additive (E-120).

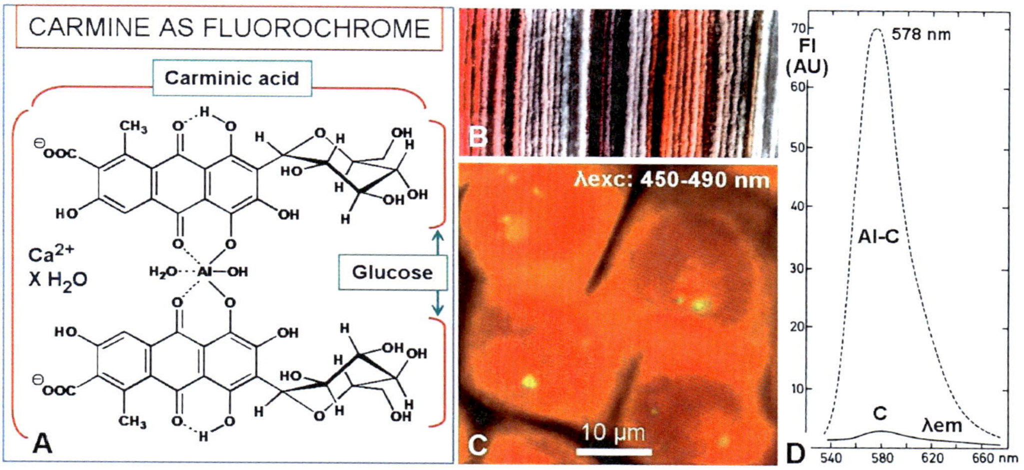

Fig. (8.6). **A**: Chemical structure of carmine with the chromophore (carminic acid), Al^{3+} and Ca^{2+} cations, and glucose units. **B**: Example of different colors in wool fibers pre-mordanted with distinct metals and post-stained with carmine. **C**: Acetic carmine squash of *Allium cepa* roots, used to make the preparation and as counterstain for FISH (rDNA; FITC, green; courtesy of J.L. Bella). **D**: Emission spectra of free carminic acid (C, 0.1 mg/mL), and its complex with $AlCl_3$ (Al/C). λexc: 480 nm.

After staining fixed cells with dilute carmine solutions (0.1 mg/mL), compact chromatin from interphase nuclei, mitotic and meiotic chromosomes, and kinetoplast DNA show high red-orange emission under 436 or 450-490 nm exciting light. When using more concentrated carmine solutions (50 mg/mL in 45% acetic acid, as for routine bright-field staining), the chromatin appears subdued red whereas the cytoplasm and nucleoli are bright orange, which is useful for plant cells squashing and FISH counterstaining (Fig. **8.6C**).

Aqueous carmine solutions show λab: 495 nm and λem: 578 nm (10 and 5 µg/mL in 0.1 and 0.05% acetic acid, respectively) [32], and carminic acid (CI 75470) produces a strongly fluorescent complex with aluminum ions (Fig. **8.6D**). Interestingly, conjugates of carminic acid with proteins have a photodynamic action that generates ROS with biological consequences [33]. Kermesic acid (CI 75460) is a natural dye similar to carminic acid and forms the lake "Venice scarlet" with a metallic mordant.

Alizarin and purpurin are natural anthraquinone dyes (now synthetic dyes) formerly isolated from madder, an extract of the roots of *Rubia tinctorum* and allied species (Fig. **8.7**). They were early used as textil dyes in combination with metal mordanting. The compound (alizarin)$_2$-Al^{3+} has been applied in the textil industry as "Turkey red" [34]. As occurs with other anthraquinone-aluminum complexes, Turkey red is fluorescent. Alizarin, quinizarin and purpurin are chelating agents that detect di-and trivalent metal cations (*i.e.* Be, Mg, Ca, Al, Zn, Ga, In, La, Pb, *etc.*), and most of these complexes are also fluorescent [3, 4, 12]. Anthrapurpurin, chrysazin, quinalizarin, anthracene blue SWR, and alizarin cyanin 2R and BBS are similar chelating dyes, and in apolar solvents, some anthraquinones (*i.e.* quinizarin) shows yellow-green emission.

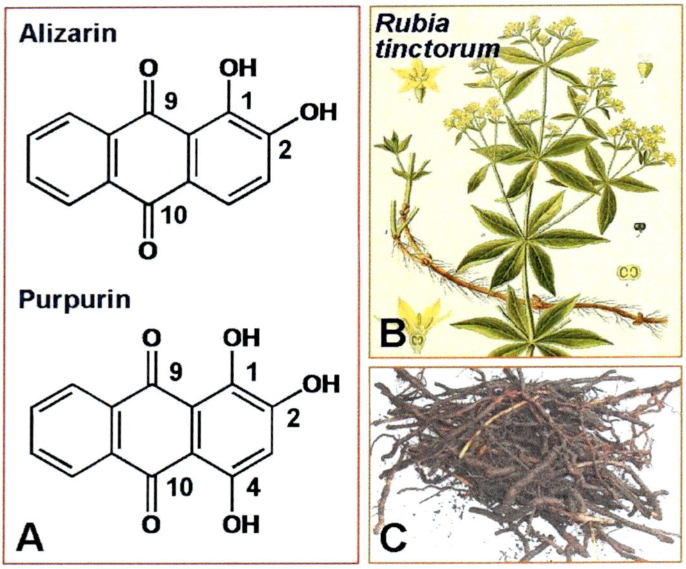

Fig. (8.7). **A**: Chemical structure of alizarin and purpurin with numbered atoms. **B**, **C**: The plant *Rubia tinctorum* and its dark red roots, respectively, from which alizarin and purpurin are obtained.

Alizarin (1,2-dihydroxy-anthraquinone, CI 58000), its sulfonated derivative alizarin red S (CI 58005), quinizarin (1,4-dihydroxy-anthraquinone, CI 58059), and purpurin (1,2,4-trihydroxy-anthraquinone, CI 58205) are easily solubilized in boiling saturated alum solutions giving preformed lakes, which under green excitation induce bright red fluorescence in nucleic acids-containing structures, mainly chromatin DNA [12]. The sulfonated dye nuclear fast red (calcium red, CI 60760; Fig. **8.8A**) is applied for the detection of Ca^{2+} and Al^{3+} and forms a bright red fluorescing complex with Al^{+3} with high affinity for chromatin (Fig. **8.8B, C**).

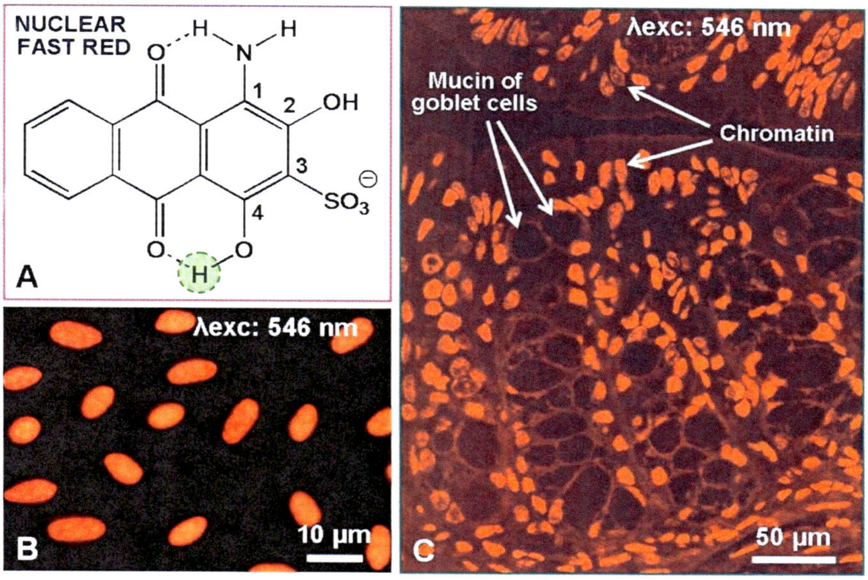

Fig. (8.8). **A**: Chemical structure of nuclear fast red. Hydrogen (green circle) is replaced by Al ion in the complex. **B**: Methanol-fixed chicken blood smear stained for 10 min with the preformed nuclear fast red-aluminum complex (0.01 mg/mL in saturated alum). **C**: Paraffin section of formaldehyde-fixed mouse large intestine after 10 min staining with the preformed nuclear fast red-aluminum complex.

The DNA-binding and antitumoral anthracycline antibiotics (*i.e.* daunomycin and doxorubicin) contain a fluorescent anthraquinone chromophore that intercalate into DNA [35, 36]. The chromophore of daunomycin is a tetracyclic aglycone (hydroxylated derivative of anthraquinone with an additional cyclohexane ring) bound to an aminosugar (Fig. **8.9A**). In the presence of Al^{3+}, daunomycin forms a strongly fluorescent complex [37] with hyper-emission and red shift (Fig. **8.9B**). The increase of red fluorescence in methanol-fixed cells is easily observed when comparing a microscopical area with nuclei stained by daunomycin alone and another area subjected to a postreatment with an alum solution (Fig. **8.9C, D**).

Possible fluorescent aluminum complexes of other anthraquinone dyes and drugs would be worth trying. They include the dyes rhein, DRAQ5, the cathartic drugs frangulin and danthron, the antineoplastic drug mitoxantrone, antibiotics such as resistomycin, pluramycins, altromycins, *etc*. Other anthraquinone-like ligands for formation or modulation of fluorescent complexes with Al^{3+}, Be^{2+}, Mg^{2+}, Ca^{2+},

Zn^{2+}, *etc.*, should be also investigated. They include the xanthone dye gentisin, the naphthoquinone and pyronequinone antibiotics actinorhodine, collinomycin, and thermorubin, metabolites such as kinurenic acid, the photosensitizing and antiretroviral agent hypericin, toxic hydroxy-xanthone metabolites secalonic acids, tetracyclines and quinolone antibiotics, antitumoral naphtho-γ-pyrone mycotoxins, quinone antibacterials and antivirals such as rubromycins, *etc.*

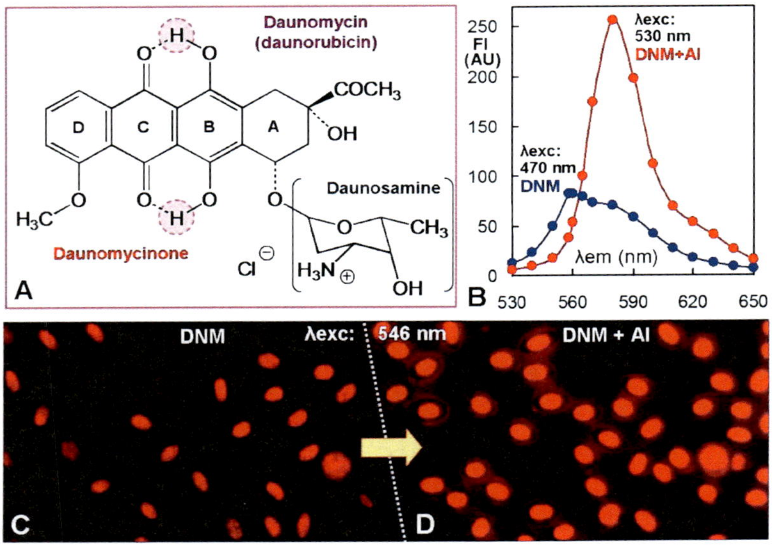

Fig. (8.9). Fluorescence of the daunomycin-aluminum complex. **A**: Chemical structure of daunomycin. Circles indicate hydrogens replaceable by metals. **B**: Emission curves of daunomycin (DNM, 0.2 mg/mL) and its Al^{3+} complex (DNM + Al; 24 h with 2.5 mg/mL aluminum sulfate). The emission peak of DNM+Al shifts from 560 to 580 nm and is 3 times higher than DNM alone. **C** and **D**: Methanol-fixed chicken blood smear stained for 10 min with 0.2 mg/mL DNM (C), followed by 24 h treatment with 2.5 mg/mL aluminum sulfate. The arrow indicates the orientation of washing after Al^{3+} treatment.

3.7. Azo Dyes and Other Reagents

The application of chromium and cooper lakes with some *o*-hydroxy-mono-azo dyes in the textil industry is well known [4]. They are lightfast, strongly colored but non-fluorescent dyes. On the contrary, several 2,2'-hydroxy-mono-azo dyes form complexes with aluminum ions that shows orange or red fluorescence, examples being anthracene chrome red A (CI 14290), acid alizarin violet N (CI 15670), palatine chrome black 6BN (CI 15705), salicin black EAG (CI 15710), diamond black PV (CI 16500), eriochrome red B (CI 18760), and flazo orange [4]. In these cases, the chelating site is accomplished by hydroxyl groups ortho to both ends of the azo bond. Some aluminum complexes of mono-azo mordant dyes are used in solution as new dye lasers.

Aluminum lakes of these *o*-hydroxy-mono-azo dyes have been applied as counterstains to tissue sections after FITC immunofluorescence to induce red-orange emission, thus avoiding the greenish autofluorescence [38]. A preformed flazo orange-Al complex induces red orange fluorescence in chromatin of fixed cells under 546 nm excitation [12]. In addition, some *o*-hydroxy-mono-azo dyes are applied for detection of magnesium, an example being the reagent magneson, which gives a bright blue color in the presence of Mg^{2+} at alkaline pH. The similar reagent lumomagneson (the sodium salt of 2-hydroxy-3-sulfo-5-chlorophen-lazo-1-barbituric acid) (Fig. **8.10A**) has been proposed for the fluorometric assay of this ion in aqueous solutions.

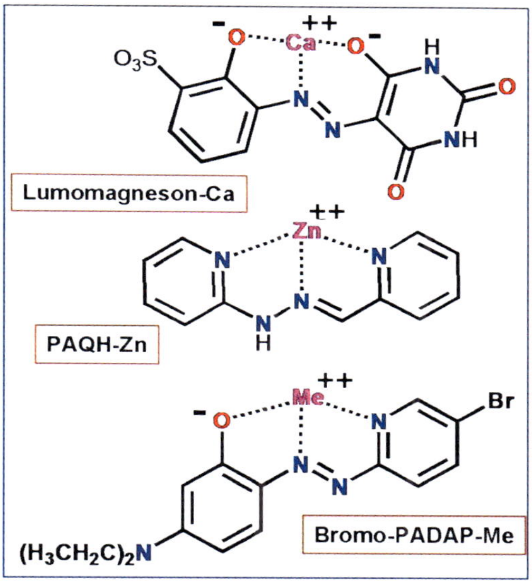

Fig. (8.10). Chemical structure of some hydroxy-mono-azo dyes complexed with divalent metals.

Lumomagneson is also a non-toxic, red fluorescent label of mineral deposits in the bone *in vivo*, which involve both Mg^{2+} and Ca^{2+} [39]. The azo-thiazole dye thiazole yellow G (CI 19590; titan yellow) is another reagent that produces red color with Mg^{2+}. It has been also applied directly as vital fluorochrome and for revealing dermal elastic fibers under UV excitation [4].

Some compounds with similar molecular geometries and chelating capacity should be considered regarding fluorescence reactions. The non-fluorescent reagent picolinealdehyde-2-quinolinyl-hydrazone (PAQH) acts as a non-selective chelating agent toward many metal ions, giving colored products with Fe, Co, Ni, Cu, Zn, and Pd, but only the zinc complex is fluorescent and has been used in the fluorometric determination of Zn^{2+} (Fig. **8.10B**). The $Zn(PAQH)Cl_2$ complex shows λexc: 478 nm and λem: 535 nm [40]. The ligand PAQH should also complex with Ru^{2+} ions to give a luminescent product similar to that of Rubipy and Ru-formazan [41]. The yellow azo dye 2-(5-bromo-2-pyridylao)-5-(diethylamino)phenol (bromo-PADAP, Fig. **8.10C**) is a chromogenic chelating reagent that forms colored complexes with a wide variety of metal ions (mainly Cu, Cd, and Pb), and has histochemical applications [8, 42]. The reagent is used for the determination of uranium(VI). Luminescent products with Mg^{2+}, Ca^{2+}, Al^{3+}, Zn^{2+}, and Ru^{2+} should be also expected.

Finally, a number of tri- and multidentate ligands are now used to complex with luminescent lanthanide ions, which are better detected with time-resolved methods (see Chapters 6.3 and 9.2.3). Examples are dipicolinic acid (2,6-pyridinecarboxylic acid), salicylaldehyde-bipyridine, diethylene-triamino-penta-acetic acid, thenoyl-trifluoro-acetone, and tris(methylene-diamino-bipyridine) cryptate, mainly used to form complexes with terbium and europium ions [43, 44]. However, Tb^{3+} and Eu^{3+} alone also give luminescent reactions useful in biochemical and physiological studies [45, 46]. Although these lanthanides generally require excitation at about 310-320 nm, some chelates allow the use of the microscopically available 365 nm (*i.e.* [salicylaldehyde]$_3$-bipyridine-Eu, thenoyl-trifluoro-acetonate-Eu, *etc.*) [47, 48].

4. FLUORESCENT PH AND ION INDICATORS

Changes in the light absorption and emission of dyes and fluorochromes can occur depending on the pH and presence of certain ions. Therefore they are used to assess the intracellular pH and specific ions. Methods to visualize pH conditions and metal cations within live cells are now highly relevant in cell and molecular biology. General references to be consulted are available [5, 7, 49 - 56].

4.1. pH Indicators

Fluorescent pH indicators are probes that show definite changes in fluorescence with changes in pH. Some of them are used in chemistry as colored ones. The principle is the same, the indicator changing in color emission or intensity as it changes from the ionized to the unionized form. One of the oldest pH indicators is neutral red (CI 50040), which has been applied to reveal intracellular pH values [57]. It is a relatively non-toxic vital dye and a sensitive indicator in the pH range 6-8, with the protonated acid form showing λab: 530 nm and the free base λab: 450 nm.

The observation of proteins with pI values at different pH by pyranine is an illustrative example of the use of pyrene-sulfonates as fluorescent pH probes [58].

Pyranine (CI 59040) is the sodium salt of hydroxypyrene-trisulfonic acid (HPTS; λem: 520 nm). Different emission intensity and localization are observed in specific cell structures revealing proteins with distinct isoelectric points (Fig. **8.11**) [59]. Pyranine is applied as anionic fluorochrome in microscopy (see Fig. **13.4C, D**). Industrially, it is mainly used in office supplies such as inks for fluorescent markers, as well as in laundry and cosmetic products.

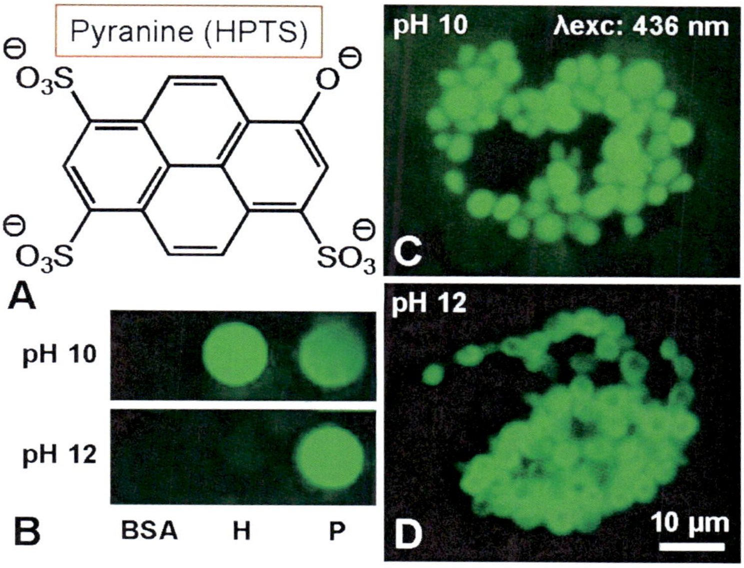

Fig. (8.11). **A**: Chemical structure of pyranine (HPTS). **B**: Fluorescence of nitrocellulose blots of bovine serum albumin (BSA, fraction V), histone (H) and protamine sulfate (P, from herring, grade III) stained with 1 µg/mL HPTS at pH 10 and pH 12, and observed in a transilluminator under UV light. **C** and **D**: methanol-fixed horse eosinophils treated with 10 µg/mL HPTS at pH 10 and pH 12 for 5 min. Green fluorescence is observed in the whole cytoplasmic granules stained at pH 10, wereas a selectively fluorescent anular region occurs in granules stained at pH 12.

Fluorescent indicators find utility for titration of opaque, highly turbid or colored media, but some fluorochromes are not suitable since their changes in emission is too gradual. The assessment of intracellular pH (H^+ concentration) is commonly performed using vital probes such as substituted fluoresceins and rhodamines. The techniques of pH detection have been reviewed [6, 55].

The use of fluorescent pH probes affords greater sensitivity than simple color indicators, and both are based on the formation of dye species with varying optical properties due to equilibrium between protonation-deprotonation. Therefore, because of the more sensitivity and simplicity of work with fluorescent dyes they are currently preferred to absorption dyes. Fluorescent pH probes operate by the occurrence of large shifts in the λem of the indicator in acidic and basic media, and they must be ratiometric. This means that the ratio of emission peaks corresponding to acidic and basic conditions can be applied to measure pH if both emissions are quantified.

Some fluorochromes derived from fluorescein or those that combine parts of both fluorescein and rhodamine are being successfuly used as pH indicators. They are semi-naphtho-fluorescein (SNAF) and semi-naphtho-rhodofluorescein (SNARF) [51]. For SNARF carboxylic acid, a typical calibration uses a dual-emission ratio with λem1: 580 nm ± 10 nm (acid form) and λem2: 640 nm ± 10 nm (basic form), as well as a fixed dual-excitation with 488 nm and 514 nm. The pH-dependent spectral shifts allow calibration of the pH response in terms of the ratio R of fluorescence (F) intensities (R: F λem1/F λem2, measured at the two excitation wavelengths). Dual emissions and dual excitations are more convenient for flow cytometry and microscopy, respectively.

The usual way of cell internalization of the otherwise impermeant dyes is based on the strategy of incubation with membrane-permeant precursors (acetate or acetoxy-methyl esters, AM) that are enzymatically converted inside the cell into the proper indicator. Both the pH and ion indicators work on the simple principle that its absorption, excitation or emission properties change significantly when the dye is protonated or binds to a particular ion. The two forms of the dye (free and bound) can then be spectroscopically distinguished, their concentration ratio determined and, *via* a standard calibration curve, converted into actual pH or metal ion concentration. This approach is called the "ratio method" or "ratio imaging", because the map of local concentration derives from an image pair. In the case of pH indicators they must have calibration curves independent of buffer composition.

An illustrative example of the procedure is shown using the simple pH indicator fluorescein (Fig. **8.12**). The membrane-permeant, colorless and non-fluorescent precursor fluorescein diacetate penetrates the cell, and after cleavage by non-specific esterases the colored and fluorescent free dye becomes trapped within live cells. To obtain the pH topography (pH maps), two images must be recorded using selective excitation for the acid and basic forms of the dye. After computerized digital-image processing of both pictures and calculation of emission differences, these values can be converted into a local pH map using a standard calibration curve.

An interesting fluorescent pH probe is the small molecule 2,3-dicyanohydroquinone (2,3-dicyano-1,4-dihydroxybenzene, DCDHB; CBN: 12) [60]. It can be excited at a single wavelength and the emission shifts from 450 to 483 nm as the pH increases, which produces ionization of one hydroxyl group. To be permeant to living cells, the more hydrophobic diacetyl derivative is used, which after esterase action gives DCDHB.

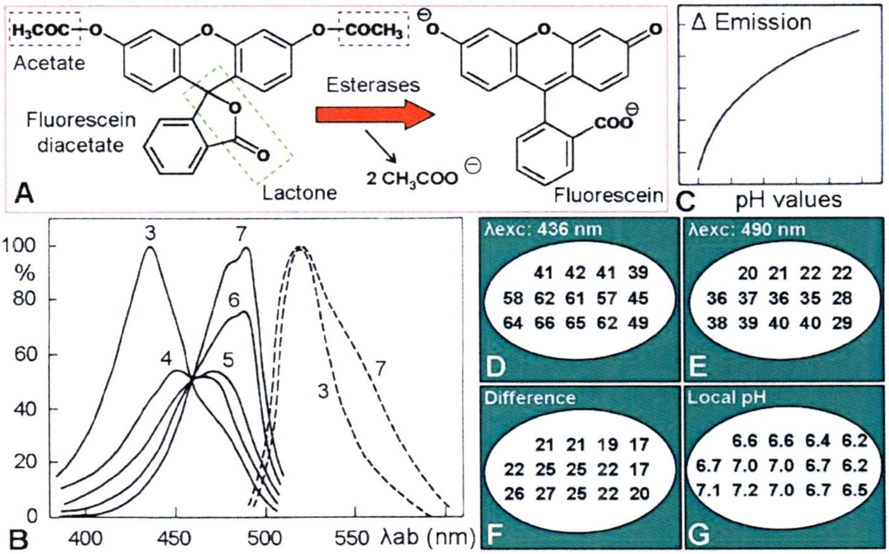

Fig. (8.12). **A**: Conversion of fluorescein diacetate (permeant and non-fluorescent) to fluorescein by non-specific cytoplasm esterases. **B**: Normalized absorption (—) and emission (- - -) curves of fluorescein excited at different pH values (between 3 and 7). **C**-**G**: Determination of H^+ ion concentration by ratio imaging. Densitometric maps of fluorescein emission excited at the indicated pH (**D**, **E**) are used to obtain the difference map (**F**), and then the local pH values after conversion from a standard pH-curve (**C**).

Complementary to carboxylic SNARF AM, a novel ratiometric pH indicator with two xanthene (fluorescein) fluorophores bound to a central Bodipy probe works on a pH-dependent energy-transfer equilibrium between xanthene (green emission at pH > 6.5) and Bodipy (red emission at pH < 6.0) under unique excitation at 488 nm [61]. The bicolor-fluorescing xanthene-Bodipy complex can be conjugated to proteins and used for studying pH processes in cell organelles (*i.e.* endosomes).

4.2. Ion Indicators

Cyclic ionophores such as chelating crown ethers are suitable to complex with metal cations, an example being the 18-crown-6 K+ salt (Fig. **8.13A**). As occurs with some antibiotics (*i.e.* enniatins, valinomycin), they allow the incorporation of inorganic cations into living cells. Dyes and ionophores can be combined to form ion indicators. In the case of alkali and alkaline-earth metal complexes with chromo(fluoro)-ionophores (where the dyes are mesomeric systems), the positive cation charge influences the electron donor heteroatom (O, N) and the electronic disturbance propagates through the whole system resulting in spectral changes [5].

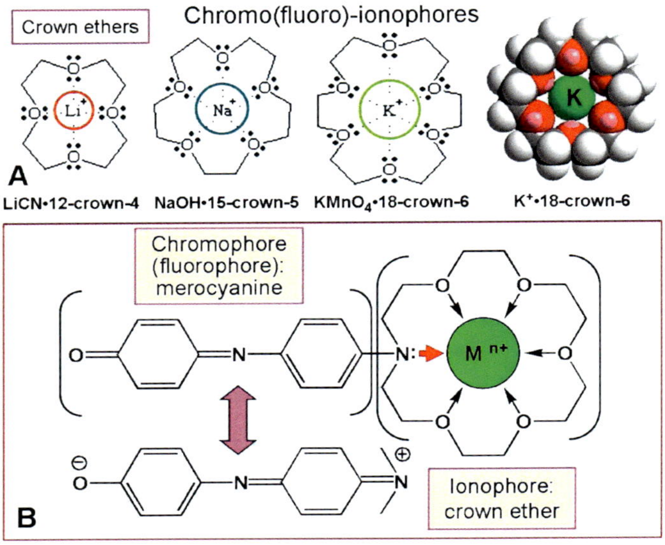

Fig. (8.13). **A**: Cryptate crown ethers of different size, containing Li^+, Na^+ and K^+ ions. Black dots and pink circles correspond to lone electron pairs. At right, representation of the crown ether compound K^+18-crown-6 in atomic volume. **B**: Phenol blue bound to a crown ether forming a chromophore-ionophore combination.

The linkage of the ionophore at the (δ^-)-polarized chromophore is advantageous for cations with high charge density. Chemical design of chromo(fluoro)-ionophores has allowed the synthesis and application of numerous metal ion indicators based on the combination of dyes with crown ethers [5, 51]. An example of chromo-ionophore is shown in Fig. (**8.13B**), where the indicated chromophore corresponds to the isomeric dye phenol blue (see Fig. **3.9B**).

A large number of dyes display changes in absorption or fluorescence upon binding a particular ion. Such dyes may be employed as indicators of the intracellular composition in live cells, and using bright-field or fluorescence microscopy associated with digital image processing, quantitative values of concentration of various ions can be obtained [52]. The "menu" of commercially offered chemicals has become really impressive and includes dyes and fluorescent probes for the determination of free cation and anion levels: Ca^{2+} (fura-2, quin2), Mg^{2+} (Mag-fura-2, Mag-indo-1), Na^+ (SBFI), K^+ (PBFI), Cl^- (SPQ), *etc.* On account of the rather hydrophilic character of ion indicators, they must be administered as the more lipophilic and permeant acetoxy-methyl (AM) esters, which generate the active ion probe after the action of intracellular esterases.

The colocalization of fluorescent metal cation indicators and organelle probes should be considered with caution, because often both of them are designed on the same fluorophore or on different fluorophores that label the same organelle (see [54]). Under this condition, the fluorophore of both ion and organelle probes will share QSAR parameters for selective organelle labeling (see Chapter 15.2.1). Accordingly, it is expected that simultaneous treatment of cells with the mitochondrial probe MitoFluor green and the Ca^{2+} indicator Rhod-2 AM originate the same mitochondrial image, although the later reveals the content of Ca^{2+} in this organelle. Likewise, cells treated with CoroNa red for Na^{+} detection and the mitocondrial probe MitoTracker green FM shows colocalization of both signals, on account of having fluorophores with the same organelle selectivity.

4.3. Calcium

Calcium probes derive mainly from developments of the basic chemical skeleton of the well known chelating agents EDTA, EGTA, and BAPTA (Fig. **8.14A-D**).

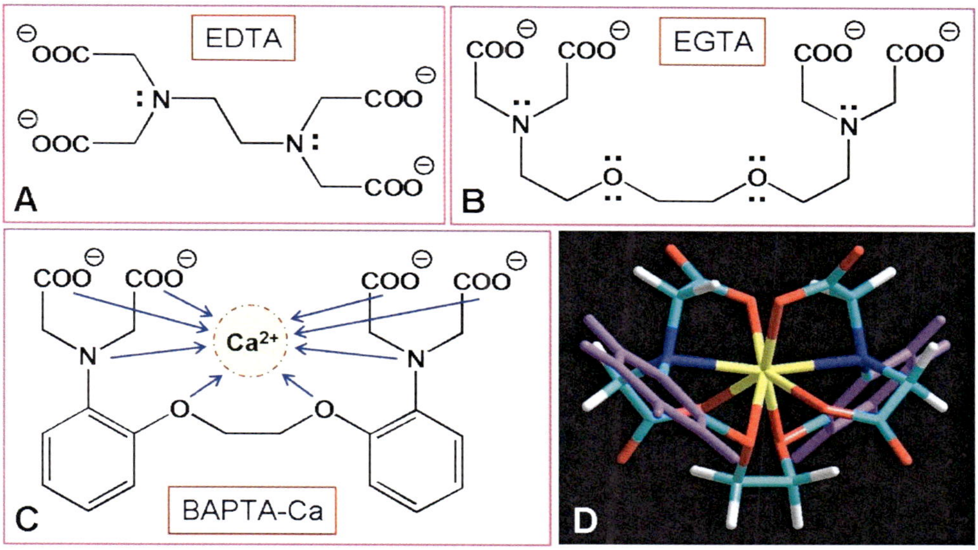

Fig. (8.14). Basic chemical structure of metal cation indicators, showing the chelating reagents EDTA (**A**), EGTA (**B**), BAPTA-Ca (**C**, coordinative bonds correspond to blue arrows), and the molecular structure (tubes) of the octahedral BAPTA-Ca complex (**D**; color code are H: white, C: cyan, O: red, N: blue, Ca: yellow, benzene rings: violet).

From the BAPTA design, and in addition to previous quinoline derivatives (quin2, which also binds strongly Mg and Zn ions), several Ca^{2+} probes were developed [51, 62], showing higher UV absorption and visible emission suitable for microscopical observation and flow cytometry. They were derived from BAPTA bound to stilbene (stil-1, stil-2), indole (indo-1), and benzofurane (fura-1, fura-2, fura-3). An important advantage of the new dyes over quin2 is their better selectivity for Ca^{2+} over divalent cations such as Mg^{2+}, Mn^{2+}, Zn^{2+}, and Fe^{2+}. Obviously, fluorescente is quenched by chelation of divalent paramagnetic metals.

Some of the commonly used fluorescent calcium indicators are shown in Fig. (**8.15A-D**), together with the structure of AM derivatives. Fura-2 is one of the most popular Ca probes, showing λab: 365 nm and λem: 512 nm (in water), and λab: 335 nm and λem: 505 nm (with Ca^{2+}).

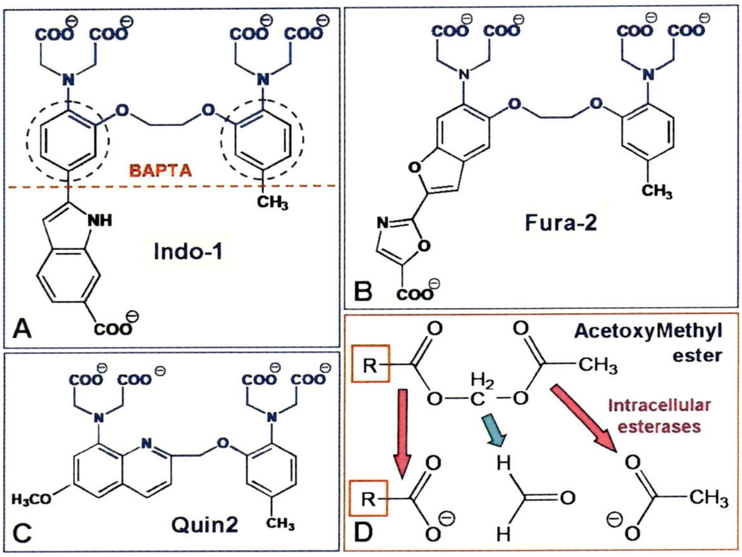

Fig. (8.15). Chemical structure of fluorescent Ca^{2+} indicators. **A**: Indo-1, showing the original EGTA skeleton (in blue) and the BAPTA parental compound. **B**: Fura-2. **C**: Quin2. **D**: The acetoxy-methyl ester component of ion indicators (R), and its degradation pathway by intracellular esterases (violet arrows) to release the active chelating probes.

Fura red (Fig. **8.16A**) and calcium green are also frequently used. Ca^{2+} indicators are often used to assess physiological kinetic processes. An example (Fig. **8.16B**) is the visualization of the free calcium wave induced by fertilization of oocytes using microinjection of dextran (MW: 10,000) labeled with calcium green-1 (λab: 506 nm, λem: 531 nm in water). At present, commercial probes such as Fluoforte® AM Calcium dyes are brighter than Fluo-3 AM and Fluo-4 AM probes for assessing Ca^{2+}.

On the other hand, it is well known that changes in the concentration of cytosolic free calcium play a key role in the regulation of physiological and pathological processes in animal, plant and bacterial cells. Some natural free Ca^{2+} indicators are known, and an interesting example is the photoprotein aequorin obtained from species of the bioluminescent jellyfish *Aequorea*, which emits blue light when Ca^{2+} is added. Aequorin is composed of a single polypeptide (apoaequorin) and a tightly bound imidazopyrazine chromophore (coelenterazine), which is closely related to *Cypripedine* and *Renilla* luciferins [63]. Aequorin has been engineered with the endoplasmic reticulum (ER)-targeting sequences of calreticulin, so that it locates efficiently in the ER of live cells. After incubating cells with coelenterazine, the reconstituted photoprotein was able to record concentration and changes of Ca^{2+} in ER [64].

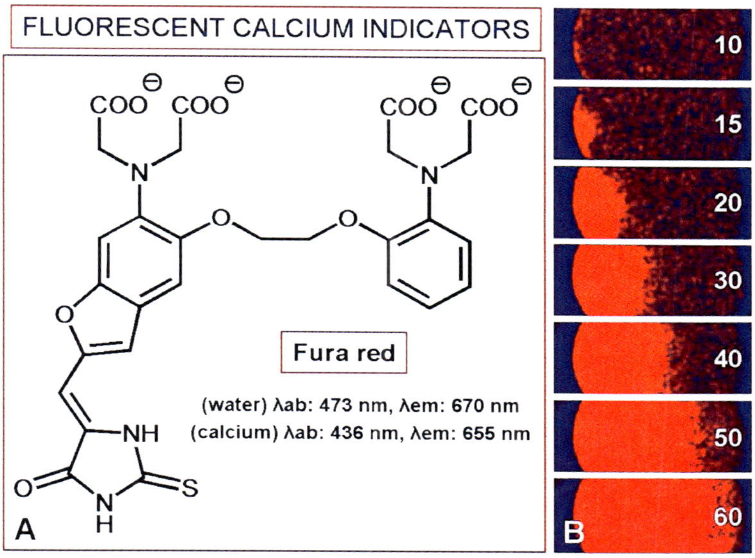

Fig. (8.16). **A**: Structure of the calcium indicator fura red showing spectral features. **B**: Fertilization-induced calcium wave (in red) from a starfish oocyte (*Pisaster ochraceus*) microinjected with dextran (MW: 10,000) labeled with calcium green-1, and observed as selected pseudocolor images taken at the indicated times after fertilization (in seconds).

4.4. Sodium, Magnesium, Zinc, and Phosphate

Examples of chelating fluoro-ionophores for the small sodium ion are CoroNa red and SBFI, which are fluorescent probes with 7 electron donors, and therefore the chelating cavity is smaller than that with 8 donors. The structure of their complexes with Na ions is shown in Fig. (**8.17A, B**). Some Ca^{2+} probes (mainly quin2) also form complexes with Mg^{2+} and Zn^{2+}. The use of magneson and lumomagneson for detecting Mg^{2+} has been already mentioned (section 3.7).

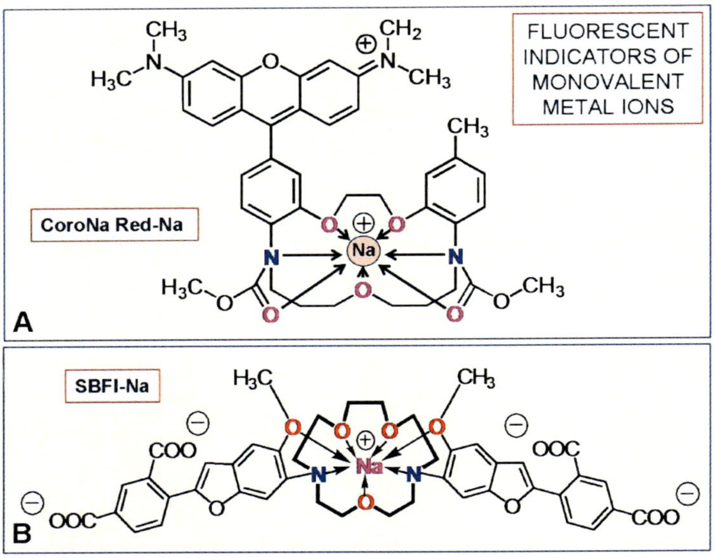

Fig. (8.17). **A**, **B**: Chemical structure of CoroNa red-Na and SBFI-Na complexes, respectively, showing hepta-coordinative bonds to the sodium ion (arrows).

Regarding zinc, histochemical methods using zincon, dithizone and rhodizonic acid in bright-field microscopy are well known [4]. Fluorescent methods for Zn^{2+} localization and determination with 8-hydroxyquinoline and derivatives have been early described [21, 65]. Zinc has also been detected by using the already described reagent PAQH [40]. Likewise, the emission intensity of 9,10-bis(--dimethylamino-ethyl-methylamino-methyl)-anthracene increases over 1000 fold in the presence of zinc chloride [66]. This mechanism is a chelate-enhanced fluorescence. When coordinated to a metal ion, the nitrogen lone electrons of the bidentate ethylenediamino chelating group become involved in bonding and are unable to donate them to the excited state of anthracene. Thus emission increases because quenching is avoided. The same occurs at low pH when the amines become protonated. On the other hand, Zinquin (2-methyl-8-*p*-toluen--sulphonamido-6-quinolyloxy) acetic acid is a vital probe for free Zn ions, the complex Zinquin-Zn^{2+} showing λexc: 350-360 nm and λem: 485 nm [67].

The chemistry and properties of a large number of fluorescent Zn^{2+} indicators such as TQEN, TPA analogues, cyclen, as well as PyPO, triazolyl, and terpyridine derivatives have been extensively revised. An interesting approach is the use of organelle-targeting Zn^{2+} probes selective for lysosomes, mitochondria, and plasma membrane [56]. This is the case for lysosome-targeting with tertiary amines that are lysosomotropic tags, and mitochondria-targeting with lipophilic alkyl-triphenylphosphonium (Ph_3P^+) (*i.e.* the ubiquinone derivative mitoQ [56, 68].

In the case of determination of phosphate, a new and interesting approach based on coordination chemistry involves the use of dinuclear dipicolylamine-zinc complexes bound to an anthracene ring as selective binding site for phosphate anions under neutral aqueous conditions [69]. When this phosphate receptor is occupied by inorganic phosphate or the phosphate group from peptides with phosphorylated tyrosine, the fluorescence increases 4- or 5-fold (exc: 380 nm, em: 413 and 435 nm).

5. CALCIFICATION MARKERS

Fluorescent calcification (mineralization) markers are often applied in anatomical, embryological, physiological, and pathological studies on bone and teeth, administered either alone or in a sequential mode to reveal detailed ossification kinetics under fluorescence microscopy.

5.1. Fluorescent Labels

In addition to compounds with hydroxy and ketone groups in *peri* position such as alizarin red S and morin, the chelating dyes calcein blue, calcein, tetracycline, xylenol orange, alizarin complexon, and hematoporphyrin are well known and widely applied chelating probes for studying calcification *in vivo* [70 - 75]. They induce strong blue, green, pale green, orange, orange-red and red fluorescence, respectively, under UV or blue exciting light (see Fig. **1.13B**). Lumomagneson can be also used [39].

It must be taken into account that only ionic calcium (not as insoluble precipitate) will be visualized. Likewise, due to the action on the whole organism, some synthetic dyes can display toxic effects. Although it is often claimed that these ossification markers recognize Ca^{2+}, it seems possible that several so-called calcium-specific labels also react with Mg^{2+}, and therefore they would reveal global mineral deposits in bones. Figs. (**8.18** and **8.19A, B**) show several commonly used calcification markers. Some approximate spectral parameters are the following:

	λab (nm, water)	**λem (nm, bone)**
Calcein blue (coumarin)	360	500
Calcein (xanthene)	490	520
Xylenol orange (triphenylmethane)	490	610
Alizarin complexon (anthraquinone)	580	625

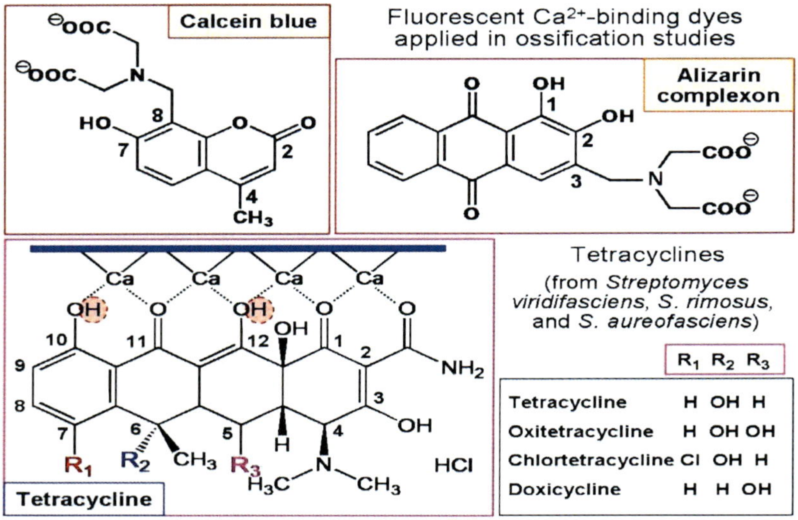

Fig. (8.18). Chemical structure of the calcification markers calcein blue (coumarin), alizarin complexon (anthraquinone), and tetracycline, the later showing the binding mechanism to calcium ions still exposed to chelating dyes during the ossification process. The blue line indicates tissue structures carrying available calcium.

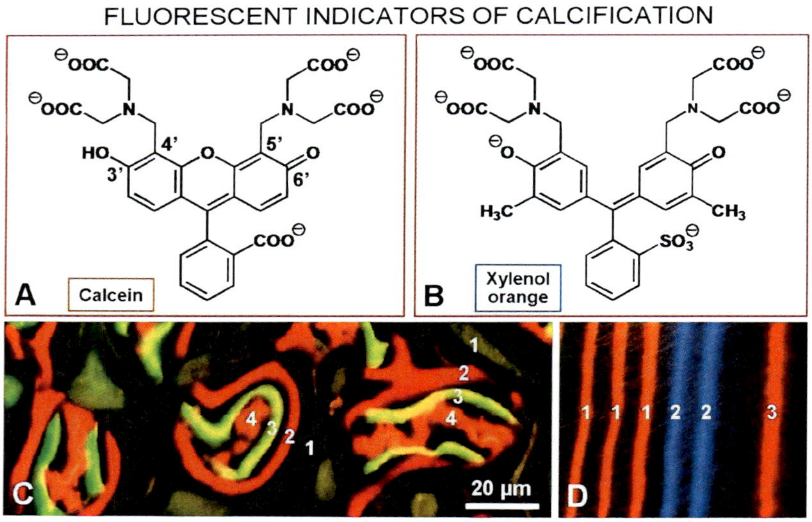

Fig. (8.19). **A**, **B**: Chemical structures of calcification markers. **C**, **D**: Polychromatic fluorescence during the endochondral osteogenesis (**C**) and odontogenesis (**D**). **C**: Tibia bone of a young rabbit showing labeling of calcium deposits from outside (1, older marker) to inside (4, latter marker). Several osteons with four-color fluorescent lamellae are visible. 1: tetracycline (pale green), 2: alizarin complexon (red), 3: calcein (green), 4: xylenol orange (orange). **D**: Labeling *in vivo* of rabbit teeth by fluorescent markers observed under UV excitation. 1: xylenol orange, 2: calcein blue, 3: alizarin complexon.

Obviously, as chelating agents these calcification markers can also detect other metal cations, but those non-physiological have scarce importance in biological studies. Thus, in addition to Ca, alizarin complexon also detect Cs, Zr, lanthanides, *etc.*

In the case of tetracyclines (aureomycine, terramycine, *etc.*) the keto-enol system allows chelation of metal cations (Fig. **8.18**). Therefore tetracyclines are used under UV excitation as green-yellow fluorescing markers of calcification, as well as mitocondrial probes and DNA-binding aluminum complexes.

5.2. Polychromatic Fluorescence

The use of fluorescent probes for Ca ions with different emission colors are simple and sensitive procedures to visualize bone and teeth calcification in living organisms allowing polychromatic fluorescence [75 - 78]. Therefore, they compare very favorably with the less sensitive and more time-consuming alcian blue-alizarin red method for studies on skeletal development. Bone formation and remodeling *in vivo* can be assessed by polychrome labeling using three or four color-contrasting calcium-binding fluorescent dyes.

Examples of microscopical calcification processes labeled by polychromatic fluorescence are illustrated in Fig. (**8.19**). New techniques of polychromatic sequential labeling of ossification followed by spectral image analysis have improved the discrimination between different calcification markers, which often overlap partially their emission spectra [79, 80].

CONFLICT OF INTEREST

The author (editor) declares no conflict of interest, financial or otherwise.

ACKNOWLEDGEMENTS

This work was supported by a grant from the Ministerio de Economía y Competitividad (CTQ2013-48767-C3-3-R), Spain.

REFERENCES

[1] Feigl F. Spot Test in Inorganic Analysis. New York: Elsevier Publishing Co. 1958.

[2] Thompson SW. Selected Histochemical and Histopathological Methods. Springfield, Illinois: Charles C. Thomas 1966.

[3] Pearse AG. Histochemistry Theoretical and Applied. 3rd ed. Edinburgh, London, New York: Churchill Livingstone 1972; 2.

[4] Lillie RD. Conn's Biological Stains. 9th ed., Baltimore: Williams and Wilkins 1977.

[5] Löhr HG, Vögtle F. Chromo- and fluoroionophores. A new class of dye reagents. Acc Chem Res 1985; 18: 65-72.

[http://dx.doi.org/10.1021/ar00111a001]

[6] Kotyk A, Slavik J, Eds. Intracellular pH and its Measurement. Boca Raton, FL: CRC Press 1989.

[7] Whitaker M. Fluorescence Imaging of Living Cells.Cell Biology: A Laboratory Handbook. New York, London: Academic Press 1994; Vol. 2.

[8] Horobin RW, Kiernan JA, Eds. Conn's Biological Stains A Handbook of Dyes, Stains and Fluorochromes for Use in Biology and Medicine. 10th ed., Oxford, UK: BIOS Scientific Publishers 2002.

[9] Stockert JC. Reactive Staining Reagents and Fluorescent Labels.Conn's Biological Stains A Handbook of Dyes, Stains and Fluorochromes for Use in Biology and Medicine. 10th ed. Oxford: Bios Scientific Publishers 2002; pp. 77-88.

[10] Scott JE. A new classification of cationic dyes based on their affinity for ionized groups of polyanions. Biochem J 1966; 99: 3-4.

[11] Scott JE. Affinity, competition and specific interactions in the biochemistry and histochemistry of polyelectrolytes. Biochem Soc Trans 1973; 1: 787-806.
[http://dx.doi.org/10.1042/bst0010787]

[12] Del Castillo P, Llorente AR, Gómez A, Gosálvez J, Goyanes VJ, Stockert JC. New fluorescence reactions in DNA cytochemistry. 2. Microscopic and spectroscopic studies on fluorescent aluminum complexes. Anal Quant Cytol Histol 1990; 12(1): 11-20.
[PMID: 2302307]

[13] Staple PH. Staining of tissue polycations by anionic dyes in salt solutions with special reference to gingiva. J Histochem Cytochem 1970; 18(6): 424-38.
[http://dx.doi.org/10.1177/18.6.424] [PMID: 4193244]

[14] Trigoso CI, Del Castillo P, Stockert JC. Influence of inorganic salts on the staining reaction of eosinophil leukocyte granules by anionic dyes. Acta Histochem 1992; 93(1): 313-8.
[http://dx.doi.org/10.1016/S0065-1281(11)80230-8] [PMID: 1382350]

[15] Stockert JC. Cytochemistry of nucleic acids: binding mechanisms of dyes and fluorochromes. Microsc Electron Biol Celular 1985; 9(2): 89-131.
[PMID: 2431250]

[16] Gutiérrez-Gonzálvez MG, Armas-Portela R, Stockert JC. Differential staining of biological structures by ruthenium red. J Microsc 1987; 145(Pt 3): 333-40.
[PMID: 2438415]

[17] Mello ML. Toluidine blue binding capacity of heterochromatin and euchromatin of *Triatoma infestans* Klug. Histochemistry 1980; 69(2): 181-7.
[http://dx.doi.org/10.1007/BF00533135] [PMID: 6161909]

[18] Marshall PN, Horobin RW. The mechanism of action of "mordant" dyes--a study using preformed metal complexes. Histochemie 1973; 35(4): 361-71.
[http://dx.doi.org/10.1007/BF00310675] [PMID: 4126143]

[19] Stockert JC, Armas-Portela R. Selective staining reaction of the cationic complex tris(1,10-phenanthroline)-iron(II) with cell and tissue polyanions. Biocell 1995; 19: 43-8.

[20] Udenfriend S. Fluorescence Assay in Biology and Medicine. New York, London: Academic Press 1962.

[21] Beltrán JL, Compañó R, Izquierdo A, *et al.* Simultaneous determination of zinc and beryllium by synchronous spectrofluorimetry. Appl Fluor Technol 1991; 3(6): 6-13.

[22] Kushi Y, Fernando Q. The crystal and molecular structure of bis(2-methyl-8-quinolinolato) aluminum(III)-μ-oxo-bis (2-methyl-8-quinolinolato) aluminum(III). J Am Chem Soc 1970; 92: 91-6.
[http://dx.doi.org/10.1021/ja00704a015]

[23] Llorente AR, Del Castillo P, Stockert JC. Aluminium binding to chromatin DNA as revealed by formation of fluorescent complexes with 8-hydroxyquinoline and other ligands. J Microsc 1989; 155(Pt 2): 227-30.
[http://dx.doi.org/10.1111/j.1365-2818.1989.tb02885.x] [PMID: 2507787]

[24] Bertolesi GE, Trigoso CI, Espada J, Stockert JC. Cytochemical application of tris (2,2′-bipyridine) ruthenium (II): fluorescence reaction with sulfated polyanions of mast cell granules. J Histochem Cytochem 1995; 43(5): 537-43.
[http://dx.doi.org/10.1177/43.5.7730592] [PMID: 7730592]

[25] Bertolesi GE, Lauría de Cidre L, Stockert JC. Fluorescence of mast cells using aluminium / tris ruthenium II staining procedures. Eur Microsc Anal 1999; 61: 5-7.

[26] Malinin GI. Permanent fluorescent staining of nucleic acids in isolated cells. J Histochem Cytochem 1978; 26(11): 1018-20.
[http://dx.doi.org/10.1177/26.11.722051] [PMID: 722051]

[27] Cowden RR, Curtis SK. Microfluorometric investigations of chromatin structure. II. Mordant fluorochroming with ions that complex with morin. Histochemistry 1981; 72(3): 391-400.
[http://dx.doi.org/10.1007/BF00501781] [PMID: 6795151]

[28] Stockert JC, Colman OD, Cañete M. Fluorescence reaction of leukocyte granules by morin. Acta Histochem Suppl 1985; 31 (Suppl.): 243-52.
[PMID: 2410949]

[29] Frohne D. Anatomisch-mikrochemische Drogenanalyse. Stuttgart: Georg Thieme Verlag 1974.

[30] Kiernan JA. Chromoxane cyanine R. I. Physical and chemical properties of the dye and of some of its iron complexes. J Microsc 1984; 134(Pt 1): 13-23.
[http://dx.doi.org/10.1111/j.1365-2818.1984.tb00500.x] [PMID: 6201615]

[31] Meloan SN, Valentine LS, Puchtler H. On the structure of carminic acid and carmine. Histochemie 1971; 27(2): 87-95.
[http://dx.doi.org/10.1007/BF00284950] [PMID: 4106026]

[32] Stockert JC, Llorente AR, Del Castillo P, Gómez A. Chromatin fluorescence after carmine staining. Stain Technol 1990; 65(6): 299-302.
[http://dx.doi.org/10.3109/10520299009105621] [PMID: 2080525]

[33] Lev-Goldman V, Mester B, Ben-Aroya N, *et al.* Conjugates of gonadotropin releasing hormone (GnRH) with carminic acid: Synthesis, generation of reactive oxygen species (ROS) and biological evaluation. Bioorg Med Chem 2008; 16(14): 6789-98.
[http://dx.doi.org/10.1016/j.bmc.2008.05.057] [PMID: 18571926]

[34] Kiel EG, Heertjes PM. Metal complexes of alizarin. I. The structure of the calcium-aluminium lake of alizarin. J Soc Dyers Colorists 1963; 79: 21-7.
[http://dx.doi.org/10.1111/j.1478-4408.1963.tb02507.x]

[35] Budavari S, Ed. The Merck Index. An Encyclopedia of Chemicals, Drugs and Biologicals. 11th ed.. Rahway, NJ: Merck and Co. 1989.

[36] Lown WJ. Discovery and development of anthracycline antitumor antibiotics. Chem Soc Rev 1993; 22: 165-76.
[http://dx.doi.org/10.1039/cs9932200165]

[37] Juarranz A, Villanueva A, Cañete M, Polo S, Domínguez V, Stockert JC. Microscopical and spectroscopic studies on the fluorescence of a daunomycin-aluminum complex. Histochem J 1999; 31(3): 201-7.
[http://dx.doi.org/10.1023/A:1003549500211] [PMID: 10421420]

[38] Hall CT, Hansen PA. Chelated azo dyes as counterstains in fluorescent antibody technic. Zentralbl Bakteriol 1962; 184: 548-54.

[39] Coutelier L. Le lumomagnéson: marqueur fluorescent de l'os. Experientia 1973; 29(2): 192-3. [http://dx.doi.org/10.1007/BF01945470] [PMID: 4692765]

[40] Jensen RE, Pflaum RT. Fluorometric determination of zinc. Anal Chem 1966; 38: 1268-9. [http://dx.doi.org/10.1021/ac60241a043]

[41] Stockert JC, Abasolo MI. Inaccurate chemical structures of dyes and fluorochromes found in the literature can be problematic for teaching and research. Biotech Histochem 2011; 86(1): 52-60. [http://dx.doi.org/10.3109/10520295.2010.489428] [PMID: 20500122]

[42] Sumi Y, Inoue T, Muraki T, Suzuki T. A highly sensitive chelator for metal staining, bromopyridylazo-diethylaminophenol. Stain Technol 1983; 58(6): 325-8. [http://dx.doi.org/10.3109/10520298309066807] [PMID: 6206612]

[43] Soini E, Hemmilä I. Fluoroimmunoassay: present status and key problems. Clin Chem 1979; 25(3): 353-61. [PMID: 400437]

[44] Prat O, Lopez E, Mathis G. Europium(III) cryptate: a fluorescent label for the detection of DNA hybrids on solid support. Anal Biochem 1991; 195(2): 283-9. [http://dx.doi.org/10.1016/0003-2697(91)90331-M] [PMID: 1750681]

[45] Axelrod D, Klein MP. Fluorescent Ca^{+2} analogs in nerve: rare-earth ions. Biochem Biophys Res Commun 1974; 57(3): 927-33. [http://dx.doi.org/10.1016/0006-291X(74)90634-2] [PMID: 4827842]

[46] Kayne MS, Cohn M. Enhancement of Tb(III) and Eu(III) fluorescence in complexes with *Escherichia coli* tRNA. Biochemistry 1974; 13(20): 4159-65. [http://dx.doi.org/10.1021/bi00717a014] [PMID: 4606518]

[47] Dyer DL, Mori K. Fluorescent nuclear staining with europium thenolytrifluoroacetonate. J Histochem Cytochem 1969; 17(11): 755-6. [http://dx.doi.org/10.1177/17.11.755] [PMID: 4195389]

[48] Sevéus L, Väisälä M, Hemmilä I, Kojola H, Roomans GM, Soini E. Use of fluorescent europium chelates as labels in microscopy allows glutaraldehyde fixation and permanent mounting and leads to reduced autofluorescence and good long-term stability. Microsc Res Tech 1994; 28(2): 149-54. [http://dx.doi.org/10.1002/jemt.1070280206] [PMID: 8054664]

[49] Bishop E. Indicators. New York: Pergamon Press 1972.

[50] Caswell AH. Methods of measuring intracellular calcium. Int Rev Cytol 1979; 56: 145-81. [http://dx.doi.org/10.1016/S0074-7696(08)61822-7] [PMID: 378885]

[51] Tsien RY. Fluorescent indicators of ion concentrations. Methods Cell Biol 1989; 30: 127-56. [http://dx.doi.org/10.1016/S0091-679X(08)60978-4] [PMID: 2538708]

[52] Slavik J. Optical ion probes for cytological applications. Appl Fluor Technol 1990; 2(1): 1-8.

[53] Gillies RJ, Lynch RM. Frontiers in the measurement of cell and tissue pH. Novartis Found Symp 2001; 240: 7-19. [http://dx.doi.org/10.1002/0470868716.ch2] [PMID: 11727938]

[54] Haugland RP, Ed. Handbook of Fluorescent Probes and Research Products. 9th ed., Eugene, OR, USA: Molecular Probes 2002.

[55] Wainwright M. The Use of Dyes and Fluorochromes as Indicators.Conn's Biological Stains A Handbook of Dyes, Stains and Fluorochromes for Use in Biology and Medicine. 10th ed. Oxford: Bios Scientific Publishers 2002; pp. 89-99.

[56] Zhu L, Yuan Z, Simmons JT, Sreenath K. Zn(II)-coordination modulated ligand photophysical processes - the development of fluorescent indicators for imaging biological Zn(II) ions. RSC Advances 2014; 4(39): 20398-440.

[http://dx.doi.org/10.1039/C4RA00354C] [PMID: 25071933]

[57] LaManna JC, McCracken KA. The use of neutral red as an intracellular pH indicator in rat brain cortex *in vivo*. Anal Biochem 1984; 142(1): 117-25. [http://dx.doi.org/10.1016/0003-2697(84)90525-6] [PMID: 6517307]

[58] Koller E. Pyrene-sulfonates: an interesting class of fluorescent probes. Appl Fluor Technol 1989; 1(2): 13-4.

[59] Trigoso CI, Espada J, Stockert JC. Fluorescence of eosinophil leucocyte granules induced by 1-hydroxy-3,6,8-pyrenetrisulfonate. Visualization of differences in protein isoelectric points. Histochem Cell Biol 1995; 104(1): 69-73. [http://dx.doi.org/10.1007/BF01464788] [PMID: 7584562]

[60] Kurtz I, Balaban RS. Fluorescence emission spectroscopy of 1,4-dihydroxyphthalonitrile. A method for determining intracellular pH in cultured cells. Biophys J 1985; 48(3): 499-508. [http://dx.doi.org/10.1016/S0006-3495(85)83805-4] [PMID: 4041540]

[61] Han J, Loudet A, Barhoumi R, Burghardt RC, Burgess K. A ratiometric pH reporter for imaging protein-dye conjugates in living cells. J Am Chem Soc 2009; 131(5): 1642-3. [http://dx.doi.org/10.1021/ja8073374] [PMID: 19146412]

[62] Grynkiewicz G, Poenie M, Tsien RY. A new generation of Ca^{2+} indicators with greatly improved fluorescence properties. J Biol Chem 1985; 260(6): 3440-50. [PMID: 3838314]

[63] Shimomura O, Johnson FH, Morise H. Mechanism of the luminescent intramolecular reaction of aequorin. Biochemistry 1974; 13(16): 3278-86. [http://dx.doi.org/10.1021/bi00713a016] [PMID: 4152180]

[64] Kendall JM, Badminton MN, Dormer RL, Campbell AK. Changes in free calcium in the endoplasmic reticulum of living cells detected using targeted aequorin. Anal Biochem 1994; 221(1): 173-81. [http://dx.doi.org/10.1006/abio.1994.1394] [PMID: 7985790]

[65] Smith GL, Jenkins RA, Gough JF. A fluorescent method for the detection and localization of zinc in human granulocytes. J Histochem Cytochem 1969; 17(11): 749-50. [http://dx.doi.org/10.1177/17.11.749] [PMID: 5386301]

[66] Huston ME, Haider KW, Czarnik AW. Chelation-enhanced fluorescence in 9,10-bis(TMEDA--anthracene. J Am Chem Soc 1988; 110: 4460-2. [http://dx.doi.org/10.1021/ja00221a083]

[67] Zalewski PD, Forbes IJ, Betts WH. Correlation of apoptosis with change in intracellular labile Zn(II) using zinquin [(2-methyl-8-*p*-toluenesulphonamido-6-quinolyloxy)acetic acid], a new specific fluorescent probe for Zn(II). Biochem J 1993; 296(Pt 2): 403-8. [http://dx.doi.org/10.1042/bj2960403] [PMID: 8257431]

[68] Kelso GF, Porteous CM, Coulter CV, *et al.* Selective targeting of a redox-active ubiquinone to mitochondria within cells: antioxidant and antiapoptotic properties. J Biol Chem 2001; 276(7): 4588-96. [http://dx.doi.org/10.1074/jbc.M009093200] [PMID: 11092892]

[69] Ojida A, Mito-Oka Y, Inoue MA, Hamachi I. First artificial receptors and chemosensors toward phosphorylated peptide in aqueous solution. J Am Chem Soc 2002; 124(22): 6256-8. [http://dx.doi.org/10.1021/ja025761b] [PMID: 12033851]

[70] Coutelier L. A new fluorescent labelling agent in the study of osteogenesis: hematoporphyrin. Rev Belg Pathol Med Exp 1964; 30: 369-75. [PMID: 14269118]

[71] Suzuki HK, Mathews A. Two-color fluorescent labeling of mineralizing tissues with tetracycline and 2,4-bis[N,N′-di-(carbomethyl)aminomethyl] fluorescein. Stain Technol 1966; 41(1): 57-60. [http://dx.doi.org/10.3109/10520296609116280] [PMID: 4161482]

[72] Rahn BA, Perren SM. Calcein blue as a fluorescent label in bone. Experientia 1970; 26(5): 519-20. [http://dx.doi.org/10.1007/BF01898484] [PMID: 5444837]

[73] Rahn BA, Perren SM. Xylenol orange, a fluorochrome useful in polychrome sequential labeling of calcifying tissues. Stain Technol 1971; 46(3): 125-9. [http://dx.doi.org/10.3109/10520297109067836] [PMID: 4105405]

[74] Rahn BA, Perren SM. [Alizarin complexon-fluorochrome for bone and dentine labeling]. Experientia 1972; 28(2): 180. [http://dx.doi.org/10.1007/BF01935743] [PMID: 4111994]

[75] Du SJ, Frenkel V, Kindschi G, Zohar Y. Visualizing normal and defective bone development in zebrafish embryos using the fluorescent chromophore calcein. Dev Biol 2001; 238(2): 239-46. [http://dx.doi.org/10.1006/dbio.2001.0390] [PMID: 11784007]

[76] Solheim T. Pluricolor fluorescent labeling of mineralizing tissue. Scand J Dent Res 1974; 82(1): 19-27. [PMID: 4132925]

[77] Aaron JE, Makins NB, Francis RM, Peacock M. Staining of the calcification front in human bone using contrasting fluorochromes *in vitro*. J Histochem Cytochem 1984; 32(12): 1251-61. [http://dx.doi.org/10.1177/32.12.6209330] [PMID: 6209330]

[78] Burt-Pichat B, Follet H, Toulemonde G, Arlot M, Delmas P, Chapurlat R. Methodological approach for the detection of both microdamage and fluorochrome labels in ewe bone and human trabecular bone. J Bone Miner Metab 2011; 29(6): 756-64. [http://dx.doi.org/10.1007/s00774-011-0291-7] [PMID: 21748462]

[79] Pautke C, Vogt S, Tischer T, *et al.* Polychrome labeling of bone with seven different fluorochromes: enhancing fluorochrome discrimination by spectral image analysis. Bone 2005; 37(4): 441-5. [http://dx.doi.org/10.1016/j.bone.2005.05.008] [PMID: 16046209]

[80] Pautke C, Tischer T, Vogt S, *et al.* New advances in fluorochrome sequential labelling of teeth using seven different fluorochromes and spectral image analysis. J Anat 2007; 210(1): 117-21. [http://dx.doi.org/10.1111/j.1469-7580.2006.00660.x] [PMID: 17229289]

CHAPTER 9

Luminescent Transition Metal Probes

Abstract: In this chapter, the fundamentals in regard to the use of transition metal complexes (TMCs) as luminescent probes will be presented. These metal complexes (MCs) between organic ligands and transition metal cations present many interesting properties that make them very suitable for a number of biomedical applications. First, a brief introduction to TMCs and their photophysics and photochemistry will be provided. Then, some particular applications of selected transition metals (TMs) in the field of fluorescence microscopy will be presented on a case-per-case basis. The relevant role played by lanthanide elements and their complexes in fluorescence microscopy, especially time-resolved, will be introduced. Some uses of actinide MCs will be presented as well.

Keywords: Actinide, d-d transitions, Denticity, f-f transitions, Lanthanide, Ligand, Ligand-to-metal charge transfer, Luminescence, Macrocycle, Metal-t--ligand charge transfer, Time-resolved fluorescence microscopy, Transition metal, Transition metal complex.

1. TRANSITION METAL COMPLEX

A TMC results from the chemical interaction of a metal cation and an inorganic/organic ligand. Commonly, one metal cation interacts with more than one ligand but two or more metal atoms can be involved in the TMC. First we will see which transition metal elements most usually take part in the MCs and then we will present some of the most common ligands.

The TMs comprise the most extensive group of chemical elements among several others in which the periodic table (see Fig. **9.1A**) can be divided attending to common features (*e.g.*, alkali metals, halogens, noble gases, *etc.*). The periodic table is arranged in accordance with the filling of the available electronic shells (*s*, *p*, *d* and *f* in particular). This provides a basis to classify the chemical elements in regard to their electronic structure (see Fig. **9.1B**) [1]. TMs are divided into the d-block and f-block after the corresponding shell that is being filled. Furthermore, the d-block is itself made up of 3 subgroups: 3d metals, 4d metals and 5d metals (see Fig. **9.1C**). Correspondingly, the f-block is subdivided into the 4f metals, commonly known as the Lanthanides, and the 5f metals, known as the Actinides (see Fig. **9.1D**). There are important differences between the characteristics of

Juan Carlos Stockert & Alfonso Blázquez-Castro

d-block TMs and the f-block TMs [2]. In what follows, d-block TMs with important roles in TMCs will be briefly presented first, and then, lanthanides and actinides will be explained in the subsequent sections. Finally, the ligands making up the TMCs will be introduced before pondering over into the photophysics and photochemistry of these chemical compounds.

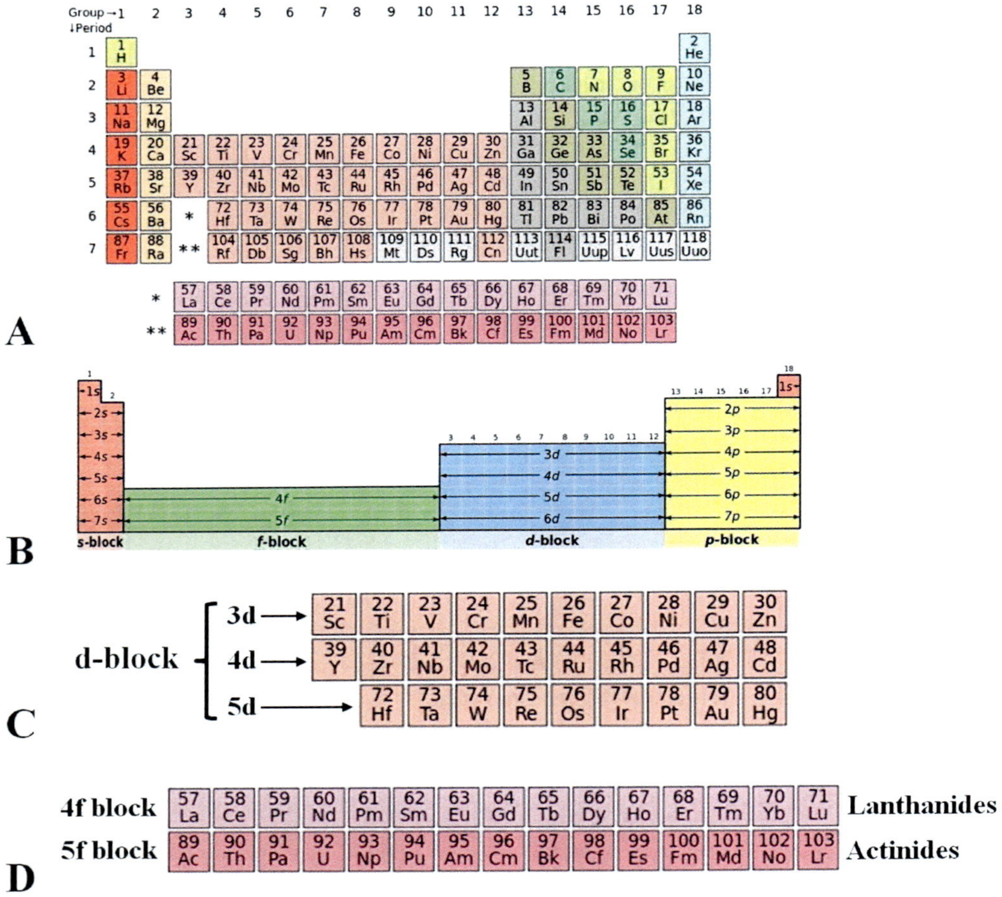

Fig. (9.1). A: Periodic table of the elements. The elements are arranged according to their atomic charge. The elements can be classified by groups (rows) or periods (columns) (reproduced from DePiep, https://commons.wikimedia.org/wiki/File:Periodic_table_(polyatomic). svg#). **B:** Alternative extended depiction of the periodic table showing the blocks in which it can be divided attending to the sequential filling of electronic shells s, p, d and f. TMs compromise the d-block (highlighted in blue) and f-block (highlighted in green) (reproduced from DePiep, https://commons.wikimedia.org/wiki/File:Periodic_table_blocks_ spdf_(32_column).svg#). **C:** Expanded d-block showing in more detail the elements making the 3d, 4d and 5d subgroups. **D:** Expanded f-block depiction in which the lanthanides (4f) and actinides (5f) subgroups can be identified.

1.1. d-block Transition Metals

The elements in the d-block represent the most common metals that are found in nature. These elements are relatively light, especially in regard to 3d metals, making them abundant in nature. Most TMCs found in nature and in biological systems display a 3d TM as the cation in the TMC. The elements in the d-block have their d orbital sequentially filled with electrons (some exceptions exist in regard to the corresponding s-orbital, although this is outside the scope of this chapter). When these metals are oxidized, they usually lose any electrons in the s orbital and at least one electron in the d orbital. In this way, the d orbital displays an incomplete filling for the cations. Depending on which d orbital is being filled, d-block TMs belong to the 3d, 4d or 5d period. Briefly, we will now introduce each of these subgroups.

3d Transition Metals

This subgroup includes elements from Scandium (Sc) to Zinc (Zn). These are the lightest TMs and, correspondingly, they are the most abundant ones. Oxidation states vary along the subgroup with the extreme representatives (Scandium and Zinc) showing only the +3 and +2 states respectively, while members in the middle of the subgroup display a richer variability in oxidation states (*e.g.*, Manganese features oxidation states from -3 to +7). This subgroup includes some of the more important TMs taking part in TMCs like Chromium (Cr), Iron (Fe) and Copper (Cu).

4d Transition Metals

This subgroup encompasses the TMs from Yttrium (Y) to Cadmium (Cd). As their atomic weights are larger than 3d members, they are less common in nature. Indeed, this group features the first unstable radioactive element of the periodic table, Technetium (Tc). No stable isotopes of this element are known. Again elements in the middle of the subgroup display a broader availability of oxidation states which reflects a richer TMC chemistry. This especially is true for the elements like Molybdenum (Mo), Ruthenium (Ru), Rhodium (Rh) and Palladium (Pd). As shown below, Ru is one of the most important TMs in regard to the quantity of complexes, it is capable of forming and the wide range of photophysical and photochemical behaviours that TMs display.

5d Transition Metals

The 5d subgroup represents the heaviest and more uncommon TMs to be found in nature (all the 6d TMs are extremely radioactive and have been artificially created in the laboratory). The subgroup is heavier than the lanthanides (4f subgroup) and includes the best known representatives of the noble metals. It encompasses from Hafnium (Hf) to Mercury (Hg). Notable members in regard to TMCs are Rhenium (Re), Osmium (Os), Iridium (Ir), Platinum (Pt) and Gold (Au). Among these, Ir and Pt play a central role in TMCs as will be explained later in this chapter.

1.2. 4f-block Transition Metals: Lanthanides

The 4f block TMs are collectively known as the lanthanides or internal transition elements. The first member of the subgroup is Lanthanum (La) which gives the name to the whole subgroup. The lanthanides display a very interesting feature: they have a partly filled 4f electronic shell that is lower in energy than the 5s and 5p shells [2]. This means that, on average, an electron in the 4f shell is closer to the nucleus than a 5s or 5p electron. From this, it follows that 4f electrons are much less prone to interact with the environment. Therefore, these elements, under different oxidation states, have electrons within the 4f shell that are quite impervious to modulation by the surroundings because of the efficient "burial" within deeper electronic shells. We will see how this screening has a determinant role in explaining the fascinating spectroscopic characteristics of lanthanides and the TMCs based on them. Because of this relative isolation from the environment, even taking into account the sequential filling of the 4f orbital, lanthanides show a very similar chemistry among themselves with small variation in physico-chemical properties as one compares the different elements.

1.3. 5f-block Transition Metals: Actinides

The actinides are a part of the internal transition elements. In their case, it is the 5f orbital; the one that is sequentially filled along the chemical period. Therefore, the actinides show a similar chemistry among themselves, although much more diverse than that of lanthanides because of the effect their larger nuclei have on the electrons making up the different atomic orbitals. Apart from this, all of them are unstable and radioactive, and therefore, great precautions should be applied when working with these elements. Among the actinides, only Thorium (Th) and Uranium (U) have isotopes with long enough lifetimes as to make the radioactivity levels relatively low in their surroundings [3]. Following this, the most employed actinide chemical compounds and metal complexes are those derived from the uranium oxide cation known as uranyl $[UO_2]^{2+}$. Uranyl forms a large variety of complexes with different ligands, with a rich spectral diversity [4]. The use of uranyl salts and complexes will be briefly explained below in sections 2.4 and 3.5. Due to their radioactive nature and the advances in desirable chemical and photochemical properties of other TMCs, the use of actinide MCs in fluorescence microscopy is nowadays minimum at best.

1.4. Ligands

Ligands provide the immediate environment surrounding the TM and with which it interacts to form the TMC. The term ligand encompasses a large array of different molecules, from simple and inorganic (H_2O, CO) to complex and organic (organic macrocycles, nitrogen compounds, *etc.*). Fig. **(9.2A-C)** shows an ensemble of some ligands that can be found as a part of TMCs. It is the interaction between the TM and the ligand that produces the particular characteristics of each TMC. Therefore, the same TM forming a complex with different ligands results in very different TMCs. For example, Cu(II) in aqueous solution is complexed by water molecules. This MC displays a light blue colour.

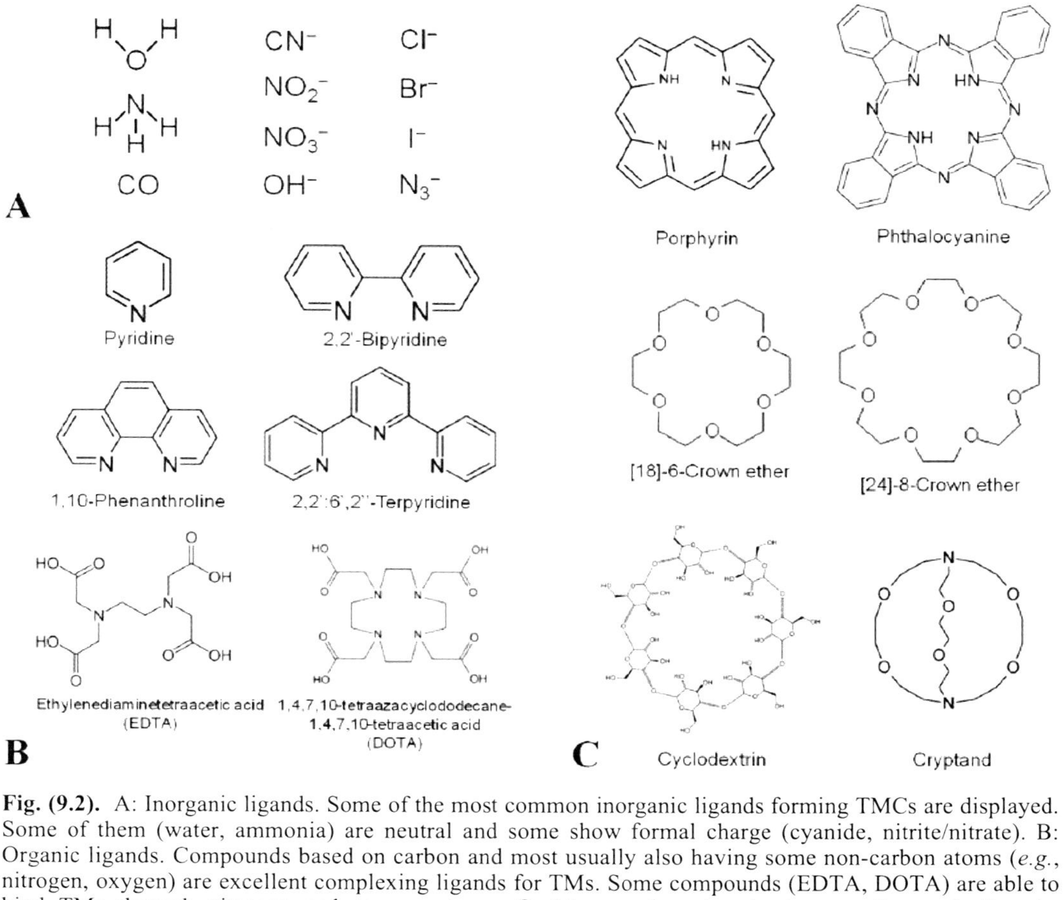

Fig. (9.2). A: Inorganic ligands. Some of the most common inorganic ligands forming TMCs are displayed. Some of them (water, ammonia) are neutral and some show formal charge (cyanide, nitrite/nitrate). B: Organic ligands. Compounds based on carbon and most usually also having some non-carbon atoms (*e.g.*, nitrogen, oxygen) are excellent complexing ligands for TMs. Some compounds (EDTA, DOTA) are able to bind TMs through nitrogen and oxygen atoms. C: Macrocycles. A subcategory of organic ligands, macrocycles represent ring-like or cavity-like molecules capable of complexing metals within their volumes. Some notable members of this category are displayed as examples.

However, adding ammonia (NH_3) to this solution brings a drastic change as ammonia molecules displace water molecules in the MC. The Cu(II) complex with ammonia now shows a deep blue/indigo colour. If carbonate is added to the solution instead of ammonia, then the colour of the obtained solution becomes cyan (blue-green) and a precipitate evolves quickly as the MC is sparingly soluble in water.

The variety of ligands is nearly as large as the molecules present there [2]. To broadly introduce a working classification, ligands will be divided in three general groups: inorganic, organic and macrocycles. Many other ways in which to present ligands are possible and this particular one has been chosen for the sake of simplicity to better understand the photophysics/photochemistry of TMCs to be presented afterwards.

Inorganic Ligands

In general, inorganic ligands are small and simple molecules. Some examples are displayed in Fig. (**9.2A**). As can be seen, solvents act as the ligand for metal cations when there is no stronger ligand present. This ligand role of a solvent makes the metal or metal cation soluble in the said solvent. As such, H_2O is one of the most important ligands. Ammonia also acts as an important ligand. Many anions act as counter ions and ligands to TM cations, such as nitrite, nitrate, azide and the reduced halogens, to name a few.

Organic Ligands

The organic ligands show a much larger variety in regard to the molecules that can act as ligands (see Fig. **9.2B**). Some are small and simple molecules, like simple acids (oxalate, formate). Others are more complex like the polyacidic molecules; examples of which are ethylenediaminetetraacetic acid –EDTA- and 1,4,7,10-tetraazacyclododecane-1,4,7,10-tetraacetic acid –DOTA-, already mentioned as an octadentate ligand. Most usually this kind of ligands serves as an alternate to nitrogen atoms with carboxylic groups to increase the amount of bonds and the strength of ligand. Also included in this ligand group are the very important aromatic nitrogen heterocycles. Common examples of these are pyridine, 2,2'-bipyridine, 1,10-phenanthroline and 2,2';6',2"-terpyridine. It can be readily appreciated that nitrogen plays an essential role for these molecules to act as ligands. They are in fact the non-bonding atoms of nitrogen the ones that bond to the metal cation and make possible the establishment of the TMC. Organic molecules where two nitrogen atoms, linked by carbon atoms, bind the TM (*e.g.* 2,2'-bipyridine or 1,10-phenanthroline) are very important ligands for TMCs. This kind of ligands is generically known as diimines and many luminescent complexes presented in the rest of the chapter belong to this chemical description.

Macrocycles

Macrocycles are introduced here as a subclass of organic ligands. When an organic molecule forms a closed structure, this brings about a ring-like (two-dimensional) or cage-like (three-dimensional) spatial arrangement. If binding atoms are adequately located on the "inside" of these structures, metal cations can bond within this internal space. As such, macrocycles are ligands which show this closed molecular structure (see Fig. **9.2C**). Again nitrogen atoms play a very important role for this kind of ligands but oxygen is a very common ligand atom as well. Different kinds of macrocycles are known both 2-D (porphyrins, phthalocyanines, chlorins, porphycenes, crown ethers, cyclens, *etc.*) and 3-D (cyclodextrins, cryptands, clathrochelates, cryptophanes, *etc.*). Most usually, a single macrocycle molecule interacts and binds a cation. But, especially for 2-D macrocycles, some metal cations can form ligands with two macrocyles, one above and one below, effectively be "sandwiched" between them. The chemistry and photochemistry of macrocycle TMCs are very important and rich (Chlorophyll, essential to bring about photosynthesis, is a macrocycle complexing magnesium, for example).

After introducing both the TMs and the ligands in a general way, it's comes the time to show some example of a TMC. Fig. (**9.3A**) presents two such examples. In the upper part, a generic aqua (water ligand) TMC can be seen. The TM sits at the centre of the complex, while six water molecules surround the metal cation complexing it. The water molecules bond the cation through their oxygen atoms establishing the coordination bonds. In the lower part, the prototypical ligand EDTA is shown binding a metal cation. A single EDTA molecule is capable of establishing six coordination bonds, two through its nitrogen atoms and four as a result of the carboxylic moieties. For this reason, EDTA is a very efficient chelating (complex-forming) agent.

In coordination chemistry, it is very usual to use the concept of denticity, which deserves explanation. The denticity of a ligand is the number of atoms per molecule directly interacting with the complexed metal [1]. Some molecules are able to establish a single interaction with a TM, while others can establish multiple bonds (the compound DOTA, see Fig. (**9.2B**), is able to form up to eight complexing bonds). Some examples of different ligands highlighting their denticity (red markers) are presented in Fig. (**9.3B**). Attending to the number of bonds a particular molecule can establish as a ligand, it is said to be monodentate (single bond), bidentate (two bonds), tridentate (three bonds), tetradentate (four bonds), *etc.* Ligands shown in Fig. (**9.3B**) display their denticity below their molecular formula.

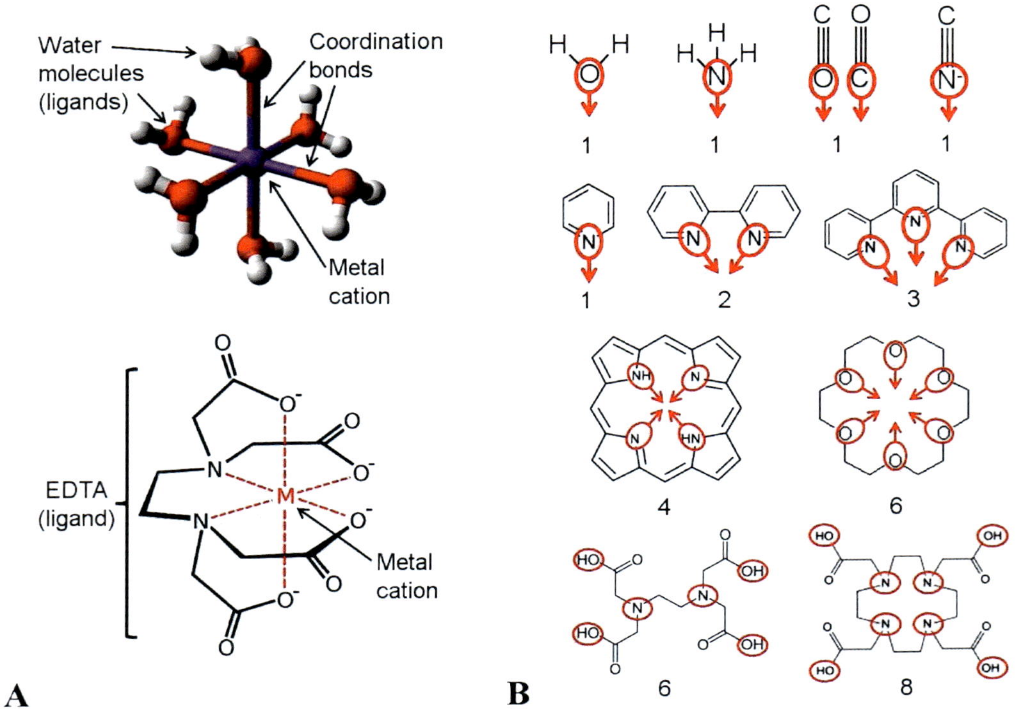

Fig. (9.3). A: Two examples of TMCs. The upper one depicts a central metal cation surrounded by six water molecules. The water molecules are complexing or coordinating the metal ion. The electrons in the oxygen atoms interact with the metal to bind it (coordination bonds) (reproduced from Ben Mills - https://commons.wikimedia.org/wiki/File:Hexaaquasodium-3D-balls.png#). In the lower example a single EDTA molecule coordinates a metal atom (red M) through six coordination bonds. Four coordination bonds are due to the carboxylic moieties and the remaining two due to the nitrogen atoms in the molecule backbone (Reproduced from Smokefootderivative work: Chamberlain2007 (talk) - https://commons.wikimedia.org/wiki/File:Metal-EDTA.svg#). B: Atoms involved in the coordination of metals in different ligands. These atoms are highlighted with a red circle and an arrow in the general direction of their coordination preference. The number below each molecule indicates the number of coordination bonds it can establish (*i.e.*, denticity). Note that for carbon monoxide (CO) both the oxygen and the carbon can act as the ligand atom, depending on the particular metal atom the molecule is encountering.

As a final point, we must stress that a huge amount of molecules can act as ligands. Therefore the previously introduced molecules are just a very small range of the possible ligands, although some of the most common/important ones have been depicted.

2. PHOTOPHYSICS AND PHOTOCHEMISTRY OF TMCS

To understand and make better use of TMCs in fluorescence microscopy and photodetection techniques, a brief introduction to their general photophysics and photochemistry seems appropriate. For deeper and more specialized texts, the reader is advised to consult the corresponding bibliography [5 - 8]. The photophysical and photochemical fundamentals already presented in chapter 2, are fully applicable to TMCs. An important point to keep in mind is that TMCs display particular and new photophysics/photochemistry in comparison to the isolated TM and the isolated ligand or ligands. New phenomena appear due to the intimate interaction that takes place between the TM and the ligands [9]. This is even true for the f-block TM, although with pertinent modifications, as will be explained below (sections 2.3 and 2.4) [2].

2.1. Optical Transitions

The interaction between the TM cation and the ligands brings new features to the TMC. Isolated, both of them have their own electronic energy level arrangement. However, when brought together in the TMC, the electronic energy levels are reshaped and split to accommodate the new chemical situation [10 - 12]. For the moment, we are going to limit the descriptions to the TM belonging to the d-block. Most of the features explained in what follows will apply to the f-block TMs as well. In Fig. (**9.4A**) the rearrangement of the electronic levels can be seen.

On one hand, the TM provides electronic levels corresponding to the s, p and d shells (left side). On the other, the ligand provides electronic levels according to its σ, π and non-bonding particular arrangement (see Chapter 3.2) (right side of the figure). When the TM and ligand make a bond, their electronic levels interact with each other. This forces them to increase or decrease in energy to fit in the new arrangement (centre). Therefore, different optical transitions appear that are absent when considering the TM and the ligand alone. Some previous optical transitions can be preserved, this depending on each particular case. This depiction of electronic energy levels and the transitions among them can be considered a Jablonski diagram. This is equivalent to the Jablonski diagram first introduced in chapter 2.

Now Fig. (**9.4B**) shows an example in which Ni(II) forms a TMC with water ($[Ni(H_2O)_6]^{2+}$) and the corresponding absorption spectrum is shown (upper figure). The absorption spectrum is such that the solution has a light green colour (see example in inset). In the same figure, (lower part) Ni(II) is forms a TMC with ammonia ($[Ni(NH_3)_6]^{2+}$) instead of water. The absorption bands change and the solution's colour changes to deep blue (example in inset). These two examples show how different ligands, water or ammonia, modulate the optical spectrum of the TMC in different ways while the cation remains the same in both cases.

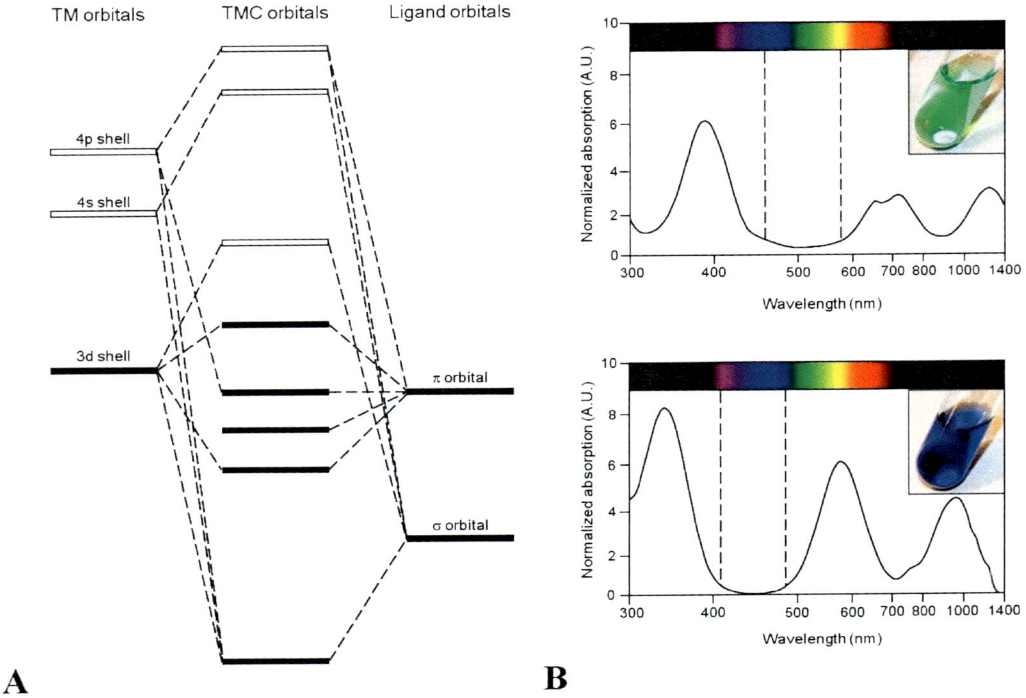

Fig. (9.4). A: Scheme of electronic level rearrangement in TMC. On the left the TM electron orbitals are displayed when not forming part of a TMC. On the right the ligand molecular orbitals displayed also on its own. When TM and ligand interact to form the TMC their electronic levels restructure themselves, splitting and hybridizing or repulsing following quantum mechanics and electrical rules. The complex notation for the TMC electronic levels has been avoided for simplicity. Occupied orbitals are shown as black boxes while empty orbitals appear as white boxes. B: Change in the optical properties of TMCs as a function of ligand. In this examples Ni(II) is forming a TMC with water (upper panel) or with ammonia (lower panel). With water the absorption spectrum is lowest in the green region whereas with ammonia it is in the blue one (see absorption curves). The regions of least absorbance in each case lay within the two broken lines in each panel (the coloured strip above each curve is a help to the eye). The insets show an example of the coloured solutions of each TMC to compare their colours (reproduced from LHcheM - https://commons.wikimedia.org/wiki/File:Color_of_various_Ni(II)_complexes_in_aqueous_solution.jpg#).

Once the TMC has been established, its optical transitions can be included within four general groups. This classification attends to the initial and final electronic states involved in the absorption process. The initial and final electronic state can belong to the TM or to the ligand. As a starting point, the following applies to both the d-block and f-block TMs, although some peculiarities apply to the lanthanides and actinides. This will be elaborated below in the corresponding section.

d-d Transitions

In these optical transitions, both electronic levels involved in the transition belong to the TM atom. They can be thought of as an internal metal-metal transition. The Jablonski diagram corresponding to this case is shown in Fig. (**9.5A**). An electron occupying the d shell absorbs a photon and transits to a higher energy level that still corresponds to the d shell. These d-d transitions are weak because the electron spin usually changes and this is forbidden [2]. However, the ligands change the symmetry and spin-orbit coupling, giving rise to a finite absorption probability. Depending on the TMC geometry (octahedral, square planar, *etc.*) the d-d splitting and transition probability change. They are usually broad because of the environmental effect on them due to the ligands. After the transition takes place, there is usually a fair efficient coupling to vibrational modes which efficiently deactivate the excited state through a non-radiative mechanism. Because of this, d-d transitions show a low quantum yield in fluorescence/phosphorescence in general.

Metal-to-Ligand Charge Transfer

These transitions are denoted as MLCT. The absorption of a photon promotes an electron from one of the TM orbitals to an unoccupied orbital in one of the ligands (see Fig. **9.5B**). As a consequence, there is a net transfer of charge from the TM to the ligand. This is why, these transitions are denoted as charge transfer. After photon absorption, the TM is oxidized (loses an electron) and the ligand is reduced (gains an electron). The complex usually rearranges to fit the new electronic distribution. Depending on each particular complex, there can be luminescence, photoisomerization, photochemistry or heat production after the light exposure. Many TMCs produce light emission after light absorption when MLCT bands are involved. These are interesting from the point of view of fluorescence microscopy applications [12 - 14].

Ligand-to-Metal Charge Transfer

In these kinds of transitions, LMCT, an electron in an occupied ligand orbital absorbs a photon and is transferred to an unoccupied d-shell orbital of the TM. It is the symmetrical transition to MLCT (see Fig. **9.5C**). In the TMC undergoing this transition, the ligand is oxidized and the TM is reduced. The same general features that have been introduced for MLCT can be applied to LMCT. However, depending on the particular TMC involved, the light wavelengths involved in each transition will vary (compare Fig. **9.5B, C**).

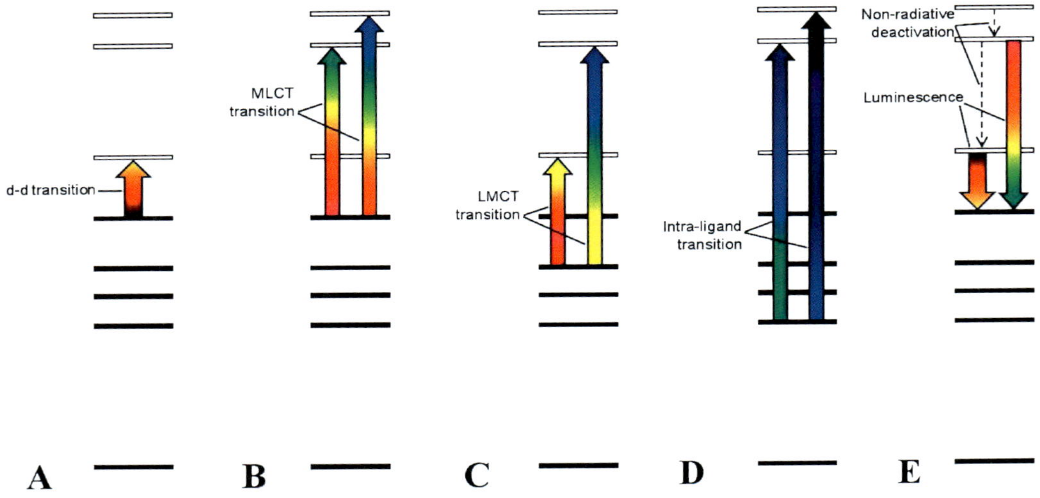

Fig. (9.5). Optical transitions in TMCs. **A**: Metal to metal electronic transitions or d-d transitions. Both the initial and final orbitals involved in the optical excitation belong to the TM. Nevertheless the wavelengths involved can be quite different from the d-d transitions of the cation in vacuum due to the interaction with the ligands (see Fig. **9.4 A**). **B**: Metal to ligand charge transfer (MLCT) transitions. The electron excited by a photon comes from one of the TM d orbitals. It fills one of the ligands empty orbitals after photon absorption. Therefore charge is transferred from the TM to one of the ligands. **C**: Ligand to metal charge transfer (LMCT) transitions. In this case, the excited electron moves from a ligand orbital to one of the TM empty orbitals. Charge moves from the ligand to the metal. **D**: Intra-ligand transitions. The initial and final orbitals involved in the transition belong to the ligand. In general d-d transitions have the smaller excitation energies, MLCT and LMCT have more or less similar excitation energies and ligand to ligand transitions are the most energetic (colour code of transition arrows depict this tendency in the figures). **E**: Scheme of a radiative de-excitation transition (luminescence) after ligand to ligand absorption. There can be non-radiative transitions (broken arrows) that populate other orbitals from the initially involved in the optical transition (analogous to internal transitions and intersystem crossings). In the scheme a LMCT transition (larger arrow) and a d-d transition (shorter arrow) have been represented. Both or only one of them can occur in real cases, depending on the particular nature of the TMC involved. Other transitions not depicted are equally possible. Occupied orbitals are shown as black boxes while empty orbitals appear as white boxes.

Intra-ligand Transitions

The fourth available option for light absorption is that of an electron starting from an occupied ligand orbital and ending in an empty orbital on the same ligand. Broadly speaking, this is the usual absorption now modulated by the interaction and complexing of the TM. Due to this the absorption and emission bands can be shifted in regard to the bands in the isolated ligand molecule (see Fig. **9.5D**).

The establishment of new bands (MLCT and LMCT) and the change of previous ones (d-d and intra-ligand transitions) is a characteristic feature of TMCs. Taking advantage of this provides means to, not only use TMCs as luminescent probes, but to detect certain metal cations through their MCs (see Chapter 8 for more information in this regard).

2.2. Luminescence

Any of the four types of electronic transitions referred above can lead to light emission from one of the populated excited states. This follows the same general process explained in chapter 2. Some of the possible radiative transitions are displayed in Fig. (**9.5E**). It is important to note that these are only some of the options, shown for the sake of simplicity. The presence of the TM in the complex largely increases the probability of intersystem crossing (ISC) of the excited states involved [15, 16]. This is due to the high spin-orbit coupling parameter of heavy atoms. The probability of ISC becomes larger with the heavier TM involved. Given that, in this situation, it becomes harder to distinguish between singlet and triplet states; the more general term of luminescence is employed for light emission by TMCs [15]. For d-block TMCs the main emissive states are those in which TM and ligand share the electrons. This translates into MLCT and LMCT being *a priori* the emissive states. Intra-ligand transitions usually occur in the ultraviolet (absorption and emission). Therefore, they are not very useful from the point of view of fluorescence microscopy. On the other hand, d-d transitions tend to deactivate non-radiatively and are of limited use usually. There are some exceptions to this, like Cr(III) complexes and some Pt(II) compounds, but in general, this rule applies. As it will be shown latter (sections 2.3 and 2.4 below), metal-metal transitions (f-f transitions) are the rule instead of the exception in the lanthanides and most actinides.

A very important feature in regard to these TMCs luminescence arises because of the ISC: the radiative lifetimes are usually very long in comparison with the typical lifetimes in organic fluorophores [15, 17]. Organic fluorophores usually have fluorescence lifetimes in the range of 1-10 ns. Many TMCs have lifetimes well above the 100 ns mark and sometimes stretching into the microsecond range. As a rule of thumb, TMCs have lifetime 10-100 ns longer than organic fluorophores. Lifetimes are even longer for f-f transitions in lanthanides (see below). These long lifetimes are put to use in time-resolved fluorescence techniques (see Chapter 6.3) as they relax the necessity to work with very fast photodetectors (see Fig. **9.6A** and **B**) [18]. They are very well suited for FLIM and related techniques [19].

By using TMCs as the imaging tools, the autofluorescence of samples (mainly due to organic fluorophores) fades in a short time while, the emission from the MCs lingers much longer. An example of this is shown in Fig. (**9.6C**) where CHO cells have been incubated with a Pt(II) compound (indicated in the figure) and fluorescein [20]. The upper panel shows the cell field 0 ns after light excitation (left image), where no details are distinguishable because both the fluorescein and the Pt complex emission overlap in wavelength and time. The image on the right shows the same field after 10 ns have passed. Now the fluorescein emission disappears and the TMC emission can be clearly seen as it is the only source of light. The lower panel shows a temporal follow up (100-2900 ns) of the same cell field to exemplify the long-term fading of the Pt(II) MC. Note how, even after ~3 µs, there is still some useful luminescence from the TMC.

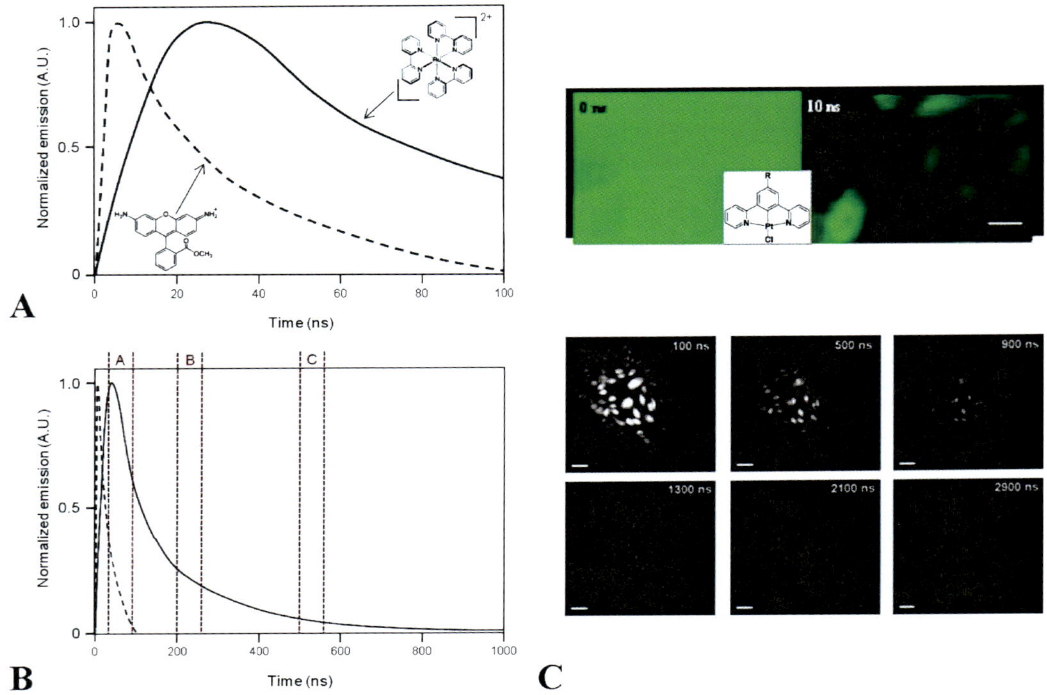

Fig. (9.6). Time-resolved luminescence with TMCs. **A**: Many TMCs are very well suited for time-resolved methods because of their intrinsic long radiative lifetimes. Measuring the emission from an organic fluorophore (rhodamine 123) and a TMC [Ru(bipy)$_3$] clearly show that the TMC emission (continuous curve) is much longer than that from the organic fluorophore (broken curve). After 100 ns fluorophore emission is almost depleted while the TMC still emits at ~ 40% of the highest intensity (chemical formulas reproduced from Yikrazuul - https://commons.wikimedia.org/wiki/File:Rhodamine_123.svg#and Organicoboist - https://commons.wikimedia.org/wiki/File:Ru(bipy)_Schematic.png#). **B**: Further stretching the sampling time it is clear how TMC luminescence can be used for time-resolved studies even when its emission spectrum overlaps other fluorophores. If light is sampled during the period A (red broken lines) both emissions can be unresolved from each other. Sampling at a later time (B) makes light from the TMC almost the unique source of light. After 500 ns (C), roughly 100 times the organic molecule lifetime, some light is still being emitted by the TMC. **C**: Time-resolved imaging of a luminescent Pt TMC (chemical formula shown in the inset) in CHO cells against a background of fluorescein. In the upper panel time-gated emission is shown after a 0 ns delay between excitation and emission (left) and after a delay of 10 ns. Clearly after 10 ns most of the fluorescei.(transitions and intersystem crossingsis still strongly emitting (scale bar: 10 µm). The lower panel shows a time series of CHO cells labelled (nuclear stain) with the Pt MC at different sampling times (upper right corner of each picture) after pulsed excitation. Although clearly dimmer there is still some emission after 2900 ns (scale bar: 50 µm). Images reproduced with permission from [20], Copyright (2008) National Academy of Sciences, U.S.A.

TMCs offer other advantages compared to the classic organic fluorophores. Those employed for fluorescence techniques are more resistant to photobleaching than usual organic molecules. Their emission is stable and long-lived. Apart from allowing a clear distinction from autofluorescence, for example, this long lifetime makes TMCs better candidates for kinetic studies. Microsecond emission times allow the excited Mc to diffuse a much longer distance which increases the kinetic information. Small changes around a common TMC lead to a broad tuning of absorption and, especially, emission. Because of the nature of the orbitals involved in the optical transitions, the Stokes shift in TMCs is quite large in general. This is more often than not an advantage in fluorescence microscopy, as the researcher is interested in the lowest spectral overlap possible to better distinguish signal from noise. The chemical synthesis and purification of these compounds are easy.

Among the disadvantages of these compounds as luminescent probes, the increased probability of photochemistry stands out. By definition, an optimal fluorophore should have a quantum yield of emission as close to 1 as possible. Any other deactivation channel can be regarded as lost energy/resources. This loss can have no deleterious effects (*e.g.* vibrational excitation and heat production) or it can have unwanted effects (photochemistry, singlet oxygen photosensitization, radical production, *etc.*) that lead to 1) photobleaching of the molecule (with its permanent loss from the pool of fluorophores) and 2) establishment of undesirable chemistry in the studied sample (as with oxidation reactions) which degrades it over time. TMCs in the excited state are, in principle, more prone to photochemistry due to their long lifetimes [21]. It is necessary to stress that the occurrence of actual photochemical reactions depends on each particular TMC and their environment. Many useful TMCs undergo internal charge transfer as they absorb a photon (MLCT and LMCT). This means that the excited electron is more localized either on one of the ligands or the TM as opposed to what happens in an excited organic fluorophore, in which the energized electron is usually delocalized around the molecule (π^* or σ^* states). This higher localization makes chemical reaction more likely. In fact, some TMCs make ideal photo-chemotherapeutic compounds precisely because of their high chemical reactivity in the excited state which makes them very cytotoxic when localized within tumoral cells [22].

Special mentioning deserves the photosensitized production of $^1O_2$ by TMCs. As has been mentioned, TMCs in the excited state display mainly a triplet character. This fact associated with the long lifetimes makes the excited TMCs good candidates to photosensitize $^1O_2$ by energy transfer. This is a disadvantage when considering a TMC as an imaging agent. Therefore, those complexes synthesized and employed as fluorophores have their Φ_Δ as low as possible. These can be achieved by fitting several parameters such as the triplet energy, redox potential, shielding of the excited state [23]. This interaction with O_2 can also be put to use to detect oxygen concentrations. By measuring light quenching, the O_2 concentration in a given sample can be assayed. Under other circumstances, such as the use of TMCs as photoactivatable chemotherapeutic agents, the ability to produce $^1O_2$ under light exposure is enhanced as high as possible in order to deliver maximum damage to the target cells. This is another very fruitful research and applicable area for TMCs [24].

As a summary, Table **9.1** presents a short list of the features of organic molecules *vs*. TMCs in the field of fluorescence microscopy. This is a generalizing list highlighting the main differences. Many exceptions can be found and should be taken into account when selecting the best option to study a certain phenomenon.

Table 9.1. General features comparison between organic fluorophores and TMCs employed in fluorescence microscopy [5, 14, 18].

Feature	**Organic fluorophores**	**TMCs**
Molecular weight	Low	High
Radiative lifetime	Short (1-10 ns)	Long (100 ns – 1μs)
Stokes shift	Small (10-100 nm)	Large (50-200 nm)
Photostability	Small to medium	Medium to large
$^1O_2$ production	Small to medium ($\Phi_\Delta \approx 0.1$-0.5)	Small to large ($\Phi_\Delta \approx 0.1$-0.9)
Dark cytotoxicity	Small	Small to large

2.3. Lanthanides

The 4f group presents particular features in regard to their photophysics. Lanthanide TMC has the same four basic types of electronic transitions explained in section 1 (MLCT, LMCT, *etc*.). Because the metal-centred electrons involved in these transitions belong to the 4f shell, the excitation energy usually ends up in one of these 4f electrons. As explained previously in the chapter, the 4f electrons spend most of their time "buried" within the 5s and 5p shells. This makes them relatively independent of interactions with the environment. Thus contrary to what happens to d-block metals, once the electron is in one of the metal-centred orbitals, their lifetimes are very long and usually the radiative pathway of deactivation is the preferred one [25]. On the other hand, it is quite difficult directly to excite these levels, and the absorption coefficient for f-f transitions is very low (much more than d-d transitions). The usual procedure to excite f-f transitions in lanthanide TMC is to excite one of the LMCT or intra-ligand transitions which then evolve naturally into an excited 4f level [26, 27].

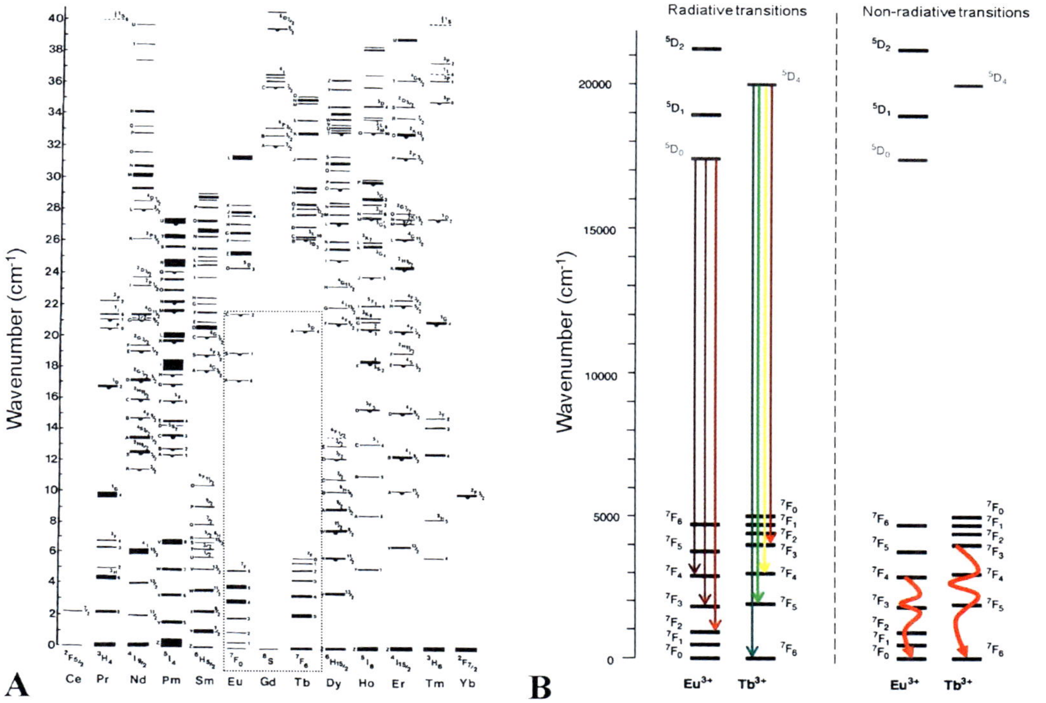

Fig. (9.7). Internal electronic transitions (f-f transitions) in lanthanides. A: The Dieke diagram displaying the electronic energy levels available for the different lanthanides in the most frequent +3 oxidation state. The levels are labelled according to the nomenclature. Levels from which radiative transitions are most frequent show a half dot on the underside. The energy scale (left axis) is expressed in wavenumbers (cm^{-1}, see Chapter 1). B: Magnified electronic levels corresponding to europium and terbium (highlighted within a broken line box in Fig. **9.7 A**). In coloured arrows the transitions from europium and terbium most frequently exploited in fluorescence microscopy. On the left the radiative transitions for Eu ($^5D_0$ to the ^7F_n manifold), in the red part of the spectrum, and for Tb ($^5D_4$ to the ^7F_n manifold), in the green, are shown. The coloured arrows closely resemble the true emission colours seen from these transitions (see Fig. **9.8 B** below for the emission spectra of these lanthanides). After light emission (right scheme) there is a non-radiative deactivation process (red wiggling arrows) within the ^7F_n manifold until the electron eventually reaches the ground state ($^7F_0$ for Eu and $^7F_6$ for Tb).

Because the most frequent oxidation state of lanthanides is +3 by far, this is the encountered state in TMCs. The 4f electronic levels for lanthanides in the +3 state are displayed in Fig. (**9.7A**). As can be appreciated, there are many available electronic levels. This scheme showing the electronic energy levels of the lanthanides is known as the Dieke diagram after Gerhard Heinrich Dieke, a physicist who studied and proposed a model for the internal electronic structure of lanthanides. It is a grotrian diagram applied specifically to lanthanides [28]. Some lanthanides display only a few f-f transitions (Ytterbium has only one) but usually they have a rich variety of levels (*e.g.* Neodymium or Dysprosium). Not all levels give rise to the radiative transitions. As a general rule, if two levels are quite close, the higher one usually deactivates non-radiatively to the lower one. This is the case, for example, for the lower levels of Europium ($^7F_6$-$^7F_1$ deactivate to the $^7F_0$ ground level) and Terbium ($^7F_0$-$^7F_5$ deactivate to the $^7F_6$ ground level) (see Fig. **9.7B**).

On the contrary, when there is a large gap between the two states, then the probability of photon emission increases greatly. Taking Europium and Terbium again as examples, radiative transitions occur for the $^5D_0$ (Eu) and $^5D_4$ (Tb) levels to the ^7F_n manifold already commented for non-radiative deactivation (see Fig. **9.7B**). It is important not to confuse the nomenclature at this point. On one hand, all of these transitions are 4f-4f, or f-f, transitions as they all take place within the 4f electronic shell (note that the "f " is small letter). However, each of these internal levels has a spectroscopic designation ($^7F_6$-$^7F_1$, $^5D_0$ or $^5D_4$) in which the letters are capital. Thus, there is a notation difference between the electronic shell/orbitals (small letters) and the spectroscopic nomenclature (capital letters).

As already noted, direct excitation of the f-f transitions is very inefficient. For this reason, lanthanide TMCs are excited in the absorption bands of the ligands to induce either intra-ligand or LMCT transitions [29]. Once this happens, the excitation energy usually ends in one of the 4f electrons (there are exceptions, of course) which finally reemit the light according to a de-excitation cascade along with the different 4f levels [30, 31]. A scheme of this typical approach is shown in Fig. (**9.8A**). In this the ligand, usually known as the antenna for its role in collecting light energy in order to transfer it to the lanthanide, is excited after absorbing a photon. Due to the strong mix of electronic states because of the heavy atom (lanthanide) effect, the excitation energy is usually transferred efficiently to one of the 4f electrons centred on the metal atom. This takes place in a time period of the order of nanoseconds or less. Afterwards, the electron decay follows the "internal" rules of the particular lanthanide atom, depending on the involved electronic levels (see Fig. **9.7A**). In this way, the difficulty of directly exciting lanthanides is bypassed by this antenna effect [32, 33].

Light emission due to f-f transitions is quite sharp. Usually, they resemble more atomic emission lines rather than molecular emission bands (see Fig. **9.8B**). This is another consequence of the relative electronic isolation of the 4f shell. In this case, one would like to excite an f-f transition, not only the low absorption coefficient must be taken into account but also a narrow excitation source should be employed precisely overlapping the absorption line. It is clear that a better suited option is to excite intra-ligand or LMCT transitions (usually broad because they involve molecular states) that then relax to one of the 4f levels of the lanthanide. This is the common approach when these TMCs are employed in fluorescence microscopy [25]. Table **9.2** introduces some of the most employed lanthanides as luminescent probes along some of their photophysical features.

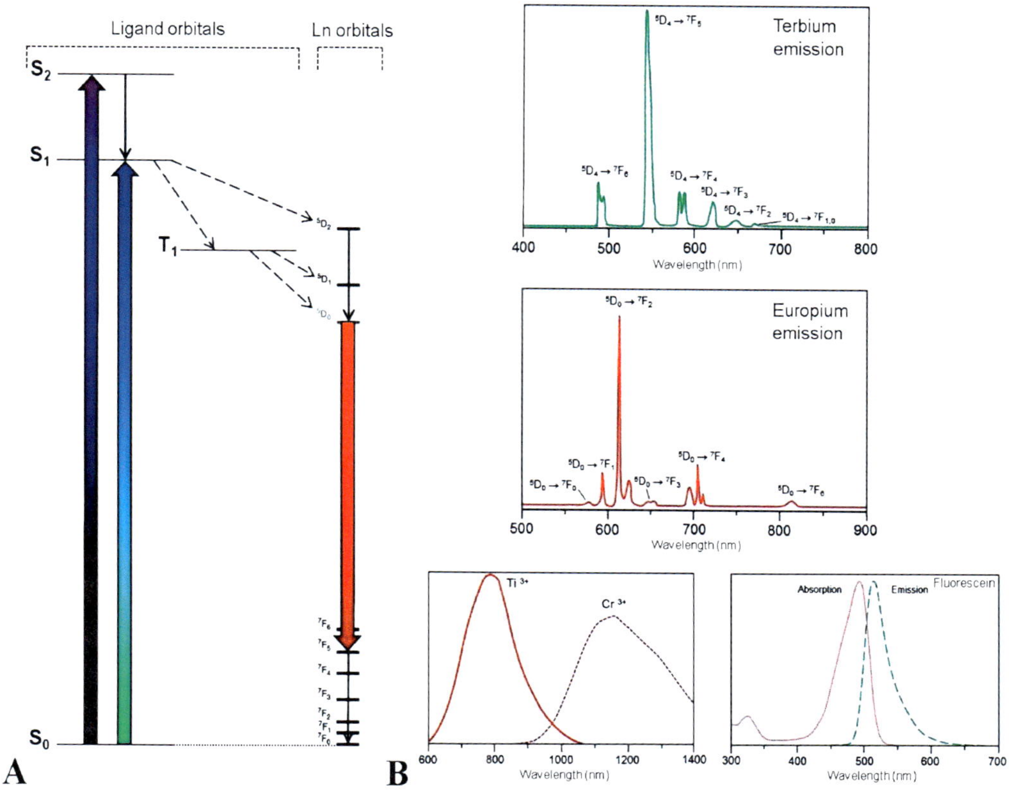

Fig. (9.8). **A**: Ligand-sensitization of lanthanide f-f emission. An allowed intra-ligand transition is exploited to indirectly excite the lanthanide ion. After photon absorption by the ligand (commonly an organic molecule) an electron transfer from the S_0 to the excited singlet manifold (S_1, S_2, *etc.*) takes place. Internal conversion populates the S_1 level after a short delay and from this the T_1 is populated by intersystem crossing after a longer time. Direct electron transfer can also fill one of the excited f levels of the lanthanide from S_1. From T_1 the electron is transferred to an f orbital. In the particular example shown the $^5D_2$, $^5D_1$ and $^5D_0$ levels of Eu are populated by the excited ligand. The $^5D_2$ and $^5D_1$ levels quickly decay non-radiatively to the $^5D_0$ level. From there the well-known red emission bands of Eu are emitted. In the scheme coloured arrows are radiative transitions, solid black arrows are internal conversions and broken arrows represent intersystem crossings. **B**: Comparison of the emission spectral bandwidths between lanthanides and other fluorophores. The two upper panels display the emission spectra from Tb (green line) and Eu (red line). The two lower panels show the emission spectra of two d-transition metals (Ti^{3+} and Cr^{3+}) on the left and the organic molecule fluorescein (both absorption and emission are shown for this) on the right. It is clear that the two lanthanides emission profiles are narrower than the others. For the lanthanides the f orbitals involved in the transitions are labelling each particular peak. Wavelength in nm is represented on the x-axis. Emission and absorption in arbitrary units represented on the y-axis.

Table 9.2. Photophysical features of interest for the commonest lanthanides employed in luminescent TMCs (modified from [30]).

Lanthanide	Radiative lifetime (ms)	Emission wavelengths (nm)	Emission colour
Europium	9.67	620	Red
Gadolinium	10.9	310	UV-B
Terbium	9.02	550	Green
Holmium	0.37	540/760	Green/NIR
Erbium	0.66	550/860/1550	Green/NIR/IR
Thulium	0.19	470	Blue
Ytterbium	1.2	980	NIR

Lanthanides in **bold** are the most employed.

It is important to remark that the same lanthanide ion can emit more than one wavelength when excited. For example, the main emission of Terbium is around 550 nm but it also emits at 490, 570, 620, 655 and 680 nm. In the same way, the indicated radiative lifetimes are examples: depending on the initial and final 4f electronic levels involved in the transition so does the lifetime change. Nevertheless, for most of the transitions of interest, the radiative lifetimes are around the millisecond.

Because of their particular characteristics and environmental isolation f-f transitions present, in general, very long lifetimes. Typical radiative lifetimes are of the order of a millisecond with some transitions going up to 10-50 ms. Of course, these are extremely long lifetimes (note that a typical radiative lifetime for an organic fluorophore is 1-10 ns, 6 orders of magnitude shorter!). These very long lifetimes make lanthanide TMCs even a better option for time-resolved fluorescence studies [26, 34]. Examples of this will be presented below for particular applications of lanthanide complexes. The very long lifetimes displayed by these elements make another unusual and very interesting phenomenon to come about: photon upconversion. This phenomenon allows the lanthanide to emit photons of higher energy than those employed to excite it. This is possible because the excited levels have such a long lifetime that there is time for a second photon to excite the lanthanide before it decays to the fundamental state. In this way, the sample can be pumped with red light and green or blue light comes out from it under the right conditions. This very interesting topic will be fully described in chapter 19.4.1, as upconversion is greatly favoured by the very long lifetimes of lanthanide ions.

Another advantage of lanthanide TMCs is that their photosensitized $^1O_2$ production is much lower when compared to d-block TMCs. Due to the screening of the 4f electrons once the excitation energy is located on them, it is relatively difficult for it to interact with the molecular oxygen. The main oxygen photosensitization source is the ligands while they stay in the excited state [24]. Once the electronic energy flows to the 4f electron, it is no longer available to excite oxygen or, at least, the excitation rate is smaller in magnitude. Nevertheless, there has been some research in employing some tailored lanthanide MCs to assess O_2 concentration [35]. This will be described later in the applications of lanthanide complexes (see section 3.4).

2.4. Actinides

Following the same trend presented before for lanthanides, actinides display f-f optical transitions due to their partly filled 5f orbitals [36]. The equivalent to the Dieke diagram for actinides is shown in Fig. (**9.9A**) [37, 38]. Note the similarities to the lanthanide data displayed in Fig. (**9.7A**). Taking advantage of this characteristic, some luminescent actinide MCs are employed in fluorescence microscopy and analytical sciences [3, 39]. However, because of their inherent unsafety in regard to their radioactivity, they tend to be displaced in these uses as new compounds are introduced. TMCs involving the actinides are included here for the sake of completeness and for historical reasons, although their use is more restricted for obvious reasons (for an excellent review on current applications of actinide TMCs see [40]).

Probably the most employed actinide in luminescent TMCs is uranium. Due to its long radioactive lifetime (> 10^9 years), uranium is relatively safe to handle in this regard. However, some of its trace isotopes have much shorter lifetimes (10^5-10^8 years) and the same is true for all of its radioactive decomposition products. This makes use of uranium and other actinide TMCs less common as other options are available. Nevertheless, there are still some applications for uranium especially as uranyl complexes ($[UO_2]^{2+}$) [41]. Some complexes of uranyl are displayed in Fig. (**9.9B**). Uranium compounds are also employed as electron dense contrast agents in electron microscopy, an area in which they are still largely used. In the following section, where applications for the different luminescent TMCs will be provided, there is a small presentation to uranyl TMCs employed as luminescent probes (see section 3.5).

The uptake and interactions of TMCs with the living cells and biological substrates are governed, obviously, by the physicochemical interactions between these entities. Therefore, not all TMCs will spontaneously enter a living cell or, if they do, they localize to the same structures. The same general chemical principles that applied in the interactions of organic molecules with biological samples are also applied to TMCs [9]. The reader is directed to chapter 15.2 for explanation of the mechanisms that govern the uptake, localization and labelling of biological structures.

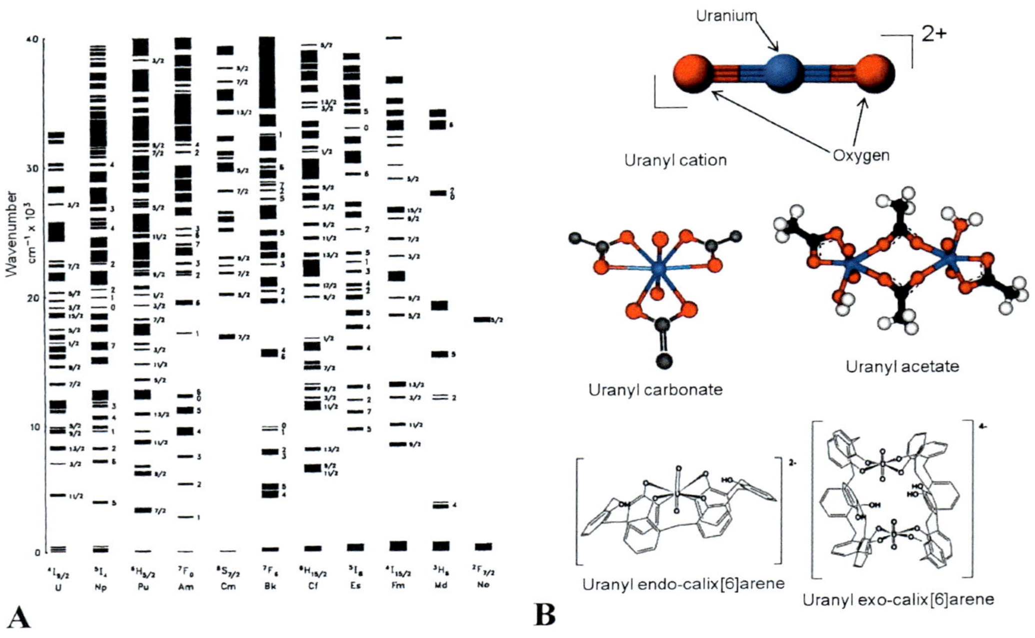

Fig. (9.9). **A**: Dieke diagram for the actinides. The f orbitals for most of the actinides are displayed in the same way as the Dieke diagram for lanthanides (see Fig. **9.7 A**). It can be seen that, in general, the actinides show a more crowded arrangement of f orbitals than the lanthanides. **B**: Ball and stick representation of the uranyl cation $[UO_2]^{2+}$ (upper drawing) showing the atomic arrangement (reproduced from Benjah-bmm27 - https://commons.wikimedia.org/wiki/File:Uranyl-3D-balls.png#). In the middle some common uranyl complexes like uranyl carbonate (reproduced from DEMETER copyright © 2009-2010 Bruce Ravel) or uranyl acetate (reproduced from Benjah-bmm27 - https://commons.wikimedia.org/wiki/File:Uranyl-acetate--D-balls.png#) largely employed in electronic microscopy too. In the lower part some more exotic uranyl complexes are also depicted (reproduced with permission from [40]).

3. BIOLOGICAL PROBES BASED ON LUMINESCENT TMCS

After presenting the main photophysical characteristics of TMCs, a variety of applied biomedical uses are going to be introduced in what follows. There are many ways in which to present these applications: attending to the particular field of application, to different methodologies, to similar physicochemical properties, *etc*. All of them have advantages and shortcomings, of course. In our case, we have opted to present practical examples attending primarily to the TM involved. Following this rationale, we will first introduce TMCs belonging to the d-block, 3d subgroup. Then TMCs belonging to the 4d and 5d subgroups will be presented. After the d-block, examples from the f block (lanthanides and actinides) will be provided. It is worth mentioning that luminescent applications of TMCs have grown exponentially in the last decade or so. Therefore relevant examples will be presented but this is by no means a comprehensive compilation on the field. For reviews on this subject, consult the following references [11 - 15, 42].

3.1. 3d-block

We start presenting those luminescent TMCs of interest for fluorescence microscopy in which the TM that forms part of the complex belongs to the 3d group. Commonly TMCs employed as luminescent probes have heavy TMs. Therefore, as we will see shortly, it is more usual to employ TMCs having 4d or 5d block metals or lanthanides. Nevertheless, there are some examples of 3d-block luminescent or photoactive TMCs.

Chromium (Cr)

Applications of TMCs based on Cr are scarce. The luminescence band is broad and is located in the far red to near infrared region (650-800 nm), while the excitation usually requires short wavelength blue. However, they are being employed as photonucleases: compounds with the ability of inducing DNA damage when illuminated and with a greater or lesser specificity as to the kind of lesion induced [16, 24].

Manganese (Mn)

Mn-based TMCs are not employed as luminescent probes. They find some use as photochemical systems for the light-driven release of NO (nitric oxide) and CO (carbon monoxide) [43, 44].

Iron (Fe)

As it happens with Mn, iron-containing MCs are not adequate as luminescent probes. In most cases, this is because of the paramagnetic nature of the iron cations that tends to favour the non-radiative deactivation of the excited complex over light emission. Nevertheless, some of them display photochemical activity and are being exploited as photonucleases in biological experiments and as a therapeutical approach [22, 43, 45].

Cobalt (Co) and Nickel (Ni)

Co and Ni TMCs are not employed in fluorescence microscopy. Again, there are uses for them in photochemotherapy due to the damaging photoaction of some of them on DNA. This feature often precludes their use as labels since they induce damage to the studied cells. Some Co TMCs have been used as luminescent labels for DNA detection [46, 47].

Copper (Cu)

Cu displays a paramagnetic character, as Fe, that favours non-radiative deactivation of the excited state. When in the +1 oxidation state, Cu(I), light emission is enhanced. Nevertheless, these complexes tend to show geometrical rearrangements in the excited state that again may hinder photon emission [16]. Some Cu MCs find applications in direct DNA photocleavage too [45] and others in ROS photogeneration [24]. The mechanism, it is believed, starts from a Cu(II) complex. After photon absorption, the Cu(II) engages in electron transfer from one of the ligands, being reduced to Cu(I). Now with a Cu(I) cation, the MC can interact with O_2 to produce ROS.

Zinc (Zn)

Zn is the second most abundant transition metal in the human body. In biological systems, it appears as Zn(II). Most Zn(II) is retained by proteins and is defined as immobilized. A minor fraction is labile or mobile, and it is this fraction that plays a very important role in physiological and pathological processes. For example, under oxidative stress situations, thiol groups get oxidized and release chelated Zn(II) to the medium. Then Zn(II) acts as a transducer of cellular stress. For this and other reasons luminescent tools for Zn(II) detection are important. The most employed approach is to provide ligands to the studied biological system that increases or changes (*e.g.* lifetime) their luminescent properties upon binding Zn(II) [12, 13]. Taking advantage of this approach, Zn(II) concentrations are studied in many subcellular structures and domains, like the mitochondria, lysosomes, plasmatic membrane, Golgi apparatus, endoplasmic reticulum and nucleus. Zn(II) has also been studied and measured in tissues and whole living animals, like zebrafish and mice. For an excellent and recent review on this issue see [48].

An example of the different intracellular Zn(II) pools in HC11 mouse mammary epithelial cells is shown in Fig. (**9.10A**) [49]. In the upper image, Zn(II) in intracellular vesicles is highlighted by the green emission of FluoZin-3 probe. Notice the punctuated granular perinuclear Zn(II) signal in the picture. The bottom image shows the Zn(II) emission from the mitochondria obtained with RhodZin-3 probe. In contrast to the vesicular signal, the mitochondrial emission displays a clear peripheral distribution and the emitting regions are elongated rod-like structures, in agreement with the mitochondrial morphology. Zn(II) dynamics have been found to correlate with cell fate too (see Fig. **9.10B**) [50]. By making use of a chelating agent that increases its emission when complexing Zn(II), Kimura *et al.* showed that free Zn(II) increases in HeLa cells undergoing apoptosis. This effect on intracellular Zn(II) appears at the early apoptotic stage (panels A-D) and peaks when the prototypical apoptotic blebbing is present (panels E and F). The Zn(II) signal dims strongly or disappears at the final stages of apoptosis (panels G-J).

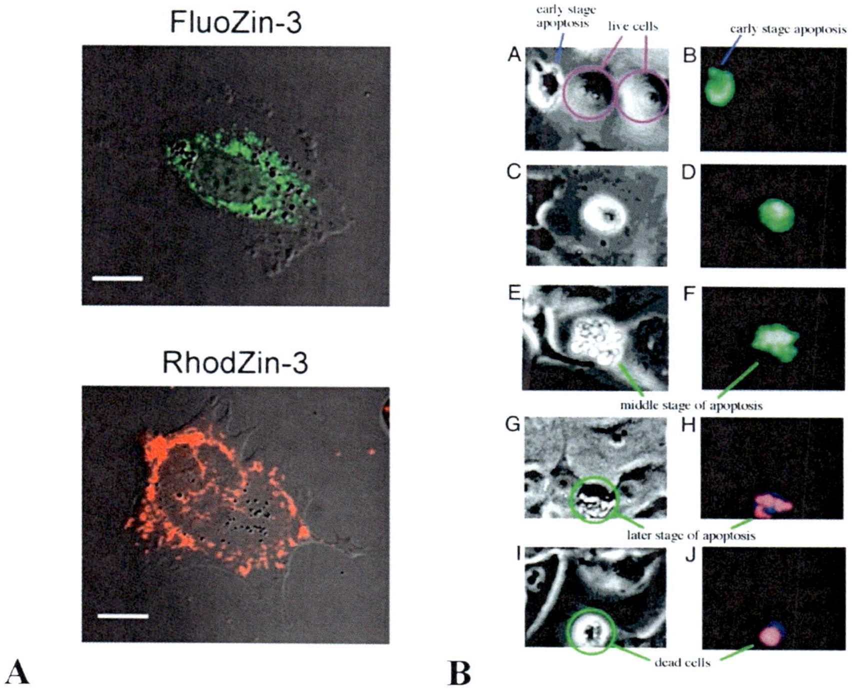

Fig. (9.10). **A**: Intracellular localization of Zn(II) in mouse mammary cells (HC11). The upper image shows the Zn(II) distribution in the vesicular component of the epithelial cell (FluoZin-3). This Zn(II) correlates with a perinuclear localization in rounded structures. In contrast the lower image shows the Zn(II) localization as revealed with RhodZin-3 that is selective for mitochondria. Indeed the highlighted structures display the rod-like shape typical of mitochondria. Also note the more cytoplasmic distribution of Zn(II) in comparison with the upper image. Scale bars: 5 μm. Reproduced with permission from [49]. **B**: Zn(II) dynamics during HeLa apoptosis. Cells were incubated with the Zn(II) complexing agent shown at the upper right corner. When it complexes Zn(II) its emission increases notably. Left panels (A-I) show bright field cell images while right panels (B-J) show the corresponding fluorescence pictures. In A-B the early apoptotic cell to the left can be detected in fluorescence due to its increased cytoplasmic Zn(II) as revealed by the complexing probe. High levels of Zn(II) can be detected through the intermediate apoptotic process too (C-F). However at late apoptosis stage Zn(II)concentration has decreased significantly (G-J) while the plasmatic membrane selective permeability has been compromised and the vital dye propidium iodine is staining the nucleic acids and fluorescing (pink areas in H and J). Reproduced with permission from [50]. Copyright (2003) National Academy of Sciences, U.S.A.

3.2. 4d-block

TMs from the 4d-block show a much richer variety of luminescent complexes for use in fluorescence microscopy than 3d-block metals. This is mainly due to the increase in atomic mass which favours electronic spin-orbit coupling and intersystem crossing. This effect leads to a more efficient radiative deactivation pathway in contrast to TMs from the 3d-block which favours non-radiative de-excitation. These comments are broad and the reader should always keep in mind that many exceptions exist to these general rules. Some of the most employed 4d TMs in regard to luminescent probes will be presented in what follows.

Zirconium (Zr)

Only certain metallocenes of this metal have stable enough excited states to allow photon emission *vs.* vibrational deactivation [51]. Nevertheless, at least one Zr TMC has been successfully employed to image living cells [52]. The authors report cellular uptake, low cytotoxicity and perinuclear localization inside the living cell. Due to the current use of a Zr isotope (^{89}Zr) for PET, these kinds of TMCs are proposed as double imaging agents.

Molybdenum (Mo)

As it happened with TMs from the 3d-block, Mo is not a good candidate for luminescent complex probes. However, it has found uses as a photoactivatable source of CO [43].

Ruthenium (Ru)

This element is one of the most employed TMs for the design of luminescent probes [12, 53]. TMCs formed with this metal offer long radiative lifetimes, large Stokes shift in emission, photostability and good biocompatibility in cellular systems. Its study and application to photophysical and photochemical studies trace back to several decades [54].

Fig. (**9.11A**) shows some examples of Ru-TMCs employed for luminescent imaging and photochemical studies. These compounds are usually excited in the near UV and blue regions of the visible spectrum (350 nm $\leq \lambda \leq$ 450 nm). The emission, due to the large Stokes shift, occurs mainly between 580-650 nm.

Ru-TMCs are widely employed to image different cellular compartments and structures, both in the living cells and in the fixed ones. Ru complexes are important nuclear DNA/chromatin markers (see Chapter 15.4.1). Most often, one or more of the complexing ligands show DNA binding properties. This feature stabilizes the interaction between DNA/chromatin and the luminescent complex. Once the interaction takes place, the rotational and solvent-dependent non-radiative de-excitation channels are greatly diminished. As a consequence, the light emission increases after DNA/chromatin binding of these compounds. An example of chromatin fluorescent labelling is shown in Fig. (**9.11B**). A smear of

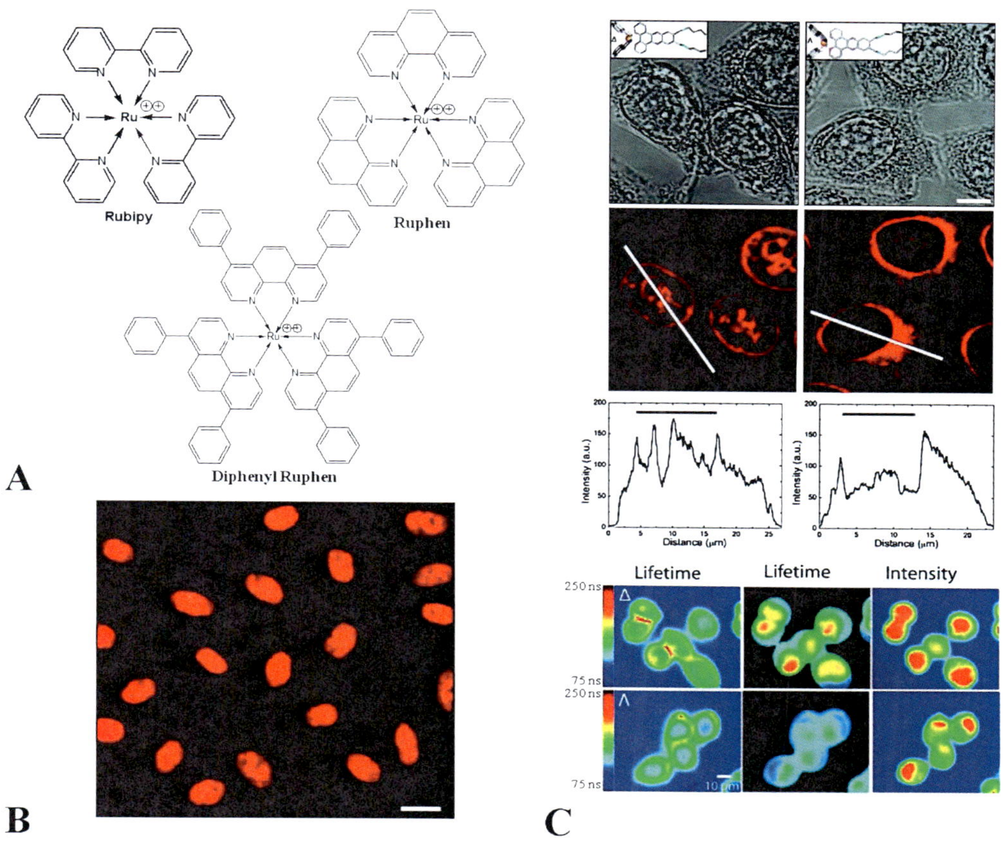

Fig. (9.11). **A**: Examples of some Ru TMCs structures. The complexes are ruthenium(II) tris(2,2'-bipyridine) (Rubipy), ruthenium(II) tris(1,10-phenanthroline) (Ruphen), and ruthenium(II) tris(4,7-diphenyl-1,-0-phenanthroline) (diphenyl Ruphen). Although represented as planar compounds the ligands arrange themselves in a 3D spherical structure around the ruthenium atom "core". **B**: Chicken red blood cell nuclei stained with Ruphen in a blood smear. The negatively-charged nucleic acids (DNA) act as a substrate for Ruphen with a positive net charge. The MC localization is revealed by exciting at 436 nm and observing the emission at 600-650 nm. Note the heterogeneity in the luminescent chromatin signal. Scale bar: 5 µm. **C**: Upper panels: Differences in luminescent properties between two isomers of a Ru complex (shown in the insets). Fixed CHO-K1 cells (bright field images at top) where incubated with either the Δ isomer or the Λ isomer. Confocal fluorescence images of these cells show that the overall light emission intensity is similar but the localization of the emission is different (scale bar 5 µm). The light intensity assessment along the indicated paths (white lines) for both samples is displayed in the middle panels. It can be seen that light comes from different cell regions. Lower panels: Luminescence lifetime measurements with FLIM in living CHO-K1 cells. The features for each isomer are shown in the upper lane (Δ isomer) or lower lane (Λ isomer). On the left is the lifetime for cells in which the Ru complex is outside the cells (note lifetime false colour scale in ns to the left). After some minutes of irradiation the complexes produce membrane photodamage and diffuse inside the cells, which modifies their radiative lifetimes (middle panels). Once inside the cells the Λ isomer displays a much shorter lifetime (blue colours) as compared to the Δ isomer (warmer colours). On the right the light intensity after permeabilization (*i.e.* equal conditions to the middle panels) is shown. Light intensity is slightly higher for the Δ isomer. Reprinted and adapted with permission from [56]. Copyright 2011 American Chemical Society.

chicken blood has been incubated with the Ru-TMC known as Ruphen (see structure in Fig. **9.11A**). The metal complex binds to the DNA due to its ligands and the luminescence yield increases appreciably (the red emission is characteristic of many Ru-TMCs). Note the granular aspect of the stained chromatin due to differences in the amount of complex bound to DNA. These differences are in part because of the more or less relaxed structure of the chromatin that allows the Ru complex to interact with the DNA. Going further with this approach, Gill *et al.* could follow changes in the chromatin structure of the living cells throughout the cell cycle by analysing the luminescence characteristics of a Ru TMC [55].

Ru-TMCs have been employed to label other subcellular structures and elements too. An interesting example is the use of two isomers (enantiomers) of a Ru-TMC for the determination of different domains in the living cells by Svensson and collaborators [56]. Chinese hamster ovary cells (CHO-K1) were stained with the two isomers. First, the fixed cells were studied (see Fig. **9.11C**). It was shown that the light intensity was roughly the same but that the two isomers stained different cell structures (Fig. **9.11C**, upper panels). The Δ isomer showed a higher affinity for the nucleus but much less for the nucleoli, while the Λ isomer stained preferentially the perinuclear region and the nucleoli. The radiative features also changed in the living cells exposed to the complexes (see Fig. **9.11C**, lower panels). Attending to the radiative lifetime, the two isomers behaved as equivalents when outside a cell. After some time of exciting light exposure (405 nm), the complexes produced photodamage and compromised the plasmatic membrane integrity, therefore diffusing within the still living cell. Now interacting with the internal cellular substrates, it was readily seen that the isomers lifetimes changed appreciably. The Δ isomer shows a longer lifetime on average while the Λ isomer displayed a shorter lifetime. Once inside the cells, the Δ isomer also displayed a somewhat stronger light emission.

Ru-TMCs have been employed as fluorescent protein labels [53]. This has been accomplished through two strategies. In the first case, covalent bonds are established between one or more amino acid side chains of the protein and one of the ligands of the complex. Typical reactions include the isothiocyanate, N-hydroxysuccinimidyl ester, aldehydes, or iodoacetamide reactive group on the ligand reacting to primary amines or sulfhydryls groups on the protein. These modifications make proteins luminescent in virtue of the attached TMC. The second strategy is to label the protein with a Ru complex that interacts non-covalently with the protein. This is usually done by introducing a hydrophobic moiety into one or more of the ligands or by attaching a functional molecule to the ligands, like an enzyme substrate or a cofactor. This last approach provides the higher amount of specificity on the detected/labelled protein.

Ru-TMCs have been employed to detect other biological substrates, like polysaccharides. The complex Rubipy (see Fig. **9.11A**) has been extensively employed to highlight the presence of negatively charged polysaccharides like heparin in specialized cell types, like mast cells. An example is shown in Chapter 11 (see Fig. **11.18**). Peritoneal fluid obtained from mouse has been incubated with aluminium-morin and afterwards with Rubipy. Now under blue light excitation at 450-490 nm, the differential emission from aluminium-morin (540 nm) and from Rubipy (600-650 nm) can be appreciated in lymphocytes (green cells) and mast cells (red cells), respectively. The orange-red emission from mast cells is due to the interaction between Rubipy and heparin stored in basophilic granules.

Ru-TMCs are being used as triplet sensitizer compounds. By using the long-lived T_1 state of these complexes, the direct excitation of triplet states of organic molecules can be achieved through collisions of the excited complexes and the target molecule [57]. Uses for this triplet sensitization process are many in photochemistry and in triplet-triplet annihilation upconversion techniques (see Chapter 19.4.3). Perhaps, one of the greatest shortcomings of Ru-TMCs, precisely because of their triplet sensitizing ability, in biological imaging, is their high quantum yield for $^1O_2$ production. This is understandable as O_2 is usually abundant in biological environments. Because $^1O_2$ is a ROS, its presence during imaging must be avoided as much as possible (see Chapter 2.5.4 in regard to its role in photobleaching and undesirable sample photochemistry). Nevertheless, this feature is taken as an advantage in order to damage certain cells or tissues (like cancer) and therefore, these compounds also find important applications in the so called photodynamic therapy and photochemotherapy [23, 24].

Rhodium (Rh)

Although sharing many of the properties of its two neighbours, Ru and Pd, in the periodic table, TMCs with rhodium as the complexed metal are not employed broadly in fluorescence microscopy. It is worth mentioning that Rh-TMCs have been used as labels for specific nucleic acid sequences (TGCA sequence) and to detect double strand mismatches [21]. A Rh complex has also found its use as β-amyloid peptide aggregation. The analysis of the complex effect upon the protein was obtained in part by its luminescence [58]. Additionally, Rh-TMCs are employed to photocleave DNA [59] and as prospective photodynamic agents [24].

Palladium (Pd)

Palladium complexes are not in widespread use for fluorescent microscopy. The particular characteristics of the Pd(II) cation that takes part in these complexes do not give rise to particularly attractive luminescent features. Nevertheless, some complexes made from Pd(II) and large macrocycles find uses as NIR probes [16].

Silver (Ag)

Luminescent complexes of Ag are not employed in fluorescence microscopy. However, Ag as a metal is a very important component of luminescent nanoparticles. See Chapter 18.3 in relation to the uses of Ag and other noble metals as imaging probes in the form of nanoparticles.

3.3. 5d-block

Several elements of the 5d-block are employed in the structure of luminescent TMCs as probes in fluorescence microscopy. Following the same trend as 4d elements, elements in this group favour radiative de-excitation when forming part of many MCs. These metals have very high atomic numbers (higher than the lanthanides) and in fact, are almost the heaviest stable elements in the periodic table. This feature makes MCs having 5d-block elements to show a very high spin-orbit coupling parameter. This also prolongs the excited state lifetime considerably. For these reasons and their photostability, many MCs with 5d elements are the preferred option to synthesise luminescent probes [53].

As it has been done with the previous groups, some members of the 5d-block that make up TMCs with interest in fluorescence microscopy are briefly introduced in what follows. So many applications have been found, and are continuously being proposed, for TMCs with 5d block elements (in particular rhenium, osmium, iridium, platinum and gold) that simply presenting them within the frame of this chapter is impossible. For those readers interested in expanding the knowledge in this area, we recommend the following comprehensive recent reviews on the theme [12, 14, 15, 17, 45, 51, 60]. Therefore, only a few selected examples will be presented for 5d block TMCs.

Hafnium (Hf)

As with zirconium (see above 4d-block), Hf has currently few examples of luminescent MCs for biomedical applications. Nevertheless, research is starting to give its fruits in regards with these two elements and some particular Hf metallocenes display adequate characteristics as luminescent probes [51, 52].

Rhenium (Re)

This heavy element is much more employed in luminescent complexes. It displays adequate characteristics to make efficient luminescent TMCs when it is in the first oxidation state. Following this, most luminescent complexes make use of Re(I). Some examples of currently employed Re-TMCs are displayed in Fig. (**9.12A**).

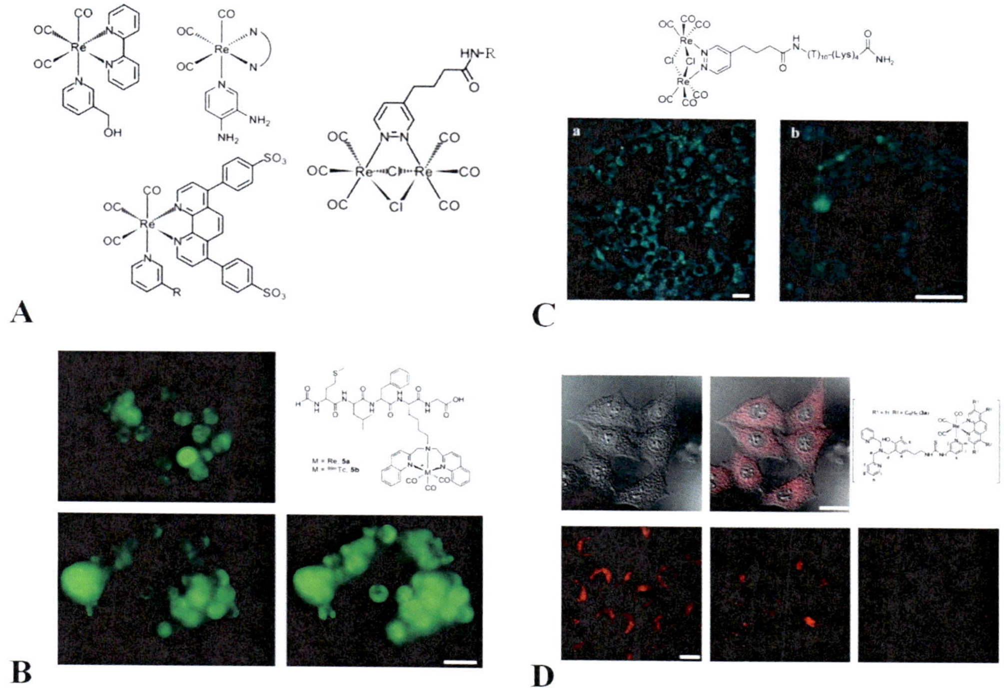

Fig. (9.12). **A**: Some examples of rhenium(I) containing complexes. Some complexes can be further functionalized by adding adequate chemical groups to the ligands (R chain). **B**: The rhenium complex displayed on the right is selective for FPR expressed by neutrophils. In the upper left panel neutrophils are labelled with the complex at 1 nM concentration (360 nm excitation light). The two lower panels show the same neutrophil population stained with the Re complex (left, 50 nM) and a selective FPR-recognizing peptide labelled with fluorescein. The two luminescent compounds highlight the same cell regions. Scale bar: 10 µm. Reprinted and adapted with permission from [61]. Copyright 2004 American Chemical Society. **C**: Re(I) complex functionalized with an attached PNA including 10 thymine bases and 4 lysine residues to enhance hydrophilicity and cellular uptake (upper panel). HEK-293 cells were incubated with this complex (left lower panel a) which stained endosome-like structures (higher magnification in right lower panel b). The internalization in endosome-like compartments seems to be a general trend for Lys-containing molecules. Complex excitation was done at 770 nm employing a two-photon absorption setup. Scale bars: 20 µm (a) and 10 µm (b). Reprinted and adapted with permission from [63]. Copyright 2012 American Chemical Society. **D**: Example of Re(I) complex for Zn^{2+} and Cd^{2+} detection within living cells. The complex is shown on the upper right panel. The left side of the complex is able to bind Zn^{2+} or Cd^{2+} then increasing the emission from the molecule. The upper left (bright field) and middle (overlay bright field and fluorescence) panels show HeLa cells incubated with the complex. The lower panels show the confocal-excited emission from cells exposed to Zn^{2+} (left), Cd^{2+} (middle) and control cells without heavy metal (right). Scale bars: 25 µm. Reprinted and adapted with permission from reference [64]. Copyright 2009 American Chemical Society.

Rhenium complexes find a wide number of applications. As labelling agents, they can mark specific membranous cell structures depending on their ligands nature. An example is given in Fig. (**9.12B**) where one of the first employed Re complexes is shown labelling a human neutrophil receptor within a population of lymphocytes [61]. The organic ligand has been functionalized with a moiety that recognizes the formyl peptide receptor (FPR). This is expressed in neutrophils. The image on the upper left shows labelling by the complex at 1 nM concentration. The two lower images show the same cells labelled with the complex (50 nM, left) and with a fluorescein-containing specific peptide for the FPR (right). The labelling overlap is notorious. This Re complex is also proposed to bind radioactive technetium-99 as a prospective selective radiosensitizing compound for neutrophil-related disorders.

Through adequate modification of one of the complex ligands, the complexes can be made selective for certain structures/organelles. For example, introducing a chloromethyl group into one of the ligands makes a Re complex a mitochondrial marking compound [62]. Re complexes have also been functionalized with biotin or with isothiocyanate to establish linkages to certain proteins. It has been functionalised with PEG to increase their biocompatibility. Aromatic ligands provide a mechanism for Re complexes to intercalate and mark DNA [45]. A very interesting approach is to link Re complexes with PNAs (peptide nucleic acids). PNAs interact with nucleic acids depending on their base sequence due to base-base recognition and pairing. In this way, it has been reported that these compounds can label different cellular structures depending on the particular PNA sequence (see Fig. **9.12C**) [63].

Re-TMCs can be employed to detect small molecules and ions in the living systems [12]. Reactive nitrogen species like nitric oxide (NO) has been successfully detected employing a Re complex. With a suitably modified Re complex to target the mitochondria (see previous paragraph) the compound can engage in chemical reaction with glutathione (GSH); the reaction that allows GSH determination. Re complexes can be employed to detect some metallic ions in the living cells, like Zn^{2+} or Cd^{2+} (see Fig. **9.12D**) where these two cations are detected in live HeLa cells [64]. The approach employed in this example is to use a Re complex whose light emission is enhanced when one of the ligands chelates Zn^{2+} or Cd^{2+}. Therefore, the complex displays some emission that is enhanced as heavy metal concentration increases. The complex shows adequate cell uptake although it displayed cytotoxicity and so its use would probably be constrained to short term experiments.

Osmium (Os)

Osmium TMCs are less employed for fluorescent imaging purposes than other 5d-block metals. In general, their emission spectra is further red-shifted into the NIR as compared to the Ru-TMCs (Os and Ru belong to the 8 periodic table group, see Fig. **9.1A**). Some applications have been reported for DNA detection *in* *vitro* in the NIR spectral region [65]. Given its ability to effectively photosensitize O_2 to its first excited state ($^1O_2$) Os-TMCs have a potential to be employed as oxygen probes and photochemotherapeutic agents [24]. Depending on the ligand's ability to interact with nucleic acids some Os compounds are able to photocleveage DNA effectively too.

Iridium (Ir)

This transition metal is one of the most employed for luminescent probes design. It finds uses mainly in its Ir(III) valence state and depending on the ligands making up the complex, photophysical and photochemical features can be fine-tuned [51]. The emission spectrum can be tuned throughout most of the visible range and radiative lifetimes range from nanoseconds to milliseconds (which make some of these compounds attractive for time-resolved microscopy). Ir-TMCs show good cellular uptake and are most usually lipophilic in their behaviour.

Compounds of this metal have found many different applications in fluorescence microscopy. The biological labelling can be due to the Ir-complex natural lipophilicity (usually localising to the perinuclear area) or the ligands can be tailored/functionalized to recognize a particular organelle, cell structure or protein/substrate. Changing the lipophilicity character of the ligands allows them to label the mitochondria (higher lipophilicity) *vs.* the plasma membrane (lower lipophilicity), for example [53]. Recently, mitochondrial dynamics have been studied with high detail due to the use of Ir-TMCs [66]. These complexes have been reported to show a high photostability and very good mitochondrial localization characteristics. As for other cellular structures, some dipyridoquinoxaline complexes of Ir(III) do nicely as nucleoli-labelling agents in live and fixed mammalian cells (see examples in Fig. **9.13A**) [67]. Other Ir(III) complexes, with dendritic polypyridine ligands, have been successfully applied to stain the Golgi apparatus in the living cells (see Fig. **9.13B**) [68]. The Golgi labelling was cross-checked with a selective golgin-97 antibody. The uptake of the more complex and bulkier dendritic complexes seems to be energy-dependent which offers the possibility to study metabolic processes related to the Golgi apparatus. At the same time, these more complex dendrimers provide a better Golgi apparatus selectivity as seen from the better localization in the pictures.

Some Ir complexes have been used to detect histidine molecules with high specificity. This includes the detection of His residues on proteins too. Ir-TMCs can be put to use in the luminescent detection of dissolved ions or small molecules like alkali cations, alkaline earth cations, heavy metals (silver, cadmium, mercury, lead) [69]. Given the very efficient energy transfer between

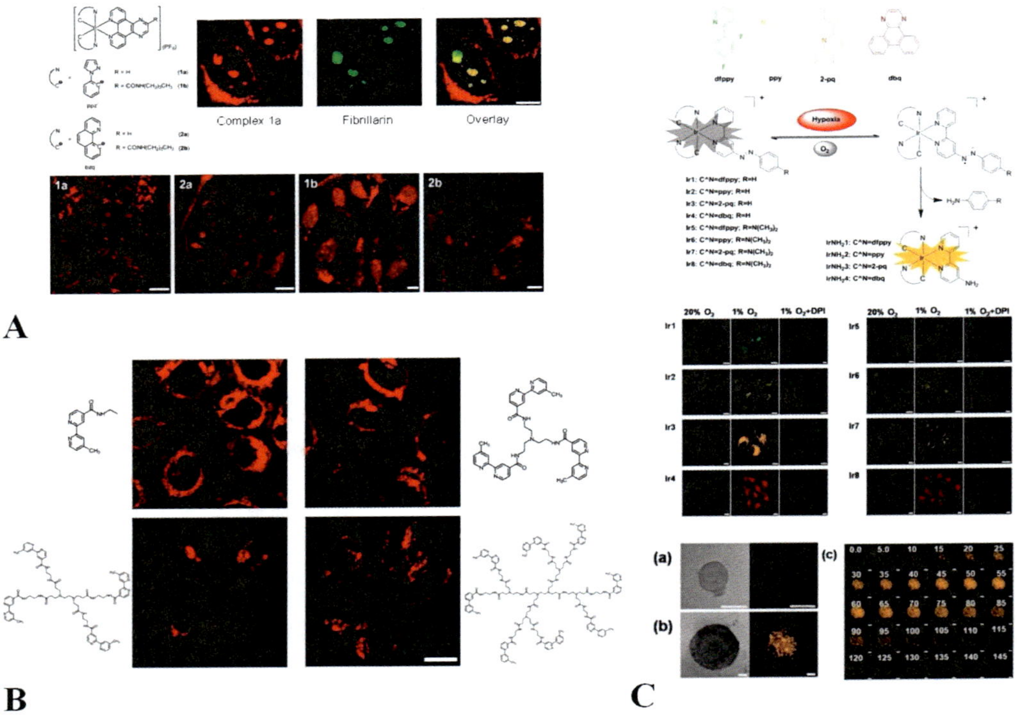

Fig. (9.13). **A**: Ir(III) complexes for labelling nuclei and nucleoli. Variants of a Ir(III) complex (1a-2b, upper left panel) were employed to stain fixed MCDK cells (upper right panels). The first image shows the labelling of nuclei by complex 1a. The signal from the nucleolar-specific fibrillarin protein (as revealed with an antibody) is shown next while the overlap of both signals is displayed last (right). Scale bar: 10 µm. In the lower four panels the luminescent signal from living MDCK cells incubated for 3 hours with complexes 1a-2b is shown. Complex 2a provides a nice nucleolar mark. Scale bars: 10 µm. Reprinted and adapted with permission from reference [67]. Copyright 2010 American Chemical Society. **B**: Ir(III) complexes labelling Golgi apparatus in living HeLa cells. The cells were incubated with the corresponding Ir(III) complex for 2 h and analysed with a confocal laser system (λ_{exc}= 405 nm). For each picture the corresponding ligands complexing Ir(III) are shown on the image side. The increase in ligand complexity correlates with a more selective Golgi localization as judged from the images resolution. Scale bar: 10 µm. Reprinted and adapted with permission from reference [68]. Copyright 2010 American Chemical Society. **C**: Ir(III) complexes to detect pO_2 levels in HeLa cells and spheroids. The Ir(III) complexes structures are displayed in the upper panel along with the activation mechanism. When O_2 concentration lowers, the azo (N=N) group-containing ligand is reduced in the cellular environment. This reduction brings about a chemical reaction that makes the complex luminescent. In the middle panel the hypoxia level is determined in HeLa cells using the Ir1-Ir8 complexes (code meaning in the upper panel). Under normoxic conditions (20% O_2) there is no signal. Under 1% O_2 the cells light up correspondingly. If diphenyliodonium (DPI), an electron-trapping compound, is used there is no signal even under hypoxic conditions. Scale bars: 10 µm. In the lower panels the negligible signal from a normoxic small spheroid is in contrast with the clear luminescence from the center (hypoxic) of a large spheroid (panels a and b). In panel c a z-scan from the top to the bottom of a large spheroid shows the 3D profile of the activated complex signal. Scale bars: 100 µm. Reproduced with permission from [70].

the most heavy metal complexes and O_2, Ir(III) complexes have also been employed to measure O_2 concentration in different media [23]. An interesting and different approach for the assessment of oxygen partial pressure (pO_2) within living cells and tumour spheroids is provided by Sun *et al.* [70]. The O_2 detecting mechanism is explained in Fig. (**9.13C**). Instead of relying on the O_2 ability to directly quench heavy metal complexes excited electronic states, the ligands compete for chemical reduction with O_2. Under high pO_2, the complex cannot compete for the electrons with oxygen and remains in a non-luminescent state. Under hypoxic conditions, the complex starts to collect electrons and gets reduced. This leads to a chemical reaction by which the radiative deactivation pathway of the complex opens. Therefore, in a hypoxic environment, the complex emits light when properly excited. This approach is exemplified both in living HeLa cells and tumour spheroids (see Fig. **9.13C**).

Platinum (Pt)

Along with ruthenium and iridium, platinum is perhaps the most favoured transition metal to formulate luminescent complexes for use in fluorescence microscopy and other luminescent analysis techniques. This noble metal is mostly employed in its Pt(II) oxidation state for luminescent complex synthesis. This brings an interesting feature in regard to luminescent Pt(II) TMCs as these complexes adopt a square planar geometry instead of the more usual octahedral geometry for most other TMCs (see complex examples in Fig. **9.14A-C**) [15]. Because of this, and not going into details, tridentate pyridine-based ligands are largely employed in these complexes as they properly arrange the electronic levels to afford adequate luminescent characteristics. The fact that Pt(II) TMCs display this unusual square planar geometry makes them interact differently with many substrates in comparison to other heavy metal TMCs.

Many different platinum compounds have been employed to label different cellular structures. For a comprehensive and nicely tabulated data review on the matter, the reader is directed to the work by Mauro *et al.* [71]. Briefly Pt(II) complexes have been used to visualize the cell nucleus, nucleoli, mitochondria, endoplasmic reticulum, cytoplasm, lysosomes and plasma membrane. This means that, generally speaking, all cellular organelles can be visualized using proper luminescent Pt complexes. These Pt(II) complexes provide reliable and photostable markers that, depending on the ligands features, show selectivity for different substrates.

As with other TMCs, it is not the goal of this chapter to thoroughly present a list of all the possible uses of Pt-TMCs. We refer the interested reader to the very large available bibliography on this subject, part of which can be found at the end

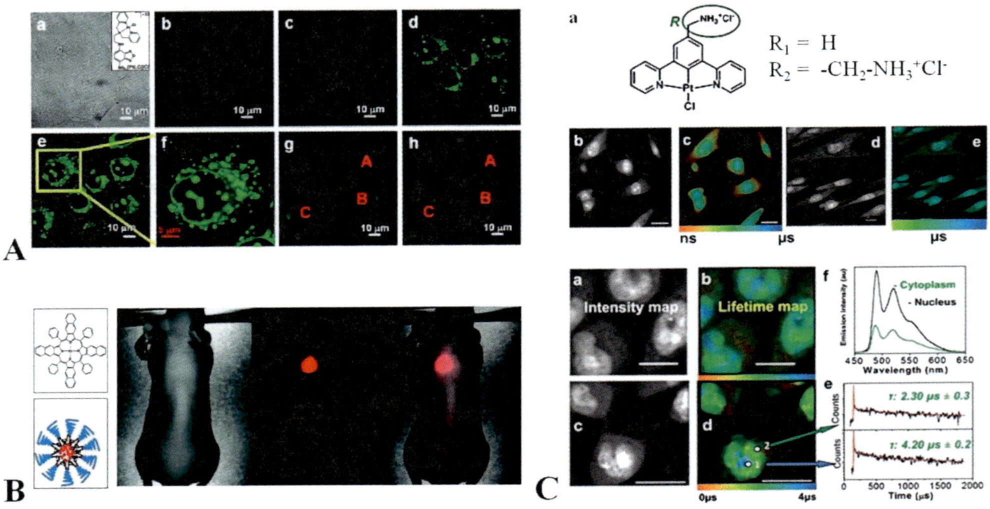

Fig. (9.14). **A**: Pt(II) antitumoral luminescent complex accumulation in living MCF-7 cells. Panel a shows a bright field image of the assessed cells. Panels b-e show the increase in intracellular complex uptake at several times (30 min, 3 h, 4 h and 5 h). Panel f is a magnified portion of panel e to better appreciate the subcellular complex localization (mainly located at the perinuclear region, the nuclear membrane and the nucleoli). Panels g and h show the complex clearance from the cells once they are incubated with clean fresh PBS after 15 min and 60 min respectively. The luminescent signal quickly disappears. Spots labelled A, B and C (red) where employed by the authors to measure the time-dependent loss of luminescent signal. Scale bars indicated on each picture. Reprinted and adapted with permission from reference [72]. Copyright 2011 American Chemical Society. **B**: Example of use of a Pt complex, Pt(II)-tetraphenyltetranaphthoporphyrin (Pt(TPNP)), for mouse tumour detection. Pt(TPNP) was incorporated within micelles (Pt(TPNP) structure and mixture scheme on the left) and administrated to tumour-bearing mice. After 24 h the mouse was imaged both bright field (left image) and by deep red light (650-700 nm) which excited the NIR emission (800-1000 nm) from the Pt macrocycle (middle image). The red spot indicates the tumour location. An overlay image mixing both modalities, bright field and fluorescence, is shown at the right clearly highlighting the tumour's anatomical position. Reprinted and adapted with permission from reference [73]. Copyright 2009 American Chemical Society. **C**: Upper panels: Pt(II) complexes for FLIM. The indicated water-soluble Pt complex (upper left panel a, complex-R_2) is taken by living cells. The less hydrophilic complex obtained when R is a hydrogen (complex-R_1) was also employed in the experiments. The displayed cells (upper panels b and c) are CHO-K1 cells incubated with complex-R_2. Panel b shows the greyscale light emission intensity from the complex. Panel c shows the emission lifetime of the complex as a false colour gradient (colour code below image). Panels d and e show HDF cells also labelled with complex-R_2. In both cases, CHO-K1 and HDF cells, it can be seen that radiative lifetimes are much longer in the nucleus as compared to the cytoplasm. The excitation source was a pulsed laser at 760 nm exciting the complex by a two-photon process (see Chapter 19.2.2). Scale bars: 20 µm. Lower panels: CHO-K1 cells incubated with complex-R_2 (hydrophilic, panels a and b) or complex-R_1 (hydrophobic, panels c-f). Panels a and c show the light intensity while panels b and d display the emission lifetime for the same cells (false colour gradient, see scale below pictures). Panel e compares the emission lifetime decay curves from two cell locations (indicated white spots 1 and 2) corresponding to the nucleus (1) and the cytoplasm (2). It is readily seen that lifetime is almost double within the nucleus. Panel f shows the emission spectra from the nucleus and the cytoplasm, both in intensity and in wavelength range. Scale bars: 20 µm. Reproduced with permission from [19]. Published by The Royal Society of Chemistry.

of this chapter. In what follows, we present a few interesting examples of the uses for Pt(II) complexes in fluorescence detection. Wu *et al.* have recently reported the biological and luminescent properties of a Pt(II) antitumoral complex, both *in vitro* and *in vivo* [72]. Some of their results in cell cultures are presented in Fig. (**9.14A**). The NIR signal from a Pt(II)-tetraphenyltetranaphthoporphyrin (Pt(TPNP)) in an *in vivo* tumour model (mouse) is shown in Fig. (**9.14B**) [73]. The very hydrophobic Pt(TPNP) was incorporated in phospholipid micelles which were then injected into the mouse bloodstream. The images show the NIR emission from the Pt complex-micelle mixture when the animals were excited with red light. This illustrates nicely the selective uptake of the complex-micelles by the tumoral mass.

Another interesting example of overlap between topics presented in this book is shown in Fig. (**9.14C**). In this case, two Pt complexes are put to use to measure radiative lifetimes in different cellular domains by using the FLIM technique (see Chapter 6.3). By modifying part of the ligand, the authors were able to control the hydrophilicity of the complexes. This allowed the complexes to accumulate differentially within the living cells attending to their physico-chemical properties (see Chapter 15.2 for the QSAR model). In the examples presented, there is a very clear differential behaviour between the complexes in the cytoplasm (shorter lifetimes) and the nucleus (longer lifetimes) [19, 74].

Gold (Au)

Gold is probably the heaviest element applied routinely to synthesize luminescent complexes and probes. The use of this metal in this area is relatively recent . Gold is employed mainly in its Au(I) oxidation state to form TMCs [51, 75]. Nevertheless, some luminescent complexes display a Au(III) cation [45]. Many ligands employed to form luminescent Au(I) MCs make use of organic phosphorous-containing molecules (see Fig. **9.15A**). Other ligands include the more usual nitrogen organic molecule configuration. It is quite usual for gold TMCs to form bimetallic or oligo-metallic complexes, where two or more gold atoms are included within the same complex (see Fig. **9.15A**) [76]. Au(I) complexes have been reported to label lysosomes and mitochondria in the living cells [77]. However, the compound reported by Barnard *et al.* localizes preferentially within the lysosomes (this was confirmed with an independent lysosome stain). Other Au(I) complexes localize in the cell nucleus. Bagowski *et al.* describe the nuclear localization of a series of naphthalimide-derivatives complexing Au(I) [78]. These compounds showed antitumoral activity aside from their luminescence. In any case, luminescent Au TMCs are not in such widespread use or diversity as other transition metals presented in the previous sections.

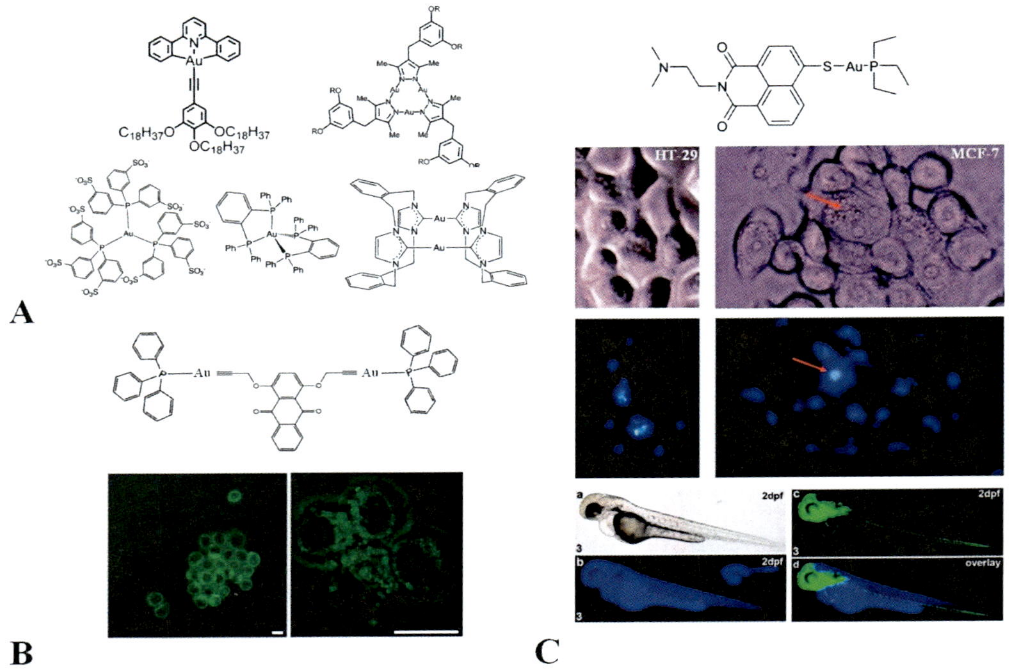

Fig. (9.15). **A**: Examples of some luminescent Au-TMCs structures routinely employed in fluorescence techniques. Reproduced with permission from [76]. Copyright 2015 American Chemical Society. Note that some of the complexes include more than one gold atom bridging the ligands. **B**: Substituted-anthraquinone Au(I) triphenylphosphine complex imaging of MCF-7 cells. The complex structure is depicted in the upper panel. Note that this compound has two gold atoms per complex molecule. Fluorescence images (lower panels) of MCF-7 colonies. The cells were incubated at 4°C to avoid active cell processes so that the complex diffusion features were clearer. The excitation source was a laser confocal system emitting at 405 nm with the microscopic system collecting the complex emission between 530 and 580 nm. On the left a general image of a cell colony showing a brighter emission pattern from the cell periphery. On the right a close up of four cells that allows better discrimination of the light signal. No particular structure is labelled strongly although it seems that the perinuclear area accumulates more complex. The nucleus does not show a signal. Scale bars: 20 μm. Reprinted and adapted with permission from reference [79]. Copyright 2012 American Chemical Society. **C**: Au(I) phosphine-naphthalimide complex assessed *in vitro* and *in vivo*. The complex employed (structure in top panel) features antiproliferative and antiangiogenic properties that make it a candidate for tumour treatment. In the middle panels the complex localization in cell cultures is shown: in HT-29 cells (left panels) and in MCF-7 cells (right panels). Bright field images of the cells are shown first. Structural changes on MCF-7 appearing after incubation is highlighted by the red arrow. The luminescence signal is displayed below the bright field pictures. The intracellular complex accumulation is clear and, in the case of MCF-7 cells, it correlates with the cellular structural changes observed in bright field (red arrow). Excitation was done at 390 nm while emission was registered at 460 nm. The compound was also tested *in vivo* in zebrafish embryos (2dpf= 2 days post-fertilization). The bottom panels show the complex localization: a) bright field embryo image, b) complex emission (inset: control embryo displaying yolk sac autofluorescence), c) emission from GFPs expressed by the embryo, d) overlay of b and c. Reprinted and adapted with permission from reference [80]. Copyright 2009 American Chemical Society.

Two practical examples of Au TMCs in fluorescence microscopy are presented in what follows. In the first one, a series of substituted anthraquinones complexing gold atoms was studied to check their feasibility for living cell imaging in microscopy [79]. The fluorescence images obtained after incubation of MCF-7 cells with one of the complexes (structure detailed in the figure) are shown in Fig. (**9.15B**). The tested compound showed a good cellular uptake after 30 min. The authors reported a widespread cytoplasmic distribution with no particular structure or organelle displaying an enhanced accumulation. The cell nuclei seem to be spared from the compound. The images were obtained by a laser confocal system.

The second presented example employs a Au(I) complex of phosphine and naphthalimide (see structure in Fig. **9.15C**). The complex is intended for use as an antiproliferative and antiangiogenic agent [80]. The cell affects and localization were assessed with two cell lines: HT-29 and MCF-7. The complex exerted a more severe action on MCF-7 cells. The luminescence signal is highly correlated with places of structural change in MCF-7. *In vivo* studies to evaluate the complex antiangiogenic action were done in zebrafish embryos. The complex was incorporated to the living organisms (see Fig. **9.15C**). These examples highlight the usefulness of luminescent TMCs that can have additional desirable features like antitumor activity.

Gold, as it has already been stated for silver above, is also employed directly as a metal to produce luminescent nanoparticles as opposed to metallic cations in TMCs. See Chapter 18.3 in relation to the uses of metallic Au as a component of luminescent solid-state markers.

3.4. 4f-block

The lanthanides form a very interesting and useful family of TMCs in relation to fluorescence microscopy applications [81 - 83]. As has been explained above, once a 4f electron has been excited, it usually has a very long radiative lifetime (~ ms) which makes them perfect probes for time-resolved fluorescence studies. A very particular use of lanthanides, the emission of short wavelength light after excitation with long wavelength photons, known as upconversion will be mainly introduced in Chapter 19.4.1. Another variant of the upconversion phenomenon, in relation to MCs which share one d-block metal and a lanthanide, will be briefly explained in Chapter 19.4.2.

Many lanthanides have emission lines or sharp bands in the visible range. However, the most widely employed lanthanides in fluorescence microscopy have been terbium(III) and europium(III). The former has several emission lines between the blue-green and orange (480-650 nm), while the later shows emission between 590 nm and 700 nm (see Fig. **9.7B** and **9.8B** for the emission spectra of these metals). Additionally, their 4f excited states are less sensitive to vibrational perturbations from the environment therefore, increasing the radiative yield [51]. Luminescent complexes of these two ions show a low cytotoxicity in the concentration range of 100-200 μM, which are 1-2 orders of magnitude higher than the usual incubation conditions for cell imaging. TMCs of Eu(III) and Tb(III) make good labelling agents for lysosomes, mitochondria and cell nucleus. This selectivity depends mainly on the physico-chemical characteristics of the ligands.

Europium (Eu)

Some practical examples of these two 4f TMCs and their uses will be next presented. Europium(III) will be introduced first. This 4f-block metal displays a characteristic red light emission. It is widely used in biological imaging [84]. In Fig. (**9.16A**), three examples of Eu(III) TMCs employed in cell imaging and bioanalysis are pictured. Two of them show the molecular mechanisms by which the Eu(III) emission increases o diminishes in response to the concentration of some co-solute (in the examples K^+ or H^+) [85]. The interaction with the co-solute promotes a molecular change that either promotes or decreases the radiative deactivation of the Eu(III) cation.

An interesting example is presented in Fig. (**9.16B**) where a short-lived ROS, singlet oxygen ($^1O_2$), is detected by a clever use of chemical reactions [35]. When there is no $^1O_{2,}$ the UV excitation of the ligand antenna leads to the excitation energy, ending in the methyl-anthracene moiety at the top of the molecular structure. The non-radiative deactivation pathway is highly favoured under this circumstance and there is no efficient Eu(III) excitation. In the presence of $^1O_{2,}$ this reactive molecule forms an endoperoxide with methyl-anthracene distorting its usually planar structure and disrupting its electronic conjugation. UV excitation of this particular chemical structure favours energy ending on the Eu(III) atom promoting a 4f electron to the excited state. When this electron relaxes back to the ground state it emits s-red photon. A practical example of living cells is also presented in Fig. (**9.16B**) where increasing amounts of photosensitized $^1O_2$ are quantified by using the depicted Eu(III) complex.

Eu(III) complexes are being designed and employed as subcellular luminescent tag probes [81, 84]. This is exemplified in Fig. (**9.16C**) that presents some europium TMCs that localize selectively in different cellular structures depending on their functional chemical groups. One of these compounds localizes initially in the mitochondria of living cells but then relocalizes to the endosome/lysosome compartments for sufficient long times (24 h). This highlights the importance of these complexes to study internal kinetics and metabolism in the living cells.

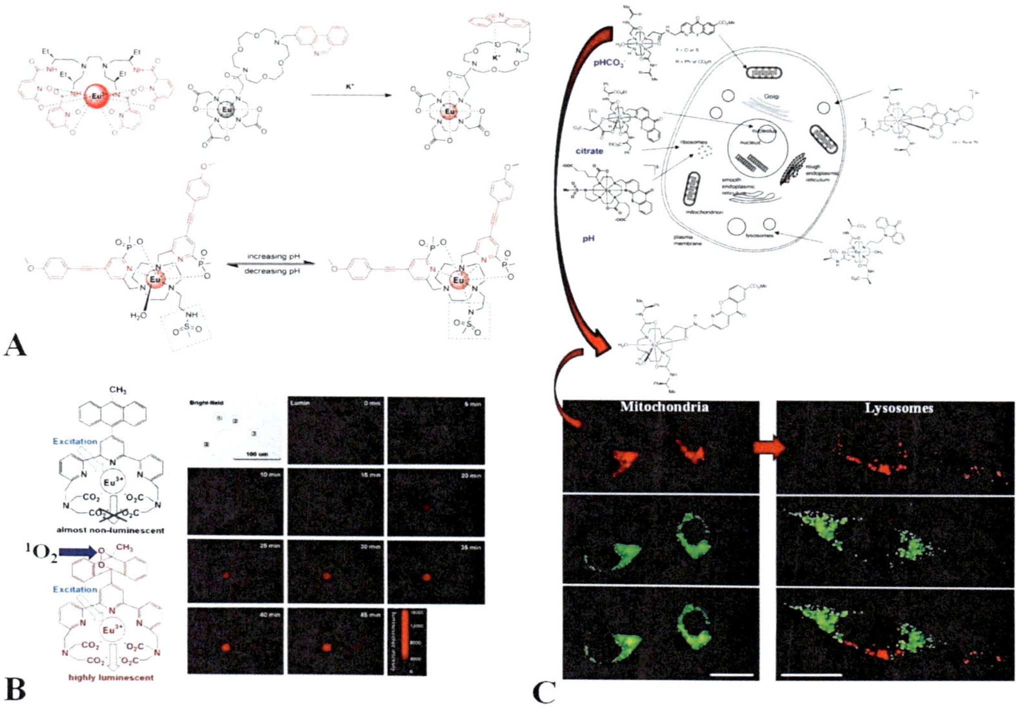

Fig. (9.16). **A**: Some europium(III) complexes employed in bioanalytical techniques and cell imaging. The complex on the top right increases its light emission when coronene-like moiety binds a K^+ cation. The complex at the bottom decreases its emission under acidic (low pH) conditions. It has been employed to measure the endoplasmic reticulum pH *in vitro*. Reprinted and adapted with permission from reference [85]. Copyright 2014 American Chemical Society. **B**: Eu(III) complex as an $^1O_2$ detection probe. In the presence of $^1O_2$ the ligand forms an endoperoxide (red arrow) that increases the energy transfer to the Eu(III) atom when UV light excites the whole complex. There is an increase in the red emission from Eu(III) in the presence of $^1O_2$. On the right panel an example of $^1O_2$ detection in living HeLa cells. The cells were incubated with the $^1O_2$ probe and a $^1O_2$ photosensitizer (TMPyP) and then irradiated for the displayed amount of time with light exciting TMPyP. It is clearly seen how the probe emission increases in direct relation to the amount of TMPyP excitation and, in correlation, $^1O_2$ production. Reprinted and adapted with permission from reference [35]. Copyright 2006 American Chemical Society. **C**: Some common cellular targets for Eu(III) and Tb(III) fluorescent label complexes (top panel). Note that the complexing motif is practically the same in these examples and that selectivity is provided by side groups attached to the complexing motif. One of the complexes (curved red arrows) is an early-time mitochondrial label that relocalizes on a longer time frame to the lysosomes (short red arrow). This is shown in living CHO cells incubated with the Eu(III) complex (bottom panels). On the left the early time (4 h) complex localization and on the right the late time (24 h) labelling. The upper panels show the complex red emission, the middle panels the mitochondrial marking by Mitotracker Green™ and the lower panels show the overlay of the previous images. It can be seen that both signals, red and green, co-localize at 4 h but do not at 24 h. Scale bars: 20 μm. Reprinted and adapted with permission from reference [81]. Copyright 2009 American Chemical Society.

Terbium (Tb)

Terbium, as Tb(III), is also one of the most employed 4f elements for luminescent

TMCs [81, 85, 86]. Its characteristic emission is a bright green colour due to its most intense band around 550 nm. Because of this, it is possible to employ Eu(III) and Tb(III) complexes at the same time on the same sample and exciting with a single light source in the UV or deep blue (350-450 nm). The differential emission from each lanthanide (550 nm *vs*. 650 nm) allows for very good signal discrimination. Three complexes that include Tb(III) as the metal centre are shown in Fig. (**9.17A**). Not only cellular structures, but also different substrates (*e.g.* physiological cations, heavy metals, organic cofactors, *etc.*) can be assessed through the use of these compounds.

A more concrete example is shown in Fig. (**9.17B**). A large complex with 15 Tb(III) cations (structure in the figure) was employed to assess the cell uptake kinetics [87]. The complex uptake was compared between the complex and the simple terbium salt Tb(III)Cl_3. It is very clear from the images that the organic complex was internalized in a much more efficient way than the much smaller but inorganic terbium chloride. This is probably due to screening of the Tb(III) positive charges by the complex. Once again, we see how ligands provide most of the functionality of the probes, while the metals act merely as the internal light source.

Complex ligands can be tailored in other ways too, for example increasing their non-linear absorption coefficient in order to enhance their multi-photon excitation (see Chapter 19.2). This has been done by Law *et al.* to excite Tb(III) complexes with a pulsed Ti:sapphire laser [88]. In this example (see Fig. **9.17C**), the ligand binding Tb(III) was designed to increase its absorption cross-section for photons at 800 nm. After absorbing three of these 800 nm photons in a very fast sequence (almost instantaneously), the complex "behaves" as if it has absorbed one photon of three times the energy of each of the IR photons ($\approx$ 266 nm). Afterwards, everything happens in the same way as if a 266 nm UV photon has been absorbed. The ligand is excited and transfers this energy to the Tb(III) f electrons. Ultimately an f-f transition takes place most probably with the emission of a green photon (550 nm). The cell images shown in Fig. (**9.17C**) were obtained with a confocal 800 nm laser system in different cell lines. Differential uptake numbers and subcellular location can be appreciated in the pictures. Law *et al.* have also reported a very nice assessment of chromatin structure changes during the normal cell cycle employing precisely a Tb(III) complex that closely interacts with chromatin [89].

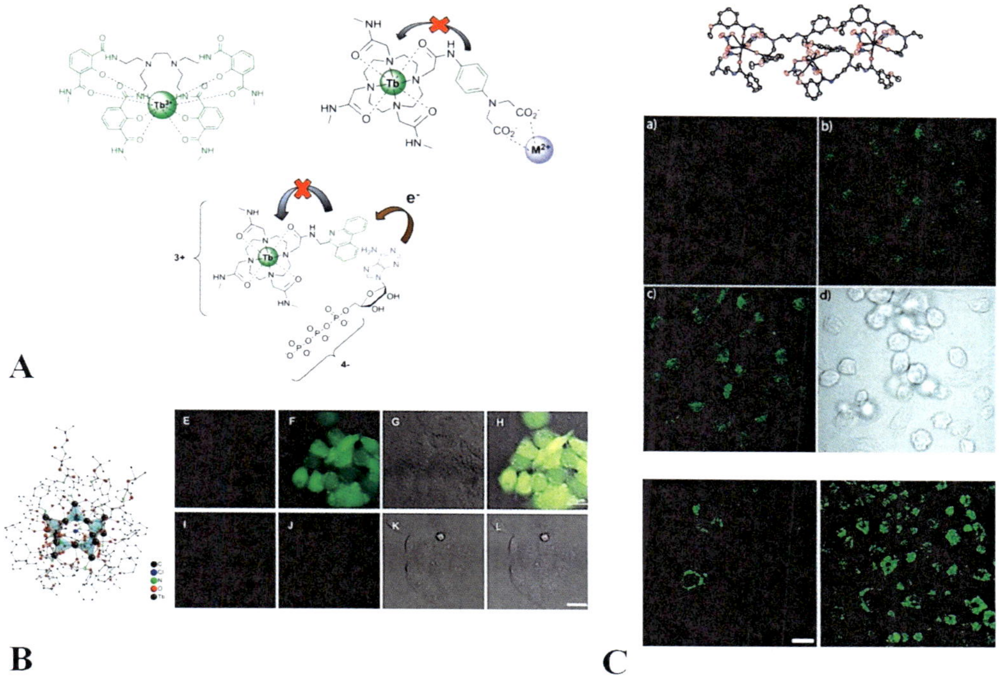

Fig. (9.17). **A**: Three examples of the chemical structures of terbium(III) complexes used in biological imaging. Side groups provide functionality, as shown for the detection of heavy metals –M^{2+}- (top right) or ATP (bottom middle). Binding the heavy metal or the ATP molecule interferes with the complex geometry or electron kinetics, both resulting in a net luminescence decrease. Reprinted and adapted with permission from reference [85]. Copyright 2014 American Chemical Society. **B**: Multi Tb(III) complex for cell imaging . The complex 3D structure is shown on the left in a ball-and-stick model (see atom colour code). A total of 15 Tb(III) cations form part of this complex. Hydrogen atoms have been omitted for clarity. On the right several cell pictures are displayed (original source labelling for each panel). First row (E-H) shows HeLa cells incubated with the complex. Second row (I-L) are HeLa cells incubated with Tb(III)Cl_3 as a control. First vertical lane (E and I) show light in the blue range, second lane (F and J) show green, third lane (G and K) are bright field images of the studied cells and last lane (H and L) are the overlay of the three previous lanes. The cellular uptake of the complex is clear in comparison to Tb(III)Cl_3. Scale bar: 25 μm. Reprinted and adapted with permission from reference [87]. Copyright 2013 American Chemical Society. **C**: Tb(III) complexes are also designed for multiphoton cell imaging. In the top panel the structure of one of these complexes is shown. The middle panels show the temporal evolution of the complex uptake by HONE1 cells (human nasopharyngeal carcinoma). Pictures show cells after 3 min (a), 45 min (b), 60 min (c) and a bright field image of the culture (d). Excitation was provided by a pulsed Ti:sapphire laser emitting at 800 nm and complex emission was registered between 480-615 nm. The complex is emitting visible light after it has absorbed three IR photons. At the bottom panels the uptake of the same complex in A549 (left) and HeLa (right) after 60 min is shown. The excitation conditions are the same as with HONE1 cells. It can be clearly notice the differences in uptake between the two cell lines: HeLa incorporates a higher amount in most of the cells and the complex emits not only from the perinuclear region (like A549) but from other punctuated cytoplasmic structures. Scale bar: 25 μm. Reprinted and adapted with permission from reference [88]. Copyright 2008 American Chemical Society.

Other 4f-block Elements

Although europium and terbium have been historically the most employed (and still are) lanthanides in the biosciences, other lanthanides are finding more use in biological imaging. Especially for those lanthanides with emission lines in the NIR as infrared imaging is becoming more widespread in recent times. Some of the more promising NIR-emitting lanthanides are Praseodymium (Pr), Neodymium (Nd), Samarium (Sm), Dysprosium (Dy), Holmium (Ho), Erbium (Er), Thulium (Tm) and, particularly promising, Ytterbium (Yb) [90, 91]. The reason for employing these NIR-emitting lanthanides in TMCs is because light in the red and NIR spectrum (650-1100 nm) penetrates better through biological tissue, the so-called biological optical window, than other shorter or longer wavelengths. This is because there is less absorption and less scattering of the light field. Optical sensor technologies have also improved very much in the last few years in regard to collecting NIR light. The use of these NIR-emitting probes offers promising applications in the field of disease diagnostics and imaging. The emission band associated with each one of these lanthanides in TMCs is listed in Table **9.3**.

Table 9.3. Principal lanthanides employed in NIR-emitting TMCs (modified from [90]).

Lanthanide	**NIR emission wavelengths (nm)**
Praseodymium	**1010-1040** / 1440
Neodymium	**870-920** / **1060-1090** / 1320-1390
Samarium	**880** / **1020-1040** / 1160-1170
Dysprosium	1280-1340
Holmium	**910** / **965-990** / 1160-1190 / 1480-1510
Erbium	**650-670** / **800** / **980** / 1540-1550
Thulium	**780-805**
Ytterbium	**980**

Wavelengths in **bold** are within the biological optical window.

Ytterbium is an interesting example of development of these NIR emitting complexes. Among the lanthanides Yb(III) has only one 4f electronic excited state (see Fig. **9.7A**, rightmost element). Radiative transitions between this excited state ($^2F_{5/2}$) and the ground state ($^2F_{7/2}$) have a wavelength around 980 nm (the wavelength varies slightly depending on the ligands complexing the Yb(III)). Therefore no matter what wavelength is employed to excite the ligand (UV-VIS or multi-photon) the emission from the complex will always be around 980 nm.

A practical use of one Yb(III) complex, in which the ytterbium atom is complexed between a modified porphyrin and a cobalt-phosphate complex (see structure in Fig. **9.18A**), is presented next [92]. By functionalizing the porphyrin with a rhodamine B molecule, the complex shows a very high specificity for the mitochondrial compartment in living HeLa cells (see Fig. **9.18B**). In the figure the complex Yb-1 is without the rhodamine B group while Yb-2 is the one functionalized with rhodamine B.

Both complexes show a good signal under single photon (430 nm) and two-photon (860 nm) excitation, with the registered emission including the signal from the porphyrin and the Yb(III) cation. However, Yb-1 displays a perinuclear distribution, while Yb-2 is localized in tubular-like structures. Cross checking with a mitochondrial probe showed an acceptable level of mitochondrial co-localization for Yb-2 (see Fig. **9.18C**). These compounds show the additional advantage that they are water-soluble. The strategy of complexing NIR-emitting lanthanides with macrocycles (*e.g.* porphyrins, *etc.*) seems very promising as the recent literature shows [91, 93, 94].

3.5. 5f-block

Finally some comments in regard to the heaviest and radioactive, elements are presented. Among the actinides, the most employed has been uranium as a part of the uranyl cation (see section 2.4 above and Fig. **9.9 B**). In this form, uranium is in its U(VI) oxidation state. Complexes of the uranyl cation have been employed in optical microscopy but especially in electron microscopy, due to the very high atomic weight of U that makes it a very good contrasting agent. Uranyl complexes and salts bind efficiently to nucleic acids, like DNA, and to a lesser extent to proteins. The interaction site can then be analysed by electron microscopy and by fluorescence microscopy due to the U(VI) emission in the green [3, 95]. The actinides display a larger amount of stable oxidation states than the lanthanides. Apart from the U(VI) present in uranyl salts, other luminescent actinide cations are Pa(IV), U(IV), Np(V) and Np(VI), and Am, Cm, Bk, Cf and Es all in the (III) state. These cations emit in the visible and NIR when properly excited, with lifetimes not as long as lanthanides, but nevertheless well into the microsecond range in most cases [39]. In any case, most current uses of luminescent actinides have to do with inorganic analytical determination and not with fluorescence microscopy of biological samples.

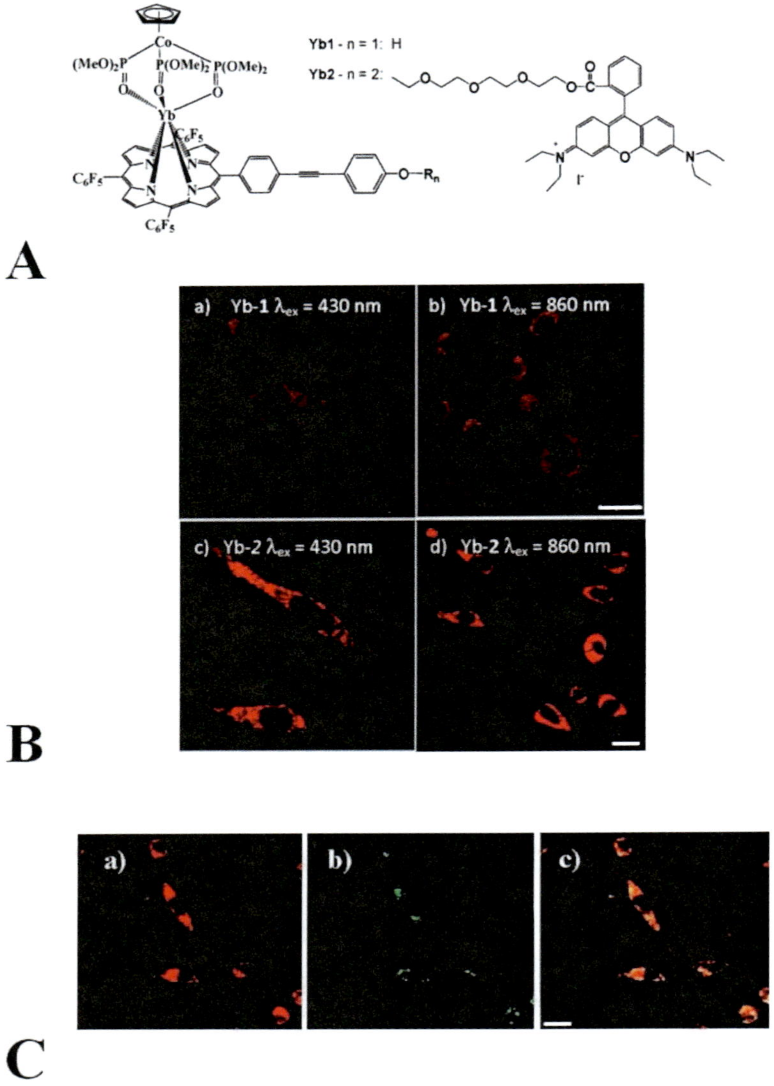

Fig. (9.18). **A**: Structure of the two Yb(III) complexes employed by Zhang *et al.* in their mitochondrial labelling experiments. The two side groups employed are shown to the right of the general formula. Yb-2 is functionalized with a rhodamine B molecule. **B**: Comparison of Yb-1 and Yb-2 cellular localization in HeLa and photoexcitation performance. Panels a and b show the subcellular distribution of Yb-1 under single-photon (430 nm) or two-photon (860 nm) excitation. Panels c and d show the corresponding emission of Yb-2 under one-photon or two-photon excitation (emission registered in the red and NIR ranges). Scale bar: 25 μm. **C**: Confirmation of the mitochondrial localization of Yb-2 in HeLa cells. Panel a shows Yb-2 emission, panel b emission from mitochondrial probe M7514 (Invitrogen), and panel c is the overlay of a and b. Excitation was done at 430 nm for both fluorophores. Scale bar: 50 μm. All three figures (A-C) reprinted and adapted with permission from reference [92]. Copyright 2011 American Chemical Society.

Elements from the 5f-block have been included both for historical and for the sake of being comprehensive. All these elements are radioactive and present a health hazard. Except U and Th, both with very long half-lives, the rest of the actinides show very high disintegration rates. Even U and Th give rise to steady levels of other daughter isotopes with much higher radioactive activity than the parent nuclides. Not only are these elements a hazard for scientists, they can also be deleterious for the sample studied, more so in the case of living biological systems.

Based on the above discussion, we discourage the use of radioactive fluorescent markers because they do not offer, in our view, any particular advantage and because of the almost inexhaustible availability of alternative (safer) compounds, especially in regard to the use of lanthanide compounds.

CONFLICT OF INTEREST

The author (editor) declares no conflict of interest, financial or otherwise.

ACKNOWLEDGEMENTS

This work was supported by the Aarhus Institute of Advanced Studies (AIAS).

REFERENCES

[1] Rayner-Canham G, Overton T. Descriptive inorganic chemistry. 5th ed., New York: WH Freeman and Company 2010.

[2] Jones CJ. d- and f-block chemistry. Cambridge, UK: The Royal Society of Chemistry 2002.

[3] Szabó Z, Toraishi T, Vallet V, *et al.* Solution coordination chemistry of actinides: Thermodynamics, structure and reaction mechanisms. Coord Chem Rev 2006; 250: 784-815. [http://dx.doi.org/10.1016/j.ccr.2005.10.005]

[4] Denning RG. Electronic structure and bonding in actinyl ions and their analogs. J Phys Chem 2007; 111: 4125-43. [http://dx.doi.org/10.1021/jp071061n]

[5] Roundhill DM. Photochemistry and photophysics of metal complexes. New York, NY: Springer Science+Business Media 1994. [http://dx.doi.org/10.1007/978-1-4899-1495-8]

[6] Montalti M, Credi A, Prodi L, *et al.* Handbook of Photochemistry. 3rd ed., Boca Raton, FL: CRC Press, Taylor & Francis Group 2006.

[7] Balzani V, Campagna S, Eds. Photochemistry and photophysics of coordination compounds I Topics in Current Chemistry. Heidelberg: Springer-Verlag Berlin Heidelberg 2007a; Vol. 280.

[8] Balzani V, Campagna S, Eds. Photochemistry and photophysics of coordination compounds II Topics in Current Chemistry. Heidelberg: Springer-Verlag Berlin Heidelberg 2007b; Vol. 281.

[9] Coogan M. Ligand design in d-block optical imaging agents and sensors. In: Storr T, Ed. Ligand design in medicinal inorganic chemistry. Chichester: John Wiley & Sons, Ltd. 2014. [http://dx.doi.org/10.1002/9781118697191.ch4]

[10] Balzani V, Juris A, Venturi M. Luminescent and redox-active polynuclear transition metal complexes. Chem Rev 1996; 96: 759-833.
[http://dx.doi.org/10.1021/cr941154y]

[11] Keefe MH, Benkstein KD, Hupp JT. Luminescent sensor molecules based on coordinated metals: a review of recent developments. Coord Chem Rev 2000; 205: 201-28.
[http://dx.doi.org/10.1016/S0010-8545(00)00240-X]

[12] Lo KK, Li SP. Utilization of the photophysical and photochemical properties of phosphorescent transition metal complexes in the development of photofunctional cellular sensors, imaging reagents, and cytotoxic agents. RSC Advances 2014; 4: 10560-85.
[http://dx.doi.org/10.1039/c3ra47611a]

[13] Liu Z, He W, Guo Z. Metal coordination in photoluminescent sensing. Chem Soc Rev 2013; 42: 1568-600.
[http://dx.doi.org/10.1039/c2cs35363f]

[14] Lo KK, Ed. Luminescent and photoactive transition metal complexes as biomolecular probes and cellular reagents. Berlin, Heidelberg: Springer-Verlag 2015.
[http://dx.doi.org/10.1007/978-3-662-46718-3]

[15] Baggaley E, Weinstein JA, Gareth Williams JA. Lighting the way to see inside the live cell with luminescent transition metal complexes. Coord Chem Rev 2012; 256: 1762-85.
[http://dx.doi.org/10.1016/j.ccr.2012.03.018]

[16] Xiang H, Cheng J, Ma X, *et al.* Near-infrared phosphorescence: materials and applications. Chem Soc Rev 2013; 42: 6128-85.
[http://dx.doi.org/10.1039/c3cs60029g]

[17] Zhao Q, Huang C, Li F. Phosphorescent heavy-metal complexes for bioimaging. Chem Soc Rev 2011; 40: 2508-24.
[http://dx.doi.org/10.1039/c0cs00114g]

[18] Lakowicz JR. Principles of fluorescence spectroscopy. 3rd ed., New York: Springer Science + Business Media LLC 2010.

[19] Baggaley E, Botchway SW, Haycock JW, *et al.* Long-lived metal complexes open up microsecond lifetime imaging microscopy under multiphoton excitation: from FLIM to PLIM and beyond. Chem Sci (Camb) 2014; 5: 879-86.
[http://dx.doi.org/10.1039/C3SC51875B]

[20] Botchway SW, Charnley M, Haycock JW, *et al.* Time-resolved and two-photon emission imaging microscopy of live cells with inert platinum complexes. Proc Natl Acad Sci USA 2008; 105: 16071-6.
[http://dx.doi.org/10.1073/pnas.0804071105]

[21] Gill MR, Thomas JA. Ruthenium(II) polypyridyl complexes and DNA-from structural probes to cellular imaging and therapeutics. Chem Soc Rev 2012; 41: 3179-92.
[http://dx.doi.org/10.1039/c2cs15299a]

[22] Smith NA, Sadler PJ. Photoactivatable metal complexes: from theory to applications in biotechnology and medicine. Phil Trans R Soc A 2013; 371: 20120519.
[http://dx.doi.org/10.1098/rsta.2012.0519]

[23] Ruggi A, van Leeuwen FW, Velders AH. Interaction of dioxygen with the electronic excited states of Ir(III) and Ru(II) complexes: Principles and biomedical applications. Coord Chem Rev 2011; 255: 2542-54.
[http://dx.doi.org/10.1016/j.ccr.2011.05.012]

[24] Stacey OJ, Pope SJ. New avenues in the design and potential application of metal complexes for photodynamic therapy. RSC Advances 2013; 3: 25550-64.
[http://dx.doi.org/10.1039/c3ra45219k]

[25] Werts MH. Making sense of lanthanide luminescence. Sci Prog 2005; 88: 101-31. [http://dx.doi.org/10.3184/003685005783238435]

[26] Faulkner S, Pope SJ, Burton-Pye BP. Lanthanide complexes for luminescence imaging applications. Appl Spectrosc Rev 2005; 40: 1-31. [http://dx.doi.org/10.1081/ASR-200038308]

[27] Leonard JP, Nolan CB, Stomeo F, *et al.* Photochemistry and photophysics of coordination compounds: Lanthanides. Top Curr Chem 2007; 281: 1-43. [http://dx.doi.org/10.1007/128_2007_142]

[28] Bünzli JC. Benefiting from the unique properties of lanthanide ions. Acc Chem Res 2006; 39: 53-61. [http://dx.doi.org/10.1021/ar0400894]

[29] O'Neill ES, New EJ. Luminescent lanthanoid probes. In: Storr T, Ed. Ligand design in medicinal inorganic chemistry. Chichester: John Wiley & Sons, Ltd. 2014. [http://dx.doi.org/10.1002/9781118697191.ch5]

[30] Bünzli JC, Piguet C. Taking advantage of luminescent lanthanide ions. Chem Soc Rev 2005; 34: 1048-77. [http://dx.doi.org/10.1039/b406082m]

[31] Vuojola J, Soukka T. Luminescent lanthanide reporters: new concepts for use in bioanalytical applications. Methods Appl Fluoresc 2014; 2: 012001. [http://dx.doi.org/10.1088/2050-6120/2/1/012001]

[32] Wang X, Chang H, Xie J, *et al.* Recent developments in lanthanide-based luminescent probes. Coord Chem Rev 2014; 273-274: 201-12. [http://dx.doi.org/10.1016/j.ccr.2014.02.001]

[33] Bünzli JC. On the design of highly luminescent lanthanide complexes. Coord Chem Rev 2015; 293-294: 19-47. [http://dx.doi.org/10.1016/j.ccr.2014.10.013]

[34] Hanaoka K, Kikuchi K, Kobayashi S, *et al.* Time-resolved long-lived luminescence imaging method employing luminescent lanthanide probes with a new microscopy system. J Am Chem Soc 2007; 129: 13502-9. [http://dx.doi.org/10.1021/ja073392j]

[35] Song B, Wang G, Tan M, *et al.* A Europium(III) complex as an efficient singlet oxygen luminescence probe. J Am Chem Soc 2006; 128: 13442-50. [http://dx.doi.org/10.1021/ja062990f]

[36] Edelstein NM. Comparison of the electronic structure of the lanthanides and actinides. J Alloys Compd 1995; 223: 197-203. [http://dx.doi.org/10.1016/0925-8388(94)09003-3]

[37] Carnall WT. A systematic analysis of the spectra of trivalent actinide chlorides in D 3h site symmetry. J Chem Phys 1992; 96: 8713-26. [http://dx.doi.org/10.1063/1.462278]

[38] Liu G, Beitz JV. Optical spectra and electronic structure. In: Morss LR, Edelstein NM, Fuger J, Eds. The chemistry of the actinide and transactinide elements. Netherlands: Springer Science + Business Media BV 2006; pp. 2013-111. [http://dx.doi.org/10.1007/1-4020-3598-5_18]

[39] Natrajan LS. Developments in the photophysics and photochemistry of actinide ions and their coordination compounds. Coord Chem Rev 2012; 256: 1583-603. [http://dx.doi.org/10.1016/j.ccr.2012.03.029]

[40] Kharissova OV, Méndez-Rojas MA, Kharisov BI, *et al.* Metal complexes containing natural and artificial radioactive elements and their applications. Molecules 2014; 19: 10755-802.

[http://dx.doi.org/10.3390/molecules190810755]

[41] Song J, Gao X, Wang ZN, *et al.* Multifunctional uranyl hybrid materials: Structural diversities as a function of pH, luminescence with potential nitrobenzene sensing, and photoelectric behavior as p-type semiconductors. Inorg Chem 2015; 54: 9046-59. [http://dx.doi.org/10.1021/acs.inorgchem.5b01364]

[42] Licandro E, Panigati M, Salmain M, *et al.* Organo metallic bioprobes for cellular imaging (Ch. 11, pp. 339-392). In: Jaouen G, Salmain M, Eds. Bioorganometallic Chemistry Applications in drug discovery, biocatalysis, and imaging. Weinheim,: Wiley-VCH Verlag GmbH & Co. KGaA 2015.

[43] Heinemann SH, Hoshi T, Westerhausen M, *et al.* Carbon monoxide – physiology, detection and controlled release. Chem Commun (Camb) 2014; 50: 3644-60. [http://dx.doi.org/10.1039/C3CC49196J]

[44] Pierri AE, Muizzi DA, Ostrowski AD, *et al.* Photo-controlled release of NO and CO with inorganic and organometallic complexes. In: Lo KK, Ed. Luminescent and photoactive transition metal complexes as biomolecular probes and cellular reagents. Berlin, Heidelberg: Springer-Verlag 2015.

[45] Ma DL, He HZ, Leung KH, *et al.* Bioactive luminescent transition-metal complexes for biomedical applications. Angew Chem Int Ed 2013; 52: 7666-82. [http://dx.doi.org/10.1002/anie.201208414]

[46] Zhang QL, Liu JH, Ren XZ, *et al.* A functionalized cobalt(III) mixed-polypyridyl complex as a newly designed DNA molecular light switch. J Inorg Biochem 2003; 95: 194-8. [http://dx.doi.org/10.1016/S0162-0134(03)00122-3]

[47] Arslantas A, Devrim AK, Necefoglu H. Dna-binding study of tetraaqua-bis(p-nitrobenzoato)cobalt(II) dehydrate complex: $[Co(H_2O)4(p\text{-}NO_2C_6H_4COO)_2]\cdot 2H_2O$. Int J Mol Sci 2007; 8: 564-71. [http://dx.doi.org/10.3390/i8070564]

[48] Chen Y, Bai Y, Han Z, *et al.* Photoluminescent imaging of Zn^{2+} in living systems. Chem Soc Rev 2015; 44: 4517-46. [http://dx.doi.org/10.1039/C5CS00005J]

[49] McCormick N, Velasquez V, Finney L, *et al.* X-Ray fluorescence microscopy reveals accumulation and secretion of discrete intracellular zinc pools in the lactating mouse mammary gland. PLoS One 2010; 5: e11078. [http://dx.doi.org/10.1371/journal.pone.0011078]

[50] Kimura E, Aoki S, Kikuta E, *et al.* A macrocyclic zinc(II) fluorophore as a detector of apoptosis. Proc Natl Acad Sci USA 2003; 100: 3731-6. [http://dx.doi.org/10.1073/pnas.0637275100]

[51] Coogan MP, Fernández-Moreira V. Progress with, and prospects for, metal complexes in cell imaging. Chem Commun (Camb) 2014; 50: 384-99. [http://dx.doi.org/10.1039/C3CC45229H]

[52] Pritchard VE, Thorp-Greenwood FL, Balasingham RG, *et al.* Simple polyphenyl zirconium and hafnium metallocene room-temperature lumophores for cell imaging. Organometallics 2013; 32: 3566-9. [http://dx.doi.org/10.1021/om400212y]

[53] Lo KK, Choi AW, Law WH. Applications of luminescent inorganic and organometallic transition metal complexes as biomolecular and cellular probes. Dalton Trans 2012; 41: 6021-47. [http://dx.doi.org/10.1039/c2dt11892k]

[54] Juris A, Balzani V, Barigelletti F, *et al.* Ru(II) polypyridine complexes: Photophysics, photochemistry, electrochemistry, and chemiluminescence. Coord Chem Rev 1988; 84: 85-277. [http://dx.doi.org/10.1016/0010-8545(88)80032-8]

[55] Gill MR, Garcia-Lara J, Foster SJ, *et al.* A ruthenium(II) polypyridyl complex for direct imaging of DNA structure in living cells. Nat Chem 2009; 1: 662-7.

[http://dx.doi.org/10.1038/nchem.406]

[56] Svensson FR, Abrahamsson M, Strömberg N, *et al.* Ruthenium(II) complex enantiomers as cellular probes for diastereomeric interactions in confocal and fluorescence lifetime imaging microscopy. J Phys Chem 2011; 2: 397-401.

[57] Zhao J, Wu W, Sun J, *et al.* Triplet photosensitizers: from molecular design to applications. Chem Soc Rev 2013; 42: 5323-51.
[http://dx.doi.org/10.1039/c3cs35531d]

[58] Man BY, Chan HM, Leung CH, *et al.* Group 9 metal-based inhibitors of β-amyloid (1-40) fibrillation as potential therapeutic agents for Alzheimer's disease. Chem Sci (Camb) 2011; 2: 917-21.
[http://dx.doi.org/10.1039/c0sc00636j]

[59] Metcalfe C, Thomas JA. Kinetically inert transition metal complexes that reversibly bind to DNA. Chem Soc Rev 2003; 32: 215-24.
[http://dx.doi.org/10.1039/b201945k]

[60] Thorp-Greenwood FL. An introduction to organometallic complexes in fluorescence cell imaging: Current applications and future prospects. Organometallics 2012; 31: 5686-92.
[http://dx.doi.org/10.1021/om3004477]

[61] Stephenson KA, Banerjee SR, Besanger T, *et al.* Bridging the gap between *in vitro* and *in vivo* imaging: Isostructural Re and ^{99m}Tc complexes for correlating fluorescence and radioimaging studies. J Am Chem Soc 2004; 126: 8598-9.
[http://dx.doi.org/10.1021/ja047751b]

[62] Louie MW, Liu HW, Lam MH, *et al.* Luminescent Rhenium(I) polypyridine complexes appended with an α-D-glucose moiety as novel biomolecular and cellular probes. Chemistry 2011; 30: 8304-8.
[http://dx.doi.org/10.1002/chem.201101399]

[63] Mari C, Panigati M, D'Alfonso L, *et al.* Luminescent conjugates between dinuclear rhenium complexes and peptide nucleic acids (PNA): Synthesis, photophysical characterization, and cell uptake. Organometallics 2012; 31: 5918-28.
[http://dx.doi.org/10.1021/om300451 5]

[64] Louie MW, Liu HW, Lam MH, *et al.* Novel luminescent tricarbonylrhenium(I) polypyridine tyramine-derived dipicolylamine complexes as sensors for Zinc(II) and Cadmium(II) ions. Organometallics 2009; 28: 4297-307.
[http://dx.doi.org/10.1021/om900311s]

[65] Trovato E, Di Pietro ML, Puntoriero F. Shining a new light on an old game-An OsII-based near-IR light switch. Eur J Inorg Chem 2012; 25: 3984-8.
[http://dx.doi.org/10.1002/ejic.201200787]

[66] Jin C, Liu J, Chen Y, *et al.* Cyclometalated iridium(III) complexes as two-photon phosphorescent probes for specific mitochondrial dynamics tracking in living cells. Chemistry 2015; 21: 12000-10.
[http://dx.doi.org/10.1002/chem.201501882]

[67] Zhang KY, Li SP, Zhu N, *et al.* Structure, photophysical and electrochemical properties, biomolecular inteactions, and intracellular uptake of luminescent cyclometalated iridium(III) dipyridoquinoxaline complexes. Inorg Chem 2010; 49: 2530-40.
[http://dx.doi.org/10.1021/ic902465b]

[68] Zhang KY, Liu HW, Fong TT, *et al.* Luminescent dendritic cyclometalated iridium(III) polypyridine complexes: Synthesis, emission behavior, and biological properties. Inorg Chem 2010; 49: 5432-43.
[http://dx.doi.org/10.1021/ic902443e]

[69] Guerchais V, Fillaut JL. Sensory luminescent iridium(III) and platinum(II) complexes for cation recognition. Coord Chem Rev 2011; 255: 2448-57.
[http://dx.doi.org/10.1016/j.ccr.2011.04.006]

[70] Sun L, Li G, Chen X, *et al.* Azo-based iridium(III) complexes as multicolor phosphorescent probes to

detect hypoxia in 3D multicellular tumor spheroids. Sci Rep 2015; 5: 14837. [http://dx.doi.org/10.1038/srep14837]

[71] Mauro M, Aliprandi A, Septiadi D, *et al.* When self-assembly meets biology: luminescent platinum complexes for imaging applications. Chem Soc Rev 2014; 43: 4144-66. [http://dx.doi.org/10.1039/C3CS60453E]

[72] Wu S, Zhu C, Zhang C, *et al. In vitro* and *in vivo* fluorescent imaging of a monofunctional chelated platinum complex excitable using visible light. Inorg Chem 2011; 50: 11847-9. [http://dx.doi.org/10.1021/ic201506y]

[73] Kumar R, Ohulchanskyy TY, Roy I, *et al.* Near-infrared phosphorescent polymeric nanomicelles: Efficient optical probes for tumor imaging and detection. ACS Appl Mater Interfaces 2009; 1: 1474-81. [http://dx.doi.org/10.1021/am9001293]

[74] Baggaley E, Weinstein JA, Williams JA. Time-resolved emission imaging microscopy using phosphorescent metal complexes: Taking FLIM and PLIM to new lengths. In: Lo KK, Ed. Luminescent and photoactive transition metal complexes as biomolecular probes and cellular reagents. Berlin, Heidelberg: Springer-Verlag 2015.

[75] Langdon-Jones EE, Pope SJ. Recent developments in gold(I) coordination chemistry: luminescence properties and bioimaging opportunities. Chem Commun (Camb) 2014; 50: 10343-54. [http://dx.doi.org/10.1039/C4CC03259D]

[76] Yam VW, Au VK, Leung SY. Light-emitting self-assembled materials based on d8 and d10 transition metal complexes. Chem Rev 2015; 115: 7589-728. [http://dx.doi.org/10.1021/acs.chemrev.5b00074]

[77] Barnard PJ, Wedlock LE, Baker MV, *et al.* Luminescence studies of the intracellular distribution of a dinuclear gold(I) N-heterocyclic carbine complex. Angew Chem Int Ed 2006; 45: 5966-70. [http://dx.doi.org/10.1002/anie.200601526]

[78] Bagowski CP, You Y, Scheffler H, *et al.* Naphthalimide gold(I) phosphine complexes a anticancer metallodrugs. Dalton Trans 2009; 10799-805. [http://dx.doi.org/10.1039/b912378d]

[79] Balasingham RG, Williams CF, Mottram HJ, *et al.* Gold(I) complexes derived from alkynyloxy-substituted anthraquinones: Syntheses, luminescence, preliminary cytotoxicity, and cell imaging. Organometallics 2012; 31: 5835-43. [http://dx.doi.org/10.1021/om300475y]

[80] Ott I, Qian X, Xu Y, *et al.* A gold(I) phosphine complex containing a naphthalimide ligand functions as a TrxR inhibiting antiproliferative agent and angiogenesis inhibitor. J Med Chem 2009; 52: 763-70. [http://dx.doi.org/10.1021/jm8012135]

[81] Montgomery CP, Murray BS, New EJ, *et al.* Cell-penetrating metal complexes optical probes: Targeted and responsive systems based on lanthanide luminescence. Acc Chem Res 2009; 42: 925-37. [http://dx.doi.org/10.1021/ar800174z]

[82] Laurent S, Vander Elst L, Muller RN. Lanthanide complexes for magnetic resonance and optical molecular imaging. Q J Nucl Med Mol Imaging 2009; 53: 586-603.

[83] Kobayashi H, Longmire MR, Ogawa M, *et al.* Rational chemical design of the next generation of molecular imaging probes based on physics and biology: mixing modalities, colors and signals. Chem Soc Rev 2011; 40: 4626-48. [http://dx.doi.org/10.1039/c1cs15077d]

[84] Butler SJ, Lamarque L, Pal R, *et al.* EuroTracker dyes: highly emissive europium complexes as alternative organelle stains for live cell imaging. Chem Sci (Camb) 2014; 5: 1750-6. [http://dx.doi.org/10.1039/c3sc53056f]

[85] Heffern MC, Matosziuk LM, Meade TJ. Lanthanide probes for bioresponsive imaging. Chem Rev

2014; 114: 4496-539.
[http://dx.doi.org/10.1021/cr400477t]

[86] Rajendran M, Yapici E, Miller LW. Lanthanide-based imaging of protein-protein interactions in live cells. Inorg Chem 2014; 53: 1839-53.
[http://dx.doi.org/10.1021/ic4018739]

[87] Thielemann DT, Wagner AT, Rösch E, *et al.* Luminescent cell-penetrating pentadecanuclear lanthanide clusters. J Am Chem Soc 2013; 135: 7454-7.
[http://dx.doi.org/10.1021/ja403539t]

[88] Law GL, Wong KL, Man CW, *et al.* Emissive terbium probe for multiphoton *in vitro* cell imaging. J Am Chem Soc 2008; 130: 3714-5.
[http://dx.doi.org/10.1021/ja710418d]

[89] Law GL, Man CW, Parker D, *et al.* Observation of the selective staining of chromosomal DNA in dividing cells using a luminescent terbium(III) complex. Chem Commun (Camb) 2010; 46: 2391-3.
[http://dx.doi.org/10.1039/b924679g]

[90] Bünzli JC, Eliseeva SV. Lanthanide NIR luminescence for telecommunications, bioanalyses and solar energy conversion. J Rare Earths 2010; 28: 824-42.
[http://dx.doi.org/10.1016/S1002-0721(09)60208-8]

[91] Zhu X. Design and synthesis of near-infrared emissive lanthanide complexes based on macrocyclic ligands. Eur J Inorg Chem 2011; 4651-74.
[http://dx.doi.org/10.1002/ejic.201100481]

[92] Zhang T, Zhu X, Cheng CC, *et al.* Water-soluble mitochondria-specific ytterbium complex with impressive NIR emission. J Am Chem Soc 2011; 133: 20120-2.
[http://dx.doi.org/10.1021/ja207689k]

[93] Bulach V, Sguerra F, Hosseini MW. Porphyrin lanthanide complexes for NIR emission. Coord Chem Rev 2012; 256: 1468-78.
[http://dx.doi.org/10.1016/j.ccr.2012.02.027]

[94] He H. Near-infrared emitting lanthanide complexes of porphyrin and BODIPY dyes. Coord Chem Rev 2014; 273-274: 87-99.
[http://dx.doi.org/10.1016/j.ccr.2013.11.006]

[95] Zobel CR, Beer M. Electron stains I. Chemical studies on the interaction of DNA with uranyl salts. J Biophys Biochem Cytol 1961; 10: 335-46.
[http://dx.doi.org/10.1083/jcb.10.3.335]

CHAPTER 10

Lipids

Abstract: In this chapter, the uses of most known fluorescent compounds for the microscopic visualization of lipids are described. Particular emphasis is given to the nature of substrates, fixing requirements, dye solubility, and examples of lipophilic stains. Probes for visualizing membranous or lipidic structures in live cells will be also described in Chapter 15.5. Hydrophobic azo and anthraquinone dyes are applied industrially to color fats, paints, plastic resins, rubbers, varnishes and waxes. Some of these dyes are employed to detect lipid droplets and fat-rich components of tissues. Several fat-soluble dyes, particularly bisazo Sudan dyes and Nile red have been routinely used in visualizing lipids [1-3]. Relevant natural and synthetic polyenes such as fluorescent antifungal antibiotics and unsaturated fatty acids will be also described.

Keywords: Carotenoids, Ceramides, Chlorophyll, Diphenyl-hexatrienes, Filipin III-cholesterol, Lipid bilayers, Lipid droplets, Lipochromatic reaction, Nile red, Oil red O, Oxidative deamination, Polyenes, Pentaene fatty acids, Saffron, Solvent dyes, Sudan dyes.

1. LIPID COMPONENTS AND STRUCTURES

Lipids are fundamental components of cell membranes (*i.e.* plasma membrane and organelle membranes), and can be classified in the following way:

1. Neutral lipids such as triglycerides (neutral fats and oils) and waxes.
2. Phospholipids (lecithins, cephalins).
3. Sphingolipids (ceramides, sphingomyelins, cerebrosides, gangliosides, sulfatides).
4. Steroids, containing the polycyclic ring cyclopentano-perhydrophenanthrene (sterane) such as sterols (cholesterol, ergosterol), sex and adrenal hormones.
5. Other compounds (terpenes, carotenoids, tocopherols, *etc*).

In the case of triglycerides, phospholipids, and sphingolipids, glycerol and sphingosine alcohols are esterified with palmitic, stearic, oleic, linoleic, and other fatty acids. These lipids show a hydrophilic head and hydrophobic tail configuration, and form membrane bilayers or limit lipid droplets as monolayers.

Juan Carlos Stockert & Alfonso Blázquez-Castro

On account to the occurrence of double bonds, partially unsaturated (conjugated) fatty acids such as α-eleostearic acid, α(*cis*)-parinaric acid, β(*trans*)-parinaric acid), and steroids (estradiol, estriol, equilin, dehydroergosterol) are fluorescent (Fig. **10.1**) [3 - 6]. However, limitations for the microscopical use of natural unsaturated polyenes and sterols are the very short λexc (below 360 nm) and λem, although they have been applied in biochemistry as probes of lipid-protein interactions and energy-transfer processes, the fatty acids being an acceptor of tryptophan emission [7]. On the other hand, there are sterols (estradiol, cholesterol) labeled with fluorescein, NBD, and anthracene fluorophores [6].

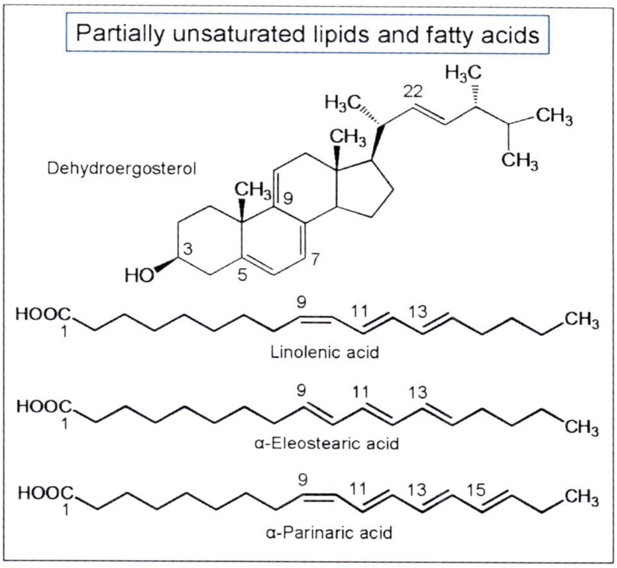

Fig. (10.1). Chemical structure of some partially unsaturated lipids and fatty acids. Substituted positions and double bonds are numbered. Fluorescent compounds are shown (dehydroergosterol, α-eleostearic acid and α-parinaric acid), but only the latter (with 4 double bonds) could be microscopically useful.

α-Parinaric acid (9,11,13,15-octadecatetraenoic acid) was isolated from seeds of the Fiji Island makita tree, *Parinari glaberrimum*, and its fluorescence at 430 nm excited at 320 nm strongly increases on binding to hydrophobic sites in proteins. Among other uses, it has been applied to examine the *cis*:*trans* ratios of unsaturated fatty acids in lipid bilayers, and as lipid probe when bound to phospholipids [4]. Interestingly, α-eleostearic acid is a linolenic acid isomer (octadecatrienoic acid) with a conjugated triene and tumor suppression activity [8]. Polyene fatty acids with longer conjugated chains and excitation above 360 nm are more useful for microscopical applications and they will be described in section 6.

Lipid histochemistry has been often reviewed [3, 9, 10]. Staining methods based on oil-soluble dyes that demonstrate lipids by virtue of their physical properties have been also called "histophysical" procedures. Fluorescence methods for lipids involve (1) direct microscopic observation of autofluorescence due to natural fluorophores (polyenes) or to oxidation products of lipids (lipofucsins) in phagosomes, lysosomes, and residual bodies, and (2) induced fluorescence (Nile red, oil red O, and other lipophilic fluorochromes). Schematic staining mechanisms of phospholipid membranes by either charged, neutral or mixed dyes and fluorochromes are illustrated in Fig. (**10.2**).

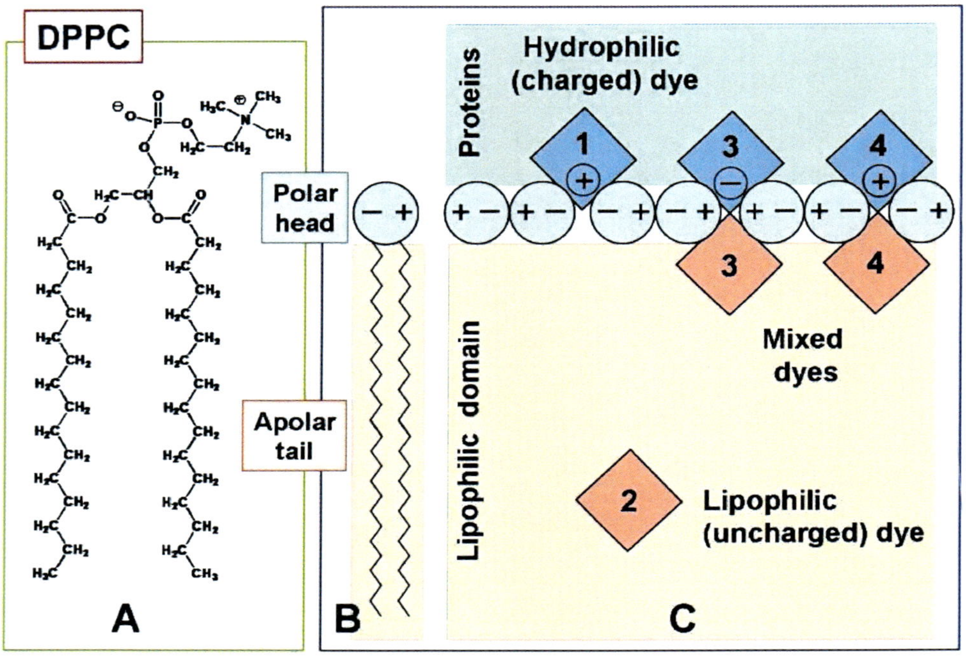

Fig. (10.2). Phospholipids and dye binding. **A**: Chemical structure of dipalmitoyl-phospatidyl-choline (DPPC). **B**: Simplified view of a phospholipid as head (blue)-tail (pink) structure. **C**: Part of a phospholipid bilayer showing binding of charged (blue squares), neutral (pink squares), and mixed dyes. Positive and negative charges are shown. Numbered dyes are 1: Nile blue, 2: Nile red, 3-3: Luxol fast blue MBS, 4-4: $DiIC_{18}(3)$.

Lipid-rich model substrates for studies using fluorescence microscopy are represented by numerous cell and tissue components, such as lipid droplets and membranous organelles, adipocytes, foam cells, atheroma plaques, *etc*. The fat body of *Drosophila* larvae is a continuous tissue mass used as test material because of the large amount of lipid inclusions [11]. Multivesicular bodies and myelin figures resulting from lysosome activity are other lipid-rich cell structures. Lipid droplets from cultured cells (particularly from adipocytes and ovarian granulosa cells) are other useful substrates [12 - 14].

The lipid droplet is now considered an important cell organelle, which is made of a hydrophobic lipid ester core surrounded by a phospholipid monolayer. Staining with lipophilic dyes in alcohols (*i.e.* 70% ethanol), and mounting in glycerol cause gross deformation of lipid droplets [15]. Fresh oil emulsions can be prepared by thoroughly mixing 1 ml commercial sunflower or olive oil and 9 ml culture medium. The treatment of cell cultures with oil emulsions for several hours results in a massive incorporation of lipid droplets into live cells, which can be visualized with fluorescent lipophilic probes, (*i.e.* chlorophyll, see Fig. **4.10**), or after fixation with 4% paraformaldehyde in PBS for 5 min or formaldehyde vapor for 1-2 min [16], and staining with appropriate dyes [13].

2. FIXING REQUISITES AND DYE SOLUBILITY

The use of lipophilic fluorochromes requires a compromise between two extremes: (1) avoiding lipid extraction from cells and tissues by its solubility in fixatives and organic solvents, and (2) allowing dissolution of lipophilic stains. Therefore, several co-solutes are often used to preserve lipids and to dissolve dyes. Lipids are generally unaffected by chemical fixation, exceptions being extraction by organic fixatives, or modifications by formation of calcium salts with phosphate groups, or cyclic ester complexes of Os, Cr, Mn, and Pd with double bonds of unsaturated fatty acids, followed by deposition of hydrated metal oxides or metals. Br and Hg also react with lipid double bonds.

The addition of Ca^{2+} ions in the fixative (*i.e.* formaldehyde, dichromate) improves preservation of tissue lipids by forming insoluble salts with phospholipids. Likewise, some cationic dyes such as malachite green and crystal violet retain and stain cellular phospholipids when added to glutaraldehyde [17, 18]. Digitonin (glycoside of a plant sterol) can also be added to fixatives to preserve cholesterol by making an insoluble sterol adduct. Detailed chemistry of lipid fixation is described in [10, 19].

Regarding solubility of dyes, several solvents, *i.e.* ethylene glycol, hydroalcoholic media, isopropanol, propylene glycol, and triethylphosphate, have been used for dissolving lipid stains, particularly Sudan dyes [20, 21]. These lipid dyes can be first prepared in absolute ethanol or n-heptane and diluted with propylene glycol [13]. Nile red can be used first dissolved in acetone and then in glycerol-water, or more practically, applied such as generated within the original Nile blue solution.

An advantage of the glycols compared to alcoholic solutions is that with the former, no extraction of lipids occurs [22]. Unfortunately, because as most lipophilic fluorochromes are soluble in alcohols and glycols, mounting media containing hydroxyl and glycol groups (*i.e.* aqueous media with glycerol and/or polyvinyl alcohol, such as Mowiol or Vectashield) should be not used. Therefore, observation and photography of preparations must be made soon after mounting in water and sealing the edges of coverslips with nail varnish.

Solubility and staining mechanism of lipophilic dyes are illustrated in Fig. (**10.3A-C**), which shows solubilization by formation of sandwich complexes with the parental cationic dyes, or directly in alcohols or lipids. Solubilizing co-solutes such as caffeine have been also employed. A benzo[*a*]pyrene method (20 mg/L in water saturated with caffeine) was very early used to reveal lipids by fluorescence microscopy [23], but because of important disadvantages (strong reactivity and carcinogenic risk of the hydrocarbon) it is no longer used. It is noteworthy that benzopyrene was also applied as a fluorescent vital probe for visualization of lipids and lysosomes in living cells [24]. Under UV excitation, a bright emission of nuclei and chromosomes is apparent, which was attributed to phospholipid components [25].

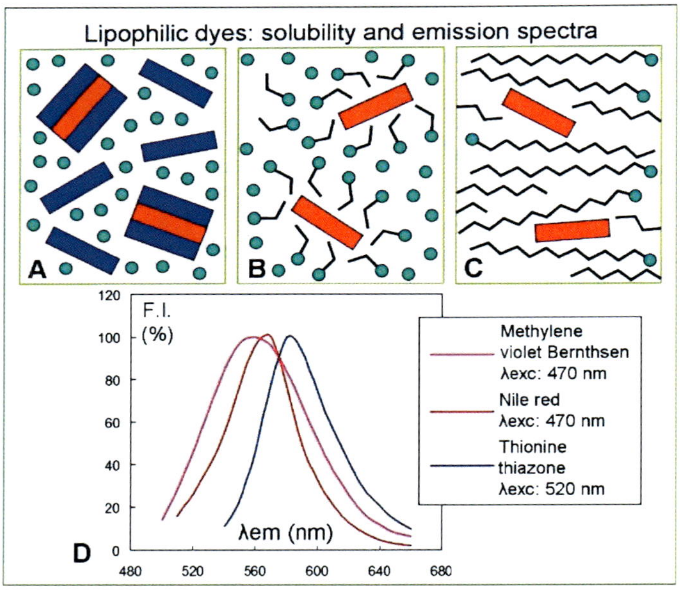

Fig. (10.3). A-C: Lipophilic dyes (red rectangles) are solubilized by complex formation with the parental cationic dyes (blue rectangles) (**A**), by solubility in alcohols (ethylene glycol, propylene glycol) (**B**), or through hydrophobic forces into lipids (**C**). Alcohols and lipids are represented as zigzag chains with polar groups (blue circles; also water). (**D**), Normalized fluorescence of some thiazone and oxazone dyes in sunflower oil.

3. LIPOCHROMATIC REACTIONS

Since long ago, it is known that general staining by some thiazine and oxazine dyes (*i.e.* methylene blue and Nile blue) is improved by the presence of thiazone and oxazone impurities (Bernthsen's methylene violet and Nile red, respectively) [1, 2]. Emission spectra of these contaminants are illustrated in Fig. (**10.3D**). Satisfactory staining in cytology, histology, hematology, and parasitology is just due to these thiazones and oxazones, which are generated by a spontaneous oxidative demethylation and deamination, in which the amino group is replaced hydrolytically by oxygen.

An example is thionine oxidation (Fig. **10.4**). In this case, thionine (Lauth's violet, CI 52000; λab: 600 and 560 nm) generates a red, non-ionic and fluorescent thiazone derivative that can be used to stain lipids. In addition, other modifications of thionine are also known. Photoreduction of thionine gives a leucobase (λab: 390 nm; λem: 440 nm), whereas at alkaline pH the uncharged red imine of thionine (free base) is formed (λab: 500 nm; λem: 590 nm).

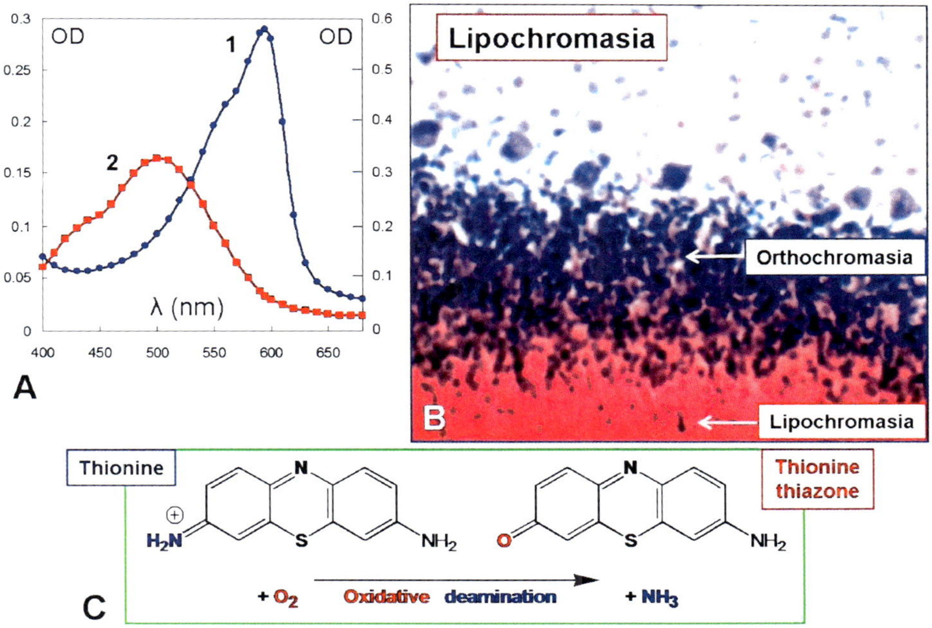

Fig. (10.4). Lipochromatic staining reaction. **A**: Absorption spectra of thionine and its thiazone. 1: aqueous 5 x 10^{-4} M thionine heated at 90 °C for 10 min with 0.1 mg/mL K_2CO_3, extracted with xylene (1:1, v/v), and diluted 1:1 with water. 2: xylene fraction diluted 1:1 with xylene. OD values at left and right corresponds to the curves 2 and 1, respectively. **B**: Frozen formaldehyde-fixed mouse cerebellum section stained with alkali-treated and heated but unextracted 0.1 mg/mL thionine solution. **C**: Oxidative deamination of thionine (hydrophilic blue cation) to thionine-derived thiazone (red and lipophilic dye).

A new concept in the stainability of lipid substrates is the "lipochromatic" reaction. This effect, formerly named "allochromasia", has useful applications in a similar way as the ortho- and metachromatic reactions. Surprisingly, the red color of myelin stained with thionine in neural tissue has been referred as "metachromatic" [27, 28], but now it is clear that the red dye is not a thiazine but a thiazone compound. The uncharged thiazone and imine free base of thiazine dyes seems to be responsible for red staining of lipid components in cells and tissues.

Thus lipophilic and fluorescent compounds can be formed from the oxidative deamination of several thiazine and oxazine dyes (thionine, toluidine blue, methylene blue, brilliant cresyl blue, Nile blue, cresyl violet), giving the corresponding thiazone and oxazone dyes [2, 11, 29]. Aged or alkaline heated thiazine solutions contain variable amounts of thiazones, such as occurs for polychrome methylene blue (Unna's blue, containing variable amounts of oxidized methylene blue derivatives) [26]. In adequate solvents, the lipophilic dye Bernthsen's methylene violet (CI 52041) might be of value as a lipid fluorochrome in a similar way as Nile red. Instead of spontaneous oxidation by aging, fluorescent oxazones and thiazones form rapidly when fresh aqueous solutions of the cationic dyes are treated with 0.1 mg/mL sodium carbonate and heated at 90 °C for 2 h.

It is interesting to note that regarding phenol blue (see Chapter 3.2.2), a very similar structure appears in dyes such as Bernthsen's methylene violet, Nile red and other thiazones and oxazones. In addition to Nile red and brilliant cresyl blue oxazone, other compounds such as orcein (amino-orcein), ommochromes (the eye pigments from insects [30]), and some antibiotics (actinomycin D, cinnabarine) are also fluorescent oxazones.

4. OXIDATIVE DEAMINATION OF DYES: NILE RED

Nile blue is a cationic oxazine dye introduced early for the cytochemical detection of lipids, showing phospholipids and neutral fats in blue and light red, respectively. The formation of Nile red oxazone from Nile blue is a relevant example of oxidative deamination (Fig. **10.5**). As indicated by Green [1], Nile red dye has been little more than a laboratory curiosity for many years but now its application as fluorescent lipid stain has increasing importance. The same occurs in the case of cresyl red (the oxazone derivative of cresyl violet) and the oxazone of brilliant cresyl blue [1, 29]. In addition to the fluorescent Nile red oxazone, Nile blue forms a lipophilic non-fluorescent orange-red imine free base at alkaline pH [26]. Although the oxazine dyes, cresyl violet and Nile blue, are claimed to be non-fluorescent, it is known that when purified to remove the red oxazone contaminants, they are stable fluorescent lasing dyes with useful emission at 630 and 670-740 nm, respectively [1].

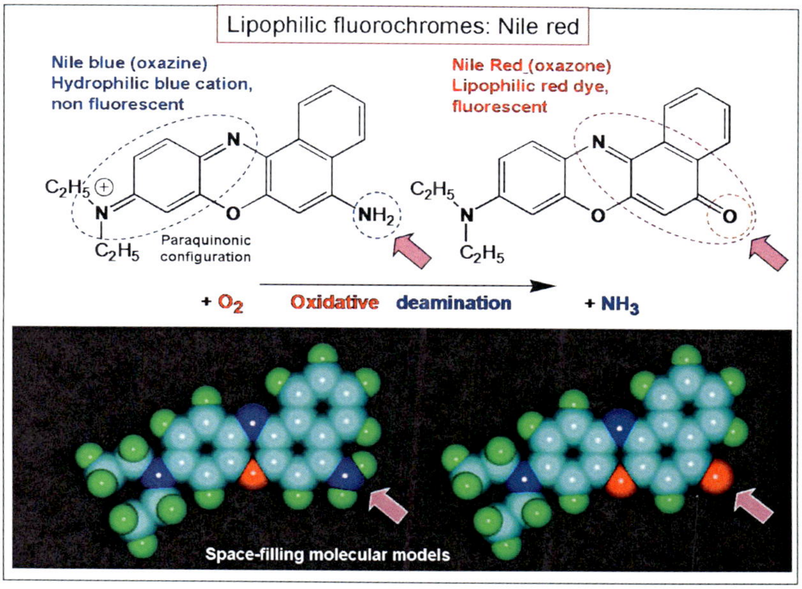

Fig. (10.5). Nile blue and Nile red produced by oxidative deamination, shown as chemicals structures (top), and space-filling models (bottom).

Nile red is one of the most important fluorescent probes for lipids in living and fixed cells, frozen sections, polyacrylamide electrophoresis of proteins, and lipid detection in both chromatography and flow cytometry [2, 11, 31 - 33]. The emission of Nile red is affected by the polarity of the solvent, likely due to an equilibrium between uncharged and zwitterion forms. In strongly non-polar solvents such as heptane it occurs at about 530 nm whereas emission at 630 nm is found with more polar solvents. Nile red dissolved in sunflower oil shows emission at 570 nm. It has been reported that the emission of Nile red shifts from red to yellow according to the degree of hydrophobicity of lipids. Polar lipids (*i.e.* phospholipids, mostly present in membranes), show red fluorescence whereas neutral lipids (esterified cholesterol and triglycerides), which are present in lipid droplets, have yellow emission [34]. The red/yellow emission ratio enables discrimination of different polar (charged) and neutral lipids.

Nile red allows visualizing lipid droplets with a high degree of resolution and brightness. After staining with aged Nile blue, oil emulsion, *Drosophila* fat body, HeLa cells, and Ehrlich tumor cells show lipid droplets with a strong fluorescence induced by Nile red (Figs. **10.6-10.8**).

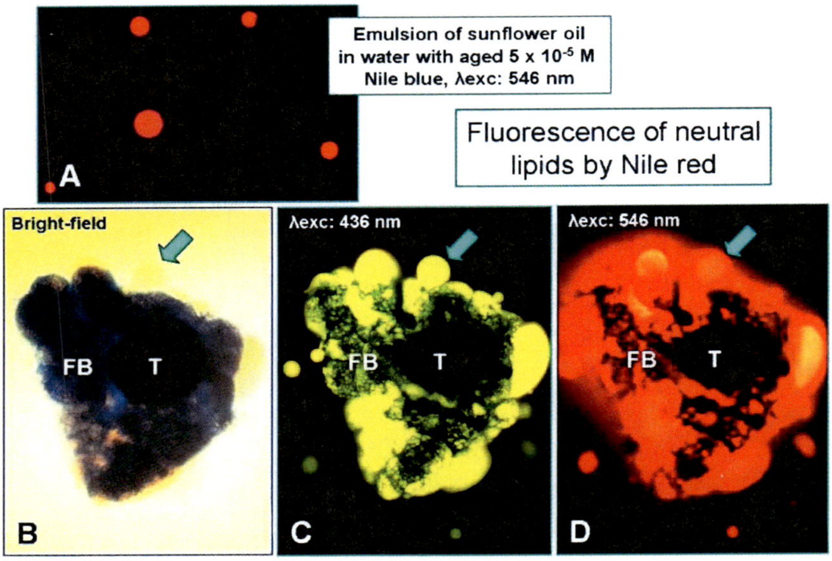

Fig. (10.6). A: Nile red fluorescence of lipid droplets from an oil/water emulsion. **B-D**: Formaldehyde-fixed testis (T) and surrounding fat body (FB) of a *Drosophila hydei* larva stained with aged 10^{-4} M Nile blue, and observed by bright-field (**B**) and fluorescence microscopy (**C**, **D**), showing lipid droplets (arrows).

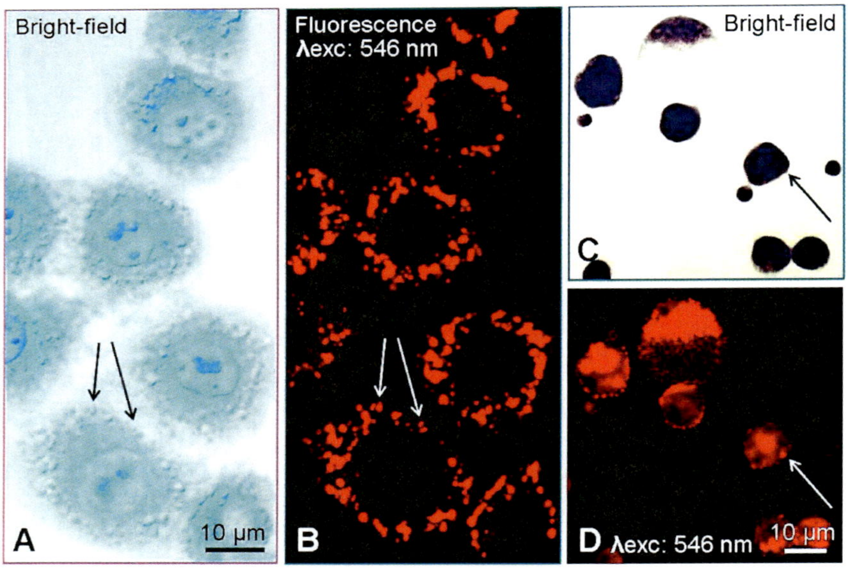

Fig. (10.7). Fluorescence of neutral lipids by Nile red. **A**, **B**: Live HeLa cells treated with a 10:1 DMEM culture medium-sunflower oil emulsion for 24 h, fixed in 4% formaldehyde, stained with aged 10^{-4} M Nile blue for 10 min, and mounted in water). **C**, **D**: Formaldehyde-fixed Ehrlich ascites tumor cells stained with the same Nile blue solution. Arrows indicate lipid droplets.

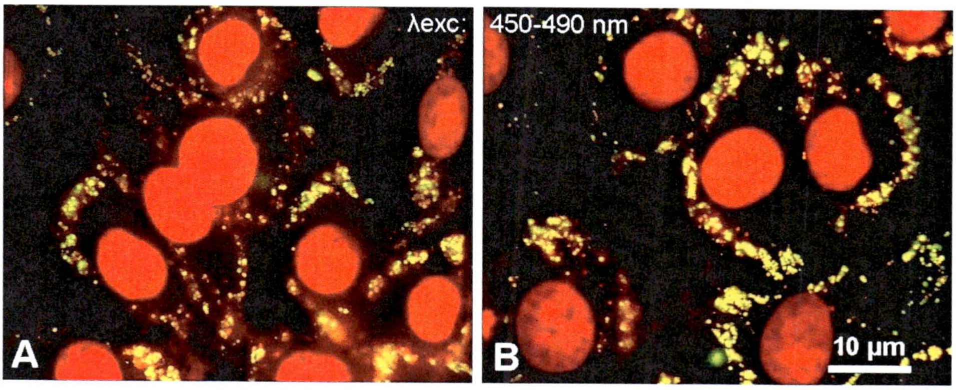

Fig. (10.8). **A**, **B**: Bovine granulosa cells stained by aged 10^{-4} M Nile blue, showing lipid droplets as yellow-green fluorescing dots. Propidium iodide counterstaining (red nuclei) (courtesy of D.M. Lombardo).

Yellow-green or red emission (using blue or green excitation, respectively) is observed also in fat drops that are released from the disrupted fat body. Also lipids in formaldehyde-fixed frozen sections can be demonstrated by means of Nile red fluorescence (Fig. **10.9**). It is worth noting that anionic polysaccharides (*i.e.* mast cell granules) stain metachromatically red by aged Nile blue, which must be not confused with the color of Nile red.

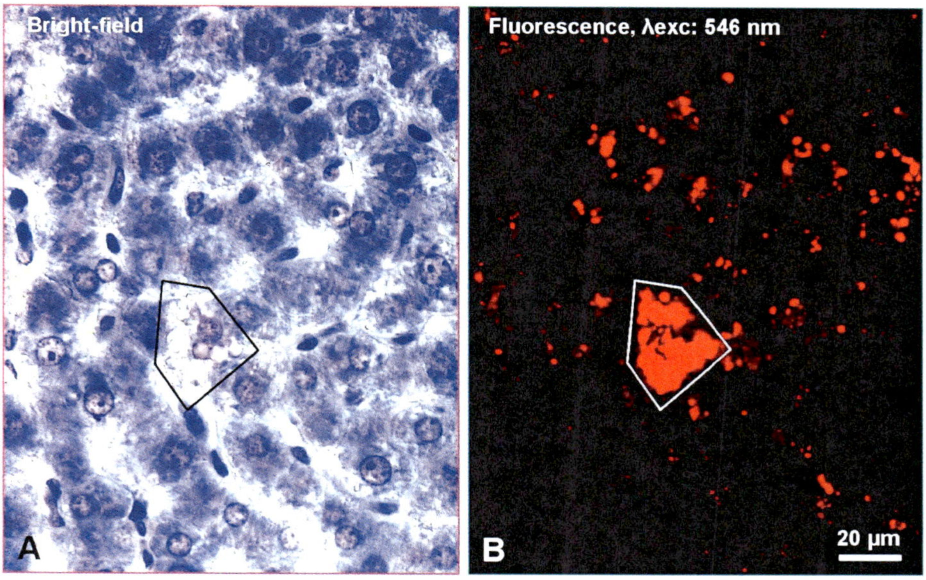

Fig. (10.9). Formaldehyde-fixed, frozen mouse liver section stained with aged 10^{-4} M Nile blue. **A**: Bright-field image showing chromatin, nucleoli, and basophilic cytoplasm. **B**: The same field showing Nile red fluorescence of lipid droplets, mainly accumulated in Kupffer cells (polygonal mark).

5. OTHER LIPOPHILIC STAINS

In order to visualize lipids, classic "solvent" dyes are widely used. Solvent dyes are colored, non-polar compounds that dissolve in lipids and render them visible under the light microscope [10]. Because colored substances are not necessarily dyes, Baker [35] preferred to call them "lysochromes". These compounds stain fats because they are more soluble in fats than in the original solvent. The first widely used solvent dyes for fat were Sudan III (CI 26100) and Sudan IV (CI 26105). Later, Sudan black (CI 26150) and the more hydrophobic oil red O (CI 26125) were successfully applied. At present, fluorescent properties of Sudan dyes attract increasing attention for cytochemical studies. Sudan III gives a red fluorescence in lipids [15]. Oil red O (1-[2,5-dimethyl-4-{2,5-dimthyl-phenylazo}phenylazo]-2-naphthol), commonly used for lipid detection in bright-field microscopy [36], is now applied for fluorescence visualization of lipids (Fig. **10.10**), allowing also application of immunofluorescence techniques to the same cell or tissue sample [14, 21].

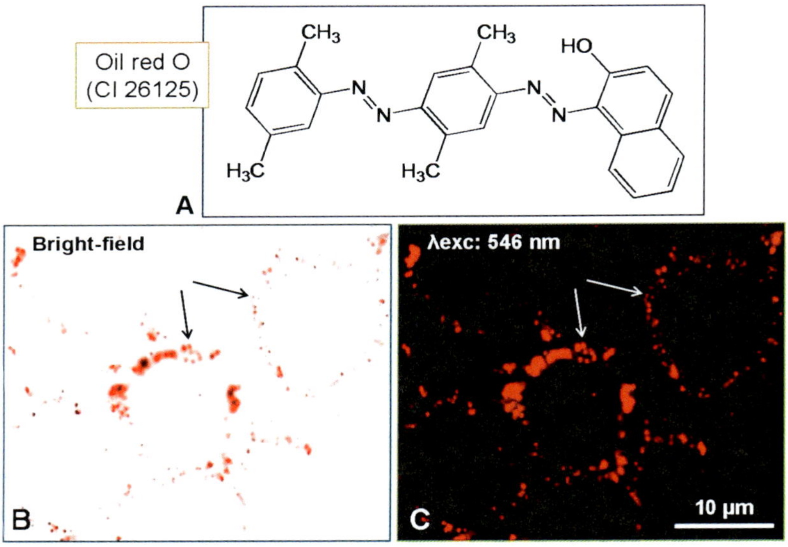

Fig. (10.10). Staining and fluorescence of neutral lipids by the bisazo dye, oil red O. **A**: Chemical structure of oil red O. **B**, **C**: Bovine granulosa cells stained by oil red O (0.1 mg/mL in ethanol/propylene glycol, 1:10, v/v), and observed either in bright-field **(A)** or fluorescent microscopy **(B)**. Lipid droplets appear as red dots (arrows) (courtesy of D.M. Lombardo).

It is worth to note that before the application of Sudan dyes for lipid staining, several compounds were used for this purpose [26]. Quinoline blue was the first aniline dye used as a fat stain. The azamethine dye indophenol blue (CI 49700), the product of early histochemical techniques for cytochrome c oxidase is a lipophilic, non-ionic dye also used as a fat stain, like fat blue Z (CI 49705). All these compounds should be fluorescent in rigid media. Porphyrins have a marked affinity for lipids and those lacking paramagnetic metals endow both lipid tissue components and serum lipoproteins with bright red fluorescence [37]. The lipophilicity and fluorescence of chlorophyll and chlorines is also known, and protoporphyrin IX has been used as a fluorescent stain for lipids.

Likewise, cyanines ($DiOC_6(3)$, DiI, merocyanine 540), Bodipy compounds, anthraquinones, and polymethines are examples of widely used fluorophores for lipid visualization [2, 3, 26]. The yellow neutral red free base (NR^0) can be also used as a lipid stain [38]. The acridine dye phosphine (CI 46045) has been employed as a fluorescent stain for lipids and hydroxybutyrate granules in bacteria [26]. The lipophilic BAO derivative, 2,5-bis(4-diethylaminophenyl)-1, 3,4-oxadiazole has been employed as a lipid fluorochrome [13]. Berberine is a fluorescent alkaloid from *Hydrastis canadensis* L. with antibacterial and antimalarial activity (see Chapter 12.4.4), which also interacts with lipid bilayers [39]. In addition to the N-quaternary compound, tautomeric forms such as the pseudobase and the imino aldehyde are also present (Fig. **10.11**) [40, 41]. These uncharged, lipophilic forms could be responsible for the fluorescent reaction of lipid droplets.

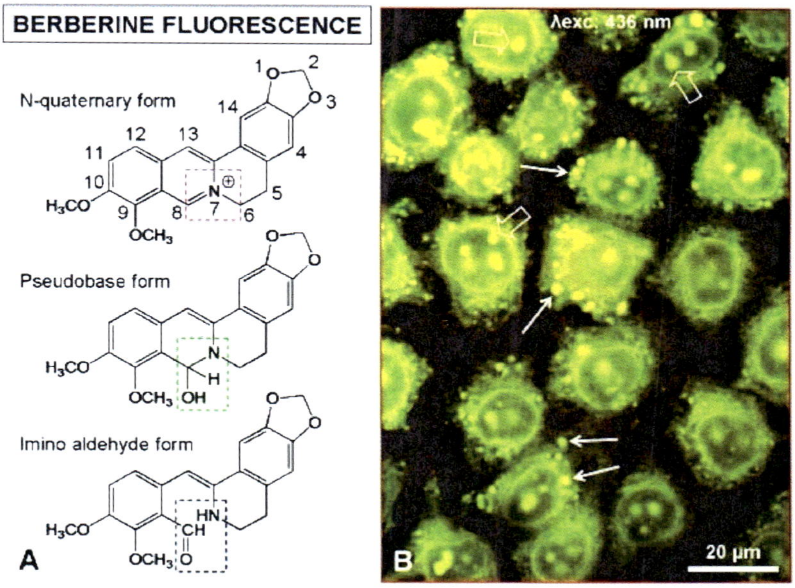

Fig. (10.11). **A**: Chemical structure of berberine tautomers. **B**: HeLa cells treated with sunflower oil emulsion, fixed in formaldehyde, and stained with 10^{-4} M berberine sulfate. Small and large arrows indicate lipid droplets and nucleoli, respectively. Threads in the cytoplasm correspond to mitochondria.

The fluorophore dipyrromethene-boron difluoride (Bodipy) is a new heterocyclic compound widely applied in fluorescence studies. Hydrophobic Bodipy dyes are now employed for lipid detection, either in the form of the fluorophore alone, or esters with fatty acid or cholesterol, examples being Bodipy FL C12 [12] and Bodipy 493/503 [14] (Fig. **10.12**). It is worth noting that Bodipy 493/503 has a green emission under blue exciting light, but when excited by green light after previous blue excitation, the dye shows a red fluorescence [42], which seems due to the generation of a Bodipy dimer. Long term observation must be avoided because the dye fades rapidly. At present, another derivative, Bodipy LD 540 (Fig. **10.12**), appears as a promising lipophilic fluorochrome for microscopical studies [43].

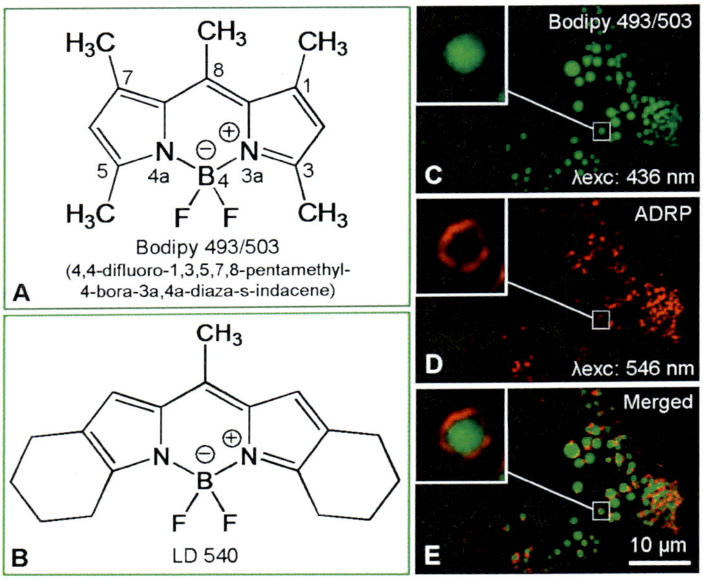

Fig. (10.12). Chemical structures of Bodipy 493/503 (**A**) and LD 540 (**B**). **C-E**: Double labeling of lipid droplets by 2 μg/mL Bodipy 493/503 and adipocyte differentiation-related protein (ADRP) in Huh7 cells. Anti-ADRP antibody and Cy3-conjugated IgG were applied after fixation with 3% formaldehyde/0.025% glutaraldehyde and treatment with 0.01% digitonin before lipid staining. Labeling of lipid droplets (green) and ADRP around them (red) is observed. Squares show details at higher magnification (courtesy of M. Suzuki and T. Fujimoto).

Some fluorescent lipid probes can be also used as lipophilic fluorochromes for cell staining (*i.e.* chlorophyll, see Chapter 4.2). In the case of NBD-ceramide, the Golgi probe is selectively trapped even in fixed cells [44]. Likewise, NBD-phospholipid probes are useful to detect phospholipidosis in living or fixed cultured cells [45 - 47].

On the other hand, tetrazolium salts are easily reduced to intensely colored and water-insoluble formazans. MTT (3-(4,5-dimethylthiazol-2-yl)-2,5-diphenyl- 2H-tetrazolium bromide) is widely used as a viability assay for bacteria and cultured cells [2]. On account of its lipophilic character, the formazan product accumulates into lipid droplets [48, 49]. MTT formazan deposits in lipid droplets appear violet blue under bright-field, whereas the Golgi apparatus shows a weak bluish color. Although it is believed that tetrazolium salts give non-fluorescent formazans, that of MTT shows red fluorescence under green exciting light (see Chapter 14.2.2 and Figs. **14.5** and **14.6**). A similar reaction of lipid droplets and Golgi apparatus is observed after fluorescent labeling with cholesterol-Bodipy FL_{12} ester [12].

6. POLYENES

Fluorescent properties and uses of several natural and synthetic lipid components will be described here. As already mentioned, compounds with only 4 conjugated double bonds require too short excitation wavelengths, but those with 5-7 or more conjugated double bonds are easily excitable using the UV source available in current fluorescence microscopes. However, at present there are new excitation sources for short excitations. Light emitting diodes (LEDs), are compact and efficient light sources based on the electroluminescence of semiconductor materials (see Chapter 6.1.1). Exciting LEDs at 260, 300, and 340 nm are now available (http://www.thorlabs.com).

6.1. Natural Polyenes

Retinol (all *trans*-retinol, vitamin A, Fig. **10.13**) is a fluorescent natural probe for membrane "fluidity". Its emission yield is higher in phospholipid membranes with cholesterol as well as when the probe is rigidly bound to bovine serum albumin [50]. Retinol is converted to 11-*cis*-retinal, the active photoreceptor chromophore. Likewise, other carotenoids show the expected fluorescence properties, some examples being phytofluene, neurosporene, lycopene and β-carotene, which have 5, 9, 11 and 11 conjugated double bonds, respectively. Canthaxanthin (4,4'-diox--β-carotene, CI 40850; Fig. **10.13**) has been used as lipophilic fluorochrome [13].

Although saffron is also a natural polyene compound (namely, a glycoside of the dicarboxylic acid crocetin, see Chapter 3.2.1), the dye induces selective fluorescence reactions of cationic proteins that are rather derived from its anionic hydrolytic products [51]. Azafrin, from roots of *Escobedia scabrifolia*, is a carboxylic carotenoid dye similar to saffron, and on account of its 7 double bonds, it should be useful as fluorochrome. Likewise, the carotenoid-carboxylic acid dye bixin (with 9 double bonds, from seeds of *Bixa orellana*) could have fluorescent applications; other xanthophylls (*i.e.* astaxanthin, capsanthin, cryptoxanthin, lycophyll, violaxanthin) should be also studied.

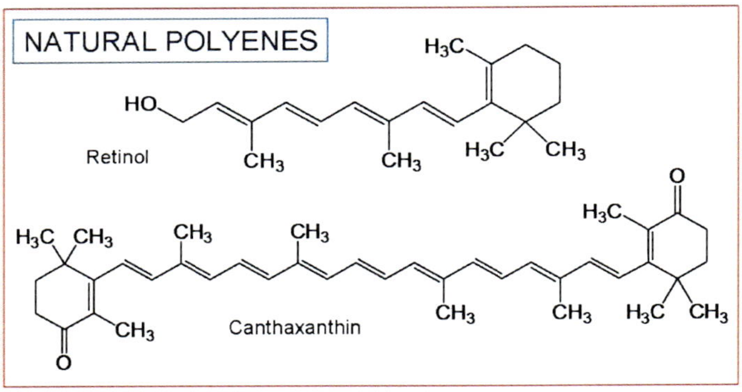

Fig. (10.13). Chemical structure of the natural fluorescent polyenes, all *trans*-retinol and canthaxanthin.

On the other hand, staphyloxanthin is a carotenoid pigment produced by some strains of *Staphylococcus aureus*, and is responsible for the characteristic golden color that gives *S. aureus* its species name. The pigment also acts as a virulence factor. It has an antioxidant action that helps the bacteria evade death by ROS produced by the host immune system [52]. Based on similar pathways for biosynthesis of staphyloxanthin and cholesterol, an anti-cholesterolemic drug was shown to block *S. aureus* pigmentation and infection disease [53]. *S. aureus* strains with (SH1000) and without (RN6390) functional *rsbU* genes show high and negligible amount of staphyloxanthin, respectively. The fluorescence of the pigment can be microscopically observed in SH1000 (Fig. **10.14**), but not in RN6390.

Regarding other lipidic compounds, sterols are specific substrates that can be studied by fluorescence techniques. They generally contain four rings, with -OH groups at C3 that provide the only hydrophilic component, allowing the proper orientation of sterol molecules in the membrane. They have a double bond at C5 and methyl groups at C10 and C13. Cholesterol is the major sterol found in the membranes of all vertebrates. Major plant sterols, sitosterol and stigmasterol, have ring structures identical to that of cholesterol, differing slightly in the side chain. The fungal sterol, ergosterol, differs from cholesterol by the presence of double bonds at C7 and C22, and by an additional methyl group at C24.

In this context, macrolide polyene antibiotics mainly produced by *Streptomyces* represent a highly selective sterol-binding class of fluorescent compounds of great interest and very useful biomedical applications. Most important polyene antifungal antibiotics are shown below, with source, number of conjugated double bonds, and presence or absence of aminosugars (from [41]):

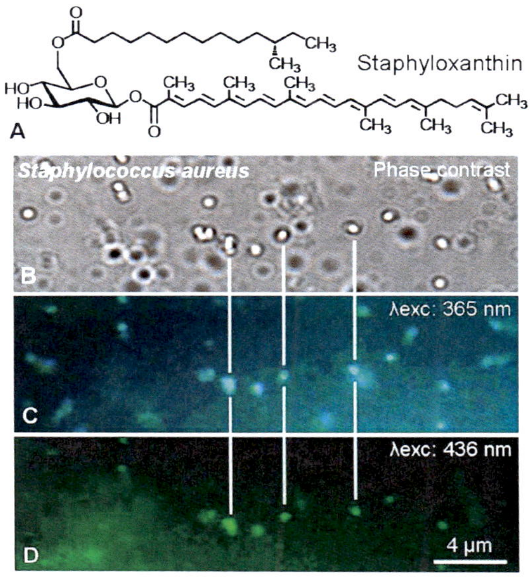

Fig. (10.14). Fluorescence of staphyloxanthin. **A**: Chemical structure of the pigment. **B-D**: *Staphylococcus aureus* (pigmented strain SH1000) grown overnight in tryptic soy broth and observed by phase contrast (**B**), and under UV (**C**) and blue (**D**) exciting light, showing blue and green fluorescing bacteria (courtesy of L. Gándara and A. Casas).

Antibiotic	**Source**	**Conjugated bonds**	**Aminosugar**
Natamycin	*S. natalensis*	4	yes
Nystatin A_1	*S. noursei*	4	yes
Tetrin	*S. illinois*	4	yes
Rimocidin	*S. rimosus*	4	yes
Filipin III	*S. filipenensis*	5	no
Fungichromin	*S. cellulosae*	5	no
Amphotericin B	*S. nodosus*	7	yes
Candicidin	*S. viridoflavus*	7	yes
Candidin	*S. griseus*	7	yes
Partricin	*S. aureofaciens*	7	yes
Perimycin A	*S. coelicolor*	7	yes

Polyene antibiotics have aminosugars (mycosamine in the case of amphotericin B, natamycin, and nystatin A_1), which allow a specific hydrogen bond between the amino group and the hydroxyl of sterols. In addition, hydrophobic forces are responsible for binding of polyenes with sterols, such as occurs in the case of filipin III-cholesterol (Fig. **10.15**). In this context, filipin was early used as a cholesterol probe [54]. Extracellular filipin-positive areas have been also demonstrated in atherosclerotic lesions. Tissue sections stained with both filipin (detecting unesterified cholesterol) and oil red O (detecting esterified cholesterol) show that extracellular filipin-positive areas do not stain with oil red O [55]. Although ergosterol is also visualized in the fungal plasma membrane by using filipin III [56] (Fig. **10.15**), amphotericin B and natamycin have higher affinity for ergosterol.

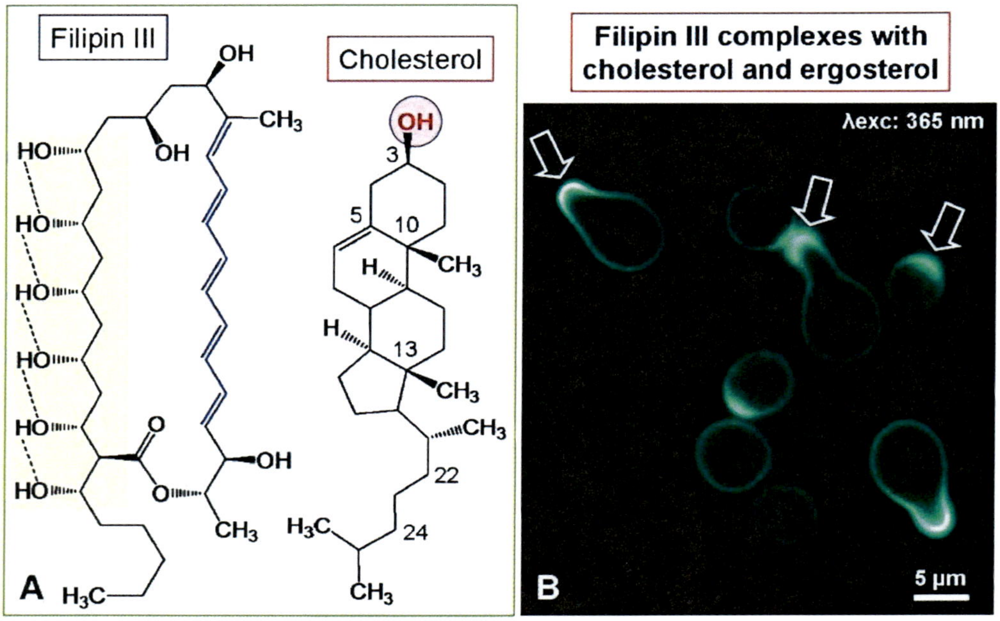

Fig. (10.15). **A**: Chemical structure of filipin III and cholesterol (pink), showing hydrophilic (yellow) and polyene (blue) chains, numbered positions, and the -OH group (red) of the sterol at C3. Filipin III shows a chain of -OH···O- hydrogen bonds. **B**: Germinating conidia of living *Penicillium discolor* treated with 15 μM filipin III for 30 s. Arrows show the location of germ tube formation with increased ergosterol content (courtesy of J. Dijksterhuis).

On account of their higher conjugation degree, filipin III, amphotericin B and perimycin A are yellow compounds and good fluorochromes for microscopical studies under UV excitation. Spectroscopic and electron microscope studies have shown strong interactions of polyene antibiotics and membrane sterols [57]. They are very selective for the visualization of sterols, the most used being filipin III, with λexc and λem at about 360 and 480 nm, respectively. Spectroscopic studies show that affinity for cholesterol binding is filipin III > amphotericin B > nystatin A_1.

Models of polyene antibiotic-sterol complexes within cell membranes have been published [58 - 60]. The flat ring systems of polyenes and sterols interact with 1:1 stoichiometry. In addition to the hydrophobic surface, polyenes contain a hydrophilic edge (polyhydroxy chain) that forms a central channel (hydrated pore) in the amphotericin B-ergosterol double octamer. In the case of filipin III-cholesterol complex, clusters instead pores are formed within the plasma membrane with polyhydroxy chains packed together. Both pores and clusters lead to membrane disruption and cell death.

6.2. Synthetic Polyenes

Synthetic compounds correspond to aromatic polyenes and polyene fatty acids. In the first case they are based on the linear diphenyl-hexatriene (DPH, Fig. **10.16**), which is used as a fluorescent probe [2, 61]. Cationic (DPH-TMA), anionic (DPH-propionic acid), and neutral compounds (DPH-propionyl cholesterol ester) are also DPH derivatives employed to label lipid bilayers [2, 3, 62, 63]. It is important to note that properties and distribution of natural lipids can be affected by fluorochrome labeling. The most frequently used tags, NBD and Bodipy, strongly influence the properties of lipids, yielding analogue molecules with quite different characteristics.

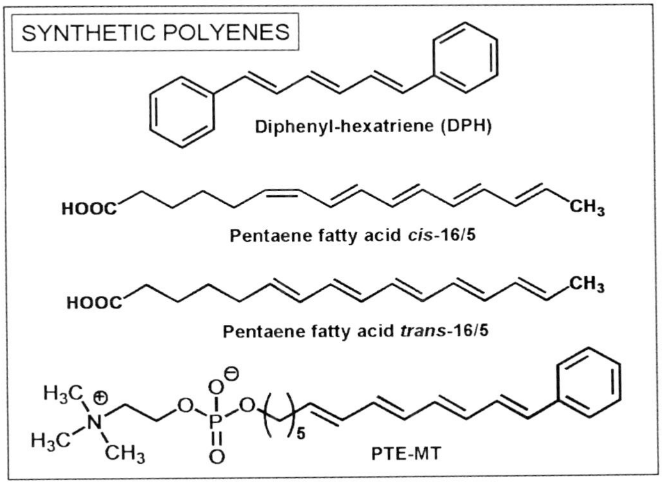

Fig. (10.16). Chemical structure of synthetic polyenes: DPH, two pentaene fatty acid chains (*cis* or *trans* configuration (c, t), followed by the number of C atoms and double bonds), and PTE-MT (all *trans*-1-phenyltrideca-6,8,10,12-tetraenyl-phosphocholine).

Therefore, fluorescent polyene derivatives of fat acids (polyene-lipids containing five conjugated double bonds) have been proposed as a new type of lipid tags. Polyene-lipids exhibit a unique structural similarity to natural lipids, resulting in minimal effects on the lipid properties. Ceramides and sphingomyelins labeled with polyene fatty acids have been successfully applied to visualize lipids in living cells [64]. The pentaene fatty acids c16:5 and t16:5 (Fig. **10.16**) were used as fluorescent labels to study the biosynthesis and storage of triglycerides, which accumulate into lipid droplets [65].

In this context and using the polyene design, a fluorescent ω-phenyltetraene analogue of the ether phospholipid edelfosine has been synthesized, showing λexc: 360 nm and λexc: 442 nm [66]. This approach is particularly interesting, since edelfosine is a synthetic antitumoral lipid that accumulates selectively in cancer cells, and therefore its fluorescent derivative is useful for the labeling and differentiation between normal and tumor cells [66]. Likewise, fluorescent phenylpolyene analogues (*i.e.* PTE-MT) of the antitumoral and leishmanicidal drug miltefosine (hexadecylphosphocholine) (Fig. **10.16**) are biologically active and induce homogeneous blue emission within *Leishmania* parasites [67].

CONFLICT OF INTEREST

The author (editor) declares no conflict of interest, financial or otherwise.

ACKNOWLEDGEMENTS

This work was supported by a grant from the Ministerio de Economía y Competitividad (CTQ2013-48767-C3-3-R), Spain.

REFERENCES

[1] Green FJ. The Sigma-Aldrich Handbook of Stains, Dyes and Indicators. Milwaukee, Wisconsin: Aldrich Chemical Company 1990.

[2] Horobin RW, Kiernan JA, Eds. Conn's Biological Stains A Handbook of Dyes, Stains and Fluorochromes for Use in Biology and Medicine. 10th ed., Oxford, UK: BIOS Scientific Publishers 2002.

[3] Horobin RW, Stockert JC. Uptake and localization mechanisms of fluorescent and colored lipid probes. 1. Physicochemistry of probe uptake and localization, and the use of QSAR models for selectivity prediction. Biotech Histochem 2011; 86(6): 379-93. [http://dx.doi.org/10.3109/10520295.2010.515489] [PMID: 20818918]

[4] Sklar LA, Hudson BS, Simoni RD. Conjugated polyene fatty acids as fluorescent probes: synthetic phospholipid membrane studies. Biochemistry 1977; 16(5): 819-28. [http://dx.doi.org/10.1021/bi00624a002] [PMID: 843518]

[5] Sklar LA, Dratz EA. Analysis of membrane bilayer asymmetry using parinaric acid fluorescent probes. FEBS Lett 1980; 118(2): 308-10. [http://dx.doi.org/10.1016/0014-5793(80)80245-6] [PMID: 7418878]

[6] Schroeder F. Fluorescent sterols: probe molecules of membrane structure and function. Prog Lipid Res 1984; 23(2): 97-113. [http://dx.doi.org/10.1016/0163-7827(84)90009-2] [PMID: 6387716]

[7] Sklar LA, Hudson BS, Simoni RD. Conjugated polyene fatty acids as fluorescent probes: binding to

bovine serum albumin. Biochemistry 1977; 16(23): 5100-8.
[http://dx.doi.org/10.1021/bi00642a024] [PMID: 911814]

[8] Tsuzuki T, Tokuyama Y, Igarashi M, Miyazawa T. Tumor growth suppression by α-eleostearic acid, a linolenic acid isomer with a conjugated triene system, *via* lipid peroxidation. Carcinogenesis 2004; 25(8): 1417-25.
[http://dx.doi.org/10.1093/carcin/bgh109] [PMID: 14963014]

[9] Bayliss-High O. Lipid Histochemistry. Oxford: Oxford University Press 1984.

[10] Kiernan JA. Histological and Histochemical Methods: Theory and Practice. Oxford, New York: Pergamon Press 1990.

[11] Cañete M, Hazen MJ, Stockert JC. Nile blue sulfate staining for demonstration of lipids in fluorescence microscopy. Acta Histochem Cytochem 1983; 16: 286-8.
[http://dx.doi.org/10.1267/ahc.16.286]

[12] Reaven E, Tsai L, Azhar S. Intracellular events in the "selective" transport of lipoprotein-derived cholesteryl esters. J Biol Chem 1996; 271(27): 16208-17.
[http://dx.doi.org/10.1074/jbc.271.27.16208] [PMID: 8663101]

[13] Stockert JC, Abasolo MI, Blázquez-Castro A, Horobin RW, Revilla M, Lombardo DM. Selective labeling of lipid droplets in aldehyde fixed cell monolayers by lipophilic fluorochromes. Biotech Histochem 2010; 85(5): 277-83.
[http://dx.doi.org/10.3109/10520290903196183] [PMID: 20560872]

[14] Suzuki M, Shinohara Y, Fujimoto T. Histochemical detection of lipid droplets in cultured cells. Methods Mol Biol 2013; 931: 483-91.
[http://dx.doi.org/10.1007/978-1-62703-056-4_25] [PMID: 23027019]

[15] Fujimoto T, Ohsaki Y, Cheng J, Suzuki M, Shinohara Y. Lipid droplets: a classic organelle with new outfits. Histochem Cell Biol 2008; 130(2): 263-79.
[http://dx.doi.org/10.1007/s00418-008-0449-0] [PMID: 18546013]

[16] Cañete M, Juarranz A, López-Nieva P, Alonso-Torcal C, Villanueva A, Stockert JC. Fixation and permanent mounting of fluorescent probes after vital labelling of cultured cells. Acta Histochem 2001; 103(2): 117-26.
[http://dx.doi.org/10.1078/0065-1281-00594] [PMID: 11368093]

[17] Lawton JR. An investigation of the fixation and staining of lipids by a combination of malachite green or other triphenylmethane dyes with glutaraldehyde. J Microsc 1989; 154(Pt 1): 83-92.
[http://dx.doi.org/10.1111/j.1365-2818.1989.tb00570.x] [PMID: 2473211]

[18] Horobin RW. A numerical approach to understanding fixative action: being a re-analysis of the fixation of lipids by the dye-glutaraldehyde system. J Microsc 1989; 154: 93-6.
[http://dx.doi.org/10.1111/j.1365-2818.1989.tb00571.x]

[19] Hayat MA. Fixation for Electron Microscopy. New York, London: Academic Press 1981.

[20] Chiffelle TL, Putt FA. Propylene and ethylene glycol as solvents for Sudan IV and Sudan black B. Stain Technol 1951; 26(1): 51-6.
[http://dx.doi.org/10.3109/10520295109113178] [PMID: 14809507]

[21] Koopman R, Schaart G, Hesselink MK. Optimisation of oil red O staining permits combination with immunofluorescence and automated quantification of lipids. Histochem Cell Biol 2001; 116(1): 63-8.
[PMID: 11479724]

[22] Carson FL. Histotechnology A Self-Instructional Text. Chicago: American Society of Clinical Pathologists Press 1993.

[23] Berg NO. A histological study of masked lipids; stainability, distribution and functional variations. Acta Pathol Microbiol Scand Suppl 1951; 90 (Suppl.): 1-192.
[PMID: 14894234]

[24] Allison AC, Mallucci L. Uptake of hydrocarbon carcinogens by lysosomes. Nature 1964; 203: 1024-7. [http://dx.doi.org/10.1038/2031024b0] [PMID: 14223066]

[25] Gahan PB. Plant Histochemistry and Cytochemistry An Introduction. London, New York: Academic Press 1984.

[26] Lillie RD. Conn's Biological Stains. 9th ed., Baltimore: Williams and Wilkins 1977.

[27] Donovick PJ. A metachromatic stain for neural tissue. Stain Technol 1974; 49(1): 49-51. [http://dx.doi.org/10.3109/10520297409116937] [PMID: 4131625]

[28] Eichler VB, Taylor PW. A new metachromatic stain technique for paraffin-embedded neural tissue using thionin. J Microsc 1976; 108(1): 97-9. [http://dx.doi.org/10.1111/j.1365-2818.1976.tb01083.x] [PMID: 64611]

[29] Bozzo L, Vidal BdeC. Nile blue sulphate and brilliant cresyl blue in fluorescence methods for the histochemical demonstration of lipids. Ann Histochim 1968; 13(3): 177-83. [PMID: 5717661]

[30] Gupta AK, Ed. The Complete Book on Natural Dyes and Pigments National Institute of Industrial Research. Delhi: Asia Pacific Business Press 2005.

[31] Fowler SD, Greenspan P. Application of Nile red, a fluorescent hydrophobic probe, for the detection of neutral lipid deposits in tissue sections: comparison with oil red O. J Histochem Cytochem 1985; 33(8): 833-6. [http://dx.doi.org/10.1177/33.8.4020099] [PMID: 4020099]

[32] Greenspan P, Mayer EP, Fowler SD. Nile red: a selective fluorescent stain for intracellular lipid droplets. J Cell Biol 1985; 100(3): 965-73. [http://dx.doi.org/10.1083/jcb.100.3.965] [PMID: 3972906]

[33] Alba FJ, Bermudez A, Bartolome S, Daban JR. Detection of five nanograms of protein by two-minute nile red staining of unfixed SDS gels. Biotechniques 1996; 21(4): 625-6. [PMID: 8891212]

[34] Diaz G, Melis M, Batetta B, Angius F, Falchi AM. Hydrophobic characterization of intracellular lipids *in situ* by Nile Red red/yellow emission ratio. Micron 2008; 39(7): 819-24. [http://dx.doi.org/10.1016/j.micron.2008.01.001] [PMID: 18329888]

[35] Baker JR. Principles of Biological Microtechnique: a Study of Fixation and Dyeing. London: Methuen 1958.

[36] Wang H, Wang H, Xiong W, *et al.* Evaluation on the phagocytosis of apoptotic spermatogenic cells by Sertoli cells *in vitro* through detecting lipid droplet formation by Oil Red O staining. Reproduction 2006; 132(3): 485-92. [http://dx.doi.org/10.1530/rep.1.01213] [PMID: 16940289]

[37] Udenfriend S. Fluorescence Assay in Biology and Medicine. New York, London: Academic Press 1962.

[38] Gurr E. Synthetic Dyes in Biology, Medicine and Chemistry. New York, London: Academic Press 1971.

[39] Mikes V, Dadák V. Berberine derivatives as cationic fluorescent probes for the investigation of the energized state of mitochondria. Biochim Biophys Acta 1983; 723(2): 231-9. [http://dx.doi.org/10.1016/0005-2728(83)90122-6] [PMID: 6849903]

[40] Hahn FE, Ciak J. Berberine.Antibiotics. Berlin: Springer 1975; Vol. 3: pp. 577-84.

[41] Budavari S, Ed. The Merck Index. An Encyclopedia of Chemicals, Drugs and Biologicals. 11th.. Rahway, NJ: Merck and Co. 1989.

[42] Ohsaki Y, Shinohara Y, Suzuki M, Fujimoto T. A pitfall in using BODIPY dyes to label lipid droplets for fluorescence microscopy. Histochem Cell Biol 2010; 133(4): 477-80.

[http://dx.doi.org/10.1007/s00418-010-0678-x] [PMID: 20191286]

[43] Spandl J, White DJ, Peychl J, Thiele C. Live cell multicolor imaging of lipid droplets with a new dye, LD540. Traffic 2009; 10(11): 1579-84.
[http://dx.doi.org/10.1111/j.1600-0854.2009.00980.x] [PMID: 19765264]

[44] Pagano RE, Sepanski MA, Martin OC. Molecular trapping of a fluorescent ceramide analogue at the Golgi apparatus of fixed cells: interaction with endogenous lipids provides a trans-Golgi marker for both light and electron microscopy. J Cell Biol 1989; 109(5): 2067-79.
[http://dx.doi.org/10.1083/jcb.109.5.2067] [PMID: 2478562]

[45] Fujimura H, Dekura E, Kurabe M, Shimazu N, Koitabashi M, Toriumi W. Cell-based fluorescence assay for evaluation of new-drugs potential for phospholipidosis in an early stage of drug development. Exp Toxicol Pathol 2007; 58(6): 375-82.
[http://dx.doi.org/10.1016/j.etp.2007.01.004] [PMID: 17408938]

[46] Bhandari N, Figueroa DJ, Lawrence JW, Gerhold DL. Phospholipidosis assay in HepG2 cells and rat or rhesus hepatocytes using phospholipid probe NBD-PE. Assay Drug Dev Technol 2008; 6(3): 407-19.
[http://dx.doi.org/10.1089/adt.2007.119] [PMID: 18537465]

[47] Hope-Roberts M, Wainwright M, Horobin RW. A nitrobenzofuran-conjugated phosphatidylcholine (C_{12}-NBD-PC) as a stain for membrane lamellae for both microscopic imaging and spectrofluorimetry. Biotech Histochem 2008; 83(1): 25-8.
[http://dx.doi.org/10.1080/10520290801886224] [PMID: 18568674]

[48] Diaz G, Melis M, Musin A, Piludu M, Piras M, Falchi AM. Localization of MTT formazan in lipid droplets. An alternative hypothesis about the nature of formazan granules and aggregates. Eur J Histochem 2007; 51(3): 213-8.
[PMID: 17921117]

[49] Stockert JC, Blázquez-Castro A, Cañete M, Horobin RW, Villanueva A. MTT assay for cell viability: Intracellular localization of the formazan product is in lipid droplets. Acta Histochem 2012; 114(8): 785-96.
[http://dx.doi.org/10.1016/j.acthis.2012.01.006] [PMID: 22341561]

[50] Radda GK, Smith DS. Retinol: A fluorescent probe for membrane lipids. FEBS Lett 1970; 9(5): 287-9.
[http://dx.doi.org/10.1016/0014-5793(70)80379-9] [PMID: 11947694]

[51] Trigoso CI, Stockert JC. Fluorescence of the natural dye saffron: selective reaction with eosinophil leucocyte granules. Histochem Cell Biol 1995; 104(1): 75-7.
[http://dx.doi.org/10.1007/BF01464789] [PMID: 7584563]

[52] Clauditz A, Resch A, Wieland KP, Peschel A, Götz F. Staphyloxanthin plays a role in the fitness of *Staphylococcus aureus* and its ability to cope with oxidative stress. Infect Immun 2006; 74(8): 4950-3.
[http://dx.doi.org/10.1128/IAI.00204-06] [PMID: 16861688]

[53] Liu CI, Liu GY, Song Y, *et al.* A cholesterol biosynthesis inhibitor blocks *Staphylococcus aureus* virulence. Science 2008; 319(5868): 1391-4.
[http://dx.doi.org/10.1126/science.1153018] [PMID: 18276850]

[54] Börnig H, Geyer G. Staining of cholesterol with the fluorescent antibiotic "filipin". Acta Histochem 1974; 50(1): 110-5.
[PMID: 4140671]

[55] Kruth HS. Filipin-positive, oil red O-negative particles in atherosclerotic lesions induced by cholesterol feeding. Lab Invest 1984; 50(1): 87-93.
[PMID: 6694353]

[56] Van Leeuwen MR, Smant W, de Boer W, Dijksterhuis J. Filipin is a reliable *in situ* marker of ergosterol in the plasma membrane of germinating conidia (spores) of *Penicillium discolor* and stains intensively at the site of germ tube formation. J Microbiol Methods 2008; 74(2-3): 64-73.

[http://dx.doi.org/10.1016/j.mimet.2008.04.001] [PMID: 18485505]

[57] Bittman R, Chen WC, Anderson OR. Interaction of filipin 3 and amphotericin B with lecithin-sterol vesicles and cellular membranes. Spectral and electron microscope studies. Biochemistry 1974; 13(7): 1364-73.
[http://dx.doi.org/10.1021/bi00704a009] [PMID: 4132123]

[58] de Kruijff B, Demel RA. Polyene antibiotic-sterol interactions in membranes of *Acholeplasma laidlawii* cells and lecithin liposomes. 3. Molecular structure of the polyene antibiotic-cholesterol complexes. Biochim Biophys Acta 1974; 339(1): 57-70.
[http://dx.doi.org/10.1016/0005-2736(74)90332-0] [PMID: 4546885]

[59] Lopes SC, Goormaghtigh E, Cabral BJ, Castanho MA. Filipin orientation revealed by linear dichroism. Implication for a model of action. J Am Chem Soc 2004; 126(17): 5396-402.
[http://dx.doi.org/10.1021/ja031782+] [PMID: 15113211]

[60] Gray KC, Palacios DS, Dailey I, *et al.* Amphotericin primarily kills yeast by simply binding ergosterol. Proc Natl Acad Sci USA 2012; 109(7): 2234-9.
[http://dx.doi.org/10.1073/pnas.1117280109] [PMID: 22308411]

[61] van Ginkel G. Diphenyl hexatrienes as fluorescent membrane probes. Appl Fluorescence Technol 1989; 1(6): 1-8.

[62] Prendergast FG, Haugland RP, Callahan PJ. 1-[4-(Trimethylamino)phenyl]-6-phenylhexa-1,3,5-triene: synthesis, fluorescence properties, and use as a fluorescence probe of lipid bilayers. Biochemistry 1981; 20(26): 7333-8.
[http://dx.doi.org/10.1021/bi00529a002] [PMID: 7326228]

[63] Kessel D. Characterization of photodynamic and sonodynamic cytotoxicity by fluorescent probes. SPIE 1993; 1881: pp. 275-84.

[64] Kuerschner L, Ejsing CS, Ekroos K, Shevchenko A, Anderson KI, Thiele C. Polyene-lipids: a new tool to image lipids. Nat Methods 2005; 2(1): 39-45.
[http://dx.doi.org/10.1038/nmeth728] [PMID: 15782159]

[65] Kuerschner L, Moessinger C, Thiele C. Imaging of lipid biosynthesis: how a neutral lipid enters lipid droplets. Traffic 2008; 9(3): 338-52.
[http://dx.doi.org/10.1111/j.1600-0854.2007.00689.x] [PMID: 18088320]

[66] Quesada E, Delgado J, Gajate C, Mollinedo F, Acuña AU, Amat-Guerri F. Fluorescent phenylpolyene analogues of the ether phospholipid edelfosine for the selective labeling of cancer cells. J Med Chem 2004; 47(22): 5333-5.
[http://dx.doi.org/10.1021/jm049808a] [PMID: 15481970]

[67] Saugar JM, Delgado J, Hornillos V, *et al.* Synthesis and biological evaluation of fluorescent leishmanicidal analogues of hexadecylphosphocholine (miltefosine) as probes of antiparasite mechanisms. J Med Chem 2007; 50(24): 5994-6003.
[http://dx.doi.org/10.1021/jm070595+] [PMID: 17973359]

CHAPTER 11

Polysaccharides

Abstract: Although the term carbohydrate is a general concept and applies to all the sugars and their derivatives, the name polysaccharides will be preferred as they are the common macromolecular substrates to be recognized in microscopy. Carbohydrates may be classified on account of their chemistry, histochemical responses in diagnostic pathology, or distribution and function in nature [1]. Some issues are central to this chapter: the use of bisazo dyes and fluorescent brighteners, chlorotriazines and PAS reaction, ionic binding and metachromatic reactions. Staining and fluorescence methods applied for histochemistry of polysaccharides have been reviewed [1-4], and some of them, in special glycoproteins (*i.e.* Congo red for amyloid), will be described in Chapter 13.11.3. More specific procedures involving the use of lectins can be found in Chapter 4.5.2.

Keywords: Acid glycosaminoglycans, Acridines, Aryl-oxazoles, Boronic acids, Brightening agents, Carbocyanines, Cell walls, Chloro-triazines, Congo red, Curcumin, Lignin, PAS reaction, Ru(II)-bipyridine, Sirofluor, Trypan blue.

1. TYPES OF POLYSACCHARIDES AND HISTOCHEMICAL REACTIONS

As a logical consequence of carbohydrate diversity, a rather confusing terminology resulted from different classifications and nomenclature rules. The most common names will be used here, attempting to simplify the carbohydrate description for microscopical studies. The most frequent monosaccharide units in the biomedical field are hexoses (*i.e.* glucose, galactose, mannose, fucose), and pentoses (*i.e.* ribose, xylose). In addition to polysaccharides, carbohydrates are present in glycolipids, nucleic acids and proteins. Proteoglycans are long polysaccharide chains covalently attached to a small protein core, whereas glycoproteins are proteins bearing several covalently bound oligosaccharide chains. Linear or branched neutral polysaccharides include glucose polymers (glycogen, starch, cellulose, chitin, callose, *etc.*). The structure of cellulose and chitin chains is shown in Fig. (**11.1**).

Other polysaccharides have amino sugars and hexose derivatives such as uronic acids or sulfate esters (chitin, pectin, heparin, chondroitin-sulfates, dermatan-sulfates, hyaluronic acid, amyloid). They are currently referred to as acid

Juan Carlos Stockert & Alfonso Blázquez-Castro

glycosaminoglycans (GAGs) and are often bound to polypeptides or proteins. The names acid (anionic) GAGs and acid polysaccharides are often employed to denote the same type of polymers, containing hexose or amino-hexose units and mild or strong anionic groups.

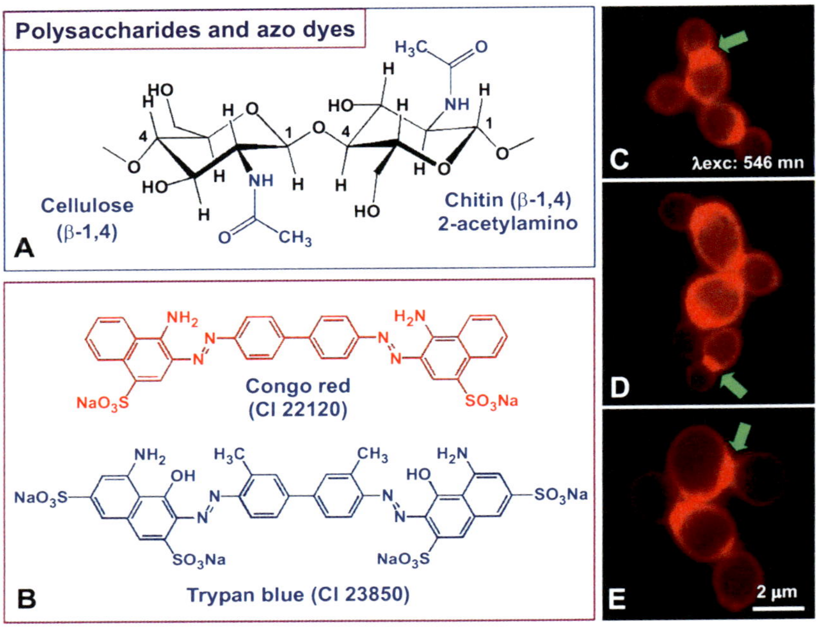

Fig. (11.1). Chemical structures of cellulose and chitin (**A**), and the bisazo dyes Congo red and trypan blue (**B**). In cellulose, the 2-acetylamino group is replaced for a -OH group. **C**-**E**: *Candida famata* cells stained 5 min with 0.2% trypan blue in PBS, showing the red fluorescence of the chitin wall and thickening of budding scars (arrows) (courtesy of A. Juarranz).

From the histochemical point of view, polysaccharides can be classified as neutral or acid polymers, which are differentially stained by a number of selective reactions (linear bisazo dyes and brightener agents, metachromatic dyes, PAS procedure, labeled lectins and antibodies). Hydroxyls are the most abundant groups in neutral polysaccharides, and they are targets for covalent binding of reactive chloro-triazines and boronic acids. Anionic polysaccharides have also carboxyl, sulfate and acetylamino groups. All these groups in the biopolymer can be used for reacting with dyes or reagents to give staining or fluorescent reactions, binding mechanisms being H bonding, ionic forces, dye aggregation, or covalent bonds in the case of chromo- and fluorogenic reactions.

In order to better preserve acid polysaccharides, cationic dyes and surfactants (*i.e.*, cetylpyridinium chloride, CPC) [2, 5] can be added to the fixative to precipitate and render them insoluble, thus improving microscopical morphology. However, caution is necessary in the histochemical analysis because after this procedure the availability of anionic groups will be reduced or abolished.

2. BISAZO DYES AND BRIGHTENING AGENTS

Classical bisazo dyes such as Congo red, trypan blue, and its isomeric Evans blue are linear chromophores that bind to polysaccharides and proteins. Industrial applications include dyeing of cotton, paper, leather, silk and wool products. Trypan blue and Evans blue are also employed in dye-exclusion assays for cell viability [4, 6] (see Chapter 16.1.1). Microscopic fluorescence is strongly enhanced in polysaccharides stained by linear and flexible fluorochromes due to the high rigidity imposed upon binding, which greatly increases the emission yield.

Although direct staining with these dyes are commonly applied for bright-field microscopy, their interesting emission properties (see Chapters 5 and 15) can be profited by using them in fluorescence microscopy to reveal selectively the chitin wall in budding yeast (Fig. **11.1**), and cellulose in plant tissues (Fig. **11.2**). Proper alignment of H donors and acceptors for H bonding between polysaccharide substrates and these direct dyes, as well as van der Waals forces (dipolar interactions) between $O^{\delta-}$ and $H^{\delta+}$ atoms, are responsible for selective binding.

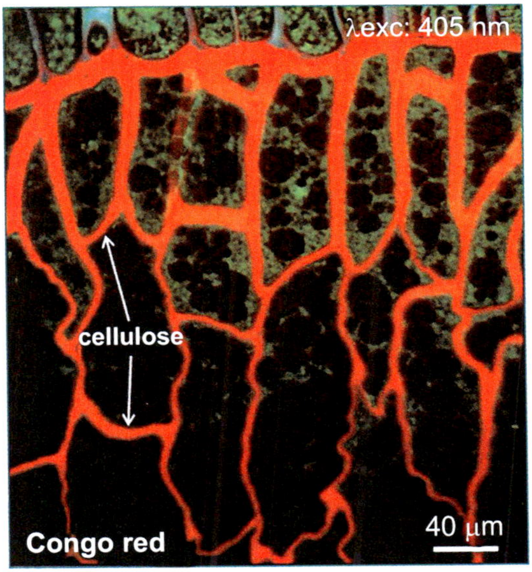

Fig. (11.2). Selective fluorescence of cellulose. Section of an oat grain stained by Congo red showing the fluorescence of cellulose walls (β-glycans) under blue exciting light. Greenish emission corresponds to autofluorescence.

Other linear dyes are also useful for visualizing neutral polysaccharides. The natural dye, curcumin, has scarcely been applied for microscopical staining, and its fluorescence reaction has been overlooked [7, 8]. Plant tissues fixed in glutaraldehyde, embedded in Epon, and stained with curcumin shows green fluorescence in chromatin, cellulose and callose walls, but no emission is found in starch granules. When glutaraldehyde fixation is followed by osmium tetroxide, the callose walls are selectively visualized (Fig. **11.3**). On account of the abundance of -OH groups, some anthraquinones dyes such as carmine bind selectively to glycogen [9], but this reaction has been overlooked in fluorescence microscopy.

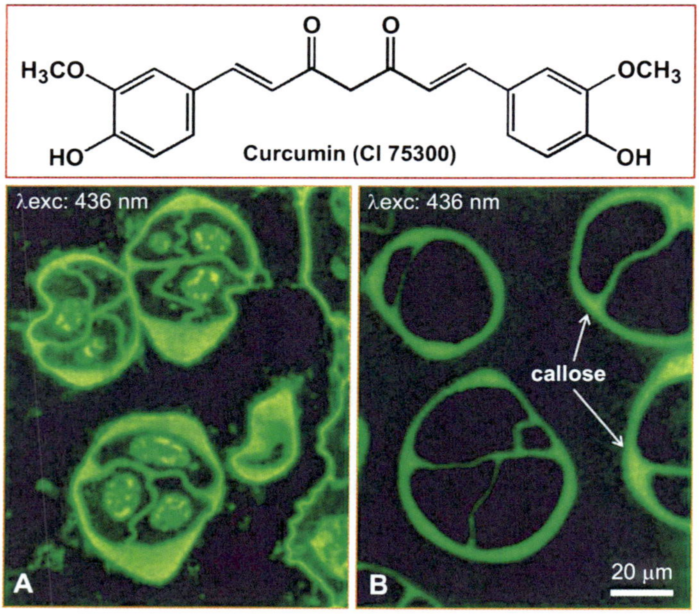

Fig. (11.3). Selective fluorescence of callose walls. Semi-thin Epon section of *Scilla peruviana* anthers after fixation with glutaraldehyde alone (**A**) and glutaraldehyde-osmium tetroxide (**B**), stained with curcumin (0.1 mg/mL in 5% ethanol, 30 min). In early tetrads chromatin fluorescence is abolished after osmium tetroxide fixation (**B**) but the fluorescence of callose walls remains unaffected.

Brightening agents or fluorescent brighteners (FBs) [10] are colorless compounds that fluoresce in blue under UV excitation. The development and industrial use of FBs (mainly based on linear stilbene derivatives that bind to cellulose, as well as on coumarin compounds) forwarded their applications in fluorescence microscopy to reveal polysaccharides. They are currently used to visualize cellulose and callose in plant tissues [6], FB 9 and FB 30 being the most known (Fig. **11.4**). The latter has been also employed for detection of collagen [11].

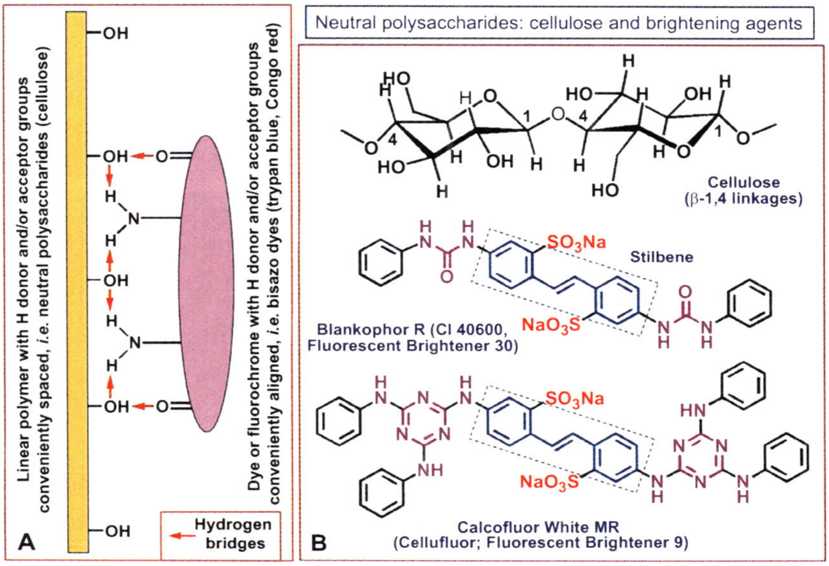

Fig. (11.4). **A**: Binding mechanism of direct dyes, showing hydrogen bonding patterns. **B**: Chemical structure of cellulose and stilbene (blue) brightening agents, with ureido and triamino-triazine groups (violet), and sodium sulfonate (red).

Calcofluor white MR (FB 9) has found application in the detection of chitin-rich scars of yeast [12] and for diagnosis of *Pneumocystis carinii* [13]. Since FB 9 is no longer produced, the analogous FB 28 (Calcofluor white M2R, CI 40622) might be a suitable substitute. Two amino phenyl groups at both ends of FB 30 are replaced for two $-N(CH_2CH_2OH)_2$ groups in FB 28. Other stilbene FBs are tinopal CBS (FB 351) and uvitex 2B (FB 362). In addition to visualization of cellulose and chitin walls, FBs are used to demonstrate protozoal and worm parasites, parasitic fungi of cereals, *etc.*

Important FB agents are now industrially applied on account that they make more white cellulosic, silk, and polyacrylamide fabrics. FBs are water soluble, and have fastness to light, chlorine and washing. Likewise, incorporation of FBs into paper, detergents, cosmetics, and soaps accomplish the same end. The rationale of their use in textiles is as follows (Fig. **11.5**): (1) With time, oxidation due to washing processes generates double bonds that absorb violet light giving a yellow color, which is considered to derive from dirtiness; (2) used as additives, UV- and violet-absorbing FBs fluoresce in bright blue, which corrects the "dirty" yellow color and produces the expected white of "cleanness".

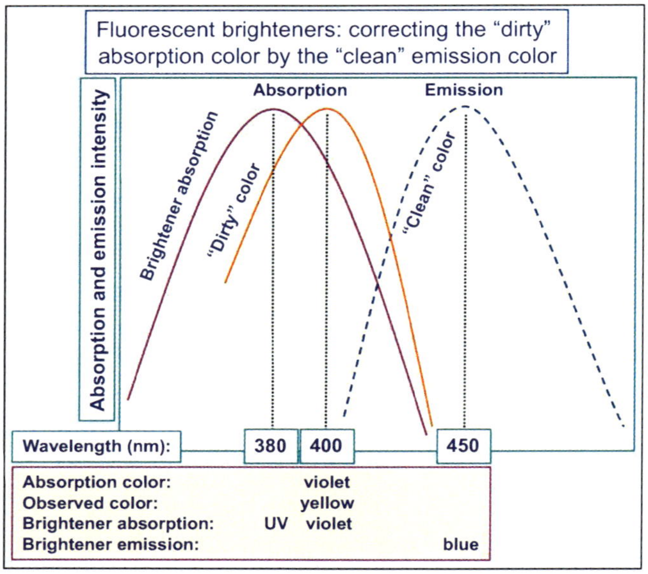

Fig. (11.5). Mechanism of spectral correction of fabrics achieved by means of fluorescent brighteners, showing ideal curves of absorption and emission.

Likewise, 4,4'-bis(2-benzoxazolyl)-stilbene is a hydrophobic FB for incorporation to plastics and should be useful as a fat fluorochrome. Interestingly, natural stilbenes such as rhapontin, from *Rheum raphonticum*, and the antioxidant polydatin (resveratrol-3-glucoside, from *Polygonum cuspidatum*) have a bright blue fluorescence that should be useful as natural FB agents. It is worth noting that some stilbene derivatives (*i.e.* diethyl-stilbestrol) have estrogenic and carcinogenic activity.

The triphenylmethane dye, aniline blue (CI 42780; cotton blue, methyl blue), is the most widely used fluorochrome in botany, inducing green fluorescence in sieve plates, cell walls and yeasts. These structures contain the β-1,3-glucan known as callose [14, 15]. The dye also binds strongly to primary cell walls and lignified walls, but in this latter case the unspecific binding can be prevented by prior staining with toluidine blue. Today, it is known that fluorescence properties of aniline blue are due to the occurrence of a contaminant dye or impurity, which was isolated and named sirofluor (4,4'-[carbonylbis{benzene-4,1-diyl}bis-{imino}] bisbenzene sulfonate disodium salt) [16, 17].

Sirofluor is a yellow and anionic vital dye that binds selectively to callose, giving bright green emission under blue excitation. At present, aniline blue still remains useful for callose visualization in fluorescence microscopy (Fig. **11.6**). Applied on blood smears, aniline blue induces blue fluorescence in eosinophil granules under

UV light. An intriguing explanation was early suggested based on the reduction of the non-fluorescent dye to a fluorescent leuco-compound by -CH=CH- groups from unsaturated lipids [18]. However, the green emission in the absence of reducing substrates is not due to this mechanism but to the occurrence of the contaminant sirofluor.

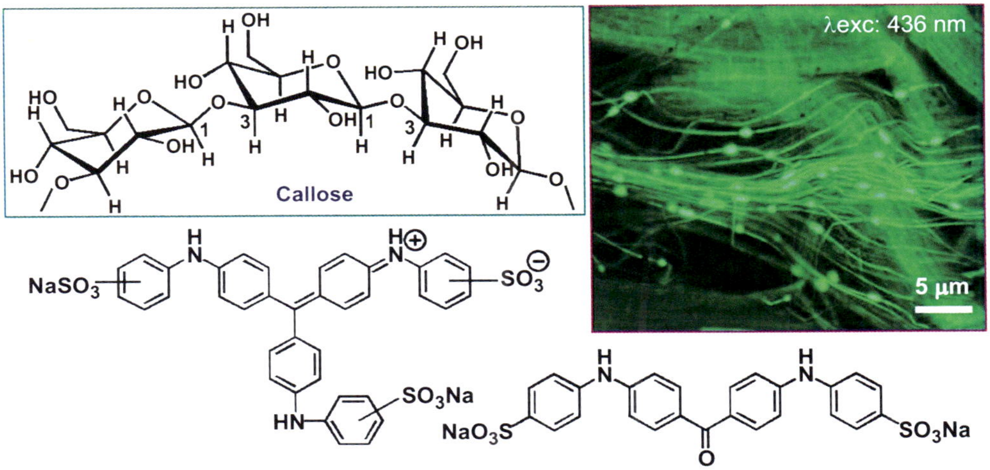

Aniline blue (CI 42780) Sirofluor (fluorescent contaminant)

Fig. (11.6). Selective visualization of callose. Chemical structure of callose, aniline blue, and sirofluor. The picture shows pollen tubes from *Tabebuia pulcherrima* x *T. heptaphylla* stained with aniline blue. Callose walls and thickerings fluoresce in green (courtesy of G.R. Facciuto).

3. REACTIVE DYES AND FLUOROCHROMES

The periodic acid-Schiff (PAS) method is an important example of a chromo- and fluorogenic reaction in polysaccharide histochemistry. In this method, oxidation of glycols by periodic acid (HIO_4) or sodium periodate ($NaIO_4$) originates aldehyde groups that are then revealed by the Schiff's reagent (Fig. **11.7**). The classical Schiff's reagent (leuco-fuchsine) is a combination of basic fuchsine and sulfurous acid, and is traditionally used for the detection of aldehydes (see Chapter 12.2.1). Although two aldehyde groups from the same glucose units are often shown to be bound to the Schiff's reagent, stereochemical studies reveal that this conformation is unlikely.

A PAS positive reaction occurs when leuco-fuchsine generates a recolored (reddish-violet) covalently-bound product, which also shows red fluorescence under green excitation [19]. When observed in bright-field microscopy, adequate nuclear counterstaining can be obtained by using methyl green. However, under both UV and blue light excitation demethylation of this dye occurs with formation of the red-fluorescing derivative crystal violet, which makes unsuitable methyl green counterstaining.

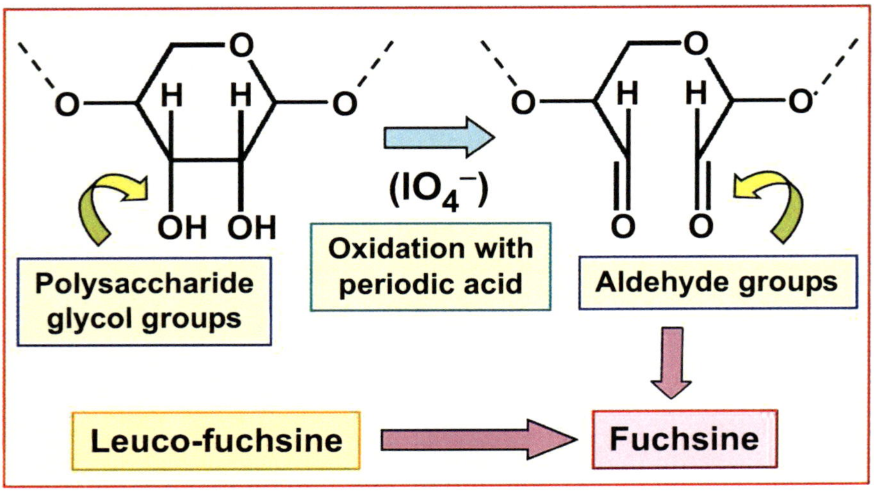

Fig. (11.7). Chromo-and fluorogenic PAS reactions for detection of polysaccharides. After mild oxidation of glycols with periodic acid or sodium periodate, aldehyde groups are demonstrated by recoloration of the Schiff's reagent (leuco-fuchsine).

On the other hand, there are other fluorescent Schiff-type ("pseudo-Schiff") reagents that are combinations of sulfurous acid and a cationic fluorochrome (see Chapter 12.2.1), the most suitable being auramine O and acriflavine, yielding a fluorescent product with aldehyde groups. Other reagents that give fluorescent products are compounds carrying hydrazine (-NH-NH_2) or hydrazide (-CO-NH-NH_2) groups. The best known are dansyl-hydrazine, salicyl-hydrazide, and 2-hydroxy-3-naphthoic acid hydrazide [1]. After Zn^{2+} or Al^{3+} chelation the product of aldehydes with the two latter reagents should be fluorescent.

Free aldehyde groups are largely absent from tissue components, although some exceptions are elastin, lignin, air-oxidized unsaturated lipids, or after glutaraldehyde fixation. On account of their content of vicinal -OH groups, glycolipids such as cerebrosides and gangliosides, as well as phosphatidyl-inositols, also give staining and fluorescence PAS positive reactions [1]. Likewise, periodic acid generates aldehydes from sphingosine, sialic acid residues and -CH=CH- groups. The same occurs after OsO_4 fixation of lipids and nucleic acids, in which -CH=CH- groups of unsaturated fatty acids and pyrimidines are converted to glycols that yield aldehydes upon mild oxidation (see Chapters 7.9 and 12.2.2).

In addition to the formation of covalent linkages between basic fuchsine and polysaccharides following the PAS reaction, there are also other reactive dye mechanisms with histochemical applications. In all these cases, dyes or fluorochromes covalently bound to substrates strongly increase their resistance to possible extraction by washing. Numerous reactive dyes have a chloro-triazine group that is able to attach to polysaccharide hydroxyls (Fig. **11.8**). In this way, poly-hydroxy or poly-amine substrates can be visualized in bright-field or fluorescence microscopy. To this respect, the reagent 1-ethoxy-4-(dichlo-o-s-triazinyl)-naphthalene (EDTN) gives a fluorescent product in substrates containing aliphatic -OH groups (λexc: 350 nm, λem: 420 nm) [20].

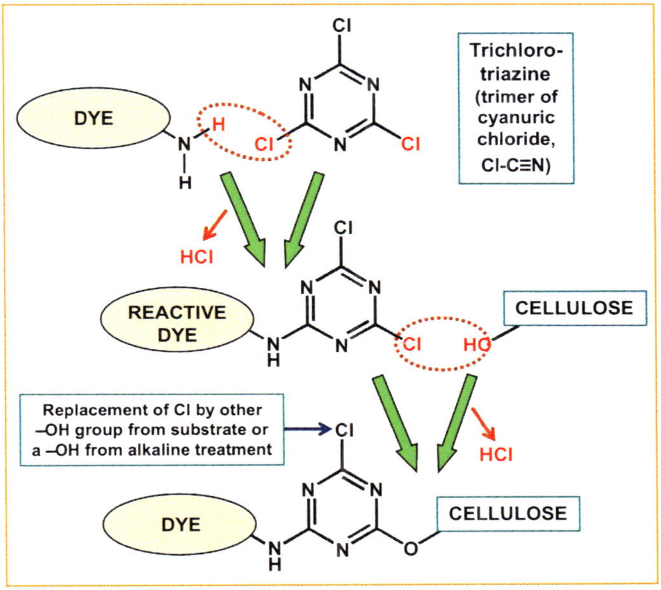

Fig. (11.8). Use of reactive dyes to demonstrate hydroxyl groups from polysaccharides. Trichloro-triazine is first bound to an adequate group of the dye, and then a remaining chloro-triazine group from the reactive dye is bound to the substrate, in all cases with elimination of HCl.

Due to its high affinity for diols, boronic acids have generated much interest as a recognition motif for the synthesis of fluorescent sensors and reporters for carbohydrates. Thus high selectivity for polysaccharide demonstration may be obtained by using boronic esters labeled with fluorescent tags [21, 22], some of which can be used to stain living cells. Carbohydrates with *cis* diols (*i.e.* mannose, gulose, allose, talose, rhamnose, ribose) are visualized with "FluoroBoras" (containing dansyl, pyrene, and quinoline fluorophores), according to the reaction mechanism illustrated in Fig. (**11.9**). Also fluorescent boronolectins can be applied in carbohydrate histochemistry [23].

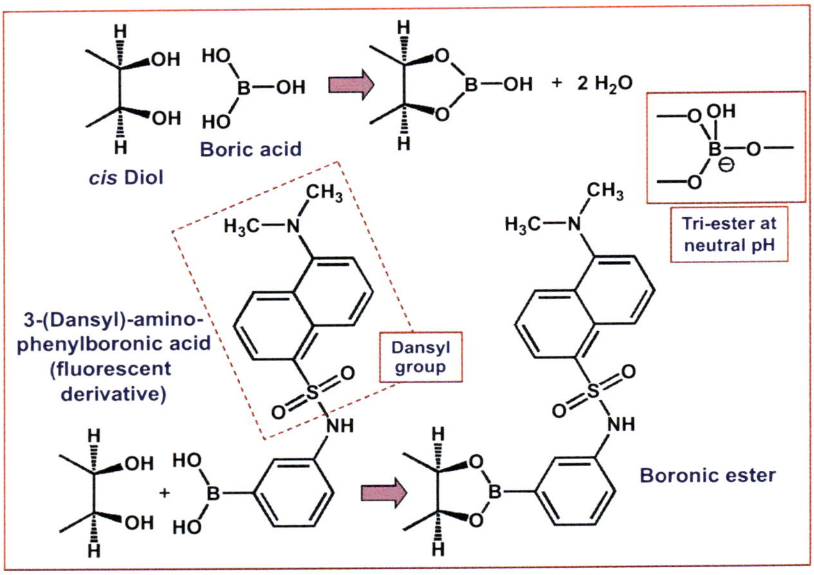

Fig. (11.9). Reaction mechanism of formation of boric and boronic acid esters with glycol groups from polysaccharides. A fluorescent dansyl-boronic acid derivative is illustrated below.

4. CATIONIC DYES

4.1. Acridines

Important staining methods for acid glycosaminoglycans (GAGs) involve the use of cationic organic dyes, fluorochromes, and metal complexes with high affinity for sulfate and carboxylate groups. Cationic dyes give orthochromatic staining and fluorescence reactions with neutral polysaccharides [24], but the most relevant reactions are the metachromatic staining induced by some planar or non-rigid fluorophores on acid GAGs [11, 25, 26]. Thus, acridines (acridine orange, acriflavine, and flavophosphine) have been applied since long ago for detection of anionic polysaccharides. Metachromatic fluorescence of the mucin of goblet cells by acridine orange is a good example of selective metachromatic reaction (Fig. **11.10A**), which contrasts with the orthochromatic emission of chromatin.

Regarding acid GAGs, there are possibilities of differential blocking reactions for specific anionic groups with inorganic cations (see Chapter 8.1). According to Heath [27], solutions of cationic dyes (about 10^{-3} M toluidine blue or neutral red) in 0.1-0.5 M aluminum sulfate induce selective staining of sulfated GAGs. Al^{3+} ions can be also used before the dye: they bind to carboxyl and phosphate groups, leaving sulfates as the only anionic groups free to combine with the cationic dye, resulting in metachromatic staining and fluorescence reactions. An illustrative example of the competitive influence of aluminum cations on the metachromasia of acid GAGs stained with toluidine blue is shown in Fig. (**11.10B**). The same results are obtained by using acridine orange instead thiazine dyes.

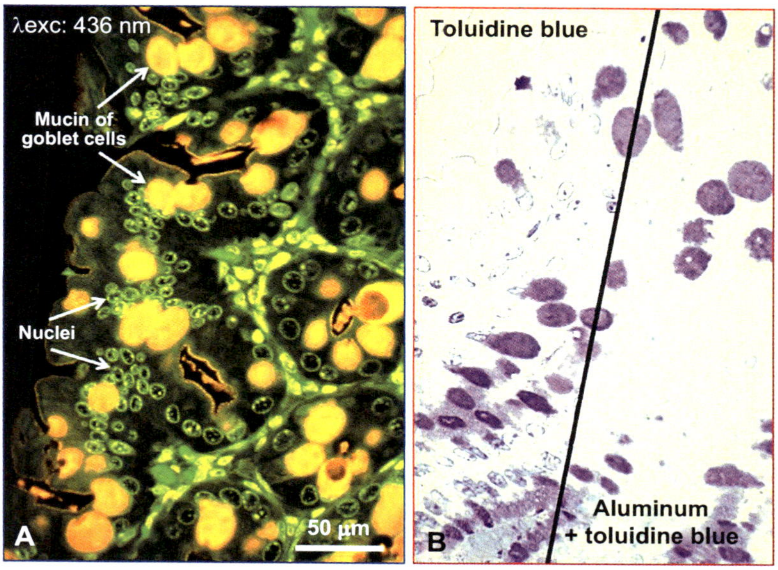

Fig. (11.10). Metachromatic reaction of acid polysaccharides. Mouse large intestine (glutaraldehyde, Epon) stained for 5 min with 10^{-2} M acridine orange (**A**), and treated with 2% alum and then stained with 0.1 mg/mL toluidine blue at pH 5 (**B**), showing orthochromatic (nuclei) and metachromatic (mucin) staining and fluorescence. Al^{+3} ions abolish the orthochromatic staining of nuclei.

On the other hand, the selective reaction of barium ions with sulfate groups allows recognizing sulfated GAGs by formation of a colored complex of Ba^{2+} with sodium rhodizonate [28]. In connection with the content of sulfate residues and regarding the type of polysaccharide substrate, it is important to know the basic structure of a representative anionic GAG (proteoglycan) such as heparin (Fig. **11.11**), which shows tetra-anionic disaccharide units, trisaccharide links (galactose-galactose-xylose), and a polypeptide core $(\text{serine-glycine})_{10-15}$.

Heparin containing mast cell granules are highly metachromatic. Fluorescence metachromasia was found in rat peritoneal mast cell granules with intracellular concentration of acridine orange as low as 5 x 10^{-16} mole per cell [29]. It is important to note that microscopical ortho- and metachromatic fluorescence reactions are related to the fluorochrome concentration. When stained at very low acridine orange concentration, both chromatin and mast cell granules are green. Likewise, at very high dye concentration they appear red. Only at suitable intermediate concentrations these cell components show an appropriate color differentiation (Fig. **11.12**).

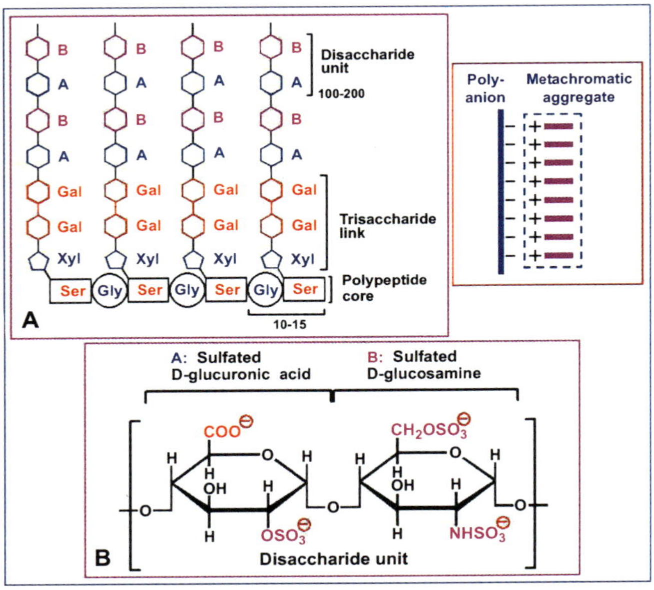

Fig. (11.11). **A**: Chemical structure of the heparin proteoglycan showing disaccharide units, trisaccharide links, and polypeptide core. **B**: Chemical structure of disaccharide units with anionic groups. **C**: Scheme of interactions between cationic dyes and a polyanionic macromolecule.

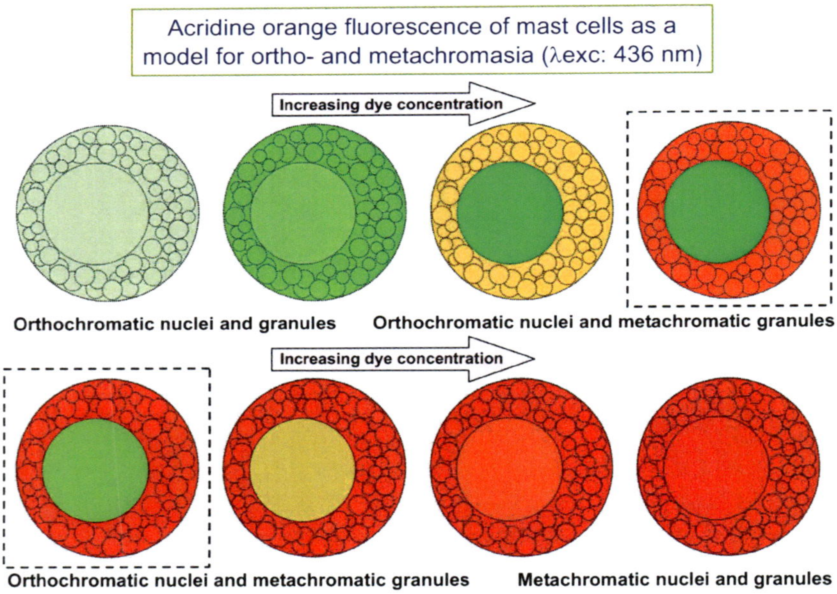

Fig. (11.12). Ortho- and metachromatic gradient shown by mast cells after staining with increasing concentrations of acridine orange. Dashed squares indicate the best differential signal between orthochromatic nuclei and metachromatic granules.

4.2. Carbocyanines, Indoles, Aryl-Oxazoles, and Other Dyes

Not only thiazines and acridines give metachromatic reactions with acid GAGs but also several linear and non-rigid fluorochromes. Although it is a poorly known feature, cationic fluorochromes used for DNA staining can also stain anionic polysaccharides. In this sense, linear cationic carbocyanines such as $DiOC_1(3)$ induces a strong metachromasia in heparin-containing granules of mast cells (Fig. **11.13**).

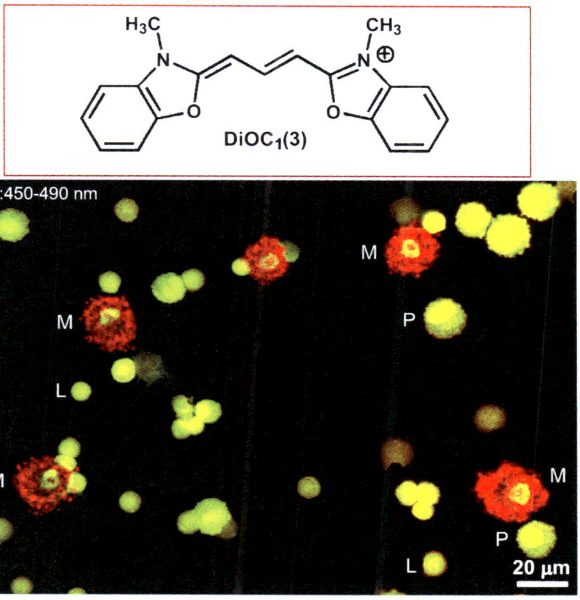

Fig. (11.13). Chemical structure of $DiOC_1(3)$ (3,3'-dimethyl-oxacarbocyanine) and cells from mouse peritoneal fluid fixed in methanol and stained by $DiOC_1(3)$ (20 µg/mL, 5 min). Green: orthochromatic emission of nucleic acids. Red: metachromatic emission of heparin. M: mast cells; P: macrophages; L: lymphocytes.

In other cases, on account of the spectral position of their emission maxima some fluorochromes do not offer striking color changes but only scarce differences of shade. Thus Grossgebauer [30] has noticed a metachromatic fluorescence reaction of yeast cell walls after staining with DAPI. The same results can be observed in the wall of *Candida famata* stained with Hoechst 33258, which shows a metachromatic shift from blue to cyan or blue-green (Fig. **11.14A-C**). In addition, the same metachromatic (red-shifted) fluorescence occurs in substrates such as chondroitin sulfate-containing cartilage matrix after DAPI saining (Fig. **11.14D**), which shows a clear difference between ortho- and metachromatic fluorescence.

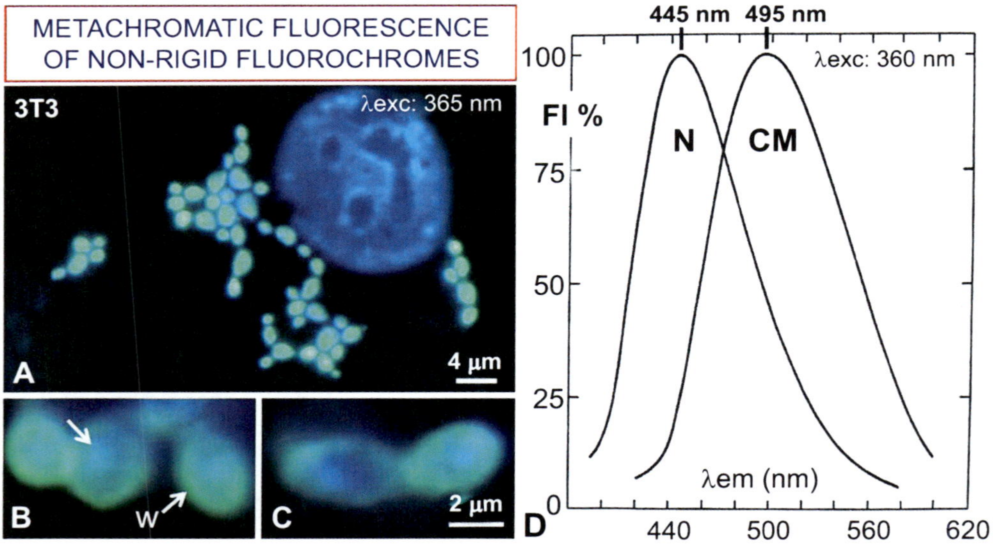

Fig. (11.14). **A**: Methanol-fixed cultures of 3T3 cells infected with *Candida famata* and stained with Hoechst 33258 (5 μg/mL, 3 min). **B**, **C**: Greater magnification of yeasts showing ortho- (blue) and metachromatic (blue-green) fluorescence of nuclei (N) and cell walls (W), respectively (courtesy of M. Pacheco and A. Juarranz). **D**: Fluorescence spectra of chicken erythrocyte nuclei (N) and formaldehyde-fixed frozen section of cartilage matrix (CM) from chicken xiphoides stained with DAPI at pH 8 (N: 10^{-6} M, 5 min, and CM: 10^{-4} M, 2 h). The emission peaks in nm are shown at the top.

On the other hand, aryl-oxazoles are well known compounds used in scintillation counting methods [31]. N-quaternary scintillators derivatives are water soluble and have been applied as lasing dyes [32] and in microscopical studies [33]. The oxazolium derivative mPyPO has been also used for visualization of acid GAGs [3] as shown in the case of heparin-containing mast cells (Fig. **11.15**).

The well known scintillators POPOP and dimethyl-POPOP can be methylated to produce the cationic and water soluble N-quaternary compounds. Staining with the quaternary dimethyl-POPOP derivative (Q-dmPOPOP) induces a bright blue fluorescence in mast cell granules [33], and chromatin DNA [34]. When preceded by the conventional hematoxylin-eosin (H&E) histological staining and under UV excitation, Q-dmPOPOP allows a striking differential cell visualization in paraffin sections giving blue emission in mast cells, whereas eosin Y induces yellow-green fluorescence in eosinophil leucocytes and reticular fibers [3]. Thus detailed analysis of mast cells can be performed in pathologic conditions such as skin mastocytosis, where the simultaneous occurrence of mast cells and eosinophils makes it difficult to recognize these cell types after routine H&E staining. When sections are additionally stained with Q-dmPOPOP or mPyPO, a clear distribution of cells and reticular fibers is observed (Fig. **11.16**).

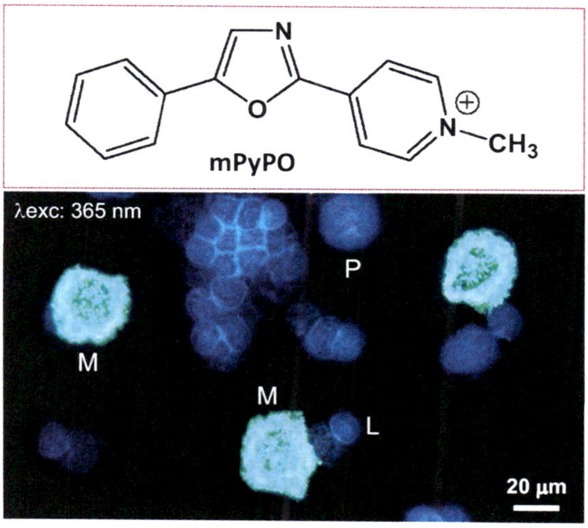

Fig. (11.15). Metachromatic fluorescence of mast cells granules from mouse peritoneal exudate by the aryl-oxazolium dye, mPyPO. Observe the scarce orthochromatic (light blue) color of basophilic nucleic acids components in macrophages (P) and lymphocytes (L), and the metachromasia (cyan) of heparin-containing mast cells granules (M).

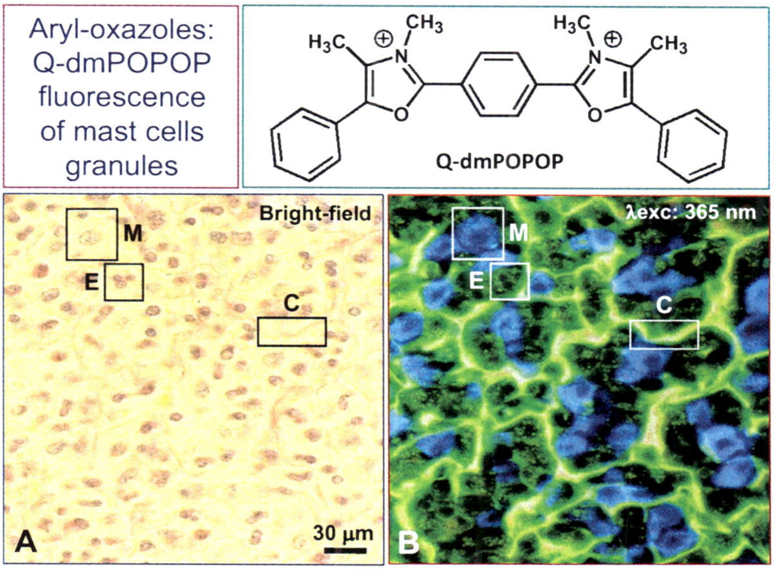

Fig. (11.16). **A, B**: Formaldehyde-fixed paraffin section of a dog skin mastocytoma stained with H&E followed by Q-dmPOPOP (10 µg/mL). Both images correspond to the same microscopic field. M: mast cells, E: eosinophils, C: collagen fibers.

Alcian yellow (CI 12840) is a cationic bis-benzothiazole monoazo dye (Fig. **11.17A-C**). As expected from its two tetramethyl-thiouronium groups, it has high affinity for polyanionic cell and tissue components such as acid GAGs and nucleic acids [35], giving fluorescence reactions with these substrates. Binding to nucleic acids may be avoided by previous treatment with Al^{3+} ions or staining with aluminum-hematoxylin. Ultrafluar® objectives are more convenient than Neofluar® ones for visualization of dyes requiring short UV excitation. On account of its alkaline-sensitive thiouronium groups, short treatment of Alcian yellow stained substrates with 1 N NaOH should precipitate the insoluble chromophore as occurs for the phthalocyanine Monastral fast blue after Alcian blue 8GS staining of acid polysaccharides [36].

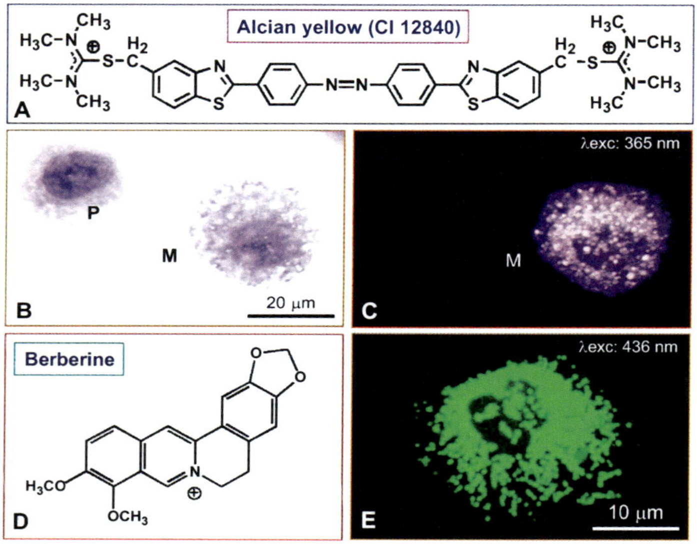

Fig. (11.17). **A**, **D**: Chemical structure of Alcian yellow and berberine. **B**, **C**: Bright-field (B) and fluorescence (C) micrographs of formaldehyde-fixed cells from mouse peritoneum treated with Mayer's aluminum hematoxylin followed by 0.2% Alcian yellow, and observed through an Ultrafluar objective. P: macrophage. M: mast cell. Only the emission of heparin-containing mast cell granules is seen. **E**: Methanol-fixed mast cell from mouse peritoneum, treated with alum (5%, 5 min), and stained with berberine (25 μg/mL, 5 min). A strong and selective fluorescence of mast cell granules is observed.

On the other hand, taking into account that the copper-containing phthalocyanine-like dye, cuprolinic blue, stains selectively acid GAGs [37], its red-fluorescing zinc analogue (zinc(II)-tetramethyl-2,3-pyridino-porphyrazinium [38]) should be also useful for histochemical applications in fluorescence microscopy.

Berberine is a natural and cationic isoquinoline alkaloid with strong fluorescence and high affinity for acid GAGs. It has been widely applied in microscopic methods to visualize and quantify the heparin content of mast cell granules [39 - 42]. The dye also induces bright fluorescence in chromatin and basophilic cytoplasm [43], but the previous treatment with aluminum ions allows achieving a high fluorescence selectivity in mast cell granules (Fig. **11.17D**, **E**).

4.3. Ruthenium Complexes

Several complexes between inorganic cations and different ligands have interesting chemical and spectral properties, and most of them has been overlooked or scarcely explored as staining and fluorescence reagents. Stable and deeply colored complexes between some metals (Fe, Ru) and planar, bidentate ligands (bipyridine, *o*-phenanthroline) are known [44], and sometimes applied as in the case of Fe(II)-phenanthroline [45] and Ru(II)-bipyridine [46, 47]. This latter complex (Rubipy) contains a diimino ligand group bound to the Ru(II) atom (see Chapter 9.3.2), which after energy absorption generates the triplet-state excited molecule Ru(III) $(bipyridine)_3^-$ that decays by a phosphorescence process (erroneously referred to as fluorescence).

Luminescent Ru complexes are stable, show a large Stokes shift and almost no fading, which make them very suitable as both direct fluorochromes and labeling agents. The cationic dye Rubipy, induces red-orange emission in chromatin (see Fig. **9.11B**) and sulfated GAGs but the reaction of chromatin can be abolished using a previous treatment with aluminum ions. In contrast, blocking sulfate groups of acid GAGs with Ba^{2+} hinders Rubipy binding. A bicolor image of mast cells that distinguishes them from lymphocytes and macrophages is easily obtained with the sequence aluminum-Rubipy-morin (Fig. **11.18**), in which the Al^{3+}-morin complex gives a green fluorescence to polyphosphates (nucleic acids) whereas polysulfates (heparin) appear with red-orange emission due to Rubipy.

Combined bright-field and fluorescence microscopy can be achieved using previous staining of nuclei and basophilic cytoplasm with aluminum-hematoxylin followed by Rubipy. This aluminum-hematoxylin/Rubipy sequence allows recognizing cells under bright-field illumination, whereas mast cells appear with the characteristic Rubipy luminescence (Fig. **11.19**). The large overlapping between Al^{3+}-hematoxylin absorption and Rubipy emission explains the fact that substrates with high affinity for Al^{3+} ions such as chromatin and basophilic cytoplasm do not show luminescent signals. Quantitative studies of heparin in mast cells from normal and tumor-bearing mice have been carried out using Rubipy [48]. The ruthenium signal from Rubipy stained mast cells can be also recorded by means of X-ray microanalysis in scanning electron microscopy [49].

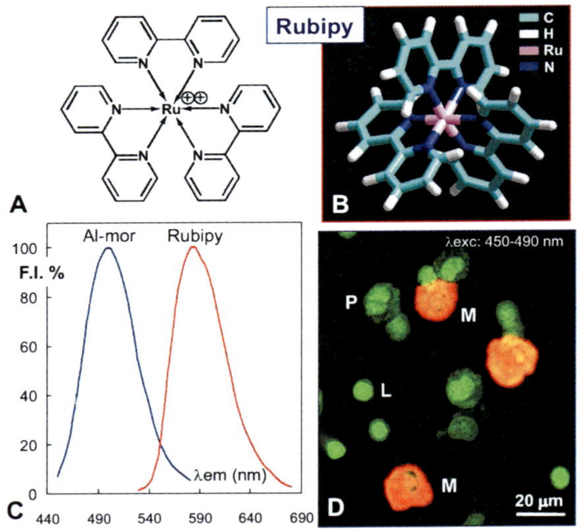

Fig. (11.18). Chemical structure (**A**) and molecular configuration (**B**) of tris(2,2'-bipyridine) ruthenium(II). **C**: Normalized spectra of aqueous aluminum-morin complex (Al-mor: 1 mg/mL alum + 10 μg/mL morin), and 25 μg/mL Rubipy under excitation at 440 and 460 nm, respectively. **D**: Methanol-fixed cells from mouse peritoneum exudate treated sequentially with alum (5%, 5 min), Rubipy (25 μg/mL, 5 min) and morin (saturated solution, 5 min). Green: Al-morin complex (polyphosphates). Red: Rubipy (polysulfates). M: mast cells. P: macrophages. L: lymphocytes.

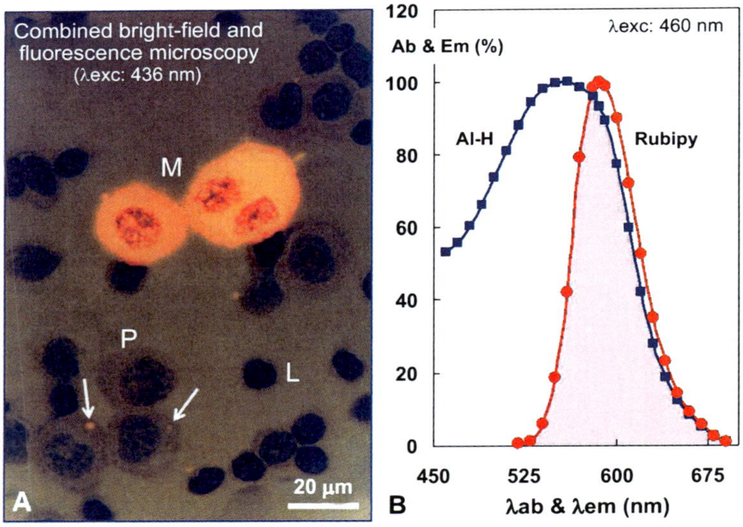

Fig. (11.19). **A**: Methanol-fixed cell smear from mouse peritoneum exudate, stained with Mayer's aluminum hematoxylin (15 min) followed by Rubipy (25 μg/mL, 5 min). M: mast cells; L: lymphocytes; P: macrophages. Arrows indicate heparin granules within macrophages. **B**: Normalized absorption spectrum of Mayers's aluminum-hematoxylin (Al-H, diluted with water, 1:16, v/v) and emission spectrum of 25 μg/mL Rubipy. The overlapping between both spectra corresponds to the shaded surface.

CONFLICT OF INTEREST

The author (editor) declares no conflict of interest, financial or otherwise.

ACKNOWLEDGEMENTS

This work was supported by a grant from the Ministerio de Economía y Competitividad (CTQ2013-48767-C3-3-R), Spain.

REFERENCES

[1] Kiernan JA. Histological and Histochemical Methods: Theory and Practice. Oxford, New York: Pergamon Press 1990.

[2] Szirmai JA. Quantitative approaches in the histochemistry of polysaccharides. J Histochem Cytochem 1963; 11: 24-34. [http://dx.doi.org/10.1177/11.1.24]

[3] Stockert JC. Cytochemistry of mast cells: new fluorescent methods selective for sulfated glycosaminoglycans. Acta Histochem 2000; 102(3): 259-72. [http://dx.doi.org/10.1078/S0065-1281(04)70034-3] [PMID: 10990064]

[4] Horobin RW, Kiernan JA, Eds. Conn's Biological Stains A Handbook of Dyes, Stains and Fluorochromes for Use in Biology and Medicine. 10th ed., Oxford, UK: BIOS Scientific Publishers 2002.

[5] Scott JE. Affinity, competition and specific interactions in the biochemistry and histochemistry of polyelectrolytes. Biochem Soc Trans 1973; 1: 787-806. [http://dx.doi.org/10.1042/bst0010787]

[6] Green FJ. The Sigma-Aldrich Handbook of Stains, Dyes and Indicators. Milwaukee, Wisconsin: Aldrich Chemical Company 1990.

[7] Stockert JC, Del Castillo P, Gómez A, Llorente AR. Fluorescence reaction of chromatin by curcumin. Z Natforsch C J Biosci 1989; 44(3-4): 327-9. [PMID: 2663005]

[8] Stockert JC, Del Castillo P, Testillano PS, Risueño MC. Fluorescence of plastic embedded tissue sections after curcumin staining. Stain Technol 1989; 64(4): 207-9. [http://dx.doi.org/10.3109/10520298909107001] [PMID: 2482998]

[9] Horobin RW, Murgatroyd LB. The staining of glycogen with Best's Carmine and similar hydrogen bonding dyes. A mechanistic study. Histochem J 1971; 3(1): 1-9. [http://dx.doi.org/10.1007/BF01686501] [PMID: 4106594]

[10] Abrahart EN. Dyes and their Intermediates. Oxford, London: Pergamon Press 1968.

[11] Süveges I, Módis L. A new fluorescence microscopic technic for the investigation of the cornea. Acta Histochem 1970; 35(1): 85-9. [PMID: 4191490]

[12] Chant J, Pringle JR. Patterns of bud-site selection in the yeast Saccharomyces cerevisiae. J Cell Biol 1995; 129(3): 751-65. [http://dx.doi.org/10.1083/jcb.129.3.751] [PMID: 7730409]

[13] Baselski VS, Robison MK, Pifer LW, Woods DR. Rapid detection of *Pneumocystis carinii* in bronchoalveolar lavage samples by using Cellufluor staining. J Clin Microbiol 1990; 28(2): 393-4. [PMID: 1690218]

[14] Smith MM, McCully ME. A critical evaluation of the specificity of aniline blue induced fluorescence.

Protoplasma 1978; 95: 229-54.
[http://dx.doi.org/10.1007/BF01294453]

[15] von Sengbusch P, Hechler J, Müller U. Molecular architecture of fungal cell walls. An approach by use of fluorescent markers. Eur J Cell Biol 1983; 30(2): 305-12.
[PMID: 11596507]

[16] Evans NA, Hoyne PA. A fluorochrome from aniline blue: structure, synthesis and fluorescence properties. Aust J Chem 1982; 35: 2571-5.
[http://dx.doi.org/10.1071/CH9822571]

[17] Stone BA, Evans NA, Bonig I, Clarke AE. The application of Sirofluor, chemically defined fluorochrome from aniline blue for the histochemical detection of callose. Protoplasma 1984; 122: 191-5.
[http://dx.doi.org/10.1007/BF01281696]

[18] Brzezinski DK. Der Chemismus der Anilinblau−Fluoreszenz der eosinophilen Leukozytengranula. Wiss Z Univ Greisfswald 1964; 13: pp. (Math.-nat. R.1)87-90.

[19] Changaris DG, Combs J, Severs WB. A microfluorescent PAS method for the quantitative demonstration of cytoplasmic 1,2-glycols. Histochemistry 1977; 52(1): 1-15.
[http://dx.doi.org/10.1007/BF00495809] [PMID: 863749]

[20] Rhys Williams AT. Fluorescence Detection in Liquid Chromatography. Buckinghamshire, UK: Perkin-Elmer FL publication 1980.

[21] Gallop PM, Paz MA, Henson E. Boradeption: a new procedure for transferring water-insoluble agents across cell membranes. Science 1982; 217(4555): 166-9.
[http://dx.doi.org/10.1126/science.6178158] [PMID: 6178158]

[22] Yang W, Yan J, Springsteen G, Deeter S, Wang B. A novel type of fluorescent boronic acid that shows large fluorescence intensity changes upon binding with a carbohydrate in aqueous solution at physiological pH. Bioorg Med Chem Lett 2003; 13(6): 1019-22.
[http://dx.doi.org/10.1016/S0960-894X(03)00086-6] [PMID: 12643902]

[23] Yan J, Fang H, Wang B. Boronolectins and fluorescent boronolectins: an examination of the detailed chemistry issues important for the design. Med Res Rev 2005; 25(5): 490-520.
[http://dx.doi.org/10.1002/med.20038] [PMID: 16025498]

[24] Armas-Portela R, Stockert JC. A reaction mechanism for the orthochromatic staining of polysaccharides: monomerizing interactions with basic dyes. Cell Mol Biol 1986; 32(4): 449-53.
[PMID: 3742556]

[25] Saunders AM. Histochemical identification of acid mucopolysaccharides with acridine orange. J Histochem Cytochem 1964; 12: 164-70.
[http://dx.doi.org/10.1177/12.3.164] [PMID: 14187320]

[26] Yamada K. A histochemical study on the acriflavine staining of mucopolysaccharide-containing tissues. Acta Histochem 1970; 35(1): 90-101.
[PMID: 4191491]

[27] Heath ID. Observations on a highly specific method for the histochemical detection of sulphated mucopolysaccharides, and its possible mechanisms. Q J Microsc Sci 1962; 103: 457-75.

[28] Stockert JC, Ibáñez N, Trigoso CI, Cañete M, Tato A. A barium method for the cytochemical detection of sulfated glycosaminoglycans in mast cells and basophilic leukocytes. Acta Histochem 1999; 101(4): 397-408.
[http://dx.doi.org/10.1016/S0065-1281(99)80040-3] [PMID: 10611928]

[29] Love LD. Fluorescence microscopy of viable mast cells stained with different concentrations of acridine orange. Histochemistry 1979; 62(2): 221-5.
[http://dx.doi.org/10.1007/BF00493322] [PMID: 500403]

[30] Grossgebauer K. Staining of acid mucopolysaccharides appearing on and in various cell types by DAPI. Microsc Acta 1979; 82(3): 291-3.
[PMID: 93685]

[31] Birks JB. The Theory and Practice of Scintillation Counting. New York: Macmillan 1964.

[32] Lee LA, Robb RA. Water soluble blue-green lasing dyes for flashlamp-pumped dye lasers. IEEE J Quantum Electron 1980; QE-16: 777-84.
[http://dx.doi.org/10.1109/JQE.1980.1070567]

[33] Espada J, Bertolesi GE, Trigoso CI, Gamallo C, Stockert JC. Fluorescence of mast cell granules in paraffin sections and cell smears induced by an N-quaternary oxazole scintillator. Histochem J 1995; 27(4): 318-22.
[http://dx.doi.org/10.1007/BF00398974] [PMID: 7635764]

[34] Stockert JC, Pelling C, Espada J. New cationic fluorochromes from diaryloxazole scintillators: fluorescence of chromatin DNA induced by N-quaternary POPOP derivatives. Acta Histochem 1997; 99(2): 195-205.
[http://dx.doi.org/10.1016/S0065-1281(97)80042-6] [PMID: 9248577]

[35] Stockert JC, Del Castillo P, Armas-Portela R. Alcian yellow as a fluorescent dye. Acta Histochem 1989; 85(1): 59-63.
[http://dx.doi.org/10.1016/S0065-1281(89)80099-6] [PMID: 2496573]

[36] Juarranz A, Stockert JC. Monastral fast blue. Cytochemical properties of a reaction product from Alcian blue stained chromatin. Acta Histochem 1982; 70(1): 130-4.
[http://dx.doi.org/10.1016/S0065-1281(82)80106-2] [PMID: 6179120]

[37] Juarranz A, Ferrer JM, Tato A, Cañete M, Stockert JC. Metachromatic staining and electron dense reaction of glycosaminoglycans by means of cuprolinic blue. Histochem J 1987; 19(1): 1-6.
[http://dx.doi.org/10.1007/BF01675286] [PMID: 3583811]

[38] Tempesti TC, Stockert JC, Durantini EN. Photosensitization ability of a water soluble zinc(II)tetramethyltetrapyridinoporphyrazinium salt in aqueous solution and biomimetic reverse micelles medium. J Phys Chem B 2008; 112(49): 15701-7.
[http://dx.doi.org/10.1021/jp808094q] [PMID: 19053687]

[39] Hahn von Dorsche H, Opitz M. Beitrag zur Kenntnis der selektiven Fluorochromierung von Mastzellen. Acta Histochem 1970; 36(1): 74-86.
[PMID: 4193858]

[40] Enerbäck L. Berberine sulphate binding to mast cell polyanions: a cytofluorometric method for the quantitation of heparin. Histochemistry 1974; 42(4): 301-13.
[http://dx.doi.org/10.1007/BF00492678] [PMID: 4448705]

[41] Hals E. Some methods for fluorochromation and staining of rat mast cells with bacic dyes. Scand J Dent Res 1970; 78(4): 301-10.
[PMID: 4097884]

[42] Dimlich RV, Meineke HA, Reilly FD, McCuskey RS. The fluorescent staining of heparin in mast cells using berberine sulfate: compatibility with paraformaldehyde or *o*-phthalaldehyde induced fluorescence and metachromasia. Stain Technol 1980; 55(4): 217-23.
[http://dx.doi.org/10.3109/10520298009067243] [PMID: 6160648]

[43] Molero ML, Stockert JC. Fluorescence reaction of chromatin and basophilic cytoplasm by berberine sulfate. Cell Mol Biol Incl Cyto Enzymol 1981; 27(5): 523-5.
[PMID: 7326706]

[44] McWhinnie WR, Miller JD. The chemistry of complexes containing 2,2′-bipyridyl, 1,10-phenanthroline, or 2,2′,6′,2”-terpyridyl as ligands. Adv Inorg Chem Radiochem 1969; 12: 135-215.
[http://dx.doi.org/10.1016/S0065-2792(08)60049-7]

[45] Stockert JC, Armas-Portela R. Selective staining reaction of the cationic complex tris (1,10-phenanthroline)-iron(II) with cell and tissue polyanions. Biocell 1995; 19: 43-8.

[46] Bertolesi GE, Trigoso CI, Espada J, Stockert JC. Cytochemical application of tris (2,2′-bipyridine) ruthenium (II): fluorescence reaction with sulfated polyanions of mast cell granules. J Histochem Cytochem 1995; 43(5): 537-43.
[http://dx.doi.org/10.1177/43.5.7730592] [PMID: 7730592]

[47] Bertolesi GE, Lauría de Cidre L, Stockert JC. Fluorescence of mast cells using aluminium/tris ruthenium II staining procedures. Eur Microsc Anal 1999; 61: 5-7.

[48] Bertolesi G, Stockert J, Decidre L. Microfluorometry and image analysis of peritoneal mast cells reveal differences in the morphology and content of sulfated polyanions between normal and tumor-bearing mice. Int J Oncol 1997; 11(6): 1221-5.
[PMID: 21528326]

[49] Stockert JC, Bertolesi GE. Analytical SEM of cell polyanions stained with tris(2,2′-bipyridine)ruthenium(II). Micron 1996; 27(3-4): 177-9.
[http://dx.doi.org/10.1016/0968-4328(96)00032-7] [PMID: 8953237]

CHAPTER 12

Nucleic Acids

Abstract: On account of the effects on cell processes, binding of small ligands with nucleic acids (NAs) has important implications in biomedical sciences [1-4]. Several fluorescent ligands allow visualizing the microscopical site of interaction. Specific DNA sequences may also be visualized by intercalating and minor groove-binding drugs. Detailed descriptions on NAs structure and function can be found in the literature [5-10]. Microscopical assays for DNA damage are described in Chapter 16.2. Detection of NAs by FISH is considered in Chapter 17 (see also Chapter 4.4). Consequences of binding modes of small ligands with NAs have been reviewed [11-14]. Structure-based design strategies have yielded new DNA-binding agents with clinical promise (*i.e.* hairpin polyamides and bis-intercalators [1, 3, 15]. Recognition of DNA sequences by fluorescent oligopeptides is a new approach. Examples are dansyl-oligopeptides from the *lac* repressor (containing sequences 19-32 and 53-71) that bind specifically to the *lac* operator DNA [16]. Computer-assisted molecular docking is increasingly applied for the design of molecules to target therapeutically relevant NAs structures. New analytical methods (*i.e.* electrospray ionization mass spectrometry [17]) allow assessment of the binding affinity of ligands to NAs.

Keywords: Anthracyclines, Auramine O, Banding methods, Benzothiazoles, Berberine, Bis-benzimidazoles, Carbocyanines, Cationic porphyrins, Centromeric heterochromatin, Chromomycin A_3, DAPI, Distamycin A, Feulgen reaction, Intercalation, Minor groove binding, kDNA, Schiff's reagent, Telomeres.

1. GENERAL STRUCTURE AND REACTIVITY

Nucleic acids (NAs) contain specific units (nucleotides) that are formed by sugars (ribose and deoxyribose), phosphate, and bases (nucleobases: purines and pyrimidines). For an excellent review concerning NAs structure Saenger's book should be consulted [5]. The general organization of NAs and the pH-dependent ionization of their components is shown in Fig. (**12.1**).

In addition to acid and alkaline hydrolysis, specific reagents react with NAs components: aldehydes (exposed $-NH_2$ gives Schiff's bases and $-CH_2-$ bridges), nitrous acid (diazotization of $-NH_2$); hydroxylamine, halogens, bisulfite, OsO_4 (addition to pyrimidine C5 and/or C6); alkylating agents and metal ions (binding to purine N1, N6 and N7, and to pyrimidine N3 and N4), and periodic oxidation of the glycol from a terminal ribose.

Juan Carlos Stockert & Alfonso Blázquez-Castro

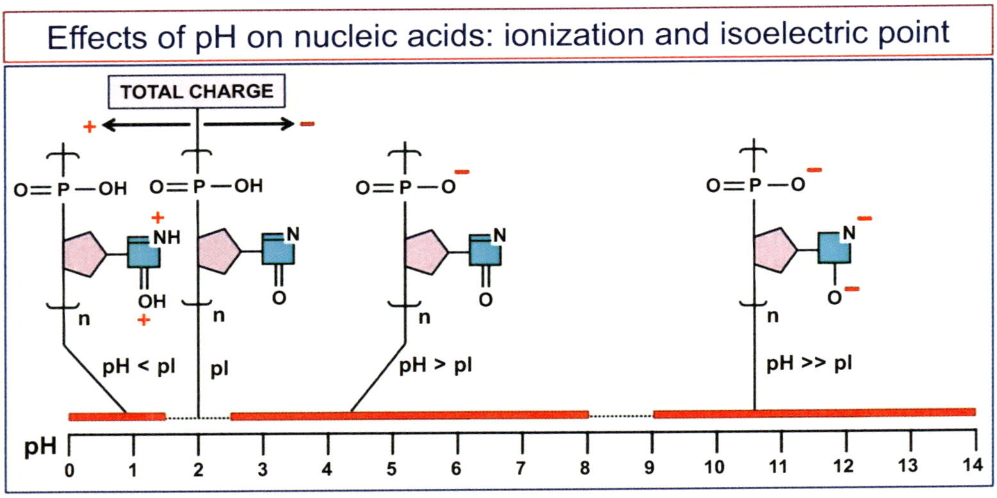

Fig. (12.1). General organization of nucleic acids, showing the pH-dependent ionization of their components. Vertical lines: sugar-phosphate chains; pink pentagons: sugars; cyan squares: bases. Polynucleotides appear as polyanions and polycations above and below their isoelectric point (pI), respectively.

On account of their different constituents, NAs can be revealed by the reactivity of pentose (Feulgen), phosphate (basophilia, metachromasia), and bases (UV absorption, intercalation, minor groove binding). This sequence will be followed in this chapter. Regarding cytochemical demonstration of bases, absorption at 260 nm (ultraviolet microscopy) or emission at 360 and 440 nm (see Fig. **7.7**) requires expensive instruments, and so in practice, they are not used. Cytochemical methods can be applied to obtain some evidence indicative of the binding mode. Several competing procedures for ionic, intercalating and minor groove binding are known, and they will be referred in the corresponding sections.

Four categories of NAs-ligand complexes may be identified. (1) Covalent interactions of bases with Pt, mitomycin C, hedamycin, psoralens, *etc.*; and binding with Schiff's reagent after acid hydrolysis. (2) Non-specific electrostatic interaction of inorganic or organic cations with phosphates. (3) Intercalative binding, in which a planar aromatic group is inserted between the base pairs of an NA duplex, increasing base pair separation from 3.4 Å to 6.8 Å. (4) Groove binding (generally restricted to B-DNA minor groove), occurring by means of H bonds, van der Waals and electrostatic forces.

Cell biologists are increasingly interested in the detection of NAs in the live cells. Fixing procedures modify the structure of biomolecules, and more direct binding modes of fluorochromes occur in fixed cells and tissues. Semi-permeable membranes and differential pH gradients in living cells are relevant factors that guide the uptake and localization of fluorescent probes. Accordingly, compounds that bind selectively to NAs *in silico* may not be able to reveal them *in vivo* (see Chapter 15.1), and microscopical staining results depend on whether the cells are dead or live.

2. FLUOROGENIC METHODS

2.1. Feulgen-Rossenbeck Reaction

One of the most important chromo- and fluorogenic methods (see Chapter 7.2.4) is the Feulgen-Rossenbeck reaction for DNA described in 1924 [18]. Unfortunately, mention of the second author is commonly omitted, so this abbreviated form will be used here for simplicity. This reaction is based on the generation of aldehyde groups by acid hydrolysis, and their detection using Schiff's reagent, which is a solution of pararosaniline or basic fuchsine decolored by sulfurous anhydride. Reviews concerning Schiff's reagent have been published [19 - 22].

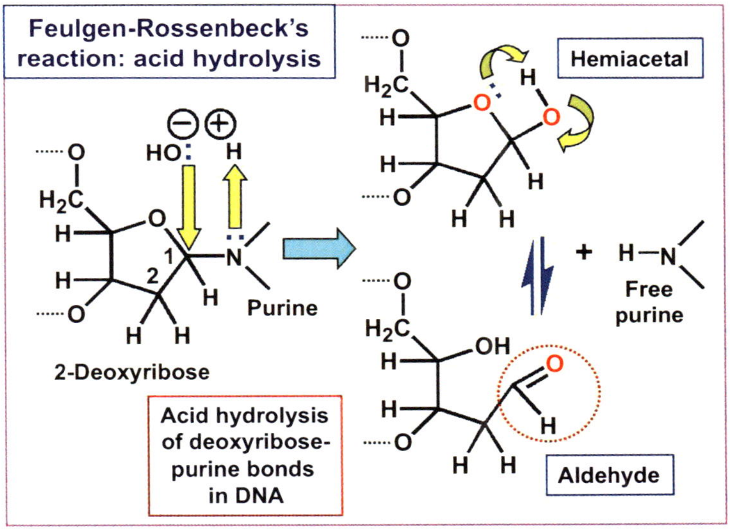

Fig. (12.2). Feulgen reaction: scheme showing acid hydrolysis and reactive aldehyde groups.

The mechanism of acid hydrolysis and formation of aldehyde groups are illustrated in Fig. (**12.2**). The acid treatment of DNA (usually with HCl) protonates the imidazole ring of purines and cleaves their N-glycosidic bonds generating the poly-aldehyde apurinic acid. The deoxyribose (in equilibrium between the closed furanose hemiacetal and the open aldehyde form) then reacts with Schiff's reagent to give the recolored dye. Although often overlooked, the alkaline hydrolysis produces a specific breakage of phosphodiester bonds in RNA.

Basic fuchsine contains several triamino-phenylmethane dyes of different degrees of methylation, from non- (pararosaniline, CI 42500), mono- (rosaniline CI 42510), di- (magenta II), and tri- (new fuchsine CI 42520) [22]. Interestingly, Schiff reported the preparation of this reagent in 1866, but the triphenyl-methane structure of these dyes was not known until 1878. Pararosaniline can be considered as involving various forms of resonance, namely paraquinone and carbonium structures. The central C is the most electropositive atom, and thus is the target for the nucleophilic attack of bisulfite or hydroxyl anions (Fig. **12.3A**). The molecular structure of pararosaniline and its sulfonate leuco-derivative (Schiff's reagent) can be seen in Fig. (**12.3B, C**), which shows the three aromatic rings twisted by about 30 ° in a three-bladed propeller-shaped conformation.

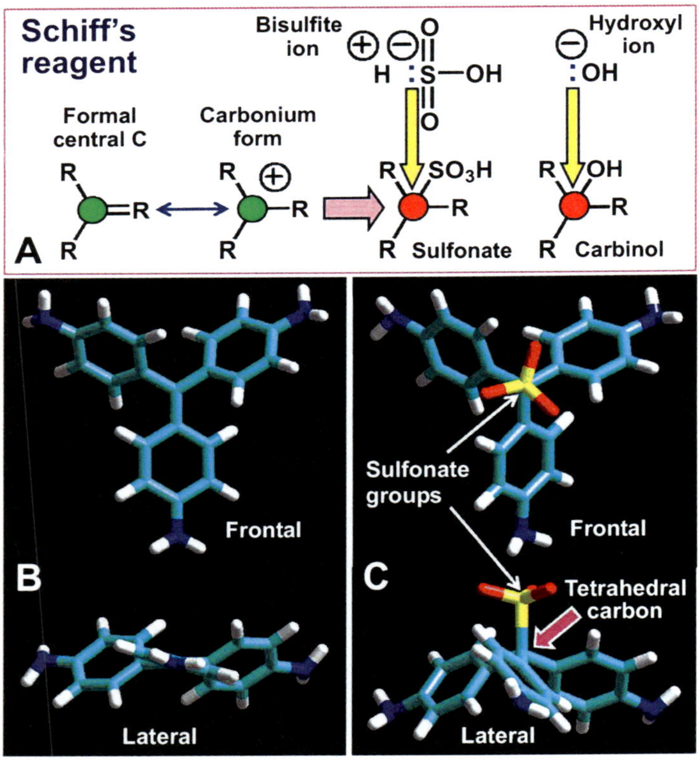

Fig. (12.3). Formation of Schiff's reagent. **A:** Nucleophilic attack of bisulfite or hydroxyl ions on the central planar carbon (green) of pararosaniline, forming a sulfonate or carbinol leuco-dye with a tetrahedral carbon (red). R: aniline residues. **B, C:** Frontal (top) and lateral view (bottom) of pararosaniline **(B)** and leuco-dye **(C)**. Note the conversion of the planar and conjugated central C of the dye to the tetrahedral C of the colorless Schiff's reagent.

Formation of the leuco-dye is usually achieved by treatment with metabisulfite (pyrosulfite, $S_2O_5^{2-}$), sulfite (SO_3^{2-}), bisulfite (HSO_3^-), thiosulfate ($S_2O_3^{2-}$), dithionite ($S_2O_4^{2-}$), or thionyl chloride ($SOCl_2$). Understanding the decoloration process requires a knowledge of the chemistry of the sulfur dioxide (SO_2)-water system [23]. In aqueous solution, SO_2 hydrates and forms sulfurous acid, which releases H^+ ions and dissociates to bisulfite and sulfite ions:

$$SO_2 + H_2O \leftrightarrow \; H_2SO_3 \; \leftrightarrow HSO_3^- \; \leftrightarrow SO_3^{2-}$$

The equilibrium between these species depends on the pH, and the products are only related to the conditions of the final solution, and not to the type of sulfur compound used to prepare it. At pH 7.2, 25 °C, and with low salt, bisulfite and sulfite are present in equal amounts. Bisulfite itself exists in two tautomeric forms ($HOSO_2^- \leftrightarrow HSO_3^-$), with HO-S(IV)$O_2^-$ predominating below 0.1 M, and H-S(VI)O_3^- above 0.3 M. Bisulfite can further dimerize at higher concentration giving metabisulfite ($2\ HSO_3^- \leftrightarrow S_2O_5^{2-} + H_2O$). Thus, the term bisulfite should be used to indicate all ionic forms from the SO_2-H_2O system, regardless of the preparative method.

The process by which Schiff's reagent reacts with an aldehyde (R-CHO) has been explained by three theories. (a) According to the original Schiff's view (1865-1867), the product would be an imine or azomethine (-N=CH-R). (b) On the contrary, Prud'homme (1900) suggested that bisulfite participates in the final dye product as an alkyl-sulfonate (-NH-CHR-SO_3^-). (c) Wieland and Scheuing (1921) proposed that the dye product would be an N-sulfinic acid derivative (-NH-SO_2-CHOH-R). The accepted mechanism of the reaction between Schiff's reagent and aldehyde group based on Prud'homme view is shown in Fig. (**12.4**).

The structure of the red-violet alkyl-sulfonate product from the Schiff-aldehyde adduct has now been unequivocally clarified [24], corresponding to a mixture of 2:1 and 1:1 aldehyde-dye adducts; the former with high yielding *in vitro* and the later showing more relevance for biological staining.

Several triarylmethane dyes (*i.e.* basic fuchsine, rosaniline, crystal violet, fast green FCF) are fluorescent (λexc: ~ 545 nm, λem: ~ 625 nm, either bound to substrates or in rigid media, see Fig. **5.11**), as is the reaction product of Schiff's reagent. This fluorogenic Feulgen reaction has been applied to identify DNA in bacteria, cell smears and tissue sections [25 - 28].

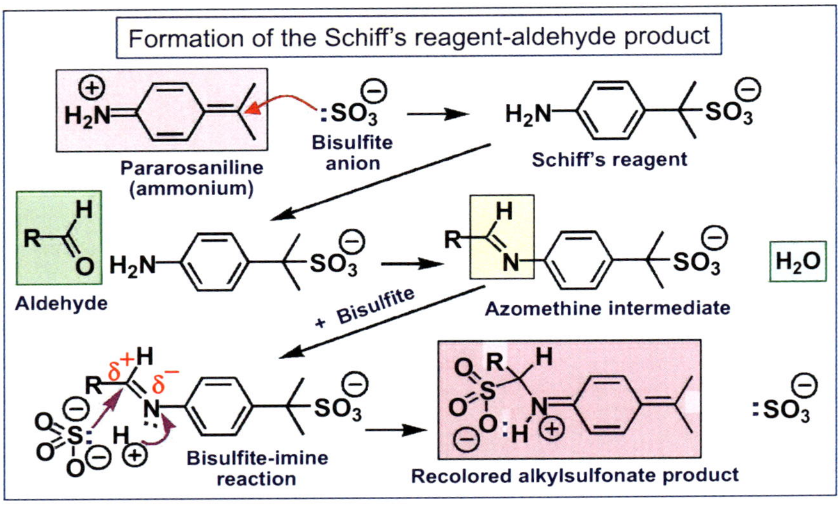

Fig. (12.4). Chromogenic reaction of Schiff's reagent with an aldehyde group, and formation of a recolored alkylsulfonate product with possible H bonding between NH and SO_3 groups. In the scheme only part of the chromophore is shown.

Fig. (**12.5**) illustrates some exemplar staining and fluorescence reactions. In autoimmune diseases, such as systemic lupus erythematosus [29], highly basophilic bodies (DNA or hematoxylin bodies) are often present in several tissues. These structures are also seen in lymph nodes after organ transplants. In both the cases, they are easily revealed by the Feulgen reaction. DNA in plant chromatin, and in Balbiani rings derived from polytene chromosomes, is also clearly visualized. The high sensitivity of Feulgen fluorescence allows the detection of small amounts of DNA (*e.g.*, bacterial nucleoids, fungal nuclei, extrachromosomal rDNA in micronucleoli of amphibian oocytes, *etc*).

Free aldehydes are largely absent from tissues, although some exceptions are known. Thus, acetal (plasmalogen) phospholipids, collagen, elastin, lignin in xylem vessels, mast cell granules, peroxidized fatty acids, and senile plaques (amyloid deposits) in Alzheimer's disease, react with the Schiff's reagent [20, 22, 30, 31]. It must be noted that after glutaraldehyde fixation, proteins often show false Feulgen positive reactions (see Fig. **7.20**).

For counterstaining of fuchsine stained materials in bright-field microscopy, the most usual dyes are light green (LG) and methyl green (MG), although misleading results were found when used in fluorescence microscopy (see Fig. **7.5**). It has been claimed that MG is not fluorescent and quenches the emission of intercalating and minor-groove fluorochromes [32]. However, a photo-induced red emission in chromatin DNA appears after MG staining followed by 3-5 min excitation under 365, 405 or 436 nm. At the same time, the blue-green color of nuclei observed in bright-field microscopy changes to violet. Microphotometric analysis shows an absorption spectrum identical to that of crystal violet, indicating the loss of a methyl group from MG.

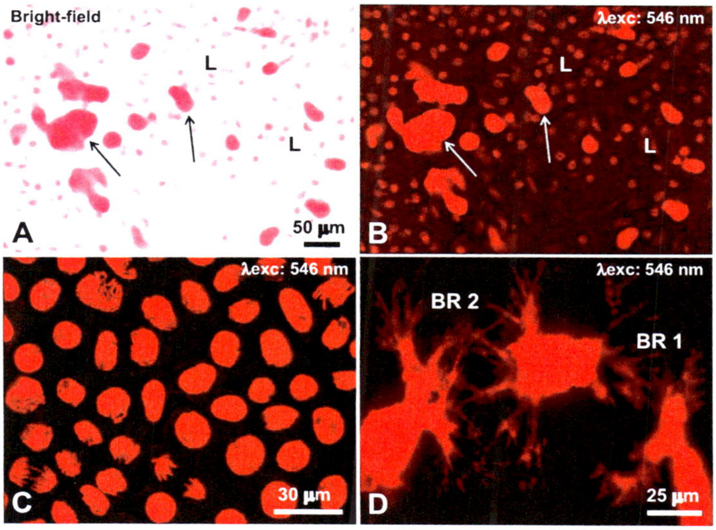

Fig. (12.5). **A, B:** Feulgen-stained paraffin section of a human lymph node after organ transplant under bright-field illumination (A) and fluorescence microscopy (B), showing large DNA bodies (arrows). L: lymphocytes. **C:** Carnoy-fixed and squashed *Allium cepa* root, with positive Feulgen reaction of chromatin DNA. **D:** Chromatin branching in Balbiani rings (BR1 and BR2) from a Carnoy-fixed IV polytene chromosome of *Chironomus pallidivittatus* after Feulgen staining.

There are fluorescent Schiff-type ("pseudo-Schiff") reagents that are combinations of sulfurous acid and a cationic fluorochrome, the most suitable being acridine yellow, acriflavine, auramine O, neutral red, proflavine, safranine O, *etc*., which yield fluorescent products with aldehydes [21, 22, 33, 34]. 2,5-Bis(aminophenyl)-1,3,4-oxadiazole (BAO) treated with sulfurous acid has also been also employed as Schiff-like reagent [35, 36].

Direct reactions of aldehyde and carbonyl groups with hydrazine or hydrazide derivatives are also used, according to the reaction:

$$-NH-NH_2 + O=C< \rightarrow -NH-N=C< + H_2O$$

Salicyloyl hydrazide has been employed to yield a blue fluorescing hydrazone (CBN: 12) under UV light. 2-Hydroxy-3-naphthoic acid hydrazide also gives a fluorescent product with aldehydes [20], which should show increased fluorescence after Zn^{2+} and Al^{3+} chelation or diazonium coupling. A formazan-like, blue-green tetraaza-pentamethine cyanine dye is formed on reaction of N-methyl-benzothiazolone-hydrazone (MBTH) with an aldehyde group [37]. MBTH is an extremely sensitive reagent for aldehydes, *i.e.* about 100 times more sensitive than Schiff's reagent, and is used for the detection of free aldehydes in tissues [30]. The fluorescence of the cyanine product should be useful in microscopy. Dansyl-hydrazine and 9-hydrazino-acridine also react with aldehydes and ketones to form fluorescent products.

In addition, other fluorometric methods for aldehydes have been described. Under acid conditions, 4,5-dimethoxy-1,2-diaminobenzene (DDB) reacts with aromatic aldehydes to give a fluorescent benzimidazole product (λexc: ~ 340 nm, λem: ~ 400 nm) [38]. A similar reaction occurs with 1,2-diaminonaphthalene. 3,3'-Diaminobenzidine (DAB) reacts with aldehydes from oxidized myelin lipids in nervous tissue giving greenish fluorescence under 460-495 nm exciting light [39]. Likewise, a fluorescent benzimidazole derivative is expected to form when DAB is applied after HCl hydrolysis of DNA.

A sensitive fluorometric assay for DNA is provided by the reaction with 3,5-diaminobenzoic acid after acid-catalyzed depurination [40], giving a fluorescent product with aldedydes of the form $-CH_2-CHO$ (λexc: 410 nm, λem: 510 nm), which has been overlooked for microscopical work. Another fluorogenic reaction uses malonaldehyde, yielding a fluorescent product with DNA (λexc: 390 nm, λem: 460 nm) from the amino-iminopropene group (-N=CH-CH=CH-NH-) bound to adenine and guanine [41]. Interestingly, acid-induced DNA aldehydes could by revealed by reaction with indole compounds (*i.e.* tryptamin, tryptophan), giving carboline or indolylmethine fluorophores, but these methods only showed analytical applications using colorimetric assays [42, 43].

2.2. Osmium Tetroxide Reaction

Oxidation of C-C double bonds in aromatic amino acids, catechols, pyrimidines, and unsaturated lipids, also gives Schiff positive results. After osmium tetroxide (OsO_4) fixation, double bonds form a *cis*-Os(VI) cyclic ester ("osmate") that hydrolyses in aqueous media yielding a *cis*-diol (glycol) group and OsO_2 (Fig. **12.6A**), which appears as a negatively-charged, polymeric dark and electron dense precipitate.

Treatment of OsO_4 fixed materials with *p*-phenylenediamine (PPD) also generates dark oxidized polymers (Bandrowski's base, phenazine dyes), increasing the ochre-brown color of the Os reaction. The glycol group is detectable by the PAS reaction (see Figs. **7.20** and **11.7**), yielding the expected colored and fluorescent PAS product (Fig. **12.6B-D**). Reoxidation by H_2O_2 or periodic acid eliminates Os staining but does not interfere with the fluorogenic reaction.

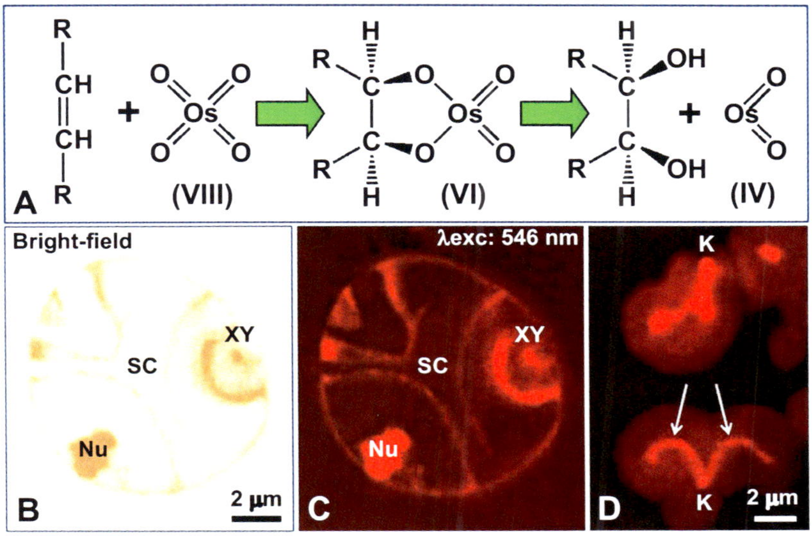

Fig. (12.6). **A**: Reactivity of the double bond with OsO_4. Numbers indicate the oxidation degree (valency) of Os. **B**-**D**: Os-PPD images of Epon semithin sections of mouse (B, C) and *Chorthippus jucundus* testes (D), either alone (B) or followed by the PAS reaction (C, D). Cells show nucleolus (Nu), synaptonemal complexes (SC), the XY pair with thick lateral elements in a primary spermatocyte (B, C), kinetochores (K) and chromatid cores (arrows) in first meiotic metaphase (D, courtesy of J.S. Rufas).

Accordingly, single-stranded RNA and DNA can be selectively detected by light and electron microscopy [44, 45]. Nucleoli, prenucleolar bodies and basophilic cytoplasm, but not native chromatin, appear positive. OsO_4 reacts very selectively with the 5,6 double bond of pyrimidines, yielding the *cis*-5,6-dihydroxy derivative; this electrophilic attack on the π orbital is necessarily out-of-plane, and thus, prevented by normal base stacking in NAs duplexes. Thymine reacts 3.5, 10, and 45 times faster than 5-methylcytosine, uracil, and cytosine, respectively. It must be noted that a strong Os-PPD reaction in chromatin results after HCl hydrolysis, due to the unstacked pyrimidines in the apurinic DNA. As OsO_4 only reacts when base stacking is disrupted, the reagent is really a probe for pyrimidine unstacking, which occurs in DNA 4-way junctions, B-Z junctions, and poly(dA-dT).

In meiotic cells, the synaptonemal complex, chromatid core, and kinetochore [45, 46] show selective staining after Os-PPD (Fig. **12.6B**) or fluorescence after Os-PPD-PAS (Fig. **12.6C-D**). The chromatid core (scaffold) and the synaptonemal complex seem to be an anchorage protein structure for chromatin fibers in the "loop-scaffold" chromosome model. AT-rich DNA tracts termed as "scaffold associated regions" (SARs, about 70% AT, 200-1000 bp long, and highly reactive toward OsO_4) are attached to the scaffold. Since ribonucleoproteins are also present in chromatid cores, kinetochores, and synaptonemal complexes, RNA could be also involved in the osmiophilia of these structures.

3. IONIC BINDING

NAs phosphate groups are obvious sites for interaction with metal and organic cations. Ionic (external) binding of cationic dyes and fluorochromes is responsible for the basophilia of NAs-containing structures. Therefore, ionic interactions of dyes with NAs can be blocked by using pre- or post-treatment of microscopical preparations with salt solutions containing Na^+, K^+, Mg^{2+}, Al^{3+}, *etc.*, which interfere with the monomeric or aggregated binding to polyphosphates. Na, K, and Mg ions are used in the "critical electrolyte concentration" (CEC) method (see Figs. **7.3** and **8.2**).

3.1. Metal Cations and Fluorescent Lakes

Monovalent (Li, Na, K, Rb, Cs) and divalent cations (Mg, Ca, Mn) bind to NAs primarily as counterions to the phosphate group (see Chapter 8.1). There are also base-specific interactions with cations such as Ag^+, Cu^{2+}, Zn^{2+}, Pt^{2+}, Hg^{2+}, *etc.* Cs^+ cations located in the minor groove are coordinated to O2 and N3 of thymine and adenine, respectively [47]. Some metal cations (*i.e.* Cu, Cd, Cr, Hg, Ni, V, Pb, Zn) are involved in oxidative DNA damage [48]. The selective binding of metal cations to DNA and RNA has been the subject of several reviews [49 - 52].

Be, Tl, Tb, Eu, and U salts give native luminescence that is dependent on conditions such as their physical state, type of complex and solvent, *etc.* [53] (see Chapter 9.2). Luminescent Ru-, Ir- and Os-probes have been designed for NAs labeling in live cells. Luminescence of a guanine-Tb^{3+} complex has been applied in studies on NAs structure [54]. Uranyl salts at pH 2 exist as the uranyl cation, UO_2^{2+}. At higher pH values other products are present (*i.e.* $UO_2(OH)^+$, and mainly $[UO_2(OH)_2UO_2]^{2+}$). UO_2^{2+} is luminescent and binds in the DNA minor groove. The photocleavage of its DNA complex (λexc: 420 nm) shows that the minor groove width and/or negative electrostatic potencial are sensed by UO_2^{2+} [55, 56]. Microscopical applications of the green emission of UO_2^{2+} seem to have been overlooked.

Metal cations, after binding to NAs, can be revealed by formation of a fluorescent complex with an appropriate ligand (see Chapter 8.3). Postformed and preformed complexes (lakes) are employed for this purpose. Peri-hydroxylated anthraquinone derivatives are excellent metal complexing ligands, inducing improved fluorescence reactions. The greater emission is related to the enhanced resonance of the chromophore after metal chelation. Al^{3+} is one of the most appropriate cations to form fluorescent complexes with high affinity for phosphate groups of chromatin, basophilic cytoplasm, and polyphospate granules [14, 51, 57]. Examples are the preformed lakes Al-alizarin red, Al-curcumin, Al-8-hydroxyquinoline, Al-morin, Al-nitrosonaphthol, Al-nuclear fast red (Al-NFR), Al-purpurin, *etc.* [58, 59]. For illustrative examples of fluorescent aluminum lakes applied on biological samples to reveal NAs-containing structures see Figs. (**8.4-8.9**).

3.2. Cationic Fluorochromes

There are fundamentally three binding modes of cationic fluorochromes to DNA, namely ionic attraction, intercalation and minor-groove insertion. Fig. (**12.7**) shows some properties and consequences of these interactions.

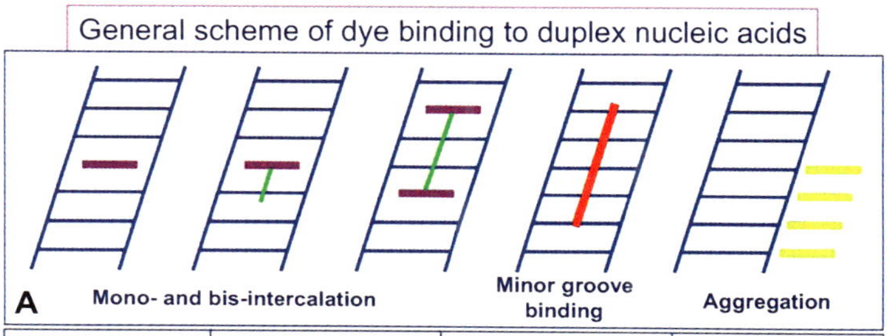

B			
Binding mode	Intercalation	External	External
Chromophore status	Monomer	Monomer or dimer	Aggregated
Spectral change (λab)	Bathochromic	Bathochromic	Hypsochromic
Color reaction	Orthocromatic	Orthochromatic	Metachromatic
Linear dichroism	⊥ > ‖	⊥ < ‖	⊥ > ‖

Fig. (12.7). **A**: Schematic representation of the principal binding modes of cationic dyes to nucleic acid duplexes, with ligands or parts of them in different colors: intercalators (violet), lateral chains and linkers (green), minor-groove binders (red), and externally aggregated dyes (yellow). **B**: Properties and consequences of the interaction of duplexes with these ligands, as revealed by microscopic and spectroscopic studies.

The intercalative and/or minor groove binding of several ligands can be evaluated in model RNA and DNA polymers by crystallography, NMR, spectroscopy, thermal melting and viscometric analysis. Polarized fluorescence is a simple method to evaluate the orientation of dyes bound to NAs. DNA fibers stained by intercalators or minor groove binders show polarized emission when recorded through a polarizer filter either parallel (E||) or perpendicular (E⊥) to the DNA axis, which is indicative of the binding mode (Fig. **12.8A-C**).

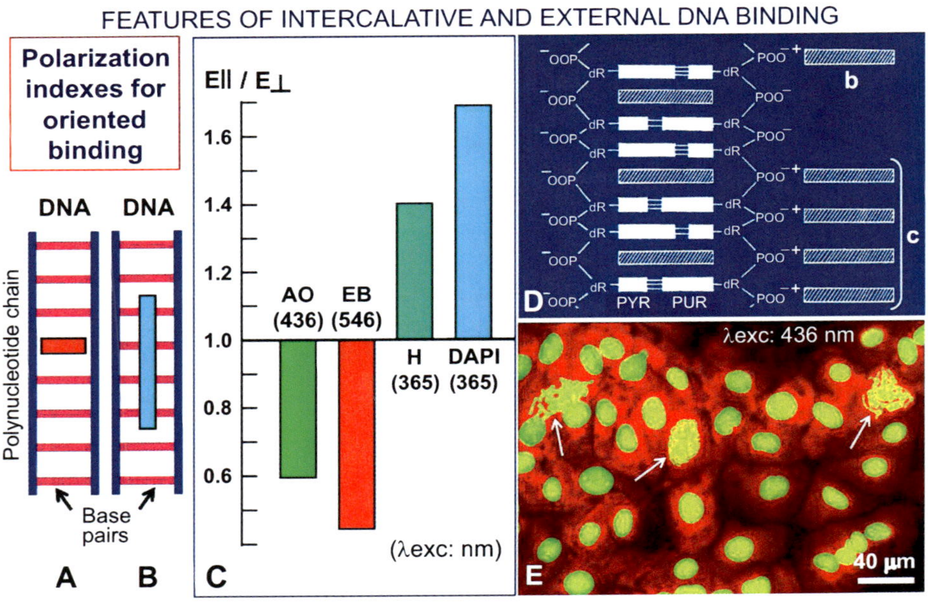

Fig. (12.8). Intercalation (**A**, red) and minor groove (**B**, cyan) binding to a DNA duplex. **C**: Polarized fluorescence of oriented DNA fibers stained with acridine orange (AO), ethidium bromide (EB), Hoechst 33258 (H), and DAPI. DNA gels and fluorochromes were mixed and stretched on glass slides, and the emission intensity assessed through a polarized filter, either parallel ($E_{\|}$) or perpendicular ($E_{\perp}$) to the fiber axis; λexc is indicated. **D**: Intercalative (a) and external binding of planar dyes as monomers (b) or aggregates (c) to a NA duplex. POO: phosphate, dR: deoxyribose, PUR, PYR: purines and pyrimidines. **E**: Methanol-fixed fibroblasts of *Pleurodeles waltlii* stained with 0.1 mg/mL AO, showing cytoplasm and nuclei with red (RNA) and green (DNA) fluorescence, respectively. Arrows: mitotic nuclei.

$E_{\|}/E_{\perp}$ indexes higher or lower than 1 reflect, respectively, the orientation of fluorophores either parallel (minor groove binders) or perpendicular (intercalators) to the DNA axis [60, 61], respectively. Studies on the DNA topography in chromatin fibers have been published [62].

Numerous dyes and fluorochromes have been applied to visualize NAs-containing structures. Acridine orange (AO), ethidium bromide (EB), DAPI, Hoechst 33258, and many other dyes are known examples [14, 61, 63]. Likewise, ortho- and metachromatic staining and fluorescence of chromatin and basophilic cytoplasm by cationic dyes (*i.e.* toluidine blue, carbocyanines) are also useful [64, 65] (see Figs. **1.6**, **1.8A**, **5.9**, **11.10** and **11.13**). The external binding of thiazine and acridine dyes that induce the metachromasy of NAs involves dye stacking, possibly on the crests limiting the minor groove in the case of DNA [66]. Other fluorochromes are used to reveal the basophilia of NAs, with emphasis on RNA visualization. Fluoro Nissl green has been proposed as a fluorescent stain for neuronal Nissl bodies and is formed from *m*-phenylene-diamine [67]. As the proposed compound seems to correspond to an azine fluorophore, other azine dyes may also be useful (*i.e.* neutral red, phenosafranine, safranine O [68, 69]).

Several cationic triphenylmetyhane dyes are relevant in NAs cytochemistry; some of them inhibit replication and displace other biologically active compounds. Methyl green (MG) is widely used for selective DNA staining [22], and inserts into the major groove of AT sequences [70]. Other tissue components (*i.e.* lignin, mast cell granules, mucin) are also stained. Due to binding preference to AT-rich DNA regions, this dye is used as counterstain in enhanced banding methods [32]. After irradiation with UV or violet-blue light, the non-fluorescent heptamethylated MG becomes the hexamethylated and red fluorescing crystal violet (see section 2.1). Crystal violet and pyronine Y interact with NAs and have been applied for visualization of Nissl bodies [71, 72].

Cresyl violet (CI 42555) is widely used in neuroanatomical studies to reveal Nissl bodies, and can be used as a red fluorescent stain (λexc: 546 nm, λem: 630 nm) for Nissl bodies, neuroglia nuclei and nucleoli [73]. The procedure requires only minor modifications of routine bright-field cresyl violet staining, the most significant being a low concentration (5 μg/mL) of the stain. In fixed cell smears, nuclei are colorless but fluoresce red after staining with thionine or toluidine blue at concentrations of 10^{-5}-10^{-6} M [74]. Heating alkaline dye solutions produces thiazones with red emission. The occurrence of poorly polar thiazones or red free bases could be responsible for these fluorescence reactions, possibly involving DNA intercalation.

Staining with AO followed by MG counterstaining induces a selective fluorescence of plant and animal nucleoli highlighted against the quenched fluorescence of the chromatin [75]. In addition to the ortho-metachromatic fluorescence of acridines, there are fluorochrome pairs that reveal RNA and DNA in different emission colors, generally requiring double excitation. Examples are pyronine Y/DAPI (see Fig. **5.17A**), propidium iodide/Hoechst 33258 (RNA/DNA), and NeuroTracer 640/660 deep-red fluorescent Nissl stain/DAPI (RNA/DNA).

4. INTERCALATION

A schematic representation of the interaction between duplex NAs and typical tricyclic planar dyes is illustrated in Fig. (**12.8D**), showing intercalation and external binding of dye monomers and aggregates, which result in ortho- and metachromatic fluorescence, respectively (Fig. **12.8E**). On account of their binding mode, intercalating compounds display important biological activities, namely antiproliferative effects, inhibition of replication and transcription, mutagenicity, *etc*.

4.1. Geometrical Parameters

The intercalation model was proposed by Lerman [76] for the binding of acridines to DNA, based on hydrodynamic, optical and chemical evidence. According to this model, a gap with the thickness of an aromatic intercalator (~ 3.4 Å) is formed between two successive base pairs in the duplex. The inserted planar ligand is stabilized by hydrophobic stacking forces resulting from the overlap of π-electrons from the chromophore and base pairs.

Polynucleotide elongation and helix unwinding are required in order to enable the intercalative binding mode. The chains adjacent to the intercalation site also unwind slightly to relieve backbone strain. The propagation of this distortion provides a rationale for the fact that intercalation saturates at a ratio of one ligand per two base bairs, a situation known as the "nearest neighbor exclusion principle". The unwinding induced by intercalation causes the relaxation of negative turns of the supercoiled circular DNA, whereas elongation induces higher viscosity of linear DNA. An axial view of the base pairs C·G/G·C, and AO intercalation into this miniduplex can be seen in Fig. (**12.9A, B**).

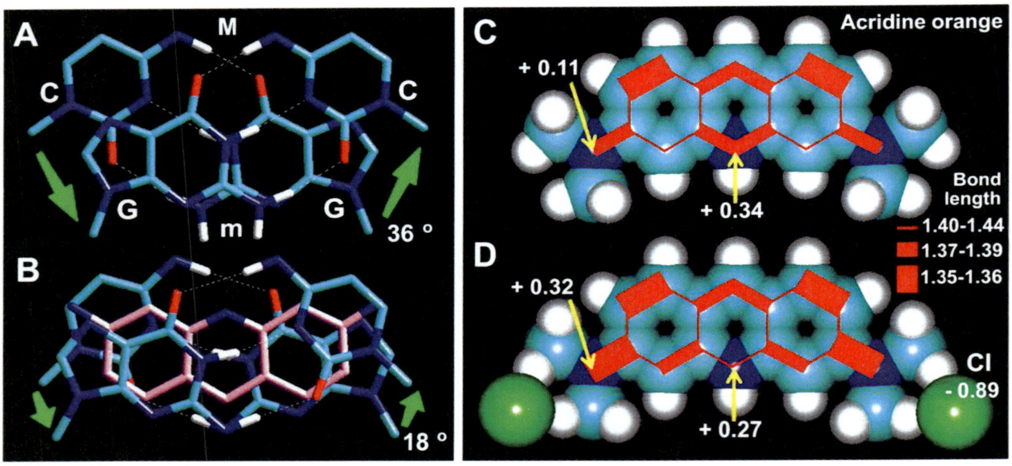

Fig. (12.9). **A**: Base pairs (G·C/C·G) from a B-DNA duplex, showing base overlapping and 36 ° winding angle. M and m: major and minor groove, respectively. Dashed lines are H bonds. **B**: AO ring (pink) intercalated between G·C/C·G with an unwinding angle of 18 °. **C**, **D**: Double bond character of AO either calculated *in vacuo* (C), or in the presence of two Cl^- ions (D). Bond lengths (in Å) are shown as lines with different width. Partial charges of Cl, central and lateral N atoms are indicated. Observe the paraquinone conformation in (D).

The electronic configuration of the dye changes when it is bound to an anionic substrate shows different bond lengths and charge distributions (Fig. **12.9C, D**). The orthoquinone structure of AO becomes a paraquinone form when the dye is subjected to strong electrostatic forces, as occurs in intercalated AO with its $(CH_3)_2N$- groups facing the negative charges (*i.e.* phosphates).

Regarding other geometrical aspects, there are three types of intercalating chromophores: (1) planar and rigid (AO, pyronine Y), (2) non-planar and flexible (thioflavine T, thiazole orange, oxazole yellow), and (3) non-planar and rigid (berberine). Computer-assisted studies on the suitable geometry of tri- and tetracyclic DNA intercalating chromophores have been published [77]. On account of the torsional freedom of one or two bonds, flexible intercalators often have increased emission when intercalated. Best known examples are thiazole orange (TO) and oxazole yellow (YO) [78, 79].

Another type of interaction is DNA "hemi-intercalation" of antibacterial quinolones (*i.e.* norfloxacin, nalidixic acid, *etc.*), which occurs within a bubble of distorted and/or partially denatured DNA induced by DNA gyrase [80, 81]. Quinolones interact directly with DNA, and Mg^{2+} ions are necessary for binding [82], and neutralizing the negative charge of the ionized carboxyl group of ligands. Mg^{2+} complexes of 1:1 quinolone-Mg with DNA are fluorescent (λexc: 330 nm, λem: 423 nm) [82], but no microscopical applications seem to have been described.

In addition to DNA and RNA duplexes, some model substrates (*i.e.* "molecular tweezers" [83] and cyclic diguanylic acid [84] are host targets for intercalation *in silico*. On account of their largely insensible binding at high ionic strength, competing agents such as the planar, uncharged, colorless and non-fluorescent caffeine and *o*-phenanthroline are employed to study the hydrophobic intercalative process at the microscopical level [85]. After washing with these competitors (or staining in their presence), the reduction or absence of emission indicate intercalative binding.

4.2. Quantitative Parameters

To understand the quantitative dye-DNA interactions, cytological molar ratios (m dye/m DNA) using the EB intercalator are shown in Fig. (**12.10**). A prerequisite of this approach is a knowledge of the molar concentration of DNA in suitable model systems such as methanol-fixed chicken blood smears [14]. The genome of the chicken erythrocyte contains DNA with 10^{12} D. Taking into account that $0.65 \cdot 10^{12}$ D corresponds to 10^{-12} g DNA ($9.13 \cdot 10^{8}$ base pairs), and that 3 mg/mL is equivalent to $9 \cdot 10^{-3}$ M (in bases), then a ratio of one nucleus/mL would correspond to a DNA concentration of $4.59 \cdot 10^{-9}$ M (in bases). Circular areas (5 mm in diameter) from homogeneous smears reveal about $4.2 \cdot 10^{4}$ nuclei; therefore, adding 20 µL solvent on this area, the molar DNA concentration can be taken as about $0.5 \cdot 10^{-5}$ M (in bases).

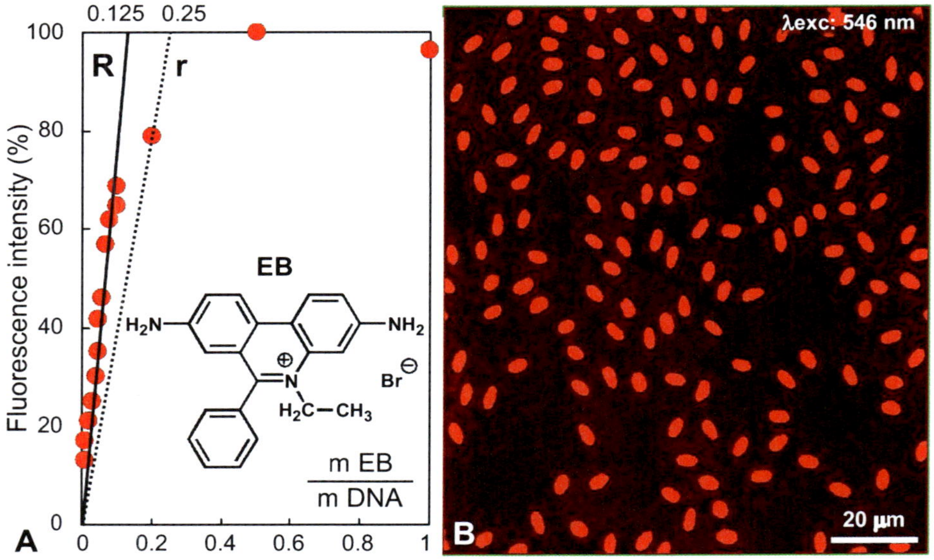

Fig. (12.10). A: Correlation between the fluorescence of chromatin DNA stained by ethidium bromide (EB), and different values of the molar ratio R (total EB/DNA bases, intercept of R at 0.125), measured for a known amount of chicken erythrocyte nuclei. The linearity of R is only observed for low molar ratios. The dotted line is the expected correlation for the molar ratio r (bound EB/DNA bases, intercept at 0.25; values not shown). **B**: Fluorescence of chicken erythrocyte nuclei after 10^{-6} M EB staining.

After EB staining at concentrations from $0.5 \cdot 10^{-7}$ M to $0.5 \cdot 10^{-5}$ M, chromatin DNA emission increases linearly only at very low EB concentrations (Fig. **12.10A**). Using "r" as the molar ratio bound dye/DNA, and recording the emission intensity (I_{em}), the expression I_{em}: f (r) is valid for DNA-EB complexes either *in silico* [86], and in chromatin from cell suspensions [87]. It is accepted that at the highest DNA concentration and lowest dye concentration all ligand molecules are bound. When the simplified value "R" (total dye/DNA *in situ*) is used, the total dye amount is greater than the bound dye, and therefore R < r. Although the intercept at R: 0.125 represents an overestimation of the fluorescent (intercalative) binding of EB to fixed chromatin DNA, the expression I_{em}: f (R) may be extended to the microscopic field and taken as indicative of the binding process [14].

At increasing dye concentration, external interactions with NAs start before intercalating sites are fully occupied. Thus I_{em} increases more slowly because of the binding of dye aggregates to external sites on DNA, and a considerable quenching of fluorescence occurs at higher r (and R) values. In this context, fluorescence based in the intercalative binding alone could only be observed by using values of R < 0.1 (*i.e.* dye < $5 \cdot 10^{-6}$ M in 0.2 mL for microscopical areas containing about $3 \cdot 10^5$ nuclei, Fig. **12.10B**). It must be noted that EB fluorescence is not specific for nucleic acid duplexes, since reactivity with collagen, keratin, and other substrates is also found.

The concentration-dependent effect is clearly observed in Figs. (**12.11** and **12.12**), which illustrate the intercalative and external binding kinetics of pyronine Y and safranine O. The relation of emission intensity with dye concentration is apparent: at high concentration a preferential staining of RNA structures is seen with bright-field illumination, whereas under fluorescence microscopy, chromatin shows the highest emission [88]. The same occurs for safranine O [68]. The orange fluorescence of pyronine Y (λem: 580 nm, see Fig. **5.17A**) has been used to reveal RNA in cell cycle analysis by flow cytometry [89]. The spectral response of pyronine Y bound to DNA and heparin is illustrated in Fig. (**5.15**).

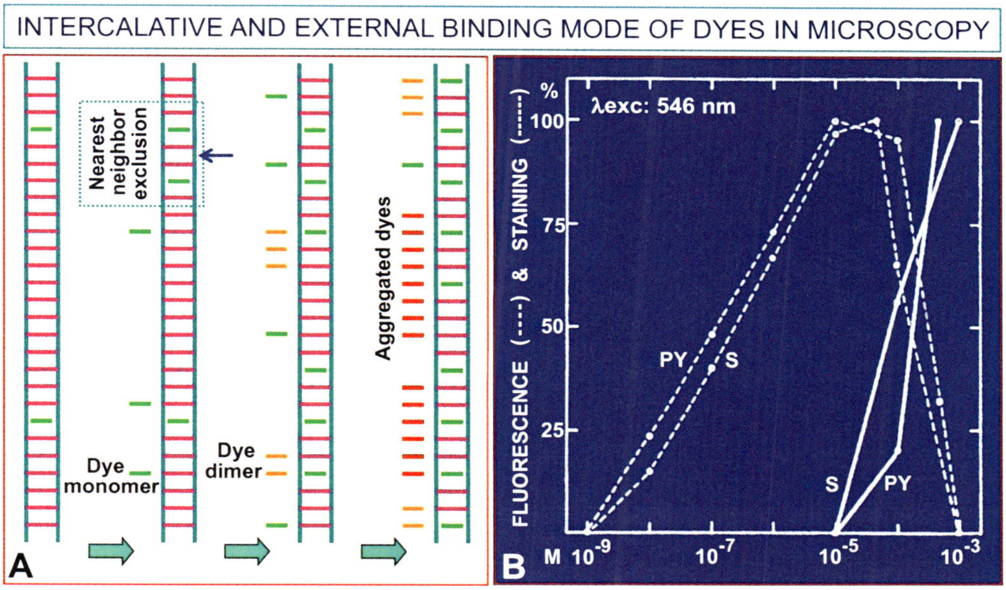

Fig. (12.11). **A**: Scheme of the kinetics of intercalation and binding of dye monomers and aggregates depending on the increasing dye concentration. **B**: Intensity of fluorescence and staining of methanol-fixed chicken erythrocyte nuclei treated for 5 min with increasing molar concentrations of pyronine Y (PY) and safranine O (S).

4.3. Applications of Acridine Fluorochromes

Cytochemical studies on the chromosome banding pattern were initiated by the pioneer work of Caspersson's group [90] using quinacrine (atebrine, mepacrine) and its alkylating bis-chloroethyl derivative, quinacrine mustard. With these dyes diprotonated species predominate at pH 7. As quinacrine dyes show fluorescence quenching at GC-rich sites, they reveal selectively AT-rich regions [91], which are now known as Q bands.

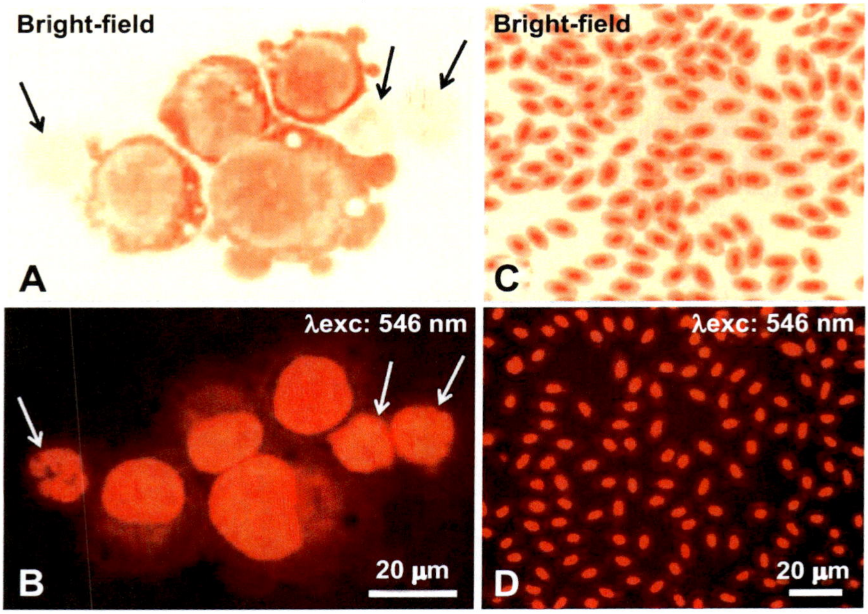

Fig. (12.12). A, B: Methanol-fixed Ehrlich ascites tumor cells stained with 0.5 mg/mL pyronine Y and observed under bright-field (A), or fluorescence microscopy (B). Arrows show leucocyte nuclei scarcely visible in A but well recognized in B. **C, D**: Methanol-fixed chicken erythrocyte nuclei stained by 10^{-3} M (C) or 10^{-5} M (D) safranine O, showing a concentration-dependent reaction.

The differential ortho- and metachromatic fluorescence along mitotic and meiotic chromosomes was achieved through DNA denaturation-reassociation followed by AO staining [92, 93]. Thus the highly repetitive DNA in the centromeric heterochromatin from mouse metaphase chromosomes, containing $([dA]_{4-6} \cdot [dT]_{4-6})_n$ sequences, could be visualized after denaturation-renaturation of the fast- and slow-reassociating highly repetitive and non-repetitive sequences, respectively (Fig. **12.13**).

As is the case with other acridines, acriflavine has high affinity for NAs by intercalation, and also binds to acid polysaccharides, giving strong fluorescence reactions [34]. An acriflavine-phosphotungstate (A-PTA) 3:1 complex has been employed for visualization of nucleic acids and acid polysaccharides using both fluorescence and electron microscopy [94]. Likewise, on account of a monomeric intercalative-like binding, AO gives a bright green emission with lignin (see Chapter 5.3.3).

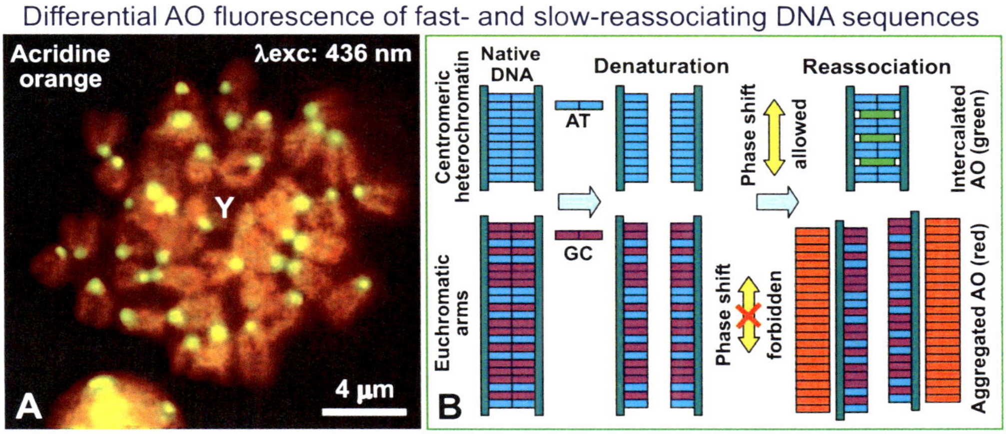

Fig. (12.13). **A**: First cytochemical visualization of the highly repetitive DNA in the centromeric heterochromatin from mouse metaphase chromosomes using denaturation (2 x SSC at 100 °C + 10% formamide, 2 min), renaturation (2 x SSC at 66 °C, 30 s + 4% formaldehyde, 1 min), and 0.125 mg/mL AO staining [92]. Y: chromosome Y. **B**: Schematic mechanism of the green-red fluorescence induced by DNA denaturation and differential renaturation of the fast- and slow-reassociating highly repetitive and non-repetitive sequences, respectively.

4.4. Alkaloids

β-Carboline alkaloids are pyrido-indole fluorophores with a wide spectrum of biological activities. The use of simple β-carbolines as cationic fluorochromes has been rather overlooked. Harmine (from *Peganum harmala*) binds to DNA by intercalation, and induces blue-white emission of nuclei in fixed cells [95]. Harmine fluorescence of nuclei in live plant cells has been reported [96]. The 3,4-dihydro-β-carbolines harmalol and harmaline also show fluorescence. Other derivatives of the harmine ring are corynanthine, flavopereirine, reserpine, rutecarpine, sempervirine, serpentine, yohimbine, *etc*., some of which are psycho-active drugs. Several derivatives are fluorescent and intercalating compounds but they have been little explored for microscopical applications [97, 98].

Berberine (from *Berberis vulgaris* and *Hydrastis canadensis*, CI 75160; Fig. **12.14A**), is a fluorescent isoquinoline alkaloid with anti-bacterial, anti-fungal, anti-protozoal, anti-tumoral, and photodynamic activity [99, 100]. It interacts with DNA duplexes by intercalation [99, 101], and externally to single-stranded NAs [102]. Berberine also binds to acid glycosaminoglycans and lipid bilayers (see Chapters 10.5 and 11.4.2), inducing mitochondrial damage and apoptosis (see Chapter 15.5.4). Other isoquinolines are similar to berberine (*i.e.* chelerythrine, columbamine, copticine, coralyne, palmatine, sanguinarine), and may also be useful in microscopic studies. Free bases of these compounds are colorless but quaternary salts are yellow and fluorescent. In fixed materials, berberine induces fluorescence of NAs-containing structures, lipid droplets (see Fig. **10.11**), and mast cell granules (see Fig. **11.17D, E**). High fluorescence is observed in RNA components (nucleoli, basophilic cytoplasm), but after acid fixation or HCl hydrolysis chromatin DNA has brighter fluorescence [103]. The fluorescent staining of nuclei by berberine is shown in Fig. (**12.14D**).

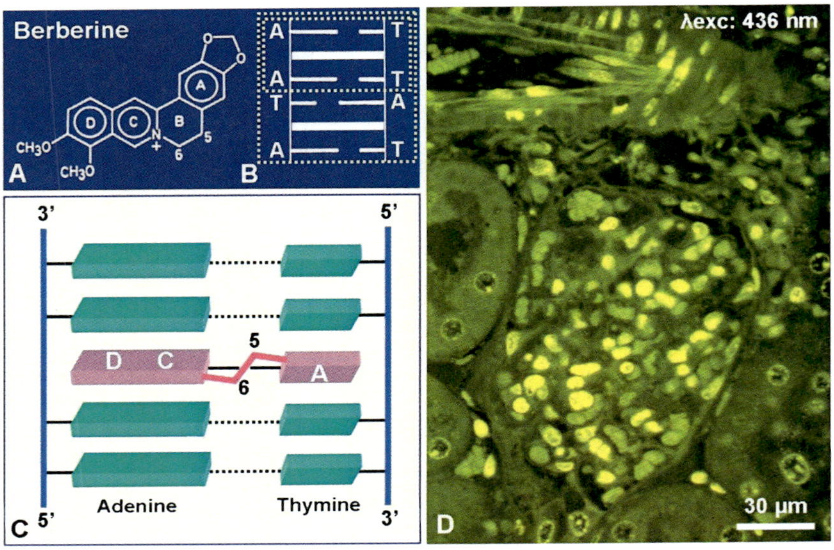

Fig. (12.14). **A**, **B**: Chemical structure of berberine and simplified scheme of its intercalation into the DNA sequences AA·TT (small rectangle, r: 0.125) and AATA·TTAT (large rectangle, r: 0.25). **C**: Model of the preferential intercalative binding showing the saturated bridge C5-C6 and the complementary geometry between the ACD rings of berberine and the propeller twist of AAAA·TTTT base pairs. **D**: Glomerulus of mouse kidney (glutaraldehyde, Epon), stained by 10^{-4} M berberine sulfate (pH 9.2), showing nuclear emission.

Intercalation of berberine into DNA and preference for AT regions is illustrated in Fig. (**12.14B, C**). Two binding modes saturate at r: 0.1 and r: 0.25. As berberine strongly interacts with purines, the statistical distribution of about 50% purine-purine sequences could result first in intercalative binding each 8 (r: 0.125) and then each 4 bases (r: 0.25). As the two aromatic rings of berberine are tilted [104] and AT sequences have high propeller twist, there is a close fitting between the intercalator and its binding site [14]. Accordingly, selective yellow-green emission (λexc: 370 nm, λem: 550 nm) is observed in the AT-rich mouse centromeric heterochromatin [105].

The pyrido-carbazole alkaloid ellipticine (from *Ochrosia elliptica*) and derivatives are fluorescent anticancer agents causing cell cycle arrest and apoptosis [106]. Mechanisms of anti-tumor, cytotoxic and mutagenic activities of ellipticine are

DNA intercalation, covalent binding to DNA after activation by cytochrome P450 and/or peroxidases, and inhibition of DNA topoisomerase II. A related olivacine alkaloid (from *Aspidosperma olivaceum*) is also fluorescent. Fluorescence microscopy of an ellipticine 1-aminopropylamine derivative shows accumulation in the nuclei of live HeLa cells. The distamycin-ellipticine hybrid is preferentially located in the nucleoli [107].

4.5. Psoralens

Psoralens are furocoumarins from *Rutaceae*, *Umbelliferae*, *Leguminosae,* and *Moraceae* plants, used as phytoalexins in a defensive response against fungi and insects. Psoralen and its 5-methoxy, 8-methoxy, and 4,5',8-trimethyl derivatives (also angelicin) bind to DNA by intercalation, and are phototherapeutic drugs for treatment of mycosis fungoides, psoriasis, and vitiligo. Psoralen is fluorescent, as is its intercalation complex. AMT (4'-aminomethyl-4,5',8-trimethyl-psoralen, λexc: 336 nm, λem: 450 nm) shows hypoemission when intercalated into DNA. The 4',5' (furan)-cyclobutane adduct with pyrimidine bases (mainly thymine) is fluorescent, but on account of the reduction in the number of conjugated double bonds, the 3,4 (pyrone)-cyclobutane adduct and the double photoadduct are not fluorescent [108]. These adducts can be photo-reversed by λexc: 260 nm.

4.6. Anthraquinones

The antitumor drug mitoxantrone is a dihydroxy-diaminoalkyl substituted anthraquinone derivative that intercalates into DNA [109, 110]. This drug and other dyes such as the fluorescent DRAQ5 (also used as a vital probe, see Chapter 15.4.1), have strong absorption in the red spectral region. The miniduplex d(CpG) has been used to study intercalative binding of mitoxantrone and similar drugs by computerized model building [111]. Other anthraquinone derivatives are the anthracyclines and the pluramycin family of antitumor antibiotics, which contains an anthrapyran-4,7,12-trione chromophore. Hedamycin achieves biological activity by intercalating between T·A/G·C bases and alkylating the DNA duplex with an epoxide group that reacts with N7 of guanine [112]. Fluorescence and microscopical use of pluramycins have not yet been described.

4.7. Porphyrins and Phthalocyanines

Cationic porphyrins and phthalocyanines bind to NAs. *Meso*-tetra(4-N-methylpyridyl)-porphine (TMPyP) intercalates into GC base pairs of DNA, and binds to AT sites on the minor groove, inducing fluorescence in chromatin DNA [113]. The similar photoactive porphyrin *meso*-tetra(4-*N,N,N*-trimethylanilinium)-porphine (TMAP) is a fluorescent outside DNA binder with external self-stacking [114]. Both porphyrins have been used for chromatin counterstaining [115]. For additional features of cationic porphyrin binding to NAs see Chapter 15.4.1.

Cuprolinic blue (cooper quinolinic phthalocyanine) stains selectively single-stranded RNA, and has been used in electrophoresis and electron microscopy [116]. Hydrophobic interactions between the dye and RNA bases seem to account for this selective staining. The fluorescent zinc derivative of this dye, zinc(II)-tetramethyl-2,3-pyridino-porphirazinium [117], should be useful for RNA visualization in fluorescence microscopy. Intercalation of these macrocycles into three-stranded RNA regions could explain the preference for this substrate [14].

4.8. Naphthalimides

Fluorescent naphthalimides are used as industrial brighteners, anionic dyes, and probes for detection of proteins and cell interactions (see Chapters 13.2.1 and 16.5.1). Cationic naphthalimides are also drugs with high antitumor activity that bind to DNA by intercalation, and inhibit replication, topoisomerase II and transcription activity. They possess excellent properties as DNA fluorochromes [118], particularly the 3-amino-naphthalimide derivative FA-142 (amonafide), which induces green-yellow fluorescence in chromatin DNA (λexc: 346/408 nm, λem: 570 nm), although other polyanions also appear fluorescent (Fig. **12.15A**).

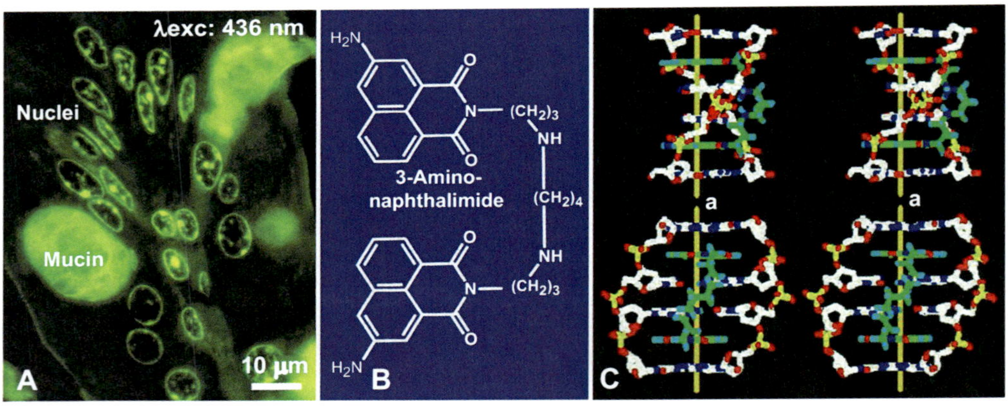

Fig. (12.15). **A**. Semithin section of mouse large intestine (glutaraldehyde, Epon) stained with the naphthalimide FA-142 (50 µg/mL). **B**: Chemical structure of bis(3-aminonaphthalimide)-spermine (BANI-343). **C**: Stereopair model of BANI-343 bis-intercalated into $d(GC)_4 \cdot d(CG)_4$. Side (up) and front (bottom) views are showed. Ligand C and H atoms are green and light blue, respectively. a: DNA axis.

Likewise, symmetric naphthalene-diimide derivatives with alkylamino substituents at each imide position have shown a threading intercalation mode with preference for GC sequences [119]. No applications in fluorescence microscopy have been found for these intercalators. However, tetrasubstituted naphthalene-diimides have proved useful for fluorescent labeling of nucleoli.

4.9. Bis-Intercalation

Bis-intercalation occurs when two aromatic chromophores separated by a link of 10.2 Å become inserted into a DNA duplex, which is the case for some antibiotics and synthetic drugs with antitumor activity. Quinoxaline antibiotics echinomycin and triostin A (from *Streptomyces* S-2-210, resembling *S. aureus*) bind to DNA by bis-intercalation and recognize GC/CG base pairs [120]. The emission of the quinoxaline and quinoline rings from these antibiotics seems to have found no applications in fluorescence microscopy. The antitumor antibiotic carzinophilin A (from *Streptomyces sahachiroi*) bis-intercalates into DNA inducing interstrand cross-linking by means of its two alkylating aziridines [121]. The emission of the naphthalene ring (λexc: 346 nm, λem: 427 nm), and its 200% increase when bound to DNA, should allow microscopical applications. Other bis-intercalators are ditercalinium [122, 123], and the quinolinic luzopeptin A [124].

In contrast, the synthetic bis-intercalating fluorochromes ethidium homodimer, TOTO and YOYO, have wide applications for DNA imaging. Members of the cyanine dimer dyes (TOTO family) are impermeant with live cells, and show high affinity for NAs. The flexible bridge between YOYO and TOTO fluorophores has 11 atoms; that of ethidium homodimer has 10 atoms. In all cases, a linker with this length is required for effective bis-intercalation. As YO (oxazole yellow) and TO (thiazole orange) have torsional freedom, a dramatic increase of emission occurs after DNA intercalation [79]. Advantages of TOTO and YOYO are high selectivity for DNA, minimal fluorescence fading, and there being no need to wash off excess stain [78, 125, 126]. The related TO-PRO family of cyanine monomers are generally impermeant to live cells, but YO-PRO-1 is very popular as a fluorescent probe for chromatin DNA in live cells (see Chapter 16.1.1).

Bis-naphthalimides are also antitumor agents [127]. Using the spermine bridge, bis-intercalating naphthalimides are obtained, which induce bright emission in chromatin DNA. Fig. (**12.15B, C**) shows the structure of bis(3-amino-naphthalimide)-spermine (BANI-343), and a model of bis-intercalated DNA miniduplex. Likewise, bis-intercalating spermine aryl-pyrrolones are strongly fluorescent and could be used in DNA cytochemistry.

Poly-intercalators based on the naphthalene diimide ring linked by peptide bonds have been suggested for DNA binding [128]. They show strong preference for GC over AT sequences, and intercalate from the major groove and not in a threading manner. The emission of the poly-intercalating chromophore have been not used in microscopy. An interesting concept is bifunctional side-by-side intercalation, in which a bis-intercalator with a rigid linker can intercalate into two adjacent DNA duplexes [129].

4.10. Intercalating Antibiotics

The antitumoral antibiotic actinomycin D binds selectively to DNA by intercalation of the oxazone ring between G·C/C·G base pairs, while the two pentapeptide rings lie in the minor groove and form H bonds with G. Although actinomycin D is somewhat fluorescent, its 7-amino derivative (7AAD) shows enhanced emission and has been used as a fluorochrome for GC-rich DNA in banding pattern [32], cytochemistry [130, 131], and flow cytometric analysis [89, 132]. Spectral features of the 7AAD-DNA complex (λexc: 488 or 543 nm; λem: 655 nm), allow the use of other simultaneous probes (*e.g.*, FITC, pyronine Y). Bis-actinomycin complexes also binds strongly to GC-rich DNA [16].

DNA-binding anthracycline antibiotics contain an anthraquinone chromophore such as daunomycin (DNM, see Fig. **8.9A**), and doxorubicin (adriamycin, DOX), both from *Streptomyces peucetius*. Antitumoral anthracyclines intercalate into DNA [7] inducing genotoxic effects related to inhibition of DNA topoisomerases, and DNA cleavage by free radicals by a Fenton reaction based on an anthracycline-Fe complex [133, 134]. Anthracyclines are fluorescent drugs, a property applied in flow cytometry, vital nuclear labeling, and Q-type banding [125, 135].

DNM and DOX intercalate only into DNA and not into RNA, as they require a B-type duplex and recognize G·C/C·G sites [7, 136]. DNM and DOX are useful DNA probes for live cells (see Chapter 15.4.1). Fluorescence intensity values of DNM are 100% free in solution, ~ 40% bound to AT duplexes and ~ 5% with GC duplexes. This differential quenching effect allows to visualize AT-rich G/Q bands in great detail [125]. After DNM or DOX staining of chromatin, Al^{3+} ions increase the red emission, but the site of metal binding in the complex is not known (for chemical structures and images see Figs. **8.9** and **15.8**). As the chelating oxygens O5/O6 and O11/O12 lie within the intercalating cavity, they are not readily available for metal ion binding [137 - 139].

Some other DNA-intercalating anthracycline drugs are fluorescent and have antitumor action. Examples are the trisaccharide-containing aclacinomycin A (from *S. galilaeus*), carminomycin (carubicin, a demethylated DNM derivative from *Actinomadura carminata*), nogalamycin (from *S. nogalater*), and the rhodomycins (from *S. purpurascens*, with two relatively symmetrical aminosugars). The synthetic DNM derivative MAR70 (with an additional hexose at C4') forms a formaldehyde adduct with 2-aminopurine [139]. Certain new bis-intercalating anthracyclines also bind with high affinity to DNA [1, 134].

5. MINOR GROOVE BINDING

NAs grooves are helicoidal ramps defined by phosphate-sugar units and the edges of base pairs. In antiparallel duplexes, two grooves (major and minor) of different width and depth characterize the A, B, and Z-forms of DNA. In B-DNA, the minor groove is the most commonly used binding site for small ligands, with the cyclic oxygen of deoxyribose and H atoms at C1', C4', and C5' forming the minor groove wall. O_2, N_3, and H_2 atoms form the electronegative floor of the narrow minor groove at AT sequences (Fig. **12.16**). In contrast, the geometry of the RNA minor groove is very different to that in DNA, the former being more shallow, wide, and electropositive [5]. Therefore, suitable interactions that occur with ligands in DNA minor groove are not possible with RNA.

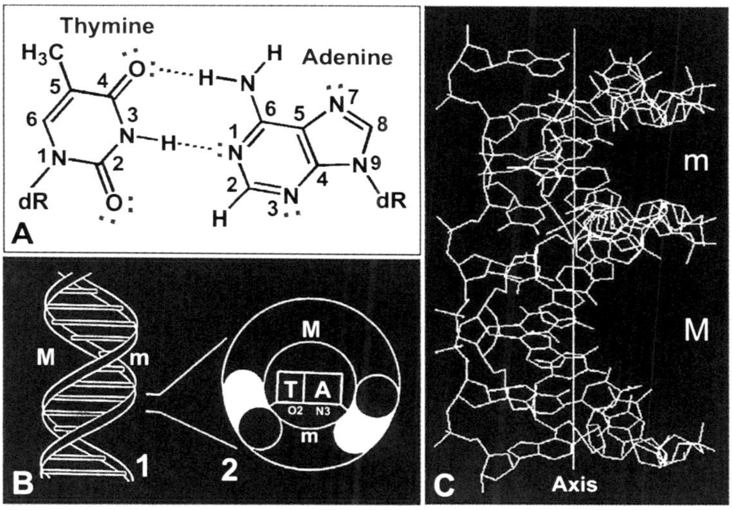

Fig. (12.16). A: AT base pair, showing atom numbering and donor electron pairs (blue dots). dR: deoxyribose. **B:** Schematic anatomy of the B-DNA grooves. 1: Side view of the major (M) and minor (m) groove. 2: Axial view of the phosphate-deoxyribose chains (curved white cylinders), major and minor grooves, and a TA base pair with electron donor atoms O_2 and N_3. **C:** Skeletal view of a $(dA)_{12}\cdot(dT)_{12}$ segment tilted 35 ° from top to the observer. The different width of M and m grooves is clearly seen.

An ordered row of water molecules (hydration spine) is bound by H bonds to O_2 and N_3, and after H_2O displacement, these atoms can also bind to metal cations (*e.g.*, Na, K, Cs) or organic cations (*e.g.*, bowed dyes and drugs). Molecular aspects [140, 141], and relevant biological activities (antiprotozoal, antitumoral, antiviral, bactericide, genotoxic and mutagenic) are often associated with minor groove binding [4, 6, 142].

Insertion of ligands into the DNA minor groove resembles the formation of inclusion complexes, and therefore host molecules are useful model systems for studying this binding mode. The interior of cyclodextrins shows an environment similar to the DNA minor groove, and is useful in spectroscopic studies [63]. Depending on the guest compound, H bonding, van der Waals, and hydrophobic forces contribute to the binding. Fig. (**5.13**) illustrates geometrical correlations between some minor groove binders and α-cyclodextrins. Likewise, AT-rich DNA substrates are useful models to analyze the possible minor groove binding of ligands. Examples are C bands from the mouse and broad bean, kDNA from *Leishmania* and *Typanosoma*, satellite DNA of crabs, *Plasmodium falciparum*, *etc*.

5.1. Pyrrole-Carboxamide Antibiotics

Distamycin A (DA) and netropsin are pyrrole-carboxamide antibiotics from *Streptomyces distallicus* and *S. netropsis*, respectively. The binding mechanism involves a close fitting between the bowed ligand and the convex floor of AT-sequences on the DNA minor groove through H bonds, van der Waals contacts, and electrostatic forces [140, 143] (Fig. **12.17A**). The figure shows the structural motif "6-atoms-in-a-row" (blue lines) that are commonly found in AT-rich minor groove ligands.

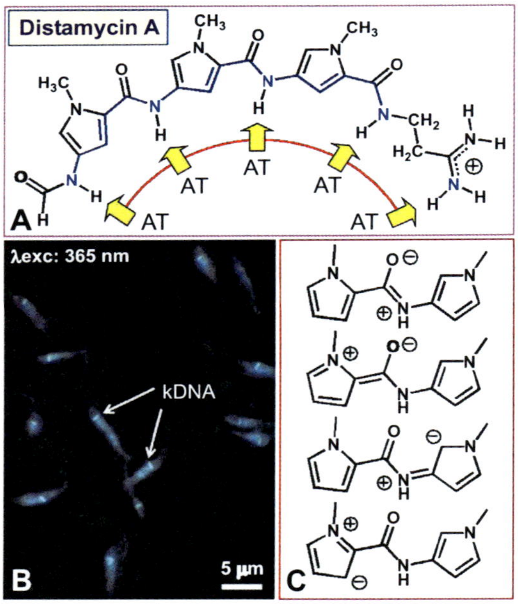

Fig. (12.17). A: Chemical structure of distamycin A showing H bonding (yellow arrows) to AT base pairs on the DNA minor groove (red line). **B:** Methanol-fixed *Trypanosoma cruzi* epimastigotes stained by 10 µg/mL distamycin A, showing the blue-white fluorescence of kinetoplasts (kDNA). **C:** Resonance hybrids of the N-methylpyrrole-carboxamide units of distamycin A allowing electronic coupling.

It is currently assumed that DA does not fluoresce, and this accounts for its use in cytogenetics as a counterstain to improve chromosome banding [144, 145]. However, DA is fluorescent and its emission strongly increases in the presence of AT-rich DNA (λexc: 365 nm, λem: 445 nm) [146]. An example is shown in Fig. (**12.17B**). Resonance hybrids of carboxamide bridges [147] can account for the electronic coupling between methylpyrrole units (Fig. **12.17C**), and this feature could explain the emission of DA.

Studies with the dodecamer duplex $d(CGCAAATTTGCG)_2$ show that at low drug/DNA ratios (< 1) there exists only one DA per duplex, but at higher ratios (> 2) a new binding mode occurs, in which two antiparallel and face-to-face DA molecules bind simultaneously to the minor groove [148]. In this case, the three-center H bond seen in 1:1 drug/DNA complexes is not possible in 2:1 complexes, and H bonds are not bifurcated. Synthetic ligands based on the DA architecture also bind in the minor groove as dimers [15, 149]. Instead the formamide group of DA, dansyl derivatives have been used in minor-groove binding studies to DNA [16].

On account of its guanidine group, arginine selectively interacts with the DNA minor groove [150], and could be applied as competing agent to prevent binding of fluorochromes at this site. Binding of arginines to narrow minor grooves is a known mode for protein–DNA recognition. This readout mechanism exploits the fact that narrow minor grooves have a strong negative electrostatic potential. The nucleosome core particle offers a prominent example of this effect. Minor-groove narrowing is often associated with the presence of A-tracts [150].

In addition to arginine, linear and non-fluorescent cationic compounds containing amidine or guanidine groups could be expected to compete for binding to the DNA minor groove. Washing microscopic preparations with 10 mM phenformin (1-phenethyl-biguanide) abolishes the emission of minor groove binders, whereas 1 M NaCl and $MgCl_2$ have no effect [118]. Since chlorhexidine, the anti-malarial paludrine, streptomycin, and poly-arginine (*i.e.* protamines) also contain guanidine groups, they should be useful as minor groove competing agents.

5.2. Isohelicity

Based on the molecular architecture of pyrrole-carboxamides, methods for predictive isohelical binding of ligands to the DNA minor groove have been described [151], also allowing the recognition of base pairs sequences [15, 149]. Simple methods to assess the curvature of both the DNA minor groove and ligands, to allow the comparison of geometrical fit and the possibility of H bonding, have been published [152, 153].

In Fig. (**12.18A**) the duplex $d(A)_{12}{\cdot}d(T)_{12}$ shows the mid line on the floor of the minor groove, which corresponds to the sequence of H_2 atoms along consecutive adenines. Helicity patterns of the duplex $d(A)_{12}{\cdot}d(T)_{12}$ can be seen in Fig. (**12.18B**), with H_2 curve in green. The white HL curve indicates the maximum proximity for hydrogens of the ligand. Taking into account the fact that numerous minor groove ligands form H bonds with thymine O_2 and adenine N_3, a new curve (HB, similar to HL) can be generated, where H donors are available for H bonding. Different aromatic units can be designed and assembled to satisfy the isohelicity, also allowing H bonds either for AT or GC base pairs according to an "8-atoms-in-a-row" pattern [153]. An example of isohelical binding model between the benzoxazole-amine (BOA) octamer and AT sequences of DNA minor-groove is illustrated in Fig. (**12.18C**).

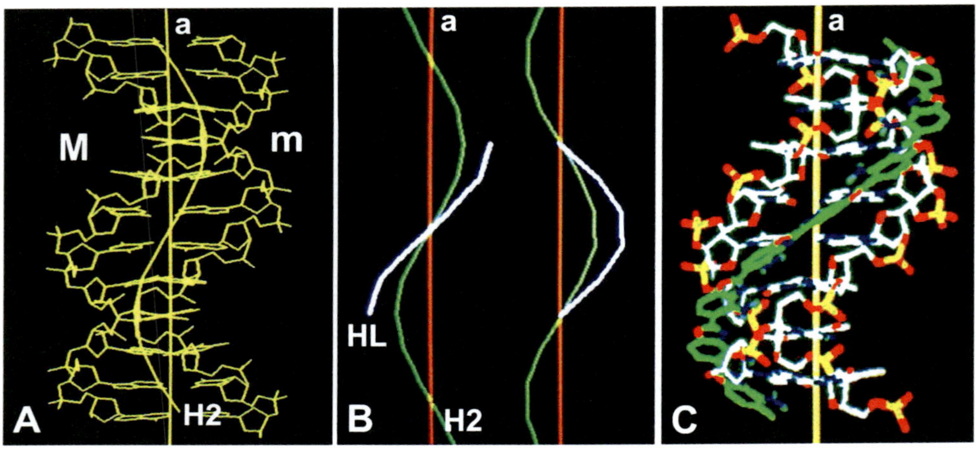

Fig. (12.18). Isohelical binding. **A:** The duplex $(dA)_{12}{\cdot}(dT)_{12}$ showing the major (M) and minor (m) groove, and the H_2 helical curve, formed by consecutive H_2 from adenines and representing the mid line on the convex floor of the minor groove. **B:** Helicity patterns in the duplex $d(A)_{12}{\cdot}d(T)_{12}$ in a frontal (left) and lateral view (right). Green line: H_2 curve. White line: HL curve as the position of maximum proximity to hydrogens of the ligand (L), which can also form H bonds. **C:** Model of isohelical minor-groove binding of a benzoxazole-amine (BOA) octamer to $(dA)_8{\cdot}(dT)_8$. C atoms of BOA are seen in green. Vertical lines (a) are the axis of the duplexes.

5.3. Benzimidazoles

The bis-benzimidazoles Hoechst 33258, Hoechst 33342 and nuclear yellow (Hoechst S-769121) are the best known and most widely used fluorochromes in this group (Fig. **12.19A**). Under UV excitation, they induce a bright white-blue emission of chromatin DNA, with applications in cell biology and cytogenetics [154 - 156]. Crystallographic studies have revealed the binding of Hoechst 33258 (H-33258) to AT sequences on the minor groove of B-DNA [157], where the dye makes two bifurcated H bonds and extends over three consecutive AT base pairs

[140, 142]. These fluorochromes bind selectively to this site at neutral pH (when protonated at N4), but at pH 2 (with protonated N4, N3 y N3') also induce fluorescence in RNA-containing structures. Human metaphase chromosomes stained with nuclear yellow and counterstained with DA or methyl green show the AT-rich heterochromatin in chromosomes 1, 9, 15, 16, and Y (Fig. **12.19B**), but mouse centromeric heterochromatin can be directly visualized with nuclear yellow.

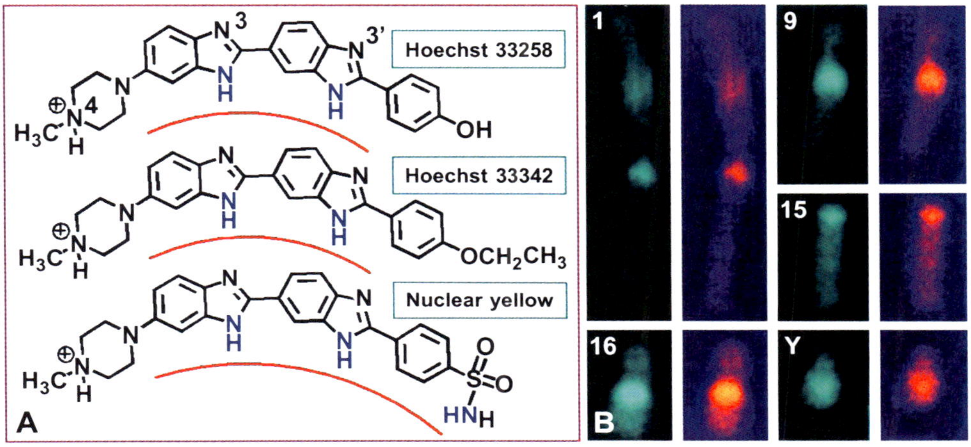

Fig. (12.19). A: Bis-benzimidazole fluorochromes located on the minor groove floor (red curves) showing H bonds between >NH groups (blue) and O_2 and N_3 atoms from AT base pairs. **B:** Human chromosomes 1, 9, 15, 16, and Y after staining with 5 μg/mL nuclear yellow followed by 0.1 mg/mL distamycin A and UV excitation, showing the AT-rich heterochromatic regions (at left), and a pseudocolor equidensity image (LUT: fire, at right).

The emission color of minor groove ligands depends on their concentration and the available substrates. Using H-33258, a polyanionic structure such as the yeast cell wall shows a clear metachromatic shift (Fig. **12.20A**), similarly to that of DAPI [158]. Likewise, when H-33258 is used at low (10^{-6} M) or high (10^{-5} M) concentration, chicken erythrocyte nuclei show emission peaks at 445 nm and 480 nm, respectively (Fig. **12.20B**). A temperature-induced hyper-chromism at 260 nm due to disaggregation that interferes with DNA melting has been described for H-33258 [159].

Several binding modes have been described for this dye. Edge-to-edge dimers of H-33258 have been suggested for DNA minor groove binding, but the H bonding geometry makes them unlikely. In contrast, antiparallel face-to-face dimers could insert into the AT-rich minor groove [160] (Fig. **12.20C**). In the case of the AT-rich satellite DNA of live mouse cells, treatment with AT binders such DA and

H-33258 induces a severe undercondensation of the centric heterochromatin, which is related to the structural curvature of A tracts [161]. H-33258 is also used as photosensitizer after BrdU labeling (see Fig. **4.16A**).

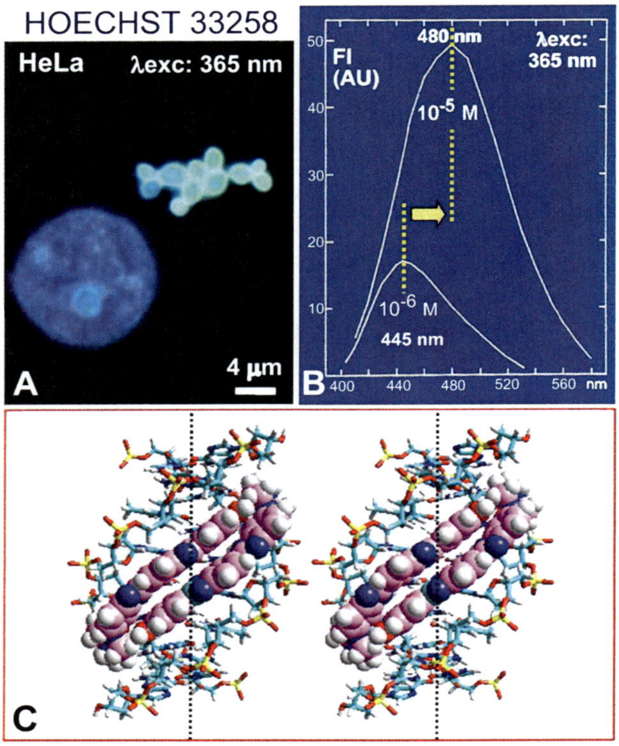

Fig. (12.20). A: Methanol-fixed HeLa cell infected with *Candida famata* and stained with 5 µg/mL Hoechst 33258, showing orthochromatic (blue) and metachromatic (blue-green) fluorescence of nuclei and yeast cell walls, respectively (courtesy of M. Pacheco and A. Juarranz). **B:** Emission spectra of nuclei stained with 10^{-5} M and 10^{-6} M Hoechst 33258. Observe the red shift. C: Model of an antiparallel face-to-face dimer of Hoechst 33258 inserted into the minor groove of the oligomer $d(GCTATAGC)_2$. Dotted line: oligomer axis.

5.4. Aromatic Diamidines

Berenil (diminazene), Dann 488/45, Dann 488/48, DAPI, diamidine-biphenyl-imidazoles, DIPI, 2-hydroxy-stilbamidine, M&B938, pentamidine, phenamidine, propamidine, *etc*., are aromatic diamidines, a class of dicationic DNA minor-groove binders with a long history of clinical success as anti-protozoal agents [3, 162, 163]. Similarly to Hoechst dyes, aromatic diamidines were originally synthesized as trypanocides and not as dyes.

4',6-Diamidine-2-phenylindole·2HCl (DAPI) was synthesized by Dann *et al.* and has been widely used as a DNA fluorochrome in biomedical studies [164, 165]. In the presence of DNA and mainly poly(dA-dT) and poly(dA)·poly(dT) duplexes, the emission of DAPI is strongly enhanced, inducing a bright white-blue fluorescence of chromatin DNA (λexc: 344 nm; λem: 449 nm). Early applications of DAPI included the visualization of mitochondrial DNA [166], chloroplast DNA [167], and mycoplasma contamination in cell cultures [168]. It has been reported that in addition to binding into the DNA minor groove, DAPI can also intercalate into RNA and polyA·polyU duplexes [169]. As both binding modes involve dye rigidity and increased emission, it is intriguing that no microscopical fluorescence is found in RNA components.

With the exception of DAPI, the use of other aromatic diamidines is not frequent in DNA cytochemistry and chromosome banding. However, berenil, Dann 288/45, Dann 288/48, and DIPI, have been employed for these purposes (see Fig. **7.15B-D**) [63, 170]. 2-Hydroxy-stilbamidine is a known fluorochrome for selective staining of DNA, although mucosubstances and elastic fibers are also stained [171]. 2-Hydroxy-stilbamidine (fluoro-gold) and nuclear yellow are retrograde axonal tracers (see Chapter 16.5.2).

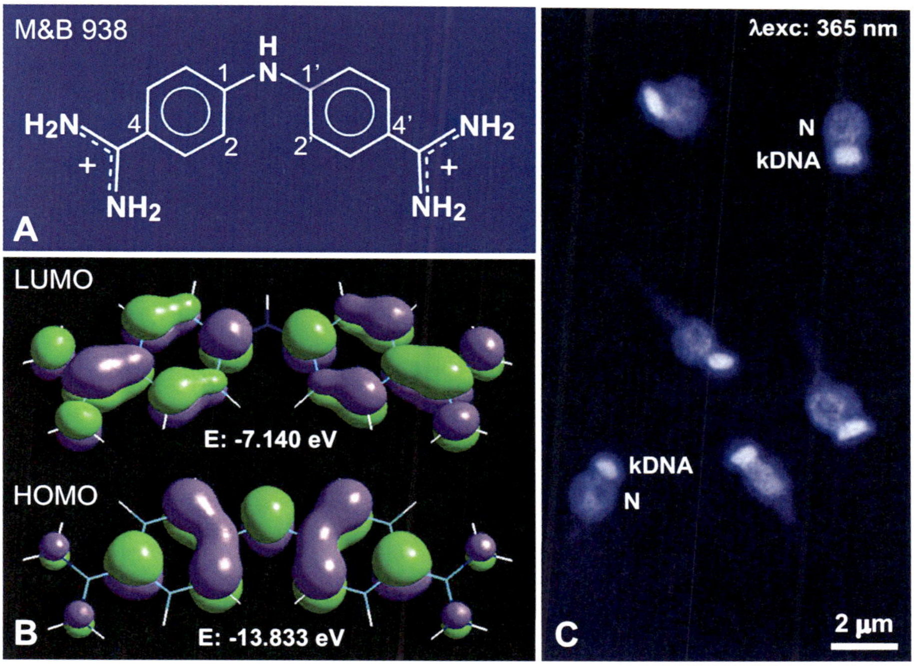

Fig. (12.21). **A**: Chemical structure of the dication 4,4'-diamidino-diphenylamine (M&B 938). **B**: HOMO and LUMO orbitals of M&B 938, showing the energy levels and the paraquinone electronic structure of LUMO. **C**: Methanol-fixed *T. cruzi* epimastigotes stained with 1 µg/mL M&B 938, showing fluorescence of kDNA and nuclei (N).

Like pentamidine and berenil, the simple bis-amidine trypanocidal drug M&B 938 (4,4'-diamidino-diphenylamine [172], Fig. **12.21A, B**), binds to the minor groove at AT sequences. It is highly fluorescent and has been used in DNA cytochemistry [118, 170, 173]. Chromatin DNA shows a bright white-blue emission at neutral or alkaline pH, whereas other polyanions (RNA, heparin) remain non-fluorescent, although they fluoresce at acid pH. Particularly intense is the emission of kDNA (Fig. **12.21C**).

A comparison between cationic DNA-binding fluorochromes with and without potential to form H bonds has shown that only the former reveal C bands [170]. This correlation between C banding and H-bonding for M&B 938 and auramine O is illustrated in Fig. **(12.22)** (see also Fig. **7.19A**).

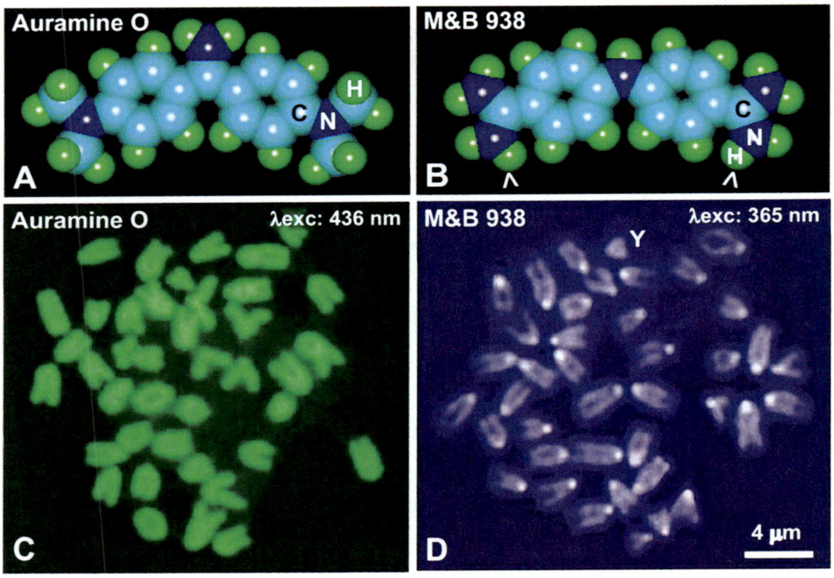

Fig. (12.22). Correlation between C-bands and H-bonds. **A**, **B**: Molecular structure of auramine O and M&B 938, showing similar shape, size, and possible H acceptors (arrowheads). **C**, **D**: Fluorescence of metaphase chromosomes from mouse bone marrow stained with 5 µg/mL auramine O (C) and 1 µg/mL M&B 938 (D). Highly fluorescent AT-rich centromeric heterochromatin and Y chromosome can be seen in D.

5.5. Benzothiazoles

Great interest is currently being devoted to benzothiazole fluorochromes. Thioflavine T (ThT, CI 49005) has been widely stated to contain a single benzothiazole (BT) unit, but analytical work [174, 175] shows that commercial lots of ThT are mixtures of the mono-, di-, and tri-BT species (see Fig. **13.7A, B**). Possibly the mono-BT is the dye fraction that intercalates into DNA [176], and labels nucleoli and mitochondria in live cells. In addition to the detection of amyloid (see Chapter 13.1.4), microscopical applications of ThT includes the visualization of basophilic cytoplasm, chromatin, granules from mast cells, Paneth cells, and juxtaglomerular cells, kinetoplast DNA (Fig. **12.23A**), nucleoli, *etc.* [14, 63, 177]. Analyzed on oriented DNA fibers, the polarized emission of thioflavine T is $\| > \perp$, which indicates minor grove binding [61], possibly due to the di- and tri-BT components.

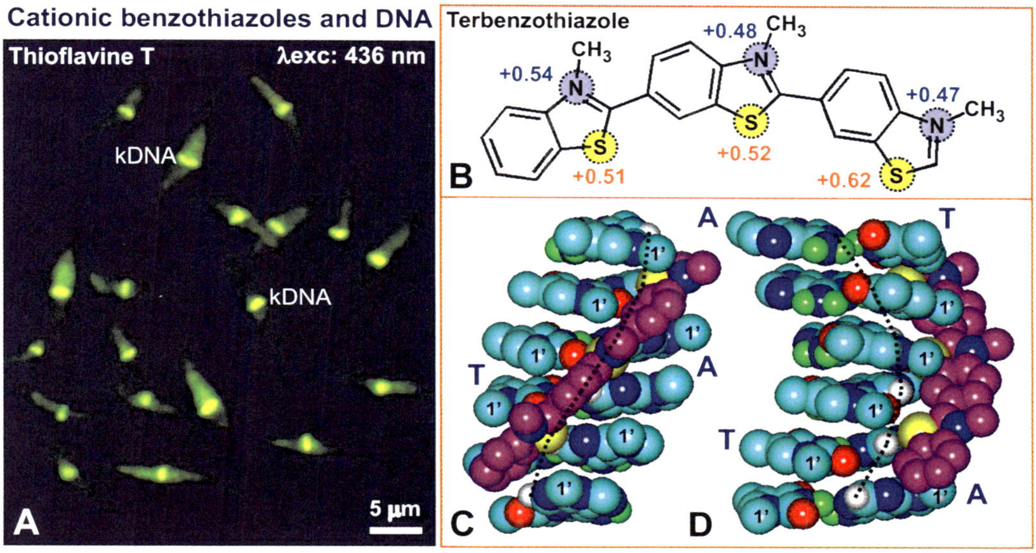

Fig. (12.23). A: Methanol-fixed *T. cruzi* epimastigotes stained with 5 µg/mL thioflavine T showing selective fluorescence of kDNA. **B:** Terbenzothiazole (tri-BT) with partial charges on N and S atoms. C, **D:** Frontal (C) and lateral (D) view of a space-filling model of $d(A)_6 \cdot d(T)_6$ (propeller twist angle: 22 °), with tri-BT (violet carbons, torsion angle of BT units: 20 °) on the minor groove. Only base pairs (AT), H bonds (green) and adenine H2 (white) are shown. Deoxyribose C1' atoms are indicated. The dotted line between consecutive H^2 is the midline of the electronegative minor groove floor.

Although commonly the formal positive charge in cationic benzothiazoles is located on the N atom, calculated partial charges show that S atoms also have a high δ^+. Because of the molecular asymmetry, partial charges between N and S atoms are not equivalent (Fig. **12.23B**). A model for the insertion of terbenzothiazole (tri-BT) into the negative charge rich AT-rich minor groove is shown in Fig. (**12.23C, D**). Benzothiazole rings occur in 4-hydroxystyrene-benzothiazole (new fluorescent substrate for peroxidases), the metabolite luciferin (substrate for bioluminescence), and thiazole orange (TO). In other fluorochromes such as the asymmetric cyanine BEBO, the structure of the known intercalating cyanine dye BO has been extended with a benzothiazole group that binds in the minor groove of AT-rich DNA [178, 179], supporting the potential of the benzothiazole group as a minor groove recognition moiety. BEBO has a crescent shape, and upon binding to DNA the fluorescence increases about 200-fold, which is the largest increase reported for a minor groove fluorochrome.

In addition, symmetrical cyanine dyes containing two benzothiazole units bind to the AT-rich minor groove either as monomers or dimers [180, 181]. A trimethine-benzothiazole-quinoline cyanine dye, TO-PRO-3, forms three complexes with DNA, complex II being inserted in the minor groove [182]. TO-PRO-3 is one of the best nuclear fluorochromes for confocal laser scanning microscopy in 3-D [183].

5.6. Other Linear Ligands

Other linear agents with pharmacological activity against *Trypanosoma* and *Babesia* diseases are bis(6-quinolyl)urea, imidocarb, quinapyramine (antrycide), *etc*. They are suitable for inserting into the DNA minor groove, and could be fluorescent. The same is expected for linear derivatives of Ehrlich's reagent (DMBA) with aromatic amino groups (see Chapter 4.6.2). Likewise, a similar binding mode could occur for imidazoline-terephthalanilides, which are fluorescent antitumor drugs. Although not using H bonds, arc-shaped fluorochromes could have preference for minor groove binding (*i.e.* BAO, pyrvinium, Q-dmPOPOP) [63, 118, 184, 185]. Curcumin (with interesting biomedical properties, see Chapter 3.4.1) induces fluorescence in kinetoplast and chromatin DNA (see Fig. **3.19**). In the presence of DNA (4.5 x 10^{-5} M in bases), the emission at 548 nm of 5 μg/mL curcumin (λexc: 440 nm) increases 3.3 times and shifts to 532 nm.

5.7. Bifunctional Minor Groove Binding

Bifunctional minor groove binding is a new interaction of some ligands with two DNA duplexes. Bis-imidazoline agents such as 4,4'-bis(imidazolinylamino--diphenylamine (CD27) are good examples of minor groove ligands [186]. They could be useful against the human pathogens *Plasmodium falciparum*, *T. cruzi*, and *Trichomonas vaginalis*, which contain very AT rich genomes. In addition to minor-groove binding to AT-tracts, two CD27 molecules are able to cross link the phosphates of two oblique DNA duplexes.

A similar cross-linking occurs between pentamidine bound to AT sequences and neighboring DNA duplexes [187]. It is likely that other drugs (*i.e.* berenil, which can form dimers by H bonds [188]) show similar interactions when bound to appropriate DNA tracts. The formation of cross-links may be a feature related to the biological action of these drugs used as therapeutic agents in AIDS-related pneumonia induced by the unicellular fungus *Pneumocystis carinii*, leishmaniasis, and trypanosomiasis.

In *Leishmania* and *Trypanosoma* parasites, kinetoplast DNA (kDNA), which is located within the single mitochondrion, has abundant oligo (dA)-oligo (dT) tracts, and appears organized as a mesokaryotic dinoflagellate chromosome, consisting of a network of catenated mini- and maxicircles, the later representing the informational mitochondrial DNA. There are 20-50 copies of maxicircles (about 40 kb each) and about 10^4 copies of about 1 kb (*Leishmania, Trypanosoma*) to 3 kb (*Crithidia*) AT-rich minicircles with a strong structural curvature [189 - 192], which is straightened by binding to DA and H-33258. Thus the minor groove of minicircles is a selective target for drug interaction, which is in part driven by the accumulation of cations in the mitochondria due to its membrane potential [193].

On the other hand, cationic porphyrins (*i.e.* alkyl-pyridinium derivatives) are fluorescent, photosensitizing, and bind DNA [113, 194]. *Meso*-tetra(4-N-methylpyridyl)-porphine (TMPyP) binds externally to AT sites [195]. Molecular modeling reveals binding to the AT-rich minor groove [196] with affinity about 200 times higher than intercalative binding to GC sites. Also the fluorescent ZnTMPyP complex binds preferentially to AT sites in the DNA minor groove [8].

Microscopical studies using TMPyP show the typical fluorescence pattern of minor groove binders, with a bright signal in chromatin DNA and no emission in basophilic cytoplasm (see Figs. **7.4A** and **15.9B**). TMPyP could also bind in a bifunctional way to the minor groove of two DNA duplexes, either orthogonal or conveniently tilted (Fig. **12.24A, B**). In addition to chromatin DNA, TMPyP binds to AT-rich kDNA from *T. cruzi* giving strong red fluorescence (Fig. **12.24C**). Bifunctional minor groove binding sites would be (1) interlocked AT-rich minicircles in kDNA, (2) supercoiled DNA molecules, and (3) linker DNA between nucleosomes in the crossed-linker model of the 30 nm chromatin fiber. An ideal fluorescent drug for bifunctional minor groove binding should be *meso*-tetra-amidinophenyl-porphine (TAPP). In this case, two amidine groups could bind to AT sites in one DNA duplex, remaining available the other two amidine groups for binding to AT sites in the second DNA duplex (Fig. **12.24D**). *Meso*-tetra-*p*-aminophenyl-porphine would show a similar binding mode.

5.8. Minor Groove-Binding Antibiotics

Chromomycin A_3 (CMA, from *Streptomyces griseus*, Fig. **12.25A**) is an antibiotic with antibacterial, antifungal, and antitumor activities, which also serves as a fluorescent DNA stain and vital probe [197, 198] (see Chapter 15.4.1). CMA blocks macromolecular synthesis by a specific, reversible interaction with DNA in the presence of bivalent metal ions, particularly Mg^{2+}. Binding to DNA minor groove at the level of GC [199] mediates an efficient competitive inhibition of DNA gyrase and significantly affects topoisomerase II activity. A model of interaction of CMA with a DNA duplex is shown in Fig. (**12.25B**). In cytogenetic studies, highly selective fluorescence of GC-rich sites is achieved by CMA (Fig. **12.25C**). Mithramycin (from *S. argillaceus*) and olivomycin A (from *S. olivoreticuli*) are also fluorescent antineoplastic antibiotics similar to CMA, and have been used for nuclear staining in the presence of Mg^{2+} [200, 201].

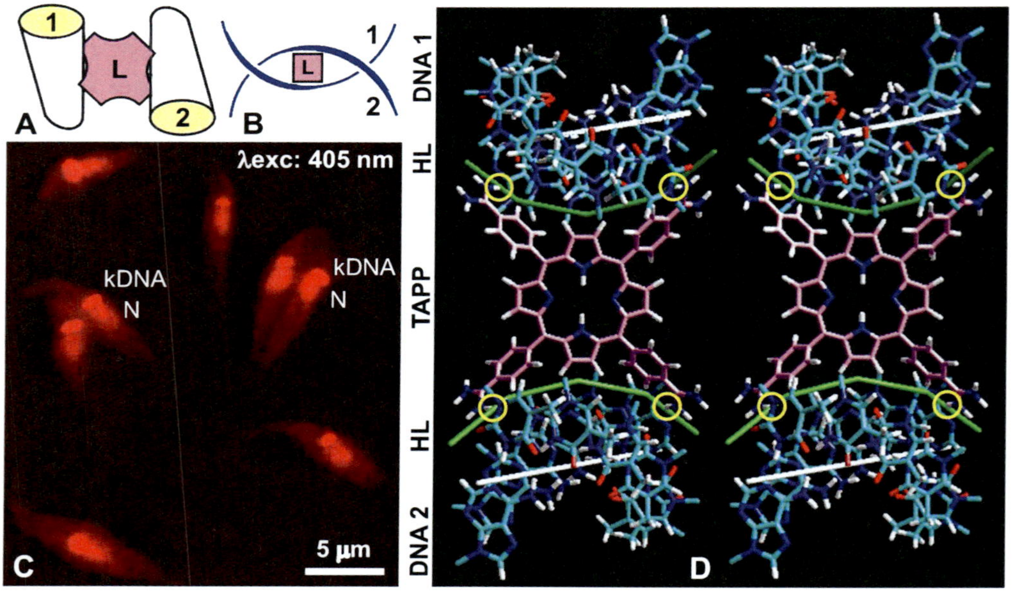

Fig. (12.24). Binding of cationic porphyrins to DNA minor groove. **A:** Schematic view of a ligand (L) bound to two DNA duplexes (1 and 2). **B:** Local structure of two catenated DNA minicircles (1 and 2) bound to L. **C:** Methanol-fixed *T. cruzi* epimastigotes stained with $5 \cdot 10^{-6}$ M TMPyP, showing the fluorescence of kDNA and nuclei (N). **D:** Stereopair of a model for TAPP binding to the minor groove of two $d(A)_6 \cdot d(T)_6$ duplexes. The ligand (TAPP, violet), HL (green curve), H bonding (yellow circles), and helical axis (white line) of DNA 1 and DNA 2 (tilted 37.5 ° to the plane of porphine ring) are indicated.

6. SPECIAL NUCLEIC ACIDS STRUCTURES

Regarding special structures of NAs, fluorescent methods based on high-affinity binding can be applied to analyze topographical features related to the structural and functional organization of DNA. Chromatin DNA is a complex structure, involving nucleosomal and supra-nucleosomal organization, regulation of the transcription machinery, topographic positioning of genes, *etc*. Reviews on the functional nuclear and chromosome structure have been published [202 - 204]. The organization of 10 nm chromatin fibers into 30 nm fibers is not yet well known, although the crossed-linker model has considerable acceptance [205 - 207]. Side-by-side intercalation and minor groove binding could selectively reveal parallel and orthogonal DNA duplexes, respectively, and thus visualize features of DNA organization (*i.e.* catenated minicircles in kDNA, consecutive turns of nucleosomal DNA, crossed-linkers, loop anchorage to the chromosome scaffold, supercoiled circular DNA, 3- and 4-way junctions, *etc*).

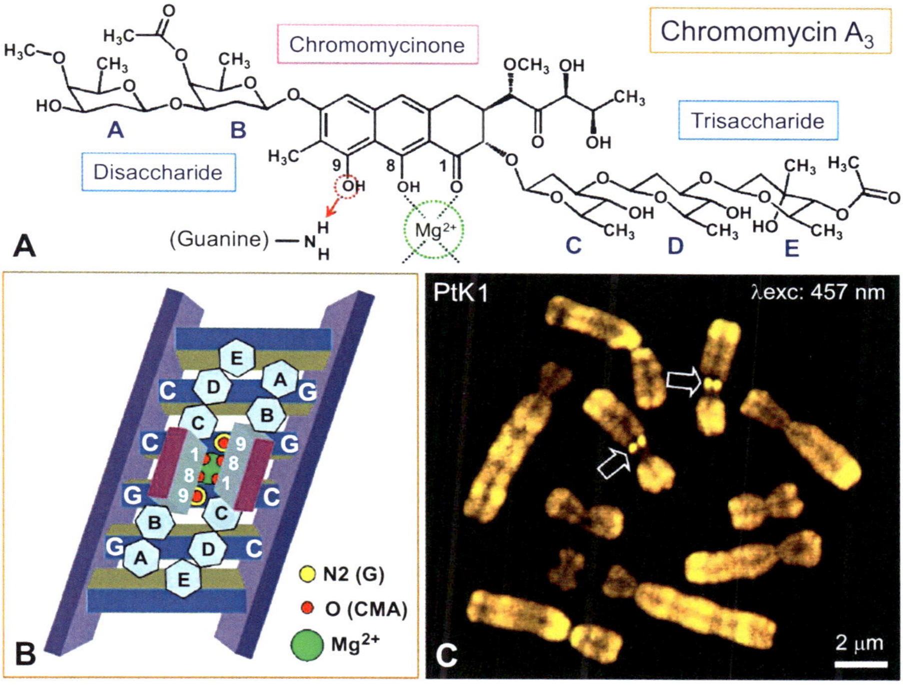

Fig. (12.25). A: Chemical structure of chromomycin A_3 (CMA) showing the chromomycinone ring, the di- (A, B) and tri-saccharide (C, D, E) chains, and the specific interactions with Mg^{2+} and the amino group of guanine (N2). B: Diagram of the $(CMA)_2$-Mg^{2+} dimer (antiparallel head-tail) bound to the sequence GGCC of B-DNA by H-bonding between O9 of CMA and guanine N2. C: Metaphase chromosomes from a PtK1cell stained with CMA and DA, showing bright GC-rich bands and the NOR-associated heterochromatin (arrows).

6.1. Aptamers

Sites for high affinity binding of ligands to NAs after *in vitro* selection ("aptamers", from the Greek "aptos": to fit [208]) can be revealed by fluorescence methods. When dissolved in viscous solvents or bound to substrates, malachite green (MG, with a very low Φ_F: ~ 10^{-5}) shows a strongly increased emission and has been employed as a fluorescent probe for detection of specific DNA or RNA sequences. MG aptamers (MGAs) are RNA molecules that have very high affinity to MG [209]. Upon binding to these RNA regions, the dye increases its fluorescence by > 2000-fold [210]. As the low amount of MGAs requires high instrumental sensitivity to be detected, microscopical uses of MG fluorescence have been overlooked.

6.2. Multistranded Polynucleotides

DNA and RNA polynucleotides can form multistranded structures (triplexes and tetraplexes, using Watson-Crick and/or Hoogsteen pairing), and possible biological roles include chromosome structure, genetic recombination, and use as possible antisense drugs and targets for anticancer agents [211 - 213]. The occurrence of triplex DNA in the chromosome structure has been suggested [214]. Interestingly, some dyes bind better to TAT triplexes than to TA duplexes (*e.g.* cresyl violet, Darrow red, ellipticine, ethidium bromide, oxazine 170, phenosafranine, safranine O, *etc.*) [212]. A benzo-pyrido-indole (BePI) drug intercalates preferentially into DNA triplexes [215].

In addition to DNA duplex intercalation, coralyne and related compounds intercalate into DNA triplexes formed from oligopurines-oligopyrimidines [212, 216]. A similar binding mode to triple helix regions of RNA could occur with berberine, which induces a selective fluorescence in RNA-containing structures. Interestingly, sanguinarine (from *Sanguinaria canadensis*), another isoquinoline alkaloid with different conformation to that of coralyne, binds selectively to T·A·T base triads, but in contrast to coralyne it does not bind to protonated $C{\cdot}G{\cdot}C^{+}$ triplex. An explanation is that the charged N^{+} of coralyne faces the electronegative minor groove, whereas in sanguinarine the N^{+} faces the electropositive major groove [212]. Uses of these alkaloids in fluorescence microscopy should be investigated.

In addition to duplex intercalation and minor groove binding, some cationic porphyrins show strong interactions with guanine quadruplexes (popular targets for drug binding [217]). Human telomeres comprise tandem repeats of the DNA motif (TTAGGG). The terminal 150–250 nucleotides at the 3'-ends are single-stranded, but are protected from G-quadruplex formation by a single-stranded DNA binding protein (hPOT1), permitting the regulation of telomerase action in cancer cells, and thereby maintaining telomere length. Loss of hPOT1 deprotects telomeres and initiates DNA damage-response and cell death. Small ligands that stabilize a G-quadruplex from the single-stranded TTAGGG compete with hPOT1 and also initiate this response [217]. G-quadruplex formation at the single-strand overhang may itself be a DNA damage signal.

Some fluorescent ligands such as acridine derivatives AS1410 and BRACO-19, quarfloxin, and substituted naphthalene-diimides bind to G-quadruplexes. The same occurs for coronene and perylene derivatives [218]. TMPyP intercalates into

G-tetraplexes, with stabilization of telomere DNA, telomerase inhibition, and cell death [219, 220]. Likewise, the enantiomer Λ-Ru(phenanthroline)$_2$(p-DMNP)$^{2+}$ also binds selectively to G-quadruplexes (see Chapter 15.4.1). Phthalocyanines and similar large fluorophores should be suitable to bind G-quadruplexes [14].

6.3. Three- and Four-Way Junctions

DNA junctions are branched structures in which several duplexes converge at one point. The best known structure is the 4-way junction, a key intermediate in homologous recombination (Holliday junction, see [213]). More simple and abundant are 3-way Y-shaped structures also involved in biological functions, which bind to the porphyrin TMAP (causing DNA photodamage) [221]. 3-Way junctions can be recognized by a tetracationic helicate complex with three bis-pyridylimines bound to two Fe^{2+} ions, which closely fits into the trigonal hydrophobic cavity of the junction [222]. Replacement of Fe^{2+} by Ru^{2+} should yield a luminiscent helicate with biochemical and biological applications.

6.4. Chromosome Banding Patterns

Chromosome banding patterns are related to the chromosome structure [125, 203], and were first reported many years ago, following the discovery of appropriate fluorochromes and preparative treatments. Most results are due to differences between intercalation and external binding of aggregates. Reviews on mechanisms of chromosome banding have been published [10, 32, 145, 223 - 225]. Bright-field banding is mainly due to external binding of aggregated thiazines or dye complexes (Giemsa stain) to chromatin DNA [226]. Fluorescent banding is mainly due to differential intercalation (DNM, quinacrine) or minor groove binding (CMA, DAPI, Hoechst dyes, *etc*).

Several types of banding are commonly described: (1) C bands correspond to constitutive (centromeric and intercalar) heterochromatin, generally with AT-rich DNA. (2) Giemsa (G) and quinacrine (Q) bands also have AT-rich DNA. (3) Chromatid axis (scaffold) with very high AT content from the SARs, related to the coiled (Q/G bands) or uncoiled (reverse, R bands) scaffold organization [125]. Accordingly, AT-rich regions correspond to Q/G bands, whereas R bands are GC-rich regions (*e.g.* CMA bands). Counterstaining-enhancement of banding is now a common procedure [32, 144, 227, 228]. Fluorescence methods to visualize replication pattern, harlequin chromosomes and sister chromatid exchanges have been described in Chapter 4.4.3.

CONFLICT OF INTEREST

The author (editor) declares no conflict of interest, financial or otherwise.

ACKNOWLEDGEMENTS

This work was supported by a grant from the Ministerio de Economía y Competitividad (CTQ2013-48767-C3-3-R), Spain.

REFERENCES

[1] Chaires JB. Drug--DNA interactions. Curr Opin Struct Biol 1998; 8(3): 314-20. [http://dx.doi.org/10.1016/S0959-440X(98)80064-X] [PMID: 9666327]

[2] Pindur U, Jansen M, Lemster T. Advances in DNA-ligands with groove binding, intercalating and/or alkylating activity: chemistry, DNA-binding and biology. Curr Med Chem 2005; 12(24): 2805-47. [http://dx.doi.org/10.2174/092986705774454698] [PMID: 16305474]

[3] Sheng J, Gan J, Huang Z. Structure-based DNA-targeting strategies with small molecule ligands for drug discovery. Med Res Rev 2013; 33(5): 1119-73. [http://dx.doi.org/10.1002/med.21278] [PMID: 23633219]

[4] Kang JS, Meier JL, Dervan PB. Design of sequence-specific DNA binding molecules for DNA methyltransferase inhibition. J Am Chem Soc 2014; 136(9): 3687-94. [http://dx.doi.org/10.1021/ja500211z] [PMID: 24502234]

[5] Saenger W. Principles of Nucleic Acid Structure. New York, Berlin: Springer-Verlag 1984. [http://dx.doi.org/10.1007/978-1-4612-5190-3]

[6] Zimmer C, Wähnert U. Nonintercalating DNA-binding ligands: specificity of the interaction and their use as tools in biophysical, biochemical and biological investigations of the genetic material. Prog Biophys Mol Biol 1986; 47(1): 31-112. [http://dx.doi.org/10.1016/0079-6107(86)90005-2] [PMID: 2422697]

[7] Wang AH. Interactions between antitumor drugs and DNA. Nucleic Acids Mol Biol 1987; 1: 53-69. [http://dx.doi.org/10.1007/978-3-642-46596-3_3]

[8] Wilson WD, Ratmeyer L, Zhao M, Strekowski L, Boykin D. The search for structure-specific nucleic acid-interactive drugs: effects of compound structure on RNA *versus* DNA interaction strength. Biochemistry 1993; 32(15): 4098-104. [http://dx.doi.org/10.1021/bi00066a035] [PMID: 7682441]

[9] Blackburn GM, Gait MJ. Nucleic Acids in Chemistry and Biology. Oxford, New York, Tokio: Oxford University Press 1990.

[10] Bickmore W, Craig J. Chromosome Bands: Pattern in the Genome. Berlin: Springer-Verlag 1997.

[11] Gabbay EJ, Wilson WD. Intercalating agents as probes of chromatin structure. Methods Cell Biol 1978; 18: 351-84. [http://dx.doi.org/10.1016/S0091-679X(08)60146-6] [PMID: 355793]

[12] Sobell HM, Reddy BS, Bhandary KK, Jain SC, Sakore TD, Seshadri TP. Conformational flexibility in DNA structure as revealed by structural studies of drug intercalation and its broader implications in understanding the organization of DNA in chromatin. Cold Spring Harb Symp Quant Biol 1978; 42(Pt 1): 87-102. [http://dx.doi.org/10.1101/SQB.1978.042.01.010] [PMID: 277367]

[13] Sobell HM, Sakore TD, Jain SC, *et al.* β-kinked DNA--a structure that gives rise to drug intercalation and DNA breathing--and its wider significance in determining the premelting and melting behavior of DNA. Cold Spring Harb Symp Quant Biol 1983; 47(Pt 1): 293-314. [http://dx.doi.org/10.1101/SQB.1983.047.01.035] [PMID: 6190609]

[14] Stockert JC. Cytochemistry of nucleic acids: binding mechanisms of dyes and fluorochromes. Microsc Electron Biol Celular 1985; 9(2): 89-131. [PMID: 2431250]

[15] White S, Szewczyk JW, Turner JM, Baird EE, Dervan PB. Recognition of the four Watson-Crick base pairs in the DNA minor groove by synthetic ligands. Nature 1998; 391(6666): 468-71. [http://dx.doi.org/10.1038/35106] [PMID: 9461213]

[16] Gursky GV, Zasedatelev AS, Zhuze AL, *et al.* Synthetic sequence-specific ligands. Cold Spring Harb Symp Quant Biol 1983; 47(Pt 1): 367-78. [http://dx.doi.org/10.1101/SQB.1983.047.01.043] [PMID: 6305558]

[17] Laughlin S, Wilson WD. May the best molecule win: Competition ESI mass spectrometry. Int J Mol Sci 2015; 16(10): 24506-31. [http://dx.doi.org/10.3390/ijms161024506] [PMID: 26501262]

[18] Feulgen R, Rossenbeck H. Mikroskopisch-chemischer Nachweis einer Nucleinsäure von Typus der Thymonucleinsäure und die darauf beruhende elektive Färbung von Zellkernen in mikroskopischen Präparaten. Hoppe Seylers Z Physiol Chem 1924; 135: 203-48. [http://dx.doi.org/10.1515/bchm2.1924.135.5-6.203]

[19] Kasten FH. The chemistry of Schiff's reagent. Int Rev Cytol 1960; 10: 1-100.

[20] Kiernan JA. Histological and Histochemical Methods: Theory and Practice. 5th ed., Banbury, UK: Scion Publishing 2015.

[21] Stockert JC. Reactive Staining Reagents and Fluorescent Labels.Conn's Biological Stains A Handbook of Dyes, Stains and Fluorochromes for Use in Biology and Medicine. 10th ed. Oxford: Bios Scientific Publishers 2002; pp. 77-88.

[22] Horobin RW, Kiernan JA, Eds. Conn's Biological Stains A Handbook of Dyes, Stains and Fluorochromes for Use in Biology and Medicine. 10th ed., Oxford, UK: BIOS Scientific Publishers 2002.

[23] Shapiro R. Genetic effects of bisulfite (sulfur dioxide). Mutat Res 1977; 39(2): 149-75. [http://dx.doi.org/10.1016/0165-1110(77)90020-3] [PMID: 405577]

[24] Robins JH, Abrams GD, Pincock JA. The structure of Schiff reagent aldehyde adducts and the mechanism of the Schiff reaction as determined by nuclear magnetic resonance spectroscopy. Can J Chem 1980; 58: 339-47. [http://dx.doi.org/10.1139/v80-055]

[25] Böhm N, Sprenger E. Fluorescence cytophotometry: a valuable method for the quantitative determination of nuclear Feulgen-DNA. Histochemie 1968; 16(2): 100-18. [PMID: 5751084]

[26] Fujita S. DNA cytofluorometry on large and small cell nuclei stained with pararosaniline Feulgen. Histochemie 1973; 36(3): 193-9. [http://dx.doi.org/10.1007/BF00306309] [PMID: 4757990]

[27] Thiébaud CH. The intra-nucleolar localization of amplified rDNA in *Xenopus laevis* oocytes. Chromosoma 1979; 73(1): 29-36. [http://dx.doi.org/10.1007/BF00294841] [PMID: 90588]

[28] Oud JL, Mans A, Brakenhoff GJ, van Der Voort HT, van Spronsen EA, Nanninga N. Three-dimensional chromosome arrangement of *Crepis capillaris* in mitotic prophase and anaphase as studied by confocal scanning laser microscopy. J Cell Sci 1989; 92(Pt 3): 329-39. [PMID: 2592439]

[29] Cohen AH. Hematoxylin bodies: their appearance in renal tubular epithelium. Hum Pathol 1975; 6(6): 741-4. [http://dx.doi.org/10.1016/S0046-8177(75)80082-7] [PMID: 1183996]

[30] Davis RP, Janis R. Free aldehydic groups in collagen and other tissue components. Nature 1966; 210(5033): 318-9. [http://dx.doi.org/10.1038/210318a0] [PMID: 5954577]

[31] Kobayashi K, Miyazu K, Fukutani Y, Nakamura I, Yamaguchi N. Gallyas-Schiff stain for senile plaques. Biotech Histochem 1992; 67(5): 256-60. [http://dx.doi.org/10.3109/10520299209110033] [PMID: 1300146]

[32] Sahar E, Latt SA. Energy transfer and binding competition between dyes used to enhance staining differentiation in metaphase chromosomes. Chromosoma 1980; 79(1): 1-28. [http://dx.doi.org/10.1007/BF00328469] [PMID: 6156806]

[33] Kasten FH. Schiff-type reagents in cytochemistry. I. Theoretical and practical considerations. Z Zellforsch Mikrosk Anat Histochem 1959; 1: 466-509. [PMID: 14404683]

[34] Roth D, London M, Manjon M. Binding specificity and affinity of acriflavine for nucleic acids. Stain Technol 1967; 42(3): 125-32. [http://dx.doi.org/10.3109/10520296709114994] [PMID: 6033653]

[35] Ruch F. Principles and Some Applications of Cytofluorometry.Introduction to Quantitative Cytochemistry. New York: Academic Press 1970; Vol. 2: p. 431.

[36] Weste SM, Penington DG. Fluorometric measurement of deoxyribonucleic acid in bone marrow cells. The measurement of megakaryocyte deoxyribonucleic acid. J Histochem Cytochem 1972; 20(8): 627-33. [http://dx.doi.org/10.1177/20.8.627] [PMID: 4114595]

[37] Sawicki E, Hauser TR, Stanley TW, Elbert W. The 3-methyl-2-benzothiazolone hydrazone test. Sensitive new method for the detection, rapid estimation, and determination of aliphatic aldehydes. Anal Chem 1961; 33: 93-6. [http://dx.doi.org/10.1021/ac60169a028]

[38] Nakamura M, Toda M, Saito H, Okhura Y. Fluorimetric determination of aromatic aldehydes with 4,5-dimethoxy-1,2-diaminobenzene. Anal Chim Acta 1982; 134: 39-45. [http://dx.doi.org/10.1016/S0003-2670(01)84175-1]

[39] von Bohlen und Halbach O, Kiernan JA. Diaminobenzidine induces fluorescence in nervous tissue and provides intrinsic counterstaining of sections prepared for peroxidase histochemistry. Biotech Histochem 1999; 74(5): 236-43. [http://dx.doi.org/10.3109/10520299909034659] [PMID: 10711503]

[40] Gurney T Jr, Gurney EG. DABA fluorescence assay for submicrogram aounts of DNA. Methods Mol Biol 1985; 2: 5-11. [PMID: 21374166]

[41] Reiss U, Tappel AL, Chio KS. DNA-malonaldehyde reaction: formation of fluorescent products. Biochem Biophys Res Commun 1972; 48(4): 921-6. [http://dx.doi.org/10.1016/0006-291X(72)90696-1] [PMID: 4636653]

[42] Ceriotti G. Determination of nucleic acids in animal tissues. J Biol Chem 1955; 214(1): 59-70. [PMID: 14367363]

[43] Silvestroni L, Frajese G, Malandrino F, Murgiano S, Rocco A, Isidori A. Determinazione microchimica del DNA negli spermatozoi umani di soggetti fertili. Folia Endocrinol 1974; 27(6pt2): 706-17. [PMID: 4480667]

[44] Stockert JC. Osmium tetroxide/*p*-phenylenediamine staining of nucleoli and Balbiani Rings in *Chironomus* salivary glands. Histochemistry 1977; 53(1): 43-56. [http://dx.doi.org/10.1007/BF00511209] [PMID: 71287]

[45] Antonio C, González-García JM, Page J, Suja JA, Stockert JC, Rufas JS. The osmium tetroxide---phenylenediamine procedure reveals the chromatid cores and kinetochores of meiotic chromosomes by light and electron microscopy. J Histochem Cytochem 1996; 44(11): 1279-88. [http://dx.doi.org/10.1177/44.11.8918903] [PMID: 8918903]

[46] Esponda P, Stockert JC. Localization of the synaptonemal complex under the light microscope. Chromosoma 1978; 68(1): 83-90.
[http://dx.doi.org/10.1007/BF00330374] [PMID: 80311]

[47] Bartenev VN, Golovamov EuI , Kapitonova KA, Mokulskii MA, Volkova LI, Skuratovskii IYa . Structure of the B DNA cationic shell as revealed by an X-ray diffraction study of CsDNA. Sequence-specific cationic stabilization of B form DNA. J Mol Biol 1983; 169(1): 217-34.
[http://dx.doi.org/10.1016/S0022-2836(83)80181-8] [PMID: 6620381]

[48] Kawanishi S, Hiraku Y, Murata M, Oikawa S. The role of metals in site-specific DNA damage with reference to carcinogenesis. Free Radic Biol Med 2002; 32(9): 822-32.
[http://dx.doi.org/10.1016/S0891-5849(02)00779-7] [PMID: 11978484]

[49] Wacker WE, Vallee BL. Nucleic acids and metals. I. Chromium, manganese, nickel, iron, and other metals in ribonucleic acid from diverse biological sources. J Biol Chem 1959; 234: 3257-62.

[50] Sissoëff I, Grisvard J, Guillé E. Studies on metal ions-DNA interactions: specific behaviour of reiterative DNA sequences. Prog Biophys Mol Biol 1976; 31(2): 165-99.
[PMID: 185659]

[51] Karlik SJ, Eichhorn GL, Lewis PN, Crapper DR. Interaction of aluminum species with deoxyribonucleic acid. Biochemistry 1980; 19(26): 5991-8.
[http://dx.doi.org/10.1021/bi00567a008] [PMID: 7470444]

[52] Chiu TK, Dickerson RE. 1 A crystal structures of B-DNA reveal sequence-specific binding and groove-specific bending of DNA by magnesium and calcium. J Mol Biol 2000; 301(4): 915-45.
[http://dx.doi.org/10.1006/jmbi.2000.4012] [PMID: 10966796]

[53] Udenfriend S. Fluorescence Assay in Biology and Medicine. New York, London: Academic Press 1962.

[54] Ringer DP, Burchett S, Kizer DE. Use of terbium (III) fluorescence enhancement to selectively monitor DNA and RNA guanine residues and their alteration by chemical modification. Biochemistry 1978; 17(22): 4818-25.
[http://dx.doi.org/10.1021/bi00615a032] [PMID: 215192]

[55] Nielsen PE, Møllegaard NE, Jeppesen C. DNA conformational analysis in solution by uranyl mediated photocleavage. Nucleic Acids Res 1990; 18(13): 3847-51.
[http://dx.doi.org/10.1093/nar/18.13.3847] [PMID: 2374711]

[56] Sönnichsen SH, Nielsen PE. Enhanced uranyl photocleavage across the minor groove of all $(A/T)_4$ sequences indicates a similar narrow minor groove conformation. J Mol Recognit 1996; 9(3): 219-27.
[http://dx.doi.org/10.1002/(SICI)1099-1352(199605)9:3<219::AID-JMR242>3.0.CO;2-S] [PMID: 8938594]

[57] Pettersson A, Kunst L, Bergman B, Roomans GM. Accumulation of aluminium by *Anabaena cylindrica* into polyphosphate granules and cell walls: An X-ray energy-dispersive microanalysis study. J Gen Microbiol 1985; 131: 2545-8.

[58] Llorente AR, Del Castillo P, Stockert JC. Aluminium binding to chromatin DNA as revealed by formation of fluorescent complexes with 8-hydroxyquinoline and other ligands. J Microsc 1989; 155(Pt 2): 227-30.
[http://dx.doi.org/10.1111/j.1365-2818.1989.tb02885.x] [PMID: 2507787]

[59] Del Castillo P, Llorente AR, Gómez A, Gosálvez J, Goyanes VJ, Stockert JC. New fluorescence reactions in DNA cytochemistry. 2. Microscopic and spectroscopic studies on fluorescent aluminum complexes. Anal Quant Cytol Histol 1990; 12(1): 11-20.
[PMID: 2302307]

[60] Stockert JC, Del Castillo P. Linear dichroism and polarized fluorescence of dye-complexed DNA fibers. Histochemistry 1989; 91(3): 263-4.
[http://dx.doi.org/10.1007/BF00490142] [PMID: 2470703]

[61] Del Castillo P, Horobin RW, Blázquez-Castro A, Stockert JC. Binding of cationic dyes to DNA: distinguishing intercalation and groove binding mechanisms using simple experimental and numerical models. Biotech Histochem 2010; 85(4): 247-56. [http://dx.doi.org/10.3109/10520290903149620] [PMID: 20560871]

[62] Stockert JC, Del Castillo P, Romero JB, *et al.* Orientation of nucleosomes in the 30 nm chromatin fiber as revealed by microscopic studies of linear dichroism and polarized fluorescence. Biol Zent Bl 1990; 109: 471-80.

[63] Stockert JC, Del Castillo P, Llorente AR, Rasskin DM, Romero JB, Gómez A. New fluorescence reactions in DNA cytochemistry. 1. Microscopic and spectroscopic studies on nonrigid fluorochromes. Anal Quant Cytol Histol 1990; 12(1): 1-10. [PMID: 2302306]

[64] Kasten FH. Cytochemical studies with acridine orange and the influence of dye contaminants in the staining of nucleic acids. Int Rev Cytol 1967; 21: 141-202. [http://dx.doi.org/10.1016/S0074-7696(08)60814-1] [PMID: 5338238]

[65] Stockert JC. Cytochemistry of mast cells: new fluorescent methods selective for sulfated glycosaminoglycans. Acta Histochem 2000; 102(3): 259-72. [http://dx.doi.org/10.1078/S0065-1281(04)70034-3] [PMID: 10990064]

[66] Nordén B, Tjerneld F. Structure of methylene blue-DNA complexes studied by linear and circular dichroism spectroscopy. Biopolymers 1982; 21(9): 1713-34. [http://dx.doi.org/10.1002/bip.360210904] [PMID: 7126754]

[67] Quinn B. "Fluoro Nissl green": Purification and functional characterization of a novel Nissl stain for neuroanatomic technique. Neurosci Lett 1989; 15: 309.

[68] Espelosin RH, Eraso P, Stockert JC. Observations on the chromatin staining and fluorescence by safranine O. Microsc Acta 1981; 84(3): 255-9. [PMID: 6165882]

[69] Espelosin RH, Stockert JC. Neutral red fluorescence of chromatin: specificity and binding mechanism. Z Naturforsch C Biosci 1982; 37(1-2): 139-41. [PMID: 7064508]

[70] Kim SK, Nordén B. Methyl green. A DNA major-groove binding drug. FEBS Lett 1993; 315(1): 61-4. [http://dx.doi.org/10.1016/0014-5793(93)81133-K] [PMID: 8416812]

[71] Wakelin LP, Adams A, Hunter C, Waring MJ. Interaction of crystal violet with nucleic acids. Biochemistry 1981; 20(20): 5779-87. [http://dx.doi.org/10.1021/bi00523a021] [PMID: 6170329]

[72] Terr LI. Demonstration of lipofuscin and Nissl bodies in crystal violet stained sections using a fluorescence technique or pyronin Y stain. Stain Technol 1986; 61(5): 309-13. [http://dx.doi.org/10.3109/10520298609109958] [PMID: 2431521]

[73] Alvarez-Buylla A, Ling CY, Kirn JR. Cresyl violet: a red fluorescent Nissl stain. J Neurosci Methods 1990; 33(2-3): 129-33. [http://dx.doi.org/10.1016/0165-0270(90)90016-9] [PMID: 2232864]

[74] Cañete M, Armas-Portela R, Stockert JC. Cytochemical mechanisms of chromatin fluorescence and staining reactions by thionin solutions. Microsc Acta 1983; 87(1): 35-8. [PMID: 6186894]

[75] Franklin AL, Filion WG. Acridine orange-methyl green fluorescent staining of nucleoli. Stain Technol 1981; 56(6): 343-8. [http://dx.doi.org/10.3109/10520298109067341] [PMID: 6176050]

[76] Lerman LS. The structure of the DNA-acridine complex. Proc Natl Acad Sci USA 1963; 49: 94-102. [http://dx.doi.org/10.1073/pnas.49.1.94] [PMID: 13929834]

[77] Miller KJ, Newlin DD. Interactions of molecules with nucleic acids. VI. Computer design of chromophoric intercalating agents. Biopolymers 1982; 21(3): 633-52. [http://dx.doi.org/10.1002/bip.360210311] [PMID: 7066476]

[78] Glazer AN, Rye HS. Stable dye-DNA intercalation complexes as reagents for high-sensitivity fluorescence detection. Nature 1992; 359(6398): 859-61. [http://dx.doi.org/10.1038/359859a0] [PMID: 1436062]

[79] Prodhomme S, Demaret JP, Vinogradov S, Asseline U, Morin-Allory L, Vigny P. A theoretical and experimental study of two thiazole orange derivatives with single- and double-stranded oligonucleotides, polydeoxyribonucleotides and DNA. J Photochem Photobiol B 1999; 53(1-3): 60-9. [http://dx.doi.org/10.1016/S1011-1344(99)00127-X] [PMID: 10672530]

[80] Siegmund K, Maheshwary S, Narayanan S, *et al.* Molecular details of quinolone-DNA interactions: solution structure of an unusually stable DNA duplex with covalently linked nalidixic acid residues and non-covalent complexes derived from it. Nucleic Acids Res 2005; 33(15): 4838-48. [http://dx.doi.org/10.1093/nar/gki795] [PMID: 16126848]

[81] Mustaev A, Malik M, Zhao X, *et al.* Fluoroquinolone-gyrase-DNA complexes: two modes of drug binding. J Biol Chem 2014; 289(18): 12300-12. [http://dx.doi.org/10.1074/jbc.M113.529164] [PMID: 24497635]

[82] Palù G, Valisena S, Ciarrocchi G, Gatto B, Palumbo M. Quinolone binding to DNA is mediated by magnesium ions. Proc Natl Acad Sci USA 1992; 89(20): 9671-5. [http://dx.doi.org/10.1073/pnas.89.20.9671] [PMID: 1409681]

[83] Chen CW, Whitlock HW. Molecular tweezers: A simple model of bifunctional intercalation. J Am Chem Soc 1978; 100: 4921-2. [http://dx.doi.org/10.1021/ja00483a063]

[84] Liaw YC, Gao YG, Robinson H, *et al.* Cyclic diguanylic acid behaves as a host molecule for planar intercalators. FEBS Lett 1990; 264(2): 223-7. [http://dx.doi.org/10.1016/0014-5793(90)80253-F] [PMID: 2162785]

[85] Stockert JC. Monomerizing effect of caffeine, *o*-phenanthroline, and tannin on cationic dyes: a model system to analyze spectral characteristics of the intercalative binding to nucleic acids. Acta Histochem 1989; 87(1): 33-42. [http://dx.doi.org/10.1016/S0065-1281(89)80028-5] [PMID: 2513697]

[86] LePecq JB, Paoletti C. A fluorescent complex between ethidium bromide and nucleic acids. Physical-chemical characterization. J Mol Biol 1967; 27(1): 87-106. [http://dx.doi.org/10.1016/0022-2836(67)90353-1] [PMID: 6033613]

[87] Eisenhut M, Choné B. Studies of the binding of ethidium bromide to cells before and after enzyme treatment. Histochemistry 1979; 62(2): 207-20. [http://dx.doi.org/10.1007/BF00493321] [PMID: 387675]

[88] Armas-Portela R, Stockert JC. Chromatin fluorescence by pyronin staining. Experientia 1979; 35(12): 1663-4. [http://dx.doi.org/10.1007/BF01953256] [PMID: 520493]

[89] Toba K, Winton EF, Koike T, Shibata A. Simultaneous three-color analysis of the surface phenotype and DNA-RNA quantitation using 7-amino-actinomycin D and pyronin Y. J Immunol Methods 1995; 182(2): 193-207. [http://dx.doi.org/10.1016/0022-1759(95)00050-K] [PMID: 7540639]

[90] Caspersson T, Zech L, Johansson C, Modest EJ. Identification of human chromosomes by DNA-binding fluorescent agents. Chromosoma 1970; 30(2): 215-27. [http://dx.doi.org/10.1007/BF00282002] [PMID: 4193398]

[91] Weisblum B, De Haseth PL. Quinacrine, a chromosome stain specific for deoxyadenylate-deoxythymidylaterich regions in DNA. Proc Natl Acad Sci USA 1972; 69(3): 629-32.

[http://dx.doi.org/10.1073/pnas.69.3.629] [PMID: 4111050]

[92] Stockert JC, Lisanti JA. Acridine-orange differential fluorescence of fast- and slow-reassociating chromosomal DNA after *in situ* DNA denaturation and reassociation. Chromosoma 1972; 37(2): 117-30.
[http://dx.doi.org/10.1007/BF00284934] [PMID: 4114487]

[93] Stockert JC. Meiotic association of X and Y mouse chromosomes as revealed by acridine orange fluorescence after DNA denaturation and differential renaturation. Exp Cell Res 1972; 74(1): 279-80.
[http://dx.doi.org/10.1016/0014-4827(72)90506-X] [PMID: 4116306]

[94] Chan-Curtis V, Belt WD, Ladoulis CT. Cytochemical localization of nucleic acids by an acriflavine-phosphotungstate complex for fluorescence microscopy and electron microscopy. J Histochem Cytochem 1970; 18(9): 609-27.
[http://dx.doi.org/10.1177/18.9.609] [PMID: 4097868]

[95] Molero ML, Hazen MJ, Pérez Gorroño AI, Stockert JC. Simple beta-carboline alkaloids as nucleic acids fluorochromes. Acta Histochem 1995; 97(2): 165-73.
[http://dx.doi.org/10.1016/S0065-1281(11)80094-2] [PMID: 7660734]

[96] Hazen MJ, Gutiérrez-Gonzálvez MG. UV-mediated toxic bioactivity of harmine in the meristematic cells of *Allium cepa*. Mutagenesis 1988; 3(4): 333-5.
[http://dx.doi.org/10.1093/mutage/3.4.333] [PMID: 3062322]

[97] Burns VW. Studies on the fluorophore sempervirene and its complexes with DNA. Biophys J 1974; 14(3): 189-99.
[http://dx.doi.org/10.1016/S0006-3495(74)85906-0] [PMID: 4207290]

[98] Beljanski M, Crochet S. The anticancer agent pb-100 concentrates in the nucleus and nucleoli of human glioblastoma cells but does not enter normal astrocytes. Int J Oncol 1995; 7(1): 81-5.
[PMID: 21552810]

[99] Hahn FE, Ciak J. Berberine.Antibiotics. Berlin, Heidelberg, New York: Springer-Verlag 1975; Vol. 3: pp. 577-84.

[100] Molero ML, Hazen MJ, Stockert JC. Photodynamic effect of berberine sulfate on the growth rate of *Allium cepa* roots. J Plant Physiol 1985; 120: 91-4.
[http://dx.doi.org/10.1016/S0176-1617(85)80126-7]

[101] Kuo CL, Chou CC, Yung BY. Berberine complexes with DNA in the berberine-induced apoptosis in human leukemic HL-60 cells. Cancer Lett 1995; 93(2): 193-200.
[http://dx.doi.org/10.1016/0304-3835(95)03809-B] [PMID: 7621428]

[102] Yamagishi H. Interaction between nucleic acids and berberine sulfate. J Cell Biol 1962; 15: 589-92.
[http://dx.doi.org/10.1083/jcb.15.3.589] [PMID: 14002271]

[103] Molero ML, Stockert JC. Fluorescence reaction of chromatin and basophilic cytoplasm by berberine sulfate. Cell Mol Biol Incl Cyto Enzymol 1981; 27(5): 523-5.
[PMID: 7326706]

[104] Ababi BE, Moss DS, Palmer RA. Molecular structure of an ancient folk medicine: berberine hydrogen sulfate. J Cryst Spectr Res 1984; 14: 269-81.
[http://dx.doi.org/10.1007/BF01161164]

[105] Moutschen J, Degraeve N, Moutschen-Dahmen M. Chromosome fluorescence with berberine. Cytobiologie 1973; 8: 112-7.

[106] Stiborová M, Frei E. Ellipticines as DNA-targeted chemotherapeutics. Curr Med Chem 2014; 21(5): 575-91.
[http://dx.doi.org/10.2174/09298673113206660272] [PMID: 24059226]

[107] Bailly C, Leclere V, Pommery N, *et al.* Binding to DNA, cellular uptake and biological activity of a distamycin-ellipticine hybrid molecule. Anticancer Drug Des 1993; 8(2): 145-64.

[PMID: 8494604]

[108] Hearst JE, Isaacs ST, Kanne D, Rapoport H, Straub K. The reaction of the psoralens with deoxyribonucleic acid. Q Rev Biophys 1984; 17(1): 1-44.
[http://dx.doi.org/10.1017/S0033583500005242] [PMID: 6385057]

[109] Rosenberg LS, Carvlin MJ, Krugh TR. The antitumor agent mitoxantrone binds cooperatively to DNA: evidence for heterogeneity in DNA conformation. Biochemistry 1986; 25(5): 1002-8.
[http://dx.doi.org/10.1021/bi00353a008] [PMID: 2870736]

[110] Denny WA, Wakelin LP. Kinetics of the binding of mitoxantrone, ametantrone and analogues to DNA: relationship with binding mode and anti-tumour activity. Anticancer Drug Des 1990; 5(2): 189-200.
[PMID: 2357264]

[111] Islam SA, Neidle S, Gandecha BM, Partridge M, Patterson LH, Brown JR. Comparative computer graphics and solution studies of the DNA interaction of substituted anthraquinones based on doxorubicin and mitoxantrone. J Med Chem 1985; 28(7): 857-64.
[http://dx.doi.org/10.1021/jm00145a003] [PMID: 4009608]

[112] Hansen M, Yun S, Hurley L. Hedamycin intercalates the DNA helix and, through carbohydrate-mediated recognition in the minor groove, directs N7-alkylation of guanine in the major groove in a sequence-specific manner. Chem Biol 1995; 2(4): 229-40.
[http://dx.doi.org/10.1016/1074-5521(95)90273-2] [PMID: 9383425]

[113] Villanueva A, Cañete M, Juarranz A, Stockert JC. Cytochemical study of the fluorescence reaction of chromatin by the porphyrin derivative meso-tetra (4-N-methylpyridyl) porphine. Basic Appl Histochem 1987; 31(1): 9-16.
[PMID: 3593176]

[114] Fiel RJ. Porphyrin-nucleic acid interactions: a review. J Biomol Struct Dyn 1989; 6(6): 1259-74.
[http://dx.doi.org/10.1080/07391102.1989.10506549] [PMID: 2684219]

[115] Juarranz A, Villanueva A, Cañete M, Stockert JC. Fluorescent porphyrin counterstaining of chromatin DNA in conjunction with immunofluorescence methods using FITC-labelled antibodies. J Microsc 1996; 182(Pt 1): 46-9.
[http://dx.doi.org/10.1046/j.1365-2818.1996.114394.x] [PMID: 8632446]

[116] Tas J, Mendelson D, Noorden CJ. Cuprolinic Blue: a specific dye for single-stranded RNA in the presence of magnesium chloride. I. Fundamental aspects. Histochem J 1983; 15(8): 801-14.
[http://dx.doi.org/10.1007/BF01003343] [PMID: 6194140]

[117] Tempesti TC, Stockert JC, Durantini EN. Photosensitization ability of a water soluble zinc(II)tetramethyltetrapyridinoporphyrazinium salt in aqueous solution and biomimetic reverse micelles medium. J Phys Chem B 2008; 112(49): 15701-7.
[http://dx.doi.org/10.1021/jp808094q] [PMID: 19053687]

[118] Stockert JC, Cañete M, Villanueva A, Juarranz A, Trigoso CI, Braña MF. Fluorescence of chromatin DNA induced by antitumoral naphthalimides. Z Natforsch C J Biosci 1997; 52(5-6): 408-11.
[PMID: 9232898]

[119] Tanious FA, Yen SF, Wilson WD. Kinetic and equilibrium analysis of a threading intercalation mode: DNA sequence and ion effects. Biochemistry 1991; 30(7): 1813-9.
[http://dx.doi.org/10.1021/bi00221a013] [PMID: 1993195]

[120] Ughetto G, Wang AH, Quigley GJ, van der Marel GA, van Boom JH, Rich A. A comparison of the structure of echinomycin and triostin A complexed to a DNA fragment. Nucleic Acids Res 1985; 13(7): 2305-23.
[http://dx.doi.org/10.1093/nar/13.7.2305] [PMID: 4000957]

[121] Lown JW, Hanstock CC. Structure and function of the antitumor antibiotic carzinophilin A: the first natural intercalative bisalkylator. J Am Chem Soc 1982; 104: 3213-4.

[http://dx.doi.org/10.1021/ja00375a047]

[122] Delepierre M, Dinh TH, Roques BP. Bisintercalation of ditercalinium into a d(CpGpApTpCpG)2 minihelix: a 1H- and 31P-NMR study. Biopolymers 1989; 28(12): 2115-42. [http://dx.doi.org/10.1002/bip.360281207] [PMID: 2605314]

[123] Gao Q, Williams LD, Egli M, *et al.* Drug-induced DNA repair: X-ray structure of a DNA-ditercalinium complex. Proc Natl Acad Sci USA 1991; 88(6): 2422-6. [http://dx.doi.org/10.1073/pnas.88.6.2422] [PMID: 2006181]

[124] Huang CH, Mirabelli CK, Mong S, Crooke ST. Intermolecular cross-linking of DNA through bifunctional intercalation of an antitumor antibiotic, luzopeptin A (BBM-928A). Cancer Res 1983; 43(6): 2718-24. [PMID: 6303566]

[125] Saitoh Y, Laemmli UK. Metaphase chromosome structure: bands arise from a differential folding path of the highly AT-rich scaffold. Cell 1994; 76(4): 609-22. [http://dx.doi.org/10.1016/0092-8674(94)90502-9] [PMID: 7510215]

[126] Tekola P, Baak JP, Beliën JA, Brugghe J. Highly sensitive, specific, and stable new fluorescent DNA stains for confocal laser microscopy and image processing of normal paraffin sections. Cytometry 1994; 17(3): 191-5. [http://dx.doi.org/10.1002/cyto.990170302] [PMID: 7531632]

[127] Braña MF, Castellano JM, Morán M, *et al.* Bis-naphthalimides: a new class of antitumor agents. Anticancer Drug Des 1993; 8(4): 257-68. [PMID: 8240655]

[128] Lokey RS, Kwok Y, Guelev V, *et al.* A new class of polyintercalating molecules. J Am Chem Soc 1997; 119: 7202-10. [http://dx.doi.org/10.1021/ja9706108]

[129] Carpenter ML, Lowe G, Cook PR. The structure of 4-way DNA junctions: specific binding of bis-intercalators with rigid linkers. Nucleic Acids Res 1996; 24(9): 1594-601. [http://dx.doi.org/10.1093/nar/24.9.1594] [PMID: 8649974]

[130] Gill JE, Jotz MM, Young SG, Modest EJ, Sengupta SK. 7-Amino-actinomycin D as a cytochemical probe. I. Spectral properties. J Histochem Cytochem 1975; 23(11): 793-9. [http://dx.doi.org/10.1177/23.11.1194669] [PMID: 1194669]

[131] Stepanova NG, Nikitin SM, Valeeva FS, Kartasheva ON, Zhuze AL, Zelenin AV. Application of 7-amino-actinomycin D for the fluorescence microscopical analysis of DNA in cells and polytene chromosomes. Histochem J 1985; 17(2): 131-42. [http://dx.doi.org/10.1007/BF01003213] [PMID: 4019246]

[132] Rabinovitch PS, Torres RM, Engel D. Simultaneous cell cycle analysis and two-color surface immunofluorescence using 7-amino-actinomycin D and single laser excitation: applications to study of cell activation and the cell cycle of murine Ly-1 B cells. J Immunol 1986; 136(8): 2769-75. [PMID: 2420866]

[133] Zweier JL. Iron-mediated formation of an oxidized adriamycin free radical. Biochim Biophys Acta 1985; 839(2): 209-13. [http://dx.doi.org/10.1016/0304-4165(85)90038-8] [PMID: 2985121]

[134] Minotti G, Menna P, Salvatorelli E, Cairo G, Gianni L. Anthracyclines: molecular advances and pharmacologic developments in antitumor activity and cardiotoxicity. Pharmacol Rev 2004; 56(2): 185-229. [http://dx.doi.org/10.1124/pr.56.2.6] [PMID: 15169927]

[135] Johnston FP, Jorgenson KF, Lin CC, van de Sande JH. Interaction of anthracyclines with DNA and chromosomes. Chromosoma 1978; 68(2): 115-29. [http://dx.doi.org/10.1007/BF00287144] [PMID: 80312]

[136] Quigley GJ, Wang AH, Ughetto G, van der Marel G, van Boom JH, Rich A. Molecular structure of an anticancer drug-DNA complex: daunomycin plus d(CpGpTpApCpG). Proc Natl Acad Sci USA 1980; 77(12): 7204-8.
[http://dx.doi.org/10.1073/pnas.77.12.7204] [PMID: 6938965]

[137] Spinelli M, Dabrowiak JC. Interaction of copper(II) ions with the daunomycin-calf thymus deoxyribonucleic acid complex. Biochemistry 1982; 21(23): 5862-70.
[http://dx.doi.org/10.1021/bi00266a021] [PMID: 7150531]

[138] Beraldo H, Garnier-Suillerot A, Tosi L, Lavelle F. Iron(III)-adriamycin and Iron(III)-daunorubicin complexes: physicochemical characteristics, interaction with DNA, and antitumor activity. Biochemistry 1985; 24(2): 284-9.
[http://dx.doi.org/10.1021/bi00323a007] [PMID: 2983753]

[139] Gao YG, Sriram M, Wang AH. Crystallographic studies of metal ion-DNA interactions: different binding modes of cobalt(II), copper(II) and barium(II) to N^7 of guanines in Z-DNA and a drug-DNA complex. Nucleic Acids Res 1993; 21(17): 4093-101.
[http://dx.doi.org/10.1093/nar/21.17.4093] [PMID: 8371984]

[140] Kopka ML, Larsen TA. Netropsin and the Lexitropsins. The Search for Sequence-Specific Minor Groove-Binding Ligands.Nucleic Acid Targeted Drug Design. New York, Basel: Marcel Dekker 1992; pp. 303-74.

[141] Kahne D. Strategies for the design of minor groove binders: a re-evaluation based on the emergence of site-selective carbohydrate binders. Chem Biol 1995; 2(1): 7-12.
[http://dx.doi.org/10.1016/1074-5521(95)90075-6] [PMID: 9383398]

[142] Geierstanger BH, Wemmer DE. Complexes of the minor groove of DNA. Annu Rev Biophys Biomol Struct 1995; 24: 463-93.
[http://dx.doi.org/10.1146/annurev.bb.24.060195.002335] [PMID: 7663124]

[143] Kopka ML, Yoon C, Goodsell D, Pjura P, Dickerson RE. The molecular origin of DNA-drug specificity in netropsin and distamycin. Proc Natl Acad Sci USA 1985; 82(5): 1376-80.
[http://dx.doi.org/10.1073/pnas.82.5.1376] [PMID: 2983343]

[144] Schweizer D. Counterstain-enhanced chromosome banding. Hum Genet 1981; 57(1): 1-14.
[PMID: 6167505]

[145] Holmquist GP, Motara MA. The Magic of Cytogenetic Technology. Cytogenetics. Berlin, Heidelberg, New York: Springer 1987; pp. 30-47.
[http://dx.doi.org/10.1007/978-3-642-72802-0_2]

[146] Stockert JC, Del Castillo P, Bella JL. DNA-induced distamycin A fluorescence. Histochemistry 1990; 94(1): 45-7.
[http://dx.doi.org/10.1007/BF00266788] [PMID: 2190951]

[147] Zimmer C. Effects of the antibiotics netropsin and distamycin A on the structure and function of nucleic acids. Prog Nucleic Acid Res Mol Biol 1975; 15(0): 285-318.
[http://dx.doi.org/10.1016/S0079-6603(08)60122-1] [PMID: 1094496]

[148] Pelton JG, Wemmer DF. Binding modes of distamycin A with $d(CGCAAATTTGCG)_2$ determined by two-dimensional NMR. J Am Chem Soc 1990; 112: 1393-9.
[http://dx.doi.org/10.1021/ja00160a016]

[149] Mrksich M, Dervan PB. Antiparallel side-by-side heterodimer for sequence-specific recognition in the minor groove of DNA by a distamycin/1-methylimidazole-2-carboxamide-netropsin pair. J Am Chem Soc 1993; 115: 2572-6.
[http://dx.doi.org/10.1021/ja00060a004]

[150] Rohs R, West SM, Sosinsky A, Liu P, Mann RS, Honig B. The role of DNA shape in protein-DNA recognition. Nature 2009; 461(7268): 1248-53.
[http://dx.doi.org/10.1038/nature08473] [PMID: 19865164]

[151] Goodsell D, Dickerson RE. Isohelical analysis of DNA groove-binding drugs. J Med Chem 1986; 29(5): 727-33.
[http://dx.doi.org/10.1021/jm00155a023] [PMID: 2422377]

[152] Zasedatelev AS. Geometrical correlations useful for design of sequence-specific DNA narrow groove binding ligands. FEBS Lett 1991; 281(1-2): 209-11.
[http://dx.doi.org/10.1016/0014-5793(91)80395-J] [PMID: 2015896]

[153] Stockert JC. Predictive binding geometry of ligands to DNA minor groove: isohelicity and hydrogen-bonding pattern. Methods Mol Biol 2014; 1094: 1-12.
[http://dx.doi.org/10.1007/978-1-62703-706-8_1] [PMID: 24162975]

[154] Raposa T, Natarajan AT. Fluorescence banding pattern of human and mouse chromosomes with a benzimidazol derivative (Hoechst 33258). Humangenetik 1974; 21(3): 221-6.
[PMID: 4136266]

[155] Comings DE. Mechanisms of chromosome banding. VIII. Hoechst 33258-DNA interaction. Chromosoma 1975; 52(3): 229-43.
[http://dx.doi.org/10.1007/BF00332113] [PMID: 1183300]

[156] Pinna-Senn E, Lisanti JA, Ortiz MI, *et al.* Specific heterochromatic banding of metaphase chromosomes using nuclear yellow. Biotech Histochem 2000; 75(3): 132-40.
[http://dx.doi.org/10.3109/10520290009066491] [PMID: 10950175]

[157] Pjura PE, Grzeskowiak K, Dickerson RE. Binding of Hoechst 33258 to the minor groove of B-DNA. J Mol Biol 1987; 197(2): 257-71.
[http://dx.doi.org/10.1016/0022-2836(87)90123-9] [PMID: 2445998]

[158] Grossgebauer K. Staining of acid mucopolysaccharides appearing on and in various cell types by DAPI. Microsc Acta 1979; 82(3): 291-3.
[PMID: 93685]

[159] Kaushik M, Kukreti S. Temperature induced hyperchromism exhibited by Hoechst 33258: evidence of drug aggregation from UV-melting method. Spectrochim Acta A Mol Biomol Spectrosc 2003; 59(13): 3123-9.
[http://dx.doi.org/10.1016/S1386-1425(03)00115-X] [PMID: 14583287]

[160] Stockert JC, Abasolo MI. Inaccurate chemical structures of dyes and fluorochromes found in the literature can be problematic for teaching and research. Biotech Histochem 2011; 86(1): 52-60.
[http://dx.doi.org/10.3109/10520295.2010.489428] [PMID: 20500122]

[161] Radic MZ, Lundgren K, Hamkalo BA. Curvature of mouse satellite DNA and condensation of heterochromatin. Cell 1987; 50(7): 1101-8.
[http://dx.doi.org/10.1016/0092-8674(87)90176-0] [PMID: 2441880]

[162] Soeiro MN, De Souza EM, Stephens CE, Boykin DW. Aromatic diamidines as antiparasitic agents. Expert Opin Investig Drugs 2005; 14(8): 957-72.
[http://dx.doi.org/10.1517/13543784.14.8.957] [PMID: 16050790]

[163] Paine MF, Wang MZ, Generaux CN, *et al.* Diamidines for human African trypanosomiasis. Curr Opin Investig Drugs 2010; 11(8): 876-83.
[PMID: 20721830]

[164] Kapuściński J. DAPI: a DNA-specific fluorescent probe. Biotech Histochem 1995; 70(5): 220-33.
[http://dx.doi.org/10.3109/10520299509108199] [PMID: 8580206]

[165] Kahn E, Frouin F, Souchier C, *et al.* Confocal multilaser focusing and single-laser characterization of ultraviolet excitable stains of cellular preparations. Cytometry 2000; 40(1): 42-9.
[http://dx.doi.org/10.1002/(SICI)1097-0320(20000501)40:1<42::AID-CYTO6>3.0.CO;2-T] [PMID: 10754516]

[166] Williamson DH, Fennell DJ. Visualization of yeast mitochondrial DNA with the fluorescent stain

"DAPI". Methods Enzymol 1979; 56: 728-33.
[http://dx.doi.org/10.1016/0076-6879(79)56065-0] [PMID: 459889]

[167] Coleman AW. The fate of chloroplast DNA during cell fusion, zygote maturation and zygote germination in *Chlamydomonas reinhardi* as revealed by DAPI staining. Exp Cell Res 1984; 152(2): 528-40.
[http://dx.doi.org/10.1016/0014-4827(84)90655-4] [PMID: 6539223]

[168] Kapuściński J, Skoczylas B. Simple and rapid fluorimetric method for DNA microassay. Anal Biochem 1977; 83(1): 252-7.
[http://dx.doi.org/10.1016/0003-2697(77)90533-4] [PMID: 21598]

[169] Tanious FA, Veal JM, Buczak H, Ratmeyer LS, Wilson WD. DAPI (4′,6-diamidino-2-phenylindole) binds differently to DNA and RNA: minor-groove binding at AT sites and intercalation at AU sites. Biochemistry 1992; 31(12): 3103-12.
[http://dx.doi.org/10.1021/bi00127a010] [PMID: 1372825]

[170] Stockert JC, Pinna-Senn E, Bella JL, Lisanti JA. DNA-binding fluorochromes: correlation between C-banding of mouse metaphase chromosomes and hydrogen bonding to adenine-thymine base pairs. Acta Histochem 2005; 106(6): 413-20.
[http://dx.doi.org/10.1016/j.acthis.2004.09.001] [PMID: 15707650]

[171] Murgatroyd LB. 2-hydroxystilbamidine isethionate: a new fluorochrome for use in general pathology. I. The selective staining of DNA, mucosubstances and elastic fibres. Histochemistry 1982; 74(1): 107-14.
[http://dx.doi.org/10.1007/BF00495056] [PMID: 6177666]

[172] Newton BA. Berenil: a Trypanocide with Selective Activity against Extranuclear DNA.Antibiotics: Mechanisms of Action of Antimicrobial and Antitumor Agents. Berlin, Heidelberg, New York: Springer-Verlag 1975.
[http://dx.doi.org/10.1007/978-3-642-46304-4_4]

[173] Tring FC, Murgatroyd LB. A new method for identifying parakeratosis in human epidermis. Dermatologica 1976; 153: 357-62.
[http://dx.doi.org/10.1159/000251147]

[174] Kelényi G. On the histochemistry of azo group-free thiazole dyes. J Histochem Cytochem 1967; 15(3): 172-80.
[http://dx.doi.org/10.1177/15.3.172] [PMID: 4166508]

[175] Stiller D, Katenkamp D, Thoss K. [Staining mechanism of thioflavin T with special reference to the localization of amyloid]. Acta Histochem 1972; 42(2): 234-45.
[PMID: 4115678]

[176] Cañete M, Villanueva A, Juarranz A, Stockert JC. A study of interaction of thioflavine T with DNA: evidence for intercalation. Cell Mol Biol 1987; 33(2): 191-9.
[PMID: 3607826]

[177] Cañete M, Stockert JC. Fluorescencia de estructuras celulares por tioflavina T: reacción con la cromatina. Citología 1985; 7: 55-60.

[178] Karlsson HJ, Lincoln P, Westman G. Synthesis and DNA binding studies of a new asymmetric cyanine dye binding in the minor groove of [poly(dA-dT)]$_2$. Bioorg Med Chem 2003; 11(6): 1035-40.
[http://dx.doi.org/10.1016/S0968-0896(02)00463-7] [PMID: 12614890]

[179] Bengtsson M, Karlsson HJ, Westman G, Kubista M. A new minor groove binding asymmetric cyanine reporter dye for real-time PCR. Nucleic Acids Res 2003; 31(8): e45.
[http://dx.doi.org/10.1093/nar/gng045] [PMID: 12682380]

[180] Seifert JL, Connor RE, Kushon SA, *et al.* Spontaneous assembly of helical cyanine dye aggregates on DNA nanotemplates. J Am Chem Soc 1999; 121: 2987-95.
[http://dx.doi.org/10.1021/ja984279j]

[181] Mikheikin AL, Zhuze AL, Zasedatelev AS. Binding of symmetrical cyanine dyes into the DNA minor groove. J Biomol Struct Dyn 2000; 18(1): 59-72. [http://dx.doi.org/10.1080/07391102.2000.10506647] [PMID: 11021652]

[182] Sovenyhazy KM, Bordelon JA, Petty JT. Spectroscopic studies of the multiple binding modes of a trimethine-bridged cyanine dye with DNA. Nucleic Acids Res 2003; 31(10): 2561-9. [http://dx.doi.org/10.1093/nar/gkg363] [PMID: 12736305]

[183] Ploeger LS, Dullens HF, Huisman A, van Diest PJ. Fluorescent stains for quantification of DNA by confocal laser scanning microscopy in 3-D. Biotech Histochem 2008; 83(2): 63-9. [http://dx.doi.org/10.1080/10520290802127586] [PMID: 18568680]

[184] Stockert JC. 2,5-Bis(4-aminophenyl)-1,3,4-oxadiazol: fluorescence reaction of chromatin and basophilic cytoplasm. Acta Histochem Cytochem 1983; 16: 66-9. [http://dx.doi.org/10.1267/ahc.16.66]

[185] Stockert JC, Trigoso CI, Llorente AR, Del Castillo P. DNA fluorescence induced by polymethine cation pyrvinium binding. Histochem J 1991; 23(11-12): 548-52. [http://dx.doi.org/10.1007/BF01041182] [PMID: 1791161]

[186] Acosta-Reyes FJ, Dardonville C, de Koning HP, Natto M, Subirana JA, Campos JL. In and out of the minor groove: interaction of an AT-rich DNA with the drug CD27. Acta Crystallogr D Biol Crystallogr 2014; 70(Pt 6): 1614-21. [http://dx.doi.org/10.1107/S139900471400697X] [PMID: 24914972]

[187] Moreno T, Pous J, Subirana JA, Campos JL. Coiled-coil conformation of a pentamidine-DNA complex. Acta Crystallogr D Biol Crystallogr 2010; 66(Pt 3): 251-7. [http://dx.doi.org/10.1107/S0907444909055693] [PMID: 20179336]

[188] Roig A, Peña MP, Vicente F, Stockert JC. Voltammetric behavior of berenil. J Pharm Sci 1993; 82(3): 251-3. [http://dx.doi.org/10.1002/jps.2600820306] [PMID: 8450416]

[189] Englund PT. The replication of kinetoplast DNA networks in *Crithidia fasciculata*. Cell 1978; 14(1): 157-68. [http://dx.doi.org/10.1016/0092-8674(78)90310-0] [PMID: 667931]

[190] Borst P, Hoeijmakers JH. Kinetoplast DNA. Plasmid 1979; 2(1): 20-40. [http://dx.doi.org/10.1016/0147-619X(79)90003-9] [PMID: 382201]

[191] Jensen RE, Englund PT. Network news: the replication of kinetoplast DNA. Annu Rev Microbiol 2012; 66: 473-91. [http://dx.doi.org/10.1146/annurev-micro-092611-150057] [PMID: 22994497]

[192] Marini JC, Effron PN, Goodman TC, *et al.* Physical characterization of a kinetoplast DNA fragment with unusual properties. J Biol Chem 1984; 259(14): 8974-9. [PMID: 6086618]

[193] Ibrahim HM, Al-Salabi MI, El Sabbagh N, *et al.* Symmetrical choline-derived dications display strong anti-kinetoplastid activity. J Antimicrob Chemother 2011; 66(1): 111-25. [http://dx.doi.org/10.1093/jac/dkq401] [PMID: 21078603]

[194] Colombo LL, Vanzulli SI, Villanueva A, Cañete M, Juarranz A, Stockert JC. Long-term regression of the murine mammary adenocarcinoma, LM3, by repeated photodynamic treatments using *meso*-tetra (4-N-methylpyridinium) porphine. Int J Oncol 2005; 27(4): 1053-9. [PMID: 16142323]

[195] Strickland JA, Marzilli LG, Gay KM, Wilson WD. Porphyrin and metalloporphyrin binding to DNA polymers: rate and equilibrium binding studies. Biochemistry 1988; 27(24): 8870-8. [http://dx.doi.org/10.1021/bi00424a027] [PMID: 3242614]

[196] Hui XW, Gresh N, Pullman B. Modelling of the binding specificity in the interactions of cationic

porphyrins with DNA. Nucleic Acids Res 1990; 18(5): 1109-14.
[http://dx.doi.org/10.1093/nar/18.5.1109] [PMID: 2320411]

[197] Jensen RH. Chromomycin A_3 as a fluorescent probe for flow cytometry of human gynecologic samples. J Histochem Cytochem 1977; 25(7): 573-9.
[http://dx.doi.org/10.1177/25.7.70448] [PMID: 70448]

[198] Barros MA, Patton JL. Genome evolution in pocket gophers (genus Thomomys). III. Fluorochrome-revealed heterochromatin heterogeneity. Chromosoma 1985; 92(5): 337-43.
[http://dx.doi.org/10.1007/BF00327464] [PMID: 4053789]

[199] Gao XL, Mirau P, Patel DJ. Structure refinement of the chromomycin dimer-DNA oligomer complex in solution. J Mol Biol 1992; 223(1): 259-79.
[http://dx.doi.org/10.1016/0022-2836(92)90730-8] [PMID: 1731073]

[200] Slater ML. Rapid nuclear staining method for *Saccharomyces cerevisiae.* J Bacteriol 1976; 126(3): 1339-41.
[PMID: 59726]

[201] Johannisson E, Thorell B. Mithramycin fluorescence for quantitative determination of deoxyribonucleic acid in single cells. J Histochem Cytochem 1977; 25(2): 122-8.
[http://dx.doi.org/10.1177/25.2.138705] [PMID: 138705]

[202] Hancock R, Boulikas T. Functional organization in the nucleus. Int Rev Cytol 1982; 79: 165-214.
[http://dx.doi.org/10.1016/S0074-7696(08)61674-5] [PMID: 6185451]

[203] Stack SM, Anderson LK. A model for chromosome structure during the mitotic and meiotic cell cycles. Chromosome Res 2001; 9(3): 175-98.
[http://dx.doi.org/10.1023/A:1016690802570] [PMID: 11330393]

[204] Espada J, Esteller M. DNA methylation and the functional organization of the nuclear compartment. Semin Cell Dev Biol 2010; 21(2): 238-46.
[http://dx.doi.org/10.1016/j.semcdb.2009.10.006] [PMID: 19892028]

[205] Bordas J, Perez-Grau L, Koch MH, Vega MC, Nave C. The superstructure of chromatin and its condensation mechanism. II. Theoretical analysis of the X-ray scattering patterns and model calculations. Eur Biophys J 1986; 13(3): 175-85.
[http://dx.doi.org/10.1007/BF00542561] [PMID: 3956446]

[206] Schalch T, Duda S, Sargent DF, Richmond TJ. X-ray structure of a tetranucleosome and its implications for the chromatin fibre. Nature 2005; 436(7047): 138-41.
[http://dx.doi.org/10.1038/nature03686] [PMID: 16001076]

[207] Wu C, Bassett A, Travers A. A variable topology for the 30-nm chromatin fibre. EMBO Rep 2007; 8(12): 1129-34.
[http://dx.doi.org/10.1038/sj.embor.7401115] [PMID: 18059311]

[208] Joyce GF. *In vitro* evolution of nucleic acids. Curr Opin Struct Biol 1994; 4: 331-6.
[http://dx.doi.org/10.1016/S0959-440X(94)90100-7] [PMID: 11539574]

[209] Kolpashchikov DM. Binary malachite green aptamer for fluorescent detection of nucleic acids. J Am Chem Soc 2005; 127(36): 12442-3.
[http://dx.doi.org/10.1021/ja0529788] [PMID: 16144363]

[210] Babendure JR, Adams SR, Tsien RY. Aptamers switch on fluorescence of triphenylmethane dyes. J Am Chem Soc 2003; 125(48): 14716-7.
[http://dx.doi.org/10.1021/ja037994o] [PMID: 14640641]

[211] Morgan AR. Do multistranded polynucleotides have a biological role? Trends Biochem Sci 1979; 244-9.
[http://dx.doi.org/10.1016/0968-0004(79)90112-9]

[212] Latimer LJ, Payton N, Forsyth G, Lee JS. The binding of analogues of coralyne and related

heterocyclics to DNA triplexes. Biochem Cell Biol 1995; 73(1-2): 11-8. [http://dx.doi.org/10.1139/o95-002] [PMID: 7662309]

[213] Stockert JC. Stereochemistry of general genetic recombination: paranemic duplexes and duplex-tetraplex junctions as possible DNA structures for homologous pairing, strand exchange and hybrid resolution. Biol Zentralbl 1992; 111: 150-63.

[214] Agazie YM, Lee JS, Burkholder GD. Characterization of a new monoclonal antibody to triplex DNA and immunofluorescent staining of mammalian chromosomes. J Biol Chem 1994; 269(9): 7019-23. [PMID: 7509814]

[215] Mergny JL, Duval-Valentin G, Nguyen CH, *et al.* Triple helix-specific ligands. Science 1992; 256(5064): 1681-4. [http://dx.doi.org/10.1126/science.256.5064.1681] [PMID: 1609278]

[216] Polak M, Hud NV. Complete disproportionation of duplex poly(dT)*poly(dA) into triplex poly(dT)*poly(dA)*poly(dT) and poly(dA) by coralyne. Nucleic Acids Res 2002; 30(4): 983-92. [http://dx.doi.org/10.1093/nar/30.4.983] [PMID: 11842110]

[217] Neidle S. Human telomeric G-quadruplex: the current status of telomeric G-quadruplexes as therapeutic targets in human cancer. FEBS J 2010; 277(5): 1118-25. [http://dx.doi.org/10.1111/j.1742-4658.2009.07463.x] [PMID: 19951354]

[218] Casagrande V, Alvino A, Bianco A, Ortaggi G, Franceschin M. Study of binding affinity and selectivity of perylene and coronene derivatives towards duplex and quadruplex DNA by ESI-MS. J Mass Spectrom 2009; 44(4): 530-40. [http://dx.doi.org/10.1002/jms.1529] [PMID: 19034888]

[219] Izbicka E, Wheelhouse RT, Raymond E, *et al.* Effects of cationic porphyrins as G-quadruplex interactive agents in human tumor cells. Cancer Res 1999; 59(3): 639-44. [PMID: 9973212]

[220] Shammas MA, Shmookler Reis RJ, Akiyama M, *et al.* Telomerase inhibition and cell growth arrest by G-quadruplex interactive agent in multiple myeloma. Mol Cancer Ther 2003; 2(9): 825-33. [PMID: 14555701]

[221] Nussbaum JM, Newport ME, Mackie M, Leontis NB. Structure-specific binding and photosensitized cleavage of branched DNA three-way junction complexes by cationic porphyrins. Photochem Photobiol 1994; 59(5): 515-28. [http://dx.doi.org/10.1111/j.1751-1097.1994.tb02977.x] [PMID: 8041806]

[222] Oleksy A, Blanco AG, Boer R, *et al.* Molecular recognition of a three-way DNA junction by a metallosupramolecular helicate. Angew Chem Int Ed Engl 2006; 45(8): 1227-31. [http://dx.doi.org/10.1002/anie.200503822] [PMID: 16463312]

[223] Comings DE. Mechanisms of chromosome banding and implications for chromosome structure. Annu Rev Genet 1978; 12: 25-46. [http://dx.doi.org/10.1146/annurev.ge.12.120178.000325] [PMID: 85431]

[224] Schnedl W, Dann O, Schweizer D. Effects of counterstaining with DNA binding drugs on fluorescent banding patterns of human and mammalian chromosomes. Eur J Cell Biol 1980; 20(3): 290-6. [PMID: 6153615]

[225] Sumner AT. Chromosome Banding. London: Unwin Hyman 1990.

[226] Stockert JC, Blázquez-Castro A, Horobin RW. Identifying different types of chromatin using Giemsa staining. Methods Mol Biol 2014; 1094: 25-38. [http://dx.doi.org/10.1007/978-1-62703-706-8_3] [PMID: 24162977]

[227] Schweizer D, Loidl J, Hamilton B. Heterochromatin and the Phenomenon of Chromosome Banding.Structure and Function of Eukaryotic Chromosomes. Berlin, Heidelberg: Springer-Verlag 1987; pp. 235-54. [http://dx.doi.org/10.1007/978-3-540-47783-9_15]

[228] Stockert JC, Trigoso CI, Cuéllar T, Bella JL, Lisanti JA. A new fluorescence reaction in DNA cytochemistry: microscopic and spectroscopic studies on the aromatic diamidino compound M&B 938. J Histochem Cytochem 1997; 45(1): 97-105.
[http://dx.doi.org/10.1177/002215549704500113] [PMID: 9010474]

CHAPTER 13

Proteins

Abstract: At present, several histochemical methods for proteins using fluorescent dyes and reagents are available. Some of them utilize simple fluorochromes, whereas others involve preferential fluorogenic reactions with amino acids. Ionic, hydrophobic, and covalent interactions occur between protein groups and fluorochromes. In addition to these classical demonstrations of proteins, more selective methods such as immunofluorescence and detection of enzymatic activity are now widely used in microscopical studies. Also other macromolecules such as nucleic acids, polysaccharides, *etc*., can be visualized by these methods. Detection of macromolecules is also possible with haptens (*i.e.* dinitrophenol, biotin, digoxigenin) that are recognizable by suitable ligands (antibodies, streptavidin), or using labeled lectins, toxins, oligonucleotides, *etc.* (see Chapter 4). Fluorescent reactions for enzymatic activity and immunofluorescence are described in Chapters 14 and 17.

Keywords: Amino groups, Amyloid deposits, Aromatic amino acids, Carboxyl groups, Catecholamines, Dansyl chloride, Fast blue B, Fast red TR, Furanones, Indolamines, Isatin, Isoelectric point, Mercury orange, NBD, Phenanthrene-quinone, Sulfhydryls.

1. GENERAL STAINING AND IONIC REACTIONS

Despite the intrinsic value of the direct use of fluorochromes and reagents for detection of amino acids in protein histochemistry, at present they are not often applied in fluorescence microscopy. This is unfortunate because interesting structural features and chemical information can be obtained using these classical methods [1]. A representation of general fluorescence reactions for proteins is shown in Fig. (**13.1**) (several binding modes are indicated; for simplicity, some of these methods are omitted).

Proteins have both basic and acid amino acids and thus these macromolecules are amphoteric compounds. As the interactions of charged groups with also charged dyes are ionic in nature, the influence of pH conditions on the binding reaction is relevant [2]. Introductory concepts regarding ionic binding mechanisms of dyes and fluorochromes are the following.

Juan Carlos Stockert & Alfonso Blázquez-Castro

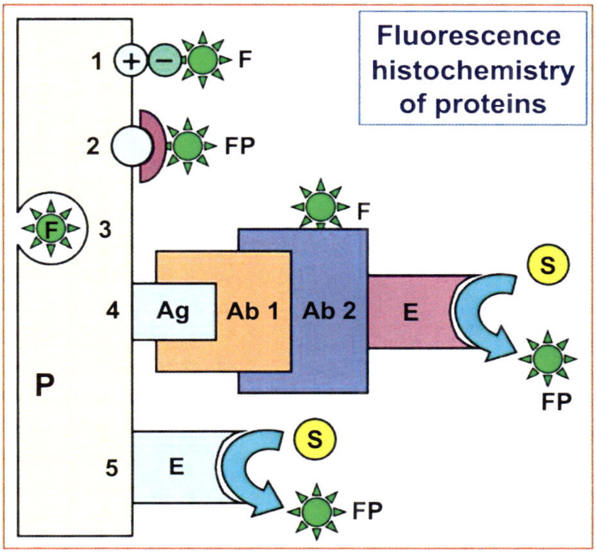

Fig. (13.1). Commonly used methods in the fluorescence histochemistry of a protein (P). Fluorophores (F) and fluorescent products (FP) are indicated. 1: Ionic interactions. 2: Selective fluorogenic reactions with amino acids (R). 3: F located within hydrophobic pockets. 4: Immunofluorescence. An antigen (Ag) is recognized by the primary antibody (Ab 1) and then by a secondary one (Ab 2) labeled with F or a suitable enzyme (peroxidase, alkaline phosphatase) that produces an insoluble FP from a non-fluorescent substrate (S). 5: The enzymatic activity of P can be revealed by the formation of FP.

1.1. Isoelectric Point and Ionic Binding

Ionic binding (electrostatic interaction) is a widespread form of linkage between dyes and biological substrates. Forces involved in this binding mode occur between ionizable and therefore charged groups. The influence of pH on ionization and ionic binding is exemplified in Fig. (**13.2**), which shows the isoelectric point (pI) and the progressive increase of electric charge of the nucleus and cytoplasm at different pH values, as well as the corresponding greater basophilia and acidophilia of these structures.

These pH-pI staining correlations have important practical consequences. When using a cationic dye or fluorochrome for nuclei at a pH below the nuclear pI, no staining will occur. On the contrary, anionic dyes or fluorochromes will stain the cytoplasm at pH values lower than its pI. Therefore, it must be taken into account that the pH at which staining reactions takes place will strongly influence quali- and quantitatively the microscopical results.

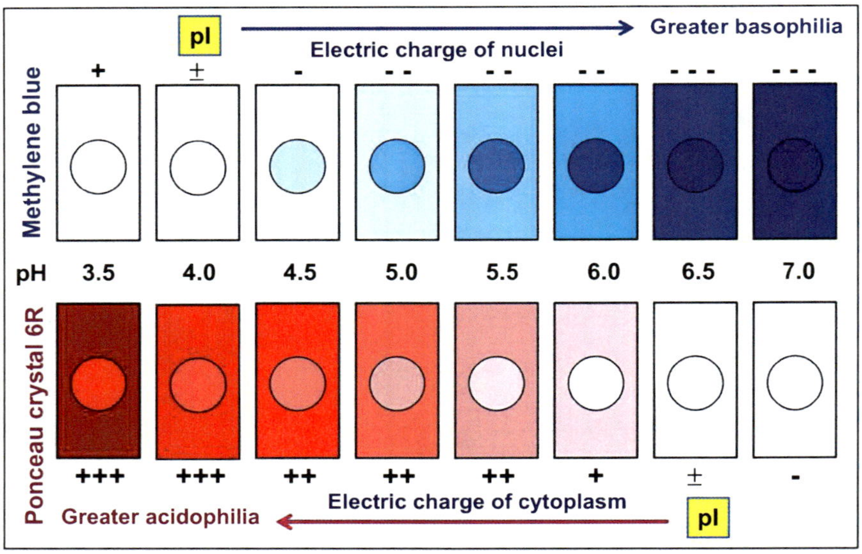

Fig. (13.2). Influence of the isoelectric point (pI) of cell components on their staining reaction at different pH values. Observe the opposite effect of pH on the staining reaction by cationic (methylene blue) and anionic (ponceau crystal 6R) dyes.

1.2. Applications of Fluorochromes in Protein Histochemistry

Selective fluorescence reactions for proteins include detection of various cell and tissue components by means of general cytochemical reactions, mainly based on anionic fluorochromes. Examples are cationic proteins from eosinophil leukocyte granules, keratohyalin granules, *etc*., which can be easily visualized by anionic fluorochromes (*i.e.* eosin Y, phloxin B, erythrosin, fluorescein, fluorone red, morin, hematoxylin, azocarmine GX, tartrazine, titan yellow, nuclear fast red, fast green FCF, bisazo and pyrene dyes, benzothiazoles, *etc.*) [3 - 5].

Anionic dyes are useful for the quantitative demonstration of proteins only if the staining intensity is proportional to the local protein concentration. The popular triphenylmethane dye Coomassie brilliant blue R-250 (CI 42660) binds stoichiometrically to proteins in histological sections and polyacrylamide gels [1] and should be useful as fluorescent dye. The anionic NIR dye indocyanine green (cardio green) forms non-covalent fluorescent complexes with proteins with λexc and λem near 780 and 820 nm, respectively. This "pseudo-fluorogenic" property has been utilized for protein assessment by capillary electrophoresis with diode laser-induced fluorescence detection. The excess of unbound dye does not interfere [6].

Anionic fluorescent dyes are of different type according to the nature of the acid group. Thus they can be phenolic, carboxylic or sulfonated dyes. Flavonols, neoflavones, and hydroxy-anthraquinones, are examples of phenolic dyes. The fluorescence of the polyphenolic dye morin, commonly used as saturated aqueous solution for blood smears [7], is illustrated in Fig. (**13.3**). It is worth to note that similar resonant forms of anionic hydroxy-quinone fluorophores occur for several compounds such as coumarins (esculin), xanthenes (eosin Y), flavonols (morin) and neoflavones (hematoxylin). Selective fluorescence of acidophilic structures is also achieved with the carboxylic derivative of natural dye saffron [8].

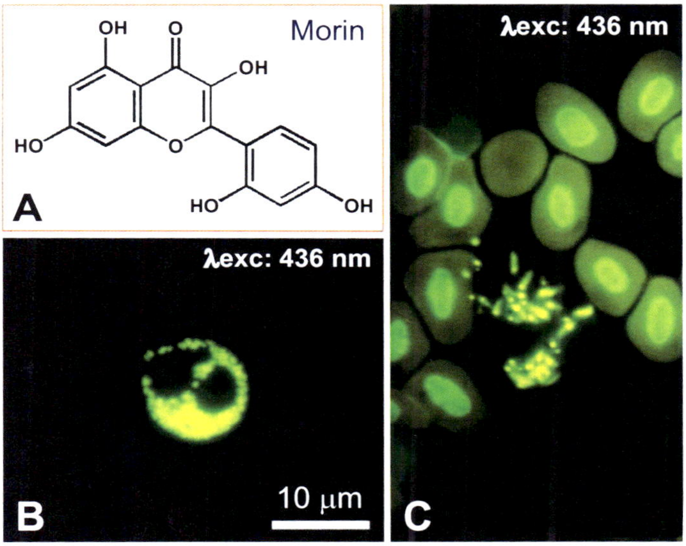

Fig. (13.3). Flavonol fluorocromes for cationic proteins. **A**: Chemical structure of morin. **B**: Methanol-fixed eosinophil leucocytes of pig blood after morin staining. **C**: Heterophil leucocytes from a methanol-fixed chicken blood smear stained with morin (saturated aqueous solution, 5 min). Observe the bright cores within the fusiform granules of heterophils.

Typical phenolic-carboxylic dyes are hydroxy-xanthenes such as fluorescein, eosin Y, phloxin B, erythrosin, *etc*., which are fluorescent dyes useful for the visualization of proteins. Eosinophil granules and other cell and tissue structures such as microtubules and reticular fibers appear selectively fluorescent by eosin Y [9 - 11] (Fig. **13.4A, B**). The hydroxylated and sulfonated pyrene derivative HPTS is strongly fluorescent (Fig. **13.4C, D**), allowing to reveal the isoelectric point of proteins [12] (see Fig. **8.11**). The anionic naphthalimide brilliant sulfoflavine FF (acid yellow 7, CI 56205; λexc: 365-450 nm, λem: 520 nm) has been used for cytofluorometric determinations of proteins [13].

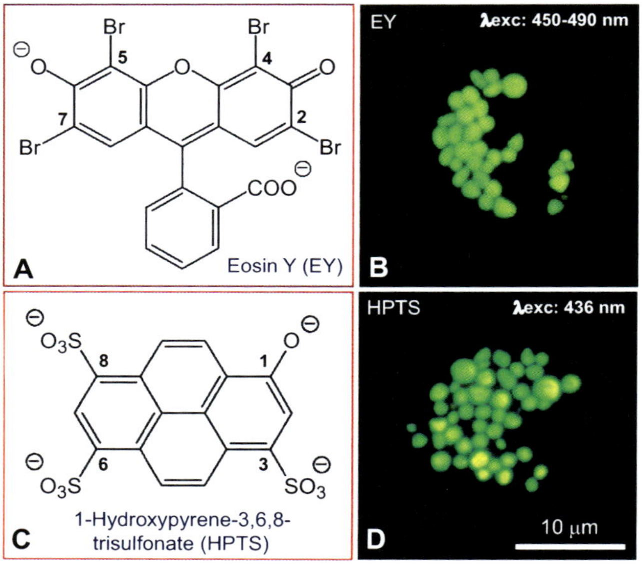

Fig. (13.4). Acid fluorochromes and eosinophil leucocytes. **A, B:** Chemical structure of eosin Y (CI 45380), and fluorescence reaction of cationic proteins from horse eosinophil granules visualized by EY (0.1 mg/mL, 5 min). **C, D:** Chemical structure of 1-hydroxypyrene-3,6,8-trisulfonate (HPTS, pyranine, CI 59040) and fluorescence of horse eosinophil granules by HPTS (10 μg/mL, pH 10, 5 min).

In addition to anionic triphenylmethane dyes such as acid fuchsine, fast green FCF, light green SF, isamine blue, aniline blue WS, yellow corallin, Coomassie brilliant blue R-250, and similar metal chelating dyes (*i.e.* xylenol orange) [14, 15], the pH indicator bromophenol blue (tetrabromophenol-sulfonphthalein) was used in early classical techniques for staining protein amino groups [14, 16]. The sultone (uncharged) form is colorless, whereas the sodium salt is deep blue [17] and should be fluorescent when rigidly bound to basic substrates.

Some fluorochromes greatly increase their fluorescent yield after binding to macromolecular substrates, typically proteins. In addition to triphenylmethanes, other dyes such as carbocyanines, coupled tetrazonium, and anionic bisazo dyes (trypan blue, Evans blue, Congo red) [4, 18 - 20] are excellent examples of this behavior (see Fig. **5.10**). The influence of solvent viscosity or rigid dye binding on the fluorescence yield is described in Chapter 5.3.1.

1.3. Bisazo Dyes

Regarding azo and bisazo dyes it may be surprising that their fluorescence has received little attention in microscopical work. This could be due to some misleading statements in the literature, which claims that the azo group destroys fluorescence (see Puchtler *et al.* [19]). However, the fluorescence of bound bisazo dyes is well known to textile chemists. Non-fluorescent azo or bisazo dyes can be rendered fluorescent by cooling, dissolving into viscous solvents, and binding or adsorption to suitable substrates, solid phases, and rigid media. Thus, in addition to Congo red, trypan blue and Evans blue, a great variety of bisazo dyes (acilan red S2B, benzopurpurin 4B, benzo fast scarlet YA and 4BS, Congo corinth, levanol brilliant red BB, levanol scarlet Y, *etc.*) shows strong emission when bound to protein components of cells and tissues.

Trypan blue (CI 23850), its isomer Evans blue (CI 23860), trypan red (CI 22850), and Congo red (CI 22120) are widely employed in microscopical methods as dyes, vital stains, and fluorescence markers [4]. Trypan blue is the prototype stain in dye exclusion assays to differentiate between living and dead cells (see Chapter 16.1.1). It is worth noting that trypan blue and Evans blue are teratogenic dyes, a feature attributed to their binding to proteins [17, 18]. Excitation curves and microscopic fluorescence of the acidophilic cytoplasm of chicken erythrocytes stained with trypan blue are seen in Fig. (**13.5**).

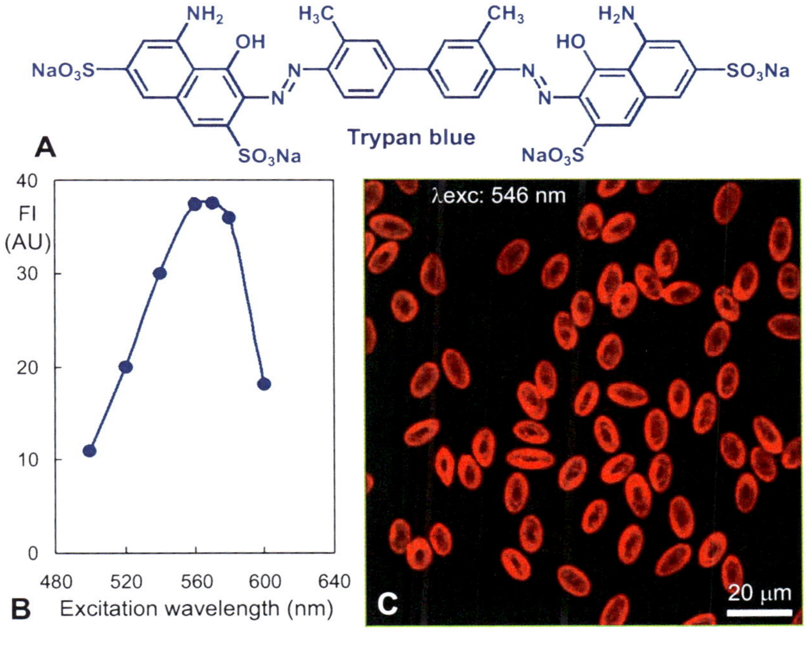

Fig. (13.5). A: Chemical structure of trypan blue. **B:** Response of 5 μg/mL trypan blue in the presence of 1 mg/mL trypsin to different excitation wavelengths. The best excitation is at 560-570 nm, and all the emission curves have their peaks at 650 nm. **B:** Methanol-fixed chicken blood smear showing the red fluorescence of erythrocyte cytoplasm after staining with trypan blue (0.1 mg/mL, 5 min).

When trypan blue binds to proteins, notably albumin, its very weak red fluorescence strongly increases, and the same occurs in the case of polysaccharides and other polymeric substrates. Regarding applications, it was shown that when injected subcutaneously to quails, highly diluted solutions of trypan blue can be applied without teratogenic effects, as tracer for exogenous yolk uptake and migration into oocytes, where the dye is visualized by its deep red fluorescence [21]. Likewise, Evans blue is used as protein fluorochrome (λab: 546 nm, λem: 680 nm), and both dyes have been applied in studies on protein transport from the blood circulation to tissues. Mixed with bovine albumin, Evans blue was the first fluorescent neuronal tracer to be used to study retrograde axonal transport [22]. In addition to demonstration of proteins (Fig. **13.6**), fluorescent applications of Congo red are numerous.

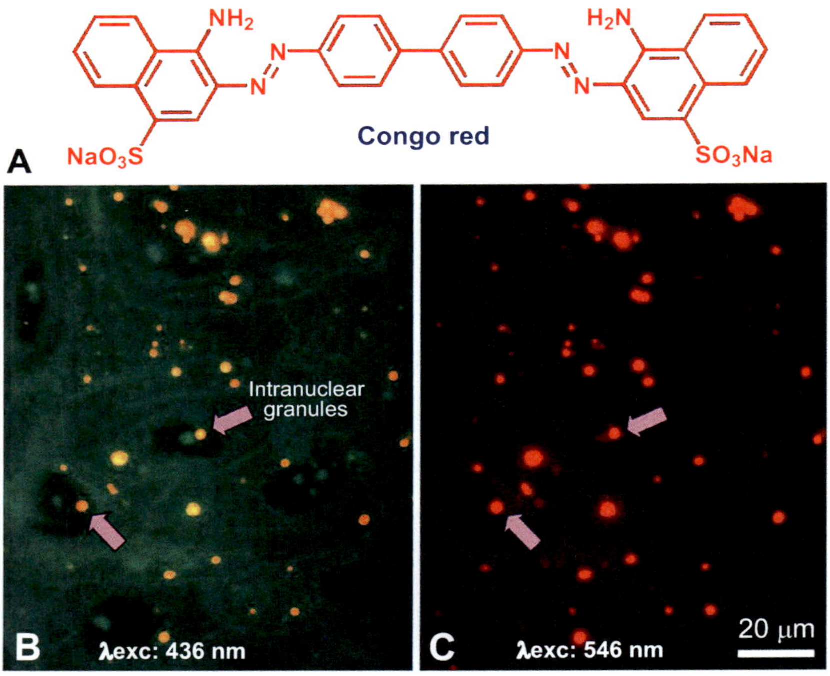

Fig. (13.6). A: Chemical structure of Congo red. **B**, **C:** The same field of a semithin section of mouse tongue (glutaraldehyde, Epon) stained with Congo red (0.1 mg/mL, 30 min), showing the fluorescence of acidophilic keratohyalin granules and the influence of the excitation and barrier filter on the emission color.

Congo red has been used for detection of smooth muscle cells and elastic fibers [23], cellulose fibers [24], chitin in growing fungal cell walls [25], amyloid deposits and Alzheimer's plaques [26]. In these cases selectivity and sensitivity are improved by observing the green birefringence under polarized light [4, 27].

Regarding amyloid staining, only direct cotton dyes reveal amyloid selectively; acid (sulfonated) dyes stain all tissue structures uniformly. Similarly to cellulose staining, histochemical studies indicate the occurrence of non-ionic hydrogen bonds between hydroxyl groups of amyloid and amino, hydroxy and azo groups of these bisazo dyes [28], but hydrophobic and van der Waals forces also take place in the binding mechanism. In addition to histological staining, bisazo dyes have been used in other biomedical fields [4, 17, 29, 30]. Trypan blue had veterinary applications against pyroplasmosis, Evans blue was used for blood volume studies, trypan red as anti-trypanosomal drug and against herpes relapses, and Congo red as hemostatic agent.

1.4. Benzothiazole Fluorochromes

As occurs with many dyes used as reagents in biomedical research, benzothiazole dyes are heterogeneous in composition, although authors often fail to realize this. Benzothiazole dyes are directly formed by heating *p*-toluidine and sulfur, and they can contain several benzothiazole units (Fig. **13.7A**). The amount of mono-, di- y tri-benzothiazoles as base dyes is related to the temperature reached in the reaction, and after sulfonation and/or N-methylation the anionic (primuline, CI 49000), cationic (thioflavine T, CI 49005), and amphoteric (thioflavine S, CI 49010) derivatives are obtained (Fig. **13.7B**). However, in current publications, primuline, thioflavine T, and thioflavine S are shown as containing a single benzothiazole unit [31].

Fluorescent visualization of cationic proteins by these benzothiazole dyes is illustrated in Fig. (**13.7C, D**). Primuline, thiazole yellow, thioflavine S and T are highly appreciated for detecting amyloid deposits and Alzheimer's plaques that fluoresce in yellow-green [32, 33]. As thioflavine T also binds to nucleic acids [34, 35] (see Chapter 12.5.5), the emission of nuclei can be quenched by previous staining with aluminum-hematoxylin. Acid glycosaminoglycans, pancreatic zymogen granules, Paneth cells, cartilage matrix, and juxtaglomerular granules are other cell and tissue structures are fluorescent after thioflavine T staining. Primuline has been also applied as retrograde neuronal tracer. A tetrameric benzothiazole derivative of primuline isolated from the commercial product is an efficient inhibitor of the hepatitis C virus helicase [31], which strengthens the potential of dyes and fluorochromes as pharmacological agents.

Amyloid fibrils are protein misfolding structures that involve a β-sheet pattern and are associated with various pathologies such as Alzheimer's disease, Creustzfeldt-Jacob's disease, or type II diabetes. At the concentration used to monitor amyloid deposits, thioflavine T occurs as micelles (~3 nm in diameter) that are responsible for enhanced fluorescence upon binding to amyloid fibrils [36]. Recently, polarization-resolved fluorescence microscopy has been employed to analyze structural details of amyloid fibrils alter thioflavine T and Congo red staining, providing useful information on the orientation of protein β-sheets by using these dyes as structural reporters of the insulin amyloid fibrils [37].

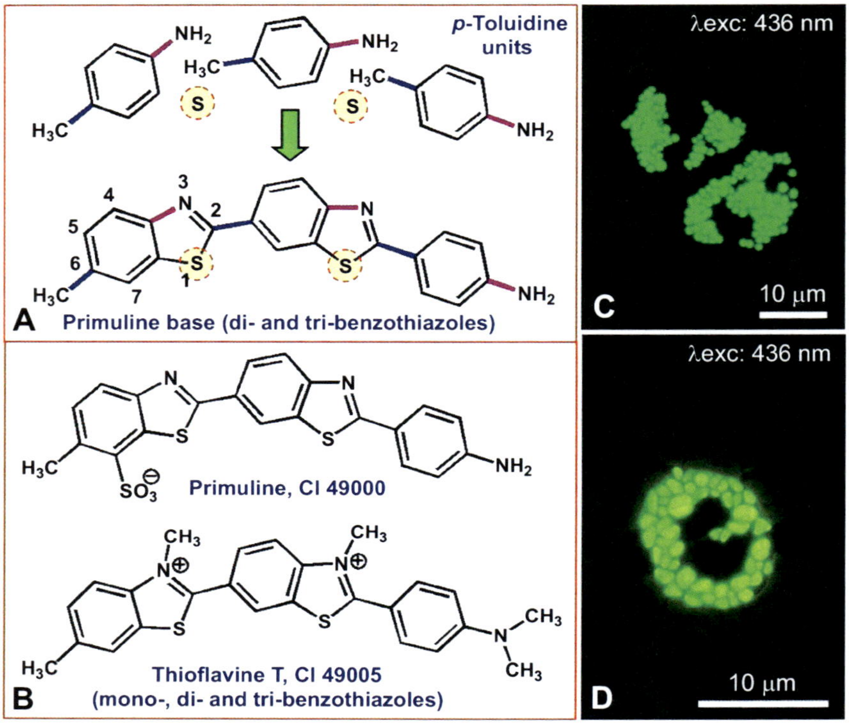

Fig. (13.7). Synthesis **(A)** and examples of benzothiazole fluorochromes **(B)**, made from p-toluidine and sulfur. Numbering of the benzothiazole ring is indicated. **C, D:** Methanol-fixed horse eosinophils stained with the anionic primuline (2.5 μg/mL, 5 min) and thioflavine S (5 μg/mL, 5 min), respectively.

1.5. Hydrophobic and Reactive Fluorochromes

The fluorescent amphiphilic probe, 1-anilino-8-naphthalene sulfonate (ANS) binds selectively to hydrophobic domains in proteins, and has been applied as fluorochrome in studies on oocyte inclusions [38], and hydrophobic proteins such as elastin [39]. It must be noted that ANS increase its fluorescence yield from Φ_F: 0.0 free in water to Φ_F: 0.9 when bound to proteins. Chemical structure and an image of ANS fluorescence of elastic fibers is shown in Fig. (**13.8A, B**). On the other hand, tetraphenyl-porphine sulfonate ($TPPS_4$) and basic fuchsine have been also used for the selective localization of elastic fibers under fluorescence microscopy [40, 41].

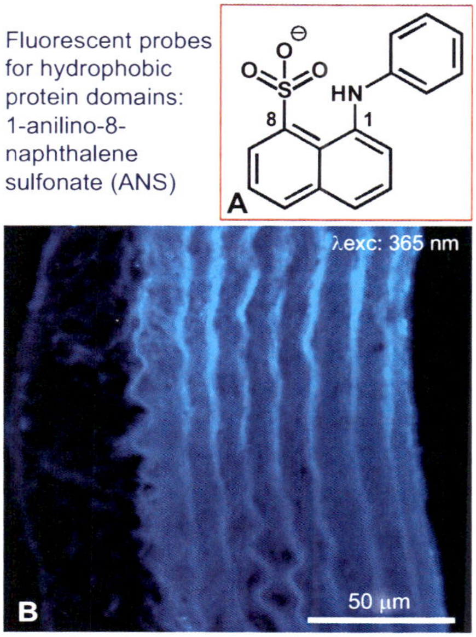

Fig. (13.8). A: Chemical structure of ANS (1-anilino-8-naphthalene sulfonate). **B:** Paraformaldehyde-fixed, paraffin section of pig aorta stained with ANS, showing selective fluorescent binding to hydrophobic regions of elastin (courtesy of B.C. Vidal).

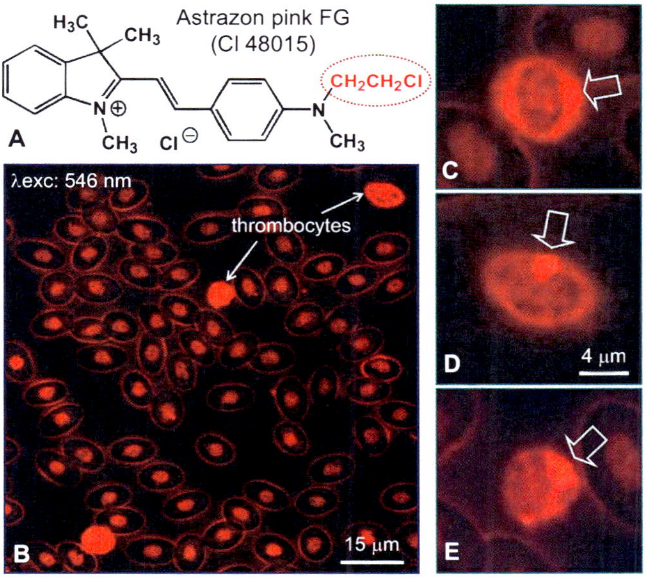

Fig. (13.9). Selective visualization of blood cell structures. **A:** Astrazon pink FG. The chloroethylamino residue (marked in red) corresponds to the reactive group. **B-E:** Smears of chicken blood fixed in methanol and stained with 0.2% methyl green followed by 1 mg/mL astrazon pink FG. Observe the attenuated nuclear emission and the high fluorescence of specific granules (arrows) in thrombocytes.

Astrazon pink FG (CI 48015) is a reactive carbocyanine dye that contains the alkylating chloro-ethylamino group (Fig. **13.9A**), giving red fluorescence mainly to protein structures and to some extend to nuclei under green excitation. The specific granule of chicken thrombocytes (called "ruby body", as described after Giemsa staining) appears highly fluorescent when treated with astrazon pink FG (Fig. **13.9B-E**). Other reactive fluorochromes (*i.e.* FITC, TRITC) that give selective fluorescence of protein amino groups are well known but scarcely employed for direct biological staining.

2. SELECTIVE FLUOROGENIC REACTIONS FOR AMINO ACIDS

Amino acids are characterized by side chains with distinct chemical groups, examples being amino, carboxyl, guanidine, sulfhydryl, and aromatic groups. They can be histochemically visualized.

2.1. Amino Groups

Amino groups from lysine and terminal $-NH_2$ can be demonstrated by using selective reagents such as dansyl chloride (5-[dimethylamino]-1-naphthalene-sulfonyl chloride), and NBD-Cl (4-chloro-7-nitro-benzo-2,1,3-oxadiazole chloride) (Fig. **13.10A, B**), which are widely used for fluorescent labeling of amino groups in proteins and lipids. Although dansyl chloride is the most used reagent [42], bansyl chloride gives similar results. However, these reagents also react with secondary amino groups and phenolic hydroxyls [43].

Dansyl-amino products have λexc: 350 nm and λem: 500-550 nm, depending upon the type of solvent. As the reaction eliminates HCl, the pH must be kept constant by addition of sodium bicarbonate or carbonate. In the case of NBD-Cl, the reaction takes place with primary and secondary aliphatic amines to produce intensely fluorescent products showing λexc: 480 nm and λem: 510-560 nm.

Furanone reagents such as fluorescamine (4-phenylspiro-[furan-2(3H)]- 1-phthalan-3,3-dione) (Fig. **13.10C**) and MDPF (2-methoxy-2,4-diphenyl-3 (2H)-furanone) have been used in fluorescence microscopy to reveal primary amino groups [44, 45], for instance in cationic proteins of eosinophilic granules (Fig. **13.10D**). The excess of reagent in water is rapidly hydrolyzed to a non-fluorescent product. It is currently assumed that both reagents give pyrrolinone products in which the C2 atom is saturated (tetrahedral, corresponding to the pseudo-base generated at alkaline pH) (Fig. **13.10E**). However, the true fluorophore is obviously the planar pyrrolone product [46, 47] (Fig. **13.10F**). The chemical structure of this pseudo-base is also similar to the pseudo-base of ethidium bromide and other well known non-fluorescent pseudo-bases.

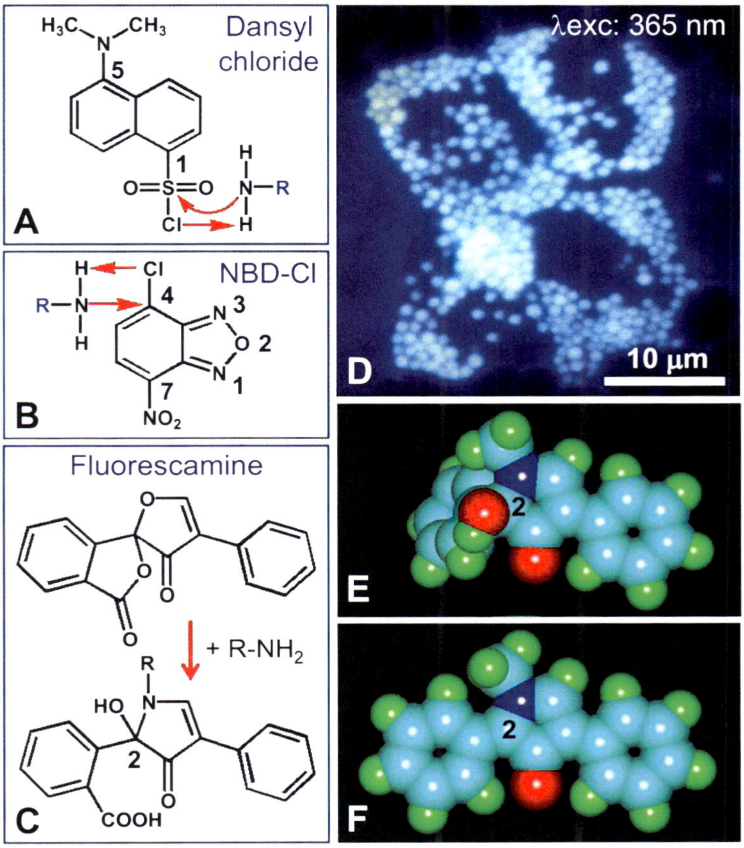

Fig. (13.10). Fluorogenic reactions for amino groups. **A**, **B:** Chemical structure of dansyl chloride and NBD-Cl, respectively, and their reaction with -NH_2 groups. Relevant atom numbering is shown. **C:** Fluorescamine (Fluram®) and the pyrrolinone (pseudo-base) reaction product with an amino group showing the tetrahedral C_2 atom. **D:** Horse eosinophil leucocytes after the fluorogenic reaction of fluorescamine for amino groups. **E**, **F:** Pseudo-base methylamino-derivative of MDPF **(E)** and the true fluorophore (cationic pyrrolone, F), showing tetrahedral and planar C_2, respectively.

Aromatic dialdehydes also react with amino groups (Fig. **13.11A**). In alkaline solution and in the presence of reducing agents, *o*-phthalaldehyde (OPA) gives a fluorescent product with -NH_2 of amino acids and amino sugars, showing λexc: 340 nm and λem: 455 nm. Likewise, other even more sensitive aromatic dialdehyde reagents are useful to reveal amino groups. Examples are naphthalene-2,3-dialdehyde (NDA), and 3-(4-carboxybenzoyl)-quinoline-2-carboxaldehyde (CBQCA), which give highly fluorescent pyrrole-naphthalene and pyrrole-quinoline reaction products (Fig. **13.11B**), with the sticking sensitivity limits of 10^{-17} y 10^{-18} M, respectively, under λexc: 440-450 nm [48].

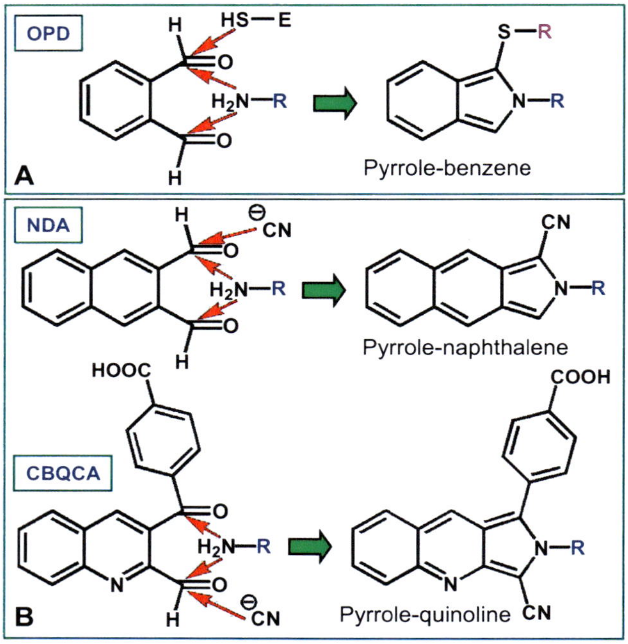

Fig. (13.11). A: Reaction of OPA (*o*-phthalaldehyde) with a -NH_2 group in the presence of 2-mercapto-ethanol (E) to give the fluorescent pyrrole-benzene product. **B:** Reactions of the aromatic dialdehydes NDA and CBQCA with -NH_2 groups in the presence of cyanide ions to give highly fluorescent pyrrole-naphthalene and pyrrole-quinoline products.

Naphthalic anhydride and some derivatives have been applied for revealing aliphatic amino groups in fluorescence histochemistry of proteins, which give the naphthalimide fluorophore [49], with λexc: 440 nm and λem: 560 nm (Fig. **13.12A, B**). The fluorogenic reaction is easily applicable to microscopical preparations, giving an intense yellow-green emission in protein components (Fig. **13.12B**). Also antitumoral amino-naphthalimides are fluorescent and its intercalative binding to DNA has been described [50].

The azomethine (imine or Schiff's base) reaction product between the -NH_2 group and salicylaldehyde has green fluorescence [2]. Likewise, 2-hydroxy-3-naphthaldehyde condenses with -NH_2 giving an azomethine. This product could complex to Al^{3+} producing a fluorescent compound. Likewise, after coupling the naphthol with a diazonium salt the product should be suitable for use in fluorescence microscopy. FITC, TRITC, and the sulfonylchloride derivative of sulforhodamine 101 (Texas red) have found wide application for labeling of amino groups (see Chapter 4.1.1). Recently, aminomethyl-coumarin acetic acid succinimidyl ester and the fluorescent orange 550 reactive dye have been also applied for coupling with amino groups.

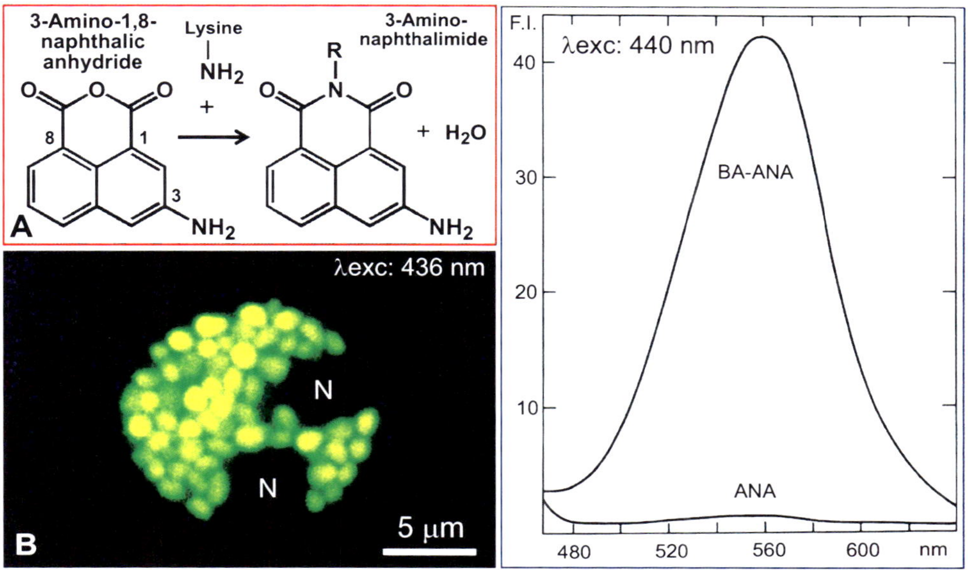

Fig. (13.12). A: Fluorogenic reaction of 3-amino-1,8-naphthalic anhydride (ANA) with an amino group to give the fluorescent amino-naphthalimide. **B:** Methanol-fixed horse eosinophil after reaction with ANA, showing the emission of acidophilic granules. N: nuclear lobes. **C:** Emission curves of ANA and its reaction product with butyl-amine (BA-ANA).

2.2. Carboxyl Groups

These groups occur in aspartic and glutamic amino acids. When both the C-terminal and the carboxylic residues of aspartic and glutamic acids (as well as other carboxyl substrates) are treated with acetic anhydride in pyridine at 60 ºC, ketone reaction products are formed. They react with organic hydrazines (*i.e.* dansylhydrazine) and hydrazides (*i.e.* salicylhydrazide, 2-hydroxy-3-naphthoic acid hydrazide) to give colored or fluorescent compounds. Whitish yellow emission has been described under blue excitation [14]. Staining or emission can be intensified after coupling with diazonium salts. In addition to ketones, similar reactions occur with aldehyde groups (see Chapter 12.2.1). Direct fluorescence of these adducts (or after chelation with Zn^{2+} or Al^{3+} ions) can be useful for histochemical detection [1, 2].

2.3. Guanidino Groups

From chemical interpretations of the classical Sakaguchi's test for arginine, an aromatic quinone reaction mechanism was suggested [1, 14]. The reagent 1,2-naphthoquinone-4-sulfonic acid condenses with the guanidino group to give a red naphthalene-imidazole product. In addition, the condensation of arginine with *o*-phthalaldehyde gives a fluorescent product (λexc: 355 nm and λem: 470 nm) [18].

A fluorescent method with phenanthrene-quinone for the histochemical demonstration of arginine residues in tissues (Fig. **13.13A**) has been proposed by Magun and Kelly [51]. The hematological application of this fluorescence reaction for guanidino groups [5] allows detecting arginine residues in blood cells, mainly cationic proteins from eosinophil leucocyte granules (Fig. **13.13B**). It is worth to note that a striking photo-enhancement of emission occurs upon continuous UV irradiation. In addition, an interesting possibility would be to apply this reaction to label the amidino or guanidino groups of other biomolecules (*i.e.* distamycin A, streptomycin, *etc*).

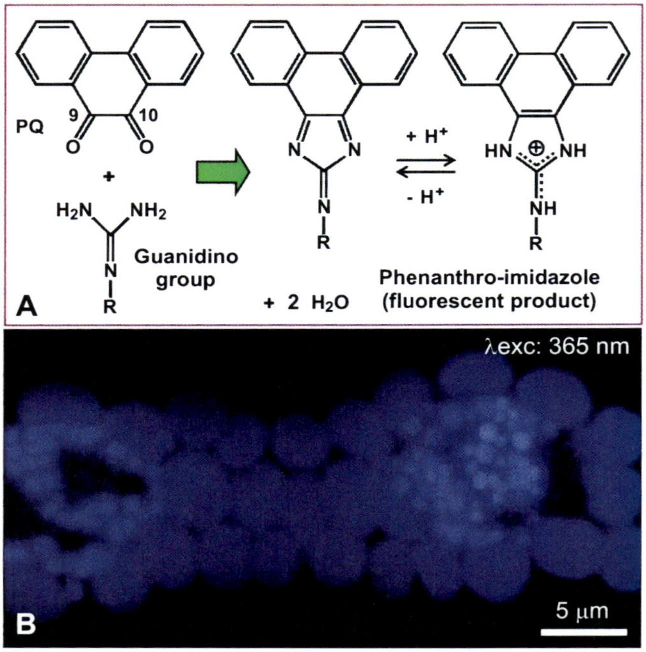

Fig. (13.13). A: Reaction of 9,10-phenanthrene-quinone (PQ) with the guanidino group. **B:** Methanol-fixed horse eosinophils showing the fluorescence reaction after treatment with 0.1% PQ in alkaline ethanolic solution, 30 min), and 5 min continuous UV excitation.

2.4. Aromatic Amino Acids

Aromatic amino acids are tryptophan, tyrosine and histidine. Although their intrinsic emission is suitable for quantitative assays purposes *in vitro* (*i.e.* λem: 350 nm and 310 nm for tryptophan and tyrosine, respectively), its application in fluorescence microscopy is unlikely. Aldehyde condensation of indole and phenol residues from biogenic amines gives the known isoquinoline and β-carboline derivatives (see section 3).

The most popular histochemical method for tryptophan is the reaction with with *p*-dimethylamino-benzaldehyde (Ehrlich's reagent) [1, 2]. Colored compounds can result from the direct benzylidene reaction using the reactive C2 atom from the indole ring with formation of an azomethine product, or condensation to form a β-carboline derivative when cyclization is followed by oxidation. In both cases the reaction products are fluorescent. Also the xanthydrol reaction for tryptophan [2] should give a fluorescent product.

Isatin (indole-2,3-dione) easily condenses with compounds having a reactive methylene group. Upon binding to indoxyl, isatin gives the dye indirubin (CI 73200) [29]. In ethanolic solvent, isatin reacts very selectively with indol derivatives such as tryptophan, forming fluorescent indigo or indirubin products (Fig. **13.14A, B**), and no other amino acid gives a similar fluorogenic reaction [5, 52]. Treatment of microscopical substrates with an ethanolic solution of isatin results in a bright green fluorescence of tryptophan-containing proteins (Fig. **13.14C**).

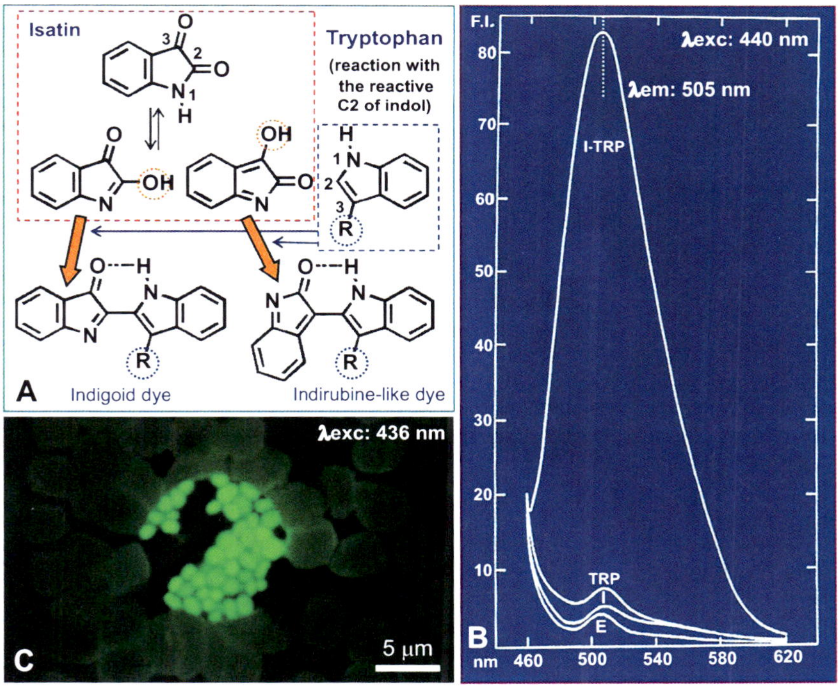

Fig. (13.14). A: Reaction of ethanolic isatin with tryptophan, and formulation of possible fluorescent products (indigo and/or indirubin derivatives). **B:** Emission spectra of compounds corresponding to the ethanolic isatin reaction: E: ethanol, I: isatin in ethanol (50 μg/mL), TRP: tryptophan (saturated), I-TRP: the reaction product. **C:** Methanol-fixed horse eosinophil showing the emission of specific granules after the ethanolic isatin reaction.

Diazonium and tetrazonium salts easily react with aromatic amino acids and derivatives giving colored products [1, 2]. The tetrazonium salt, fast blue B, has been applied to reveal aromatic amino acids in proteins using fluorescence microscopy, the product being a bisazo dye [20] (Fig. **13.15A, B**). This reaction mainly occurs with the indole group of tryptophan (*ortho* to the N atom), the phenol group of tyrosine (*ortho* to -OH), and the imidazole ring of histidine (between the two N atoms). The strongly colored tetrazonium product *in silico* (*i.e.* fast blue B followed by 1-naphthol, λexc: 520-535 nm, λem: 660 nm in 80% glycerol), does not have sulfonate groups as in common bisazo dyes, and then it must be dissolved in ethanol-glycerol mixtures to make spectral studies. Interestingly, diazonium salts alone (*i.e.* fast red TR) bind to the reactive C2 of tryptophan and derivatives (tryptamine, serotonine) and generate the corresponding fluorescent azo dyes.

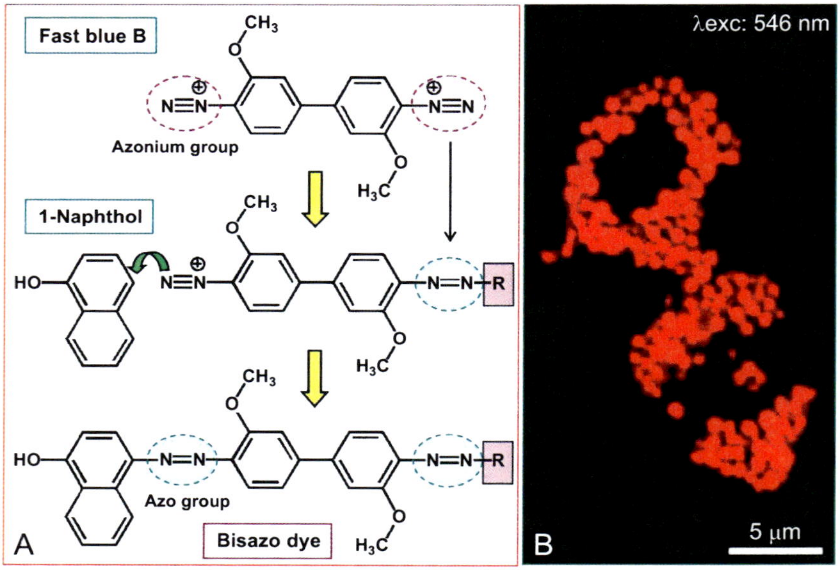

Fig. (13.15). A: Formation of a bisazo dye *in situ* involving aromatic residues (R) from tryptophan, tyrosine, or histidine amino acids, and the tetrazonium salt fast blue B. When the second diazonium group remains uncoupled to R, it is blocked with 1-naphthol. **B:** Fluorescence of methanol-fixed horse eosinophils after the fast blue B/1-naphthol reaction.

N-bromo-succinimide (NBS) selectively induces fluorescence in protein components under UV excitation. Cytoplasmic granules of mammalian eosinophils (Fig. **13.16**) and chicken heterophils appear with a striking white-blue fluorescence [53], which rapidly increases in the first minutes (see Chapter 7.5).

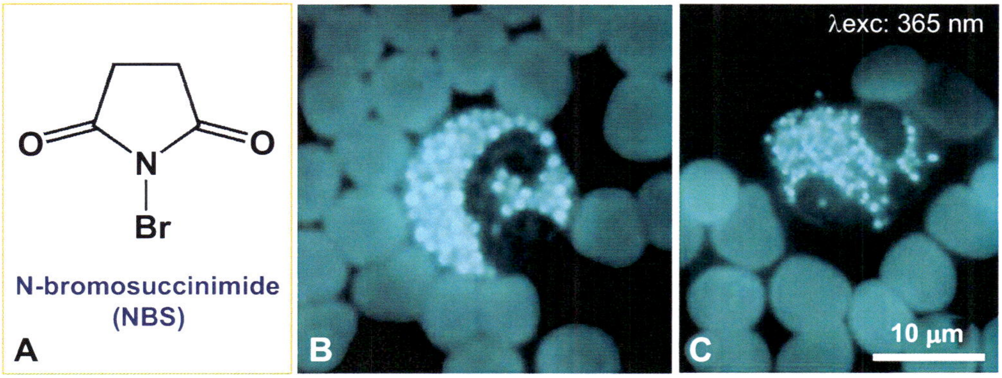

Fig. (13.16). A: N-bromosuccinimide (NBS), and fluorogenic reaction with horse (B) and human (C) eosinophil leucocytes fixed in methanol, treated 10 min with 10 mM NBS, and observed after 1 min of continuous UV excitation.

NBS generates the strongly oxidizing reagent hypobromous acid (BrOH), causing dehydrogenation and formation of double bonds. In proteins, the aromatic amino acids tryptophan, tyrosine, and histidine are modified by NBS treatment giving highly fluorescent products (λexc: 365 nm, λem: 450 nm) [54].

For the histochemical recognition of phenols, especially tyrosine, the Millon's test (1849) and Meyer's reaction (1864) have been routinely applied. Both are based on the red color produced when tyrosine is heated with mercury ions in the presence of nitrite in acid solution. It is accepted that in these tests and other similar procedures, the reaction proceeds in two steps: first a nitroso-phenol is formed by the substitution of -NO for -H in *ortho* position to the phenol -OH, and then mercury forms a new ring involving -NO and -OH groups giving the red color. Mercuric sulfate and sodium nitrite are used in the most popular method for tyrosine [55]. It is interesting to note that other adequate cations (*i.e.* aluminum) form green fluorescing complexes with nitroso-phenol (see Chapter 8.3.1). A similar nitroso-naphthol complex with Fe^{3+} ions corresponds to the dye naphthol green B (CI 10020). Likewise, a strongly fluorescent product (λexc: 460 nm, λem: 570 nm) was found after reaction of tyrosine with 1-nitroso-2-naphthol [2, 18].

Also the phenol group of tyrosine reacts with dinitro-fluorobenzene and after reduction, diazotization, and naphthol coupling a final azo dye is formed. In a similar way, nitrosation of tyrosine, conversion to a diazonium salt and coupling to naphthol gives an azo dye. In both cases, fluorescence should be expected for these final products.

2.5. Sulfhydryls

To maintain the availability of sulfhydryl groups, fixation with mercury or lead salts, and oxidation by reagents or atmospheric oxygen must be avoided [1]. Then these groups can be adequately demonstrated using organic mercurials as well as bromo, iodoacetamide, and maleimide dye derivatives.

The fluorescent mercurial azo dye, mercury orange (*p*-chloromercuri-phenylazo-2-naphthol) (Fig. **13.17A**), is a commonly used reagent to reveal the thiol group (-SH) [2, 56, 57]. It can be applied in both bright-field and fluorescence microscopy and is considered a selective dye for thiols, although nucleic acid bases also react to some extent.

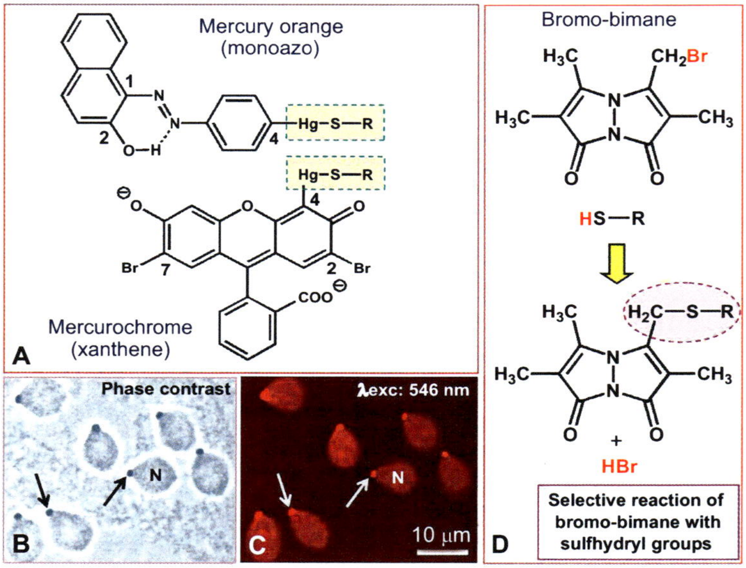

Fig. (13.17). A: Chemical structure of the reaction product of the fluorescent mercurials, mercury orange and mercurochrome, with thiol groups. Rectangles indicate the linkage. **B, C:** Centriolar adjuncts from ethanol-acetic acid-fixed spermatids of *Arcyptera fusca* stained with mercurochrome (25 µg/mL, 15 min), and observed by phase contrast (B) and fluorescence (C). **D:** Detection of -SH groups by monobromo-bimane.

Mercurochrome (2,7-dibromo-4-hydroxy-mercurifluorescein) (Fig. **13.17A**) is a mercurial xanthene dye [58] that has been used for detection of thiols by fluorescence and electron microscopy [59]. Green and red emissions are observed under excitation at 436 and 546 nm, respectively. It is known that the centriolar adjunct of grasshopper spermatids is a strongly argentophilic paranuclear body with a high content of thiol groups, and this structure gives a selective reaction with mercurochrome [60] (Fig. **13.17B, C**). Other cell components are also visualized after mercurochrome staining. In the case of chicken blood, granules of heterophil leucocytes, the ruby body of thrombocytes, and a weakly autofluorescent body in erythrocyte cytoplasm show selective fluorescence, which is resistant to extraction by 2M NaCl and then does not represent simple ionic binding. Likewise, fluorescein mercuric acetate has been employed as a fluorescent mercurial.

Monobromo-bimane (mBrB) (Fig. **13.17D**), and dibromo-bimane (dBrB) are essentially non-fluorescent compounds that react easily with thiol groups to give fluorescent products with λexc: 385-390 nm, λem: 477-484 nm, and Φ_F from 0.18 to 0.33, depending on the substrate [61, 62]. The reagents give positive results in membrane proteins and glutathione (GSH) in living cells and under physiological conditions (*i.e.* 80 μM, 90 s), but the disulfide GSSG does not react with mBrB. Some collateral reactivity with ethanolamine-containing phospholipids can not be excluded.

In addition, iodoacetamide and maleimide derivatives of some coumarin and amino-xanthene dyes are useful for -SH detection. Examples are the recent maleimides of Atto 425, 465, 495, 525, 565, 590, 610 and 647. Bodipy derivatives such as Bodipy FL N-2 aminoethylmaleimide and Bodipy 499/508 give highly fluorescent products with thiols. Also naphthalene and pyrene derivatives have been employed to reveal thiols. Examples are 6-acryloyl-2-dimethylamino-naphthalene and N-(1-pyrenyl)-maleimide. The later is non-fluorescent and gives fluorescent excimers when bound to -SH. Among other applications, N-methyl and N-ethyl pseudoisocyanines have been employed for detection of cysteic acid groups after oxidation of insulin cystine residues. At present, new fluorophores for detection of cysteine and homocysteine are available from commercial sources.

3. FLUOROGENIC REACTIONS FOR BIOGENIC AMINES AND NEUROTRANSMITTERS

Biogenic amines such as histamine and the catecholamines adrenalin, noradrenalin, dopamine and tyramine, and the indolamines serotonin and melatonin, are decarboxylated amino acid derivatives. They are low molecular weight metabolites found as freely soluble neurotransmitters and within secretory granules of endocrine or similar cells. This is the case of catecholamine-containing chromaffin cells in adrenal medulla, serotonin-containing mast cells, and argentaffin and enterochromaffin cells in the intestinal mucosa, where binding to a protein or carbohydrate matrix greatly reduces their diffusion.

Neurotransmitters can be detected through chemical reactions forming highly fluorescent products (isoquinolines from catecholamines and β-carbolines from indolamines). It is well known that according to the Pictet-Spengler-type condensation, aldehydes react with the -NH_2 group of biogenic amines followed by cyclization to produce new and red-shifted chromophores [63 - 65]. They have interesting fluorescence properties allowing to recognize chatecolamines from indolamines (Fig. **13.18A-C**). These neurotransmitters are readily converted into fluorescent products by treatment with gaseous formaldehyde under mild conditions, and reaction has been found to be catalyzed by some amino acids and peptides [66], which is relevant regarding induction of autofluorescence after aldehyde fixation. Some less oxidized isoquinoline and β-carboline products could also contribute to the fluorogenic reactions. The importance of the fluorescence histochemistry of biogenic amines is now widely accepted [1, 67, 68].

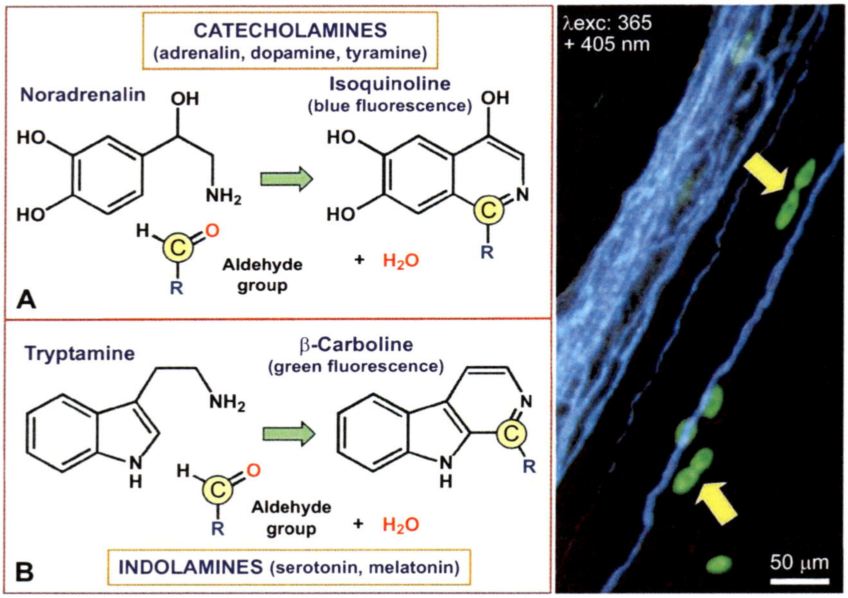

Fig. (13.18). Fluorogenic reactions for neurotransmitters. **A, B:** Aldehyde-induced reactions for catecholamines (A) and indolamines (B), showing the formation of blue-fluorescing isoquinolines and green-fluorescing β-carbolines. **C:** Rat mesentery fixed with formaldehyde vapor. Fluorescence of catecholamines in nerves (blue) and indolamines (serotonin) in mast cells (green) can be seen.

In the highly sensitive ALFA method used mainly for freeze-dried tissues and cryostat sections, previous treatment of the samples with Al^{3+} ions followed by formaldehyde generates an increased fluorescence [1]. Other aldehydes (*i.e.* acetaldehyde, glutaraldehyde, glyoxal, glyoxylic acid) can also be applied for this aldehyde-induced fluorescence. A variable degree of autofluorescence is obtained after any type of aldehyde fixation. Glyoxylic acid is a good alternative to formaldehyde because its fluorescence yield is higher and freeze-drying methodology is not necessary (perfusion is allowed).

CONFLICT OF INTEREST

The author (editor) declares no conflict of interest, financial or otherwise.

ACKNOWLEDGEMENTS

This work was supported by a grant from the Ministerio de Economía y Competitividad (CTQ2013-48767-C3-3-R), Spain.

REFERENCES

[1] Kiernan JA. Histological and Histochemical Methods: Theory and Practice. Oxford, New York: Pergamon Press 1990.

[2] Pearse AG. Histochemistry Theoretical and Applied. 3rd ed. Edinburgh, London, New York: Churchill Livingstone 1968; 1.

[3] Bobadilla JR, Gutiérrez-Gonzálvez MG, Armas-Portela R, Stockert JC. 9-Fenil-2,3,7-trihidroxi- 6-fluorona como colorante y fluorocromo. Acta Bioquim Clin Latinoam 1988; 22: 419-25.

[4] Horobin RW, Kiernan JA, Eds. Conn's Biological Stains A Handbook of Dyes, Stains and Fluorochromes for Use in Biology and Medicine. 10th ed., Oxford, UK: BIOS Scientific Publishers 2002.

[5] Stockert JC. The horse eosinophil as a model leucocyte for morphological and cytochemical studies. Braz J Morphol Sci 2005; 22: 73-84.

[6] McCorquodale EM, Colyer CL. Indocyanine green as a noncovalent, pseudofluorogenic label for protein determination by capillary electrophoresis. Electrophoresis 2001; 22(12): 2403-8. [http://dx.doi.org/10.1002/1522-2683(200107)22:12<2403::AID-ELPS2403>3.0.CO;2-B] [PMID: 11519943]

[7] Stockert JC, Colman OD, Cañete M. Fluorescence reaction of leukocyte granules by morin. Acta Histochem Suppl 1985; 31 (Suppl.): 243-52. [PMID: 2410949]

[8] Trigoso CI, Stockert JC. Fluorescence of the natural dye saffron: selective reaction with eosinophil leucocyte granules. Histochem Cell Biol 1995; 104(1): 75-7. [http://dx.doi.org/10.1007/BF01464789] [PMID: 7584563]

[9] Espada J, Valverde P, Stockert JC. Selective fluorescence of eosinophilic structures in grasshopper and mammalian testis stained with haematoxylin-eosin. Histochemistry 1993; 99(5): 385-90. [http://dx.doi.org/10.1007/BF00717051] [PMID: 7687595]

[10] Espada J, Stockert JC, Gamallo C. Fluorescencia selectiva de estructuras eosinofílicas en secciones teñidas con hematoxilina-eosina: mecanismos histoquímicos y aplicaciones histológicas. Patologia (Madr) 1994; 27: 335-41.

[11] Stockert JC. Cytochemistry of mast cells: new fluorescent methods selective for sulfated glycosaminoglycans. Acta Histochem 2000; 102(3): 259-72. [http://dx.doi.org/10.1078/S0065-1281(04)70034-3] [PMID: 10990064]

[12] Trigoso CI, Espada J, Stockert JC. Fluorescence of eosinophil leucocyte granules induced by 1-hydroxy-3,6,8-pyrenetrisulfonate. Visualization of differences in protein isoelectric points. Histochem Cell Biol 1995; 104(1): 69-73. [http://dx.doi.org/10.1007/BF01464788] [PMID: 7584562]

[13] Leemann U, Ruch F. Cytofluorometric determination of basic and total proteins with sulfaflavine. J Histochem Cytochem 1972; 20(9): 659-71. [http://dx.doi.org/10.1177/20.9.659] [PMID: 4116453]

[14] Lillie RD. Conn's Biological Stains. 9th ed., Baltimore: Williams and Wilkins 1977.

[15] Stockert JC, Espada J, Trigoso CI, Del Castillo P. Fluorescence reaction of the anionic triarylmethane dye fast green FCF. Appl Fluor Technol 1991; 3(6): 14-5.

[16] Chapman DM. Dichromatism of bromphenol blue, with an improvement in the mercuric bromphenol blue technic for protein. Stain Technol 1975; 50(1): 25-30. [http://dx.doi.org/10.3109/10520297509117027] [PMID: 46633]

[17] Green FJ. The Sigma-Aldrich Handbook of Stains, Dyes and Indicators. Milwaukee, Wisconsin: Aldrich Chemical Company 1990.

[18] Udenfriend S. Fluorescence Assay in Biology and Medicine. New York, London: Academic Press 1962.

[19] Puchtler H, Sweat F, Gropp S. An investigation into the relation between structure and fluorescence of azo dyes. J R Microsc Soc 1967; 87(3): 309-28. [http://dx.doi.org/10.1111/j.1365-2818.1967.tb04513.x] [PMID: 5582815]

[20] Espada J, Stockert JC. Fluorescence of bisazo dye reaction products from the coupled tetrazonium method for proteins. Acta Histochem 1994; 96(3): 315-24. [http://dx.doi.org/10.1016/S0065-1281(11)80042-5] [PMID: 7531933]

[21] Harrisson F, Callebaut M, Vakaet L. Microspectrographic analysis of trypan blue-induced fluorescence in oocytes of the Japanese quail. Histochemistry 1981; 72(4): 563-78. [http://dx.doi.org/10.1007/BF00493276] [PMID: 7298390]

[22] Kristensson K, Olsson Y, Sjöstrand J. Axonal uptake and retrograde transport of exogenous proteins in the hypoglossal nerve. Brain Res 1971; 32(2): 399-406. [http://dx.doi.org/10.1016/0006-8993(71)90332-5] [PMID: 4109164]

[23] Hospelhorn AC, Faris B, Mogayzel PJ Jr, Tan OT, Franzblau C. Congo red binding of elastin in aortic smooth muscle cell cultures. J Histochem Cytochem 1988; 36(10): 1353-8. [http://dx.doi.org/10.1177/36.10.3418110] [PMID: 3418110]

[24] Verbelen JP, Stickens D. *In vivo* determination of fibril orientation in plant cell walls with polarization CSLM. J Microsc 1995; 177: 1-6. [http://dx.doi.org/10.1111/j.1365-2818.1995.tb03528.x]

[25] Matsuoka H, Yang HC, Homma T, *et al.* Use of Congo red as a microscopic fluorescence indicator of hyphal growth. Appl Microbiol Biotechnol 1995; 43: 102-8. [http://dx.doi.org/10.1007/BF00170630]

[26] Puchtler H, Sweat F. Congo red as a stain for fluorescence microscopy of amyloid. J Histochem Cytochem 1965; 13(8): 693-4. [http://dx.doi.org/10.1177/13.8.693] [PMID: 4160077]

[27] Howie AJ, Brewer DB. Optical properties of amyloid stained by Congo red: history and mechanisms. Micron 2009; 40(3): 285-301. [http://dx.doi.org/10.1016/j.micron.2008.10.002] [PMID: 19019688]

[28] Puchtler H, Sweat F, Kuhns JG. On the binding of direct cotton dyes by amyloid. J Histochem Cytochem 1964; 12: 900-7. [http://dx.doi.org/10.1177/12.12.900] [PMID: 14249351]

[29] Miall S, Miall LM. A New Dictionary of Chemistry. London, New York, Toronto: Longmans, Green and Co. 1953.

[30] Budavari S, Ed. The Merck Index An Encyclopedia of Chemicals, Drugs and Biologicals. 11th ed.,

Rahway, NJ: Merck and Co. 1989.

[31] Li K, Frankowski KJ, Belon CA, *et al.* Optimization of potent hepatitis C virus NS3 helicase inhibitors isolated from the yellow dyes thioflavine S and primuline. J Med Chem 2012; 55(7): 3319-30. [http://dx.doi.org/10.1021/jm300021v] [PMID: 22409723]

[32] Vassar PS, Culling CF. Fluorescent stains, with special reference to amyloid and connective tissues. Arch Pathol 1959; 68: 487-98. [PMID: 13841452]

[33] Puchtler H, Sweat Waldrop F, Meloan SN. Application of thiazole dyes to amyloid under conditions of direct cotton dyeing: correlation of histochemical and chemical data. Histochemistry 1983; 77(4): 431-45. [http://dx.doi.org/10.1007/BF00495799] [PMID: 6190786]

[34] Cañete M, Villanueva A, Juarranz A, Stockert JC. A study of interaction of thioflavine T with DNA: evidence for intercalation. Cell Mol Biol 1987; 33(2): 191-9. [PMID: 3607826]

[35] Stockert JC, Del Castillo P, Llorente AR, Rasskin DM, Romero JB, Gómez A. New fluorescence reactions in DNA cytochemistry. 1. Microscopic and spectroscopic studies on nonrigid fluorochromes. Anal Quant Cytol Histol 1990; 12(1): 1-10. [PMID: 2302306]

[36] Khurana R, Coleman C, Ionescu-Zanetti C, *et al.* Mechanism of thioflavin T binding to amyloid fibrils. J Struct Biol 2005; 151(3): 229-38. [http://dx.doi.org/10.1016/j.jsb.2005.06.006] [PMID: 16125973]

[37] Duboisset J, Ferrand P, He W, Wang X, Rigneault H, Brasselet S. Thioflavine-T and Congo Red reveal the polymorphism of insulin amyloid fibrils when probed by polarization-resolved fluorescence microscopy. J Phys Chem B 2013; 117(3): 784-8. [http://dx.doi.org/10.1021/jp309528f] [PMID: 23289901]

[38] Cowden RR, Curtis SK. Use of a fluorescence probe for hydrophobic groups, anilinonaphthalene sulphonic acid, in the supravital study of unusual slug oocyte nucleoli. Histochem J 1974; 6(4): 447-50. [http://dx.doi.org/10.1007/BF01012435] [PMID: 4859583]

[39] De Campos Vidal B. The use of the fluorescent probe 8-anilinonaphthalene sulfate (ANS) for collagen and elastin histochemistry. J Histochem Cytochem 1978; 26(3): 196-201. [http://dx.doi.org/10.1177/26.3.632557] [PMID: 632557]

[40] Winkelman J, Spicer SS. Staining with tetraphenyl-porphinesulfonate *in vivo*. Stain Technol 1962; 37: 303-5. [http://dx.doi.org/10.3109/10520296209114490] [PMID: 14007507]

[41] Pihlman K, Linder E. Fluorescence microscopical visualization of elastic fibres using basic fuchsin. Histochemistry 1983; 79(2): 157-65. [http://dx.doi.org/10.1007/BF00489778] [PMID: 6196330]

[42] Rosselet A, Ruch F. Cytofluorometric determination of lysine with dansylchloride. J Histochem Cytochem 1968; 16(7): 459-66. [http://dx.doi.org/10.1177/16.7.459] [PMID: 4176026]

[43] Rhys Williams AT. Fluorescence Detection in Liquid Chromatography. Buckinghamshire, UK: Perkin-Elmer FL publication 1980.

[44] Cuéllar T, Gosálvez J, Del Castillo P, Stockert JC. Fluram induces species-dependent C and G bands in mammalian chromosomes, revealing heterogeneous distribution of chromosomal proteins. Genome 1991; 34(5): 772-6. [http://dx.doi.org/10.1139/g91-120] [PMID: 1955156]

[45] Stockert JC, Trigoso CI. Fluorescence of eosinophil leukocyte granules induced by the fluorogenic

reagent 2-methoxy-2,4-diphenyl-3 (2H)-furanone. Blood Cells 1993; 19(2): 423-30. [PMID: 8312573]

[46] Stockert JC, Blázquez A, Galaz S, Juarranz A. A mechanism for the fluorogenic reaction of amino groups with fluorescamine and MDPF. Acta Histochem 2008; 110(4): 333-40. [http://dx.doi.org/10.1016/j.acthis.2007.10.011] [PMID: 18272208]

[47] Stockert JC, Abasolo MI. Inaccurate chemical structures of dyes and fluorochromes found in the literature can be problematic for teaching and research. Biotech Histochem 2011; 86(1): 52-60. [http://dx.doi.org/10.3109/10520295.2010.489428] [PMID: 20500122]

[48] Sugden JK. Photochemistry of dyes and fluorochromes used in biology and medicine: some physicochemical background and current applications. Biotech Histochem 2004; 79(2): 71-90. [http://dx.doi.org/10.1080/10520290412331292400] [PMID: 15513709]

[49] Stockert JC, Trigoso CI, Braña MF. A new fluorescence reaction in protein cytochemistry: formation of naphthalimide fluorophores from primary amino groups and 1,8-naphthalic anhydride derivatives. Eur J Histochem 1994; 38(1): 29-39. [PMID: 7517729]

[50] Stockert JC, Cañete M, Villanueva A, Juarranz A, Trigoso CI, Braña MF. Fluorescence of chromatin DNA induced by antitumoral naphthalimides. Z Natforsch C J Biosci 1997; 52(5-6): 408-11. [PMID: 9232898]

[51] Magun BE, Kelly JW. A new fluorescent method with phenanthrenequinone for the histochemical demonstration of arginine residues in tissues. J Histochem Cytochem 1969; 17(12): 821-7. [http://dx.doi.org/10.1177/17.12.821] [PMID: 5367018]

[52] Trigoso CI, Ibañez N, Stockert JC. A specific fluorogenic reaction for tryptophan residues using isatin in organic solvents. J Histochem Cytochem 1993; 41(10): 1557-61. [http://dx.doi.org/10.1177/41.10.8245414] [PMID: 8245414]

[53] Martínez-Ramírez A, Bella JL, Stockert JC. Fluorogenic reaction of blood cells induced by N-bromosuccinimide. Micron 2002; 33(4): 399-402. [http://dx.doi.org/10.1016/S0968-4328(01)00030-0] [PMID: 11814878]

[54] Brand L, Shaltiel S. Modification of histidine residues leading to the appearance of visible fluorescence. Biochim Biophys Acta 1964; 88: 338-51. [PMID: 14249842]

[55] Baker JR. The histochemical recognition of phenols, especially tyrosine. Q J Microsc Sci 1956; 97: 161-4.

[56] Horobin RW, Flemming L. Is Mercury Orange a selective stain for thiols? Histochem J 1982; 14(6): 1003-6. [http://dx.doi.org/10.1007/BF01005240] [PMID: 6757192]

[57] Cowden RR, Curtis SK. Further comments on the selectivity of Mercury Orange for protein thiols. Histochem J 1984; 16(11): 1237-41. [http://dx.doi.org/10.1007/BF01003447] [PMID: 6210274]

[58] Cowden RR, Curtis SK. Demonstration of protein-bound sulfhydryl and disulfide groups with fluorescent mercurials. Histochemie 1970; 22(3): 247-55. [http://dx.doi.org/10.1007/BF00306102] [PMID: 5450233]

[59] Ferrer JM, González Garrigues M, Stockert JC. Mercurochrome: a fluorescent and electron opaque dye. Microsc Acta 1983; 87(3): 293-300. [PMID: 6195509]

[60] de la Torre J, Gosálvez J, Stockert JC. Cytochemical observations on the centriolar adjunct of grasshopper spermatids in relation to its silver stainability. J Microsc 1990; 159: 109-12. [http://dx.doi.org/10.1111/j.1365-2818.1990.tb03024.x]

[61] Kosower NS, Kosower EM, Newton GL, Ranney HM. Bimane fluorescent labels: labeling of normal human red cells under physiological conditions. Proc Natl Acad Sci USA 1979; 76(7): 3382-6. [http://dx.doi.org/10.1073/pnas.76.7.3382] [PMID: 291011]

[62] Mirabelli F, Vairetti M, Bellomo G. Fluorescence analysis of intracellular soluble and protein thiols using bimane derivatives. Appl Fluor Technol 1992; 4(2 & 3): 9-13.

[63] Tramezzani JH, Chiocchio S, Wassermann GF. A technique for light and electron microscopic identification of adrenalin- and noradrenalin-storing cells. J Histochem Cytochem 1964; 12: 890-9. [http://dx.doi.org/10.1177/12.12.890] [PMID: 14249350]

[64] Gardner HA, Silver MD, McLean S, Strom-Gundersen EK. The chemical basis of the argentaffin reaction applied to glutaraldehyde fixed tissue. Histochemie 1971; 28(2): 95-8. [PMID: 4945434]

[65] Björklund A, Falck B, Lindvall O, Svensson LA. New aspects on reaction mechanisms in the formaldehyde histofluorescence method for monoamines. J Histochem Cytochem 1973; 21(1): 17-25. [http://dx.doi.org/10.1177/21.1.17] [PMID: 4694535]

[66] Corrodi H, Hillarp NA. Fluoreszenzmethoden zur histochemischen Sichtbarmachung von Monoaminen. 2. [1][1]) Identifizierung des fluoreszierenden Produktes aus Dopamin und Formaldehyd. Helv Chim Acta 1964; 47: 911-8. [http://dx.doi.org/10.1002/hlca.19640470326]

[67] Jonsson G. Quantitation of fluorescence of biogenic monoamines demonstrated with the formaldehyde flourescence method. Prog Histochem Cytochem 1971; 2(4): ix-35. [http://dx.doi.org/10.1016/S0079-6336(71)80001-3] [PMID: 23195750]

[68] Krinke G, Hess R. The value of the fluorescence histochemistry of biogenic amines in neurotoxicology. Histochem J 1981; 13(5): 849-63. [http://dx.doi.org/10.1007/BF01003294] [PMID: 6170615]

CHAPTER 14

Enzymes

Abstract: In recent years fluorescence techniques have become increasingly popular for the estimation and localization of enzymatic activities. Basic methodological concepts on enzyme histochemistry can be found in standard books and reviews [1 - 6]. Enzyme activity studies may be carried out to assess the localization of enzymes within cells and tissues, and to determine the effects of pH, ionic strength, temperature, presence of inhibitors, *etc*., on the rate of an enzyme-catalyzed reaction. In addition, enzymes can be applied as histochemical reagents to remove macromolecular substrates, examples being RNases, DNases, amylases, *etc*., or to identify substrates after their fluorescent labeling. Chromo- and fluorogenic reactions are usually employed to reveal enzymatic activity. A relevant issue to avoid diffusion artifacts or false localization is that the colored or fluorescent product of a given enzymatic reaction must have very scarce or null water solubility. However, insoluble or crystalline precipitates should be not confused with specific cell structures. Some examples will illustrate these points.

Keywords: AEC, Arylsulfatases, Benzothiazole-hydroxystyrene, CARD, Caspases, Cathepsins, Dehydrogenases, ELF, Esterases, Fluorogenic substrates, Glycosidases, Indigogenic methods, Peroxidases, Phosphatases, Tyramide signal amplification.

1. GENERAL ASPECTS OF ENZYMATIC HISTOCHEMISTRY

The fluorescence histochemistry of enzymatic activity is a biomedical field of great interest. Quantitative measuring of enzymatic activity is usually necessary, and therefore stoichiometric reactions are required. In fluorometric techniques it is also important that the product of an enzymatic reaction absorbs and fluoresces differently from the substrate and the background.

A schematic representation of the use of chromogenic reactions to demonstrate enzymatic activity is shown in Fig. (**14.1**), exemplified by known indigogenic methods. Indoxyl esters and glycosides are useful chromogenic substrates for detecting enzyme activities in histochemistry, biochemistry and bacteriology.

Most used substrates are 5-bromo-, 5-bromo-6-chloro- and 5-bromo-4-chloro-indoxyl derivatives. Pairs of diffusible indoxyl molecules released by the acti-

Juan Carlos Stockert & Alfonso Blázquez-Castro

vity of carboxylic esterases, phosphatases, phosphodiesterases, arylsulfatase, and glycosidases react rapidly with oxygen to yield insoluble blue or purple indigo derivatives. As leuco-indigo compounds are strongly fluorescent [7], this property is exploited in a variety of sensitive assays for hydrolases [8]. A similar strategy, but based on compounds such as naphthol esters followed by diazonium salts, is also valid for fluorogenic reactions.

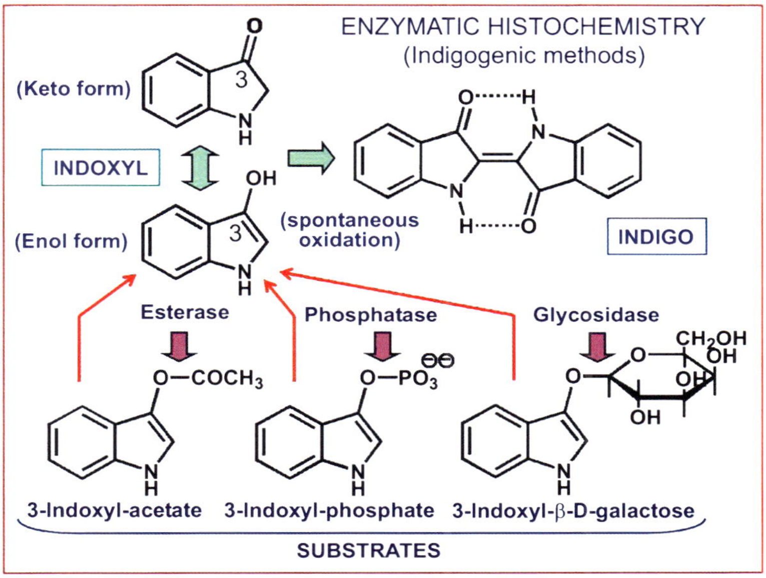

Fig. (14.1). Structure and oxidation of indoxyl and demonstration of the activity of some hydrolytic enzymes based on indigogenic (chromogenic) methods.

Most common enzymatic activities correspond to dehydrogenases, esterases, glycosidases, peroxidases, phosphatases, and proteases. Enzymatic histochemistry is based on the development of a color or fluorescence reaction induced by a given enzyme. In addition to the example of indigogenic methods shown in Fig. (**14.1**), the well known procedure for demonstrating succinic dehydrogenase clearly illustrates the strategy of using suitable hydrogen acceptors in chromogenic reactions (Fig. **14.2**). Obviously, other substrates can be employed in bright-field microscopy to generate colored products as in the case of peroxidases (*i.e.* benzidine, diaminobenzidine, tetramethylbenzidine, *o*-dianisidine, amino-ethylcarbazole, *etc.*). Unfortunately, no systematic studies on the possibility of using the final colored products in fluorescence microscopy have been performed.

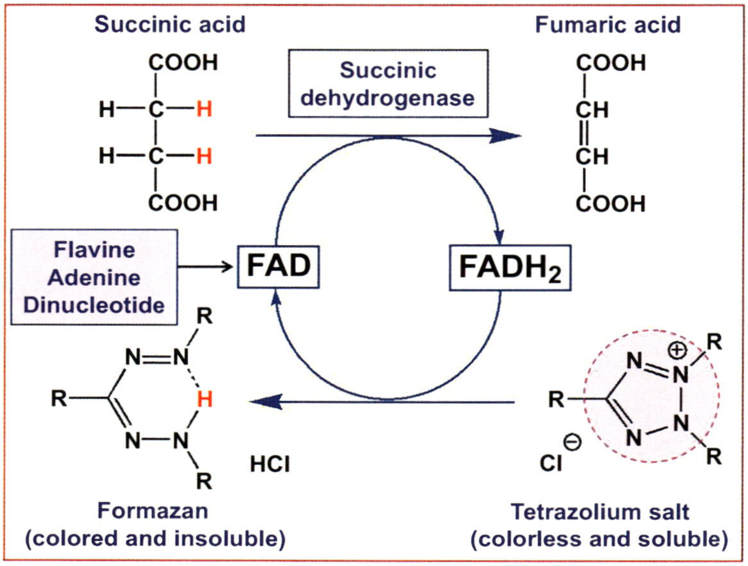

Fig. (14.2). Simplified diagram of the demonstration of succinic dehydrogenase activity by reduction of a tetrazolium salt to a colored formazan.

2. FLUORESCENCE METHODS FOR ENZYMATIC ACTIVITY

The fundament for the fluorescent detection of enzymatic activity is the use of suitable substrates that give rise to fluorescent reaction products only in the presence of a particular enzyme. Fluorogenic substrates based on 4-methyl-umbelliferone exhibit high sensitivity because of the intense blue fluorescence of this compound upon its release by the enzyme action. Esterase, lipase, glycosidase, phosphatase and arylsulfatase activities are the common enzymes that are revealed by fluorogenic 4-methyl-umbelliferone substrates.

In addition to classic (FITC, TRITC) or recent (Bodipy, Cy3, Cy5) reactive fluorochromes for labeling, suitable enzymes (alkaline phosphatase, peroxidase, β-galactosidase) can be employed to label secondary antibodies, probes and metabolites to reveal the reaction product in an amplified way. This corresponds to a fluorescent labeling amplification based on enzymatic activity. Lanthanide luminescence has been also applied in enzyme-amplified methods, mainly using Eu^{3+} and Tb^{3+} complexed with salicylic acid derivatives [9].

On the other hand, the reduction of tetrazolium salts and other compounds by dehydrogenases of living cells to fluorescent products has found wide applications in viability and cytotoxicity assays. In this Chapter only the most usual fluorescence methods for microscopical detection of enzymatic activity will be described.

2.1. Esterases and Hydrolases

Esterases are hydrolases that cleavage the ester bond, and a well known example is the hydrolysis of fluorescein diacetate (Fig. **14.3**) and calcein AM, which are useful probes for assessing cell viability, living cells showing bright green fluorescence (see Chapter 16.1.2). An illustrative case of the hydrolytic cleavage of carbohydrates by glycosidases is the use of the fluorogenic substrate 4-methyl-umbelliferyl-β-D-galactopyranoside to demonstrate the presence of the corresponding β-galactosidase activity [10]. Other fluorogenic substrates based on 1- or 2-naphthol are also used for the histochemical detection of hydrolytic activities.

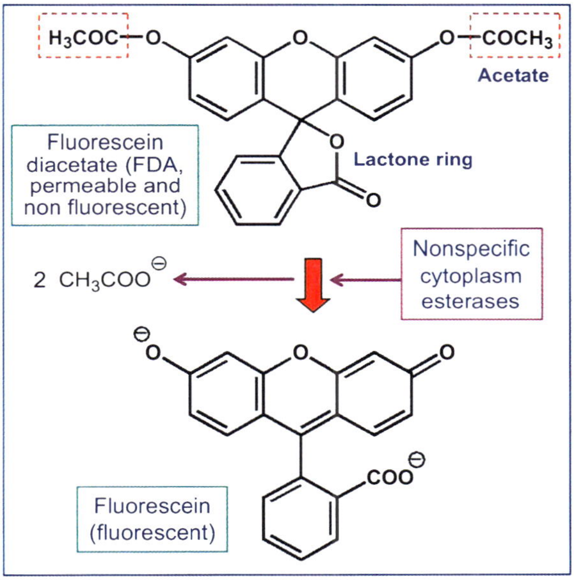

Fig. (14.3). Fluorogenic substrates for esterases. Conversion of non-fluorescent and cell permeable fluorescein diacetate into fluorescein by esterases within living cells.

A widely used biomarker associated with the senescent phenotype is the increase of β-galactosidase, which can be detected by histochemical staining of cultured cell populations and aging tissues, using the substrate 5-bromo-4-chloro- 3-indolyl-β-D-galactoside [11, 12]. Resorufin butyrate and indoxyl acetate have been employed as fluorogenic and chromogenic substrates for the visualization of cholinesterase activity.

2.2. Dehydrogenases

In the case of detection of dehydrogenase activity, the most important and widely used method is the reduction of a tetrazolium salt in the presence of the corresponding substrate (see Fig. **14.2**). The histochemical application of tetrazolium salts has been extensively reviewed [6, 13, 14]. Key dehydrogenases in the cell metabolism that can be studied and measured are glucose-3- and 6-phosphate dehydrogenase, lactate, glutamate, isocitrate and succinate dehydrogenases.

Tetrazolium salts are colorless redox indicators that are converted to intensely colored and water-insoluble formazans upon reduction. It is generally claimed that tetrazolium salts give non-fluorescent formazans, and that only the formazan of cyanotolyl tetrazolium chloride (CTC) (Fig. **14.4A**) shows fluorescence properties (λexc: 380-440 nm, λem: 625 nm) in the solid state [15, 16]. This feature has found application in fluorescence microscopy of cells, flow cytometry, and viability assays of bacteria in biofilms [17, 18]. However, other formazans also proved to be fluorescent.

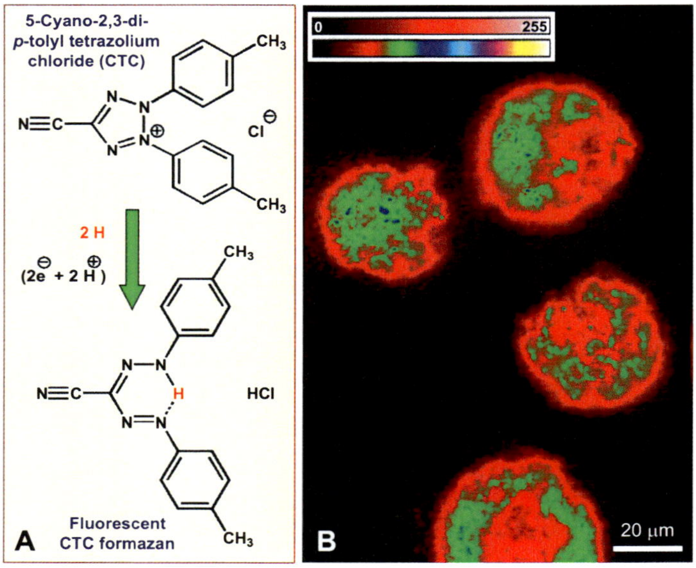

Fig. (14.4). **A**: Structure of CTC and reduction reaction leading to the formazan product. **B**: Ehrlich ascites tumor cells incubated with CTC and succinic acid for the demonstration of succinic dehydrogenase, showing the fluorescent formazan under blue excitation. The picture is a pseudocolor image processed with the ImageJ software (LUT: 6-shades). Although the enzyme is localized in mitochondria, the reaction product appears as green granules corresponding to insoluble CTC formazan deposits and resembling mitochondria.

Unfortunately, crystalline formazan deposits can be confused with organelles. Red fluorescing CTC formazan precipitates are mainly localized on the surface of living or briefly prefixed cells, bearing some resemblance in size and shape to mitochondria (Fig. **14.4B**). Tetrazolium salts are reduced by dehydrogenases and reducing agents (ascorbate, glutathione, *etc*.), and it is also expected that reduction to insoluble formazan can take place mostly under the influence of membrane-bound redox enzymes that are known to exist on the cell surface.

MTT (3-[4,5-dimethylthiazol-2-yl]-2,5-diphenyl-2H-tetrazolium bromide) is widely used as a viability assay for bacteria and cultured cells [6] (Fig. **14.5A**). Biochemical work reporting on the use of MTT has assumed that it is reduced by the activity of the mitochondrial enzyme succinic dehydrogenase [19]. Consequently, the site of formation of the formazan dye was claimed to be the mitochondria. However, reduction to MTT formazan is mainly due to the reducing power of the cytoplasm in living cells, particularly based on NAD(P)H. On account of its lipophilic character, the formazan accumulates into lipid components [20, 21]. After MTT viability assay, lipid droplets appear violet blue under bright-field, whereas the Golgi apparatus shows a weak bluish color (Fig. **14.5B**) (see Chapter 16.1.2).

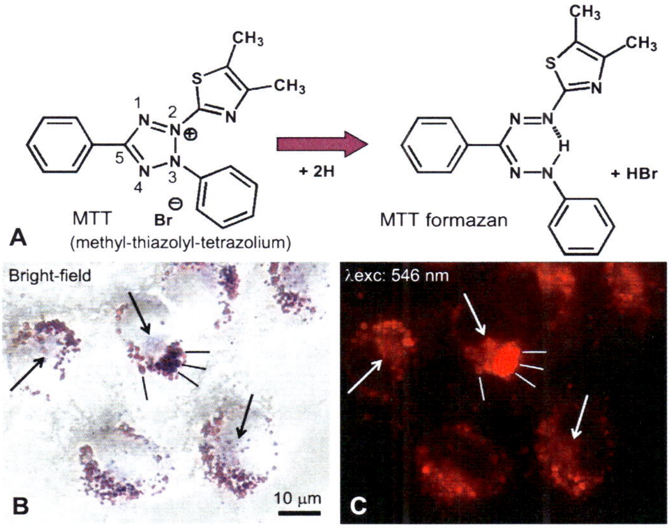

Fig. (14.5). A: Reduction of MTT to its formazan product. **B, C:** Uptake of the lipophilic MTT formazan into lipid droplets and Golgi apparatus of living HeLa cells treated with sunflower oil emulsion followed by the MTT viability assay (0.05 mg/mL MTT in culture medium, 3 h at 37 ºC), observed in bright-field (B) and fluorescence microscopy (C). Lines: lipid droplets. Arrows: Golgi apparatus.

When observed under 546 nm excitation, red fluorescing MTT formazan within lipid droplets is observed (Fig. **14.5C**). In addition, Golgi membranes also appear somewhat fluorescent in living cells, but the emission intensity becomes strikingly increased after formaldehyde fixation (Fig. **14.5D, E**). This intriguing effect can be explained by the influence of solvent viscosity on the emission of a non-rigid fluorochrome (Fig. **14.6 A-C**), the highest emission intensity being observed upon rigid binding to substrates.

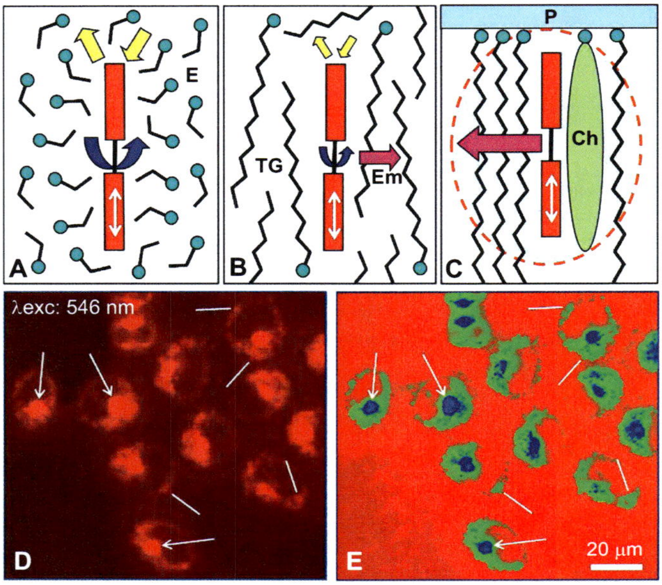

Fig. (14.6). Interaction of MTT formazan (red rectangles) with solvents (E: ethanol, TG: triglycerides) and binding sites. Violet arrows represent emission. Decay mechanisms (arrows) are shown as bond rotation (curved blue), solvent collision, (yellow), and atomic vibration (white). The dye is shown in ethanol **(A)**, oil **(B)**, and Golgi membranes **(C)** with cholesterol (Ch). Circles: polar heads. P: membrane proteins. Formaldehyde fixation makes more rigid the binding sites (ellipse in **C**). **D:** HeLa cells treated with sunflower oil emulsion, MTT, and then formaldehyde fixation. E: pseudocolor gradient of (D) (brightness range: blue > green > red). Lines: lipid droplets. Arrows: Golgi apparatus.

This interpretation is supported by the analysis of MTT formazan emission in the solid state. After reduction by ascorbic acid *in silico*, a massive precipitate of microscopic needle-shaped and red fluorescing MTT formazan crystals is obtained (Fig. **14.7A, B**). The same crystal pattern is often found on and between cells in some cultures. X-ray crystallography of a bromo derivative of formazan [22] reveals an almost planar structure stacked at 60 ° along the crystal axis, with hydrogen bond between N1···H − N**5**. Molecular modeling of the similar compound 1,3,5-triphenylformazan illustrates this packing mode in the solid state (Fig. **14.7C**), and explains the fluorescence of rigid formazans in a crystalline matrix.

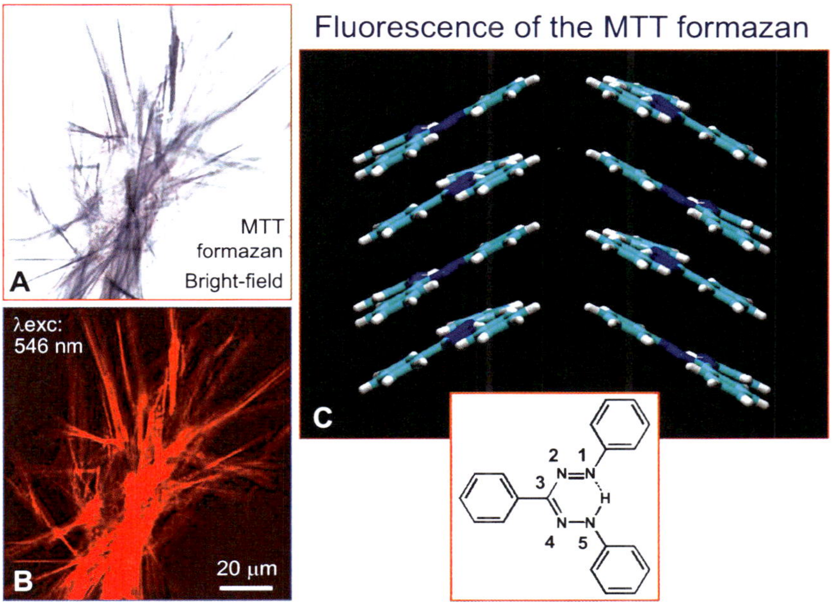

Fig. (14.7). MTT formazan crystals observed under bright-field **(A)** and fluorescence microscopy **(B)**. The molecular model of the original compound 1,3,5-triphenylformazan **(C)** according to X-ray crystallography of its bromo derivative [22] shows an extensive stacking in the crystal structure.

As occurs with MTT formazan, a rigid environment could be responsible for the possible fluorescence of other soluble formazan products such as, among others, MTS (sulfonated MTT derivative), and the reagent zincon that forms colored complexes with zinc and magnesium. When rigid, the linear formazan-like tetraaza-pentamethine cyanine dye [23, 24] should be also fluorescent. On the other hand, metal chelation of water-insoluble formazans greatly reduces the size of precipitates. Although the commonly used metal cations Cu^{2+}, Co^{2+}, and Ni^{2+} abolish fluorescence, the complex with Ru^{2+} strikingly increases the luminescence of MTT formazan on account of the formation of a metal-to-ligand-charge-transfer complex [25].

There are also useful redox dye systems to reveal reducing conditions. Resazurin is the N-oxide derivative of an oxazone dye that reduces to resorufin by dehydrogenase activity. It is used for cell viability and proliferation assays as 1% solution in PBS or Hanks for several hours (kit TOX-8, Sigma). Likewise Alamar blue® is a trade mark product (solution of resazurin and stabilizing agents, kit BUFO12) promoted for cytotoxicity and cell proliferation assays by color and fluorescence (www.bionova.es, www.ab-direct.com/alamarBlue), instead MTT, ^{3}H-TdR, *etc.* Resazurin is a non-fluorescent, blue dye (λab: 598 nm in water), and its reduced product is a red-violet dye (λexc: 546 nm, λem: 585 nm in water) (Fig.

14.8A, C). However, it must be noted that upon more reduction, resorufin converts in the colorless N-hydroresorufin.

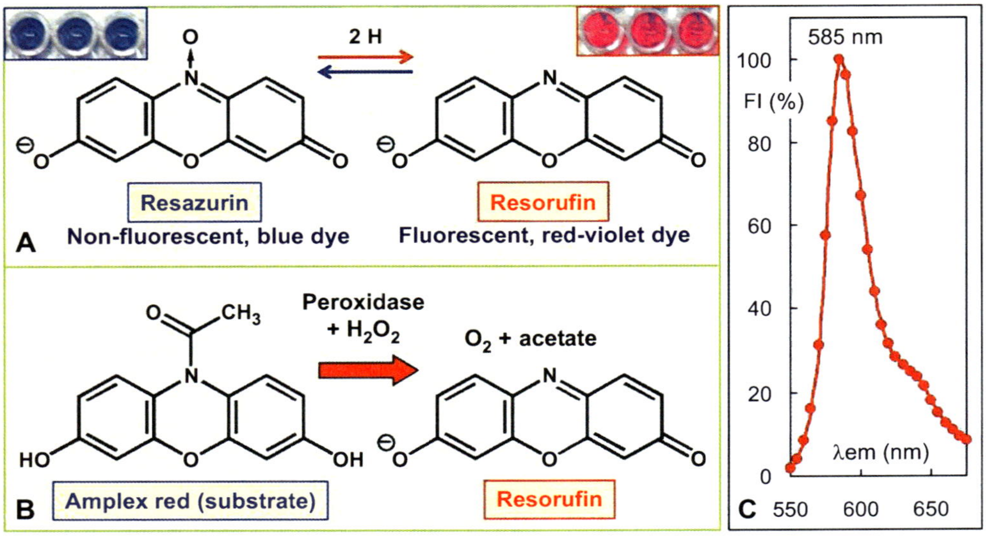

Fig. (14.8). A: Dehydrogenase activity. Redox reaction of resazurin-resorufin. Inserts show the color of wells containing the dyes. **B:** Detection of peroxidase activity by amplex red. **C:** Resorufin fluorescence under 546 nm exciting light in water at pH 9 (emission peak at 585 nm).

2.3. Peroxidases

Peroxidases catalyze the oxidation of various compounds by H_2O_2. In addition to endogenous peroxidases, horseradish peroxidase obtained from the root of *Armoracia rusticana* is widely applied as intravital axonally transported tracer and enzymatic reagent for immunohistochemistry. Well known examples of chromogenic reactions for demonstration of peroxidase activity use benzidine, 3,3'-diaminobenzidine (DAB), 3,5,3',5'-tetramethylbenzidine (TMB), and 3-amino-9-ethylcarbazole (AEC) [4, 26, 27]. The corresponding reaction products are non-fluorescent, although experimental designs to convert them in fluorescent products have been described or can be envisaged.

The oxidation of DAB generates brown and insoluble polymeric products by oxidative polymerization and cyclization. They contain quinonoid phenazine and indamine chromophores, which can complex with metals (Ni^{2+}, Co^{2+}) and react with osmium tetroxide [2, 4]. Oxidized benzidine, in its stabilized blue form, also appears highly birefringent [28]. When plastic semithin sections stained by immunoperoxidase-DAB methods are exposed to ultraviolet light under the fluorescence microscope, positive cells develop a strong yellowish fluorescence within 2-4 min, allowing the possibility of visualizing even reaction products which are barely identified by bright-field microscopy [29]. This process of fluorescent photo-conversion is possibly due to cleavage of oxidized DAB polymers giving small fluorophores. Endogenous peroxidases may also be visualized using this procedure.

Homovanillic acid has been proposed as a fluorometric substrate for peroxidase, which in the presence of H_2O_2 yields a fluorescent dimeric product (λexc: 315 nm, λem: 425 nm) [30]. The assay is also useful for other oxidative enzymes (*i.e.* glucose oxidase, xanthine oxidase). After homovanillic acid reaction, the product can be precipitated as a complex of lead and rhodamine (B or 6G), and visualized under fluorescence or electron microscopy [31]. The similar reagent, 3-(4-hydroxyphenyl)propionic acid (phloretic acid), has been also proposed as fluorogenic substrate for oxidative enzyme assays in analytical biochemistry [32].

Carbazole and its derivative AEC are colorless reagents that show strong blue fluorescence under UV excitation. In the case of AEC, oxidation would produce a red and insoluble 3,3'-azo dye precipitate. Polymeric imine dyes from AEC [33] could be formed through a Wurster's red radical cation [34]. Some of the AEC oxidation products should be fluorescent. A recent review covers chromogenic and fluorogenic approaches for the histochemical detection of peroxidase activity [35]. On the other hand, it would be expected that in the presence of H_2O_2, endogenous peroxidase activity in peroxisomes could be microscopically visualized by luminol without light excitation through a chemiluminescence reaction (λem: 425 and 485 nm). Forensic investigations also employ the pseudoperoxidase activity of hemoglobin to detect blood traces using luminol + H_2O_2.

In addition, the oxazone dye derivative, N-acetoxy-resorufin or amplex red (Fig. **14.8B**), is a stable fluorogenic substrate employed to reveal peroxidase activity giving the fluorescent resorufin product [36]. Amplex red can be also used in coupled enzymatic assays. For instance, oxidation of glucose by glucose oxidase generates H_2O_2, which is then coupled with the conversion of amplex red to resorufin. It must be taken into account that in the presence of light (mainly UV) self photo-oxidation of amplex red also results in resorufin [37].

An important method such as tyramide signal amplification (TSA) involves peroxidase activity on a biotinyl tyramide (BT) substrate. A biotinylated antibody is recognized by streptavidin-horseradish peroxidase conjugate that oxidizes the phenol group of the tyramide to a reactive free radical, which then attaches to aromatic residues of near macromolecules (Fig. **14.9A-C**). Another round of streptavidin-peroxidase (or streptavidin-alkaline phosphatase) in the presence of suitable chromo- or fluorogenic substrates generates the amplified final product. The conventional TSA method with BT and peroxidase is illustrated in Fig. (**14.9D, E**), showing the visualization of the EBV antigen in Hodgkin lymphoma cells using the sequence anti-EBV biotinyl-IgG, streptavidin-peroxidase, BT, streptavidin-alkaline phosphatase, naphthol AS-MX phosphate, and fast red TR.

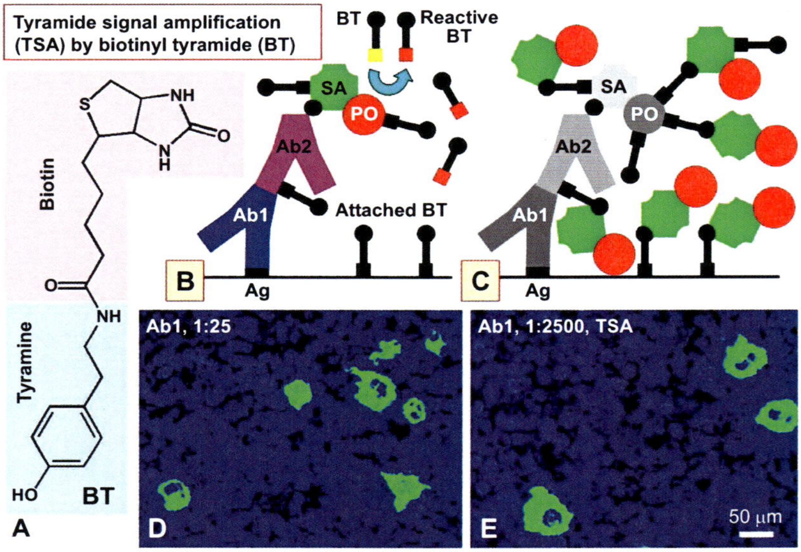

Fig. (14.9). **A**: Structure of biotinyl tyramide (BT). **B**, **C**: Tyramide signal amplification. Biotin: black circles; tyrosine: yellow squares. BT is oxidized by peroxidase (PO) bound to streptavidin (SA), with the production of reactive BT (red squares), and BT (black squares) attached to near macromolecules (B), followed by binding of additional SA-PO in the next TSA round. **D**, **E**: Visualization of the EBV antigen in Hodgkin lymphoma cells by using anti-EBV biotinyl-IgG (Ab1) at the indicated dilution. Alkaline phosphatase: AP. Image processing (LUT: rainbow-RGB). **D**: control image: anti-EBV biotinyl-IgG, SA-AP, naphthol AS-MX phosphate, fast red TR. **E**: TSA image: anti-EBV biotinyl-IgG, SA-PO, BT, SA-AP, naphthol AS-MX phosphate, fast red TR.

By contrast to a wide variety of well established chromogenic techniques, fluorescent labels for detecting peroxidase activity have been only confined to a few substrates. In the CARD (catalyzed reporter deposition) approach, a tyramine-fluorochrome couple is directly used, an example being Alexa 546-tyramide. However, in this case there is electronic isolation between the fluorescent label and the anchoring group.

In contrast, novel fluorescent substrates of peroxidases that are derived from heterocyclic 4-hydroxy-styrenes have been described as useful tools for detecting endogenous and exogenous targets in fixed cells and tissues. In the original method introduced by Krieg *et al.* [38], reporting fluorescent label and anchoring function are no longer isolated, and both have a full π-conjugated merocyanine-like system (*i.e.* benzothiazole-hydroxystyrene, Fig. **14.10A**). Excellent localization, high fluorescence intensity and photostability, and low background emission are achieved by using these 4-hydroxy-styryl derivatives (Fig. **14.10B, C**), which are promising substrates for fluorescent detection of endogenous and immuno-bound peroxidase activity in fixed cells and tissues. Structure-staining relationships for fluorescent 4-hydroxy-styryl derivatives has been discussed [35]. It would be interesting to investigate the possibility that the phenol group of isolated GFP or similar fluorophores could also form a free radical for use in conjugated CARD methods.

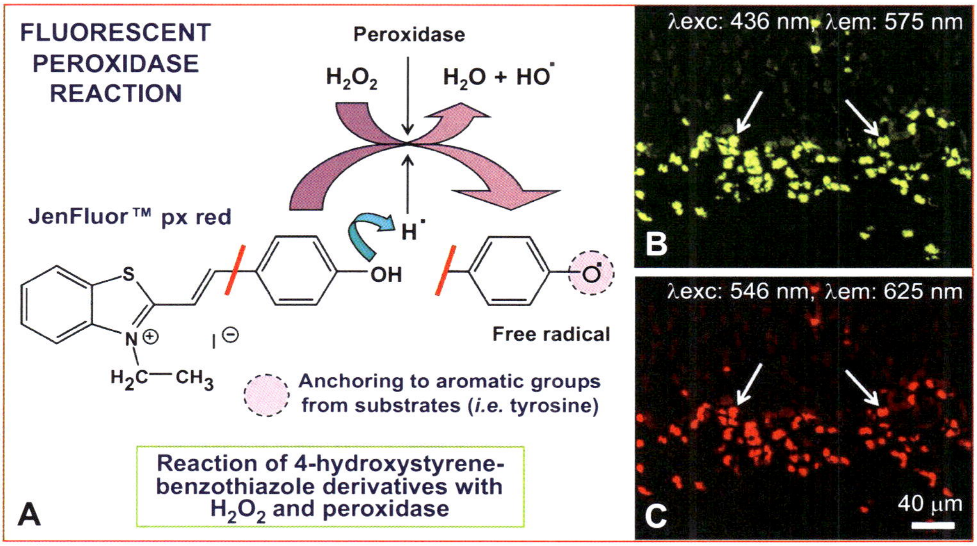

Fig. (14.10). **A**: Structure and activation mechanism of JenFluor™ px red to give a reactive free radical. **B**, **C**: Images of endogenous peroxidase activity in eosinophilic granulocytes from rat small intestine (cryostat 10 µm sections fixed in 2% formaldehyde) subjected to JenFluor™ px red observed under 436 and 546 nm exciting light (courtesy of R. Krieg). Note the different emission colors depending on the λexc.

2.4. Phosphatases

Classic reactions for phosphatases involve the use of naphthol-phosphate derivatives followed by copulation with diazonium salts [3, 6]. In this respect, it has been reported that the colored naphthol-azo products of the diazonium salts fast red TR and fast violet LB, used for the demonstration of phosphatase activity, show red fluorescence [39, 40]. The reaction has been also applied for nucleic acid hybridization using an alkaline phosphatase-fast red fluorescent procedure [41]. The reaction products of theses azo methods for phosphatases were also shown to be birefringent dyes [28].

This procedure allows the demonstration of phosphatase activity in both fixed and living cells [42]. The reaction mechanism between naphthol AS-MX phosphate and the diazonium salt fast red TR is illustrated in Fig. (**14.11A**), and the red fluorescence from the activity of acid phosphatase in lysosomes can be clearly observed in Fig. (**14.11B**). The final product between naphthol AS-MX and fast red TR has a CBN value of 31 and shows λexc: 500-540 nm, and λem: 560 nm. When this reaction product is excited at 546 nm and its fluorescence recorded in the spectrofluorimeter through the corresponding microscopic barrier filter (LP520), the emission peak shifts to 595 nm (see also Fig. **7.6C**).

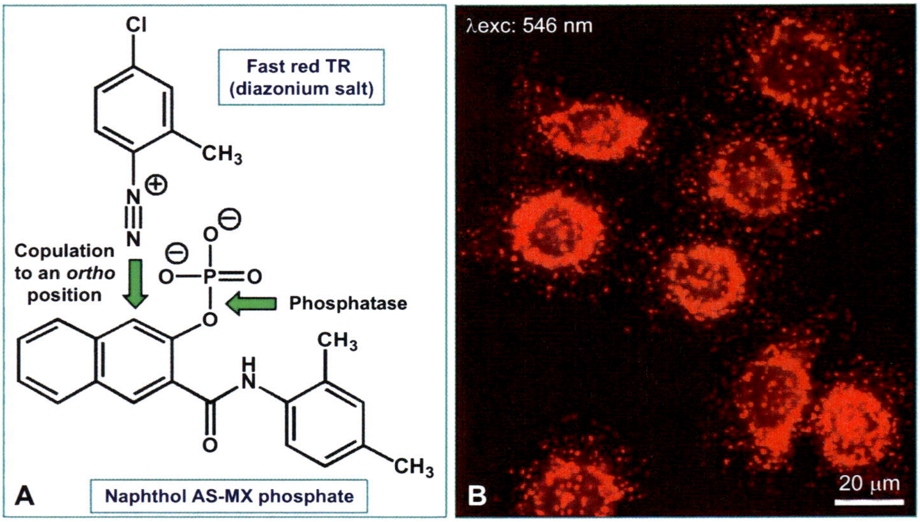

Fig. (14.11). **A**: Phosphatase activity. Reaction mechanism between naphthol AS-MX phosphate and the diazonium salt fast red TR. **B**: Fluorescence of acid phosphatase activity in methanol-fixed MCA3D mouse keratinocytes, induced by naphthol AS-MX phosphate (0.4 mg/mL, pH 5.2, 15 min), followed by fast red TR (0.2 mg/mL, pH 8.5, 30 min).

Interestingly, the procedure above can be also applied to reveal acid phosphatase activity and lysosome labeling in live cells, using 10 μg/mL naphthol AS-MX phosphate and 20 μg/mL fast red TR in the culture medium for 15 min at 37 °C [42]. The formation of an aluminum complex with the phenolic and carbonyl oxygens of the naphthol could improve the fluorescence. The same strategy can be applied to detect arylsulfatase activity (also a lysosome marker enzyme) using naphthol AS sulfate and fast red TR. In addition to fast red TR, tetrazonium salts such as fast blue black could be also used for coupling with two free naphthols (from naphthol-phosphate) giving a fluorescent product (see Chapter 13.2.4).

Phosphate derivatives of coumarin are also useful substrates for revealing the phosphatase activity. 6,8-Difluoro-4-methylumbelliferyl phosphate is the best general substrate for fluorescence-based assays of tyrosine or serine/threonine protein phosphatases. The hydrolysis product (6,8-difluoro-7-hydroxy-4-methylcoumarin) shows Φ_F: 0.89 and strong fluorescence at acidic, neutral or alkaline pH (λexc: 358 nm, λem: 455 nm). 3-Indolyl phosphate derivatives can be also used as substrates, with oxidation of indoxyl to blue and insoluble indigoid dyes, which are often coupled with reduction of nitroblue tetrazolium chloride to a blue formazan precipitate.

An interesting fluorescence reaction for demonstration of phosphatase activity uses the fluorogenic ELF (Enzyme Labeled Fluorescence) substrate, which can be also applied for enzymatic signal amplification. Phosphatase activity using ELF is illustrated in Fig. (**14.12A**). Microscopical applications are seen in Fig. (**14.12B**), where ELF forms a fluorescent photostable precipitate in the site of enzymatic activity (amplified immunofluorescence of epidermal growing factor receptors in epithelial cells using alkaline phosphatase signal). Another examples are the streptavidin-alkaline phosphatase detected by ELF after anti-tubulin IgG followed by biotinyl-IgG to demonstrate microtubules, or after FISH with biotinylated complementary oligonucleotides to visualize cytoplasmic mRNA.

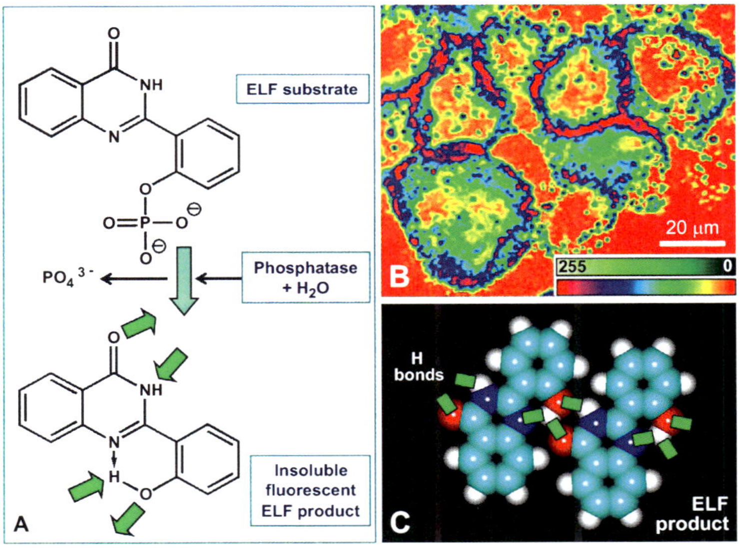

Fig. (14.12). **A:** Phosphatase reaction on ELF substrate and the fluorescent product. Green arrows show possible hydrogen bonds. **B:** Detection of plasma membrane receptors of epidermal growing factor (EGF) in A431 human cells after incubation with EGF-biotin and then with streptavidin-alkaline phosphatase and ELF. The fluorescent image under blue excitation was converted to pseudocolor using ImageJ (LUT: spectrum). **C:** Molecular structure of the ELF product showing extended hydrogen bonding (green bars).

It must be noted that although water insolubility of a given reaction product is desirable for microscopical purposes, dye precipitation can often result in a strong distortion of images. The ELF product appears with granular (possibly microcrystalline) aspect that does not correspond to true cell structures. Regarding this issue, it is at first sight intriguing the high water insolubility of some relatively simple dyes. This is the case of indigo and quinacrone dyes, linked together by H bonds between NH and CO groups in the crystal lattice [43]. The same seems to occur for the insoluble ELF product, which would form a polymeric array due to H bonding. Molecular modeling indicate that polymeric ELF complexes can be formed through extensive H bonds with 2.6 - 2.8 Å length (Fig. **14.12C**).

2.5. Proteases

Proteolytic enzymes (briefly, proteases) catalyze the hydrolysis of peptide bonds between amino acids in polypeptides and proteins. They can be peptidases (exopeptidases, breaking terminal amino acids) and proteinases (endopeptidases, which break internal peptide bonds). According to the nature of the active site, proteolytic enzymes are classified in thiol proteases (papain, cathepsins B and C), carboxyl proteases (pepsin, rennin), serine proteases (chymotrypsin, trypsin, elastases, thrombin), and metalloproteases (collagenases, requiring either Zn^{2+}, Mg^{2+}, Ca^{2+}, Fe^{2+}, Co^{2+}, *etc*). Some proteases are markers of specific organelles such as cathepsin D and elastase for lysosomes.

Methods for the histochemical visualization of proteases are generally based on the use of naphthylamide substrates (such as N-benzoylarginine-2-naphthylamide) [4]. Released naphthols then react with tetrazonium salts (*i.e.* fast blue B) giving a bisazo dye that forms a chelated complex with Cu^{2+} ions. The final product is a bluish-purple and lipid insoluble dye [2]. Selective fluorescence methods have been also developed for these enzymes, for instance using an AEC-labeled substrate such as peptidyl-3-amido-9-ethylcarbazole, which generates the free AEC (λexc: 360 nm, λem: 460 nm) [44, 45]. This is a sensitive fluorogenic substrate for demonstration of protease activity (*i.e.* trypsin, elastase, plasmin, cathepsins, *etc*).

However, new sensitive, selective and pH-independent bisamide-derivatives of rhodamine 110 (Rh110) are now widely used fluorogenic substrates for assaying proteases, and several amino acid- and peptide-derivatives of Rh110 are available. The appearance of caspases in the protease scenario of apoptosis has strongly promoted the design of new fluorogenic substrates. Caspases are cysteinyl aspartate-specific proteases that are involved in the proteolytic cascade of apoptosis, affecting nuclear lamins, actin, adhesion proteins, *etc*. There are several types of these proteases (*i.e.* initiator and effector caspases), but the most commonly demonstrated is the effector caspase-3 that contains the recognition site Asp-Glu-Val-Asp. Labeling with rhodamine 110-bis-L-aspartate-amide (substrates of caspase-3 and -7) induces a green fluorescence in the cytosol when caspases are active (Fig. **14.13A, B**). Nuclear counterstaining of normal cells can be performed with Hoechst 33342. In addition to rhodamine 110, other peptide-bound fluorophores are used for detection of caspases-3 and -8 (*i.e.* 7-amino-4-methylcoumarin).

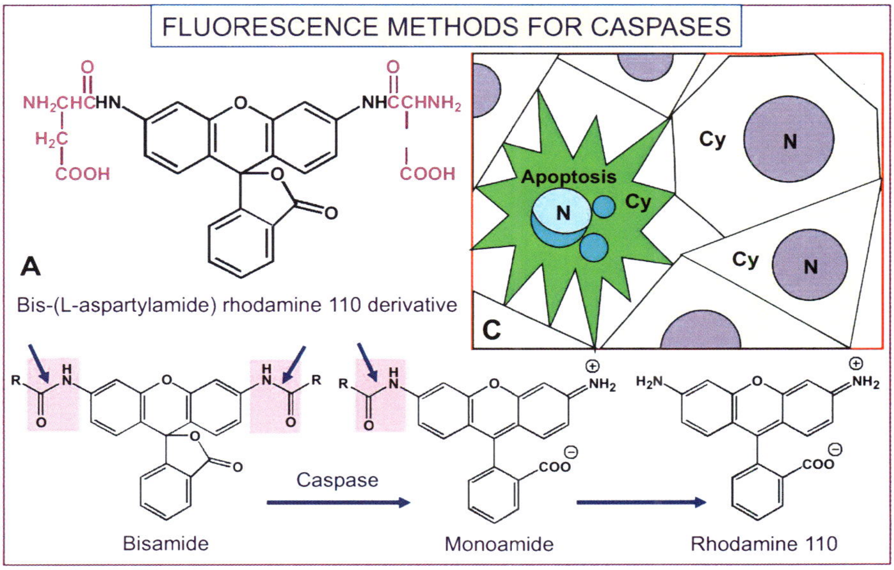

Fig. (14.13). **A**: Substrate for caspase-3 and -7 activity. Structure of the bis-(L-aspartylamide) rhodamine 110 derivative. **B**: Cleavage of peptide bonds by these caspases showing different steps. **C**: Schematic view of the apoptotic caspase activity induced on cultured cells after damaging H_2O_2 treatment, using rhodamine 110-bis-L-aspartate-amide (cytosol, green). Hoechst 33342 counterstaining (nuclei, blue).

CONFLICT OF INTEREST

The author (editor) declares no conflict of interest, financial or otherwise.

ACKNOWLEDGEMENTS

This work was supported by a grant from the Ministerio de Economía y Competitividad (CTQ2013-48767-C3-3-R), Spain.

REFERENCES

[1] Thompson SW. Selected Histochemical and Histopathological Methods. Springfield, Illinois: Charles C. Thomas 1966.

[2] Pearse AG. Histochemistry Theoretical and Applied. 3rd Ed. Edinburgh, London, New York: Churchill Livingstone 1972; 2.

[3] Lillie RD. Conn's Biological Stains. 9th ed., Baltimore: Williams and Wilkins 1977.

[4] Kiernan JA. Histological and Histochemical Methods: Theory and Practice. Oxford, New York: Pergamon Press 1990.

[5] Haugland RP. Handbook of Fluorescent Probes and Research Products. 9th ed., Eugene, OR: Molecular Probes, Inc. 2002.

[6] Horobin RW, Kiernan JA, Eds. Conn's Biological Stains A Handbook of Dyes, Stains and Fluorochromes for Use in Biology and Medicine. 10th ed., Oxford, UK: BIOS Scientific Publishers 2002.

[7] Stockert JC, Trigoso CI. Selective fluorescence reaction of indigocarmine stained eosinophil leucocyte granules induced by alkaline reduction of the bound dye to its leuco derivative. Acta Histochem 1994; 96(1): 8-14.
[http://dx.doi.org/10.1016/S0065-1281(11)80003-6] [PMID: 7518177]

[8] Kiernan JA. Indigogenic substrates for detection and localization of enzymes. Biotech Histochem 2007; 82(2): 73-103.
[http://dx.doi.org/10.1080/10520290701375278] [PMID: 17577701]

[9] Evangelista RA, Pollak A, Templeton EF. Enzyme-amplified lanthanide luminescence for enzyme detection in bioanalytical assays. Anal Biochem 1991; 197(1): 213-24.
[http://dx.doi.org/10.1016/0003-2697(91)90381-3] [PMID: 1952068]

[10] Kämpfer P, Kulies I, Dott W. Fluorogenic substrates for differentiation of gram-negative nonfermentative and oxidase-positive fermentative bacteria. J Clin Microbiol 1992; 30(6): 1402-6. [PMID: 1624555]

[11] Dimri GP, LeeX , Basile G, *et al.* A biomarker that identifies senescent human cells in culture and in aging skin *in vivo*. Proc Natl Acad Sci USA 1995; 92(20): 9363-7.
[http://dx.doi.org/10.1073/pnas.92.20.9363] [PMID: 7568133]

[12] Itahana K, Campisi J, Dimri GP. Methods to detect biomarkers of cellular senescence: the senescence-associated beta-galactosidase assay. Methods Mol Biol 2007; 371: 21-31.
[http://dx.doi.org/10.1007/978-1-59745-361-5_3] [PMID: 17634571]

[13] Stoward PJ, Altman FP, Seidler E. Principles of Oxidoreductase Histochemistry.Histochemistry Theoretical and Applied. 4th Ed. Edinburgh: Churchill Livingstone 1991; 3.

[14] Seidler E. The tetrazolium-formazan system: design and histochemistry. Prog Histochem Cytochem 1991; 24(1): 1-86.
[http://dx.doi.org/10.1016/S0079-6336(11)80060-4] [PMID: 1712976]

[15] Severin E, Stellmach J, Nachtigal HM. Fluorometric assay of redox activity in cells. Anal Chim Acta 1985; 170: 341-6.
[http://dx.doi.org/10.1016/S0003-2670(00)81760-2]

[16] Stellmach J, Severin E. A fluorescent redox dye. Influence of several substrates and electron carriers on the tetrazolium salt-formazan reaction of Ehrlich ascites tumour cells. Histochem J 1987; 19(1): 21-6.
[http://dx.doi.org/10.1007/BF01675289] [PMID: 3583812]

[17] Severin E, Seidler E, Huang CJ. Flow cytometric assay for specific dehydrogenases as demonstrated with human tumor cell lines. Appl Fluor Technol 1992; 4(5 & 6): 3-7.

[18] Rodriguez GG, Phipps D, Ishiguro K, Ridgway HF. Use of a fluorescent redox probe for direct visualization of actively respiring bacteria. Appl Environ Microbiol 1992; 58(6): 1801-8.
[PMID: 1622256]

[19] Slater TF, Sawyer B, Straeuli U. Studies on succinate-tetrazolium reductase systems. III. Points of coupling of four different tetrazolium salts. Biochim Biophys Acta 1963; 77: 383-93. [http://dx.doi.org/10.1016/0006-3002(63)90513-4] [PMID: 14089413]

[20] Diaz G, Melis M, Musin A, Piludu M, Piras M, Falchi AM. Localization of MTT formazan in lipid droplets. An alternative hypothesis about the nature of formazan granules and aggregates. Eur J Histochem 2007; 51(3): 213-8. [PMID: 17921117]

[21] Stockert JC, Blázquez-Castro A, Cañete M, Horobin RW, Villanueva A. MTT assay for cell viability: Intracellular localization of the formazan product is in lipid droplets. Acta Histochem 2012; 114(8): 785-96. [http://dx.doi.org/10.1016/j.acthis.2012.01.006] [PMID: 22341561]

[22] Tunc T, Yildirim LT. Synthesis, crystal structure and spectroscopic studies of 1-(p-bromphenol)-3, 5-diphenylformazan. Open Cryst J 2010; 3: 54-8. [http://dx.doi.org/10.2174/1874846501003010054]

[23] Sawicki E, Hauser TR, Stanley TW, Elbert W. The 3-methyl-2-benzothiazolone hydrazone test. Sensitive new method for the detection, rapid estimation, and determination of aliphatic aldehydes. Anal Chem 1961; 33: 93-6. [http://dx.doi.org/10.1021/ac60169a028]

[24] Hünig S, Balli H, Breither E, *et al.* Heterocyclic azo dyes by oxidative coupling. Angew Chem Int Ed Engl 1962; 1: 640-6. [http://dx.doi.org/10.1002/anie.196206401]

[25] Stockert JC, Abasolo MI. Inaccurate chemical structures of dyes and fluorochromes found in the literature can be problematic for teaching and research. Biotech Histochem 2011; 86(1): 52-60. [http://dx.doi.org/10.3109/10520295.2010.489428] [PMID: 20500122]

[26] Burstone MS. Histochemical demonstration of cytochrome oxidase with new amine reagents. J Histochem Cytochem 1960; 8: 63-70. [http://dx.doi.org/10.1177/8.1.63] [PMID: 13806313]

[27] Litwin JA. Histochemistry and cytochemistry of 3,3′-diaminobenzidine. A review. Folia Histochem Cytochem (Krakow) 1979; 17(1): 3-28. [PMID: 220157]

[28] Kaplow LS. Birefringence of peroxidase and azo dye stains. J Histochem Cytochem 1970; 18(7): 515-7. [http://dx.doi.org/10.1177/18.7.515] [PMID: 4194118]

[29] Grube D. Immunoperoxidase methods: increased efficiency using fluorescence microscopy for 3,3-diaminobenzidine (DAB) stained semithin sections. Histochemistry 1980; 70(1): 19-22. [http://dx.doi.org/10.1007/BF00508841] [PMID: 6161911]

[30] Guilbault GG, Brignac P Jr, Zimmer M. Homovanillic acid as a fluorometric substrate for oxidative enzymes. Analytical applications of the peroxidase, glucose oxidase, and xanthine oxidase systems. Anal Chem 1968; 40(1): 190-6. [http://dx.doi.org/10.1021/ac60257a002] [PMID: 5634544]

[31] Papadimitriou JM, Van Duijn P, Brederoo P, Streefkerk JG. A new method for the cytochemical demonstration of peroxidase for light, fluorescence and electron microscopy. J Histochem Cytochem 1976; 24(1): 82-90. [http://dx.doi.org/10.1177/24.1.768373] [PMID: 768373]

[32] Matsumoto T, Furuta T, Nimura Y, Suzuki O. 3-(p-hydroxyphenyl)propionic acid as a new fluorogenic reagent for amine oxidase assays. Anal Biochem 1984; 138(1): 133-6. [http://dx.doi.org/10.1016/0003-2697(84)90780-2] [PMID: 6428269]

[33] Krieg R, Halbhuber KJ, Oehring H. Novel chromogenic substrates with metal chelating properties for

the histochemical detection of peroxidasic activity, derived from 3-amino-9-ethylcarbazole (AEC) and 3,6-diamino-9-ethylcarbazole. Cell Mol Biol 2000; 46(7): 1191-212.
[PMID: 11075949]

[34] Koretz K, Leman J, Brandt I, Möller P. Metachromasia of 3-amino-9-ethylcarbazole (AEC) and its prevention in immunoperoxidase techniques. Histochemistry 1987; 86(5): 471-8.
[http://dx.doi.org/10.1007/BF00500619] [PMID: 2438258]

[35] Krieg R, Halbhuber KJ. Detection of endogenous and immuno-bound peroxidase--the status quo in histochemistry. Prog Histochem Cytochem 2010; 45(2): 81-139.
[http://dx.doi.org/10.1016/j.proghi.2009.11.001] [PMID: 20488278]

[36] Zhou M, Diwu Z, Panchuk-Voloshina N, Haugland RP. A stable nonfluorescent derivative of resorufin for the fluorometric determination of trace hydrogen peroxide: applications in detecting the activity of phagocyte NADPH oxidase and other oxidases. Anal Biochem 1997; 253(2): 162-8.
[http://dx.doi.org/10.1006/abio.1997.2391] [PMID: 9367498]

[37] Summers FA, Zhao B, Ganini D, Mason RP. Photooxidation of Amplex Red to resorufin: implications of exposing the Amplex Red assay to light. Methods Enzymol 2013; 526: 1-17.
[http://dx.doi.org/10.1016/B978-0-12-405883-5.00001-6] [PMID: 23791091]

[38] Krieg R, Eitner A, Günther W, Halbhuber KJ. Optimization of heterocyclic 4-hydroxystyryl derivatives for histological localization of endogenous and immunobound peroxidase activity. Biotech Histochem 2007; 82(4-5): 235-62.
[http://dx.doi.org/10.1080/10520290701714013] [PMID: 18074270]

[39] Ziomek CA, Lepire ML, Torres I. A highly fluorescent simultaneous azo dye technique for demonstration of nonspecific alkaline phosphatase activity. J Histochem Cytochem 1990; 38(3): 437-42.
[http://dx.doi.org/10.1177/38.3.1689343] [PMID: 1689343]

[40] Kamalia N, McCulloch CA, Tenenbaum HC, Limeback H. Direct flow cytometric quantification of alkaline phosphatase activity in rat bone marrow stromal cells. J Histochem Cytochem 1992; 40(7): 1059-65.
[http://dx.doi.org/10.1177/40.7.1376742] [PMID: 1376742]

[41] Speel EJ, Schutte B, Wiegant J, Ramaekers FC, Hopman AH. A novel fluorescence detection method for *in situ* hybridization, based on the alkaline phosphatase-fast red reaction. J Histochem Cytochem 1992; 40(9): 1299-308.
[http://dx.doi.org/10.1177/40.9.1506667] [PMID: 1506667]

[42] Espada J, Horobin RW, Stockert JC. Fluorescent cytochemistry of acid phosphatase and demonstration of fluid-phase endocytosis using an azo dye method. Histochem Cell Biol 1997; 108(6): 481-7.
[http://dx.doi.org/10.1007/s004180050188] [PMID: 9450630]

[43] Zollinger H. Color Chemistry. Syntheses, Properties, and Applications of Organic Dyes and Pigments. 3rd Ed., Zürich, Weinheim: Wiley-VCH, VHCA 2003.

[44] Monsigny M, Kieda C, Maillet T. Assay for proteolytic activity using a new fluorogenic substrate (peptidyl-3-amino-9-ethyl-carbazole); quantitative determination of lipopolysaccharide at the level of one picogram. EMBO J 1982; 1(3): 303-6.
[PMID: 7188184]

[45] Harnois-Pontoni M, Monsigny M, Mayer R. Hydrosoluble fluorogenic substrates for plasmin. Anal Biochem 1991; 193(2): 248-55.
[http://dx.doi.org/10.1016/0003-2697(91)90017-N] [PMID: 1831333]

CHAPTER 15

Vital Probes 1. Cell Organelles

Abstract: The visualization of cell structures by fluorescent vital probes to study cell organelles, cell viability and signaling processes has great importance in biomedical sciences. Main interest is devoted to the uptake, accumulation and prediction of localization of drugs and xenobiotics in cultured cells. Mechanistic aspects and predictive rules are studied using numerical parameters, which conform the quantitative structure-activity relationships (QSAR) for vital probes. The uptake of probes into cells is the first step for labeling. Several mechanisms of cell uptake and accumulation in cell organelles are known, the most relevant being passive diffusion and active endocytosis (*i.e.* adsorptive and fluid-phase, binding to receptors). In addition to theoretical predictions, direct cytotoxic effects, photodynamic action, and fading must be taken into account in the microscopical evaluation of uptake and localization of fluorescent probes.

Keywords: Acridine orange, Amphiphilicity, Autofluorescence, Benzothiazoles, Chromatin DNA, Colocalization, $DiOC_1(3)$, Golgi apparatus, Hoechst dyes, Hydroethidine, Hydrophobic index, Log P, Lysosomes, Merocyanine 540, Mitochondria, Nile red, QSAR, Rhodamine 123, Trypan blue.

1. MECHANISMS OF UPTAKE AND ACCUMULATION

A relevant question is: "Where do dyes and probes go, inside living cells and tisues?" [1], and it can be extended to additional questions: "Who needs to know where probes go and why?"; "Which probes are of particular concern?"; "Which cells and structures are labeled? And in what context?", and "How are probes distributed within cells?" General articles and reviews dealing with uptake and organelle labeling should be consulted [2 - 11]. Interesting historical accounts on vital dyes such as methylene blue and neutral red are those of Chang [12] and Barbosa and Peters [13].

A schematic description of differences between live and fixed cells observed after acridine orange (AO) staining can be seen in Fig. (**15.1**), which shows the mechanisms of cell uptake and localization selectivity. Obviously, the localization of fluorescence signals after staining fixed (dead) or living cells differs considerably. Generally this occurs irrespectively of the type of fluorescent dye used for staining. The chemical category of dyes does not have relation with the

Juan Carlos Stockert & Alfonso Blázquez-Castro

labeling selectivity in living cells. Thus some acridines accumulate in nuclei, other in lysosomes, and other in mitochondria. Numerous cationic fluorochromes are know to bind DNA *in silico* and to stain nuclei and chromosomes in fixed cells and tissues, but only few of them are able to label selectively the chromatin DNA in live cells.

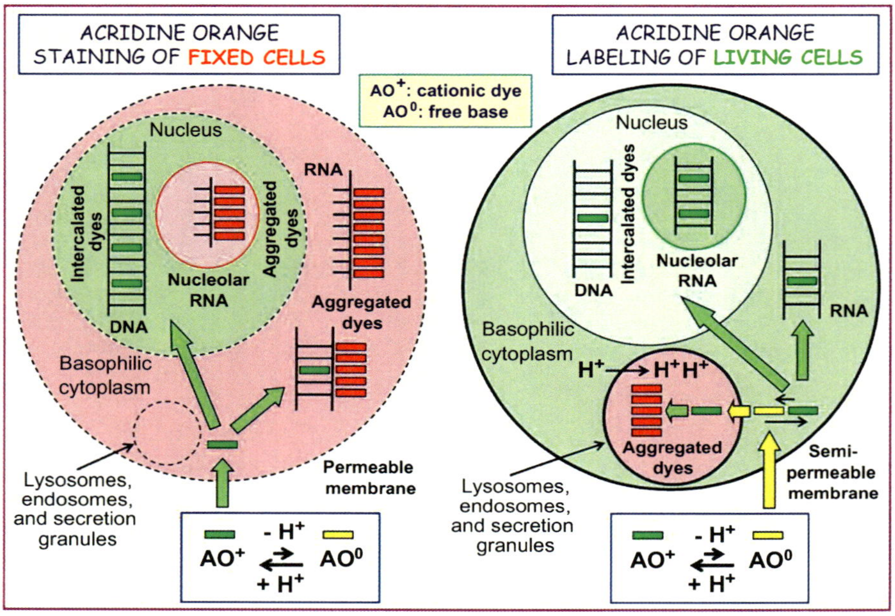

Fig. (15.1). Differences between fixed **(A)** and living cells **(B)** subjected to staining or labeling with acridine orange (AO). The schemes show the AO uptake and localization mechanisms related to the equilibrium between the protonated (cationic) and deprotonated dye (free base, yellow). AO monomers and aggregates are shown as green and red bars, respectively.

The metachromatic (red-orange) and orthochromatic (green) emission after AO appear very different in fixed (or dead) and living cells (Fig. **15.2A, B**) (see also Fig. **1.9**). After AO labeling, the cytoplasm and nucleoli of live cells show green fluorescence, and also the chromatin, but in lower degree. This feature represents intercalative binding to RNA and DNA at very low AO concentrations. Of course, the emission intensity of chromatin can vary according to the cell type. AO also accumulates in live cell lysosomes giving a metachromatic emission, but after photodamage or fixation, nuclei become the most conspicuous orthochromatic fluorescing structures [14].

This case of AO is illustrative, because of its photosensitizing properties that led to cell damage and death after blue irradiation, with striking changes in dye localization [14, 15]. The frequently mentioned process of "relocalization" of photosensitizers after irradiation can be explained by this mechanism of converting "live cell labeling" to "dead cell staining".

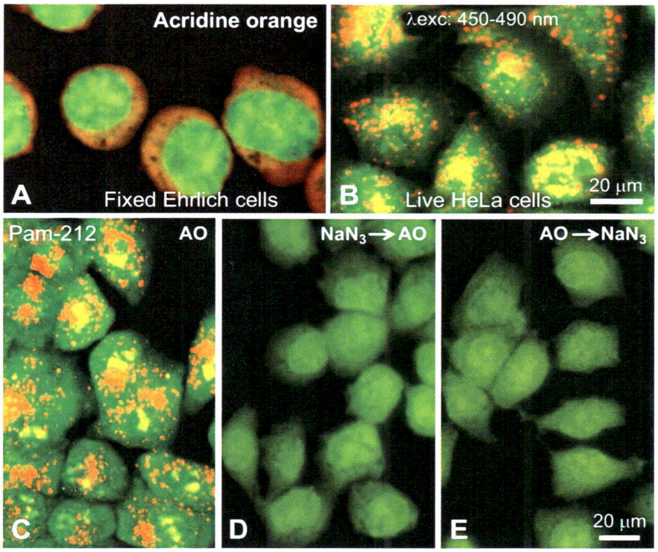

Fig. (15.2). Differential fluorescence of dead (fixed) cells *vs.* live cells by acridine orange (AO). **A:** Methanol-fixed smear of Ehrlich ascites tumor cells stained with 50 µg/mL AO for 5 min, showing orthochromatic (green) and metachromatic (red-orange) emission in nuclei and basophilic cytoplasm, respectively. **B:** Living HeLa cells labeled with 5 µg/mL AO for 15 min showing metachromatic lysosomes. **C-E:** Influence of lysosomal pH on AO labeling. Mouse Pam212 keratinocytes treated with (C) 2 µg/mL AO in DMEM, 15 min; (D) PBS (15 min) followed by NaN_3 (10 mM in PBS, 30 min), and then AO labeling, and (E) AO followed by NaN_3. In D and E the lysosomal red-orange signal is lacking and cells show a difusse green emission.

Fluorescence labeling of live cells first requires the uptake of probes and then their selective localization within organelles. In contrast to the protonated (cationic) form of AO, the deprotonated and uncharged (free base) form can diffuse through cellular membranes. In the aqueous medium inside the cell the pH-dependent equilibrium between cation and free base is rapidly re-established. Simple diffusion of the free base accounts for cell uptake, but accumulation into lysosomes is an energy-dependent process (Figs. **15.1B**, and **15.2C-E**). The inhibition of the lysosomal proton pump (*i.e.* due to low temperature or action of metabolic poisons such as sodium azide, NaN_3) blocks energy-dependent cell processes on account of the depletion of ATP. Therefore, maintenance of the acid pH within lysosomes is inhibited [10, 16, 17], and the red labeling of lysosomes by AO is abolished (Fig. **15.2C-E**). The leakage of the dye out the lysosomes can also occur by photodamage [18], which inactivates the proton pump and causes deprotonation of AO due to the rise in lysosomal pH.

2. QUANTITATIVE STRUCTURE-ACTIVITY RELATIONSHIPS

The possible relations between chemical structure and localization of fluorescent probes can be assessed using numerical parameters, and various physico-chemical characteristics are involved in models regarding the selective localization of fluorescent labels [1].

The novelty of this approach is the parallel use of imaging of live cells and numerical description of the physico-chemistry of probes. Cell biologists image live cells, but do not use numerical modeling. Pharmacologists use structure-activity relations, but do not image living cells [19]. The quantitative structure–activity relationships (QSAR, in this case, "activity" relates to "localization" of probes, which can be considered the result of uptake and accumulation activities) specify the physico-chemical properties of probes that accumulate preferentially within live cells, which are microscopically observed by staining or fluorescence.

QSAR models have been introduced in biological staining mainly on account of the work of Horobin [20, 21], in relation to histochemical and chromosome banding studies [21 - 23]. QSAR prediction rules have been recently reviewed [11, 24, 25].

Most used parameters for QSAR of fluorescent probes are the following:

- Electric charge (Z).
- Cation/free base or cation/pseudo-base equilibrium.
- Ionization constant, acid-base strength (pK_a).
- Conjugated bond number (CBN).
- Number of conjugated bonds in the largest fragment (LCF).
- Hydrophilicity-lipophilicity (log P, HI).
- Amphiphilic index (AI).
- Head group size (HGS).
- Head group hydrophilicity (HGH).
- Substituent bulk (SB).
- Ratio between molecular width and length (W/L).

Other factors such as the type of uptake (pasive diffusion, active transport, delivery media, *etc.*) should be also taken into account.

2.1. Numerical Parameters

The QSAR approach requires describing relevant physico-chemical characteristics numerically. In some cases this is straightforward, *e.g.* with overall (nominal) electric charge (Z), which will vary from n^+ through zero to n^-, where n is an integer, and relative molecular mass (MW: molecular weight), which is suitable to estimate the whole molecular size.

The acid–base strength (pK_a) is the negative logarithm of the equilibrium constant for the free base/protonated base reaction. This pH-dependent equilibrium between a cationic dye and the free base or pseudo-base corresponds to the alkaline deprotonation of an amino group or hydroxylation of a carbon conjugated with a quaternary (ammonium) nitrogen (see Fig. **3.8**). It is important to note that in addition to probes, also the ionization constant of chemical groups from the substrate must be taken into account. Examples are the different pK_a values for some amino acids (aspartic and glutamic acid: ~4.7; histidine: ~6.5; lysine: ~10.2; arginine: ~12).

The size of the conjugated systems is easily modeled using the conjugated bond number (CBN) and the largest conjugated fragment (LCF) parameters, which are estimated by inspection of chemical structures. Thus the sizes of the overall, and largest planar portion, of probes' aromatic systems (with conjugated substituents) are defined by the CBN and LCF parameters, respectively (Fig. **15.3**), with the LCF/CBN ratio providing a measure of the planarity of the aromatic component of the molecule. The molecular area of a conjugated system (in $Å^2$) is also proportional to its LCF value [26].

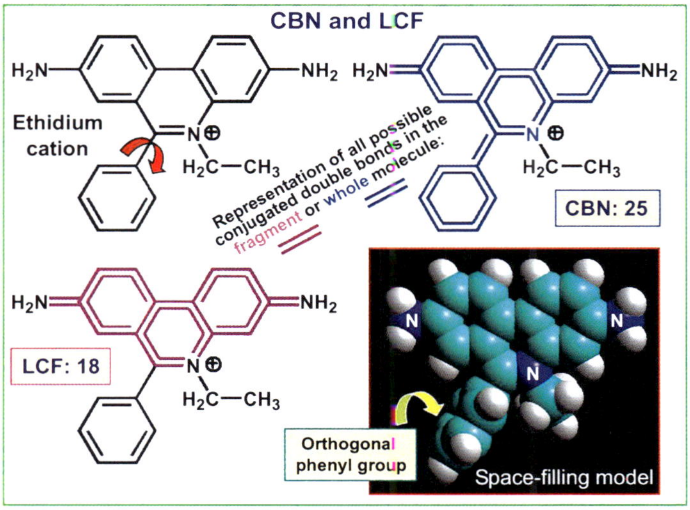

Fig. (15.3). Estimation of the aromatic size of fluorophores: CBN and LCF values for ethidium bromide. CBN (Conjugated Bond Number) corresponds to the total number of conjugated bonds. LCF (Largest Conjugated Fragment) is the number of conjugated bonds in the largest fragment. Red arrow indicates the rotation of orthogonal phenyl group. A space-filling model of ethidium is also shown.

Following pharmacological usage, the hydrophobic-hydrophilic balance of a probe (for both the cation and free or pseudo-base species) is assessed by the logarithm of the experimental water–octanol partition coefficient (log P) or the more recent computational hydrophobic index (HI). Theoretically, the log P can be estimated by using the fractional Hansch π values [27, 28], which are combined to assess the hydrophobic-hydrophilic character of a molecule. The influence of neighboring chemical groups, expressed numerically by using the Hammett σ values should be also considered. The result is the calculated log P, with negative and positive values corresponding to increasing hydrophilicity and lipophilicity, respectively.

Other calculation of log P was reported by Black and Mould [29] using similar "hydrophobic fragmental constants". To illustrate differences between log P values of both a free base and the protonated cationic compoud, imidazole and neutral red (NR) will be useful. The log P of both imidazole free base and imidazolium cation are shown in Fig. (**15.4A**).

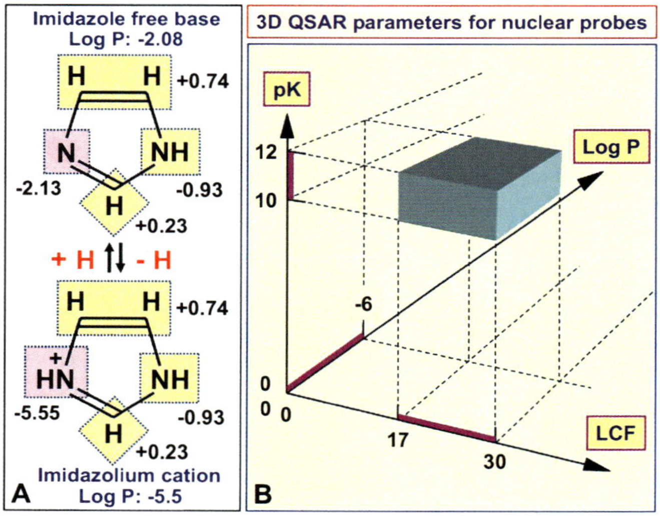

Fig. (15.4). A: Calculation of log P of the imidazole free base (top) and imidazolium cation (bottom) by using the fragmental values of Black and Mould [29]. Observe the increase of hydrophilia in the charged (protonated) molecule. **B:** 3D representation of parameters required for labeling of chromatin DNA, satisfied as shown by the blue parallelepiped.

Likewise, the moderately hydrophilic neutral red cation (NR^+) and the lipophilic neutral red free base (NR^0) have a log P of−1.1 and +1.9, respectively. Obviously, the equilibrium between the ratio of protonated compound and free base depends on the pH.

Log P can be also calculated using the QSAR properties tool of modeling software (*e.g.* ChemDraw, HyperChem), but unfortunately the log P thus assessed does not accommodate metals and charged atoms. In contrast, the hydrophobic index (HI) [30] allows the handling of charged molecules. HI is easily obtained after energy minimization of molecules by the semi-empirical AM1 or PM3 methods. The formula is HI: 1/(sum of the absolute value of partial charge on each atom/number of atoms). In the case of lipophilic fluorochromes, the hydrophilic-lipophilic parameters (log P and HI) are highly correlated, with a linear fitting R^2: 0.93 [31]. Therefore, HI values are more suitable for assessing the hydrophobic-hydrophilic balance than log P values.

Molecules containing distinct lipophilic and hydrophilic domains are termed amphipathic (amphiphilic). Amphiphilicity is the tendency of molecules with both lipophilic and hydrophilic domains to orient so that each molecular region is associated with an environment similar to its own [32]. Therefore, amphiphilic probes orient within membranes so that the lipophilic domain is within the region of acyl chains and the hydrophilic domain is near the polar head groups. A lipophilic, but non-amphiphilic probe, will dissolve in lipidic structures with no particular orientation. Although no rigorous rules are available to model amphiphilic properties, they may be estimated using the log P value of the lipophilic domain. Although not a ratio, this parameter is termed the amphiphilic index (AI), and corresponds to the nominal lipophilicity of the largest lipophilic domain of amphipathic compounds [32, 33]. Significant or selective binding to fluid membranes are usually only seen when $AI > 3.5$; non-selective insertion into membranes occurs when $AI > 5$, and strong binding to membranes at $AI > 8$.

In addition, other numerical parameters are important [11, 32]. The hydrophilic domain in an amphiphile is termed the head group, even if a probe is not a surfactant, fatty acid or phospholipid. The size of the head group can be modeled loosely by summing the relative atomic masses of all atoms in the head group to yield a head group size (HGS) parameter. The HGS value tends to co-vary with other significant factors such as head group hydrophilicity (HGH). Another parameter is the substituent bulk (SB) that corresponds to the overall size of groups appended to an aromatic core. SB can be assessed by summing the atomic weights of the atoms of the non-conjugated bulk substituents.

On the other hand, to assess the shapes of probes and regarding comparisons between reactangular and rod-shape DNA binding fluorophores, the simplistic length (L) and width (W) values can be evaluated by counting the number of conjugated bonds along length and width of each probe in 2D molecular diagrams. Then the W/L ratio provides information about possible intercalation ($W/L > 0.33$) and minor groove binding mode ($W/L < 0.33$) (see Chapter 12.3.2).

Several numerical parameters such as Z, MW, CBN, LCF, AI, HGS, *etc.*, can be obtained by inspection of chemical structures given in standard sources [9, 34, 35]. Obviously, when chemical structures are not disclosed for reasons of commercial secrecy, no numerical analysis can be made. This is unfortunate, because understanding the uptake and selective accumulation of probes is only possible when their chemical structure and properties are known.

2.2. Prediction Rules

QSAR rules are routinely applied in biomedical sciences (*e.g.* physiology, pharmacology), and are also useful to investigate histochemical staining mechanisms [20, 21, 23, 36], localization of fluorescent probes and photosensitizers within organelles [10, 11, 17, 32, 37, 38], and uptake of compounds into tumor cells [39]. QSAR rules have been applied for prediction of uptake and accumulation of probes in the cytosol, lysosomes, mitochondria, endoplasmic reticulum, Golgi apparatus, lipid droplets, nuclei, and plasma membrane [6, 24, 26, 31, 40 - 42].

Although a single QSAR parameter can roughly indicate the localization, more precise and reliable predictions are achieved by using several parameters simultaneously. In the first case, compounds with log P between -5 and 0 are expected to localize in both nuclei and lysosomes, whereas those with log P in the range 0 to +5 should label both endoplasmic reticulum and mitochondria [26]. The use of additional physico-chemical properties allows multiparametric predictions (Fig. **15.4B**). QSAR models used to predict intracellular localization are here shown:

Cell Sites for Which Models are Available		**Source of the Models**
Cytosol		[43]
Endoplasmic reticulum		[41]
Golgi apparatus		[24, 44]
Lipid droplets		[32, 42]
Lysosomes		[19]
Mitochondria	(cationic probes)	[45]
	(anionic probes)	[40]
Nuclei (chromatin and nucleolus)		[11, 26]
Plasma membrane		[32, 37]

Fig. (**15.5**) represents a summary of QSAR predictions for the localization of probes within several organelles [1, 24, 25]. An "expert" system running as a computer software, and QSAR predictions for multicellular organisms are interesting developments [1, 6]. At present, there is available software to make QSAR predictions for pharmacophores [46].

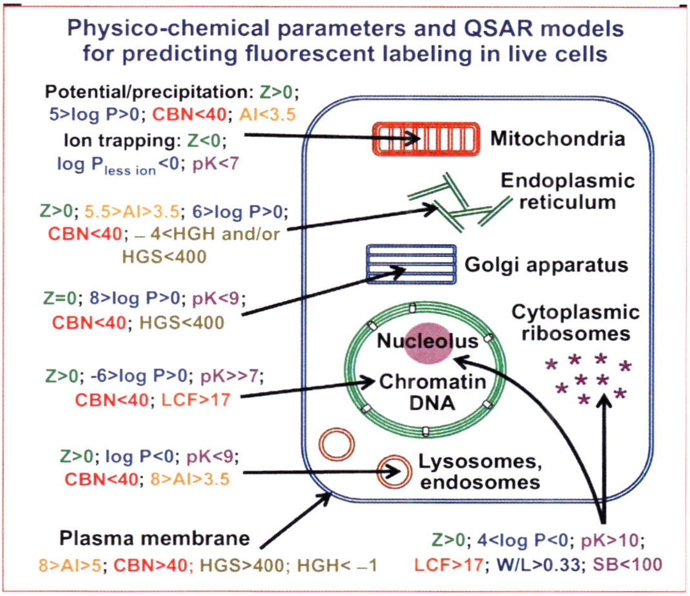

Fig. (15.5). General summary of vital probes localization according to QSAR prediction rules.

3. FLUORESCENT LABELING OF CELL ORGANELLES

Regarding practical aspects of organelle labeling [7, 14, 16, 47, 48], cell cultures are routinely grown on coverslips using common media, supplemented with 10% (v/v) of fetal calf serum, antibiotics and L-glutamine, at 37 °C in a humidified atmosphere with 5% CO_2. Usual conditions of labeling involve the treatment with solutions of vital probes in PBS, sterilized by filtration through a 0.22 µm Millipore filter.

Dye concentrations of 1-5 mg/mL (about 2-10 mM) are commonly employed to stain fixed cells and tissue sections for bright-field microscopy. However, fluorescent probes for live cells are generally used at concentrations three orders of magnitude lower (*i.e.* 1-5 µg/mL). Higher concentrations are not necessary and they could induce quenching or cytotoxicity. Labeling time is variable, but in most cases incubations of 1-10 min are used. After probe treatment, cells are washed 2-3 times with PBS, postincubated in culture medium for 15 min, mounted in PBS and observed under a suitable exciting light. In addition to reactive probes that form covalent bonds with organelles [49], fixation with formaldehyde vapor is a convenient and simple procedure to retain unchanged the localization of several probes [50].

Although vital probes are commonly used in aqueous solution, fluorescent drugs have frequently low water solubility and therefore they must be delivered within carriers such as phospholipid liposomes, Cremophor EL (polyethoxylated castor oil) micelles, LDL or HDL conjugates, nanoparticles, *etc*. Obviously, when cells are treated with drugs within delivery media, QSAR predictions could be not entirely correct.

At present, it is a common practice the multiple labeling of cell organelles by using probes with contrasting emission colors, examples being the simultaneous visualization of mitochondria, lysosomes and nuclei by labeling with MitoFluor Far Red 680, LysoTracker Green DND-26, and Hoechst 33342, respectively, following multiple exposure. A similar approach is the labeling of organelles with fixable probes followed by immunofluorescence labeling. Likewise, two signals from the same organelle in living cells can be recorded (*i.e.* by simultaneous labeling with a mitochondria probe and a Ca^{2+} indicator), but it is worth noting that in this case the ion indicator must also localize in mitochondria.

On the other hand, confirmation of accumulation of a given fluorescent compound often requires the use of already known selective probes. The overlay of the two images will result in colocalization. There are topographic requisites for a true colocalization: fluorescence signals should come from only one structure in order that the overlapping colors indicate true colocalization (Fig. **15.6A**). Otherwise, a false colocalization will be obtained (Fig. **15.6B**). An example is illustrated in Fig. (**15.6C-E**), where lysosomes of HeLa cells appear labeled with the photosensitizer zinc(II)-*meso*-tetraphenyl-porphine (ZnTPP) [51], followed by labeling with LysoTracker® Green. Red and green signals from ZnTPP and LysoTracker, respectively, show clear colocalization. As ZnTPP is somewhat more hydrophilic than TPP, lysosome rather plasma membrane localization is favored (see Section 5.1).

Influence of kinetics on probe targeting is another not well recognized factor. Following routinely used protocols for probe localization, failures to obtain the expected results are regarded as either technical errors or artifacts. When a wider range of incubation times are explored, however, marked differences in localization between the shorter and longer times can occur. Such kinetic effects are illustrated by $DiOC_6(3)$ or hexylrhodamine B to label the endoplasmic reticulum (ER). The initial uptake is into mitochondria and accumulation in the ER is a secondary step [52].

Small modifications of the chemical structure suffice to produce a completely different labeling. In the strong base acridine orange (AO), the N10 atom is almost completely protonated at pH 7, and the cationic dye is accumulated in lysosomes (see Fig. **15.1B**). Alkylation of AO leads to the related 10-alky--acridinium dyes; alkyl groups from ethyl to nonyl decrease the lysosome fluorescence whereas increase mitochondria emission [53]. The nonyl-AO derivative is a selective mitochondrial probe for living cells, but it stains DNA and RNA in fixed materials.

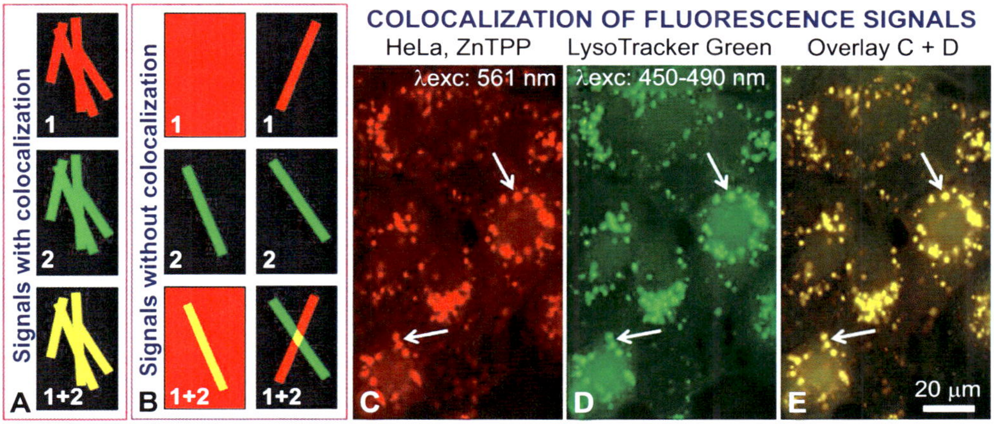

Fig. (15.6). A, B: Topographic requisites for a true colocalization: The same structure with different colors (1, 2) corresponds to a true colocalizattion (A, 1+2), whereas differently shaped signals originate false colocalizations (B). **C-E:** Lysosomes from HeLa cells (arrows) after 24 h incubation with 10 μM ZnTPP encapsulated in folate-targeted multilamellar liposomes (C) to be recognized by folate receptors [51], and then labeled for 30 min with 50 nM LysoTracker® Green DND-26 (D). The overlay of images (E) shows true colocalization (yellow signal).

The type of delivery can influence the uptake of probes and drugs. When lipophilic compounds are incorporated into liposomes, specific mechanisms of cell internalization can occur in addition to simple membrane fusion. For example, when zinc(II)-phthalocyanine (ZnPc) is delivered into dipalmitoyl-phosphatidyl-choline (DPPC) liposomes, accumulation in A-549 cells occurs by clathrin-mediated endocytosis [54, 55]. In other cases, the type of delivery does not appear a decisive factor: the same lysosomal localization is observed when ZnTPP is administered either in DPPC liposomes or in PEG-coated poly(D,L-lactide-co-glycolide) nanoparticles [56].

4. NUCLEIC ACIDS-CONTAINING STRUCTURES

Numerous techniques have been recently introduced for nucleic acids (NAs) labeling and imaging in live cells (*i.e.* chromatin, nucleoli), giving rise to a field of increasing interest in microscopical studies. Progresses in this field have been revised [4, 11, 57 - 59]. As NAs probes often have biological activity, the subject has relevant implications in cell biology and cancer research [11, 60, 61]. Some vital probes for NAs are selective fluorochromes for staining fixed cells (see Chapter 12.3.2). Several cationic mitochondrial probes (*i.e.* $DiOC_1(3)$ also induce brigth emission in kinetoplast DNA from living *T. cruzi*, see section 5.4). Although not using a DNA probe, chromatin visualization in live cells can be achieved by targeting histones (*i.e.* the fusion protein histone H1-GFP, see Fig. **4.19B**), and site- or sequence-specific DNA-binding proteins labeled with GFP [61].

4.1. Chromatin DNA

Cationic DNA probes are widely used in fluorescence microscopy. The dication 4',6-diamidino-2-phenylindole (DAPI) binds to the DNA minor groove at AT sequences and is applied as a vital nuclear probe [62], but it stains better fixed material than live cells [63], and its use for DNA labeling has been rather limited.

The labeling of nuclei (chromatin DNA) by minor-groove binding Hoechst (H) fluorochromes is dependent on their log P and cell uptake. As H-33258 is more hydrophilic than H-33342, the uptake of the later is greater and induces stronger labeling [64]. Therefore, H-33342 is the most suitable probe for DNA labeling in live cells using fluorescence microscopy and flow cytometry [4, 57, 65]. After intravenous or intraperitoneal injection of H-33342 fluorescent nuclei were found in several organs from 1 h to 28 days later [66], which resisted conventional histological procedures. Chromatin DNA labeling in living embryos has also been attempted by H-33342 [67].

Differences between intercalating (rectangular shaped) and minor groove binding (rod shaped) probes must be noted [11]. The former (*i.e.* AO) gives a weak signal in chromatin DNA but generally stronger fluorescence in RNA-containing nucleoli and basophilic cytoplasm. No metachromatic lysosome labeling occurs using a low probe concentration and short incubation time. In contrast, DNA minor groove binders (*i.e.* H-33342) induces highly selective fluorescence in chromatin DNA (Fig. **15.7A, B**).

Regarding intercalation, two diverse reports suggest a possible explanation for the puzzles noted above. AO *in silico* is known to intercalate into DNA [68, 69], as in fixed cells. However this is justly not the case for chromatin of live cells, as DNA elongation and unwinding to permit dye intercalation is inhibited when DNA is organized as nucleosomes (Fig. **15.7C, D**). However, when Φ_F is high and R is very low, a small amount of suitable ligands could intercalate into chromatin DNA giving a still serviceable emission. Since fluorescence of ribosomal RNA does occur *in vivo* with AO, live cells show fluorescent nucleoli and cytoplasm, with very weak emission of chromatin [70]. In contrast, minor groove ligands do not cause DNA elongation or unwinding. Since histones and other proteins associated to chromatin DNA let off the minor groove [71], this site remains relatively free for binding to specific probes.

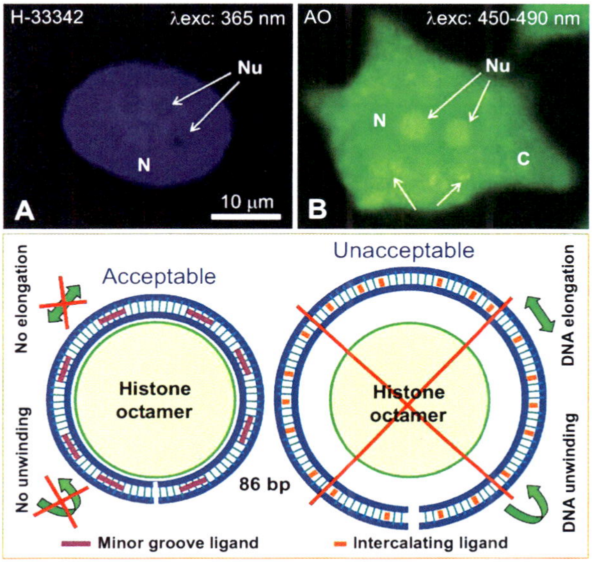

Fig. (15.7). A: Nuclear (DNA) labeling of a HeLa cell treated 5 min with 2.5 μg/mL H-33342. **B:** Nucleolar and cytoplasmic labeling of an A-549 cell treated 0.5 min with 0.5 μg/mL acridine orange (AO). **C, D:** Binding modes of minor groove (H-33342, C) and intercalating (AO, D) probes with DNA in the nucleosomes of live cells, showing the limitation of intercalative binding due to demands for DNA elongation and unwinding. Probe ligands (L) are bound to a single turn of nucleosomal DNA with 86 base pairs (bp) and adenine–thymine clusters, giving energetically acceptable minor groove binding (C), and unacceptable intercalation (D). Binding ratios (r: saturation limit, L/bp, and R: saturation ratio shown) are (C) r: 0.2 (1L/5 bp), R: ≈ 0.1, and (D) r: 0.5 (1L/2 bp), R: ≈ 0.25.

Intercalating antitumoral antibiotics are also DNA probes for living cells. Consider in particular the anthracycline drugs daunomycin (DNM) and doxorubicin (DOX), which intercalate into DNA [72], and induce chromatin fluorescence [73, 74] following QSAR predictions. Anthracyclines comprise an important class of anticancer drugs, with a metal-chelating hydroxy-anthraquinone chromophore that intercalates only into DNA and not into RNA, as they require a B-type duplex. Biochemical consequences of anthracycline intercalation are inhibition of topoisomerase II activity, which is the main factor responsible for antitumoral effects.

In the case of DOX (Fig. **15.8A**), highly selective red emission is found in chromatin DNA from live cells, whereas RNA from nucleoli and basophilic cytoplasm shows no fluorescence (Fig. **15.8B**). As occurs in the case of DNM (see Chapter 8.3.6), post-treatment of DOX labeled cells with Al^{3+} ions increases emission of chromatin DNA (Fig. **15.8C**). Esterification at C14 of DOX with a valerate residue leads to changed localization, favoring accumulation in the perinuclear cytoplasm rather than in the nucleus [75].

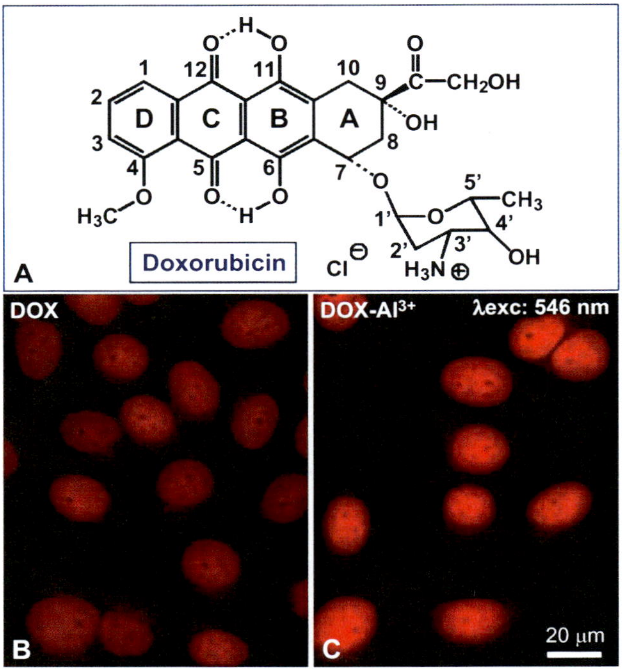

Fig. (15.8). A: Chemical structures of doxorubicin (DOX) with atom numbering. **B, C:** Live HeLa cells treated with 0.3 μM DOX for 24 h (B), and similar DOX-labeled HeLa cells after fixation with formaldehyde vapor for 5 min, and treatment with 2 mg/mL $Al_2(SO_4)_3$ (C). Nucleoli and cytoplasm show no fluorescence.

Interestingly, DOX and formaldehyde generated by iron-mediated free radical reactions react to give a dimer (DOXFORM) in which two anthracycline molecules bind together with three CH_2 groups, two forming oxazolidine rings and one binding the oxazolidines together at their amino groups [75]. DOXFORM can hydrolyze to give an active monomeric metabolite in which the carbon of formaldehyde is recovered as a Schiff's base at the amino sugar group. The anthracycline-FORM conjugate has attracted interest because when binding occurs at the 5'-NGC-3' sequence, the drug intercalates between N and G, covalently binds to the 2-amino group of a G-base on one strand in the minor groove using the Schiff's base, and binds to the G-base on the opposing strand using H bonds. Such an unusual combination of intercalation, covalent- and H-bonding is referred to as the virtual cross-linking of DNA by anthracyclines [75].

Other fluorescent DNA binding antibiotics are often used for chromatin labeling in living cells. Chromomycin A_3 (CMA) (see Chapter 12.5.8), and mithramycin (from *Streptomyces plicatus*) are anticancer antibiotics that in the presence of Mg^{2+} binds to the DNA minor groove and inhibits transcription and protein synthesis. Fluorescent DNA labeling in living plant cells is achieved by using 50 µg/mL mithramycin and 5 mM $MgCl_2$ under blue light excitation [63]. In addition, CMA, mithramycin and olivomycin complexed with Mg^{2+} or Zn^{2+} have been used for DNA fluorescence in flow cytometry [76].

The anthraquinone dye DRAQ5 (deep red-fluorescing bisalkylamino-anthraquinone number 5) induces selective nuclear fluorescence of live cells [77]. Althrough the anthraquinone fluorophore is only weakly fluorescent [78], the combination of low Φ_F but high AT-rich DNA affinity explains the favorable signal-to-noise profile of nuclear fluorescence [79]. Molecular modeling suggests that some anthraquinone derivatives (*i.e.* 1,4- and 1,8-$NH(CH_2)_2NH(CH_2CH_3)_2^+$) intercalate into DNA, with both substituents in the same groove, while the substituted 1,5-derivative intercalates in a threading mode, with a side chain in each groove [80].

However, the microscopical emission of DRAQ5 corresponds to DNA minor groove binders such as H-33342, but not to intercalators. A molecular DRAQ5-DNA model involving binding to the minor groove has been suggested [79], with breaking of the H bonds of one AT base pair, base displacement out the DNA backbone, and stacking of DRAQ5 aromatic ring with the two unpaired bases. Alternatively, a head-to-tail dimer of the ligand could also bind to the minor groove, explaining the low (quenched) Φ_F of the probe, and the selectivity for certain base pairs by H bonding. A 2:1 Mg^{2+} complex would be well suited to bind to the DNA minor groove and to recognize AT base pairs. A similar situation occurs for the 2:1 CMA-Mg^{2+} complex, which recognizes GGCC sequences [81] (see Chapter 12.4.8).

Cationic porphyrins are also used for DNA labeling of live cells. *Meso*-tetra(4-*N*-methyl-pyridinium) porphine (TMPyP) (Fig. **15.9A**) binds with high selectivity to the minor groove of consecutive AT sequences [82 - 84]. The localization of TMPyP is detected by fluorescence microscopy, and most studies indicate that the chromatin is selectively labeled [85 - 88]. Chromatin DNA labeling by 10^{-7} M TMPyP for 4 h is shown in Fig. (**15.9B**), but localization in lysosomes has been also described using 10^{-6} M for 1 h [89]. According to QSAR rules some cationic probes can label both nuclei and lysosomes [26]. In most cases of shared labeling (N + L), an initial localization in nuclei is generally followed by lysosome labeling (N → L). From 13 N + L probes, only one showed the inverse L → N uptake sequence. Labeling of chromatin DNA is also achieved by intratumoral injection *in vivo* of TMPyP and the outside DNA binder TMAP (*meso*-tetra(4-*N,N,N*-trimethylanilinium) porphine) [10, 90]. Both TMPyP and TMAP are also fluorochromes for DNA in fixed cells [91, 92].

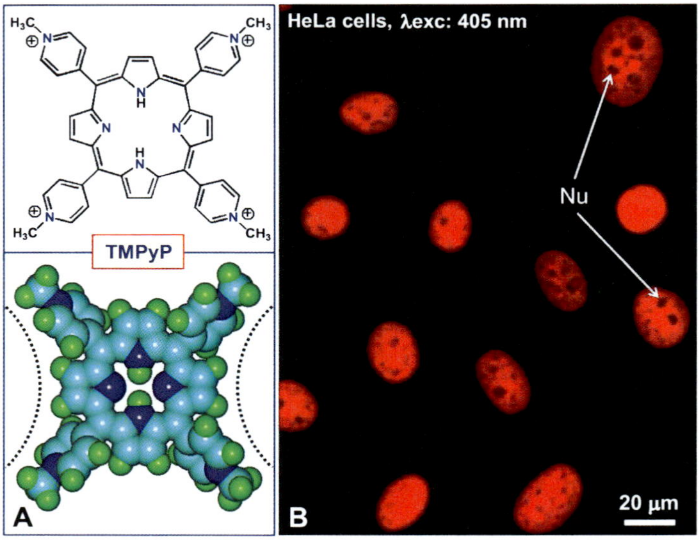

Fig. (15.9). A: Chemical (top) and molecular structure (bottom) of the cationic porphyrin TMPyP, showing the four twisted pyridyl groups bound to the porphine ring. The dotted curved lines represent the middle point on the floor of the minor groove at the level of AT sequences [216]. **B:** HeLa cells labeled with TMPyP (10^{-7} M, 4 h). Selective chromatin DNA labeling can be seen; nucleoli appear as dark round areas within nuclei. This image pattern corresponds with that of H-33342 and similar minor groove ligands.

Although luminescent ruthenium complexes as Ru(bipyridine)$_3^{2+}$ and Ru(phenanthroline)$_3^{2+}$ (see Chapter 9.3.2) do not enter living cells [93], other less hydrophilic Ru polypyridyl complexes are permeable and can label chromatin DNA. Recently, the dinuclear Ru(II) complexes [(phen)$_2$Ru(tpphz)Ru(phen)$_2$]$^{4+}$ and [(bpy)$_2$Ru(tpphz)Ru(bpy)$_2$]$^{4+}$ (where phen: 1,10-phenanthroline; bpy: bipyridine; tpphz: tetrapyrido-phenazine) as cellular DNA stains have been investigated [94, 95].

These complexes are non-emissive in water but display intense metal-to-ligand charge-transfer (MLCT) luminescence when bound to DNA, a "light switch" effect used to study the interaction of Ru complexes with DNA. Electron excitation occurs from the Ru atom to a π* orbital on the tpphz ligand. The ^{1}MLCT excited state then decays *via* intersystem crossing to a ^{3}MLCT excited state localized on the phenazine N atoms. Water quenches the excited state by H bonding to these N atoms, but when tpphz is shielded from water the luminescence is activated and the complex acts as it would be in organic solvents [95]. Live cells are permeable to $(phen)_2Ru(tpphz)Ru(phen)_2$ (log P: − 0.96), showing DNA labeling (λexc: 458 nm; λem: 680 nm) and colocalization with H-33342 [94]. In contrast, the bpy analogue (log P:−1.61) is not a vital probe but a dead cell stain for DNA that colocalizes with the non-membrane-permeable DNA intercalator, propidium iodide. $Ru(phenanthroline)_2$-phenanthroimidazol--dimethylaniline derivatives were recently found to label chromatin DNA in live cells, the enantiomer Λ-$Ru(phen)_2(p\text{-}DMNP)^{2+}$ showing selective binding to G-quadruplex DNA [96].

Some styryl-pyridinium derivatives are also fluorescent probes for DNA [97], the compound A41B2 showing the best properties. Both 2,4,5-trimethoxy groups and the adamantyl residue are important for the specific interaction with DNA but no binding mechanism was described. In addition, a linear thiazole-coumarin (TC) cationic dye has been reported to intercalate between AT and GC base pairs showing intense red fluorescence only in AT regions, because a charge transfer complex with guanine abolishes the emission [98]. No emission of TC in presence of single-stranded DNA, RNA and proteins was found. Although an intercalative mechanism has been proposed, binding of the slightly curved TC dye to AT tracts on the DNA minor groove should be not discarded. Applicacions for TC are DNA labeling and staining in live and fixed cells, flow cytometry, DNA electrophoresis, *etc*. The dye SYTO 13 allows visualizing chromatin DNA in live cells, but unfortunately, the chemical structure of SYTO dyes is not known and then no predictions can be put forward.

4.2. Nucleoli and Basophilic Cytoplasm

For more than 60 years, RNA has been detectable in fixed cells and tissues by relatively specific staining methods. More recently, it has become possible to study RNA in unfixed, live cells. Likewise, nucleolus and RNA-targeted chemotherapy have now increasing interest [99]. The treatment with cationic rectangle-shaped fluorochromes induces fluorescence in nucleoli and basophilic cytoplasm according to the QSAR prediction rules (Fig. **15.5**). This occurs typically with AO labeling (Fig. **15.7**).

Ethidium bromide (EB) intercalates into DNA and RNA duplexes. An excellent review on the chemistry and present uses of this and other phenantridine drugs has been published [100]. Although chromatin DNA in live cells can be labeled by EB [45], this labeling has been shown to be erratic, and fluorescent nucleoli and basophilic cytoplam are commonly found. As pseudo-base/cation interconversions are rather slow, this may explain why EB, although having a moderately lipophilic pseudo-base, stains nuclei slowly and in a concentration-dependent way [26]. However hydroethidine, the reduced non-ionized form of EB, is used when oxidized to label chromatin DNA [101, 102]. Hydroethidine has a log P: 3.2, and therefore readily enters the live cell, within which it is oxidized to the hydrophilic ethidium cation that accumulates in the nucleus.

When the hydrophilicity is high (*i.e.* propidium iodide (PI, log $P_{dication}$:−5.1; log $P_{uncharged}$:−1.4), the probe is not permeant and only stains dead cells. In the case of EB (log P_{cation}:−1.6; log $P_{uncharged}$: 3.2), sufficient probe enters the cells and labels RNA-containing structures [103, 104] (Fig. **15.10**). Chromatin DNA generally appears poorly labeled because of its nucleosomal organization. It must be taken into account that a great amount of RNA in living cells occurs as double- and triple-stranded structures (*i.e.* ribosomal and transfer RNA [105 - 110]), which are relevant targets for intercalating probes.

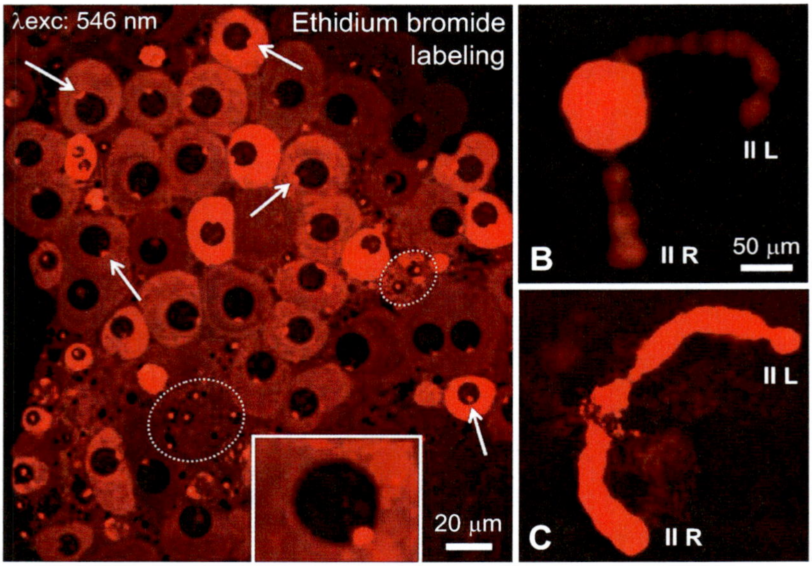

Fig. (15.10). A: Ethidium bromide (EB) labeling of live primary spermatocytes and nurse cells (encircled) from a *Drosophyla hydei* testis incubated in 5 μg/mL EB for 5 min. Spermatocyte nucleoli attached to the nuclear envelope (arrows and insert), nurse cell nucleoli, and basophilic cytoplasm are intensely fluorescent. **B, C:** EB labeling of salivary gland chromosomes of *Chironomus pallidivittatus* 4th instar larvae. **B:** Isolated chromosome II from a living nucleus previously treated with 5 μg/mL EB for 5 min. **C:** Acetic acid-fixed chromosome II stained with EB as in A. Observe the differential emission of chromatin DNA and nucleolar RNA. Right and left arms of the nucleolar chromosome are indicated (II R and II L).

The cationic thioflavine T (ThT, CI 49005) is an interesting fluorochrome that is useful to see differences between labeling and staining processes on live and fixed cells, respectively, also in relation to the heterogeneity of dyes. Commercial samples of the dye contain mono-, di- y tri-benzothiazole (BT) units, although in current publications, ThT is shown with only one BT unit. The mono- and di-BT dyes, ThT1 and ThT2, respectively, have different log P values (ThT1: +1.9, ThT2: -2.7), and because of its greater lipophilia only ThT1 can enter live cells and label nucleoli and mitochondria (Fig. **15.11**). ThT1 is a planar "tea tray"-shaped dye that can intercalate into RNA and DNA [111, 112], which explains the fluorescent labeling of nucleolar RNA and kinetoplast DNA in living cells. Nuclear chromatin does not fluoresce, possibly due to limitations imposed by its nucleosomal organization, as occurs with AO [11], but the non-nucleosomal kDNA in *T. cruzi* mitochondria allows its selective labeling by ThT1 [113] (see section 5.4). Similarly, SYTO RNA Select green-fluorescent cell stain is a probe of unrevealed structure that like ThT1 also label nucleoli and mitochondria in live cells.

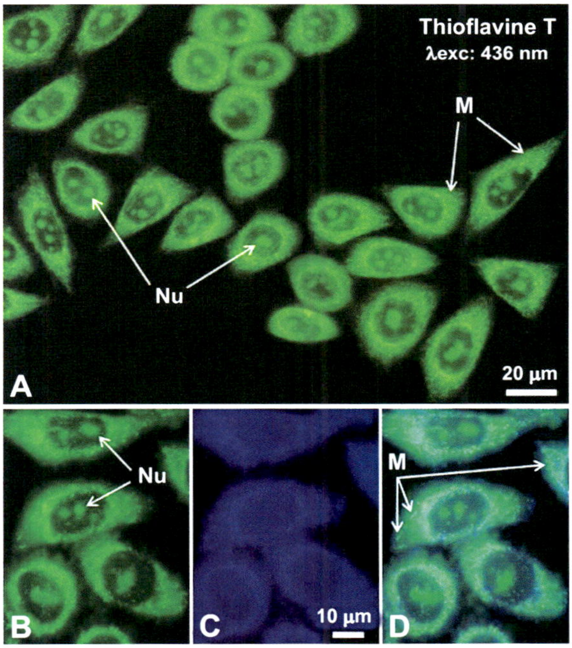

Fig. (15.11). Living HeLa cells incubated with 1 µg/mL ThT in PBS for 3 min, showing green emission of mitochondria and nucleoli (**A**, **B**; λexc: 436 nm), blue autofluorescence of mitochondria (**C**; λexc: 365 nm), and cyan colocalization of both signals in mitochondria **(D)** (courtesy of A. Lázaro and A. Villanueva). A low and diffuse green emission in the cytoplasm is likely due to ribosomes.

Both the zwitterion and methochloride derivative of the alkaloid sempervirine (from *Gelsemium sempervirens*) are fluorescent (λexc: 348 and 387 nm, λem: 437 nm) and bind to DNA and RNA. They enter live cells inducing blue fluorescence in the cytoplasm [114]. Likewise, the anticancer agent flavopereirine (from *Strychnos melinoniana*; PB-100, with a chromophore similar to sempervirine) enter only live tumoral cells and the blue-fluorescing zwitterion appears selectively accumulated in the nucleoli [115]. Binding to purine bases of ribopolymers has been described.

On the other hand, the antitumor fluoroquinoline drug quarfloxin (CX-3543) and substituted naphthalene diimides are fluorescent agents that bind preferentially to G-quadruplex and localize selectively in nucleoli of live cells. Quarfloxin disrupts interaction of rDNA G-quadruplexes with nucleolin, inhibits rRNA synthesis and induces apoptosis [116]. Tri- and tetra-dimethylamino-propyl-naphthalene diimides have exceptional affinity for telomeric G-quadruplexes, inhibit telomerase, and show antiproliferative effects [117].

An europium complex selectively binds to live cell nucleoli, allowing to visualize either the red Eu^{3+} luminescence (λexc: 380-405 nm) or the green emission of the 1-azathioxanthone antenna chromophore (λexc: 450 nm) by fluorescence microscopy [118]. Binding mechanism to RNA is not known but it is tempting to speculate that intercalation of the planar 1-azathioxanthone ring (similarly to the drug hycanthone) could be the factor responsible for RNA labeling.

Luminescent cyclometalated iridium(III)-dipyridoquinoxaline complexes (nearly non-emissive in aqueous medium) are permeable to living mammalian cells and accumulate in the cytoplasm and nucleoli giving strong orange luminescence [119]. Upon photoexcitation d(Ir)→π*(L), the Ir-diimine complexes display intense and long-lived orange ^{3}MLCT emission in aprotic organic solvents at room temperature. Despite the relatively strong DNA-induced emission of these complexes intercalated *in silico*, chromatin DNA is not labeled possibly because intercalation is hindered by the nucleosomal organization.

Intriguingly, the selective labeling of nucleoli and basophilic cytoplasm by these Ir-complexes does not depend on interaction with RNA, but possibly binding to hydrophobic pockets of proteins [119]. The hydrophobic probe 8-anilino-1-naphthalene-sulfonate also shows distinct nucleolar staining [120, 121]. Likewise, the lipophilic dye Nile red labels the nucleolus in addition to the expected oil droplets [122]. These results indicate that nucleoli can be imaged using hydrophobic probes with low charge, and strong environment-sensitive emission properties.

A dicationic pyridyl-vinyl-carbazole dye (2,7-9E-BHVC) has been recently proposed as fluorescent probe for RNA labeling [123]. Although the dye is formulated in an extended (rod-shaped) form, a more compact structure suitable for intercalation is possible. Labeling with the commerial probes SYTO 9 (log P:−0.93, membrane-permeable nucleic acid dye) and SYTO RNA Select Green-fluorescent stain are also suitable to visualize RNA components of live cells, such as nucleoli and basophilic cytoplasm. The same seems to occur in the case of the isoquinoline alkaloids berberine and sanguinarine, the former only when the mitochondrial transmembrane potential is abolished [124]. Fluorescently labeled nucleolar targeting peptides can be also used for selective nucleolar visualization [99].

In addition, new techniques based on molecular biological methods are available for imaging of RNA, and for studies on gene expression. They include micro-injection of fluorescent RNA into living cells, use of molecular beacons, energy transfer and/or quenching between fluorescent oligonucleotides, fluorogenic compounds or GFP-fused RNA binding proteins to label aptamer tags in RNA, *etc*., and have been recently reviewed [58 - 60, 125].

5. MEMBRANOUS STRUCTURES

The basic pattern of biomembranes corresponds to a lipid bilayer associated to trans-membrane and surface proteins. In the case of probes that combine a lipid metabolite with a fluorophore, numerical QSAR parameters can be assessed and prediction rules applied. Recent reviews have been published [24, 25].

NBD- and Bodipy-tagged phospholipids, ceramides and sphingomyelins are most frequently used fluorescent probes for living cells, although FITC, TRITC, anthracene and pyrene fluorophores are also employed [8, 9]. Fluorescent labeling of phosphatidyl-ethanolamine with furanones (*i.e.* MDPF) seems to have been overlooked. Likewise, synthetic lipophilic fluorochromes such as carbocyanines are also employed to reveal membranous structures [32]. Interestingly, more physiological tagged lipids using fluorescent polyene fatty acids (pentaene-ceramides and -sphingomyelins) have been synthesized and used on live cells [126] (see Chapter 10.6.2). Lipophilic and uncharged fluorescent drugs accumulate preferentially in lipid domains of membranes. Examples are hydrophobic macrocycles such as photosensitizing porphyrins, phthalocyanines, and porphycenes [10, 127, 128].

5.1. Plasma Membrane

Although the main feature of the plasma membrane is its lipid bilayer structure, compelling evidence now exists that in eukaryotic cells the phospholipids are asymmetrically distributed between the two lipid leaflets. In this non-uniform enzyme- and energy-dependent distribution, phosphatidyl-ethanolamine (PE), phosphatidyl-serine (PS), phosphatidyl-inositol (PI), and phosphatidic acid are mainly located in the inner leaflet, whereas phosphatidyl-choline (PC), sphingomyelin (SM), and glycosphingolipids are concentrated in the outer leaflet [129, 130].

Typical fluorescent probes for plasma membrane have a polar head and a lipohilic chain. Fluorescein, carbocyanin (*i.e.* DiI), Bodipy, NBD, anthracene, and amphiphilic styrene derivatives (*i.e.* FM 1-43 and di-8-ANEPPS) bound to fatty acids or alkyl chains are the most useful markers for the plasma membrane [9, 32]. Some of these lipophilic probes have been used for visualizing lipid components of the complex cell wall from mycobacteria in fluorescence microscopy [131]. Plasma membrane labels find also applications for cell tracking (reviewed in [8, 32, 132, 133]). To this respect, the plasma membrane is frequently labeled with FM dyes, for example, FM 4-64. This probe is used to label selectively the plasma membrane of eukaryotic cells [134], but if the staining time is extended, endosomal membranes formed by internalization are also labeled.

The anionic amphiphile merocyanine 540 (MC540) (Fig. **15.12A**) is a fluorescent dye that localizes primarily in the plasma membrane of mammalian cells [135]. The dye may serve not only as a fluorophore for probing membrane structure [136], but also for inducing lethal photodamage of neoplastic cells [137, 138]. Unsaturated phospholipids and cholesterol are readily peroxidized by MC540. The fluorescence of the plasma membrane induced by MC540 (λexc: ~530 nm, λem: ~580 nm) is illustrated in Fig. (**15.12B, C**). Other examples such as filipin labeling of cholesterol and ergosterol present in the plasma membrane have been already described in Chapter 10.6.1.

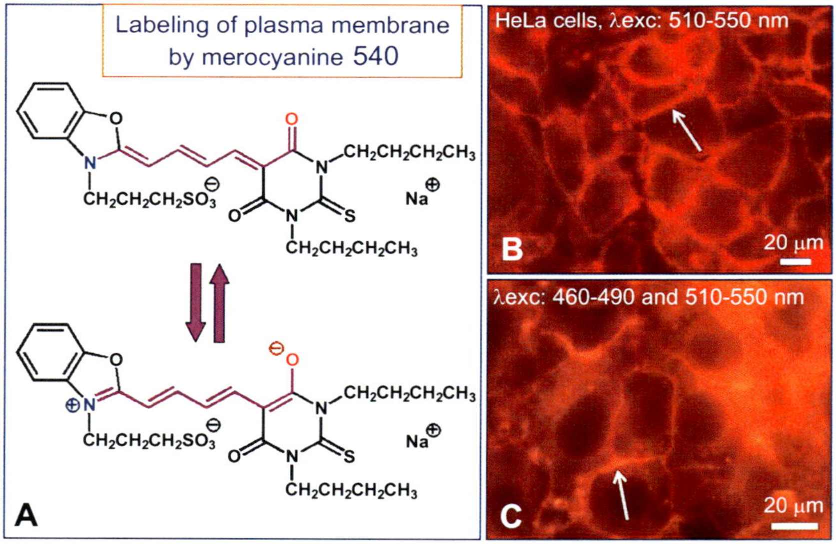

Fig. (15.12). A: Chemical structure of merocyanine 540 and its mesomeric forms. **B, C:** Fluorescence micrographs of the plasma membrane of HeLa cells (arrows) labeled by merocyanine 540 (10 µM, 60 min), under green or blue-green double excitation (courtesy of A. Casas).

Regarding porphyrin macrocycles, *meso*-tetraphenyl-porphyrin (TPP; 5,10,15,20-tetraphenyl-21H,23H-porphine, the simplest aryl derivative of the porphine ring), can be used as a fluorescent probe for lipid detection. A strong emission (λem: 640 nm) occurs when TPP is solubilized in DPPC liposomes, which labels the plasma membrane [139]. After 6 h incubation with TPP-liposomes, a continuous and/or punctate red fluorescence is found on the cell surface. Some internalization of TPP also occurs, since after prolonged incubation (18 h) red granules are found in the cytoplasm. A monocationic porphyrin photosensitizer was also found to label the plasma membrane [140]. With suitable lateral residues (*i.e.* four palmitoyl chains), fluorescent porphycenes should also interact with lipid bilayers.

5.2. Endoplasmic Reticulum

The endoplasmic reticulum (ER) is a central membranous organelle integrated by the rough (ribosomal) ER and the smooth ER. The functions of ER appear closely related to other organelles and resulted in the concept of "Golgi-associated ER from which lysosomes arise" (GERL). In addition, ER is related to endosomes, phagosomes, secretion granules (zymogen), and lipid droplets [141]. ER is the site of synthesis of membrane and secreted proteins, as well as the main calcium store. It is known that the impediment of cells to properly fold and modify secretory and transmembrane proteins in the ER leads to a buildup of misfolded proteins, a condition that is called ER stress [142], which can be revealed by using selective fluorescent probes such as thioflavine T [143].

Considering the number of available probes for organelles, there are very few probes for ER. They include carbocyanines, metal coordination complexes, and the ER-tracker series, which consist of Bodipy conjugated to glibenclamide, a drug which targets ATP-sensitive K^+ channels often found in the ER. Some carbocyanines have been employed for visualizing ER [144, 145], the most popular being $DiOC_6(3)$. This lipophilic cationic dye, permeable to the plasma membrane, is a general membrane probe, and then it is not specific for the ER. In live cells exposed to low dye concentration, $DiOC_6(3)$ accumulates only in mitochondria due to the high negative membrane potential of this organelle. With prolonged incubation time or higher concentration, the dye labels the ER [52, 146].

When the technique is successful, the ER is recognized by its characteristic shape and continuity. When not, it is often because ER can not be distinguished from other organelles, or because it has a dense 3D organization that prevents detailed visualization. Therefore, best results are obtained with cells that have regions where ER shows a 2D organization (*i.e.* large thin peripheral areas). Photoactivity of $DiOC_6(3)$ also induces a selective damage to microtubules [147].

Regarding QSAR, routine ER probes are amphipathic, lipophilic cations with moderate-sized conjugated systems [41]. The lipophilic character permits probe uptake by passive diffusion without non-specific accumulation in other biomembranes. The amphipathic character favors uptake into the ER, perhaps owing to lipids with high amount of zwitterionic head-groups.

In the last few years, ER targeting has been also shown for luminescent coordination complexes of Eu(III), Tb(III), and Zn(II) [148 - 151]. In addition, luminescent Pt(II) complexes containing bidentate N-heterocyclic carbene ligands selectively label ER, induce ER stress and cell apoptosis. Some of them show phototoxicity to cancer cells [152]. Recently, the dinuclear Ru(II) complex $[(Ru(DIP)_2)_2(tpphz)]^{4+}$ (DIP: 4,7-diphenyl-1,10-phenanthroline and tpphz: tetrapyridophenazine) was found to display an increase in MLCT luminescence on addition of liposomes. This lipophilic Ru complex targets the ER (λexc: 450 nm; λem: 620 nm), and is an imaging agent for live cells [153].

5.3. Lysosomes

Selective fluorescence labeling of endocytic compartments such as endocytosis vesicles, endosomes, fagosomes, *etc.*, has been described [16, 154, 155]. As most of them share cytochemical characteristics with lysosomes, the later can also appear labeled. Neutral red (NR) was one of the earliest fluorochromes to reveal lysosomes and at present it remains as a useful probe [156]. After treatment with 20 μg/mL NR and green light excitation, red fluorescing lysosomes are clearly seen (Fig. **15.13A-C**). After observation for some minutes, nuclei and cytoplasm regions begin to show a diffuse red emission (possibly, damaged cells). Very often and specially in round cells, a great amount of lysosomes appears as a granular cloud surrounding the Golgi apparatus and then the precise localization can be confused [157].

The ability to label lysosomes by AO in living cells is well known, giving a metachromatic image of the organelle (Fig. **15.13D**), which depends on its acidic pH. As occurs with other lysosomal probes (acridine yellow, Bismarck brown, brilliant cresyl blue, euchrisine, iodine green, neocyanine, phosphine, proflavine, and thionine [26]), AO is an N → L probe. Metachromatic labeling of lysosomes by AO requires a suitable time and concentration to enter the organelle and to aggregate. Lysosome probes such as AO and Nile blue are photoactive and have been used to induce tumor cell death [18, 158, 159]. ZnTPP (Fig. **15.6**) also labels lysosomes. Widely used lysosome probes are the LysoTracker® family of dyes, including blue-, green-, red-, and deep red-fluorescing dyes (Fig. **15.14**).

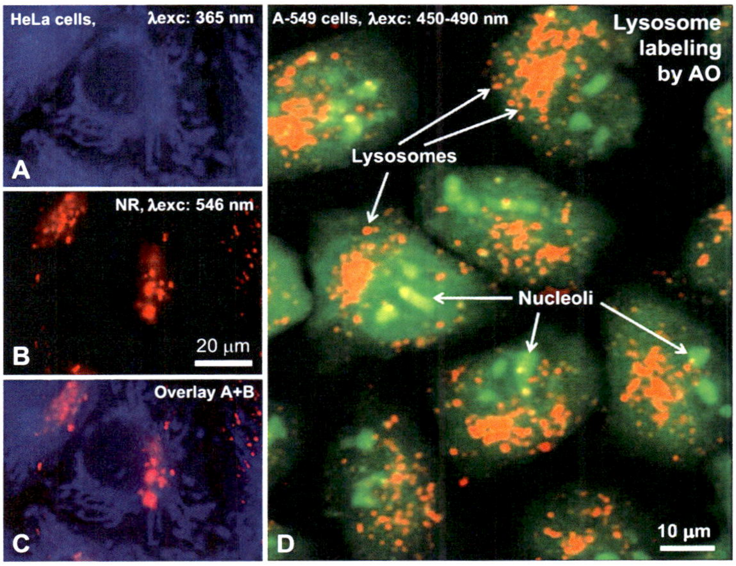

Fig. (15.13). A-C: Fluorescence images of living HeLa cells stained with 3 μg/mL neutral red (NR) for 15 min and observed under UV and blue excitation. **C:** overlay, showing red fluorescing lysosomes and blue autofluorescent mitochondria. **D:** Live human lung adenocarcinoma A-549 cells labeled with acridine orange (AO, 5 μg/mL, 15 min), washed in PBS and further incubated in DMEM for 15 min before observation. Red-orange metachromatic emission of lysosomes is clearly seen.

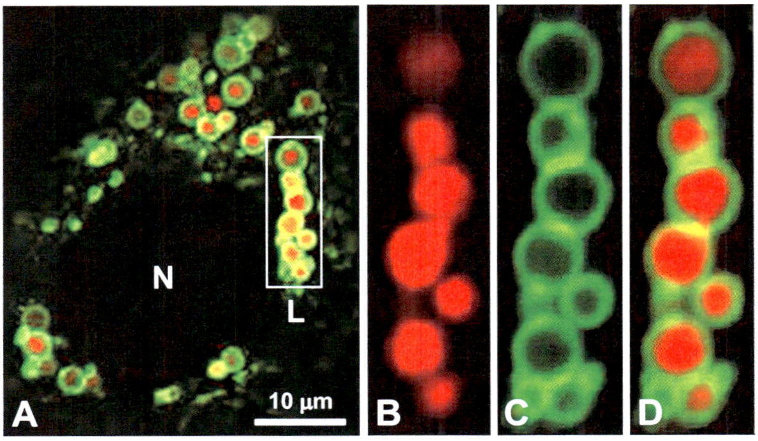

Fig. (15.14). Lysosome labeling. **A:** U_2OS cells transfected with CellLight® Lysosomes-GFP, BacMam 2.0 (green), labeled with LysoTracker® Deep Red (red), and observed using FITC/GFP and red excitation filtres (N: nucleus, L: lysosomes). **B-D:** Enlargement of the rectangular area in (A) showing LysoTracker® Deep Red localized in the lumen of the lysosome (B), membrane labeling by CellLight® Lysosome-GFP (C), and image overlay (D). The label in (C) is a fusion construct of Lamp1 (lysosomal-associated membrane protein 1) and GFP packaged in the insect baculovirus [217], which targets lysosome membranes (Reprinted from BioProbes® 69, Journal of Cell Biology Applications 2013, with permission of Thermo Fisher Scientific Inc.).

There are three main mechanisms of fluorescent lysosome labeling [16]:

1. With membrane permeant, acidotropic weak bases (AO, NR): entry into cells and then into lysosomes is simple diffusion. This requires moderate lipophilia: too lipophilic probes become trapped within the plasma membrane, and too hydrophilic probes remain in the aqueous extracellular medium. Consider NR with a pK: 6.9. The cationic dye (NR^+) with log P: − 1.1 is membrane impermeant. Under physiological conditions, about half NR is present as the permeant free base (NR^0) with log P: 1.9. Once inside the cell the equilibrium NR^+/NR^0 is re-established according to the local pH. Thiazine dyes, proflavine, quinacrine, chloroquine, 2-hydroxy-stilbamidine, *etc.*, show this uptake mechanism [19, 160].
2. Non-permeant probes can also enter the cell by adsorptive pinocytosis through strong adsorption to membrane proteins, partitioning into membrane lipids, internalization of the plasma membrane, and localization in lysosomes, examples being trypan blue, ovalbumin-Texas red, *etc.*
3. A third mode of entry for lysosomal probes is fluid-phase pinocytosis. Probes are not membrane permeant and remain in the aqueous extracellular fluid; they do not adsorp onto, or partition into, the plasma membrane. Examples are indigocarmine, Lucifer yellow CH, FITC- or TRITC-dextran, the styryl dye FM 4-64, *etc.* Another case of lysosome labeling by fluid-phase endocytosis is the demonstration of acid phosphatase by simultaneous treatment with naphthol AS-MX phosphate and fast red TR [17].

Fluorescent lysosome labeling by thiazine dyes has been described in live cells. Although thiazines are listed as non-fluorescent dyes under commonly available excitations, emission of thionine, toluidine blue and methylene blue was found in lysosomes [19], but the nature of fluorophores is not clear. Cytoplasmic granules (probably lysosomes and lysosome-derived bodies) from ascitic sarcoma 37 cells treated with toluidine blue show bright fluorescence during exposure from 0.5 to 10 min to UV excitation (changing emission from yellow to blue) [161]. Other dyes such as thionine, methylene blue and azur A also induce fluorescence to these granules [162], but as they are positive to staining reactions with oil red O, sudan black, and OsO_4, the participation of lipid droplets can not be excluded.

Thiazine dyes such as methylene blue and derivatives are known to be fluorescent (far red emission at about 685-695 nm under red excitation at 665 nm, see [163, 164]), but it is difficult that this feature can be seen in fluorescence microscopy using available excitations. Light absorption of most thiazine and oxazine dyes is very high from 580 to 660 nm, and thus emission in this spectral region should be severely quenched. Within live cells, other derivatives can also be formed during irradiation (see Chapter 10.3). Thiazines show significant photoreduction [165], and thus reduced thiazines could be responsible for unexpected microscopic emission at short wavelengths.

On the other hand, secretion (zymogen) granules (*i.e.* from pancreatic and parotid acinar cells) are acidic compartments due to a pH gradient generated by a proton pump. Metachromatic AO emission of zymogen granules is reversibly dissipated by protonophores (methylamine, NH_4Cl, monensin), but recovery of acidity requires ATP [166, 167]. Other membranous organelles are peroxisomes and glyoxysomes, which are involved in specific metabolic pathways: oxidase and peroxidase activity (peroxisomes), and glyoxylic acid cycle (glyoxysomes). Peroxisomes can be recognized by immunostaining of fixed cells with anti-PMP70 IgG followed by a fluorescent secondary IgG.

Finally, a greenish autofluorescence of lysosomes has been early reported, possibly due to flavines and lipofuscin pigments (see Chapter 7.3). Autofluorescent lysosomal residues (lipofuscin granules, age pigments) are universal biomarkers of cellular aging. An amphotheric quaternary amine that arises as a Schiff-base reaction from ethanolamine and retinaldehyde [168, 169] is a relevant lipofuscin fluorophore. Cytoplasmic granules from eosinophil leucocytes (an organelle that contains lysosome and peroxisome marker enzymes), also show a characteristic autofluorecence (Fig. **15.15A, B**).

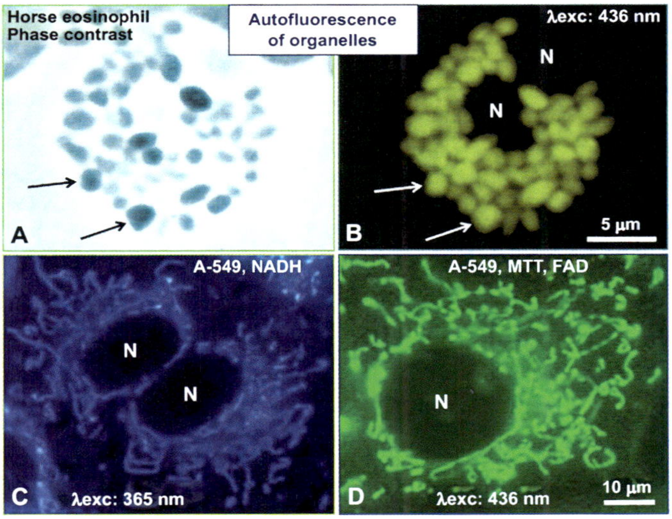

Fig. (15.15). A: Phase-contrast image of an unstained horse eosinophil with bilobed nucleus (N, N) and specific granules in the cytoplasm. **B:** The same cell showing greenish autofluorescence of eosinophil granules (arrows). **C:** Mitochondria in live A-549 cells showing blue autofluorescence due to the endogenous fluorophore, NAD(P)H. **D:** A-549 cell treated with 50 µg/mL MTT for 1.5 h, showing green autofluorescence of mitochondria due to the increase of FAD [42].

5.4. Mitochondria

Mitochondria autofluorescence is a well known feature of this organelle [170 - 172]. Reduced nicotinamide-adenine dinucleotides (NADH, NADPH), and oxidized flavine nucleotides (FAD, FMN) are the most important fluorophores responsible for autofluorescence of mitochondria. The λexc/λem for NADH and NADPH is ~340 nm/~460 nm, and for FAD and FMN is ~450 nm/~520-540 nm. The mitochondrial bioenergetics was early probed by monitoring the fluorescence of NADH [47, 173]. At present it is accepted that NAD(P)H is the main mitochondrial fluorophore responsible for the blue autofluorescence [172, 174].

When excited at 365 nm, blue autofluorescent mitochondria is clearly seen (Fig. **15.15C**) [42, 90, 175]. MTT viability assay induces considerable depletion of blue-fluorescing NADH, and significant increase of green FAD emission in mitochondria (Fig. **15.15D**). Thus, the consumption of reducing power in MTT treated cells results in an inversion of the coenzyme-dependant mitochondrial autofluorescence [42], which is an example of how image-based data are directly related to biochemical processes. Oxidative stress and ROS hyper-production generate a rapid decrease of mitochondrial NAD(P)H fluorescence followed by slow recovery [174]. Cell death can result when the coenzyme depletion induced by oxidative stress reaches severe damaging levels.

Early application of Janus green B staining was made for visualization of living mitochondria in bright-field microscopy, based on the mantenaince of the oxidized (colored) dye within mitochondria and its reduction to a colorless derivative in the cytoplasm. Numerous fluorescent, cationic probes for mitochondria have been described. Examples are astrazone red 6B, the styryl dye DASPMI, JC-1, orcein, quinacrine, rhodamine B, rhodamine 3B, rhodamine 6G, rhodamine 123, $DiOC_n(3)$ (with n from 2 to 6), safranine O, pinacyanol, tetracyclines, berberine, *etc*. [47, 176 - 178]. All of them are lipophilic probes with delocalized positive charge ($pK_a > 8.2$), and fulfill the QSAR requirements for localization in mitochondria [45].

In addition, several non-cationic mitochondrial probes (lipophilic, weak acids) such as anthraquinone, flavonol, neoflavone, and hydroxy-xanthene derivatives, have been also described, examples being alizarin, brazilin, eosin Y, fluorescein, gallein, hematein, hematoxylin, erythrosin, phloxin B, purpurin, quinalizarin, morin, umbelliferone, *etc*. [40]. In this case, somewhat higher concentration and incubation time must be used to label mitochondria.

One of the earliest mitochondrial probes is rhodamine 123 (Rh123) [47, 179, 180] (Fig. **15.16**). Mitochondria are clearly seen as green fluorescing threads and granules. Rh123 does not show cytotoxic effects but it has some phototoxicity

(Φ_Δ: ~0.01). In comparison, tetrabromo-Rh123 is a much more effective $^1O_2$ generator (Φ_Δ: 0.7). Further advances in mitochondria labels are the development of reactive probes such as MitoTracker red CMXRos carrying a thiol-reactive chloromethyl group that reacts with proteins and forms a covalent bond [181]. Then labeled cells can be fixed and subjected to additional procedures (*i.e.* immunofluorescence, FISH, electron microscopy).

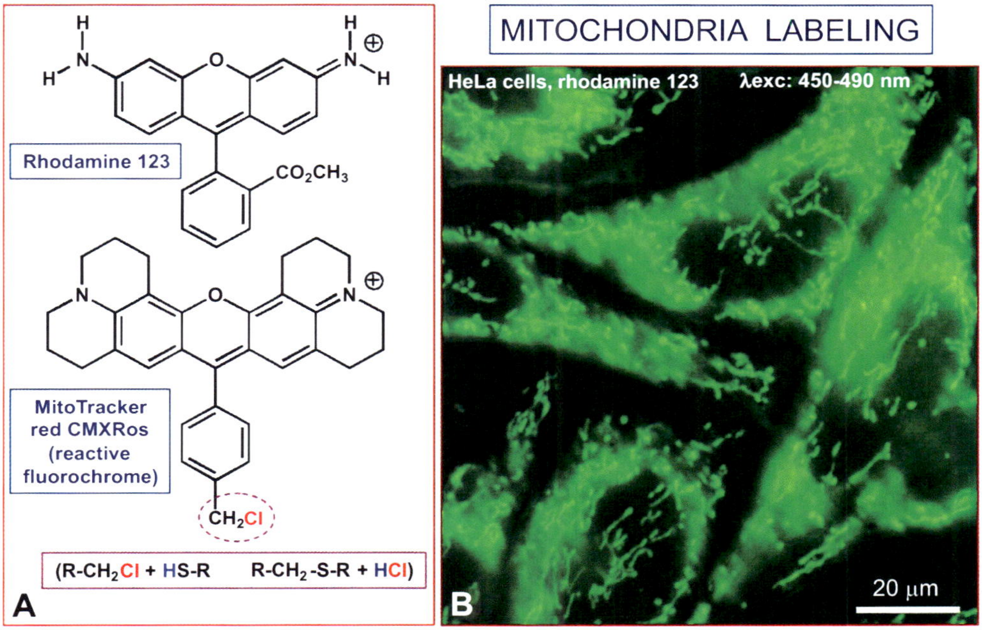

Fig. (15.16). Rhodamines commonly used for mitochondria labeling. **A:** Chemical structure of rhodamine 123 and the reactive probe MitoTracker Red CMXRos for covalent binding to -SH groups of proteins. **B:** Mitochondria of HeLa cells labeled with rhodamine 123 (6.5 μg/mL, 5 min).

On the other hand, as some mitocondrial probes and ion indicators are based in similar fluorophores, caution must be taken in the interpretation of colocalization results. For instance, the Ca^{2+} and Na^+ indicators Rhod-2 AM and CoroNa Red have the same xanthene fluorophore and structural features as the mitochondrial probes MitoFluor Green and MitoTracker Green FM. Thus, the QSAR models predict that these ion indicators could accumulate within mitochondria, irrespective of the presence of Ca or Na ions [11].

Most of the N-n-alkyl-AO derivatives are excellent fluorescent probes for mitochondria, mainly nonyl-AO (NAO) [53, 182]. Zimmermann's group [183] has suggested a mechanism of mitochondria labeling that involves the formation of an insoluble salt between lipophilic dye cations and the anionic phosphatidic acid derivative cardiolipin (used for syphilis diagnosis), a marker of the inner mitochondrial membrane that is selectively labeled with 10-alkyl AO [184]. The probable explanation for this fact is that owing to its unusual stereochemistry, NAO forms stable complexes with cardiolipin [185].

Mitochondria-selective interactions of cationic probes is dependent on the high trans-membrane potential (Δ_Ψ, inside negative) maintained by mitochondria [47, 178]. Dissipation of this potential by ionophores or inhibition of electron transport prevents the selective mitochondrial labeling. A special type of aggregates (multimers) of carbocyanine dyes was early identified by Jelly in 1937, and named J-aggregates. The most suitable dye system for visualizing mitochondrial Δ_Ψ is JC-1 and its J-aggregates, which display green and red fluorescence related to low and high membrane polarization, respectively [186]. The abbreviated name of JC-1 stands for the first J-aggregate-forming cationic dye that is membrane potential-sensitive, and correspond to 5,5',6,6'-tetrachloro-1,1',3,3'-tetraethyl-benzimidazolyl-carbocyanine iodide.

At present, a great importance is devoted to the implication of mitochondria in physiological and pharmacological studies [43, 187, 188]. This organelle is related to apoptosis and is a particular target for anti-cancer drugs [184, 189]. Berberine accumulation in mitochondria [177] shows great interest because the alkaloid induces fragmentation and dysfunction of this organelle leading to apoptosis [124], likely by blocking the adenine nucleotide translocator [190]. Interestingly, when the mitochondrial electric potential Δ_Ψ is collapsed, berberine fluorescence of mitochondria is abolished and it occurs in nucleoli and basophilic cytoplasm [124]. Mitochondria-targeted probes such as berberine derivatives with alkyl chains and conjugates with plastoquinones [191, 192] have revealed great potencial to analyze the physiology of this organelle, which is relevant for new approaches in cancer treatment [193, 194].

Fluorescent mitochondrial probes are often good photosensitizers for photodynamic therapy of cancer [195]. As endogenously biotinylated proteins occurs in mitochondria from eukaryotic cells, this organelle can be detected by methods using labeled streptavidin. Mitochondria are also detected in live cells by means of fluorescent fusion proteins (see Fig. **4.20**).

Fluorescent cationic probes such as pyronine Y and $DiOC_1(3)$ are used to visualize selectively mitochondria but not nuclei in live cells, although in fixed cells they stain chromatin DNA [14, 196]. A clear example of mitochondria labeling by $DiOC_1(3)$ is illustrated in Fig. (**15.17**).

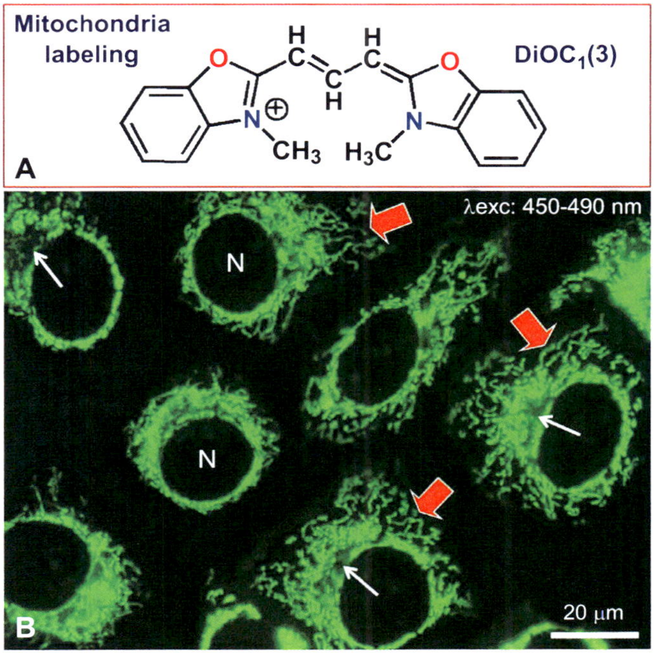

Fig. (15.17). A: Chemical structure of $DiOC_1(3)$. **B:** Living 3T3 murine fibroblasts treated with $DiOC_1(3)$ (15 μg/mL, 1 min), showing green mitochondria (red arrows). White arrows and N indicate the unlabeled Golgi region and nuclei, respectively.

Some protozoa (*i.e.* from the genus *Trypanosoma*, *Leishmania*, *Crithidia*, *etc*.) contain an unusual mitochondrial DNA body, the kinetoplast DNA (kDNA) [197] (see Chapter 12.5.7), which can be labeled by cationic mitochondria probes. The treatment of live *T. cruzi* epimastigotes with $DiOC_1(3)$ induces a bright green emission in kDNA under blue excitation (Fig. **15.18**). Also the mitochondrial probe Rh123 gives a bright fluorescence in kDNA [198], as well as other mitochondrial probes such as pyronine Y, berberine, pinacyanol and thioflavine T, which induce a selective emission in kDNA of live epimastigotes. Obviously, cationic probes must first penetrate the outer and inner membrane of the mitochondria for kDNA labeling.

On the other hand, after crossing the mitochondrial membranes, small amounts of acridines (AO), phenanthridines (EB), and aromatic diamidines (DAPI) induce abnormalities in this organelle from yeast, protozoa, and mammalian cells [199]. The trypanocidal drug EB binds to kDNA of *T. cruzi* and inhibits kDNA replication, inducing morphological changes and a dyskinetoplastic condition [200, 201], without adverse effects on nuclear DNA synthesis. This could be due to limitations for DNA binding based on the nucleosomal organization of chromatin.

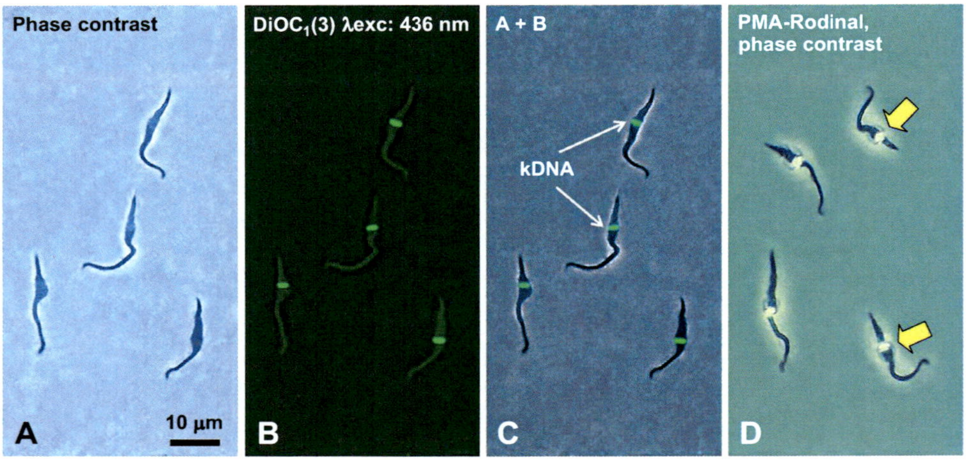

Fig. (15.18). A, B: Living *Trypanosoma cruzi* epimastigotes treated with 5 µg/mL $DiOC_1(3)$ for 0.5 min, fixed with formaldehyde vapor to immobilize the cells, and observed under phase contrast (A) and fluorescence microscopy (B). **C:** Merged A and B images showing kDNA. **D:** Methanol-fixed *T. cruzi* epimastigotes subjected to 2% phosphomolybdic acid followed by photographic developer (Rodinal®, Agfa) [218] to form molybdenum blue. Observe the high refringence and swelling of kDNA (yellow arrows).

5.5. Golgi Apparatus

Since the pioner work of Lipsky and Pagano [157] on fluorescent labeling of the Golgi apparatus in live cells, numerous investigations have been carried out. The golden standard probe for Golgi labeling is a fluorescent sphingolipid, namely NBD-C_6-ceramide, which is a widely applied vital probe [202, 203]. Its selectivity was early confirmed by colocalization with thiamine pyrophosphatase activity, and immunofluorescence against specific Golgi proteins [157]. It is worth to note that the Golgi apparatus of fixed cells can also be stained with this compound [204]. From the morphological point of view, the paranuclear localization and reticulo-granular appearance of the Golgi apparatus are also diagnostic criteria.

The mechanism by which the NBD-labeled ceramide enters and accumulate within cells is thought to be a transfer of the lipophilic probe to almost all intracellular membranes, followed by selective trapping in the Golgi apparatus due to the interaction with Golgi lipids. Among 27 fluorescent Golgi probes showing varied chemistry, five distinct classes of compound were described [44], including (1) weakly basic, (2) weakly acids (both lipophilic), (3) markers of adsorptive and (4) fluid-phase pinocytosis, and (5) lipid-partitioning probes. As occurs with NBD-ceramide, other Golgi probes require membrane recycling, ending up in the Golgi apparatus after 20-30 min postincubation of cells in the culture medium.

Chemical structure and labeling of the Golgi apparatus by NBD-C_6-ceramide is illustrated in Fig. (**15.19**) The green emission of NBD fluorophore (λexc: 460 nm, λem: 530 nm) permits the easy detection of this organelle in living cells, although some fluorescent background is commonly found in the cytoplasm. This probe is routinely administrated within liposomes or complexed with bovine serum albumin, but its direct used after dilution in DMSO and then in culture medium gives often good results. In all cases, postincubation for 30 min is required. When colocalization are made using UV excitation, the blue autofluorescence of mitochondria can be removed [90].

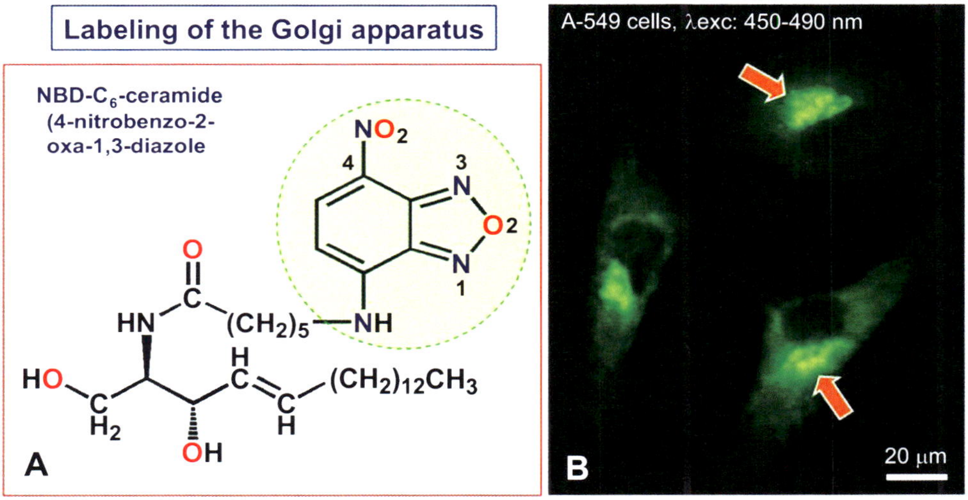

Fig. (15.19). A: Chemical structure of NBD-C_6-ceramide. The fluorophore has been highlighted in green. **B:** Golgi apparatus of A-549 cells (orange arrows) labeled by NBD-C_6-ceramide (25 μg/mL in DMEM, 30 min), followed by further incubation in fresh DMEM for 30 min (courtesy of M. Cañete and A. Villanueva).

On the other hand, both the Golgi apparatus and lipid droplets from cultured granulosa cells are labeled by treatment with high density lipoprotein (HDL)-Bodipy FI-C_{12}-cholesterol ester [205]. Chlorophyll-labeled oil micelles also allow detection of the Golgi apparatus and lipid droplets in living cells (see Fig. **4.10**). Labeled lectins and antibodies are used for detection of the Golgi apparatus in fixed cells.

A smart "off–on"-specific fluorescence probe (ANQ-IMC-6) for the Golgi apparatus of cancer cells has been recently designed, based on the use of a selective cyclooxygenase-2 (COX-2) inhibitor (IMC) linked to a fluorophore (ANQ) [206]. COX-2 is a biomarker of virtually all cancer cell lines and localizes in the Golgi apparatus. Free in aqueous solution, ANQ-IMC-6 mainly exists in a folded conformation where probe fluorescence is quenched through photoinduced electron transfer between the fluorophore ANQ and the indomethacin recognition group IMC. Fluorescence is turned on (λexc: 457 nm, λem: 547 nm) when ANQ-IMC-6 is forced to adopt the unfolded state following binding to COX-2 in the Golgi apparatus of cancer cells, offering an important tool for early cancer diagnosis.

5.6. Lipid Droplets

The lipid droplet is now considered an important cell organelle, which is made of a hydrophobic lipid core (mainly formed by triglycerides) surrounded by a phospholipid monolayer, representing a lipid storage compartment of the cell. Fluorescent lipophilic probes are commonly used to visualize lipid droplets in live cells [32]. Several of them can be employed after fixation [31].

Lipid droplets appear labeled after treatment with HDL-Bodipy FI-C_{12}-cholesterol ester [205], as well as after incubation with chlorophyll-oil micelles (see Fig. **4.10**). Lipophilic Bodipy dyes are now employed to reveal lipid droplets [207, 208] (see Fig. **10.12**). Likewise, due to its lipophilic nature, the formazan from reduction of MTT by live cells remains trapped within lipidic substrates, typically lipid droplets and the Golgi apparatus, giving staining and fluorescence reactions (see Figs. **14.5** and **14.6**).

5.7. Localization of Drugs

The possibility of using fluorescence imaging to reveal the uptake and localization of xenobiotics and drugs in live cells is an important biomedical application (*i.e.* toxic agents, antipsichotic drugs, photosensitizers, anticancer agents, *etc.*) [10, 33, 41, 64]. Cultured cells treated with zinc(II)-phthalocyanine (ZnPc) show the red fluorescence localized in the Golgi apparatus (Fig. **15.20**).

The signal colocalizes with the green NBD-C_6-ceramide emission that identify this organelle, and contrasts with the blue autofluorescence of mitochondria [90, 175, 209]. Some red granules dispersed in the cytoplasm do not colocalize with NBD-C_6-ceramide, suggesting that this granular component could be endosomes, lipid droplets, or ZnPc-containing liposomes.

Camptothecin is an alkaloid isolated from the stem wood of the Chinese tree, *Camptotheca acuminata* (happy tree, cancer tree) with antileukemic and antitumor activity. Camptothecin binds to DNA inducing photo-activated chain breaks [210], and like other derivatives, it intercalates into DNA in the ternary topoisomerase I (topo I)-DNA-drug complex [211 - 213]. Topo I is responsible for relaxation of supercoiled DNA during replication. The camptothecin derivative, irinotecan (CPT-11, Fig. **15.21A, B**) is converted *in vivo* into its most active metabolite SN-38 by esterase cleavage, which stabilizes the resulting ternary complex and inhibites DNA synthesis.

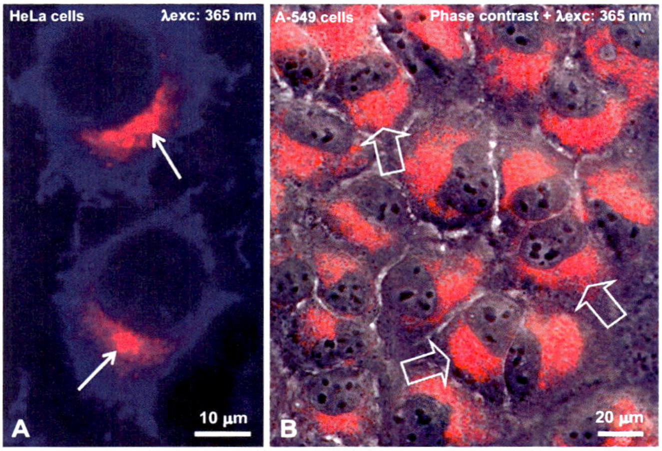

Fig. (15.20). A: HeLa cells treated for 18 h with 10^{-6} M ZnPc in DPPC liposomes. The Golgi apparatus is labeled in red by ZnPc (arrows). The blue autofluorescence of mitochondria is due to NAD(P)H. **B:** A-549 cells treated for 24 h with 10^{-6} M ZnPc in DPPC liposomes. A massive accumulation of the red-fluorescing drug is found in the Golgi apparatus (arrows) and in numerous vesicles, which likely are liposomes, lipid droplets or lysosomes. The image is an overlay of a phase contrast picture and the red ZnPc signal isolated from the original fluorescence micrograph under UV excitation.

On account of its blue fluorescence under UV excitation, the localization of irinotecan is visualized in the endosome-lysosome compartment (Fig. **15.21C, D**). From this site irinotecan should release SN-38 to reach the nucleus in order to exert its cytotoxic effect as topo I poison. Other derivatives such as topotecan show different localizations: chromatin in MCF-7 cells [214], and mitochondria in HT-29 cells [215].

Finally, a precursor metabolite may be delivered into cells to form another metabolite that in turn is localized by fluorescence microscopy. This is the case of the precursor 5-aminolevulinic acid (5-ALA) that forms the physiological photosensitizer protoporphyrin IX (PpIX). 5-ALA and succinyl-coenzyme A produce first porphobilinogen and then PpIX, the later being localized in mitochondria and also diffusely in the cytoplasm and between confluent cells (Fig. **15.22**).

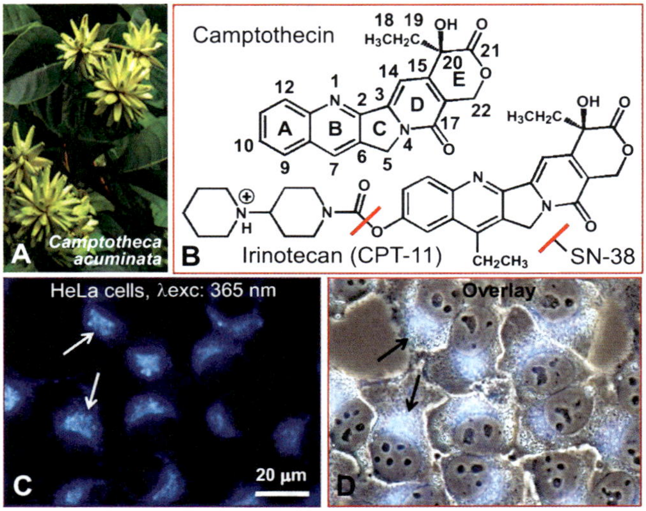

Fig. (15.21). A: The Chinese tree *Camptotheca acuminata*. **B:** Chemical structure of camptothecin (numbered) and irinotecan. The red line indicate the esterase breackage to form the active metabolite SN-38 (7-ethyl-10-hydroxy-camptothecin). **C, D:** Fluorescence image (C) and phase contrast overlay (D) of HeLa cells treated with 100 µM irinotecan in DPPC liposomes for 24 h. Observe the massive accumulation of the drug in lysosomes surrounding the Golgi region (arrows) (courtesy of A. Villanueva and A. Casadó).

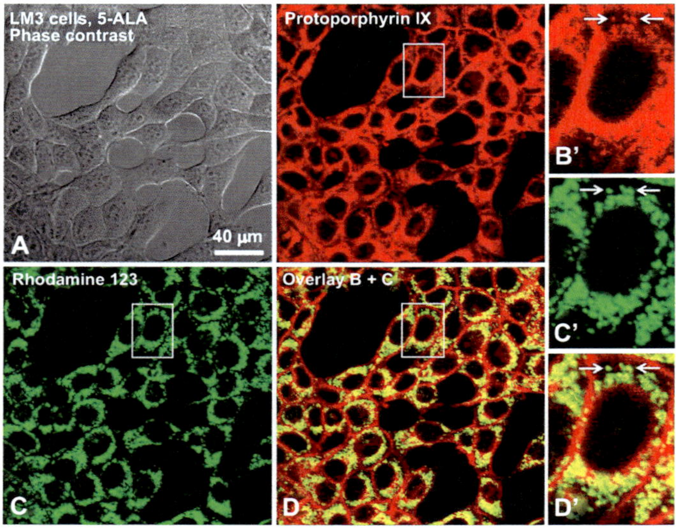

Fig. (15.22). A-D: LM3 cells treated with 5-aminolevulinic acid (5-ALA) to induce the formation of protoporphyrin IX **(B)**, followed by labeling with rhodamine 123 **(C)**, and image overlay **(D)**, showing colocalization of the particulate red signal (mitochondria) but not of the diffuse red signal in cytoplasm. B'-D': Zoom of images corresponding to rectangles in **B-D**. Arrows indicate mitochondria (courtesy of A. Casas).

CONFLICT OF INTEREST

The author (editor) declares no conflict of interest, financial or otherwise.

ACKNOWLEDGEMENTS

This work was supported by a grant from the Ministerio de Economía y Competitividad (CTQ2013-48767-C3-3-R), Spain.

REFERENCES

[1] Horobin RW. Where do Dyes Go, Inside Living Cells and Tissues?Colour Science. Leeds: Leeds University 1999; Vol. 1: pp. 248-58.

[2] Beddart GS, West MA, Eds. Fluorescent Probes. New York, London: Academic Press 1981.

[3] Waggoner AS. Fluorescent Probes for Analysis of Cell Structure, Function, and Health by Flow and Imaging Cytometry.Applications of Fluorescence in the Biomedical Sciences. New York: Alan R. Liss 1986; pp. 3-28.

[4] Arndt-Jovin DJ, Jovin TM. Fluorescence labeling and microscopy of DNA. Methods Cell Biol 1989; 30: 417-48.
[http://dx.doi.org/10.1016/S0091-679X(08)60989-9] [PMID: 2467179]

[5] Kashiwasake M, Imai K. Interaction of fluorescent labels with cells and its analytical potential. Organelle specific staining. Appl Fluor Technol 1991; 3(1): 11-27.

[6] Horobin RW, Rashid-Doubell F. Fluorescent Probes. Where do they Go, and Why?Biotechnology Applications of Microinjection, Microscopic Imaging, and Fluorescence. New York: Plenum Press 1993.
[http://dx.doi.org/10.1007/978-1-4615-2828-9_8]

[7] Whitaker M. Fluorescence Imaging in Living Cells.Cell Biology A Laboratory Handbook. New York, London: Academic Press 1994; pp. 37-43.

[8] Haugland RP. Handbook of Fluorescent Probes and Research Products. 9th ed., Eugene, OR: Molecular Probes, Inc. 2002.

[9] Horobin RW, Kiernan JA, Eds. Conn's Biological Stains A Handbook of Dyes, Stains and Fluorochromes for Use in Biology and Medicine. 10th ed., Oxford, UK: BIOS Scientific Publishers 2002.

[10] Stockert JC, Juarranz A, Villanueva A, *et al.* Photodynamic therapy: selective uptake of photosensitizing drugs into tumor cells. Curr Top Pharmacol 2004; 8: 185-217.

[11] Horobin RW, Stockert JC, Rashid-Doubell F. Uptake and localisation of small-molecule fluorescent probes in living cells: a critical appraisal of QSAR models and a case study concerning probes for DNA and RNA. Histochem Cell Biol 2013; 139(5): 623-37.
[http://dx.doi.org/10.1007/s00418-013-1090-0] [PMID: 23542926]

[12] Chang HC. The so-called neutral red "vacuome" and the Golgi apparatus. Anat Rec 1935; 62: 95-103.
[http://dx.doi.org/10.1002/ar.1090620108]

[13] Barbosa P, Peters TM. The effects of vital dyes on living organisms with special reference to methylene blue and neutral red. Histochem J 1971; 3(1): 71-93.
[http://dx.doi.org/10.1007/BF01686508] [PMID: 4106597]

[14] Alvarez M, Villanueva A, Acedo P, Cañete M, Stockert JC. Cell death causes relocalization of photosensitizing fluorescent probes. Acta Histochem 2011; 113(3): 363-8.
[http://dx.doi.org/10.1016/j.acthis.2010.01.008] [PMID: 20138336]

[15] Wolf MK, Aronson SB. Growth fluorescence and metachromasy of cells cultured in the presence of acridine orange. J Histochem Cytochem 1961; 9: 22-9.
[http://dx.doi.org/10.1177/9.1.22] [PMID: 14007945]

[16] Swanson J. Fluorescent labeling of endocytic compartments. Methods Cell Biol 1989; 29: 137-51.
[http://dx.doi.org/10.1016/S0091-679X(08)60192-2] [PMID: 2464119]

[17] Espada J, Horobin RW, Stockert JC. Fluorescent cytochemistry of acid phosphatase and demonstration of fluid-phase endocytosis using an azo dye method. Histochem Cell Biol 1997; 108(6): 481-7.
[http://dx.doi.org/10.1007/s004180050188] [PMID: 9450630]

[18] Zdolsek JM, Olsson GM, Brunk UT. Photooxidative damage to lysosomes of cultured macrophages by acridine orange. Photochem Photobiol 1990; 51(1): 67-76.
[http://dx.doi.org/10.1111/j.1751-1097.1990.tb01685.x] [PMID: 2304980]

[19] Rashid F, Horobin RW, Williams MA. Predicting the behaviour and selectivity of fluorescent probes for lysosomes and related structures by means of structure-activity models. Histochem J 1991; 23(10): 450-9.
[http://dx.doi.org/10.1007/BF01041375] [PMID: 1743993]

[20] Horobin RW. Structure-staining relationships in histochemistry and biological staining. Part 3. Some comments on the intentional and artifactual staining of lipids. Acta Histochem Suppl 1981; 24 (Suppl.): 237-46.
[PMID: 6165042]

[21] Horobin RW. Histochemistry An Explanatory Outline of Histochemistry and Biophysical Staining. Stuttgart, New York: Gustav Fischer 1982.

[22] Curtis D, Horobin RW. Chromosome banding: specification of structural features of dyes giving rise to G-banding. Histochem J 1982; 14(6): 911-28.
[http://dx.doi.org/10.1007/BF01005233] [PMID: 6184333]

[23] Juarranz A, Horobin RW, Proctor GB. Prediction of *in situ* fluorescence of histochemical reagents using a structure-staining correlation procedure. Histochemistry 1986; 84(4-6): 426-31.
[http://dx.doi.org/10.1007/BF00482974] [PMID: 3721913]

[24] Horobin RW, Stockert JC, Rashid-Doubell F. Uptake and localization mechanisms of fluorescent and colored lipid probes. Part 2. QSAR models that predict localization of fluorescent probes used to identify ("specifically stain") various biomembranes and membranous organelles. Biotech Histochem 2015; 90(4): 241-54.
[http://dx.doi.org/10.3109/10520295.2015.1005129] [PMID: 25801297]

[25] Horobin RW, Stockert JC, Rashid-Doubell F. Uptake and localization mechanisms of fluorescent and colored lipid probes. Part 3. Protocols for predicting intracellular localization of lipid probes using QSAR models. Biotech Histochem 2015; 90(4): 255-63.
[http://dx.doi.org/10.3109/10520295.2015.1006680] [PMID: 25801295]

[26] Horobin RW, Stockert JC, Rashid-Doubell F. Fluorescent cationic probes for nuclei of living cells: why are they selective? A quantitative structure-activity relations analysis. Histochem Cell Biol 2006; 126(2): 165-75.
[http://dx.doi.org/10.1007/s00418-006-0156-7] [PMID: 16463179]

[27] Hansch C, Leo A. Substituent Constants for Correlation Analysis in Chemistry and Biology. New York: Wiley 1979.

[28] Hansch C, Leo A. Exploring QSAR: Fundamentals and Applications in Chemistry and Biology. Washington, DC: American Chemical Society 1995; p. 160.

[29] Black SD, Mould DR. Development of hydrophobicity parameters to analyze proteins which bear post- or cotranslational modifications. Anal Biochem 1991; 193(1): 72-82.
[http://dx.doi.org/10.1016/0003-2697(91)90045-U] [PMID: 2042744]

[30] Dapson RW. Dye-tissue interactions: mechanisms, quantification and bonding parameters for dyes used in biological staining. Biotech Histochem 2005; 80(2): 49-72. [http://dx.doi.org/10.1080/10520290500219982] [PMID: 16195171]

[31] Stockert JC, Abasolo MI, Blázquez-Castro A, Horobin RW, Revilla M, Lombardo DM. Selective labeling of lipid droplets in aldehyde fixed cell monolayers by lipophilic fluorochromes. Biotech Histochem 2010; 85(5): 277-83. [http://dx.doi.org/10.3109/10520290903196183] [PMID: 20560872]

[32] Horobin RW, Stockert JC. Uptake and localization mechanisms of fluorescent and colored lipid probes. 1. Physicochemistry of probe uptake and localization, and the use of QSAR models for selectivity prediction. Biotech Histochem 2011; 86(6): 379-93. [http://dx.doi.org/10.3109/10520295.2010.515489] [PMID: 20818918]

[33] Delieu JM, Horobin RW, Duguid JK. Exploring the relationship of drug-induced neutrophil immaturity & haematological toxicity to drug chemistry using quantitative structure-activity models. Med Chem 2009; 5(1): 7-14. [http://dx.doi.org/10.2174/157340609787049307] [PMID: 19149645]

[34] Gurr E. Synthetic Dyes in Biology, Medicine and Chemistry. London, New York: Academic Press 1971.

[35] Green FJ. The Sigma-Aldrich Handbook of Stains, Dyes and Indicators. Milwaukee, Wisconsin: Aldrich Chemical Company 1990.

[36] Horobin RW, Proctor GB. A numerical method of selecting stains for epoxy-embedded tissues. Trans Roy Microsc Soc 1990; 1: 199-202.

[37] Horobin RW. Uptake, distribution, and accumulation of dyes and fluorescent probes within living cells. A structure-activity modelling approach. Adv Colour Sci Technol 2001; 4: 101-7.

[38] Tejedor-Estrada R, Nonell S, Teixidó J, *et al.* An artificial neural network model for predicting the subcellular localization of photosensitisers for photodynamic therapy of solid tumours. Curr Med Chem 2012; 19(15): 2472-82. [http://dx.doi.org/10.2174/092986712800269290] [PMID: 22420336]

[39] Trapp S, Horobin RW. A predictive model for the selective accumulation of chemicals in tumor cells. Eur Biophys J 2005; 34(7): 959-66. [http://dx.doi.org/10.1007/s00249-005-0472-1] [PMID: 15895221]

[40] Rashid F, Horobin RW. Accumulation of fluorescent non-cationic probes in mitochondria of cultured cells: observations, a proposed mechanism, and some implications. J Microsc 1991; 163(Pt 2): 233-41. [http://dx.doi.org/10.1111/j.1365-2818.1991.tb03175.x] [PMID: 1719208]

[41] Colston J, Horobin RW, Rashid-Doubell F, Pediani J, Johal KK. Why fluorescent probes for endoplasmic reticulum are selective: an experimental and QSAR-modelling study. Biotech Histochem 2003; 78(6): 323-32. [http://dx.doi.org/10.1080/10520290310001646659] [PMID: 15473580]

[42] Stockert JC, Blázquez-Castro A, Cañete M, Horobin RW, Villanueva A. MTT assay for cell viability: Intracellular localization of the formazan product is in lipid droplets. Acta Histochem 2012; 114(8): 785-96. [http://dx.doi.org/10.1016/j.acthis.2012.01.006] [PMID: 22341561]

[43] D'Souza GG, Cheng SM, Boddapati SV, Horobin RW, Weissig V. Nanocarrier-assisted sub-cellular targeting to the site of mitochondria improves the pro-apoptotic activity of paclitaxel. J Drug Target 2008; 16(7): 578-85. [http://dx.doi.org/10.1080/10611860802228855] [PMID: 18686127]

[44] Rashid-Doubell F, Horobin RW. Selection of Fluorescent Golgi Complex Probes Using Structure-Activity Models.Biotechnology Applications of Microinjection, Microscopic Imaging, and Fluorescence. New York: Plenum Press 1993; pp. 73-8.

[http://dx.doi.org/10.1007/978-1-4615-2828-9_10]

[45] Rashid F, Horobin RW. Interaction of molecular probes with living cells and tissues. Part 2. A structure-activity analysis of mitochondrial staining by cationic probes, and a discussion of the synergistic nature of image-based and biochemical approaches. Histochemistry 1990; 94(3): 303-8. [PMID: 1698190]

[46] Agrafiotis DK, Bandyopadhyay D, Wegner JK, Vlijmen Hv. Recent advances in chemoinformatics. J Chem Inf Model 2007; 47(4): 1279-93. [http://dx.doi.org/10.1021/ci700059g] [PMID: 17511441]

[47] Chen LB. Fluorescent labeling of mitochondria. Methods Cell Biol 1989; 29: 103-23. [http://dx.doi.org/10.1016/S0091-679X(08)60190-9] [PMID: 2643756]

[48] Celis JE, Dejgaard K. Vital Staining of Mitochondria with Rhodamine 123 and of Acidic Organelles with Acridine Orange.Cell Biology: A Laboratory Handbook. New York, London: Academic Press 1994; pp. 394-6.

[49] Poot M, Zhang YZ, Krämer JA, *et al.* Analysis of mitochondrial morphology and function with novel fixable fluorescent stains. J Histochem Cytochem 1996; 44(12): 1363-72. [http://dx.doi.org/10.1177/44.12.8985128] [PMID: 8985128]

[50] Cañete M, Juarranz A, López-Nieva P, Alonso-Torcal C, Villanueva A, Stockert JC. Fixation and permanent mounting of fluorescent probes after vital labelling of cultured cells. Acta Histochem 2001; 103(2): 117-26. [http://dx.doi.org/10.1078/0065-1281-00594] [PMID: 11368093]

[51] García-Díaz M, Nonell S, Villanueva A, *et al.* Do folate-receptor targeted liposomal photosensitizers enhance photodynamic therapy selectivity? Biochim Biophys Acta, Membranes 1808; 1808: 1063-71.

[52] Terasaki M. Probes for Endoplasmic Reticulum.Fluorescent and Luminescent Probes for Biological Activity. London: Academic Press 1993; pp. 120-3.

[53] Zimmermann HW. Physicochemical and cytochemical investigations on the binding of ethidium and acridine dyes to DNA and to organelles in living cells. Angew Chem Int Ed Engl 1986; 25: 115-30. [http://dx.doi.org/10.1002/anie.198601153]

[54] Soriano J, Stockert JC, Villanueva A, Cañete M. Cell uptake of Zn(II)-phthalocyanine-containing liposomes by clathrin-mediated endocytosis. Histochem Cell Biol 2010; 133(4): 449-54. [http://dx.doi.org/10.1007/s00418-010-0679-9] [PMID: 20191285]

[55] Soriano J, Villanueva A, Stockert JC, Cañete M. Vehiculization determines the endocytic internalization mechanism of Zn(II)-phthalocyanine. Histochem Cell Biol 2013; 139(1): 149-60. [http://dx.doi.org/10.1007/s00418-012-1012-6] [PMID: 22899479]

[56] Boix-Garriga E, Acedo P, Casadó A, *et al.* Poly(D,L-lactide-co-glycolide) nanoparticles as delivery agents for photodynamic therapy: enhancing singlet oxygen release and phototoxicity by surface PEG coating. Nanotechnology 2015; 26: 365104 (16pp).

[57] Pederson T. The molecular cytology of gene expression: fluorescent RNA as both a stain and tracer *in vivo*. Eur J Histochem 2004; 48(1): 57-64. [PMID: 15145776]

[58] Dirks RW, Tanke HJ. Advances in fluorescent tracking of nucleic acids in living cells. Biotechniques 2006; 40(4): 489-96. [http://dx.doi.org/10.2144/000112121] [PMID: 16629396]

[59] Armitage BA. Imaging of RNA in live cells. Curr Opin Chem Biol 2011; 15(6): 806-12. [http://dx.doi.org/10.1016/j.cbpa.2011.10.006] [PMID: 22055496]

[60] Hurley LH. DNA and its associated processes as targets for cancer therapy. Nat Rev Cancer 2002; 2(3): 188-200. [http://dx.doi.org/10.1038/nrc749] [PMID: 11990855]

[61] Boutorine AS, Novopashina DS, Krasheninina OA, Nozeret K, Venyaminova AG. Fluorescent probes for nucleic acid visualization in fixed and live cells. Molecules 2013; 18(12): 15357-97. [http://dx.doi.org/10.3390/molecules181215357] [PMID: 24335616]

[62] Coleman AW, Goff LJ. Applications of fluorochromes to pollen biology. I. Mithramycin and 4′,6-diamidino-2-phenylindole (DAPI) as vital stains and for quantitation of nuclear DNA. Stain Technol 1985; 60(3): 145-54. [http://dx.doi.org/10.3109/10520298509113905] [PMID: 2411019]

[63] Bach PH, Reynolds CH, Brant S. Fluorescent Probes for the Assessment of the Subcellular Effects of Chemicals and Disease Processes *in vivo* and *in vitro*.Biotechnology Applications of Microinjection, Microscopic Imaging and Fluorescence. New York, London: Plenum Press 1993; pp. 79-97. [http://dx.doi.org/10.1007/978-1-4615-2828-9_11]

[64] Hardin JA, Sherr DH, DeMaria M, Lopez PA. A simple fluorescence method for surface antigen phenotyping of lymphocytes undergoing DNA fragmentation. J Immunol Methods 1992; 154(1): 99-107. [http://dx.doi.org/10.1016/0022-1759(92)90217-H] [PMID: 1401949]

[65] Gregoire M, Hernandez-Verdun D, Bouteille M. Visualization of chromatin distribution in living PTO cells by Hoechst 33342 fluorescent staining. Exp Cell Res 1984; 152(1): 38-46. [http://dx.doi.org/10.1016/0014-4827(84)90228-3] [PMID: 6201374]

[66] Sawicki W, Moskalewski S. Hoechst 33342 staining coupled with conventional histological technique. Stain Technol 1989; 64(4): 191-6. [http://dx.doi.org/10.3109/10520298909106998] [PMID: 2482997]

[67] Debey P, Adenot P, Geze M, Besombes D. Following chromatin in living embryos. Appl Fluor Technol 1991; 3(2): 17-22.

[68] Darzynkiewicz Z, Traganos F, Kapuscinski J, Staiano-Coico L, Melamed MR. Accessibility of DNA *in situ* to various fluorochromes: relationship to chromatin changes during erythroid differentiation of Friend leukemia cells. Cytometry 1984; 5(4): 355-63. [http://dx.doi.org/10.1002/cyto.990050411] [PMID: 6468176]

[69] Lyles MB, Cameron IL. Interactions of the DNA intercalator acridine orange, with itself, with caffeine, and with double stranded DNA. Biophys Chem 2002; 96(1): 53-76. [http://dx.doi.org/10.1016/S0301-4622(02)00036-4] [PMID: 11975993]

[70] Delic J, Coppey J, Magdelenat H, Coppey-Moisan M. Impossibility of acridine orange intercalation in nuclear DNA of the living cell. Exp Cell Res 1991; 194(1): 147-53. [http://dx.doi.org/10.1016/0014-4827(91)90144-J] [PMID: 2015848]

[71] Von Hippel PH, McGhee JD. DNA-protein interactions. Annu Rev Biochem 1972; 41(10): 231-300. [http://dx.doi.org/10.1146/annurev.bi.41.070172.001311] [PMID: 4570958]

[72] Wang AH. Interactions between antitumor drugs and DNA. Nucleic Acids Mol Biol 1987; 1: 53-69. [http://dx.doi.org/10.1007/978-3-642-46596-3_3]

[73] Di Marco A, Arcamone F, Zunino F. Daunomycin (Daunorubicin) and Adriamycin and Structural Analogues: Biological Activity and Mechanism of Action. Antibiotics. Berlin, New York: Springer-Verlag 1975; Vol. 3: pp. 101-28. [http://dx.doi.org/10.1007/978-3-642-46304-4_8]

[74] Egorin MJ, Clawson RE, Cohen JL, Ross LA, Bachur NR. Cytofluorescence localization of anthracycline antibiotics. Cancer Res 1980; 40(12): 4669-76. [PMID: 6934029]

[75] Minotti G, Menna P, Salvatorelli E, Cairo G, Gianni L. Anthracyclines: molecular advances and pharmacologic developments in antitumor activity and cardiotoxicity. Pharmacol Rev 2004; 56(2): 185-229. [http://dx.doi.org/10.1124/pr.56.2.6] [PMID: 15169927]

[76] Crissman HA, Stevenson AP, Orlicky DJ, Kissane RJ. Detailed studies on the application of three fluorescent antibiotics for DNA staining in flow cytometry. Stain Technol 1978; 53(6): 321-30. [http://dx.doi.org/10.3109/10520297809111954] [PMID: 89719]

[77] Smith PJ, Blunt N, Wiltshire M, *et al.* Characteristics of a novel deep red/infrared fluorescent cell-permeant DNA probe, DRAQ5, in intact human cells analyzed by flow cytometry, confocal and multiphoton microscopy. Cytometry 2000; 40(4): 280-91. [http://dx.doi.org/10.1002/1097-0320(20000801)40:4<280::AID-CYTO4>3.0.CO;2-7] [PMID: 10918279]

[78] Bell DH. Characterization of the fluorescence of the antitumor agent, mitoxantrone. Biochim Biophys Acta 1988; 949(1): 132-7. [http://dx.doi.org/10.1016/0167-4781(88)90063-2] [PMID: 3334848]

[79] Njoh KL, Patterson LH, Zloh M, *et al.* Spectral analysis of the DNA targeting bisalkylaminoanthraquinone DRAQ5 in intact living cells. Cytometry A 2006; 69(8): 805-14. [http://dx.doi.org/10.1002/cyto.a.20308] [PMID: 16969814]

[80] Islam SA, Neidle S, Gandecha BM, Partridge M, Patterson LH, Brown JR. Comparative computer graphics and solution studies of the DNA interaction of substituted anthraquinones based on doxorubicin and mitoxantrone. J Med Chem 1985; 28(7): 857-64. [http://dx.doi.org/10.1021/jm00145a003] [PMID: 4009608]

[81] Gao XL, Mirau P, Patel DJ. Structure refinement of the chromomycin dimer-DNA oligomer complex in solution. J Mol Biol 1992; 223(1): 259-79. [http://dx.doi.org/10.1016/0022-2836(92)90730-8] [PMID: 1731073]

[82] Strickland JA, Marzilli LG, Gay KM, Wilson WD. Porphyrin and metalloporphyrin binding to DNA polymers: rate and equilibrium binding studies. Biochemistry 1988; 27(24): 8870-8. [http://dx.doi.org/10.1021/bi00424a027] [PMID: 3242614]

[83] Fiel RJ. Porphyrin-nucleic acid interactions: a review. J Biomol Struct Dyn 1989; 6(6): 1259-74. [http://dx.doi.org/10.1080/07391102.1989.10506549] [PMID: 2684219]

[84] Hui XW, Gresh N, Pullman B. Modelling of the binding specificity in the interactions of cationic porphyrins with DNA. Nucleic Acids Res 1990; 18(5): 1109-14. [http://dx.doi.org/10.1093/nar/18.5.1109] [PMID: 2320411]

[85] Villanueva A, Juarranz A, Diaz V, Gomez J, Cañete M. Photodynamic effects of a cationic mesosubstituted porphyrin in cell cultures. Anticancer Drug Des 1992; 7(4): 297-303. [PMID: 1510799]

[86] Georgiou GN, Ahmet MT, Houlton A, Silver J, Cherry RJ. Measurement of the rate of uptake and subcellular localization of porphyrins in cells using fluorescence digital imaging microscopy. Photochem Photobiol 1994; 59(4): 419-22. [http://dx.doi.org/10.1111/j.1751-1097.1994.tb05058.x] [PMID: 8022883]

[87] Snyder JW, Lambert JD, Ogilby PR. 5,10,15,20-tetrakis(*N*-methyl-4-pyridyl)-21*H*,23*H*-porphine (TMPyP) as a sensitizer for singlet oxygen imaging in cells: characterizing the irradiation-dependent behavior of TMPyP in a single cell. Photochem Photobiol 2006; 82(1): 177-84. [http://dx.doi.org/10.1562/2005-05-30-RA-553] [PMID: 16351325]

[88] Hatz S, Lambert JD, Ogilby PR. Measuring the lifetime of singlet oxygen in a single cell: addressing the issue of cell viability. Photochem Photobiol Sci 2007; 6(10): 1106-16. [http://dx.doi.org/10.1039/b707313e] [PMID: 17914485]

[89] Villanueva A, Stockert JC, Cañete M, Acedo P. A new protocol in photodynamic therapy: enhanced tumour cell death by combining two different photosensitizers. Photochem Photobiol Sci 2010; 9(3): 295-7. [http://dx.doi.org/10.1039/b9pp00153k] [PMID: 20221454]

[90] Stockert JC, Vanzulli SI, Cañete M, *et al.* Regression of the murine LM3 tumor by repeated

photodynamic therapy with *meso*-tetrakis (4-*N,N,N*-trimethylanilinium) porphine. J Porphyr Phthalocyanines 2009; 13: 560-6.
[http://dx.doi.org/10.1142/S1088424609000577]

[91] Villanueva A, Cañete M, Juarranz A, Stockert JC. Cytochemical study of the fluorescence reaction of chromatin by the porphyrin derivative meso-tetra (4-N-methylpyridyl) porphine. Basic Appl Histochem 1987; 31(1): 9-16.
[PMID: 3593176]

[92] Juarranz A, Villanueva A, Cañete M, Stockert JC. Fluorescent porphyrin counterstaining of chromatin DNA in conjunction with immunofluorescence methods using FITC-labelled antibodies. J Microsc 1996; 182(Pt 1): 46-9.
[http://dx.doi.org/10.1046/j.1365-2818.1996.114394.x] [PMID: 8632446]

[93] Dobrucki JW. Interaction of oxygen-sensitive luminescent probes $Ru(phen)_3^{2+}$ and $Ru(bipy)_3^{2+}$ with animal and plant cells *in vitro*. Mechanism of phototoxicity and conditions for non-invasive oxygen measurements. J Photochem Photobiol B 2001; 65(2-3): 136-44.
[http://dx.doi.org/10.1016/S1011-1344(01)00257-3] [PMID: 11809371]

[94] Gill MR, Garcia-Lara J, Foster SJ, Smythe C, Battaglia G, Thomas JA. A ruthenium(II) polypyridyl complex for direct imaging of DNA structure in living cells. Nat Chem 2009; 1(8): 662-7.
[http://dx.doi.org/10.1038/nchem.406] [PMID: 21378959]

[95] Gill MR, Thomas JA. Ruthenium(II) polypyridyl complexes and DNA--from structural probes to cellular imaging and therapeutics. Chem Soc Rev 2012; 41(8): 3179-92.
[http://dx.doi.org/10.1039/c2cs15299a] [PMID: 22314926]

[96] Sun D, Liu Y, Yu Q, Liu D, Zhou Y, Liu J. Selective nuclei accumulation of ruthenium(II) complex enantiomers that target G-quadruplex DNA. J Inorg Biochem 2015; 150: 90-9.
[http://dx.doi.org/10.1016/j.jinorgbio.2015.04.003] [PMID: 25911217]

[97] Lee JW, Jung M, Rosania GR, Chang YT. Development of novel cell-permeable DNA sensitive dyes using combinatorial synthesis and cell-based screening. Chem Commun (Camb) 2003; 7(15): 1852-3.
[http://dx.doi.org/10.1039/b303960a] [PMID: 12931999]

[98] Narayanaswamy N, Kumar M, Das S, *et al.* A thiazole coumarin (TC) turn-on fluorescence probe for AT-base pair detection and multipurpose applications in different biological systems. Sci Rep 2014; 4: 6476.
[http://dx.doi.org/10.1038/srep06476] [PMID: 25252596]

[99] Pickard AJ, Bierbach U. The cell's nucleolus: an emerging target for chemotherapeutic intervention. ChemMedChem 2013; 8(9): 1441-9.
[http://dx.doi.org/10.1002/cmdc.201300262] [PMID: 23881648]

[100] Tumir LM, Radić Stojković M, Piantanida I. Come-back of phenanthridine and phenanthridinium derivatives in the 21st century. Beilstein J Org Chem 2014; 10: 2930-54.
[http://dx.doi.org/10.3762/bjoc.10.312] [PMID: 25550761]

[101] Bucana C, Saiki I, Nayar R. Uptake and accumulation of the vital dye hydroethidine in neoplastic cells. J Histochem Cytochem 1986; 34(9): 1109-15.
[http://dx.doi.org/10.1177/34.9.2426339] [PMID: 2426339]

[102] Kapraun DF, Hinson TK, Limus AJ. Karyology and cytophotometric estimation of inter- and intraspecific nuclear-DNA variation in four species of Porphyra (Rhodophyta). Phycologia 1991; 30: 458-66.
[http://dx.doi.org/10.2216/i0031-8884-30-5-458.1]

[103] Burns VW. Location and molecular characteristics of fluorescent complexes of ethidium bromide in the cell. Exp Cell Res 1972; 75(1): 200-6.
[http://dx.doi.org/10.1016/0014-4827(72)90537-X] [PMID: 4564470]

[104] Stockert JC. Ethidium bromide fluorescence in living and fixed cells. Naturwissenschaften 1974;

61(8): 363-4.
[http://dx.doi.org/10.1007/BF00600306] [PMID: 4851915]

[105] Kallenbach NR, Berman HM. RNA structure. Q Rev Biophys 1977; 10(2): 138-236. [http://dx.doi.org/10.1017/S0033583500000202] [PMID: 333501]

[106] Woese CR, Gutell R, Gupta R, Noller HF. Detailed analysis of the higher-order structure of 16S-like ribosomal ribonucleic acids. Microbiol Rev 1983; 47(4): 621-69. [PMID: 6363901]

[107] Saenger W. Principles of Nucleic Acid Structure. New York, Berlin, Heidelberg, Tokyo: Springer-Verlag 1984; pp. 331-49. [http://dx.doi.org/10.1007/978-1-4612-5190-3_15]

[108] Stern S, Weiser B, Noller HF. Model for the three-dimensional folding of 16 S ribosomal RNA. J Mol Biol 1988; 204(2): 447-81. [http://dx.doi.org/10.1016/0022-2836(88)90588-8] [PMID: 2464693]

[109] Tinoco I Jr, Kieft JS, Wyatt JR. The ion core in RNA folding. Nat Struct Biol 1997; 4(7): 509-12. [http://dx.doi.org/10.1038/nsb0797-509] [PMID: 9228937]

[110] Moore PB. Recent RNA structures. Curr Opin Struct Biol 1993; 3: 340-4. [http://dx.doi.org/10.1016/S0959-440X(05)80104-6]

[111] Cañete M, Stockert JC. Fluorescencia de estructuras celulares por tioflavina T: reacción con la cromatina. Citología (Barcelona) 1985; 7: 55-60.

[112] Cañete M, Villanueva A, Juarranz A, Stockert JC. A study of interaction of thioflavine T with DNA: evidence for intercalation. Cell Mol Biol 1987; 33(2): 191-9. [PMID: 3607826]

[113] Stockert JC, Del Castillo P, Llorente AR, Rasskin DM, Romero JB, Gómez A. New fluorescence reactions in DNA cytochemistry. 1. Microscopic and spectroscopic studies on nonrigid fluorochromes. Anal Quant Cytol Histol 1990; 12(1): 1-10. [PMID: 2302306]

[114] Burns VW. Studies on the fluorophore sempervirene and its complexes with DNA. Biophys J 1974; 14(3): 189-99. [http://dx.doi.org/10.1016/S0006-3495(74)85906-0] [PMID: 4207290]

[115] Beljanski M, Crochet S. The anticancer agent pb-100 concentrates in the nucleus and nucleoli of human glioblastoma cells but does not enter normal astrocytes. Int J Oncol 1995; 7(1): 81-5. [PMID: 21552810]

[116] Drygin D, Siddiqui-Jain A, O'Brien S, *et al.* Anticancer activity of CX-3543: a direct inhibitor of rRNA biogenesis. Cancer Res 2009; 69(19): 7653-61. [http://dx.doi.org/10.1158/0008-5472.CAN-09-1304] [PMID: 19738048]

[117] Cuenca F, Greciano O, Gunaratnam M, *et al.* Tri- and tetra-substituted naphthalene diimides as potent G-quadruplex ligands. Bioorg Med Chem Lett 2008; 18(5): 1668-73. [http://dx.doi.org/10.1016/j.bmcl.2008.01.050] [PMID: 18243701]

[118] Yu J, Parker D, Pal R, Poole RA, Cann MJ. A europium complex that selectively stains nucleoli of cells. J Am Chem Soc 2006; 128(7): 2294-9. [http://dx.doi.org/10.1021/ja056303g] [PMID: 16478184]

[119] Zhang KY, Li SP, Zhu N, *et al.* Structure, photophysical and electrochemical properties, biomolecular interactions, and intracellular uptake of luminescent cyclometalated iridium(III) dipyridoquinoxaline complexes. Inorg Chem 2010; 49(5): 2530-40. [http://dx.doi.org/10.1021/ic902465b] [PMID: 20131874]

[120] Dyckman J, Weltman JK. A morphological analysis of binding of a hydrophobic probe to cells. J Cell Biol 1970; 45(1): 192-7.

[http://dx.doi.org/10.1083/jcb.45.1.192] [PMID: 5458996]

[121] Cowden RR, Curtis SK. Use of a fluorescence probe for hydrophobic groups, anilinonaphthalene sulphonic acid, in the supravital study of unusual slug oocyte nucleoli. Histochem J 1974; 6(4): 447-50.
[http://dx.doi.org/10.1007/BF01012435] [PMID: 4859583]

[122] Diaz G, Melis M, Batetta B, Angius F, Falchi AM. Hydrophobic characterization of intracellular lipids *in situ* by Nile Red red/yellow emission ratio. Micron 2008; 39(7): 819-24.
[http://dx.doi.org/10.1016/j.micron.2008.01.001] [PMID: 18329888]

[123] Liu X, Sun Y, Zhang Y, *et al.* A 2,7-carbazole-based dicationic salt for fluorescence detection of nucleic acids and two-photon fluorescence imaging of RNA in nucleoli and cytoplasm. Org Biomol Chem 2011; 9(10): 3615-8.
[http://dx.doi.org/10.1039/c1ob05123g] [PMID: 21431187]

[124] Pereira GC, Branco AF, Matos JA, *et al.* Mitochondrially targeted effects of berberine [Natural Yellow 18, 5,6-dihydro-9,10-dimethoxybenzo(g)-1,3-benzodioxolo(5,6-a) quinolizinium] on K1735-M2 mouse melanoma cells: comparison with direct effects on isolated mitochondrial fractions. J Pharmacol Exp Ther 2007; 323(2): 636-49.
[http://dx.doi.org/10.1124/jpet.107.128017] [PMID: 17704354]

[125] Hövelmann F, Gaspar I, Ephrussi A, Seitz O. Brightness enhanced DNA FIT-probes for wash-free RNA imaging in tissue. J Am Chem Soc 2013; 135(50): 19025-32.
[http://dx.doi.org/10.1021/ja410674h] [PMID: 24295172]

[126] Kuerschner L, Ejsing CS, Ekroos K, Shevchenko A, Anderson KI, Thiele C. Polyene-lipids: a new tool to image lipids. Nat Methods 2005; 2(1): 39-45.
[http://dx.doi.org/10.1038/nmeth728] [PMID: 15782159]

[127] Stockert JC, Cañete M, Juarranz A, *et al.* Porphycenes: facts and prospects in photodynamic therapy of cancer. Curr Med Chem 2007; 14(9): 997-1026.
[http://dx.doi.org/10.2174/092986707780362934] [PMID: 17439399]

[128] Cañete M, Stockert JC, Villanueva A. Preclinical photodynamic therapy research in Spain. 3. Localization of photosensitizers and mechanisms of cell death *in vitro.* J Porphyr Phthalocyanines 2009; 13: 544-51.
[http://dx.doi.org/10.1142/S1088424609000516]

[129] Whatmore JL, Allan D. Phospholipid asymmetry in plasma membrane vesicles derived from BHK cells. Biochim Biophys Acta 1994; 1192(1): 88-94.
[http://dx.doi.org/10.1016/0005-2736(94)90146-5] [PMID: 8204655]

[130] Yamaji-Hasegawa A, Tsujimoto M. Asymmetric distribution of phospholipids in biomembranes. Biol Pharm Bull 2006; 29(8): 1547-53.
[http://dx.doi.org/10.1248/bpb.29.1547] [PMID: 16880602]

[131] Christensen H, Garton NJ, Horobin RW, Minnikin DE, Barer MR. Lipid domains of mycobacteria studied with fluorescent molecular probes. Mol Microbiol 1999; 31(5): 1561-72.
[http://dx.doi.org/10.1046/j.1365-2958.1999.01304.x] [PMID: 10200973]

[132] Edidin M. Fluorescent labeling of cell surfaces. Methods Cell Biol 1989; 29: 87-102.
[http://dx.doi.org/10.1016/S0091-679X(08)60189-2] [PMID: 2464121]

[133] Horan PK, Melnicoff MJ, Jensen BD, Slezak SE. Fluorescent cell labeling for *in vivo* and *in vitro* cell tracking. Methods Cell Biol 1990; 33: 469-90.
[http://dx.doi.org/10.1016/S0091-679X(08)60547-6] [PMID: 2084480]

[134] Janecki AJ, Janecki M, Akhter S, Donowitz M. Quantitation of plasma membrane expression of a fusion protein of Na/H exchanger NHE3 and green fluorescence protein (GFP) in living PS120 fibroblasts. J Histochem Cytochem 2000; 48(11): 1479-92.
[http://dx.doi.org/10.1177/002215540004801105] [PMID: 11036091]

[135] Valinsky JE, Easton TG, Reich E. Merocyanine 540 as a fluorescent probe of membranes: selective staining of leukemic and immature hemopoietic cells. Cell 1978; 13(3): 487-99. [http://dx.doi.org/10.1016/0092-8674(78)90322-7] [PMID: 77735]

[136] Del Buono BJ, Williamson PL, Schlegel RA. Relation between the organization of spectrin and of membrane lipids in lymphocytes. J Cell Biol 1988; 106(3): 697-703. [http://dx.doi.org/10.1083/jcb.106.3.697] [PMID: 3279051]

[137] Dougherty TJ, Gomer CJ, Henderson BW, *et al.* Photodynamic therapy. J Natl Cancer Inst 1998; 90(12): 889-905. [http://dx.doi.org/10.1093/jnci/90.12.889] [PMID: 9637138]

[138] Zareba M, Nizioleka M, Korytowski W, *et al.* Merocyanine 540-sensitized photokilling of leukemia cells: role of post-irradiation chain peroxidation of plasma membrane lipids as revealed by nitric oxide protection. Biochim Biophys Acta 2005; 1722: 51-9.

[139] Cañete M, Villanueva A, Domínguez V, Polo S, Juarranz A, Stockert JC. Meso-tetraphenylporphyrin: photosensitizing properties and cytotoxic effects on cultured tumor cells. Int J Oncol 1998; 13(3): 497-504. [PMID: 9683784]

[140] Kessel D. Relocalization of cationic porphyrins during photodynamic therapy. Photochem Photobiol Sci 2002; 1(11): 837-40. [http://dx.doi.org/10.1039/b206046a] [PMID: 12659521]

[141] Pitt D. Lysosomes and Cell Function. London, New York: Longman 1975.

[142] Oakes SA, Papa FR. The role of endoplasmic reticulum stress in human pathology. Annu Rev Pathol 2015; 10: 173-94. [http://dx.doi.org/10.1146/annurev-pathol-012513-104649] [PMID: 25387057]

[143] Beriault DR, Werstuck GH. Detection and quantification of endoplasmic reticulum stress in living cells using the fluorescent compound, Thioflavin T. Biochim Biophys Acta 2013; 1833: 2293-301.

[144] Quader H, Schnepf E. Endoplasmic reticulum and cytoplasmic streaming: fluorescence microscopical observations in adaxial epidermis cells of onion bulb scales. Protoplasma 1986; 131: 250-2. [http://dx.doi.org/10.1007/BF01282989]

[145] Terasaki M. Fluorescent labeling of endoplasmic reticulum. Methods Cell Biol 1989; 29: 125-35. [http://dx.doi.org/10.1016/S0091-679X(08)60191-0] [PMID: 2643757]

[146] Terasaki M. Labeling of the Endoplasmic Reticulum with $DiOC_6(3)$. Cell Biology: A Laboratory Handbook. New York, London: Academic Press 1994; pp. 381-6.

[147] Lee C, Wu SS, Chen LB. Photosensitization by 3,3′-dihexyloxacarbocyanine iodide: specific disruption of microtubules and inactivation of organelle motility. Cancer Res 1995; 55(10): 2063-9. [PMID: 7743503]

[148] Pandya S, Yu J, Parker D. Engineering emissive europium and terbium complexes for molecular imaging and sensing. Dalton Trans 2006; (23): 2757-66. [http://dx.doi.org/10.1039/b514637b] [PMID: 16751883]

[149] Hai Y, Chen JJ, Zhao P, *et al.* Luminescent zinc salen complexes as single and two-photon fluorescence subcellular imaging probes. Chem Commun (Camb) 2011; 47(8): 2435-7. [http://dx.doi.org/10.1039/C0CC04113K] [PMID: 21170447]

[150] McMahon BK, Pal R, Parker D. A bright and responsive europium probe for determination of pH change within the endoplasmic reticulum of living cells. Chem Commun (Camb) 2013; 49(47): 5363-5. [http://dx.doi.org/10.1039/c3cc42308e] [PMID: 23652607]

[151] Primik MF, Göschl S, Meier SM, *et al.* Dicopper(II) and dizinc(II) complexes with nonsymmetric dinucleating ligands based on indolo[3,2-c]quinolines: synthesis, structure, cytotoxicity, and

intracellular distribution. Inorg Chem 2013; 52(17): 10137-46.
[http://dx.doi.org/10.1021/ic401573d] [PMID: 23952332]

[152] Zou T, Lok CN, Fung YM, Che CM. Luminescent organoplatinum(II) complexes containing bis(N-heterocyclic carbene) ligands selectively target the endoplasmic reticulum and induce potent phototoxicity. Chem Commun (Camb) 2013; 49(47): 5423-5.
[http://dx.doi.org/10.1039/c3cc40953h] [PMID: 23515304]

[153] Gill MR, Cecchin D, Michael G, *et al.* Targeting the endoplasmic reticulum with a membrane interactive luminescent ruthenium(II) polypyridyl complex. Chem Sci (Camb) 2013; 4: 4512-9.
[http://dx.doi.org/10.1039/c3sc51725j]

[154] Racoosin EL, Swanson JA. Labeling of Endocytic Vesicles using Fluorescent Probes for Fluid-Phase Endocytosis.Cell Biology: A Laboratory Handbook. New York, London: Academic Press 1994; Vol. 2.

[155] Vida TA, Emr SD. A new vital stain for visualizing vacuolar membrane dynamics and endocytosis in yeast. J Cell Biol 1995; 128(5): 779-92.
[http://dx.doi.org/10.1083/jcb.128.5.779] [PMID: 7533169]

[156] Sousa C, Sá e Melo T, Gèze M, Gaullier JM, Mazière JC, Santus R. Solvent polarity and pH effects on the spectroscopic properties of neutral red: application to lysosomal microenvironment probing in living cells. Photochem Photobiol 1996; 63(5): 601-7.
[http://dx.doi.org/10.1111/j.1751-1097.1996.tb05662.x] [PMID: 8628751]

[157] Lipsky NG, Pagano RE. A vital stain for the Golgi apparatus. Science 1985; 228(4700): 745-7.
[http://dx.doi.org/10.1126/science.2581316] [PMID: 2581316]

[158] Lin CW, Shulok JR, Kirley SD, Cincotta L, Foley JW. Lysosomal localization and mechanism of uptake of Nile blue photosensitizers in tumor cells. Cancer Res 1991; 51(10): 2710-9.
[PMID: 2021950]

[159] Gèze M, Morlière P, Mazière JC, Smith KM, Santus R. Lysosomes, a key target of hydrophobic photosensitizers proposed for photochemotherapeutic applications. J Photochem Photobiol B 1993; 20(1): 23-35.
[http://dx.doi.org/10.1016/1011-1344(93)80128-V] [PMID: 8229466]

[160] Wessendorf MW. Fluoro-Gold: composition, and mechanism of uptake. Brain Res 1991; 553(1): 135-48.
[http://dx.doi.org/10.1016/0006-8993(91)90241-M] [PMID: 1933270]

[161] Bastos AL, Marques DV, Filipe Da Silva J, *et al.* Primary inducible fluorescence in secretory granules of tumor cells supravitally stained by a toluidine blue dye. Z Naturforsch B 1968; 23(7): 969-75.
[http://dx.doi.org/10.1515/znb-1968-0716] [PMID: 4178917]

[162] Marques D, Bastos AL. Ultraviolet inducible fluorescence of tumour cells supravitally stained with thiazine dyes. Exp Cell Res 1972; 73(2): 525-8.
[http://dx.doi.org/10.1016/0014-4827(72)90086-9] [PMID: 4115485]

[163] Hogan ME, Hayes B, Wang NC, Austin RH. Ion-induced DNA structure change in nucleosomes. Biochemistry 1986; 25(18): 5070-82.
[http://dx.doi.org/10.1021/bi00366a015] [PMID: 3768333]

[164] Peng Q, Brown SB, Moan J, *et al.* Biodistribution of a methylene blue derivative in tumor and normal tissues of rats. J Photochem Photobiol B 1993; 20(1): 63-71.
[http://dx.doi.org/10.1016/1011-1344(93)80132-S] [PMID: 8229471]

[165] Koizumi M, Obato H, Hayashi S. Photoreduction of thiazine dyes in aqueous solution. Bull Chem Soc Jpn 1964; 37: 108-17.
[http://dx.doi.org/10.1246/bcsj.37.108]

[166] Niederau C, Van Dyke RW, Scharschmidt BF, Grendell JH. Rat pancreatic zymogen granules. An actively acidified compartment. Gastroenterology 1986; 91(6): 1433-42.

[http://dx.doi.org/10.1016/0016-5085(86)90197-6] [PMID: 3021556]

[167] De Lisle RC, Steinberg R, Williams JA. Zymogen granules of mouse parotid acinar cells are acidified *in situ* in an ATP-dependent manner. Cell Tissue Res 1988; 253(1): 267-9. [http://dx.doi.org/10.1007/BF00221764] [PMID: 3416345]

[168] Eldred GE, Lasky MR. Retinal age pigments generated by self-assembling lysosomotropic detergents. Nature 1993; 361(6414): 724-6. [http://dx.doi.org/10.1038/361724a0] [PMID: 8441466]

[169] Sparrow JR, Parish CA, Hashimoto M, Nakanishi K. A2E, a lipofuscin fluorophore, in human retinal pigmented epithelial cells in culture. Invest Ophthalmol Vis Sci 1999; 40(12): 2988-95. [PMID: 10549662]

[170] Aubin JE. Autofluorescence of viable cultured mammalian cells. J Histochem Cytochem 1979; 27(1): 36-43. [http://dx.doi.org/10.1177/27.1.220325] [PMID: 220325]

[171] Benson RC, Meyer RA, Zaruba ME, McKhann GM. Cellular autofluorescence--is it due to flavins? J Histochem Cytochem 1979; 27(1): 44-8. [http://dx.doi.org/10.1177/27.1.438504] [PMID: 438504]

[172] Andersson H, Baechi T, Hoechl M, Richter C. Autofluorescence of living cells. J Microsc 1998; 191(Pt 1): 1-7. [http://dx.doi.org/10.1046/j.1365-2818.1998.00347.x] [PMID: 9723186]

[173] Chance B, Cohen P, Jobsis F, Schoener B. Intracellular oxidation-reduction states *in vivo*. Science 1962; 137(3529): 499-508. [http://dx.doi.org/10.1126/science.137.3529.499] [PMID: 13878016]

[174] Petrat F, Pindiur S, Kirsch M, de Groot H. NAD(P)H, a primary target of $^1O_2$ in mitochondria of intact cells. J Biol Chem 2003; 278(5): 3298-307. [http://dx.doi.org/10.1074/jbc.M204230200] [PMID: 12433931]

[175] Cristóbal J, Stockert JC, Villanueva A, Rello-Varona S, Juarranz A, Cañete M. Caspase-2: a possible trigger of apoptosis induced in A-549 tumor cells by ZnPc photodynamic treatment. Int J Oncol 2006; 28(5): 1057-63. [PMID: 16596221]

[176] Bereiter-Hahn J. Dimethylaminostyrylmethylpyridiniumiodine (daspmi) as a fluorescent probe for mitochondria *in situ*. Biochim Biophys Acta 1976; 423(1): 1-14. [http://dx.doi.org/10.1016/0005-2728(76)90096-7] [PMID: 764877]

[177] Borodina VM, Zelenin AV. [Fluorescence microscopy demonstration of mitochondria in tissue culture cells using berberine]. Tsitologiia 1977; 19(9): 1067-8. [PMID: 73245]

[178] Johnson LV, Walsh ML, Bockus BJ, Chen LB. Monitoring of relative mitochondrial membrane potential in living cells by fluorescence microscopy. J Cell Biol 1981; 88(3): 526-35. [http://dx.doi.org/10.1083/jcb.88.3.526] [PMID: 6783667]

[179] Johnson LV, Walsh ML, Chen LB. Localization of mitochondria in living cells with rhodamine 123. Proc Natl Acad Sci USA 1980; 77(2): 990-4. [http://dx.doi.org/10.1073/pnas.77.2.990] [PMID: 6965798]

[180] Ronot X, Benel L, Adolphe M, Mounolou JC. Mitochondrial analysis in living cells: the use of rhodamine 123 and flow cytometry. Biol Cell 1986; 57(1): 1-7. [http://dx.doi.org/10.1111/j.1768-322X.1986.tb00458.x] [PMID: 2945606]

[181] Poot M. Staining of Mitochondria. Cell Biology: A Laboratory Handbook. 2nd ed. New York, London: Academic Press 1998; Vol. 2: pp. 513-7.

[182] Septinus M, Seiffert W, Zimmermann HW. Uber hydrophobe Acridinfarbstoffe zur

Fluorochromierung von Mitochondrien in lebenden Zellen. 1. Mitteilung: Thermodynamische und spektroskopische Eigenschaften von 10-n-Alkyl-acridiniumorange-chloriden. Histochemistry 1983; 79(3): 443-56.
[http://dx.doi.org/10.1007/BF00491779] [PMID: 6197394]

[183] Röding J, Naujok A, Zimmermann HW. Effects of ethidium bromide, tetramethylethidium bromide and betaine B on the ultrastructure of HeLa cell mitochondria *in situ*. A comparative binding study. Histochemistry 1986; 85(3): 215-22.
[http://dx.doi.org/10.1007/BF00494807] [PMID: 3744904]

[184] Biasutto L, Dong LF, Zoratti M, Neuzil J. Mitochondrially targeted anti-cancer agents. Mitochondrion 2010; 10(6): 670-81.
[http://dx.doi.org/10.1016/j.mito.2010.06.004] [PMID: 20601192]

[185] Mileykovskaya E, Dowhan W, Birke RL, Zheng D, Lutterodt L, Haines TH. Cardiolipin binds nonyl acridine orange by aggregating the dye at exposed hydrophobic domains on bilayer surfaces. FEBS Lett 2001; 507(2): 187-90.
[http://dx.doi.org/10.1016/S0014-5793(01)02948-9] [PMID: 11684095]

[186] Reers M, Smith TW, Chen LB. J-aggregate formation of a carbocyanine as a quantitative fluorescent indicator of membrane potential. Biochemistry 1991; 30(18): 4480-6.
[http://dx.doi.org/10.1021/bi00232a015] [PMID: 2021638]

[187] Boitier E, Rea R, Duchen MR. Mitochondria exert a negative feedback on the propagation of intracellular Ca^{2+} waves in rat cortical astrocytes. J Cell Biol 1999; 145(4): 795-808.
[http://dx.doi.org/10.1083/jcb.145.4.795] [PMID: 10330407]

[188] Olszewska A, Szewczyk A. Mitochondria as a pharmacological target: magnum overview. IUBMB Life 2013; 65(3): 273-81.
[http://dx.doi.org/10.1002/iub.1147] [PMID: 23441041]

[189] Neuzil J, Dong LF, Rohlena J, Truksa J, Ralph SJ. Classification of mitocans, anti-cancer drugs acting on mitochondria. Mitochondrion 2013; 13(3): 199-208.
[http://dx.doi.org/10.1016/j.mito.2012.07.112] [PMID: 22846431]

[190] Pereira CV, Machado NG, Oliveira PJ. Mechanisms of berberine (natural yellow 18)-induced mitochondrial dysfunction: interaction with the adenine nucleotide translocator. Toxicol Sci 2008; 105(2): 408-17.
[http://dx.doi.org/10.1093/toxsci/kfn131] [PMID: 18599498]

[191] Mikes V, Dadák V. Berberine derivatives as cationic fluorescent probes for the investigation of the energized state of mitochondria. Biochim Biophys Acta 1983; 723(2): 231-9.
[http://dx.doi.org/10.1016/0005-2728(83)90122-6] [PMID: 6849903]

[192] Chernyak BV, Antonenko YN, Domnina LV, *et al.* Novel penetrating cations for targeting mitochondria. Curr Pharm Des 2013; 19(15): 2795-806.
[http://dx.doi.org/10.2174/1381612811319150015] [PMID: 23092317]

[193] Diogo CV, Machado NG, Barbosa IA, Serafim TL, Burgeiro A, Oliveira PJ. Berberine as a promising safe anti-cancer agent - is there a role for mitochondria? Curr Drug Targets 2011; 12(6): 850-9.
[http://dx.doi.org/10.2174/138945011795528930] [PMID: 21269266]

[194] Ortiz LM, Lombardi P, Tillhon M, Scovassi AI. Berberine, an epiphany against cancer. Molecules 2014; 19(8): 12349-67.
[http://dx.doi.org/10.3390/molecules190812349] [PMID: 25153862]

[195] Oseroff AR, Ohuoha D, Ara G, McAuliffe D, Foley J, Cincotta L. Intramitochondrial dyes allow selective *in vitro* photolysis of carcinoma cells. Proc Natl Acad Sci USA 1986; 83(24): 9729-33.
[http://dx.doi.org/10.1073/pnas.83.24.9729] [PMID: 3467335]

[196] Cowden RR, Curtis SK. Supravital experiments with Pyronin Y, a fluorochrome of mitochondria and nucleic acids. Histochemistry 1983; 77(4): 535-42.

[http://dx.doi.org/10.1007/BF00495807] [PMID: 6190789]

[197] Cosgrove WB. Why kinetoplasts? J Protozool 1973; 20(2): 191-4. [http://dx.doi.org/10.1111/j.1550-7408.1973.tb00861.x] [PMID: 4122735]

[198] Wolfson JS, McHugh GL, Swartz MN, Ng EY, Hooper DC. Visualization of the mitochondrion-kinetoplast in viable *Trypanosoma cruzi* with rhodamine 123. J Parasitol 1987; 73(4): 866-9. [http://dx.doi.org/10.2307/3282439] [PMID: 2442343]

[199] McGill M, Baur PS, Hsu TC. Ultrastructure and microchemical composition of ethidium bromide-induced intramitochondrial complexes. Exp Cell Res 1976; 99(1): 7-14. [http://dx.doi.org/10.1016/0014-4827(76)90673-X] [PMID: 1261616]

[200] Steinert M. Specific loss of kinetoplastic DNA in trypanosomatidae treated with ethidium bromide. Exp Cell Res 1969; 55(2): 248-52. [http://dx.doi.org/10.1016/0014-4827(69)90487-X] [PMID: 5772120]

[201] Affranchino JL, de Tarlovsky MN, Stoppani AO, Burgos MH. Ultrastructural and metabolic alteration in dyskinetoplastic Trypanosoma cruzi. Microsc Electron Biol Celular 1985; 9(1): 67-87. [PMID: 3916125]

[202] Pagano RE. A fluorescent derivative of ceramide: physical properties and use in studying the Golgi apparatus of animal cells. Methods Cell Biol 1989; 29: 75-85. [http://dx.doi.org/10.1016/S0091-679X(08)60188-0] [PMID: 2913469]

[203] Pagano RE, Martin OC. Use of Fluorescent Analogs of Ceramide to Study the Golgi Apparatus of Animal Cells.Cell Biology: A Laboratory Handbook. New York, London: Academic Press 1994; Vol. 2.

[204] Pagano RE, Sepanski MA, Martin OC. Molecular trapping of a fluorescent ceramide analogue at the Golgi apparatus of fixed cells: interaction with endogenous lipids provides a trans-Golgi marker for both light and electron microscopy. J Cell Biol 1989; 109(5): 2067-79. [http://dx.doi.org/10.1083/jcb.109.5.2067] [PMID: 2478562]

[205] Reaven E, Tsai L, Azhar S. Intracellular events in the "selective" transport of lipoprotein-derived cholesteryl esters. J Biol Chem 1996; 271(27): 16208-17. [http://dx.doi.org/10.1074/jbc.271.27.16208] [PMID: 8663101]

[206] Zhang H, Fan J, Wang J, Zhang S, Dou B, Peng X. An off-on COX-2-specific fluorescent probe: targeting the Golgi apparatus of cancer cells. J Am Chem Soc 2013; 135(31): 11663-9. [http://dx.doi.org/10.1021/ja4056905] [PMID: 23862760]

[207] Spandl J, White DJ, Peychl J, Thiele C. Live cell multicolor imaging of lipid droplets with a new dye, LD540. Traffic 2009; 10(11): 1579-84. [http://dx.doi.org/10.1111/j.1600-0854.2009.00980.x] [PMID: 19765264]

[208] Suzuki M, Shinohara Y, Fujimoto T. Histochemical detection of lipid droplets in cultured cells. Methods Mol Biol 2013; 931: 483-91. [http://dx.doi.org/10.1007/978-1-62703-056-4_25] [PMID: 23027019]

[209] Rello-Varona S, Stockert JC, Cañete M, Acedo P, Villanueva A. Mitotic catastrophe induced in HeLa cells by photodynamic treatment with Zn(II)-phthalocyanine. Int J Oncol 2008; 32(6): 1189-96. [http://dx.doi.org/10.3892/ijo_32_6_1189] [PMID: 18497980]

[210] Lown JW, Chen HH. Studies on the effects of the antitumor agent camptothecin and derivatives on deoxyribonucleic acid. Mechanism of the scission of deoxyribonucleic acid by photoactivated camptothecin. Biochem Pharmacol 1980; 29(6): 905-15. [http://dx.doi.org/10.1016/0006-2952(80)90221-X] [PMID: 7387708]

[211] Staker BL, Hjerrild K, Feese MD, Behnke CA, Burgin AB Jr, Stewart L. The mechanism of topoisomerase I poisoning by a camptothecin analog. Proc Natl Acad Sci USA 2002; 99(24): 15387-92. [http://dx.doi.org/10.1073/pnas.242259599] [PMID: 12426403]

[212] Marchand C, Antony S, Kohn KW, *et al.* A novel norindenoisoquinoline structure reveals a common interfacial inhibitor paradigm for ternary trapping of the topoisomerase I-DNA covalent complex. Mol Cancer Ther 2006; 5(2): 287-95.
[http://dx.doi.org/10.1158/1535-7163.MCT-05-0456] [PMID: 16505102]

[213] Huang M, Gao H, Chen Y, *et al.* Chimmitecan, a novel 9-substituted camptothecin, with improved anticancer pharmacologic profiles *in vitro* and *in vivo.* Clin Cancer Res 2007; 13(4): 1298-307.
[http://dx.doi.org/10.1158/1078-0432.CCR-06-1277] [PMID: 17287296]

[214] Errington RJ, Ameer-Beg SM, Vojnovic B, Patterson LH, Zloh M, Smith PJ. Advanced microscopy solutions for monitoring the kinetics and dynamics of drug-DNA targeting in living cells. Adv Drug Deliv Rev 2005; 57(1): 153-67.
[http://dx.doi.org/10.1016/j.addr.2004.05.005] [PMID: 15518927]

[215] Croce AC, Bottiroli G, Supino R, Favini E, Zuco V, Zunino F. Subcellular localization of the camptothecin analogues, topotecan and gimatecan. Biochem Pharmacol 2004; 67(6): 1035-45.
[http://dx.doi.org/10.1016/j.bcp.2003.10.034] [PMID: 15006540]

[216] Stockert JC. Predictive binding geometry of ligands to DNA minor groove: isohelicity and hydrogen-bonding pattern. Methods Mol Biol 2014; 1094: 1-12.
[http://dx.doi.org/10.1007/978-1-62703-706-8_1] [PMID: 24162975]

[217] Dolman NJ, Kilgore JA, Davidson MW. A review of reagents for fluorescence microscopy of cellular compartments and structures, part I: BacMam labeling and reagents for vesicular structures. Curr Protoc Cytom 2013; Chapter 12: 30.
[PMID: 23835803]

[218] Stockert JC. A simple and selective method for visualizing the kinetoplast of trypanosomes by means of phosphomolybdic acid and a photographic developer. Parasitol Res 1992; 78(3): 263-4.
[http://dx.doi.org/10.1007/BF00931738] [PMID: 1317034]

CHAPTER 16

Vital Probes 2. Cell Viability and Signaling

Abstract: In this chapter several fluorescent probes for the assessment of cell viability, types of cell death, identification of signals, cellular traffic and connections will be described. These biomedical fields have increasing interest [1-3]. The fluorescence histochemistry of enzymatic activity is currently applied in assays for cell viability, apoptotic processes, and detection of reactive oxygen species (ROS). Several probes of viability and signaling are also markers of oxidative stress and redox visualization. Numerous probes are now commercially available for studies on cell metabolism and toxicity, signal transduction, drug screening, molecular diagnosis, *etc*. Fluorescent labels and biosensors are applied to perform quali- and quantitative analysis of normal and pathological cell processes (*i.e.* sensing dyes for trans-membrane potential, ions, ROS, pH, *etc*). There exist also approaches to the direct real-time signal imaging of live-cells *in vitro* and *in vivo*, and for 2D and 3D analysis.

Keywords: Apoptosis, Autophagy, Calcein-AM, Comet assay, Genotoxicity, H_2DCFDA, Halo assay, Hexidium iodide, H-33342, Hydroethidine, Mitotic catastrophe, Necrosis, Nitric oxide, ROS, Retrograde axonal tracers, Senescence, TUNEL assay, Trypan blue.

1. CELL VIABILITY ASSAYS

Several assays for cell viability are commonly used. In addition to traditional methods, organelle probes and indirect methods are also applied for this purpose. This is the case of the mitochondrial probe rhodamine 123 (Rh123), which on account of the loss of the mitochondrial trans-membrane potential in apoptotic and necrotic cells, results in the delocalization of Rh123 signal [4].

1.1. Dye Exclusion

Most common dye exclusion tests under bright-field microscopy employ trypan blue, Congo red, and Evans blue, although other dyes (*i.e.* ruthenium red [5]) were also used on account of their high charge and hydrophilia. Since long ago, it is known that trypan blue staining indicates plasma membrane damage associated with cell death [6 - 10]. This classic dye can be used in manual, semi-automated, and automated counting methods for cell viability [11]. Trypan blue has been often applied in embryological studies, showing possible teratogenic effects and

Juan Carlos Stockert & Alfonso Blázquez-Castro

reduction of fertilization and subsequent development of mouse oocytes [12]. The emission of trypan blue (see Fig. **13.5**) can be used in flow cytometry for viability assays (λexc: 488 nm, λem: 660 nm) [13].

Nuclear staining by trypan blue has been described long ago in fixed plant tissues [14], and nuclear labeling of death cells by the trypan blue exclusion test was found in human retinal pigment epithelial cells exposed to triamcinolone [15], and in apoptotic and necrotic cells after photosensitization [16] (Fig. **16.1B-D**). In contrast with live cells, in chromatin from both necrotic and apoptotic cells, negative and positive charges of DNA and histones could become separated enough to allow different interactions with dyes. Negative charges of trypan blue would interact with positive groups of histones, whereas the convex edge of the dye could bind to DNA minor groove (Fig. **16.1A**).

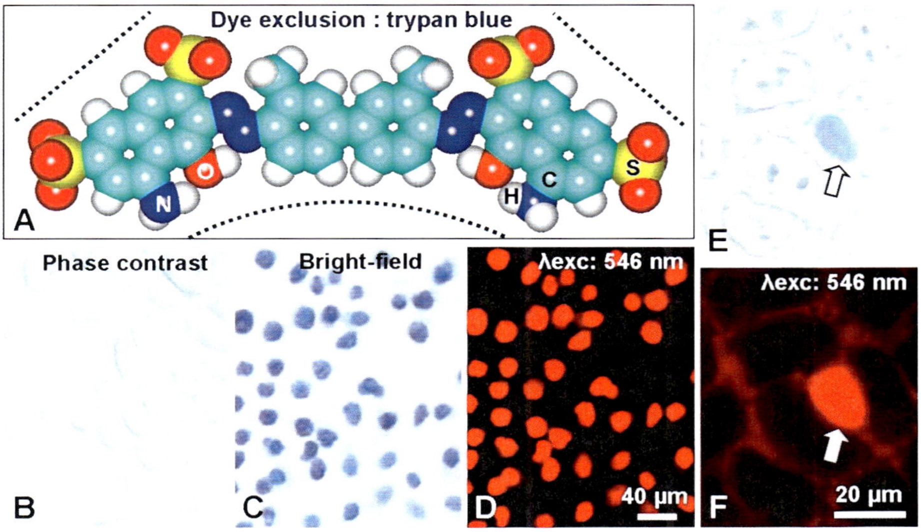

Fig. (16.1). A: Van der Waals representation of trypan blue, showing atom symbols and possible binding to histones (dotted straight lines) and to the floor of DNA minor groove (dotted curved line). **B-D:** Trypan blue assay (0.4% in PBS, 3 min) of viable (untreated) HeLa cells (B), and cells 24 h after lethal photodynamic treatment with ZnPc and red light (C, D), showing positive nuclear staining (C) and fluorescence (D). **E, F:** Images of HeLa cells after 24 h starvation in PBS, showing trypan blue staining (E) and fluorescence (F) of an apoptotic cell (arrow). Trypan blue assays were performed directly on plastic wells without trypsinization.

Likewise, the plasma membrane of live cells can be permeable or non-permeable to fluorescent vital probes. AO is one of the oldest probes applied to differentiate live cells (green) from dead cells (red) [17 - 19]. Reagents and fluorochromes that can penetrate into living cells are AO, neutral red, H-33342, fluorescein diacetate

(FDA), calcein-AM, and YO-PRO-1 [3, 20 - 22]. On the contrary, propidium iodide (PI, Fig. **16.2**), is excluded from living cells and only stains the dead ones [3, 23]. Reduced ethidium (hydroethidine) is a non-charged and lipophilic probe that after uptake and oxidation within live cells yields the fluorescent ethidium cation.

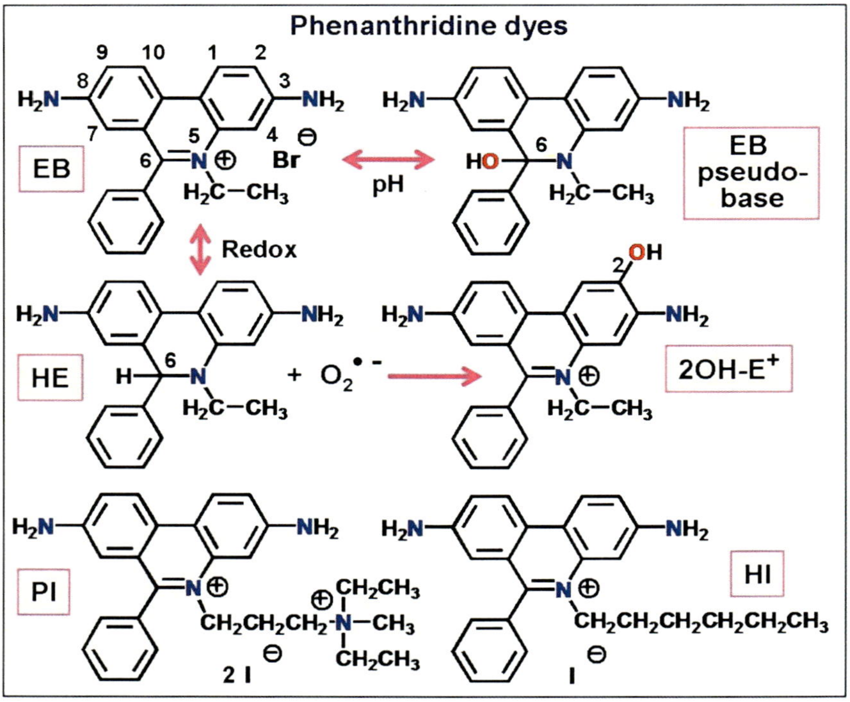

Fig. (16.2). Chemical structure of phenanthridine dyes. EB: ethidium bromide (with ring numbering), EB pseudo-base (related to pH), HE: hydroethidine (reduced form), 2OH-E$^+$: 2-hydroxy-ethidium derivative (oxidized by the radical superoxide), PI: propidium iodide, and HI: hexidium iodide.

In the case of ethidium bromide (EB, Figs. **15.3** and **16.2**), while the dye penetrates live cells very slowly, it rapidly enters dead cells. The dicationic PI (calculated values, log $P_{dication}$:−5.1; log $P_{pseudo\text{-}base}$:−1.4) only enters cells whose plasma membrane is damaged (*i.e.* dead and fixed cells), and induces red fluorescence in NAs-containing structures (particularly chromatin DNA). Therefore PI is very useful as a live/dead labeling probe [4, 24, 25]. PI penetrates a dysfunctional or compromised plasma membrane but in the case of apoptotic cells, since the membrane generally ceases to function only during the final steps of this processs, the PI assay is useful for visualizing advanced stages of apoptosis.

Obviously, two main parameters of cell viability such as intracellular esterase activity and plasma membrane integrity can be measured using two probes. By combining FDA and EB, improved viability tests allow to obtain a clear contrast between live and dead cells (green and red emission, respectively) [20]. Therefore, current procedures for fast and easy viability and cytotoxicity tests are based on two color fluorescence, which show a code such as the traffic light symbols (live cells, green; dead cells, red). They are applied for fluorescence microscopy and flow cytometry. The couples FDA/EB, AO/EB, AO/PI (Fig. **16.3A, C**), calcein-AM/ethidium homodimer-1, and H-33342/PI are commonly employed assays [22, 26 - 29]. The four first methods have the advantage that only one exciting light is required for both dyes. Various live/dead kits are commercially available (*i.e.* calcein-AM/PI, YO-PRO-1/PI, SYBR 14/PI).

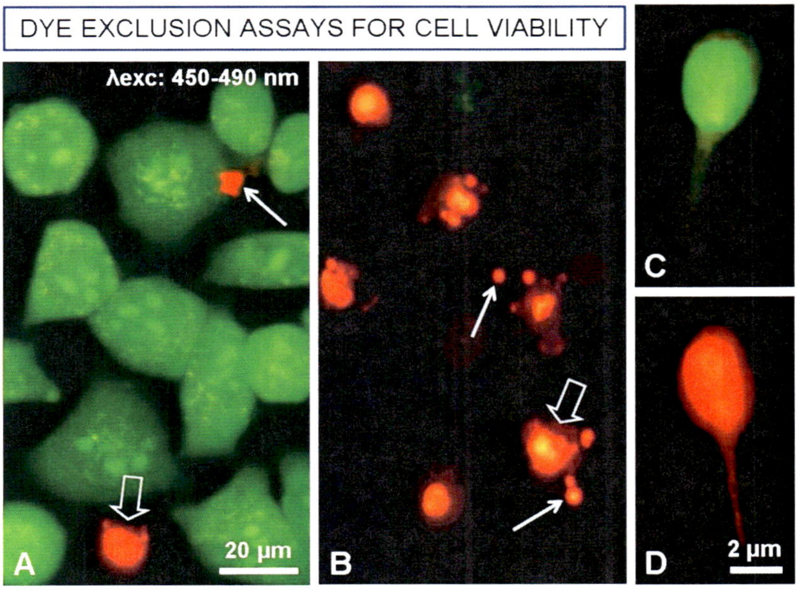

Fig. (16.3). A, B: HeLa cells after the AO/PI assay (0.5 μg/mL AO for 0.5 min and 10 μg/mL PI for 4.5 min). **A:** Cells maintained in PBS alone for 24 h to induce apoptosis. **B:** Cells subjected to sublethal photosensitizing treatment with ZnPc. Dead cells appear red; thin and large arrows indicate apoptotic bodies and apoptotic cells, respectively. Live (C) and dead (D) bovine spermatozoa revealed by a Live/Dead Sperm Viability Kit, showing live sperm (SYBR 14 nucleic acid stain, green), and dead sperm (PI, red).

A direct live/dead test for cell populations that avoids the loss of detached cells involves the use of PI fluorescence without any intermediate washing. The dye is added to cell cultures in excess and the emission is recorded in a fluorescence plate reader. As the emission of free PI (not bound to nucleic acids) is very small and can be subtracted from experimental values, the remaining red signal only represents dead cells. Then cells are fixed with a small amount of formaldehyde, and the fluorescence recorded again. The new emission value corresponds to the whole cell population. After simple calculation, the percent value of live *vs* dead cells is obtained. As positive controls, cell necrosis can be induced by LD_{100} treatments (freezing-thawing, H_2O_2). Obviously, other non-permeable fluorescent probes can be also employed. Chemical structures of some probes are illustrated in Fig. (**16.4A, C**).

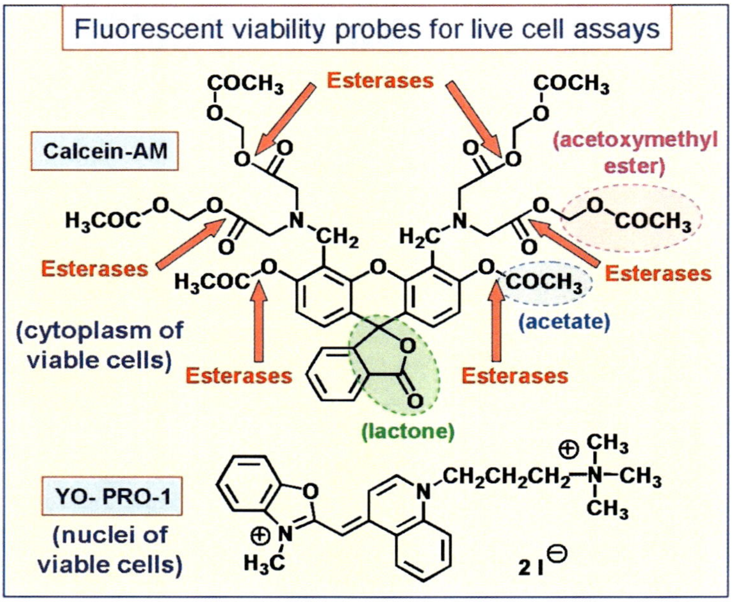

Fig. (16.4). Chemical structure of calcein-AM (calcein-diacetate-acetoxymethyl ester) and YO-PRO-1. Arrows indicate the hydrolysis of acetate and acetoxymethyl ester groups by esterases.

In the case of bacterial cells, there are several viability and Gram stain probes as commercial kits. Examples are DAPI and SYTOX green, which label live and dead bacteria, respectively. SYTO 9 stains Gram-negative bacteria green. SYTO 13 nucleic acid stain labels nuclei of live cells. On account of its more lipophilic character, the nucleic acid stain hexidium iodide (HI, Fig. **16.2**) is permeant to mammalian cells and stains red both the cytoplasm and nuclei, as well as almost all Gram-positive bacteria in the presence of Gram-negative bacteria. Upon binding to DNA, HI exhibits λexc: 518 nm, and λem: 600 nm. In addition, wheat germ agglutinin-Texas red conjugate labels red Gram-positive bacteria.

1.2. Enzymatic Assays and Redox Probes

The occurrence of intracellular esterases that hydrolyze esterified dyes is utilized in cell viability assays. Uptake of the membrane permeant and non-fluorescent FDA (see Fig. **14.3**) into live cells is one of the most used viability assays for live cells (*i.e.* mammalian, plant, and fungi cells, bacteria, protozoa, *etc.*, see [20, 30]), and is generally applied in combination with dye exclusion probes. Ester hydrolysis also occurs in the case of calcein-AM (Fig. **16.4**). Living cells accumulate FDA on account of its non-polar character, and then the probe is hydrolyzed giving the polar and green-fluorescing product fluorescein, which is retained within cells [3]. Likewise, the permeant hydroethidine (the reduced and uncharged form of ethidium) has been applied as a fluorescent probe for live-dead assays [31]. The structure of phenanthridine dyes (pseudo-base, reduced, oxidized, and cationic derivatives [32]) is shown in Fig. (**16.2**).

In the presence of appropriate substrates the reagent 3-(4,5-dimethylthiazol-2-yl)-2,5-diphenyl-tetrazolium bromide (MTT) is reduced by the corresponding dehydrogenases yielding a purple formazan product (see Fig. **14.2**). MTT assay (regarded as the golden standard), is the most popular test for cytotoxicity and viability, which is based on its reduction to the formazan in metabolically active cells [33 - 35]. It must be noted that after vital labeling of rats by intraperitoneal or subcutaneous injection of neotetrazolium and nitro blue tetrazolium, a violet formazan appears in kidney and liver lysosomes, as well as in peritoneal mast cell granules [36].

It is known that MTT is mainly reduced in the cytoplasm [37, 38], and that the formazan product is accumulated in lipid droplets [39, 40] (see Chapter 14.2.2). Therefore the MTT assay is a misleading method for testing mitochondrial activity. The main source of reducing power is the reduced nicotinamide-adenine dinucleotide (NADH) [38, 41], but other reducing intracellular compounds (*i.e.* ascorbic acid, dihydrolipoic acid, cysteine, tocopherols, glutathione, *etc.*) can be also involved.

It has been reported that the reduction of 5-cyano-2,3-di-*p*-tolyltetrazolium chloride (CTC) generates a fluorescent formazan (λexc: 380-440 nm, λem: 600 nm) [42, 43], which is suitable for microscopical studies (see Fig. **14.4**). Although the fluorescence of other tetrazolium formazans has been neglected, that of MTT can be clearly visualized under green excitation (see Figs. **14.5** and **14.6**). It is tempting to assume that as the number of free bond rotations is lower for diaryl (CTC) than for triaryl (MTT) formazans, the thermal decay would be lower in the case of CTC with increase of the Φ_F. Anyway, triaryl formazans are also efficient fluorophores in the solid state or when bound to rigid sites.

The redox indicator resazurin is the N-oxide derivative of an oxazone dye that reduces to resorufin by dehydrogenase activity. Resazurin sodium salt is employed as 1% solution in PBS or Hanks solution for several hours (kit TOX-8, Sigma) in proliferation and cytotoxicity assays (see Chapter 14.2.3 and Fig. **14.8A**). It is accepted that the conversion of resazurin to red-fluorescent resorufin occurs mostly in the mitochondria, and therefore the amount of resorufin generated can be used as indicator of metabolic activity [44 - 46]. Spectral data of resazurin and resorufin in water are following: the non-fluorescent and blue dye shows λab: 598 nm; the fluorescent, red-violet dye shas λem: 585 nm, and can be excited at 546 nm. Alamar blue® is a trade mark product (solution of resazurin and stabilizing agents [kit BUFO12]) promoted for fluorescent assays of cytotoxicity and cell proliferation (www.bionova.es, www.ab-direct.com/ alamarBlue) instead MTT, ^{3}H-TdR, *etc*.

RedoxSensor Red (2,3,4,5,6-pentafluoro-dihydro-tetramethylrosamine) is a probe whose oxidized product appears localized in relation to the cytosolic redox potencial. The probe passively enter cells and is oxidized in the cytosol to a red-fluorescent product (λexc: 540 nm, λem: 600 nm), which then accumulates in mitochondria of proliferating cells, or the probe is transported to lysosomes of quiescent cells, where it is subsequently oxidized [47].

1.3. Organelle Labeling

As described above, in the Rh123 assay the recognition of a live cell involves the visualization of mitochondria, which can be seen as granular or thread-like structures whereas the nucleus appear non-fluorescent. In "compromised" cells, Rh123 emission is diffuse and mitochondria are not resolved. In dead cells the nuclear fluorescence of Rh123 is the most striking event, likely related to the "relocalization" of probes after cell damage [48].

The neutral red (NR) uptake assay is used in cell viability and cytotoxicity studies [46, 49, 50]. It is assumed that this colorimetric assay is based on the intracellular accumulation of the dye in lysosomes [51, 52] (see Fig. **15.13B**). NR uptake occurs in live cells and is best visualized in fluorescence microscopy, but it must be noted that NAs-containing structures (*i.e.* nuclei, nucleoli, basophilic cytoplasm) are also stained in dead cells. Therefore, differences between live and dead cells using NR are rather morphological, and quantitative results in cell populations could be misleading.

Transmembrane potential changes are essential events that control basic cellular processes, and they are due to the non-equal distribution of ions (mainly K^+, Na^+, and Cl^-) on the outer and inner face of the membrane/aqueous interface. Fluorescent probes are important tools for assessing these changes regarding mitochondria and plasma membrane potentials. The most used carbocyanine dyes for potential measurements are JC-1, JC-9, $DiSC_3(5)$ and $DiOC_5(3)$ [53] (see

Chapter 15.5.4). Merocyanine 540, oxonols and styryl dyes are also used. Dye aggregation, fading, non-specific binding to macromolecules, cytotoxicity, and photodynamic damage are common drawbacks that must be taken into account.

1.4. Protein Quantification

One of the most suitable dyes to reveal protein amount is sulforhodamine B (SRB) [54], widely used for the colorimetric assay in cytotoxicity screening [46]. In contrast with anterior methods, the SRB assay measures total protein content due to proliferation, and does not rely on cell functionality [46, 54]. At present the SRB assay is the preferred high-throughput assay of the National Cancer Institute (NCI) in the USA and is the assay used in the NCI's lead compound screening programme [55]. On account of the fluorescence of rhodamine dyes, the emission of SRB should be suitable for assays with improved sensitivity.

2. TYPES OF CELL DEATH

Several types of cell death, namely apoptosis, necrosis, autophagia, mitotic catastrophe, *etc.*, have been reviewed [56 - 61]. The capacity of many chemotherapeutic agents to induce either apoptosis or necrosis in a dose-dependent mode makes necessary clear criteria to identify the type of cell death in a rapid and objective way. Now some correspondence between these types of cell death and old histopathological descriptions can be made. The term "pyknosis" reflects a rather uniform chromatin compaction equivalent to necrosis *in vitro*, whereas "karyorhexis" reflects chromatin fragmentation that would be equivalent to apoptosis. "Atypical mitosis" (tri- and multipolar spindles, *etc.*) seems to represent the so-called "mitotic catastrophe". In addition to changes in the nuclear morphology, relevant alterations in the cytoskeleton are evident in apoptotic and necrotic processes [58, 62, 63]. Toluidine blue, neutral red, H-33258, *etc.*, have been applied in morphological studies to identify the type of cell death [16, 64, 65].

2.1. Apoptosis

There is a growing interest in apoptosis from diverse fields of research, but it is specially relevant in oncology. The development of new chemotherapeutic agents such as taxoids, camptothecins and anthracyclins has provided apoptogenic therapeutic tools, based in microtubule and DNA interactions. Typical features observed in apoptotic cells are (a) condensation and fragmentation of chromatin, (b) cell shrinkage, presumably by cytoskeletal degradation, (c) exposure of phosphatidylserine at the outer surface of the plasma membrane, and (d) cytoplasmic blebbing and formation of apoptotic bodies [58, 59].

Microscopic quantification of apoptosis using H-33342 and AO/EB has been described [27 - 29]. The characterization of apoptosis can be now made by fluorescence microscopy and flow cytometry (*i.e.* measuring cell parameters such as annexin V binding, mitochondrial membrane depolarization, and caspase activity). Finally, permeabilization of the plasma membrane occurs as a late event during apoptosis *in vitro*. General articles and reviews concerning apoptosis should be consulted [66 - 69].

The analysis of nuclear morphology and TUNEL assay are widely used to identify apoptotic cells (Fig. **16.5**), in which the condensation-fragmentation of chromatin is the most conspicuous feature. The ladder pattern of DNA in agarose electrophoresis is another biochemical feature of apoptosis. The disorganization of actin microfilaments (Fig. **16.6**), and loss of E-cadherin mediated cell-cell adhesion are also characteristic changes during apoptosis [63, 70], as well as mitochondrial membrane depolarization as revealed by the mitochondria labels JC-1 and JC-9, which are probes for the trans-membrane potential (Δ_Ψ).

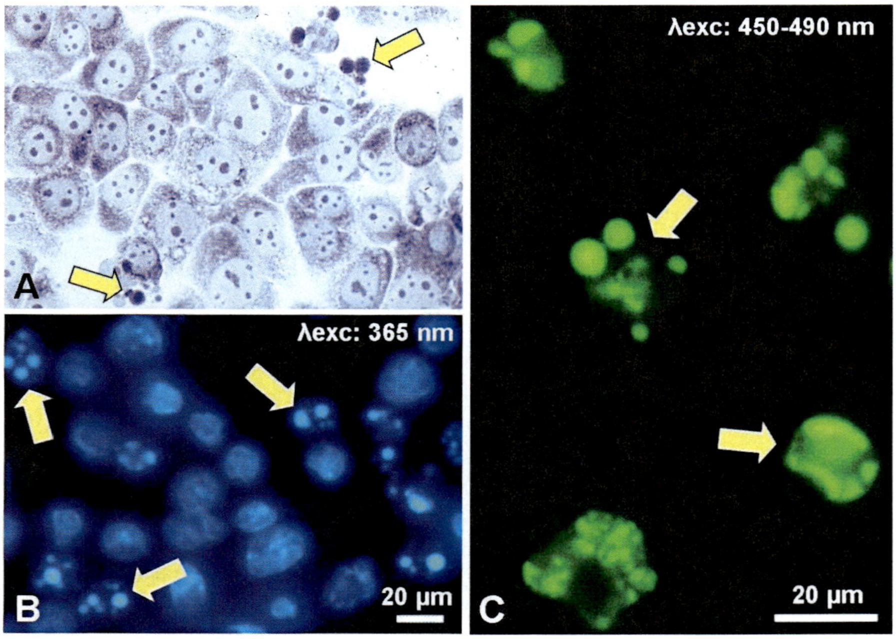

Fig. (16.5). Apoptotic morphology and TUNEL assay in cell cultures. **A-C:** Apoptosis in Pam212 cells induced by sublethal ZnPc photodynamic treatments observed after toluidine blue (**A**), H-33258 (**B**), and TUNEL assay using fluorescein-12-dUTP (**C**). Arrows indicate apoptotic cells.

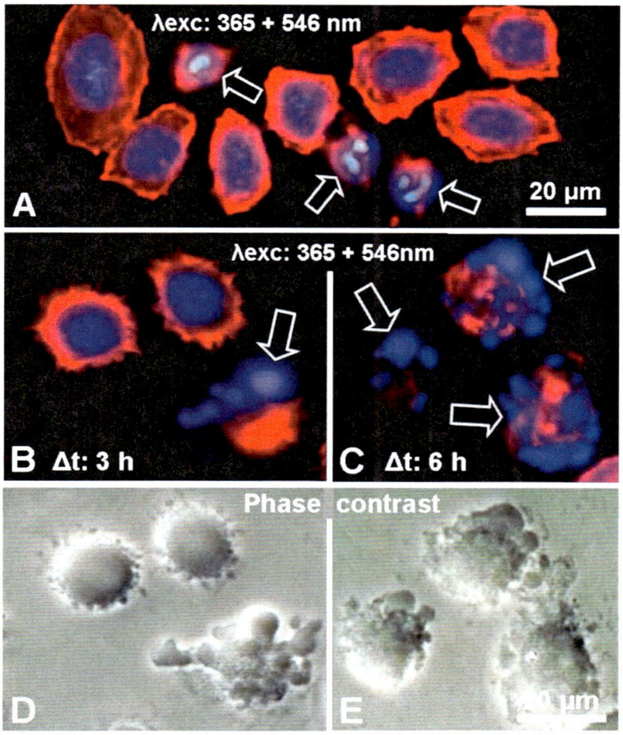

Fig. (16.6). Apoptosis and actin cytoskeleton. **A:** HeLa cells 3 h after sublethal photodynamic treatment with ZnPc, labeled by TRITC-phalloidin and DAPI. **B-E:** HeLa cells 3 h (B, D), and 6 h (C, E) after photodynamic treatment with the tricationic porphyrin Py_3MeO-TBPo, labeled by TRITC-phalloidin and Hoechst 33258. Apoptotic cells are indicated by arrows.

Apoptotic translocation (externalization) of phosphatidylserine (PS) to the external side of the plasma membrane is one of the first indicators of apoposis, which can be revealed by treatment with fluorescently labeled annexin V (AV, anticoagulant human protein that binds PS with high affinity in presence of Ca^{2+}) [71]. This assay is represented in Fig. (**16.7**), which shows the binding of AV-FITC or AV-Alexa Fluor 488, followed by counterstaining with PI, and observation using λexc: 450-490 nm.

But AV has several drawbacks: 1) it is an expensive protein, 2) it requires extracellular Ca^{2+} for binding to externalized phosphatidylserine, and 3) it reacts with membranes that contain by-products of lipid peroxidation. A low molecular-weight fluorescent probe that mimics AV is the zinc-dipicolylamine-anthracene compound (PSS-380, λexc: 380 nm, λem: 440 nm), which acts as a fluorescent sensor of apoptotic cells [72]. In addition to morphological criteria, translocation of cytochrome c from mitochondria to cytosol, translocation of Bax from the cytosol to mitochondria (Fig. **16.8A-E**), and occurrence of caspases are well known events used in fluorescence methods for apoptosis detection.

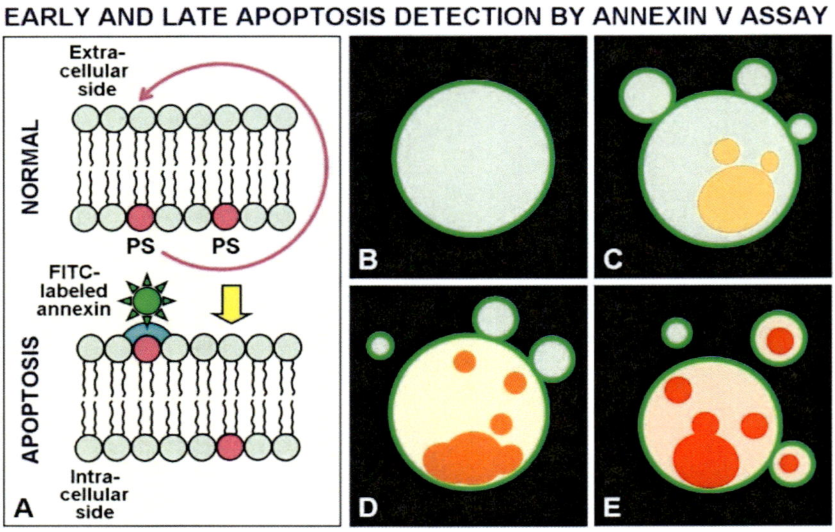

Fig. (16.7). Schematic assay for annexin V. **A:** Diagram of the lipid bilayer with serine phospholipids (PS, violet) and their translocation from the internal to the external side during apoptosis. **B-E:** Sequence of annexin V (FITC)- and DNA (PI)-labeled cells during early (**B**), mid (**C**, **D**) and late (**E**) apoptosis. Green signal: early apoptosis (**B**); green and red signals: late apoptosis (**E**).

Caspases are cysteine-activated, aspartate-directed proteases involved in proteolytic cascade of apoptosis. Caspase-3, -7, and -2 are visualized using suitable fluorogenic substrates (see Chapter 14.2.5, and Fig. **14.13**). The cell permeable bis-L-aspartic acid-(Rh110)-amide substrate (Rh110 flanked by two aspartic acid residues) can be used for detection of caspase-3 and -7 [73]. Most applied methods for caspase-3 involve the recognition site Asp-Glu-Val-Asp, either labeled with rhodamine 110, or 7-amino-4-methyl-coumarin [10], yielding green and blue fluorescence, respectively. The recognition site of the initiator caspase-8 is Ile-Glu-Thr-Asp, and it can also be labeled. At present, caspase detection by immunofluorescence microscopy is also used [74]. Photodynamic-induced apoptosis by ZnPc and *m*-tetra(hydroxyphenyl) porphycene involves the immediate activation of caspase-2 in the Golgi apparatus of HeLa and A-549 tumor cells (Fig. **16.8F-H**) [75, 76].

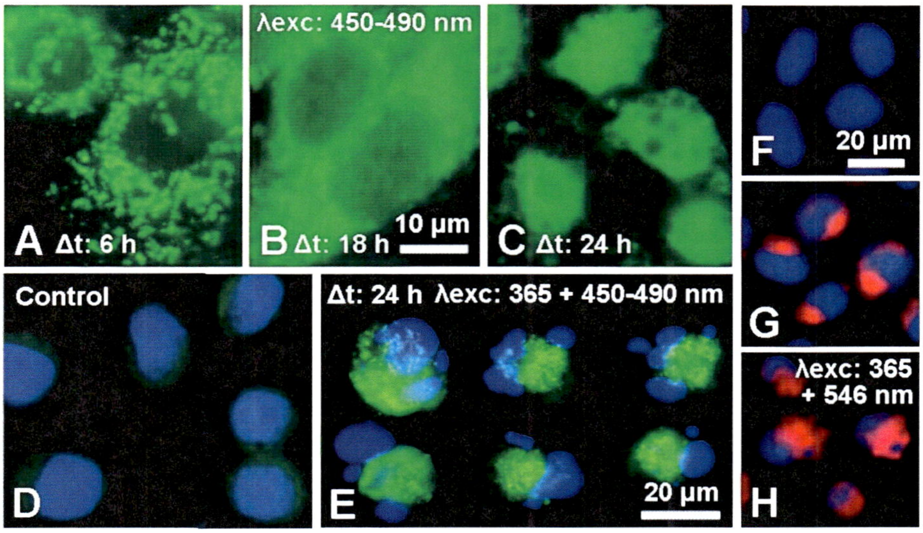

Fig. (16.8). Apoptotic cells: **A-C:** Release of cytochrome c from mitochondria to the cytosol of Pam212 cells 6, 18, and 24 h, respectively, after ZnPc photosensitization; anti-cytochrome c IgG, and IgG-FITC. **D**, **E:** Translocation of Bax (Bcl-2, pro-apoptotic) from cytosol to mitochondria, induced in HeLa cells by Py_3MeO-TBPo photosensitization; anti-Bax IgG, IgG-FITC, and H-33258. Observe the apoptotic morphology (**E**). **F**-**H**: HeLa cells (F, control) subjected to *m*-THPPo photosensitization processed 30 min (**G**) and 3 h later (**H**); anti-cleaved caspase-2 IgG, IgG-Texas red, and H-33258. Observe the red signal in the Golgi apparatus.

Clearance of apoptotic cells, also termed efferocytosis, is mediated by amateur and professional phagocytes [77], but the process is not often seen in cell cultures. Interestingly, homologous phagocytosis of apoptotic remants in cultured ovarian granulosa cells can be detected using current hematoxylin, DAPI, and TUNEL staining (Fig. **16.9**), which reveals both apoptotic cells and bodies engulfed by granulosa cells [69].

2.2. Necrosis

Highly damaging xenobiotics, chemicals, drugs, and physical conditions often induce cell necrosis. In cultured cells exposed to lethal photosensitization, typical features are (a) loss of plasma membrane integrity and rapid influx of Na^+, Ca^{2+}, and water, resulting in cytoplasmic swelling, (b) occurrence of tubulin-positive bubbles, (c) release of cell components in the extracellular space, (d) nuclear condensation with swelling of the perinuclear space, and (e) loss of the F-actin and microtubule network [58, 78, 79]. Immediate plasma membrane bubbles and late nuclear pyknosis are characteristic events shown in Fig. (**16.10**).

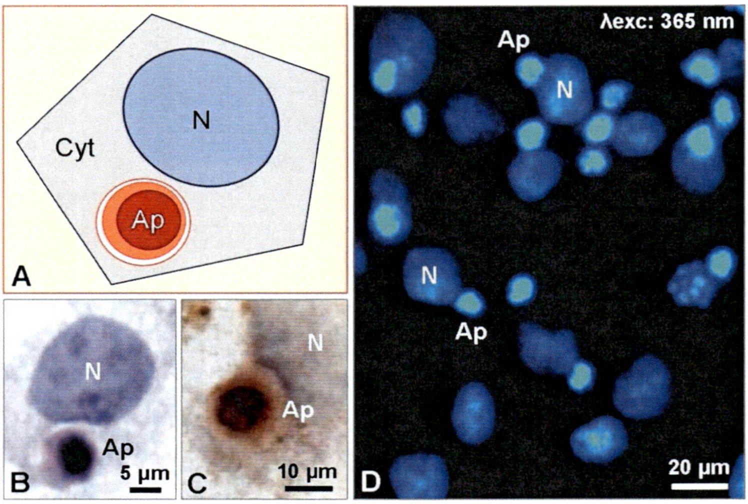

Fig. (16.9). A: Diagram of homologous phagocytosis of apoptotic bodies (Ap), showing cytoplasm (Cyt) and nucleus (N) of the phagocytic cell. **B-D:** Apoptosis from bovine granulosa cells (BGC-1) treated with 100 nM leuprolide acetate for 24 h, fixed in methanol, and observed after Mayer´s hematoxylin (B), TUNEL assay (peroxidase-DAB) and hematoxylin counterstaining (C), and 5 μg/mL DAPI (D), showing nuclei (N) and apoptotic bodies (Ap).

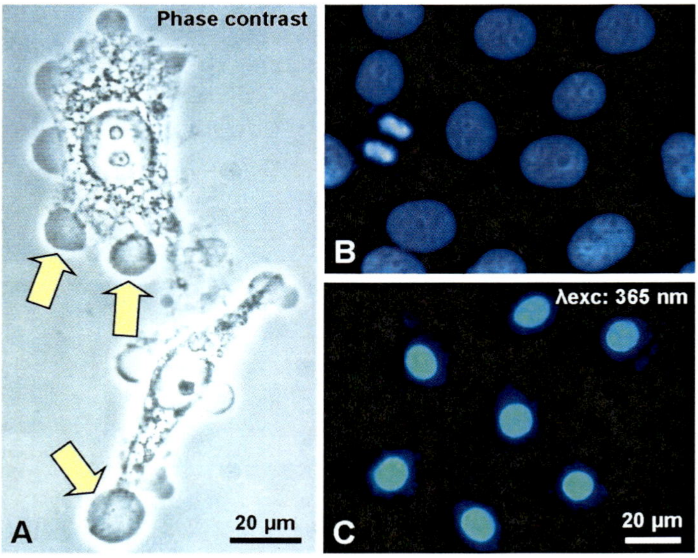

Fig. (16.10). Necrotic cells. **A:** HeLa cells after lethal phodynamic treatment with 5 μM TPPo, showing immediate surface bubbles (arrows). **B**, **C:** H-33258 staining of A-549 cells (untreated, B) and necrotic cells 5 h after lethal photodynamic treatment (C) with 5 x 10^{-7} M PdTPPo, showing severe chromatin condensation (pyknosis).

An interesting feature of necrotic cells is the frequent intranuclear accumulation of tubulins [58]. After lethal phosensitization with ZnPc, immuno-detection of α- or β-tubulin reveals a highly selective signal in the nuclear remnants and dispersed chromatin droplets (Fig. **16.11**). Intranuclear tubulin occurs in necrotic cells but not in apoptotic ones, and in contrast with the early appearance of tubulin in surface bubbles, nuclear localization of tubulin is a late event. This intriguing process has been described in guinea pig lung mast cells during anaphylactic shock [80]. Using immunofluorescence and electron microscopy, α-tubulin and microtubules were also found in nuclei of lethally damaged cisplatin-treated C6 glioma astrocytes in culture [81].

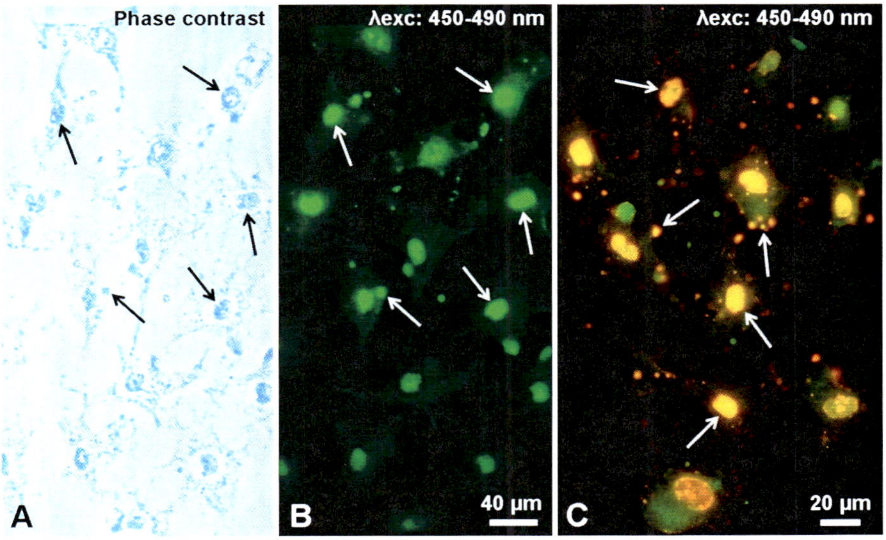

Fig. (16.11). Nuclear localization of α-tubulin in HeLa cells. **A:** Necrotic cells after lethal photosensitization with 5 x 10^{-6} M ZnPc, Δt: 24 h. **B:** The same cells observed after FITC-immunofluorescence for α-tubulin, with positive nuclei and micronuclei (arrows). **C:** Similar cells labeled for α-tubulin with IgG-FITC and counterstained with the cationic porphyrin TMPyP, showing yellow colocalization of α-tubulin and DNA signals (arrows).

In contrast to the uptake of dyes into dead cells, the necrotic release of the soluble cytosolic enzyme glucose-6-phosphate dehydrogenase or lactate dehydrogenase (LDH) in the culture medium after membrane injury of damaged and dead cells can be detected by reduction of resazurin to resorufin. A fluorometric method of coupling the NADH to the electron acceptor resazurin in the presence of lactate and the cofactor phenazine methosulfate, yielding fluorescent resorufin, has been used for LDH measuring [82]. High levels of lactate dehydrogenase occur in acute and chronic leukemia, in relapse and carcinomatosis.

2.3. Other Types of Cell Death

Mitotic Catastrophe

Mitotic catastrophe is a recently characterized cell death process intimately related with defective mitosis behaviour causing cell cycle arrest, spindle alterations, aneuploidy, and finally cell death [83]. This process is also linked with genomic instability, tumor progression, drug resistance, and senescence. Mitotic catastrophe typically results from apoptogenic treatments (*i.e.* sublethal photosensitization, podophyllotoxin derivatives such as etoposide, *etc.*) [60, 84]. Mitotic (metaphase) arrest is followed by aneuploidy and cell death by an apoptotic process. Morphological features of mitotic catastrophe can be easily recognized by anti-tubulin immunofluorescence and staining of chromatin DNA (*i.e.*, DAPI, Hoechst dyes, *etc.*).

Autophagy

The appearance of autophagic vacuoles is related to processes of either cell survival or cell death. Autophagy (from the Greek, "to eat oneself") is primarily a pro-survival rather than a pro-death mechanism. It denotes the process of self-cannibalization through a lysosomal degradation pathway, and corresponds to the cell's major regulated mechanism for degrading long-lived proteins and the only known pathway for degrading organelles. The basic processes of autophagy have been reviewed [85, 86].

In mammalian cells, two major catabolic mechanisms (the ubiquitin-proteasome system and the autophagy pathway) degrade short- and long-lived proteins, toxic protein aggregates and damaged organelles. Misregulation of these processes is associated with a number of diseases, from cancers to cardiac myopathies and neurodegenerative disorders. Autophagic vacuoles or autophagosomes containing various organelles are morphological characteristics of cells during autophagy [87]. The methionine analog homopropargyl-glycine (HPG) contain an alkyne group that can be detected with a fluorescent Alexa Fluor® azide through a click reaction (see Chapter 4.4.3). HPG is incorporated into proteins undergoing translation during the labeling pulse, and after a chase period, HPG-containing proteins are revealed by the azide dye [88]. In adition, "nucleophagy" appears as a new process of cell death [89], although it appears less relevant.

Regulated Necrosis

Regulated necrosis (necroptosis) induces a cell morphology that does not correspond to any classical apoptotic or necrotic process. Cells treated with ZnPc for 3 h, irradiated, and observed 24 h later show nuclei with the nuclear envelope apparently intact, and chromatin condensed in the form of small spots. This "granulated nuclei" morphology represents the process of necroptosis [90].

2.4. Senescence

Cellular senescence occurs when cells move from an actively dividing to a non-dividing state, yet remaining metabolically active. This is due for accruing a fixed number of cell divisions, or when cells are subjected to a variety of damaging factors. Senescent cells (as well as quiescent cells) do not undergo DNA synthesis, and are associated with several age-related diseases. Although senescent cells may remain viable for a long time, finally cell death usually occurs. Senescence-associated β-galactosidase (located in lysosomes, optimal pH: 6.0) is a suitable biomarker of senescence. β-Galactosidase staining can be accomplished by using the substrate 5-bromo-4-chloro-3-indolyl-β-D-galactoside, which generates the halogenated indoxyl and then the insoluble indigo derivative. A senescent cells staining kit is available for the histochemical detection of these cells, which is based on this procedure [91, 92]. Fluorogenic methods for revealing this enzyme are mentioned in Chapter 14.2.1.

3. GENOTOXICITY

Several physical and chemical agents can produce DNA damage, consisting in modification or replacement of bases, breackage of polynucleotide chains, *etc*.. Mutagenesis and inhibitory effects on replication and transcription, leading to cell death, result from DNA damage, and therefore methods to assess genotoxicity are of increasing importance. Oxidative DNA damage is of course the most relevant. Although there are useful tests to reveal DNA damage (*i.e.* presence of 8-hydroxyguanine, FISH and flow cytometry), most used microscopical methods are TUNEL, and chromatin dispersion techniques such as comet and halo assays.

3.1. TUNEL Assay

The TUNEL method (terminal deoxynucleotidyl transferase-mediated dUTP nick-end labelling, also named ISEL assay, *in situ* end labeling) recognizes internucleosomal DNA breaks induced by DNases typical of apoptotic cells [9, 93] (Figs. **16.5C**, and **16.9C**). TUNEL assays utilize generally fluorescein-dUTP, but also other labeled nucleotides such as biotinylated- and BrdU-dUTP can be used, showing selective reaction in DNA from condensed and/or fragmented chromatin of apoptotic cells. Oligobiotinyl and Br derivatives are revealed by streptavidin-peroxidase and AlexaFluor 488-labeled anti-BrdU-IgG, respectively.

3.2. Comet Assay

The single-cell micro-gel electrophoresis (SCGE) or "comet" assay is a reliable and sensitive technique for measuring DNA breakage in individual cells. As many of the designed antitumor drugs target genomic DNA, the comet assay is a valuable tool to monitor DNA damage. The development, protocols, and applications of this popular genotoxicity assay have been reviewed [94 - 99]. At present, comet assay associated to FISH is an interesting approach for analyzing DNA damage and its intranuclear localization [100, 101] (see Chapter 17.3).

Cells or nuclei must be in suspension, and then fresh blood or hemopoietic cells are specially suited. They are mixed with low melting point agarose at 37 °C and smeared on a microscopic slide. Nuclear lysis (with DNA denaturation and unwinding) is achieved by alkaline pH and detergent. Slides are then subjected to electrophoresis, stained by EB or PI, and observed under fluorescence microscopy (Fig. **16.12**).

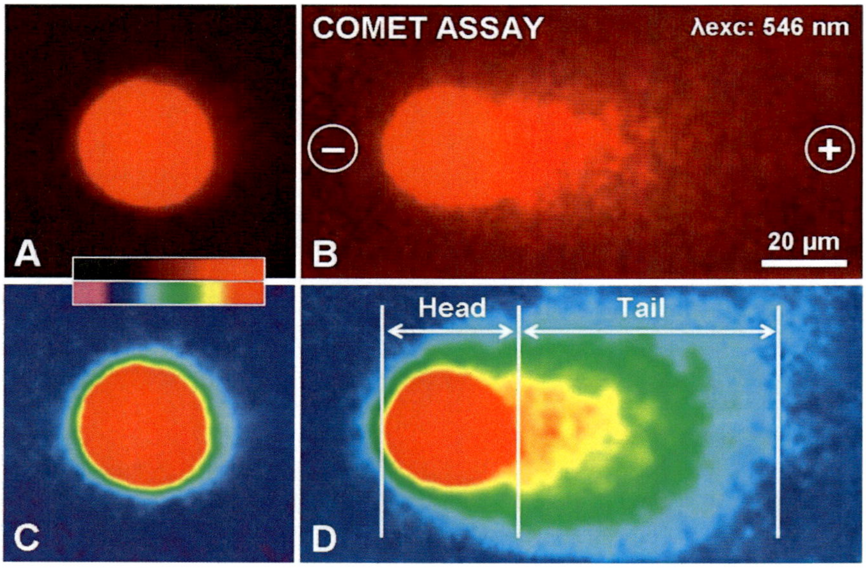

Fig. (16.12). Comet assay in HeLa cells stained with ethidium bromide. **A:** Control (untreated) cells. **B:** Cells after 5 Gy X-rays, subjected to alkaline (0.3 M NaOH) electrophoresis. Positive and negative symbols are anode and cathode, respectively. C, D: Pseudo-color images of pictures A and B, respectively, showing the splitting of the red gradient from head and tail into color equidensities. The red-rainbow code is indicated.

Negatively charged DNA fragments migrates to the anode (+), small chains faster than large ones, whereas intact (native) DNA does not migrate at all. Nuclei from normal cells appear round, but nuclei with DNA strand breaks appear like comets.

Quantitative evaluations on the degree of DNA breakage can be performed by using specific software for the geometrical analysis of comets. SYBR Green I and Giemsa staining have been also applied [102].

3.3. Halo Assay

After DNA breakage, treatments of cell nuclei with strong saline or alkaline solutions induce chromatin dispersion generating nuclear halos that surround nuclear remnants and can be visualized at microscopical level. These simple methods represent a rapid assay in genotoxicity studies. The fast halo (alkaline) assay assesses the extend to which DNA strand breakage occurs in a single cell as a result of genotoxic treatment [103, 104]. In this test, control and treated cell populations are first immobilized whithin a thin layer of low melting point agarose and then subjected to a strongly alkaline lysis (0.3 M NaOH for 30 min). Both compact and dispersed chromatin are revealed by using EB or PI, and the radial diffusion of DNA fragments can be easily observed and quantified as a fluorescent halo surrounding the nucleus (Fig **16.13**). Giemsa or Wright staining and bright-field microscopy are also possible [105].

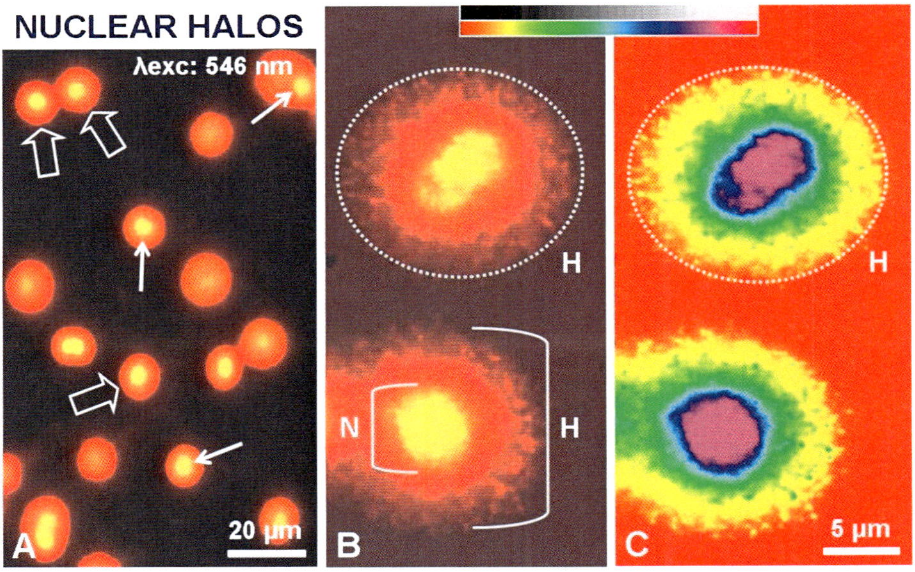

Fig. (16.13). A, B: Nucleated erythrocytes from chicken blood subjected to 0.88 mM H_2O_2 for 15 min, mixed with 2% low melting point agarose, and smeared on glass slides. When solidified, smears were treated with 0.3 M NaOH for 15 min, air-dried, and stained with 10 µg/mL EB. Observe the nuclei (small arrows, N) and nuclear halos (large arrows, H). **C:** Pseudo-color image of B, showing the splitting of the gray gradient into color areas (LUT: spectrum). Top: gray-color code.

A simplified fast halo method and a mechanistic model for the differential formation of alkaline or saline (reverse) nuclear halos after DNA breakage have been described [105]. By using image processing and analysis, this simplified nuclear halo assay can be used as a rapid, reliable and inexpensive procedure to detect and quantify DNA breakage induced by ionizing radiation and chemical agents. The nuclear area (N), halo plus nucleus area (H), and the N/H index can be easily determined, showing values of 1 for normal (non-damaged) nuclei or lower for damaged nuclei with chromatin halos. Halo procedures are more suitable than comet assays, which are time-consuming and require more sophisticated instrumental.

4. CELL SIGNALING

Several cell signals (inducing both survival or damaging effects) are generated as response to stimuli or changes in the environmental conditions of cells and tissues, and methods have been designed to detect these signaling molecules. The excessive accumulation of reactive oxygen species (ROS) and free radicals, and their negative impact on cell homeostasis, collectively known as oxidative stress, are implicated in the pathogenesis of aging, cancer, neurodegeneration, cardiovascular diseases, and inflammation. Likewise, nitric oxide (NO) is involved in diverse physiological processes, including smooth muscle relaxation, platelet inhibition, neurotransmission, immune regulation, *etc.* Recent reviews on the assessment of ROS and NO signaling have been published [106 - 108].

4.1. ROS Detection

The original fluorescent assay to detect ROS and oxidative stress is based on the reduced non-fluorescent fluorescein derivative, dihydrofluorescein (also named fluorescin, which must be not confused with fluorescein). In the presence of ROS, dihydrofluorescein is oxidized and regenerates the highly fluorescent dye, fluorescein. At present, dihydro-dichloro-fluorescein diacetate (H_2DCFDA) is often used to reveal oxidative stress in cells and tissues (Fig. **16.14**). In cultured cells or whole organisms exposed to oxidative stress, appearance of ROS is revealed by oxidation of H_2DCFDA, which produces the fluorescent DCF (dichlorofluorescein) [108 - 112]. Halogen atoms makes the dichloro derivative more resistant to oxidation and fading due to ROS. As an example, treatment of alfalfa plantlets with agents such as L-buthionine sulfoximine (BSO, an inhibitor of glutathione synthesis), as well as cadmium poisoning increases the cell peroxidation level, which can be demonstrated using H_2DCFDA (Fig. **16.15**).

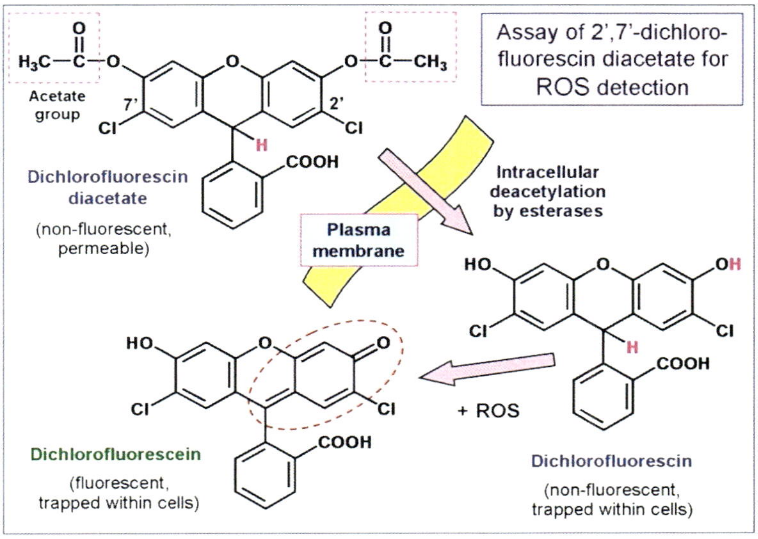

Fig. (16.14). Assay of 2',7'-dichloro-dyhydrofluorescein diacetate for ROS detection, showing the action of esterases on acetate groups, and the oxidation of the product to give the final fluorophore, fluorescein.

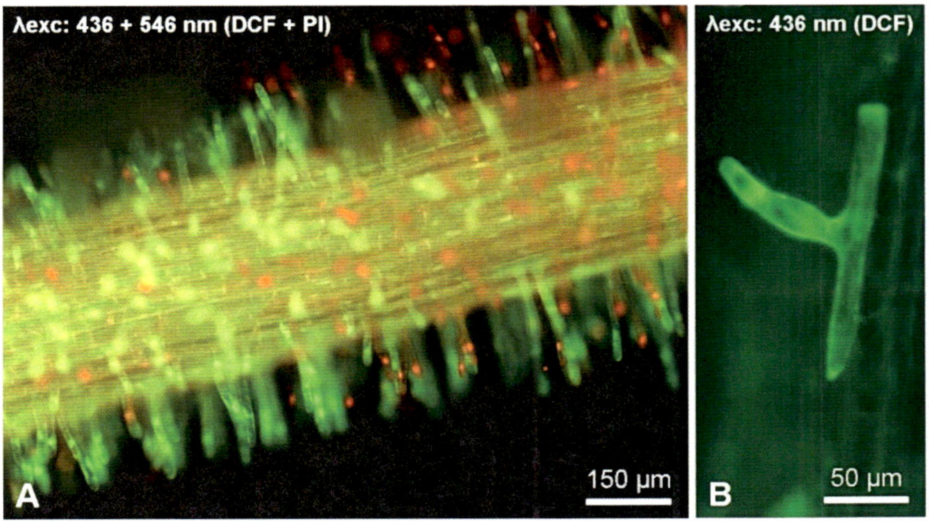

Fig. (16.15). Alfalfa (*Medicago sativa*) roots grown between slide and coverslip, treated with 1 mM L-buthionine sulfoximine (or 20 μM $CdSO_4$) for 24 h. After further 24 h incubation in culture medium, roots were treated with 10 μM H_2DCFDA and 25 μM propidium iodide for 15 min. The root segment corresponds to root hairs nearly finishing elongation and shows dead cells (red nuclei) (courtesy of C. Ortega-Villasante).

Aminophenyl- and hydroxyphenyl-fluorescein are other probes for ROS detection [113], which appear more selective and stable than H_2DCFDA, also allowing the visualization of the hypochlorite anion (ClO^-) generated in activated neutrophils. Likewise, Singlet Oxygen Sensor Green® (SOSG) is a commercial probe (possibly a fluorescein-anthracene derivative) for $^1O_2$ [114]. Disadvantages are that it becomes fluorescent at alkaline pH, and itself induces photoproduction of $^1O_2$ [115]. Detection of mitochondrial ROS can be made using the non-fluorescent dihydro-rhodamine 123, which accumulates in the mitochondria where it is oxidized to the fluorescent rhodamine 123 [10]. The same occurs in the case of dihydro-fluorescein diacetate [116]. In addition, cell-permeant reduced and non-fluorescent CellROX® ROS sensors show strong fluorescence when oxidized in nuclei, cytoplasm or mitochondria. Dihydro-calcein-AM is also used in flow cytometry as ROS indicator [117].

Genetically encoded fluorescent probes are unique tools to study ROS production in live cells. Current recombinant redox biosensors have green emission (*i.e.* roGFP2-Orp1, and the HyPer family reporters) [118]. HyPer consists of a circularly permuted YFP (cpYFP) molecule inserted into the regulatory domain of the *E. coli* H_2O_2 sensing protein OxyR [119], whose activity relies on two cysteine residues, Cys199 and Cys208. When the OxyR domain is exposed to H_2O_2 with disulfide formation, an intramolecular change occurs shifting the λexc of the attached cpYFP from 420 to 500 nm, while the emission peak at 516 nm remains unchanged. Due to this H_2O_2-induced change in λexc, transgenic HyPer is expressed as a ratiometric and reversible biosensor for targeting organelles (*i.e.* mitochondria, peroxisomes, and chloroplasts). Likewise, HyPerRed is a new red fluorescent sensor that uses OxyR and cpRFP for H_2O_2 detection [120].

On the other hand, the strong reducing agent 3,3′-diaminobenzidine (DAB) is often used as chromogenic substrate in bright-field microscopy for assessing oxidative processes, giving dark brown and insoluble polymeric products [2, 3, 121]. Spontaneous production of ROS in mitochondria and mast cell granules of frozen tissue sections [122, 123], and photodynamic production of ROS [124] can be also visualized by oxidation of DAB. Likewise, photoconversion of fluorescent signals into colored and osmiophilic DAB precipitates has been applied for immunofluorescence and neuroanatomical studies (see Chapter 7.6).

It is worth to note that under oxidative conditions, the oxazine Amplex red generates the fluorescent resorufin (λexc: 563 nm; λem: 587 nm) (see Fig. **14.8**). In the case of lucigenin (10,10'-dimethyl-9,9'-bis-acridinium), its oxidation gives rise to two molecules of 10-methylacridone together with chemiluminescence at 470 nm, which has been used to detect the presence of superoxide anion radical in isolated mitochondria [10]. The probe is also fluorescent (λab: 455 nm, and λem:

505, both in water). Although lucigenin is mainly applied in non-imaging technologies, it has been used in fluorescence microscopy [125].

The redox-sensitive probe, hydroethidine (HE; the reduced form of ethidium, E^+) is used for detection of ROS in live cultured cells [126]. It is commonly assumed that the intracellular reaction between HE and ROS results in the formation of ethidium (E^+), which binds to DNA with emission enhancement (λexc: 480 nm; λem: 590–610 nm). However, a highly selective demonstration of the superoxide anion radical (O_2•-) can be achieved by means of HE, which gives the yellow fluorescent product, 2-hydroxy-ethidium (2OH-E^+, λexc: 480 nm; λem: 567 nm) (Fig. **16.2**) [112, 127 - 129].

Oxidation of HE is not accomplished by $^1O_2$, HO•, H_2O_2 or nitrogen radicals. Normally, the major source of O_2•- appears to be the mitochondria. Thus after oxidation of HE, 2OH-E^+ fluorescence in dendrites is punctate, blocked by CN^-, and colocalizes with mitochondria [130]. It must be noted that HE also has a blue fluorescence (λexc: 355 nm, λem: 420 nm) that is equivalent to that of benzidine (λem: 400 nm [131]). This feature results from the benzidine nature of the HE chromophore. Regarding inhibitors, sodium azide (NaN_3), histidine, β-carotene, and dimethylfuran prevent the oxidative stress generated by $^1O_2$. Care must be taken by using some of these drugs (*i.e.* NaN_3 inhibites ATP production, myeloperoxidase and catalase activity, and is potentially mutagenic [Owais and Kleinhofs 1988]).

The peroxidation of membrane lipids can be caused by drugs, xenobiotics, heavy metals, and photodynamic action, by inducing the formation of aldehyde and carbonyl groups from unsaturated fatty acids. As lipoperoxidation products could play a pathogenic role in human diseases (*i.e.* ischemia, inflammation, cancer, atherosclerosis, *etc.*), fluorescence methods for the histochemical detection of aldehyde and carbonyl compounds in tissue sections and isolated cells have been developed. They use Schiff's reagent, 2-hydroxy-3-naphthoic acid hydrazide/fast blue B, and 7-diethylamino-3-(2-benzothiazolyl)coumarin sulfonyl-hydrazide [132, 133].

Malondialdehyde produced by lipoperoxidation is a genotoxic agent and a known biomarker of oxidative stress. Its condensation with thiobarbituric acid gives a fluorescent product [134]. MitoFOX Green is a commercial probe for lipid peroxidation in mitochondria. In the food industry, lipid peroxidation is an important factor that affects food quality, nutrition, safety, and consumers' acceptance, and both the recognition of lipid peroxides and strategies to improve the stability of food are relevant.

4.2. NO Detection

In the current perspective, nitrogen monoxide (NO), identified with the free radical nitric oxide (NO•), is generated enzymatically by NO synthases and implicated in a number of biological functions. The broader chemistry of NO involves an array of interrelated and reactive redox forms: nitrosonium cation (NO^+), nitric oxide radical (NO•), and nitroxyl anion (NO^-) [135]. Neutral NO• has a single electron in its $2p$-π antibonding orbital; the removal of this electron forms NO^+, and the addition of an electron forms NO^-. These forms show similarities (but also differences) with analogous diatomic molecules such as O_2.

The reagents *o*-diamino-benzene and 2,3-diamino-naphthalene can be used to detect NO, giving fluorescent benzo- and naphtho-triazole (azoimide) products (Fig. **16.16**), in the later case with λexc: ~ 340 nm and λem: ~ 410 nm. Some triazole compounds are also constituents of fluorescent brighteners. Likewise, diamino-benzidine (DAB) could be an adequate probe to form a fluorescent bis-benzo-triazole derivative upon binding to NO.

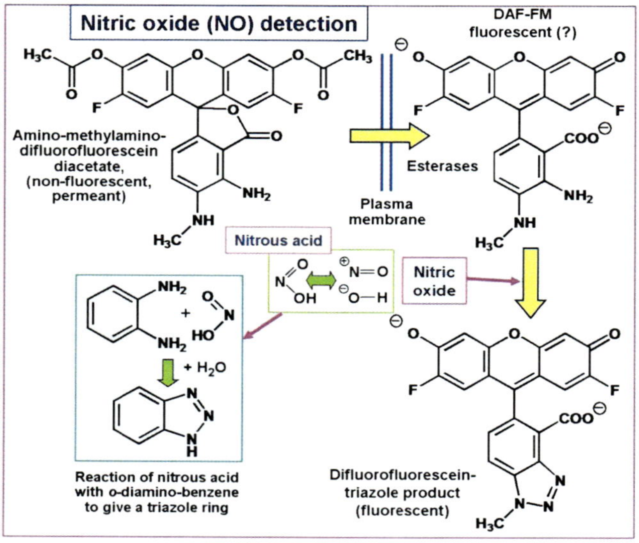

Fig. (16.16). Detection of nitric oxide by means of reaction with *ortho*-diamino-aryl probes (*i.e.* amino-methylamino-difluorofluorescein diacetate) to yield fluorescent triazole products. Observe the intermediate formation of nitrous acid.

At the microscopical level, the reagent *o*-diamino-fluorescein diacetate (DAF-DA) is used as a cell permeable probe for fluorescent detection of NO in living cells [112, 136]. The probe is hydrolyzed by esterases giving the parent DAF, which reacts with NO to form the fluorescent triazole derivative. The same occurs using the similar probe amino-methylamino-difluorofluorescein diacetate (DAF-FM-DA) (Fig. **16.16**), through a reaction with the oxidized NO^+ cation. The green fluorescing product can be visualized with standard fluorescein exciting filters (λexc: 450-490 nm). However, it would be expected that after esterase hydrolysis, the green emission of the DAF intermediate (reported as weakly fluorescent [108]) could interfere with that of the triazole product. This is a feature that deserves further investigation.

5. LABELING OF CELL POPULATIONS

Fluoresent labeling methods of cell populations for identification of cells in whole organisms are valuable in studies of cell traffic, motility, migration, transplant, *etc*., corresponding mainly to methods for phagocytosis, cell tracking, and neuronal tracing.

5.1. Phagocytosis and Cell Tracking

Complementary to the old labeling methods for the reticulo-endothelial system using China ink and bisazo dyes, at present there are new fluorescent probes of phagocytosis and tracking of macrophages. Most used are fluorescent latex particles of different size (see Chapter 18.4.1). FluoSpheres are commercially available fluorescent microspheres of different colors that are suitable for microscopy and flow cytometry studies. Commercial Green-Fluorescing *E. coli* BioParticles are also internalized by phagocytosis and assessed selectively in a microplate reader after quenching the extracellular signal from non-internalized bacteria by the addition of trypan blue to the culture medium. Likewise, the Fc OxyBURST Green reagent allows to monitor the oxidative burst in phagocytic cells. It consists of bovine serum albumin (BSA) covalently linked to dichloro-dihydro-fluorescein, and complexed with anti-BSA IgG antibodies. When this immune complex is internalized within phagocytic cells and subsequently oxidized, it developes green fluorescence [137].

Fluorescent probes for plasma membrane that remain bound long time at the uptake site can be employed to label isolated cell types, and observed in organs and tissues after reinoculation into whole animals [138]. Most used fluorescent probes for cell tracking are the vital dyes PKH 67 (green), and PKH 26 (red). Although finally internalized, no exchange of labels must occur after mixture of differentially labeled cell populations. Some of the plasma membrane probes are also used for neuronal and axonal tracing.

Lucifer dyes are highly fluorescent 4-amino-naphthalimides which are readily visible in living cells at concentrations and levels of illumination at which they are non-toxic [139]. Lucifer dyes are also markers of fluid-phase pinocytosis, and they are suitable for visualizing the retrograde axonal transport, and for covalent labeling of macromolecules. Intercellular connections by gap junctions are used for transmision of metabolic information between cells (*i.e.* signaling and regulating processes by ions, metabolites, *etc.*). In this context, the carbonyl-reactive Lucifer yellow CH with a hydrazine group (λexc: 438 nm, λem: 536 nm) is useful in marking nerve cells and gap junctions. This dye and the similar Lucifer yellow VS (with a vinyl-sulfone group) [3] are related to the acid fluorochrome brilliant sulfoflavine FF (CI 56205). Other fluorescent polar tracers such as pyrene derivatives can be used to track movement of fluids in whole organisms, mainly solute pathways in living plants [140].

5.2. Retrograde Axonal Tracers

The bisazo dyes, trypan blue and Evans blue have ben applied as retrograde axonal tracers, but at present new fluorochromes such as some diamidino-vinyl derivatives are more suitable. The cationic fluorochromes, fast blue (diamidino 253/50; diamidino-benzofuranoyl-vinyl-indole; λab: 372 nm λem: 400 nm, in water), true blue (diamidine-vinyl-bisbenzofurane), granular blue (*p*-amidin--phenyl-6-carboxamide-2-indolyl-phenyl ether), and fluoro-gold are used as retrograde axonal tracers in neuroanatomical studies [3, 141] (Fig. **16.17**). Diamidino yellow seems useful as axonal tracer [142], but unfortunately its chemical structure is poorly known. The bis-benzimidazole dye nuclear yellow is also suitable as retrograde fluorescent tracer.

Since its introduction in 1986 [143], fluoro-gold has gained popularity as retrograde axonal tracer due to its bright yellow emission in cytoplasm and lysosomes, good filling of dendritic processes, and compatibility with immunohistochemistry. It is now assumed that this dye corresponds to hydroxy-stilbamidine (2-hydroxy-4,4'-diamidinostilbene) [144]. Hydroxy-stilbamidine, like chloroquine and many other fluorescent weak bases, can cross cell membranes in their uncharged form, whereas the protonated form are trapped within the acidic interior of endosomes and lysosomes [144]. In addition, several carbocyanines such as DiO, DiI, and DiD ($DiOC_{18}(3)$, $DiIC_{18}(3)$, and $DiIC_{18}(5)$, with green, red-orange and far red emission, respectively) label the plasma membrane and have been used for cell tracking as well as for neuronal labeling (both retrograde and anterograde tracing) [145, 146]. Axonal labeling with biotinylated dextran-Texas red allows to identify neuronal bodies in different nervous nuclei using fluorescence microscopy and streptavidin-peroxidase-DAB by bright-field optics.

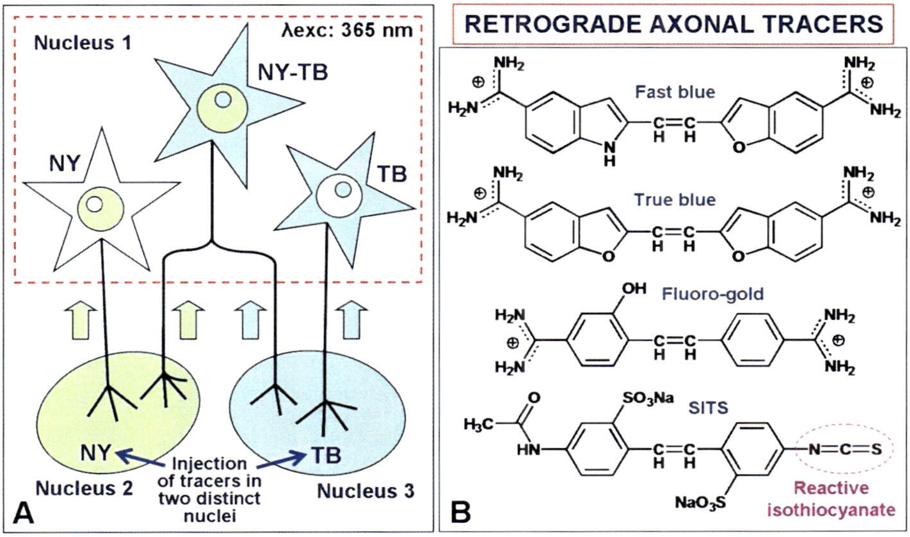

Fig. (16.17). A: Diagram depicting the neuronal labeling by fluorescent retrograde tracers and the use of double-labeling with different probes: nuclear mark by NY (nuclear yellow, green-yellow) and cytoplasmic-nucleolar mark by TB (true blue, blue). Neuronal bodies with bifurcated axons are visualized simultaneously. **B:** Structure of retrograde tracers fast blue, true blue, and fluoro-gold. SITS is shown for comparative purposes.

SITS (4-acetamido-4'-isothiocyanato-stilbene-2,2'disulfonate; λexc: 336 nm, λem: 436 nm) is an anionic fluorochrome sometimes used for detection of cationic proteins, but aqueous solutions are unstable [3]. Samples of SITS used as neuronal tracers are often of uncertain composition [147], but contrary to earlier statements, pure SITS is not retrogradely transported in neuronal axons. The transported agent is a red-orange impurity which arises occasionally during the preparation of SITS. A possible structure for this red-orange compound has been proposed: namely a dimeric azo or azoxy condensation product [148].

CONFLICT OF INTEREST

The author (editor) declares no conflict of interest, financial or otherwise.

ACKNOWLEDGEMENTS

This work was supported by a grant from the Ministerio de Economía y Competitividad (CTQ2013-48767-C3-3-R), Spain.

REFERENCES

[1] Waggoner AS. Fluorescent Probes for Analysis of Cell Structure, Function, and Health by Flow and

Imaging Cytometry.Applications of Fluorescence in the Biomedical Sciences. New York: Alan R. Liss 1986; pp. 3-28.

[2] Kiernan JA. Histological and Histochemical Methods: Theory and Practice. Oxford, New York: Pergamon Press 1999.

[3] Horobin RW, Kiernan JA, Eds. Conn's Biological Stains A Handbook of Dyes, Stains and Fluorochromes for Use in Biology and Medicine. 10th ed., Oxford, UK: BIOS Scientific Publishers 2002.

[4] Hatz S, Lambert JD, Ogilby PR. Measuring the lifetime of singlet oxygen in a single cell: addressing the issue of cell viability. Photochem Photobiol Sci 2007; 6(10): 1106-16. [http://dx.doi.org/10.1039/b707313e] [PMID: 17914485]

[5] Mickwitz CU, Zipper J, Mauersberger B, Axt J, Heder G. [The usefulness of ruthenium red in dye exclusion test on cultured cells (author's transl)]. Exp Pathol (Jena) 1977; 14(1-2): 24-32. [http://dx.doi.org/10.1016/S0014-4908(77)80088-4] [PMID: 23306]

[6] Brus I, Glass GB. Estimation of cytotoxic injury to gastric parietal cells by trypan blue exclusion test followed by hematoxylin-eosin counterstaining of fixed smears. Stain Technol 1973; 48(3): 127-32. [http://dx.doi.org/10.3109/10520297309116600] [PMID: 4122194]

[7] Hagino S, Itagaki H, Kato S, Kobayashi T, Tanaka M. Quantitative evaluation to predict the eye irritancy of chemicals: Modification of chorioallantoic membrane test by using trypan blue. Toxicol *In Vitro* 1991; 5(4): 301-4. [http://dx.doi.org/10.1016/0887-2333(91)90006-Y] [PMID: 20732030]

[8] Kovács A, Foote RH. Viability and acrosome staining of bull, boar and rabbit spermatozoa. Biotech Histochem 1992; 67(3): 119-24. [http://dx.doi.org/10.3109/10520299209110020] [PMID: 1377502]

[9] Perry SW, Epstein LG, Gelbard HA. Simultaneous *in situ* detection of apoptosis and necrosis in monolayer cultures by TUNEL and trypan blue staining. Biotechniques 1997; 22(6): 1102-6. [PMID: 9187759]

[10] Li Y, Stansbury KH, Zhu H, Trush MA. Biochemical characterization of lucigenin (Bis--methylacridinium) as a chemiluminescent probe for detecting intramitochondrial superoxide anion radical production. Biochem Biophys Res Commun 1999; 262(1): 80-7. [http://dx.doi.org/10.1006/bbrc.1999.1174] [PMID: 10448072]

[11] Cadena-Herrera D, Esparza-De Lara JE, Ramírez-Ibañez ND, *et al.* Validation of three viable-cell counting methods: Manual, semi-automated, and automated. Biotechnol Rep (Amst) 2015; 7: 9-16. [http://dx.doi.org/10.1016/j.btre.2015.04.004] [PMID: 28626709]

[12] Nijs M, Van Blerk M, Van Steirteghem AC. Activation of mouse oocytes, fertilization and development of mouse embryos *in vitro* after staining with trypan blue or fluorescein diacetate. Biotech Histochem 1992; 67(6): 351-5. [http://dx.doi.org/10.3109/10520299209110048] [PMID: 1489837]

[13] Avelar-Freitas BA, Almeida VG, Pinto MC, *et al.* Trypan blue exclusion assay by flow cytometry. Braz J Med Biol Res 2014; 47(4): 307-15. [http://dx.doi.org/10.1590/1414-431X20143437] [PMID: 24652322]

[14] Hoffmeister ER. Trypan blue as a nuclear stain for plant material. Stain Technol 1953; 28(6): 309-10. [http://dx.doi.org/10.3109/10520295309105562] [PMID: 13113461]

[15] Chang YS, Wu CL, Tseng SH, Kuo PY, Tseng SY. Cytotoxicity of triamcinolone acetonide on human retinal pigment epithelial cells. Invest Ophthalmol Vis Sci 2007; 48(6): 2792-8. [http://dx.doi.org/10.1167/iovs.06-1146] [PMID: 17525214]

[16] Ruiz-González R, Acedo P, Sánchez-García D, *et al.* Efficient induction of apoptosis in HeLa cells by a novel cationic porphycene photosensitizer. Eur J Med Chem 2013; 63: 401-14. [http://dx.doi.org/10.1016/j.ejmech.2013.02.028] [PMID: 23517729]

[17] Wolf MK, Aronson SB. Growth fluorescence and metachromasy of cells cultured in the presence of acridine orange. J Histochem Cytochem 1961; 9: 22-9.
[http://dx.doi.org/10.1177/9.1.22] [PMID: 14007945]

[18] Schwartz DS, Larsh HW, Bartels PA. Enumerative fluorescent vital staining of live and dead pathogenic yeast cells. Stain Technol 1977; 52(4): 203-10.
[http://dx.doi.org/10.3109/10520297709116776] [PMID: 71766]

[19] Söderström K-O , Parvinen L-M , Parvinen M. Early detection of cell damage by supravital acridine orange staining. Experientia 1977; 33(2): 265-6.
[http://dx.doi.org/10.1007/BF02124103] [PMID: 66156]

[20] Correa B. Fluorescent viability test of fungi cells. Appl Fluor Technol 1991; 3(1): 8-10.

[21] Lee YJ, Shacter E. Oxidative stress inhibits apoptosis in human lymphoma cells. J Biol Chem 1999; 274(28): 19792-8.
[http://dx.doi.org/10.1074/jbc.274.28.19792] [PMID: 10391922]

[22] Cañete M, Juarranz A, López-Nieva P, Alonso-Torcal C, Villanueva A, Stockert JC. Fixation and permanent mounting of fluorescent probes after vital labelling of cultured cells. Acta Histochem 2001; 103(2): 117-26.
[http://dx.doi.org/10.1078/0065-1281-00594] [PMID: 11368093]

[23] Edidin M. A rapid, quantitative fluorescence assay for cell damage by cytotoxic antibodies. J Immunol 1970; 104(5): 1303-6.
[PMID: 5444341]

[24] Beletsky IP, Umansky SR. A new assay for cell death. J Immunol Methods 1990; 134(2): 201-5.
[http://dx.doi.org/10.1016/0022-1759(90)90381-5] [PMID: 2254666]

[25] Nicoletti I, Migliorati G, Pagliacci MC, Grignani F, Riccardi C. A rapid and simple method for measuring thymocyte apoptosis by propidium iodide staining and flow cytometry. J Immunol Methods 1991; 139(2): 271-9.
[http://dx.doi.org/10.1016/0022-1759(91)90198-O] [PMID: 1710634]

[26] Ross DD, Joneckis CC, Ordóñez JV, *et al.* Estimation of cell survival by flow cytometric quantification of fluorescein diacetate/propidium iodide viable cell number. Cancer Res 1989; 49(14): 3776-82.
[PMID: 2736519]

[27] Stapper NJ, Stuschke M, Sak A, Stüben G. Radiation-induced apoptosis in human sarcoma and glioma cell lines. Int J Cancer 1995; 62(1): 58-62.
[http://dx.doi.org/10.1002/ijc.2910620112] [PMID: 7601568]

[28] Banda NK, Satterfield WC, Dunlap A, *et al.* Lack of gp120-induced anergy and apoptosis in chimpanzees is correlated with resistance to AIDS. Apoptosis 1996; 1: 49-62.
[http://dx.doi.org/10.1007/BF00142078]

[29] Vowels BR, Yoo EK, Gasparro FP. Kinetic analysis of apoptosis induction in human cell lines by UVA and 8-MOP. Photochem Photobiol 1996; 63(5): 572-6.
[http://dx.doi.org/10.1111/j.1751-1097.1996.tb05658.x] [PMID: 8628747]

[30] Widholm JM. The use of fluorescein diacetate and phenosafranine for determining viability of cultured plant cells. Stain Technol 1972; 47(4): 189-94.
[http://dx.doi.org/10.3109/10520297209116483] [PMID: 4113995]

[31] Bucana C, Saiki I, Nayar R. Uptake and accumulation of the vital dye hydroethidine in neoplastic cells. J Histochem Cytochem 1986; 34(9): 1109-15.
[http://dx.doi.org/10.1177/34.9.2426339] [PMID: 2426339]

[32] Wainwright M. Dyes, trypanosomiasis and DNA: a historical and critical review. Biotech Histochem 2010; 85(6): 341-54.

[http://dx.doi.org/10.3109/10520290903297528] [PMID: 21080764]

[33] Mosmann T. Rapid colorimetric assay for cellular growth and survival: application to proliferation and cytotoxicity assays. J Immunol Methods 1983; 65(1-2): 55-63.
[http://dx.doi.org/10.1016/0022-1759(83)90303-4] [PMID: 6606682]

[34] Carmichael J, DeGraff WG, Gazdar AF, Minna JD, Mitchell JB. Evaluation of a tetrazolium-based semiautomated colorimetric assay: assessment of chemosensitivity testing. Cancer Res 1987; 47(4): 936-42.
[PMID: 3802100]

[35] Merlin JL, Azzi S, Lignon D, Ramacci C, Zeghari N, Guillemin F. MTT assays allow quick and reliable measurement of the response of human tumour cells to photodynamic therapy. Eur J Cancer 1992; 28A(8-9): 1452-8.
[http://dx.doi.org/10.1016/0959-8049(92)90542-A] [PMID: 1387543]

[36] Koenig H. Intravital staining of lysosomes and mast cell granules by tetrazolium salts. J Histochem Cytochem 1965; 13: 411-3.
[http://dx.doi.org/10.1177/13.5.411] [PMID: 14344262]

[37] Liu Y, Peterson DA, Kimura H, Schubert D. Mechanism of cellular 3-(4,5-dimethylthiazol-2-yl--2,5-diphenyltetrazolium bromide (MTT) reduction. J Neurochem 1997; 69(2): 581-93.
[http://dx.doi.org/10.1046/j.1471-4159.1997.69020581.x] [PMID: 9231715]

[38] Berridge MV, Herst PM, Tan AS. Tetrazolium dyes as tools in cell biology: new insights into their cellular reduction. Biotechnol Annu Rev 2005; 11: 127-52.
[http://dx.doi.org/10.1016/S1387-2656(05)11004-7] [PMID: 16216776]

[39] Diaz G, Melis M, Musin A, Piludu M, Piras M, Falchi AM. Localization of MTT formazan in lipid droplets. An alternative hypothesis about the nature of formazan granules and aggregates. Eur J Histochem 2007; 51(3): 213-8.
[PMID: 17921117]

[40] Stockert JC, Blázquez-Castro A, Cañete M, Horobin RW, Villanueva A. MTT assay for cell viability: Intracellular localization of the formazan product is in lipid droplets. Acta Histochem 2012; 114(8): 785-96.
[http://dx.doi.org/10.1016/j.acthis.2012.01.006] [PMID: 22341561]

[41] Winterbourn CC, Hampton MB. Thiol chemistry and specificity in redox signaling. Free Radic Biol Med 2008; 45(5): 549-61.
[http://dx.doi.org/10.1016/j.freeradbiomed.2008.05.004] [PMID: 18544350]

[42] Stellmach J, Severin E. A fluorescent redox dye. Influence of several substrates and electron carriers on the tetrazolium salt-formazan reaction of Ehrlich ascites tumour cells. Histochem J 1987; 19(1): 21-6.
[http://dx.doi.org/10.1007/BF01675289] [PMID: 3583812]

[43] Severin E, Seidler E, Huang CJ. Flow cytometric assay for specific dehydrogenases as demonstrated with human tumor cell lines. Appl Fluor Technol 1992; 4(5 & 6): 3-7.

[44] Goegan P, Johnson G, *VincentR* . Effects of serum protein and colloid on the Alamar Blue assay in cell cultures. Toxicol In vitro 1995; 9(3): 257-66.
[http://dx.doi.org/10.1016/0887-2333(95)00004-R] [PMID: 20650087]

[45] Al-Nasiry S, Geusens N, Hanssens M, Luyten C, Pijnenborg R. The use of Alamar Blue assay for quantitative analysis of viability, migration and invasion of choriocarcinoma cells. Hum Reprod 2007; 22(5): 1304-9.
[http://dx.doi.org/10.1093/humrep/dem011] [PMID: 17307808]

[46] van Tonder A, Joubert AM, Cromarty AD. Limitations of the 3-(4,5-dimethylthiazol-2-yl--2,5-diphenyl-2H-tetrazolium bromide (MTT) assay when compared to three commonly used cell enumeration assays. BMC Res Notes 2015; 8: 47.

[http://dx.doi.org/10.1186/s13104-015-1000-8] [PMID: 25884200]

[47] Chen CS, Gee KR. Redox-dependent trafficking of 2,3,4,5, 6-pentafluorodihydrotetramethylrosamine, a novel fluorogenic indicator of cellular oxidative activity. Free Radic Biol Med 2000; 28(8): 1266-78. [http://dx.doi.org/10.1016/S0891-5849(00)00265-3] [PMID: 10889457]

[48] Alvarez M, Villanueva A, Acedo P, Cañete M, Stockert JC. Cell death causes relocalization of photosensitizing fluorescent probes. Acta Histochem 2011; 113(3): 363-8. [http://dx.doi.org/10.1016/j.acthis.2010.01.008] [PMID: 20138336]

[49] Zhang SZ, Lipsky MM, Trump BF, Hsu IC. Neutral red (NR) assay for cell viability and xenobiotic-induced cytotoxicity in primary cultures of human and rat hepatocytes. Cell Biol Toxicol 1990; 6(2): 219-34. [http://dx.doi.org/10.1007/BF00249595] [PMID: 2113829]

[50] Repetto G, del Peso A, Zurita JL. Neutral red uptake assay for the estimation of cell viability/cytotoxicity. Nat Protoc 2008; 3(7): 1125-31. [http://dx.doi.org/10.1038/nprot.2008.75] [PMID: 18600217]

[51] Borenfreund E, Puerner JA. Toxicity determined *in vitro* by morphological alterations and neutral red absorption. Toxicol Lett 1985; 24(2-3): 119-24. [http://dx.doi.org/10.1016/0378-4274(85)90046-3] [PMID: 3983963]

[52] Lasarow RM, Isseroff RR, Gomez EC. Quantitative *in vitro* assessment of phototoxicity by a fibroblast-neutral red assay. J Invest Dermatol 1992; 98(5): 725-9. [http://dx.doi.org/10.1111/1523-1747.ep12499927] [PMID: 1569321]

[53] Malinski T, Kiechle F. Fluorescence measurement of membrane potential changes. Appl Fluor Technol 1991; 3(1): 1-7.

[54] Vichai V, Kirtikara K. Sulforhodamine B colorimetric assay for cytotoxicity screening. Nat Protoc 2006; 1(3): 1112-6. [http://dx.doi.org/10.1038/nprot.2006.179] [PMID: 17406391]

[55] National Cancer Institute Screening Services. http://www.cancer.gov2014.

[56] Ruiz MC, Michelangeli F, Ludert JE, *et al.* Fluorimetric quantification of cell death in monolayer cultures and cell suspensions. J Biochem Biophys Methods 1991; 23(3): 237-48. [http://dx.doi.org/10.1016/0165-022X(91)90016-P] [PMID: 1779095]

[57] Nieminen AL. Apoptosis and necrosis in health and disease: role of mitochondria. Int Rev Cytol 2003; 224: 29-55. [http://dx.doi.org/10.1016/S0074-7696(05)24002-0] [PMID: 12722948]

[58] Villanueva A, Vidania R, Stockert JC, *et al.* Photodynamic Effects on Cultured Tumor Cells: Cytoskeleton Alterations and Cell Death Mechanisms.Handbook of Photochemistry and Photobiology. California: AmericanScientific Publishers 2003; Vol. 4: pp. 79-117.

[59] Rello S, Stockert JC, Moreno V, *et al.* Morphological criteria to distinguish cell death induced by apoptotic and necrotic treatments. Apoptosis 2005; 10(1): 201-8. [http://dx.doi.org/10.1007/s10495-005-6075-6] [PMID: 15711936]

[60] Rello-Varona S, Stockert JC, Cañete M, Acedo P, Villanueva A. Mitotic catastrophe induced in HeLa cells by photodynamic treatment with Zn(II)-phthalocyanine. Int J Oncol 2008; 32(6): 1189-96. [http://dx.doi.org/10.3892/ijo_32_6_1189] [PMID: 18497980]

[61] Su Z, Yang Z, Xu Y, Chen Y, Yu Q. Apoptosis, autophagy, necroptosis, and cancer metastasis. Mol Cancer 2015; 14: 48. [http://dx.doi.org/10.1186/s12943-015-0321-5] [PMID: 25743109]

[62] Juarranz A, Villanueva A, Cañete M, Stockert JC. Damage to microtubules of cultured cells induced by photodynamic treatments. Res Adv Photochem Photobiol 2000; 1: 129-36.

[63] Juarranz A, Espada J, Stockert JC, *et al.* Photodamage induced by Zinc(II)-phthalocyanine to

microtubules, actin, α-actinin and keratin of HeLa cells. Photochem Photobiol 2001; 73(3): 283-9. [http://dx.doi.org/10.1562/0031-8655(2001)0730283PIBZIP2.0.CO2] [PMID: 11281025]

[64] Acedo P, Stockert JC, Cañete M, Villanueva A. Two combined photosensitizers: a goal for more effective photodynamic therapy of cancer. Cell Death Dis 2014; 5: e1122. [http://dx.doi.org/10.1038/cddis.2014.77] [PMID: 24625981]

[65] Boix-Garriga E, Acedo P, Casadó A, *et al.* Poly(D, L-lactide-co-glycolide) nanoparticles as delivery agents for photodynamic therapy: enhancing singlet oxygen release and photototoxicity by surface PEG coating. Nanotechnology 2015; 26(36): 365104. [http://dx.doi.org/10.1088/0957-4484/26/36/365104] [PMID: 26293792]

[66] Telford WG, King LE, Fraker PJ. Flow cytometric analysis of apoptosis: A review. Appl Fluor Technol 1992; 4(5 & 6): 12-7.

[67] Verhaegen S. Microscopical study of cell death *via* apoptosis. Eur Microsc Anal 1998; 51: 31-3.

[68] Oleinick NL, Morris RL, Belichenko I. The role of apoptosis in response to photodynamic therapy: what, where, why, and how. Photochem Photobiol Sci 2002; 1(1): 1-21. [http://dx.doi.org/10.1039/b108586g] [PMID: 12659143]

[69] Carou MC, Cruzans PR, Maruri A, Stockert JC, Lombardo DM. Apoptosis in ovarian granulosa cells of cattle: morphological features and clearance by homologous phagocytosis. Acta Histochem 2015; 117(1): 92-103. [http://dx.doi.org/10.1016/j.acthis.2014.11.006] [PMID: 25511683]

[70] Galaz S, Espada J, Stockert JC, *et al.* Loss of E-cadherin mediated cell-cell adhesion as an early trigger of apoptosis induced by photodynamic treatment. J Cell Physiol 2005; 205(1): 86-96. [http://dx.doi.org/10.1002/jcp.20374] [PMID: 15880654]

[71] van Engeland M, Nieland LJ, Ramaekers FC, Schutte B, Reutelingsperger CP. Annexin V-affinity assay: a review on an apoptosis detection system based on phosphatidylserine exposure. Cytometry 1998; 31(1): 1-9. [http://dx.doi.org/10.1002/(SICI)1097-0320(19980101)31:1<1::AID-CYTO1>3.0.CO;2-R] [PMID: 9450519]

[72] Koulov AV, Stucker KA, Lakshmi C, Robinson JP, Smith BD. Detection of apoptotic cells using a synthetic fluorescent sensor for membrane surfaces that contain phosphatidylserine. Cell Death Differ 2003; 10(12): 1357-9. [http://dx.doi.org/10.1038/sj.cdd.4401315] [PMID: 12970674]

[73] Hug H, Los M, Hirt W, Debatin KM. Rhodamine 110-linked amino acids and peptides as substrates to measure caspase activity upon apoptosis induction in intact cells. Biochemistry 1999; 38(42): 13906-11. [http://dx.doi.org/10.1021/bi99133951] [PMID: 10529236]

[74] Joshi SG, Sahni SK. Immunofluorescent detection of activation of initiator caspases-8 and -9 during pharmacologically induced apoptosis of cultured HeLa and endothelial cells. Histochem Cell Biol 2003; 119(6): 463-8. [http://dx.doi.org/10.1007/s00418-003-0536-1] [PMID: 12764618]

[75] Cristóbal J, Stockert JC, Villanueva A, Rello-Varona S, Juarranz A, Cañete M. Caspase-2: a possible trigger of apoptosis induced in A-549 tumor cells by ZnPc photodynamic treatment. Int J Oncol 2006; 28(5): 1057-63. [PMID: 16596221]

[76] Soriano J, García-Díaz M, Mora M, *et al.* Liposomal temocene (m-THPPo) photodynamic treatment induces cell death by mitochondria-independent apoptosis. Biophys Biochim Acta (General Subjects) 2013; 1830: 4611-20.

[77] Ablin J, Verbovetski I, Trahtemberg U, Metzger S, Mevorach D. Quinidine and procainamide inhibit murine macrophage uptake of apoptotic and necrotic cells: a novel contributing mechanism of drug-

induced-lupus. Apoptosis 2005; 10(5): 1009-18.
[http://dx.doi.org/10.1007/s10495-005-1189-4] [PMID: 16151636]

[78] Stockert JC, Juarranz A, Villanueva A, *et al.* Photodynamic therapy: selective uptake of photosensitizing drugs into tumor cells. Curr Top Pharmacol 2004; 8: 185-217.

[79] Cañete M, Ortega C, Gavaldá A, *et al.* Necrotic cell death induced by photodynamic treatment of human lung adenocarcinoma A-549 cells with palladium(II)-tetraphenylporphycene. Int J Oncol 2004; 24(5): 1221-8.
[PMID: 15067345]

[80] Borgers M, De Brabander M, Van Reempts J, Awouters F, Jacob WA. Intranuclear microtubules in lung mast cells of guinea pigs in anaphylactic shock. Lab Invest 1977; 37(1): 1-8.
[PMID: 69052]

[81] Krajcí D, Mares V, Lisá V, Bottone MG, Pellicciari C. Intranuclear microtubules are hallmarks of an unusual form of cell death in cisplatin-treated C6 glioma cells. Histochem Cell Biol 2006; 125(1-2): 183-91.
[http://dx.doi.org/10.1007/s00418-005-0094-9] [PMID: 16283354]

[82] Guilbault GG, Kramer DN. Fluorometric procedure for measuring the activity of dehydrogenases. Anal Chem 1965; 37: 1219-21.
[http://dx.doi.org/10.1021/ac60229a011] [PMID: 14342709]

[83] Galluzzi L, Maiuri MC, Vitale I, *et al.* Cell death modalities: classification and pathophysiological implications. Cell Death Differ 2007; 14(7): 1237-43.
[http://dx.doi.org/10.1038/sj.cdd.4402148] [PMID: 17431418]

[84] Rello-Varona S, Gámez A, Moreno V, *et al.* Metaphase arrest and cell death induced by etoposide on HeLa cells. Int J Biochem Cell Biol 2006; 38(12): 2183-95.
[http://dx.doi.org/10.1016/j.biocel.2006.06.013] [PMID: 16931106]

[85] Levine B, Yuan J. Autophagy in cell death: an innocent convict? J Clin Invest 2005; 115(10): 2679-88.
[http://dx.doi.org/10.1172/JCI26390] [PMID: 16200202]

[86] Baehrecke EH. Autophagy: dual roles in life and death? Nat Rev Mol Cell Biol 2005; 6(6): 505-10.
[http://dx.doi.org/10.1038/nrm1666] [PMID: 15928714]

[87] Tsujimoto Y, Shimizu S. Another way to die: autophagic programmed cell death. Cell Death Differ 2005; 12 (Suppl. 2): 1528-34.
[http://dx.doi.org/10.1038/sj.cdd.4401777] [PMID: 16247500]

[88] Narita M, Young AR, Arakawa S, *et al.* Spatial coupling of mTOR and autophagy augments secretory phenotypes. Science 2011; 332(6032): 966-70.
[http://dx.doi.org/10.1126/science.1205407] [PMID: 21512002]

[89] Mijaljica D, Prescott M, Devenish RJ. The intricacy of nuclear membrane dynamics during nucleophagy. Nucleus 2010; 1(3): 213-23.
[http://dx.doi.org/10.4161/nucl.11738] [PMID: 21327066]

[90] Soriano J, Villanueva A, Stockert JC, Cañete M. Regulated necrosis in HeLa cells induced by ZnPc photodynamic treatment: a new nuclear morphology. Int J Mol Sci 2014; 15(12): 22772-85.
[http://dx.doi.org/10.3390/ijms151222772] [PMID: 25501332]

[91] Dimri GP, Lee X, Basile G, *et al.* A biomarker that identifies senescent human cells in culture and in aging skin *in vivo*. Proc Natl Acad Sci USA 1995; 92(20): 9363-7.
[http://dx.doi.org/10.1073/pnas.92.20.9363] [PMID: 7568133]

[92] Itahana K, Campisi J, Dimri GP. Methods to detect biomarkers of cellular senescence: the senescence-associated beta-galactosidase assay. Methods Mol Biol 2007; 371: 21-31.
[http://dx.doi.org/10.1007/978-1-59745-361-5_3] [PMID: 17634571]

[93] Chohan KR, Griffin JT, Lafromboise M, De Jonge CJ, Carrell DT. Comparison of chromatin assays

for DNA fragmentation evaluation in human sperm. J Androl 2006; 27(1): 53-9. [http://dx.doi.org/10.2164/jandrol.05068] [PMID: 16400078]

[94] McKelvey-Martin VJ, Green MH, Schmezer P, Pool-Zobel BL, De Méo MP, Collins A. The single cell gel electrophoresis assay (comet assay): a European review. Mutat Res 1993; 288(1): 47-63. [http://dx.doi.org/10.1016/0027-5107(93)90207-V] [PMID: 7686265]

[95] Hooghe RJ, Verschaeve L, Vankerkom J. DNA comets for monitoring pollution. Eur Microsc Anal 1995; 38: 19-20.

[96] Fairbairn DW, Olive PL, O'Neill KL. The comet assay: a comprehensive review. Mutat Res 1995; 339(1): 37-59. [http://dx.doi.org/10.1016/0165-1110(94)00013-3] [PMID: 7877644]

[97] Speit G, Hartmann A. The comet assay (single-cell gel test). A sensitive genotoxicity test for the detection of DNA damage and repair. Methods Mol Biol 1999; 113: 203-12. [PMID: 10443422]

[98] Collins AR. The comet assay for DNA damage and repair: principles, applications, and limitations. Mol Biotechnol 2004; 26(3): 249-61. [http://dx.doi.org/10.1385/MB:26:3:249] [PMID: 15004294]

[99] Wood DK, Weingeist DM, Bhatia SN, Engelward BP. Single cell trapping and DNA damage analysis using microwell arrays. Proc Natl Acad Sci USA 2010; 107(22): 10008-13. [http://dx.doi.org/10.1073/pnas.1004056107] [PMID: 20534572]

[100] Mladinic M, Zeljezic D. Modification of comet-FISH technique by using temperature instead of chemical denaturation. MethodsX 2014; 1: 162-7. [http://dx.doi.org/10.1016/j.mex.2014.08.010] [PMID: 26150949]

[101] Glei M, Schlörmann W. Analysis of DNA damage and repair by comet fluorescence *in situ* hybridization (Comet-FISH). Methods Mol Biol 2014; 1094: 39-48. [http://dx.doi.org/10.1007/978-1-62703-706-8_4] [PMID: 24162978]

[102] Osipov A, Arkhangelskaya E, Vinokurov A, Smetanina N, Zhavoronkov A, Klokov D. DNA comet Giemsa staining for conventional bright-field microscopy. Int J Mol Sci 2014; 15(4): 6086-95. [http://dx.doi.org/10.3390/ijms15046086] [PMID: 24727376]

[103] Sestili P, Martinelli C, Stocchi V. The fast halo assay: an improved method to quantify genomic DNA strand breakage at the single-cell level. Mutat Res 2006; 607(2): 205-14. [http://dx.doi.org/10.1016/j.mrgentox.2006.04.018] [PMID: 16766222]

[104] Sestili P, Fimognari C. Alkaline nuclear dispersion assays for the determination of DNA damage at the single cell level. Methods Mol Biol 2014; 1094: 49-70. [http://dx.doi.org/10.1007/978-1-62703-706-8_5] [PMID: 24162979]

[105] Galaz-Leiva S, Pérez-Rodríguez G, Blázquez-Castro A, Stockert JC. A simplified chromatin dispersion (nuclear halo) assay for detecting DNA breakage induced by ionizing radiation and chemical agents. Biotech Histochem 2012; 87(3): 208-17. [http://dx.doi.org/10.3109/10520295.2011.604163] [PMID: 21916782]

[106] Soh N. Recent advances in fluorescent probes for the detection of reactive oxygen species. Anal Bioanal Chem 2006; 386(3): 532-43. [http://dx.doi.org/10.1007/s00216-006-0366-9] [PMID: 16609844]

[107] Wardman P. Fluorescent and luminescent probes for measurement of oxidative and nitrosative species in cells and tissues: progress, pitfalls, and prospects. Free Radic Biol Med 2007; 43(7): 995-1022. [http://dx.doi.org/10.1016/j.freeradbiomed.2007.06.026] [PMID: 17761297]

[108] Xie YJ, Shen WB. *In vivo* imaging of nitric oxide and reactive oxygen species using laser scanning confocal microscopy. Methods Mol Biol 2012; 913: 191-200. [PMID: 22895760]

[109] Huang X, Frenkel K, Klein CB, Costa M. Nickel induces increased oxidants in intact cultured mammalian cells as detected by dichlorofluorescein fluorescence. Toxicol Appl Pharmacol 1993; 120(1): 29-36.
[http://dx.doi.org/10.1006/taap.1993.1083] [PMID: 8511780]

[110] Bethke PC, Jones RL. Cell death of barley aleurone protoplasts is mediated by reactive oxygen species. Plant J 2001; 25(1): 19-29.
[http://dx.doi.org/10.1046/j.1365-313x.2001.00930.x] [PMID: 11169179]

[111] Ortega-Villasante C, Rellán-Alvarez R, Del Campo FF, Carpena-Ruiz RO, Hernández LE. Cellular damage induced by cadmium and mercury in *Medicago sativa*. J Exp Bot 2005; 56(418): 2239-51.
[http://dx.doi.org/10.1093/jxb/eri223] [PMID: 15996984]

[112] Sandalio LM, Rodríguez-Serrano M, Romero-Puertas MC, Del Río LA. Imaging of reactive oxygen species and nitric oxide *in vivo* in plant tissues. Methods Enzymol 2008; 440: 397-409.
[http://dx.doi.org/10.1016/S0076-6879(07)00825-7] [PMID: 18423232]

[113] Setsukinai K, Urano Y, Kakinuma K, Majima HJ, Nagano T. Development of novel fluorescence probes that can reliably detect reactive oxygen species and distinguish specific species. J Biol Chem 2003; 278(5): 3170-5.
[http://dx.doi.org/10.1074/jbc.M209264200] [PMID: 12419811]

[114] Flors C, Fryer MJ, Waring J, *et al.* Imaging the production of singlet oxygen *in vivo* using a new fluorescent sensor, Singlet Oxygen Sensor Green. J Exp Bot 2006; 57(8): 1725-34.
[http://dx.doi.org/10.1093/jxb/erj181] [PMID: 16595576]

[115] Ragàs X, Jiménez-Banzo A, Sánchez-García D, Batllori X, Nonell S. Singlet oxygen photosensitisation by the fluorescent probe Singlet Oxygen Sensor Green. Chem Commun (Camb) 2009; (20): 2920-2.
[http://dx.doi.org/10.1039/b822776d] [PMID: 19436910]

[116] Diaz G, Liu S, Isola R, Diana A, Falchi AM. Mitochondrial localization of reactive oxygen species by dihydrofluorescein probes. Histochem Cell Biol 2003; 120(4): 319-25.
[http://dx.doi.org/10.1007/s00418-003-0566-8] [PMID: 14574587]

[117] Rohnstock A, Lehmann L. Evaluation of the probe dihydrocalcein acetoxymethylester as an indicator of reactive oxygen species formation and comparison with oxidative DNA base modification determined by modified alkaline elution technique. Toxicol In vitro 2007; 21(8): 1552-62.
[http://dx.doi.org/10.1016/j.tiv.2007.05.001] [PMID: 17574384]

[118] Hernández-Barrera A, Velarde-Buendía A, Zepeda I, *et al.* Hyper, a hydrogen peroxide sensor, indicates the sensitivity of the *Arabidopsis* root elongation zone to aluminum treatment. Sensors (Basel) 2015; 15(1): 855-67.
[http://dx.doi.org/10.3390/s150100855] [PMID: 25569758]

[119] Belousov VV, Fradkov AF, Lukyanov KA, *et al.* Genetically encoded fluorescent indicator for intracellular hydrogen peroxide. Nat Methods 2006; 3(4): 281-6.
[http://dx.doi.org/10.1038/nmeth866] [PMID: 16554833]

[120] Ermakova YG, Bilan DS, Matlashov ME, *et al.* Red fluorescent genetically encoded indicator for intracellular hydrogen peroxide. Nat Commun 2014; 5: 5222.
[http://dx.doi.org/10.1038/ncomms6222] [PMID: 25330925]

[121] Stoward PJ, Pearse AG. Histochemistry, Theoretical and Applied.Enzyme Histochemistry. 4th ed. Edinburgh: Churchill-Livingstone 1991; 3.

[122] Kerver ED, Vogels IM, Bosch KS, Vreeling-Sindelárová H, Van den Munckhof RJ, Frederiks WM. *In situ* detection of spontaneous superoxide anion and singlet oxygen production by mitochondria in rat liver and small intestine. Histochem J 1997; 29(3): 229-37.
[http://dx.doi.org/10.1023/A:1026453926517] [PMID: 9472385]

[123] Frederiks WM, Bosch KS, Vreeling-Sindelárová HA. *In situ* detection of constitutive superoxide anion

production in granules of mast cells. Histochem J 1997; 29(4): 287-91. [http://dx.doi.org/10.1023/A:1026470430151] [PMID: 9184843]

[124] Stockert JC, Blázquez-Castro A. Establishing the subcellular localization of photodynamically-induced ROS using 3,3′-diaminobenzidine: A methodological proposal, with a proof-of-concept demonstration. Methods 2016; 109: 175-9. [http://dx.doi.org/10.1016/j.ymeth.2016.04.031] [PMID: 27154745]

[125] Braakman I, Pijning T, Verest O, Weert B, Meijer DK, Groothuis GM. Vesicular uptake system for the cation lucigenin in the rat hepatocyte. Mol Pharmacol 1989; 36(4): 537-42. [PMID: 2811856]

[126] Vit JP, Moustacchi E, Rosselli F. ATM protein is required for radiation-induced apoptosis and acts before mitochondrial collapse. Int J Radiat Biol 2000; 76(6): 841-51. [http://dx.doi.org/10.1080/09553000050029002] [PMID: 10902739]

[127] Benov L, Sztejnberg L, Fridovich I. Critical evaluation of the use of hydroethidine as a measure of superoxide anion radical. Free Radic Biol Med 1998; 25(7): 826-31. [http://dx.doi.org/10.1016/S0891-5849(98)00163-4] [PMID: 9823548]

[128] Zhao H, Kalivendi S, Zhang H, *et al.* Superoxide reacts with hydroethidine but forms a fluorescent product that is distinctly different from ethidium: potential implications in intracellular fluorescence detection of superoxide. Free Radic Biol Med 2003; 34(11): 1359-68. [http://dx.doi.org/10.1016/S0891-5849(03)00142-4] [PMID: 12757846]

[129] Zhao H, Joseph J, Fales HM, *et al.* Detection and characterization of the product of hydroethidine and intracellular superoxide by HPLC and limitations of fluorescence. Proc Natl Acad Sci USA 2005; 102(16): 5727-32. [http://dx.doi.org/10.1073/pnas.0501719102] [PMID: 15824309]

[130] Bindokas VP, Jordán J, Lee CC, Miller RJ. Superoxide production in rat hippocampal neurons: selective imaging with hydroethidine. J Neurosci 1996; 16(4): 1324-36. [PMID: 8778284]

[131] Berlman IB. Handbook of Fluorescence Spectra of Aromatic Molecules. 2nd ed., New York, London: Academic Press 1971.

[132] Pompella A. The use of fluorescent hydrazides for the detection of carbonyl products of lipid peroxidation in tissues. Appl Fluor Technol 1991; 3(1): 30-2.

[133] Pompella A, Comporti M, Müller K, *et al.* Confocal laser scanning fluorescence microscopy of DABCSH-labeled protein carbonyls in isolated cells. Appl Fluor Technol 1992; 4(5 & 6): 8-11.

[134] Del Rio D, Stewart AJ, Pellegrini N. A review of recent studies on malondialdehyde as toxic molecule and biological marker of oxidative stress. Nutr Metab Cardiovasc Dis 2005; 15(4): 316-28. [http://dx.doi.org/10.1016/j.numecd.2005.05.003] [PMID: 16054557]

[135] Stamler JS, Singel DJ, Loscalzo J. Biochemistry of nitric oxide and its redox-activated forms. Science 1992; 258(5090): 1898-902. [http://dx.doi.org/10.1126/science.1281928] [PMID: 1281928]

[136] Beltrán B, Quintero M, García-Zaragozá E, O'Connor E, Esplugues JV, Moncada S. Inhibition of mitochondrial respiration by endogenous nitric oxide: a critical step in Fas signaling. Proc Natl Acad Sci USA 2002; 99(13): 8892-7. [http://dx.doi.org/10.1073/pnas.092259799] [PMID: 12077295]

[137] Ryan TC, Weil GJ, Newburger PE, Haugland R, Simons ER. Measurement of superoxide release in the phagovacuoles of immune complex-stimulated human neutrophils. J Immunol Methods 1990; 130(2): 223-33. [http://dx.doi.org/10.1016/0022-1759(90)90052-W] [PMID: 2165099]

[138] Horan PK, Melnicoff MJ, Jensen BD, Slezak SE. Fluorescent cell labeling for *in vivo* and *in vitro* cell tracking. Methods Cell Biol 1990; 33: 469-90.

[http://dx.doi.org/10.1016/S0091-679X(08)60547-6] [PMID: 2084480]

[139] Stewart WW. Lucifer dyes--highly fluorescent dyes for biological tracing. Nature 1981; 292(5818): 17-21.
[http://dx.doi.org/10.1038/292017a0] [PMID: 6168915]

[140] Gisel A, Barella S, Hempel FD, Zambryski PC. Temporal and spatial regulation of symplastic trafficking during development in Arabidopsis thaliana apices. Development 1999; 126(9): 1879-89.
[PMID: 10101122]

[141] Köbbert C, Apps R, Bechmann I, Lanciego JL, Mey J, Thanos S. Current concepts in neuroanatomical tracing. Prog Neurobiol 2000; 62(4): 327-51.
[http://dx.doi.org/10.1016/S0301-0082(00)00019-8] [PMID: 10856608]

[142] Payne JN. Comparisons between the use of true blue and diamidino yellow as retrograde fluorescent tracers. Exp Brain Res 1987; 68(3): 631-42.
[http://dx.doi.org/10.1007/BF00249806] [PMID: 3691732]

[143] Schmued LC, Fallon JH. Fluoro-Gold: a new fluorescent retrograde axonal tracer with numerous unique properties. Brain Res 1986; 377(1): 147-54.
[http://dx.doi.org/10.1016/0006-8993(86)91199-6] [PMID: 2425899]

[144] Wessendorf MW. Fluoro-Gold: composition, and mechanism of uptake. Brain Res 1991; 553(1): 135-48.
[http://dx.doi.org/10.1016/0006-8993(91)90241-M] [PMID: 1933270]

[145] Honig MG, Hume RI. Fluorescent carbocyanine dyes allow living neurons of identified origin to be studied in long-term cultures. J Cell Biol 1986; 103(1): 171-87.
[http://dx.doi.org/10.1083/jcb.103.1.171] [PMID: 2424918]

[146] Honig MG, Hume RI. Dil and diO: versatile fluorescent dyes for neuronal labelling and pathway tracing. Trends Neurosci 1989; 12(9): 333-335, 340-341.
[http://dx.doi.org/10.1016/0166-2236(89)90040-4] [PMID: 2480673]

[147] Payne JN, Lawes IN, Proctor GB, Horobin RW. Variation between different samples of SITS with respect to axonal transport and toxicity. Neurosci Lett 1983; 42(3): 229-34.
[http://dx.doi.org/10.1016/0304-3940(83)90266-5] [PMID: 6198609]

[148] Horobin RW, Payne JN, Jakobsen P. Histochemical implications of the chemical and biological properties of SITS and some related compounds. J Microsc 1987; 146(Pt 1): 87-96.
[http://dx.doi.org/10.1111/j.1365-2818.1987.tb01329.x] [PMID: 2439692]

CHAPTER 17

Immunofluorescence and FISH

Abstract: This chapter presents and summarizes two very important techniques in biomedical research, immunofluorescence and fluorescence *in situ* hybridization (FISH), which relay completely in the use of fluorescent methods to achieve their results. Although the particular molecules involved are different both methods make use of the same approach: the detection of a specific molecular sequence or moiety by a specific molecule (antibody or nucleic acid sequence) that has a fluorophore attached to it. This fluorophore permits the detection of the specific target. An improvement over fluorophore-tagged antibodies has been the introduction of fluorescent proteins, generically known as GFPs (green fluorescent proteins) or FPs (fluorescent proteins). Here we will elaborate a little bit further on their practical applications, as one can think of FPs as a "subtype" of immunofluorescence.

Strictly speaking immunofluorescence and FISH provide no new concept to the principles of fluorescence. They rely on using a more or less classical fluorophore and the real breakthrough was the use of the exquisite specificity of antibodies and nucleobase complementarity to recognize certain substrates. However, immunofluorescence represents probably the most widespread employed technique in fluorescence microscopy. And FISH is only a step behind if at all in regards as its use as a diagnostic and research tool in genetics and medicine. Therefore we will introduce the basics of both techniques and some of their most promising recent applications in this chapter.

Keywords: Anti-fading agent, Antibody, Antigen, Chromobody, Chromosome painting (M-FISH), Comet-FISH, Counterstaining dye, Direct immuno-fluorescence, Enzyme-linked immunosorbent assay (ELISA), Fluorescence *in situ* hybridization (FISH), Fluorescent protein, Fluorophore-tagged antibody, Genetic analysis, Immunofluorescence, Indirect immunofluorescence, Nanobody, Western blot.

1. IMMUNOFLUORESCENCE

In immunofluorescence the technique takes advantage of the very high specificity of an antibody (Ab) to detect a particular molecular pattern or moiety. An Ab is a type of protein produced by immune cells in vertebrates with essential roles in the defence systems of these organisms. A depiction of an Ab molecular structure is shown in Fig. (**17.1A**).

Juan Carlos Stockert & Alfonso Blázquez-Castro

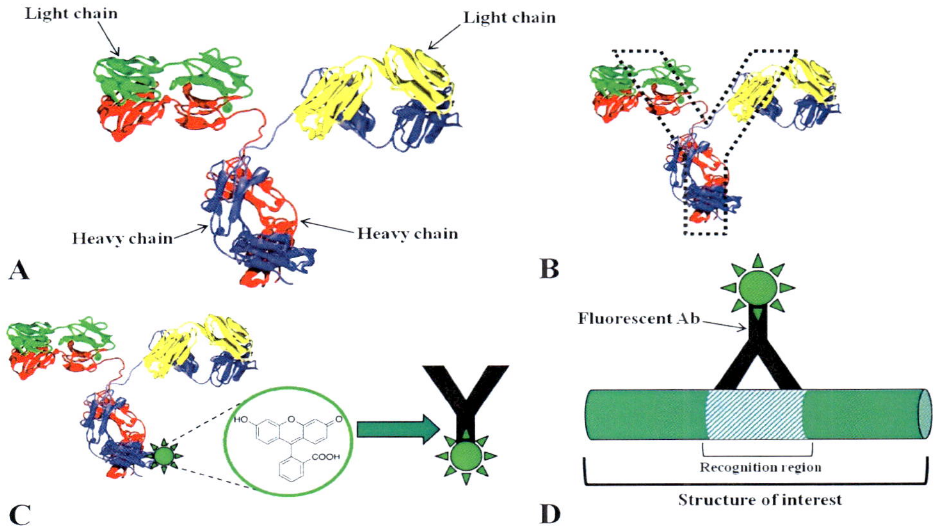

Fig. (17.1). A: Depiction of an antibody (Ab) structure. Abs are proteins made of two heavy chains and two light chains (indicated in the cartoon). The two light chains associate with each other (forming the Ab's "stem" or "trunk") and then each heavy chain associates with one of the light chains (the "stalks"). It is these regions that recognize antigens by molecular interactions. (Figure reproduced with permission from TimVickers - https://commons.wikimedia.org/w/ index.php?curid=3434214). **B**: Schematics of Ab representation as a "Y" letter relating the letter form with the Ab structure. **C**: Simple representation of Ab functionalization to produce a fluorescent Ab. In this example the Ab is labelled with a fluorescein moiety. The fluorophore is usually localized in the heavy chain's interaction region, far from the antigen recognizing sites to avoid interferences. A fluorescent Ab can be depicted in a very simple way as a "Y" with a "shining light "attached to the stem, as displayed in the figure. **D**: Fluorescent Ab detection mechanism. The labelled Ab recognizes its epitope for which it has a very high specificity and binding constant. This recognition region (hatched colour) is part of a larger structure (protein, nucleic acid, *etc.*) that is to be analysed and imaged. In this cartoon the structure to be recognized can be one of the tubulin subunits making up one of the components of the cytoskeleton.

An Ab structure loosely resembles the letter "Y", so Abs are often represented as Y structures (see Fig. **17.1B**). This accepted simplification will also be employed in this chapter to represent Abs. Abs are made of heavy and light proteins, or chains. Each Ab is a dimer of two heavy and two light chains. In Fig. (**17.1A**) the two heavy chains are represented in blue and red, while the light chains appear as yellow and green. The stem of the Y (red + blue) is conserved and determines the interactions of the Ab with different cell receptors to mediate the immune response. The two "horns" at the upper end of the Y are the hypervariable regions (green + red and yellow + blue). These regions are in charge of recognizing particular molecular moieties, known as antigens, and each Ab responds to a unique antigen.

With such a high specificity for recognizing a molecule or part of a molecule, it is straightforward to functionalize an Ab with some kind of luminescent tag to detect the entity of interest [1, 2]. So an Ab can be functionalized with one or more fluorophores chemically-linked to the molecular structure of the Ab (Fig. **17.1C**). Wherever the Ab recognizes and attaches to its antigen, a light signal will be obtained under appropriate excitation conditions (Fig. **17.1D**). Abs employed in immunofluorescence can be of two types: polyclonal or monoclonal. Polyclonal Abs is a mixture of Abs that recognizes the same antigen but do so at slightly different molecular moieties. For example, polyclonal Abs against tubulin subunits will recognize the same subunit but will bind to different amino acids/regions on the subunit (see Fig. **17.2A**). Monoclonal Abs, on the other side, are obtained from a single population of immune cells and therefore they all recognize the same molecular motif (see Fig. **17.2B**). In this sense the signal obtained from monoclonal Abs is more reliable as to what structure is being exactly recognized. However, they are more difficult to obtain and, as a consequence, more expensive.

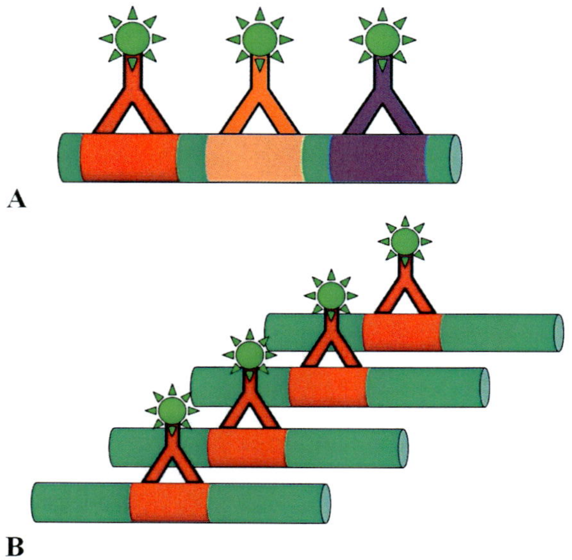

Fig. (17.2). A: Polyclonal Abs recognizing slightly different epitopes (different colours) in the same structure of interest. Depending on how close are the recognized regions for each Ab the final visual signal can be more or less resolved. **B**: Monoclonal Abs they all recognize the same epitope (red area). However only one Ab will label each structure of interest and it is necessary to have more Abs to obtain a strong signal. If the structure to be visualized is scarce employing monoclonal Abs can lead to very low signal-to-noise ratio.

Immunofluorescence can be considered as an evolution of immunohistochemistry [3]. Whereas immunohistochemistry was developed first and makes use of differences in light transmission (sample staining) to highlight structures of interest, immunofluorescence relies on differential light emission, not absorption, to present such structures. Therefore these two techniques can be seen as complementing one to the other: a bright background with darker coloured structures (immunohistochemistry) *vs*. a dark background with brighter coloured structures (immunofluorescence). The protocol followed in both techniques is indeed quite similar. The sample is fixed with some solvent (methanol), chemical product (formaldehyde) or physical procedure (freezing) that preserves the structure to be analysed. Ideally the fixing procedure does not alter the sample although, in the practice, this is utopic. However current protocols are thought to be as neutral as possible. Then the sample must be permeabilized to allow the Abs to enter the fixed cell. In some cases the fixing agent (*e.g.* methanol) acts as a permeabilizing agent too. In other cases some kind of permeabilizing agent (detergents) must be employed. Once permeabilized the sample is exposed –incubated- with the recognizing Ab. In immunofluorescence the Ab is tagged with a fluorescent marker. Afterwards the sample is taken to a fluorescence microscope to be imaged.

Some examples of immunofluorescent labelling in cultured cells are shown in Fig. (**17.3**). The green structures in Figs. (**17.3A, B**) is the microtubule cytoskeleton network in human fibroblasts (**A**) and immortalized human keratinocytes –HaCaT- (**B**).

It is clearly appreciated the differences in cellular architecture between these two cell types. The Ab employed to visualize these structures recognizes the α subunit of the tubulin dimer. Therefore the images show the microtubules. A secondary Ab that recognizes the first Ab (see below for the indirect immunofluorescence approach) is tagged with an Alexa 488 fluorescent molecule. This commercial dye is excited in the blue and emits in the green, the reason why the microtubule network appears as a green structure. Complementary Figs. (**17.3C, D**) show another component of the cytoskeleton, in this case the actin microfilaments. The procedure was a little bit different as phalloidin, a small oligopeptide produced by some fungi with very high affinity for filamentous actin, was employed to recognize the actin network. The phalloidin molecule is commercialized with a tetra-methyl-rhodamine molecule attached to it. This rhodamine (see Chapter 4.1.3) is preferentially excited in the green and then emits in the red. This is beautifully exemplified in the images shown for 3T3 murine fibroblasts (**C**) and HaCaT (**D**). Note that, as it was the case with the microtubules, the actin cytoskeleton arrangement is different between these two cell types.

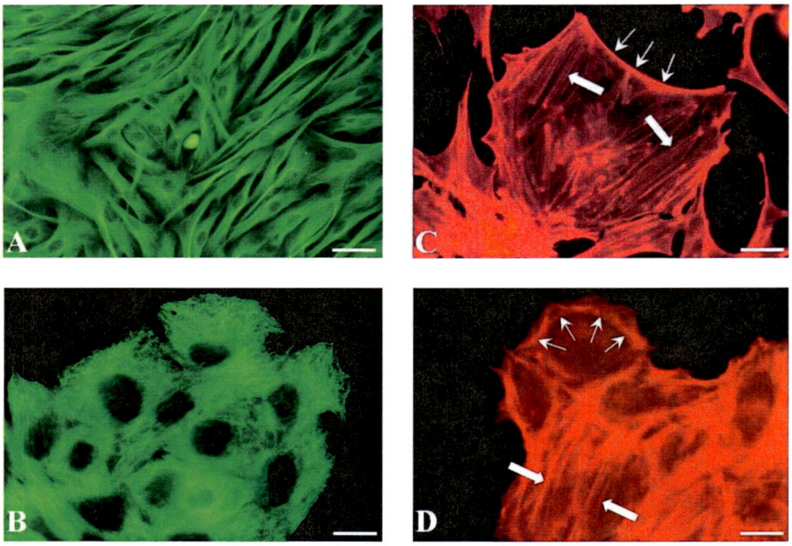

Fig. (17.3). Examples of Ab labelling of cytoskeletal features in different cell types. **A**: Tubulin labelling in human fibroblasts with a primary Ab recognizing the α subunit of the tubulin dimer. The secondary Ab is functionalized with Alexa Fluor 488 that provides the bright green emission (520-550 nm) registered in the picture. Note the elongated (acicular) arrangement of the tubulin network in most fibroblasts. The dimmer circular spot at the centre of each cell is the "hole" left by the nucleus as it does not contain tubulin. At the centre of the image a fibroblast undergoing mitosis can be seen as the brightest rounded small structure. Scale bar: 50 μm. **B**: Tubulin labelling of HaCaT cells (same conditions as in "**A**"). The tubulin cytoskeleton appears with an overall rounded structure while the individual tubules form a dense crisscrossed network, making each cell look like a "birds nest" with the central hole marking the position of the nucleus. Scale bar: 15 μm. **C**: Actin fluorescent staining with phalloidin tagged with rhodamine (red emission 580-630 nm). The imaged cell is a 3T3 murine fibroblast. The actin cytoskeleton is forming stress fibres (arrows) in the basal cell side that increase the mechanical resilience of the cells. At the same time some cortical actin (arrowheads) is localized to the inner plasma membrane reinforcing it and helping the cell maintain its shape. Scale bar: 5 μm. **D**: Labelling of actin in HaCaT cells under the same conditions as in "**C**". Notice the less prominent stress fibres (arrows) but the much more intense signal from the cortical actin (arrowheads). HaCaT cells are more densely packed than fibroblasts with tighter cell-cell unions. Scale bar: 20 μm.

Many different fluorophores can be attached or linked to Abs. Generally speaking all the different luminescent sources presented in this book have been or can be employed as Ab fluorophores. Even in the case of nano- (*e.g.* metallic particles) or micro-particles (*e.g.* latex beads), Abs are in fact linked to the particle, in a certain reverse of roles. In this case we talk about particle coating with Abs. Nevertheless their detection capacity remains. Since its inception more than 50 years ago the classical fluorophores conjugated to Abs have been organic fluorescent molecules. These have been presented broadly in Chapter 4. Most typical Abs employed are labelled with fluorescein or rhodamine, or any of their derivatives. These molecules show a very good compromise between brightness, adequate absorption/emission spectral ranges, adequate Stokes shift to avoid exciting light, radiative lifetime, low photobleaching rate and high quantum yield of fluorescence (thereby low quantum yield for other deactivation channels including photochemistry) [4, 5]. A useful and clarifying chart is presented in Fig. (**17.4**) where the normalized molecular fluorescent brightness is plotted *versus* the maximum in the absorption band for many different fluorescent molecules routinely employed in fluorescence microscopy [4]. The colour of each molecule's structure represents the emission colour of it. Note that not all the displayed molecules are employed as fluorophores to tag Abs. Some of them are counterstaining dyes (like DAPI, see below), fluorescent dyes that label cell structures due to their particular characteristics (Cy3 and Cy7 for membranes) or they are metabolites that produce some of the cell autofluorescence (NADH and FMN). Some companies have introduced new fluorescent molecules (*e.g.* Alexa Fluor commercialized by Molecular Probes) with improved parameters, especially in respect to the photobleaching rate, that permit better image acquisition [6].

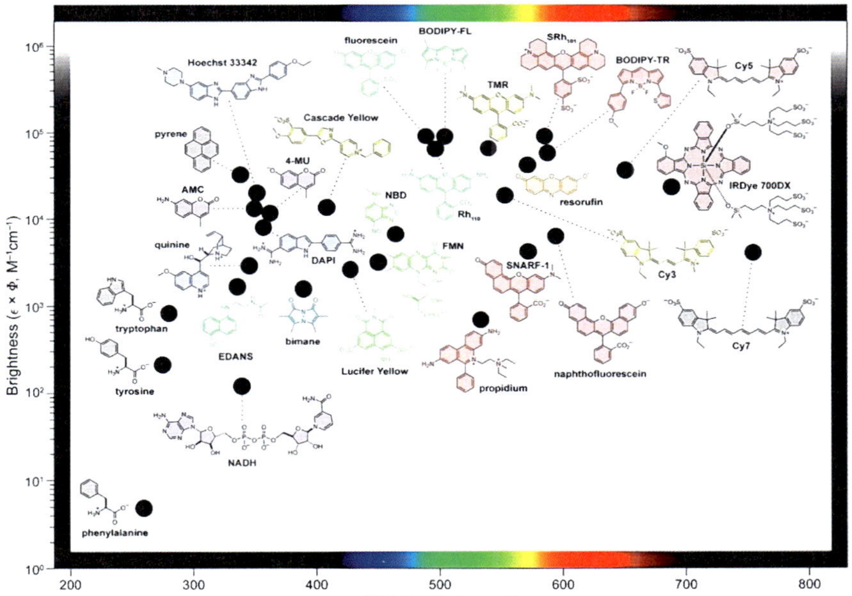

Fig. (17.4). Chart of organic fluorophores plotting a normalized brightness parameter *vs.* the maximum wavelength of the absorption band. The normalized brightness was calculated by multiplying the fluorophore's molar absorption (how efficient it is in collecting photons) by the fluorescence quantum yield (how efficient it is emitting light once it has absorbed a photon). Each fluorophore is depicted by its chemical formula. The colour assigned corresponds to the emission colour the molecule provides when excited. Black spots indicate the plot point where each fluorophore's features map. In this plot the molecules on the top emit more light as compared to molecules at the bottom. The molecules absorb more to the red part of the spectrum and also emit more to the red when moving from left to right. Reprinted and adapted with permission from reference [4]. Copyright 2008 American Chemical Society.

Other fluorophores have been employed to label Abs. In principle many if not all of the luminescent molecules/particles described in this book can be employed for such a task. For example, Lin *et al.* have employed a immunofluorescence strategy with a Ru(II) transition metal complex to enhance light emission and signal stability (see Chapter 9.3.2) [7]. The strategy was employed in an ELISA-like fluorescent assay (see later section 1.4) but the principles can be applied to classical immunofluorescence studies. In a similar trend other researchers have very recently reported the use of a terbium(III) complex linked to an Ab [8]. In addition to the very stable emission from lanthanides these kind of Abs can have immense impact in time-resolved microscopy due to the very long radiative lifetimes of lanthanides (see Chapter 9.2.3). We will present more examples of Ab conjugation to other fluorophores (solid-state) in Chapter 18.

The classical procedure to label an Ab is to have a fluorescent molecule chemically attached to the Ab by means of a covalent bond to one of the amino acids. This makes a permanent reaction and wherever the Ab goes the fluorophore will be "dragged" on. However, recently some issues have been raised in regards as to the adequacy of cellular components detection in fixed cells in comparison to live cells [9]. The rationale behind this argument is that the very process of fixing, permeating and incubating the biological samples to reveal with Abs modifies to a lesser or greater extent them. It is important for the microscopist to understand that what is studied under the microscope is never exactly the unperturbed biological entity. This argument is not without support so complementary and alternative assays are being proposed and tested to better study biological samples. One of these approaches is to rely on advanced labelling strategies [10]. Some of these proposed strategies are briefly presented. First is the use of intrinsic labelling, something routinely done for many decades. With intrinsic labelling the cell is incubated in the presence of a fluorophore that then interacts with some cellular structure. The fluorophore localization will depend both on the physico-chemical properties of the molecule and the cellular status (live, dead, metabolic level, *etc.*) at the moment of interaction (see Chapters 15 and 16). A classic example of this is the fluorescent labelling of live cells by acridine orange (see Chapter 15.1).

A second strategy is to label the cell directly, by making use of fluorescent proteins (FPs). These are genetically-encoded and transfected into a live cell that then expresses it at the, hopefully, right place and moment. No approach is without shortcomings and the use of FPs can lead to inappropriate protein expression due to their chimeric nature (incorrect expression pattern, function, localization, *etc.*). More will be presented below about FPs (section 2 below) and the reader can also consult Chapter 4.6 for additional information.

Then different tag-ligand (TL) systems have been proposed too. These strategies try to take advantage of both small fluorescent molecules and protein engineering. A ligand protein is expressed in the cell that displays an enhanced interaction with a small fluorophore. It is a plug-and-socket strategy in which both partners interact within the cell. Affinity-based TL systems take advantage of strong recognition-interaction between the fluorophore and the protein/oligopeptide. This situation is similar to that of a protein interacting with its substrate or a cofactor. Self-labelling TLs use the enzymatic activity of the protein to be tagged to "uncap" a fluorophore that is in a silenced mode because it is chemically interacting with the protein substrate. Once the enzyme catalyses the substrate reaction, the fluorophore is able to interact with the enzyme itself labelling it. This approach reduces greatly the background fluorescence because the "silenced" fluorophore is non-emissive. Finally ligase-based TL takes advantage of ligases enzymes present in the studied cell that link the fluorophore to a target protein/biomolecule recognized by the ligase. These approaches are interesting and deserve a deep follow up to check their promising outcomes.

Returning to the classical use of labelled Abs we will explain the differences between direct and indirect immunofluorescence so that the reader can choose the most advantageous in regards to his or her necessities.

1.1. Direct Immunofluorescence

In direct immunofluorescence, also known as primary immunofluorescence, the Ab labelled with a fluorophore directly recognizes the structure of interest (see Fig. **17.5A**). This approach is more straightforward than secondary immunofluorescence (see below) as the moiety to be studied is directly recognized. It saves time, steps and can be cheaper than indirect immune because it only employs one Ab. It also reduces background signal as there is less non-specific spurious binding to other structures (false positives) as compared to indirect immune.

However direct immune increases the chance of false negatives as there is no amplification step or steps and the target signal can be very low or below the detection threshold. For this reason direct immune is best indicated in analysis where a high amount of protein/structure of interest is anticipated in the studied sample. The use of a monoclonal Ab for direct immune can provide very specific results but at the risk of obtaining a very faint signal. On the other hand using polyclonal Ab increases the chances of detection (binding to different parts of the structure is a kind of signal amplification) but it increases the chances of cross-recognition with unrelated targets (false positives).

Fig. (17.5). **A**: In direct immunofluorescence the Ab that directly recognizes the structure of interest is labelled with a fluorophore. In the example shown three monoclonal Abs each bind one of the structures and provide a light signal in virtue of their fluorophores. **B**: In indirect immunofluorescence the sample is first exposed to a recognizing Ab (primary) that is not tagged with any fluorophore. After washing, the sample is left with a "latent" image, waiting to be "developed". A second step ("developing") is done by exposing the sample to another Ab (secondary) that recognizes the first primary Ab already present in the sample. The secondary carries the fluorophore. Now the structure of interest can be imaged thanks to the light signal emitted by the secondary Ab. Note that more than one secondary Ab can bind to the primary therefore providing signal amplification in this way.

1.2. Indirect Immunofluorescence

The indirect immunofluorescence approach, or secondary immunofluorescence, enhances the signal-to-noise ratio by making use of one (usually) or more amplification steps. First an unlabelled Ab (primary Ab) that recognizes the structure of interest is employed to "mark" the sample. This mark gives no signal as the Ab is non-fluorescent. The sample is washed to reduce as much as possible the amount of non-binding primary. Then the amplification step comes that, at the same time, is a revealing step. A second Ab (secondary Ab) that is fluorescent is employed. This secondary recognizes and binds to the primary Ab structure *not* the cellular target (see Fig. **17.5B**). By having several secondary Abs labelling each primary Ab a single recognized site at the cellular target provides many light signals. This is the most important advantage of the indirect immune approach that allows for an important increase in the detection signal. For this reason indirect immune is more resistant to bleaching than direct detection. On the other hand, the secondary can provide a false positive by binding to some molecule unrelated to the primary. This technique can produce a higher unspecific background fluorescence that interferes with the real signal under certain circumstances (inadequate washing the secondary). The indirect immune is more time consuming, makes use of more steps and can be more expensive as more reagents (Abs and others) are employed.

1.3. Counterstain Dyes and Anti-fading Agents

After completing the immune-detection steps it is often highly desirable either to subject the sample to another round of immune-detection with another Ab to study several structures/molecules at the same time, or to label those structures with complementary fluorescent labels, counterstain dyes, that provide additional information. Other compounds can prevent or at least delay the onset of bleaching, quenching and other undesirable reactions that diminish the fluorophore's signal. These compounds are known as anti-fading agents. The role of counterstain dyes and anti-fading agents will be briefly introduced in what follows.

Counterstaining Dyes

Counterstaining is a widely employed procedure in immunohistochemistry to increase the contrast between differential structures observed in bright field. The same rationale can be applied to counterstaining in immunofluorescence or fluorescence microscopy in general. These dyes are fluorophores that by one mechanism or another provide additional information on top of that provided by the immunofluorescence itself. This information can be related to the highlight of cellular structures, reporting on the vital status of the cell or to physiological processes such as ionic potential, metabolism, *etc.*

One of the most employed counterstains in immunofluorescence is that of that of the nucleus. This is achieved by incubating the sample with a compound that presents affinity for nucleic acids. Among these DAPI or the Hoechst group of dyes are prominent (see Chapter 12.5). These dyes allow for the nucleus localization once the sample is viewed in the microscope. Two examples of this double marking are shown in Fig. (**17.6**) in HaCaT cells. The top pictures (**A**-**D**) show the fluorescent signal from microtubules as revealed with an anti-α tubulin primary Ab followed with a secondary Ab against the primary. The secondary carries Alexa 488 as the fluorophore. The middle pictures (**B**-**E**) show the same field as the corresponding upper image but now the excitation light is changed to the UV (365 nm). The cells have been incubated with Hoechst 33258 that binds to the minor groove of the DNA. The bottom pictures (**C**-**F**)) show the overlay of the previous ones to visualize the nuclei position within the tubulin cytoskeleton.

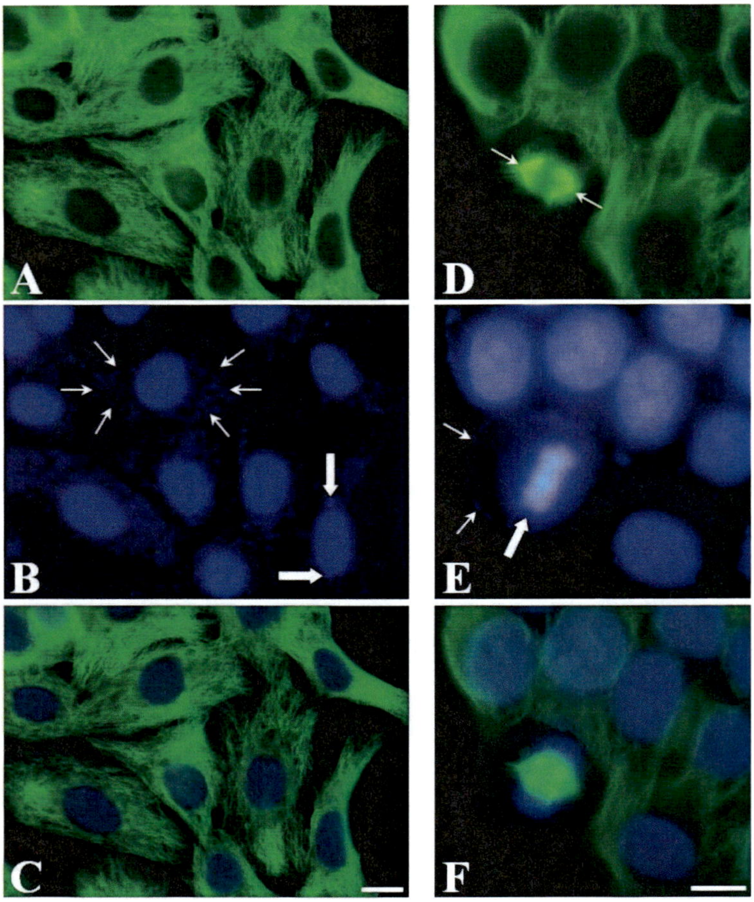

Fig. (17.6). Immunofluorescence and counterstaining in fixed HaCaT cells. **A**: Tubulin cytoskeleton imaged through indirect immunofluorescence. The secondary Ab was tagged with Alexa Fluor 488 and emits in the green. **B**: Counterstaining of the same cell field as in "A" with Hoechst 33258, a minor groove-binding dye. Excitation in the near UV (365 nm) makes the dye emit in the blue (450-480 nm). All nuclei in the picture are interphasic (rounded). Note that one cell shows two micronuclei (arrows) which appear as a consequence of structural chromatin damage. Also notice the staining of mitochondria (arrowheads) because of Hoechst 33258 interacts with mitochondrial DNA too. **C**: Overlay of images **A** and **B**. **D**: Tubulin imaging at higher magnification than in **A**. A mitotic (metaphase) cell is displayed at the centre left. Note how the whole tubulin network is recruited to form the mitotic bundles (arrowheads) that arrange and distribute the chromosomes during cell division. **E**: Counterstain Hoechst 33258 image of D to show the localization of chromatin. The mitotic cell displays a prominent chromosomic metaphasic plate (arrow) which is the direct consequence of the chromosomes interaction with the tubulin bundles shown in D. All other nuclei appear rounded and smoother as they are in interphase. Notice how the mitochondria (arrowheads) arrange in a ring-like structure at the very edge of the dividing cell. Probably they localize just below the plasma membrane. **F**: Overlay of D and E to better appreciate the interaction between the mitotic bundles and the chromosomal plate. Scale bars: 10 µm.

Other counterstain exist that can be employed to label organelles like lysosomes (acridine orange), mitochondria (rhodamine 123) or membrane lipids (cyanines, BODIPY dyes). In case of working with living cells or samples, counterstaining with vital dyes (propidium iodide) can also be employed to assess vital status (see Chapter 16.1.1). An additional example is provided by sulforhodamine 101 which is employed as a counterstain to discriminate between astrocytes/oligodendrocytes (positive staining) and neurons (negative staining) when studying neuronal physiology with fluorescent potential probes [11, 12].

Anti-fading Agents

These agents are a series of chemical compounds whose physical and/or chemical characteristics diminish the rate at which sample fading/quenching takes place [13] (see Chapter 7.4). Fluorescent signal loss occurs because non-radiative deactivation pathways take over the radiative process (see Chapter 2.5 for an explanation of these mechanisms). Fluorescence loss can take place temporally (quenching) or on a permanent basis (fading). The most common cause of quenching is by transfer of excitation energy from the excited fluorophore to another molecule. This is a physical mechanism. Two common causes of quenching are the presence of water and a fluid medium. Water, being a very light molecule with two hydrogen atoms, presents relatively high vibrational and rotational frequencies. These act as sinks for the electronic energy stored in an excited fluorophore. In this regard one strategy to reduce quenching is to reduce the water content of a sample (see mounting below). However this goes into conflict with the necessity of water when assessing living biological samples, so reducing the water content is not always possible. Another path for an excited fluorophore to lose its energy is internal conversion. Usually this takes place with internal rearrangement of the molecule's structure. A viscous medium opposes such rearrangements by increasing friction at the molecular level. Using viscous or solid matrices for an immunofluorescent sample reduces the loss of signal.

In regards to fading photochemistry is the most common cause. This photochemical reaction changes the fluorophore in a permanent way, destroying it in many instances. Most current fluorophores are specifically designed to avoid intramolecular changes and internal photochemistry. The most common cause of photochemistry that leads to fading is the presence of molecular oxygen (see Chapter 2.5.4). When O_2 is present in the sample it can readily react with an excited fluorophore producing ROS. These species are capable of permanently oxidize the fluorophore and the sample. Once the fluorophore is damaged it cannot longer emit light. One procedure to eliminate O_2 is to prevent its diffusion into the medium (see mounting below). Another complementary strategy is to incorporate some compound in the sample's medium that protects from O_2. These are known as anti-fading agents [13, 14]. Some of them, like sodium azide (NaN_3), physically interact with an excited oxygen molecule and quench it by transforming the excitation energy into vibrations and heat. This does not eliminate the O_2 but at least minimizes the potential of ROS to oxidize the fluorophore. Other anti-fading compounds chemically react with the ROS to avoid fluorophore damage: they are sacrificial agents. They are molecules with a high reducing potential, commonly known as antioxidants. Some examples are ascorbate, 1,4-diazabicyclo[2.2.2]octane (DABCO), para-phenylenediamine (PPD), 6-hydroxy-2,5,7,8-tetramethylchroman-2-carboxylic acid (Trolox, TX), n-propyl gallate, mercaptoethylamine and cyclo-octatetraene [13]. If the O_2 availability is low or negligible, due to consumption and/or very low diffusion, these agents can consume all the O_2 in the sample, thereby eliminating O_2 as a fading agent.

Mounting is the final procedure during the sample preparation for fluorescence microscopy. It involves having the sample in a more or less permanent platform (for example between a slide and a coverslip) embedded with a mounting medium. Mounting can be seen as an anti-fading procedure for two reasons: the mounting medium usually has anti-fading properties and diffusion of chemical agents, like O_2, is diminished or avoided by mounting. Mounting media can be aqueous or non-aqueous. Aqueous mounting, with Vectashield for example, are more sample friendly because water is the natural solvent in biology. This media usually employ one or a combination of those anti-fading reagents listed before. Also they are made viscous on purpose to reduce the chemical rearranging. On the other hand, water acts as a quenching agent itself, as already mentioned. The viscosity also helps to reduce O_2 diffusion into the sample.

Non-aqueous mounting media (*e.g.* DePeX) are hydrocarbon-based compounds. They provide a less physiological environment as most of the water is eliminated from the sample. However, the mounting is defined as permanent because part of the mounting medium evaporates quickly leaving a solid matrix (resin-like) that encases the sample. This makes the viscosity practically highest. So these media provide two ways of eliminating quenching mechanisms: they displace water and provide a solid matrix. This solid state also negates any O_2 diffusion inside the sample once it has been established, usually within minutes of preparing the sample. Non-aqueous mounting offers a "permanent" mounting although fluorescence signal can degrade over time due to other unavoidable processes. As a bottom line here, we would like to call the reader's attention to the fact that these advantages of the solid state in regards to minimising luminescence loss is one of the reasons solid-state probes are being so widely employed in fluorescence microscopy currently. We will elaborate on this topic in Chapter 18.

A common misconception is that aqueous-based mounting media are necessary for the correct visualization and preservation of immunofluorescent samples. This is incorrect and, in fact, these samples benefit from a non-aqueous permanent mounting, for the reasons considered in the previous paragraph. A practical example is provided in Fig. (**17.7**) where HCT-116 human colon carcinoma cells have been subject to an immuno against β-catenin (a protein involved both in cell-cell adhesion and in nuclear transcription) and counterstained with DAPI to visualize the nucleus [15]. Two samples were mounted with Mowiol (aqueous) and DePeX (non-aqueous) mounting media. The images clearly show that DePeX mounting offers a much brighter and resolved images than those obtained from Mowiol-mounted cells. Even more, the DePeX-mounted samples displayed an almost unchanged brightness even after one year storage at room temperature. This is a lapse of time unthinkable when using aqueous media.

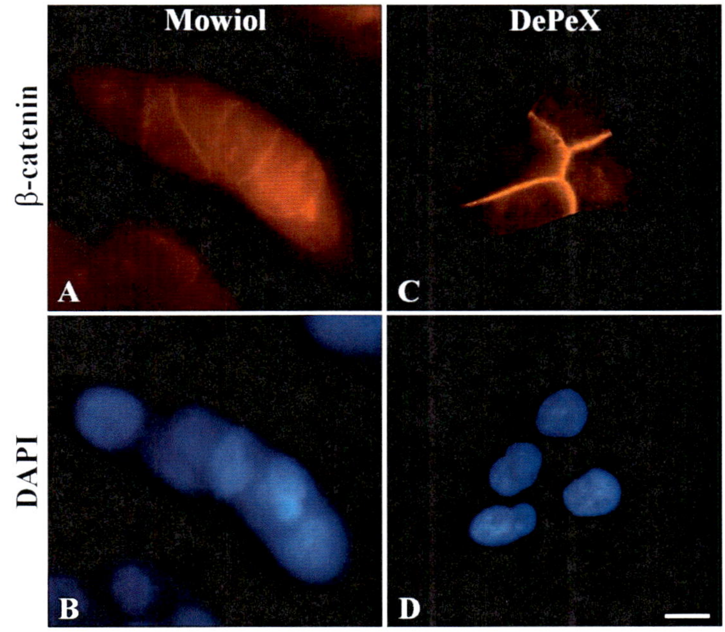

Fig. (17.7). Differential results of alternative mounting media, aqueous-based Mowiol (A-B) or hydrophobic DePeX (C-D) in immunofluorescence done on HCT-116 human colon carcinoma cells. **A**: β catenin detection in fixed cells. It is quickly realized that the image is fuzzy and details are lost. **B**: DAPI counterstaining image that also presents a poorly defined nuclear structure and localization. **C**: In contrast to A, this image shows a much better resolved image where β catenin can be detected both at the intercellular junctions (stronger signal between cells) and in the cytoplasmic domain (diffuse emission within the cells). The image resolution is much better even when the same microscope was employed to obtain all these images. **D:** In regards to nuclear imaging with DAPI using DePeX to mount greatly increases the resolving power too. In comparison with B the nuclei displayed in D are more defined and some details can be resolved. Scale bar: 10 μm. Reproduced with permission from reference [15].

1.4. New Uses and Related Techniques

The use of Abs for immunofluorescence is something totally established: a cornerstone of fluorescence microscopy. These techniques have been employed for decades so we will not elaborate more on them as literally thousands of scientific papers, works and books have been published on the field. We will highlight two new uses of immunofluorescence, or more correctly, expansions upon the general mechanistic. These two topics are the establishment of a proteome atlas and the use of nanobodies.

Proteome Studies

Since the study of the human genome started more than two decades ago, researchers have tried to obtain not only the complete genomes sequences of different organisms but also other "libraries" that provide information to the diverse components that make up living organisms. Among these the proteome, the range of proteins expressed by an organism under certain circumstances, is of utmost importance as it provides a much more dynamic picture than the genome. Towards this goal the use of immunofluorescence has resulted in great amounts of information in relation to the proteome especially in regards to the relationship between protein expression levels and cellular spatial localization [16 - 19]. An example of such an approach is provided in Fig. (**17.8**) [20]. In this case the authors imaged 4005 proteins in different cellular compartments and organelles of three different cell lines in order to establish a location-function relationship. The proteins location was obtained by immunofluorescence with the corresponding Abs and the use of a confocal laser system to obtain the fluorescence images. Some examples are shown in Fig. (**17.8A**). After obtaining all the pertinent data computer programs were employed to obtain networks of compartment-protein frequency for all the experiments (see Fig. **17.8B**). This allows a straightforward analysis of protein function, location and the involved organelles/compartments. Automated approaches to this kind of analysis are already providing huge amounts of information in relation to protein expression patterns in human cultured cells [21].

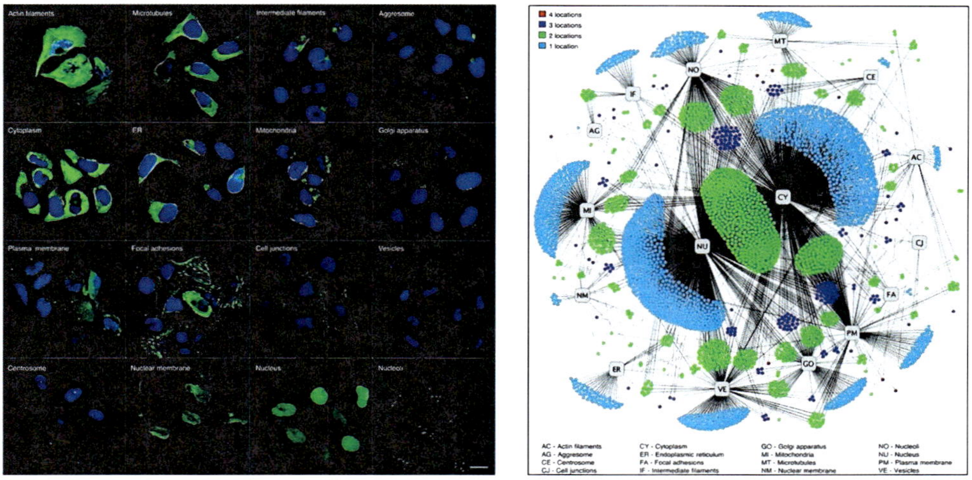

A **B**

Fig. (17.8). Immunofluorescence as a tool for proteomic libraries. A: Example of 16 different organelles and cellular structures targeted with different Abs for proteins implicated in the function/structure of these cell components. The cells have been counterstained with DAPI to better visualize the nuclei. Scale bar: 10 µm. B: Plot displaying the relationship between any one of the 3223 assessed proteins and their cell localization in U-2 OS osteosarcoma cell line. Each protein appears in the plot as a coloured dot. The colour provides information as to in how many cellular compartments the protein appeared (maximum 4 different locations). The lines join each protein (dot) with the particular compartment(s) were it was found. Each compartment is denoted as a grey box with a letter code. The letter code is explained in the legend at the bottom of the figure. This plot provides some insight as to the huge interrelation protein network that makes a cell such a dynamic system. Reprinted and adapted with permission from reference [20]. Copyright 2011 American Chemical Society.

Nanobodies

A recent incorporation to the immunofluorescence range of techniques is the use of nanobodies (Nbs). Nbs or single-domain antibodies are fragments of a full Ab. The fragment of interest is the variable region of the light and heavy chains (the very "tip" of the two horns in the Y depiction, see Fig. **17.9A**). Alternatively heavy chain fragments from camelid or shark Abs have been very successfully employed to produce Nbs. These fragments display the specificity for antigen recognition but their molecular weight is one order of magnitude lower than the whole Ab (10-15 kDa *vs*. 150 kDa) [22, 23].

This very low molecular weight makes Nbs ideal for living cell penetration and for mass production in bacteria, among other advantages. By attaching an organic fluorophore, as it is the usual procedure with Abs, one obtains fluorescent Nbs (see Fig. **17.9B**). However, the preferred strategy with Nbs is to express them as a chimeric protein with a GFP moiety (see Fig. **17.9C**). These highly substrate-specific and photostable proteins are known as chromobodies (Cbs) [24]. Cbs are finding many applications in research and diagnostics, and Nbs in therapeutics too. An example of Cb use to detect vimentin and relate this protein levels to epithelial-mesenchymal transition is shown in Fig. (**17.10**) [25]. Vimentin is a protein that belongs to the intermediate filaments group of cytoskeleton proteins.

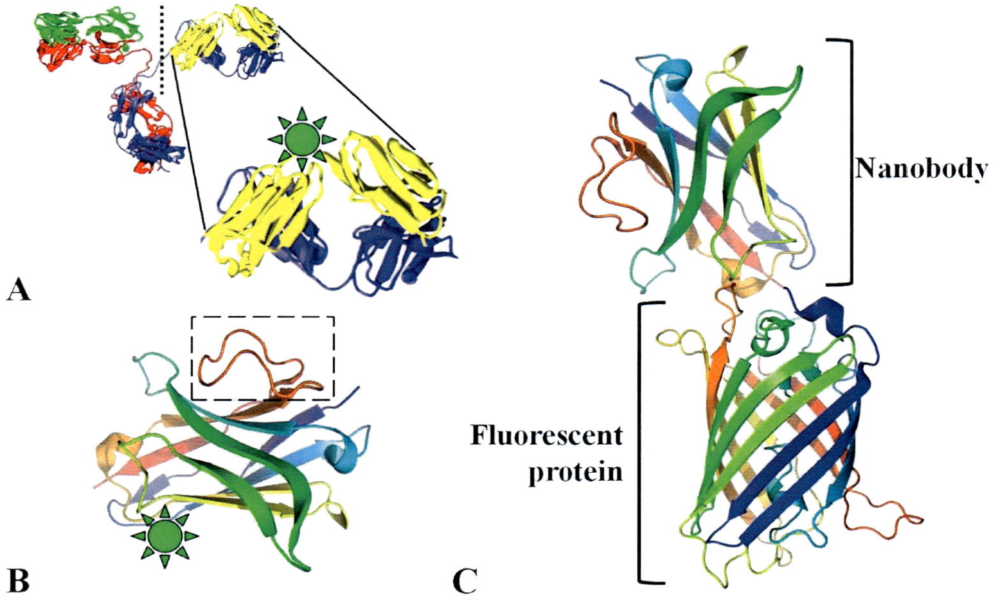

Fig. (17.9). A: Basic approach to Nbs structure. The hyper-variable Ab region, where light and heavy chains interact to recognize the epitope, can be clipped from a whole Ab. This light fragment can be functionalized with a fluorophore in the same way as a full Ab but with the advantage of lower size and molecular weight (Ab structure reproduced with permission from TimVickers -https://commons.wikimedia.org/w/index.php?curid=3434214). **B:** Current experimental approach to Nbs. Fragments from camelid or shark immunoglobulins are expressed in bacteria. The structure shown corresponds to a Nb obtained from a camelid. The epitope-recognizing region is the orange "ribbon" at the top of the structure (dashed box). The Nb can be functionalized with a fluorophore as schematically shown in the figure (structure reproduced with permission from S. Jähnichen - https://commons.wikimedia.org/w/index.php?curid=7318216). C: Chromobody structure. Cbs are fusion proteins of Nbs with a GFP or, generically, any FPs. The displayed structure shows the Nb region at the top and a FP at the bottom. Although this approach increases the size and molecular weight of the structure the FP moiety provides many fluorescent advantages as compared to classical organic fluorophores (FP structure reproduced with permission from Zephyris https://commons.wikimedia.org/w/index.php?curid=2718822).

It is considered a hallmark of cancer cell transformation in epithelial tissues. Epithelial cells undergo the epithelial-mesenchymal transition in this transformation process, with increased levels and reorganisation of vimentin. By making use Cbs that recognize vimentin the authors were able to detect vimentin in a very detailed way in living cells (see Fig. **17.10A**). Going further, this work also reports making use of these Cbs to study vimentin kinetics within living cells by employing FRAP techniques (see Chapter 6.6.1) on the Cbs (see Fig. **17.10B**). This can provide potentially fundamental information about cell transformation to better understand cancer development and design strategies against it. Additional examples dealing with the use of Nbs and Cbs can be found in Chapters 18.2.1 (conjugated to Qdots) and 20 (use for super-resolution microscopy).

Fig. (17.10). A: Selection of different Cbs for vimentin detection in HeLa cells. At the top the amino acid sequences defining each one of the eight Cbs assessed for vimentin detection (Cb code at the left). At the bottom HeLa cells images for vimentin with the eight Cbs. Additionally cells expressed generic GFP as a control. Also a positive immunofluorescence control for vimentin detection was done using an Ab specific for α vimentin. It can be seen that VB3-CB, VB6-CB and VG1-CB are the three Cbs with a better vimentin detection profile (compare with α-VIM-IgG). Scale bar: 20 μm. **B:** Time-resolved FRAP studies of vimentin kinetics in living HeLa cells. (Top) Cells expressing Cbs VB6-CB, VB3-CB or vimentin-GFP (GFP-VIM) were subjected to a photobleaching pulse in the highlighted cellular area (white box). The fluorescent signal recovery after photobleaching is shown for 10 and 25 s after the pulse. VB6-CB showed the best recovery profile followed by VB3-CB. GFP-VIM displayed a poor recovery after photobleaching. Scale bars: 10 μm. (Bottom) Kinetics plot of fluorescence recovery after photobleaching for the three proteins showed in the images. This provides quantitative data supporting the best result for VB6-CB closely followed by VB3-CB. The plot confirms that GFP-VIM shows negligible recovery after photobleaching. Reproduced with permission from reference [25].

Now some established biological research techniques that also make use of immunofluorescence will be briefly introduced.

Flow Cytometry

Flow cytometry was introduced in Chapter 6.8. This technique can be considered an evolution of immunofluorescence as it usually makes use of labelled antibodies to recognize a sample's feature of interest. The difference stands in that in immunofluorescence the assessment is made upon a sample standing on a substrate, like a coverslip. In flow cytometry the substrate is a liquid flow that carries the sample. Each technique offers advantages and shortcomings, of course. Immunofluorescence allows for a much better structural analysis of the sample. The particular location of a given target can be determined. Flow cytometry allows for some structural features to be analysed but in a very limited way. On the other hand flow cytometry provides a much faster signal processing when compared to immunofluorescence (1,000-100,000 faster). So it is up to the researcher to choose which technique better suits his or her necessities as a function of the studied parameters and the magnitudes considered.

Enzyme-linked immunosorbent assay (ELISA)

In its classical use ELISA employs Abs immobilized on a substrate (usually at the well bottom of a microtiter plate) to recognize a given substrate of interest (other ELISA approaches exist but we will not discuss the details). Once the substrate has been recognized it binds and attaches to the Ab. After washing, a second Ab that also recognizes the substrate is provided. This second Ab is modified to produce some chemical reaction, usually functionalizing it with an enzyme. The enzymatic reaction produces a coloured chemical product whose concentration can then be measured through optical absorption. This approach is, therefore, a chromogenic one: colour is produced and measured as a function of initial substrate concentration.

It follows from this experimental setup that is perfectly possible to functionalize the "active" recognizing Ab with a fluorophore and then measure the brightness level in the sample. The procedure is very similar to an immunofluorescence but instead of assessing the sample under the microscope it is analysed for its global light emission. This fluorescent approach to ELISA is providing very successful outcomes in regards to analysis of very small volumes of analyte through microfluidics [26, 27]. In addition to the very high sensitivity of this fluorescence technique, microfluidics ELISA allows for truly enormous throughput as a vast numbers of microchannels can perform analysis at the same time in parallel [28, 29]. The measurement of fluorescent light lends itself to analysis with other techniques already presented. For example, Saraheimo *et al.* have recently reported the use a fluorescent ELISA enhanced by use of time-resolved FRET. This allowed them to measure samples not only on a substrate but also in homogenous solution [30].

Western Blot

The most employed systems to reveal western blots rely on a colorimetric reactions by an enzyme linked to the secondary Ab or chemiluminescent reactions also driven by the secondary Ab. However it is becoming more and more usual to employ fluorescent secondary Abs, as in fluorescence microscopy, to highlight the protein bands in the Western, both in basic research and in clinical diagnosis [31, 32]. This has the advantage as compared to chemiluminescence in that light emission is steady, as long as the sample is adequately excited, and does not diminishes over time.

2. FLUORESCENT PROTEINS

The proposal and realization of fluorescent proteins (FPs) for detection purposes in Biology and Medicine have revolutionized these fields. Their impact has been recognized by awarding it with Nobel Prize in Chemistry in 2008. The basic idea started with the use of a green fluorescent protein (GFP) isolated from the jellyfish *Aequorea victoria*. More complete information in regards to the basic structure and related photophysics and photochemistry of GFP and FPs has been introduced in Chapter 4.6. Here we will briefly present some applications of FPs in relation to new concepts being developed in the biomedical field.

As stated at the Chapter's introduction FPs can be thought as a type of primary-secondary fluorescent Ab in regards to the substrate recognized by the chimeric FP. Perhaps the most cited asset in regards to the use of FPs is their adequacy to measure parameters in living samples (cells and tissues, and even whole organisms) [33, 34]. In this sense Lavis has already proposed FPs as one of the fundamental labelling strategies to be considered for studies with living cells [10]. The correlation between function and subcellular location, classically assessed by Ab immunofluorescence in fixed samples, is changing its paradigm to the use of FPs in living cells, which is seen as the next step in fluorescence microscopy research [9]. This offers three advantages over classical immunofluorescence: it still correlates function-location but in a much more physiological way, the object of analysis is alive (less artefactual and perturbative approach), and it opens the possibility to study kinetics in a truly continuous process (think of the difference between still pictures and cinema). Therefore FPs use, very especially in regards to dynamical and proteomics studies, is indeed revolutioning cell biology [18, 35]. FPs have also been designed in combination with Nbs (see chromobodies above). The combination of the very low molecular weight Nbs with the excellent photodetection characteristics of FPs (see Fig. **17.9C**) make this emerging technique really promising [24].

Alternative FPs can be designed that do not reproduce the classical GFP from *A. Victoria*, and that are amenable to genetic and metabolic control in living cells. An example of this was recently reported by Chatterjee *et al.* where the intrinsically fluorescent amino acid 3-(6-acetylnaphthalen-2-ylamn)-2-aminopropanoic acid (Anap) was incorporated into proteins expressed by a mammalian cell under the genetic control of the TAG codon [36]. This approach offers the advantage that only a single amino acid has to be included in the transformed protein and not a whole protein as with chimeric FPs. The FP "label" can in many instances perturb or even abolish the physiological role of the tagged protein. Anap was incorporated into a histone for nuclear localization (see Fig. **17.11A**) or endoplasmic reticulum and Golgi proteins (see Fig. **17.11B**). Comparison (see the figures) with an equivalent FP (mCherry) as a positive control shows the very promising results of this approach for cell studies.

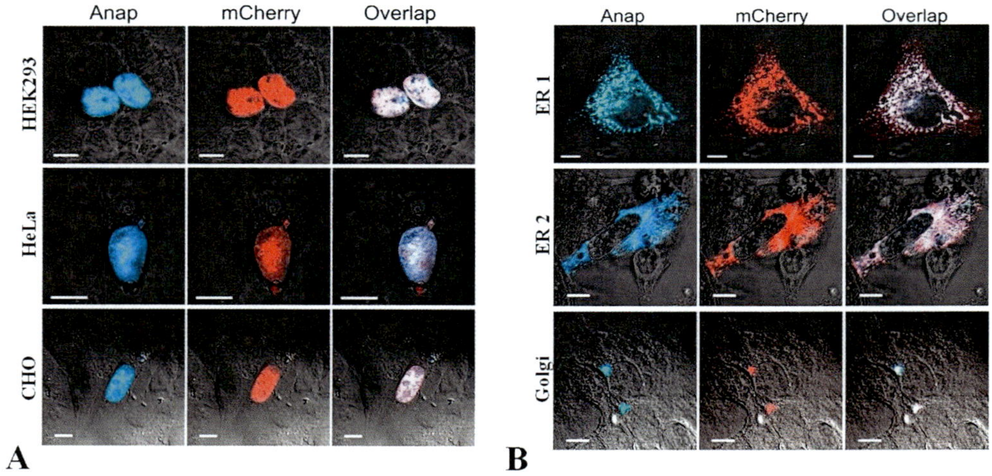

Fig. (17.11). Use of a fluorescent artificial amino acid (Anap) incorporated in a genetically-controlled way into different living cells. **A:** Fluorescent localization of Histone H3 by two complementary approaches. In the first column position number 59 in H3 sequence has been replaced by the fluorescent amino acid Anap. Under UV excitation (360 nm) the H3 location is revealed by Anap emission at 470-510 nm. In the second column the same H3 protein has, additionally to the Anap amino acid, a red FP (mCherry) fused to it. Under 540 nm excitation mCherry emits at 580-610 nm allowing H3 localization at the nucleus too. The third column is an overlay of the two previous images to check the signals overlap. Images obtained by confocal laser excitation. Scale bars: 10 µm. **B:** Same procedure as in A but now labelling proteins localizing at the endoplasmic reticulum (ER) or the Golgi apparatus. ER 1 (first row) refers to a variant of ER resident protein Grp94 with Anap incorporated at position 127. ER 2 (second row) refers to another variant of Grp94 with Anap at position 7. Both ER variants have an mCherry FP fused at the C-terminal and both were expressed in HeLa cells. In the third row the Golgi-resident protein Galt1had Anap incorporated at position 3 and fused with mCherry. In all three examples it can be seen the very good co-localization of both signals. Imaging conditions same as in A. Scale bars: 10 µm. Reprinted and adapted with permission from reference [36]. Copyright 2013 American Chemical Society.

We would like to include here a brief comment in regards to other biophotonic techniques currently available thanks to the use of photo-reacting proteins. The newly developed field of Optogenetics is based precisely on the expression of ionic membrane channels that open or close in response to light [33]. This opens the possibility to directly control an excitable (neuron, muscle cell) or non-excitable (control of Ca^{2+} fluxes, for example) cell at will [37]. Even the optogenetic control of gene expression is starting to be at hand by these photoresponsive channel proteins [38]. Finally another group of photoactive proteins are able to generate ROS when illuminated, a sort of photosensitizing proteins which can be employed in the emerging field of redox biology [39]. All these new tools at the researcher's disposal are promoting a revolution in Biology and Medicine.

3. FLUORESCENCE *IN SITU* HYBRIDIZATION (FISH)

The technique of fluorescence *in situ* hybridization (FISH) employs the complementary recognition between nucleobases to detect and analyse gene localization and genetic expression levels [40 - 44]. It can make use of RNA or DNA probes to identify very specific sequences on mRNA, lncRNA, miRNA and DNA. The generic steps of this technique are displayed in Fig. (**17.12A**) and will be explained next. First a genetic probe of known sequence and affinity must be obtained by molecular biology techniques. This is the genetic sequence that will recognize the target sequence in the sample (it is the equivalent to a primary Ab in immunofluorescence). This probe is tagged with a fluorophore of choice. The genetic material (chromatin, DNA or RNA) is extracted from the sample and undergoes some kind of physical or chemical treatment that denaturalizes it. Then it is incubated with the probe. After some time the target and the probe hybridize in a very specific way. After washing to eliminate any free probe and low specificity unions with non-target sequences, the sample is mounted and assessed under a fluorescence device, usually a microscope.

It is customary, although not essential for the technique, to counterstain the sample after hybridization with a general nuclear dye like DAPI or any of the Hoechst stains. This provides a certain background signal from the genetic material (chromosome, comet) on top of which the brighter probe signal is localized. The deep blue emission from the chromatin in Fig. (**17.12B, C**) is due to nuclear counterstaining. Throughout the years since its implementation FISH has evolved in many fields for many specific applications. Therefore this generic almost schematic protocol introduces the basic steps. There are many scientific reviews and book chapters focused on FISH protocols within specific fields [43, 45 - 48].

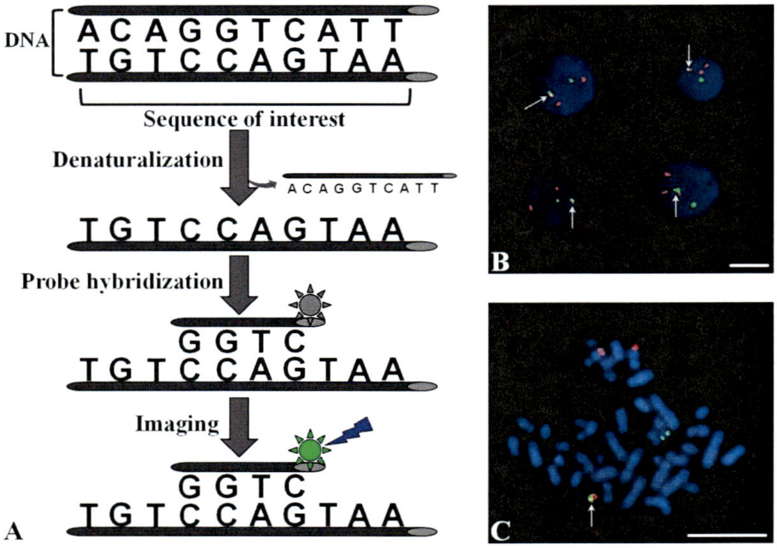

Fig. (17.12). A: Simple schematics of the FISH experimental procedure. After selecting a particular sequence of interest (a DNA example is shown but RNA works in the same way) the genetic material is gently denaturalized to eliminate the complementary sequence to the one of interest. After this step the genetic material is incubated with the genetic probe. The sequence and probe hybridize and establish a relatively strong hydrogen bonding. Non-selective partial bonding can take place but after a washing step (not shown) this false recognitions disappear. Afterwards the sample can be excited with adequate light to obtain a fluorescent signal from the fluorophore attached to the probe (imaging). **B:** Image of FISH on interphase lymphocyte nuclei. In these nuclei two genes on two different chromosomes (9 and 22) have been labelled. The red probe recognizes a sequence in chromosome 22 while the green probe recognizes chromosome 9. It can be appreciated that there is a mixed green+red signal, giving a yellowish colour, at the some points (arrowheads). This is the result of a translocation of one chromosome fragment to another chromosome. In this case the two genes involved in the translocation and detected by FISH are ABL 1-chromosome 9 (green) and BCR-chromosome 22 (red). This particular translocation (Philadelphia chromosome) is involved in leukemia development (reproduced with permission from https://commons.wikimedia.org/w/ index.php?curid=585574). Scale bar: 10 µm. C: Same example as in B but FISH was done on metaphasic chromosomes instead. The red and green signals are clearly seen as is the fusion chromosome (arrowhead) with a combination of green, red and yellow (reproduced with permission from https://commons.wikimedia.org/w/ index.php?curid=254833). Scale bar: 10 µm. In both B and C chromatin counterstaining was done with DAPI (blue background).

Although several reports and proposals that can be englobed within the concept of FISH were proposed and tested in the 60s and 70s, the technique as such was first employed in 1982 to map genes in the fruit fly [49]. At that time image was revealed by making use an indirect technique very employed during the first years of FISH use. In short the genetic probe was conjugated to a biotin molecule. Once hybridization took place the biotinylated probe was recognized by a primary Ab that bonded to biotin. A secondary fluorescent Ab, recognizing the first Ab, was then employed to produce the image. It can be seen that this approach is a certain variation and evolution of common immunofluorescence. The next step was to directly tag the genetic probe with a fluorophore to avoid the intermediate labelling steps by Abs. This was done at the beginning of the 90s [50]. It offered the advantages of simplifying the experimental protocol, increasing the sensitivity and decreasing the background, it could be combined with detection through biotinylated/digoxinylated probes, and it was open to further signal amplification by employing Abs that recognize the fluorescein moiety on the probe. At present a whole range of fluorophores, including most of those presented in this book like organic molecules, transition metal complexes or luminescent nanoparticles, are being employed in conjugation with genetic probes for FISH. Fig. (**17.13**) shows a list of some fluorophores routinely employed for FISH detection [51].

The field in which FISH finds applications is very broad. One of the most classical uses is the identification of particular sequences in chromosomes (see Fig. **17.12B, C**). This allows the direct karyotyping of the chromosome set and assess if there are any changes in the chromosome set. An example of such a chromosome analysis is shown in Fig. (**17.14A**) where some of the chromosomes from a human sperm cell are displayed [52]. In this case particular case the sperm cell derives from an infertile subject and the technique allows a fast detection of numerical abnormalities (super numerary chromosomes) in the karyotype (see figure caption for details). Even further, by combining FISH with 3D confocal imaging it is possible to localize chromosomes within the nuclear volume with great detail (see Fig. **17.14B**).

The technique known as chromosome painting (M-FISH) is a further development on the FISH theme. It consists on labelling different chromosome regions (on the same or different chromosomes) with different probes [43, 53]. Each probe has a different emission colour. By analysing the sample's emission spectrum it is possible to assign a specific colour or combination thereof to each chromosome. Usually the coloured image is artificially enhanced by using false colours to improve resolution. Chromosome painting can easily provide information in regards to structural chromosomal defects (see Fig. **17.15**). In the figure example a cell exposed to a radioactive source is then analysed for cytogenetic damage through chromosome painting [54]. Any colour mixing between chromosomes as well as missing parts provides information as to structural rearrangements taking place after the treatment. Similar studies have employed radioactive sources and FISH to analyse new genetic variants in plants [55].

Chemical family	Fluorophore	Functionalization	λ_{ABS} (nm)	λ_{EMS} (nm)	ϕ_{FL}	Remarks	Suppliers
Cyanine (Cy®) dyes	Cy®3	SE, CA, strep, mal, azide, alkyne, NH_2, UTP, CTP, GTP, ATP, pCp, dUTP, dCTP	550	570	0.15	Broad range of applications, less bright than Alexa Fluor® 546	Invitrogen, Jena Bioscience, GE Healthcare Life Sciences, Lumiprobe, ...
	Cy®5		649	670	0.27	Broad range of applications, less bright than Alexa Fluor® 647	
Cyanine dimers	YOYO-1	N/A	491	509	0.38 0.25	Photocleavage of NA and bleaching reported	Molecular Probes™
	TOTO-3		642	660	0.06	Bleaches fast (20-30s)	
Ethidium bromide		N/A	285	605	0.20	Especially used for gel staining, but suspected mutagenicity	Invitrogen, Sigma-Aldrich, Bio-Rad, ...
Alexa Fluor®	Alexa Fluor®488	SE, CA, strep, mal, azide, alkyne, UTP, dUTP	495	519	0.92	Fluorescence output unmatched by any other spectrally similar dye	Molecular Probes®
	Alexa Fluor®546		556	573	0.79	Cy®3 and TM-Rhodamine substitute	
	Alexa Fluor®594	SE, CA, strep, mal, azide, alkyne, dUTP	590	617	0.66	Bright and photostable	
	Alexa Fluor®647	SE, CA, strep, mal, azide, alkyne, ATP, dUTP, dCTP	650	668	0.33	brighter substitute for Cy®5	
ATTO dyes	ATTO 488	SE, CA, strep, mal, azide, alkyne, NH_2, biotin, UTP, ATP, dUTP, dCTP	501	523	0.80	Very high photostability	ATTO-TEC, Sigma-Aldrich, Jena Bioscience
	ATTO 647N		644	669	0.65	Cy®5 substitute, high thermal and photostability, exceptionally high stability towards atmospheric ozone	
Fluorescein	FITC (isothiocyanate)	SE, mal, azide, alkyne, UTP, CTP, ATP, dUTP, dCTP, dGTP, dATP	494	518	0.86 (0.1 N NaOH)	pH sensitive, prone to photobleaching	Thermo Scientific™, Sigma-Aldrich, Jena Bioscience, ...
	5-FAM (carboxylic acid)	SE, azide, alkyne, strep, NH_2, dUTP, dCTP	494	518	0.90	SE form more stable bond with amines compared to FITC-SE	Molecular Probes™, Sigma-Aldrich, Jena Bioscience, Lumiprobe, ...
TM-Rhodamine	TRITC (isothiocyanate)	SE, strep, mal, azide, dUTP, dCTP	555	580	0.10	Prone to aggregation	Thermo Scientific™, Sigma-Aldrich, ...
	5-TAMRA (carboxylic acid)	SE, strep, mal, azide, alkyne, NH_2, dUTP, dCTP	555	580	0.10	More stable than TRITC	Molecular Probes™, Sigma-Aldrich, AnaSpec.
CX-Rhodamine	5-ROX	SE, mal, azide, alkyne	576	591	0.70 (PO_4 buffer, pH 9)	Unstable compared to other rhodamines	Molecular Probes™, Sigma-Aldrich, Jena Bioscience, AnaSpec.
Fluorescent proteins	EGFP	N/A	488	509	0.60	Most used, brighter variants available, no problems with photostability	Thermo Scientific™, Sigma-Aldrich, BioVision
	mCherry	N/A	587	610	0.22	Superior photostability when compared to other red proteins	BioVision, Abcam®
Quantum dots	CdSe-ZnS	CA, strep, mal, azide, alkyne (cyclooctyne), NH_2, biotin			0.20-0.50	Bright, photostable, broad excitation, tunable narrow emission, but concerns about cell toxicity	Molecular Probes™, Sigma-Aldrich, Mesolight
Lanthanide chelate	Eu(III)-chelates	SE, CA, strep, mal, azide, alkyne, NH_2, biotin	376	616	~1	Narrow emission peaks, long fluorescence lifetimes, high photostability but often needs antenna to transfer absorbed excitation energy to the lanthanide	PerkinElmer,
	Tb(III)-chelates		340	545			Invitrogen, PerkinElmer
Transient metal complexes	Pt(II), Ir(III), Ru(II), Re(I)based	SE, CA, strep, mal, azide, alkyne, NH_2, biotin, dUTP				Long fluorescence lifetimes interesting for FLIM, tunable excitation/emission, high photostability, good quantum yield	Sigma-Aldrich

Fig. (17.13). Chart of some of the most employed fluorophores in FISH. Along with the fluorophore and its chemical family or group, the usual functionalization group employed for genetic probe conjugation is shown. The maximum of the absorption (λ_{ABS}) and emission (λ_{EMS}) bands, and the fluorescence quantum yield (ϕ_{FL}) are provided too. Some useful remarks and commercial suppliers can be found last. Abbreviations: SE: succinimidyl ester; CA: carboxylic acid; strep: streptavidin; mal: maleimide; NH_2: amine; FLIM: fluorescence-lifetime imaging microscopy. Reprinted and adapted with permission from reference [51]. Copyright 2016 American Chemical Society.

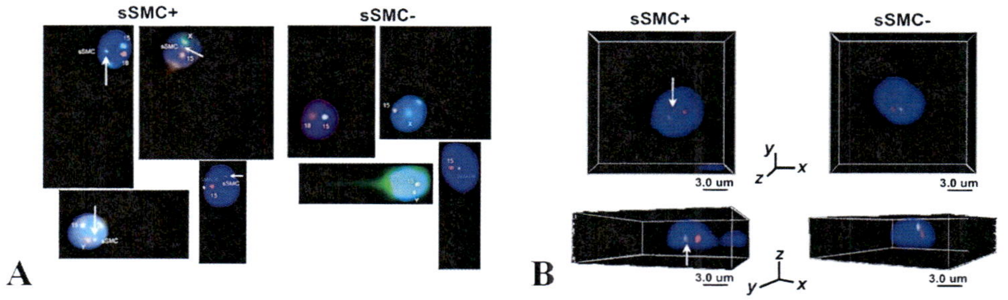

Fig. (17.14). A: Example of chromosome detection by FISH. In this case chromosomes 15, 18, X and a supernumerary-sSMC- (chromosomal numerical abnormality) were detected in spermatozoa cells subjected to FISH for those chromosomes. On the left four examples of spermatozoa with the additional chromosome (highlighted by arrowheads). On the right spermatozoa with the normal chromosomal set. **B:** Fluorescent detection of chromosomes (as in A) by FISH but taking advantage of a confocal setup to localize the chromosomes within the volume of the spermatozoa head. The supernumerary chromosome is indicated by the arrowheads. Reproduced with permission from reference [52].

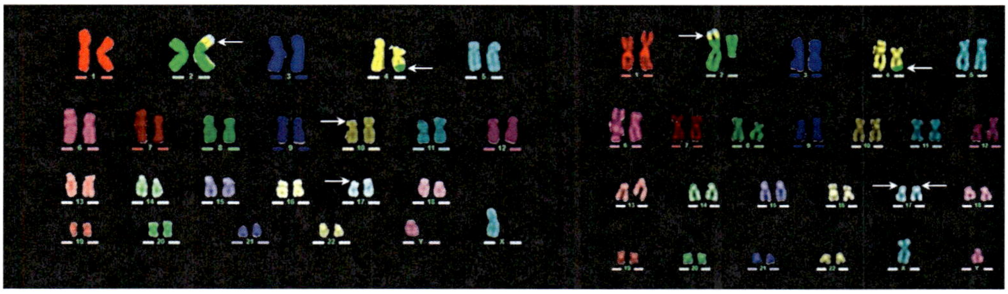

Fig. (17.15). Chromosome painting employed in the analysis of chromosomal structural aberrations after radiation exposure. Human bone marrow progenitor cells, implicated in maintaining immune cells, were exposed, on average, to a single alpha particle. After that single deleterious event the cells were analysed for chromosomal integrity making use of chromosomal painting (M-FISH). Two sets of chromosomes are presented in the figure. Both present translocations between chromosomes 2, 4, 10 and 17 (arrowheads). Notice how aberrant chromosomes display more than one colour banding. Intact chromosomes, at least at the major structural level, display a single colour (Reproduced with permission from reference [54]).

This very precise analysis of the chromosomal set by painting is put to use in the identification of species and subspecies [56]. An impressive assortment of maize chromosomes obtained by chromosome painting is shown in Fig. (**17.16**). Repetitive DNA sequences were chosen to identify each one of the 10 chromosomes present in maize [57]. This particular analysis was done to establish specific genetic variants of maize in order to propose the technique as a standard for plant genetic identification and selection. As can be seen in the figure differences in sequence frequency are immediately apparent. This method has the potential to be automatized for plant selection.

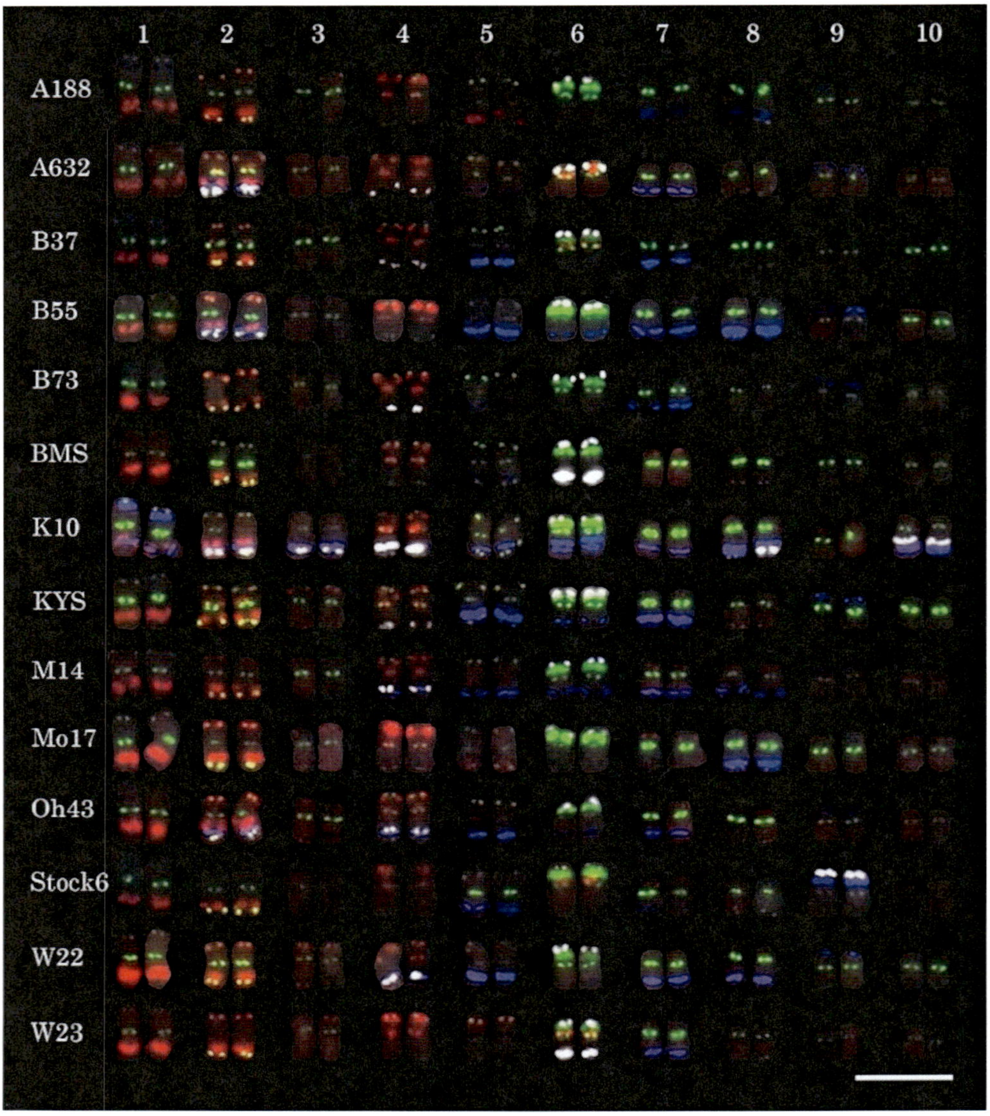

Fig. (17.16). Example of chromosome painting to identify the 10 somatic chromosomes in root-tip cells from maize plants. Probes were designed from different tandemly repeated DNA sequences and tagged with different fluorophores to obtain a colour range for identification. The analysis was done on 14 different maize variants (left column A188-W23) across the 10 chromosomes (top lane). Differences in the appearance and frequency of the DNA sequences can be appreciated by comparing the existence and intensity of colours (*e.g.* compare A188 and K10 variants on chromosome 1). Scale bar: 10 μm. Reproduced with permission from reference [57]. Copyright (2004) National Academy of Sciences, U.S.A.

A variation of this use of FISH-chromosome painting is the analysis of intra-species differences in the genetic makeup. This can be related to genetic changes at the gene level, but also when chromosome rearrangements take place or even the appearance of whole new chromosomes (supernumerary chromosomes or B chromosomes). An example of such a case is shown in Fig. (**17.17**) where supernumerary chromosome analysis was undertaken in the fish species *Astyanax paranae* (a species of freshwater fish) [58].

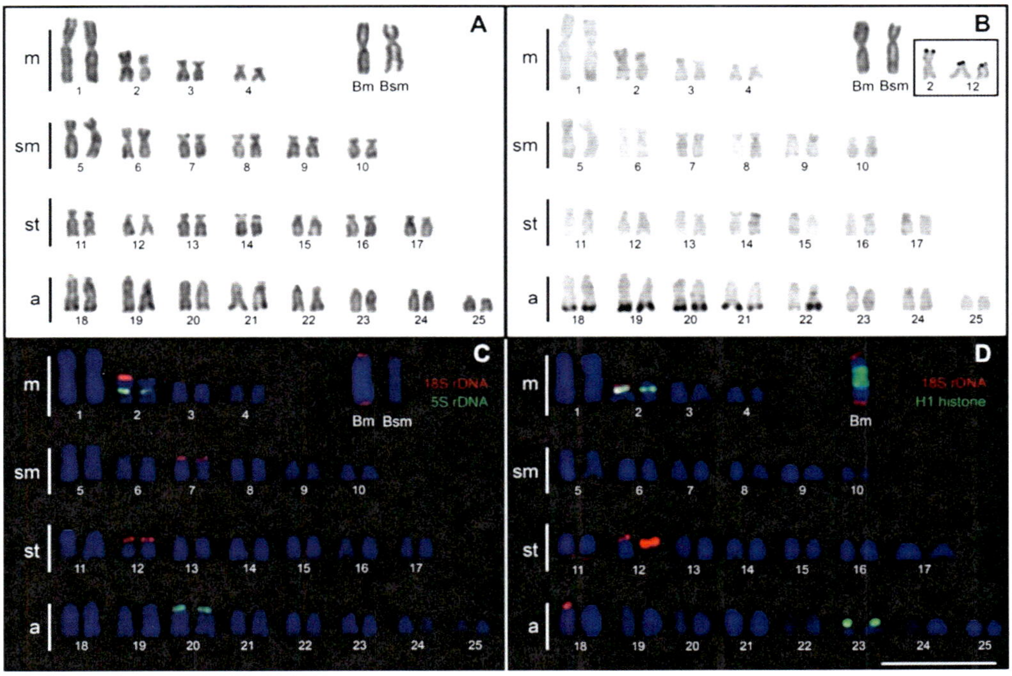

Fig. (17.17). Analysis of B chromosomes (supernumerary) in the fish species *Astyanax paranae*. **A:** Conventional bright field staining of chromosomal set. This shows 25 somatic chromosomes and 2 supernumerary (Bm and Bsm) chromosomes. **B:** C-banding staining of the chromosomal set. **C:** M-FISH to detect sequences related to 18S rRNA (red) and 5S rRNA (green). Note that both B chromosomes show duplications of 18S rRNA genes at the chromosomal tips. **D:** Same procedure (M-FISH) to detect 18S rRNA and H1 histone in a different sample. In this case extraordinary copies of both genes are localized to the Bm chromosome. These chromosomal differences can be used to assess subspeciation within a species. m=metacentric, sm=sub-metacentric, st=subtelocentric, a=acrocentric. Scale bar: 10 μm. Reproduced with permission from reference [58].

Chromosome painting allows comparison with classical cytogenetic staining and banding techniques too. This is also shown in the figure (see the figure caption for more details). The analysis and identification of species is routinely done not only in eukaryotes but also in prokaryotes through the use of FISH [48, 59, 60]. Starting from very low amounts of biological material it is possible to detect and allocate bacterial organisms with a very low probability of error. One approach is to use probes for bacterial rRNA to detect the amount and species present in a sample. This can be combined with confocal microscopy assessment to obtain information not only in regards to the absence or presence of a particular microorganism but also into the spatial structure of the community. In their interesting review Wagner *et al.* provide many troubleshooting strategies to handle specific problems in regards to FISH applied to microbial detection [59]. FISH reaches such a level of detection that single cell representatives in a microbial community can be detected through rRNA hybridization. This combined with the possibility of employing flow cytometry methods largely increases the amount of sample assessed in a given time [60]. These studies permit the establishment of microorganism/bacterial communities in such different environments as sea or fresh waters, soils, sediments, biofilms, in symbiosis or parasitism of other organisms, in patients suffering infection, *etc.* Therefore FISH is a very important tool not only at an ecological but also at the clinical level.

In regards to clinical applications and given the high sensitivity of FISH for the detection of cytogenetic aberrations, including numerical disorders like aneuploidies (trisomies, tetrasomies) and structural defects like deletions, duplications, translocations, inversions, insertions, *etc.*, it has indeed been employed for such a goal since its implementation [44, 61, 62]. FISH is an established routine prenatal diagnosis technique. It provides results in a time period of 24 h relying on uncultured amniocyte interphase analysis. Dedicated assays exist to detect aneuploidies of chromosomes 13, 18, 21, X and Y, the most common observed in human pregnancy. FISH is employed in preimplantation genetic diagnosis too. Single-cell analysis can be done on polar bodies, blastomere or trophoblast cells in cases of knowledge or suspicion of chromosomal or gene disorders in one or both progenitors [43].

One of the largest application areas of FISH in diagnosis and prognosis is in relation to cancer detection [45, 63, 64]. This is quite straightforward as numerical aberrations and structural abnormalities are directly involved in the initiation and progression of many cancers. Coming back to a previous example, it can be seen in Fig. (**17.15**) that a single alpha particle, one modality of radioactive decay, traversing a cell is capable of introducing severe structural chromosomal rearrangements [54]. These can lead to several outcomes, one of which can be tumour development. Furthermore, the authors of this work reported that the chromosomal abnormalities where inherited by successive generations of cells starting with the irradiated cell. In a world with an increase in aged population the risk of suffering cancer is high. FISH diagnostic techniques provide a very valuable tool in the hand of the clinics to deal with this range of diseases.

Following this trend more efforts are focused on faster and more accurate diagnostics. For example, it has been recently reported a microfluidic FISH platform to provide quick circulating tumour cell analysis [65]. To combine microfluidics with FISH dramatically reduces detection time and cuts costs down. It also lends itself to automatization which further increases the efficiency. Given that microfluidics FISH can work on single-cell samples a whole new wave of research is dedicated to promote this research tool [66, 67].

Another combination that has provided very fruitful results is the so called comet-FISH approach. The comet assay has been explained in detail in the previous chapter (see Chapter 16.3.2). Briefly the cells under study after a given treatment are lysed and subjected to an electric field in order to induce an electrophoretic migration of the genetic material. When a treatment produces a fragmentation of this genetic material the migration is faster and more spread in comparison to a negative treatment or a control. When imaged the damaged cells display a "comet-like" structure with a "head" (cell body) and a "tail" (electrophoretically-stretched genetic material). The extent of the tail provides a direct measure of the genetic damage and fragmentation. In the combined comet-FISH method the chromatin tail is produced first and then a selective hybridization (FISH) step is done on the sample [68, 69]. This greatly increases the amount of information obtained by the method. Where the comet for itself only provides semi-quantitative information in regards to chromatin damage, the comet-FISH produces additional information as it reports if some specific sequences (those detected by the probes) have been damaged. Therefore the probed sequences can appear on the comet's head (no damage), on the tail (damaged), at both sites (differential damage assuming the sample comes from a diploid organism), *etc.* (see Fig. **17.18A** for a schematic explanation). Examples of this technique application to human lymphocytes with different degrees of chromosomal damage are shown in Fig. (**17.18B-D**) [70]. In this case the authors report an enhanced version of comet-FISH through the use of a thermal denaturation step instead of the more common chemical denaturation. In the experiments two probes were employed: a p53 gene probe (red Texas red signal) and a chromosome 17 centromeric region probe (green fluorescein signal). The whole chromatin was counterstained with DAPI (deep blue background). This example gives an idea of the information to be gained by using a comet-FISH assay.

FISH is employed to study chromosomal domains and their dynamics too. Chromosomal domains are functional regions that compromise genes related to a common cellular function. Chromosomal domains can be integrated by regions from different chromosomes (this is usually the case) that are "brought together" to increase efficiency and promote coordination for gene expression related to a certain cellular function. FISH can map different genes in interphasic nuclei to measure the interaction level among different genes. This domain visualization has been done for the full complement of chromosomes in human cells (see Fig. **17.19**) [71]. The 24 chromosomes (22 autosomes + X and Y sexual chromosomes) were labelled and imaged in a quiescent (G_0) fibroblast. The results provide the spatial arrangement of these chromosomes within the cell nucleus. The authors reported not only the 3D arrangement of the chromosomes but also some of the interrelations that take place among them. For example, the smaller a chromosome is the closer it tends to be to the nucleus centre. And the other way, larger chromosomes locate in the nuclear periphery. Those chromosomes that code for the nucleolar organizing regions (NORs) localize to the core of the nucleus, in agreement with the observation that nucleoli are found far from the nuclear membrane and close to the nucleus centre in living cells.

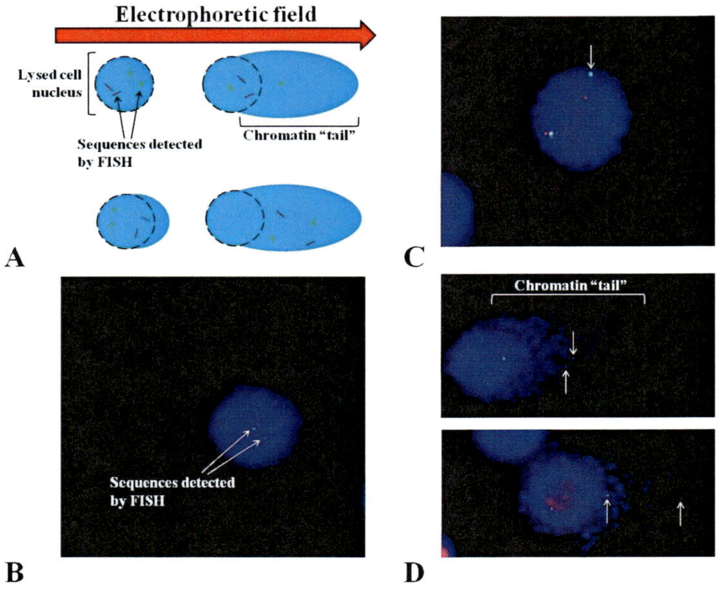

Fig. (17.18). A: Simple schematics of the different outcomes for a comet-FISH assay. The lysed nucleus can display no damage (no tail) and therefore the chromosomes sequences detected by FISH (green and red) map within the nucleus "head" (top left). The treatment may slightly damage the nucleus (bottom left). In this case a small tail develops. A larger damage results in a well-developed "tail" (top right) in which some chromosomal sequences can appear (green signal in tail). Finally when there is widespread nuclear damage the "tail" is prominent and most usually all FISH-detected regions appear within it (bottom right). **B-D:** Examples of different degrees of chromosomal damage that correlate with the schematics shown in A. The damage increases from none **(B)** through mild **(C)** to large **(D)**. The sequences detected by FISH (damaged ones indicated by arrowheads in **C** and **D**) include a centromeric sequence for chromosome 17 (green) and the p53 gene sequence (red). The blue background signal is due to DAPI counterstaining. The reddish background in the nucleus at D bottom is due to non-specific hybridization. Reproduced with permission from reference [70].

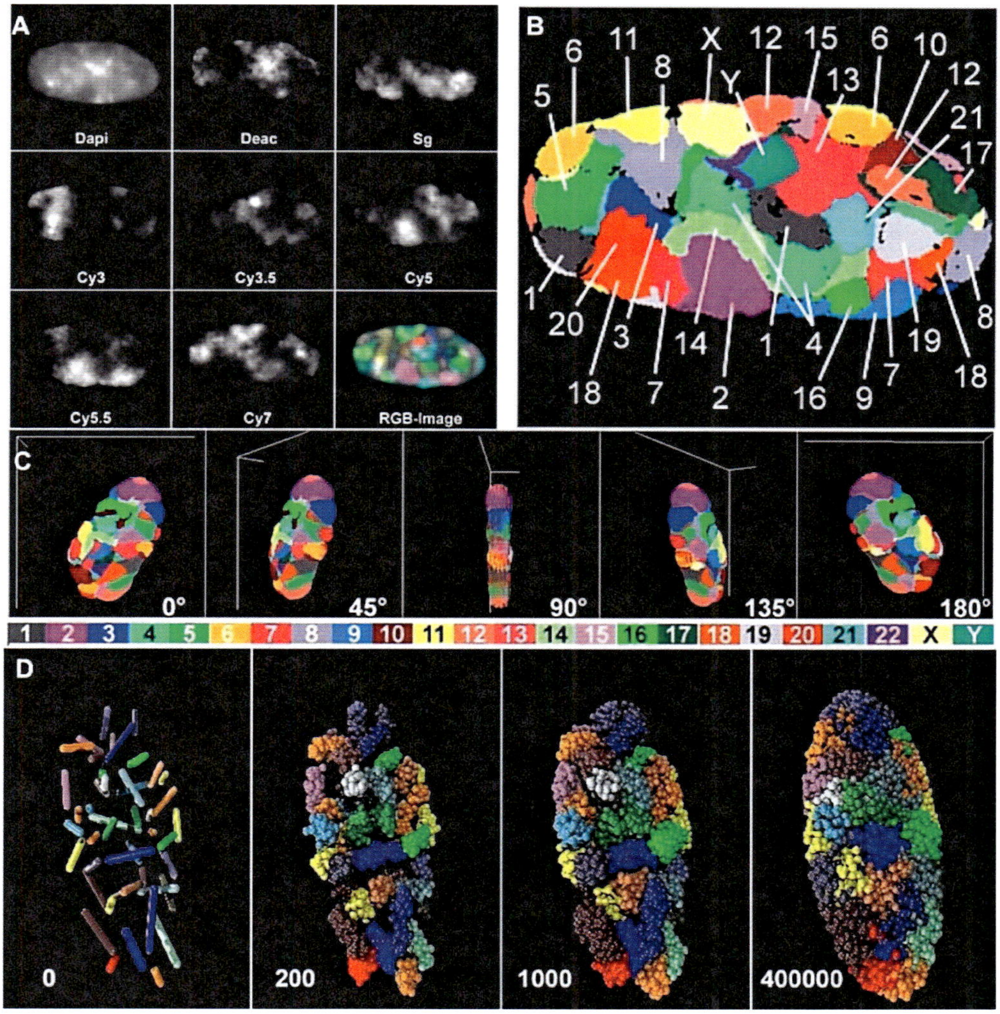

Fig. (17.19). A: Fluorescent signals (greyscale) obtained from the different fluorophores employed for the FISH probes (DAPI-non-specific counterstaining-; Deac: diethylaminocoumarin; Sg: Spectrum green; Cy3-Cy7: Cyanine dyes). The RGB reconstructed image corresponds to the compound colour image of the 24 labelled chromosomes. **B:** False colour representation of the chromosomal domains in a mid-plane nuclear reconstruction based on data obtained from FISH. The numbers indicate each one of the chromosomes. **C:** 3-D nucleus reconstruction viewed from different angles. The false colour code assigned to each chromosome is shown at the bottom. **D:** Simulation of the decondensing process in a fibroblast nucleus after mitosis and into interphase. Step 0 shows the 46 chromatids redistributed around the nucleus right after telophase. Steps 200-400000 show a further decondensation of the chromatin until the typical interphasic structure is achieved. False colours indicate the different chromosomes. Reproduced with permission from reference [71].

When FISH is combined with super-resolution microscopic techniques it is possible to obtain information of chromosomal and genetic domains with unprecedented detail [72]. One such application has been recently reported by Beliveau *et al.* who obtained very high spatial resolution of chromosomal structures employing Stochastic Optical Reconstruction Microscopy (STORM, see Chapter 20.4.1) [73]. The researchers designed probes for specific sequences of the *Drosophila melanogaster* genome. These probes where employed for FISH and then the labelled chromosomal structures were interrogated.

Some of the obtained results are displayed in Fig. (**17.20**). It is immediately appreciated that the differences in resolving power between the conventional and super-resolution techniques are enormous. Where only a few pixels can be resolved with the conventional approach really detailed structures emerge with STORM. Even when imaging structures below 100 nm in size appreciable details can still be discerned (see Fig. **17.20**). The combination of FISH specificity and super-resolution resolving power make a very strong approach to study relationships between structure and function at the chromatin level.

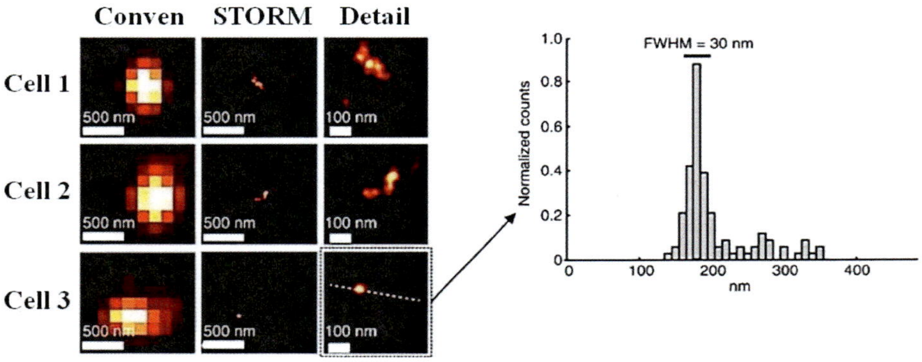

Fig. (17.20). Super-resolution images obtained by STORM of very short (4.9 kb) chromosomal sequences in *Drosophila melanogaster* cells. Three individual examples are shown (Cell 1-3). The first vertical lane shows the resolution obtained with a conventional confocal microscope (Conven). The middle lane illustrates the STORM super-resolution images (STORM). Compare the resolved detail with the left lane and notice the magnification level is the same. On the right lane a further detailed image of each example that allows for fine structural details of the assessed area (Detail). The light intensity was evaluated along the displayed dotted line for Cell 3-Detail and plotted in the graph to the right. Very fine detail (□30-50 nm) can be achieved in this way as shown (reproduced with permission from reference [73]).

CONFLICT OF INTEREST

The author (editor) declares no conflict of interest, financial or otherwise.

ACKNOWLEDGEMENTS

This work was supported by the Aarhus Institute of Advanced Studies (AIAS).

REFERENCES

[1] Lichtman JW, Conchello JA. Fluorescence microscopy. Nat Methods 2005; 2(12): 910-9. [http://dx.doi.org/10.1038/nmeth817] [PMID: 16299476]

[2] Cox G. Optical imaging techniques in cell biology. 2nd ed., Boca Raton, FL: CRC Press, Taylor & Francis, LLC 2012.

[3] Schacht V, Kern JS. Basics of immunohistochemistry. J Invest Dermatol 2015; 135(3): 1-4. [http://dx.doi.org/10.1038/jid.2014.541] [PMID: 25666678]

[4] Lavis LD, Raines RT. Bright ideas for chemical biology. ACS Chem Biol 2008; 3(3): 142-55. [http://dx.doi.org/10.1021/cb700248m] [PMID: 18355003]

[5] Kobayashi H, Longmire MR, Ogawa M, Choyke PL. Rational chemical design of the next generation of molecular imaging probes based on physics and biology: mixing modalities, colors and signals. Chem Soc Rev 2011; 40(9): 4626-48. [http://dx.doi.org/10.1039/c1cs15077d] [PMID: 21607237]

[6] Panchuk-Voloshina N, Haugland RP, Bishop-Stewart J, *et al.* Alexa dyes, a series of new fluorescent dyes that yield exceptionally bright, photostable conjugates. J Histochem Cytochem 1999; 47(9): 1179-88. [http://dx.doi.org/10.1177/002215549904700910] [PMID: 10449539]

[7] Lin JT, Chen PC, Goud TV, *et al.* A sulfhydryl-reactive ruthenium (II) complex and its conjugation to protein G as a universal reagent for fluorescent immunoassays. PLoS One 2012; 7(4): e36086. [http://dx.doi.org/10.1371/journal.pone.0036086] [PMID: 22563441]

[8] O'Malley WI, Abdelkader EH, Aulsebrook ML, *et al.* Luminescent alkyne-bearing terbium(III) complexes and their application to bioorthogonal protein labeling. Inorg Chem 2016; 55(4): 1674-82. [http://dx.doi.org/10.1021/acs.inorgchem.5b02605] [PMID: 26821062]

[9] Schnell U, Dijk F, Sjollema KA, Giepmans BN. Immunolabeling artifacts and the need for live-cell imaging. Nat Methods 2012; 9(2): 152-8. [http://dx.doi.org/10.1038/nmeth.1855] [PMID: 22290187]

[10] Lavis LD. Histochemistry: live and in color. J Histochem Cytochem 2011; 59(2): 139-45. [http://dx.doi.org/10.1369/0022155410395760] [PMID: 21339179]

[11] Nimmerjahn A, Kirchhoff F, Kerr JN, Helmchen F. Sulforhodamine 101 as a specific marker of astroglia in the neocortex *in vivo.* Nat Methods 2004; 1(1): 31-7. [http://dx.doi.org/10.1038/nmeth706] [PMID: 15782150]

[12] Hill RA, Grutzendler J. *In vivo* imaging of oligodendrocytes with sulforhodamine 101. Nat Methods 2014; 11(11): 1081-2. [http://dx.doi.org/10.1038/nmeth.3140] [PMID: 25357236]

[13] Cordes T, Maiser A, Steinhauer C, Schermelleh L, Tinnefeld P. Mechanisms and advancement of antifading agents for fluorescence microscopy and single-molecule spectroscopy. Phys Chem Chem Phys 2011; 13(14): 6699-709. [http://dx.doi.org/10.1039/c0cp01919d] [PMID: 21311807]

[14] Florijn RJ, Slats J, Tanke HJ, Raap AK. Analysis of antifading reagents for fluorescence microscopy. Cytometry 1995; 19(2): 177-82. [http://dx.doi.org/10.1002/cyto.990190213] [PMID: 7743897]

[15] Espada J, Juarranz A, Galaz S, *et al.* Non-aqueous permanent mounting for immunofluorescence microscopy. Histochem Cell Biol 2005; 123(3): 329-34.

[http://dx.doi.org/10.1007/s00418-005-0769-2] [PMID: 15856278]

[16] Barbe L, Lundberg E, Oksvold P, *et al.* Toward a confocal subcellular atlas of the human proteome. Mol Cell Proteomics 2008; 7(3): 499-508.
[http://dx.doi.org/10.1074/mcp.M700325-MCP200] [PMID: 18029348]

[17] Stadler C, Hjelmare M, Neumann B, *et al.* Systematic validation of antibody binding and protein subcellular localization using siRNA and confocal microscopy. J Proteomics 2012; 75(7): 2236-51.
[http://dx.doi.org/10.1016/j.jprot.2012.01.030] [PMID: 22361696]

[18] Stadler C, Rexhepaj E, Singan VR, *et al.* Immunofluorescence and fluorescent-protein tagging show high correlation for protein localization in mammalian cells. Nat Methods 2013; 10(4): 315-23.
[http://dx.doi.org/10.1038/nmeth.2377] [PMID: 23435261]

[19] Stadler C, Fagerberg L, Sivertsson Å, *et al.* RNA- and antibody-based profiling of the human proteome with focus on chromosome 19. J Proteome Res 2014; 13(4): 2019-27.
[http://dx.doi.org/10.1021/pr401156g] [PMID: 24579871]

[20] Fagerberg L, Stadler C, Skogs M, *et al.* Mapping the subcellular protein distribution in three human cell lines. J Proteome Res 2011; 10(8): 3766-77.
[http://dx.doi.org/10.1021/pr200379a] [PMID: 21675716]

[21] Li J, Newberg JY, Uhlén M, Lundberg E, Murphy RF. Automated analysis and reannotation of subcellular locations in confocal images from the Human Protein Atlas. PLoS One 2012; 7(11): e50514.
[http://dx.doi.org/10.1371/journal.pone.0050514] [PMID: 23226299]

[22] Harmsen MM, De Haard HJ. Properties, production, and applications of camelid single-domain antibody fragments. Appl Microbiol Biotechnol 2007; 77(1): 13-22.
[http://dx.doi.org/10.1007/s00253-007-1142-2] [PMID: 17704915]

[23] Muyldermans S. Nanobodies: natural single-domain antibodies. Annu Rev Biochem 2013; 82: 775-97.
[http://dx.doi.org/10.1146/annurev-biochem-063011-092449] [PMID: 23495938]

[24] Rothbauer U, Zolghadr K, Tillib S, *et al.* Targeting and tracing antigens in live cells with fluorescent nanobodies. Nat Methods 2006; 3(11): 887-9.
[http://dx.doi.org/10.1038/nmeth953] [PMID: 17060912]

[25] Maier J, Traenkle B, Rothbauer U. Real-time analysis of epithelial-mesenchymal transition using fluorescent single-domain antibodies. Sci Rep 2015; 5: 13402.
[http://dx.doi.org/10.1038/srep13402] [PMID: 26292717]

[26] Eteshola E, Balberg M. Microfluidic ELISA: on-chip fluorescence imaging. Biomed Microdevices 2004; 6(1): 7-9.
[http://dx.doi.org/10.1023/B:BMMD.0000013360.65653.c2] [PMID: 15307439]

[27] Darain F, Yager P, Gan KL, Tjin SC. On-chip detection of myoglobin based on fluorescence. Biosens Bioelectron 2009; 24(6): 1744-50.
[http://dx.doi.org/10.1016/j.bios.2008.09.004] [PMID: 18945609]

[28] Herrmann M, Veres T, Tabrizian M. Enzymatically-generated fluorescent detection in micro-channels with internal magnetic mixing for the development of parallel microfluidic ELISA. Lab Chip 2006; 6(4): 555-60.
[http://dx.doi.org/10.1039/b516031f] [PMID: 16572219]

[29] Zhan M, Chingozha L, Lu H. Enabling systems biology approaches through microfabricated systems. Anal Chem 2013; 85(19): 8882-94.
[http://dx.doi.org/10.1021/ac401472y] [PMID: 23984862]

[30] Saraheimo S, Hepojoki J, Nurmi V, *et al.* Time-resolved FRET -based approach for antibody detection - a new serodiagnostic concept. PLoS One 2013; 8(5): e62739.
[http://dx.doi.org/10.1371/journal.pone.0062739] [PMID: 23667515]

[31] Zellner M, Babeluk R, Diestinger M, Pirchegger P, Skeledzic S, Oehler R. Fluorescence-based Western blotting for quantitation of protein biomarkers in clinical samples. Electrophoresis 2008; 29(17): 3621-7.
[http://dx.doi.org/10.1002/elps.200700935] [PMID: 18803224]

[32] Eaton SL, Hurtado ML, Oldknow KJ, *et al.* A guide to modern quantitative fluorescent western blotting with troubleshooting strategies. J Vis Exp 2014; 93(93): e52099.
[PMID: 25490604]

[33] Chudakov DM, Matz MV, Lukyanov S, Lukyanov KA. Fluorescent proteins and their applications in imaging living cells and tissues. Physiol Rev 2010; 90(3): 1103-63.
[http://dx.doi.org/10.1152/physrev.00038.2009] [PMID: 20664080]

[34] Remington SJ. Green fluorescent protein: a perspective. Protein Sci 2011; 20(9): 1509-19.
[http://dx.doi.org/10.1002/pro.684] [PMID: 21714025]

[35] Liebel U, Starkuviene V, Erfle H, *et al.* A microscope-based screening platform for large-scale functional protein analysis in intact cells. FEBS Lett 2003; 554(3): 394-8.
[http://dx.doi.org/10.1016/S0014-5793(03)01197-9] [PMID: 14623100]

[36] Chatterjee A, Guo J, Lee HS, Schultz PG. A genetically encoded fluorescent probe in mammalian cells. J Am Chem Soc 2013; 135(34): 12540-3.
[http://dx.doi.org/10.1021/ja4059553] [PMID: 23924161]

[37] Deisseroth K. Optogenetics: 10 years of microbial opsins in neuroscience. Nat Neurosci 2015; 18(9): 1213-25.
[http://dx.doi.org/10.1038/nn.4091] [PMID: 26308982]

[38] Müller K, Naumann S, Weber W, Zurbriggen MD. Optogenetics for gene expression in mammalian cells. Biol Chem 2015; 396(2): 145-52.
[http://dx.doi.org/10.1515/hsz-2014-0199] [PMID: 25153239]

[39] Wojtovich AP, Foster TH. Optogenetic control of ROS production. Redox Biol 2014; 2: 368-76.
[http://dx.doi.org/10.1016/j.redox.2014.01.019] [PMID: 24563855]

[40] Price CM. Fluorescence *in situ* hybridization. Blood Rev 1993; 7(2): 127-34.
[http://dx.doi.org/10.1016/S0268-960X(05)80023-2] [PMID: 8369661]

[41] Swiger RR, Tucker JD. Fluorescence *in situ* hybridization: a brief review. Environ Mol Mutagen 1996; 27(4): 245-54.
[http://dx.doi.org/10.1002/(SICI)1098-2280(1996)27:4<245::AID-EM1>3.0.CO;2-C] [PMID: 8665869]

[42] Levsky JM, Singer RH. Fluorescence *in situ* hybridization: past, present and future. J Cell Sci 2003; 116(Pt 14): 2833-8.
[http://dx.doi.org/10.1242/jcs.00633] [PMID: 12808017]

[43] Tsuchiya KD. Fluorescence *in situ* hybridization. Clin Lab Med 2011; 31(4): 525-542, vii-viii.
[http://dx.doi.org/10.1016/j.cll.2011.08.011] [PMID: 22118735]

[44] Durmaz AA, Karaca E, Demkow U, *et al.* Evolution of genetic techniques: Past, present, and beyond. Biomed Res Int 2015; 2015: 461524.

[45] Campbell LJ, Ed. Cancer cytogenetics. Methods and protocols.Methods in Molecular Biology. New York, NY: Springer Science+Business Media LLC 2011; 730.
[http://dx.doi.org/10.1007/978-1-61779-074-4]

[46] Levenson R, Beechem J, McNamara G. Spectral imaging in preclinical research and clinical pathology. Anal Cell Pathol (Amst) 2012; 35(5-6): 339-61.
[http://dx.doi.org/10.1155/2012/904828] [PMID: 22475632]

[47] Stockert JC, Espada J, Blázquez-Castro A. Functional analysis of DNA and chromatin.Methods in Molecular Biology. New York, NY: Springer Science+Business Media LLC 2014; 1049.

[http://dx.doi.org/10.1007/978-1-62703-706-8]

[48] Pita M, Orellana J, Martinez-Rodriguez P, *et al.* FISH methods in cytogenetics studies. Methods in Molecular Biology New York, NY: Springer Science+Business Media LLC 2014; 1049. [http://dx.doi.org/10.1007/978-1-62703-706-8_10]

[49] Langer-Safer PR, Levine M, Ward DC. Immunological method for mapping genes on Drosophila polytene chromosomes. Proc Natl Acad Sci USA 1982; 79(14): 4381-5. [http://dx.doi.org/10.1073/pnas.79.14.4381] [PMID: 6812046]

[50] Wiegant J, Ried T, Nederlof PM, van der Ploeg M, Tanke HJ, Raap AK. *In situ* hybridization with fluoresceinated DNA. Nucleic Acids Res 1991; 19(12): 3237-41. [http://dx.doi.org/10.1093/nar/19.12.3237] [PMID: 2062640]

[51] Rombouts K, Braeckmans K, Remaut K. Fluorescent labelling of plasmid DNA and mRNA: Gains and losses of current labelling strategies. Bioconjug Chem 2016; 27(2): 280-97. [http://dx.doi.org/10.1021/acs.bioconjchem.5b00579] [PMID: 26670733]

[52] Olszewska M, Wanowska E, Kishore A, *et al.* Genetic dosage and position effect of small supernumerary marker chromosome (sSMC) in human sperm nuclei in infertile male patient. Sci Rep 2015; 5: 17408. [http://dx.doi.org/10.1038/srep17408] [PMID: 26616419]

[53] Speicher MR, Gwyn Ballard S, Ward DC. Karyotyping human chromosomes by combinatorial multi-fluor FISH. Nat Genet 1996; 12(4): 368-75. [http://dx.doi.org/10.1038/ng0496-368] [PMID: 8630489]

[54] Sumption N, Goodhead DT, Anderson RM. Alpha-particle-induced complex chromosome exchanges transmitted through extra-thymic lymphopoiesis *in vitro* show evidence of emerging genomic instability. PLoS One 2015; 10(8): e0134046. [http://dx.doi.org/10.1371/journal.pone.0134046] [PMID: 26252014]

[55] Zhang J, Jiang Y, Guo Y, *et al.* Identification of novel chromosomal aberrations induced by ^{60}Co--irradiation in wheat-*Dasypyrum villosum* lines. Int J Mol Sci 2015; 16(12): 29787-96. [http://dx.doi.org/10.3390/ijms161226134] [PMID: 26694350]

[56] Beliveau BJ, Joyce EF, Apostolopoulos N, *et al.* Versatile design and synthesis platform for visualizing genomes with Oligopaint FISH probes. Proc Natl Acad Sci USA 2012; 109(52): 21301-6. [http://dx.doi.org/10.1073/pnas.1213818110] [PMID: 23236188]

[57] Kato A, Lamb JC, Birchler JA. Chromosome painting using repetitive DNA sequences as probes for somatic chromosome identification in maize. Proc Natl Acad Sci USA 2004; 101(37): 13554-9. [http://dx.doi.org/10.1073/pnas.0403659101] [PMID: 15342909]

[58] Silva DM, Pansonato-Alves JC, Utsunomia R, *et al.* Delimiting the origin of a B chromosome by FISH mapping, chromosome painting and DNA sequence analysis in Astyanax paranae (Teleostei, Characiformes). PLoS One 2014; 9(4): e94896. [http://dx.doi.org/10.1371/journal.pone.0094896] [PMID: 24736529]

[59] Wagner M, Horn M, Daims H. Fluorescence *in situ* hybridisation for the identification and characterisation of prokaryotes. Curr Opin Microbiol 2003; 6(3): 302-9. [http://dx.doi.org/10.1016/S1369-5274(03)00054-7] [PMID: 12831908]

[60] Amann R, Fuchs BM. Single-cell identification in microbial communities by improved fluorescence *in situ* hybridization techniques. Nat Rev Microbiol 2008; 6(5): 339-48. [http://dx.doi.org/10.1038/nrmicro1888] [PMID: 18414500]

[61] King W, Proffitt J, Morrison L, Piper J, Lane D, Seelig S. The role of fluorescence *in situ* hybridization technologies in molecular diagnostics and disease management. Mol Diagn 2000; 5(4): 309-19. [http://dx.doi.org/10.2165/00066982-200005040-00009] [PMID: 11172495]

[62] Kwasny D, Vedarethinam I, Shah P, *et al.* Advanced microtechnologies for detection of chromosome

abnormalities by fluorescent *in situ* hybridization. Biomed Microdevices 2012; 14(3): 453-60. [http://dx.doi.org/10.1007/s10544-011-9622-7] [PMID: 22222279]

[63] Bartlett JM. Roulston JE, Bartlett JM. Fluorescence in situ hybridization. Molecular diagnosis of cancerMethods in molecular medicine. Totowa, New Jersey: Humana Press Inc 2004; 97.

[64] Kearny L, Shipley J. Fluorescence in situ hybridization for cancer-related studies.Methods in Molecular Biology: Metastasis research protocols. New York, NY: Springer Science+Business Media LLC 2012.
[http://dx.doi.org/10.1007/978-1-61779-854-2_10]

[65] Sato K. Microdevice in cellular pathology: Microfluidic platforms for fluorescence *in situ* hybridization and analysis of circulating tumor cells. Anal Sci 2015; 31(9): 867-73. [http://dx.doi.org/10.2116/analsci.31.867] [PMID: 26353951]

[66] Sieben VJ, Debes-Marun CS, Pilarski LM, Backhouse CJ. An integrated microfluidic chip for chromosome enumeration using fluorescence *in situ* hybridization. Lab Chip 2008; 8(12): 2151-6. [http://dx.doi.org/10.1039/b812443d] [PMID: 19023479]

[67] Kurz CM. Towards a cellular multi-parameter analysis platform: Fluorescence *in situ* hybridization (FISH) on microhole-array chips. 33rd Annual International Conference of the IEEE EMBS. 8408-11.

[68] Glei M, Schlörmann W. Analysis of DNA damage and repair by comet fluorescence *in situ* hybridization (comet-FISH). Methods in Molecular Biology: Functional analysis of DNA and chromatin. New York, NY: Springer Science+Business Media LLC 2014.

[69] Spivak G. New developments in comet-FISH. Mutagenesis 2015; 30(1): 5-9.
[http://dx.doi.org/10.1093/mutage/geu036] [PMID: 25527722]

[70] Mladinic M, Zeljezic D. Modification of comet-FISH technique by using temperature instead of chemical denaturation. MethodsX 2014; 1: 162-7.
[http://dx.doi.org/10.1016/j.mex.2014.08.010] [PMID: 26150949]

[71] Bolzer A, Kreth G, Solovei I, *et al.* Three-dimensional maps of all chromosomes in human male fibroblast nuclei and prometaphase rosettes. PLoS Biol 2005; 3(5): e157.
[http://dx.doi.org/10.1371/journal.pbio.0030157] [PMID: 15839726]

[72] Markaki Y, Smeets D, Cremer M, *et al.* Sousa AA, Kruhlak MJ. Fluorescence *in situ* hybridization applications for super-resolution 3D structured illumination microscopy. Methods in Molecular Biology. New York, NY: Springer Science+Business Media LLC 2013; 950.

[73] Beliveau BJ, Boettiger AN, Avendaño MS, *et al.* Single-molecule super-resolution imaging of chromosomes and *in situ* haplotype visualization using Oligopaint FISH probes. Nat Commun 2015; 6: 7147.
[http://dx.doi.org/10.1038/ncomms8147] [PMID: 25962338]

CHAPTER 18

Luminescent Solid-State Markers

Abstract: In this chapter, we will introduce the fundamentals and the main types of solid-state fluorophores in common use for fluorescence microscopy. The photophysics underlying the absorption and emission processes are universal and shared with other fluorophores already presented (*e.g.* organic molecules, see Chapters 2 and 3). However, fluorescent/luminescent solid state matrices present their own peculiarities, and it is convenient to introduce them in a generic way. As the reader will notice progressesing through the chapter, there is no common chemical basis for the presented solid-state fluorophores. The common theme is precisely solid state structures composed of a relatively large number of atoms (10^6-10^{12} atoms/structure) as compared to the low number of atoms (10^1-10^3 atoms/structure) of the fluorophores already presented in other chapters. The solid state is broadly divided into three categories according to the (increasing) electrical conductivity: insulators (dielectrics), semiconductors and conductors (metals). There are luminescent representatives within each of these categories. Therefore, we will begin the chapter by explaining the physical basis of electrical conductivity, which is directly related to the optical properties of the material, and then move forward to introduce the different subtypes of luminescent solid-state markers. To understand the electrical/optical properties of solids we will first present the band structure model.

Keywords: Band structure model, Carbon nanoparticle (C-dot), Conduction band, Conductor, Dielectric, Electroluminescence, Electron, Energy gap (Eg), Hole, Insulator, Lanthanide-doped nanoparticle, Light emitting diode (LED), Metal, Nanodiamond, Nanoparticle, Photoluminescence, Quantum dot (Qdot), Semiconductor, Silica nanoparticle (Si-dot), Solid-state physics, Valence band.

1. BAND STRUCTURE MODEL

Condensed matter (solids and liquids) is characterised by strong atomic and molecular interactions and, as a consequence, small interatomic and intermolecular distances. These interactions are much stronger in solids, and we will focus on them for the rest of the chapter. This small interatomic distance distorts the electronic clouds from each of the atoms/molecules making up the solid. The distortion is a direct consequence of the so-called Pauli exclusion principle that states that electrons cannot occupy the same energy levels in the same spatial region [1]. Thus if electrons have the same quantum numbers (*i.e.*

Juan Carlos Stockert & Alfonso Blázquez-Castro

they have the same energy "content"), then they "repulse" one another when they are brought close enough. A scheme of this phenomenon is shown in Fig. (**18.1A**). One can think of electrons as small magnets. When far apart nothing happens, but when we try to bring them together (assuming that we try to couple the same pole in both magnets), repulsion sets in that grows and becomes stronger as the distance between the two magnets decreases. Electrons "solve" this problem when forced together, as in solids, by assuming energy levels that are either higher or lower in energy than the average.

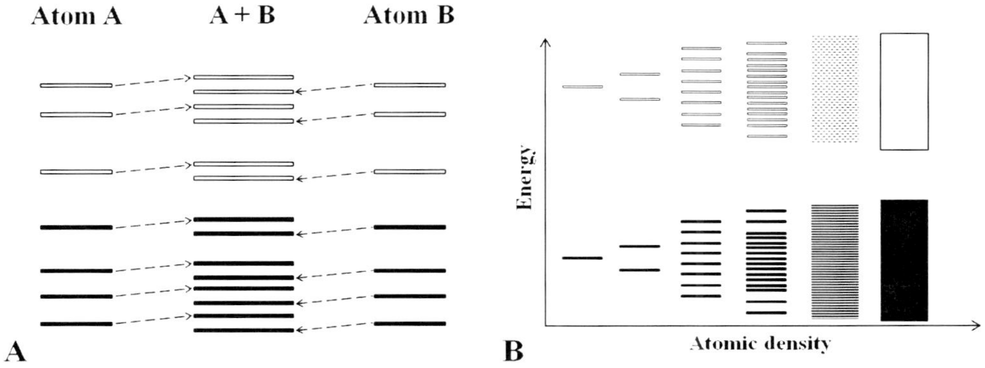

Fig. (18.1). Pauli Exclusion Principle. **A:** When relatively isolated (gas phase) atoms **A** and **B** from the same element display the same electronic energy levels. When brought close enough, level repulsion sets in as a consequence of the fermionic character of electrons. The different levels rearrange, increasing or decreasing their energies, to accommodate in the new situation. Filled levels are represented as black lines and empty levels as white lines. **B:** Simplified scheme for electronic band establishment. Initially (left) there is a filled electronic level (black) and an empty one (white) higher in energy. As atoms are packed in a small volume, the atomic density increases, and the electronic levels fill positions at different energies (middle). Finally, as solid state is reached, the levels stack so closely in energy that they form an energy continuum and combine forming bands (right). Filled levels are represented in black and empty ones in white.

This process has already been presented to explain the electronic energy levels and optical properties of transition metal complexes (see Chapter 9.2). In Fig. (**9.4A**), the rearrangement of energy levels due to the interaction and repulsion of the transition metal and the ligands electrons was introduced. This can be understood as an intermediate case between the isolated atoms and small molecules and the solid state.

If more electrons are condensed in a small volume, this trend continues and amplifies. The electrons then tend to fill more energy levels, so densely packed that it is no longer possible to discern a particular electronic level: the levels merge into the electronic bands (see Fig. **18.1B**). This is the origin of the band structure in solids. Band diagrams are analogous to Jablonski diagrams but applied to the solid state (Jablonski diagrams were introduced in Chapter 2.2.2). Moving upwards in the diagram means that the system is increasing its energy, while moving downwards means the energy is released by the system. Following this, photon absorption will move an electron upwards while photon emission will displace it downwards. The same general photophysical principles introduced in chapter 2 are also applied to the band model. Some effects are enhanced in solids while others are diminished, and still others are modified; however, the overall processes are the same.

What is the physical meaning of the band structure? One has to imagine this structure spreading in three dimensions. The convention, as it happened with Jablonski diagrams and potential energy curves, is to represent energy in the vertical axis while the horizontal axis can have different meanings depending on what the depiction intends (usually interatomic distances). However, as explained with regard to the potential energy curves, it is not possible to represent three dimensional volumes and energy values at the same time. This would call for a four dimensional diagram. Therefore, bands represent a "flattened" version of the 3D structure.

1.1. Valence and Conduction Bands

This band model permits us to understand and predict the electrical and optical behaviour of the different solids. One of the principal divisions for solid state matter is between conductors (metals) and insulators (non-metals or dielectrics). A particular solid behaves as a conductor or an insulator as a direct consequence of the way in which its electronic bands arrange [1, 2]. Let us first present the conductor case, and then move to dielectrics.

The establishment of electronic bands in a metal, the prototypical conductor, is shown in Fig. (**18.2A**). Starting from isolated atoms with their defined electronic orbitals, the condensation into a solid lattice leads to the apparition of bands. From the scheme it can be seen that the different bands overlap, both the filled and the empty ones. In the solid there exists a continuum between filled and empty states. It follows that a very small increase in energy is necessary to bring an electron above the filled states, where it is free to move in the solid matrix. This small amount of energy can come from many different sources like heat, light, electric fields, *etc*. For this reason metals are very good conductors of heat and electricity, because it takes an almost negligible amount of work to move the electrons around.

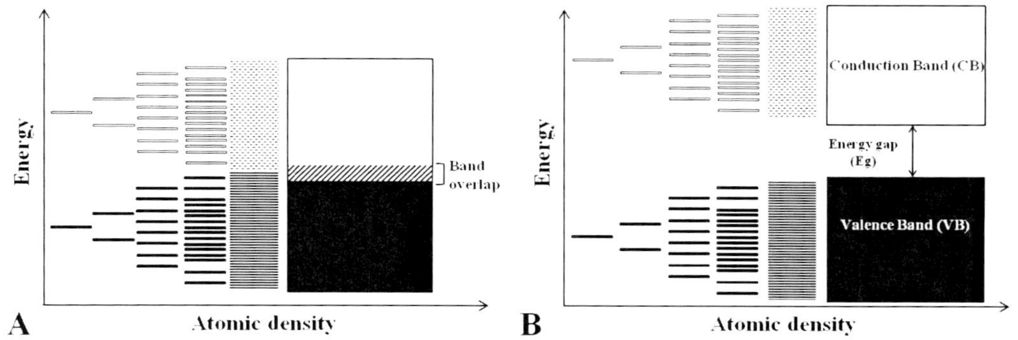

Fig. (18.2). Conductors (metals) and dielectrics (insulators) **A:** Band inception in conductors. The initial energy gap between electronic levels in isolated atoms (left) is small. As atoms condense into a solid the Pauli Exclusion Principle makes the energy levels spread in energy (middle). This leads to the final overlap of empty and filled states (right). Due to this reason, electrons in conductors can easily move through the empty states, and conduct heat and electricity very efficiently. **B:** Band arrangement in dielectrics. The same general process takes place in insulators, but the initial energy gap between filled and empty orbitals is much larger, even at the isolated atom level (left). By the time a solid is formed there is still a certain energy difference between filled (Valence Band -VB-) and empty levels (Conduction Band -CB-). This difference is called the energy gap (Eg), and corresponds to the minimum energy to be applied to an electron in order for it to be promoted from the VB to the CB.

The situation is different for a dielectric. In this case the broadening of the electronic levels to form the bands leads to a final structure where empty and filled bands do not overlap (see Fig. **18.2B**). Under these circumstances the electrons are fixed in place, unable to move because they are surrounded by the rest of the electrons in the solid. The lower energy-filled band is denoted as the valence band (VB) because it is made up of the outermost (valence) electrons of the material. The higher energy empty band is known as the conduction band (CB) because an electron that reaches this band is able to move across the solid in a very similar way to electrons in a conductor. With this scheme in mind, conductors can be thought of in terms of overlapping between the CB and the VB, so there is no gap between them.

Slightly increasing the energy of one electron in the VB of a dielectric does not allow it to reach the CB, as was the case for conductors. A minimum amount of energy must be provided to an electron to promote it from the VB to the CB (see Fig. **18.2B**). This amount of energy that "bridges" the bands is known as the energy gap (Eg). If the sample is illuminated with photons whose energy is less than Eg, then the absorption is very low if at all. An electron absorbing such a photon cannot reach a stable position at the CB, so it immediately reemits the energy. However, once photons with a minimum energy equal to Eg interact with the sample, they are absorbed because the electrons can now reach the CB across the gap. In fact, complicated phenomena can take place during light absorption in solids. Our intention here is to present these events in a simplified, operative way. We direct readers interested in a more profound description and explanation of solid state physics and photophysics to the bibliography already presented at the beginning of this section. Given that Eg is conventionally establish as larger than 4 eV, this means that only UV light with wavelength shorter than ≈ 300 nm (hν > 4 eV) can photoexcite electrons from the VB to the CB. Therefore, dielectrics do not absorb visible light (energy per photon between 1.8 and 3.1 eV) and appear transparent to the eye.

Once an electron reaches the CB, it can only stay on it for a certain period of time. On being promoted to the CB the electron leaves behind an uncompensated positive charge, belonging to an atomic nucleus in the VB. This positive charge – which is the absence of an electron – is known as a hole. As charges of opposite sign, electrons and holes attract each other. Thus, the excited electron is metastable while dwelling in the CB. After some characteristic time that depends on the particular material a CB electron and a VB hole will recombine and neutralize. This recombination can lead to photon emission, acoustic/mechanical wave generation, or simply to heat production depending on the compound involved and other circumstances (temperature, pressure, external electric or magnetic fields, *etc*.). These processes of electron excitation and decay between bands are analogous to the ground state and excited potential energy curves explained in Chapter 2.4.8.

The prototypical examples of conductors and insulators have been briefly presented. However, there is a very important class of materials where Eg is relatively small. These materials are known as semiconductors, and their band structure will be introduced next.

1.2. Semiconductors

Semiconductors display features intermediate between metals and insulators. Strictly speaking, they are dielectrics as they present an Eg between bands. However, this Eg is small (usually between 0 and 4 eV; see Fig. **18.3A**) in comparison to typical insulators.

In this energy range, even background thermal energy at room temperature pumps some electrons to the CB. Consequently, semiconductors show a certain baseline degree of electronic conduction that is intermediate between true conductors and dielectrics. Due to their relatively small Eg, semiconductors absorb strongly in the visible range and appear coloured or black to the eye. Semiconductors are chemically made from some pure elements or from mixtures of other elements. Pure element semiconductors can be obtained from elements in the group 14: carbon, silicon and germanium. These elements have 4 electrons in their outermost orbital. Other semiconductors made from elemental mixtures are composed of element combinations that also produce 4 electrons. For example, by mixing gallium and arsenic we obtain gallium arsenide, one of the most industrially-employed semiconductors. The combinations usually take place between group 13 and group 15 elements, and group 12 and group 16 elements. Almost any of these combinations produces a semiconductor. As a rule of thumb, the heavier the initial elements entering the composition, the smaller Eg will be (*e.g.* InSb has an Eg of only 0.17 eV while AlP has Eg=2.45 eV). Other combinations exist, but we will not go into detail here because most semiconductors employed as fluorescent probes conform to the examples described.

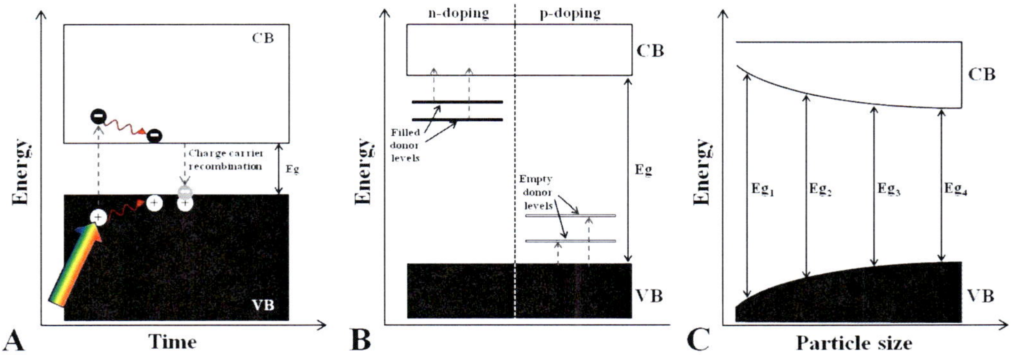

Fig. (18.3). Charge carrier excitation in dielectrics (insulators) and semiconductors. **A:** In the scheme shown a photon (coloured arrow) is absorbed by an electron in the valence band –VB-, and its energy is employed to pump the electron to the conduction band –CB- (negative black dot). The electron excitation leaves an uncompensated positive charge in the VB, a hole (positive white spot). Within a very short time (~picoseconds), the electron and the hole reemit part of the energy as lattice vibrations (red wiggly arrows). Sometime later the two charge carriers, electron and hole, recombine due to their mutual electrostatic attraction. The evolved energy in this process is equal to the energy gap (Eg). This energy can radiate as a photon, acusto-mechanical waves, thermal vibrations, *etc.* depending on the particular circumstances. **B:** Doping effect on dielectrics and semiconductors. The addition of an electron-rich impurity leads to n-doping (left). The impurity provides some additional electron-rich levels that localize below the CB. Electron excitation from these levels (black lines) to the CB is accomplished with much less energy than from the VB. For p-doping (right) an electron-deficient impurity is chosen. This provides empty electronic levels (white lines) above the VB. Again, pumping electrons from the VB to these impurity levels requires much less energy than overcoming the full Eg. **C:** Quantum confinement effect. A decrease in the particle size results in less electrons being constrained within the same volume. This partly relieves the Pauli Exclusion Principle, and the bands become thinner and more distant in energy terms. Thus Eg becomes larger for smaller particles ($Eg_1>Eg_2>Eg_3>Eg_4$).

The small Eg in semiconductors provides great plasticity in their electronic behaviour, and they represent a cornerstone in modern technology (electronics, photonics, computing, *etc.*). Because we are interested in explaining the photonics and uses of semiconductors in fluorescence microscopy, we will focus on how photonic processes can be modulated in them [3 - 5]. This modulation is mainly achieved by manipulating Eg. In general terms, there are two ways to change Eg in a given semiconductor: impurity doping and particle size control.

Doping consists in the introduction of another minor element (impurity) within the semiconductor, so that either an excess or a defect of electrons is introduced in the lattice [2]. If the result is an increase in the available electrons, it is called n-doping. If there is a defect of electrons (which can be seen as an excess of holes), it is called p-doping. The effect that doping has on the band structure is shown in Fig. (**18.3B**). In the case of n-doping, additional levels or thin bands are established close to the CB. Electrons can be promoted from these levels to the CB, but much less energy is necessary because the Eg is smaller than the canonical semiconductor Eg. The same principle applies in the case of p-doping, but the electronic levels are closer to the VB in this case, and they are empty. Therefore, electron excitation can take place from the VB to one of these acceptor levels, because Eg is again much smaller than the intrinsic Eg of the semiconductor matrix (see Fig. **18.3B**). By controlling the amount and oxidation state of the impurities, the width and position of the energy levels within the Eg can be finely tuned. This, in consequence, modulates the electrical and optical properties of the doped semiconductor.

The second approach to modulate the properties of a semiconductor is by controlling its size. This approach, semiconductor size control, can be applied when working with very small (nanometric) semiconductor crystals known as quantum dots (Qdots). The origin of the modulation is due to the quantum confinement effect [6, 7]. This effect can be understood if we recall how bands are established in the first place (see Fig. **18.1B**). It was shown that, as the solid particle gets larger, the repulsion between electrons generates the bands. The quantum confinement effect is this same process, but acting in reverse: as the particle gets smaller, fewer electrons are available to produce the bands, and Eg shifts accordingly (see Fig. **18.3C**). As can be concluded from the scheme, the smaller the particle is, the larger Eg. Thus, Qdots made up from the same semiconductor material can have different optical properties, depending exclusively on their sizes. Size reduction is one of the most employed techniques to finely control semiconductor properties. This size control can take place in one dimension (giving rise to very thin plates known as quantum wells), two dimensions (producing nanometric-section "threads" known as quantum wires), or three dimensions (Qdots, semiconductor particles in which any spatial dimension is always less than 100 nm) [1]. In principle, nothing bans the simultaneous control of size and doping in a semiconductor, and indeed this is usually the case for practical applications [8]. In regards to their use in fluorescence microscopy,

Qdots are currently the most employed luminescent solid structures. More information and practical examples will be provided below (see section 2.1).

1.3. Solid-state Luminescence

Once the principles of band structure have been introduced, we will present the basics of photonics in solid-state to understand the assets and disadvantages of these materials in fluorescence microscopy. We will start with the process of light absorption, and then move on to light emission and other deactivation pathways.

Dielectric and Semiconductor Luminescence

As stated before, there are no significant differences between light absorption in atoms and molecules, and by a solid-state matrix once the band structure is taken into account. We will present absorption in a material with a band gap (Eg > 0 eV, *i.e.* a semiconductor or a dielectric), and then explain what happens in a conductor [1]. To be absorbed, a photon must have energy content equal or higher than Eg, otherwise an electron cannot be successfully promoted to the CB (see Fig. **18.4A**).

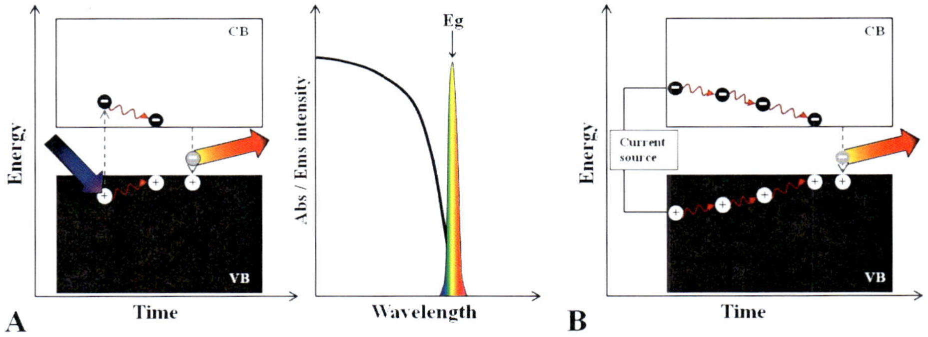

Fig. (18.4). Luminescence in dielectrics and semiconductors. A: Photoluminescence. The process starts with photon absorption, promotion of an electron to the CB, and hole creation at the VB (left scheme). After charge carrier thermalization (red arrows), the electron-hole system recombines re-emitting a photon with energy equal to the energy gap (hv = Eg). Absorption/emission spectra (right scheme) displaying the relationship between the absorption spectrum of a dielectric/semiconductor (thick black line) and the emission spectrum, highly peaked around a wavelength satisfying the equality hv = Eg. B: Electroluminescence. An electrical source establishes an electric potential across the material. This potential induces a current that flows inside the material. The current injects electrons into the CB, and draws them from the VB, which is equivalent to injecting holes into the VB. To guarantee efficient charge carrier flow, the voltage provided should be higher than Eg. Therefore, there is some thermalization of the injected carriers (red arrows). After this, the carriers radiatively recombine emitting photons in the processes.

If the energy requirement is fulfilled, then an electron will absorb the photon and transfer to the CB leaving a hole in the VB. It must be mentioned that there is also a momentum requirement for photon absorption, but this is beyond the purpose of this book and it will not be considered here. If the absorbed energy moves the electron high into the CB, it will lose part of this energy as mechanical and/or thermal emissions in a very short time (~ 10^{-12} s), until it reaches the bottom of the CB. This is because energy levels within the CB are not quantized (the electron can take or give any amount of energy within the band), but when it reaches the band bottom it must emit a defined amount of energy (= Eg) to move back to the VB. This process is analogous to the vibrational deexcitation that takes place in a molecule after light absorption, if it is excited to a vibrational level above the v=0 of the excited energy curve (see Chapter 2.4.6 and Fig. **2.9**).

After coming to a rest at the bottom of the CB, the electron will spend some time there before moving back to the VB. This state is said to be metastable. The lifetime depends on each particular material, and other variables like physical fields or impurities. When returning to the VB the electron loses energy that reappears as a photon. Because the radiative transition takes place between the limits of the band gap, the energy associated with photon emission is constant and the emission spectrum is quite narrow. Other non-radiative mechanisms can take place and reduce light emission from the solid, like photochemistry or just heat production. But in general, dielectrics and semiconductors are less prone to this non-radiative deactivation channels than organic molecules, as most of their emitting units are within the volume of the solid and, therefore, shielded from external influences. Also, the rigid lattice helps to eliminate molecular movements (in particular rotations) as a deactivation channel. These make semiconductors very good candidates for stable, reliable, narrow-emission fluorophores [7].

The excitation process just explained takes place when a semiconductor or dielectric is exposed to light. This process is known as photoluminescence, because the excitation source for the emitted light are photons. This is the typical situation with this kind of fluorophores in fluorescence microscopy. However, these materials can be excited very efficiently by an electrical current to produce light. This complementary excitation approach is known as electroluminescence, and its elemental action mechanism is shown in Fig. (**18.4B**) [1]. Instead of creating electron–hole pairs *in situ*, as photons do, an electric source is connected to the material. The flowing current provides electrons to the CB, while driving them away from the VB. When electrons are displaced from the VB holes are, by definition, created. The consequence of current flow is to create a population of electrons and holes that produce photons when undergoing recombination. The light production mechanism is equivalent in photo- and electroluminescence; it is the excitation mechanism that differs. The light emission by LEDs (introduced in Chapter 6.1.1) is a nice and widespread example of electroluminescence. The process generating light in LEDs is the one schematically displayed in Fig. (**18.4B**).

Conductor Luminescence

Light emission by conductors (metals) has its own peculiarities, because of the overlap between the VB and the CB. As can be seen in Fig. (**18.5A**) the CB and VB in conductors overlap (alternatively, in some cases the band is only partly filled with electrons; the end result is the same for our purposes). This means that there is no gap (Eg=0) for the electrons to cross, and in consequence nothing delays the recombination of an electron with a hole.

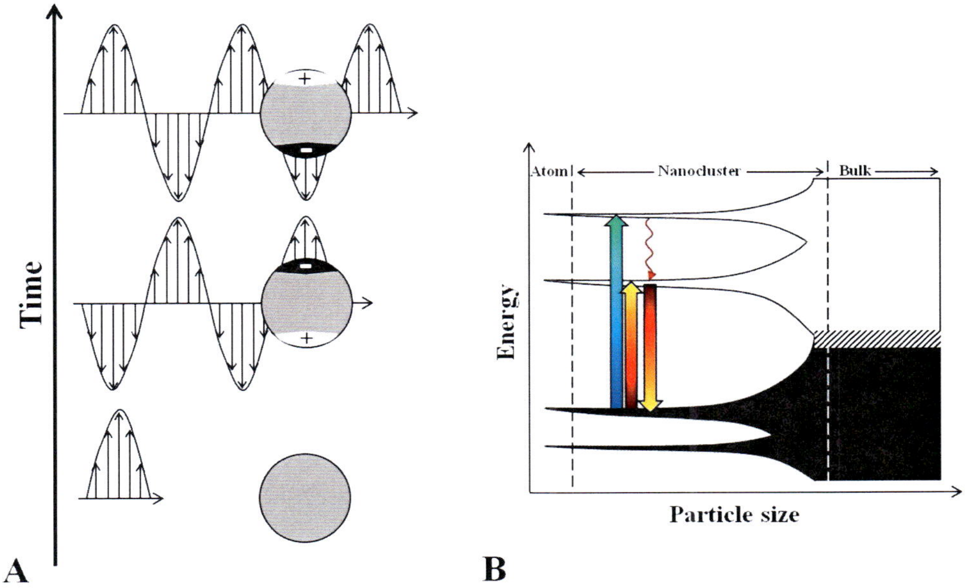

Fig. (18.5). Photonic processes in conductors. **A:** Plasmon resonance. A metallic nanoparticle (grey circle) is exposed to a light beam (electromagnetic wave). Only the electric field oscillation of the wave is shown for clarity (upwards/downwards pointing arrows). Before interaction (bottom) the electrons are homogenously distributed through the nanoparticle. When the wave starts interacting (middle) with the conductor, the electrons tend to locally accumulate at the top of the nanoparticle (negative black spot) driven by the wave's electric field. This produces local electron depletion at the bottom of the nanoparticle, which can be interpreted as a local increase in holes (positive white spot). After some time (top), the electric field changes polarity, which polarizes the nanoparticle in the opposite way (negative bottom and positive top). This chain of polarizations continues as long as the light excites the nanoparticle. **B:** Size dependence of metal luminescence. For very small metal particles, between "bulk" size (above 20 nm) and isolated atoms, there is a size range (0.33-2 nm) where the CB (white) and the VB (black), that overlap at larger dimensions, split into defined bands. Radiative transitions (coloured arrows) can take place between these bands. Non-radiative transitions (red wiggly arrow) can also take place, usually producing heat as a by-product of photon excitation.

In fact, the amount of these "quasi-free" electrons is so high in metals that they react in a collective way to light excitation. As light is an electromagnetic wave, the electrons will move according to the electric field vector of the light: it is as if very short electric currents were applied to the metal. These collective motions of the electron cloud are known as plasmon resonances, and some light frequencies are more efficient at exciting these plasmons than others [1, 9]. These excitations have very short lifetimes, and its most usual outcome is vibrational excitation of the solid lattice (*i.e.* heat). This is represented by the very efficient non-radiative deactivation of the excited levels after light absorption (see Fig. **18.5A**).

The plasmon resonance phenomenon is very important on solid-state photonics, but we will not pursue it further here as it quenches solid-state light emission in most situations, instead of promoting it. However, plasmon resonance plays a critical role in the process of plasmon-enhanced luminescence, although in this case the plasmon increases the light emission of some external fluorophore located very close to the metal structure (*e.g.* an organic dye) through electric field effects [9].

The plasmon effect, due to the very high density of electrons in metals, operates even at very small dimensions when one is dealing with metallic nanoparticles (down to 5-10 nm). So it seemed for many years that metal nanoparticles were useless as fluorophores. Research in the last 15 years has demonstrated that this conception was wrong, and now it is possible to produce luminescent metallic nanoparticles. This field is in its initial steps and the precise mechanism for light emission is not perfectly established. In any case, we will present the basics known at this moment [10 - 13].

Efficient light emission by metallic nanoparticles is obtained principally from noble metals (gold, silver, platinum and copper) [14]. The fundamental parameter for reliable light emission is a very small particle size. And we really mean this, because luminescent noble metal nanoparticles range in size from 20 nm down to 0.33 nm (few atoms clusters). Three different mechanisms, related to three different size regions, have been established to explain light emission by these particles [15]. We will start from the largest particles, and then move down in size to the atomic clusters in order to explain the light emission mechanisms.

In the particle size range from 20 nm down to 2 nm the plasmon resonance still occurs. However by careful synthesis, it is possible to obtain multicrystalline nanoparticles composed of different size nanodomains. Some nanodomains within the nanoparticle are small enough (below 2 nm) to be luminescent (see below). The whole nanoparticle, however, still displays plasmon resonances. So, there is a coupling of both phenomena, and there is energy transfer from the continuum of electronic states in the larger domains to the discrete levels in the smaller ones [15]. Radiative transitions between these discrete levels provide the mechanism of luminescence. This is the so-called grain size effect.

The second mechanism, the valence state effect/surface ligand effect, appears in the 1-5 nm size range. There are still many electronic states available in the nanoparticle, but the plasmon resonance approximation is no longer valid. Nanoparticles of this size are commonly synthesized through chemical interactions with organic ligands, especially sulfur-containing compounds. Therefore the surface of these metallic nanoparticles is enriched in +1 oxidation state atoms of the metal (this applies generally to gold nanoparticles), and closely located reduced sulfur atoms. This closeness provides a mixing of the metal and non-metal orbitals, and the establishment of discrete electronic energy levels. These are capable of efficiently absorbing/emitting light.

Finally, the particle size effect sets in at the smallest range in size [15 - 17]. The nanoparticles in this range span from 2 nm down to atomic size, as this effect has been observed even for gold clusters composed of only 3 atoms (Au_3). The luminescence is a consequence of the definitive breakdown of continuous bands into discrete levels, because of the decreasing number of available electrons in the nanoparticle/cluster. This is represented in Fig. (**18.5B**), where it can be seen how the continuous energy bands finally split into defined levels with a decrease in size. The metal cluster, usually composed of between 3 and 50-100 atoms, can be approximated to a very large "supra-atom" (cited in the literature as multielectronic artificial atoms), with an internal "core" of atomic nuclei and an outer electronic shell made by shared "band" electrons. This electronic shell displays quantized electronic levels, the energy of which depends on the cluster composition. For larger clusters (*e.g.* Au_{55}), the absorption and emission spectra are in the near infrared. For Au_3, the absorption takes place at 280-300 nm and emission at ~ 310 nm. The emission band is narrow and the radiative lifetimes are in the nanosecond range. These luminescent clusters are very photostable, reliable and, being made of noble metals, are minimally toxic for biological systems. Very small (~ 1 nm) noble metal nanoparticles seem a very promising group of fluorophores, probably called to displace other luminescent solid-state materials like Qdots in the near future.

In the rest of the chapter, we will present many of the currently employed solid-state markers in fluorescence microscopy. Some practical applications will also be outlined. As with many other technologies and materials, already introduced in this book, the field of solid-state markers is evolving at a very fast pace. To avoid presenting a too long chapter to the reader, the different classes of materials will be explained in short. The main goal of this chapter is that the reader becomes familiar with the photonic mechanisms behind the absorption and emission of light presented in this first section for different materials. We refer the interested reader to the selected bibliography for further information on each area.

2. SEMICONDUCTOR LUMINESCENT MARKERS

The first group of solid-state luminescent markers that we will review is that of semiconductors. These display a defined energy gap between a CB and a VB, although the Eg is small (less than 4 eV). The most important and widespread in use luminescent markers are Qdots, although other structures have been recently incorporated to this group. We will start by introducing Qdots.

2.1. Quantum Dots (Qdots)

As explained above (see section 1.2) Qdots are nanometric structures (1-100 nm) made from different semiconductor materials [3, 4, 7]. Many different elemental mixtures produce semiconducting materials, but we will describe here only the most common Qdots compositions. Fig. (**18.6A**) shows part of the periodic table of the elements, highlighting groups 12-16. Semiconductors can be made from single-element particles (group 14), as for carbon, silicon and germanium, which can then be tuned in their properties through impurity doping. Alternatively, most semiconductors are synthesized by mixing elements from other groups in different proportions [1]. The most frequent combinations involve elements from groups [13 + 15] and from groups [12 + 16]. Following this, compounds like BN, AlP, GaAs, InAs or InSb are semiconductors of the [13 + 15] kind. Likewise, ZnO, ZnS, ZnSe, CdS, CdSe or CdTe are very important semiconductors of the [12 + 16] kind. All these semiconductors can be further modified in their properties, mainly electronic/photonic ones, by adequate doping and/or particle size control.

For fluorescence microscopy applications, Qdot luminescence is usually tailored through particle size control (see quantum confinement effect in section 1.2 above). Although a single Eg value is provided for many semiconductors in different application fields, the consequences of the quantum confinement are unavoidable for Qdots in fluorescence microscopy. Therefore, it is more correct to provide a range of Eg values for different nanoparticle sizes of the same semiconductor. A selection of emission wavelengths for different semiconductors, depending on the nanoparticle size, is shown in Fig. (**18.6B**) [5]. For example, cadmium telluride (CdTe) emits in the range 550 to 750 nm. The precise emission wavelength depends exclusively on the particle size, being further to the blue the smaller it is. Thus the smallest CdTe nanoparticles emit at 550 nm. The same principle applies to the rest of semiconductors.

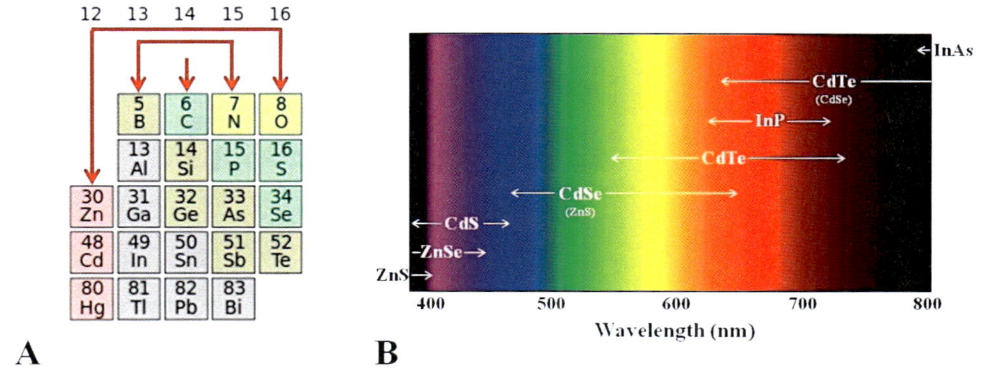

Fig. (18.6). Semiconductors. **A:** Most common elemental combinations resulting in semiconducting compounds. The first three elements of the group 14 are intrinsic semiconductors. Tin and lead can form semiconducting alloys with other elements of groups 13 and 15. Any combination of groups 13 and 15 or groups 12 and 16 (red arrows) results in a semiconductor. The heavier the elements making the semiconductor, the smaller Eg and closer to metallic behaviour (reproduced with permission from DePiep, https://commons.wikimedia.org/wiki/File: Periodic_table_(polyatomic). svg#). **B:** Chromatic plot displaying the range of light emission for different semiconductors as a function of the nanoparticle size. The smaller the particle size, the shorter the wavelength of the emitted light. Some semiconductors vary their emission as a consequence of being encased in a shell of another semiconductor (compare CdTe with CdTe coated with CdSe). Some Qdots, emitting in the near infrared (beyond 750 nm), find widespread application thanks to the sensitivity of some electronic cameras in that spectral region (spectral data taken from [5]).

Qdots are commonly composed of a luminescent core and a protective outer shell. This coat can be made of another closely related semiconductor, for example CdSe coated with ZnS, in order to alleviate lattice mismatch between core and shell. Because many useful Qdots in fluorescence microscopy have Cd in their composition, which is a very toxic heavy metal, the shell prevents metal leaking too. An example of a CdSe-ZnS Qdot is shown in Fig. (**18.7A**). A real demonstration of the quantum confinement effect on these Qdots is also displayed. Most often, Qdots are functionalized in one way or another to make them selectively interact with a biological target. Many different molecules can adsorb at the Qdot surface, and provide the desired functionalization (see Fig. **18.7B** for some examples) [18, 19].

As to the advantages presented by these fluorophores, they provide a very high photostability (very low photobleaching) in comparison to common organic molecules. The emission brightness is very high, and the spectral profile narrow. The absorption/emission spectra can be size-tailored as already described, and the optical absorption is very large [7, 19]. In regards to two-photon excitation Qdots present high extinction coefficients, much larger than most molecules. The emission Stokes shift is large, making luminescent light detection easier in regards to the excitation light. Due to their long lifetimes, Qdots make good fluorophores for time-resolved and multiplexing studies.

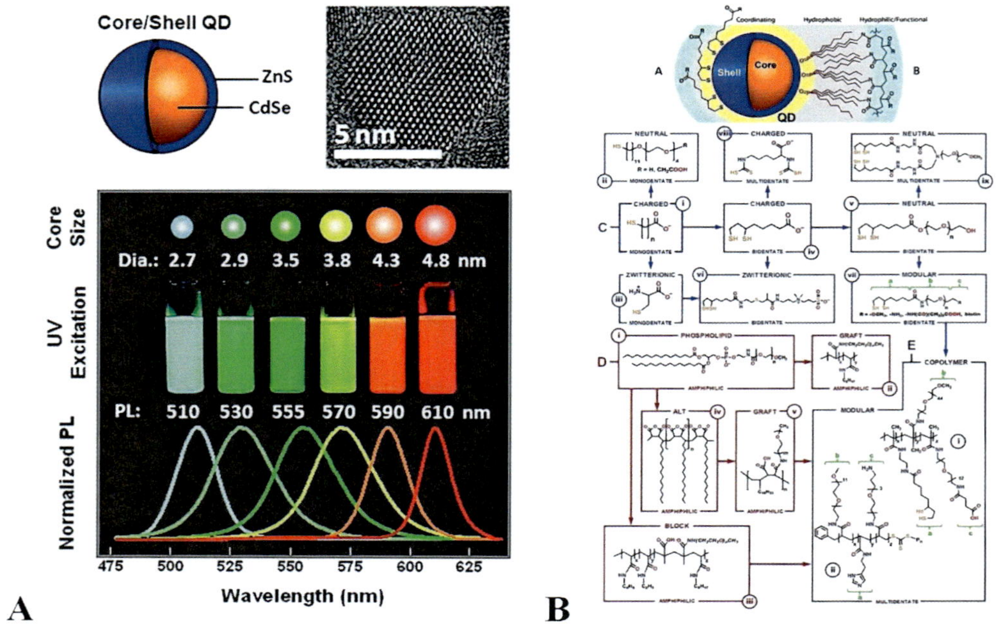

Fig. (18.7). Qdot structure and functionalization. **A:** Typical structural Qdot arrangement employed in fluorescence microscopy (top left). This is a core/shell layout. The core (CdSe in this example) is the luminescent unit. The shell (ZnS) plays two roles: isolating the luminescent core from the environment, and reducing the chances of core elements leaking to the biological milieu. An electron microscopy picture a Qdot (top right) shows the very small physical dimensions of these systems. Notice the regular crystalline core structure. Example of the quantum confinement effect in CdSe Qdots (bottom). As Qdot size becomes smaller (from 4.8 to 2.7 nm), the emission spectrum blue-shifts from 610 nm to 510 nm. The UV excitation spectrum is the same in all cases. Note the relative narrowness of the emission spectra for the different sized Qdots (30-40 nm). **B:** Selection of some organic molecules employed to functionalize Qdots. Through this selection, the Qdot charge or hydrophilicity, for example, can be tuned according to the research necessities. Qdots admit a very large arrangement of surface-modifying molecules. Reprinted with permission from [18]. Copyright (2011) American Chemical Society.

Qdots display some drawbacks too [19]. These are related basically to their large size (in molecular terms) and their leaking of toxic components. Due to their size (10-100 nm), their bioavailability can be compromised or hindered in some biological systems. This size effect can make them interfere with some physiological systems (kidney filtration), especially in animal models [20]. Also, it is still not clear how long it takes for a biological system to clear up the Qdots after incubation, which can take days, weeks or even years. Nanoparticle wearing can lead to the release of toxic elements from it, most notoriously cadmium and arsenic. Considering the extremely low amount of semiconductor being introduced in the system, even the total breakdown of the Qdots would result, in most cases, in negligible toxicity but further studies are needed in this sense. Some Qdots surface can act as catalysts, promoting undesirable side reactions with deleterious biological outcomes (*e.g.*, ROS production) [20]. From the photonic perspective, Qdots display sometimes the so-called "blink" effect where the particle's emission suddenly disappears, transiently or permanently, while is excited. The reasons behind this "blinking" are still matter of debate, but photo- or thermally-driven structural changes seem to be behind the phenomenon.

Qdots are routinely employed to image cells by fluorescence microscopy, but they can also be put to other, complementary, uses. This is shown in Fig. (**18.8A**), where CdSe Qdots have small interference RNA (siRNA) molecules attached to their surface by an amphiphilic organic compound [21]. Additionally, the siRNA molecules are tagged with fluorescein (FITC).

Human breast adenocarcinoma cells were incubated with these Qdots, and the temporal uptake sequence is displayed in the figure. The Qdots emit in red and the free siRNA in green. During the first 60 min almost the only signal registered is a red one, *i.e.* emission from the Qdots. Afterwards, siRNA detaches from the Qdots and starts to emit (green signal). Note that, by t = 6 h, two very different intracellular light patterns are registered: a red punctuate emission from the Qdots (reported as within endosomes by the authors), and a green diffuse cytoplasmic + nuclear emission due to siRNA. Another use of Qdots in cell microscopy is shown in Fig. (**18.8B**). In this case, modified Qdots were used as a fluorophore to measure pH by FLIM (see Chapter 6.3) [22]. CdSe/ZnS core/shell Qdots had their surface coated with mercaptopropionic acid (MPA). MPA shows a pKa around pH 7. Depending on its protonation state, MPA modifies the photonic properties of the Qdots which show a radiative lifetime increase from ~8 ns at pH 5.7 to over 16 ns at pH 8. By measuring the emission lifetime, it is possible to obtain the pH around the Qdots. Note that in the cell images provided at the top of Fig. (**18.8B**), the colour is a false colour scale related to the emission lifetime (warmer colours for longer lifetimes, see scale to the right), and it is not a real change in the emitted light wavelength (although a small wavelength shift does indeed take place under different pH). This is a very instructive example of fine-tuning Qdot photonic properties to measure parameters in living cells.

Qdots allow for measurement of different cellular parameters, including detection of ROS/RNS production [23], uptake processes (endocytosis, phagocytosis), cell metabolism/distribution by Qdot surface modification, intracellular ions, pO_2, drug and gene delivery assessment, and single particle tracking in relation to cell cytoskeleton and internal trafficking, to name a few [7]. In recent years, the use of

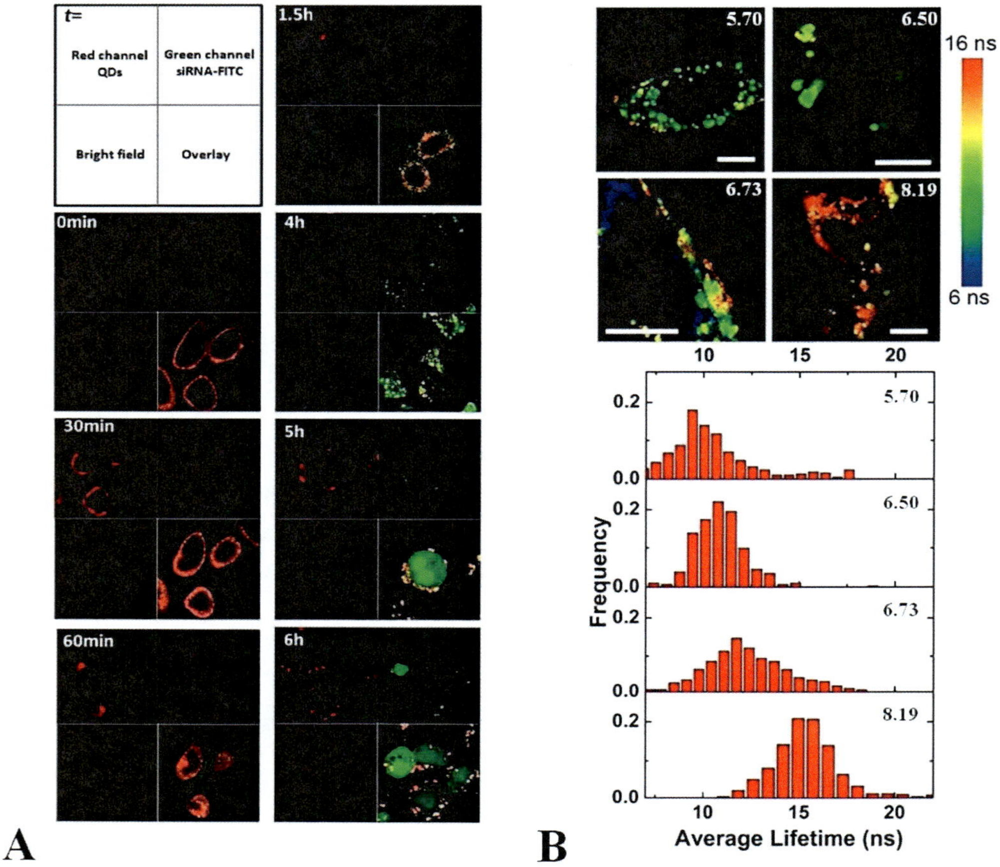

Fig. (18.8). Qdots in cell microscopy. A: Cell uptake kinetics of Qdots. Human breast adenocarcinoma cells (SK-BR-3) were incubated with CdSe Qdots decorated with FITC-tagged siRNA for the Her-2 gene product. The images were obtained from three channels (see upper left square for details): red emission from Qdots, green emission from fluorescein (siRNA), and bright field. An overlay of the three channels is also provided for each time frame. At t= 0 min the Qdots-siRNA is located on the plasmatic membrane. As time progresses, there is an increase in the intracellular Qdot mark (see red channel images). Green emission from FITC-siRNA is initially quenched due to association with Qdot surface. From t= 1.5 h onwards it can be seen how the green signal increases with time as a consequence of siRNA detaching from the Qdots surface. Reprinted with permission from [21]. Copyright (2008) American Chemical Society. B: Detection of intracellular pH with Qdots employing FLIM. CHO-k1 hamster ovary cells were incubated with CdSe/ZnS Qdots functionalized with mercaptopropionic acid (MPA). The Qdots radiative lifetime is sensitive to the environmental pH, becoming longer the more alkaline the pH is. Cell pictures (top panel) at different pH (pH value indicated on the upper right corner) indeed show different radiative lifetimes (see false colour scale on right). Under mild acidic conditions (pH=5.70) the lifetime is around 8-9 nanoseconds, while at pH 8.19 lifetime increases to 16 nanoseconds. A sample (bottom panel) of measured radiative lifetime frequencies for different incubated cells as a function of pH is also provided. Scale bars: 10 μm. Reprinted with permission from [22]. Copyright (2013) American Chemical Society.

Qdots in combination with super-resolution techniques is providing a wealth of new information (see Chapter 20). The use of Qdots with new technologies is providing new tools for the microscopist. In this sense, the combination of very small Qdots with nanobodies (see Chapter 17.1.4) allows for nanometric resolution (7-8 nm) of cellular motor protein function, as recently reported by Wang *et al.* [24].

A current trend showing much success is Qdot doping with transition metal (TM) ions [25]. This strategy leads to an enhancement of light emission, photostability and light tuning. This is a variation of the impurity doping strategy already mentioned at the beginning of the chapter. Incorporation of TM cations, especially Mn^{2+}, leads to a very fine-tuning of the emission spectrum with almost negligible self-absorption due to Stokes shift. Doping with Cu sounds very promising as to obtain white light-emitting nanoparticles due to the arrangement of d-bands of Cu atoms within the semiconductor lattice.

All in all, semiconductor Qdots represent one of the most successful approaches to very efficient fluorophores for fluorescence microscopy. However, other alternatives exist in relation to solid-state luminescent nanoparticles, and the most important of them will be introduced below.

2.2. Carbon and Silicon Nanoparticles

Carbon and silicon nanoparticles could, in principle, be classified within the semiconductor Qdot group (previous section 2.1). We have decided to present them in a different section for two reasons. First, carbon is not considered a semiconductor as such. However, its chemistry is so rich that many of its allotropes do display a semiconductor character. Second reason is that, until recently, silicon luminescent nanoparticles were not available to the scientific community. Silicon is an intrinsic semiconductor, but it is an indirect gap one, which classically has translated into very low radiative quantum yields. We will start with carbon luminescent nanoparticles, and then proceed to introduce silicon Qdots.

Carbon is a fascinating element, capable of forming atomic bonds with almost any other known element. This "bond-friendly" feature extends also to other carbon atoms. As a consequence, carbon can form many different structural compounds (allotropes) without intervention from other elements. These allotropes display particular characteristics, different from the other allotropes. Historically, carbon has been known to form two allotropes: a dielectric known as diamond and a conductor known as graphite. However, nowadays the range of carbon allotropes known is huge [26]. In this section we will talk about luminescent "semiconductor" (graphite-like) carbon nanoparticles, those based on structures with a prominent sp^2 electronic hybridization and abundant delocalized π electrons (see Chapter 3). Luminescent carbon compounds with dielectric characteristics (nanodiamonds) will be treated later (see section 4.2 below).

The "semiconductor" carbon compounds are characterized by a dominant sp^2 electronic orbital hybridization, which promote carbon atoms to form sheets (graphite-like) in which each carbon is bonded to another three carbons in the same lattice plane. A delocalized π electron "cloud" forms above and below the plane of this sheet. This is the basic 2D arrangement that, when made from a single layer, is known as graphene. When this sheet is rolled upon itself to form cylindrical (1D) or spherical (3D) structures, very interesting nanoparticles can be obtained, like nanotubes or fullerenes. A display of some of the most common nanometric carbon structures is shown in Fig. (**18.9A**) [26].

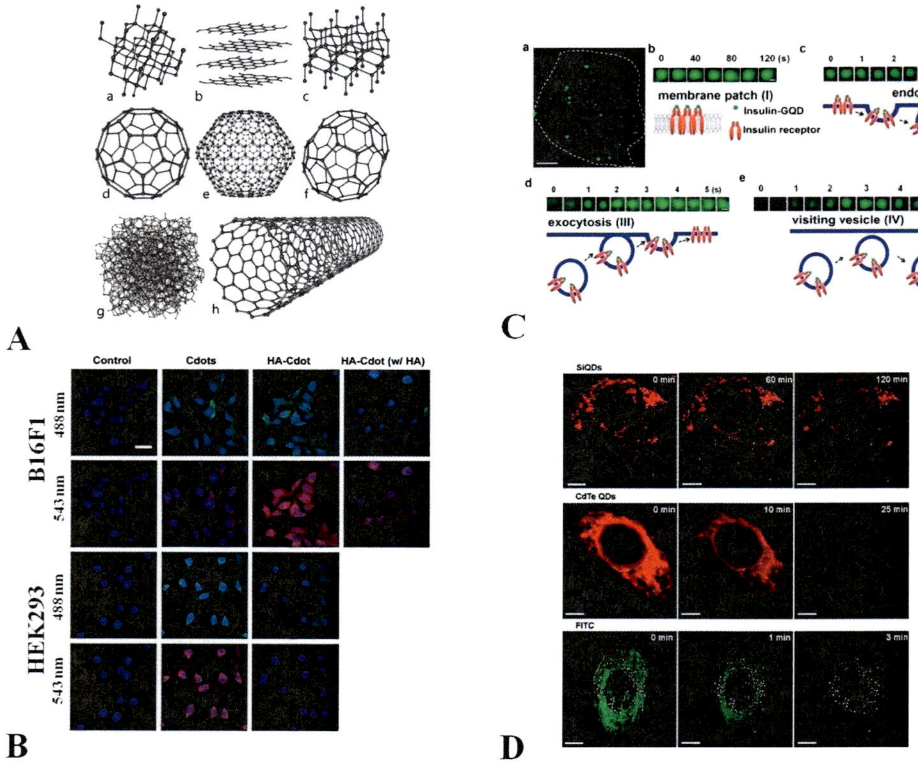

Fig. (18.9). Carbon and silicon Qdots in cells. **A:** Some carbon allotropic structures commonly employed in luminescent solid-state carbon markers. (a) Diamond, (b) graphite (4 graphene layers displayed), (c) lonsdaleite (hexagonal diamond), (d) C_{60} fullerene, (e) C_{540} fullerene, (f) C_{70} fullerene, (g) amorphous carbon, and (h) single-wall carbon nanotube. Reproduced with permission from Michael Ströck (mstroeck), https://commons.wikimedia.org/w/ index.php?curid=584786. **B:** Uptake of hyaluronic acid-coated C-dots

(HA-C-dots) by B16F1 (mouse melanoma cells) and HEK293 (human embryonic kidney cells). C-dots were excited by a confocal laser system at two wavelengths: 488 nm and 543 nm. Control cells were not exposed to C-dots, and only the deep blue emission from counterstaining DAPI dye is observed (left lane). Cells incubated with C-dots only display some internal signal. However, when C-dots were functionalized with HA, their uptake was largely increased but only for the B16F1 cell line that expresses HA receptors. HEK293 in fact show a somewhat decreased HA-C-dot uptake. Under conditions of excess free HA (100x) in comparison to HA-C-dots B16F1, cells reduce drastically their incorporation of the nanoparticles due to competitive HA receptor saturation (two pictures on the right). Scale bar: 30 μm. Reprinted and adapted with permission from [32]. Copyright (2012) American Chemical Society. **C:** Insulin cell receptor dynamics assessed by graphene C-dots. In (a) a 3T3-L1 murine adipocyte incubated for 1 h with graphene C-dots decorated with insulin is shown under laser confocal TIRFM excitation (λ= 488 nm). Notice the clustered appearance of the C-dot signal (cell outlined by the slashed line). Scale bar: 5 μm. In (b) – (e) four different dynamic insulin receptor related vesicle behaviour examples are presented. Each frame corresponds to the dynamics of just one cluster as the ones shown in (a). A sketch of the underlying behaviour in each case is depicted below the emission sequences. The time is indicated above the sequences in seconds. Scale bars: 200 nm. Reprinted with permission from [33]. Copyright (2013) American Chemical Society. **D:** Si-dots display a remarkable resistance to photobleaching. HeLa cells labelled with Si-dots (top row), CdTe Qdots (middle row) or fluorescein (FITC, bottom row) were excited with a laser confocal system continuously, and the diminishing light emission was registered at different times (time displayed at the upper right corner of each image). Clearly, fluorescein bleaches in a matter of minutes and CdTe Qdots in less than half an hour. In contrast, Si-dots signal shows almost no decay during the first hour and starts to bleach somewhat after 2 h. Scale bars: 5 μm. Reprinted and adapted with permission from [36]. Copyright (2011) American Chemical Society.

Very large graphene-like structures tend to efficiently absorb light, but deactivate quickly through non-radiative channels. Thus, luminescent graphene-like carbon nanoparticles (carbon-dots or C-dots) are synthesized to be of very small dimensions, so that quantum confinement effects favour radiative processes over non-radiative deactivation [27 - 29]. Additionally, certain structural modifications, like carbon-oxygen bonds, create distortions in the lattice that favour light emission [27]. We will not go into further detail, because the basic electronic processes that take place in the solid-state also apply to C-dots. We encourage the interested reader to consult the referred literature on this rapidly expanding field.

Probably, the main advantage of C-dots and carbon-based luminescent nanomaterials in fluorescence microscopy of living systems is their biocompatibility [29, 30]. Being made of carbon, they seem to be much better tolerated by organisms than other solid-state heavier elements fluorophores. As we already mentioned, Qdots may leak some heavy metals after a long exposure. Additionally, C-dots offer a better clearance profile from organisms in comparison to other markers. However, there is extensive literature supporting a deleterious effect of very small carbon particles from combustion engines and other machines. Thus, more studies are necessary to assess the safety of these carbon nanostructures.

C-dots are finding many applications in different areas, like chemical and biochemical sensing, nanoreactors and energy harvesting, and drug delivery and photodynamic therapy, to name some [30, 31]. In bioimaging, their photoluminescent and biocompatible properties make these compounds serious challengers of Qdots as solid-state luminescent markers [27, 29 - 31]. They display enormous photostability and resistance to photobleaching. Therefore, they do not suffer from the blinking instability that is more common with Qdots. However, for the moment, their emission spectrum seems quite restricted to the short wavelength visible (blue and green), where most biological autofluorescence takes place. On the other hand, C-dots are interesting agents to obtain upconversion emission, in which shorter wavelength light is emitted under long wavelength excitation (see Chapter 19.4).

Two practical applications of C-dots for fluorescence bioimaging are presented next. In Fig. (**18.9B**) two different cell lines (B16F1 and HEK293) are imaged after incubation with C-dots and hyaluronic acid-coated C-dots (HA-C-dots). C-dots entered the cells *via* passive mechanism. Meanwhile, HA-C-dots were incorporated through a HA receptor-dependent active mechanism [32]. The functionalization of these nanoparticles, as with other similar structures presented in this chapter, provides a very robust analysis tool to study cellular processes. A very interesting use of C-dots was reported by Zheng *et al.* in regards to the membrane dynamics of insulin receptors in living adipocytes [33]. The authors employed the TIRFM technique (see Chapter 6.5) to study insulin receptors located at the plasma membrane, by attaching insulin to the graphene C-dots surface. Incubating the cells with these functionalized C-dots allowed the uptake dynamic to be observed (see Fig. **18.9C**). The experiments clearly showed membrane clusters/microdomains, where insulin receptors were located. Taking advantage of the temporal resolution of TIRFM, it was shown that these microdomains undergo lateral movement alongside the plasma membrane, and "in-and-out" dipping movements between the cell surface and the immediately underlying cytoplasm. The study was made possible partly because of the almost negligible graphene C-dot photobleaching.

Silicon is the next element after carbon in group 14 of the periodic table. They share many characteristics. However, silicon is included in the semiconductor group, unlike carbon of which only certain allotropes are considered so. Being a semiconductor, it would be expected for silicon Qdots (Si-dots) to be widely available and employed. As it turns out, silicon is an indirect gap semiconductor, which greatly precludes its use as a luminescent semiconductor. In a simplistic way, it can be said that indirect gap semiconductors require the simultaneous emission of a phonon (quantized lattice vibration) and a photon for light emission to occur (for further details on indirect gap semiconductors see [1]). This double process is improbable, thus non-radiative deactivation pathways domain in silicon. However, it has been shown that silicon nanoparticles below ~4 nm display quantum confinement effects that boost radiative deactivation. Due to this feature, their adequate biocompatibility, emission in the near infrared and very high resistance to photobleaching, Si-dots are gaining popularity as solid-state fluorophores for bioimaging and fluorescence microscopy [34, 35]. An example of the impressive photostability of Si-dots has been reported by He *et al.* (see Fig. **18.9D**) [36]. In the figure, it can be clearly seen how Si-dots keep emitting light even after 120 min of continuous excitation by a confocal laser system. CdTe Qdots display a noticeable fading after 10 min, and are almost completely bleached after 25 min. Fluorescein is severely bleached after 1 min of laser exposure, and the signal has nearly disappear after 3 min.

2.3. Semiconducting Polymers (P-dots)

Finally, we would like to mention semiconducting organic polymers (P-dots), sometimes called conjugated polymers, as another type of semiconductor luminescent compounds [4, 13]. They are polymers composed of dye molecular units, or a polymeric backbone decorated with dye units. Each unit is a fluorophore in itself, but when several or many units are put together, the overall photophysical behaviour changes with respect to the isolated units. The delocalized π electrons, which are directly related to the photonic molecular features, become even more delocalized over the polymer (molecular wires) [37]. As a consequence, the polymer behaves in some aspects more as a Qdot than as a molecule (collective electronic effects become very important). Depending on the number of subunits (dye monomers), the absorption coefficients can be very high ($\sim 10^6$ $M^{-1}cm^{-1}$) and the emission can be properly tailored by adding/subtracting functional side groups. Given the very high electronic concentration, non-linear optical effects are also very important with these compounds [38]. They can be functionalized for different bioimaging roles, and are quite biologically-friendly. Most P-dots present sizes in the 10-100 nm range. But most of them do not extent in the three dimensions to the same length. They usually are either long needle-like 1D structures, or 2D sheet-like complexes. The smallest P-dots fill the gap between small organic dye molecules, introduced at the beginning of the book (see Chapter 3.3.1), and large dendrimer structures.

3. METALLIC LUMINESCENT MARKERS

This kind of solid-state markers are relatively new in the bioimaging and fluorescence microscopy field. Due to chemical and photochemical reasons, efficient metallic nanoparticles are made of noble metals, most frequently gold and silver. Other metal nanoparticles made of copper or platinum have been reported in the literature [12]. The mechanisms leading to the particular photonic properties of these nanoparticles were explained in detail previously (see section

1.3 above). They offer excellent photostability, tuneable absorption and emission spectra (depends directly on cluster size and surface functionalising molecules), and chemical/biological inertness [4, 13, 37]. Their biological interactions are mainly determined by their functionalization, by surface molecular moieties that interact with particular cell/tissue targets. Because of the synthetic approach to obtain these very small size nanoparticles (it was previously stated that size must be smaller than 2-3 nm for these markers to efficiently emit light), "scaffold" molecular structures are necessary to avoid precipitation. Frequently employed "scaffold" molecules include small organic ligand molecules, polymers, dendrimers, oligopeptides and nucleic acids (single strand DNA). Once synthesized, this scaffold can be disposed of, or it can secondarily increase the nanoparticle's preference for some target/substrate. Next, we will introduce the two most employed metal nanoparticles in fluorescence microscopy: gold and silver clusters.

3.1. Gold Luminescent Nanoparticles

Gold nanoparticles or nanoclusters are the preferred format currently for use of luminescent metal nanoparticles for fluorescence microscopy. Depending on the molecules attached to the nanoparticle surface, it will show preference for certain cell types or cellular structures. For example, folic acid-coated Au nanoparticles have a clear preference for rapidly-replicating tumoral cells [16]. Other researchers have employed streptavidin-decorated Au nanoparticles to bind and detect biotin in fixed cells, as a proof of concept for detecting important biological cofactors [39]. Only when Au nanoparticles were functionalized with streptavidin, there was a clear fluorescent signal coming from the HepG2 cells employed as experimental model (see Fig. **18.10A**). Another example, Au nanoparticles with triphenylphosphonium (TPP), a cation that tends to localise at the mitochondria, permit fluorescent imaging of these organelles in living cells. It has been recently reported that ROS and antioxidants can be detected in cultured cells using adequately modified Au nanoclusters (see Fig. **18.10B**) [40]. Due to their extremely small size (~1-2 nm), Au nanoclusters make perfect luminescent probes to enter the cell nucleus and assess nuclear functions and structures [17].

These luminescent markers have been successfully employed as FLIM probes, due to their long radiative lifetimes and large Stokes shift in emission [16, 17]. Radiative lifetimes of these gold nanoclusters show a dependence on their environment. As such, it is possible to take advantage of this feature to assess subcellular microdomains with these markers. For example, it has been reported that there are subtle differences in temperature within HeLa cells by measuring the emission lifetimes. As these lifetimes were calibrated before in cuvette, it was possible to know the temperatures within a living cell [41]. Additionally, Au nanoparticles display very high two-photon absorption cross sections. This feature makes them an adequate option in non-linear bioimaging [16, 17].

Gold nanoparticles have been successfully employed in the upconversion of light, taking advantage of the surface plasmon resonance. By exciting intensively the plasmon resonance, the associated electric field can enhance non-linear optical processes by many orders of magnitude, due to symmetry breaking and electric-field enhancement of non-linear coefficients.

3.2. Silver Luminescent Nanoparticles

Silver (Ag) nanoparticles/nanoclusters share many of the features that Au nanoparticles display: very small size, extremely high photostability, size-dependent photonic characteristics, *etc.* [42]. The most employed molecular scaffold for Ag nanoparticle growing and stabilization is DNA. The emission ranges from deep blue (450 nm) all the way to the near infrared (800-900 nm). Perhaps, one of the most distinguishing properties of Ag nanoparticles is their relatively short lifetime (~ 0.2-4 ns), which is between one and two orders of magnitude shorter than typical Au nanoparticles. They also show very high two photon absorption cross sections.

Apart from their uses in other areas like chemical sensing (*e.g.* to detect metal cations), Ag nanoclusters are finding more and more use in bioimaging and fluorescence microscopy [42]. Although most commonly stabilized and shuttled by using DNA, other carriers are employed, like oligopeptides and proteins. Depending on the enveloping carrier, the Ag nanoparticles mark different cell structures, like the nucleus and the nucleoli. Functionalising these kind of metallic nanoparticles with biotin, streptavidin or antibodies provides a way to obtain robust and biocompatible luminescent markers, both for fixed samples and for living material.

A somewhat different, but very interesting and successful, approach to Ag luminescent biomarkers has been reported by Zhang *et al.* [10]. They synthesized larger two-component nanoparticles: a silica core (40 nm) coated with a silver shell (10 nm). Doping the silica core with three different Ruthenium luminescent complexes, three different nanoparticles were obtained. The interesting point is that the three luminescent Ru complexes initially have roughly equivalent radiative lifetimes. It is by being encapsulated within the silver metal shell, that the lifetimes increase to different extents for the Ru compounds (10, 45 and 200 ns). An example of the three different emission lifetimes is provided in Fig. (**18.10C**). These doped-silica/silver nanoparticles were employed to label living cells for a time-resolved fluorescence analysis (see Fig. **18.10D**). The cell surface was biotinylated artificially, and then exposed to avidin-coated silica-silver nanoparticles. Cells were then fixed by simple air drying. The emission lifetimes as recorded are shown in the figure, where it is clear that the three types of nanoparticles displayed much longer emissions than the cell autofluorescence.

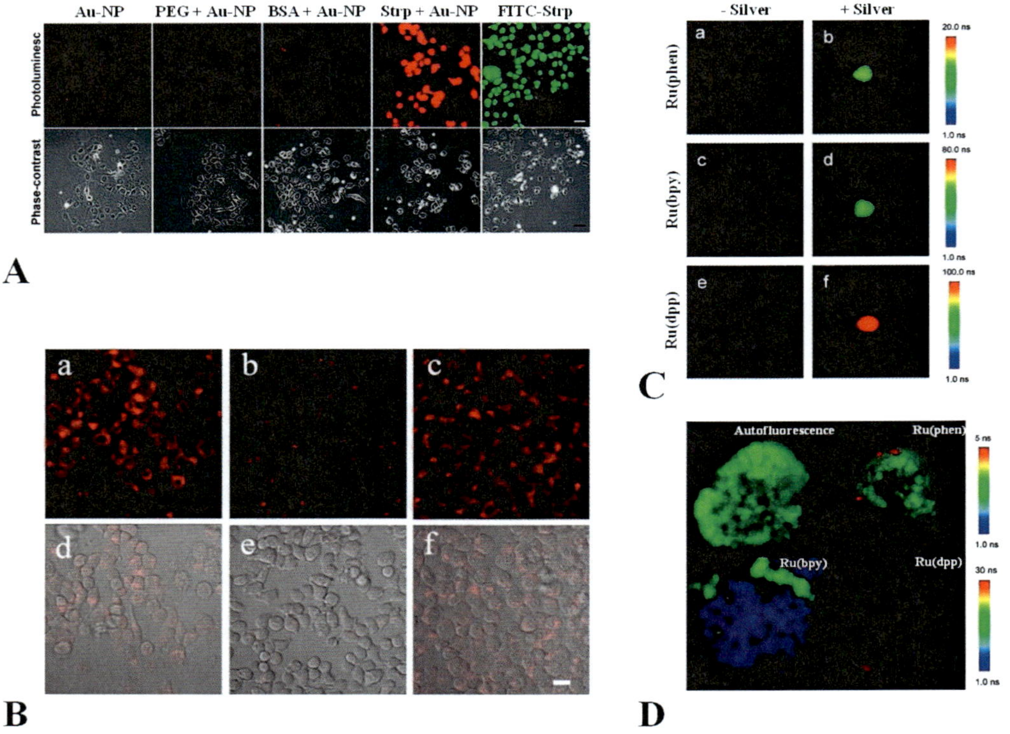

Fig. (18.10). Noble metal (gold and silver) nanoparticles for fluorescence microscopy. **A:** Biotin labelling inside fixed $HepG_2$ human hepatoma cells. The top row shows fluorescence images, and the bottom row the corresponding phase contrast field. From left to right, the cells were exposed to plain Au nanoparticles (Au-NP), Au-NP functionalized with polyethyleneglycol (PEG), bovine serum albumin (BSA) or streptavidin (Strp). As a positive control for biotin detection, fluorescein-tagged streptavidin (FITC-Strp) images were also obtained. Although there is some level of baseline noise in the first three pictures, it can be seen that the biotin signal is quite clear for the Strp + Au-NP and FITC-Strp. Images were obtained with a 488 nm excitation confocal laser system. Scale bars: 50 µm. Reprinted and adapted with permission from [39]. Copyright (2009) American Chemical Society. **B:** ROS/antioxidant detection in living cells. HeLa cells were incubated with Au nanoclusters for 2 h. The cell uptake is readily seen, because of the strong fluorescent red emission (a). If cells were exposed afterwards to hypochlorite, a strong ROS, the Au nanocluster emission is quenched immediately (b). When cells were incubated with ascorbic acid before exposure to hypochlorite, then the light signal did not diminish (c). This proofs that the Au nanoclusters are sensitive to their redox environment, and such parameter can be measured in living cells. Images (d)-(f) show the bright field pictures corresponding to (a)-(c) respectively. Scale bar: 20 µm. Reprinted with permission from [40]. Copyright (2014) American Chemical Society. **C:** Radiative lifetime modulation of Ruthenium (II) complexes in silver shell over silica core nanoparticles. Three Ru(II) complexes (Ru(phen), Ru(bpy) and Ru(dpp)) were introduced into silica nanoparticles. Images (a), (c) and (e) show the false colour emission lifetime on these uncoated particles, which is too short to map in the scale (< 1 ns). When a 10 nm thick silver layer coated the core, the emission lifetimes greatly increase for the three complexes, but in different amounts for each one. This is indicated in images (b), (d) and (f). Note that the false colour time codes show different scales. **D:** Differences in the nanoparticles lifetime on a cell surface. Human lymphoblastoid CEM-SS cells were "coated" with biotin through a chemical reaction. Then Ru(II) complex containing nanoparticles (silica/silver) with avidin on their surfaces were added to the cell medium. After 1 h incubation, the cells were air dried and the fluorescence lifetime images were obtained in a confocal laser microscope (λ_{exc}= 470 nm). The different lifetimes are coded in false colour (time scales to the right). Note that the time scales are not equivalent. Both **C** and **D** reprinted and adapted with permission from [10]. Copyright (2011) American Chemical Society.

Taking into account all the positive features shown by noble metal nanoclusters, they are probably one of the most important contenders for the next generation of solid-state luminescent markers.

4. DIELECTRIC LUMINESCENT MARKERS

Finally, we will briefly introduce some dielectric (insulator) luminescent nanoparticles. Some are very established markers (*e.g.* dye-doped polystyrene or latex beads), while others are new additions to the microscopist toolbox, like nanodiamonds. We will provide less information in this section, as in fact the nanoparticle itself does not emit. This is in contrast to the semiconductor and conductor nanoparticles introduced in previous sections, given that the lattice itself is the luminescent agent in those cases. The dielectric lattice offers a better suited environment for radiative decay to take place, by providing a rigid solid matrix, and confining the fluorophore so that it does not undergo photochemistry. The nanoparticle does modulate the light emission process, but it is not the emitter itself. In most cases, organic dyes are the fluorophore of choice encased within the nanoparticle. Other fluorophores, however, are not uncommon as we will see.

4.1. Polymer Particles

Polymeric plastic particles have been used in fluorescence microscopy and other fluorescence related areas for decades [13]. They are usually made of polystyrene, although other plastics can enter in the composition, like polyacrylonitrile or poly-(vinyl butyral). These plastics are characterized by their hydrophobicity and their chemical inertness. They do not emit light by themselves, but rely on internal or surface doping with secondary fluorophores to achieve this property. Due to the particle rigidness and absence of, or very limited, diffusion, the doping fluorophores find a better suited environment for radiative transitions. For this reason they also have a relatively low photobleaching rate. The most common doping fluorophores are small organic molecules. But other options have been employed, like transition metals (*e.g.* Eu(III) or Pt(II)), or even smaller Qdots embedded within the plastic matrix. The particular photophysical features of a given particle depend mainly on the doping fluorophore. As such, plastic particles with emission ranges from the near UV (390 nm) all the way to the near IR (~800 nm) are in common use.

They are very well tolerated by biological systems (cultured cells, organisms), and they can be functionalized according to their target with different surface moieties (like biotin or streptavidin, for example). They range in size from very large micrometric particles (above 10 μm) to truly nanometric ones (around 10 nm). Common uses for these plastic particles are very diverse, and include blood or other fluid flows, study of circulation systems, evaluation of phagocytosis and intracellular trafficking, cellular tracing (see Chapter 16.5), study of cell receptor function, study of molecular motors, cell imaging and quantification (both fluorescence microscopy and flow cytometry), among many others.

A second kind of polymeric particles is represented by the generic group of polymeric gels [13]. These are hydrophilic polymers based on organic backbones. Common materials forming these gels are polyethylene glycols, polyacrylamide, polyurethanes or poly-hydroxyethyl methacrylamide. They present a less rigid lattice (porous) than plastic particles, and their photonic performance as labelling agents is considered poorer than that of plastics. However, this drawback is an asset when considering their role as drug-delivering agents. As it happens with plastic particles, these gel materials do not emit light by themselves, but rather act as a carrier/encasing agent for a fluorophore. The fluorophore can be covalently linked to the gel monomers, or be retained by ionic bonding. This kind of materials tends to interact more with their surroundings than plastic particles, swelling/shrinking for example under the right pH, temperature, hydration, *etc.* conditions.

We will not discuss further this kind of polymeric particles, as they do not constitute real fluorophores but fulfil the role of carriers for other emitting species, the description of which can be found in previous chapters of this book.

4.2. Nanodiamonds

Carbon nanodiamonds (NDs) are a subtype of carbon nanoparticle. They have not been included as C-dots (section 2.2 above) because NDs are not a semiconductor material, but rather an insulator. The bonding structure in NDs is different from that in semiconductor graphene-like C-dots explained before. Where in C-dots sp^2 hybridization of the electronic levels takes place, in NDs it is the sp^3 hybridization that occurs. The essential difference is that, in NDs, the bonding between carbon atoms occurs in the three dimensions and there are no delocalized π electrons in the lattice. All electrons are σ in character, and form part of atom-atom bonds [26]. As a consequence, the material is a dielectric, only absorbing in the UV part of the spectrum. Indeed, the band gap is around 5.5 eV, which translates into an absorption edge wavelength of ~225 nm. NDs, especially the larger ones, exhibit physical characteristics very similar to bulk diamond, like hardness, chemical inertness, high thermal conductivity, *etc.*

NDs range in size between 1 and 20 nm. The smaller the ND, the less diamond-like are its properties, as a higher proportion of its carbon atom constituents are at the particle's surface, and not in the bulk. At the surface, the carbon atoms are usually bonding other types of atoms like hydrogen, nitrogen or oxygen [26]. Additionally, at the surface there is a transition from a sp^3 to a sp^2 hybridization, as the particle's structure "closes upon itself". Thus, NDs surfaces are much more similar to graphene layers than to diamond. This surface reactivity offers the possibility to functionalize the NDs with different organic or inorganic molecules, to better suit the ND sensing function.

Luminescence in NDs traces back to the presence of point defects within the lattice [27, 35]. This defects, or dislocations, involve an impurity nitrogen atom incorporated into the particle. Usually the starting concentration of nitrogen impurities is adequate, but there are no structural defects. The usual procedure to obtain luminescent NDs is to subject them to heavy particle bombardment in a particle accelerator. The high energy particles impact the lattice, and produce defects. Afterwards, the NDs go through a thermal annealing step. Under high temperatures the created defects migrate through the structure until they encounter a nitrogen atom. The nitrogen stabilizes the defect (vacancy), and the defect-nitrogen form what is called a N-V centre: a colour centre that is the luminescent unit. This is analogous to the colour centres in other dielectric materials, also known as F-centres (from *Farbe* meaning "colour" in German), where defects in the solid matrix trap electrons or ions and give rise to atomic-size luminescent centres [1].

The most common type of N-V centre is composed of one vacancy and one nitrogen atom. The absorption spectrum is broad in the visible and the emission is broad too, covering the 600-750 nm region. The emission quantum yield is ~1.0, with a radiative lifetime of 12 ns. The photostability is exceptional, and blinking is a rare event. It is possible to obtain NDs emitting in the green (around 504 nm), with two nitrogen atoms per vacancy (N-V-N centre). This works also in the blue (three nitrogen atoms surrounding a vacancy, N3 centre), that displays a weaker luminescence. Other impurities have been successfully incorporated into NDs to produce luminescence. This is the case of silicon (Si-V) and nickel. Si-V centres emit in the near IR in the region of 740 nm. Ni-V centres emit at 710 nm and they show another emission doublet at 880-885 nm. Without doubt, more luminescent impurity centres will be reported in the near future as research proceeds in this area. Because of the much localized emission centre in NDs, the usual quantum confinement effects observed in other nanometric solid fluorophores is very weak or absent in these materials. As such, there is no absorption/emission spectral fine tuning through the particle's size as in Qdots, for example. The emission wavelength options is more limited (for the moment) when employing NDs. This is partly offset by the very high photostability and absence of toxicity they display [35].

In regards to bioimaging and fluorescence microscopy, NDs have been employed both at the cell level up to the whole organism. As with other nanoparticles, the particular interaction taking place between the ND and the biological target is mostly mediated by some surface-functionalizing molecule (cofactor, antibody, *etc.*). NDs have been successfully employed to image cell uptake (both targeted to specific receptors and non-targeted), analyse internalization mechanisms, drug and xenobiotic cell delivery, vesicle dynamics (endo- and exocytosis), and subcellular nanothermometry. An illustrative example of energy-dependent endocytosis in NIH 3T3 murine fibroblasts assessed by using NDs is shown in Fig. (**18.11A**) [43]. The researchers employed to different cationic coatings for the NDs, polyallylamine (PAH) and polyethylenimine (PEI). Fibroblasts were incubated with the modified NDs under three conditions: 37 °, 4 °, or 37 ° plus sodium azide (NaN_3) to inhibit mitochondrial ATP production. The NDs emit red light (more details in the figure caption). It is readily seen that cells internalize both types of NDs at physiological temperature, but fail to do so at 4 ° due to the almost total energetic depletion at this low temperature. On the other hand, at 37 ° plus inhibiting ATP production by NaN_3, ND-PAH uptake is very low while ND-PEI is still quite high. In summary, ND established the different uptake mechanism for the two different organic coatings.

NDs have been employed to image biological processes in whole organisms. Mohan *et al.* reported using bare NDs to assess their dynamics and toxicity in a *C. elegans* model [44]. The animals were either fed the NDs or had them microinjected. No toxicity, as assessed with different assays, was observed in any case. Microinjecting NDs into the gonads led to ND accumulation in the worm embryos (see Fig. **18.11B**). The NDs remained within the embryos even after hatching, but there was no deleterious effect registered at all. Therefore, NDs seem a very interesting option for very long, even intergenerational, biological follow up experiments. The authors noted the very high photostability and brightness of the employed NDs in this study.

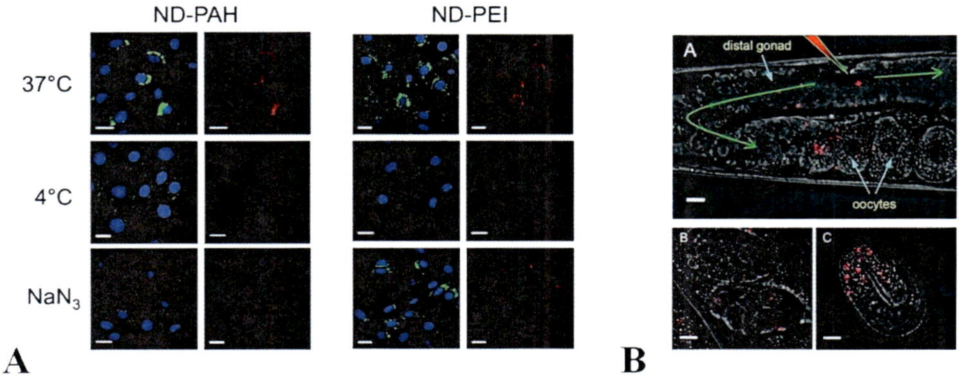

Fig. (18.11). Nanodiamonds (NDs) in fluorescence microscopy. **A:** Cell uptake of NDs. Two different cationic organic coatings (prospective siRNA cell delivery agents), polyallylamine (PAH) and polyethylenimine (PEI), were employed to coat the same type of NDs. The endocytic cell uptake was studied at different temperatures (37°, 4°) and with NaN_3 incubation, in a NIH 3T3 murine fibroblast model with 4 h incubation time. Additionally, the ND coatings had FITC covalently bonded, as an extra control (green emission) to guarantee that the coating molecules were at all times attached to the NDs (red emission). Counterstaining was done with DAPI (blue emission) for the cell nuclei. It can be seen that uptake was as expected at 37 ° for both coatings (upper panels). At 4 ° neither class of NDs is internalized (middle panels), due to temperature-inhibited cellular metabolism. At 37°, with NaN_3 in the cell medium (selectively inhibiting the mitochondria), ND-PAH uptake was greatly inhibited, but this was not the case with ND-PEI. This can point to partially different internalization mechanisms. Images were obtained with a confocal laser microscope. Scale bars: 20 µm. Reproduced with permission from [43]. **B:** NDs dynamics in *C. elegans*. NDs (bright red spots) were microinjected into an adult animal (red "needle" in A). After 30 min the NDs have dispersed (green arrows) within the gonad and into the oocytes. The NDs are internalized into the early stage worm embryos (B), which nevertheless continue their normal development. A late stage embryo (C) shows most of the NDs in its intestinal track, its development undisturbed by the nanoparticles. Images are overlays of epifluorescence and phase-contrast pictures. Scale bars: 10 µm. Reproduced and adapted with permission from [44]. Copyright (2010) American Chemical Society.

4.3. Silica Nanoparticles

Luminescent nanoparticles based on silica (SiO_2) have been used for quite a long time. The matrix itself is not luminescent, and its light emission depends on a true fluorophore being embedded within the nanoparticle's structure [13]. These markers are basically made up of common glass (with different particular compositions), which then are functionalized according to the target to be imaged. Most commonly, these nanoparticles are of the mesoporous type, meaning that they have many nanometric scale pores traversing the material. This greatly increases the surface to volume ratio for loading the particle with fluorophores and/or chemical compounds. Size can be largely tuned according to the necessities (from a few nanometers up to several microns). These particles are well tolerated by biological systems. These nanoparticles are usually loaded with organic fluorophores. Due to the solid-state environment, they offer many advantages in comparison to liquid solution: radiative decay enhancement (due to rigidity), decreased photobleaching (diminished O_2 diffusion), higher dye concentrations, *etc.* [45]. Luminescent transition metal complexes or lanthanide cations can be loaded into these nanostructures (see section 4.4 below). It has been very recently reported that ultrabright silica nanoparticles loaded with rhodamine 6G can achieve at least 100x brightness levels as compared to semiconductor Qdots [45]. This is explained by the extremely high dye concentration within the nanoparticle's pores (locally close to 100 mM), almost impossible to obtain in bulk solution.

The excitation and emission features depend on the internal fluorophore carried by the nanoparticle. In principle, the whole optical range from the UV to the NIR is open to these structures. A recent publication reports ultrabright emission in the NIR (peak emission at 815 nm) when mesoporous nanometric silica particles were loaded with the LS277 fluorophore (a member of the carbocyanine dye group) [46]. NIR-emitting compounds are gaining widespread use, as they emit in the biological window providing a tool to study deep tissues in thick biological samples (*i.e.* tissues or organisms). By making use of adequate synthetic approaches, like introducing hydrophobic sites within the mesoporous network to prevent dye leaking, these ultrabright long-lasting silica nanoparticles are now available [47].

4.4. Lanthanide-doped Dielectrics

Lanthanides have been incorporated into many solid-state dielectric materials. Given that the f-f transitions characteristic of the 3+ oxidation state of these metals are quite independent of the surrounding medium, the matrix does not usually enhance or diminish the photophysical features of these metals. However, it can be very interesting to have a solid-state marker doped with these metals to study flows, cell dynamics and other biological processes. Lanthanide(III)-doped nanoparticles are frequently employed for time-resolved measurements, given the very long radiative lifetimes of these species [13]. Codoping with two or more lanthanides allows for multicolour emission with the same excitation source. Apart from silica, many other materials have been doped with lanthanides like fluorides, hydroxyapatites, oxides (vanadates, iron oxides, *etc.*), titanium dioxide or zirconia (ZrO_2) to name some.

In addition to their role in time-resolved measurements and their almost complete photophysical "inertness", lanthanide-doped nanoparticles have a definitive outstanding function in the field of light upconversion. Due to their electronic level arrangement and very long excited lifetimes, it is relatively easy to obtain photons of higher energy (shorter wavelength) when pumping lanthanides with an adequate light source. This is an anti-Stokes process, in which light of shorter wavelength than the excitation light is emitted by the sample. As this phenomenon falls completely within the scope of the non-linear optics field, it will be treated in detail in the next chapter (see Chapter 19.4.1). We stress here that solid rigid environments greatly enhance the upconversion process by shutting down non-radiative deactivation pathways. We refer the interested reader to the next chapter for more information on the upconversion process.

4.5. Other Materials

To finish this chapter, we would like to mention a few other solid-state markers that are gaining popularity within the research and clinical communities. This provides an overview of upcoming materials, which can be part of the next generation of tools in the biomedical field. As has been mentioned in regards to lanthanide(III) nanoparticle doping, many materials, not only silica, have been employed to synthesise luminescent markers. Examples include many different oxides and ceramics (TiO_2, ZrO_2, *etc.*) and other solid compounds (*e.g.* fluorides, sulphides, carbides) [13].

Recently, ZnO has been proposed as a luminescent material to produce biological markers [48]. ZnO is a wide gap semiconductor. Its absorption spectrum is located in the UV. But, as it happened with nanodiamonds, defects and dislocations can provide traps for excited electrons. Therefore, there is photoluminescence in the visible for this material. The mechanisms have not been definitively established yet. ZnO nanoparticles find application in FLIM microscopy, due to their long radiative lifetimes, which can be longer than 1000 ns. They are also a material of election for two-photon excitation (see Chapter 19.2.2). Apart from their imaging possibilities, antibacterial and drug delivery properties have been described for them.

Nanoparticles combining magnetic and luminescent properties have been described [49]. They have dual composition: a magnetic manganite-based core and a silica dye-doped shell. This type of markers is very popular at present, as part of the so-called multimodal approach. This multimodal approach tries to merge several properties in one particle. It can be optical imaging and drug delivery, optical imaging combined with X-ray opaque materials or, as is the case with the manganite particles, combining luminescent imaging with magnetic detection/manipulation. Indeed, the general trend in the biomedical field is to implement and use these combinatorial tracers to better suit clinical diagnosis/therapy (theranostics).

Also within this multimodal paradigm, we can find nanoscintillators [50]. These nanoparticles, usually made of oxides like garnets, have heavy metals (commonly lanthanides) incorporated into the lattice. When exposed to X-rays, the emitting metal absorbs part of this ionizing radiation, and reemits a fraction of it as light in the optical (mainly visible) range. Because of the heavy element presence, these materials are X-ray opaque. Therefore, they should be adequate to obtain both radiographies and fluorescent images. Nanoscintillators are actively proposed as therapeutic agents. Their radioluminescence can be employed to excite photosensitizing compounds, inducing additional deleterious effects on diseased tissues exposed to X-rays in the presence of these materials [51].

CONFLICT OF INTEREST

The author (editor) declares no conflict of interest, financial or otherwise.

ACKNOWLEDGEMENTS

This work was supported by the Aarhus Institute of Advanced Studies (AIAS).

REFERENCES

[1] Fox M. Optical properties of solids. Oxford: Oxford University Press 2001.

[2] Band YB. Light and Matter Electromagnetism, Optics, Spectroscopy and Lasers. Chichester: John Wiley & Sons Ltd 2006.

[3] Kobayashi H, Longmire MR, Ogawa M, Choyke PL. Rational chemical design of the next generation of molecular imaging probes based on physics and biology: mixing modalities, colors and signals. Chem Soc Rev 2011; 40(9): 4626-48. [http://dx.doi.org/10.1039/c1cs15077d] [PMID: 21607237]

[4] Demchenko AP. Nanoparticles and nanocomposites for fluorescence sensing and imaging. Methods Appl Fluoresc 2013; 1: 022001. [http://dx.doi.org/10.1088/2050-6120/1/2/022001]

[5] Yao J, Yang M, Duan Y. Chemistry, biology, and medicine of fluorescent nanomaterials and related systems: new insights into biosensing, bioimaging, genomics, diagnostics, and therapy. Chem Rev 2014; 114(12): 6130-78. [http://dx.doi.org/10.1021/cr200359p] [PMID: 24779710]

[6] Lakowicz JR. Principles of fluorescence spectroscopy. 3rd ed., New York: Springer Science + Business Media LLC 2010.

[7] Wegner KD, Hildebrandt N. Quantum dots: bright and versatile *in vitro* and *in vivo* fluorescence imaging biosensors. Chem Soc Rev 2015; 44(14): 4792-834. [http://dx.doi.org/10.1039/C4CS00532E] [PMID: 25777768]

[8] Wang C, Gao X, Su X. *In vitro* and *in vivo* imaging with quantum dots. Anal Bioanal Chem 2010; 397(4): 1397-415. [http://dx.doi.org/10.1007/s00216-010-3481-6] [PMID: 20174786]

[9] Halas NJ, Lal S, Chang WS, Link S, Nordlander P. Plasmons in strongly coupled metallic nanostructures. Chem Rev 2011; 111(6): 3913-61. [http://dx.doi.org/10.1021/cr200061k] [PMID: 21542636]

[10] Zhang J, Fu Y, Lakowicz JR. Fluorescent metal nanoshell: Lifetime-tunable molecular probes in fluorescent cell imaging. J Phys Chem C Nanomater Interfaces 2011; 115(15): 7255-60. [http://dx.doi.org/10.1021/jp111475y] [PMID: 21743823]

[11] Zhou C, Yang S, Liu J, Yu M, Zheng J. Luminescent gold nanoparticles: a new class of nanoprobes

for biomedical imaging. Exp Biol Med (Maywood) 2013; 238(11): 1199-209. [http://dx.doi.org/10.1177/1535370213505825] [PMID: 24108449]

[12] Mathew A, Pradeep T. Noble metal clusters: Applications in energy, environment, and biology. Part Part Syst Charact 2014; 31: 1017-53. [http://dx.doi.org/10.1002/ppsc.201400033]

[13] Wolfbeis OS. An overview of nanoparticles commonly used in fluorescent bioimaging. Chem Soc Rev 2015; 44(14): 4743-68. [http://dx.doi.org/10.1039/C4CS00392F] [PMID: 25620543]

[14] Zheng J, Nicovich PR, Dickson RM. Highly fluorescent noble-metal quantum dots. Annu Rev Phys Chem 2007; 58: 409-31. [http://dx.doi.org/10.1146/annurev.physchem.58.032806.104546] [PMID: 17105412]

[15] Zheng J, Zhou C, Yu M, Liu J. Different sized luminescent gold nanoparticles. Nanoscale 2012; 4(14): 4073-83. [http://dx.doi.org/10.1039/c2nr31192e] [PMID: 22706895]

[16] Chen LY, Wang CW, Yuan Z, Chang HT. Fluorescent gold nanoclusters: recent advances in sensing and imaging. Anal Chem 2015; 87(1): 216-29. [http://dx.doi.org/10.1021/ac503636j] [PMID: 25275676]

[17] Qu X, Li Y, Li L, *et al.* Fluorescent gold nanoclusters: Synthesis and recent biological application. J Nanomater 2015; 784097.

[18] Algar WR, Susumu K, Delehanty JB, Medintz IL. Semiconductor quantum dots in bioanalysis: crossing the valley of death. Anal Chem 2011; 83(23): 8826-37. [http://dx.doi.org/10.1021/ac201331r] [PMID: 21928771]

[19] Bilan R, Fleury F, Nabiev I, Sukhanova A. Quantum dot surface chemistry and functionalization for cell targeting and imaging. Bioconjug Chem 2015; 26(4): 609-24. [http://dx.doi.org/10.1021/acs.bioconjchem.5b00069] [PMID: 25710410]

[20] Zhang Y, Bai Y, Jia J, *et al.* Perturbation of physiological systems by nanoparticles. Chem Soc Rev 2014; 43(10): 3762-809. [http://dx.doi.org/10.1039/C3CS60338E] [PMID: 24647382]

[21] Qi L, Gao X. Quantum dot-amphipol nanocomplex for intracellular delivery and real-time imaging of siRNA. ACS Nano 2008; 2(7): 1403-10. [http://dx.doi.org/10.1021/nn800280r] [PMID: 19206308]

[22] Orte A, Álvarez-Pez JM, Ruedas-Rama MJ. Fluorescence lifetime imaging microscopy for the detection of intracellular pH with quantum dot nanosensors. ACS Nano 2013; 7(7): 6387-95. [http://dx.doi.org/10.1021/nn402581q] [PMID: 23808971]

[23] Adegoke O, Forbes PB. Challenges and advances in quantum dot fluorescent probes to detect reactive oxygen and nitrogen species: a review. Anal Chim Acta 2015; 862: 1-13. [http://dx.doi.org/10.1016/j.aca.2014.08.036] [PMID: 25682423]

[24] Wang Y, Cai E, Rosenkranz T, *et al.* Small quantum dots conjugated to nanobodies as immunofluorescence probes for nanometric microscopy. Bioconjug Chem 2014; 25(12): 2205-11. [http://dx.doi.org/10.1021/bc5004179] [PMID: 25397889]

[25] Pradhan N, Sarma DD. Advances in light-emitting doped semiconductor nanocrystals. J Phys Chem Lett 2011; 2: 2818-26. [http://dx.doi.org/10.1021/jz201132s]

[26] Georgakilas V, Perman JA, Tucek J, Zboril R. Broad family of carbon nanoallotropes: classification, chemistry, and applications of fullerenes, carbon dots, nanotubes, graphene, nanodiamonds, and combined superstructures. Chem Rev 2015; 115(11): 4744-822. [http://dx.doi.org/10.1021/cr500304f] [PMID: 26012488]

[27] Demchenko AP, Dekaliuk MO. Novel fluorescent carbonic nanomaterials for sensing and imaging. Methods Appl Fluoresc 2013; 1: 042001.
[http://dx.doi.org/10.1088/2050-6120/1/4/042001]

[28] Wang Z, Zeng H, Sun L. Graphene quantum dots: Versatile photoluminescence for energy, biomedical, and environmental applications. J Mater Chem C Mater Opt Electron Devices 2015; 3: 1157-65.
[http://dx.doi.org/10.1039/C4TC02536A]

[29] Zheng XT, Ananthanarayanan A, Luo KQ, Chen P. Glowing graphene quantum dots and carbon dots: properties, syntheses, and biological applications. Small 2015; 11(14): 1620-36.
[http://dx.doi.org/10.1002/smll.201402648] [PMID: 25521301]

[30] Wang J, Qiu J. A review of carbon dots in biological applications. J Mater Sci 2016; 51: 4728-38.
[http://dx.doi.org/10.1007/s10853-016-9797-7]

[31] Lim SY, Shen W, Gao Z. Carbon quantum dots and their applications. Chem Soc Rev 2015; 44(1): 362-81.
[http://dx.doi.org/10.1039/C4CS00269E] [PMID: 25316556]

[32] Goh EJ, Kim KS, Kim YR, *et al.* Bioimaging of hyaluronic acid derivatives using nanosized carbon dots. Biomacromolecules 2012; 13(8): 2554-61.
[http://dx.doi.org/10.1021/bm300796q] [PMID: 22804331]

[33] Zheng XT, Than A, Ananthanaraya A, Kim DH, Chen P. Graphene quantum dots as universal fluorophores and their use in revealing regulated trafficking of insulin receptors in adipocytes. ACS Nano 2013; 7(7): 6278-86.
[http://dx.doi.org/10.1021/nn4023137] [PMID: 23799995]

[34] Chinnathambi S, Chen S, Ganesan S, Hanagata N. Silicon quantum dots for biological applications. Adv Healthc Mater 2014; 3(1): 10-29.
[http://dx.doi.org/10.1002/adhm.201300157] [PMID: 23949967]

[35] Montalti M, Cantelli A, Battistelli G. Nanodiamonds and silicon quantum dots: ultrastable and biocompatible luminescent nanoprobes for long-term bioimaging. Chem Soc Rev 2015; 44(14): 4853-921.
[http://dx.doi.org/10.1039/C4CS00486H] [PMID: 26051500]

[36] He Y, Zhong Y, Peng F, *et al.* One-pot microwave synthesis of water-dispersible, ultraphoto- and pH-stable, and highly fluorescent silicon quantum dots. J Am Chem Soc 2011; 133(36): 14192-5.
[http://dx.doi.org/10.1021/ja2048804] [PMID: 21848339]

[37] Xu H, Li Q, Wang L, *et al.* Nanoscale optical probes for cellular imaging. Chem Soc Rev 2014; 43(8): 2650-61.
[http://dx.doi.org/10.1039/c3cs60309a] [PMID: 24394966]

[38] Massey M, Wu M, Conroy EM, Algar WR. Mind your P's and Q's: the coming of age of semiconducting polymer dots and semiconductor quantum dots in biological applications. Curr Opin Biotechnol 2015; 34: 30-40.
[http://dx.doi.org/10.1016/j.copbio.2014.11.006] [PMID: 25481436]

[39] Lin CA, Yang TY, Lee CH, *et al.* Synthesis, characterization, and bioconjugation of fluorescent gold nanoclusters toward biological labeling applications. ACS Nano 2009; 3(2): 395-401.
[http://dx.doi.org/10.1021/nn800632j] [PMID: 19236077]

[40] Hu L, Deng L, Alsaiari S, Zhang D, Khashab NM. "Light-on" sensing of antioxidants using gold nanoclusters. Anal Chem 2014; 86(10): 4989-94.
[http://dx.doi.org/10.1021/ac500528m] [PMID: 24785707]

[41] Shang L, Stockmar F, Azadfar N, Nienhaus GU. Intracellular thermometry by using fluorescent gold nanoclusters. Angew Chem Int Ed Engl 2013; 52(42): 11154-7.
[http://dx.doi.org/10.1002/anie.201306366] [PMID: 24039076]

[42] Choi S, Dickson RM, Yu J. Developing luminescent silver nanodots for biological applications. Chem Soc Rev 2012; 41(5): 1867-91.
[http://dx.doi.org/10.1039/C1CS15226B] [PMID: 22076614]

[43] Alhaddad A, Durieu C, Dantelle G, *et al.* Influence of the internalization pathway on the efficacy of siRNA delivery by cationic fluorescent nanodiamonds in the Ewing sarcoma cell model. PLoS One 2012; 7(12): e52207.
[http://dx.doi.org/10.1371/journal.pone.0052207] [PMID: 23284935]

[44] Mohan N, Chen CS, Hsieh HH, Wu YC, Chang HC. *In vivo* imaging and toxicity assessments of fluorescent nanodiamonds in Caenorhabditis elegans. Nano Lett 2010; 10(9): 3692-9.
[http://dx.doi.org/10.1021/nl1021909] [PMID: 20677785]

[45] Kalaparthi V, Palantavida S, Sokolov I. The nature of ultrabrightness of nanoporous fluorescent particles with physically encapsulated fluorescent dyes. J Mater Chem C Mater Opt Electron Devices 2016; 4: 2197-210.
[http://dx.doi.org/10.1039/C5TC04221F]

[46] Palantavida S, Tang R, Sudlow GP, *et al.* Ultrabright NIR fluorescent mesoporous silica nanoparticles. J Mater Chem B Mater Biol Med 2014; 2: 3107-14.
[http://dx.doi.org/10.1039/C4TB00287C]

[47] Cho EB, Volkov DO, Sokolov I. Ultrabright fluorescent mesoporous silica nanoparticles. Small 2010; 6(20): 2314-9.
[http://dx.doi.org/10.1002/smll.201001337] [PMID: 20859948]

[48] Zhang ZY, Xiong HM. Photoluminescent ZnO nanoparticles and their biological applications. Materials (Basel) 2015; 8: 3101-27.
[http://dx.doi.org/10.3390/ma8063101]

[49] Kačenka M, Kaman O, Kikerlová S, *et al.* Fluorescent magnetic nanoparticles for cell labeling: flux synthesis of manganite particles and novel functionalization of silica shell. J Colloid Interface Sci 2015; 447: 97-106.
[http://dx.doi.org/10.1016/j.jcis.2015.01.071] [PMID: 25702866]

[50] Jung JY, Hirata GA, Gundiah G, *et al.* Identification and development of nanoscintillators for biotechnology applications. J Lumin 2014; 154: 569-77.
[http://dx.doi.org/10.1016/j.jlumin.2014.05.040]

[51] Liu Y, Chen W, Wang S, *et al.* Investigation of water-soluble x-ray luminescence nanoparticles for photodynamic activation. Appl Phys Lett 2008; 92: 043901.
[http://dx.doi.org/10.1063/1.2835701]

CHAPTER 19

Non-linear Optics

Abstract: This chapter deals with some of the most fascinating topics in fluorescence microscopy and optics, in general: non-linear optics. All the optical phenomena explained in the previous chapters, with some particular exceptions, have dealt with linear optics; this means that a single photon is involved in the optical process during and after being absorbed. However, it is possible to have a multiphoton optical response, one that depends on more than one photon interacting with the system "at the same time". This leads to a whole new range of optical phenomena; many of them unobservable at the relatively low light intensities with characteristic of linear optics. Some of the non-linear optical phenomena provide very interesting and useful applications in fluorescence microscopy. In this chapter, we will introduce and explain the main non-linear optical processes employed in fluorescence microscopy, enumerate their advantages and drawbacks, and highlight their importance in the life sciences.

Keywords: Biphotonic microscopy, Coherent anti-Stokes Raman scattering (CARS), Excited state absorption, Harmonic microscopy, Lanthanide upconversion, Multiphoton microscopy, Non-linear optics, Photon upconversion (UC), Raman microscopy, Second harmonic generation, Stimulated Raman scattering (SRS), Third harmonic generation, Triplet-triplet annihilation upconversion, Triplet state, Virtual state.

1. NON-LINEAR OPTICS

Optical effects are commonly linear. If we consider about a fluorescence signal for example, this means that increasing the excitation light to double will accordingly bring a double increase in the fluorescent emission intensity. At the molecular level this is straightforward to understand. Assuming a fluorophore with a fluorescence quantum yield close to 1 for every excitation photon absorbed, the fluorophore will emit a single corresponding fluorescent photon. Increasing the number of exciting photons (*i.e.* the light energy) will increase the number of re-emitted photons in the same ratio. This is the linear regime characterized by low light intensities: a single photon is involved in the optical process after being absorbed. Put another way, the outcome of light absorption (fluorescence, phosphorescence, photochemistry, *etc.*) increases proportionally (*i.e.* linearly) with light intensity (*i.e.* photon flux or number of photons). Yet a

Juan Carlos Stockert & Alfonso Blázquez-Castro

third way to define this range is that the optical response of the system depends linearly on the magnitude of the electric field that supports the electromagnetic wave interacting with the system. Linear optics is the range of our common everyday perception of light-driven processes (*e.g.* vision, photosynthesis, *etc.*).

If the light intensity is increased high enough a whole new range of optical phenomena enters into the scene [1]. In truth, it can be said, that a vast realm of optical processes, much broader in scope than linear optics, becomes accessible when high enough light intensities interact with matter. The key concept here is light intensity: when enough photons reach a target in a certain amount of time the probability that the system absorbs or interacts with two or more photons at the same time becomes non-negligible. Thus, the optical response of the system to the electric field of the incoming electromagnetic wave becomes non-linear because the equations describing this response have power terms (squared, cubed, *etc.*) that are substantial in comparison to the linear term [2, 3]. This simplified equation clarifies this concept:

$$OR = (c_1 \cdot E^1) + (c_2 \cdot E^2) + (c_3 \cdot E^3) + \ldots\ldots + (c_n \cdot E^n) \tag{1}$$

The term on the left describes the optical response (OR, *e.g.* absorption or fluorescence) of the system. The terms on the right describe the electromagnetic wave electric field magnitude (E) that is directly related to the light intensity. Each electric field term on the right part of the equation (E^1, E^2, *etc.*) pairs with its corresponding coefficient (c_1, c_2, *etc.*). The c_1 coefficient is the linear coefficient (note that it multiplies E because E^1 = E). For low light intensities, *e.g.* low E values, the linear approximation is obtained:

$$OR = c_1 \cdot E \tag{2}$$

We can neglect the non-linear terms for reasons that are explained below. Equation **(2)**, for example, is derived based on the Lambert-Beer law that defines a linear attenuation of light intensity as it traverses an absorbing medium [4]:

$$Abs = \sigma l \cdot I, \tag{3}$$

where the light absorbed (Abs) depends on: the absorption coefficient or cross section (σ, specific for each material or mixture), the absorbing medium thickness (l), and the light intensity (I). This linear relationship remains true as long as the light intensity (I) is low enough. All the rest of the coefficients in equation **(1)** are the non-linear coefficients (they multiply E to powers larger than 1, that is E^2, E^3, *etc.*). More often than not, these non-linear coefficients are extremely small. In the optical range, the coefficient c_1 is in the order of magnitude larger than c_2, coefficient c_2 is in the order of magnitude larger than c_3, and so on. The precise value of the different coefficients depends on the particular material being exposed to the light and the wavelength of this light.

It can readily be seen that at low light intensities, when E is small, all coefficients beyond c_1 make any non-linear contribution negligible, as they are quite small in magnitude. Therefore, the interaction can be described by the classical linear theory of optics. However, when the light intensity increases, so too does the value of E. There will be a threshold intensity range when the E^2 term starts to make a significant contribution to the optical response of the system, even when multiplied by a very small c_2. This is the onset of non-linear optics. The light intensity necessary to observe such non-linear optical effects is highly variable and depends on the material, the wavelength, polarization, presence/absence of external electric/magnetic fields, *etc*. As a rule of thumb the light intensities needed to start observing these phenomena are around 10^4-10^5 W/cm^2 — at least one million times more intense than midday sunlight (0.1 W/cm^2). For a more rigorous and in depth treatment of non-linear optics we recommend the interested reader to the following literature [1 - 4].

It comes as no surprise that non-linear optics could practically be studied only after the advent of laser systems. Few theoretical studies of non-linear optics, as well as very few experimental results were published before 1960 (however, the work of Maria Göppert-Mayer in the early 1930s [5] deserves a special remark as the first publication pointing to the possibility of two-photon absorption). But with the arrival of lasers in the 1960s, the very high irradiances necessary to study non-linear optics were finally available for the research community to start the experimental study of these phenomena. As a curious aside, the light intensities reached by current laser systems can go as high as 10^{21} W/cm^2. Such intensities permit the direct transformation of light into an electron-positron plasma or the acceleration of subatomic particles (protons, electrons, *etc*.) to energies in the GeV range (to mention just a few applications).

There are three main features that make lasers unique light sources to study non-linear optics: 1) high light intensity, 2) coherence, and 3) large focusing. Lasers are notorious for their ability to store electromagnetic energy in the form of electronic excitation, then suddenly releasing it in a very short period of time as a light pulse. This can be thought about as a "light spring" in which energy builds up within the laser material and then, under the right conditions, is suddenly released in a very short time. This results in very high instantaneous power (up to petawatts —10^{15} W— for some lasers), although the energies involved are not necessarily very high (*e.g.* 1 joule of energy emitted over the course of few femtoseconds —10^{-15} s — has an instantaneous power of 10^{14}-10^{15} W). A schematic comparison of the outputs of a continuous wave (cw) and pulsed lasers is shown in Fig. (**19.1A**). The cw laser emits a stable stream of photons. This does not mean that cw lasers necessarily have low power; some industrial and military infrared cw lasers can reach powers close to the megawatt. However, most cw lasers employed in fluorescence microscopy rarely emit above the watt power level. A pulsed laser can be thought of as a photon bunch or cloud, where most of the photons emitted in a pulse are "piled up" close together both in time and space (see Fig. **19.1A**). As a large population of photons arrive at the sample target at the same time, the probability that more than one photon is absorbed in a very short time period is high. Thus, this pulsed intensity makes lasers perfect tools to study non-linear optics [6].

The high coherence is the result of the stimulated light emission process in lasers. Without going into detail, it can be said in a general way, that all photons involved in the laser emission are equal or almost equal. They share the same energy per photon (and thus they are all of the same wavelength), the same polarization, and the same propagation direction. This is not strictly correct in all cases, but it will serve for our purposes (for a more detailed and comprehensive explanation of lasers and their working mechanisms, see [1, 6]). The high photon coherence makes observation of non-linear phenomena easier. Finally, the possibility to focus the laser beam to very small dimensions (below a micron for visible wavelengths) is the basis to make a very large proportion of the light to traverse a tiny area/volume, therefore subjecting a relatively small population of atoms/molecules to a very large photon flux (photons/surface), which again increases the non-linear interactions. The large focusing ratio of lasers is a result of their high coherence. A representation of this is shown in Fig. (**19.1B**). When the laser output is focused by the lens the photons follow a convergent trajectory towards the focal point (the spot of maximum concentration). As they approach the focal point the same number of photons has to traverse smaller and smaller surfaces and they become more "concentrated" as a consequence. Therefore, the photon flux increases exponentially the closer it gets to the focal point (represented schematically in the figure graphs). At the focal point the photon flux is largest as the surface traversed is narrowest. For many applications, the sample is localized at or close to the focal point/plane to take advantage of this large photon flux. After the focal point the photons diverge and the photon flux rapidly diminishes. It is important to note that for a given set of conditions the laser energy is the same no matter at which point in the light cone we measure (here we assume that no absorption/scattering is taking place). The laser energy and power is conserved. However, the photon flux changes as a result of the smaller surface the light traverses on its way to the focal point. Following this argument non-linear effects that can take place at or close to the focal point do not occur far from it, even when the laser energy is the same [7]. Colloquially we can say that far from the focal point the photons are "too diluted" to produce non-linear effects.

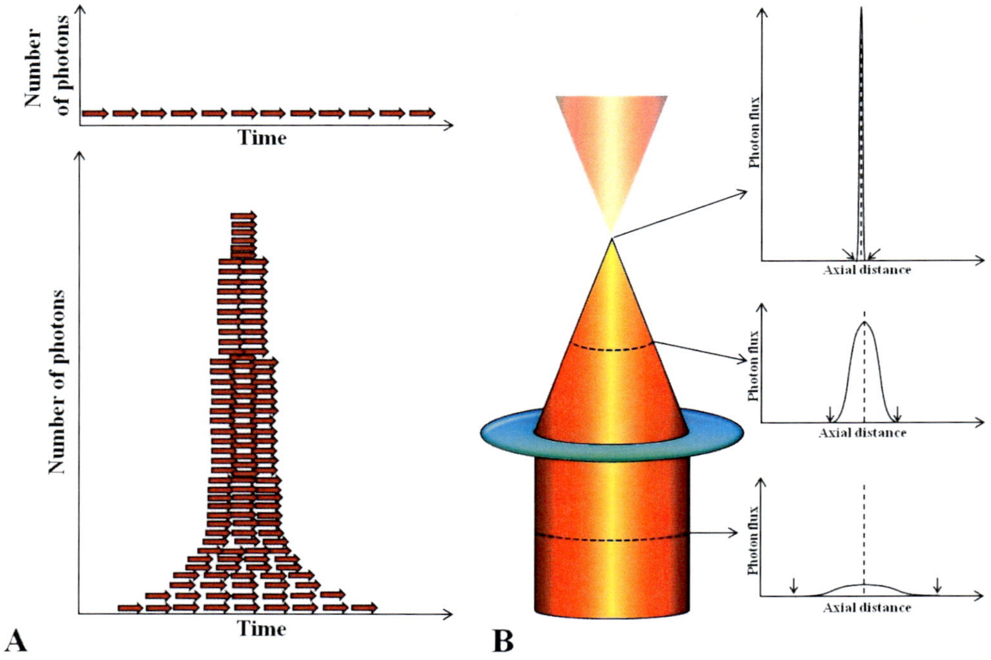

Fig. (19.1). A: Comparison of continuous wave (cw) *vs*. pulsed laser outputs. A cw laser (top panel) emits a continuous stream of photons (represented as arrows). The number of photons, and thus the energy, does not vary. The average power (energy per unit time) is usually very low as compared to pulsed lasers. A laser pulse (bottom panel) is made up of many photons grouped together within a short time. Usually the photon distribution in time is Gaussian-like (as in the cartoon) although other profiles are possible. The key point is that many photons can interact with a target at the same time, meaning that the instantaneous power (energy divided by pulse duration) can be extremely high. **B:** Effect of laser focusing. When laser light interacts with a focusing element the light flux increases due to a reduction in the surface the light travels through. Before the lens (lower section of the beam) there is a certain amount of light crossing through the surface derived from the laser beam diameter (small arrows in the graph). Usually the light flux is not homogenous but light is more concentrated at the beam axis (see photon distribution in the bottom graph). After the focusing element but before the focal point (middle section) the radial dimensions are shrinking and the photon flux increases accordingly (see middle graph). Note that the number of photons has not changed (assuming the lens did not absorb/scatter any light) but the surface is smaller than before the lens. At the focal point (top section) the photon flux is highest because the surface is smallest, almost to a point. It is at or near the focal point where non-linear effects are most likely.

We are not going to elaborate further on the fundamentals of non-linear optics as the topic is vast and can get highly complex for some issues [3]. Also, it is not appropriate to continue much further without relying too strongly on mathematical arguments, which is against the spirit of this handbook (that is, to present optical microscopy in the simplest and most intuitive way). In what follows we will introduce the physical fundamentals whenever it is necessary in a case-by-case basis for each phenomenon or technique. Our initial approach is to present and explain the non-linear optical effects from a quantum (photon) perspective. In our experience this is a much more intuitive approximation, which depends less on a deep mathematical understanding of the processes. This photonic approximation is perfectly valid, although some effects are better explained from a wave perspective. Whenever necessary we will use the wave approach.

2. MULTIPHOTON MICROSCOPY

The most widespread non-linear optical effect put to use in fluorescence microscopy is multiphotonic excitation. Indeed multiphotonic imaging and photonics has revolutionized not only fluorescence microscopy but the life sciences in general [8 - 11]. In this section we will present the fundamentals of this technique and then its technical implementations, including biphotonic microscopy and higher-order photon microscopy.

2.1. Fundamentals

From a practical point of view, the main advantage of non-linear microscopy is that the researcher decides where a given sample is transparent and where it begins to absorb light, and, in fluorescence microscopy, when it starts to emit light [10]. This seemingly magical effect derives from the already mentioned threshold for non-linear processes (see section 1 above). Because of the tight light beam focusing, the surface across which light has to cross rapidly decreases just below and above the focal point. Two- or multiphoton absorption does not appreciably take place until a relatively intense photon flux is achieved, and so the sample does not absorb light away from the focal point. Only in close proximity to this location, just below or above it, is the light intensity high enough for absorption to happen. Once absorption takes place, fluorescence, or generally speaking, luminescence, can follow depending on the particular features of the involved non-linear chromophore.

This is exemplified in Fig. (**19.2**) where typical conditions for linear and non-linear microscopy are shown in a fluorescein solution contained in a cuvette. With the low light intensities typical of one-photon imaging, the full length of the beam path results in fluorescence from the molecules (see Fig. **19.2A**). The excitation volume defined by the light cones is easily spotted due to the bright green fluorescence emission. Fluorescence intensity is largest at and around the focal point (see figure inset). An example of the extremely confined excitation spot obtained through non-linear excitation is shown in Fig. (**19.2B**). When the same fluorescein solution is illuminated with orange-red light of high enough intensity, a very different behaviour is observed. The light cones have completely disappeared and only a tiny light point (show by the arrow) can be noticed (magnified view in the inset). Only within a ~100-150 µm thickness around the focal point is the light intensity high enough to provide a biphotonic excitation of fluorescein molecules. The rest of the sample is transparent to the light. Employing higher magnification objectives to further narrow down the focal point allow near diffraction-limited submicrometric imaging. It becomes obvious with this example that non-linear excitation inherently affords very high spatial resolution without relying on additional optical elements.

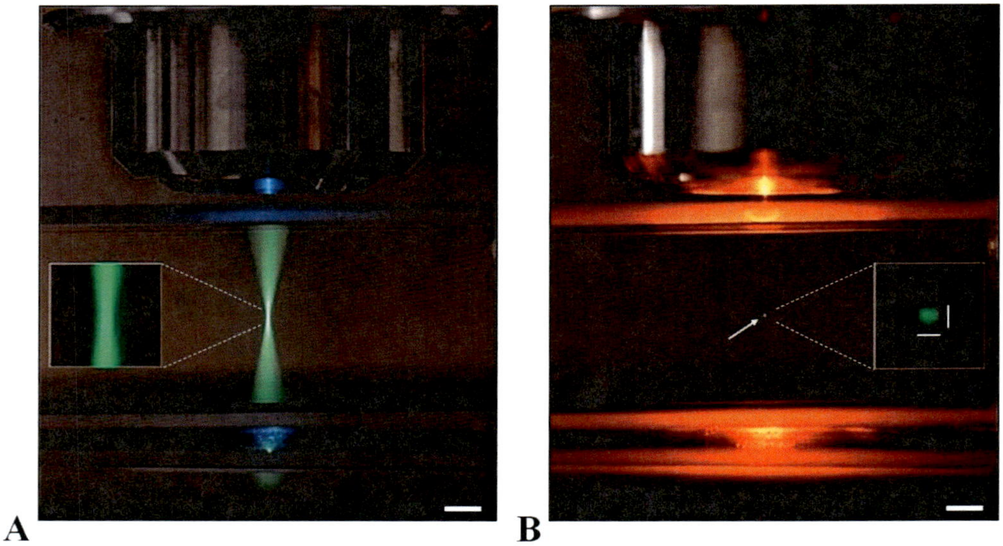

Fig. (19.2). Comparison of one-photon *vs*. two-photon fluorophore excitation. **A:** Linear excitation of a diluted fluorescein solution by 488 nm blue light focused through an objective inside a cuvette. Fluorescein green emission is observed along the full length of the beam path. Notice the converging-diverging light cones defined by the incoming exciting light. Inset: magnified image of the focal point. **B:** Two-photon excitation of the same fluorescein solution with 590 nm laser light. Notice that the light cones have completely disappeared and only a tiny emitting spot (arrow) can be observed. This spot is magnified in the inset where its dimensions can be appreciated (displayed bars are 150 µm in length). Scale bars: 2 mm. Images courtesy of Prof. Steven E. Ruzin, University of California at Berkley.

How much light is necessary for a given system to start absorbing in the non-linear regime? To answer this question, we have to introduce the GM unit named after Maria Göppert-Mayer. The GM has units of 10^{-50} cm^4 s $photon^{-1}$ for two-photon (biphotonic) absorption [7, 12]. The GM units at a certain wavelength must be known to obtain the non-linear absorption cross section of a chromophore. The light flux intensity (photons s^{-1} cm^{-2}) is also needed. The absorption coefficient (in units of cm^2) is obtained by multiplying the light flux intensity by the GM units. Absorption coefficients of the order of 10^{-20} cm^2 or higher are adequate to observe biphotonic processes. For example, let us assume we are working with a fluorophore with 1 GM at 800 nm. Exciting the fluorophore with an 800 nm laser pulse and light flux intensity of 10^{25} photons s^{-1} cm^{-2} (around 10^6 W cm^{-2}, quite mild conditions for biphotonic fluorescence microscopy) results in an absorption cross section per fluorophore of 10^{-25} cm^2. This is a low absorption value. If the fluorophore has a high fluorescence quantum yield, then it could be enough to image it. There are two ways to increase the biphotonic absorption: to use fluorophores with a higher non-linear coefficient (higher GM units), increase the light flux intensity, or both. If the fluorophore of choice now has 1000 GMs at 800 nm and the laser intensity is increased to 10^{28} photons s^{-1} cm^{-2} (~1 GW cm^{-2}) the absorption coefficient goes to 10^{-19} cm^2 which is more than enough to obtain two-photon excited fluorescence. Because increasing the laser intensity also increases the chances of exciting other undesirable chromophores or non-linear effects in the sample, the typical strategy is to employ fluorophores with as high as possible biphotonic cross sections. The GM is usually associated with two-photon processes but can be defined for other non-linear effects. For example, for three-photon processes the "cubic" GM unit has units of cm^6 s $photon^{-1}$ with 10^{-75} or lower coefficients [13].

Next we will explain the physical processes that happen during non-linear absorption and excitation. For this we will use again the Jablonski diagram (see Chapter 2.4.9). This diagram relates energy to atomic and molecular levels and it is a simple way to present energy transitions between these levels. In the Jablonski diagram (see Fig. **19.3**) we have first represented one-photon absorption, which is the kind of transition we have been dealing with in all the previous chapters. Second, there is a two-photon transition. It can be seen that there is only enough energy to bridge the gap between the ground state and the first excited state when the two photons are absorbed at the same time [7, 13].

Each photon by itself does not provide enough energy for the excitation to take place, thus the sample is transparent for one-photon transitions at the corresponding wavelength. Once the chromophore has been excited to a higher electronic level, the deactivation channels at work are the same as those operating when a single photon is absorbed. Therefore, the chromophore does not keep a "record" of the process that led it to reach the excited state. This is advantageous for fluorescence microscopy as the usual fluorescence spectrum is observed after non-linear excitation.

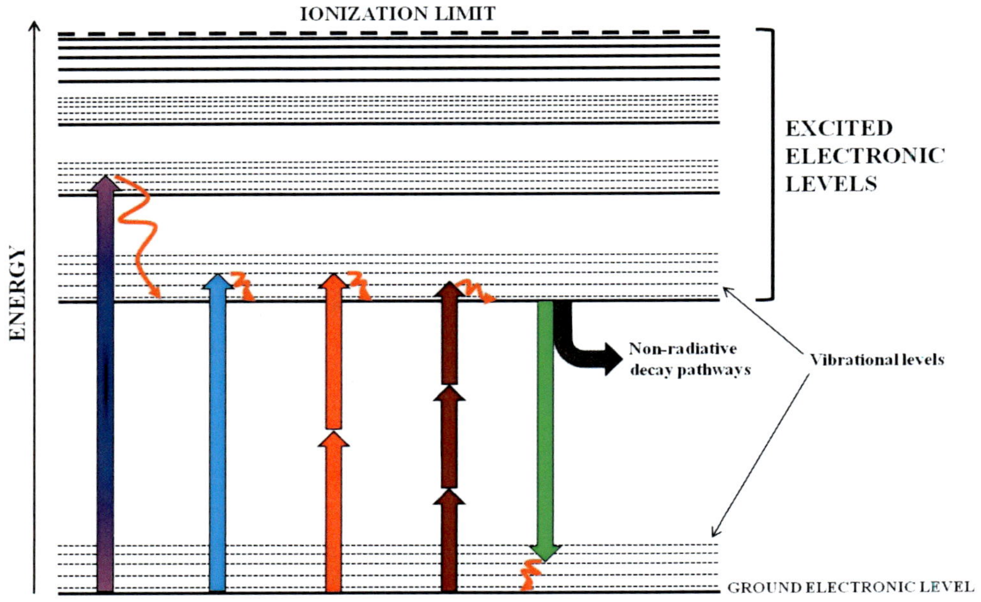

Fig. (19.3). Comparison between one-photon and multiple photons absorption processes. A Jablonski diagram is employed to easily visualize the transitions involved. A single UV photon (left) pumps the molecule from the ground state to the second excited state. Non-radiative decay (represented here as wiggly red arrows) accounts for deexcitation to the first excited state. A single blue photon excites the molecule to the first excited electronic level with some minor vibrational deexcitation. These are one-photon processes. Two near IR photons (red arrows) reaching the molecule within a very short time can be absorbed simultaneously and excite the first electronic level (two-photon or biphotonic absorption). Three IR photons can also excite a transition to the first electronic level under the right conditions. Once the molecule (or other fluorophore) has been excited the actual process that led to the excitation is no longer of importance, at least for the time scales we are dealing with in fluorescence microscopy. Consequently, our example molecule will emit a green photon no matter which one of the four possible excitation schemes has been employed. Other non-radiative deactivation channels can also occur apart from light emission, depending on the corresponding quantum yields for each process. Note that in all the cases depicted we are dealing with resonant photon absorption as real electronic levels are involved in the process.

It is important to make a remark at this point: the kind of non-linear photon absorption processes we are explaining here belong to the class of resonant non-linear optical transitions [3]. In these resonant transitions, the initial and final electronic states are real states, in the sense that electrons show a stable or metastable (long-lived) lifetime when in those states. On the other hand, there can be virtual states with a very short lifetime that do not show a stable/metastable electronic configuration. In the two-photon absorption process depicted in Fig. (**19.3**) the excited electron reaches a virtual state (dashed line) for a very short time (~femtosecond or less) after absorbing the first photon. Only after absorbing the second photon, which must reach the electron before it "decays" from this virtual level, does the electron achieve a metastable level (the first excited state in this case). Many non-linear optical applications make use of these virtual levels although they usually require much higher light irradiances than those commonly used in multiphotonic fluorescence microscopy. Nevertheless, it is convenient to make this distinction between real and virtual levels (virtual levels are crucial to understand non-linear harmonic generation and Raman microscopy for example, see sections 3 and 5 below). The use of resonant transitions is convenient as it leads to weaker light intensities necessary to produce non-linear effects.

We want to mention that the selection rules for multiphotonic transitions can be different from those involving one photon [1, 3]. As it was indicated in Chapter 2, atomic or molecular electronic levels have different quantum numbers and symmetry features. Photons carry momentum in addition to energy, and so they are able to change these properties to a certain extent. As a consequence, even when a photon provides sufficient energy to excite a particular excited level it can be that the transition is forbidden due to selection rules. The same can be applied to multi-photon transitions except that now the momentum from two or more photons has to be taken into account. Because of this it is possible to excite certain levels through a multi-photonic process that are unattainable with only one photon. The opposite is also true: some levels can be excited by one photon and not by two photons. For most common applications in non-linear microscopy, this is of no further concern, although we wanted to mention it for the sake of clarity.

Now we will introduce the most common non-linear optical technique in use for fluorescence microscopy: biphotonic microscopy. We will then briefly present higher-order microscopies.

2.2. Biphotonic Microscopy

Two-photon absorption microscopy, or biphotonic microscopy, is probably the most common non-linear methodology in use within the area of fluorescence microscopy. Aside from the general advantages of non-linear processes already mentioned in previous sections, biphotonic imaging provides two main assets: lowest intensity threshold and lowest transparency for one-photon absorption in biological samples. As it follows from the arguments presented in the first section (see section 1 above) the most easily observable non-linear effects are those in which two photons are involved. Therefore, biphotonic microscopy is the easiest to attain. As a direct consequence of this, biphotonic imaging shows the lowest light intensity threshold [7, 13, 14]. The light intensities commonly found to excite biphotonic emission are between 10^6 and 10^9 W cm^{-2}. Lower irradiances can also work (this depends critically on the fluorophore two-photon absorption coefficient as was described when introducing the GM units above) with magnitudes between 10^3 and 10^6 W cm^{-2} not being uncommon. Below the kW cm^{-2} no biphotonic response should be expected from biological samples. Irradiances higher than 10^9 W cm^{-2} usually lead other non-linear processes to manifest, usually with deleterious consequences for the sample (like multiphoton ionization and plasma creation). Pulsed lasers are the natural election for this methodology, in particular commercially-available femtosecond near-infrared Ti:sapphire lasers. These types of pulsed lasers show average powers in the mW range, guarantee instantaneous irradiances above 10^9 W cm^{-2} at the focal point, and at the same time, produce a very low thermal and optical load for the sample. Use of cw lasers for biphotonic microscopy excitation was already reported more than 20 years ago [15], but the higher powers involved, usually above 100 mW and even above 1 W, subject the sample to damaging levels of light and heat [13, 14].

The other principal advantage of biphotonic microscopy is the low absorption coefficient of most chromophores in the biological samples for one-photon transitions. This is easily understood with an example. Let's assume we want to assess the cellular binding site of a FITC-tagged antibody. Looking at the absorption/emission spectra of fluorescein (see Fig. **19.4A**), it can be seen that it absorbs light with wavelengths 540 nm and shorter. Some wavelengths will be more efficient at exciting the molecule (*e.g.* peak at 490 nm), but there is some noticeable absorption at any wavelength between 300 and 540 nm. If we focus on the peak at ~ 325 nm one-photon excitation at this wavelength is discouraged because the UV light damages the sample. But a two-photon excitation process with photons of double the 325 nm wavelength, that is 650 nm, will ensure a much more innocuous light interaction because 650 nm is in the red light spectrum. There are far fewer chromophores, both natural and artificial, that absorb at 650 nm as compared to 325 nm.

The comparison between the two photonic processes, one-photon *vs*. two-photon, is displayed in Fig. (**19.4B**). Biphotonic absorption of 650 nm light will bring the fluorescein molecule to its second excited singlet state (S_2, see Chapter 2.4.7). From there the molecules will non-radiatively (internal conversion) decay to the S_1 state releasing heat. From S_1, some percentage of the molecules will emit fluorescence light returning the molecules to the fundamental S_0 state.

Also shown in Fig. (**19.4B**) is the outcome of biphotonic absorption at 800 nm. In this case the molecule is excited to a highly excited vibrational level of the S_1 state. After losing most of this vibrational energy, the molecule will decay to the S_0 level with the corresponding quantum yields. It is preferable to excite biological samples in the so-called "biological window" which spans from ~ 700 nm to ~ 900 nm [13]. In this wavelength range, few chromophores absorb by a one-photon process and very little vibrational absorption by water and other molecules takes place, sparing heat production. Thus, biphotonic excitation in the easily attainable 700-900 nm region minimizes one-photon transitions, but still works very well to non-linearly excite most chromophores of interest. A practical example of biphotonic visualization of a fluorophore is shown in Fig. (**19.4C**). The chromatin of mice neuroblasts was made fluorescent with the SYTO 82 nuclear stain. Then the cells were excited by a pulsed laser emitting at 780 nm and the dye-chromatin complex emitted in the orange-red part of the spectrum (grey scale images in the figure). With this approach, it was possible to assess the chromatin rotation that takes place in some cells during metaphase [16].

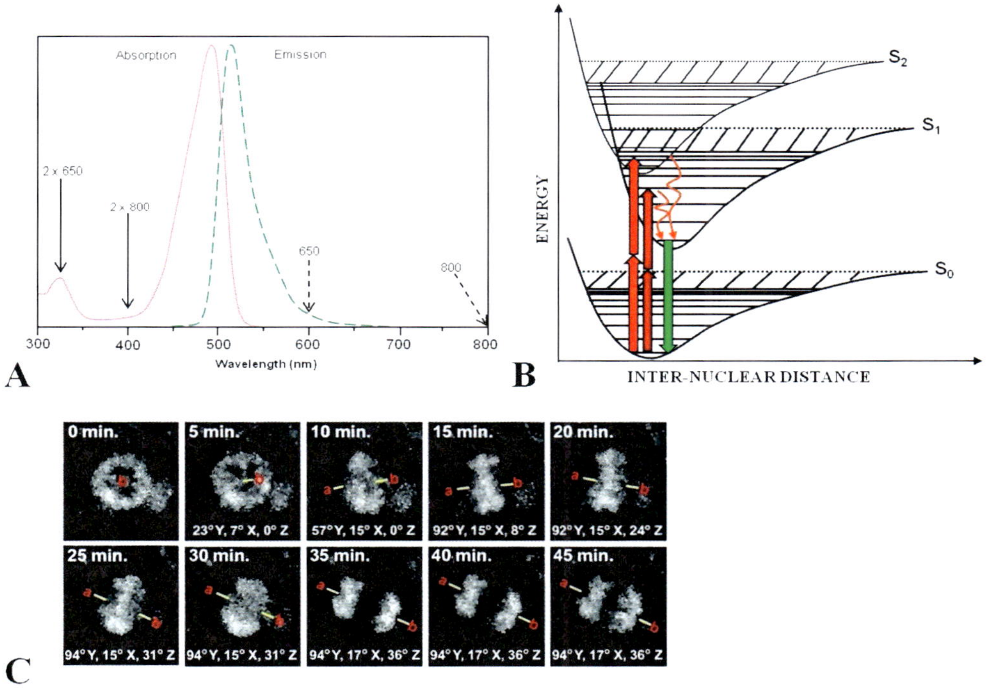

Fig. (19.4). A: Absorption (pink) and emission (green) spectra of fluorescein. Light of 650 or 800 nm wavelength (dashed arrows) is not absorbed under low intensity light because the linear absorption spectrum starts at ~540 nm. Under intense enough light excitation two photons at those wavelengths will be absorbed as they carry the equivalent energy of a 325 nm (2 x 650 nm) or 400 nm (2 x 800 nm) photon (solid arrows). **B:** Schematic potential energy curves of fluorescein illustrating the biphotonic absorption process at two different wavelengths: 650 nm (bright red upward arrows) or 800 nm (deep red upward arrows). Absorbing two 650 nm photons initially excites the molecule to the second excited singlet state (S_2, upper curve). By internal conversion the molecule decays non-radiatively (wiggly arrow) to the first excited singlet state (S_1, middle curve). Absorption of two 800 nm photons directly excites the S_1 level concomitantly with large amounts of vibrational energy. After vibrational relaxation (wiggly arrow) the molecule reaches a metastable state at the bottom of the S_1 curve. After some time it emits a fluorescence photon (green downward arrow) releasing its excess energy, occupying again the ground state (S_0, lower curve). **C:** Real-time mitotic spindle dynamics visualized by biphotonic imaging. Live mouse neuroblasts were visualized in cortical slices. The cells were stained with the chromatin dye SYTO 82. The same neuroblasts were irradiated with a 780 nm pulsed laser to obtain the displayed images. At time 0 min the metaphasic mitotic spindle is arranged with its axis parallel to the observation axis (coordinates indicated for each image). With time, the spindle axis (denoted a-b) tilts and at 15 min it has rotated 92°. Then it rotates towards the upper left part of the image (20-30 min) and finally enters anaphase (30-45 min). All the images were obtained by two-photon excitation of the SYTO 82 fluorophore. Reproduced with permission from [16], Copyright (2003) National Academy of Sciences, U.S.A.

Examples of biphotonic excitation of endogenous compounds in living cells are shown in Fig. (**19.5**). In the first example, a HeLa cell culture was incubated with 5-aminolevulinic acid (ALA), a precursor metabolite for the cytochrome synthesis route. This increased the intracellular concentration of protoporphyrin IX (PpIX), a fluorescent endogenous porphyrin. The selected cell subpopulation is imaged with a 60x objective in bright field (Fig. **19.5A**, upper panel). The cell at the centre and slightly down the image was selected for irradiation with a laser beam from a Ti:sapphire femtosecond laser emitting at 800 nm. The average laser power was 10 mW with 76 million pulses per second, each one of which had an instantaneous power of ~1 kW. The selected cell was irradiated with the focused laser, and the biphotonic excited red fluorescence from intracellular PpIX is shown in the middle panel (arrow highlighted). The emitting spot has dimensions ~0.7 x 0.9 μm. The exciting laser light intensity at the spot is $\sim10^{11}$ W cm^{-2}. An overlay image based on the two previous pictures is shown in the bottom panel. This image puts the biphotonic excitation spot in the context of the cell size.

The previous example shows that it is feasible to excite biphotonically an endogenous molecule or fluorophore in biological samples. PpIX is a naturally-occurring member of an enzymatic route (the synthesis of cytochromes and hemo group). But it can be imaged *via* biphotonic excitation. In the example provided, PpIX levels were increased artificially above physiological levels, but it is nevertheless possible to obtain two-photon excitation of this and other biomolecules at normal levels [17, 18]. Common examples of biphotonic biomolecule excitation are provided by imaging of NAD(P)H or FAD/$FADH_2$ with excitation at 700-800 nm [4]. This results in what is known as label-free non-linear microscopy because no exogenous compound is introduced in the sample to assess it. This technique is put to use to identify precancerous changes in cells and tissues in the example shown in Fig. (**19.5B**). Two *in vitro* grown skin analogues were analysed with biphotonic excitation to assess the arrangement of structures and metabolites [19]. One of the analogues was obtained from primary human keratinocytes (top image) while the other was grown from immortalized keratinocytes displaying some pretumoral features (bottom image). All data were obtained by exciting the samples with pulsed lasers either at 755 nm or 800 nm. The green signal comes from NADH, the orange signal from FAD and the blue signal from collagen (this blue signal was obtained by second harmonic generation; see section 3.2 below for details). It is readily apparent how the skin-like tissue from primary keratinocytes shows an ordered arrangement. Meanwhile tissue derived from pretumoral cells displays a more disordered distribution.

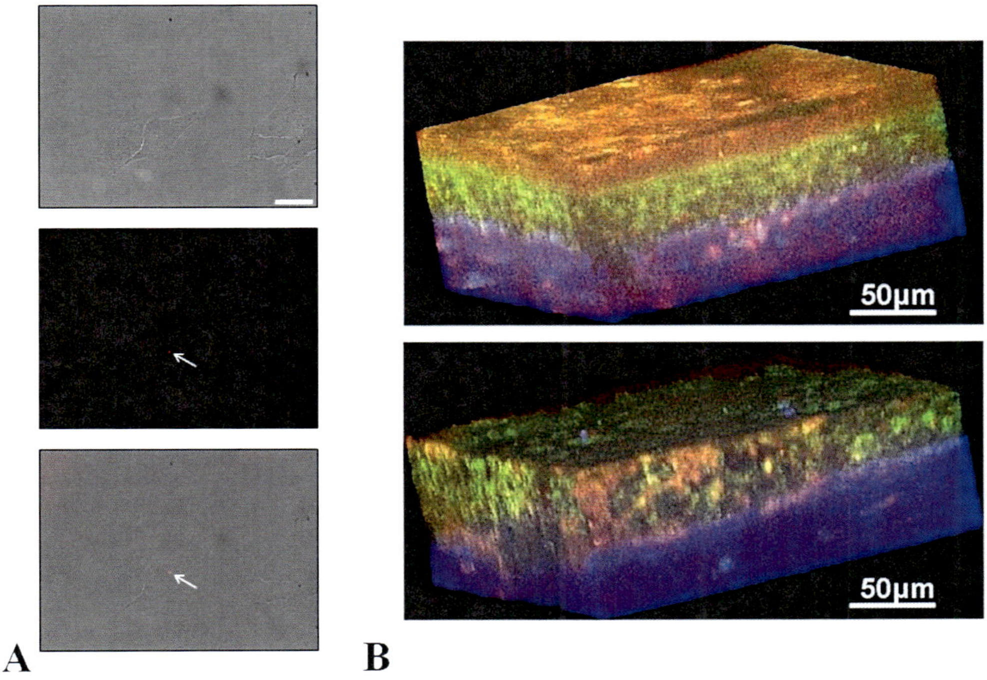

Fig. (19.5). Biphotonic cell and tissue imaging. **A:** Near-diffraction limited imaging of protoporphyrin IX (PpIX) fluorescence excited biphotonically at 800 nm with a femtosecond laser. The cell culture was incubated with 5-aminolevulinic acid to promote the synthesis of intracellular PpIX. The most centred cell (top image bright field) was irradiated with the laser to a near-diffraction limited spot (0.7 x 0.9 µm in this case) by employing a 60x objective. The fluorescence signal (middle image) is obvious against the dark background (arrow). Only at the focal point was the PpIX subjected to high enough light intensity to be two-photon excited. An overlay of the two previous images (bottom image) is provided to compare the size of the emitting spot (arrow) compared to the whole cell. Scale bar: 10 µm. **B:** Two-photon imaging of skin-like tissues derived from primary keratinocytes (top image) and pretumoral keratinocytes (bottom image). The images were obtained by scanning a pulsed laser at either 755 nm or 800 nm. The images are false colour in order to present all the fluorescent signals: NADH (green), FAD (orange) and collagen (blue). It becomes immediately clear that there is a structured gradient in the "normal" tissue (top) as compared to the "precancerous skin" (bottom). Reproduced with permission from [19].

In addition, biphotonic microscopy, and by extension higher-order processes, show an advantageous inherent confocal effect (see Chapter 6.2 for confocal microscopy) [14, 20]. This is because the biphotonic excitation volume is usually so small (given the light intensity constraint detailed at the beginning of the chapter) it behaves as if it had been limited by the pinhole element that characterizes confocal systems. In other words, as the physical dimensions of the non-linear excited volume are so reduced and because this volume is constrained by definition to the focal point, it automatically provides a confocal-like image. Two examples of this are shown in Fig. (**19.6**). In the first example (see Fig. **19.6A**), the pore-sealing ability of human dermal fibroblasts was tested in a PEG polymer matrix [21]. This has important consequences to understand the process of wound healing, for example. Both the intracellular actin cytoskeleton and the external deposition of fibronectin were studied by means of a biphotonic confocal laser system.

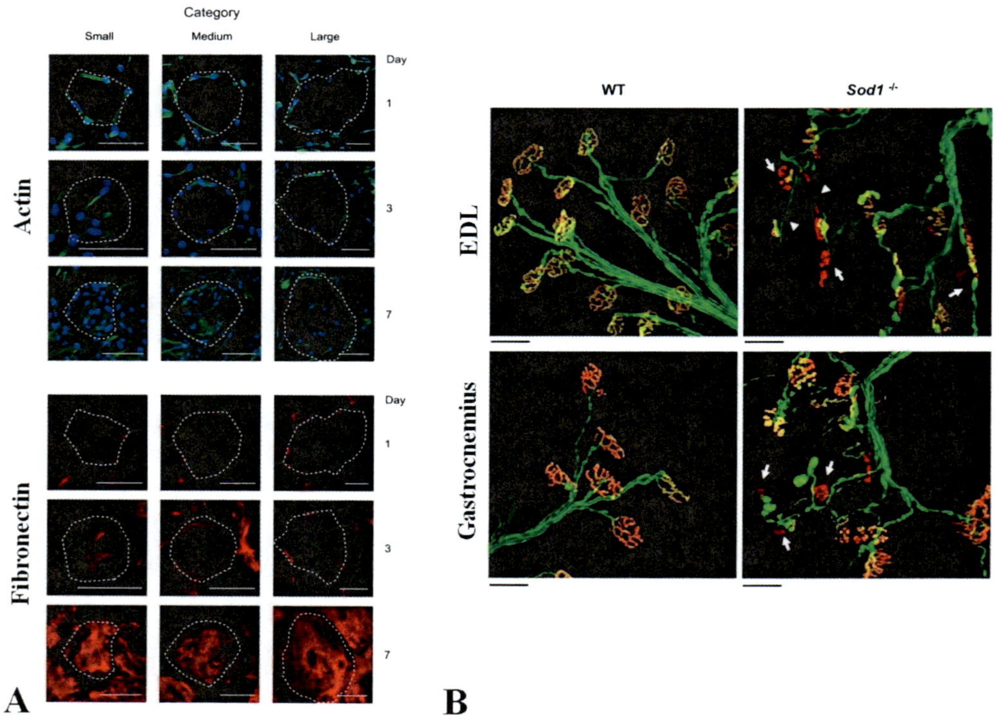

Fig. (19.6). Inherent confocal property of biphotonic excitation. **A:** Human primary dermal fibroblasts were grown in polyethylene glycol polymer with different sized pores (small, medium, large; dashed white contours) to study their rate of pore closure and structural arrangement. After the indicated time (in days) of cell growing, actin (green emission, top panel) and fibronectin (red emission, bottom panel) were detected in fixed samples by indirect immunofluorescence. Chromatin was also detected alongside actin (blue emission, top panel) by DAPI counterstaining. The excitation source was a confocal multiphoton pulsed laser emitting in the near infrared. Sections of 20 µm thickness were employed to construct the images. Scale bars: 100 µm. Reproduced with permission from [21]. **B:** Comparison of neuromuscular junctions in a model of ROS-derived neuromuscular degeneration. Neuromuscular junctions from the extensor digitorum longus (EDL) and gastrocnemius muscles from wild type mice (WT) and genetically-modified mice lacking superoxide dismutase (Sod1 $^{-/-}$), which results in higher ROS concentrations at the tissues, were imaged with a biphotonic laser confocal system. The presynaptic axons and motor plate were visualized by cell expression of a yellow fluorescent protein (YFP, green emission). Postsynaptic acetylcholine receptors were labelled with Alexa Fluor 594-tagged α-bungarotoxin (red emission). The less structured motor plates in the high-ROS model are immediately appreciated on the images for both muscle samples. In addition, partially or totally denervated neuromuscular junctions can be observed in the mutant mice (arrows) because there is no overlap between the green and red signals (the neuromuscular overlap in the controls is the sum of green plus red resulting in an orange signal). Scale bars: 50 μm. Reproduced with permission from [22].

Shi *et al.* reported important structural and functional changes at the neuromuscular junctions (see Fig. **19.6B**) in a mouse strain engineered to have higher than normal ROS levels [22]. The detailed neuromuscular junction microscopic images were obtained with a pulsed laser system providing biphotonic confocal capabilities.

This inherent confocal-like feature of non-linear excitation systems also makes these systems a competitive alternative for microscopic thick sample imaging, for example tissues or even whole organisms [23, 24]. The principle is the same: due to the restricted volume where light intensity is high enough to promote biphotonic excitation, layers at different positions from the focal point/plane are transparent and thus non-emissive. Making use of this approach, structures like the eye, retina, or the skin have been successfully imaged with great detail. But also, other deep organs and structures, like the brain in flies or in mice, have been analysed through two-photon excitation techniques [25]. In Fig. (**19.7A**) there is an example of deep skin tissue vascular irrigation during wound healing [26]. An *ex vivo* mouse skin model was imaged through biphotonic excitation at 800-825 nm. As a target for the two-photon excitation, a fluorescent probe was employed that recognized specific integrins in fibroblasts and endothelial cells activated by the wound. In panel (a) of Fig. (**19.7A**), it can be seen that sections up to 1.6 mm in depth can be imaged with micrometric detail. Two clearly differentiated regions, (1) an upper one where the wound is located, which is undergoing healing granulation between 1000 and 1600 μm, and (2) a lower non-damaged tissue where the vascular structure is clearly visible between 0 and 1000 μm. Notice the detail of the capillary network. Another example where biphotonic confocal properties have been exploited is depicted in Fig. (**19.7B-D**). The authors studied in detail the microscopic structure of a living butterfly's wing epithelium [27]. Thanks to the imaging properties of two-photon excited fluorescence it was possible to obtain high-detail pictures of the very compacted wing epithelium. Imaging this tissue with classical one-photon procedures would have resulted in very blurred images (see figure caption for more details).

As with conventional one-photon fluorescence microscopy, the methodology of two-photon excitation has also been successfully applied to flow cytometry (see Chapter 6.8 for some basics in regards to this technique). Given that lasers are the most widely employed excitation source in flow cytometers, taking a further step to employ pulsed excitation to promote biphotonic excitation was quite obvious [28]. The same assets mentioned with regard to biphotonic excitation as compared to single photon microscopy can be applied to non-linear cytometry. Under the correct conditions, much less photodamage —almost negligible— is inflicted on the sampled cells. Multicolour probing is better with biphotonic excitation, and *in vivo* flow cytometry with optical fibers is in the process of being developed. The possibility not only to image, but also to photoexcite different light-responsive compounds (*e.g.* caged compounds, photodynamic agents, multiphotonic optogenetics, *etc.*), has transcended the basic scientific research, and biphotonic (and by extension, multiphotonic methodologies) are starting to be common players in areas beyond fluorescence microscopy and in the clinical world [17, 24, 29].

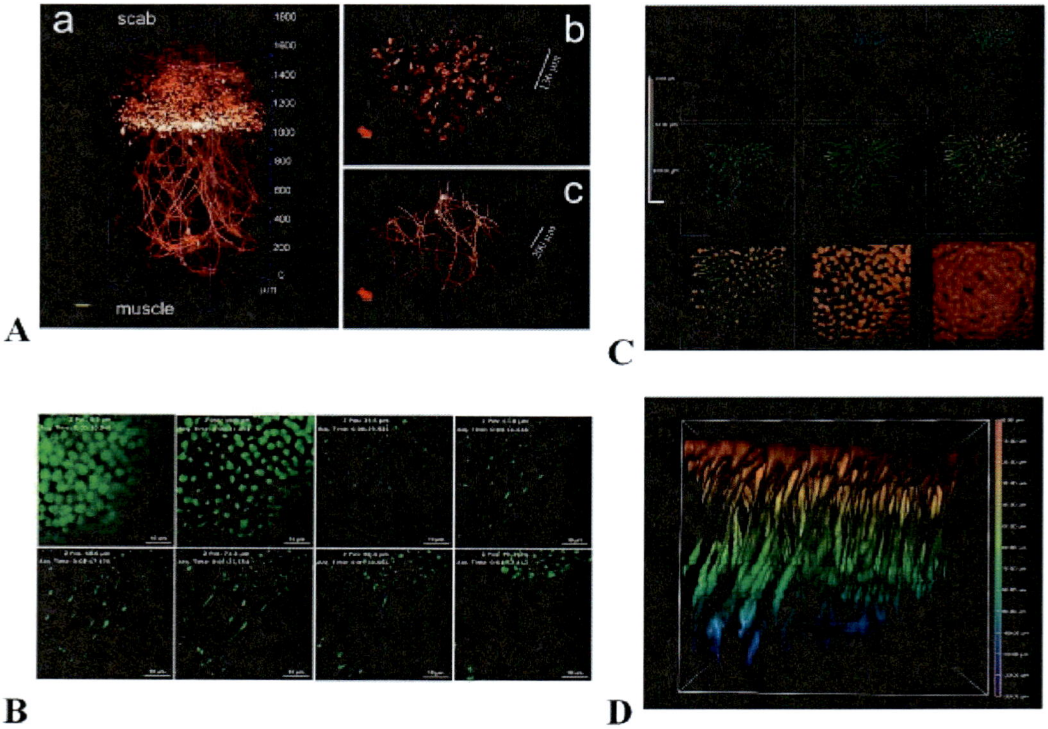

Fig. (19.7). Thick tissue and whole organism biphotonic imaging. **A:** Biphotonic imaging of the wound healing process in a mouse skin model. Wounded skin stripes were prepared *ex vivo* and then studied by two-photon microscopy at 800-825 nm to excite a probe binding a fibronectin subtype in regenerating tissues. Panel (a) shows a reconstructed 3D volume of skin with a total depth of 1.6 mm. From 0 to 1 mm the vascular network underlying the skin is imaged. From 1 to 1.6 mm the granulated skin undergoing regeneration. Panel (b) shows an image stack from 1.2 to 1.336 mm in which the granulating tissue is observed in greater detail. Most signals come from activated fibroblasts. Panel (c) shows a lower level stack of the subdermal vascular network (0.2 to 0.4 mm) also at higher magnification. Arrows indicate the excitation and light collection directions. Reproduced with permission from [26]. **B:** Biphotonic imaging of butterfly wing epithelium. Live animals were stained with CFSE (a cytoplasmic-marking derivative of fluorescein) and then imaged with a pulsed laser. Images follow a superficial to deep sequence starting at 9.5 μm (top left) and ending at 99.3 μm (bottom right). **C:** Top-viewed computer-generated false colour reconstruction of the wing epithelial structure from the images obtained through biphotonic microscopy. The reconstruction is presented from bottom (top left) to wing surface (bottom right) and spans around 130 μm (cooler colours deeper and warmer ones closer to surface). It can be appreciated that cells are filamentary at the base and then spread out and densely pack at the surface (flower-like structure). **D:** Side-viewed 3D computer-generated false colour reconstruction of the wing epithelial structure. As in (C), cool colours represent cells buried deeper and warm colours structures closer to the surface. The trumpet-like arrangement of the epithelium is clearly appreciated in this reconstruction. (B-D) Reproduced with permission from [27].

Biphotonic excitation has also been employed in combination with other techniques already introduced in this book like FLIM [30] (see Chapter 6.3) or photon correlation counting to determine excited lifetimes [31] (see Chapter 6.7). As practically all of these multiphotonic systems are based on very short laser pulses to excite the sample, it is straightforward that the excited states lifetimes and decay pathways can be studied in great detail making use of this approach. A closely related technique to biphotonic microscopy is pump-probe microscopy [32]. We will only mention this technique here. It can be implemented by attending to three different non-linear optical mechanisms: excited state absorption, stimulated emission, and ground state depletion. All three share a common factor: the fluorescent probe ground state population is largely modified by a first pump laser pulse. This pump pulse brings many fluorophores to the first (or higher) excited state. At this point the sample is probed with a second, usually much less intense, pulse and the way this probe pulse interacts with the sample is the basis to differentiate the three techniques. In excited state absorption, the probe pulse takes part of the population in the excited state and pumps it further into more excited levels. The signal can come from the emission from these higher levels or by measuring how much light of the probe pulse has been absorbed (*e.g.* as a function of delay between pump and probe pulses). In stimulated emission, the excited states are stimulated by the probe to emit before they decay spontaneously. So, the probe pulse is amplified by interacting with the sample. The amount of amplification gives a measure of how many fluorophores were in the excited state. An extension of this technique is now widely employed in the STED super-resolution modality (see Chapter 20.4.4). In ground state depletion, the probe pulse (with wavelength very similar or equal to the pump pulse) is absorbed only by those fluorophores that have returned to the ground state. Those still in the excited state do not absorb probe photons. Thus, these techniques allow the measurement of radiative and non-radiative lifetimes. A wealth of information is obtained through these methods by making use of other complementary approaches, like photothermal or photoacoustic detection [32]. These pump-probe techniques are of importance for super-resolution microscopy and will be further explored in Chapter 20.

As the reader can understand, the detailed account of these topics surpass the limits of this book. Whole books, reviews and papers are dedicated exclusively to different areas of non-linear optics, including non-linear optical imaging. We hope that the interested reader will find the selected bibliography enlightening and that it will encourage further research. We will now move forward to higher-order multiphotonic processes.

2.3. Higher-order Microscopy

Although biphotonic microscopy is currently the most employed non-linear technique in fluorescence microscopy, other higher-order (*i.e.* more than two photons involved) processes are exploited too. This is especially true for three-photon excitation and we will now discuss its advantages and uses in fluorescence microscopy.

The fundamentals of non-linear three-photon absorption are the same as for biphotonic absorption, only that three photons are simultaneously absorbed (see Fig. **19.3**). As a consequence, higher light intensities are necessary to reach the absorption threshold. Usual intensities are in the GW cm^{-2} range or higher. From our perspective, there are two main shortcomings of three-photon microscopy in comparison to two-photon: (1) as stated, higher light intensities are necessary to excite the samples, and (2) the wavelength of the photons involved must be longer (less energetic) than those photons involved in two-photon absorption. These longer wavelengths usually are in the range of 1000 – 2000 nm (for example, three 1500 nm photons carry the same energy as one 500 nm photon). This range lays in the infrared and many cell components, in particular water, can strongly absorb these photons as a one-photon process.

Nevertheless, by judiciously selecting certain wavelength regions where one-photon absorption is low, it is possible to obtain very interesting applications for three-photon microscopy. For example, it has been reported that higher tissue penetration with *in vivo* mouse brain can be accomplished by exciting fluorescein and Alexa 680 with 1280 nm photons (3-photon absorption) as compared to 775 nm (2-photon) [23]. With 775 nm, the maximum depth imaged was ~0.5 mm, wheras ~1.1 mm was possible with 1280 nm. The authors selected 1280 nm because it is a relatively non-absorbed wavelength by water.

Another very interesting application of three-photon microscopy is the excitation of endogenous chromophores that absorb in the short visible or UV wavelength ranges, when one-photon excited. The three-photon excitation of the amino acid tryptophan at 750-800 nm allows for the direct visualization and kinetics of unlabelled proteins and peptides [4]. Therefore, the researcher does not need to rely on GFP-constructs that (although absolutely useful in most instances) can sometimes modify the original protein structure/function due to the chimeric nature of the fluorescent GFP. Other endogenous chromophores like cofactors (NADH, FAD) or nucleic acid bases can be excited in a three-photon process to study biological or biochemical processes. Due to the three-photon nature of these processes even higher spatial resolution can be obtained under the right circumstances as compared to biphotonic microscopy.

Finally, we would like to briefly mention higher photonic processes with more than 3 photons involved. As the irradiances are increased further towards and above the TW cm^{-2} level, it becomes increasingly probable to excite 4-, 5- or higher-order absorption processes. Given the energy content of each individual photon in the near infrared, atoms/molecules absorbing that amount of energy (equivalent to far UV) will become highly excited, dissociated and/or ionized [33]. Therefore, these light levels are employed not for imaging, but instead for inducing ROS, low-density plasmas, plasma ablation, and therapeutic applications. For example, in single cell experiments it is possible to cut a single chromosome within a living cell with these methodologies.

3. HARMONIC MICROSCOPY

In this section we will explain the fundamentals as well as some applications of harmonic microscopy. Generally speaking, these transitions are characterized by their non-resonant nature as opposed to resonant ones that were explained in section 2 above. Nevertheless, they are part of the phenomena grouped under multiphotonic techniques. We will go into further details with the fundamentals.

3.1. Fundamentals

In harmonic emission, the irradiated sample emits light with wavelengths that are an optical harmonic of the fundamental exciting light [3]. For example, if 1064 nm fundamental laser light is employed to excite the harmonics, light of wavelengths 532 nm (2nd harmonic), 355 nm (3rd harmonic), 266 nm (4th harmonic), and so on, can be obtained. This process is schematically represented in Fig. (**19.8A**). One of the key points of this process of harmonic generation is that it is non-resonant. The involved excited states are virtual: no resonant excited state is reached [1]. For this reason, the emitted light wavelengths are directly dependent on the exciting light wavelength. On the contrary, when resonant multiphotonic absorption takes place it is the atom or molecule that dictates which wavelengths are emitted, irrespective of the wavelength employed to excite it.

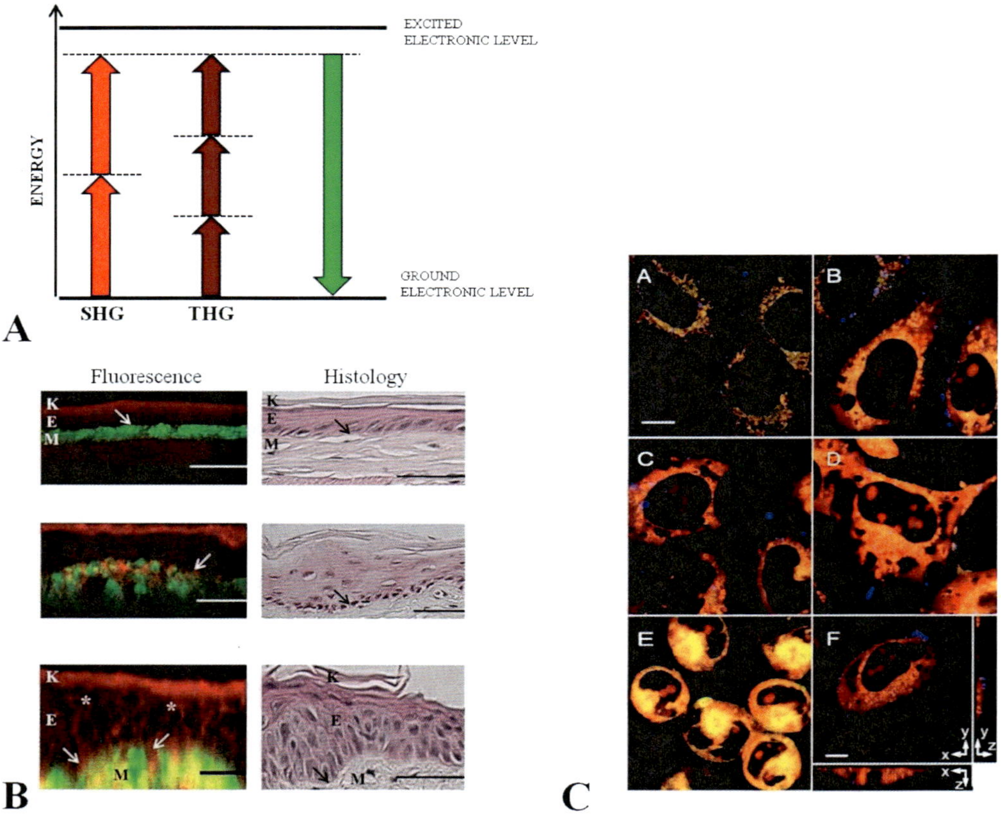

Fig. (19.8). Harmonic generation in fluorescence microscopy. **A:** General scheme of the harmonic generation process. Two photons (red) interact with the sample through two virtual levels. The two exciting photons annihilate to create a single photon (green) with double energy. This is the case for second harmonic generation (SHG). If three photons (deep red) interact to generate a higher energy (green) photon the process is third harmonic generation (THG). Real electronic levels are represented as solid lines and virtual ones as dashed lines. Vibrational levels have not been depicted to avoid confusion. **B:** Tissue structure analysis in an oral mucosa dysplasia model in mice. Normal tissue (top), mild dysplastic (middle) and severely dysplastic (bottom) were imaged with pulsed laser excitation. The biphotonic tissue autofluorescence (red signal) and the SHG by the skin collagen (green signal) can clearly be distinguished in the images (multiphotonic fluorescence, left lane). The normal tissue shows a regular and highly defined structure. As dysplasia progresses the tissue is less structured. Histological tissue samples where obtained as controls for each condition (right lane). The differences in histology are patent and support the fluorescence findings. K: keratinizing layer; E: epithelial layer; M: lamina propria; arrows; epithelial-connective tissue interface; * enlarged dysplastic nuclei. Scale bars: 40 µm. Reproduced with permission from [36] **C:** SHG from different doubling materials incorporated into BEAS-2B cells (transformed human bronchial epithelium) as PEG-stabilized nanoparticles: (A) $BaTiO_3$; (B) $KNbO_3$; (C) $LiNbO_3$; (D) KTP; and (E) ZnO. The nanoparticles stay mainly outside the nucleus although some nucleolar marking can be seen. ZnO (E) resulted in cell toxicity as revealed by the rounded morphology. (F) Three different planes of the same cell loaded with $KNbO_3$ imaged by SHG. Indeed, some SHG material is located within the nucleus although most of it is outside. Scale bars: 10 µm. Reprinted with permission from [40]. Copyright (2012) American Chemical Society.

As can be understood from the Jablonski diagram for harmonic generation (see Fig. **19.8A**), it is important that the harmonic produced is non-resonant with any allowed transitions. Otherwise the emitted light will be reabsorbed by the sample and no image can be obtained. Another important requirement for harmonic generation is structural asymmetry of the material interacting with the light. The mechanism underlying harmonic generation is a wave phenomenon. When a monochromatic wave (single frequency) moves through a material, the interaction matter-wave spontaneously generates harmonics of the fundamental wave [3]. This applies not only to optical electromagnetic waves but it is a general phenomenon in physics related to all kind of waves. This means that when 1064 nm light moves through matter, some 532 nm light is generated. However due to wave propagation and symmetry arguments, the harmonic waves generated interfere destructively among themselves. The end result is that extremely low harmonic, or no harmonic at all, is generated under usual circumstances. For efficient harmonic generation, some asymmetry in the material substrate at the atomic level is necessary. This allows for waves propagating in certain directions to interfere constructively and emit harmonics. It is possible to obtain harmonic emission only by having the light propagating in strictly defined directions. This is known as phase matching because the phase velocities of the fundamental and the harmonic of interest within the material must be the same in a certain direction to interfere constructively.

The phase matching condition is essential for harmonic generation. But there is also an energetic argument. For every photon at the second harmonic, for example, two fundamental photons must annihilate to provide the corresponding energy. As a consequence, optical harmonic generation is a non-linear process that depends on the light intensity raised to the power of the harmonic being produced. Second harmonic generation has a squared dependence on fundamental light intensity (I^2), third harmonic a cubic dependence (I^3) and so on. It is possible to generate harmonics in any material as long as the substrate has some kind of asymmetry. But crystalline solid-state materials are the material of election for this kind of non-linear process. They provide asymmetry in their crystalline lattice (some of them), high density of atoms that increase the harmonic yield, and adequate resilience and heat-dissipating properties that are also of interest.

The most employed technique in biological studies is second harmonic generation (SHG). It is the easiest to obtain at relatively low powers and a wider range of materials comply with the symmetry requisites for SHG than for other harmonics. Third harmonic generation (THG) is nevertheless also employed, although less frequently. We will mainly focus on SHG bioimaging applications. Two approaches make use of SHG for microscopic imaging. One relying on SHG from endogenous biological structures (label-free SHG), and another that introduces some kind of doubling material (usually crystalline nanoparticles) providing SHG when interacting with laser light. We will introduce both methodologies in greater detail below.

3.2. Harmonic Generation in Biological Materials

The main advantage of this imaging approach (label-free harmonic microscopy) is its label-free nature, which makes it possible to study biological samples with minimum interference [34]. The light intensities are of the same order of magnitude as with multiphoton-excited fluorescence. The clear shortcoming is that only certain crystalline or asymmetric biological structures will produce harmonics efficiently. As a consequence, a narrower range of biological structures can be directly imaged with this methodology. As mentioned, SHG is the most employed harmonic generation system for imaging. Some SHG-generating endogenous biological structures are the microtubules (*i.e.* in principle, all eukaryotic cells could be imaged), collagen, muscles (myosin chains), starch-containing structures (granules), polysaccharides and chloroplasts [34, 35]. Due to the breaking of symmetry at interfaces between two different media, even when these are symmetric or homogeneous, it is possible to find adequate conditions for SHG at these locations. A very important example of biological interfaces assessed with SHG microscopy are lipid membranes and proteins/structures embedded in them [35].

An example of SHG microscopy put to use is shown in Fig. (**19.8B**). The authors investigated the differences in structure between normal oral tissue, mild dysplastic tissue, and severe dysplasia [36]. By making use of non-linear optical excitation, the biphotonically-excited cell autofluorescence (red emission) and the second harmonic derived from collagen (green emission) were imaged at the same time with the same laser source. The position and arrangement of the different optical signals correlated with histochemical analysis of the same samples (see Fig. **19.8B** for more details). In light of these results, it is possible to provide diagnostic information through a non-invasive optical method.

It is also possible, and has been realized in practice, to excite the third harmonic (THG) from endogenous structures [34, 35, 37]. However, THG requires higher light intensity than SHG and either long wavelength infrared photons are employed to excite a visible harmonic (with the disadvantage of infrared absorption by the sample) or near infrared/visible photons excite the third harmonic, which is then in the UV (which can damage the sample). In any case, THG is used much less than SHG.

3.3. Harmonic Generation in Artificial Materials

Harmonics can also be generated in non-biological materials that are introduced into a biological sample to provide a light signal. As it was mentioned in the fundamentals above, crystalline solids have been the material of election for SHG and other harmonic phenomena almost since the beginning of non-linear optical interactions after the discovery of lasers. Many different crystals are adequate for producing optical harmonics. Some of the most employed are oxides like lithium niobate ($LiNbO_3$), potassium niobate and tantalate ($KNbO_3$ and $KTaO_3$), potassium dihydrogen phosphate (KH_2PO_4), β-barium borate (BaB_2O_4), as well as many others. They display the required structural asymmetry mentioned in the fundamentals as well as the high non-linear optical coefficients that enhance the non-linear response to a given light intensity [3, 6].

It is possible to design SHG and THG agents capable of labelling biological substrates by synthesizing micro- or nanoparticles of these non-linear materials, and correctly functionalising them [38, 39]. An example of cell labelling with nanoparticles of different non-linear materials is shown in Fig. (**19.8C**) [40]. These kinds of nanoparticles have been employed to measure dynamical features within cells. For example, Macias-Romero *et al.* reported the measurement of real-time translational kinetics of $KNbO_3$ nanoparticles of ~100 nm in HeLa cells [41]. Not only was translational movement measured; the rotational movements imparted on the nanoparticles could also be measured thanks to the change in second harmonic signal depending on the angle between the fixed impinging laser light and the rotating crystals. This kind of ratiometric results can result in harmonic-generating particles to fill a specific niche in fluorescence microscopy. Protocols for labelling whole organisms with SHG materials have been published recently, too [42]. As these materials can be grouped with semiconductors or dielectrics in regards to their interactions with biological samples, we refer the reader to general discussions of this topic in Chapter 18.1.

4. PHOTON UPCONVERSION

Photon upconversion (UC) is a group of techniques that favour the emission of photons of shorter wavelength over those that are employed to excite the sample (*i.e.* anti-Stokes emission). Because at least two low energy photons are necessary to produce a high energy photon (energy conservation principle), the processes have a non-linear dependence on exciting light intensity. But the mechanisms underlying UC are fundamentally different from those already introduced in this chapter for multiphotonic absorption and harmonic generation. In UC, the photons absorbed do not necessarily have to interact with the upconverting material at the same time (or nearly the same time). Rather, the energy of the first absorbed photon is stored in the medium until a second photon (or more) is absorbed. Then, part of the total energy stored is reemitted by the material as anti-Stokes light. Three different UC mechanisms will be introduced in this section. We believe that UC applied to fluorescence microscopy is one of the most promising new methodologies to be incorporated into this field. This is primarily because the light intensities necessary to excite the anti-Stokes emission are many orders of magnitude lower than those needed for multiphotonic microscopy as well as harmonic generation. UC exciting light intensities are around 1-100 W cm^{-2}, roughly 3-6 orders of magnitude less intense than other non-linear imaging techniques. Reports exist for successful UC exciting intensities of 100 mW cm^{-2} and for triplet-triplet annihilation UC (see 4.3 below), even down to less than 1 mW cm^{-2}. Such low light intensities guarantee negligible sample photodamage during assessment, and low-power cw lasers (or even non-laser light sources) can be employed for excitation.

Now we will introduce the three techniques for UC. Although related in their mechanistic background, each one has its own peculiarities which we will present.

4.1. Lanthanide-based Upconversion

The most classical UC technique makes use of lanthanide-doped materials. The fundamentals and proposal for this technique were established in the late 1960s [43]. This UC technique is based on energy transfers between lanthanide ions. It was introduced previously (see Chapters 9.2.3 and 18.4.4), that the lanthanide elements, in particular in the +3 oxidation state, present very long excited state lifetimes. This is related to the partial screening of the f shell electrons (see Chapter 9.1.2). Due to this peculiarity, these metal ions have lifetimes in the range from microseconds (10^{-6} s) to tens of milliseconds (10^{-1} s). These excited state lifetimes are anomalously long. For example, organic molecular fluorophores (*e.g.* fluorescein or rhodamine) in contrast have lifetimes in the nanosecond range (10^{-9} – 10^{-8} s). Two basic mechanisms allow for UC in lanthanide-doped materials: excited state absorption and excitation energy transfer [44 - 47]. Both mechanisms depend on the very long excited state lifetimes and are not mutually exclusive: they can occur at the same time in a sample.

Excited state absorption is straightforward to understand. A first photon is absorbed by one lanthanide cation promoting an electron from the ground state to some excited state. Due to the very long lifetime, there is a chance that a second photon arrives before decay, and it is absorbed by the excited centre. This sequential two-photon absorption will pump the electron to an even higher excited state (see Fig. **19.9A**). From there, the electron can radiatively decay to the ground state (or to some lower excited state) emitting a photon with more energy (*i.e.*

shorter wavelength) than the originally absorbed photons (anti-Stokes emission) [43, 46, 47]. Because the second photon must arrive before the excited centre decays, the light flux must still be relatively intense although much less than with the previous techniques discussed.

Excitation energy transfer involves the migration of the electronic excitation energy among different cations and their electronic levels. Once a photon has been absorbed, its energy can be transferred between energy levels. Several different ways of energy migration and redistribution are presented in Fig. (**19.9B**) [48]. By achieving a certain level of excitation, the population of excited centres is such that by mere energy redistribution some centres end up in higher energy states that then decay radiatively. This light is made of higher energy (*i.e.* upconverted) photons.

Because both processes, excited state absorption and excitation energy transfer, depend on the same electronic levels, they usually take place at the same time. The ratio between them depends on the particular lanthanides taking part on the sample's composition and on the light flux. Usually excitation energy transfer is favoured at low light intensities. In any case, the important endpoint is the wavelengths of the emitted light. And this depends directly on which particular lanthanides are present in the marker. The most commonly employed lanthanides in UC for bioimaging are Neodymium, Europium, Terbium, Holmium, Erbium and Thulium. All of them can be excited directly with infrared or visible light to obtain shorter wavelength visible or even UV emission light (see Table 9.2 in Chapter 9 for some emission wavelengths typical of lanthanide UC too). An example of some of the electronic levels involved in UC in several lanthanides along with their emission wavelengths are displayed in Fig. (**19.9C**) [48].

For many applications, the UC markers (nanoparticles in most cases) are co-doped with Ytterbium (Yb). Yb features a single f shell excited level (see Fig. **9.7A** in Chapter 9) with an absorption/emission wavelength ~980 nm (1.26 eV, 10,000 cm^{-1}). Yb has a particularly intense absorption at 980 nm for a lanthanide. As a consequence, it is a perfect candidate to absorb light and transfer the excitation energy to other lanthanides co-doping the same matrix (UC sensitizer). At 980 nm, most biological samples are still relatively transparent, although 980 nm is coincident with an overtone absorption peak of the water molecule. Fortunately, the light intensities to observe UC in Yb-doped UC materials are usually far below the threshold for photothermal damage. Fig. (**19.9D**) shows an example of Yb co-doped materials, and their UC emission, which depends on which particular lanthanide acts as receptor of the energy initially absorbed by Yb^{3+} ions [49]. In the examples presented, Er and Tm are the two emitting lanthanides. Notice that Er^{3+} has two emission wavelengths in the visible, one at

530-570 nm (green) and another at 640-680 nm (red). Tm^{3+} has a main emission band at 450-500 nm (blue) although some emission occurs at ~650 nm (not shown in the examples) and there is a large emission band at 760-840 nm in the near infrared. The figure also depicts a scheme of the proposed energy transfer and migration taking place after light absorption by Yb^{3+} ions [50].

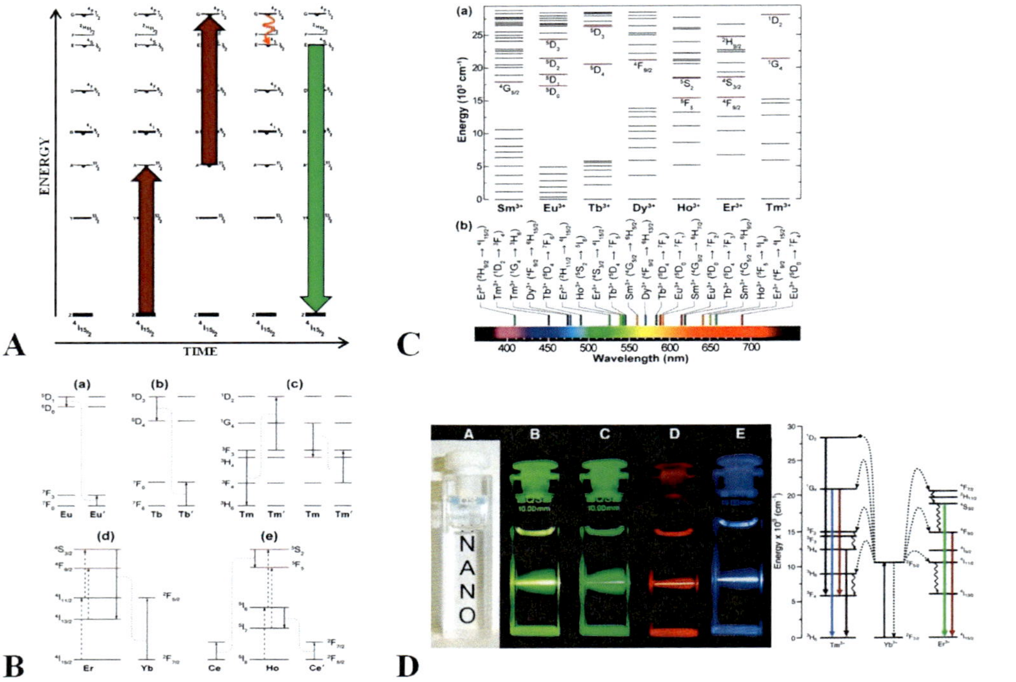

Fig. (19.9). Upconversion (UC) mechanisms in lanthanide-doped materials. **A:** Scheme of the excited state light absorption process. For this scheme the electronic energy levels of Er^{3+} have been depicted to show a real example (left). A first infrared (~980 nm) photon excites the ion from the fundamental $^4I_{15/2}$ state to the excited state $^4I_{11/2}$. This excited level has a lifetime of around 8 milliseconds. Sometime later, before decay takes place, a second 980 nm photon is absorbed by the excited centre further pumping the electron to a higher electronic level, the $^4F_{7/2}$ in the example. In a short time (nanoseconds) this level decays non-radiatively (wiggly arrow) to the $^4S_{3/2}$ state. On an average time of ~ 1 millisecond this level emits a green (550 nm) photon of upconverted nature. **B:** Different excitation energy transfer examples. When an excited electron decays (downward arrows) it non-radiatively transfers its energy to another close electron (upward arrows) which can be in the ground or in an excited level. These transfers are resonant (connecting dashed lines); the energy difference between the levels involved must be roughly the same. (a) transfer between Eu ions, (b) transfer between Tb ions, (c) two different transfers between Tm ions, (d) transfer between Er and Yb ions, (e) two possible transfers between Ho and Ce ions. In (d) and (e) the upward dashed arrows represent photon absorption. Reprinted with permission from [48]. Copyright (2014) American Chemical Society. **C:** Some of the commonest lanthanide ions employed in UC materials. (a) Emitting levels of UC photons are highlighted in red. Other electronic levels are in black. (b) Wavelengths associated with transitions from the indicated levels for the selected lanthanides. Reprinted with permission from [48]. Copyright (2014) American Chemical Society. D: Example of UC emission. Er-Yb co-doped nanoparticles

(A-D) and Tm-Yb co-doped nanoparticles (E) were excited with a 977 nm laser (1.3 kW cm^{-2}). (A) shows a photograph of the cuvette, (B) shows the total UC luminescence under laser excitation, (C) is the same as (B) but filtering the red emission and leaving the green light, (D) is the same as (B) but filtering the green emission and leaving the red component, (E) shows the blue emission from Tm ions. To the right a scheme showing Yb absorption of 977 nm photons (upward black arrow), its excitation energy transfer to either Er or Tm ions (dotted arrows), and the radiative (coloured arrows) or non-radiative (downward black and wiggly arrows) decay of these ions. Images reprinted with permission from [49]. Copyright (2006) American Chemical Society. Scheme reprinted with permission from [50]. Copyright (2007) American Chemical Society.

The most common approach for using lanthanide UC in bioimaging and fluorescence microscopy is to use lanthanide-doped micro- or nanoparticles. The solid-state environment offers the most advantages for UC to take place (for the same reasons mentioned in Chapter 18.1.3 with regards to luminescent decays in general). Thulium, featuring blue and near infrared emissions, and erbium, emitting both green and red, are most commonly employed today. Their efficient interaction with the Ytterbium single excited f shell level makes materials excited at 980 nm and emitting in the visible a perfect combination for fluorescence microscopy uses [51]. These nanoparticles show negligible levels of cytotoxicity in cell cultures or organic toxicity in organisms. In addition to their inertness the light intensities required to excite the UC can be reached with almost any modern confocal laser system commercially available today (exciting in the red and recording the blue or green signal). Given the near infrared excitation of these systems (mostly at ~980 nm but other wavelengths can work perfectly well, like 800 nm), tricolour emission in the blue-green-red, or even more colours, could be obtained theoretically [44, 48].

Several examples of these types of UC nanomaterials employed in biological assessments are presented in Fig. (**19.10**).

The first example makes use of mesoporous UC nanoparticles (Yb+Er) that incorporate doxorubicin in specific particle pores. This drug quenches the green upconverted emission from Er but does not interfere with the red emission from the same ion. Therefore, by knowing the emission ratio between green and red it is possible to measure the amount of drug in close proximity to the emitting part of the nanoparticle. When doxorubicin-loaded nanoparticles were incorporated by living HeLa cells, it was possible to establish the drug-releasing kinetics by measuring the green:red emission ratio (see Fig. **19.10A**) [52]. Working with a different, more complex, biological model, Lim *et al.* studied the dynamics in the digestive tract in the worm *C. elegans* (see Fig. **19.10B**) [53]. UC nanoparticles were ingested by the worms. Afterwards nutrients were added to the medium and the researchers measured the time it took for the nanoparticles to clear up from the digestive tract. This is a particularly good example of the innocuous nature of these fluorescent agents both in regards to dark toxicity and when exposed to exciting light for long periods of time. Recently Hu *et al.* have reported the targeted incorporation of UC nanoparticles to the bone structures of mice [54]. The UC nanoparticles were functionalized with 17β-estradiol, a hormone that targets bony tissues *in vivo*. After several hours, the animals were sacrificed and their bone structures analysed by illuminating them at 980 nm. In this case, the UC elements were Tm+Yb. The 800 nm strong Tm UC emission was registered with impressive results (see Fig. **19.10C**). This 800 nm wavelength is more than adequate for imaging thick biological samples as it is in the biological window. As it is clear from the images, whole biological structures can be imaged in great detail by making use of UC procedures.

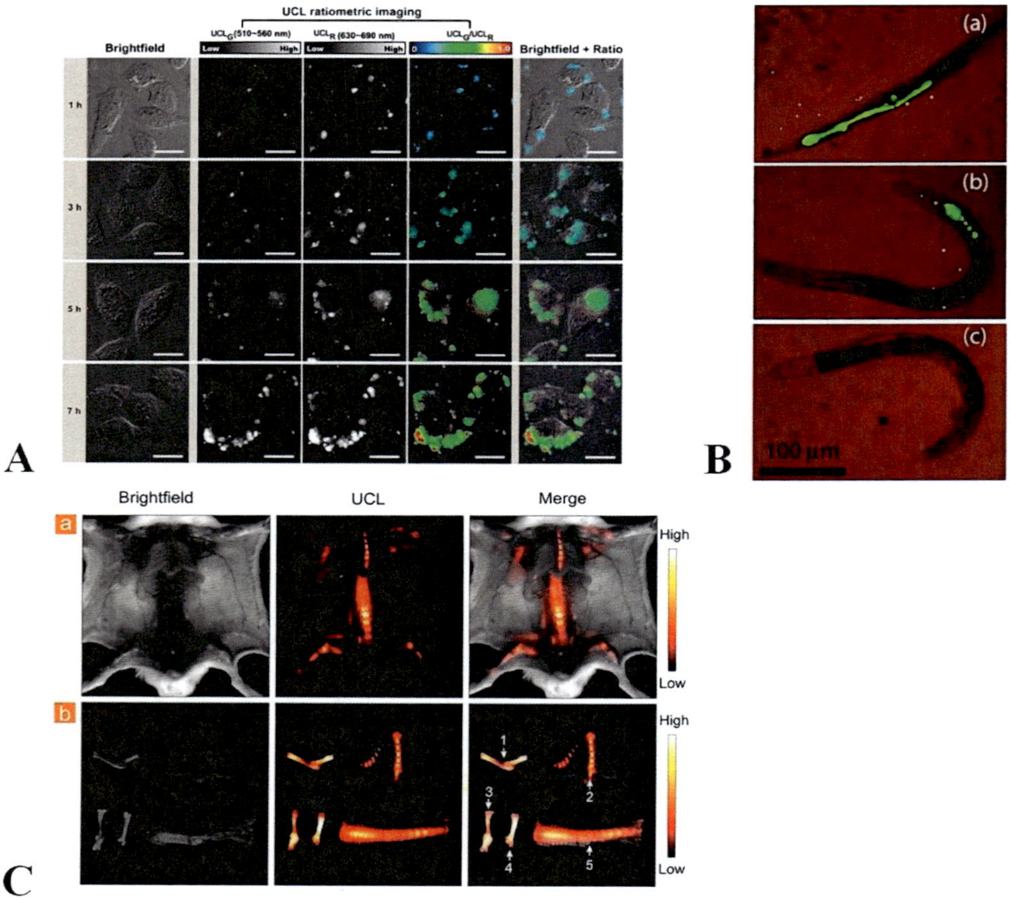

Fig. (19.10). UC imaging in biological samples. **A:** Ratiometric use of Er UC to detect drug release from nanoparticles. HeLa cells were exposed to UC nanoparticles carrying doxorubicin. This compound quenches the green UC emission from Er but not the red emission. As the drug is released the green emission recovers while the red one stays at the same intensity. The cells are shown at 1 h, 3 h, 5 h and 7 h after nanoparticle addition. The images are (from left to right) bright field, green emission, red emission, ratio green:red (false colour), and overlay of bright field and emission ratio. Inspecting the ratio images it is clear how the green emission recovers with time as the drug is released (warmer colours, see colour scale above pictures). The UC was excited by a laser emitting at 980 nm. Scale bars: 20 μm. Reprinted with permission from [52]. Copyright (2015) American Chemical Society. **B:** Study of the digestive tract in living *C. elegans* with UC Er/Yb-doped nanoparticles. *C. elegans* worms were exposed to UC nanoparticles for 3 h. Afterwards food was included in the dishes and the worms started feeding. Immediately after adding food (a) the nanoparticle UC emission is considerable inside the imaged worms. After 1 h (b) a substantial decrease in the internal nanoparticle load is observed. By 2 h (c) almost all signal is lost. Worms were imaged in both bright field and with an IR LED laser to excite UC. The images shown are an overlay of bright field and epifluorescence pictures. Reprinted with permission from [53]. Copyright (2006) American Chemical Society. C: Uptake of 17β-estradiol by mouse bones *in vivo*. 17β-estradiol-loaded UC nanoparticles ($NaYF_4:Tm^{3+}$) were injected into healthy mice. After 6 h the animals were sacrificed and the organs removed to obtain better image resolution. The skeletons (a, top row) were illuminated with a 980 nm cw laser. The prominent UC emission of Tm at 800 nm was registered and transformed to a false colour intensity gradient. The nanoparticle concentration is directly related to the light intensity (UCL column). Some bone pieces (b, bottom row) were dissected and imaged with the same laser system. The obtained results indicate the efficient deposition of 17β-estradiol-loaded nanoparticles in the adult bone tissue. Reprinted with permission from [54]. Copyright (2015) American Chemical Society.

As it was already mentioned in the previous chapter, luminescent solid-state markers often need to be functionalized and bioconjugated to adequately fulfil their imaging function [55]. Almost universally, the particle's surface is decorated with some kind of organic moiety that affords adequate solubility in the aqueous living medium and confers electrical charge properties to be adequately uptaken by biological targets. UC nanoparticles have been functionalized by attaching to them amino acids, peptides, proteins, nucleobases, nucleic acids and other biomolecules or organic molecules. These modifications warrant the adequate recognition of the selected target. Activation by using 980 nm to excite Yb^{3+} centres that then promote the UC mechanism does not usually result in target overheating. But alternatives have been suggested and are in the process of being tested. Recently, it has been proposed to include Nd^{3+} ions to an Yb^{3+} plus upconverting lanthanide material to bypass the 980 nm wavelength [56]. Nd^{3+} ions absorb strongly at 808 nm while water does not. Once an 808 nm photon is absorbed, the Nd centre will transfer the energy to an Yb centre, with the classical UC cascade taking place afterwards. This is an interesting approach to avoid excessive sample heating, of special concern when working with living systems.

Lanthanide-based UC, particularly in conjunction with nanoparticle carriers, is finding uses in many basic and applied areas, not only in research [57 - 59]. Of course, given their characteristic very long excited state lifetimes, they find use in time-resolved studies, like FLIM microscopy (see Chapters 6.3 and 9.2.3). These UC agents are becoming very important and robust tools in clinical imaging. Due to the possibility to both excite and emit in the near infrared —within the biological window— these agents allow reasonable resolution in obtaining images of different structures or processes at the whole organism level. One of the most obvious applications in this regard is for tumour imaging. But specific organ targeting for other reasons can be obtained as well as imaging of dynamic biological processes like tissue blood perfusion. In recent years, the development and application of the so-called multimodal imaging agents or multi-imaging agents has seen an impressive advancement. This approach aims to unify, within a single agent, two or more imaging modalities to better define and locate a structure of interest (*e.g.* a tumour). One of the most successful combinations is the use of agents with both optical and magnetic imaging properties. In this area lanthanide-doped UC elements are at the vanguard because some lanthanide ions display interesting magnetic characteristics too [51, 55, 57].

Apart from their use in fluorescent microscopic techniques, lanthanide-based UC are also employed in very sensitive assays as luminescent reporters [57, 58]. This involves ELISA-like techniques where binding of the target compound usually leads to an increase in UC light emission, or to a change in UC light wavelength (by inducing or quenching FRET, for example). In this regard, lanthanide-doped materials are employed most frequently with antibodies or nucleic acid sequences that provide the very high selectivity for target detection. Employing these elements, a large array of biomolecules can be detected with high sensitivity. Metallic ions, both from light elements (*e.g.* Mg^{2+} or Ca^{2+}) and from heavy polluting elements (*e.g.* Hg, Pb, *etc.*) have been detected in very low concentrations thanks to the use of UC sensors.

Lanthanide-based UC agents are also on the way to become clinical agents for the treatment of cancer [60]. As already mentioned, lanthanide-doped nanocarriers can be used to measure the drug delivery rate within a tissue or organ (see Fig. **19.10A**). Employing similar strategies of light emission quenching, these agents have also been employed to assess delivery of nucleic acids (DNA, siRNA) in genetic therapy. In addition to imaging, UC light has been employed in the photo-uncaging of different drugs and products from the nanoparticles. This allows light-controlled release of a therapeutic product only at the site of illumination. Photodynamic therapy driven *in situ* by UC light from lanthanide-doped materials is a promising approach to deliver short-wavelength light while exciting with long-wavelength light from external sources. Many good photodynamic sensitizers have not been employed for decades because they are excited in the blue or green region of the spectrum, which does not adequately penetrate tissues. But creating this light within the target structure through UC by illuminating with near infrared light in the biological window has afforded very successful results so far. UC materials have been employed also as photothermal agents. Subjecting these materials to ionizing radiation increases the radiation dose locally. This enhancement seems to be related with the production of short-wavelength UV light when the UC material is excited by the radiation. This effect lowers the burden for surrounding tissues and better delimits the treated volume. Additionally some lanthanides display very high cross sections for neutron capture. This is put to use in neutron capture therapy to, again, better control the spatial resolution and dose of the therapeutic modality.

4.2. Transition Metal Complex-based Upconversion

In transition metal complex-based upconversion, a judiciously selected d-block metal cation forms part of a complex (see Chapter 9 for transition metal complexes) which also binds a lanthanide cation (selected according to its absorption/emission features, lifetime, *etc.*) [61]. The role of the d-block metal is to act as an antenna to capture light; this is because d-block metal containing-complexes display very broad absorption bands (see Fig. **9.4B**). When a photon is absorbed, its energy transforms into electronic excitation energy by one of the possible transition types already explained (d-d, LMCT, MLCT, *etc.*; see Fig. **9.5** in Chapter 9). The long excited state lifetimes usually featured by transition metal complexes makes it more probable for the electronic energy to internally (*i.e.* through resonance energy transfer) flow to the lanthanide that also makes part of the complex [62]. The energy finally "funnels" into one of the lanthanide f shell electrons. As already mentioned, lanthanide f-f radiative transitions are exceedingly long in time. Consequently, most of the energy originally provided by the absorbed photon is stored in the lanthanide for some time. If the photon flux is high enough (*i.e.* enough photons reach the metal complex per unit time), there is a non-negligible probability that the d-block metal centre absorbs a second photon and transfers its energy to the lanthanide *before* the lanthanide emits light. In this way the already excited f electron receives a second energy packet that pumps it to a higher excited f shell level. It is from this higher level that the upconverted photon is emitted with more energy compared to each of the absorbed photons [63, 64]. The emitted light is anti-Stokes shifted in respect to the absorbed light because its wavelength is shorter than that of the excitation photons.

This approach can be seen as the molecular equivalent of the more classical UC method based on lanthanide-doped solid particles. Although it is an emerging and promising field in itself, we will not explore it further. We wanted to mention it for the sake of completeness.

4.3. Triplet-triplet Annihilation Upconversion

Our final incursion into the UC methodologies will be the introduction and explanation of triplet-triplet annihilation upconversion (T-T UC). The same general principle of energy storage after long wavelength photon absorption, followed by energy release as a shorter wavelength anti-Stokes quantum, applies to T-T UC [65, 66]. The key component in T-T UC is long-lived triplet excited states [67]. In its most basic arrangement, the system consists of a two component mixture. One component is the emitter (Emt) and the other component is generically called the sensitizer (Sn). The Sn absorbs long wavelength light and has a very high intersystem crossing quantum yield (see Chapter 2.5.6 for quantum yields). Therefore, it usually ends in the first excited triplet state. At this point it is important to note that triplet states show relatively long lifetimes (~microseconds) which allow for some diffusion after the light absorption event. As we will now see, diffusion is an essential part of this UC approach.

The T-T UC process is made possible through the following steps (see Fig. **19.11A** for details).

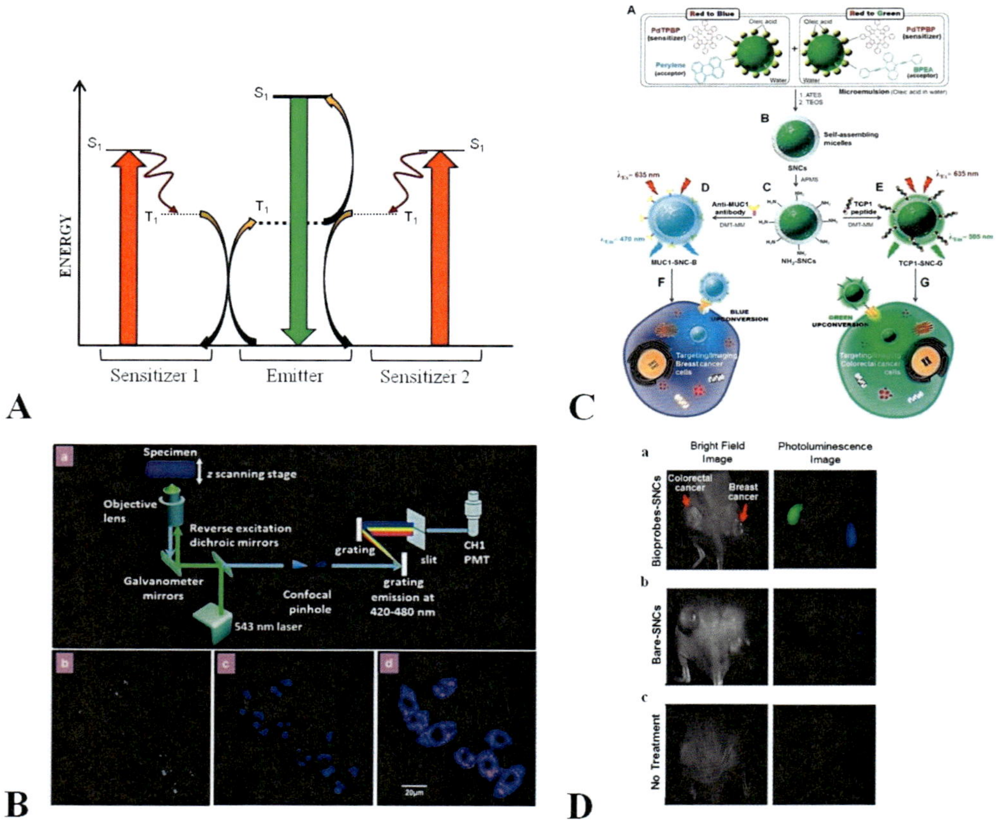

Fig. (19.11). T-T UC for bioimaging. **A:** Mechanistic scheme of T-T UC. A sensitizer agent (left) absorbs a low-energy photon (red arrow) and is excited to the first singlet state (S_1). After some time, it decays (wiggly arrow) to the triplet state (T_1). Upon collision with an emitter, the sensitizer transfers its excitation energy to the former (curved arrows), so that the sensitizer returns to the ground state while the emitter is pumped directly to the T_1 state. The process of photon absorption and sensitizer excitation is repeated (right), and when the excited sensitizer collides with the T_1 emitter there is a second energy exchange. The sensitizer returns to the ground state and the emitter is further excited to the S_1 state. From there, the emitter agent decays, radiatively emitting an upconverted high-energy photon (green arrow). **B:** T-T UC for living cell visualization. (a) Cartoon of the experimental setup to observe T-T UC in cell cultures. (b) UC images from silica nanoparticles doped with a palladium metal complex (sensitizer) + diphenylanthracene (emitter) taken up by living Hela cells. The excitation source was a cw laser emitting at 543 nm. (c) Conventional fluorescence from the emitter excited at 405 nm. (d) Overlay of bright field cell images, panel (b) and panel (c). The UC emission has been given a false colour (pinkish) to distinguish it from the conventional fluorescence. Reprinted and adapted with permission from [68]. Copyright (2012) American Chemical Society. **C:** Cartoon presenting the two types of silica nanoparticles employed to differentially target and label breast and colorectal cancer cells (see D below). The sensitizer molecule is the same in both cases: a palladium benzoporphyrin derivative (PdTPBP) that absorbs 635 nm light. To image breast cancer cells, the emitter is perylene (peak emission 470 nm). For colorectal cancer cells the emitter is an anthracene derivative (BPEA, peak emission 505 nm). The nanoparticles were functionalized with antibody anti-MUC1 (recognizing breast cancer cells) or TCP1 peptide (for colorectal cancer cells). **D:** Detection of breast cancer or colorectal cancer cells using two different types of T-T UC nanoparticles (see C above). Mice bearing both types of tumour were injected with a nanoparticle mix. After 16 h, the mice were imaged in a fluorescence chamber. It can be seen that functionalized nanoparticles (a) correctly recognized and labelled their target tumour type. Non-functionalized nanoparticles (b) still localize to the tumoral tissue (probably due to higher vascular leakiness), but colour differences and, especially emission intensities, are diminished. Control mice (c) not subjected to the nanoparticles showed no fluorescence emission. C and D reprinted with permission from [69]. Copyright (2016) American Chemical Society.

First a Sn (usually a metal complex but can also be other type of chromophore) absorbs a photon, becomes excited and undergoes intersystem crossing to the first triplet state T_1. This T_1 Sn* collides with an Emt agent (this usually, but not exclusively, is an organic molecule) and transfers its excitation energy to it. The Emt* goes directly to its T_1 state which is also very long-lived. Note that triplet states can be populated directly through collisions. The T_1 excitation energies of each partner are selected as to allow such spontaneous energy transfer from Sn* to Emt* to proceed as efficiently as possible. Under adequate light flux conditions, the excited T_1 Emt* has a high chance of colliding with another T_1 Sn* that has been excited by a second photon. Collision of Sn* and Emt*, both in the T_1 state, deactivates Sn and pumps Emt** to the first singlet (S_1) state. Radiative transitions (fluorescence) are much more likely from the S_1 state to the ground state (S_0) than is triplet to singlet deactivation (phosphorescence). The end result is that a large part of the energy of two absorbed photons is released as one upconverted photon with a shorter wavelength.

Examples of practical uses of TT-UC in bioimaging are shown in Fig. (**19.11**). In the first example, a cw laser confocal system is employed to excite TT-UC in silica nanoparticles that have been incorporated by living HeLa cells (see Fig. **19.11B**) [68]. A Pd-octaethylporphyrin metal complex was employed as Sn, in conjunction with 9,10-diphenylanthracene as Emt. Thanks to the blue (420-480 nm) UC light emission, the images taken with 543 nm excitation light (panel (b)) clearly show the location of the nanoparticles within the cell. In Fig. (**19.11C, D**), T-T UC is employed to distinguish between two different tumour types implanted in the same mouse thanks to selective-targeting nanoparticles that display different UC colours [69]. The Sn (palladium(II) tetraphenyltetrabenzoporphyrin) is the same for both nanoparticles, and it absorbs in the red spectral region. The Emt is different for each nanoparticle type: one carries perylene (emitting in the blue), and the other carries 9,10-bis-(phenylethynyl) anthracene (emitting in the green). As a result of the different nanoparticle functionalization, blue-emitting particles detected breast cancer cells, while green-emitting particles detected colorectal cancer cells. The two different tumours are clearly differentiated in the fluorescence images under the same red light excitation (see Fig. **19.11D**).

Currently, T-T UC is a young but established imaging and photochemical technique [47, 66]. More and more applications are being reported, not only in the imaging field, but also to sensitize photochemical reactions through low-energy photons that otherwise would not take place. The UC process itself is not qualitatively different in T-T UC from the other processes presented previously, nor does it afford a particularly efficient mechanism of obtaining anti-Stokes photons if compared with other non-linear methodologies. However, we see a tremendous advantage in T-T UC: the extremely low light intensities necessary for the UC to be excited. For many applications, this is below 1 W cm^{-2}, although some publications report UC emission below 1 mW cm^{-2} excitation intensity. This is unparalleled by any other technique introduced in this chapter. Of course, less intense light means a lower chance of perturbing the system under study. Also, T-T UC combinations working with photons in the red or near infrared (700-850 nm) are now available, taking advantage of the biological window to reach the structure of interest [70, 71]. Following this, detection of even a few visible UC photons can be done with excellent reliability given the extremely sensitive photodetectors now existing.

In our opinion, the most important shortcoming with regard to T-T UC use is that the triplet-relying UC system is highly sensitive to the presence of dissolved O_2 in the medium. This is easy to understand, as ground state $^3O_2$ efficiently interacts with excited triplet states, deactivating them. At the same time, the deactivating $^3O_2$ absorbs the excitation energy and is pumped to the singlet state ($^1O_2$). As it was already explained in Chapter 2.5.4, $^1O_2$ is a reactive oxygen species, responsible for much of the chromophore photobleaching action in oxygenated systems (*e.g.* cell cultures or living tissues/samples). Much research is being dedicated to find T-T UC systems impervious to or, at least, less sensible to O_2 action. Encapsulating the T-T UC system within an anoxic matrix (*e.g.* nanoparticle) is a very promising approach. On the other hand, a use that takes advantage of this O_2 triplet-deactivating feature is the measurement of partial O_2

pressure through a T-T UC system. As O_2 pressure/concentration increases, the T-T UC system emits less and less UC light. Adequate calibration allows accurate O_2 concentration assessment [72].

5. RAMAN MICROSCOPY

Raman microscopy is based upon the Raman scattering phenomenon (see Chapter 5.1). Briefly, when light of a certain wavelength is scattered, there is a chance that the outgoing photon has less energy (Stokes-shift) than the incoming photon. The energy difference appears as molecular vibrational excitation. This is an example of inelastic light scattering [73]. Raman microscopy can use this classical light scattering effect, but most commonly it is the non-linear stimulated Raman scattering (SRS) that is employed (thanks to the use of lasers). Qualitatively, SRS and normal Raman scattering are equivalent. However, while Raman scattering relies on a spontaneous, uncontrolled, and inelastic scattering event to take place, SRS "forces" this event to happen through a stimulated process that requires illuminating the sample with both exciting and Stokes-shifted photons (see Fig. **19.12A**).

For example, let's assume that we excite a water sample with 532 nm green light. One of the Raman-active vibrational levels of water is 0.43 eV above the fundamental level. Therefore, exciting at 532 nm (2.33 eV) we will find a Stokes-shifted Raman band at ~ 652 nm (1.9 eV = 2.33 - 0.43 eV). Now, if we excite water both at 532 and 652 nm, we will observe that light at 532 nm gets depleted while light at 652 nm becomes amplified through the non-linear SRS. By measuring the increase in the Raman line/band, imaging of specific vibrational modes within the sample can be obtained [74]. Properly speaking, SRS is not a non-linear process, as it depends linearly on the excitation light intensity; however, because it depends on a stimulated light emission process, we have found it adequate to include SRS in this chapter.

In particular, for bioimaging and fluorescence microscopy, SRS offers some advantages. It is a label-free technique because endogenous biomolecules can be directly visualized with it [74, 75]. However, this does not ban the use of exogenous molecules to enhance the scattering response [76]. It requires relatively low laser intensities and can be exploited with cw lasers [77]. It is a very flexible methodology, as it can detect many different molecules by simply changing the light wavelengths involved in the SRS process. The only required information is the vibrational energy difference of a particular molecule to be imaged with high spectral resolution. Nevertheless, it is important to note that not all molecular vibrational modes are Raman-active modes. There are selection rules that allow or forbid particular molecular vibrations to be excited through Raman scattering

[73]. Use of Raman microscopy requires previous knowledge of those Raman-active modes in the molecule to be imaged. Resolution is quite good, in particular when the SRS is combined with a confocal setup. Obviously, the resolving power will depend on the wavelength of the Stokes-shifted light (the light carrying information about the particular molecule). Assuming SRS emission in the visible, lateral resolutions below the micron and depth resolutions in the 1-2 microns range are attainable.

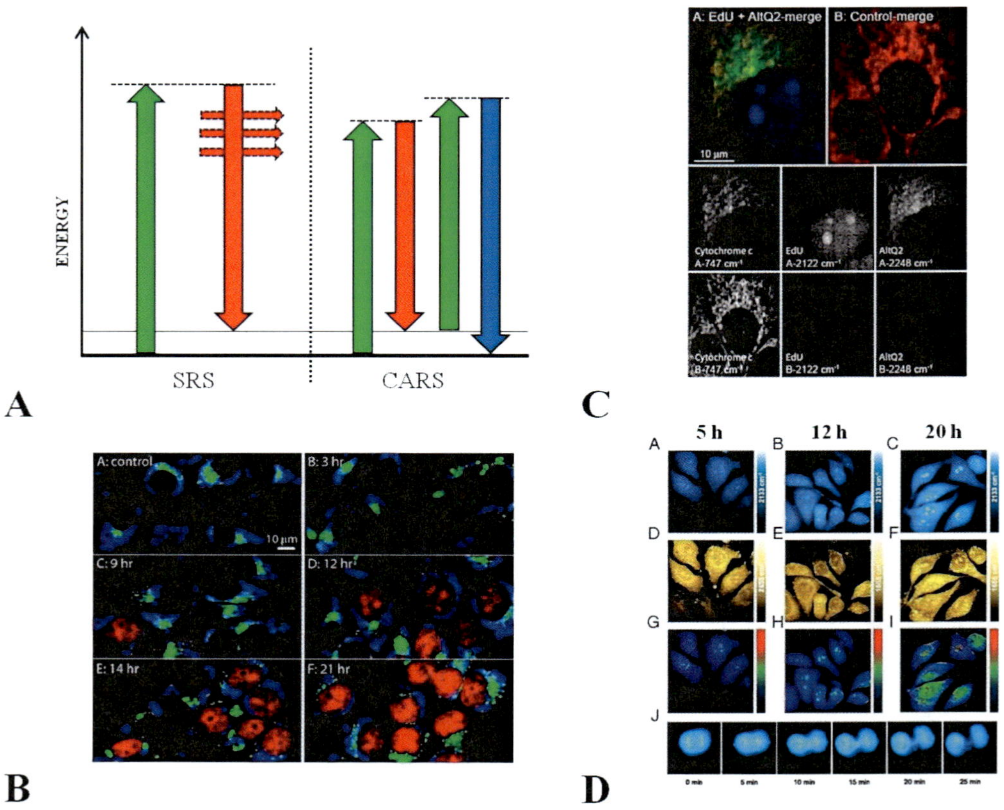

Fig. (19.12). Raman scattering in bioimaging. **A:** Scheme of the principal events taking place in stimulated Raman scattering (SRS) and coherent anti-Stokes Raman scattering (CARS). In SRS the sample is illuminated with light (green arrow) that transiently excites it to a virtual non-resonant level (dashed line). At the same time, longer wavelength light (small red arrows) interacts with this transiently excited state to stimulate emission of another long wavelength photon (red downward arrow) that brings the molecule to a real vibrationally-excited state (thin line). In CARS, two short wavelength photons (green arrows) transiently interact with a long wavelength photon (red arrow) through virtual levels (dashed lines) to stimulate the emission of a high energy anti-Stokes shifted photon (blue arrow). Notice that the molecule is not excited at the end of the process, contrary to what happens in SRS where the molecule is left in a vibrationally-excited state after the light interaction. **B:** Kinetics of EdU incorporation in living HeLa cells under culture conditions. The images presented are false colour-transformed to show SRS signals at 749 cm^{-1} (blue, cytochrome c), 2123 cm^{-1} (red, EdU) and 2849 cm^{-1} (green, lipids). In the images, the blue signal indicates mitochondrial localization, and green highlights the Golgi-endoplasmic reticulum. With time (3-21 h), the Raman-active nucleic acid analogue EdU is incorporated into the chromatin and, therefore the nuclei appear red as time passes. Reprinted with permission from [78]. Copyright (2011) American Chemical Society. **C:** Raman multi-signal from three different intracellular compounds in living HeLa cells. The top images show a false colour overlay of the Raman signals obtained (left experimental cell, right control cell). The greyscale panels show the obtained intensity Raman signals for the experimental cell (middle images) and the control one (bottom images). The signal from cytochrome c is observed in both cells. EdU and AltQ2 signals are observed only in incubated cells. EdU clearly localizes in the nucleus. AltQ2 overlaps in its distribution with cytochrome c. This is expected as both AltQ2 and cytochrome c should localize in the mitochondrial compartment. Reprinted with permission from [79]. Copyright (2012) American Chemical Society. **D:***De novo* protein synthesis in living Hela cells. HeLa cells were incubated for the indicated times with a deuterated leucine analogue which displays a Raman-active signal at 2133 cm^{-1} (A-C). The signal intensity increases with time. Imaging at 1655 cm^{-1} (total protein levels due to amide bond vibration) indicates that total protein levels are steady (D-F). The ratio between total protein content and new protein was calculated and is represented as false colour images (G-I) in which warmer colours indicate a relative abundance of newly synthesized protein. Sequence of images (J) of a dividing HeLa cell that has been exposed to the deuterated amino acid for 20 h. The ana-telophase clearly indicates that the procedure does not interfere with normal cell metabolism. Additionally, some structures can be clearly visualized with this methodology like the chromosomal spindles. Reproduced with permission from [80].

Some examples of bioimaging with SRS will follow. In Fig. (**19.12B**) a uridine nucleic acid analogue (5-ethynyl-2'-deoxyuridine (EdU), with a triple carbon-carbon (alkyne) bond) is employed to obtain Raman images of nucleic acid incorporation [78]. The alkyne bond displays a Raman active signal at 2123 cm^{-1}, in a spectral region where few other molecules are active, in particular biological molecules. Living HeLa cells in culture were incubated with EdU and the incorporation signal was measured as a function of time. At time 0 h (control), there is no EdU signal. As time passes, there is a clear increase in EdU signal (red false colour in the images). Two other vibrations, corresponding to cytochrome c and lipids, were also imaged as Raman controls. The same research group reported a two-colour alkyne signal from two different alkyne-modified biomolecules within the same cell [79]. The researchers employed EdU and a coenzyme Q derivative displaying an alkyne moiety (AltQ2) to image HeLa cells (see Fig. **19.12C**). In the control cell only, the endogenous Raman signal from cytochrome c is clearly observed. In the treated cells the signal from AltQ2 shows an overlapping emission with that coming from cytochrome c. This is expected as AltQ2 is a derivative of coenzyme Q, which is located in the mitochondria under physiological circumstances. EdU localizes only within the nucleus, which again is expected as it gets incorporated into the DNA molecule.

Another example of multi-pseudocolour assessment of cellular parameters is provided in Fig. (**19.12D**). In these impressive images, the levels of *de novo* synthesized protein are assessed in living HeLa cells [80, 81]. The cell culture was incubated with a deuterated analogue of the amino acid leucine. The bond C-D (carbon-deuterium) is present under normal conditions in negligible amounts.

Therefore, cells incubated with this deuterated analogue will incorporate it to new proteins. The C-D bond displays a Raman-active signal at 2133 cm^{-1}, a spectral region usually "dark" for cells in normal conditions. The time sequence shows an increase in the signal intensity over time. When the total amount of protein was also measured (employing the 1655 cm^{-1} signal for the imide bond), it was verified that it is maintained in time. Establishing a ratio between signals at 2133 cm^{-1} and 1655 cm^{-1} (images G-I in Fig. **19.12D**) clearly indicates that new protein was being synthesized between 5 and 20 h incubation period. The methodology seems not to interfere with normal metabolism, as cells were imaged dividing without problems after subjecting them to the technique. Indeed, the SRS imaging was employed to obtain the images of a dividing cell.

We would like to briefly address the Raman-based technique of coherent anti-Stokes Raman scattering (CARS). This spectroscopic and imaging technique truly depends on non-linear optical interactions, as it requires four photons to interact at the same time (see Fig. **19.12A**). The process can be understood as a double Raman scattering taking place simultaneously on the same molecule [3]. A photon (called fundamental here) is scattered by a Raman-active bond producing a Stokes shifted photon of longer wavelength, and the molecule temporally occupies an excited vibrational level. Immediately, a second fundamental photon is scattered by the vibrationally-excited molecule, but this time the vibrational energy is added to the outgoing photon. This photon is anti-Stokes shifted in regards to the two fundamental photons. The experimental setup usually provides fundamental photons and Raman-shifted photons at the same time. Through non-linear interactions, two fundamental photons are annihilated and one Stokes and one anti-Stokes shifted are created. The total energy in the system is conserved as the energy originally contained in the two fundamental photons is shared by the Raman-shifted photons. CARS offers the advantage that anti-Stokes shifted photons are emitted, which we have seen is particularly adequate to get rid of autofluorescence and other perturbing signals that appear on the Stokes side of the exciting spectrum [82]. Additionally, CARS provides a coherent signal, as coherence is preserved in the photons system. This provides better resolving power with confocal-like properties. On the other hand, it requires more astringent optical conditions as compared with Raman microscopy or SRS.

Non-linear optics, and its applications to optical imaging, is a very wide-ranging research area. We have attempted to provide the reader with some basic notions of this research area. However, we acknowledge that it is impossible to cover such a broad topic in a single book chapter. The literature cited provides a good starting point for the interested reader in furthering his or her knowledge of this exciting topic.

CONFLICT OF INTEREST

The author (editor) declares no conflict of interest, financial or otherwise.

ACKNOWLEDGEMENTS

This work was supported by the Aarhus Institute of Advanced Studies (AIAS).

REFERENCES

[1] Band YB. Light and Matter Electromagnetism, Optics, Spectroscopy and Lasers. Chichester: John Wiley & Sons Ltd 2006.

[2] Fox M. Optical properties of solids. Oxford: Oxford University Press 2001.

[3] Shen YR. The principles of nonlinear optics. New York: Wiley-Interscience NY 2002.

[4] Lakowicz JR. Principles of fluorescence spectroscopy. 3rd ed., New York: Springer Science + Business Media LLC 2010.

[5] Göppert-Mayer M. Über elementarakte mit zwei quantensprüngen. Ann Phys Leipzig 1931; 9: 273-94. [http://dx.doi.org/10.1002/andp.19314010303]

[6] Silfvast WT. Laser Fundamentals. 2nd ed., Cambridge: Cambridge University Press 2004. [http://dx.doi.org/10.1017/CBO9780511616426]

[7] Diaspro A, Chirico G, Collini M. Two-photon fluorescence excitation and related techniques in biological microscopy. Q Rev Biophys 2005; 38(2): 97-166. [http://dx.doi.org/10.1017/S0033583505004129] [PMID: 16478566]

[8] Nature Milestones-Light Microscopy. Nature milestones 2009.

[9] Nature Milestones-Photons. Nature milestones 2010.

[10] Ustione A, Piston DW. A simple introduction to multiphoton microscopy. J Microsc 2011; 243(3): 221-6. [http://dx.doi.org/10.1111/j.1365-2818.2011.03532.x] [PMID: 21777244]

[11] Garmire E. Nonlinear optics in daily life. Opt Express 2013; 21(25): 30532-44. [http://dx.doi.org/10.1364/OE.21.030532] [PMID: 24514630]

[12] Yao S, Belfield KD. Two-photon fluorescent probes for bioimaging. Eur J Org Chem 2012; 3199-217. [http://dx.doi.org/10.1002/ejoc.201200281]

[13] König K. Multiphoton microscopy in life sciences. J Microsc 2000; 200(Pt 2): 83-104. [http://dx.doi.org/10.1046/j.1365-2818.2000.00738.x] [PMID: 11106949]

[14] Diaspro A, Bianchini P, Vicidomini G, Faretta M, Ramoino P, Usai C. Multi-photon excitation microscopy. Biomed Eng Online 2006; 5: 36. [http://dx.doi.org/10.1186/1475-925X-5-36] [PMID: 16756664]

[15] Hänninen PE, Soini E, Hell SW. Continuous wave excitation two-photon fluorescence microscopy. J Microsc 1994; 176: 222-5. [http://dx.doi.org/10.1111/j.1365-2818.1994.tb03518.x]

[16] Haydar TF, Ang E Jr, Rakic P. Mitotic spindle rotation and mode of cell division in the developing telencephalon. Proc Natl Acad Sci USA 2003; 100(5): 2890-5. [http://dx.doi.org/10.1073/pnas.0437969100] [PMID: 12589023]

[17] Cicchi R, Pavone FS. Multimodal nonlinear microscopy: A powerful label-free method for supporting standard diagnostics on biological tissues. J Innov Opt Health Sci 2014; 7: 1330008. [http://dx.doi.org/10.1142/S1793545813300085]

[18] Uckermann O, Galli R, Beiermeister R, *et al.* Endogenous two-photon excited fluorescence provides label-free visualization of the inflammatory response in the rodent spinal cord. BioMed Res Int 2015; 2015: 859084.
[http://dx.doi.org/10.1155/2015/859084] [PMID: 26355949]

[19] Levitt JM, McLaughlin-Drubin ME, Münger K, Georgakoudi I. Automated biochemical, morphological, and organizational assessment of precancerous changes from endogenous two-photon fluorescence images. PLoS One 2011; 6(9): e24765.
[http://dx.doi.org/10.1371/journal.pone.0024765] [PMID: 21931846]

[20] Zipfel WR, Williams RM, Webb WW. Nonlinear magic: multiphoton microscopy in the biosciences. Nat Biotechnol 2003; 21(11): 1369-77.
[http://dx.doi.org/10.1038/nbt899] [PMID: 14595365]

[21] Joly P, Duda GN, Schöne M, *et al.* Geometry-driven cell organization determines tissue growths in scaffold pores: consequences for fibronectin organization. PLoS One 2013; 8(9): e73545.
[http://dx.doi.org/10.1371/journal.pone.0073545] [PMID: 24039979]

[22] Shi Y, Ivannikov MV, Walsh ME, *et al.* The lack of CuZnSOD leads to impaired neurotransmitter release, neuromuscular junction destabilization and reduced muscle strength in mice. PLoS One 2014; 9(6): e100834.
[http://dx.doi.org/10.1371/journal.pone.0100834] [PMID: 24971750]

[23] Kobat D, Durst ME, Nishimura N, Wong AW, Schaffer CB, Xu C. Deep tissue multiphoton microscopy using longer wavelength excitation. Opt Express 2009; 17(16): 13354-64.
[http://dx.doi.org/10.1364/OE.17.013354] [PMID: 19654740]

[24] Wang BG, König K, Halbhuber KJ. Two-photon microscopy of deep intravital tissues and its merits in clinical research. J Microsc 2010; 238(1): 1-20.
[http://dx.doi.org/10.1111/j.1365-2818.2009.03330.x] [PMID: 20384833]

[25] Hoover EE, Squier JA. Advances in multiphoton microscopy technology. Nat Photonics 2013; 7(2): 93-101.
[http://dx.doi.org/10.1038/nphoton.2012.361] [PMID: 24307915]

[26] Yanez CO, Morales AR, Yue X, *et al.* Deep vascular imaging in wounds by two-photon fluorescence microscopy. PLoS One 2013; 8(7): e67559.
[http://dx.doi.org/10.1371/journal.pone.0067559] [PMID: 23844028]

[27] Ohno Y, Otaki JM. Live cell imaging of butterfly pupal and larval wings *in vivo.* PLoS One 2015; 10(6): e0128332.
[http://dx.doi.org/10.1371/journal.pone.0128332] [PMID: 26107809]

[28] Tkaczyk ER, Tkaczyk AH. Multiphoton flow cytometry strategies and applications. Cytometry A 2011; 79(10): 775-88.
[http://dx.doi.org/10.1002/cyto.a.21110] [PMID: 21796772]

[29] Dal Maschio M, Difato F, Beltramo R, Blau A, Benfenati F, Fellin T. Simultaneous two-photon imaging and photo-stimulation with structured light illumination. Opt Express 2010; 18(18): 18720-31.
[http://dx.doi.org/10.1364/OE.18.018720] [PMID: 20940765]

[30] Jahn K, Hille C. Asante Calcium Green and Asante Calcium Red--novel calcium indicators for two-photon fluorescence lifetime imaging. PLoS One 2014; 9(8): e105334.
[http://dx.doi.org/10.1371/journal.pone.0105334] [PMID: 25140519]

[31] Kumar S, Dunsby C, De Beule PA, *et al.* Multifocal multiphoton excitation and time correlated single photon counting detection for 3-D fluorescence lifetime imaging. Opt Express 2007; 15(20): 12548-61.
[http://dx.doi.org/10.1364/OE.15.012548] [PMID: 19550524]

[32] Wei L, Min W. Pump-probe optical microscopy for imaging nonfluorescent chromophores. Anal

Bioanal Chem 2012; 403(8): 2197-202.
[http://dx.doi.org/10.1007/s00216-012-5890-1] [PMID: 22411535]

[33] Vogel A, Noack J, Hüttman G, *et al.* Mechanisms of femtosecond laser nanosurgery of cells and tissues. Appl Phys B 2005; 81: 1015-47.
[http://dx.doi.org/10.1007/s00340-005-2036-6]

[34] Tong L, Cheng JX. Label-free imaging through nonlinear optical signals. Mater Today 2011; 14: 264-73.
[http://dx.doi.org/10.1016/S1369-7021(11)70141-9]

[35] Carriles R, Schafer DN, Sheetz KE, *et al.* Invited review article: Imaging techniques for harmonic and multiphoton absorption fluorescence microscopy. Rev Sci Instrum 2009; 80(8): 081101.
[http://dx.doi.org/10.1063/1.3184828] [PMID: 19725639]

[36] Pal R, Yang J, Ortiz D, *et al.* *In-vivo* nonlinear optical microscopy (NLOM) of epithelial-connective tissue interface (ECTI) reveals quantitative measures of neoplasia in hamster oral mucosa. PLoS One 2015; 10(1): e0116754.
[http://dx.doi.org/10.1371/journal.pone.0116754] [PMID: 25633927]

[37] Yelin D, Silberberg Y. Third-harmonic microscopy in biology. Eur Microsc Anal 2000; 68: 13-5.

[38] Rogov A, Irondelle M, Ramos Gomes F, *et al.* Simultaneous multiharmonic imaging of nanoparticles in tissues for increased selectivity. ACS Photonics 2015; 2: 1416-22.
[http://dx.doi.org/10.1021/acsphotonics.5b00289]

[39] Li J, Qiu J, Guo W, *et al.* Cellular internalization of $LiNbO_3$ nanocrystals for second harmonic imaging and the effects on stem cell differentiation. Nanoscale 2016; 8(14): 7416-22.
[http://dx.doi.org/10.1039/C6NR00785F] [PMID: 27001708]

[40] Staedler D, Magouroux T, Hadji R, *et al.* Harmonic nanocrystals for biolabeling: a survey of optical properties and biocompatibility. ACS Nano 2012; 6(3): 2542-9.
[http://dx.doi.org/10.1021/nn204990n] [PMID: 22324660]

[41] Macias-Romero C, Didier ME, Zubkovs V, *et al.* Probing rotational and translational diffusion of nanodoublers in living cells on microsecond time scales. Nano Lett 2014; 14(5): 2552-7.
[http://dx.doi.org/10.1021/nl500356u] [PMID: 24735468]

[42] Čulić-Viskota J, Dempsey WP, Fraser SE, Pantazis P. Surface functionalization of barium titanate SHG nanoprobes for *in vivo* imaging in zebrafish. Nat Protoc 2012; 7(9): 1618-33.
[http://dx.doi.org/10.1038/nprot.2012.087] [PMID: 22899331]

[43] Auzel F. Upconversion and anti-Stokes processes with f and d ions in solids. Chem Rev 2004; 104(1): 139-73.
[http://dx.doi.org/10.1021/cr020357g] [PMID: 14719973]

[44] Cheng L, Wang C, Liu Z. Upconversion nanoparticles and their composite nanostructures for biomedical imaging and cancer therapy. Nanoscale 2013; 5(1): 23-37.
[http://dx.doi.org/10.1039/C2NR32311G] [PMID: 23135546]

[45] Liu Y, Tu D, Zhu H, Ma E, Chen X. Lanthanide-doped luminescent nano-bioprobes: from fundamentals to biodetection. Nanoscale 2013; 5(4): 1369-84.
[http://dx.doi.org/10.1039/C2NR33239F] [PMID: 23223801]

[46] Prodi L, Rampazzo E, Rastrelli F, Speghini A, Zaccheroni N. Imaging agents based on lanthanide doped nanoparticles. Chem Soc Rev 2015; 44(14): 4922-52.
[http://dx.doi.org/10.1039/C4CS00394B] [PMID: 26090530]

[47] Zhou J, Liu Q, Feng W, Sun Y, Li F. Upconversion luminescent materials: advances and applications. Chem Rev 2015; 115(1): 395-465.
[http://dx.doi.org/10.1021/cr400478f] [PMID: 25492128]

[48] Wang F, Liu X. Multicolor tuning of lanthanide-doped nanoparticles by single wavelength excitation.

Acc Chem Res 2014; 47(4): 1378-85.
[http://dx.doi.org/10.1021/ar5000067] [PMID: 24611606]

[49] Boyer JC, Vetrone F, Cuccia LA, Capobianco JA. Synthesis of colloidal upconverting $NaYF_4$ nanocrystals doped with Er^{3+}, Yb^{3+} and Tm^{3+}, Yb^{3+} *via* thermal decomposition of lanthanide trifluoroacetate precursors. J Am Chem Soc 2006; 128(23): 7444-5.
[http://dx.doi.org/10.1021/ja061848b] [PMID: 16756290]

[50] Boyer JC, Cuccia LA, Capobianco JA. Synthesis of colloidal upconverting $NaYF_4$: Er^{3+}/Yb^{3+} and Tm^{3+}/Yb^{3+} monodisperse nanocrystals. Nano Lett 2007; 7(3): 847-52.
[http://dx.doi.org/10.1021/nl070235+] [PMID: 17302461]

[51] Dong H, Du SR, Zheng XY, *et al.* Lanthanide nanoparticles: From design toward bioimaging and therapy. Chem Rev 2015; 115(19): 10725-815.
[http://dx.doi.org/10.1021/acs.chemrev.5b00091] [PMID: 26151155]

[52] Li K, Su Q, Yuan W, *et al.* Ratiometric monitoring of intracellular drug release by an upconversion drug delivery nanosystem. ACS Appl Mater Interfaces 2015; 7(22): 12278-86.
[http://dx.doi.org/10.1021/acsami.5b03204] [PMID: 25975535]

[53] Lim SF, Riehn R, Ryu WS, *et al. In vivo* and scanning electron microscopy imaging of up-converting nanophosphors in *Caenorhabditis elegans.* Nano Lett 2006; 6(2): 169-74.
[http://dx.doi.org/10.1021/nl0519175] [PMID: 16464029]

[54] Hu Y, Li J, Zhu X, *et al.* 17β-estradiol-loaded PEGlyated upconversion nanoparticles as a bone-targeted drug nanocarrier. ACS Appl Mater Interfaces 2015; 7(29): 15803-11.
[http://dx.doi.org/10.1021/acsami.5b02831] [PMID: 26133323]

[55] DaCosta MV, Doughan S, Han Y, Krull UJ. Lanthanide upconversion nanoparticles and applications in bioassays and bioimaging: a review. Anal Chim Acta 2014; 832: 1-33.
[http://dx.doi.org/10.1016/j.aca.2014.04.030] [PMID: 24890691]

[56] Wang YF, Liu GY, Sun LD, Xiao JW, Zhou JC, Yan CH. Nd($^{3+}$)-sensitized upconversion nanophosphors: efficient *in vivo* bioimaging probes with minimized heating effect. ACS Nano 2013; 7(8): 7200-6.
[http://dx.doi.org/10.1021/nn402601d] [PMID: 23869772]

[57] Wang F, Banerjee D, Liu Y, Chen X, Liu X. Upconversion nanoparticles in biological labeling, imaging, and therapy. Analyst (Lond) 2010; 135(8): 1839-54.
[http://dx.doi.org/10.1039/c0an00144a] [PMID: 20485777]

[58] Chen J, Zhao JX. Upconversion nanomaterials: synthesis, mechanism, and applications in sensing. Sensors (Basel) 2012; 12(3): 2414-35.
[http://dx.doi.org/10.3390/s120302414] [PMID: 22736958]

[59] Liu Q, Feng W, Yang T, Yi T, Li F. Upconversion luminescence imaging of cells and small animals. Nat Protoc 2013; 8(10): 2033-44.
[http://dx.doi.org/10.1038/nprot.2013.114] [PMID: 24071909]

[60] Zhang Y, Wei W, Das GK, *et al.* Engineering lanthanide-based materials for nanomedicine. J Photochem Photobiol Photochem Rev 2014; 20: 71-96.
[http://dx.doi.org/10.1016/j.jphotochemrev.2014.06.001]

[61] Aboshyan-Sorgho L, Cantuel M, Petoud S, *et al.* Optical sensitization and upconversion in discrete polynuclear chromium-lanthanide complexes. Coord Chem Rev 2012; 256: 1644-63.
[http://dx.doi.org/10.1016/j.ccr.2011.12.013]

[62] Suffren Y, Zare D, Eliseeva SV, *et al.* Near-infrared to visible light-upconversion in molecules: From dream to reality. J Phys Chem C 2013; 117: 26957-63.
[http://dx.doi.org/10.1021/jp4107519]

[63] Aboshyan-Sorgho L, Besnard C, Pattison P, *et al.* Near-infrared→visible light upconversion in a molecular trinuclear d-f-d complex. Angew Chem Int Ed Engl 2011; 50(18): 4108-12.

[http://dx.doi.org/10.1002/anie.201100095] [PMID: 21462286]

[64] Aboshyan-Sorgho L, Nozary H, Aebischer A, *et al.* Optimizing millisecond time scale near-infrared emission in polynuclear chrome(III)-lanthanide(III) complexes. J Am Chem Soc 2012; 134(30): 12675-84.
[http://dx.doi.org/10.1021/ja304009b] [PMID: 22725838]

[65] Singh-Rachford TN, Castellano FN. Photon upconversion based on sensitized triplet-triplet annihilation. Coord Chem Rev 2010; 254: 2560-73.
[http://dx.doi.org/10.1016/j.ccr.2010.01.003]

[66] Monguzzi A, Tubino R, Hoseinkhani S, Campione M, Meinardi F. Low power, non-coherent sensitized photon up-conversion: modelling and perspectives. Phys Chem Chem Phys 2012; 14(13): 4322-32.
[http://dx.doi.org/10.1039/c2cp23900k] [PMID: 22370856]

[67] Zhao J, Wu W, Sun J, Guo S. Triplet photosensitizers: from molecular design to applications. Chem Soc Rev 2013; 42(12): 5323-51.
[http://dx.doi.org/10.1039/c3cs35531d] [PMID: 23450221]

[68] Liu Q, Yang T, Feng W, Li F. Blue-emissive upconversion nanoparticles for low-power-excited bioimaging *in vivo*. J Am Chem Soc 2012; 134(11): 5390-7.
[http://dx.doi.org/10.1021/ja3003638] [PMID: 22369318]

[69] Kwon OS, Song HS, Conde J, Kim HI, Artzi N, Kim JH. Dual-color emissive upconversion nanocapsules for differential cancer bioimaging *in vivo*. ACS Nano 2016; 10(1): 1512-21.
[http://dx.doi.org/10.1021/acsnano.5b07075] [PMID: 26727423]

[70] Wohnhaas C, Mailänder V, Dröge M, *et al.* Triplet-triplet annihilation upconversion based nanocapsules for bioimaging under excitation by red and deep-red light. Macromol Biosci 2013; 13(10): 1422-30.
[http://dx.doi.org/10.1002/mabi.201300149] [PMID: 23868857]

[71] Amemori S, Yanai N, Kimizuka N. Metallonaphthalocyanines as triplet sensitizers for near-infrared photon upconversion beyond 850 nm. Phys Chem Chem Phys 2015; 17(35): 22557-60.
[http://dx.doi.org/10.1039/C5CP02733K] [PMID: 26270770]

[72] Borisov SM, Larndorfer C, Klimant I. Triplet-triplet annihilation-based anti-Stokes oxygen sensing materials with a very broad dynamic range. Adv Funct Mater 2012; 22: 4360-8.
[http://dx.doi.org/10.1002/adfm.201200794]

[73] Svanberg S. Atomic and molecular spectroscopy Basic aspects and practical applications. 4th ed., Berlin, Heidelberg: Springer-Verlag 2004.

[74] Min W, Freudiger CW, Lu S, Xie XS. Coherent nonlinear optical imaging: beyond fluorescence microscopy. Annu Rev Phys Chem 2011; 62: 507-30.
[http://dx.doi.org/10.1146/annurev.physchem.012809.103512] [PMID: 21453061]

[75] Palonpon AF, Ando J, Yamakoshi H, *et al.* Raman and SERS microscopy for molecular imaging of live cells. Nat Protoc 2013; 8(4): 677-92.
[http://dx.doi.org/10.1038/nprot.2013.030] [PMID: 23471112]

[76] Hong S, Lin L, Xiao M, Chen X. Live-cell bioorthogonal Raman imaging. Curr Opin Chem Biol 2015; 24: 91-6.
[http://dx.doi.org/10.1016/j.cbpa.2014.10.018] [PMID: 25461727]

[77] Hu CR, Slipchenko MN, Wang P, *et al.* Stimulated Raman scattering imaging by continuous-wave laser excitation. Opt Lett 2013; 38(9): 1479-81.
[http://dx.doi.org/10.1364/OL.38.001479] [PMID: 23632524]

[78] Yamakoshi H, Dodo K, Okada M, *et al.* Imaging of EdU, an alkyne-tagged cell proliferation probe, by Raman microscopy. J Am Chem Soc 2011; 133(16): 6102-5.
[http://dx.doi.org/10.1021/ja108404p] [PMID: 21443184]

[79] Yamakoshi H, Dodo K, Palonpon A, *et al.* Alkyne-tag Raman imaging for visualization of mobile small molecules in live cells. J Am Chem Soc 2012; 134(51): 20681-9. [http://dx.doi.org/10.1021/ja308529n] [PMID: 23198907]

[80] Wei L, Yu Y, Shen Y, Wang MC, Min W. Vibrational imaging of newly synthesized proteins in live cells by stimulated Raman scattering microscopy. Proc Natl Acad Sci USA 2013; 110(28): 11226-31. [http://dx.doi.org/10.1073/pnas.1303768110] [PMID: 23798434]

[81] Wei L, Hu F, Shen Y, *et al.* Live-cell imaging of alkyne-tagged small biomolecules by stimulated Raman scattering. Nat Methods 2014; 11(4): 410-2. [http://dx.doi.org/10.1038/nmeth.2878] [PMID: 24584195]

[82] Rodriguez LG, Lockett SJ, Holtom GR. Coherent anti-stokes Raman scattering microscopy: a biological review. Cytometry A 2006; 69(8): 779-91. [http://dx.doi.org/10.1002/cyto.a.20299] [PMID: 16752420]

CHAPTER 20

Super-Resolution Microscopy

Abstract: In this final chapter, the fundamentals and main super-resolution microscopy (SRM) techniques currently in use will be presented. Super-resolution allows imaging of structures that are not usually resolvable with optical frequencies in the visible range. There are different strategies to be able to do so and these will be introduced in this chapter. This field of SRM is relatively young yet it providies huge amount of information on biological process and structures. This is done with experimental setups that are much more benign to biological samples in comparison to other "super-resolution" techniques like electron microscopy. In the chapter, the fundamentals and techniques will be presented in a simple and intuitive way. This is a field currently experiencing a huge development and new advancements are guaranteed in the upcoming years.

Keywords: 4Pi microscopy, Abbe limit, Airy disc, Confocal microscopy, Diffraction limit, I^5M microscopy, Moiré fringes, Photoactivated Localization Microscopy (PALM), Point spread function (PSF), Rayleigh criterion, Resolution, REversible Saturable OpticaL Fluorescence Transitions (RESOLFT), Saturated-SIM (S-SIM), Scanning near-field optical microscope (SNOM), STimulated Emission Depletion (STED), STochastic Optical Reconstruction Microscopy (STORM), Structured Illumination Microscopy (SIM), Total internal reflection fluorescence microscopy (TIRFM).

1. THE DIFFRACTION LIMIT

To appreciate the value of SRM, first one has to understand what are the limits of microscopy in the optical range. Put in another way: how small can I see with visible light? Physicists in the late 19th century already answered this question. Ernst Abbe established in 1873 that there is a minimum size for an observation spot. The spot size depends on the wavelength of radiation employed to image the spot and the amount of radiation collected [1, 2]. To calculate the spot size the following equation is employed:

$$\text{Spot size} = \lambda / 2 \cdot n \cdot \sin\alpha = \lambda / 2 \cdot NA, \tag{1}$$

Juan Carlos Stockert & Alfonso Blázquez-Castro

where λ is the light wavelength, n is the medium's refractive index and α is the half-angle of the light cone collected by the imaging objective. As ($n \cdot \sin\alpha$) is the numerical aperture (NA) of the optical system; the equation can be written in this way as well (see Chapter 6.1.3 for NA). When substituting for typical microscopic imaging conditions (500 nm wavelength light and NA ~1.3-1.5), the spot dimensions are around 200-250 nm [3, 4]. This is with regard to the lateral dimensions. The imaging depth is larger due to light emission and geometrical issues and results in minimum spot "thickness" given by;

$$\text{Spot size (z-axis)} = 2 \cdot n \cdot \lambda / NA^2 \qquad \textbf{(2)}$$

or 500-600 nm under the same conditions as the previous example [5].

These imaging limitations come from two parameters: light wavelength and collecting optics. When light is emitted by a fluorophore, it tends to "spread out" from its molecular source. This means that even when the fluorophore is usually a few nanometres in size, the emitted light has a much larger effective size (roughly the size of its wavelength). On the other hand, it is quite intuitive that the larger the amount of light we can collect from an emitting unit, the better we can pinpoint its three dimensional position. But again there is a limit to the amount of light an objective can capture. The ideal case would be to use a spherical objective retrieving all light emitted in a sphere surrounding the fluorophore. Although this "spherical objective" is not physically realizable, some super-resolution approaches try to get as spherical as possible to increase light collection (see section 2 below).

Thus there is a limitation to accurately localize a single fluorophore. Now comes the problem of distinguishing two or more fluorophores that are closely located to each other. Under most fluorescence microscopy situations, the structure/s labelled show fluorophores closely packaged in a space. There is a difficulty in distinguishing two emitting units when they are relatively close to each other. To understand why, we will first introduce the concept of the Airy disc. The Airy disc is a light intensity pattern that results from the diffractive wave nature of light in conjunction with the imaging optics of an objective [5, 6]. At the moment of imaging, the light wave interferes with itself due to diffraction and creates a "brightness pattern" (see Fig. **20.1A**).

This pattern displays a central high-brightness spot and a series of lower-brightness circles surrounding the central area (see Fig. **20.1B** for brightness *vs.* xy plane plot). If we graph the brightness through the medial plot plane (see Fig. **20.1C**), we notice that there is a minimum brightness; as we move away from the emitting object (arrow), followed by a small increase in brightness further away

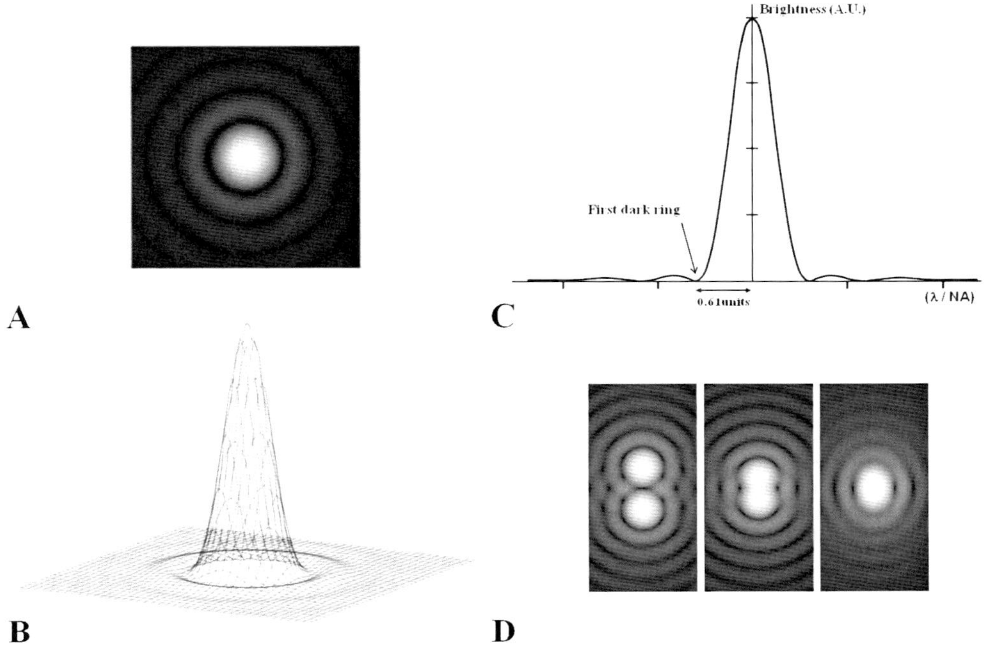

Fig. (20.1). Airy disc. **A:** Diffraction pattern creating an Airy disc. There is a central bright region (proper Airy disc) surrounded by successive dark and bright rings (Airy rings). Reproduced with permission from Sakurambo-https://commons.wikimedia.org/w/index. php?curid=18640280. **B:** Three dimensional plot of an Airy disc brightness. This is the 3D representation of the structure shown in (A). Note the ripple-like structure (Airy rings) surrounding the central peak (Airy disc). Reproduced with permission from Sakurambo-https://en.wikipedia.org/w/index.php?curid= 18371548. **C:** Brightness *vs*. radius plot of an Airy disc. Brightness is normalized to the highest value at radius=0. Radius is expressed as units of (λ / NA). The first dark ring (brightness = 0) is located at 0.61 radius units from the centre (arrow). **D:** Rayleigh criterion to discriminate two objects. When two objects emit (or scatter) light they can be discerned clearly when the centre of their respective Airy discs falls outside the first dark ring (left panel). If their peak brightness is located at the first dark ring of the other structure (middle panel) discrimination becomes difficult and the structure is blurry. If the brightness peak is within the other structure's Airy disc it is no longer feasible to discern the two structures properly (right panel). Reproduced with permission from Spencer Bliven-https://commons. wikimedia. org/w/ index.php?curid=31456019.

(first Airy ring). This pattern of dark and bright rings (Airy rings) continues as we move away from the object. The radius of the Airy disc in the xy plane from the brightness peak to the first dark ring is given by:

$$\text{Airy disc (xy-plane)} = 0.61 \cdot \lambda / \text{NA} \quad \textbf{(3)}$$

Due to two different emitting structures of the image, we must be able to recognize the Airy disc of one object from the disc of the other. Lord Rayleigh proposed an argument (Rayleigh criterion, 1879) for such a distinction: to distinguish two objects, the maximum in the intensity of an Airy disc of the first object must fall within the first dark circular area of the second object's Airy disc or further apart [3, 7]. This is exemplified in Fig. (**20.1D**) where two emitting spots are easily discernible when there is some separation between them (left panel). However, as they come closer, it becomes increasingly difficult to assess if there are two or only one object (middle panel). When the Rayleigh criterion is not satisfied (right panel), the two objects appear as one.

With this background in mind, there are two obvious approaches to enhance the optical resolution: employ shorter light wavelengths and/or increase the collecting efficiency (NA) of the system. Both have been implemented for many decades. In this context, the shorter wavelength light refers to the light emitted by the fluorophore, not the light employed to excite it. So to increase the resolution, an option is to employ compounds emitting in the near UV. But this usually requires them to be excited in the deep UV (due to the Stokes shift, see Chapter 2.4). Exciting biological samples in the UV is always discouraged as these wavelengths produce undesirable and damaging photochemical effects on them. Increasing the NA can be done by employing larger objectives and/or increasing the imaging medium refractive index. There are limitations as to how big an objective can be or how refractive the medium can behave; however,by optimizing these parameters, a lateral resolution of 150-200 nm and a depth resolution of 400-500 nm can be obtained. But this is as far as it can get.

For image at a smaller scale with fluorophores emitting visible light, other approaches are necessary. These are known as super-resolution techniques. We will describe the most common SRM techniques currently in use. They have been grouped according to their general approach in four categories: increasing collection efficiency, near-field techniques, far-field techniques, and computational strategies. Other criteria are frequently employed to describe SRM techniques, but we think our approach is intuitive and in line with this book's scope.

2. LIGHT COLLECTION ENHANCEMENT

As established in the previous section, increasing the amount of light collected by the optical system allows for an increase in resolution. This is because more light waves are collected, which then allows image reconstruction. When light is emitted or diffracted from an object, it carries information about details of the object (form, size, *etc*.). This information is codified in what is called spatial frequencies. Simple objects (*e.g.* a circle) have few details and show low spatial frequencies. Complex objects (cell or tissue) have many details and feature a full display of spatial frequencies from low to high. When an objective collects light from an object, more light waves are necessary to reconstruct the high spatial frequencies (*i.e.* the fine details). If the objective only collects a small amount of light (narrow light cone), it is missing part of the codified information by "letting go" most of the light. So to accurately reconstruct a complex object, as much light coming from it as possible must be retrieved in order to obtain the higher spatial frequencies (this again brings in the spherical objective argument). Some approaches rely on collecting as much light as possible or collecting high-quality waves to reconstruct the observed object. These microscopic techniques will be briefly described next.

2.1. Confocal

The confocal microscope described in Chapter 6.2 does, theoretically, provide some increase in resolving power [3 - 5]. As a quick reminder, confocal microscopes work by providing excitation light through a very narrow pinhole (or by a laser system) and then collecting the fluorescence light through a second pinhole. This restricts the light that reaches the photodetector to a single focal plane, rejecting light coming from other planes (above and below). Due to this confocal arrangement, arbitrarily small excitation and detection pinholes can reduce the light diffracting from the observed point. This theoretically should afford a 2x increase in the resolving power. But because the detection pinhole in fact collects only a very small light cone from the studied point, many high spatial frequencies do not reach the detector. As a consequence, some improvements in resolution can be made with a confocal system by fine tuning the detection pinhole diameter. But this improvement is marginal. The main asset of confocal systems is still the rejection of out-of-focus light in the z-axis.

2.2. 4Pi

This imaging approach tries to excite and detect emission light from as many angles as possible around the region of interest. As stated before, a spherical objective -collecting light at every point around the imaged spot- is the optimum imaging device. With this "perfect" objective, featuring the largest possible NA, all light waves could be recovered for a perfect reconstruction of all spatial frequencies. However this objective is not physically realizable. A setup that closely resembles a spherical objective can indeed be designed by using two very large NA objectives facing one another at a very close distance. In this way, much of the surface surrounding the spot is "covered" by one of the two objectives [8]. In fact, the name of this microscopic imaging modality derives from the $4 \cdot \pi$ coefficient to obtain the surface of a sphere ($S_{sphere} = 4 \cdot \pi\, r^2$), alluding to the almost complete coverage of the spherical surface around the object studied.

By carefully aligning both objectives so that they share the same geometrical focus, constructive light interference can be obtained that enhances resolution [3, 9, 10]. If, in addition, a coherent laser source is employed in a confocal arrangement, the resolution is higher than the Abbe limit. In most 4Pi microscopes, the laser excitation is non-linear through a two-photon absorption process (see Chapter 19.2) that further enhances the resolving power because of the inverse dependence of Airy disc size on the non-linear exponent [9]. All in all, resolutions around 100 nm in the lateral dimensions (xy plane) and also around 100 nm in axial direction (z axis) are now routinely accessible by means of commercial 4Pi microscopes.

2.3. I^5M

This technique is basically the incoherent equivalent of the 4Pi setup [3, 9, 10]. It takes advantage of two closely located opposing objectives with as large as possible NA. Both are focused carefully to the same focal plane/spot. The paths travelled by the light beams towards the two objectives are finely adjusted to be equal. The light from both objectives arrives then at the photodetector and interferes, producing a pattern that provides high resolution imaging. These two objective arrangements are known as Image Interference Microscopy (I^2M) [11]. The excitation source is an incoherent one (arc discharge lamp, *etc.*; see Chapter 6.1.1). Now if the excitation light is made to travel through two equivalent light paths and then focused on the sample using the two microscopic objectives, then a constructive excitation pattern occurs at the focal spot. This also increases resolution. This approach has been called Incoherent Interference Illumination Microscopy (I^3M). Then it becomes obvious that, by making use of both interference excitation and interference detection, resolution is twice enhanced. This double interference approach with incoherent light source has become known as I^5M [11]. Using an incoherent source still limits the lateral resolution to around 200 nm (xy plane). But the enhancement in axial detail is quite great (z-axis ~100 nm). Given the objective closeness necessary to take advantage of this setup, one of the most obvious drawbacks is that only very thin sample can be imaged with this methodology.

2.4. Superlenses and Hyperlenses

These strategies aim at collecting a sizeable fraction of the light that escapes with more traditional optics. Note that uncollected light means that high spatial frequencies (and hence, details) are lost for assessment. Therefore, the goal is to capture the evanescent waves (see Chapter 6.5 and Fig. **6.20A** for a description of evanescent waves) emitted by the studied object in the far-field, even when these waves show an exponential decay with distance from their source [9]. In this sense, these approaches share similarities with near-field techniques that will be treated in the next section (see section 3). To do so, these approaches take advantage of certain materials and/or structures that behave as better light-collecting optical elements than classical objectives made of dielectric materials (glasses).

Superlenses are made of materials that show a negative refractive index at optical frequencies (usually very thin metallic coatings with nanometric thickness). These materials resonate with evanescent waves and concentrate them instead of transforming them into propagating waves. This most commonly occurs as the evanescent waves couple with plasmon resonances of the metal (see Chapter 18.1.3 for more on plasmon resonances).

Hyperlenses work on the opposite principle; they enhance the conversion of evanescent waves to propagating ones. Thus even when a very weak evanescent wave is collected, it can be employed to contribute to object imaging, increasing the range of spatial frequencies. They are constructed of metamaterials, materials controlled both in composition and structure to the nanometric level. This allows for very fine tuning of the dielectric properties and, as a consequence, of the optical properties too. The refraction index is carefully tailored inside the material, taking negative values at some locations for some optical wavelengths. As a result, compensating the natural diffraction with negative refractive index, the diffraction limit can be reduced to extremely low values (theoretically almost to 0). In practice, imaging has been obtained to the dimensions of ~100 nm.

3. NEAR-FIELD SRM TECHNIQUES

These techniques take advantage of the much relaxed constraint for the diffraction limit in the optical near-field to obtain super-resolution images. When a fluorophore emits light, most of the previous arguments with regard to the diffraction limit and Airy disc (see section 1 above) apply only in the optical far-field. The near-field space can be broadly defined as that immediate region in space surrounding the emitting (or generally imaged) object to a distance substantially smaller than the wavelength of the light being emitted (or diffracted/absorbed). For example, the near-field region of a fluorescein molecule or a Qdot emitting at 500 nm would be the immediate surroundings to a distance no further away than 200-300 nm (roughly half wavelength is a good rule of thumb). In this near-field region, evanescent electromagnetic waves still contain a substantial portion of the energy emitted. We explained in Chapter 6 (see discussion on TIRFM, Chapter 6.5) that evanescent waves decay exponentially with the distance from the emission point. In the visible range after travelling 100-200 nm evanescent waves decay almost completely and ordinary propagation waves begin to "spread out", which can be seen as the start of diffraction in the far-field. Thus, techniques that can detect and register near-field evanescent waves before diffraction sets in can circumvent the diffraction limit as they are, by definition, unaffected by it. Among these techniques, two find use in fluorescence microscopy: total internal reflection fluorescence (TIRF) and scanning near-field optical microscopy (SNOM) [3, 4, 9].

3.1. TIRFM

This is the same technique that was already introduced in Chapter 6.5 as total internal reflection fluorescence microscopy (TIRFM). It registers the evanescent waves confined to a very thin volume close to a surface (see Fig. **6.20A**). This surface reflects incoming light by a total internal reflection process and, in doing so, excites the evanescent waves at the other side of the reflecting surface (where the substrate to be studied is located). This technique provides below-diffraction limit imaging but only in the z-axis (depth). Its advantage is that it is a totally contrasted methodology for studying living or fixed biological samples. Its main uses in biology are related to membrane, cell motility or extracellular matrix research (see Fig. **6.21**). It can also be employed to study certain microbiological samples.

3.2. SNOM

The scanning near-field optical microscope (SNOM) employs a special tip to image the sample's surface at very close distances, commonly less than 100 nm. The tip is usually the sharp end of an optical fiber shaped as a truncated cone at the end. This tip is frequently metal coated (*e.g.* gold) to enhance evanescent wave coupling to the fiber through plasmon resonances. Excitation light can be guided through the optical fiber to the sample or provided by an independent source. The fluorescence is collected by the tip in the near-field, thereby below the diffraction limit. Details at a scale of 10-20 nm have been obtained employing SNOM. The distance to the sample, a critical issue in this type of microscopy, is continuously monitored and adjusted through piezoelectric controls. Although providing extremely accurate details of the scanned sample, many technical and implementation issues (closeness to sample, extremely thin volume sampling, biologically-aggressive conditions, *etc.*) limit the practical application of this microscopy modality.

Although these near-field techniques, TIRF and SNOM, are in use in biology, they feature serious trade-offs from a practical point of view, especially SNOM. TIRF is biologically-friendly to a great extent but only offers super-resolution in a single dimension. Despite this, TIRF provides a wealth of information in many different research areas. As for SNOM, the necessity of extremely close scanning to obtain the super-resolution information makes it an aggressive technique with regards to biological substrates. Additionally, as happens with TIRF, SNOM is practically limited to surface assessment. For these reasons, other super-resolution techniques, capable of three dimensional imaging, have been developed. These techniques can be defined by their use of far-field electromagnetic waves. Therefore they are, in principle, under the limitation of the diffraction limit. We will now introduce these far-field super-resolution techniques and how they bypass this limit.

4. FAR-FIELD SRM TECHNIQUES

These techniques focus on imaging at the far field when diffraction phenomena have set in. However, there are clever ways to circumvent the diffraction limit to a certain extent. Usually the smaller the imaging spot dimensions, the longer it takes for the super-resolution technique to obtain the image. This trade off can have important consequences when studying living samples. Therefore, the researcher must decide under which experimental conditions, it is better to obtain more physiological results with less resolution, or better resolution but taking longer, sometimes stressing imaging periods to obtain information.

Far-field techniques can be broadly divided between linear and non-linear techniques. Linear techniques employ linear optical processes to achieve the super-resolution data. Non-linear techniques take advantage of non-linear optical processes to provide an image with super-resolution. This is because the resolving power of an optical process depends on the larger exponent of the optical phenomena involved [3]. If, for example, a super-resolution technique uses two-photon absorption to image a sample, the resolving power is double of the equivalent linear process. Under usual two-photon absorption, this non-linear resolution enhancement is offset by increase in the wavelength of the exciting photons to the far red or near infrared. Therefore, the increase in exciting wavelength usually compensates for the non-linear effect upon resolution.

In this section, we will first present common linear SRM techniques and then introduce those commonly employed non-linear techniques. As stated at the beginning of the chapter, the field of SRM is the most actively evolving in the area of fluorescence microscopy and interested readers should refresh periodically their bibliographical sources to keep updated with the latest developments.

SRM Linear Techniques

4.1. Single Molecule Localization (PALM/STORM)

These techniques rely on the same general process to achieve super-resolution: the stochastic excitation of a very small ensemble of fluorophores in the region to be imaged. The acronyms stand for Photoactivated Localization Microscopy (PALM) or STochastic Optical Reconstruction Microscopy (STORM), but despite the different names, they share the same working mechanism of single molecule localization. In the case of PALM, the fluorophore employed is some kind of fluorescent protein (*e.g.* GFP). In STORM, the fluorophore is an organic dye or a luminescent nanoparticle like a Qdot.

The mechanism supporting PALM/STORM is the following. We have stated before that even a nanometric size emitting fluorophore produces a relatively large light source whose size is defined by the Airy disc. This Airy disc is sometimes referred to as the point spread function (PSF) when the emitting source is point-like. One approach to obtain super-resolution images is to reduce the PSF to subdiffraction values. In single molecule localization technique, this is achieved by applying a mathematical fit function to the emitting spot. When the diffraction-limited light spot is imaged, the most probable localization of the fluorophore can be calculated using a Gaussian function [12]. The more photons detected, the more accurately this probabilistic fit will coincide with the real fluorophore localization. The accuracy in the PSF centroid position scales with the number of detected photons according to the following formula:

$$\text{Centroid position (xy-plane)} = \sigma / \sqrt{N_{ph}}, \tag{4}$$

where σ is the standard deviation of the PSF position and N_{ph} is the number of detected emission photons [13].

The problem in establishing the fluorophore position is that, under normal imaging conditions, the PSFs of many sources overlap in the imaging field. To be able to use the mathematical fit function, we must be sure that the PSF, the Airy disc, comes from a single molecule [3]. One way to reduce the amount of emitting sources is to use a very low density of fluorophores. But for practical reasons, this is rarely achievable due to the high fluorophore concentration even at very high dilution factors.

Another approach to "dilute" the fluorophores and achieve a very low concentration of emitting entities is to turn their emission "on" and "off" in a random way (stochastic) [4 - 7, 9]. Thus, the vast majority of a fluorophore population is in a "dark" state while light excitation is taking place, and only a very small minority is susceptible of being excited by photon absorption and subsequent fluorescence emission (see Fig. **20.2**).

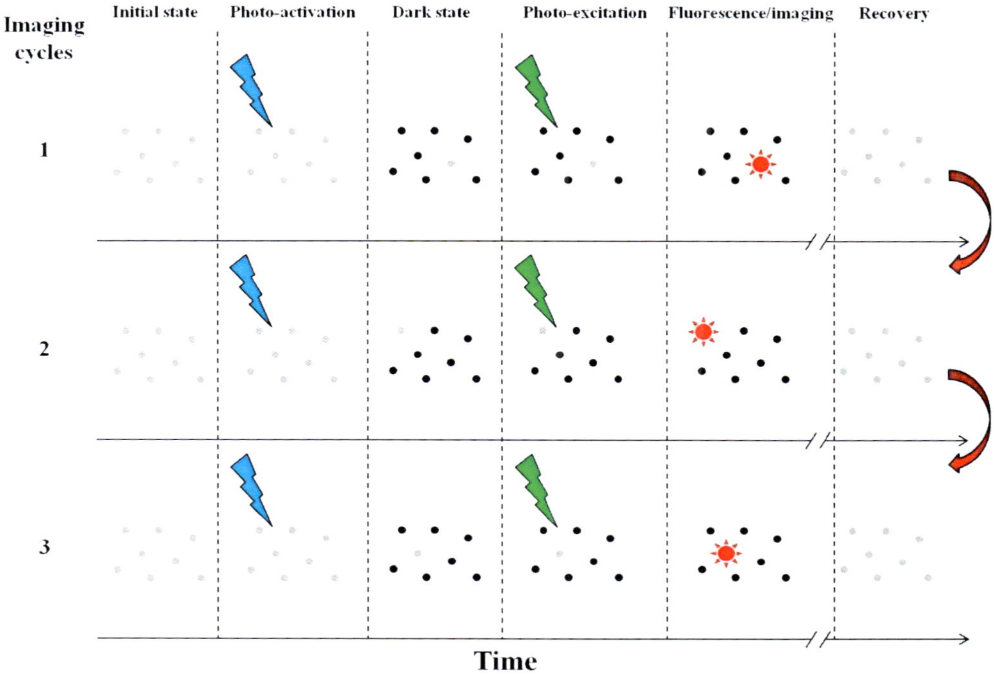

Fig. (20.2). Mechanism of single molecule localization (PALM/STORM). In the scheme 3 imaging cycles are shown. Each cycle starts with an unperturbed population of fluorophores (grey circles). First a photo-activating (also called photo-switching) pulse is provided (blue arrow). The vast majority of fluorophores in the irradiated spot absorb light and undergo a transition to a dark state (black circles). These dark fluorophores no longer respond to the usual fluorescence excitation light. After a short time the field is illuminated with fluorescence excitation light (green arrow) that only a few non-dark fluorophores can absorb (isolated grey circle). These few fluorophores become excited and after some time decay, emitting a fluorescence photon (red emission) that is imaged as usual. Each fluorescent spot is surrounded by a large population of non-emitting dark fluorophores. Depending on the lifetime of the dark state, the "photo-excitation" and "Fluorescence/imaging" steps can be repeated up to several thousand times in order to obtain as many photons from the isolated fluorophores as possible. Eventually dark fluorophores recover and become photo-excitable once again, setting the starting point for another imaging cycle (curved arrows).

The typical process starts with a certain population of fluorophores in the ground state. A photo-activating or photo-switching light pulse drives the vast majority of fluorophores to an excited state from which some kind of change takes place that renders this population unresponsive to further light excitation (dark state) for a relatively long time (from microseconds to seconds). A second light pulse, usually at a different wavelength, now excites the few responsive fluorophores. After a very short time they decay, emitting a fluorescence photon that is recorded and employed in the calculation of the fluorophore's position [7, 14].

Although shown in a simplified way in Fig. (**20.2**), each fluorophore capable of being excited during the one cycle usually does so many times before the dark state population recovers. This leads to more photons being emitted during this period and better chances of accurately localizing the source (see equation **(4)**). Indeed fluorophores to be employed in PALM/STORM are required to show high absorbance at the second pulse wavelength and a high fluorescence quantum yield [13].

During the last few years, many different strategies have been developed to achieve a highly efficient ON - OFF behaviour. Almost all of them are based on some kind of photoreaction taking place from the excited state populated after the first excitation pulse (see Fig. **20.2**) [10, 15 - 17]. Some make the product more fluorescent or induce a conformational change that decreases the molecule´s absorption or fluorescence quantum yield. Other strategies induce a molecular change whereby the emission wavelength changes afterwards for a certain period of time. The fluorophores can reach an excited state with a very long lifetime (metastable) that either does not absorb subsequent light of a certain wavelength or does not emit fluorescence light. Photoisomerism or resonance energy transfer between different parts of the fluorophore or molecular couples is widely exploited to achieve these effects. Chemical reactions from the excited state after the first priming pulse can be tailored so that the products behave in the dark state. A commonly employed strategy is to make the excited fluorophore react with some reduced reagent (less commonly with an oxidant like dissolved O_2) to kinetically and energetically drive the photo-reaction [18, 19]. All of the changes briefly introduced so far can be reversible or irreversible. In principle, reversible changes provide more imaging cycles but at the cost of less control over the actual dark states population. Irreversible changes (photobleaching, see Chapter 2.5.4) give rise to a permanent dark state population but deplete the total fluorophore pool in time.

Some practical examples of super-resolution images obtained with PALM/STORM are shown in Fig. (**20.3**). The first example is the extracellular matrix synthesized by human fibroblasts imaged with STORM (see Fig. **20.3A**) [20]. The fibronectin in the extracellular matrix was recognized with an antibody. The image depth (z-axis) is indicated in false colour (red indicates closer to coverslip and blue more apical). Note the high detail of the fibronectin deposits especially in the detailed image which comes to below 100 nm.

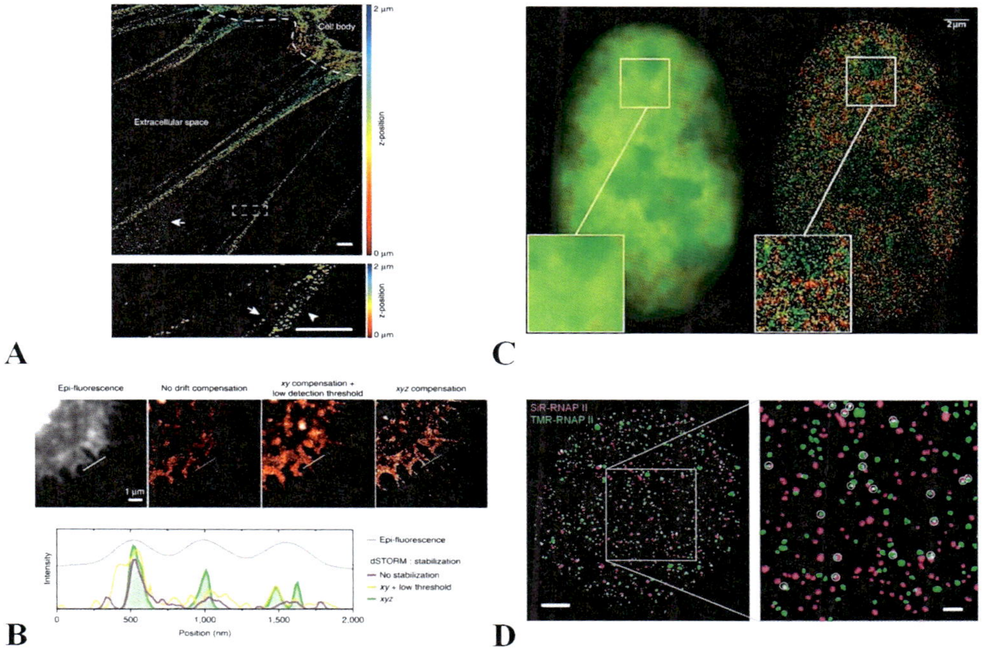

Fig. (20.3). Use of PALM/STORM techniques to visualize cellular structures. **A:** STORM images of fibronectin deposited in the extracellular matrix by human fibroblasts (top panel) prior to fixation and imaging. The fibronectin was localized by employing a specific antibody. Fibronectin is deposited in nano-deposits (arrow). The dashed line indicates the limits of a nearby fibroblast. The depth (z-axis) is indicated by a false colour scale (indicated on the right). The highlighted area (dashed box) is shown in the bottom panel. Two fibrils, a thin one (arrow) and a thick one (arrowhead) are indicated. Scale bars: 1 µm. Reproduced with permission from [20]. **B:** Resolution differences between widefield epifluorescence and STORM super-resolution. F-actin was detected in fixed CHO cells employing fluorescent antibodies. The top panels show the different imaging approaches. The three STORM images (orange-coloured) have different degrees of drift compensation that further enhances image quality. The bottom panel shows the light intensity along the four lines drawn on the pictures. Much more detail is obtained with the STORM images (violet-yellow-green lines) than with the widefield (grey). Reproduced with permission from [21]. **C:** Detailed nuclear protein structure in bone cancer cell. Fluorescent fusion proteins Histone 2A-RFP and chromatin remodelling Snf2H-GFP were imaged with PALM. Under widefield fluorescence imaging (left panel) some subtle differences in colour can be noticed but the image is quite featureless. Under SRM (right panel) the same nucleus appears in a greatly improved detailed way, with clear localization of the two proteins. The white boxes indicate the magnified nuclear regions at the bottom of the images. Notice the widespread distribution of histone 2A while Snf2H appears only in certain nuclear regions. Reproduced with permission from Andy Nestl - https://commons.wikimedia.org/w/index.php?curid=12693849. **D:** Detailed imaging of RNA polymerase II nuclear distribution by SRM-STORM. One of the proteins of the RNA Polymerase II complex was labelled with a silicon rhodamine probe (SiR-RNAP II, cyan) and another of the proteins with a tetramethylrhodamine probe (TMR-RNAP II, green). In this way the mono-colour unassembled proteins could be imaged. When the RNA Pol II complex is actively transcribing, SiR-RNAP II and SiR-RNAP II are recruited and appear together as a bicolour particle or spot (white circles, higher magnification in the inset to the right). In this way active RNA transcription was imaged with great detail within the nucleus of living human osteosarcoma cells. Scale bar: 2 µm, 500 nm (inset). Reproduced with permission from [23].

Bon *et al.* compared the resolving power of STORM with normal epifluorescence and also designed drift-controlling devices to increase the imaging resolution [21]. Their pictures clearly indicate the enhancement in detail obtained with STORM (see Fig. **20.3B**). As will be commented (see below), it is very important to control any movement of the sample, including any drift. With drift control, much more defined images can be obtained in conjunction with SRM.

The nuclear localization of two proteins, Histone 2A and Snf2H, has been imaged in great detail with a bicolour variant of PALM [22]. The proteins were fluorescent chimeric constructs, emitting in the green and in the red (see Fig. **20.3C**). Widefield images allow a very low level of detail with regard to the intranuclear protein distribution (left panel). Under super-resolution imaging, the localization of each protein is sharply defined in the imaged plane. Histone 2A is distributed in the whole nucleus (green spots). Although relatively abundant, the chromatin remodelling protein Snf2H (red spots) is clearly localized only to certain nuclear domains. Other researchers have reported about the RNA transcription process studied in living cells with SRM [23]. The RNA Polymerase II complex is composed of several different proteins. The researchers labelled two of those proteins with two different dye fluorophores, silicon rhodamine and tetramethylrhodamine, that emit at different wavelengths. Making use of STORM, the precise localization of the two proteins could be defined within the living nucleus of an osteosarcoma cell in culture (see Fig. **20.3D**). Furthermore, by identifying whenever the emission from the two proteins was coming from the same structure (white encircled spots on right panel), it could be concluded that such a structure was an assembled RNA Pol II complex. This assembled complex is formed when active RNA transcription takes place; therefore, the spatial localization of these dual-coloured structures correlates with the nuclear region where transcription occurs. Note that the level of detail in the right panel image is around 50-100 nm.

Due to the multiple images (sometimes thousands) that must be obtained to accurately calculate each fluorophore position, the possibility of imaging living systems is restricted. Any cellular movement interferes with the image acquisition mechanism of PALM/STORM. Therefore, only slow processes can be adequately registered with this technique [18]. Even when large numbers of images must be obtained, the acquisition rate is relatively fast, but image processing is lengthy. This makes this technique unfriendly for work with living samples but very adequate for assessing fixed biological samples or for analysing inanimate materials. Recently it has also been possible to increase the resolving power in the z-axis and to obtain 3D images due to the implementation of biphotonic excitation systems (that constrain the excitation plane, see Chapter 19.2.1) or by the use of astigmatic lenses that limit the excitation plane depth too [13].

4.2. SIM

The technique of Structured Illumination Microscopy (SIM) is based on codifying part of the high spatial frequencies (*i.e.* details) from the sample in an interference pattern [4, 5, 7, 16]. By providing an interference pattern with a very high density of interference lines, part of the high spatial frequencies of the sample remains in the subsequent fluorescent emission and can be recovered by adequate means. This is because the exciting interference pattern interferes itself with the sample and "encodes" detailed information in the secondary emission pattern. These secondary interference patterns are known as Moiré fringes. Given that the detailed features of the initial exciting patterns are precisely known, by comparing many emission images information can be extracted to a high resolution level. A scheme of the procedure is provided in Fig. (**20.4**). First, the exciting laser light is made to interfere with itself in a very controlled way to produce an exciting interference pattern (blue-black patterns on the left of the scheme). This excitation light illuminates the sample (previously labelled with some fluorophore for a particular structure) and excites fluorescence. Due to the exciting interference pattern, the emitted fluorescence also shows a pattern (green-black patterns in the scheme). It is precisely this emission pattern that "encodes" part of the detailed structure of the sample. This image is saved and the excitation pattern is changed in a controlled way, for example by rotating it or changing its phase. A second interference fluorescent image is obtained and is again saved for latter assessment. This process of excitation and image acquisition goes on until enough images are obtained to proceed to the super-resolution analysis. Taking into account the particular features of each excitation pattern, a computer program is capable of extracting a highly detailed image of the sample (reconstruction step in the scheme).

Through SIM, an increase in the xy-plane resolution of 2x (~ 100 nm) can be obtained as compared to the widefield diffraction limit. Additionally, any kind of fluorophore can be employed with this technique as it does not rely on any special photophysics/photochemistry. It is possible to obtain multicolour signals (from different structures, for example) with the same excitation light or by exciting at different wavelengths [7, 16, 17]. A 3D SIM is now available which simply makes use of interference patterns in three dimensions instead of only in a plane. Of course, this increases the amount of images necessary to extract the super-resolution information (from 100 images/cell for 2D to 1000 images/cell for 3D) but volumes about eight times smaller than with widefield excitation can be probed (~ 100 x 100 x 100 nm^3). SIM does not require special conditions with regard to living samples, making it a very adequate technique for imaging processes in living cells. Due to its much shorter acquisition times, this technique is much less susceptible to artifacts and blurring from cell-scale movements.

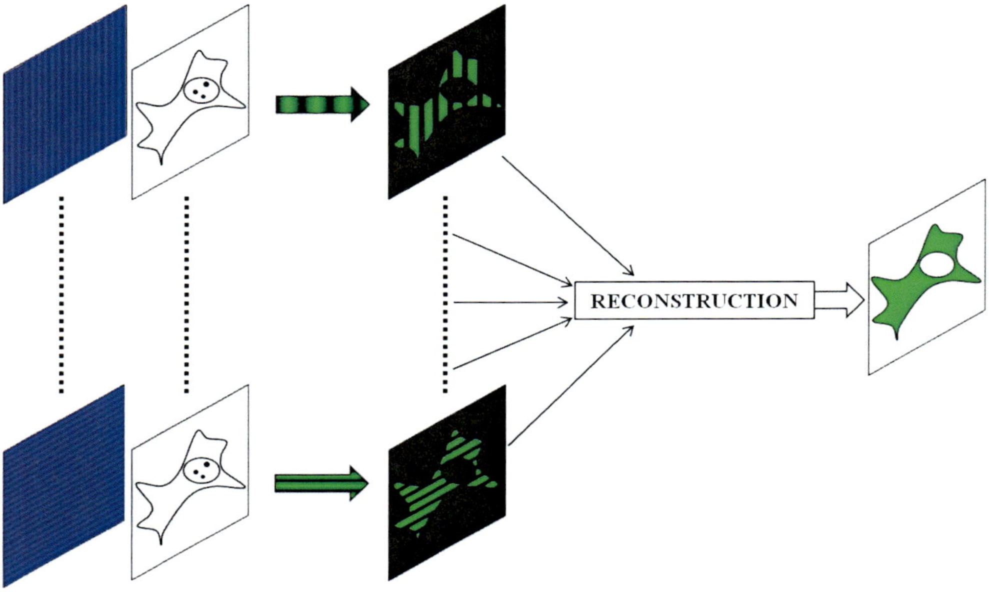

Fig. (20.4). Mechanism of super-resolution with Structured Illumination Microscopy (SIM). An interfering light pattern from an exciting laser (blue and black patterns) interacts with a biological sample (black and white cell) that has been previously labelled with some kind of fluorophore (in this scheme a cytoplasmic tag). The excited fluorophores maintain the interference pattern in their emission (green and black arrow) which leads to a "bar code" fluorescence image (green and black image). This fluorescent image contains high spatial frequencies from the sample codified in the interference pattern. The image is stored and the exciting laser interference pattern is changed (rotating the pattern or changing its phase) to obtain a series of fluorescent images, each one carrying part of the high-detail information from the sample. This image package is analyzed by a computer that reconstructs a super-resolution image of the sample (final green and white fluorescent image). For the sake of simplicity a whole cell has been shown in this scheme. However, real SIM images will focus on much smaller targets like the plasmatic membrane or intracellular structures.

Some microscopic images obtained with SIM are displayed in Fig. (**20.5**). The first example is a study of the ER structure in dendritic spine formation in living mouse hippocampal neurons [24]. Using 3D SIM, it was possible to assess the differences in ER structure when the neurons expressed a mutant variant of the Myc protein (see Fig. **20.5A**, bottom panel). In the mutant neurons, it can be seen that the ER is located more peripherally within the dendrite in comparison with the wild type neurons. SIM images of the ER were obtained with two different approaches, one employing a red fluorescent fusion protein (DsRed-ER) and the other using immunofluorescence to label an ER protein (calreticulin).

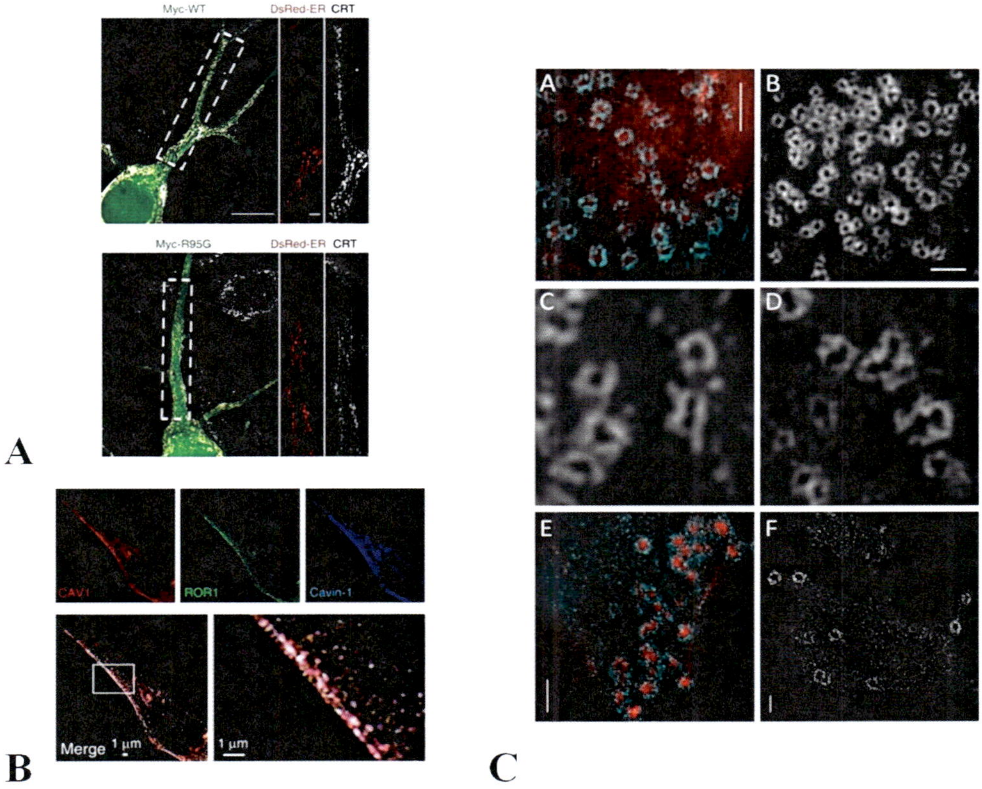

Fig. (20.5). Super-resolution images obtained by SIM. **A:** Mouse hippocampal neurons imaged with 3D-SIM. A wild type neuron (top panel) was engineered to express a red fluorescent protein (DsRed-ER) in the endoplasmic reticulum (ER). The ER fine-structure organization in a growing dendritic spine (dashed box) is shown in detail (red image). Calreticulin (CRT), an ER-localizing protein, was imaged as a positive ER control with an antibody (grey scale image). Neurons expressing a mutant Myc protein (Myc-R95G) show a subtly disorganized ER structure (bottom panel). Scale bars: 10 µm and 2 µm (detail). Reproduced with permission from [24]. **B:** Colocalization of three different proteins necessary for the establishment and maintenance of caveolae in tumor cells (NCI-H1975). Caveolin-1 (CAV1, red), a membrane kinase receptor (ROR1, green) and Cavin-1 (blue) were imaged in fixed cells with SIM and immunofluorescence. The three independent channels for each protein localization are displayed in the top panel. The merged image for the three channels (bottom panel, left) and, especially, the magnified section (white box) show (bottom panel, right) a high degree of colocalization for the three proteins (white spots). Reproduced with permission from [25]. **C:** Podosome structures studied on the membrane of THP-1 human monocytic leukaemia cells by SIM and immunofluorescence. In (A) the localization of actin (red) and vinculin (blue) is highlighted. In (B) only the vinculin rings are shown displaying a characteristic polygonal arrangement. (C,D) Vinculin rings shown at greater magnification. (E) Localization of actin (red) and paxillin (blue) in podosomes. (F) Paxillin rings show a polygonal arrangement too, but their fine structure seems more punctuate and coarse than that of vinculin. Scale bars: 2 µm. Reproduced with permission from [26].

In Fig. (**20.5B**) the colocalization of three different proteins implicated in caveolae formation and maintenance in the plasmatic membrane was studied with

SIM [25]. Each of the proteins was detected with immunofluorescence on fixed samples at different emission wavelengths (red, green and blue). By merging the three images the colour overlap can be assessed. Where all three colours overlap, the total emission will be white. This is what can be appreciated in the bottom row of images, where the colocalization in a majority of caveolae is displayed. Finally, the spatial arrangement of podosomes, plasmatic membrane structures in monocytes, was revealed in great detail by SIM [26]. The authors showed that podosomes, once thought to be rounded in their geometrical pattern, are indeed more polygonal, sometimes approaching a hexagonal shape (see Fig. **20.5C**). Also shown in the images are the differences in protein arrangement between two podosome components: vinculin (more homogeneous) and paxillin (punctuate-like).

SRM Non-Linear Techniques

4.3. S-SIM

As explained in the previous section, SIM does not allow for a very large improvement in resolution (~ 2x in comparison to widefield). However, SIM is a linear super-resolution technique which leaves room for improvement when using the same principle (interference Moiré fringes) in combination with non-linear fluorophore absorption saturation. This is the technique known as Saturated-SIM or S-SIM. The principle supporting this approach is very similar to linear SIM but the exciting light intensity is much higher to saturate the fluorophore response. Towards this end, pulsed lasers are employed to produce the exciting interference pattern that interacts with the sample. Due to the non-linear fluorescent emission of the excited fluorophores, higher spatial frequencies are produced in the final interference pattern. These are able to encode more detailed information about the sample, which translates into higher resolution [4, 5, 9]. Currently details on the range 40-50 nm are discernible using S-SIM as an imaging technique.

One of the biggest drawbacks of S-SIM is that it requires very high light intensities to reach the saturation fluorescence threshold. Obviously, this exposes the sample to very high photon fluxes, which can result in temporal or permanent photodamage/photobleaching. To try to avoid this as much as possible, S-SIM needs special fluorophores able to withstand the very intense light fluxes necessary to obtain the images [27]. Qdots and metallic nanoparticles (see Chapter 18.2 and 18.3) are also interesting candidates for non-linear super-resolution imaging [28]. In any case, the very high light intensities are not the best starting point to analyze biological, especially live, samples. Probably, as happens with other non-linear optical processes applied to biological targets, the use of femtosecond lasers is mostly appropriate for S-SIM when it is of interest to minimize the potential photodamage to the sample. S-SIM is one of the SRM techniques with a promising immediate future if the photodamage issue can be dealt with appropriately.

4.4. STED

The technique of STimulated Emission Depletion (STED) is also a non-linear super-resolution approach. It is based on exciting a relatively small population of fluorophores, followed by an almost total, controlled deactivation of those fluorophores except in a very small region at the centre of the imaged spot [4 - 7, 9]. The procedure steps are schematically shown in Fig. (**20.6**). First, a diffraction-limited laser excitation spot is illuminated to excite the fluorophores within the corresponding volume. The fluorophores show transition to the first excited electronic state. After a very short time (shorter than the fluorophore´s natural radiative lifetime) a second laser pulse, known as the depleting pulse, forces most of the excited fluorophores to emit a fluorescence photon and move back to the ground state. The key aspect of this process is that the depleting pulse takes advantage of the stimulated emission phenomenon to make the excited fluorophores emit with the same wavelength light as the depleting laser. That is the reason the technique is known as, stimulated emission depletion. The depleting laser is commonly tuned to the red part of the fluorophore´s emission band to filter it and prevent interference with the imaging system.

Additionally, the depleting pulse is tailored so as to excite a ring-shaped area that overlaps with the excitation pulse area. The centre of the depleting pulse spot is dark; no photon travels through it. In this way, a very small (much smaller than the diffraction limit) area in the centre is left unperturbed by the depleting pulse. Within this area, fluorophores are left to decay by their natural radiative lifetime. Because of the non-linearly depleted area surrounding the central spot, it is possible to reconstruct the light emitted from this volume with a detail much higher than the linear diffraction limit [15 - 17, 29]. The process can be visualized as if a custom-positioned, arbitrarily small confocal system is created every time the laser pulses excite-deexcite the sample.

STED allows for very high lateral (xy plane) resolution down to 50-70 nm. Under optimized conditions, resolutions in the 20 nm range have been reported (see Table **20.1** below). Axial (z-axis) resolution is not as good with numbers in the 100-500 nm range. However, initial biphotonic fluorophore excitation greatly improves axial resolution (see Chapter 19.2.1) and increases the resolution in this dimension to less than 100 nm [29].

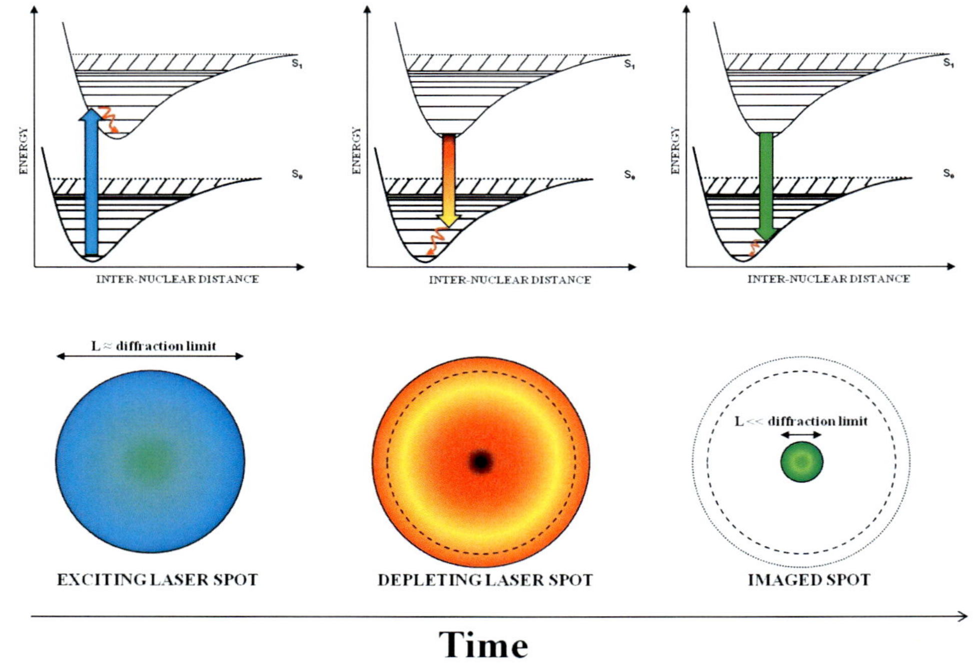

Fig. (20.6). Mechanism of super-resolution with STimulated Emission Depletion (STED). The process takes place in several consecutive steps. In the scheme the photophysical events the fluorophores undergo are displayed in the top panels. The corresponding microscopic irradiated/imaged fields are shown in the bottom panels. In the first step (left) the sample is excited with a laser pulse of diffraction-limited dimensions (blue spot). The fluorophores absorb the exciting photons (blue arrow) and decay to the first excited state (upper energy curve). Around one nanosecond later (middle) a second laser pulse (orange-red circle) irradiates a slightly larger area centred where the first pulse was located (dashed circle). This pulse is known as the depleting pulse and in fact is ring-shaped, leaving a very small central spot in darkness (black area). Fluorophores within the ring area undergo stimulated photon emission at the laser´s wavelength (orange-red arrow on top panel). Finally (right) only those fluorophores within the spared central dark spot (green area) can radiatively decay according to their spontaneous lifetime, commonly with a shorter wavelength (green arrow) than the depleting pulse laser. These spontaneous photons can be detected and imaged with subdiffraction-limited resolution.

Table 20.1. Resolving power comparison between the most commonly employed SRM modalities.

SRM modality	**Heintzmann *et al.* (2006) [3]**	**Rice (2007) [9]**	**Schermelleh *et al.* (2010) [4]**	**Yamanaka *et al.* (2014) [5]**	**Godin *et al.* (2014) [16]**	**Fornasiero *et al.* (2015) [17]**
4Pi	200(xy) 90(z)	80-190(xy) 80-190(z)				
I^5M	230(xy) 90(z)					

(Table 20.1) contd.....

SRM modality	Heintzmann *et al.* (2006) [3]	Rice (2007) [9]	Schermelleh *et al.* (2010) [4]	Yamanaka *et al.* (2014) [5]	Godin *et al.* (2014) [16]	Fornasiero *et al.* (2015) [17]
PALM/STORM	20(xy)	10(xy) 10(z)	20-50(xy) 20-75(z)	10-30(xy) 10-75(z)	50(xy)	10-30(xy) 20-100(z)
SIM	100(xy)		100-130(xy) 250-350(z)	100-130(xy) 300(z)	50-100(xy)	120(xy) 600(z)
S-SIM	50(xy)		50(xy)	50(xy)		50(xy) 350(z)
STED	20(xy) 50(z)	15-76(xy) 15-76(z)	20-100(xy) 500(z)	20-70(xy) 40-150(z)	50-70(xy)	20-60(xy) 100-600(z)

(xy) *refers to the lateral resolution*; (z) *refers to the axial resolution. All values in nm.*

The imaged spot size depends non-linearly on the depleting laser intensity above a certain threshold. This spot size is given by the following equation:

$$\text{Spot size (xy-plane)} = \Lambda / \sqrt{(1 + I_{dep} / I_{sat})} \quad \textbf{(5)}$$

where Λ is the linear classical diffraction-limited spot size (defined by equation **(1)**), I_{dep} is the light intensity of the depleting laser, and I_{sat} the light intensity (threshold) at which the fluorophore employed starts to saturate [7]. Of course I_{sat} depends not only on the specific fluorophore but also on the specific excitation wavelength employed. Using equation **(5)** it is easy to calculate the resolution improvement provided by STED. For example, if the depleting laser light is 3 times the intensity needed to reach the saturation, then the spot size will be: $\Lambda / \sqrt{(1 + 3 / 1)}$, which is: $\Lambda / \sqrt{4} = \Lambda / 2$. Therefore we can expect a 2x increase in resolution when we employ a depleting laser intensity three times above the threshold. Light intensities 100 times above the threshold will give a 10x resolution enhancement. Obviously the light intensity will depend on the sample´s resilience to photodamage. Fixed samples will be able to withstand much higher light intensities than living samples. Also, very photostable fluorophores are necessary to reach the very intense light fluxes that allow for high resolution (see equation **(5)**) [27, 30]. Fluorophores with relatively long lifetimes are desirable as this gives more flexibility for the precise timing of the depleting pulse. Additionally, the emitting agents should have a high cross section for stimulated emission in order to ensure a complete depopulation of excited states in the ring area.

In 2009 Schmidt *et al.* published some impressive images of mitochondria obtained through STED [31]. They employed an iso-STED approach, basically a 4Pi arrangement (see section 2.2 above) coupled with STED, to image features with a 30 nm resolution limit. Some very fine mitochondrial structural details, like the cristae, were visualized with this technique, something never done before with optical systems (see Fig. **20.7A**).

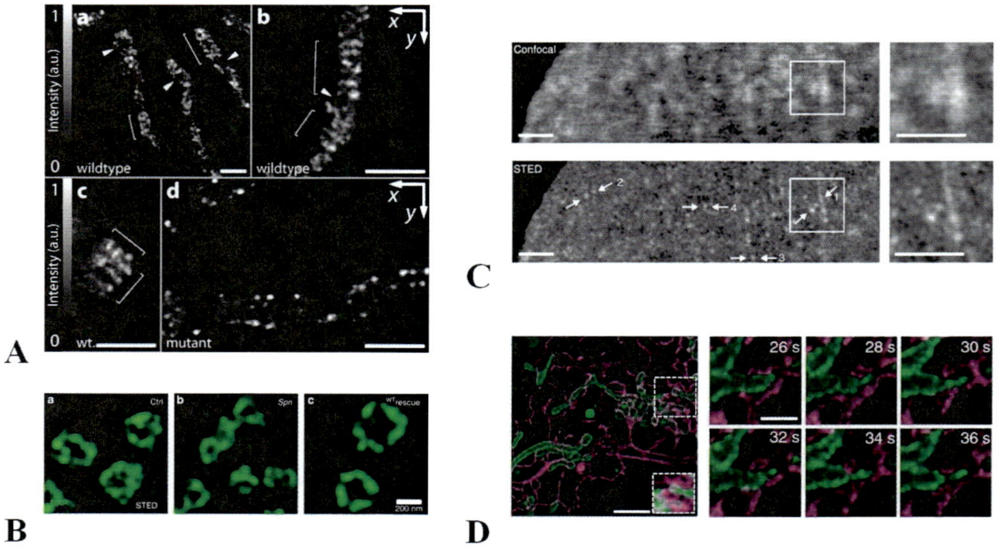

Fig. (20.7). Images obtained through STED. **A:** Mitochondrial cristae imaged with STED after labelling the ATPase complex with a fluorescent antibody in fixed PtK2 cells. (a-c) Increasing detail of mitochondrial cristae in wild type cells (brackets: cristae perpendicular to the mitochondria major axis; arrowheads: cristae-devoid regions). (d) Mitochondria of mitofilin-depleted cell. The cristae are less abundant and appear closely-clustered where they locate. Scale bars: 500 nm. Reprinted with permission from [31]. Copyright 2009 American Chemical Society. **B:** Detail of the cellular structures (BRP rings) involved in synapses at the neuromuscular junction in *Drosophila melanogaster* imaged by STED. (a) shows a control junction, (b) a junction in a mutant lacking the spinophillin (Spn) protein which helps to organize the BRP rings. Notice the smaller and less geometrical structures. (c) shows a rescue phenotype in which the BRP rings again display a normal morphology. Images from larval neuromuscular junctions labelled with immunofluorescence. Reproduced with permission from [32]. **C:** Confocal (top) and STED (bottom) images of the same chromatin stained with a far-red emitting dye (silicon-rhodamine + Hoechst). The images to the right are details of the highlighted areas (white boxes) to better appreciate the increased detail of STED as compared to confocal. The white arrows and numbers correspond to four different areas analyzed by image processing in the original paper. Images obtained from living HeLa cells. Confocal laser excitation at 633 nm. For STED excitation laser at 640 nm and depletion laser at 775 nm. Scale bars: 1 μm. Reproduced with permission from [33]. D: Two-colour STED images of endoplasmic reticulum (silicon -rhodamine, magenta) and mitochondria (rhodamine derivative, green) in living COS-7 cells (left panel). The dashed white box highlights a subarea that is imaged at different times (right panels). The same area is also displayed (bottom right corner) as imaged with confocal for comparison. The time sequence (26 s - 36 s) shows the fine-detail real time interactions between the ER and the mitochondrial network in a living cell. Note how the ER envelopes a mitochondria at the image centre (32-36 s). Note that the displayed colours are false colours. Excitation laser emitting at 595 nm and depletion laser at 775 nm. Scale bars: 2 μm (left panel) and 500 nm (time sequence). Reproduced with permission from [34].

Another example of the great detail achievable with STED is the imaging of the synaptic structures involved in muscular excitation at the neuromuscular junction in *Drosophila melanogaster* larvae (see Fig. **20.7B**). Structures well below the 100 nm can be clearly visualized (panels a and c) as well as size and structural changes in mutants for a specific protein (spinophillin) involved in the molecular scaffold necessary for the correct assembly of the synaptic structure [32]. Striking differences in the resolution provided by a confocal system and an STED system are displayed in Fig. (**20.7C**). The chromatin of a living HeLa cell was stained with a far-red emitting derivative of Hoechst + silicon rhodamine fluorophore [33]. The confocal image presents a blurry aspect (especially in the magnified area shown to the right). Meanwhile, imaging by STED provides a much better resolution (~ 100 nm).

Finally in Fig. (**20.7D**), a recent report on cellular dynamics obtained through STED is shown [34]. The authors were able to image a living COS-7 cell and record physiological movements at the organelle level by adopting a two-fluorophore stain strategy (endoplasmic reticulum and mitochondria). In the real time sequence displayed in the figure, the changes in shape and dynamic interactions between the ER and the surrounding mitochondria are clearly seen (the interactions shown in frames 32 - 36 s deserve special mention). The scale at which these dynamical events are taking place is below 100 nm. Therefore, STED and SRM, in general, are providing the frame to obtain physiologically relevant information about vital cellular events in the living samples in an unprecedented resolution and detailed level.

An evolution over STED is the technique known as REversible Saturable OpticaL Fluorescence Transitions (RESOLFT). This technique tries to reduce as much as possible the depleting laser pulse to avoid sample photodamage. It does so by employing an approach very similar to PALM/STORM, inducing a long-lifetime dark state in the ring area instead of forcing the stimulated emission, to diminish interference with emission from the central spot [5, 7]. The options to produce the dark states are the same as in PALM/STORM, like photoswitching, photobleaching, reversible photochemistry, *etc.* (see section 4.1 above). Employing RESOLFT and cw lasers to deplete the ring area, it is possible to decrease the light flux intensities from MW/cm^{-2} to kW/cm^{-2} at the focal level, which minimizes the chances of photodamaging the studied specimen [16].

Summarizing the most important SRM techniques introduced in the previous sections, Table 20.1 presents published resolution values for different SRM approaches (resolution in nanometres both for the xy-plane -lateral resolution- and for z-axis -depth resolution-). The cited references can be found in the bibliography at the end of the chapter.

Obviously SRM is not set apart from the rest of the techniques we have been introducing in this book. As already presented in some of the practical examples above (see for example Fig. **20.7D** and **20.3D**), SRM makes use of fluorescent antibodies -immunofluorescence- and fluorescent proteins to image biological targets in great detail. Another example is displayed in Fig. (**20.8**) where intimate structural details of HeLa cells have been obtained by combining a RESOLFT technique with a 4Pi microscope arrangement [35].

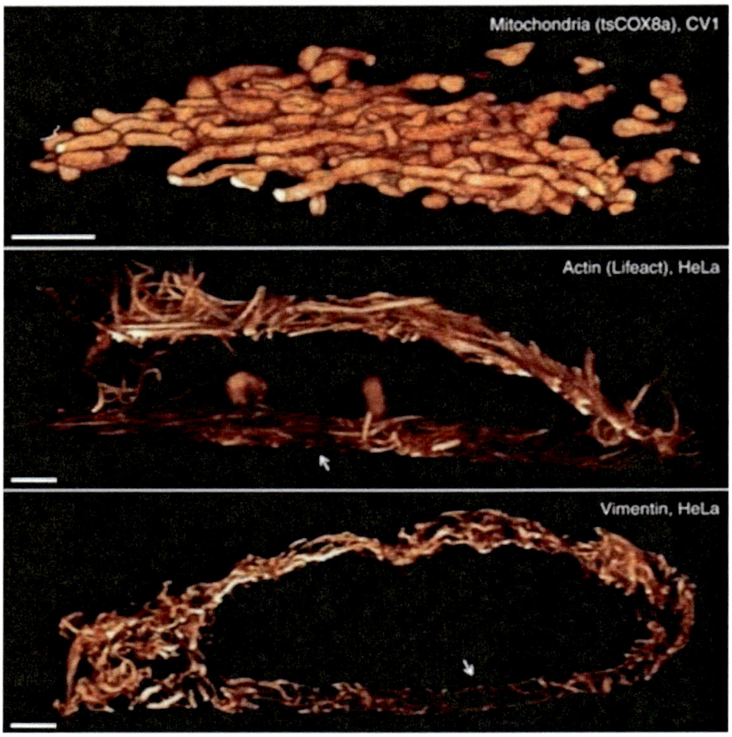

Fig. (20.8). RESOLFT technique in combination with a 4Pi microscope arrangement to image a HeLa cell. The top panel shows the mitochondrial network (fixed cell), the middle panel the actin cytoskeleton (living cell) and the bottom panel the intermediate filament cytoskeleton (vimentin, living cell). Changes in the cytoskeleton structure close to the substrate (coverslip) are highlighted (arrows in middle and bottom panels). Note that only a thin slice across the cell has been imaged, not the whole cell. The panels show computer-generated simulations based on the real cell images. Scale bars: 1 μm. Reproduced with permission from [35].

The mitochondria and two cytoskeletal components, actin and vimentin, are shown in great structural detail. In fact, the two images corresponding to the cytoskeleton were obtained from the living specimens, which proves the evolution of SRM towards less aggressive techniques capable of providing key super-resolution information about living systems. Many approaches introduced in this book can be combined with SRM with the proper adjustments, like immunofluorescence, FISH, vital probes, fluorescent proteins, *etc*. Recently, for example, nanobodies and chromobodies (see Chapter 17) have been successfully employed in combination with SRM to accurately study biological substrates and to isolate proteins [36, 37]. SRM techniques are starting to be used also in the field of microbiology [38, 39]. Bacteria, on average having a much smaller size than eukaryotic cells, are an ideal target for SRM. These microorganisms have traditionally been studied for detailed structure through electron microscopies. With SRM, the opportunity to study structure and metabolism at the same time in living prokaryotes is an open door awaiting researchers to explore this realm.

There are other SRM techniques and more, without doubt, are right now in the process of being developed. For example, there are some mathematical/computational super-resolution techniques that are not going to be introduced here [6, 9, 16]. We have judged that, since this is a general handbook about fluorescence microscopy, delving too deep into other fields is beyond the scope of this work. Nevertheless, it is important to mention that they exist to alert the interested reader to them, at least. Undoubtedly the SRM techniques introduced here are in the processes of evolving into more sample-friendly and towards a more widespread use among the biomedical community. And surely new super-resolution approaches will be proposed in the upcoming years as the field matures. SRM was awarded the 2014 Nobel Prize in Chemistry; a clear recognition of the huge impact these methodologies are already having in the scientific arena. Taking into account the very short history and the current development trend of these optical super-resolution techniques, we can only conclude that their future path and achievements are bright and not yet resolved in detail.

CONFLICT OF INTEREST

The author (editor) declares no conflict of interest, financial or otherwise.

ACKNOWLEDGEMENTS

This work was supported by the Aarhus Institute of Advanced Studies (AIAS).

REFERENCES

[1] Band YB. Light and Matter Electromagnetism, Optics, Spectroscopy and Lasers. Chichester: John Wiley & Sons Ltd 2006.

[2] Saleh BE, Teich MC. Fundamentals of Photonics. New Jersey: Wiley-Interscience 2007.

[3] Heintzmann R, Ficz G. Breaking the resolution limit in light microscopy. Brief Funct Genomics Proteomics 2006; 5(4): 289-301.

[http://dx.doi.org/10.1093/bfgp/ell036] [PMID: 17170013]

[4] Schermelleh L, Heintzmann R, Leonhardt H. A guide to super-resolution fluorescence microscopy. J Cell Biol 2010; 190(2): 165-75. [http://dx.doi.org/10.1083/jcb.201002018] [PMID: 20643879]

[5] Yamanaka M, Smith NI, Fujita K. Introduction to super-resolution microscopy. Microscopy (Oxf) 2014; 63(3): 177-92. [http://dx.doi.org/10.1093/jmicro/dfu007] [PMID: 24671128]

[6] Cox S, Jones GE. Imaging cells at the nanoscale. Int J Biochem Cell Biol 2013; 45(8): 1669-78. [http://dx.doi.org/10.1016/j.biocel.2013.05.010] [PMID: 23688552]

[7] Huang B, Babcock H, Zhuang X. Breaking the diffraction barrier: super-resolution imaging of cells. Cell 2010; 143(7): 1047-58. [http://dx.doi.org/10.1016/j.cell.2010.12.002] [PMID: 21168201]

[8] Hell S, Stelzer HK. Fundamental improvement of resolution with a 4Pi-confocal fluorescence microscope using two-photon excitation. Opt Commun 1992; 93: 277-82. [http://dx.doi.org/10.1016/0030-4018(92)90185-T]

[9] Rice JH. Beyond the diffraction limit: far-field fluorescence imaging with ultrahigh resolution. Mol Biosyst 2007; 3(11): 781-93. [http://dx.doi.org/10.1039/b705460b] [PMID: 17940661]

[10] Habuchi S. Super-resolution molecular and functional imaging of nanoscale architectures in life and materials science. Front Bioeng Biotech 2014; 2(20): 1-13. [http://dx.doi.org/10.3389/fbioe.2014.00020]

[11] Gustafsson MG, Agard DA, Sedat JW. I^5M: 3D widefield light microscopy with better than 100 nm axial resolution. J Microsc 1999; 195(Pt 1): 10-6. [http://dx.doi.org/10.1046/j.1365-2818.1999.00576.x] [PMID: 10444297]

[12] Hu D, Zhao B, Xie Y, Orr G, Li AD. Understanding super-resolution nanoscopy and its biological applications in cell imaging. Phys Chem Chem Phys 2013; 15(36): 14856-61. [http://dx.doi.org/10.1039/c3cp51629f] [PMID: 23739871]

[13] Henriques R, Mhlanga MM. PALM and STORM: what hides beyond the Rayleigh limit? Biotechnol J 2009; 4(6): 846-57. [http://dx.doi.org/10.1002/biot.200900024] [PMID: 19548237]

[14] van de Linde S, Aufmkolk S, Franke C, *et al.* Investigating cellular structures at the nanoscale with organic fluorophores. Chem Biol 2013; 20(1): 8-18. [http://dx.doi.org/10.1016/j.chembiol.2012.11.004] [PMID: 23352135]

[15] Sanderson MJ, Smith I, Parker I, Bootman MD. Fluorescence microscopy. Cold Spring Harb Protoc 2014; 2014(10): top071795. [http://dx.doi.org/10.1101/pdb.top071795] [PMID: 25275114]

[16] Godin AG, Lounis B, Cognet L. Super-resolution microscopy approaches for live cell imaging. Biophys J 2014; 107(8): 1777-84. [http://dx.doi.org/10.1016/j.bpj.2014.08.028] [PMID: 25418158]

[17] Fornasiero EF, Opazo F. Super-resolution imaging for cell biologists: concepts, applications, current challenges and developments. BioEssays 2015; 37(4): 436-51. [http://dx.doi.org/10.1002/bies.201400170] [PMID: 25581819]

[18] Henriques R, Griffiths C, Hesper Rego E, Mhlanga MM. PALM and STORM: unlocking live-cell super-resolution. Biopolymers 2011; 95(5): 322-31. [http://dx.doi.org/10.1002/bip.21586] [PMID: 21254001]

[19] Vogelsang J, Steinhauer C, Forthmann C, *et al.* Make them blink: probes for super-resolution microscopy. ChemPhysChem 2010; 11(12): 2475-90.

[http://dx.doi.org/10.1002/cphc.201000189] [PMID: 20632356]

[20] Früh SM, Schoen I, Ries J, Vogel V. Molecular architecture of native fibronectin fibrils. Nat Commun 2015; 6: 7275.
[http://dx.doi.org/10.1038/ncomms8275] [PMID: 26041410]

[21] Bon P, Bourg N, Lécart S, *et al.* Three-dimensional nanometre localization of nanoparticles to enhance super-resolution microscopy. Nat Commun 2015; 6: 7764.
[http://dx.doi.org/10.1038/ncomms8764] [PMID: 26212705]

[22] Gunkel M, Erdel F, Rippe K, *et al.* Dual color localization microscopy of cellular nanostructures. Biotechnol J 2009; 4(6): 927-38.
[http://dx.doi.org/10.1002/biot.200900005] [PMID: 19548231]

[23] Zhao ZW, Roy R, Gebhardt JC, Suter DM, Chapman AR, Xie XS. Spatial organization of RNA polymerase II inside a mammalian cell nucleus revealed by reflected light-sheet superresolution microscopy. Proc Natl Acad Sci USA 2014; 111(2): 681-6.
[http://dx.doi.org/10.1073/pnas.1318496111] [PMID: 24379392]

[24] Shih YT, Hsueh YP. VCP and ATL1 regulate endoplasmic reticulum and protein synthesis for dendritic spine formation. Nat Commun 2016; 7: 11020.
[http://dx.doi.org/10.1038/ncomms11020] [PMID: 26984393]

[25] Yamaguchi T, Lu C, Ida L, *et al.* ROR1 sustains caveolae and survival signalling as a scaffold of cavin-1 and caveolin-1. Nat Commun 2016; 7: 10060.
[http://dx.doi.org/10.1038/ncomms10060] [PMID: 26725982]

[26] Walde M, Monypenny J, Heintzmann R, Jones GE, Cox S. Vinculin binding angle in podosomes revealed by high resolution microscopy. PLoS One 2014; 9(2): e88251.
[http://dx.doi.org/10.1371/journal.pone.0088251] [PMID: 24523880]

[27] van de Linde S, Heilemann M, Sauer M. Live-cell super-resolution imaging with synthetic fluorophores. Annu Rev Phys Chem 2012; 63: 519-40.
[http://dx.doi.org/10.1146/annurev-physchem-032811-112012] [PMID: 22404589]

[28] Ilovitsh T, Danan Y, Meir R, Meiri A, Zalevsky Z. Cellular superresolved imaging of multiple markers using temporally flickering nanoparticles. Sci Rep 2015; 5: 10965.
[http://dx.doi.org/10.1038/srep10965] [PMID: 26020693]

[29] Otomo K, Hibi T, Kozawa Y, Nemoto T. STED microscopy--super-resolution bio-imaging utilizing a stimulated emission depletion. Microscopy (Oxf) 2015; 64(4): 227-36.
[http://dx.doi.org/10.1093/jmicro/dfv036] [PMID: 26152214]

[30] Fernández-Suárez M, Ting AY. Fluorescent probes for super-resolution imaging in living cells. Nat Rev Mol Cell Biol 2008; 9(12): 929-43.
[http://dx.doi.org/10.1038/nrm2531] [PMID: 19002208]

[31] Schmidt R, Wurm CA, Punge A, Egner A, Jakobs S, Hell SW. Mitochondrial cristae revealed with focused light. Nano Lett 2009; 9(6): 2508-10.
[http://dx.doi.org/10.1021/nl901398t] [PMID: 19459703]

[32] Muhammad K, Reddy-Alla S, Driller JH, *et al.* Presynaptic spinophilin tunes neurexin signalling to control active zone architecture and function. Nat Commun 2015; 6: 8362.
[http://dx.doi.org/10.1038/ncomms9362] [PMID: 26471740]

[33] Lukinavičius G, Blaukopf C, Pershagen E, *et al.* SiR-Hoechst is a far-red DNA stain for live-cell nanoscopy. Nat Commun 2015; 6: 8497.
[http://dx.doi.org/10.1038/ncomms9497] [PMID: 26423723]

[34] Bottanelli F, Kromann EB, Allgeyer ES, *et al.* Two-colour live-cell nanoscale imaging of intracellular targets. Nat Commun 2016; 7: 10778.
[http://dx.doi.org/10.1038/ncomms10778] [PMID: 26940217]

[35] Böhm U, Hell SW, Schmidt R. 4Pi-RESOLFT nanoscopy. Nat Commun 2016; 7: 10504. [http://dx.doi.org/10.1038/ncomms10504] [PMID: 26833381]

[36] Ries J, Kaplan C, Platonova E, Eghlidi H, Ewers H. A simple, versatile method for GFP-based super-resolution microscopy *via* nanobodies. Nat Methods 2012; 9(6): 582-4. [http://dx.doi.org/10.1038/nmeth.1991] [PMID: 22543348]

[37] Pleiner T, Bates M, Trakhanov S, *et al.* Nanobodies: site-specific labeling for super-resolution imaging, rapid epitope-mapping and native protein complex isolation. eLife 2015; 4: e11349. [http://dx.doi.org/10.7554/eLife.11349] [PMID: 26633879]

[38] Coltharp C, Xiao J. Superresolution microscopy for microbiology. Cell Microbiol 2012; 14(12): 1808-18. [http://dx.doi.org/10.1111/cmi.12024] [PMID: 22947061]

[39] Tuson HH, Biteen JS. Unveiling the inner workings of live bacteria using super-resolution microscopy. Anal Chem 2015; 87(1): 42-63. [http://dx.doi.org/10.1021/ac5041346] [PMID: 25380480]

SUBJECT INDEX

A

Abbe limit 687, 692
Absorbing photons, molecular population 58
Absorption 3, 13, 22, 29, 32, 33, 39, 40, 43, 46, 48, 49, 57, 70, 76, 83, 111, 119, 135, 136, 140, 142, 143, 146, 148, 151, 172, 194, 227, 254, 268, 270, 293, 294, 295, 297, 301, 326, 364, 382, 571, 592, 606, 609, 613, 617, 618, 643, 647, 648, 649, 650, 652, 653, 660, 698
  biphotonic 649, 650, 652, 660
Absorption bands 76, 77, 119, 300, 573, 673
Absorption coefficients 298, 627, 643, 649, 651
  fluorophore two-photon 651
Absorption-emission conversions 237
Absorption spectrum 41, 48, 49, 147, 148, 170, 208, 239, 291, 292, 387, 613, 633, 637
Acid GAGs 368, 369, 371, 372, 374, 375
Acid glycosaminoglycans 254, 359, 368, 399, 443
Acid hydrolysis 112, 243, 382, 383
Acidophilic substrates 240, 241
Acid polysaccharides 360, 369, 374, 398
Acridine orange (AO) 2, 11, 12, 14, 15, 62, 63, 81, 88, 106, 111, 112, 135, 141, 142, 143, 146, 147, 149, 150, 152, 225, 238, 368, 369, 370, 392, 393, 394, 395, 398, 481, 482, 483, 490, 491, 492, 493, 499, 504, 505, 506, 511, 533, 535, 574, 579
Active pixel sensor (APS) 176
Adipocyte differentiation-related protein (ADRP) 348
Agents 219, 222, 223, 253, 359, 361, 362, 363, 395, 407, 577
  brightening 359, 361, 362, 363
  competing 219, 222, 223, 253, 395, 407
  counterstain dyes and anti-fading 577
Ag nanoparticles 629
Airy disc 687, 688, 689, 690, 693, 696
Aldehydes 117, 221, 242, 310, 365, 366, 381, 385, 387, 388, 456, 553
Alexa fluorochromes 11
Alizarin 252, 257, 262, 275, 508
Alizarin complexon 275, 276, 277
Alkaline 8, 68, 69, 70, 86, 98, 187, 237, 265, 341, 342, 412, 446, 473, 475, 548, 549, 550, 552, 622
Aluminum ions 262, 264, 375
Aluminum salts 252, 254
Amines, biogenic 226, 450, 455, 456
Amino acids 103, 122, 231, 436, 437, 446, 447, 451, 456, 476, 485, 574, 671
Ammonia 5, 6, 240, 287, 288, 291, 292
Amphiphilicity 481, 487
Amphoteric dyes 1, 9
Amplification steps 575, 576
Amyloid deposits 386, 436, 442, 443
Analogues, fluorescent 115, 116, 117
Aniline 6, 364, 365, 440
Anionic groups 223, 256, 361, 368, 370
Anthracene 7, 143, 262, 274, 275, 501, 502, 675, 676
Anthracyclines 105, 381, 401, 404, 493, 494
Anthraquinones 10, 79, 232, 252, 257, 259, 261, 262, 263, 275, 276, 347, 401, 508
Anti-fading agents 233, 234, 568, 577, 579, 580
Antifading agents 219
Anti-oxidants 64, 233, 235, 260
Aperture, numerical 172, 173, 174, 199, 688
Apoprotein 119
Apoptosis 104, 306, 307, 399, 400, 476, 500, 510, 532, 534, 535, 539, 540, 542, 544
Apoptotic cells 102, 533, 534, 535, 539, 540, 541, 542, 543, 547
Applications of fluorochromes in protein histochemistry 438
Aromatic amino acids 152, 226, 227, 231, 236, 388, 436, 450, 452
Aryl-oxazoles 82, 359, 371, 372
Arylsulfatases 462

Juan Carlos Stockert & Alfonso Blázquez-Castro

Astrazon 97, 147, 445, 446
Atomic nuclei 32, 40, 41, 74, 610, 617
Atomic orbitals (AOs) 61, 66, 73, 74, 75, 286
Atomic transitions 27, 29, 162
Auramine O 14, 15, 147, 235, 241, 366, 381, 387, 412
Autocorrelation curves 207, 210, 211
Autofluorescence 79, 117, 140, 219, 223, 227, 228, 229, 230, 231, 232, 295, 297, 338, 361, 456, 481, 680
Autofluorescence intensity 187, 229, 230
Autofluorescence signal 186, 192, 212, 229, 230
Autophagy 532, 546
Auxochromes 61, 66, 68
Azo groups 77, 440, 443

B

Banding methods 381, 393
Band structure 607, 608, 610, 612, 613
Band structure model 606
Basic fuchsine 365, 367, 383, 384, 385, 444
Basophilic cytoplasm 142, 144, 345, 375, 389, 391, 392, 400, 413, 415, 483, 492, 493, 497, 498, 500, 501, 510, 538
Bathochromic shift 135, 140, 141, 149
B-diketones 81, 86, 257, 260
Benzene 7, 54, 55, 68, 70, 71, 72, 74, 75, 76, 82
  electronic configuration of 75, 76
Benzothiazole-hydroxystyrene 462, 473
Benzothiazoles 381, 412, 413, 438, 443, 481
Berberine 12, 62, 79, 347, 374, 381, 395, 399, 400, 418, 508, 511
Berenil 146, 147, 236, 237, 410, 411, 412, 414
Binding mechanisms 135, 148, 219, 221, 276, 363, 406, 443, 497, 500
Biological activities 72, 79, 96, 98, 102, 105, 393, 399, 401, 491
Biological structures 219, 222, 228, 304, 664, 670
Biological substrates 1, 10, 79, 143, 148, 219, 220, 253, 303, 311, 437, 665, 695, 711
Biology, molecular 14, 123, 266
Biomolecules 1, 2, 14, 96, 98, 99, 101, 113, 116, 117, 212, 213, 382, 450, 654, 671, 672
  labeling of 96, 113
Bio-orthogonal reactions 96, 113
Biotinyl tyramide (BT) 412, 471, 472, 498
Biphotonic biomolecule excitation 654
Biphotonic excitation 205, 648, 653, 654, 656, 657, 658, 659
Biphotonic imaging 651, 653, 658
Bipyridine 80, 81, 230, 255, 259, 288, 309, 359, 375, 376, 496
Bisazo dyes 116, 346, 359, 361, 440, 441, 443, 452, 476, 555, 556
Bis-benzimidazoles 381, 408
Bisulfite, nucleophilic attack of 384
Blood smears, methanol-fixed chicken 258, 263, 264, 395, 439, 441
Blue emission 8, 86, 103, 354, 589, 625, 635, 656, 669
Blue fluorescence 8, 85, 86, 103, 364, 515, 542, 553
Bodipy 10, 82, 99, 104, 269, 348, 353, 455, 464, 502, 503
Borate buffer 12, 65, 68, 240
Boronic acids 359, 360, 367
Bottom images 197, 306, 654, 655, 679
BO values 72
Bovine granulosa cells 345, 346, 544
Bovine serum albumin (BSA) 267, 349, 513, 555, 630
BP filters 170, 171, 172
BrdU labeling 112, 410
Bright field 313, 318, 320, 577, 622, 654, 655, 671
Bright field images 309, 318, 320, 325
Bromo-substituted DNA 111, 112, 113
BRP rings 708

C

Caged fluorophores 96, 100
Calcein-AM 532, 534, 535, 536, 537, 552
Calcification markers 252, 260, 275, 276, 277
Callose 359, 362, 364, 365

Camptothecins 105, 514, 515, 516, 539
Cancer cells 118, 354, 418, 504, 513, 514, 675
  colorectal 675
Carbocyanines 10, 11, 359, 371, 381, 392, 440, 501, 503, 556
Carbon and silicon nanoparticles 623
Carbon atoms 288, 623, 624, 632, 633
Carbon nanoparticle 606, 623, 625, 632
Carboxyl groups 259, 436, 449
Carmine 5, 62, 79, 252, 261, 362
Carotenoids 118, 336, 349
Caspases 462, 476, 477, 542
Catecholamines 230, 436, 455, 456
Cathepsins 462, 476
Cationic dyes 9, 62, 69, 141, 145, 221, 253, 254, 255, 339, 340, 342, 360, 368, 370, 390, 391, 392, 437, 485, 490, 497, 506
  parental 340
Cationic fluorochromes 112, 147, 149, 223, 366, 371, 387, 391, 399, 482, 556
Cationic porphyrins 223, 381, 401, 415, 416, 418, 495
Cationic probes 488, 495, 508, 510, 511
Cationic proteins 68, 117, 238, 242, 349, 438, 439, 440, 443, 446, 450, 557
Cell imaging 322, 323, 325, 632
Cell lines 321, 324, 325, 582, 625, 626
Cell nuclei 107, 151, 187, 203, 317, 319, 321, 322, 549, 598, 628, 635
Cell populations 151, 211, 212, 535, 536, 538, 555
Cellular structures 314, 315, 317, 322, 324, 574, 577, 583, 628, 699, 708
Cellulose 143, 359, 360, 361, 362, 363
Cell viability 361, 465, 469, 481, 532, 535, 538
Centromeric heterochromatin 241, 381, 398, 399
Ceramides 104, 105, 336, 354, 501
Charge-coupled devices (CCDs) 175, 176, 177, 184
Chelating reagents 252, 259, 271
Chemical bonds 25, 33, 34, 52, 53, 55, 136
Chemical groups 10, 79, 97, 220, 485, 486
Chemical reactions 3, 53, 56, 169, 297, 314, 316, 317, 322, 455, 586, 631, 698
Chemical structure of chromomycin A3 (CMA) 415, 416, 417, 419, 495
Chlorophyll 8, 62, 96, 107, 108, 118, 119, 228, 289, 336, 339, 347, 348
Chloroplasts 107, 228, 229, 552, 664
Chloro-triazines 359, 360
CHO cells 295, 296
Cholesterol 105, 336, 337, 339, 348, 349, 350, 352, 468, 502
Chromatic aberrations 173, 211
Chromatid cores 389, 390
Chromatids 111, 112, 599
Chromatin DNA 112, 151, 258, 263, 372, 386, 387, 396, 401, 402, 403, 408, 411, 412, 414, 415, 416, 419, 481, 482, 486, 492, 493, 494, 495, 497, 498, 500, 534, 546
Chromatin DNA emission 396, 494
Chromatin DNA labeling 486, 492, 495
Chromatin fibers 389, 392, 415, 416
Chromatin staining 254, 255
Chromo- and fluorogenic reactions 225, 226, 360, 365, 462
Chromobody 568, 584
Chromomycin A3 257, 381, 415, 417, 495
Chromophores 2, 9, 55, 56, 61, 66, 67, 68, 72, 119, 125, 150, 153, 154, 225, 261, 263, 386, 390, 394, 499, 553, 649, 652, 653, 675
Chromosomal domains 597, 599
Chromosomal set 593, 595
Chromosome banding patterns 397, 419
Chromosome painting 568, 591, 593, 594, 595
Chromosome structure 416, 418, 419
Cis-trans isomerism 96, 122
Clay minerals 135, 147
Co-doped nanoparticles 668, 669
Coherent anti-stokes Raman scattering 642, 678, 680
Collisions 37, 41, 42, 43, 44, 45, 46, 50, 52, 57, 311, 675
  molecular 43, 52
Comet-FISH 568, 597, 598
Complexes 81, 87, 89, 118, 154, 257, 259, 262, 273, 283, 285, 286, 292, 295, 297, 306, 309, 310, 311, 313, 314, 315, 316, 317, 318, 319, 320, 321, 322, 323, 324,

325, 327, 328, 375, 407, 414, 496, 500, 504, 627, 630, 631
lanthanide 302, 303
Complexing motif 323
Compound 1,3,5-triphenylformazan 468, 469
Concentrations 136, 142, 144, 222, 252, 253, 254, 370, 390
critical electrolyte 222, 252, 253, 254, 390
increasing 136, 142, 144, 370
Conduction band (CB) 80, 584, 606, 608, 609, 610, 611, 612, 613, 614, 615, 618
Conduction bands 80, 606, 608, 609, 611
Confocal approaches 182
Confocal microscope 159, 181, 183, 600, 691
Confocal microscopy 125, 165, 180, 182, 192, 596, 655, 687
Confocal systems 180, 181, 183, 184, 205, 656, 691, 709
Congo red 10, 144, 359, 360, 361, 440, 441, 442, 443, 444, 532
Conjugated bond number (CBN) 67, 81, 84, 122, 258, 268, 387, 484, 485, 488
Coordination bonds 289, 290
Correlation functions 208, 209, 210
Coumarins 8, 10, 11, 64, 79, 81, 86, 99, 275, 276, 439, 455, 475
Counterstaining dye 568, 573, 577
Coupling 96, 98, 406, 407, 448, 449, 453, 474, 545, 616
electronic 406, 407
Cristae, mitochondrial 708
Critical electrolyte concentration (CEC) 222, 252, 253, 254, 390
Cruzi epimastigotes 411, 413, 416, 511, 512
Cultured cells 101, 105, 124, 151, 186, 339, 349, 467, 477, 481, 514, 543, 550, 571, 628, 632
Curcumin 62, 86, 87, 146, 147, 257, 260, 359, 362, 414
Cyanotolyl tetrazolium chloride (CTC) 466, 537
Cyclodextrins 145, 146, 148, 289, 406
Cytochrome 124, 125, 347, 542, 543, 654, 679
Cytometer, flow 190, 211, 212, 213, 657
Cytoplasm 13, 120, 139, 200, 202, 225, 230, 261, 317, 318, 319, 347, 437, 467, 482, 492, 493, 494, 497, 499, 500, 501, 503, 507, 508, 510, 513, 514, 515, 516, 536, 537, 552, 556, 626
nucleoli and basophilic 493, 497, 500, 501, 510
Cytosol 477, 488, 538, 542, 543

D

Dansyl chloride 99, 436, 446, 447
Dark-field illumination 1, 13
Dark fluorophores 697
Dead cells 441, 498, 533, 534, 535, 536, 538, 545, 551
De-excitation mechanisms, non-radiative 50
De-excitation pathways 26, 41, 57, 58
available 41, 57, 58
Defect of electrons 612
Dehydrogenases 462, 463, 464, 465, 467
Deoxyribose 381, 383, 392, 405
Detection, lipid 343, 346, 348, 503
Detection pinholes 691
Diatomic molecules 35, 36, 40, 554
Dichroic mirrors 160, 161, 164, 170, 171, 172, 194
Diffraction limit 165, 182, 199, 687, 693, 694, 695, 705
Diphenyl-hexatrienes 336
Display, molecular excited levels 53
Dissociation energy, molecular 53
Distamycin 152, 381, 406, 409, 450
Distamycin A 152, 381, 406, 409, 450
DNA breakage 549, 550
DNA damage 381, 547, 548
DNA duplexes 392, 401, 403, 414, 415, 416
DNA elongation 492, 493
DNA labeling 109, 111, 112, 492
DNA probes 492, 493, 589
Doping fluorophores 631
Double bond index (DBI) 71, 72
Doxorubicin 263, 404, 493, 494, 669, 670

E

Ehrlich ascites tumor cells 12, 152, 466, 483
Electric field 24, 25, 597, 608, 615, 643
Electroluminescence 126, 349, 606, 613, 614
Electromagnetic field 23, 24, 57
Electromagnetic waves 22, 23, 24, 25, 26, 198, 615, 616, 643
Electron delocalization 63, 258
Electron excitation 496, 610, 611, 612
Electronic configuration 41, 71, 72, 75, 394
Electronic energy level arrangement 291
Electronic excited states 40, 326
Electronic levels 27, 30, 36, 37, 38, 40, 42, 47, 49, 199, 291, 292, 293, 299, 302, 317, 607, 609, 612, 632, 649, 650, 651, 662, 667, 668
  excited 49, 199
  first 650
  higher 649, 668
  molecular 38, 42, 651
  real 650, 662
Electronic orbitals 29, 32
Electronic shells 284, 286, 300, 617
Electronic transitions of molecular electrons 42
Electron microscopy 238, 303, 327, 389, 398, 402, 454, 471, 509, 545, 687, 711
Electron spin 43, 293
Electron transfer, photoinduced 53, 55, 154, 514
Electron transitions 79, 80
Electrophoresis 14, 402, 548
Electroporation 123, 124
Electrostatic interaction 253, 382, 437
Emission 3, 23, 26, 28, 29, 40, 42, 46, 48, 49, 50, 57, 66, 79, 81, 83, 87, 117, 119, 120, 122, 135, 137, 138, 139, 140, 141, 142, 144, 145, 146, 147, 148, 150, 152, 153, 154, 162, 164, 167, 168, 172, 184, 185, 188, 189, 190, 197, 202, 207, 223, 225, 227, 230, 236, 237, 239, 240, 252, 254, 266, 267, 268, 269, 274, 295, 296, 297, 301, 306, 307, 308, 309, 313, 314, 320, 321, 322, 323, 324, 325, 326, 327, 328, 343, 349, 361, 362, 364, 371, 374, 382, 386, 390, 395, 403, 407, 408, 411, 412, 414, 415, 443, 449, 450, 451, 468, 482, 483, 497, 500, 506, 533, 535, 538, 539, 592, 617, 618, 619, 621, 627, 628, 629, 632, 633, 636, 651, 653, 659, 661, 663, 665, 666, 667, 668, 675, 678, 696, 700, 702, 709
  anti-Stokes 665, 666, 667
  bright white-blue 408, 412
  complex 295, 320, 325
  excite biphotonic 651
  fluorescent protein 202
  harmonic 661, 663
  metachromatic 371, 482
  peak 636, 675
Emission bands 45, 137, 138, 152, 168, 294, 300, 326, 617, 705
  molecular 300
Emission colors 65, 79, 219, 225, 277, 393, 409, 442, 473
Emission colour 162, 204, 302, 573, 591
Emission filters 161, 170, 171, 172, 177, 193
Emission intensity 2, 85, 112, 143, 184, 186, 189, 205, 267, 274, 392, 396, 397, 467, 482, 675
Emission lifetimes 190, 318, 621, 628, 629, 630, 631
Emission light 171, 172, 188, 189, 193, 201, 691
Emission properties 61, 361
Emission ratio 669, 671
Emission spectrum 41, 48, 49, 50, 126, 135, 136, 142, 160, 162, 165, 167, 168, 225, 296, 315, 376, 591, 613, 614, 620, 623, 626
  continuous 160, 162
  sample's 591
Emission values 219, 536
Emission wavelengths 138, 168, 302, 618, 667, 704
Emit light 22, 26, 50, 52, 164, 185, 186, 206, 628, 631, 632, 647
Emits, molecules to 47, 51
Emitting species 185, 632

Endoplasmic reticulum 273, 306, 317, 323, 488, 490, 503, 588, 703, 708, 709
Energy 3, 23, 24, 25, 26, 27, 28, 30, 31, 32, 33, 34, 35, 36, 37, 38, 42, 44, 45, 46, 47, 48, 49, 50, 52, 54, 55, 56, 58, 74, 75, 76, 77, 84, 87, 136, 153, 164, 166, 169, 198, 199, 207, 232, 233, 286, 291, 322, 324, 579, 607, 608, 609, 610, 611, 612, 613, 614, 617, 644, 645, 646, 649, 651, 653, 660, 665, 666, 667, 668, 671, 673, 675, 677, 680, 693
  electronic 579, 673
  laser 645
  minimum 609
  molecular 33
  transition 30, 36
Energy absorption 66, 73, 375
Energy gap 606, 609, 611, 613
  initial 609
Energy levels 27, 28, 30, 31, 37, 38, 39, 47, 57, 75, 77, 291, 299, 411, 606, 607, 609, 612, 614, 617, 667, 668
  discrete electronic 617
  electronic 28, 30, 38, 39, 291, 299, 607, 668
Energy transfer 56, 57, 61, 84, 122, 126, 152, 153, 154, 297, 323, 501, 616, 666, 668
  intramolecular 153
Energy transitions 29, 649
Energy values 30, 33, 38, 608
Environments, fluorophore's 184, 185
Enzymatic activity 14, 116, 226, 436, 437, 462, 464, 475, 532, 575
Enzyme-linked immunosorbent assay 568, 586
Eosin 10, 61, 79, 81, 89, 90, 99, 106, 141, 142, 219, 225, 232, 233, 236, 238, 239, 240, 241, 372, 438, 439, 440, 508
Epidermal growth factor (EGF) 190, 191, 475
Epifluorescence microscope 160, 161, 171, 173
Erythrocyte nuclei 151, 235, 255, 372, 396, 397, 398, 409
  chicken 151, 235, 255, 372, 396, 409
  methanol-fixed chicken 397, 398
Esterases 268, 462, 463, 464, 465, 536, 551, 554
Ethanolic isatin reaction 451
Ethidium bromide (EB) 69, 147, 149, 153, 154, 223, 235, 238, 392, 396, 418, 446, 485, 497, 498, 511, 534, 535, 548, 549
Evanescent waves 199, 200, 692, 693, 694
Excitation 2, 3, 22, 28, 41, 42, 43, 55, 81, 101, 126, 136, 137, 138, 139, 140, 154, 172, 177, 185, 186, 188, 189, 190, 199, 208, 209, 219, 226, 228, 230, 233, 235, 236, 237, 263, 265, 266, 268, 296, 305, 320, 321, 322, 325, 327, 328, 337, 376, 387, 442, 454, 467, 506, 578, 616, 636, 637, 644, 648, 649, 650, 654, 657, 658, 660, 666, 667, 701
  available 506
  electronic 3, 22, 41, 43, 644
  multiphoton 199, 209
  non-linear 648, 649
  pulsed 186, 296, 658
  three-photon 660
  two-photon 328, 637, 648, 654, 657
Excitation band 136, 137
Excitation energy 32, 50, 56, 57, 153, 208, 294, 298, 300, 303, 322, 579, 580, 667, 673, 675, 676
  electronic 32, 667, 673
  exchanging electronic 50
Excitation energy transfer 579, 666, 667, 669
Excitation filters 2, 161, 170, 171, 172
Excitation light 170, 171, 172, 181, 182, 189, 193, 313, 577, 619, 637, 642, 675, 691, 692, 694, 701
Excitation modulation 188, 189
Excitation photons 642, 673
Excitation power 207, 208
Excited electron 39, 40, 43, 44, 46, 53, 54, 56, 57, 74, 294, 297, 610, 637, 650
Excited forms 75
Excited levels 43, 49, 302, 616, 659, 667, 668
Excited state absorption 642, 659, 666, 667
Excited state lifetimes, long 207, 666, 671, 673
Extension 40, 51, 52, 201, 655, 658, 659

F

False colour emission lifetime 630
Fast green FCF 79, 89, 144, 145, 223, 224, 385, 438, 440
Fast red TR 436, 452, 472, 473, 474, 506
Fatty acids 105, 228, 229, 336, 337, 339, 348, 353, 354, 366, 487, 501, 502, 553
  unsaturated 228, 229, 336, 337, 339, 366, 553
Feulgen reaction 381, 383, 386
Fiber, elastic 12, 228, 240, 265, 411, 442, 444
Fibroblasts, human 571, 572, 698, 699
Filipin III-cholesterol 336, 351, 353
Flavine-adenine-dinucleotide (FAD) 153, 507, 508, 654, 655, 661
Flavonoids 7, 64, 79, 228, 235
Flavonols 64, 65, 66, 252, 257, 259, 439, 508
Flow cytometry 120, 159, 160, 211, 212, 213, 268, 271, 343, 397, 404, 466, 492, 495, 497, 533, 535, 540, 547, 552, 585, 586, 632, 657
Fluorescamine 69, 81, 446, 447
Fluorescein 8, 61, 81, 88, 89, 98, 99, 100, 101, 104, 107, 115, 117, 118, 234, 239, 268, 269, 295, 296, 313, 337, 438, 439, 465, 502, 508, 550, 551, 572, 621, 622, 625, 627, 648, 652, 653, 659, 660, 666
  caged 100, 107
Fluorescein emission 269, 296
Fluorescein molecules 120, 648, 652, 693
Fluorescein solution 647, 648
Fluorescence 2, 3, 11, 14, 15, 17, 22, 34, 46, 47, 48, 56, 57, 58, 61, 73, 81, 82, 83, 84, 85, 88, 89, 98, 102, 116, 120, 124, 125, 126, 135, 136, 137, 138, 139, 142, 143, 144, 145, 150, 151, 152, 153, 172, 186, 189, 191, 204, 211, 219, 220, 222, 225, 226, 227, 228, 232, 233, 234, 235, 236, 238, 239, 240, 252, 253, 255, 257, 258, 264, 266, 267, 277, 268, 270, 275, 307, 313, 318, 337, 346, 347, 350, 351, 361, 362, 366, 368, 369, 371, 372, 374, 375, 388, 389, 392, 393, 396, 397, 398, 399, 400, 401, 409, 410, 412, 414, 416, 418, 439, 440, 441, 442, 444, 452, 453, 454, 456, 457, 468, 469, 471, 474, 484, 492, 493, 494, 497, 502, 506, 508, 514, 533, 536, 537, 539, 553, 568, 573, 631, 642, 643, 647, 655, 675, 694
  metachromatic 143, 150, 368, 371, 393, 398
  polychromatic 252, 276, 277
  sample's 172
Fluorescence correlation spectroscopy (FCS) 159, 205, 206, 207, 208, 209, 211
Fluorescence emission 222, 675, 696
Fluorescence excitation light 697
Fluorescence images 120, 200, 237, 241, 242, 320, 321, 505, 516, 582, 630, 676
Fluorescence *in situ* hybridization (FISH) 14, 109, 110, 114, 261, 381, 475, 509, 548, 568, 569, 571, 573, 575, 577, 579, 581, 583, 585, 587, 589, 590, 591, 592, 593, 595, 596, 597, 598, 599, 600, 711
Fluorescence labeling of live cells 483
Fluorescence lifetime 186, 188, 189, 295
Fluorescence lifetime images 631
Fluorescence lifetime imaging microscopy (FLIM) 159, 184, 295, 318, 592, 621, 622, 659
Fluorescence light 171, 173, 189, 193, 691
Fluorescence localization 159, 204
Fluorescence localization after photobleaching (FLAP) 159, 204
Fluorescence microscopes 159, 160, 170, 172, 177, 179, 180, 470, 571
Fluorescence microscopy 159, 178, 179, 184, 198, 283, 687, 694
  time-resolved 179, 184, 283
  total internal reflection 159, 178, 198, 687, 694
Fluorescence microscopy applications 293, 321, 618
Fluorescence microscopy techniques 184
Fluorescence parameters 219, 221
Fluorescence photobleaching 159, 200
Fluorescence photon 46, 51, 57, 58, 653, 697, 705
Fluorescence quantum yield 8, 14, 58, 61, 79, 119, 573, 592, 642, 649, 698

Fluorescence reactions 1, 11, 12, 68, 219, 220, 225, 235, 239, 254, 255, 258, 259, 265, 362, 368, 386, 393, 440, 450, 463, 475, 514
Fluorescence recovery 159, 201, 585
Fluorescence recovery after photobleaching (FRAP) 159, 201, 202, 203, 585
Fluorescence resonance energy transfer (FRET) 57, 154, 190
Fluorescence spectra 137, 139, 140, 142, 147, 372
Fluorescence techniques 1, 159, 297, 320, 350, 462, 586
Fluorescent azides 113, 114
Fluorescent brighteners (FBs) 359, 362, 363, 364, 554
Fluorescent complexes 252, 257, 259, 263, 390, 438
Fluorescent compounds 2, 7, 10, 82, 336, 337, 342, 350, 449
Fluorescent dyes 2, 62, 79, 81, 201, 267, 277, 436, 438, 439, 481, 502, 550, 573
Fluorescent emission 173, 192, 203, 213, 701
Fluorescent images 2, 106, 170, 219, 475, 638, 701, 702
Fluorescent labeling 96, 98, 100, 104, 105, 120, 349, 403, 446, 462, 499, 501, 512
Fluorescent light 171, 172, 173, 194, 586
Fluorescent molecules 39, 47, 49, 148, 205, 571, 573, 574
Fluorescent pH probes 252, 266, 267
Fluorescent probes 2, 10, 99, 107, 208, 212, 235, 240, 252, 270, 273, 277, 343, 353, 403, 417, 484, 488, 489, 497, 500, 501, 502, 503, 532, 537, 538, 541, 555, 611, 657
  low molecular-weight 541
Fluorescent products (FP) 125, 231, 236, 366, 367, 387, 388, 437, 446, 447, 449, 451, 453, 455, 456, 462, 464, 470, 474, 475, 553, 584, 588
Fluorescent proteins, green 82, 118, 121, 568, 587
Fluorescent proteins (FPs) 1, 62, 84, 118, 121, 123, 124, 190, 191, 197, 204, 568, 574, 584, 587, 588, 657, 696, 710, 711
Fluorescent stain 83, 347, 392
Fluorochromes 112, 145, 147, 238, 267, 371, 412, 438, 443, 444, 468, 577
  anionic 267, 438, 557
  benzothiazole 412, 443, 444
  non-rigid 145, 147, 371, 468
  photosensitizing 112, 238
Fluorogenic reactions 219, 225, 226, 360, 365, 388, 447, 448, 449, 451, 453, 455, 456, 462, 463
Fluorogenic substrates 86, 226, 462, 464, 465, 471, 472, 476, 542
Fluorophore concentration 205, 207, 208
Fluorophore emission 83
Fluorophore emission lifetime 189
Fluorophore of choice 589, 631, 649
Fluorophore photobleaching techniques 199
Fluorophore populations 204, 696, 697
Fluorophores move 206
Fluorophore-tagged antibody 568
Formaldehyde 224, 226, 229, 339, 344, 347, 399, 456, 457, 473, 494, 536, 571
Formyl peptide receptor (FPR) 313, 314
Franck-condon principle 39, 40, 41, 45, 52, 186
Frequency-domain (FD) 184, 188
Function 57, 58, 206, 207, 209, 210
  autocorrelation 206, 207, 209, 210
  electronic wave 57, 58
Furanones 226, 436, 501
Fusion protein lamina-RFP 124, 125
Fusion proteins 123, 124, 191, 584

G

Gene expression 123, 126, 501, 589, 597
Genetic analysis 568
Genetic material 589, 590, 597
Genotoxicity 532, 547, 548, 549
GFP fluorophore 121, 122, 125, 231
Giant unilamellar vesicles (GUVs) 106
Glycerol 106, 143, 144, 145, 186, 336, 340, 452
Glycols 340, 365, 366, 381, 388
Glycosidases 462, 463, 464, 465

Gold atoms 319, 320
Golgi apparatus of cancer cells 513, 514
Green emission 81, 86, 87, 152, 240, 269, 306, 348, 365, 390, 483, 500, 513, 552, 555, 622, 635, 648, 656, 657, 664, 669, 670, 671
Green fluorescent protein (GFP) 14, 82, 96, 118, 121, 122, 123, 124, 125, 126, 197, 203, 211, 320, 492, 505, 568, 584, 587, 696
Ground state depletion 659
Ground vibrational levels 45, 46, 48, 49

H

HaCaT cells 572, 577
Halo assay 532, 547, 549, 550
Harmine fluorescence of nuclei in live plant cells 399
Harmonic generation 661, 663, 665, 666
Harmonic microscopy 642, 661, 664
Head group hydrophilicity (HGH) 484, 487
Head group size (HGS) 484, 487, 488
Heavy metal complexes 317
HeLa cells 13, 124, 125, 306, 313, 316, 325, 328, 343, 347, 468, 490, 491, 493, 496, 502, 509, 515, 516, 533, 535, 541, 543, 544, 545, 548, 585, 588, 625, 628, 630, 665, 670, 679, 710
  live 314, 344, 401, 494
Hematoxylin 1, 5, 6, 10, 79, 89, 90, 137, 138, 219, 225, 226, 240, 256, 260, 438, 439, 508, 544
  oxidized 6, 10, 89, 90, 138, 225, 240, 256, 260
Hematoxylin and eosin 89, 219
Heparin 149, 242, 255, 311, 359, 369, 371, 375, 397, 412
Hexidium iodide 532, 534, 536
High density lipoprotein (HDL) 513
High-energy photons 50, 74
Histochemical applications 258, 259, 260, 266, 367, 374, 466
Histones 267, 492, 533, 588, 595, 699, 700
Hodgkin lymphoma cells 472
Hoechst dyes 410, 419, 481, 546
Human breast adenocarcinoma cells 621, 622
Human chromosomes 114, 115, 409
Hybridization 73, 109, 110, 589, 590, 598
Hydroethidine 481, 498, 532, 534, 553
Hydrophobic dyes 1, 9
Hydrophobic forces 219, 221, 340, 351, 406
Hydrophobic index 481, 486, 487
Hyper-emission 140, 142, 227, 263
Hypo-emission 142, 147, 227
Hypsochromic shift 135, 140, 141

I

Image Correlation Spectroscopy (ICS) 209, 211
Immune cells 568, 570
Immunohistochemistry 470, 556, 570, 571, 577
Indigocarmine 6, 10, 237, 238, 506
Indigogenic methods 462, 463
Indirect gap semiconductors 626
Indolamines 436, 455, 456
Initial biphotonic fluorophore excitation 705
Inorganic ligands 287, 288
Intercalates 263, 314, 401, 403, 411, 412, 492, 493, 495, 497, 499, 515
Intermolecular energy transfer 153, 154
Internal reflection, total 159, 178, 179, 198
Internal transition elements 286
Intranuclear movements of RNA molecules 100
Ionic dyes 1
Ionic strength 135, 141, 143, 395, 462
Isoelectric point 382, 436, 437, 438, 439
Isolated ligand molecule 294
Isothiocyanate 96, 97, 98, 100, 107, 238, 239, 310, 314

K

Kinetic energy 31, 32, 33, 52, 161

L

Label chromatin DNA 496, 497, 498
Label nucleoli 499
Labels nucleoli and mitochondria 413, 499
Lanthanide-doped materials 666, 668, 672
Lanthanide-doped nanoparticle 606, 636
Lanthanide ions 301, 302, 666
Lanthanides 81, 266, 277, 283, 284, 285, 286, 292, 295, 298, 299, 300, 301, 302, 303, 304, 305, 312, 321, 324, 326, 327, 574, 636, 637, 667, 673
  particular 667
Lanthanide upconversion 642
Largest conjugated fragment (LCF) 81, 484, 485, 488
Laser confocal system 320, 321, 625
Laser light 164, 165, 178, 212, 646, 648, 664, 701
Lasers 126, 160, 163, 164, 165, 166, 168, 178, 181, 182, 185, 192, 193, 201, 205, 209, 212, 644, 645, 654, 655, 657, 665, 669, 671, 677, 702, 705, 706, 707
  depleting 705, 707
Lectins 1, 96, 117, 118, 359
Lifetimes 168, 184, 186, 189, 191, 303, 309, 310, 319
  fluorophore's 184, 186, 189
  operational 168
  shorter 191, 303, 309, 310, 319
Ligand molecules 396, 628
  small organic 628
Ligand orbital, occupied 293, 294
Ligands 81, 153, 222, 283, 287, 288, 289, 314, 381, 617
  antenna 81, 153
  binding affinity of 222, 381
  organic 283, 287, 288, 289, 314, 617
Ligand-to-metal charge transfer 283, 293
Light absorption 3, 13, 63, 66, 67, 73, 80, 170, 233, 252, 266, 293, 294, 506, 609, 613, 614, 616, 642, 668
Light cones 180, 645, 647, 648, 688
Light emission 3, 81, 82, 84, 126, 167, 185, 293, 295, 299, 300, 305, 306, 308, 310, 314, 323, 574, 587, 613, 614, 616, 619, 623, 625, 626, 635, 650, 688
Light emission process 46, 168, 631
Light emitting diodes (LEDs) 14, 166, 167, 168, 192, 349, 606, 614
Light intensity 3, 182, 188, 206, 207, 309, 310, 318, 600, 642, 643, 644, 647, 648, 651, 655, 657, 660, 663, 664, 665, 666, 667, 669, 671, 676, 699, 704, 707
  higher 660, 664, 707
  low 642, 643, 644, 647, 666, 667, 676
Light sheet fluorescence microscope (LSFM) 192, 193, 194, 195, 196, 197
Light-sheet microscope 159
Lignum nephriticum 1, 7, 8
Lipid bilayers 106, 107, 336, 337, 347, 399, 501, 503, 542
Lipid droplets 107, 108, 336, 338, 339, 343, 344, 345, 346, 347, 348, 349, 354, 400, 467, 468, 488, 503, 506, 513, 514, 515, 537
  red fluorescence of 344, 345
Lipid peroxidation 231, 541, 553
Lipofuscin granules 140, 219, 228, 230, 507
Lipophilic 62, 69, 270, 315, 342, 347, 486, 487, 508, 512
Lipophilic dyes 339, 340, 341
Lipophilic fluorochromes 338, 339, 340, 348, 349, 487
Liposomes 105, 106, 224, 491, 504, 513, 515
Live cell nucleoli 500
LMCT transitions 294, 300
Lucifer dyes 555, 556
Luminescence events 33, 46, 47
Luminescence signal 320, 321
Luminescent 80, 188, 208, 305, 308, 317, 319, 321, 323, 390, 500, 606, 616, 617, 623, 624, 625, 635, 671
Luminescent complexes 259, 288, 307, 312, 317, 322
Luminescent probes 283, 294, 297, 301, 303, 305, 308, 312
Luminescent signal 316, 318, 375
Luminescent TMCs 302, 303, 304, 305, 312, 321

Lymphocytes 118, 212, 235, 311, 314, 371, 373, 375, 376, 387
Lysosome labeling 474, 495, 505, 506
fluorescent 506
Lysosomes 225, 228, 231, 274, 306, 317, 319, 322, 323, 338, 340, 474, 476, 481, 482, 483, 488, 490, 491, 495, 503, 504, 505, 506, 507, 515, 516, 538, 547, 556, 579

M

Macrocycles 283, 287, 288, 289, 327, 402
Magnetic fields 23, 24, 58, 610
Mammalian cells 117, 120, 502, 511, 536, 546, 588
Mant fluorophore 115
Mast cell granules 242, 345, 369, 372, 374, 375, 386, 393, 400, 552
Mast cells 151, 311, 370, 371, 372, 373, 374, 375, 376, 413, 456
heparin-containing granules of 151, 371
Mercurochrome 79, 89, 236, 454
Mercury orange 436, 454
Merocyanine 540 347, 481, 502, 539
Mesomeric forms 61, 67, 89, 502
Metachromasia 147, 149, 368, 373, 382
Metachromatic fluorescence reactions 369, 371
Metachromatic reactions 11, 147, 342, 359, 369, 371
Metachromatic staining 149, 255, 368, 392
Metal cations 65, 252, 254, 256, 257, 258, 259, 260, 266, 269, 277, 283, 288, 289, 290, 294, 390, 405, 629
Metal complexes (MCs) 252, 256, 257, 259, 283, 286, 294, 312, 319, 321, 368
Metallic nanoparticles 615, 616, 617, 629, 704
Metal-t--ligand charge transfer 283
Metaphase chromosomes 109, 110, 112, 113, 237, 412, 417
Methanol-fixed horse eosinophil 267, 444, 449, 450, 451, 452
Microscopy 159, 163, 165, 166, 315, 574, 642, 647, 651, 652, 655, 659, 660, 661
biphotonic 642, 647, 651, 652, 655, 659, 660, 661
three-photon 660
time-resolved 159, 163, 165, 166, 315, 574
Minor groove binders 391, 392, 406, 407, 415, 492, 495
Minor groove binding 115, 381, 382, 391, 405, 409, 414, 415, 417, 418, 419, 492, 493
Mitochondria autofluorescence 108, 230, 499, 508, 513, 514, 515
blue 108, 499, 513, 514, 515
Mitochondrial membrane depolarization 540
Mitochondrial probes 327, 492, 508, 511
Mitotic catastrophe 532, 539, 546
Molecular absorption 37, 45, 48, 49, 50
general 48
Molecular arrangements 42, 54
Molecular beacons 100, 501
Molecular change 53, 322, 698
Molecular electrons 42
Molecular emission band features 48
Molecular energy reservoirs 33
Molecular geometries 70, 122, 265
Molecular moieties 570, 628
Molecular orbitals (MOs) 39, 56, 61, 66, 72, 74, 75, 77, 78, 292
Molecular probes 10, 573
Molecular rearrangements 39, 55, 186
Molecular structure 62, 66, 68, 72, 78, 84, 88, 111, 271, 322, 384, 412, 475, 496, 568, 570, 628
Molecular transitions 27, 28, 29, 40, 50
electronic 40
Molecular vibrations 34, 35, 36, 40, 58
Molecule breaks 52
Molecule decays 74, 653
Molecule emits 41, 45, 51
excited 51
Molecule localization, single 695, 696, 697
Molecule relaxes 50, 52
Molecules 2, 3, 9, 22, 26, 28, 29, 30, 32, 33, 34, 35, 36, 37, 38, 39, 40, 41, 42, 43, 44, 45, 46, 47, 48, 49, 50, 51, 52, 53, 54, 55, 56, 57, 58, 62, 66, 71, 73, 74, 77, 84, 85, 88, 98, 116, 119, 120, 135, 142, 161, 164, 184, 186, 189, 200, 204, 205, 207,

208, 213, 232, 287, 288, 289, 290, 297, 313, 327, 328, 381, 407, 485, 486, 487, 552, 568, 570, 572, 573, 574, 576, 579, 580, 613, 614, 619, 621, 627, 628, 647, 650, 652, 653, 661, 677, 678, 679, 680, 698
hydrogen 33, 35, 39
Molecules display 32, 43
Molecules emit 42, 50
Molecules move 205
Molecule's structure 573, 579
Molecule transitions 43
Molecule transits 46, 49, 50, 51, 52, 53
Monochromatism 164, 165, 168
Monomers 142, 147, 149, 150, 392, 414
Mordant dyes 1, 9, 257, 264
MTT formazan 467, 468, 469
Multiphoton microscopy 642, 647
Multiphoton sample excitation 213

N

Nanodiamond 606, 624, 631, 632, 635, 637
NAs-containing structures 390, 391, 392, 400, 534, 538
Natural dyes 4, 5, 6, 64, 260, 262, 362
Necrosis 532, 536, 539, 543, 546
Neuromuscular junctions 181, 656, 657, 708, 709
Neutral density (ND) 177, 326, 633, 634
Neutral lipids 336, 343, 344, 346
Neutral red (NR) 62, 63, 69, 106, 150, 255, 266, 347, 368, 387, 393, 481, 486, 504, 505, 506, 533, 538, 539
Nile 1, 10, 138, 336, 338, 339, 341, 342, 343, 344, 345, 481, 504
emission of 343
Nile red 138, 336, 338, 339, 341, 342, 343, 344, 345, 481, 500
NIR emission 318, 319
NIR-emitting lanthanides 326, 327
NIR excitation 126
NIR laser emission of heptamethine pyrylium dyes 83
Nitric oxide 305, 314, 532, 550, 554
Nitrogen atoms 55, 288, 289, 290, 633
Non-linear effects 646, 649, 651, 695
Non-linear processes 165, 182, 647, 651, 652, 663, 677
Non-linear techniques 165, 194, 695
Non-radiative de-excitation 22, 50, 52, 308
Ns fluorophore emission 296
Nuclear counterstaining 98, 223, 477, 589
Nuclear localization of α-tubulin in HeLa cells 545
Nuclear morphology 539, 540
Nucleic acids-containing structures 257, 258, 263, 491
Nucleic acids labeling 96, 97, 109
Nucleolar organizing regions (NORs) 598
Nucleosomes 415, 492, 493
Nucleus centre 598
Numerical parameters 84, 481, 484, 487, 488

O

Oblique plane microscopy (OPM) 193
Oligonucleotides 1, 100, 105, 114, 436
One-photon absorption 651, 660
Optical elements 159, 160, 164, 170, 171, 173, 175, 178, 179, 180, 193
Optical excitation 51, 294
Optical fibers 163, 166, 178, 179, 198, 658, 694
Optical imaging 637, 680
Optical transitions 28, 29, 34, 36, 48, 291, 292, 293, 294, 297, 303
molecular 28
Organic fluorophores 164, 295, 296, 298, 302, 583, 635
Organic molecules 45, 56, 61, 62, 74, 185, 288, 297, 298, 301, 303, 311, 591, 606, 614, 620, 671, 675
fluorescent 45, 185
Organization, nucleosomal 498, 499, 500, 512
Orthochromatic emission 368, 371
Orthochromatic staining 368, 369
Orthogonal-plane fluorescence optical sectioning (OPFOS) 192, 193
Oxazine dyes 341, 342, 506

Oxazones 70, 341, 342, 404
Oxidation states 285, 286, 299, 306, 317, 319, 327, 612, 636, 666
Oxidative deamination 336, 341, 342, 343
Oxidative stress 233, 235, 508, 532, 550, 553
Oxygen, molecular 56, 207, 232, 303, 579
Oxygen atoms 53, 55, 86, 90, 287, 289

P

PAS reaction 243, 359, 366, 367, 388, 389
Pauli exclusion principle 606, 607, 609, 611
Pentaene fatty acids 336, 354
Peptide bonds 231, 403, 476, 477
Peptide nucleic acids (PNAs) 109, 110, 314
Perinuclear localization 307, 308
Peroxidases 117, 235, 401, 413, 437, 462, 463, 464, 470, 471, 472
Phalloidin 102, 103, 571, 572
Phalloidin-TRITC 98, 102, 103, 224, 241
Phallotoxins 96, 102, 116
Phase contrast and red fluorescence images of living HeLa cells 108
Phosphatases 462, 463, 464, 473
Phosphates 62, 67, 81, 256, 273, 275, 381, 382, 392, 395, 414
Phosphatidyl-ethanolamine (PE) 501
Phosphatidyl-inositol (PI) 238, 382, 437, 438, 498, 501, 534, 535, 541, 542, 548, 549
Phosphatidyl-serine (PS) 135, 233, 501, 541, 542
Phospholipids 104, 105, 106, 228, 229, 336, 337, 338, 339, 343, 386, 487, 501
Phosphorescence 3, 22, 46, 47, 48, 58, 80, 84, 98, 143, 227, 232, 233, 642, 675
Photoactivated localization microscopy (PALM) 195, 687, 696, 699, 700
Photoactivation 204
Photoactive materials 176
Photobiotin 100
Photobleaching 56, 159, 184, 200, 201, 203, 204, 232, 297, 311, 585, 625, 626, 627, 698, 709
  fluorescence loss in 159, 203
Photobleaching techniques 199, 201, 205
Photocathode 177
Photochemical damage 232, 233
Photochemical reactions 84, 147, 233, 297
Photoconversion 204, 238, 552
Photoconversion FLAP 204
Photodamage 234, 482, 483, 658, 707
Photodetector 180, 184, 187, 691, 692
Photodissociation 39, 52, 53
  delayed 52, 53
Photodynamic treatment, lethal 533, 544
Photoelectric effect 176, 177
Photoelectron 177
Photoexcitation 153, 500
Photoexcite electrons 610
Photofading curves of fluorochromes and fluorescent metal complexes 235
Photoinduced electron transfer (PET) 53, 54, 55, 154, 308, 514
Photoisomerization 54, 55, 293
Photoluminescence 3, 79, 126, 606, 613, 614, 637
Photoluminescent 80, 626
Photolysis, strand break induction by 113, 115
Photon absorption 39, 45, 52, 53, 55, 57, 73, 185, 293, 294, 301, 306, 608, 613, 629, 668, 675, 696
Photon absorption/emission 30, 31
Photon distribution 646
Photon emission 22, 26, 41, 42, 46, 50, 51, 52, 54, 84, 164, 167, 184, 185, 232, 300, 308, 608, 610, 614, 665
Photon energies 33, 49, 54, 55, 136, 153, 166
Photon flux 642, 645, 646, 673
  large 645
Photonics 33, 611, 613, 627, 647
Photons 3, 22, 24, 26, 27, 28, 29, 30, 31, 32, 33, 34, 37, 38, 39, 40, 41, 42, 43, 45, 46, 47, 50, 51, 52, 53, 55, 56, 57, 58, 74, 80, 84, 136, 164, 166, 177, 184, 185, 186, 293, 294, 297, 300, 321, 324, 325, 327, 573, 609, 610, 611, 613, 614, 626, 636, 642, 643, 645, 646, 647, 649, 650, 651, 652, 653, 660, 661, 662, 663, 665, 666, 667, 668, 669, 671, 673, 674, 675, 676, 678, 680, 695, 696, 697, 705, 706
  absorbed 46, 50, 58, 667, 673, 675

fundamental 663, 680
long wavelength 321, 678
low-energy 674, 676
optical 31, 33
shifted 678, 680
single 327, 642, 649, 662
upconverted 673, 675
Photonucleases 305
Photon upconversion 302, 642, 665
Photophysical features 301, 302, 636
Photophysics 116, 283, 284, 290, 298, 587, 606, 610
Photoreaction 54, 698
Photoreduction 233, 236, 506
Photosensitization, lethal 543, 545
Photosensitizers 17, 99, 105, 107, 135, 232, 233, 235, 239, 323, 410, 483, 488, 510, 514
Photostability, high 79, 315, 619, 629, 634
Photosynthesis 119, 192, 228, 289, 643
Phycobiliproteins 14, 96, 118, 119, 120, 121
Plant tissues 118, 228, 361, 362
Plasma membrane 105, 118, 199, 209, 274, 315, 317, 336, 353, 488, 501, 502, 503, 506, 533, 534, 539, 540, 541, 556, 578, 626
Plasmatic membranes 306, 307, 622, 702, 703
PNA probes 110
Point spread function (PSF) 687, 696
Polarization microscopy 1, 13
Polyanions 149, 242, 256, 382, 402, 412
Polyenes 10, 63, 68, 70, 79, 96, 105, 336, 337, 338, 349, 351, 352, 353, 354
Polymers 150, 360, 627, 628
Porphine 17, 223, 401, 415, 416, 495, 496, 503
Porphyrins 10, 79, 82, 107, 228, 289, 327, 347, 401
Potential energy curves for triatomic molecules 38
Properties 23, 26, 166, 172, 267, 292, 353, 354, 606, 607, 612, 693
lipid 353, 354
optical 23, 26, 166, 172, 267, 292, 606, 607, 612, 693
Propidium iodides 223, 497, 498, 534, 579
Protein histochemistry 436, 438
Protein labeling 99, 116
Protein levels, total 679
Proteins, synthesized 679
Protons 29, 30, 39, 40, 69, 644
Pseudo-color images 548, 549
Pulsed lasers 165, 201, 318, 645, 646, 652, 653, 654, 655, 656, 659, 704
Pyranine 252, 266, 267, 440
Pyronine 11, 62, 63, 147, 149, 150, 152, 235, 255, 393, 395, 397, 398, 404, 510, 511
interaction of 149
Pyronine Y (PY) 11, 62, 63, 147, 149, 150, 152, 235, 255, 393, 395, 397, 398, 404, 510, 511

Q

QSAR predictions 488, 489, 490, 493
Quantitative structure-activity relationships (QSAR) 61, 81, 481, 484, 504
Quantum 29, 30, 35, 38, 57, 58, 72, 81, 98, 135, 297, 633, 647, 674
Quantum confinement effect 611, 612, 618, 619, 620
Quantum dot 80, 96, 606, 612, 618

R

Radiative de-excitation 22, 50, 52, 294, 308, 312
Radiative lifetimes 295, 298, 302, 309, 310, 318, 319, 573, 617, 621, 622, 628, 629, 633
Radiative transitions 299, 300, 301, 326, 614, 615, 617, 631, 673, 675
Raman band 135, 136, 138, 139
Raman microscopy 642, 651, 677, 678, 680
Raman scatter 135, 137, 138, 140
Raman-shifted photons 680
Range, optical 25, 26, 30, 33, 636, 643, 687
Ratio imaging 252, 268, 269
Rayleigh criterion 687, 689, 690
Rayleigh-Tyndall band 135, 136
Reactive dyes 1, 9, 97, 365, 367, 448

Reactive fluorochromes 107, 117, 118, 125, 444, 446, 464
Reactive oxygen species (ROS) 219, 232, 233, 235, 262, 306, 311, 350, 532, 550, 552, 553, 580, 589, 628, 676
Red emission 8, 111, 135, 139, 152, 204, 229, 243, 269, 310, 323, 345, 393, 404, 454, 506, 535, 556, 572, 622, 635, 656, 657, 664, 669, 670, 671, 697
Red fluorescence 6, 107, 117, 119, 223, 237, 240, 259, 263, 264, 275, 343, 345, 346, 348, 349, 360, 365, 441, 442, 473, 474, 503, 510, 514, 534
Red-orange emission 225, 265, 375
Refractive index 11, 174, 175, 178, 688, 690
Relaxation time molecules 46
RESOLFT technique 710
Resolving power 581, 600, 678, 691, 692, 695, 700
Resonance, plasmon 615, 616, 629, 693, 694
Resonant forms 8, 62, 68, 70, 85, 86, 439
Resorufin 89, 469, 470, 471, 537, 538, 545
Retrograde axonal tracers 411, 532, 556
REversible saturable optical fluorescence transitions (RESOLFT) 687, 709
Rhodamine 123 296, 481, 508, 509, 516, 532, 552, 579
Rhodamine dyes 100, 539
Rhodamine fluorophores 83, 117
RNA, nucleolar 498, 499
RNA molecules 100, 417
RNA Polymerase II 699, 700
ROS/antioxidant detection in living cells 630
Rotational energy, molecular 34
Ru(II)-bipyridine 255, 359, 375

S

Salts 88, 226, 265, 267, 349, 440, 448, 449, 452, 453, 463, 464 466, 467, 473, 474 476
  diazonium 448, 449, 452, 453, 463, 473, 474
  sodium 88, 265, 267, 440
  tetrazolium 349, 464, 466, 467
  tetrazonium 226, 452, 474, 476
Scanning near-field optical microscope 687, 694
Schiff's reagent 242, 365, 366, 381, 382, 383, 384, 385, 386, 388, 553
Second harmonic generation (SHG) 642, 662, 663, 664, 665
Signal amplification 109, 117, 575, 576, 591
  tyramide 96, 100, 462, 471, 472
Silica nanoparticle 606, 630, 635, 675
Silicon nanoparticles 623, 626
Silicon Qdots 623, 624, 626
Single EDTA molecule 289, 290
Singlet states 43, 44, 45, 652, 653, 676
  second excited 652, 653
Sirofluor 359, 364, 365
Sister chromatid exchanges (SCEs) 111, 112, 419
Small ligands 145, 258, 381, 405, 418
Solid-state physics 606
Spatial frequencies, high 691, 692, 701, 702
Spatial light modulators (SLM) 183, 192, 193
Spectra 35, 48, 49, 120, 122, 135, 136, 139, 141, 142, 144, 148, 227, 376, 613, 619, 652, 653
  absorption/emission 49, 613, 619, 652
  electronic molecular 35
Spectral changes 146, 147, 148, 252, 269
Spectral features 70, 122, 123, 140, 143, 404
Spectroscopy 29, 32, 33, 38, 40, 50 159, 205, 206
  fluorescence correlation 159, 205, 206
  molecular 29, 32, 33, 38, 40, 50
Spectrum, electromagnetic 24, 25, 26
Sphingomyelins 104, 105, 336, 354, 501
Spin 22, 30, 34, 42, 43, 44, 45, 47, 74
  electronic 43, 45
SRM techniques 690, 705, 709, 711
Staining 2, 4, 11, 61, 105, 112, 138, 141, 152, 220, 223, 225, 232, 235, 241, 253, 254, 258, 261, 263, 307, 339, 341, 343, 346, 348, 360, 366, 369, 370, 371, 372, 374, 375, 386, 388, 393, 395, 397, 409, 437, 441, 443, 449, 481, 482, 484, 492, 497, 508, 546, 578
  lipid 346, 348

Staining reactions 222, 253, 254, 437, 438, 506
STimulated emission depletion (STED) 687, 705, 706, 707, 708, 709
Stimulated raman scattering (SRS) 642, 677, 678, 679, 680
STochastic optical reconstruction microscopy (STORM) 600, 687, 696, 698, 700
Streptavidin 117, 436, 472, 592, 628, 629, 630, 632
Sudan dyes 336, 339, 346
Sulfhydryls 310, 436, 454
Super-resolution techniques 175, 182, 600, 623, 687, 690, 695

T

Techniques 211, 589, 690, 693, 694, 696
  fluorescent imaging 211
  molecular biology 589
  near-field 690, 693, 694
  single molecule localization 696
Terbium 81, 266, 299, 300, 302, 321, 323, 325, 326, 574, 667
Terminal deoxynucleotidyl transferase 113, 115, 547
Tetracyclines 257, 264, 275, 276, 277, 508
Thiazines 10, 11, 62, 63, 106, 341, 342, 371, 392, 506
Thiazones 70, 340, 341, 342, 393
Thioflavine T (TT) 147, 395, 400, 412, 413, 443, 444, 498, 503, 511
Thionine 62, 63, 150, 233, 236, 255, 341, 342, 393, 504, 506
Third harmonic generation (THG) 642, 662, 663, 664
Thulium 302, 326, 667, 669
Thymine 111, 389, 390, 401
Tissues 5, 16, 83, 84, 106, 116, 122, 197, 219, 227, 228, 232, 242, 306, 311, 336, 339, 342, 382, 386, 388, 441, 442, 450, 462, 472, 473, 482, 497, 550, 555, 587, 636, 654, 655, 656, 657, 662, 672, 690
  neoplastic 16, 106
  normal 655, 662
Tissue structures 2, 13, 219, 240, 241, 259, 276, 439, 443
TMC emission 296
TMCs, lanthanide 298, 300, 302
TMCs in fluorescence microscopy and photodetection techniques 290
Toluidine 11, 62, 63, 141, 150, 254, 255, 342, 364, 368, 369, 392, 393, 506, 539, 540
Total internal reflection (TIR) 159, 178, 179, 198, 200, 687, 694
Total internal reflection fluorescence (TIRF) 159, 178, 198, 209, 687, 694, 695
Total internal reflection fluorescence microscopy (TIRFM) 159, 178, 198, 199, 200, 626, 687, 693, 694
Traditional color chemistry 61, 66
Transition metal complexes (TMCs) 283, 284, 285, 286, 287, 288, 289, 290, 291, 292, 293, 294, 295, 296, 297, 298, 299, 300, 303, 304, 305, 308, 311, 312, 317, 319, 321, 322, 324, 326, 591, 607, 673
Transition metals (TMs) 82, 96, 283, 285, 286, 287, 289, 306, 307, 308, 315, 319, 574, 607, 623, 631, 673
Transitions 27, 28, 30, 31, 32, 34, 36, 38, 39, 40, 41, 44, 45, 47, 48, 49, 66, 74, 75, 76, 77, 78, 283, 291, 293, 294, 295, 296, 298, 299, 300, 301, 302, 324, 633, 584, 636, 649, 650, 651, 652, 653, 661, 663, 668, 697, 705
  electronic 39, 40, 41, 66, 75, 76, 294, 295, 298
  epithelial-mesenchymal 584
  intra-ligand 294, 295, 298, 301
  one-photon 649, 652, 653
  photonic 27, 28, 45, 48
  resonant 650, 651
  rotational 34
Triatomic molecules 38
Triplet-triplet annihilation upconversion 311, 642, 673
Trypan 116, 144, 147, 359, 360, 361, 440, 441, 442, 443, 481, 506, 532, 533, 555, 556

Trypan blue 116, 144, 147, 359, 360, 361, 440, 441, 442, 443, 481, 506, 532, 533, 555, 556
Tryptophan 116, 119, 125, 226, 388, 450, 451, 452
Tubulin 100, 101, 545, 572, 577
Tubulin binding 101, 102
TUNEL assay 532, 540, 544, 547
Two-photon absorption process 650, 692
Two-photon fluorophore excitation 648
Tyramide signal amplification (TSA) 96, 100, 462, 471, 472
Tyrosine 119, 258, 450, 452, 453, 472, 475

U

UC emission 667, 668, 675, 676
UC light emission 672, 675
UC materials 668, 672, 673
UC nanoparticles 669, 670, 671
Uranium 266, 286, 303, 327
Uranyl complexes 303, 327
UV excitation 137, 223, 227, 230, 258, 265, 276, 277, 322, 340, 352, 362, 372, 408, 409, 452, 471, 506, 513, 515, 588

V

The very hydrophobic Pt (TPNP) 318, 319
Vibrational energy 33, 34, 35, 36, 37, 38, 41, 46, 50, 136, 652, 653, 680
Vibrational excitation 52, 297, 616
Vibrational levels 36, 37, 38, 40, 41, 47, 49, 50, 51, 614, 662
Vibrational relaxation 41, 45, 46, 47, 48, 51, 53, 653
Vimentin detection 585

W

Western blot 568, 587

X

Xanthenes 8, 10, 62, 63, 79, 99, 269, 275, 439

Made in United States
North Haven, CT
04 November 2026

21727470R00416